GREEN DESIGN, MATERIALS AND MANUFACTURING PROCESSES

T0225574

PROCEEDINGS OF THE 2ND INTERNATIONAL CONFERENCE ON SUSTAINABLE INTELLIGENT MANUFACTURING, LISBON, PORTUGAL, JUNE 26–29, 2013

Green Design, Materials and Manufacturing Processes

Editors

Helena Maria Bártolo, Paulo Jorge da Silva Bártolo,
Nuno Manuel Fernandes Alves, Artur Jorge Mateus,
Henrique Amorim Almeida, Ana Cristina Soares Lemos,
Flávio Craveiro, Carina Ramos, Igor Reis, Lina Durão &
Telma Ferreira
Centre for Rapid and Sustainable Product Development,
Polytechnic Institute of Leiria, Marinha Grande, Portugal

José Pinto Duarte, Filipa Roseta, Eduardo Castro e Costa,
Filipe Quaresma & João Paulouro Neves
Faculty of Architecture, Technical University of Lisbon, Lisbon, Portugal

CRC Press
Taylor & Francis Group
Boca Raton London New York Leiden

CRC Press is an imprint of the
Taylor & Francis Group, an **informa** business

A BALKEMA BOOK

CRC Press/Balkema is an imprint of the Taylor & Francis Group, an informa business

© 2013 Taylor & Francis Group, London, UK

Typeset by V Publishing Solutions Pvt Ltd., Chennai, India
Printed and bound in Great Britain by CPI Group (UK) Ltd, Croydon, CR0 4YY

Published by: CRC Press/Balkema
P.O. Box 11320, 2301 EH Leiden, The Netherlands
e-mail: Pub.NL@taylorandfrancis.com
www.crcpress.com – www.taylorandfrancis.com

ISBN: 978-1-138-00046-9 (Hbk)
ISBN: 978-1-315-87948-2 (eBook)

Green Design, Materials and Manufacturing Processes – Bártolo et al. (eds)
© *2013 Taylor & Francis Group, London, ISBN 978-1-138-00046-9*

Table of contents

Sustainable construction

Green Design, Materials and Manufacturing Processes – Bártolo et al. (eds)
© 2013 Taylor & Francis Group, London, ISBN 978-1-138-00046-9

Preface

Green Design, Materials and Manufacturing Processes contains papers presented at the 2nd International Conference on Sustainable Intelligent Manufacturing (SIM 2013), conjointly organized by two Portuguese institutions, the Polytechnic Institute of Leiria through its research unit of excellence, the Centre for Rapid and Sustainable Product Development, and the Faculty of Architecture of the Technical University of Lisbon. This event was held at the facilities of the Faculty of Architecture at Lisbon, Portugal, from June 26 to June 29, 2013.

The Centre for Rapid and Sustainable Product Development of the Polytechnic Institute of Leiria (CDRSP-IPL) is a FCT Research Unit of Excellence aiming at contributing to the advancement of science and technology, leading to more suitable, efficient and sustainable products, materials and processes, helping to generate added value to the industry, and promoting the awareness of the role and importance of rapid and sustainable product development in society.

The main mission of the Faculty of Architecture of the Technical University of Lisbon (FA-UTL) is to ensure the creation, development, and transmission of scientific, artistic, and technical knowledge in a socio-culturally responsible and operative manner. The research work developed at FA-UTL promotes the production of knowledge and innovation in architecture, urbanism and design. Its Research Centre for Architecture, Urbanism and Design, called CIAUD, is a FCT research unit of Excellence.

The rise of manufacturing intelligence is fuelling innovation in processes and products considering a low environmental impact over the product's lifecycle. Sustainable intelligent manufacturing is regarded as a manufacturing paradigm for the 21st century, in the move towards the next generation of manufacturing and processing technologies. On the one hand, the manufacturing industry is at a turning point in its evolution and new business opportunities are emerging. On the other hand, sustainability has become a key concern for government policies, businesses and general public. Model cities are moving forward towards novel ecosystems, combining environmental, social and economic issues in more inclusive and integrated frameworks.

This International Conference on Sustainable Intelligent Manufacturing was designed to be a major international forum for academics, researchers and industrial partners to exchange ideas in the field of sustainable intelligent manufacturing and related topics, making a significant contribution to further development of these fields. Participants came from more than 35 countries and very distinct backgrounds, such as architecture, engineering, design and economics. Such diversity was parallel to the various multidisciplinary contributions to the conference, whose subjects cover a wide range of topics like Eco Design and Innovation, Energy Efficiency, Green and Smart Manufacturing, Green Transportation, Life-Cycle Engineering, Renewable Energy Technologies, Reuse and Recycling Techniques, Smart Design, Smart Materials, Sustainable Business Models and Sustainable Construction. All participants were strongly engaged in the development of innovative solutions to solve industry problems, contributing to a more healthy and sustainable way of life.

We are deeply grateful to authors, participants, reviewers, the International Scientific Committee, Session Chairs, student helpers and administrative assistants, for contributing to the success of this conference. The conference was endorsed by:

– The Centre for Rapid and Sustainable Product Development, Polytechnic Institute of Leiria
– The CIAUD Research Centre for Architecture, Urban Planning and Design, Faculty of Architecture, Technical University of Lisbon
– The Portuguese Foundation for Science and Technology

Green Design, Materials and Manufacturing Processes – Bártolo et al. (eds)
© *2013 Taylor & Francis Group, London, ISBN 978-1-138-00046-9*

Committee members

CONFERENCE CHAIRS

Helena Bártolo	*Centre for Rapid and Sustainable Product Development, Polytecnic Institute of Leiria*
José Pinto Duarte	*Faculty of Architecture, Technical University of Lisbon*
Paulo Bártolo	*Centre for Rapid and Sustainable Product Development, Polytecnic Institute of Leiria*
Filipa Roseta	*Faculty of Architecture, Technical University of Lisbon*

KEYNOTE SPEAKERS

Branko Kolarevic	*University of Calgary, Canada*
Francesco Jovane	*Politecnico di Milano, Italy*
Gabriela Celani	*UNICAMP, Brasil*
Giuseppe D'Angelo	*FIAT Research Center, Italy*
Joost Duflou	*University of Leuven, Belgium*
Klaus Sedlbauer	*Fraunhofer-Instituts für Bauphysik IBP, Germany*
Lawrence Sass	*Massachusetts Institute of Technology, USA*
Marco Santochi	*University of Pisa, Italy*
Mario Buono	*Seconda Università degli Studi di Napoli, Italy*
Paulo Jorge Ferreira	*University of Texas at Austin, USA*
Rivka Oxman	*Technion Israel Institute of Technology, Israel*
Robert Miles Kemp	*Variate Labs, USA*

INTERNATIONAL SCIENTIFIC COMMITTEE

Alain Bernard	*École Centrale de Nantes, France*
Andres Harris	*Architectural Association School of Architecture, UK*
Antonio Frattari	*University of Trento, Italy*
António Marques	*Universidade do Porto, Portugal*
Aouad Ghassan	*University of Salford, UK*
Antje Kunze	*ETH Zürich, Switzerland*
David Hayhurst	*University of Manchester, UK*
David L.S. Hung	*Shanghai Jiao Tong University, China*
Derek Clements-Croome	*University of Reading, UK*
Dirk Uwe Sauer	*RWTH Aachen University, Germany*
Fernando Moreira da Silva	*Universidade Técnica de Lisboa, Portugal*
Geoffrey Mitchell	*Centre for Rapid and Sustainable Product Development (IPL), Portugal*
Gerhard Schmitt	*ETH Zürich, Switzerland*
Gideon Levy	*Centre for Rapid and Sustainable Product Development (IPL), Portugal*
Hans Haenlein	*Hans Haenlein Architects, UK*
Hazim B. Awbi	*University of Reading, UK*
Hojjat Adeli	*Ohio State University, USA*
Humberto Varum	*Universidade de Aveiro, Portugal*

Ian Gibson	*National University of Singapore, Singapore Centre for Rapid and Sustainable Product Development (IPL), Portugal*
Jay Yang	*Queensland University of Technology, Australia*
Jan Halatasch	*ETH Zürich, Switzerland*
Joaquim Jorge	*Universidade Técnica de Lisboa, Portugal*
Joaquim de Ciurana	*University of Girona, Spain*
John Sutherland	*Purdue University, USA*
Jorge de Brito	*Universidade Técnica de Lisboa, Portugal*
Jorge Lopes dos Santos	*Pontifícia Universidade Católica do Rio de Janeiro, Brasil*
José Bártolo	*Escola Superior de Artes e Design Matosinhos, Portugal*
Kevin Lyons	*The State University of New Jersey, USA*
Kunze Antje	*ETH Zürich, Switzerland*
Luc Laperrière	*Université du Québec à Trois-Rivières, Canada*
Luís Bragança	*Universidade do Minho, Portugal*
Luísa Caldas	*University of California, Berkeley, USA*
Maria da Graça Carvalho	*European Parliament*
Marwan Khraisheh	*Masdar Institute, United Arab Emirates*
Manuel Pinheiro	*Universidade Técnica de Lisboa, Portugal*
Mohamed Hussein	*University of Connecticut School of Business, USA*
Mohsen Aboulnaga	*University of Dubai, United Arab Emirates*
Neri Oxman	*MIT Media Laboratory, USA*
Paul Chamberlain	*Sheffield Hallam University, UK*
Paulo Lourenço	*Universidade do Minho, Portugal*
Pedro Gaspar	*Universidade Técnica de Lisboa, Portugal*
Peter Lansley	*University of Reading, UK*
Peter Lund	*Aalto University School of Science, Finland*
Rangan Banerjee	*Indian Institute of Technology, India*
Regiane Pupo	*Universidade Federal de Santa Catarina, Brasil*
Rita Almendra	*Universidade Técnica de Lisboa, Portugal*
Russell Marshal	*Loughborough University, UK*
Shengwei Wang	*Hong Kong Polytechnic University, China*
Steve Evans	*University of Cambridge, UK*
Tahar Laoui	*King Fahd University of Petroleum & Minerals, Saudi Arabia*
Thomas Bock	*Technische Universität München, Germany*
Vasco Rato	*ISCTE – IUL, Portugal*
Victor Ferreira	*Universidade de Aveiro, Portugal*
Wilfried Sihn	*Vienna University of Technology, Austria*
Winifred Ijomah	*University of Strathclyde, UK*

Eco design and innovation

New visions in the manufacturing design—jewels as multiples

Chiara Scarpitti
Design and Innovation, Second University of Studies of Naples, Naples, Italy

ABSTRACT: Clearly there is a technological humanized advancement. The user-centered design is replaced by the concept of human-centered-design. Technological processes, different in techniques, styles, philosophies and visions transform the productive geographies and place the human being at the center of a new design approach. The territory is spread with a constellation of local excellences that create synergies and unexpected winners, while new markets are opening to micro niches for highly specialized products. The production is the ultimate consequence of a way of thinking. A new figure of "design-thinker" embraces these new requirements through a difficult process of setting the entire production system up. Nodal point of the process is the ability to influence, through the project, product innovation, using new production possibilities, renewing the concept of smart manufacturing.

1 INTRODUCTION

1.1 *The productive crisis as design challenge*

We are going through a phase of transition. The contemporary scene is fragmented. The modernity has been called from time to time weak, liquid, dusty, while the world of productive design can and should be rethought in a smarter, more flexible way and responsive to the contingent reality. Disintegrated by the current economic crisis all the old landmarks, the obsolete system of values is by now overthrown.

If the track is not drawn you should learn to walk outside and in parallel, braving the risk, but with the advantage of being first in achieving the paradigm of society that is gradually forming. In this sense, a systemic crisis can be seen as an extraordinary opportunity for cultural growth and reconfiguration of territories.

In Baudrillard's "The System of Objects" the official status of the object is represented by the opposition between the unique object (transcendent reality) and replicated object (immanent reality). Significant changes in the way we produce have effectively combined these two productive plans. Everything is pure model and at the same time model of itself. We are dealing with a society of "originals", unique pieces. In such a transformed context, the concept of Industrial Design, Product Design and Craft must be re-contextualized. If the twentieth century was the golden age of the objects (A. Branzi, 2000), what are we facing in the current century? What are the new ways in which men transform and change the world?

The unstoppable multiplication of objects, information, sensory stimuli place the man of today in a new situation, especially from a communication point of view. In this sense, Dorfles uses the effective expression "Horror Pleni", with the intent to emphasize the idea of rejection, horror for an aseptic overproduction, without soul nor meaning. "What is happening to our society and to our economy is that we insert information in vending machines so that they spit out rubbish in huge quantities and at a fraction of the cost". (V. Flusser, 2007) Another significant image of this change is the cover of The Economist (April 2012), which illustrates a man who designs and manufactures directly with the computer, independently, comfortably sitting at home.

Airily the design explores the fields of information, social, new economies and once more takes on the role of critical discipline capable of exciting the international debate. "Thanks to the role of information, design objects, nowadays, are not different by the interacting communication systems, ... if the perspective is that of a progressive narrowing of the world's materiality, the production of objects will also invade the fields of immateriality and communication" (Maldonado, 2008).

The technical and aesthetic composition of artifacts becomes metaphorical model for the reconstruction of social and political world. The change we glimpse announces a transformation of production processes both material and immaterial.

1.2 *A new "culture"*

In view of the above the industrial design must necessarily be a driving force behind the modernization of production that places people at the center of design thinking through a deep path of trans-disciplinary knowledge. This new way of doing design opens the door to a "new form of culture" (Flusser, 2010), which harmoniously blends the disciplines of humanity, such as philosophy, aesthetics, anthropology along with scientific disciplines such as engineering materials and machinery.

The old creative industry is replaced by a new, more sophisticated and complex one that corresponds to the so-called "need-more" (P. Restany, 1990): an industry which hybridizes art, science, nature, poetry through an entirely humanistic approach, typical of our Mediterranean culture.

The new industry must produce a smart thinking, knowledge and explores the existential dimension of men, through an anthropological approach. The products of his labor are devices of reflection for a better future.

The focus of contemporary design moves from the Design Product to the Design Process and to a reconfiguration of the role of the factory as a place dedicated to the making of these processes. There is urgent need to investigate in a new way the concepts of seriality--production--process--designer--user.

New materials and technologies are added to a new way of thinking about the production system, the relationship between people, places, time. The encounter between designers, users and manufacturing excellence have generated hybrid processes and technological transfers, in some cases fortuitous and not always in line with the methodological programmed protocols.

There is no absolute methodological practice. Only a trans-disciplinary freedom of action may be able to generate an advancement of a certain cultural entity, able to critically reflect on those who can be the future scenarios. The innovation of a system comes out from the meeting between different knowledge and an anarchic experimental methodology free from any programme.

Key player in this modus operandi is the web 2.0. that allows the development of an open system and the aggregation among users, designers and companies. The future smart manufacturing, designed as flexible production, will be an open serial system. These systems have shared a new approach to the issue of mass production and design collective.

We are facing the passage of paradigm.

2 CASE STUDY—JEWELS AS MULTIPLES

2.1 *State of the art of the manufacturing goldsmith*

The current historical moment full of rapid technological and social changes has deeply changed the manufacturing of jewelry and the system of values which is historically linked to it. In a few years the development of new production techniques, such as chemical cutting, laser cutting, rapid prototyping, sintering in environmentally friendly materials, has effectively undermined the obsolete methods of production, shortly opened to innovation and hybridization. The overproduction of the objects and aesthetics of the globalized century brings out a strong need for product customization, and at the same time makes people feel the need for permanence and care of the object and its computerization.

In the last ten years with the development of new digital technologies we are witnessing a gradual disintegration of industrial production which was concentrated in typical territorial districts, such as Vicenza and Arezzo (print/chains), Valenza (jewelry with stones), Torre del Greco (processing coral). The majority of jewelry factories closed, sold machinery and technologies to the poorest countries or relocated production. Only the Vicenza area has increased from 1,800 units to a few hundred medium-sized factories. The brands, mostly associated with fashion industries resist with a general poor innovative production, but they are supported by a strong investment on image and focus mainly on luxury markets in emerging countries.

2.2 *An innovative model of collective enterprise*

Jewels as Multiples is a research project that was founded in 2011 by a cultural need of AGC--Association of Contemporary Jewelry--to build a bridge with ADI--Industrial Design Association--and the world of Italian manufacturing industry. Focus of the project is the theme of seriality and of new production systems and the analysis of the collaborative possibilities between designers and companies. The working group is coordinated by designer Carla Riccoboni, together with the active collaboration of various professionals, from goldsmith, designer, marketing field (Chiara Scarpitti, Eliana Negroni, Maurizio Stagni, Sara Progressi).

The project involved many other figures of Italian jewelry design and organized numerous activities in several Italian cities, such as Bologna, Naples, Milan, Vicenza, Trieste. It continues to work to build relationships with potential partners from business and cultural life, associations, universities and training centers. Its ultimate goal is the promotion of a new way to design, focused on new technologies and manufacturing processes through a collective effort.

He founded a blog as a manifesto (www.gioiellicomemultipli.tumblr.com) which will be launched during the Milan Design Week 2013 and will slowly grow with the development of the work.

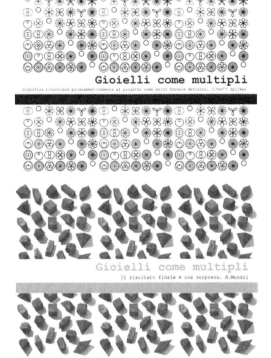

Figure 1. Graphic promotion for the launch of the project.

	PAST	FUTURE
product	amount of pieces all equals	multiplicity of unique pieces
designer	the studio is dislocated from the factory	studio = place of production
technique of production	mechanical or investment casting	digital
market	makers-wholesalers-retailer-customer	makers-customer
place of production	industrial districts	everywere

Figure 2. Framework of analysis for a new paradigm.

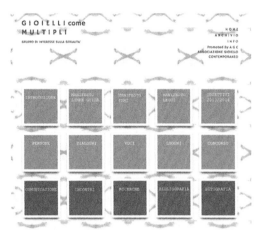

Figure 3. The website "Jewels as Multiples".

Currently the group is involved in the organization of an international competition on the theme of mass production based on new manufacturing technologies that will be declared in June 2013.

The analysis of the current modus operandi of Italian designers has enlightened the enormous amount of energy and resources to do research, promote and sell the products. In the light of these considerations, a system of shared work has spontaneously been built, which optimizes existing resources and create new opportunities for development always open to the transformation of reality. The web 2.0. has made this development possible. "The hope of an industrial creativity is to create small series aimed to economic and cultural minorities, not a spoon for everyone, but an infinite number of infinite spoons to meet individual needs" (Gaetano Pesce, 2010).

The project Jewels as Multiples is not limited to product design and communication, but handles the entire production process which includes the development, promotion and sale system. Everything has been designed thanks to the networking.

The concept of mass production, created from the first industrial revolution with the mechanization, changes its principles.

3 CONCLUSIONS

Jewels as Multiples positions itself consciously in this historical context.

On this basis, the manifesto combines the aesthetic and technical qualities with an ethical attitude that goes through all the steps in the supply chain, from the choice of materials, environmentally friendly techniques, to the recovery of collective memory, meant as codes, knowledge of a given geographical area.

The program for 2014 leaves the following insights open:

What are the innovative manufacturing systems that limit the risk of a huge loss of traditional knowledge related to handicrafts?

What are the ethical responsibilities of the designer in front of the consumption of energy and materials, which productive processes require?

What are the new forms of innovative manufacturing that include social creativity?

How manufacture companies of territory may work synergistically with cultural institutions and how can you make immediately visible and active these collaborations?

REFERENCES

ADI—Association Industrial Design, 2013. www.adi-design.org

AGC—Association Contemporary Jewellery, 2013. www.agc-it.org

AJF—Art Jewelry Forum, 2013. www.artjewelryforum.org

Baudrillard, J. 2007. Il sistema degli oggetti. Bologna: Bompiani.

De Fusco, R. 2012. Filosofia del design. Torino: Einaudi.

Dorfles, G. 2008. La (in)civiltà del rumore. Roma: Castelvecchi.

Flusser, V. 2007. Filosofia del design. Milano: Mondadori.

Gioielli come Multipli, 2013. www.gioiellicomemultipli.tumblr.com

Klimt02—Platform for Contemporary Jewellery, 2013. www.klimt02.net

Latouche, S. 2008. Breve trattato sulla decrescita serena. Torino: Bollati Boringhieri.

Maldonado, T. 2008. Disegno Industriale, un riesame. Milano: Feltrinelli.

Restany, P. 1990. Arte e Produzione. Milano: Arti Grafiche Colombo.

Green Design, Materials and Manufacturing Processes – Bártolo et al. (eds)
© 2013 Taylor & Francis Group, London, ISBN 978-1-138-00046-9

Communicating effectively in writing: Issues and strategies for engineers

L. Nazarenko
University of Applied Sciences Technikum Wien, Austria
University of Vienna, Vienna, Austria

G. Schwarz-Peaker
University of Vienna, Vienna, Austria

ABSTRACT: Sustainable intelligent manufacturing is an essential innovation for the 21st century, and advancements in this field are essential for developing new ways of doing business. The professionals who are working in this area specialize in technical fields, but not necessarily in written communication skills. Yet, without communicating these ideas effectively, both to the public and to policy makers, these innovators could hinder the transfer of these technologies. As teachers of English who specialize in written communication skills, we have worked with engineers and professionals in technical areas to improve the connection between their ideas and their ability to communicate these ideas effectively. This paper focuses on the areas to be aware of in order to write about innovations more clearly and presents specific strategies to achieve this clarity.

1 INTRODUCTION

Communicating ideas clearly is the basic purpose of writing; and when those ideas are not clear to the intended audience, then the message is not communicated. When this message relates to new ideas and innovations, then the need for clarity is even greater as when faced with new concepts, the audience is not able to draw on their existing knowledge to aid understanding. The audience of such scientific or technical texts often includes not only the general public, but also policy makers and journalists, who will in turn use the information to form messages of their own. The focus of this paper is the features of language that engineers and scientists need to be aware of when communicating their ideas in written form, in order to best reach their intended audience and achieve their intended purpose. We have compiled a list of twelve checkpoints that experts writing for a lay audience should bear in mind in order to make texts accessible and comprehensible to this non-expert group.

2 FOCUS ON AUDIENCE

Identifying a very specific audience for a particular text is the first, and perhaps most essential, step a writer has to take. Audience, in this sense, refers to the intended recipients of the message, whether spoken or written. This is not necessarily all those who *could* read the text, but rather the particular group the writer has in mind when crafting the text. For example, a proposal for funding that is written internally within an organization may be read by many people within that organization, but the intended audience will be the people who have the power to grant the required funding. Thus the way the proposal is written will need to resonate with this group. This means focusing on that particular audience, being aware of what they already know, what they need to know, why they need to know that, and what strategies are best for ensuring that this knowledge is imparted.

While this may be relatively easy to deal with in one's own organization, it becomes more difficult when dealing with people from outside. Misunderstandings can quickly lead to absurd, and potentially damaging, situations such as the case of Professor George Church from Harvard Medical School, who was faced with headlines in several newspapers claiming that he was looking for a suitable candidate to be surrogate mother for a cloned Neanderthal 'cave baby' after being misunderstood by journalists. The interesting thing here is that Professor Church was not misquoted; there was simply a lack of understanding about what he meant. His discussion of what was theoretically possible was taken as a description of what he intended to do. Journalists misunderstood his purpose, and he did not understand his audience. He would have needed to make clear that he was talking about theoretical possibilities rather than describing an intended plan of action.

- Be aware of your audience and relate all other features of text to this audience.
- Add information that this specific audience needs in order to understand what you mean.
- Leave out all unnecessary information this audience does not need so that they are not confused or overwhelmed by too much input.

3   FOCUS ON VOCABULARY CHOICES

In the same way, the choice of words can lead to misunderstanding. Engineers and scientists feel at home with the language of their field. They are familiar with shades of meaning within descriptions of technical procedures and processes and are usually competent in writing either technical descriptions or academic papers to be read by others within their field. However, this is the very language which hinders communication with non-expert audiences and which needs to be adapted in order to meet the requirements of the different recipients.

Specialized vocabulary that is restricted in meaning to certain disciplines (i.e., "insider" vocabulary) can have a different meaning to laypeople or those in a different discipline. In language studies this is referred to as "technical vocabulary," although this term to laypeople generally means "words relating to technology". Therefore, if a language teacher is writing for those outside the field, this term would have to be made clear in order for the message to be communicated. In fact, technical terms can be everyday English words that are known to people outside the field, but with a different meaning; and technical terms can vary in meaning across fields as well.

A lack of awareness of this fact is the basis of many misunderstandings about climate change. Hassol (2008) and Somerville and Hassol (2011) identify a range of words that are used in the science of climate change, and yet which have a completely different meaning for those outside the field. For example, when climate scientists use the word *manipulation*, they are referring to the processing of scientific data, whereas in the eyes of the general public manipulation means illicit tampering. It is clear that through such differences in understanding, distrust and suspicion can arise between scientists and the very people they might need to have on their side.

Likewise, the term *sustainable*, which is currently used in many areas of business and government policy, can be understood in a variety of ways in different contexts. Although most people think of sustainability as a 'good' thing, a search of the term on various websites indicates that not only is it defined differently by experts in different contexts, but that many members of the public dismiss the term as a business "buzzword" that does not really have any meaning. For example, a post on the website *Forum on Science, Technology & Innovation for Sustainable Development* asks contributors, "What does sustainability mean to you?" The answers are not only varied, but in some cases contradictory. To add to the confusion, the term could indicate a somewhat different concept depending on which area of business it is used in. For example, among many professionals in manufacturing, the term "sustainable manufacturing" is not clear. As Pojasek (2010) writes, "The term sustainable manufacturing means different things to different people. Many of these differences come from the discipline and bias of the manufacturing professional."

For professionals in any field, communicating their ideas is a high-stakes game that they do not want to lose. Therefore, complex or field-specific terminology should generally be avoided in communication with those not in the know. Keeping a record of such tricky vocabulary items could be a focused effort among colleagues at the firm or in the same field of business.

Sometimes choosing the right words is particularly essential in a business context when innovators want to seek support to develop their innovation. A *New York Times* article (Eisenberg, 2008) about an online registry for inventors focuses on the fact that the service will "provide in-person or online workshops to help inventors recast their often technical prose in jargon-free descriptions for the business and industrial customers that are expected to shop at the site." The article quotes Sandy Johnson, chief executive of the Mid-America Manufacturing Technology Center: "'A business may have a brilliant idea,' she said, "but if they need a small piece of technology to make it work, it's virtually impossible to find right now,' because of incomprehensible descriptions."

In this case, writing a description of the innovation using language that is too technical could result in a failure to bring the idea to reality.

3.1   *Checkpoints for vocabulary*

- Avoid vocabulary specific to your professional field, or if it must be used, explain what is meant by it.
- Keep a record of tricky vocabulary items as you become aware of them, or make this a focused effort among colleagues.
- If necessary, consider seeking input from professionals who provide consultations in jargon-free writing.

4 FOCUS ON STYLE

Another important factor that needs to be considered is the choice of an appropriate style. Technical descriptions and academic papers usually rely heavily not only on technical terms but on a formal scientific style. In addition, they are usually written impersonally, which means that it does not sound as if the writer is speaking directly to the reader. Of course, such a style is necessary in a professional context in order to sound justifiably authoritative. But the combination of these technical terms written in an impersonal, scientific style makes texts extremely difficult to decipher for non-experts. To ensure comprehension, the style should be changed to a more conversational, informal one by using personal pronouns (you, we) and using the active voice rather than passive. Another change to make the style less formal is the use of shorter sentences. This makes it easier for the non-expert reader to process the new information.

In addition, in a scientific style, the information is usually structured as subject/problem, methods used, result/solution. First the subject of the text is presented, which includes the basic problem it is attempting to solve. Then, specific information about the methods used to solve the problem is focused on. Finally, the result of the research, i.e., the solution of the problem, is given. This is clearly seen even in abstracts for academic or professional articles. However, for non-experts (including both journalists and policy makers), this type of text and information structure is unnecessary. A scientific paper has an accepted format that is used to disseminate new research to other scientists who may want to replicate the work, but texts for non-experts will have different purposes, such as to persuade readers to change their behaviour or to convince them that a new idea has potential. In each case it would generally be best to start with the solution to the problem, then explain what the problem was, and finally mention how the problem was solved, if necessary. A non-expert audience is not interested in the methods, but just wants to know what the end result is and why it is important.

An example of how such a scientific style can be changed for a non-expert audience is the first two sentences from an abstract by Palmas, et al. (2012) dealing with renewables in residential development:

"In recent years, there has been an increasing interest in using micro-renewable energy sources. However, planning has not yet developed methodological approaches (1) for spatially optimizing residential development according to the different renewable energy potentials and (2) for integrating objectives of optimized energy efficiency with other environmental requirements and concerns."

The basic idea expressed in these sentences is a fairly simple one, so an educated reader could interpret what is meant. However, if the audience were future homeowners, i.e., the general public, who should be convinced of the importance of developing the methodological approaches mentioned in the article, some changes would have to be made.

First of all, it would not be clear what, exactly, "micro-renewable" energy sources are, so this technical term would have to be changed. Then, the information structure could be changed so that the main idea of what needs to be done comes first. Finally, the use of the passive voice ("has been an increasing interest") could be changed to make clear who is increasingly interested. This would allow the use of personal pronouns, which would give the text a more direct, personal style:

"We have to plan new housing areas that consider the best use of space for different types of renewable energy, and make sure that these types of energy work well with other environmental requirements and concerns. This is important because home owners are becoming more interested in generating heat and electric power on a small scale for use in their own communities."

Reformulating is not a case of dumbing down content, but rather of making the message clearer and, as can be seen from this example, avoiding technical vocabulary, addressing the audience directly, shortening sentences, and changing the order in which the information is presented.

4.1 *Checkpoints for style*

- Restructure information so that the audience reads what's most important to them.
- Write in short, concise sentences.
- Use the active voice instead of the passive where possible.
- Use the personal pronouns "you" and "we" to talk directly to your audience and achieve a conversational style.

5 FOCUS ON ACHIEVING PURPOSE

The features of the text must also relate to the purpose to be achieved. Is it important merely for information to be given, or should the audience also be persuaded to do something with that information? To another specialist in the field, the information itself can be persuasive. For a layperson who has no knowledge of the subject, however, the importance of the information is not evident, so it would not likely be persuasive. In the same way, a fellow expert is interested in more detail, without the background information an outsider would need. For decision-makers, whether inside or outside the field, focusing on the "bottom line," omitting extraneous detail, is more effective.

In considering how to achieve the purpose with a specific audience, it is important to focus on which content it is necessary to include. The goal is to include enough information to make the concept clear, without including too much that will cause confusion or boredom; to inform clearly without being patronizing. To relate information about concepts with which the audience is not familiar, it is useful to use everyday examples the audience is likely to know, and to relate the unknown to the known. For example, to explain the motion of a cartesian robot to a lay audience, a reference to an office chair is used: "Move your chair front to back while moving your arm in and out." (Brumson, 2001). Similarly, in an article from *The Economist* (2008), a new type of solar panel is compared to a more familiar object: "Solyndra's approach is to coat glass tubes with CIGS and encase them in another glass tube with sealed ends. They look a bit like fluorescent-lighting tubes." By being related to something so common in everyday life, both these examples allow the non-expert audience to visualise the products. In this way a writer of technical information is able to explain concepts more effectively to an audience outside the field of expertise.

5.1 *Checkpoints for purpose*

- Determine what your purpose is; i.e., why you are relating this information to this particular audience.
- Use examples to help the audience understand, and relate those examples to everyday knowledge (i.e., relating the unknown to something known).

6 CONCLUSION

Clear writing is important to communicate ideas effectively. By taking the specific audience into account and adjusting language and structure to fit the purpose, engineers and scientists will be able to convince the general public that new research is relevant to them, encourage policy makers to introduce legislation to implement the findings of new research, and enable journalists to write about new research more accurately. In summary:

- Be aware of your audience and relate all other features of text to this audience.
- Add information that this specific audience needs in order to understand what you mean.
- Leave out all unnecessary information this audience does not need so they are not confused or overwhelmed by too much input.

- Avoid vocabulary specific to your professional field, or if it must be used, explain what is meant by it.
- Keep a record of tricky vocabulary items as you become aware of them, or make this a focused effort among colleagues.
- If necessary, consider seeking input from professionals who provide consultations in jargon-free writing.
- Restructure information so that audience reads what's most important to them.
- Write in short, concise sentences.
- Use the active voice instead of the passive where possible.
- Use the personal pronouns "you" and "we" to talk directly to your audience and achieve a conversational style.
- Determine what your purpose is; i.e., why you are relating this information to this particular audience.
- Use examples to help the audience understand, and relate those examples to everyday knowledge (i.e., relating the unknown to something known).

REFERENCES

Brumson, B. 2001. Scara vs. Cartesian Robots: Selecting the Right Type for Your Applications. *Robotics Online*.

Eisenberg, A. 2008. A Buyer's Guide to Inventions, in Plain English. *New York Times*.

Hassol, S.J. 2008. Improving How Scientists Communicate About Climate Change. *Eos* 89(11): 106–107.

Nazarenko, L. 2012. Identifying formulaic features of text type to guide engineering students in writing texts. *Challenges in Higher Education & Research* 10: 33–35.

Nazarenko, L. & Schwarz, G. 2010. A Refocused Approach to Writing Instruction: Incorporating a Focus on Genre into the Writing Process. *Writing & Pedagogy* 2.1: 103–116.

Palmas, C. et al. 2012. Renewables in residential development: an integrated GIS-based multicriteria approach for decentraized micro-renewable energy production in new settlement development: a case study of the eastern metropolitan area of Cagliari, Sardinia, Italy. *Energy, Sustainability and Society* 2:10.

Pojasek, R.B. 2010. When is sustainable manufacturing sustainable? on *MDDI Medical Device and Diagnostic Industry News Products and Suppliers*: http://www.mddionline.com.

Somerville, R.C. & Hassol, S.J. 2011. Communicating the science of climate change. *Physics Today* 64(10): 48.

The Economist 2008. Tubular sunshine: A new sort of solar panel is less fussy about where the sun shines from.

Green Design, Materials and Manufacturing Processes – Bártolo et al. (eds)
© 2013 Taylor & Francis Group, London, ISBN 978-1-138-00046-9

Sustainable design for furniture and wood products

P. Browne & P. Tobin
Galway-Mayo Institute of Technology, Galway, Ireland

ABSTRACT: A truly sustainable design for any product results from a design process that takes account of a complex mix of criteria and constraints. The sustainable requirements must be embedded in a normal customer-focused design process; otherwise the customer may not wish to buy the product. The purpose of the studies presented in this paper is to test a proposed sustainable design process for furniture and wood products small to medium enterprises (SME's) and to explore the potential for the implementation of sustainable design guidelines in the Irish furniture and wood products industry.

1 INTRODUCTION

Addressing sustainability is increasingly recognized as an important strategic issue for businesses and other organizations. Sustainable awareness can reduce your organizations climate risk, dependence on finite resources and identify your organization as an environmental leader to important stakeholder groups such as customers, shareholders, investors, governments officials and employees (Sanchez, 2009). This paper presents a qualitative view of the current practice of sustainable design in furniture and wood products SME's and proposes guidelines for the development of sustainable products, in a structured framework called 'Green Dot'.

2 REVIEW OF IRISH FURNITURE AND WOOD PRODUCTS INDUSTRY

As the scientific evidence regarding global warming has become widely accepted, so too has the realization by global businesses that resisting the move toward sustainability is no longer a viable business strategy (Tuttle & Heap, 2008). So far, sustainable design in the furniture and wood products industry in Ireland has been fragmented with much emphasis on sustainable harvesting and associated certification. The industry has not focused enough on integrating other useful tools and methodologies to provide a product design and development process that result in high-scoring sustainable products. Further to this, there is a huge margin for wastage as well as for reuse as the furniture and wood products industry predominantly uses timber and timber products.

Some work has been undertaken to identify the main issues and to quantify volumes and wastage

Table 1. Percentage waste of furniture materials used during manufacture (FIRA, 2010).

Material	% Waste
Hardwoods	40–50
Softwoods	10–15
Board Materials	5–10
Fabrics	15–20
Foams	3–4
Steel	3–4
Veneer	40–50

rates of particular raw materials by The Furniture Industry Research Association (FIRA) which gives a good indicator of waste problems in the sector. These statistics are listed in the table above.

3 RESEARCH METHODOLOGY

3.1 Overview

The research and development of sustainable design guidelines, suitable for sustainably enhancing design, manufacturing and associated activities involved researching subject areas such as 'eco' design, 'green' branding and 'green' consumerism. It also involved an examination of existing certifications and sustainable tools techniques and methodologies, national and international drivers for sustainable product development in the Irish furniture manufacturing context.

The guidelines were further developed through primary research. This consisted of a focus group and a case study. The focus group was attended by leading Irish designers, manufacturers and academics in the area of furniture and wood products. This group explored the question of 'green branding' and the viability of investing in sustainable manufacturing processes going forward. For the purposes of the case study, a 'sustainable' design process for an Irish wood products design and manufacturing company was investigated through the design and development of a 'sustainable' children's furniture range. The case study followed a typical design and development process; detailing customer design specifications, concept development and refinement and cumulating in final prototype, as well as associated engineering drawings.

3.2 *Key findings of case-study*

The case-study demonstrated difficulties reconciling sustainable goals and commercial realities. The final product was the output of a design process that included sustainable criteria, but achieved a mediocre score in typical measures such as LCA and carbon footprint. The key findings were as follows:

The concept for the case-study was to come up with an 'eco-friendly' range of children's furniture so that the company could launch themselves into the sustainable furniture market. The company believed that this 'green' initiative would give their company a unique selling point and competitive edge. The company rejected early concepts such as the 'rocking animals' as they were considered to be too brash in colour and too *'IKEA-esque'*. They also felt that the sole use of medium density fibre board (MDF) as a material, may give an inferior image to a product which was to be aimed at the high end of the furniture and wood products market.

The company originally indicated that FSC certified woods would be used where possible, this coupled with using environmentally friendly water based paints and lacquers, and using local suppliers and employees, were intended to seal the label of 'eco-friendly' to the product.

However, in real terms, as the design development progressed, the company focused more on the product aesthetics and function rather than its environmental impacts. The company decided not to use FSC certified timbers because of the high expense and lack availability of them. The company also decided to buy in small components for the product from UK as these would be much too time consuming to make. Finally, the product had originally being specified to be made entirely of FSC certified timbers, however the company felt that the muted tones of the timber would not entice children to interact with it, so it was made from MDF (to reduce costs) and sprayed in a regular coloured lacquer finish. Thus, the aesthetics

of the product out-weighed the 'eco-friendly' factor in the decision. In this case, while starting out as conscientious 'eco' enthusiasts, the company design decisions were swayed by aesthetic appearance and price.

3.3 *Establishing the focus group*

Randomly sampled groups are unlikely to hold a shared perspective on a research topic and may not even be able to generate meaningful discussions on a topic (Morgan, 1997). Therefore, participants for this focus group were not chosen at random because of the specialized nature of the furniture and wood products industry and the specific nature of the information the authors sought. Also, determining the group's composition involved seeking out strangers versus allowing acquaintances to participate together. A rule of thumb favours strangers because, although acquaintances can converse more readily, this is often dues to their ability to rely on the kind of taken-for-granted assumptions that are exactly what the researchers were trying to investigate (Morgan, 1997).

It was decided to use a 'funnel' based interview technique to facilitate the group. The focus group began with a less structured approach that emphasises free discussion and then moved toward a more structured discussion of specific questions. The funnel analogy matches an interview with a broad, open beginning and a narrower, more tightly controlled ending (Morgan, 1997). This compromise made it possible to hear the participants own perspectives in the early part of the focus group, as well as their responses to the author's specific interests in the later part of the discussion.

3.3.1 *Rational for recruitment of participants*
Eleven leaders in the field of furniture design, manufacture and consumer markets in Ireland were recruited and are detailed below in table 2.

3.3.2 *Key findings of focus group*
Leading academics, designers and furniture manufacturers attended the focus group, as they felt that 'going green' and implementing a sustainable design process could help their companies and the industry. The group felt that the manufacture of Irish furniture and wood products will continue to decline if they cannot find a unique selling point and new markets in which to sell their product.

The following conclusions were drawn from the focus group:

Growth in our economy and its subsequent slump undoubtedly put pressure on Ireland's emissions and also our ability to invest in 'green' technologies going forward.

Table 2. Categorization of focus group participants.

Participant	Age group	Gender	Occupation/ Specialization
P1	35–45	M	Head of Department of furniture technology and design institute
P2	35–45	M	Academic in furniture technology
P3	25–35	M	Academic in industrial & product design
P4	35–45	F	Managing director of furniture design and manufacturing company
P5	45–55	M	Enterprise Ireland Marketing Specialist
P6	45–55	F	Enterprise Ireland Export Specialist
P7	25–35	M	Furniture & Lighting Designer
P8	25–35	F	Researcher in energy systems
P9	55–65	M	Managing director of architectural mill-work company
P10	45–55	M	Owner of furniture design, manufacture export company
P11	55–65	M	Managing director of architectural mill-company

It is difficult for Irish manufacturer's to assess the merits of one certification or management system over another due to overlap in guidelines, voluntary versus mandatory auditing, implementation costs and stake holder benefits, etc.

Participants in the focus group had particular experience in end of life responsibility for manufacturing projects. 'End of life' is a sustainability dimension which is being increasingly built into projects where contractors have to take responsibility for end of life of plastics, packaging etc. and must reconstitute these materials or recycle them back into their processes. This can incur waste disposal costs as materials often are not sufficient quality to be used again.

The focus group felt that certifications such as ISO 14001 & ISO 9000 were not easily recognizable or valuable brands to the lay consumer. They felt that a need for a new green branding/logo would make the products instantly recognizable. They drew on the 'Guaranteed Irish' symbol as a good example of brand awareness.

The group discussed the problem of 'green washing' in the industry: the act of tagging on 'green' or

'sustainable' credentials to a product that may or may not have been designed or manufactured in a sustainable way. From this, it was concluded that without tangible and regulated benchmarks on which to gauge these 'sustainable' credentials, the consumer can end up making poor buying decisions with regard to 'sustainable' products.

The group came to the mutual agreement that they would like to meet again to discuss starting an 'Irish Wood Federation Alliance'. This would be 'Principles-based' alliance made up of core principles or guidelines to give an advantage to companies that are already implementing green techniques. It could begin as a network of Irish furniture designers and manufacturers, who all want to invest in sustainability and want to build a brand associated with value and strong sustainable principles.

The group explored the question of 'green branding' saturation in the market and if it was worth actually investing in sustainability just yet. They concluded that market for 'green' is evolving very slowly and that there is no metric or legal framework present to audit whether or not companies are producing products that really embody sustainable thought and action.

All the participants believed that developing and introducing a new certification process to incorporate a sustainable design process was a viable idea.

3.4 Review of currently available 'eco-friendly' certifications

A review of currently available independent eco-friendly certification and labelling was undertaken under the following criteria:

- The certification should have potential global applications.
- The certification whilst primarily applicable to furniture and wood products should be applicable to various different consumer products.
- The certification should be based on measurable sustainable credentials such as life cycle analysis (LCA) or carbon foot printing.

The conclusions of the review were that:

- There was a vast array of certifications to choose from and it was sometimes difficult to assess the merits of one certification or management system over another.
- Processes such as life cycle analysis (LCA), voluntary certifications such as ISO 14001 and ISO 9001 and eco-labels such as Cradle to Cradle were examined.
- Cradle to Cradle bases itself on certification of sustainable materials and manufacturing processes similar to the Building Research Establishment

which assesses the environmental effects associated with building materials over their life cycle—their extraction, processing, use and maintenance, however this is also quite similar to LCA.

- Leadership in Energy and Environmental Design (LEED) certification in America gives third-party verification that a building project meets the highest green performance and includes all stakeholders in the process providing training and continual review.
- Eco-Management and Audit Scheme (EMAS) is a voluntary environmental management instrument aimed at the management of an organization and is state audited rather than independently audited.
- It was found that there is a huge amount of choice and overlap between certifications, making it difficult to choose which methodology to implement and one which best fits an organisation.

4 DRAFT CRITERIA FOR SUSTAINABLE DESIGN GUIDELINES FOR FURNITURE AND WOOD PRODUCTS

Based on the results of the focus group discussion and the case study, seven principles or guidelines to support sustainable design were drafted:

1. End of life—recycling
2. 'Green' Credentials
3. Sustainable harvesting
4. Sustainable Material Choices
5. Sustainable Engineering
6. Education & Outreach
7. Marketing & Branding

These seven draft principles are further explored below:

4.1 End of life—recycling

4.1.1 Intent
To minimize damage to the environment by reducing waste through reuse, reconstitution and recycling.

4.1.2 Recommended performance criteria
Waste should be categorized, segregated, collected and treated accordingly, e.g., wood for wood chipping, metals and irons for smelting, paper, glass, cardboard, etc. Collection points for waste should be made available on the shop floor—these should be in proper, dry and well ventilated areas to allow for high quality recycling.

The quality of the waste affects the recyclability. Training should be provided for staff on how to manage and segregate waste and the importance and reason of doing so. Packaging should be designed specifically for the product—this will increase the chances of the packaging being recycled by the company.

4.1.3 Basis for inclusion of guideline
Manufacturers can no longer continue to encourage consumption, without beginning to take responsibility for the way in which their products will be dealt with at the end of their lives (Dowie, 1994). Implementation of waste minimization and re-use strategies (and continuous improvement in this field) can provide a competitive advantage for manufacturer's (Mohanty & Deshmukh, 1999). However, resulting value gained from a product depends on many factors such as: the types of materials recovered (e.g. non-ferrous metals have highest recycling value whereas plastics have a low value); amount of contamination on the material (this lowers the value greatly); and the current state of the recycling industry (i.e. current demand).

Participants in the focus group had particular experience in end of life responsibility for projects.

4.2 True 'green' credentials

4.2.1 Intent
To ensure true 'green' credentials of the product, to maintain customer confidence and avoid 'greenwashing', which can damage brand reputation.

4.2.2 Recommended performance criteria
From research, it was found that consumer confidence can be easily lost if consumers feel that the product is not truly 'green'. To avoid this members must: strive for integrity in all areas of procurement, acquisition, manufacture and disposal of product, constantly reviewing their processes with a 'check, plan, do, act' policy and maintain an element of transparency to all their stakeholders.

4.2.3 Basis for inclusion of guideline
The authors found that many companies are positioning themselves environmentally, with a view to obtaining competitive advantage or a unique selling point. They compete with one another to demonstrate their environmental credentials although this has led, in some instances to a term called 'green washing' where companies claim to have environmentally friendly products/services, but on closer inspection it fails to be really 'environmental' or 'sustainable' (Hickey, 2008).

Further to this, Wagner and Hansen (2005) state *'that consumers formed sceptical attitudes towards green advertising, indicating the danger of consumers avoiding purchase'* and proposed that the reason for this scepticism was *'that green claims were limited in scope by the overuse of such terms such descriptions as "environmentally friendly" and "natural"*. Therefore it is important that a product has true green credentials to attract and keep new customers.

The participants in the focus group highlighted the problem of 'green washing' and concluded that a robust certification and 'brand' of sustainability would be important going forward—as they could then compete on the brand and not solely on cost. They felt that being more sustainable was more important than being 'green', as sustainability should be built into all of their processes.

4.3 Sustainable harvesting

4.3.1 Intent
To avoid damage to the environment by using sustainably harvested materials that are extracted in sympathy with the environment.

4.3.2 Recommended performance criteria
Chain of custody certification—be able to provide evidence of an unbroken chain of custody from extraction to product.

4.3.3 Basis for inclusion of guideline
Environmental management has typically focused on managing internal environmental practices. Attention is increasingly shifting towards the management of an organization's impacts outside the boundaries of the firm, into the management of upstream and downstream activities (Holt & Ghobadian, 2009). Timber is a beautiful, natural material, but harvesting can have significant impacts on the environment (FIRA, 2010). The most important distinction is between timber harvested from old-growth forests and timber harvested from plantations. Regular harvesting from old-growth forests may destroy the habitats of small mammals and birds that rely on the larger, older trees for nesting hollows (FIRA, 2010). Forests are being cut at an astonishing rate, destroying our natural heritage and causing long-term ecological damage (FIRA, 2010). Rainforests are particularly vulnerable because of the rate at which they are being destroyed and the difficulties involved in regeneration. The United Nations Food and Agricultural Organisation (FAO) has estimated that an average of 17 million hectares of rainforest were cut down each year between 1981 and 1990 (FIRA, 2010). Designers should therefore specify the use of recycled products or timber that has a proven chain of custody back to well-managed forests (FIRA, 2010). There are two key elements to ensuring that a timber product derives

from well-managed sources. Firstly the forest of origin has to be independently certified to verify that it is being managed in accordance with the requirements of an accredited forest management standard.

Secondly, when the timber leaves the forest, it enters a 'chain of custody' system which provides independent certification of its unbroken path from the forest to the consumer, including all stages of manufacturing, transportation and distribution (TRADA, 2010).

There are four main timber certifications delivered by the following organisations:

- FSC (Forest Stewardship Council)
- PEFC (Programme for the Endorsement of Forest Certification)
- SFI (Sustainable Forestry Initiative)
- CSA (Canadian Standards Association)

4.4 Sustainable materials

4.4.1 Intent
To create sustainable products by specifying sustainable materials in product design and manufacture.

4.4.2 Recommended performance criteria
Use environmentally preferable materials where possible. Incorporate salvaged materials and specify reusable, recyclable or biodegradable materials where viable. Use local materials manufactured regionally where possible to avoid emission damage from transport. Increase use of sustainable and carbon sink materials such as timber. Using reclaimed timber can be both cheaper and more environmentally friendly than its virgin counterpart.

4.4.3 Basis for inclusion of guideline
From the case study, it was clear that the aesthetics of the product and cost outweighed the 'eco-friendly' credentials. However, the more the demand increases for products made from genuinely sustainable materials, the more widely available and cheaper they will become. Timber is especially useful in this context as a sustainable material, as it is bio-degradable and a carbon sink.

Further to this, manufactured wood products such as medium density fibreboard (MDF), particleboard and plywood have other environmental and health and safety problems that are primarily associated with the resins and glues used in production (FIRA, 2002). Inhalation of particle wood dust is a potential health risk during the manufacturing process (FIRA, 2002). Formaldehyde, which is used in traditional particleboard and fibreboard to bind the particles together, is a potential health risk through release of free formaldehyde ('off-gassing') in the workplace and after the product has been installed (FIRA, 2002). Painting or laminating the board can reduce emissions. Alternatives are being developed to replace MDF and plywood. An example is Gridcore™, a recycled fibre material manufactured in a honeycomb formation. The product can be manufactured from 100% recycled paper and cardboard without resins or adhesives. It has yet to be assessed within the context of furniture manufacturing for strengths and weaknesses (FIRA, 2002).

4.5 *Sustainable engineering*

4.5.1 *Intent*
To incorporate proven engineering techniques to lessen the carbon footprint of the product.

4.5.2 *Recommended performance criteria*
Design for deconstruction should be incorporated at engineering stage—this will allow the product to be broken down easily at end of life to allow for efficient recycling/salvaging where possible. Design for manufacture should be considered—minimizing parts and also minimizing processes means less use of resources. Use an integrated design process by involving appropriate stakeholders from each discipline and also end users. An integrated design process leads to improved communication and a systems approach to problem solving resulting in optimizing performance at the lowest cost.

4.5.3 *Basis for inclusion of guideline*
Historically market demand for cheap products has been combined with a lack of producer responsibility legislation and relatively cheap landfill disposal options in the UK and Ireland. This has deterred the sector from investing in new methods of sustainable engineering, design and production (TRADA, 2007). The application of traditional and modern engineering design tools to furniture and wood products is an area that was beyond the scope of this study. However, anecdotal evidence and the previous experience of the researchers suggests that it an area of weakness with significant potential for improvement. There is little evidence of the use of tools such as Design for Manufacture, Quality Function Deployment, and Value Analysis. The anecdotal evidence suggests that the industry is increasing its use of digital design tools, CAD/CAM, and 3D modeling. Many of these modeling tools now include tools for Life Cycle Assessment and Environmental Performance Simulation. The integration of these engineering tools in the design process also improves the manufacturing process efficiency and environmental impact. Any manufacturing process will generate environmental aspects. It is the way that these aspects are controlled or minimized that is important (FIRA, 2002).

4.6 *Education and outreach*

4.6.1 *Intent*
In order to be truly environmentally responsible, companies must go beyond monitoring their own actions and educate others. Educating others helps keep all stakeholders (employees, investors and customers) informed on the importance of sustainability, how the product/company meets its environmental obligations and why it is important to do so.

4.6.2 *Recommended performance criteria*
Constant employee training on how to design, manufacture and dispose of products responsibly. Studies have shown that when staff are trained properly that they will have more interest in the sustainable activities of the company. Outreach into local communities and schools will benefit the company's reputation but also disseminate the message of the importance of sustainability. Companies should discuss sound environmental practices with their suppliers, continually evaluating and documenting the company's practices to protect the environment. Also, they should conduct planning/ review workshops at key phases with all team members continually.

4.6.3 *Basis for inclusion of guideline*
Senior managers are often so busy in their routine fire-fighting operations and meeting daily production targets that they are left with no time to articulate strategies for improved productivity, clean manufacturing and sustainable development (Mohanty & Deshmukh, 1999). Similarly, most such reviews and activity are confined to the factory floor; yet there is often considerable scope for examination of the other areas of the business. It is difficult to expect employees to take a real interest in waste and its removal unless they are educated to understand the nature of waste and its causes (Mohanty & Deshmukh, 1999).

It was explored in Wagner and Hansen (2005) that consumers may not buy green products because they may perceive them to be of inferior quality and unable to deliver the environmental, therefore, education and outreach is important to build consumer knowledge and confidence.

4.7 Marketing and branding

4.7.1 Intent
To advertise the sustainable credentials of the products with integrity.

4.7.2 Recommended performance criteria
All sustainable certifications/awards/process information should be available for employees or customers to easily access. This creates an air of transparency and confidence in the sustainability of the product.

4.7.3 Basis for inclusion of guideline
The focus group, agreed that there was a need for a new green branding/logo would make the products instantly recognisable. The Sustainable Furnishings Council (an American organization) believes that sustainability is becoming more important to the buying public. As consumers become more educated, they seek out acceptable choices that meet their needs for style, value, and eco-responsibility (Sustainable Furnishing Council, 2010).

With a higher consumer awareness of environmental issues, many companies have adopted overtly "green" strategies often making environmental claims in their advertising campaigns with the aim of gaining an edge over their competitors (Wagner & Hansen, 2005). As described already in the 'green' credentials guideline, products need to have true 'green' credetials to attract and maintain this new market share. The furniture industry is a competitive domain and it is forcing manufacturers to be in touch with their customers' needs in order to preserve or to increase their market share (Tammela & Canen, 2008). Brands are an important and rich source of decision-making information to consumers and positive differentiation from competing brands can be achieved by constructive positioning, and can be exploited as a competitive advantage (Wagner & Hansen, 2005). Therefore, the development of schemes such as eco-labels may be the tools to prove environmental excellence within a market whilst also guaranteeing a degree of quality (FIRA, 2002). For this type of scheme to be successful public funding will be required to raise the profile of the label so that the correct marketing will benefit the individual companies who pursue this route (FIRA, 2002).

5 'GREEN DOT' SUSTAINABILITY BRANDING—A PROPOSAL

5.1 Overview of 'Green Dot' branding

The 'Green Dot' sustainable guidelines are presented graphically in figure 1.

Figure 1. Green Dot' brand logo.

This name carries strong branding imagery and would be easy to market as a label for certified sustainable products.

5.2 Potential internal benefits of implementing 'Green Dot' guidelines

- Demonstrate environmental responsibility
- Be an industry leader
- Enhance quality and investment in people and systems
- Quality of management improved
- Improved level of training
- Legal compliance is documented and can be demonstrated
- Stimulate improvements in manufacturing processes, transport, raw materials and packaging
- Cost savings from energy and waste reductions and efficiencies
- Improved employee morale
- Provides a platform for better communication and dialogue between staff and stakeholders.

5.3 Potential external benefits of 'Green Dot'

- Create a positive public image
- Gain new customers/business and satisfy existing customers
- Stay in business (in recessionary times)
- Environmental friendly products—unique selling point
- Constantly reviewed, so continual environmental performance improvement
- Legal compliance
- Increased energy and material efficiencies—resulting in possible cost savings

- Reduced pollution and environmental damage
- Develop better customer, employee and supplier relationships and communication
- Set an example for other companies in a sector

5.4 *Possible limitations of 'Green Dot'*

- Higher than usual staff training costs
- Capital expenditure may be required
- A certification/consultancy fee may have to apply to ensure company's compliance, if the certification is to be developed going forward
- Time and associated cost required in developing sustainable processes/training within the company
- Unfavourable economic climate—lack of market drivers
- Uncertainty about value/promotion of 'sustainable certification' in market place
- Lack of financial state support to implement such initiatives
- Expense of sustainable materials may be more that cheaper imported materials.

6 CONCLUSIONS

The following conclusions can be drawn from this study:

- It is difficult to assess the merits of current certification or management system over another due to overlap in guidelines, voluntary versus mandatory auditing, implementation costs and stake holder benefits, etc.
- A small but growing number of consumers are actively seeking environmentally friendly merchandise and it is hard ascertain if this is through specific product placement or personal choice, and therefore, it is also difficult to define the buying habits of an 'eco' consumer.
- Because there is no legally binding metric in place, there are many ways that manufacturers can launch a product labelled 'sustainable', 'green', 'natural', 'eco-friendly', etc., onto the market, which may in fact have low 'sustainable' integrity, thus weakening consumer confidence in 'green' products.
- Some consumers may be cautious of buying a perceived 'green' product because of a saturation of 'greening' and problems with 'green-washing' in the market place.
- Certifications such as ISO 14001 & ISO 9000 are not recognisable or valuable brands to the lay consumer. There is a need for a new green branding/logo would make the products instantly recognisable.

- There are significant difficulties reconciling sustainable goals and commercial realities.
- The output of a design process that included sustainable criteria, only achieved a mediocre score in typical measures such as life cycle analysis and carbon footprint.
- The aesthetics and cost of the product outweighed the 'eco-friendly' credentials in the design process.
- There is no comprehensive metric present to audit whether or not companies are producing products that embody true sustainability.

6.1 *Recommendations for further research*

From this study, there is an apparent gap in the market for providing an easily recognizable, independent, broad based sustainability certification for consumer products. The authors suggest that market trials, initially in the furniture and wood products area be undertaken. If the outcome of these trials indicate measurable and significant consumer support for independent sustainability branding and labelling of products then a full launch of a 'Green Dot' type certification may be viable.

REFERENCES

A.R.A. & Baksh, M.S.N., *'Case study method for new product development in engineer-to-order organizations'*, Vol. 52, No. 1, Emerald Group Publishing Limited, Johor, Malaysia, 2003.

Beard, C. & Hartmann, R., 'Naturally enterprising–eco-design, creative thinking and the greening of business products', European Business Review, Vol. 97, No. 5, pp. 237–243, 1997.

BSI, *'BS 16000:2009 Standard'*, 'Energy managment systems—Requirements with guidance for Use British Standards Institute', pdf, 2009.

Dowie, T., 'Green Design', World Class Design to Manufacture, Vol. 1, No. 4, pp. 32–38, MCB University Press, UK, 1994.

FIRA–Furniture industry research association, 'Sustainable design in the UK office furniture sector-a scoping study', UK, 2002.

Hillary, R., *'SME's and Experiences with Environmental Management Systems'*, Journal of Cleaner Production, Vol. 12, No. 6, August 2004, Pages 561–569. Retrieved March 27th from Science Direct data base.

Holliday, C., Schmidheiny, S. & Watts, P, *'Walking the Talk: The Business Case for Sustainable Development'*, Berrett-Koehler Publishers, San Francisco, CA., 2002.

Holt, D. & Ghobadian, A., *'An empirical study of green supply chain management practices amongst UK manufacturers'*, Vol. 20, No. 7, pp. 933–956, Emerald Group Publishing Limited, UK, 2009.

I.S. 393:2005, *'Energy Management Systems Technical Guideline'*, Sustainable Energy Ireland, 2006.

ISO 14001:1996, 'Environmental Management Systems—Specifications of Guidance for Use', European Committe for Standarisation, 1996.

Mohanty, R.P. & Deshmukh, S.G., *'Managing green productivity: a case study'*, Vol. 48, No. 5, pp. 165–169, Emerald Group Publishing Limited, 1999.

Morgan, D.L., *'Focus Groups as Qualitative Research'*, Second Edition, Sage Publications, Inc., California, USA, 1997.

Phau, I., & Ong, D., *'An investigation of the effects of environmental claims in promotional messages for clothing brands'*, Marketing Intelligence & Planning, Vol. 25, No. 7, pp. 772–788, Emerald Group Publishing Limited, Australia, 2007 Rahim.

Strong, C., *'The problems of translating fair trade principles into consumer purchase behaviour'*, Marketing Intelligence & Planning, pp. 32–37, Wales, UK, 1997.

Tammela, I. & Canen, A.G., *'Time-based competition and Multiculturalism: A comparative approach to the Brazilian, Danish and Finnish furniture industries'*, Vol. 46, No. 3, pp. 349–364, Finland, 2008.

Tuttle, T. & Heap, J., *'Green productivity: moving the agenda'*, International Journal of Productivity and Performance Management, Vol. 57, No. 1, pp. 93–106, Emerald Group Publishing Limited, UK, 2008.

TRADA, *'Sustainable timber sourcing: Certified timber products, Section 2/3 Sheet 58, Subject: Timber'*, June 2007, TRADA Technology, UK.

Victor, D.G. & Cullenward, D., *'Making Carbon Markets Work'*, Scientific American, December 2007.

Wagner, E.R. & Hansen, E.N., *'Innovation in large versus small companies: insights from the US wood products industry'*, Vol. 43, No. 6, pp. 837–850, Emerald Group Publishing Limited, USA, 2005.

Yin, R.K., *'Case Study Research: Design and Methods'*, 2nd edition, Sage publications, London, 1994.

Green Design, Materials and Manufacturing Processes – Bártolo et al. (eds)
© 2013 Taylor & Francis Group, London, ISBN 978-1-138-00046-9

Door's design for people with mobility impairments and service dogs

C.P. Carvalho & A.B. Magalhães
Faculty of Engineering of Oporto University, Oporto, Portugal

J.B. Pedro
National Laboratory of Civil Engineering, Lisbon, Portugal

L. de Sousa
Institute of Biomedical Sciences Abel Salazar of Oporto University, Oporto, Portugal

ABSTRACT: A door can be an obstacle when it does not meet the needs of its users. The purpose of the work was to determine the dimensional and functional requirements of the use of internal doors in public buildings by people with mobility impairments and with service dogs. This paper addresses three subjects: (i) human and canine functioning, (ii) accessibility standards, and (iii) comparison between personal and environmental components. As a result proposals for the improvement of the Portuguese accessibility standard are presented.

1 INTRODUCTION

Accessibility is characterized by a set of environmental conditions allowing its users to "reach, understand or approach something or somebody" (WHO & WB 2011). The accessibility standards set out good practices and, when compulsory, regulate the minimum requirements to be met.

Inclusive design is "The design of mainstream products and/or services that are accessible to, and usable by, as many people as reasonably possible (...) without the need for special adaptation or specialised design" (BSI 2005). The partnership of designers with extreme users (users with a severe disability or multiple disabilities) is one of many different approaches to inclusive design. By meeting the needs of extreme users, inclusive design can create mainstream design solutions that are accessible, usable and enjoyable by a broad range of users.

2 CONCEPTUALIZATION

2.1 Doors in public spaces

In buildings doors are the boundary element between two spaces enabling or conditioning the passage. Nowadays, many exterior doors in public buildings are automated. The investment to install and maintain these systems, decrease their viability in internal doors, mainly due to their greater quantity. Despite the advantages for accessibility, automated systems also have detection, timing and power cuts problems.

2.2 Portuguese accessibility standard

The Portuguese accessibility standard is mandatory and enforced by the Decree-law 163/2006. The standard sets minimum criteria to be met in public spaces, community facilities, public buildings and residential buildings. Requirements for doors accessibility are set in the standard. The standard reportedly focuses mainly on people with reduced mobility, namely wheelchair users.

2.3 Target subjects

It was estimated that there are 171,255 individuals living in Portugal with motor impairments (INE 2001), representing almost 1,7% of the total population.

The service dog (SD), like the guide dog, belongs to the assistance dog category and their function is to assist people with motor impairments in overcoming challenges in autonomous activities of life, e.g. picking up objects, switching lights and operating doors. In Portugal, the use of service dogs is regulated by the Decree-law 74/2007, which grants people the right to be accompanied by their service dog to locations, transports and facilities of public access. Presently there are four service dogs in action in Portugal and they are exclusively Labrador's Retriever breed.

2.4 Problem statement and goal

People with mobility impairments (PMI) are more likely to have more difficulties using the doors. Their structural and functional anthropometric differences give rise to gaps in the interaction with doors.

Although many PMI depend on their canine partner for door operation, dogs are not usually considered in doors design. A rope attached to the door handle is an improvised solution often adopted for doors operation by SD, but is not viable in public spaces.

The Portuguese accessibility standard focuses on wheelchair users. This approach does not necessarily address the needs of the remaining PMI or the SD.

The objectives of the work were to i) determine the dimensional and functional requirements for the use of internal doors in public buildings by PMI and SD, and ii) make proposals to improve the requirements regarding doors set by the Portuguese accessibility standard.

3 RESEARCH DESIGN

3.1 General approach

The implementation of accessibility principles may be compromised by the lack of theoretical fundaments in at least one of the three levels of knowledge (Iwarsson & Stahl 2002): standards (environmental component), human functioning (personal component), and compatibility between personal and environmental components. This study addresses the three components.

3.2 Environmental component

A comparative analysis of accessibility standards from Portugal (DL 163/2006), Australia (AS 1428.2-1992 and AS 1428.1-2009), United Kingdom (BS8300: 2009) and United States of America (ICC/ANSI A117.1-1998) was carried out. Portuguese and Australian standards are mandatory. The remaining two standards are recommendatory. Accessibility standards set the rules that shape the environment where PMI and SD have to operate.

3.3 Users component

For human functioning, anthropometric studies involving direct consultation of PMI (a total of 690 individuals) were analyzed to obtain live data.

A questionnaire was applied to determine the constraints and preferences in doors use by PMI in the Portuguese context. A sample of persons with heterogeneous mobility impairments was selected. The questionnaire collected data on how the performance was influenced by the type of door (hinged and sliding), handles and mechanical closing devices. A set of 28 questions using a Likert scale (never, rarely, sometimes, usually, always) was used and the collected data were treated statistically, using descriptive statistics.

The research about Labrador as a service dog using doors was limited mainly due to the sparse literature on the subject. Three sources of data were collected: the breed standard, an interview with a service dog educator and a study on SD opening hinged doors using the rope's method (Coppinger et al. 1998).

3.4 Compatibility between personal and environmental components

Proposals to improve the Portuguese accessibility standard were based on the comparative analysis of accessibility standards and the constraints and preferences in doors use by PMI and SD.

4 RESULTS

4.1 Comparative analysis on accessibility standards

Portuguese minimum clear width is the smallest of the four standards (from 770 mm till 850 mm). In the PT, AU and US standards the clear width of a hinged door is measured between the jamb and the door leaf open at 90 degrees. The UK measurement method allows different amplitudes in door opening and discounts the projection of door hardware (Fig. 1).

The maximum force for door opening is similar between the four standards, varying from 20 N to 22.5 N. The US and UK standards set that the maximum strength for door hardware operation should be less than those agreed for door use. The US standard does not specify any values. The UK standard sets "the torque force required to operate keys and cylinder turns should not exceed 0.5 Nm". PT and AU standards establish a unique maximum force for all door operations.

All four standards indicate that door hardware shall allow the door to be unlocked and opened with one hand and without prehension need. The

Figure 1. Measuring methods for clear width in PT, AU, US (1) and UK (2, 3, 4) accessibility standards.

Table 1. Parameters for locating doors hardware.

	(1)	(2)	(3)	(4)
AU	900–1100	900–1200	60	50
UK	800–1050	700–1300	54	72
US	865–1220	(*)	(*)	(*)
PT	800–1200	(*)	50	(*)

(*) Unspecified.
(1) Lower-upper limits for door hardware (mm).
(2) Lower-upper limits—pull handles (mm).
(3) Minimum distance from door free edge (mm).
(4) Minimum distance between devices (mm).

Figure 2. Minimum height (540 mm) of Labrador Retriever standard and maximum estimated height range (840 mm).

dimensional constrains for doors hardware location are presented in Table 1.

4.2 Anthropometric studies on population with motor impairments

The two studies analyzed included various anthropometric parameters from PMI and tested hinged doors use (Steinfeld et al. 1979, 2010). The following three main recommendations were collected from these studies.

Wider doors are better for accessibility and there is no need to set a maximum width of doors in regulations, although doors wider than 1040 mm could pose some problems (Steinfeld et al. 2010).

Grip precision is inversely proportional to device contact area. Maximum force for small devices is not expected to exceed 9 N. For operation of larger devices the desirable upper limit is 22 N. Knowing that many users have very limited or no grasping ability, whenever possible, design should promote solutions for device operation without the need for prehension (Steinfeld et al. 2010).

The upper limit of 1220 mm in lateral reach of the American standard is adapted to the majority of wheeled mobility devices (WMD) users and it could rise even higher, "but this may result in limitations for people of small stature". In turn, the 380 mm lower limit of the same standard needs to be updated to 700 mm, since lower ranges have proven unsafe for WMD users (Steinfeld et al. 2010).

4.3 Labrador's anatomy and functioning

Taking into account the ideal posture for Labrador in door operation (all four paws on the floor), the dog is hardly capable of grasping the handle at 800 mm high with its teeth in a forced extension of the neck (Fig. 2).

The snout use is not just a canine natural preference for doors operation but also a way to avoid doors deterioration caused by the use of paws (Castro Lemos 2012).

Strength demands for Labrador opening doors with a rope increases with the height differences between dog's mouth and door handle. Even without maximum force quantification, it can be said that Labrador can exert forces higher than 22 N (Coppinger et al. 1998).

4.4 Constraints and preferences in doors use by PMI in the Portuguese context

The most relevant results from the 41 respondents to the questionnaire are presented below.

The problems in reaching door handle are considerable, since half of the sample experiences "sometimes" or "usually" difficulties in reaching the handle. Moreover almost half of the sample states the use of handles for body support, at least in some situations (Table 2).

Preferred height for doors hardware is not consensus among respondents: 44% prefer a lower handle, 25% a higher handle and 24% a "well located". These differences are most likely related to reach capabilities, often constrained by mobility limitations or changes in body structures functioning such as the ones in upper and lower limbs, trunk or stature, or even the limitations imposed by their own assistive devices for mobility.

4.5 Proposals to improve the Portuguese standard

The location of door hardware between 800 mm and 1100 mm in the Portuguese standard constitutes a reach problem for Labrador since only in forced extension of the neck the dog is only able to reach the 800 mm height with its snout. Many human users also stated their preferences on lowering door hardware to resolve their own reaching problems. Meantime, other users have needs for higher devices. Therefore door hardware should present as many different height options as possible for more users benefit, i.e. multi-point solutions, and the heights range should be reduced from 800 mm to at least 700 mm in their lower

Table 2. Questionnaire results.

	MEAN*	STDDEV
I cannot reach door handles	2.46	1.142
I use the door handle to support me	2.41	1.161

*1 (Never), 2 (Rarely), 3 (Sometimes), 4 (Usually), 5 (Always).

Figure 3. Distances in door hardware (example).

limit. Ideally this improvement may extent to a full height solution.

The use of the UK measurement system for clear width, would contribute to conflict regulation between the projection of door hardware and effective width for passage. This would not only beneficiate users but would also benefit the designing of spaces, expanding project options with variations in door opening angle.

Operating doors using only the snout contact is a condition for service dog that meets the prehension need for human users when handling door hardware. Therefore, design development of door hardware solutions that exonerate prehension needs will benefit human and canine use. Following this specification, clearance considerations are required around door hardware to facilitate its use with a closed hand, a dog's snout or similar. In this matter, at least two considerations for Portuguese standard improvement can be made: (i) increasing the minimum distance from door free edge and (ii) establishing the same value for the minimum distance between devices. Preferably, this value should be 72 mm as set by the UK standard (Fig. 3).

5 CONCLUSIONS

It was concluded that SD, as an extreme user, have requirements for door design which can also benefit human users.

The suggestions to improve the minimum requirements for doors design set by the Portuguese accessibility standard would enable more inclusive practices.

The main suggestions are the following: reduce the lower limit of door hardware, change the measurement system of door clear width, set minimum clearance criteria around door devices and promote door hardware solutions with no need for grasping.

This study has some limitations. The results are adapted to the Portuguese context and based on a small sample of the population with mobility impairments.

To increase reliability of the results, the study should proceed. Further research should be based on observation, experimentation and participation of a more representative sample of the target population.

ACKNOWLEDGEMENTS

This paper presents results of a master dissertation presented at FEUP in 2012 (Carvalho 2012). Thanks are extended to all collaborations, specially ÂNIMAS, Centre for Professional Rehabilitation in Gaia (CRPG), Portuguese Disabled Association in Oporto (APD) and Rehabilitation Centre in Areosa.

REFERENCES

ANSI, American National Standards Institute. 1998. *ICC/ANSI A117.1-1998: Accessible and usable buildings and facilities.* NY, USA: International Code Council.

BSI, British Standards Institution. 2005. *British Standard 7000-6:2005. Design management systems—Managing inclusive design—Guide.*

BSI, British Standards Institution. 2009. *BS 8300: 2009: Design of buildings and their approaches to meet the needs of disable people—code of practice.* London, UK: BSI.

Carvalho, C.P. 2012. *Acessibilidade e design universal de portas—requisitos dimensionais e funcionais de utilizadores com incapacidades motoras e de cães de serviço.* Master Dissertation. Engeneiring Faculty, Oporto University.

Castro Lemos, S. 2012. *Características do Cão de Serviço e o seu modus operandi.* Ânimas: Quinta do Côvo, 28-01-2012.

Coppinger, R., Coppinger, L. & E. Skillings. 1998. "Observations on assistance dog training and use." *Journal of Applied Animal Welfare Science* no. 1(2): 133–44.

CSA, Council of Standards Australia. 1992. *AS 1428.2-1992: Design for access and mobility.* Australia: Council of Standards Australia.

CSA, Council of Standards Australia. 2009. *AS 1428.1-2009: Design for access and mobility.* Australia: Council of Standards Australia.

Decree-Law 163/2006. *Portuguese Official Journal* no. 152 (8 August 2006): 5670–5689.

Decree-Law 74/2007. *Portuguese Official Journal* no. 61 (27 March 2007): 1764–1767.

INE, Instituto Nacional de Estatística. 2002. *Relatório Censos 2001*. INE.

Iwarsson, S. & Stahl, A. 2002. "Accessibility, usability and universal design—positioning and definition of concepts describing person-environment relationships." *Disability and Rehabilitation* no. 25 (2): 57–66.

Steinfeld, Edward, Steven Schroeder & Marilyn Bishop. 1979. *Accessible buildings for people with walking and reaching limitations*. Washington, DC: Superintendent of Documents, U.S. Government Printing Office.

Steinfeld, Edward, Victor Paquet, Clive D'Souza, Caroline Joseph & Jordana Maisel. 2010. *Anthropometry of Wheeled Mobility Project—Final Report*. Buffalo, New York: Center for Inclusive Design and Environmental Accesss (IDeA Center).

WHO, World Health Organization & WB, World Bank. (2011). *World Report on Disability*. Geneva: World Health Organization.

Green Design, Materials and Manufacturing Processes – Bártolo et al. (eds)
© *2013 Taylor & Francis Group, London, ISBN 978-1-138-00046-9*

Self-building process: Living the crisis in a sustainable way

F. Pugnaloni, E. Bellu & M. Giovanelli
*Department of Civil, Building, Engineering and Architecture, Università Politecnica delle Marche,
Ancona (AN), Marche, Italy*

ABSTRACT: The aim of this research is to analyse different conscious housing models as a sustainable tool to overcome the crisis. By applying the "Arnstein's Ladder of Citizen Participation" on the different phases of the building process, it is possible to assess how these projects can involve nonprofessional manpower in the field of the low-cost building. Among these, the most significant ones are represented by the self-build experiences, in response to the quality and low-cost building growing demand, based on a participatory and inclusive design. These experiences initially depend on local resources, requiring a change in the productive building system oriented to the community design, and providing building solutions and materials easy to use, as well as technique suitable for all people.

In the end, obviously, the formative experience helps develop skills, creates identity and territorial belonging, which in turn, can consciously increase the awareness about the valorisation and the safeguard of local resources.

1 INTRODUCTION

1.1 *The current scenery: The world economic crisis*

Nowadays, the reduction of public investments in key sectors such as service industry, facilities and welfare due to the increasing global financial crisis, goes together with arising social conflict cases and the environmental emergency.

Furthermore, the socio-demographic changes, occurred during the last decades in many countries, have led to the emergence of a new middle class. The latter has been deeply affected in more sectors by the life quality worsening and by housing problems, even in the private home ownership field, as in Italy. Therefore, actions aimed to promote low cost dwellings access seem to be of particular interest.

This paper is mainly focused on self-building projects, which are seen as a solution for the quality and low-cost building growing demand, based on a participatory and inclusive design. These formative experiences helps develop skills, creates identity and territorial belonging, which in turn, can consciously increase the awareness about the valorisation and the safeguard of local resources, often not considered as common good. This shows how the world economic crisis could represent a turning point and could be conveniently turned into a resource (as the etymology of this word suggests).

These projects aim to the necessity of establishing satisfactory economies and protecting resources to preserve their use for future generations, fully applying the "sustainability" term, as described by the Brundtland Report.

Strongly linked to the resources available in the territory (materials and social wealth) the self-building process turns out to be not only a sustainable developing model but also and mainly "appropriate".

1.2 *Appropriate technologies as a response*

Therefore a model which is standardized, replicable and adaptable to any location cannot exist. Everywhere, considerable importance is given to the values identification, to the local systems characteristics and to the support in creating alliances between public and private actors, in order to identify shared and sustainable development strategies. This approach can turn potentialities into resources. For these reasons, awareness and technologies, which are seen as the main two critical factors, have to be taken into account when reshaping this model. So, in order to identify a model which is effectively appropriated to the context, increasing the popular knowledge and adopting open and transparent methodologies, becomes of essential importance.

2 TOOLS AND STRATEGIES FOR AN APPROPRIATE DEVELOPMENT MODEL

2.1 *Local resources as tools: The "unconscious sustainability" of the genius loci*

In the attempting to draw up strategies and instruments aimed to reach higher and more equally life quality levels, the promotion and the protection of local resources cannot be ignored, since a preliminary design stage through planning and urban regeneration is required.

Considering the local resources as central topics implies to start studying the specific climatic conditions of the place in question, in order to exploit renewable energy resources. In addition, it is essential to choose materials and technologies pertaining to the context, as well as to give importance to the social substrate. This kind of project approach is based on the "culture of respect", typical of the past. Nowadays this design method which is faithful "to the use, the capabilities, the resources, the pre-existences, but also to the place as recognizable domain to the human actions, to its history, to the ability to control transformations" (Trombetta 2002) can be defined as "unconsciously sustainable", motivated by the common sense and by a common good conscience.

Since the last century, the extreme industrialization and the lack of attention to the environmental ecosystem have brought to another design approach, based on a research focused on the individual well-being at the expense of the community. Further, these factors have lead to a uncontrolled resources consumption and, in particular, to a building heritage creation, based on energy dissipation models. These models are characterized by an artificial control system, through the diffusion of "active" technologies on the environment.

Therefore, when referring to the bioclimatic architecture, we have to restart from our ancestors knowledge. As a matter of fact, they used to take advantage from the typical features of the site, getting the nowadays so-called "design guidelines".

Considering this loss of awareness, the need to come back to a sustainable building method implies approaching again, not only places, but also individuals and knowledge. In this way, the concept of *genius loci* is revived as architectural design basis, as well as phenomenological approach for the study of the environment, as well as identity and place interaction. So the tradition should not be seen as an obstacle, but it needs to be reinterpreted. In fact it can represent the determining factor able to link together, and in a proper way, climate, environment, territory and users.

2.2 *A new/old living as strategy: Inclusive housing models*

Hence, when studying the potentialities of each territory, it is essential to take into consideration, as resources, along with the cultural heritage, skilled workers, vernacular crafts, and the self-construction know-how recovery, for us and for the community. (Ivan Illich 1974).

Taking active part in building the own habitat and community, belongs to the social human history. Building directly, in whole or in part, the house where a person will spend their life, nowadays is still a common procedure. This is mainly a developing countries custom, but it is also common in some northern European nations and in many states of North America.

Hive living, which used to be a specific peculiarity of the rural historical settlements in the Apennines, nowadays is rediscovered in Italy. This co-housing forms have been practiced as long as the land and the agriculture have been essential and qualifying part of the society life and economy.

Recently, few spontaneous movements, inspired to the past, sprang up with the intent of reducing the ecological footprint.

3 THE RESEARCH WORK

3.1 *Citizen participation analysis*

The research work started from the analysis of the "inclusive housing models" above mentioned, seen as local participatory forms. The "responsive housing" such as cohousing, self-building, eco-villages, etc, currently represents an efficient help for the social housing politics. In these experience, housing is considered a practical action in which the decision-making power must be handle by the users themselves (Turner & Fichter 1972). These experiences create social inclusion phenomena which give life to community, in a heterogeneous system as that of the current cities, thanks to 'social contact design' principles (Williams 2005). In addition, they bring to a reduction of the economical and environmental impact, and can be defined as valuable tools to achieve high quality and environmentally friendly standards.

These experiences, born spontaneously (bottom up) or carried out by the government (top-down), prove that, amongst citizens, there is a will to interact with the institutional bodies and the community itself. The intent is creating new landscapes, rethinking and improving the relation between the natural environment, through the stakeholders collaboration. The purpose of this actions is to develop an environment more responsive and suitable for its inhabitants and users, through a user democratization and empowerment culture (intended as acknowledgement and strengthening of the user role).

3.2 *Arnstein's Ladder of Citizen Participation. Application to the case studies*

Despite the fact that the European directives aimed to define inclusive planning are increasing, in Italy as elsewhere, they are not yet a structural component of the decision-making and formative process, but are often related to an experimental field or bottom-up. Several studies have shown that some top-down projects, considered participative, actually are unable to reach the fourth step of the Arnstein's Ladder of Citizen Participation. Obviously, in order to be considered as an effective tool to rule the territory, the participation has to be adapted to the scale of action. Moreover, it is necessary to understand in which process phase it has to be included.

The case studies aim of this research are 30 best practices, bottom-up type or top-down type. They have been divided as follow: five Italians (I) and five International (W), for each housing model (cohousing, self-building, eco-villages), over a period of 25 years. The design process of these experiences has been generally divided into 5 phases: start-up (SU), analysis (A), planning (P), construction (C), use/maintenance (UM). Each case study presents one or more of these phases, open to a participatory management. Therefore, through extracted information, it is possible to deduce which level of participation has been achieved in each phase, superimposing the Arnstein's Ladder of Citizen Participation. This observation demonstrates that the involvement of stakeholders reaches a higher level as the process goes in detail. Table 1 illustrates the results of this analysis. To simplify, the table shows the minimum, the maximum and the modal value of the levels reached in each process phase, from all 5 cases in each model.

Definitely, at the initial stage of this analysis, it is likely not to go beyond the fourth step (4. Consultation). In this case, the community knowledge and citizens awareness have to rise to the role of baseline, which today is recognized only to the technical and scientific contributions. In this first phase, Questionnaire surveys, Interviews with stakeholders and Brainstorming (Fisher et al., 1991), the Focus Group (Krueger & Kasey 2000) the Open Space Technology (Owen 1997) or effective Walking guided visits of the district (Sclavi et al., 2002), at the beginning can be suitable tools.

This process allows residents to express their opinions on a range of issues, to work together

Table 1. The Citizen Participation in the design process of "responsive housing" model.

Minimum (Modal Value) Maximum						
Process phases		SU	A	P	C	UM
Co-housing	W	2(4)8	2(6)7	5(6)6	3(6)6	6(8)8
	I	2(3)8	2(4)6	3(6)6	3(3)7	7(8)8
Self-building	W	2(2)8	5(5)5	6(6)7	8(8)8	8(8)8
	I	2(3)8	3(6)6	5(6)7	6(8)8	7(8)8
Eco-villages	W	5(8)8	5(6)6	6(6)8	6(8)8	8(8)8
	I	5(8)8	5(6)6	6(6)6	6(7)7	8(8)8

with the purpose of identifying priorities, as well as to develop an action plan for a change, in collaboration with local government agencies. It is essential to include information activities during the initial stage of fact-finding, especially when the decision taken can cause significant and adverse effects on a specific community. In most cases this is useful to prevent the so-called Nimby syndrome ("Not In My Back Yard"). In these first phases, it is important to respect local perceptions, and involve local people in setting goals and strategies. As far as the design level is concerned, the participation is opened on the basis of possible scenarios and directly on the reworked project by using specific tools (Visioning, Planning for Real (Gibson 1991), Urban Workshops), useful to make the stakeholders interact and to encourage them to change the design or the 3D scale model. All this with the purpose of building together a shared vision, starting from different requests.

Finally, it is obvious that a participatory process achieves, by its nature, the highest steps of the Arnstein's Ladder during the practical execution and direct management phases (7. Delegated power, 8. Citizen Control). This happens thanks to the fact that, in the design process, whole phases can be left to the self-management or be delegated to the users.

Among these experiences, the most important ones are those related to the self-building in which, along with the participatory planning, even the physical transformation of the own habitat, through a manual labor, is jointly managed.

3.3 *Citizen control applied to the self-building process*

It starts from the spatial features analysis, related to climatic variables, rapid space changes, presence of poor and readily available materials as usable and fair approach. This has already been at the base of some research tools, developed by the depart-

29

ment some years ago, in collaboration with some universities in the Chinese and Indo-Chinese areas. These projects are based on the heritage promotion and protection, made possible thanks to some training activities and through the involvement of local people and students. This action promotes a social architectural movement based on a Learning by doing process, through the promotion of the territory carried out by the users.

On the contrary, in Italy, the self-building process is still seen with skepticism. For many people this is considered as a spontaneous and individual experience (DIY means "do-it-yourself") without a technical and regulatory coordination. Actually, we refer to some case studies related to the field of associated group of self-builders (or *common-users*), assisted by competent technicians. As a matter of fact, few governments have recently understood the social, economic and ecological benefits of self-building and have consequently simplified the current regulation. In Italy, this still seems far to be reached, while it would be required in order to have a better housing stock on the market, also through the self-construction. Sharing physical and manual work creates exclusive bonds within the community and a sense of territorial belonging which are peculiar. The cooperative learning is essential in order to work for the common good (Comoglio 1996). The long-term sustainability of the community depends on the human and social capital development. Involving local people in controlling their own situation and managing local resources, can help them understand how to plan and influence a process work. Through the communication and the user training before and during the construction phases, these experiences are able to engender building capabilities starting from the manual labor (Learning by doing). Empowerment generates sustainable local skills which represent new territorial resources to face the crisis.

These experiences often originate to promote social inclusion amongst people. As of fact, self-builder communities include people with different ages, gender, backgrounds, racial heritage, religious and political affiliation who, planning and working together, play a fundamental role in the community building. Moreover, the self-building process helps lower the building total costs (up to 40%) providing manual work coming from common-users spare time. Furthermore, management costs are saved through self-managing and avoiding the assistance of a middle-man. In economic crisis times, self-built housing may also represent an affordable option for those who cannot afford to take a mortgage. Furthermore, it helps mitigate housing crises, where those occur. Definitively, the self-built housing development may be the key factor to move from individual and stand-alone sustainable homes to sustainable neighborhoods, and why not, to the sustainable cities of tomorrow. The following factors are seen as crucial to the success of a self-build process. The most essential factor is selecting participants on the basis of the social capital they are able to bring and assessing the degree of motivation they have. In order to achieve good results, however, the presence of a professional team, with the role of coordinating the group in all the process phases, is essential. The team must be able to manage social relations, to check budgets and to supervise the building site. It has been proved that, supervising these experiences is essential in Italy, and very often it was the line between failure and success (Cantini 2008).

For a successful close identification of professionals who are able to guarantee the quality of the process, appears to be essential to ensure a close proximity to the institutions and persons involved in it. This is something which will allow people to keep effectively under control all the possible variables of the process. In addition, the process must be "simplified" and designed to be participated during the executive phase, from the beginning. Technological and typological choices have to be suitable for the self-building process and therefore have to be "appropriate" for the inclusive design. For example, self-build and community projects might not be easily applied when it comes to the realization of blocks of high rise apartments.

3.4 *Inclusive design*

The citizen control, which is the highest rung of Arnstein's Ladder, takes place if and only if the technology and the design allow it. This is possible if the project is conceived at any stage for a common-user and if the production system provides suitable components to support it.

So, at this point, the role of the designer, who leads the self-builder community, becomes of primary importance. He, thanks to his knowledge, has to be able to simulate every aspect of the construction process directly in the design, to facilitate the users. This can only be achieved through a recovery of the know-how culture, a manual-skill oriented training, i.e. IKEA culture (Cusatelli 2006).

For a common-user, simplifying the constructive system means to base it on the use of facilitating technologies. Until now, the productive process has been able to study facilitated technologies only suitable for qualified users, while it has not yet realized the emergence of a common-user. It is interesting to note that "self-build associated and assisted" housing initiatives realized in Italy until now, are constituted by traditional construction technologies and materials, using just a streamlined assembly. The lack of architectural product suppliers for self-building makes the greatest difference between Italy and countries where self-building is regulated.

To assess the technology suitability for the self-building, it is necessary to take into account some factors of technical, economic, social, environmental and institutional order.

From a technical point of view, it is easy understandable that, facilitating the common-user means providing lightweight, flexible, repeatable components, easy to maintain, to transport and to process. For instance, wood technology is one of the most suitable technologies for the self-building method, thanks to the use of dry manufacturing as screwing and gluing. Moreover, because it can be handled and adaptable to every kind of soil, thanks to the lightness of the foundations (the Segal self-build method is a good example of this kind of technology). The two most popular methods of construction for a self build home are traditional brick and block construction and timber framed houses.

The traditional method takes longer, while timber frame self-build homes are often bought as a kit—with a precut frame, they can be erected in a much shorter time. Self-builders enthusiasm is influenced by the time table, so it is essential choosing building techniques which are as much fast as possible to use (Cantini 2008). But it would be wrong to dismiss tradition playing no part in the choice of using brick and block build. Bricks and blocks are also a more flexible tools—if a measurement is slightly not perfect, the timber frame has to go back to the manufacturer to be fixed, while an extra line of bricks can easily solve the problem. Innovation in brick and block building is moving forward. Aircrete blocks in the inner leaf of external walls increase the construction speed. Being half-weight compared to other ones, these light aggregate blocks are easy to handle and can be easily cut by hand with common tools.

In addition, for the economy of the process, these components, on the whole, have to be thought starting from the local materials and the human resources involved in the project. A self-build community enables further savings thanks to bulk purchase and to join forces for researching new affordable sustainable building materials. Cost management is also important to preserve the community involvement, and it is often a successful key factor (Cantini 2008).

As far as the environmental factor is concerned, it is important to underline that an eco-friendly technology is often easy to use. This is possible because it absorbs part of the construction complexity (Cantini 2006). Furthermore if the use of "passive solar design" is made possible by the site, the home will be not only eco-friendly, but also simpler to maintain.

With regard to the social factor, the components have to be suitable for all kind of users. This will allow people of different ages, backgrounds, racial heritage or gender, to use them.

Another essential factor is the normative one: to this day, the lack of facilitating regulations, open to new experimentations, together with a scarce political and institutional support, put a strain on the technology development.

Local governments (primarily Regions) can play an important role in creating the conditions of certainty and security for those who want to benefit from this housing access opportunity. Government authorities can put into effect regulatory and procedural facilitations, provision of resources and support measures (not only and not necessarily cheap, but information and training, technical support, guidance and participation).

The DIY construction sites do not deny the production system, but have to be seen as a laboratory in which it is possible to experiment and direct the above mentioned system, with the purpose of facilitating and accelerating the construction process.

To sum up, it is clear how the technologies evolution matter stands between standardization and execution speed and simplicity (which tends to the creation of mass-produced kit homes), and a question less devoted to the efficiency but more related to sustainability, material local availability and to site-specific solutions.

4 CONCLUSIONS

In conclusion, the research is focused on these forms of "responsive housing", considered as a sustainable solution for the low-cost and quality building growing demand. In situations where the participation and the inclusive design can then be applied, are recorded more contributions aimed to the creation of a sustainable local development.

Effectively, starting from local resources, they generate additional local resources and, above all, a *place identity,* which is extremely important when the involved users belong to different nationalities.

In order to achieve this, it is necessary that the manufacturing system interfaces with the DIY construction sites. These are seen as technological innovation and experimentation places, useful to define building products which are increasingly suitable for the common-users participation.

REFERENCES

Arnstein, S.R., 1969. A ladder of citizen participation. *Journal of the American Institute of Planners* 35(4): 216–224.

Bertoni, M. & Cantini, A. 2008. *Autocostruzione associata e assistita in Italia. Progettazione e progetto edilizio di un modello di housing sociale*. Roma: Dedalo librerie.

Comoglio, M. & Cardoso, M.A. 1996. *Insegnare e apprendere in gruppo. Il Cooperative Learning*. Roma: LAS.

Cusatelli, G. 2006. La formazione del progettista e dell'esecutore per l'autocostruzione. In Rogora A. (ed.) *La sostenibilità dell'autocostruzione nell'ERP: processi, politiche e riflessioni*. Milano: Clup Edizioni.

Fisher, R., Ury, W. & Patton, B. 1991. *Getting to yes. Negotiating agreement without giving in*, New York: Penguin book.

Gibson, T. 1991. Planning for real: The approach of the Neighbourhood Initiatives Foundation in the UK. *RRA Notes*, 11: 29–30.

Illich, I. 1974, *La convivialità*. Milano: Arnoldo Mondadori Editore.

Krueger, R.A. & Kasey, M.A. 2000. *Focus groups: a practical guide for applied research*. Thousand Oaks: SAGE.

Marcetti, C. et al. (eds) 2011. *Housing Frontline. Inclusione sociale e processi di autocostruzione e auto recupero*. Firenze: University Press.

Norberg Schulz, C. 1992. *Genius loci. Paesaggio ambiente architettura*. Milano: Electa Mondadori.

Owen, H. 1997. *Open Space Technology. A user's guide*. San Francisco: Berrett-Koheler Publishers.

Rogora, A. (ed.) 2006. *La sostenibilità dell'autocostruzione nell'ERP: processi, politiche e riflessioni*. Milano: Clup Edizioni.

Sclavi, M. et al. 2002. *Avventure urbane: progettare la città con gli abitanti*. Milano: Eleuthera.

Trombetta, C. 2002. *L'attualità del pensiero di Hassan Fathy nella cultura tecnologica contemporanea. Il luogo, l'ambiente e la qualità dell'architettura*. Catanzaro: Rubbettino editore.

Turner, J.F.C. & Fichter, R. 1972. *Freedom to Build: Dweller Control of the Housing Process*. New York: The Macmillan Company.

Wates, N. 2000. *Community Planning Handbook*. London: Earthscan.

Williams, J. 2005. Designing Neighbourhoods for Social Interaction: The case of Cohousing. *Journal of Urban Design* 10(2): 195–227.

World Commission on Environment and Development, 1987. *Our Common Future*. Oxford: Oxford University Press.

Green Design, Materials and Manufacturing Processes – Bártolo et al. (eds)
© 2013 Taylor & Francis Group, London, ISBN 978-1-138-00046-9

Eco-design concept manual as a web tool to reduce the impact of a product on the environment

Z. Tončíková
Technical University in Zvolen, Slovakia

ABSTRACT: The contribution discusses a need for an implementation of eco-design in the process of shaping the personality of designer even at the stage of his study. The objective of a paper is to introduce the Eco-design concept manual (ECM) as a new eco-design tool for the students of industrial design. The first part of a contribution contains the elaborated comprehensive overview of present ways of eco-design implementation. A methodology of the research project concerning the conditions of an efficient ECM implementation is also described here. This tool of a vocational education should easily and systematically lead students of industrial design to become able to improve environmental aspects of their design concepts and products prototypes. The last part of a paper describes the research methodology and the assessment of efficiency concerning the new eco-design tool by using the objective methods of mathematical statistics including the expected results of the project.

1 PRESENT WAYS OF ECO-DESIGN IMPLEMENTATION

Industrial designers should be considered first and the most important link in the process of implementing green products and innovations. Most companies are aware of this fact. Designers activity may be guided towards better living conditions if design research advances the practice of design (Hongo & Amirfazli, 1995).

Due to increasingly strict European and national legislation (Ecological design Act 529/2010 Coll.), consumer demands and practical requirements, schools offering Design studies will be forced to comprehensively profile their graduates also in the area of eco-design in the coming years.

Jackson (1996) and Van Weenan (1995) describe a future development as a gradual transition from acute interventions in a production to preventive measures in order to protect the environment and to shift the solution of a product impact on the environment before the production stage in the stage of the origin of an idea itself.

If both the creation and contribution of a designer are to be efficient, he has to learn to anticipate and include into a product concept the largest scale of production operations during the process of idea origin, as soon as possible. Early changes are easier and less expensive to handle than changes during the later phases (Ullman, 2003).

According to Bhamra et al. (1999), there is an understanding among a number of companies of the importance of considering environmental aspects as early as possible in the design process. There is a realization that, beyond a certain point in the design process, it is extremely difficult to alter some product features that may be crucial to environmental performance (Åkermark, 2003). Unfortunately, the earlier stages of design suffer from lack of tools and methods for efficient environmental design. Ries et al. (1999) also point out the lack of environmentally oriented methods for the earlier phases.

Including the necessary changes of the internal structure of a product is the least financially demanding just during this stage. It is clear, that there is a really actual task to try set up the conditions of a designer's education so he would be able to acquire knowledge how to apply the principles of the environmental approach to his creation directly at school.

Numerous qualitative and quantitative tools to evaluate and improve the environmental impact of products have been created since 1990s.

These functional tools, procedures and manuals are very similar. Ecodesign literature shows that many existing tools fail because they do not focus on design, but instead are aimed at strategic management or retrospective analysis of existing products (Walker, 1998). They were designed primarily for businesses and enterprises.

The beginning of the 21st century marked a change in designers' way of thinking thanks to the Cradle to Cradle concept (McDonough & Braungart, 2002). Complex approaches to implementation of eco-design into business processes were also created (Wimmer et al. 2004), as well as the concept of Biomimicry and the specific natural principle of sustainability (Benyus 1998).

The issue of sustainable product design integration has become an effective part of educational process at many foreign universities in the last decade (LENS, 2007).

However, application of existing eco-design tools suffers from one weakness—its complexity for the user. Existing methods often include advanced and complex procedures, which require a large amount of user expertise or external consultation (Lofthouse, 2006). Our research is therefore aimed at finding a solution to this problem.

2 IMPLEMENTATION OF ECO-DESIGN INTO EDUCATIONAL PROCESS IN THE FIELD "2/2/6 DESIGN" IN SLOVAKIA

In the context of university education in the field of design in Slovakia, no targeted implementation of eco-design into the creation process can be found. If even occur some projects that would-be "eco", they lack systematic availability and the methodology of testing their environmental impact on a product (Toncikova, 2012).

Available study literature is written in complicated scientific language. It is intended for students of environmental engineering and technology. Students of design in Slovakia lack essential knowledge of ecology, environmentalistics and both the environmental management and marketing. They are not able to apply knowledge concerning environmental production processes, environmental technologies and suitable selections and combinations of materials and their impacts on the environment. They are able neither to distinguish nor to use any of available eco-design tools without this essential knowledge.

The flow of information concerning the essential environmental knowledge within the academic world is very limited. And where is available, there is used not very often. Students do not know where to look for the relevant information about eco-design and sources that would be simply applicable for the quick understanding of important coherences.

So far, students have not had access to any available procedures in Slovak language which would be adapted to local national needs in the area of education or would take into account various requirements connected to the economic and social situation and tradition.

3 ECO-DESIGN CONCEPT MANUAL

Our goal was to improve the state of education in the field of design by creating an educational portal with a new comprehensive eco-design tool called ECM—Eco-design Concept Manual.

Eco-design Concept Manual provides systematic methodology for the integration of eco-design into the creation process of an industrial designer. The tool helps students and designers to anticipate and learn to include ecological criteria into their concepts as soon as possible. The manual is available at www.ekm.ekodizajn.sk

Eco-design tools and information by which we would like to improve the knowledge and skills of students are themselves relatively demanding either to understanding or application.

As our objective is to address students by this methods not only in a formal way, but also to help them with their applications, it was necessary to create a prototype of the ECM that would be well arranged, truthful, functional and positively accepted by students also from the psychological point of view. It means that we do our best to attract student's attention not only by the offered information, but also by a graphical design itself.

As to a structure of this web tool, our intention was to approach a user compatibility with touch displays, tablets and smartphones with the objective to enlarge its possible ways of use and applications.

3.1 Structure of ECM

ECM consists of three basic units. The first section includes basic information on eco-design, its terminology, environmental policy, ecology, etc.—essential information that enables designers to work independently in the following practical part. The second part consists of ECM-customized tools, concepts and methods. The third section of ECM stores information about existing technologies and materials that are suitable (or not suitable) for use in terms of environmental impact assessment. The aim is to gradually build a long-term database of ecological materials, their properties and possible applications in projects.

3.2 ECM—costumized tools, concepts and methods

The second section of this manual—the key section—includes three basic groups of eco-design tools. Student's works and designer's projects differ from case to case, so we regarded as important to provide a student with a possibility to select an eco-design method, because none of them is a universal one and suitable for the all types of assignments and projects. To declare just our proposed procedure for the only and a universal one would has been counter-productive and we rather selected the way of adaptation of existing tools for the needs of student's projects.

The first group consists of qualitative tools: LiDS Wheel, ABC analysis, E-concept spiderweb diagram, eco-design checklists.

The second group of ECM includes qualitative methods: MET matrixes (Brezet & Van Hemel, 1997), explanation of the essence of LCA analysis, free versions of software, such as GABi, Sima Pro, ECOLIZER, ECOIT, ECOINDICATOR 99, ECODESIGN PILOT.

In addition to eco-design tools we have also compiled comprehensive banks of information with examples of the following concepts: "CRADLE TO CRADLE", BIOMIMICRY and 12 steps to implement eco-design into business processes.

By having access to the ECM manual, students should be able to quickly assess, which methodology is suitable for the emerging concept of their product design, or to find a suitable method for a high quality redesign.

3.3 Used methods and solution process

In the first stage of the project, we have worked with students from the second-degree study program "Furniture design" at Technical University in Zvolen. We provided students with the information about existing eco-design tools, conceptions and methods within the new established subject of Eco-design. As there does not exist any targeted implementation of ecological procedures into a studio creation, here in Slovakia, students tested their works within particular assignments by our brand new eco-design tools week by week and provided us with a feed-back in the form of their presentations containing their comments to particular applied methods.

Based on our work with these students, we have identified key areas of information and knowledge, without which students could not understand and use any type of eco-design tools correctly. Using the collected data we developed an outline of the first section of the web portal—ecological/environmental info section.

Feedback from designers has confirmed our assumption that currently available eco-design tools are inherently designed for business processes and not for independent designers and students, whose ability to process difficult scientific information is limited by their study programs and profession. We have identified an imaginary "boundary" of students' skills, and found areas, in which students were able to work individually with the methods, as well actions, which students were

not able to perform without external consultation of the issues.

Based on the data, we have adjusted and adapted individual tools for students' needs in the second section of ECM. In parallel with the work on ECM-customized eco-design tools we also worked on the third section of the portal—materials and technologies that were analyzed in terms of their suitability and impact on the environment.

3.4 Creating teaching aids DIPO1 and DIPO2

In the last stage of the project, we verified the proposed processes by creating a custom design. Designed prototypes of products (stools DIPO) served as a unique teaching aid to explain various principles of ecological product design and to demonstrate the way of working with ECM to the students.

The philosophy of this set of teaching aids is based on two different approaches to product design.

Figure 1. Prototypes of DIPO1 and DIPO2 stools.

Table 1. List of materials for DIPO1.

Name	Pcs	Material	Weight Kg	Details
Seat	1	Oak	2.391	Solid wood
Leg	3	Oak	1.196	Solid wood
Lower block	1	Oak	0.797	Solid wood
Clamex S	3	PVC + Zn alloy	0.005	ironwork
Stabilizing ironwork	3	Steel	0.08	ironwork
Cord	1	90% cotton	0.002	–
Varnish	1	PUR	0.001	–

Table 2. List of materials for DIPO2.

Name	Pcs	Material	Weight Kg	Details
Leg1	3	Al	0.083	Tube Ø 20 × 30 mm
Leg2	3	Al	0.089	Tube Ø 20 × 30 mm
Leg3	3	Al	0.094	Tube Ø 20 × 30 mm
Foot1	3	Oak	0.015	Solid wood
Foot2	3	Oak	0.025	Solid wood
Foot3	3	Oak	0.017	Solid wood
Seat	1	Oak	0.545	Solid wood
Pattern1	1	Felt	0.004	–
Pattern2	1	Felt	0.005	–
Pattern3	1	Felt	0.004	–

The first product, DIPO1, has been designed in a "standard" way and its design philosophy includes environmental design criteria only formally. The second product, reviewed using ECM, really recognizes real ecological design criteria from early stages of the creation process. Based on a comparison of environmental impact of these products, we can explain how to work with ECM, and also illustrate the differences between the products, which might seem similar at first glance.

3.5 Steps of using ECM

The procedure of using the Eco-design Concept Guide (ECM) can be summarized into four steps:

A. Getting familiar with basic information on ecology and environmental science.
B. Getting familiar with eco-design tools, methods and concepts, which are tailored to your needs.
C. Choice of methodology suitable for the specific project.
D. Including ecological criteria into the design using this methodology and the third section "materials and technologies".

4 METHOD OF EVALUATING THE RESULTS OF USING ECM

Nationwide deployment of the Eco-design Concept Manual tool is planned for the beginning of the winter semester 2013/2014. First results will be followed by an experiment based on selective evaluation of the created procedure's effectiveness. An experimental group of design students will be created consisting of (n) = 20 to 30 students. The group of designers will be given a task within their individual studio assignment. One part of students will follow the proposed guide to improve the environmental profile of the product. The second one—control group—will not.

After an objective evaluation of the work of both groups of students, we will conduct a statistical evaluation of the experiment. We will test our hypotheses, which assume that there will be a significant difference in terms of ecological elements in design projects of individual groups. For this evaluation, we will use methods of mathematical statistics as provided by (Everitt 1986). This way we will be able to objectively evaluate the effectiveness of the educational portal.

4.1 Expected benefits of the project

As a part of a nationwide project aiming to include eco-design into works of students of design, a series of scientific and technical education seminars will be held in the fall of 2013, taking place at five universities in Slovakia which have accredited courses in design.

Based on feedback from students who have used our manual to integrate eco-design into their work, we will be able to objectively assess the improvement of environmental profile of their designs, as well as the effectiveness of the proposed system of education in this area. Projects provided by these students will also be made available on the ECM portal so that they can also be used as an inspiration for new students and designers interested in eco-design.

5 CONCLUSION

Products designed using simplified "ECM" tools are, naturally, not flawless in terms of ecological qualities. ECM cannot replace a complex lifecycle analysis. However, the mere fact that students think about environmental criteria and they try to minimize the negative impacts of their products on the environment is very important.

The ECM eco-design tool is not perfect and it has its flaws. The current version is still a pilot. However, it is currently the only tool of its kind in Slovakia or in Slovak language. In future we will try to get objective feedback on the tool from its users and it will be continuously modified to include other missing features.

REFERENCES

Åkermark, A., 2003. The Crucial Role of the Designer in EcoDesign, Doctorial Thesis, Department of Machine Design, Royal Institute of Technology, ISSN 1400-1179. 59 p.

Benyus, J., 1998, Biomimicry: Innovation Inspired by Nature. New York, Harper Perennial, 42 p.

Bharma, et al., 1999. 'Integrating environmental decisions into the product development process: part 1 the early stages' in Ecodesign '99: First Symposium on Environmentally Conscious Design and Inverse Manufacture IEEE, Tokyo, Japan, p. 329–333.

Brezet, H. & van Hemel, C., 1997. Ecodesign, A promising approach to sustainable production and consumption. Paris, Edited by UNEP.

Everitt, B. S., 1986. The analysis of Contingency Tables. London, Chapman and Hall. 128 p.

Hongo, K. & Amirfazli, A., 1995, *Scientific prediction of design implications and responsibility of design—a speculative approach to design research*, International Conference on Engineering Design (ICED), Czechoslovakian, Vol. 1, p. 9–16.

Jackson, T., 1996. Material Concerns: Pollution, *Profit and Quality of Life*, Routledge, London. 36 p.

LENS 2007, Learning Network on Sustainability Working Hypothesis, www.lens.polimi.it

Lofthouse, V., 2006. *Ecodesign tools for Designers: Defining the Requirements*, Loughborough's Institutional Repository, Journal of Cleaner Production, Volume 14, Issues 15–16, p. 1386–1395.

McDonough, W. & Braungart, M. 2002: Cradle-to-Cradle. Rethinking the Way How We Make Things, US, North Point Press, 193 p.

Ries, et al., 1999. *Barriers for a Successful Integration of Environmental Aspects in Product Design*, First International Symposium on Environmentally Conscious Design and Inverse Manufacturing, Tokyo, Japan, p. 527–532.

Tončíková, Z., 2012. Need for integration of eco-design in the process of shaping the personality of furniture designer, In.: Wood and Furniture industry in Times of Change-New Trends and Challenges (outcome of the International Conference WoodEMA, Faculty of Mass Media Communication UC, Trnava.

Ullman, D. G., 2003. *The Mechanical Design Process*, ISBN 0–07–237338–5, McGraw-Hill International Editions.

van Weenan, J. C., 1995. Towards Sustainable Product Development. Journal of Cleaner Production, 3(1–2), p. 95–100.

Walker S., 1998. Experiments in sustainable product design. The Journal of Sustainable Product Design, 7: p. 41–50.

Wimmer, et al., 2004. Ecodesign Implementation—A Systematic Guidance on Integrating Environmental Considerations into Product Development, Dordrecht, Springer, 145 p.

Green Design, Materials and Manufacturing Processes – Bártolo et al. (eds)
© *2013 Taylor & Francis Group, London, ISBN 978-1-138-00046-9*

The meaning of public spaces

M. Hanzl

Lodz University of Technology, Lodz, Poland

ABSTRACT: The paper discussed the information layer of urban public spaces presenting considerations which relate their appearance with the cultural characteristics of a given community. A classification of elements pertaining to a notion of spatial order is proposed. The epistemological approach of anthropology and urban design studies is focused around the outdoor, public realm which is represented in the urbanscapes by a void of street or square. The current paper yields some clues in this respect, providing theoretical background both for analyses of existing urbanscapes with emphasis on their cultural aspects, and for the design process.

1 INTRODUCTION

1.1 *Epistemological elucidation*

The endeavours which address the possibilities to transform the urban setting into a more sustainable environment require in depth understanding of processes how these spaces are formed through interaction of various forces and flows, acting at different intensities and speeds and under various cultural circumstances. The analyses of the above processes are addressed by several specific disciplines, among them urban design and urban morphology and anthropological studies. Outdoor spaces are delimited by a boundary consisting of facades of buildings, fences, greenery, etc or, in some cultures, non-material, socially agreed edges.

A void which constitutes a physical representation of a public realm may be considered an element providing the commensurability of the theoretical frameworks of these disciplines. The anthropological concept of walking, elaborated by Certeau (1988, p. 98) as a space of enunciation and his comparison of the usage of space and urban structures to speaking in a given language, provides a valuable asset for the analysis of the way urban spaces are created and read. A desirable harmony of urbanscapes requires congruency of the form of structures and of human behaviour. Certeau (1988, ix) discusses a concept of singularity—in this approach defined as the scientific study of relationship—that links everyday pursuits to particular circumstances.

1.2 *Meaning of public spaces*

Understanding of the 'meaning of public spaces' in this approach requires looking for relations between the urban structures and the culture of usage of space. There are three main issues, which should be considered with regard to this: physical features, including distribution, shape and size of forms defining the space, distribution and behaviour of users, which reflect social order and flows of human movement, which find their reflection in the sociometric layout of a given place. The current paper addresses the problem of classification of the above relations which is central both for the epistemology of existing and historical structures and when looking for effective normative theory in the field of urban design.

2 CLASSIFICATION OF ACTIVITIES

2.1 *General classification*

Activities which are performed in public spaces may be classified into two main categories: (1) movement and transportation and (2) interpersonal communication and social activities (Carmona et al, 2009). The mutual relations between the two components of use of public spaces is extensively explained in the classical study by Appleyard (1981), especially with regard to car traffic. Our focus is on the relations between the way space is used in its multifarious variations and the forms of public spaces themselves. We will try to analyse and review the previous research with regard to these mutual relations.

2.2 *Flows and movements*

The first category connected with transportation and movement is reflected at its basic level by street layout, mostly the elements of street profile: traffic lanes of various parameters, tram lines, pavements, cycling paths, street greenery of various character, etc. The presence of these elements of outdoor spaces means that defined mode of transportation are available. Their absence or underdevelopment as well as an abundance or overdevelopment may effectively influence users' preferences.

The human presence in social spaces may be divided into flows and concentrations: flows are connected with movement/traffic and are related to space, following the definition by Yi Fu Tuan (2001). The intensity of traffic flows and movements is a feature which was extensively described by Hillier and Hansen (1984), including also the discussion concerning the control of space. The mentioned theory provided the rules for simulations, which found their application in the well recognised Space Syntax software developed by scientists all around the world. Hillier (2009) discusses *"human system made up of movement, interaction and activity"* and its relations with the urban physical structure. Space Syntax research refers first of all to flows and patterns of human movements. The sociometric layout, as he proves, reflects the culture of usage of space appropriate for a given community, which Hillier (2009) attributes to cultural idiosyncrasy generated by conservative use of space aiming *"to reinforce existing features of society"*, in opposition to the generative one—aiming to generate co-presence and make new things happen.

2.3 Social component

The components of social and communication activities require a more elaborated classification with regard to relations with physical settings. Some considerations on urban settings of various periods as a nonverbal system of signifying elements are reviewed in an essay by Choay (1972), who gives account of the gradual reduction of meaning of contemporary urbanscapes as a result of the predominance of economical over social and related to everyday lives factors. The description of human behaviour in public spaces is addressed by the concept of situation, which is defined by anthropologists as a theatre of human activities (Perinbanayagam, 1974). The frame of situation is defined as *"the smallest viable unit of a culture that can be analyzed, taught, transmitted, and handed down as a complete entity. Frames contain linguistic, kinesics, proxemic, temporal, social, material, personality, and other components"* (Hall 1989, p. 129). Situation is classified as one of the components of meaning along with verbal communication itself and along with *"the background and preprogrammed responses of the recipient"*. Environmental contexting is supposed to involve two distinguishing processes, internal and external

(Hall 1989, p. 95). The first type engages the brain and results of past experience (internalised or programmed contexting) or the innate features of the nervous system (innate contexting). The second one consists of settings and the situation of the event (situational or environmental context). Situation provides an external context whereas other features belong to the internal contexting. The use of contextualising allows us to deal with the overload of information and makes communication much easier. One of the communication channels in cities is the form of urbanscapes. It may prove easy to grasp information about the kinds of behaviour and situations one may expect in given settings as well as give hints about required rules of conduct—*"(...) the environment is seen to consist of highly structured, improbable arrangements of objects and events which coerce behavior in accordance with their own dynamic patterning."* (Barker, 1968 after Hall 1989, p. 99).

In the field of anthropological theory these phenomena are explained with the concept of screening and filtering of sensory data. *"The architectural and urban environments that people create are expressions of this filtering-screening process"* (Hall, 1966, p. 2), which admits some things and rejects others. As a result, the experience is affected by a given *"set of culturally patterned sensory screens"*. Hall (1989, p. 87) lists at least five sets of categories of events which should be taken into account when discussing rules behind individual perception or omission: (1) subject or activity, (2) situation, (3) status in a social system, (4) former experience, (5) culture. Humphrey (1988 after Pellow 1996, p. 216) confirms the role which culture plays in the interaction between architecture and the way space is organised as both are cultural constructions.

2.4 High context versus low context cultures

Cultures may be classified according to the role of context in communication situations. Hall (1989, p. 92) emphasises the importance of the level of context pointing at its determining role versus the nature of communication and all subsequent behaviour. In high context (HC) communication most of the message is in the physical environment or is internalised in the person (Hall 1989, p. 91). It requires preprogramming, in which preparing appropriate settings is one of the necessary elements. In this kind of communication art forms are often used (Hall 1989, p. 101). According to Hall (1989, p. 102) *"Extensions that now make up most of man's world are for the most part low-context"* and could be replaced by some other inventions along with the development of technology. Currently such a shift is observed from car overusage to fascination with IT communication devices.

2.5 *"Information channels" of urban settings*

Components containing the meaning of public spaces may 'speak' in different way. Some features are obvious and result from functional conditions of given development. They may be classified as direct communication. In this group the way streets are furnished may described, like: traffic lanes of various widths and surfaces, cycling paths, pavements, greenery, bollards, etc. It is a platitude to say that there are streets which are designed to drive fast and others inviting for a walk in the shade of trees. An elementary analysis of street profile allows one to distinguish a commercial street from a residential one. The general character of public spaces and their development through the ages is addressed in manuals of history of urbanism. The relation of physical features such as size in relation to human scale, which may vary from cosy to monumental, is presented on the background of the history and culture of given period (e.g., Rasmussen, 1969).

2.6 *Indirect sphere of communication*

Another part of communication may be classified as indirect, in analogy to non verbal cues. According to Hall (1989, p. 82), *"Nonverbal systems are closely tied to ethnicity (...) they are of the essence of ethnicity."* The consistency of urban pattern, as experienced in public spaces, is a consequence of the rules of crowd behaviour constituting part of a given culture. Concentrations are static rather than dynamic, thus place related. Whenever the human flow stops for a moment concentration occurs, though interrelations require more comfortable conditions to take place, among others: time and spatial arrangement. The indirect/nonverbal part of communication affects the phenomenology of perception.

The meaning of public spaces remains in close relation with the kinesics patterns of groups of users. Kinesics is a way a person moves and handles their body (Birdwhistell, 1970, after Hall, 1989, p. 75). People specialised the language of the body making it integrated and congruent with everything they do, it is culturally determined and should be read against the given cultural background (Hall, 1989, p. 76). *"Each culture has its own characteristic manner of locomotion, sitting, standing, reclining, and gesturing."* (Hall 1989, p. 75). Hall (pp. 76–77) describes a phenomena of group movement synchronisation. He underlines the way people belonging to disparate cultures move, pointing at the presence of local rhythms and necessity *"to conform to local beat"* in order to fit in (1989, p. 79). He emphasises the role of synchronisation between people belonging to the same culture. According to Condon, who performed a comprehensive study

on the way groups of people move, the "bond" between humans should be seen as *"the result of participation within shared organizational forms"*. Further, he explains that *"(...) humans are tied to each other by hierarchies of rhythms that are culture specific and expressed through language and body movement."* (Hall 1989, p. 74)

Also a presence of a synchronisation with settings is claimed, which takes place when the urbanscape belongs to the same culture as the visitor. Then a sense of belonging may be present and a place is perceived as more attractive than when the synchronisation is lacking. Settings which are out of phase are more likely to seem alien, unordered. Synchronisation is particularly noticeable in high context cultures, where it *"functions on a high level of awareness, and is consciously valued"* (Hall 1989, p. 79). Handling synchrony is both innate and acquired as a result of the learning process and influenced by culture (Hall 1989, p. 79). When referring this theory to urbanscapes, also the settings where a child is raised affect his or her perception and kinesics behaviour. When older he or she would perceive similar spatial order as familiar, homelike. Hall suggests that human relationship to all the art forms is *"more intimate than is commonly supposed (...)"* as it is based on synchrony in which the *"(...) audience and artist are part of the same process."* (Hall 1989, p. 80)

The claim is made that the rules which govern the nonverbal communication component of human group behaviour are the same which serve the distribution of cues affecting the communicative features of public spaces, as discussed by Rapoport (1990). Following Hillier and Hanson societies vary *"in the type of physical configuration [and] in the degree in which the ordering of space appears as a conspicuous dimension of culture"* (1984, p. 4). Hillier and Hansen (1984, p. 224) ponders the method of investigation of encounters as morphic languages, concluding that the aim is to establish the way encounter systems acquire differential properties which would have different manifestations in space.

3 METHODOLOGY OF DESCRIPTION

3.1 *Urban morphology studies*

The research within the three main schools of urban morphology: (1) Conzenian, following the thoughts of M.R.G. Conzen (Whitehand, 2001), (2) Italian, continuing the tradition of Muratori (Cataldi et al, 2002) and (3) French, started by Panerai and others (Panerai et al, 2009), have so far overlooked detailed analyses of outdoor spaces. The level of detail required for description of culture related features was limited to buildings, whereas open spaces were discussed at the level of street/block layout analyses (Vernez Moudon, 1997, p. 7).

The oeuvre of morphologists who incline towards an anthropological method (Rapoport, 1990, 2003; Rykwert, 1989; Lynch, 1994) is located on the edge between the cognitive and normative approach, both providing characteristics of existing or historical cityscapes and looking for normative theory (Gauthier, Gilliland, 2006). The most elaborated methodology is provided by Rapoport (1990, pp. 106–107), who proposed a comprehensive list of cues as a basis for description, including elements of vision, sounds, smells and some social aspects: characteristics of people, activities, uses and objects present in outdoor spaces.

3.2 The notion of boundary

When discussing outdoor urban places one of the key issues is the sensation of enclosure of space and form of the edges. The notion of enclosure and its development has been extensively discussed in Hanzl (2013). Let's focus on a concept of boundary which in the anthropological approach yield a symbolic meaning. According to Prussin (1989 after Pellow 1996, p. 215), "Boundary making is a cultural process, as in 'placemaking' (…), which is organised cognitively." As Pellow (1996, p. 215) admits, in some cultures the boundaries are physically demarcated with the use of fixed features and in others these are social activities or institutional understanding which create the division.

At the same time the notion of boundary itself may be understood in various ways, following the field of studies. In urban design the most basic boundary is the one between public and private space, which may be either defined spatially or not. Even within the scope of this single discipline clear definition of boundary requires the assumption of defining criteria, e.g. accessibility or property. For the purpose of current considerations, boundary is understood as the edge of space, which represents: streets, squares, piazzas, markets. In the field of anthropology, and more generally: cultural studies, boundary means a border which has prescribed meaning which functions in the public consciousness of a given community and refers to social category placement.

3.3 Phenomenology

People perceive context as a whole (Hall 1989, p. 131). The impression is evoked by 'orientation, suggestions of movement (and) markings creating 'concentrations, directions, configurations in space.' (Böhme, 2002/2005, after Andersen 2012). Andersen (2012) points at a number of

phenomenological qualities of building facades which pertain into an overall spatial character: (1) contour, (2) shift (like shift of cornice providing dynamic transition between neighbouring facades), (3) colour palette, (4) profile, (5) relief, (6) plasticity—twists of line of construction, corrugation, (7) rhythm—of facades, windows, etc, (8) framing—as emphasis of architectural elements, (9) pattern—touchable, increasing tactile qualities. He defines after Böhme (2004, after Andersen, 2012) the smallest "sensuous experiential order" and calls it an atmosphere.

Another feature crucial for the gestalt perception of urbanscapes is profile. In order to become a true form, a street, as well as any other outdoor space, must possess a 'figural character' (Norberg-Schulz, 1963, p. 83, after Ashihara 1981, p. 142). An analyses of the changes of proportions of street profiles in various periods was undertaken by Ashihara (1981, p. 41). The approach was based on the relation of building height (H) to street width (D) and further also to the width of the facade (W) (Ashihara 1983 p. 141). The review of analytical studies with regard to enclosures is provided in Hanzl (2013), it includes the research of Wejchert (1984) and his followers.

4 CASE STUDY

This paper is part of a three year research project 'Morphological analysis of urban structures—the cultural approach. Case studies of Jewish communities in the chosen settlements of Lodz and Masovian voivodeships', funded by the Polish National Centre of Science. One of the main goals of the project is to define a methodology for the description of the idiosyncrasies of urban outdoor spaces in relation with the cultural background of the defined group. The paper concentrates on a particular stage of the project, which is to define the theoretical background enabling classification of relations between urbanscapes and human behaviour, which they embrace.

The presented classification is conceived in a way to allow the explanation of various forms of cityscapes developed from the same cultural background through adoption of habits and lifestyles connected with various ranges of cultural assimilation. It allows one to distinguish the conscious and formalised forms of equipment of public spaces from the unconscious order of space. Jewish orthodox and Hasidic culture in Poland in the period before World War II was a high context one. Along with the development of the Haskalah movement, the processes of enlightenment and secularisation led to the loss and transformation of some of the former habits and the culture itself becoming a low context (LC) one. This is visible when looking at the urban settings, which, along with changes of lifestyle and religious attitudes of their inhabitants, became more cosmopolitan.

5 CONCLUSIONS

The quest for sustainability and proper living conditions should encompass the cultural systems which affect human behaviour. This requires comprehensive research covering the relations between human behaviour in urban spaces and forms of structures. This challenge is becoming recognised, as numerous studies show (Schumacher, 2011). The description of relations between human activities and physical settings is central for the ontology for urban design. The answer to these questions engages epistemology of a few disciplines, namely anthropology, urban design and urban morphology studies.

As Humphrey (1988 after Pellow 1996, p. 216) admits, *"architecture and spatial organization have been peculiarly neglected by academic anthropology as subjects in their own right"*. An attempt to apply the anthropological approach to the description of public spaces as used in the practice of urban design has been undertaken with the void of public spaces and the physical border between public and private as a common epistemological ground. The classification of relations between human activities and physical spaces has been proposed on the background of anthropological theory, with the emphasis on the non verbal components of communication. The meaning of space is explained by contextualisation provided by the built environment as one of the parts of cultural settings which should help citizens to deal with the overload of information. Hall (1989, p. 85) refers to meaning when saying *"One of the functions of culture is to provide a highly selective screen between man and the outside world"*. A concept of singularity (Certeau, 1988, ix)—congruence of everyday pursuits to particular circumstances and synchrony between human nonverbal communication cues and physical settings yield important assets into the definition of 'spatial order'.

REFERENCES

Andersen, N.B. 2012. In search of a modus operandi for a specific urban architecture. A critical approach to the collective amnesia of urban design In *Cities in transformation Research & Design, EAAE/ARCC International Conference on Architectural Research:* 4–5. Milano: Politechnico di Milano.

Appleyard, D. 1981. *Livable streets*. Berkeley: University of California Press.

Ashihara, Y. 1981. *Exterior design in Architecture, Revised Edition*. New York, Cincinnati, Toronto, London, Melbourne: Van Nostrand Reinhold Company.

Ashihara, Y. 1983. *The Aesthetic Townscape*. Cambridge, London: The MIT Press.

Barker, R.G. 1968. *Ecological Psychology: Concepts and Methods for Studying the Environment of Human Behavior*. Stanford, California: Stanford University.

Birdwhistell, R.L. 1970. *Kinesics and Context*. Philadelphia: University of Pennsylvania Press.

Böhme, G. 2004. Atmospheres: The Connection between Music and Architecture beyond Physics in Metamorph. 9. International Architecture Exhibition. Venice: Focus.

Böhme, G. 2002/2005. Atmosphere as the Subject Matter of Architecture in Ursprung P. (ed.) Herzog & de Meuron: Natural History: 405. Montreal: Lars Müller Publishers.

Carmona, M., Heath, T., Oc, T. & Tiesdell, S. 2009. *Public Places Urban Spaces The Dimensions of Urban Design*. Oxford: Architectural Press.

Cataldi, G., Maffei, G.L. & Vaccaro, P. 2002. 'Saverio Muratori and the Italian school of planning typology', *Urban Morphology* 6 (1) 3–12.

Certeau de, M. 1988. *The Practice of Everyday Life*. Berkeley, Los Angeles, London: University of California Press.

Condon, W.S. & Sander, L.W. 1974. Neonate Movement Is Synchronized with Adult Speech: International Participation and Language Acquisition. *Science* 183 (4120).

Gauthier, P., Gilliland, J. 2006. 'Mapping urban morphology: a classification scheme for interpreting contributions to the study of urban form' *Urban Morphology* 10 (1) 41–50.

Hall, E.T. 1966. *Hidden Dimension*. Garden City, NY: Doubleday.

Hall, E.T. 1989. *Beyond Culture*. New York: Anchor Books.

Hanzl, M. 2013. 'The morphology of public spaces' *Urban Morphology* in press.

Hillier, B. & Hanson, J. 1984. *The Social Logic of Space*. Cambridge: Cambridge University Press.

Hillier, B. 2009. The genetic code for cities – is it simpler than we thought? Keynote paper for conference *Complexity Theories of Cities have come of Age,* TU Delft.

Lynch, K. 1994. *Good City Form*. Cambridge, London: The MIT Press.

Norberg-Schulz, Ch. 1963. Existence, Space and Architecture. London: Studio Vista.

Panerai, P., Depaule, J. Ch., Demorgon M. 2009. *Analyse urbaine*. Marseille: Édition Parenthèses.

Pellow, D. 1996. Concluding Thoughts in *Setting Boundaries, The Anthropology of Spatial and Social Organisation*. Westport, Connecticut, London: Bergin and Garvey.

Perinbanayagam, R.S. 1974. The Definition of the Situation: An Analysis of the Ethnomethodological and Dramaturgical View. *The Sociological Quarterly* 15: 521–541.

Prussin, L. 1989. The Architecture of Nomadism In *Housing, Culture and Design* Setha, M.L. & Chambers, E. (eds.): 141–163. Philadelphia: University of Pensylvania Press.

Rapoport, A. 1990. *The Meaning of the Built Environment. A Nonverbal Communication Approach*. Tuscon: The University of Arizona Press.

Rapoport, A. 2003. *Culture, Architecture et Design*. Paris: Collection. Infolio Éditions.

Rasmussen, S.E. 1969. *Towns and Buildings*. Cambridge: The MIT Press.

Rykwert, J. 1989. *The Idea of a Town, The Anthropology of Urban Form in Rome, Italy and the Ancient World*. Cambridge, London: The MIT Press.

Schumacher, P. 2011. *The Autopoiesis of Architecture A New Framework for Architecture*, vol. 1. Great Britain: A John Wiley and Sons, Ltd, Publication.

Tuan, Y.-F. 2003. *Space and Place: The Perspective of Experience*. Minneapolis: University of Minnesota Press.

Vernez-Moudon, A. 1997. Urban morphology as an emerging interdisciplinary field *Urban Morphology* 13–10.

Wejchert, K. 1984. *Elementy kompozycji urbanistycznej*. Warsaw: Wydawnictwo Arkady.

Whitehand, J.W.R. 2001. British urban morphology: the Conzenian tradition *Urban Morphology* 5(2), 103–109.

Green Design, Materials and Manufacturing Processes – Bártolo et al. (eds)
© *2013 Taylor & Francis Group, London, ISBN 978-1-138-00046-9*

Designing innovation with craft-evolution to develop sustainability

E. Aparo

Instituto Politécnico de Viana do Castelo, Portugal & CIAUD—The Research Centre for Architecture,
Urban Planning and Design, Portugal

L. Soares

Instituto Politécnico de Viana do Castelo, Portugal, CIAUD—The Research Centre for Architecture,
Urban Planning and Design, Portugal & ID + Research Institute for Design, Media and Culture, Portugal

H. Santos-Rodrigues

Instituto Politécnico de Viana do Castelo, Portugal & CIEO—Centre of Spatial Research and Organizations,
Algarve University, Portugal

ABSTRACT: This paper intends to recognize and value how handicraft practices can help to create a model, which studied how design schools are embracing sustainability. The research is support by the craft culture, the design-method, the cross-fertilization and the material-centered design. The main focus of this research is to realize how it is possible to explore a craft-design collaboration that may reinvent a model based on local craft design collaboration that aims to reinvent all craft industry. The paper presents a pilot study using students and straw's artisans. The outcome gained from this research serve as a reference for local craft development and design practice. This research also contributes to SIM2013 by providing a craft-design relationship process to form the basis of an approach for local craft improvement and sustainability. The paper highlights the value of the union between craft and design as a shared learning mechanism. Both sides can exchange knowledge, enhance their professional capabilities, proposed different and spontaneous meanings and created new market opportunities.

1 INTRODUCTION

Local craft is a reflection of the alliance between the individual and his environment as well as the *genius loci* of a place. Today, there are still many places characterized by the presence of handmade artifacts. Although, in Portugal the mass-production and mechanization tend for the decline of the demand of traditional craftsmanship, and the artisans sell their products at lower prices. Moreover, on one side the artisans are becoming older and fewer in numbers. On the other hand, younger generations show disinterest in craft activities. This means that those handmade singular objects need input, not only to keep survive those ancestral activities that characterize a cultural place, but also to contribute to the business opportunities that can keep it alive and render it competitive. In addition, artisans cannot relate change with the demands of the current market. As John Christopher Jones (1992) states, the method of craft-evolution tends for artisan production. It means that in this method each part of a craft object is formed by many causes and not only for one reason. But, the artisan often acts based on the knowledge that as transmitted,

respecting the process in a blindly way. He only knows how to do it as he received it as a familiarly legacy. However, the United Nations Educational intends to bring traditional crafts into ordinary life, in people's life. An intention wish comprises appealing designers to work with artisans in order to develop new products and radical thinking for

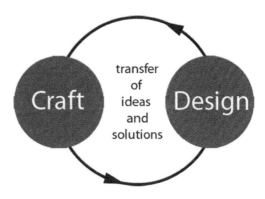

Figure 1. Craft-design alliance. Image from the authors.

new markets development. The problem raised is that some designers don't want to learn with artisans and several designers are probably more expert about production than reasonable users (Jones, 1992) and do not realize that today's client is a person and not a user. "A person's life that has become a paid-for experience". (Rifkin, 2001), and not a paid-for a product. If each person wants to live an experience, by buying a product she/he is also buying the system of values of a brand (Zurlo, 2003).

Thus, this circumstance may be an excellent and unique opportunity for artisans, designers and entrepreneur work together. That may means that companies that are created between craft-design alliance may associate their image with the values of product quality as the *Genius Loci* of a place, which is unique and singular and can not be transfer, ensuring meaning in handmade production and in design action. This idea can be support in a craft-design alliance. As Medardo Chiapponi claims, in design "often a strong innovation in a sector can be determined by the transfer of ideas and solutions from another field in which the same ideas and solutions are not more innovative, but who are already fully acquired long ago." (Chiapponi, 1999). In this research, the authors chose the proposition of changing and sharing knowledge with the artisans in order to improve intelligent sustainability and innovation, based on the awareness of crafters. It means to recognize the value added of craft work.

2 RESEARCH PROCESS

2.1 *Aim*

This paper aims to reveal an approach in which craft-evolution can be integrated in design process. It means, using craft thinking and know-how to work with the material, in this case, the rye straw. As John C. Jones sates "there is hidden in the apparent simplicity of primitive craftwork, a subtle and reliable information-system that is probably more efficient that design-by-drawing and compares in many ways with the new design methods (...)" (Jones, 1992). The designer assumes, as well, the role of a "reflective practitioner" (Schön, 1983), contributing in order to the artisan to recognize the values of sustainability. Thus, the artisan can better develop its manufacturing process.

2.2 *Related work*

Today, products with a new interpretation of traditional materials, wish are locally referenced, may gain interest in global market, such as the fashion or domestic scenarios. In Portugal, there are some pilot projects that relate de design school with the craft context (Branco, 2005) although, there is little research oriented to the reinvention of the rye straw's identity. In addition, officially, the artisans of rye straw do not exist (Fernandes, 2012).

3 PARTICIPANTS AND METHOD

Fafe is a portuguese city located in the north. According to the website of the Business Association of Minho (AIMinho), it is characterized by industry activities of small enterprises with fewer than 10 people. Some of these industries are disappearing with the crisis and persons are returning to handicraft, agriculture and emigration. Thus, the Fafe City Hall and the Ave Inter-municipal Community ask for design action. This pilot research was conducted in different phases. A first phase took place in Fafe and was the first contact between the 30 design students, the professors and 3 artisans. The next stage was focus oriented on the fieldwork, the collection and the analysis of data. Throughout the process it was necessary to return more often to Fafe. Some cases required other materials, other techniques and other artisans, providing the entrance of new partners in the project.

3.1 *Material and appropriation*

The rye crops appeared as weed infesting wheat fields, asserting itself later in land prone. Its increase is due to its great hardiness, the few requirements that require the terrain and its high resistance to cold. Besides the use in baking cobs, straw and fodder for feeding animals, this is also widely used to manufacture various types of mats and in the case of Fafe, a whole range of plaited. The braided rye straw, plaited as other materials, uses a system of wires (at least three) intersecting diagonally. It is carried out by hand by alternately passing the wires (or 'straws') above and below. The rye's cycle has the following stages (Fernandes, 2012): 1) The sowing; 2) The bleaching-stage one; 3) The harvest; 4) The cut stem straw; 5) Bleaching-stage two; 6) The choice; 7) Bleaching-stage three; 8) The staining. After this set of operations, the raw material of plaited is ready to be worked. For each type of braiding, there is a description of a technique and implementation.

3.2 *Techniques and performance*

Before performing any type of braid it is necessary to wet the straw in a minimum time of 12 hours, for the easiness of handling. This information would be essential to learn how to design the material.

Table 1. Techniques of rye straw.

Braid name	Tip of straw	N° of tips	Length of braided
Nylon	Tocos	3	150 cm × 15
Repique	Tocos	4	150 cm × 12
Trancelim	Onze	5	150 cm × 12
Tocos	Tocos	5	150 cm × 9
Peixeira	Escumilha	7	150 cm × 12
Rêlo Fino	Rêlo	7	150 cm × 12
Rendilhada*	Onze	7	150 cm × 12
Escumilha	Escumilha	11	150 cm × 9

*There is no artisan who knows how to do this technique.

There is no scientific information, about these techniques, thus the artisans shared their knowledge with the students and designers.

Some techniques are simple, such as the 'Nylon' as it has only 3 tips. Whenever the number of points, the difficulty of execution increases. Sometimes, the difficulty of execution provoked that there is nobody who knows how to make it. This happens, for example, with the braid 'rendilhada'. Concerning the performance, throughout the various strokes and any braided, it is necessary to splice strips straw. They are superimposed on the terminal that will replace one of the banks of plaited. The braids are riddled with small tips, which are the straw that were replaced. These leftovers are trimmed with scissors flush against the surface of the braid. After that, the braid becomes the feedstock.

3.3 The lacing

Previously, the artifacts of straw were sewn by hand. Today, the braids are sewn with pedal of motor sewing machine. The fundamental difference between stitching by hand or by machine is the greater speed of execution that it allows, and the greater perfection of the final product. This factor is crucial both to the artisan and to the consumer. Moreover, the difference between the handwork and the work done with the machine is focus on the effort that the artisan applies. It is much less significant in the second case. Another element to this study is that straw is sewn still wet. This information is relevant in order to improve the processability of the material, assuming a particular configuration.

4 FINDINGS

4.1 Advantages of the process

– Using various types of braids and applying three, four, five, seven or eleven straws, it is possible to carry out a number of different articles. It was the recognition of the potential of the technique.
– The market for fashion accessories and home accessories has been growing. New products under these two markets can be expanded to fulfill market demands (Turinetto, 2005).
– The potential of increasing rush-straw to other industries should be investigated to create cross-fertilization (Cappellieri, 2006) occasions for the craft community and for other potential business.
– The existence of other crafts may create partnerships and cross-materials development.
– The sale is made in the workshops directly to intermediaries. This means that the artisans are accustomed to direct commercial exchange.

4.2 Disadvantages of the process

– Older people are dedicated to this work. The dramatic aging of artisans, going almost all the age group of 60 years old, put in serious jeopardy the future of the county's most representative handicrafts. The generational legacy risks losing up.
– The weak labor income, unsusceptible to attract young apprentices.
– The competition from Chinese hats with less quality, but cheaper.

5 PROPOSITIONS FOR DESIGN

For this research the design development process was focus to increase approach to look for cross-fertilization chances. An innovation process can be organized into three different sections: (1) market pull, when the innovation process is driven by market need, (2) technology push, when the innovation process bets on resources for research and development and (3) design driven innovation, when there is a design-oriented research into new languages (Verganti, 2001). In this study, the design action was focused on creating new markets. Reinventing craft by design may produce new luxury, which authors such as Cappellieri (2005), Morace (2008), Zurlo (2003) stated as luxury ethical. This design-oriented strategy is determined by the socio-cultural context of its origin. As Bucci states "today, in the post-industrial world, we find surprisingly that large levels of aesthetic quality, of care productive of well done and innovation are almost connected to a place." (Bucci, 2003). Therefore, as propositions for this study, there were two objectives in creating new products. One was to define new markets opportunities. The other objective was to focus on internationalization.

Figure 2.   The material, the students, the artisans and the techniques of rye straw. Image from Sílvia Fernandes.

6   CASE STUDY

6.1   *The sustainable intelligent process*

The development of the pilot project was founded on a craft-design alliance, linking different players. The process can be defined as an iterative approach, with students, professors, artisans, the Fafe City Hall and the Ave Inter-municipal Community (CIM), which combined drafting, discussing, prototyping and presenting all the projects. The students' proposes were presented in the Fafe City Hall to: the artisans, the mayor of the City Hall, the secretary of the CIM, the director of the School, the media and several local retailers. As the 30 students were divided into 10 teams, after the proposal's presentation it was established that, during the final prototype, 1 member of each group would remain with one of the artisans. In addition, the 2 designers of the lecturers' team would also stay in touch with the artisans. This cross-culture communication would allow the correct understanding of both fields. If on one hand, the artisans had made contact with the logic of design, on the other hand, the designers would realize how to interpret this specific technology. With this methodology in mind, the students proposed new concepts based on previous experience in various fields and created new opportunities for innovation.

6.2   *Product development*

The artifacts made of rye straw are characterized by natural resources, by the traditional craftsmanship and the semantic competence communicated both by tactile stimulation, and by visual communication. Thus, the new product development should highlight not only the material characteristics, but also the semantic value of the rye straw, including the tactile qualities. On one hand, the new craft products must emphasize the practical function of the material, designing the material. Particularly, it means the rye straw must be designed, while still humid, and sometimes wet. On the other hand, the new craft products should highlight the aesthetic potential of straw, which qualifies traditional artifacts, particularly praising the tactile qualities of this natural material. This pilot project shows that these desirable characteristics can inspire new products with authenticity and sensory feeling, as demonstrated by fashion accessories or the packaging for gourmet products that were created during this research. The aesthetic qualities of straw were adapted to design cross-culture oriented products. For instance, endorsing links between symbols and narratives, such as a packaging for gourmet or links between the low technology and the high technology, such as a tablet cover. The first case, a packaging for gourmet entitled *Sabores com Tradição*, is related with three typical foods of the region: wine, sausages and bread. The students developed a line with two different packaging including a cultural transfer from the fashion segment. For instance, one of the projects seams a monk with a hat. The other packaging looks like a suitcase when it is closed, and becomes like a towel for picnic when it is opened, as the liner is in flax as shown in Figure 3. The collaborative teamwork functioned as a mutual learning mechanism and both sides—students and artisan—exchanged knowledge to enhance their capabilities. In this case, students realize that the project needs other materials than only the rye straw, such as cotton, wood, and flax, and this also means they need to get in touch with other artisans. The second case is a line of urban fashion products, such as a hat, a basket shop for computer and a tablet cover entitled *PureDuel* as shown in Figure 3. The idea was to emphasize the semantic competence of the rye straw, relating the traditional material with new technological products, such as, a personal computer and a tablet. Design performs as a cultural catalyst to transfer ideas from one source to another, guiding to new partnerships and to new ideas. In the specific case, students made use of abductive reasoning (Cross, 2006) and qualified the rye straw with the *Genius Loci* of Fafe. The rye straw is designed, relating the malleability of the straw with the traditional dick game of Fafe. As in Fafe only one artisan knows how to work the dick of quince, this case was also an opportunity for the survival and the sustainability of this tradition, as the stick quince needs a specific handmade work. With this idea in mind the

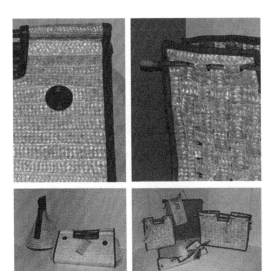

Figure 3. From left to right: Project 'Sabores com Tradição' developed by students Ana Margarida and Fábio Preto. Project 'PureDuel' developed by students Daniela Fonseca, João Rodrigo and Sara Burica. Images from the Joana Ferreira.

students developed the other projects. The eight proposals promote craft-design alliances.

Some projects exalt other craft activities, such as, filigree or embroidery. Other proposals relate the rye straw with industrial activities, such as cork industry, promoting the creation of semi-industrial goods.

7 CONCLUSIONS

The paper develops a pilot project, proposing a design development; wish was conducted by design and materialized by the craft. The competences of designers have increased and design skills are extensively applied to various markets. The topic of stimulating local manufacturing requires design knowledge. Therefore, craft-design collaboration may reinvent a local craft industry. The results of this collaboration indicate that a local craft, which import knowledge from other industries, through design, becomes a source to export new languages for new markets development. The inputs created by design in the rye straw activity permits other craft activities and other industries to reinvent themselves and to create opportunities for interdisciplinary alliances and to establish a new market territory for the craft. In this research, the product development introduced design-driven innovation as the craft-design union proposed different and spontaneous meanings to straw and created new market opportunities. It means, the research creates the basis for the development of new products. Particularly, the

artisans profited through the gradual transmission of design knowledge during the teamwork between designers and artisans. Craft-design collaboration inspires artisans to values their techniques, their culture and to fall in with further innovation. Considering the age of the artisans, more than 60 years old, the workload required to make a product was reduced, combining the straw with other materials such as wood, silver, linen or cork, which may be produced by other artisans. Students created a project interconnecting, not only the rye straw, but also other materials like pure linen, wood quince, leather, silver filigree, embroidery, wool, glass and cork. To do so, they sought for other craftsmen. Whit this action they become entrepreneurs of their own idea. From the design side, this pilot project was an occasion for professors training future designers the impact of design phenomenology (De Fusco, 1985). Thus, design schools and their students become social catalyst between the Academia and the real world. Methodologically, the present study introduced new manufacturing methods in local crafts presented in Fafe, such as new typologies and new connections for the creation of new straw artifacts. The creation of semi-industrial products is important both for crafts and for industries. Craft ensures survival of craft activities in the market, industry ensures the quality factor that handmade conveys. In the near future, the 10 proposals developed during this pilot project will be presented in Fafe to the community, the media and several local retailers. The products will also expected to be present in fairs abroad. Finally, this pilot project has challenged the community to act in a constructive way and the results of the collaboration have received attention from designers. It was also a way to explore new market opportunities for the artisan community. We hope other practitioners can apply the knowledge and experiences achieved from this study as an incentive for further research.

ACKNOWLEDGMENTS

The authors acknowledge the financial support of the Fafe City Hall and the Ave Inter-municipal Community, recognizing the support of the president of the IPVC, the students and the Fafe artisans. This paper is partially supported by the Portuguese Foundation for Science and Technology-FCT.

REFERENCES

Branco, João. 2005. Artesanato e Design: parcerias com futuro? *Mãos—Revista de Artes e Ofícios.* 27(28): 8–13.
Cappellieri, Alba. 2006. Pomellato in *Sistema Design Italia* (4)-15: 1–8.

Bucci, Ampelio. 2003. *L'impresa guidata dalle idee: Lezioni di Management Creativo dalla Moda e dal Design*. Milano: Arcipelago.

Cappellieri, Alba. 2005. Dalla faccia di Platone alla cravatta di Marinella: storie, contesti e valori di lusso. In Celaschi, Flaviano; Cappellieri, Alba; Vasile, Alessandra (coord.) Lusso versus Design. Milano: Franco Angeli, 63–94.

Chiapponi, Medardo (ed.) 1999. *Cultura sociale del prodotto*. Milano: Campi del Sapere-Feltrinelli.

De Fusco, Renato. 1985. *Storia del Design*. Bari-Roma: Laterza.

Fernandes, Sílvia. 2012. *Tecnologias tradicionais 'os entrançados de palha'*. Fafe: Naturfafe.

Jones, John Cristopher (1992) *Design Methods: seeds of human futures*. New York: John Wiley & Sons. (1st esd. 1970).

Kälviäinen, M. 2000. The significance of 'Craft' qualities in creating experiential design products. *The Design Journal* 3(3): 4–15.

Morace, Francesco. 2008. L'Accademia del valore. *Nova 24* (semestrale de Il sole 24 ore). (118). 8–9.

Rifkin, Jeremy (ed.) 2001. *A era do acesso*. Lisboa: Editorial Presenca.

Schön, D. A. (ed.) 1983. *The reflective practitioner: How professionals think in action*. New York: Basic Books.

Turinetto, Marco. 2005. *Be different: il valore attrattivo del brand-design nelle imprese moda*. Milano: POLIdesign.

Verganti, Roberto. 2001. *Le nuove sfide per l'innovazione di prodotti e servizi*. Milano: Dipartimento di Ingegneria Gestionale—Politecnico di Milano.

Zurlo, Francesco 2003. La strategia del design. *Impresa e Stato* (62) Milano: Camera di Commercio di Milano: 8–10.

Green Design, Materials and Manufacturing Processes – Bártolo et al. (eds)
© *2013 Taylor & Francis Group, London, ISBN 978-1-138-00046-9*

In direction to sustainable eco resorts

C. Alho
Auxiliar da FAUTL, Universidade de Lisboa, Lisboa, Portugal

J.C. Pina
Doutorando CIAUD-FAUTL, Universidade de Lisboa, Lisboa, Portugal

ABSTRACT: This research paper intends to show the emerging concepts in sustainable eco-resorts. The purpose of the study takes in consideration study cases in rural areas of Europe in order to define basic principles in eco-architecture and eco-urbanism to develop rural areas. The research methodology is based on case studies. The results achieved are formal and specifications in technologies and materials and the authors used the case of the most popular area at international level, in Portugal to illustrate the concepts achieved. The results show that low-technology is possible to increase good architecture and design. This evidence and conclusions allows the emerging concepts to be used in Mediterranean geographical areas.

1 INTRODUCTION

The purpose of this study takes in consideration study cases in rural areas of Europe in order to define basic principles in eco-architecture and eco-urbanism to develop rural areas, refering the most relevant research and theories based on the real world of camping and holiday parks in European rural areas.

This research, has the objective to show and illustrate the basic principles of eco-architecture and eco-urbanism and the new emerging concepts.

In the real world of camping and parks, the main idea of architects, town planners, surveyors and managers is based on proposals to contribute and resolve the economic and financial crisis with solutions supported by solar and aeolian energy, sustainable vision on architecture and urbanism, water supplies and eco-maintenance of buildings and materials.

The research uses the experiences in France (Montpellier and La Rochelle), Spain (Tarragona) and Netherlands. According to recent theories in architecture and urbanism, the principles supported by European Directives, Charts, Conventions and Recommendations refers that the emerging concepts to intervene in rural areas is based on eco-architecture and eco-urbanism.

The research uses a holistic vision and takes in consideration the antecedents that had led to modify the change of the paradigms from an industrial consuming society for a model that points in the direction of urban sustainability.

The objective of the research is to systemize a set of emergent concepts in the areas of architecture, urban plannig and tourism, in the way to establish an innovative program and, consequently, an urban and architectural proposal for a real context in the south of Portugal, more properly in Zambujeira do Mar, (Odemira, Alentejo).

2 THE RESEARCH METHODOLOGY

The methodology adopted was based on "case study" in accordance with Robert Yin (1994), one of the most appraised authors in the use of this methodology. According with Alho (2000) the research identifies a contemporary problem which intends to study emerging concepts in architecture and urbanism terms specifically in rural areas.

Based on Yin (Yin, 1994), the researcher does not have control on the data, going to use only one case study or multiple cases as form to prove and to generate knowledge for new research.

This technique, being sufficiently including, must be flexible and to use, at the same time, others techniques of research associated in order to produce final results which can be confirmed by another researcher that, following the same methodology, would arrive to identical conclusions (Hinks, 1996).

According to Denscombe (1998) case studies characteristics are:

- spotlight on instance,
- in-depth study,

- focus on relationships and processes,
- tend to be holistic and
- the case study is a naturally occurring phenomenon (Yin, 1994).

The authors take in consideration in special case study advantages (Denscombe, 1998):

- deal with the subtleties and intricacies of complex social situations,
- the use of multiple methods and multiple sources of data in order to capture the reality,
- no pressure on the researcher to impose controls or to change circumstances,
- concentrating effort on one research site,
- theory testing and building.

Finally take care with the disadvantages of case study approach (Denscombe, 1998):

- credibility of generalizations made from its findings,
- perceived as producing "soft" data,
- boundaries are difficult to define and poses difficulties in deciding sources of data,
- negotiating access to people and documents can generate ethical problems,
- the presence of the researcher can lead to the observer effect.

3 THE EMERGING CONCEPTS
   IN ECO-ARCHITECTURE
   AND DESIGN

Until the decade of 80, of XX century, the related global problems with the ozone depletion and with the climate change are a priority for a minority of scientists who did not hear its voice in the medias. The few that had attended the petroliferous crisis of years 70 forgotten the problem that the "energy crisis" raised and, each time, raise more.

Why the smashing majority of the experiences in architecture and urbanism consumes so much energy and they produce an exaggerated impact in the environment?

Based on the consideration that we do not learn the sufficient with the good practical examples or we do not systemize the technical standards which answer with severity to the sustainable principles of human being survival according to actual standards of life (Emmanuel, 2005).

According to Emmanuel (2005) the questions related with the global heating, the exhaustion of the energy resources and the bankruptcy of the speculative capitalist model meet in an impasse. Today we reach the border between the concerns of sustainable comfort and the projection of conducive secular shares to the survival of the species (Geyer-Allely, 2002).

Our generation, probably, is conditioned to prevent future cataclysms provoked for a set of factors that, cumulatively, has come to play a basic role in the constant degradation of the ecosystem conditions of the planet. Almost two decades that we are going to create knowledge capable "to support" the development model, guaranteeing the occidental ways of life to emergent countries intend in practical way.

According to possible scenes such as the considerable increase of the average level of waters of the oceans for saw of the global heating or the announced end of the fossil fuels, the world that we know would be obliged to move drastically (Stern, 2007). In such a way, we can consider that the changes, each time more visible, the climate, allied to the social changes and politics in the world provoked by the energy crisis need being equated according to a new "paradigm". In this picture, it will be necessary to review the models currently accepted for the organization of the built environment, in general, and of the architecture and urbanism in particular.

The "eco-resort" means a friendly environmental development of the area and has assumed the figure of a concept of resort whose location and destination offer a set of products, services and animation related to the environmental questions.

In a next future the "eco" will have to be transversal to all this developments. In the near future the "Eco" should be across all developments. It is no longer a cover of a concept but a levy of the hospitality market. Based on this vision it is necessary to understand the emerging concepts in eco-architecture and eco-urbanism.

Sustainability has become a widely applied Concept—so much, that the meaning lost precision and definition; today, it probably acts more like a symbol of a necessary Civilizational change, i.e. a different perception of Human activities and Values, in relation with an environment conscious attitude and accounting.

The concept of Environment was also evolving, at the same time—from an almost identity with Nature and the physical quality of its components affecting Mankind, to the perception and evaluation of the surrounding Universe, through social, economical, philosophical and cultural criteria, focused on the more subjective goals of "Quality of Life" and "Sustainable Development".

In the field of Architecture, sustainability is now also becoming mainstream; but the seeds were already there for the last decades—mainly after the oil crisis of the Seventies: Passive Solar, Bioclimatic, Green and Eco-Architecture had often claimed for the need of a better relation with Site, Physical Environment, Resources, Human scale and Cultural Diversity, pointing out the importance of local input and scale, towards a more Humane Architecture.

Governments, specially of the industrialized northern countries, have supported the climate conscious approach that some of these trends proclaimed, on a saving energy policy basis; but up till now, failed to influence the majority of architects and public opinion-besides the first buildings formal inconsequence and certain lack of quality, the Consumerist Way of Life that the industrialized World also sustained and publicized, and the civilizational blind faith on Techno scientific solutions to dominate Nature and mechanically solve problems, prevented a wide acceptance of an environmental attitude in the Architectural process.

A very representative number of architects and theorics choose the Ecological Principles as the reference to follow, in order to achieve the desired sustainability in Architecture-even here with a wide range of attitudes.

If one follows the original concept applied in the Bruntdland Report, and besides the optimal resolution of the binomial relation between Resources, Management and Quality of Life, Sustainability requires also other fundamental aspects:- Continuity, which translates better in the dynamic adaptation of a building (or urban fabric) to the continuous changing ways of Life and specially, Ethical responsibility towards next generations, to incorporate local and civilization information, seen as the essential resource to understand the Past and to provide alternative paths to build the Future.

Some authors consider Sustainable Architecture impossible, if a strict meaning is applied to the concept; in the context the definition was presented above, but it can be considered a redundancy, because a responsible Architecture should always incorporate those fundamental aspects referred, regardless of programmatically, economical, formal or other conditioning aspects in the process of Architecture design and implementation. However that is still not yet the case for the majority of the architectural approaches all over the World, and so, rather than another trend or formal style, Sustainable Architecture should stand for a basic integrative attitude to introduce in all levels of the Architectural Process.

The Bioclimatic Architecture consists of the conception of buildings having in consideration the local climate, using to advantage the available natural resources (sun, wind, vegetation) with the purpose to get through the drawing and with low energy consumption a degree of comfort raised in the use of the building.

The Bioclimatic Architecture integrates some climatic, ambient, cultural knowledge and partner—economic finding only solutions for each design.

The application of bioclimatic strategies in the buildings is essential to obtain to reduce the energy consumption and of carbon emission. A bioclimatic architecture is that one takes care of all climatic conditions in the conception of project, using passive solar systems of form to increase and energy efficiency. Not to confuse with the active solar Architecture that is associated with the use of mechanic instruments, for example, solar and photo voltaic panels, hybrid systems of cooling for evaporation, etc. Based on the roots of empiricism, the bioclimatic architecture is unproved of the one of technologies to acclimatize or to illuminate. Such constraints compelled to an efficient and inserted construction in the surrounding climate, using the local materials mainly.

The sustainable construction is defined as a constructive system that promotes interventions on the environment, adapting it to the use necessities, production and human consumption, without depleting in this intervention the natural resources.

Thus the systems of exploitation of pluvial waters, passive heating and cooling, quality of air and the water, maximization of the natural illumination as well as the use of renewed energies and impact of the used materials, are in pair and integrated in its global with the programmatically and aesthetic questions in the conception of the buildings.

As result of sustainable construction we have a building that it generates the resources rationally such as the energy, the water and the impact of the materials used in construction, taking care as a building to all estimated calculus of resistance of time, to allow the continuity of its function of shelter for a definitive use based on the waited indices of comfort.

To relate that the sustainable construction is transversal to all the concepts, styles and stylistics languages adopted and employed in the buildings. The used passive measures can modify the form but never the language of the building. As action base we have to find a good relationship of the building with the local climate searching in its selective permeability the capacity to accumulate and to absorb heat or cold, to renew air and to control the artificial light.

4 CASE STUDY

To create a playful park with two different sources of clients, first destined to a floating public who will go there only to spend the day and usufruct the varied equipment which will offer, second destination to the public that will lodge and privilege comfort and the benefits of nature in bungalows or mobile homes.

The eco-resort will function as a self-sufficient island, where the guests will find satisfaction for one varied gamma of interests and leisure, like sports, environmental and regional culture in a combination that attends de demands of different age levels.

This design project differs from its similar in camping, for the raised standard of quality that it reflects in the following areas:

4.1 Functionality

While unitary, this eco-resort intends to take care of two types of different visitors and the functionality was divided in four functional sectors, as it can be observed in the Master Plan.

4.1.1 First sector 0-lodging

The Lodging area, foresees beyond a dimensioned Parking, a generous Reception, and a Cycle-center (to rent a bicycle), which precede a Central Square, where the first contact with the resort will arrive, with a set of services and activities, and Leisure, such as Restoration area, Commerce and Entertainment complemented for units of entertainment like Amphitheatre, Mini-golf, Toys, and so on.

This Central Square will function as a distribution zone linked with the remaining areas. Its location is central and privileged in relation to the entrance in the complex and relation to the elements of bigger landscape attraction and offers several activities.

4.1.2 Second sector A-shelter

This camping leisure will be the nuclear area of the complex, and will agglutinate the biggest number of users and will offer a different concept of lodging linking with the nature. It corresponds to the camping zone, with diverse sources lodging like: tents, caravans, motorized caravans, until the most definitive solutions, such as mobile homes and bungalows.

It will enjoy a great autonomy of functioning and management of resources, and it is the sector that will make economically project viable.

This sector, on which we based our case study, is the most structural base of the complex and will be supported by a diversity of activities and services that complement a panoply of existing activities in the other sectors.

It is also responsible for the landscape and environmental development of the complex, with the creation of a biological swimming pool as well as a big tent that provides this area of a encounter area and a meeting point.

4.1.3 Third sector B-aquatic leisure

This sector corresponds to the main zone of the aquatic entertainment, where more traditional equipments are proposed, presenting more options on the sport offer. Complementing the wealth of the beautiful beaches in the proximity, this sector endows this complex with an attractive of great importance for the resort.

This sector counts on swimming pools of diverse activities (swimming pool/beach with waves, jacuzzis, lakes, islands, rivers, cascades, aquatic animation and nautical, health club and winter swimming pool, etc.), persecuting the main objective of working all the year.

4.1.4 Fourth sector C-environmental animation

This sector beyond usufruct of the natural resources will improve and rent areas of the park, with activities of preservation and environmental animation, such as sport of the nature, programs of sensibilization and environmental interpretation, activities agricultural, fifth of animals, environmental and cultural animation.

5 CONCLUSIONS

Eco-camping-resort gains a new importance in front of the new paradigm and it seems to appear in the camping leisure, two different and opposite forms in evolution.

In an upward direction, where this proposed project is integrated, expresses sophistication and the constant institutionalization, earning new contours pointing until the new classifications of "Luxury". In descending direction, the provisory lodging becomes in a form of permanent residence for a considerable part of population with modest incomes.

Thus, made these considerations of this case study, it seems us pertinent to take off the following conclusions:

- Tourism is to move, answering the new ecological and environmental concerns and also to a dramatic problem created by the development of masses tourism, that increases from Second World War until present, leaving entire cities structures that are not used during half of the year, what configures a clear problem of sustainability.
- That, will make the industry of tourism, on the XXI century, go thought substantially structuralized changes, and surely, make it more responsible, environmental and socially speaking.
- The proposal of echo-camping-resort that configured our case study is surely a new reply to a new program of tourism, looking for creative standards of quality and to propitiate a bigger contact of the customers with nature, as well as minimizing the negative impact of tourism.

- Thus, it makes sensible to conclude that the lodging in tourism of nature, in the case of eco-camping, is probably the evolution on the direction of sophistication and comfort, in the physical and architectural type of lodging, as well as in the number and quality of the leisure equipment.
- The notion of "luxury" is changing, and the close link to nature, related to unpolluted areas and harmonious natural environment is one of the great luxuries today.
- Solar and aeolian alternative energies had come to be and are in a primitive period of training development and will go to prosper and reach efficiency standards that at this moment still are considered utopian.
- As well, architecture and urbanism will follow development standards supported on "low" and "high" technology concepts. In the first case, simple and economic traditional constructive processes will be retaken and improved. In the second case, new, lighter and sophisticated materials will be developed, as well as, new equipments in order to achieve and improve technology.
- The 'economic' and 'efficacy' notions will walk along with energy efficiency, comfort and human well-being and in harmony with preservation of nature.
- Eco-camping-resort appears as a valid option to the construction of a new conventional tourist enterprise, and search to create permanent employment.
- In conclusion, "changes in society" creates new human and physical conditions on the built environment witch defines emerging concepts for eco sustainable resorts in rural areas of Europe.

REFERENCES

Alho, C. 2000. Authenticity Criteria for the Conservation of Historic Places. Ph.D. Thesis. Salford, U.K.: University of Salford.

Denscombe, M. 1998. The Good Research Guide for small-scale social research projects. Buckingham, Philadelphia: Open University Press.

Emmanuel, R. 2005. An Urban Approach to Climate Sensitive Design: Strategies for the Tropics, Spon Press-Taylor & Francis Group.

Geyer-Allely E. 2002. Sustainable consumption: an insurmountable challenge. UNEP, Industry and Environment Review, January–March.

Hinks, J. 1996. Research Methods. M.Sc. In Information Technology In Property and Construction. U.K.: Unpublished Collection, University of Salford.

Stern, N.H. 2007. "The Economics of Climate Change: The Stern Review", Great Britain Treasury Edition: illustrated, reprint, Cambridge University Press.

Yin, Robert K. 1994. Case Study Research: Design and Methods (2nd ed.), Sage Pub., California.

Green Design, Materials and Manufacturing Processes – Bártolo et al. (eds)
© 2013 Taylor & Francis Group, London, ISBN 978-1-138-00046-9

Disruptive innovation and learning to create architectural forms

L.S. Leite, A.T.C. Pereira & R. Pupo
*Programa de Pós-Graduação em Arquitetura e Urbanismo, Universidade Federal de Santa Catarina,
Santa Catarina, Brasil*

ABSTRACT: This purpose of this article is to explore the relationship between the transition process from the mass production industrial perspective to a mass customizing perspective and the necessity of generating such a teaching and learning environment on architectonic forms based on the free exploration of the digital technologies potential centered on students' abilities. Based on the convergence between Lawson's (2011) project process concepts and Christensen's (2012) disruptive innovation, didactic experiences were developed in the early phases of architecture and urbanism so as to identify the correlation between the digital technology use stimulation and the intrinsic motivation generation.

1 FROM THE INFORMATION ERA TO THE CONECTION ERA

According to The Economist (2012), there is a series of remarkable converging technologies, amongst them, the tridimensional printing and the wide range of net based services can be highlighted. As a result, the factory of the future will be focused on the mass customization; much more alike an artesian house than a Ford assembly line. It is worth pointing out that most of the jobs shall not be found on the manufacturing shop floors but in the nearby offices consisting of (designers, engineers, IT specialists) which shall demand additional skills and more and more collaborative work in order to create new products due to the barrier falling triggered by the internet.

Oliveira (2010) described the emergence of such a new era as the Connection Era, in which the higher value is not the possession of the information itself, considering that anyone is able to have access to it through search engines such as Google; rather, the value lies within the association to social relationships. In this context is the social media revolution, and according to Telles (2008), this expression has been widely used aiming to describe the phenomenon of Websites developed so as to make the collaborative content creation, social interaction and information sharing in several formats possible.

When comparing the three authors, we can catch a glimpse on which aspects the technology evolution caused profound socioeconomic changes marked by the transition from the Information Era to the Connection Era, from the mass manufacturing to the mass customization.

1.1 Disruptive innovations and learning

According to Christensen (2012), disruption is the process by which an innovation transforms the market whose services or products are complicated and very expensive where simplicity, convenience, accessibility and sustainability are the main characteristics of the sector. Additionally, it occurs in the adaptation process on digital technologies innovations as a possible influence on the learning methods modification. In the first decade of the 21st century, from Kenski's point of view, (2007) economy is globalized and volatile and formations has diluted into more and more peculiar professional demands which make widespread validity course organization harder. For the author, the use of digital technologies resources, such as simulations, telepresence and virtual reality bring a new moment to educational process. However, the most usual communication technologies on education have not provoked radical changes in the structure of the courses, especially in the articulation of contents and on teaching methods and didactics used by teachers when working with students.

Coll & Monereo (2010) report the key competences that every citizen should acquire in this new scenario, in accordance with DeSeCo—Definition and Selection of Competencies—OECD (Organization for Economic Cooperation and Development), can be grouped into three categories: being able to act autonomously; being able to interact in socially heterogeneous groups; being able to use resources and tools in an interactive manner (especially the digital medias).

Christensen (2012), when expounding on how disruptive innovation changes the way one learns, analyses the reasons why schools resist making improvements and concludes that the catalyst ingredient for a successful innovation is motivation. For the author, motivation can be intrinsic or extrinsic, and whenever there is high extrinsic motivation for learning something, school work is facilitated. Extrinsic motivation is the one that does not come from the task; it is when something has to be accessed. Intrinsic motivation takes place when there is a natural stimulus for the work, so it is inherently fun and pleasant. However, when many countries achieve stability and prosperity, extrinsic motivation vanishes and the students have more freedom to get involved in fields they regard as more pleasant, and as a consequence, schools must create intrinsically motivating teaching methods.

The author also highlights an important step so as to cause the school to be intrinsically motivating, which is customizing teaching in order to equalize it to the way that best suits the students' ability to learn. Additionally, the author says that in order to introduce the customization, it is necessary to step away from the monolithic instruction model and move forward towards a modular student centered approach, and recommends the use of software able to adapt itself to a specific kind of intelligence or learning style in the search of intrinsic motivation.

1.2 Creativity and method in the digital age

According to Kiwiatkowska (2007) and Orciuoli (2010), the fear that digital design suppresses the creativity for modifying the conception process is one of the main resistance argument against the acceptance of digital technologies. Nevertheless, the authors ratify that, much on the contrary to what some feared, digital technologies act as a creativity development element. Kiwiatowka (2007) points out that architects of the digital era are more capable to control the conception process of high complexity level forms, both when it comes to spaces geometry and advanced structural systems related to the professional that uses the traditional design; but the author clarifies that the architects choices during the creative process remain depending on aesthetic criteria coming from an idealizer's originary intuition.

Florio (2011) explains that the development phase can be more easily accomplished by computing designs with the usage of the digital tools, but points out that it is possible to anticipate Project decisions and experimenting solutions through the use of digital tools in the early stages of the project.

Pupo and Celani (2011) describe the association of the new production ways in Architecture with digital technologies joined in the project innovation, manufacturing and construction, for the tridimensional representations along with the physical model allow a greater success in the communication between the associated parts, establishing proportionalities, perspectives and project inherent functionalities, whereas a bidimensional might not make evident. The authors the possibility of an increase in the complexity of what can be built and in the possibility of experimentation with tangible examples.

Such questions lead us to Christensen's theory (2012), in his search of a modular student centered approach and highlights chiefly that the current theories on how architects and designers do their projects, Lawson (2011), must be deepened in the search of the understanding on how architecture and design students do projects, especially within their cultural and social reality context.

2 EXPERIENCE

The analysis of experiences carried out in two steps by professors Leandro Silva Leite and Nedilo Xavier Pinheiro Júnior during the second semester of 2012 with students of Formal Composition II, of the Architecture and Urbanism Course of the Universidade Católica de Santa Catarina.

Step 01: 3D digital models, developed with the software Sketchup have been made available to the students for the development of graphical analysis. The use of the digital maquette aimed at making the understanding of the composition as a whole possible, and also the training of the student's view on capturing images that best represent the conceptual element in analysis.

In order to develop the collaborative work ability, Moodle learning virtual environment was used for posting the presentations (pdf, ppt), that were shared for everyone through the tool "Fórum". After that, each work twosome analyzed and posted a work review on the work of 05 other work twosomes. When the collective construction process was over, the works were redistributed and each work twosome took the responsibility of reviewing the work developed by other twosome based on the reviews presented on Moodle platform. The result displayed strong integration among students and sensible improvement on the quality of the understanding of the concepts considered.

Step 02: Composed of the development of the architectural proposition of single-family residential unit from tridimensional composition through $3,0 \times 3,0$ m modules manipulation on a 729 m^3 fictitious terrain. The purpose of the exercise was putting into practice the application of

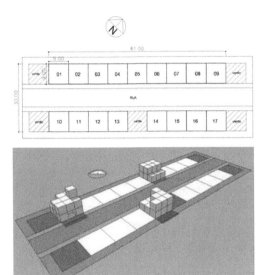

Figure 1. Proposed Distribution. Digital drawing developed by Prof. Nedilo Xavier.

the Gestalt concepts and developing the ability to comprehend environmental comfort basic concepts (lighting, insolation and ventilation), so as to integrate the formal composition development with other technical and functional needs. For that matter, a fictitious allotment was elaborated so that it was possible simulating the interaction between the housing units and the reflections on the space formal composition quality and their environmental consequences. Figure 1 shows the digital drawing developed to illustrate the proposition. The softwares used were the AutoCAD and the Sketchup. For the development of the exercise, it was regarded as a convention the elaboration of the physical volumetric model of the propositions so as to make direct interaction amongst students possible, and the use of digital resources was granted in agreement with the intrinsic motivation of each twosome.

2.1 Results

In order to verify the relevance degree of the digital tools free use students make compared to the obligatory development of a physical model, readings of the presented works were done as for the creative process (analysis, synthesis, evaluation), for identification of the interrelation of use of the tools. The mains factor verified was the development of the digital model for 85% of the students, software Sketchup and Floor Planner, the first one made available in computers of the institution and the second one found in the web as a result of initiative of one of the twosomes.

Within the creation process, the interrelation took place in several forms for each step (analysis, synthesis, evaluation). In order to understand such process, see Figure 2. In the analysis step, in which the formal conception possibilities were tested, there was balance between the use of the physical model (PM) or digital model (DM) and that was the moment in the process in which more twosomes took part using PM + DM. In the synthesis step, moment of decision on the adopted solution, there were greater use of the digital models compared to the use of the physical models and/ or joining PM + DM. See Figure 3.

In the evaluation step, graphic reading process for verification of the theoretical concepts application, there was a similar interrelation to the synthesis step; the difference is that in many cases, even when there was use of the physical model, it went through the graphic reading process due to the use of digital tools. As for the convention of elaborating a physical volumetric model of the propositions so as to make the direct interaction between propositions possible, we can notice that the interaction degree was lower than expected. Although there is a wood physical model of the implementation of the allotment, the form composition studies development were carried out in an isolated manner, between members of each twosome, rather than in a collaborative manner between twosomes. Figure 4 represents the joining of all models on the base, fact which occurred only at the moment of the delivery and presentation of exercises to the teachers.

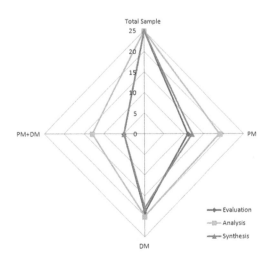

Figure 2. Graphics of the relation between the use, the total research students (25) the physical model (PM), digital model (DM), digital model + physical model (PM + DM), according to each step of the process (analysis, synthesis, evaluation).

Evaluation   Synthesis

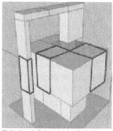

Digital Model (DM)

Analysis

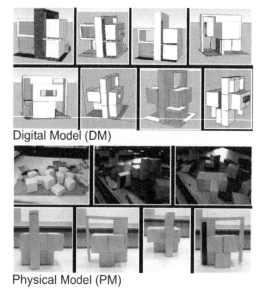

Digital Model (DM)

Physical Model (PM)

Figure 3. Picture of student work demonstrating the process of evaluation, analysis and synthesis.

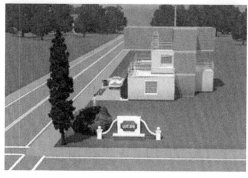

Figure 4. Images show students' work in the search of integration between the proposed models developed in physical model and digital model in the game "The Sims."

One of the twosomes, voluntarily developed a digital model of their proposition. In addition to the model in Sketchup; in the life simulation electronic game "The Sims", see Figure 4. When presenting to the other students, there was agreement that it would have been more interesting (motivating) if the allotment had been elaborated within the game, given the possibility of using avatars to navigate into the created spaces, and viewing all digital models in the same virtual environment.

When we compare perceptions on the obtained results in steps 01 and 02, we can point out the high degree of intrinsic motivation of students when it comes to the use of several digital tools, especially in the analysis and synthesis processes; both architectural references and developing propositions. As for the collaborative work and integration amongst participants and propositions, against all odds, better results were obtained using digital medias (Learning virtual environment—Moodle) rather than the use of presential physical elements.

3 CONCLUSIONS

The experiment carried out in the teaching of formal composition focused on architecture project demonstrated that establishing learning goals and the disruptive innovation stimulus made possible the understanding of which were the actual potentialities and learning needs of each student. Not forbidding nor demanding, not stipulating compulsory rules based on the fear that the new cannot achieve the same results as the old opened a fertile field for the development of the will to learn, of making sense within the apprentice's universe, of intrinsic motivation.

Such fact becomes evident when we realize that 85% of the works presented the use of digital models, which were optional, in many ways and effectively contributing to the creativity and 3D conception ability development.

Another questionable paradigm is the fear of the lack of integration between the process subjects and products generated by the so-called isolation caused by the use of digital tools, as demonstrated in the results obtained in the exercises carried out: Step 01exercise showed greater integration between the two subjects and especially between the products of the process by the use of the Moodle rather than the one carried out in Step 02 totally focused on the presential environment (classroom).

As a greater result of disruptive innovation as architectural forms teaching element we were able to obtain the development of a students' intrinsic motivations and capabilities centered teaching, as well as the exercise of autonomy ability, capability for interacting in heterogeneous groups and mostly the ability of using resources and instruments in an interactive manner, especially digital media.

REFERENCES

Chistensen, Clayton M., Horn, Michael., Johnson, Curtis W. 2012. Inovação *na sala de aula: como a inovação disruptiva muda a forma de aprender*. Porto Alegre: Bookman.

Coll, César & Monereo, Carles. 2010. *Psicologia da educação virtual: aprender e ensinar com as tecnologias da informação e comunicação*. Porto Alegre: Artmed.

Florio, Wilson. 2011. *Modelagem Paramétrica, Criatividade e Projeto: duas experiências com estudantes de arquitetura*. Gestão & Tecnologia, v. 6, n. 2, p. 43–66.

Kenski, Vani Moreira. 2007. *Educação e Tecnologias - o novo ritmo da informação*. Papirus Editora.

Kwiatkowska, Ada. 2007. A genêse das formas arquitetônicas: projetos inventivos na era virtual. In: *O lugar do projeto: no ensino e na pesquisa em arquitetura e urbanismo*. Cristiane R. Duarte, Paulo Afonso Rheingantz, Giselle Azevedo, Laís Bronstein [orgs.]. Rio de Janeiro: Contra Capa Livraria.

Lawson, Bryan. 2011. *Como arquitetos e designers pensam*. São Paulo: Oficina de textos.

Oliveira, Sidnei. 2010. *Geração Y: o nascimento de uma nova versão de líderes*. São Paulo: Integrate Editora.

Orciuoli, Affonso. 2010. *Projeto Assistido por computador: ontem, hoje e amanhã*. Revista AU, n° 197, August.

Pupo, Regiane; Celani, Gabriela. 2011. *Prototipagem rápida e fabricação digital na Arquitetura: fundamentação e formação*. Present in: Kowaltowski, Doris C.C.K.; Moreira, Daniel de Carvalho; Petreche, João R.D.; Fabricio, Márcio M. *O processo de projeto em arquitetura*. São Paulo: Oficina de Textos.

Telles, André. 2010. A Revolução das Mídias Sociais. Cases, Conceitos, Dicas e Ferramentas. São Paulo. M. Books do Brasil Editora Ltda.

The Economist. 2012. The Third Industrial Revolution. April 21, 2012. Available at: http://www.economist.com/node/ 21553017.

Green Design, Materials and Manufacturing Processes – Bártolo et al. (eds)
© *2013 Taylor & Francis Group, London, ISBN 978-1-138-00046-9*

Additive manufacturing as a social inclusion tool

G.M. Bem, R.T. Pupo & A.T.C. Pereira
Universidade Federal de Santa Catarina, Florianópolis, Santa Catarina, Brazil

ABSTRACT: This paper brings examples of elements manufactured by additive manufacturing (AM) used to improve impaired users' life. It was possible to identify cases of its application in the development of accessibility devices through a review of published papers in conferences and journals in the field of architecture, cartography, medicine and museology. Such applications are present on cartographic, urban mobility, perception of art, medicine and daily activities, related to assistive technologies. This work highlights the importance of this innovative manufacturing process in order to implement devices and initiatives towards social inclusion.

1 ACESSIBILITY AND SOCIAL INCLUSION

There was an increase in concern about global citizenship rights and social involvement of people with disabilities after the Second World War and more expressively during the 60's. In this context, the concept of inclusion refers to the possibility of social participation of these people in conditions of equality and without discrimination (Dischinger et al., 2009).

Nowadays, accessibility has become a challenge for the government and society, since it requires the elimination of architectural and urban barriers, in buildings, transport and communication (Bernardi et al., 2011).

An analysis by Sperling (2006) on the conceptual tool User Pyramid—proposed by the Ergonomic Design Group in Stockholm (Bengtzon, 1993, 1994 apud Sperling, 2006), brings a quantitative population distribution considering their performance in daily activities and interaction with the environment. There are three involved stages: 1) the bottom one is composed of people without disabilities and slight functional limitation; 2) the middle portion is occupied by users with disabilities, who need some technological support or personal assistance, and 3) the top division is represented by people with severe disabilities, which require advanced technology and staff support. This panorama shows the insertion of technology in improving impaired users lives.

Further, according to Vanderheiden (1997 apud Story, 1998) there are three ways to improve the capabilities of individuals with disabilities: (1) modifying the physical form, (2) equip the individual with tools they can use, or (3) modify the environment.

An analysis of articles, journals, theses and dissertations in fields such as cartographical, medical and architectural, showed that the Additive Manufacturing (AM) is present to satisfy the first two conditions, and also related to user interaction with art and the environment.

AM defines automated production methods, customized and personalized, whose goal is the development of small-scale prototypes for the purpose of study, review and presentation or 1:1 scale manufacturing (Pupo & Celani, 2011). AM machines can build prototypes layer-by-layer in an automated way, from a 3D digital model and it is based on solid, liquid, powder or laminated material, depending on the equipment used (Pupo & Celani, 2011). On the other hand, subtractive processes use cutters to chop or cut the material adopted through the milling machines with Computer Numerical Control (CNC) (Pupo & Celani 2011).

Some authors have adopted this term for processes that are quick but not truly AM. Everything from machining to molding is now described as a AM process (Geng, 2004). According to Celani et al. (2009), they were originally created for product design development and more recently, its use has been widely spread for any application where materialization is needed.

2 APPLICATION FIELDS

2.1 Tactile mapping

Tactile Mapping is a sub-field on a wide range of applications on accessing image-based information through touching. As a sub-division of cartography, it seeks for best ways to represent spatial information, for visual impaired users (Koch, 2012).

In principle, all maps for visually impaired users can be manufactured automatically or partially automatically. The two most used techniques are thermoform (vacuum forming) and micro-capsule (paper) and fusion. Regarding the thermoform technology, a digital tactile map is generated and automatically transformed into control commands for a CNC. Then, a negative model is generated, which is used as a mould ("master") for de vacuum-forming of the 3D-map (Koch, 2012).

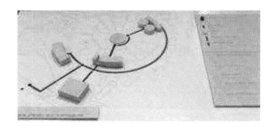

Figure 1. Tactile model. *Source*: Capeli & Bernardi, 2011.

2.2 Tactile models

Spatial orientation conditions are determined based on the characteristics of the environment that enables users to identify the spaces and its functions, enabling the development of displacement strategies and use (Dischinger et al., 2009).

To represent the environment and its understanding by visually impaired users it is necessary the use of tools and symbols that, combined with braille, make part of a reading instrument and aid mobility, such as tactile maps (Capeli & Bernardi, 2011).

In this sense, a project called "Project audible and tactile map for the central area of UNICAMP, campus developed in 2009/2010" relies on the use of two technologies of AM in making a tactile map to achieve these goals (Fig. 1).

The first process used was the laser cut in laminated sheets to represent the accessible route of the studied area. In addition to the tactile benefit obtained with the choice of material, the use of the laser cutter guaranteed accuracy and product finish (Capeli & Bernardi, 2011).

The second technique used was the 3D printing (3DP) for manufacturing braille subtitles and prototypes of existing buildings on campus. The process was developed using a 3D printer from ZCorp (310Plus), which works by layered printing process using powder as raw material. According to Capeli & Bernardi (2011), due to the chosen technique, 3D printed models showed no satisfactory final performance, requiring major adjustments and tests, which would be unfeasible for research.

However, according to Capeli & Bernardi (2011) it was clear that, compared to the handcrafted process of making tactile maps, the use of AM proved to be much more precise, easy and quick to be made. They also claim that printers that use other types of raw materials, such as plastic instead of powder, would provide a better tactile sensation obtaining better results.

2.3 Perception of Art

Because of the visual language being the more expressive form of communication in museums, users with visual disabilities face a great deal of restrictions on enjoying existing assets in museums than any other visitor. According to Sarraf (2006), despite the visually impaired persons have the right of access to museums, initiatives for inclusion of this public are still scarce.

Facing this context, a work developed at University of Technology (TU Vienna) in Vienna, Austria, presented a computer-assisted workflow that creates tactile representations of paintings, capable to be used as a learning tool in the context of guided tours in museums or galleries (Reichinger & Maierhofer, 2011).

Starting from high-resolution scan or photograph of original paintings, the process allows an artist to quickly design the desired form, and generate data suitable for AM machines to produce the physical touchable pieces. The use of Laser Cut, were implemented to reproduce the objects space layout and also to augment their depth relation. In order to produce fine details, like brush strokes, textured reliefs suitable for the sense of touch, a CNC-milling machine was used. Figure 2 shows the final stage of the process. As a result, the authors affirm that these methods mimic aspects of the visual sense, make sure that the output is quite faithful to the original paintings, and do not require special manual abilities like sculpting skills (Reichinger & Maierhofer 2011).

2.4 Assistive technology and medicine aplications

Assistive Technology (AT) was defined by "The World Health Organization" as any product, instrument, equipment or technology used to improve the functionality of an impaired user (Organização Mundial de Saúde, 2004).

In medical field, according to Milovanović & Trajanovic (2007), AM emerges in the design and development of medical devices and instrumentation corresponding on its best results of implementation. It applies from hearing aids to several surgical aid tools. Besides, it brings a great improvement to the prostheses and implants fields, which has the ability to customize and make dimensional adjustments according to users particularities.

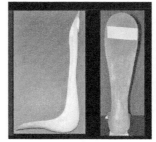
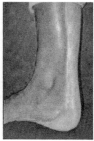

Figure 3. Flexible AM AFO. *Source*: Adapted Mavroidis et al. 2011.

Figure 2. Raffael's Madonna of the Meadow proto-typed. *Source*: Adapted Reichinger & Maierhofer, 2011.

Another field of application in medicine is related to orthoses, which, according to Mavroidis et al. (2011), are traditionally manufactured to serve a lot of users which do not guarantee optimal adjustment to the anatomy of each patient. Therefore, the authors conducted a study on the development of custom ankle-foot orthoses (AFO). The process was developed in three stages: (1) 3D scanning of the related part from user model, (2) manipulation of the mold using Computer Aided Design (CAD) software, and (3) manufacturing orthose through the process of stereolithography (SLA) (Fig. 3). Due to the raw material used, it was possible to achieve lightness and flexibility to the prototype. Moreover, in a comparative analysis to the traditionally produced models, besides the advantage of customizing the product to the anatomy of the user, there was a decrease in manufacturing time and the reduction of the cost of replacement, when necessary (Mavroidis et al. 2011).

According to Phillip (1993, apud Hurst & Tobias, 2011), the result of a survey of 227 adults with disabilities who use AT, showed that almost 33% of all devices were completely abandoned. The main factors identified were: (1) user involvement in device selection; (2) difficulty on purchasing the device; (3) device performance and (4) change in ability. In this sense, Hurst & Tobias (2011) studied that the "how-do-it-yourself" culture and tools could be applied to create, modify, or enhance AT. As a result, a new generation of affordable AM tools made it possible for individuals to build and cus-

tomize physical devices such as wheelchair accessories, prosthetics, and tools to support activities of daily living such as eating, dressing, and accessing a computer.

Lipson & Kurman (2010) state that personal-scale manufacturing tools enable people that have no special training in woodworking, metalsmithing, or embroidery to manufacture their own complex, one-of-a-kind artisan-style objects. AM gives the opportunity for users with any kind of impairment to create their own physical object, by using CNC or 3DP, for example, without the risks that is inherent to traditional manufacturing process (Hurst & Tobias, 2011).

3 CONCLUSION

Studies indicate that the process about the needs of users with disabilities is relatively recent. Fortunately, the portion of the population who seeks for solutions that provide a better life for these citizens is growing, often through government sectors, either as legislative and executive or research incentive.

Due to the benefits of its use, AM techniques can be applied in several areas. Through an analysis of articles, journals, conference proceedings and other publications in the medical field, it was identified that the use of this technology has much to contribute to social inclusion in several areas such as architectural, cartographic, engineering and technology.

Moreover, it is clear that technologies, especially AM, have a great power of insertion directly or indirectly in the development of daily activities of people with disabilities. Such applications refer to spatial orientation as an aid towards perception of accessible routes, improving mobility performance, communication regarding sound perception, the viability of surgical procedures with better planning, study and performance, custom manufacturing of assistive technologies. Besides, they allow visually impaired users to feel and understand visual elements by tactile representation.

REFERENCES

Bernardi, Núbia; Pina, Silvia A. Mikami G.; Arias, C.R. & Beltramin, R.M.G. 2011. O desenho universal no processo de projeto. In: Kowaltowski, Doris C.C. K.; Moreira, Daniel de Carvalho; Petreche, João R.D.; Fabrício, Márcio M. *O processo de projeto em Arquitetura—da teoria à tecnologia*. São Paulo: Oficina de Textos. p. 222–244.

Capeli, Giovanni Andreas & Bernardi, Núbia. 2011. *Projeto de mapa tátil e sonoro para área central do campus da UNICAMP:* desenvolvimento de recursos que auxiliem na orientação espacial de usuários com deficiência visual. Campinas.

Celani, G.; Pupo, R.; Duarte, J.P. 2009. *Technology transfer in digital prototyping by means of research laboratories: two case studies in architecture schools.* In P. Bártolo et al. (eds) Proceedings of the 4th International Conference Advanced Research on Virtual Reality and Rapid Prototyping, on Innovative Developments in Design and Manufacturing, Leiria, Portugal, October. p. 687–690. Leiden, The Netherlands: CRC Pres/Balkema.

Dischinger, M.; Bins Ely, V.H.M. & Piardi, S.M.D.G. 2009. *Promovendo acessibilidade em edifícios públicos:* programa de acessibilidade às pessoas com deficiência ou mobilidade reduzida nas edificações de uso público. Florianópolis: [s.n.].

Geng, H. 2004. *Manufacturing Engineering Handbook*, McGraw-Hill Professional, New York.

Hurst, A. & Tobias, J. 2011. *Empowering individuals with do-it-yourself assistive technology.* University of Maryland. Baltimore.

Koch, W.G. 2012. State of the art of tactile maps for visually impaired people. In: Buchroithner, M. (ed). *True 3D in cartography: autostereoscopic and solid visualization of geodata.* Lecture notes in geoinformation and cartography, Springer-Verlag. Berlin Heidelberg. p. 137–152.

Lipson, Hod & Kurman, Melban. 2010. Factory @ Home: *The Emerging Economy of Personal Manufacturing.* Report Commissioned by the Whitehouse Office of Science & Technology Policy.

Mavroidis, C.; Ranky, R.G.; Sivak, M.L.; Patritti, B.L.; DiPisa, J.; Caddle, A.; Gilhooly, K.; Govoni, L.; Sivak, S.; Lancia, M.; Drillio, R. & Bonato, P. 2011. Patient-specific ankle-foot orthoses using rapid prototyping. *Journal of neuroengineering and rehabilitation.* v. 8, n. 1.

Milovanović, Jelena; Trajanović, Miroslav. 2007. *Medical applications of rapid prototyping.* Scientific Journal Facta Universitatis. Mechanical Engineering. v. 5. n. 1. Niš. p. 79–85.

Organização Mundial de Saúde (OMS). 2004. *Classificação internacional de funcionalidade, incapacidade e saúde* (CIF). Lisboa.

Pupo, R.T. & Celani, M.G.C. 2011. Prototipagem rápida e fabricação digital na Arquitetura: fundamentação e formação. In: Kowaltowsky, Doris C.C.K.; Moreira, Daniel de Carvalho; Petreche, João R.D.; Fabrício, Márcio M. *O processo de projeto em Arquitetura—da teoria à tecnologia*. São Paulo: Oficina de Textos. p. 470–485.

Reichinger, Andreas & Maierhofer, Stefan. 2011. *High-quality tactile paintings.* ACM Journal on Computing and Cultural Heritage. Nov. Vol. 4, n. 2, Article 5.

Sarraf, V.P. 2006. *A inclusão do deficiente visual nos museus.* Musas (IPHAN), Rio de Janeiro, v. 2, n.2. p. 81–86.

Sperling, Lena. 2006. Design concepts and theory. In: Paulsson, Jan. *Universal design education.* Sweden: EIDD Sverige & NHR. p. 31–41.

Story, M.F. 1998. Maximizing usability: the principles of universal design. *Assistive technology.* v. 10. 1. p. 4–12.

Green Design, Materials and Manufacturing Processes – Bártolo et al. (eds)
© 2013 Taylor & Francis Group, London, ISBN 978-1-138-00046-9

The Olivetti factory as a paradigm of sustainable growth

F. Castanò & A. Gallo
Seconda Università degli Studi di Napoli, Aversa, Italia

ABSTRACT: The Olivetti's industries are the first example to analyze and understand the complicated dialogue between the past and the present, between culture and economy, between architectonic space and anthropized environment. The aim is to give rise to considerations about the part of modernity based on historical knowledge.

This thesis tries to stimulate new considerations regarding the "Sustainable factory planning and scheduling" topic from an historical point of view, using Adriano Olivetti's work as clever and innovative development model. In this model the product excellence is matched by oriented and continuous education, attention towards human relationships, technological research, scientific work organization, new project development and active communication methods.

1 THE ADRIANO OLIVETTI'S CONTRIBUTION TO THE DEVELOPMENT OF INNOVATION CULTURE

Halfway through the 20th century, the Italian industrial buildings within the Adriano Olivetti's *communitas* obtain in the architectonical field a continuously more defined expressiveness, which join the awareness of the functional aspects, a harmonic renovation of the environmental conditions and a reestablished human dimension.

In the same way studies on the industrial construction industry take in consideration new aspects about the field of investigation, by defining with increased attention the relationship between the factory and the territory, in the attempt of bringing this project topic to the conversation with the "environment", which started after the Second World War, when the architectural intervention in peripheral and rural locations meant the recovery of social and collective values, in contrast with the anonymity of speculation. Olivetti's ideologies were founded on active qualification, spontaneity cult, defense of environmental values and control over buildings. They shape the features of a new urban arrangement, showing ahead of time attention towards what Roberto Pane identified in 1980 as the "ecological request, many years before people started to talk about ecology" (Pane, 1980). The modern factory project includes an alternative vision of the relationship between form and function, which previously regulated the rationalistic project, in line with a research of a new quality of work, a renewed culture of the territory, a prospect of social welfare.

The initial process of rationalization of the first thirty year of the century was followed, from 1946, by the enlightened attempt by Olivetti to bring the young Italian architecture towards linguistic experimentation. It was able to join the functional request of European origin and the productive process technologies of American influence with a well-established handcrafted condition; to bend the informative principles coded by Walter Gropius, and the fordist inventions created by Albert Kahn, to the syntax of tradition and to the dialogue with nature (Darley, 2007). Thus, the Olivetti's industrial architecture focuses on the slow and gradual factory transformation in an effective production organism integrated in the territory, able to discuss with the environment, as its component. «The invention and creation of industrial landscape», as described by Eduardo Vittoria in the first number of *Città Aperta* in 1957, imposes the «homogeneous creation of entire areas intended for the production, which are directly and indirectly linked to the life of the modern human being» (Vittoria, 1957). Using a point of view that anticipated the sustainability issue, the industrial projects become more governable, by confronting the man and his habitat. The Olivetti's architects realized in Ivrea, in the "canavese" region, in Caserta and mostly in Pozzuoli the "green factories", imagined by Le Corbusier, places of integration for the production space and the natural context.

They shape the anti-industry, imagining them like some kind of "big houses" immersed in landscape full of air, light and rediscovered beauty (Tafuri, 1982). The composition scheme of the extended estate in Pozzuoli, planned by Luigi Cosenza between 1951 and 1954, denies the indus-

Figure 1. View of Olivetti factory in Pozzuoli planned by Luigi Cosenza with Pietro Porcinai for the project of the gardens (photo Charlotte Sørensen, 2012).

try idea itself, based on the traditional building type, in which all functions are focused. On the contrary, Cosenza realizes a set on urban scale, perfectly integrated in the surrounding context. The planimetry is based on a big cross with arms of different length, emphasized at the center by a squared hall for the workshop and, at the extremity of each wing, by technical and service buildings. The structure is formed by two kinds of thin round-section cement pillars, differentiated by their height, which sustain the shingles that jut out and define the estate perimeter. The collective environments, all towards the sea, are characterized by a transparent perspectives on the green. The external part was created by Pietro Porcinai, and it enters the composition through sloping floor solutions, the orderly rhythm of tree trunks, the unexpected reflections game in the central body of water, joining the plant development with the building complexity. Cosenza breaks the limitative distinction between inside and outside, by integrating the factory in the landscape. He denies its exclusive belonging to the technic world, even if it represents its best expression, and, instead, he delivers it to poetic sensibility and humanistic considerations (Cosenza, 2006).

All Olivetti's factories are born from the consideration of architecture according to which it represents a useful instrument for ethical and social revolution; to a greater extent the industrial one, lived by the working man, which is always considered with respect, because he is the main actor in the chain of progress, the uncontested protagonist of future civilization. This approach certainly was cutting-edge for that period, given the sensitivity and attention to the ethic and cultural request that only many years later would have called attention to the project policy. In effect, this is what the young director Roberto Guiducci will write in 1955 in the notes of the construction site of the factory in Pozzuoli: "Creating examples of civilisation and culture does not merely solicit the future, but determines a duty that cannot be compromised

Figure 2. Glimpse of the building and the garden (foto Charlotte Sørensen, 2012).

Figure 3. Porch entrance to the office complex open to the Gulf of Pozzuoli (photo Charlotte Sørensen, 2012).

or ignored. That's why, in building the factory and quartier, the deliberate decision was taken not to shy at any atgtempt to reach the highest goal precisely where the basic essentials are lacking; noto to be afraid of proposing the highest levels of civilisation where time-compressed backwardness bears the weight of centuries. It is here, in our opinion, that the deepest innovative sense of this "example" of Southern architecture lies" (Guiducci, 1955).

2 THE SUSTAINABILITY AND OLIVETTI

Halfway Sustainability, as defined at the end of the 80s, is characterized by a particular complexity that contributes to its value; however, at the same time it represents a limitation to the integration into current scenarios. Paolo Tamburini, thanks to his studies considering sustainable growth as an "ambitious strategy", shows how he wishes to take in consideration several phenomena able to generate ambiguity regarding this concept. On one side the ability to intervene in different situations, sometimes in contrast, like anthropic, ecological, economic, political, social and cultural ones, involves the related development and improvement of the above-mentioned systems. On the other, the contemplation of multiple directions weakens its meaning, and it brings questions concerning the real scope and purpose of sustainability, its likelihood, its being desirable and sufficient for society growth (Tamburini, 2005).

Using project creativity and technological culture, two elements that nowadays are considered essential to sustainable development, Adriano Olivetti creates an "industrial strategy" that gives value to the whole company system through the identification, in its entirely, of the potentialities to activate an evolution and both economic and social growth, thanks to the integration of different components from the world of work (Castronovo, 2012). This is possible thanks to a "progressive spirit", meaning the capacity to handle complexities, while offering in his period and to future generations a solution put before the problem itself. The Olivetti sustainable project is mostly based on the connection among an ethic concept and a development-oriented idea. This recalls the "political and cultural mediation" nowadays identified by Luca Davico as a fundamental aspect of the sustainable development, which is the ability to bring together emerging ethic sensibility and the modern request for a continuous and increasing progress (Davico, 2012). From his research it stands out that the full overlapping of the sustainability and smart growth concepts is possible when the second one combines economic, profit-driven development, to responsibility and consciousness.

Looking at the industrial architecture evolution throughout the 20th century up to now, are clear the

differences with the model created by Olivetti. This situation risks to diminish its action to an attempt to isolated progress, and to make utopian the industrial strategy he created, which is oriented to cultural, material and social elevation for the human being and the places in which he operates (Zevi, 2012). Instead, starting from the current vision of the factory of the future, smart, sustainable and innovative, some significant elements stand out; they are already present in Olivetti's model, being considered as development levers because they are the reason of its success throughout time and its realization. Among these, the intuition to centralize the temporal factor within the growth process, in order to assure that the extension of the industrial buildings on the territory, and the production and sales increase match equal interest towards the service optimization, the quality of life and of the environment, autonomous aspects compared to the production activity. One example is Ivrea, given that over thirty years it grew combining the architectural expansion to the realization of infrastructure of social impact as the Kindergarten, houses for the working class, the Service Desk, the dining hall, the Research and Experience Facility. Essentially, the "canavesi" architectures, nowadays a candidate to become Unesco World Heritage, as well as buildings in Pozzuoli and Caserta converted over time to other destinations, are even now an economic resource and cultural heritage (Olmo & Bonifazio, 1996). Of particular interest for the innovative characteristic, there is also the choice to substitute the linear and rigid hierarchical structure with a more flexible order, able to adequate to productive process transformation (Castagnoli, 2012). The constant research

Figure 5. The presence of small courtyards strengthens the relationship between architecture and nature (photo Charlotte Sørensen, 2012).

Figure 4. Detail of Olivetti factory in Pozzuoli (foto Charlotte Sørensen, 2012).

for improvements shapes over time industrial buildings identifiable as undefined places, in terms of a distance from being specialized organisms which are delimited by obsolete functional typologies. The coding of building types remains evident, but the specific functions are hidden in favor of the celebration of a poetic and spiritual environmental dimension (Branzi, 2006). To this end, the organization and the order of the job roles are defined in function of the external space, in a constant dialogue, and it grows the interest towards the achievement of living and working conditions ideal to stimulate the planning, thinking and communicating phases among people. The Olivetti development plan considers the set of relationships within architecture, entrusting growth into the hands of enlightened, aware and educated commission-based orders, and a concurrent collaboration of many different personality. Thanks to the help of technicians and intellectuals, he creates a thick, flexible and variable relationship-wise dimension, which reflects the fluidity and heterogeneity of contemporary societies and which reaches a high efficiency because of its own complexity. The interdisciplinary research and the always different experiments regarding languages and solutions tend to reach a "partial and completed" quality, which does not claim to embody an absolute result, because it represent part of a project that each time establishes its values and sense according to a specific purpose (La Rocca, 2006).

Olivetti's strength consists in his ability to renovate the whole factory concept, by giving new life to its processes, services, products and architecture, thanks to an healthy innovation. This kind of operation is usually studied from afar as positive reference model, yet here it is proposed as a continuous evolutionary process to be followed in order to move in the direction of a smart and sustainable growth, transferable in this day and age and able to carry on in the future. Through Olivetti planning idea, it is possible to keep on tracking a functional path to economic, social and cultural development for the contemporary production system. The possibility to experiment consciously, determined by the comprehension of the ongoing changes and the big issues of that period, makes even more likely to conceive new formal, cultural and technological languages able to educate the community.

Hence the need to identify the main points to lever for a sustainable planning and scheduling of the factory system; the development of a long-term growth strategy aimed to trigger real and durable changes thanks to the culture, meaning research, knowledge, education, as an instrument for social and economic growth, such as the construction and organization of the workspace as a driving force to collect, comprehend and transmit creative stimulus outwards, expressing through architectural complex the quality of the environment, of the work done inside and of the people involved; even the choice to oppose the interconnection of single specializations to the division of work and knowledge, thanks to the creation of open, flexible and versatile systems, able to foster progress through communication and sharing of knowledge and experience. Moreover Adriano Olivetti has also the ability to establish the principles behind ethical and ecological actions that flow over time. This approach transcends the utopia of spiritual elevation and it becomes concrete in a peculiar sensibility in the factories humanization by making them people-oriented and in harmony with the surrounding environment. An ethic measure for its being "fair", given its ability to satisfy even future generations.

REFERENCES

Branzi, A. 2006. Postfazione. In Francesca La Rocca, *Il tempo opaco degli oggetti: forme evolutive del design contemporaneo*. Milano: Franco Angeli.

Castagnoli, A. 2012. *Essere impresa nel mondo. L'espansione internazionale dell'Olivetti dalle origini agli anni Sessanta*. Bologna: il Mulino.

Castronovo, V. 2012. Prefazione. In Adriana Castagnoli, *Essere impresa nel mondo. L'espansione internazionale dell'Olivetti dalle origini agli anni Sessanta*. Bologna: il Mulino.

Cosenza, L. 2006. *La fabbrica Olivetti a Pozzuoli*. Giancarlo Cosenza (ed.). Napoli: Clean.

Darley, G. 2007. *Fabbriche. Origine e sviluppo dell'architettura industriale*, Bologna: Pendragon.

Davico, L. 2012. Etica e sostenibilità. *Lo sguardo—Rivista di Filosofia* 8(1): 75–76.

Guiducci, R. 1955. Appunti dal giornale del direttore dei lavori. *Casabella-Continuità* 206 (luglio-agosto): 64–74.

La Rocca F. 2006. *Il tempo opaco degli oggetti: forme evolutive del design contemporaneo*. Milano: Franco Angeli.

Olmo, C e Bonifazio, 1996. Serendipity a Ivrea. In Vittorio Gregotti e Giovanni Marzari (eds.), *Luigi Figini e Gino Pollini, Opera completa*. Milano: Electa.

Pane, R. 1980, Adriano Olivetti. *Napoli nobilissima* XIX (V-VI): 232–234.

Tafuri, M. 1982. *Storia dell'architettura italiana 1944–1985*. Torino: Einaudi.

Tamburini, P. 2005. Scenari di sostenibilità. In Giovanni Borgarello (ed.), *Condividere mondi possibili. Formazione, management di rete e sviluppo sostenibile*. Perugia: Regione Umbria.

Vittoria, E. 1957. L'invenzione di una fabbrica. *Città aperta* 1. In Giovanni Guazzo (ed.), 1995. *Eduardo Vittoria. L'utopia come laboratorio sperimentale*. Roma: Gangemi.

Zevi, L. (ed.) 2012. Le quattro stagioni. *Architetture del Made in Italy da Adriano Olivetti alla Green Economy*. Milano: Electa.

Green Design, Materials and Manufacturing Processes – Bártolo et al. (eds)
© 2013 Taylor & Francis Group, London, ISBN 978-1-138-00046-9

Avieiro Boat—its recovery as a strategic factor for a sustainable development process

C. Barbosa
IADE-U Instituto de Arte, Design e Empresa—Universitário, Portugal

ABSTRACT: The study on the avieiro boat falls within the proposal of a project started about three years ago, aimed at the application of the Avieira Culture for National Heritage.

A primary objective, to substantiate the legitimacy of applying for national heritage, will undergo scientific research focusing on all aspects of their ethnography and culture.

Beyond its anthropological values the recovery of the avieiro boat will contribute to a sustainable development process.

1 THE SCOPE

The fact that the boat is a peculiar object and an inseparable part of avieiras community life capable of being investigated in several respects, including those related to its shape and all the components that define it, as well as specific gear for the different types of fishing, but especially the relationship of life that avieiros fishermen kept with it.

If the peculiarities of their physical structure, by itself, justify a research study, first with regard to the origins and possible influences, will win, of course, a new design if they are examined in the light of anthropology and sociology, in attention to the fact that the construction of the avieiro boat considers a subdivision as if it were a domestic space and match, as a rule, the first home, a circumstance to reveal the population of fishermen in Portugal.

Moreover, fishing activities, today, with low profitability, complemented with the provision of services in the areas of leisure and discovery of nature, with reference to the principles of sustainability, put this boat in the center of a wealth-generating process by creating new opportunities for professional work assuming the respect for the Nature.

2 THE QUESTION

Avieiras communities, in an attitude of defense, chose as a rule to live far away from Lezíria populations working in the fields.

Accordingly, the next block from the Ribatejo Songbook (Alves Redol) is significant:

"I don't want to go to the field/It is very hot there

I don't want to be a rural worker/Because my love is a fisherman"

(literal/free translation)

In fact, avieiros fishermen, consider that to them, rural labor was little consistent with the motivations of their migration. They preferred to say that your "field" was the "sea" as often referred to the river Tagus. Despite the poverty and hardship that characterized his life, avieiras communities, sought to preserve a certain autonomy and independence.

As a trend they were also the subject of discrimination by local people, who looked at these communities with some suspicion.

This situation has contributed to the emergence of a very particular culture whose essence can be found in its homeland—Vieira de Leiria. The initiatives of the project which are promoting the application of the Avieira Culture for National Heritage have revealed a wealth of ethnological and anthropological values that are unique, showing at the same time that it has the features and capabilities suitable for a sustainable development process.

Among the tangible heritage, we find stand stilts buildings of houses and docks and the boat.

And there is now a combination of dynamic and wills, even political, in order to identify and recover the references of the Avieira cultural heritage.

The specific objective is the scientific and analytical study of all aspects of avieiro boat that characterize as a structural element of the *modus vivendi* of the avieiras communities, investigating the sociological and anthropological virtues that distinguish it as it manifests significant factors and identifying its culture, to integrate a sustainable development process considering the parameters that this concept implies: quality of life of social actors involved, nature conservation and economic growth.

To this end, applying the design methodologies, it will be studied ways to use the models of the boat (notably with regard to their size), reintroducing the use of the sail (replaced by motorized equipment) that will allow entrepreneurship options, environmentally friendly and respectful of ecosystems within the tourism, culture and sport.

3 THE AVIEIRAS COMMUNITIES

"Nomads of the river, like the gypsies in the land, had come from Praia da Vieira and were living apart: they were called avieiros"
(Alves Redol, in "Avieiros—literal/free translation)

There is no record of the arrival of fishermen avieiros to Borda d'Água. Indeed, Maria Adelaide Neto Salvado, writes in *Os Avieiros nos finais da década de cinquenta* (1985), that "no news were written, accurate or inaccurate, and in memory of the olds, she moves back and lies beyond a time that no one seems to know."

The sedentarization resulted in the formation of families, usually numerous, the formation of settlements and construction of villages stilts. The "tents" or "hut storage"—thus the avieiros refer to their homes—were wooden buildings covered with thatch.

The figure 1, it can be said, is an iconic image since, that in addition to showing a housing-type, reveals the shapeof the boat.

However, this happy composition contains a different and perhaps more important, factor analysis, since it leads us, unequivocally, to the inseparable relationship of the avieiros with its boat which has always been assumed as integral and essential part of their way of being and living, thus calling attention to its importance within the Avieira anthropological culture.

4 THE AVIEIRO BOAT

The boat is an integral part of Avieiros' life, being both a working tool, local dating and "home" where their sons were made and raised, coming to serve as a deathbed. It is a witness (6/7 m²) of intimacy, of anxieties, dreams, expectations, joys and sorrows.

This fishing boat is designed for the demands of the Tagus River, being carefully partitioned to match the features of a "workshop", of a "kitchen", and a "room". The sail, oars and rod are the means to navigate and perform various types of fishing and transport.

The marquee is an element with unique features that was placed in the area of "room" to serve as

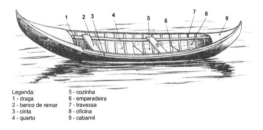

Legenda:
1 - draga
2 - banco de remar
3 - cinta
4 - quarto
5 - cozinha
6 - emparadeira
7 - travessa
8 - oficina
9 - cabarnil

Figure 2. The structure of the avieiro boat—Archive Interpretation Centre Cais da Vala (SM).

Figure 1. The Avieiro Boat and the blockhouse—in Arquitectura Popular em Portugal.

Figure 3. Avieiro boat with fisherman and fisherwoman, hard at work—Photo by Madalena de Mello Viana.

Figure 4. The marquee protecting the family—Archive Interpretation Centre Cais da Vala.

Figure 6. Povoa de Santa Iria: avieiros boats, beached in a yard, serving housing.

Figure 5. Bateira avieira exposed in the Center of Intrepertação Cais da Vala (Salvaterra de Magos).

Figure 7. Alhandra, beached boats avieiros.

a shield and shelter on rainy days or hot sun, particularly in the period that mediated between the acts of launch-ing and collect ing the fishing gear.

The bateira/saveiro or avieiro boat, served as a principle, as the first "home", to protect the couple, with or without children, and his few worldly possessions, weather and heat waves while they could not afford to build a "tent".

The text transcribed below, represents a testimony by descriptive results of in situ observations and generally ends up serving as a legend of the 1960 images that are presented below, is particularly instructive because it gives us a very objective view of the importance of the boat when used as a basis for different housing options for families avieiras.

"When they get some time in the same place, these bateiras villa, built balconies on the ground and soil, are covered by a large tarpaulin sheet, armed with cane poles like a tent, and opened at the front. The sheet then covers, in addition to drumming, where people sleep in a room next to where they cook and eat, especially on rainy days and where they arm themselves with boxes, bales, household items and

fishing tackle. These bateiras are grouped in small clusters, one after the other, right by the water. In Alhandra in Povoa de Santa Iria, etc., They line up on the yard smooth margins, surrounded by the mob of people that live there, fishermen are busy with their chores, or talking up nets, women dealing with domestic arrangements, kids playing, sleeping dogs, etc. By day the marquees remain open for that yard, having each one before it, at the time of food preparation, the cooker of iron burning charcoal. At night or when it rains, a blanket hung from the edge of the marquee in front, sealing and closeing this set. Around and inside, the Bateira is always well swept, very clean and well ordered, in an arrangement that does not lack taste, and that's fine tradition of the fisherman's central coast. These people often have two bateiras: one is where they live, and another for fishing. The villa of Bateira goes to water only when the family moves. People wandering, shifting easily from place according to the convenience of fishing, with every step you see a Bateira loaded with all the material that constitutes the 'home' of the avieiro."

---

* Veiga de Oliveira, Ernesto et al., Construções Primitivas Portugal (3ª ed.—1994), Publicações Dom Quixote, Lisboa.

In some cases, and particularly when the avieiro has only one bateira, that we can't halt, the arrangement is more simple: only the marquee is mounted on it, since the coverage of the back, behind, until the middle of the hull, which is open, arched and the sides stuck to the board, or until the bow, opening the front likea curtain, to enter.

The figure 8 shows the sense of pragmatism that has characterized the spirit of fishermen avieiros to overcome a variety of difficulties with which they were daily confronted, applying what, today, we could classify as a recycling process, giving the boat, without conditions for fishing, a new life cycle.

Naturally, we are faced with a solution that results from very specific knowledge of the structure of Bateira and deep relationship that avieiros had with this kind of means which guaranteed them survival.

Figure 8. Povoa de Santa Iria: based on a Tagus' bank an avieiro boat already unusable for fishing was used for housing.

Figure 9. Escaroupim: avieiro fisherman on the boat with pots.
* In The Avieiros from nomadic to sedentary, Municipality of Salvaterra de Magos, 2001.

Poor by nature, it seems that avieiros see in their boat and "sea" (as they relate to the river) the expression of their freedom by assuming, intimate, intrinsically, as fishermen that only in extreme situation decided to work in the fields.

Avieiros fishermen consider their speed boat ("with the wind to give it, we put ourselves in an instant wherever we want to be", as one character in the Alves Redol's novel) and versatile, allowing them access to places of the river, which can only be reached with skill and sometimes with the audacity that is in their genes.

According to avieiros testimonies collected by the author, a few months ago, the fishing arts,—component, of course, inseparable from the boat—without being particularly original, are, however, adapted as a novel way to obtain greater efficiency in chores, not being the fishermen inhibited of diving into the river where there can be more profitable with fishing nets.

Like in many other fishing communities, there are rituals accompanying the life cycle of the avieiro boat, e.g., acts related to the powers of magical protection and the values referred to the transcendence of the spirit of life (birth and death).

In this sense, the boat is an essential component of the avieira culture and as such, considered an anchor project, applying for National Heritage and Intangible Culture of UNESCO.

To accomplish this program both cultural and ethnographic initiatives are been implemented.

In what concerns the avieiro boat, there's a technological rivers training center being structured, on a sustainable entrepreneurship way, that will provide human resources (particularly the masters boat builders, housing and pier stilts, whose knowledge is essential) and the technical means to:

– The training of technicians for construction, maintenance and repair of avieiro boat, stilt piers and avieiras houses.
– The training of technicians for the preservation of the rivers' ecology and their biodiversity.
– The preparation of professionals to provide support services to all types of vessels involved in fishing, tourism and leisure.

5 CONCLUSION

The identity imprint of Avieira Culture is inscribed not only in the intangible legacy but also the tangible heritage of the communities whose essence avieiros have sought to preserve and pass on from generation to generation.

In this context, the boat is a reference characterizing avieira ethnography. The fishermen originating from Vieira de Leiria began, probably even in the first half of the nineteenth century by forming seasonal form clusters along the banks of the river Tagus, that only in early 1900, were transformed into villages with substandard housing—the "tents" or "huts". However the boat was passing most of the time, day and night drifting in search of places where they could fish, since the activity on the river, was virtually the only source of income that provided them with livelihoods and with which, consequently, they ensured the survival of each difficult day. Only the pollution of the Tagus, which drastically affected the fish species, forced the avieiros to opt for other activities, maintaining, however, even today, some that rely almost exclusively on fishing and others who play as a complement of professional occupations in which they engage.

With the improvement of ecological conditions of the river, new perspectives open up now so that the Avieiros can return to activities that best identify with themselves, in addition to fishing, can contemplate entrepreneurship initiatives in tourism relying on those who have deep knowledge of the Tagus.

And so, given its intrinsic value and unequivocal, it seems evident that the avieiro boat reveals the specific characteristics to integrate a sustainable development process, among other initiatives, as instigator of the use of large and under-exploited potential, that the river Tagus itself contains, as part of tourist activities. Indeed, taking into consideration the opinion of specialists in different areas, it is possible to structure a broad range of options in the areas of leisure and culture, including those offered by the discovery of the river with its unique landscape, fauna observation and flora in their natural habitats. Thus it might create business opportunities managed in accordance with the principles of social, environmental and economic sustainability, and, among other objectives, to contribute decisively to the improvement of living conditions of avieiras families, which still live on the Tagus, exploring a sustainable, endogenous resources of the areas of Borda d'Água.

Work of researchers has been identified, particularly in the decades of the 60 s and 70 s of last century; they were dedicated to the study of avieiros, and there was therefore some bibliographical sources that now are being released.

The avieiro boat as much as we know, was not yet the subject of a specific research work, lying, only scattered references and allusions in the plays written in documentary films.

However, it is increasingly recognized the relevance of a paper that, from an anthropological and socio-cultural point of view shows, that the boat is a primary reference avieira culture and virtues that contains material likely to contribute effectively to improving the living conditions of communities avieiras and be integrated into a process of sustainable development, understood as the concept of "Triple Bottom Line" (John Elkington, 1994) advocate the balanced interaction of the three pillars of sustainability—People | Planet | Profit—a dynamic that fits the Agenda XXI goals.

REFERENCES

Baker, S. (2006) *Sustainable Development*. New York: Routeledge.

Brito, R.S. (2009) *Palheiros de Mira*. Edição Facsimil. Figueira da Foz: Centro de Estudos do Mar.

Brandão, R. (2005) *Os Pescadores*. Lisboa: Editora Ulisseia e Editorial Verbo.

Dresner, S. (2008) *The Principles of Sustainability*, 2nd Edition. London: Earthscan.

Elkington, J. (1999) *Cannibals with Forks, the triple bottom line of 21st century business*. UK: Capstone Publishing Limited.

Enrenfeld, J.R. (2008) *Sustainability by Design*. London: Yale University Press.

Gomes, F. (1993) *Vieira de Leiria. A História, o Trabalho, a Cultura*. Lousã: Tipografia Lousanense.

Gomes, S.P. (2001) *Esteiros*.12ª Edição, Mem-Martins: Publicações Europa América.

Lopes, A. & Serrano, J. (2010) *A Reconstrução do Sagrado—Religião popular nos Avieiros da Bordad'Água*. Lisboa: Editora Âncora.

Redol, A. (n.d.) *Avieiros*, 2ª Edição, Mem-Martins: Publicações Europa América.

Rodrigues, V. (2009) *Desenvolvimento Sustentável—uma introdução crítica*. Parede: Principia.

Roosa, S.A. (2007) *Sustainable Development Handbook*. USA: The Fairmont Press, Inc.

Salvado, M.A.N. (1985) *Os avieiros, nos finais da década de cinquenta*. Castelo Branco.

Schmidt, L. et al. (2005) *Autarquias e Desenvolvimento Sustentável—Agenda 21 Local e novas estratégias ambientais*. Porto: Fronteira do Caos Editores.

Soares, M.M. (1986) *A cultura avieira. Continuidade e mudança. in Separata* do Colóquio "Santos Graça" de Etnografia Marítima.

Soares, M.M. (1995) *Varinos e Avieiros. in* "Navegando no Tejo". Lisboa: Comissão de Coordenação da Região de Lisboa e Vale do Tejo (CCRLVT).

Souto, H. (2007) *Comunidades de pesca artesanal na costa portuguesa na última década do século XX*. Lisboa: Academia de Marinha.

Green Design, Materials and Manufacturing Processes – Bártolo et al. (eds)
© 2013 Taylor & Francis Group, London, ISBN 978-1-138-00046-9

Patterns of sustainability in textile design use of traditional technologies

A. Rusu
National University of Arts, Bucharest, Romania

ABSTRACT: During the last century technological progress has begun to reveal its faults. Industry has brought many benefits to humanity but negative repercussions are becoming apparent, particularly where ethical issues are concerned. As a fiber artist and textile designer I seek to revitalize traditional solutions like manual weaving, manual printing and plant dyeing technologies. This direction in contemporary art is a reaction that has gathered together not only artists but a group of people interested in the importance of traditional patterns of sustainability, like ethnologists, anthropologists, art critics and economists. Artists cannot limit themselves to raising questions; their artworks have to make a statement towards change, otherwise they just perpetuate a society in which aesthetics is just another word for beautiful waste. It is important to revitalize green technologies and implement solutions that have been tested throughout the ages; the ritual gestures that strengthen links between humans, their artifacts and the environment.

1 CONTEXT

During the last century technological progress has begun to reveal its faults. Industry has brought many benefits to humanity but negative repercussions are becoming apparent, particularly where ethical issues are concerned. Faced with modern technology, rather than the ancient *techné*, people are reconsidering their moral relationship with technologies. The idea of sustainability first emerged in various context such as the US national energy policy, a new mandate of the World Conservation Union and United Nation Environment Program in Stockholm (Adams 2006). Researching the forty years of sustainability policies Braungart concludes: "...sustainability is a type of guilt management that helps companies feel better without actually having to make much of a stride to rectify the problem" (Braungart 2005). His solution is the "cradle to cradle" (McDonough & Braungart 2002a) theory that offers companies the alternative to design products to have another cycle or "nutrient" function after their use.

1.1 *Textile design and sustainability issues*

Textile products are major contributors to environmental problems; the situation is encountered throughout the lifecycle of the product. Nowadays, the industry is using significantly more synthetic fibers that natural ones and the chemical dyes can prove harmful to people.

Textile manufacturing is one of the main sources of water pollution. Production uses large amounts of water and energy; on average, approximately 20 gallons of water are required to produce a pound of textile product (Cao et al. 2006). The real problem in textile production begins with design education and research. According to Mackenzie (1997) design has not been taught in the context of its social and ecological impact.

Designers need a major rethinking of the educational system to stimulate creativity for a sustainable future. Higher educational system prepares students for social insertion and working in a competitive environment rather than building their inclinations for a sustainable society. In 2002 the UNESCO has declared 2005–2014 the decade for Sustainable Education Development and has designed the standards for a Sustainable Education Development but changing manufacturing systems is a rather slow process in many countries. Education for sustainability implies not only developing empathy, cooperation, but also developing skills for problem-solving and filtering decisions through a moral values framework. In addition to this it implyes new curricula and paradigm shift for teachers as well as students.

Designers first have to master a complete knowledge about materials used in their domain. In textile design, fiber and fabric properties can make the difference between the conscious decision of working with natural fibers, that become bio-nutrients after they complete their life cycle or using synthetic fibers combinations in fabrics that cannot be recycled, not even as technical nutrients.

Designers cannot limit themselves to raising questions, their concepts, use of materials and end products have to be a statement towards change.

This paper is build following my experience as a fiber artist and textile-design teacher. I am interested in the degree of adaptation of new and ancient technologies both to the requirements of the industry and the need to develop the creativity of new generation of artists/designers, for a sustainable society.

2 PATTERNS OF SUSTAINABILITY IN TEXTILE DESIGN

2.1 Traditional and new technologies

There is not an area of our world unaffected by the advances in textile research. Principles of textile science and technology merge with other research fields such as engineering, chemistry, biotechnology, material and information science.

New generation smart textiles are designed to adapt their behavior or characteristics to the circumstances: adapt their insulation function according to temperature changes, detect vital signals, change color or emit light upon defined stimuli, generate or accumulate electric energy to power medical and other electronic devices.

Smart textiles technologies enable computing; digital components and electronics are in-built (wires, conducting textile fibers, transistors, diodes and solar cells, buttons, LED mounted on woven conducing fiber networks, organic fiber transistors) (Nakad 2003; Lee B.J. & Subramanian V. 2003; Rusu, 2012a).

Fiber artists and textile designers have always used a multitude of materials and media to support their statement but now they are testing the limits of textile technology and fibers properties. Innovation, unexpected combinations and unusual materials are the hallmarks of XXI fiber art.

Textile designers turn to tradition from the point of view of new technologies, they use traditional technologies to shape new solutions, or investigate tradition from an ecological perspective. Tapestry, textile design, mixed-media, installation, textile sculptures are only a few examples of the diversity of expressions that textile art encompasses. Fiber artists understand that you cannot open yourself to innovation overlooking the cultural background of the matter with which you are working because tradition and nature are often the keepers of perfect solutions. In teaching textile art-textile design I often combine modern techniques of printing and weaving with traditional methods and emphasize the wide array of environmental friendly techniques, from block-printing, mud printing, tie and dye, and *ikat* to *batik*, fiber plant dye and manual weaving techniques like tapestry weaving (*haute-lisse* and *basse-lisse*), *karamani*, knotted and *Oltenian* weaving. In the process of learning

a technique students also gain knowledge about the history of practice and technological cultural decision made by communities to serve their needs without ignoring the environment. (Rusu, 2012b)

2.2 Study case I: Textile design education

The experience gained in textile art-textile design education and the use of new digital printing technologies in particular, determined the change of my perspective over the possibilities textile design practice offered until not too long ago. Art universities, like National University of Art in Bucharest, in which textile design is a priority understood the need for investing in new technologies for the art studios. Classic technologies, like screen printing and manual weaving are now used in combination with digital printers and digital weaving technologies.

Textile digital printing is not an end in itself and it doesn't affect creativity levels but it is an entire new framework of experimenting. My only concern focuses on the possibilities of making textile printing an environmental friendly technology. In my comparative study between screen printing technique and digital printing technique impact on environment some aspects became obvious. Some printers use reactive dyes and natural fiber fabrics but other printers use pigments and can print on synthetic fibers also. There are many factors to be considered but I think that textile printing can become a sustainable technology in studio practice. Compared with manual printing techniques, digital printing has important benefits: the small amount of dye used per square meter, up to no residual dye, moderate use of cleaning liquid, the potential of converting the technology in a soya dye based one. Other factors advocate the use of manual technologies in studio design practice: the preservation of technological gesture, the increase in creativity, the use of technological gesture instead of electricity.

Guiding a new generation of textile designers, whom experienced a wide array of technological possibilities, studied extensively fiber and fabric properties and commit to changing the mindset of consumers towards sustainable products is one of the solutions I envisage.

Textile design practice in Romania is totally dependent of studio and small design business due to the fall of once a powerful textile industry. Making policies that encourage designers, artists and artisans in developing studios and micro-businesses could mean significant decrease in the use of industrial products, uncommitted to sustainable practices. This could also lead to: the use of non-polluting manual technologies, use of organic materials, the preservation of traditional technologies, recycling fabrics in art practice, increased awareness in sustainability amongst the large population, creating premises for art students to develop skills for sustainable solutions.

2.3 *Study case II: The Maps of Time project*

In my textile design research I experience the possibilities of traditional technologies, like manual weaving and plant dyeing in becoming solutions for stressing social problems: unemployment rate, literacy and low skill rate, environmental issues in textile production.

In the past year I have worked with a group or artists/designers/ITspecialist/researchers in developing a research project on traditional technologies ("The Maps of Time. Real Communities-Virtual Worlds—Experimented Pasts". Funded by the Romanian National Authority for Scientific Research, CNCS-UEFISCDI Project Registration Code: PN-II-ID-PCE-2011-3-0245. Project manager: Prof. Dragos Gheorghiu). "The Maps of Time" is a research project exploring the ways to preserve and revitalize ancient technologies, bring them into contemporary contexts, and present them as viable and tested solutions in lifelong education for sustainability.

As a fiber artist/textile designer my responsibility in the project varies from documenting traditional textile technologies, experimenting with technology reconstructions in workshops, and creating an e-learning environment for people that want to acquire fiber art skills. In my theoretical research I questioned the revitalization of technological tradition and study various ancient weaving technologies to understand the technological decisions that contributed significantly to the humanity's development. Another reason why we are experimenting with ancient technologies is to understand their sustainable component. The patterns of strong interaction between human-environment-artifact are solutions for small communities' economic growth but also for human development issues.

The project team aims to develop an e-learning program for sustainable development of small communities. In this purpose we use visual support, expert online lessons and advice for learning textile technologies. This project is the perfect portrayal of the way artists can contribute to education, economy and sustainability.

3 REDEFINING EDUCATION

The research follows my educational and art practice. Every change begins in redefining education. In textile design education emphasize should be made on creating the premises for students to understand the properties of new industrial products and the degree of adaptation of technological

innovation to humanity's stride for sustainability. Another important change has to redefine art itself. Today, art calls for social engagement and developing projects that empower small communities, reduce poverty and engage people in lifelong skill acquiring practices.

Sustainable products, amongst which we find eco-textiles, use green technologies like manual spinning and weaving and plant dyeing. Such products don't have a major impact on environment and are compatible with the environment throughout their life cycle. The unsought sustainable character and the high degree of adaptation to the needs of society from the very beginning until today have turned fiber art and textile design technologies into solutions for small communities. Designers rediscover now traditional technologies, transform recycling in a creative process and reuse materials. In the process of matter transformation they understand the vulnerability of resources and the link between human and nature. It is of utmost importance to revitalize green technologies and implement solutions that have been tested throughout the ages; the ritual gestures that strengthen links between humans, their artifacts and the environment.

REFERENCES

Adams, W.M. 2006. The future of sustainability: re-thinking environment and development in the twenty-first century, report of the IUCN renowned thinkers meeting. Available at: http://cmsdata.iucn.org/downloads/iucn_future_of_sustainability.pdf, Accessed 15 February 2012.

Braungart, M. 2005. It's time for corporations to move beyond guilt management. *Corporate responsibility management* 1(6): 3.

Cao, H., Farr, C.A., Frey, L.V. & Gam, H. 2006. An environmental sustainability course for design and merchandising students. *Journal of family and consumer sciences* 98(2): 75–80.

Lee, B.J. & Subramanian, V. 2003. *Organic Transistors on Fiber: A first step towards electronic textiles.* Berkley: University of Berkley.

Mackenzie, D. 1997. *Green design: Design for the environment.* London: Laurence King.

McDonough, W. & Braungart, M. 2002a. *Cradle to cradle: Remaking the way we make things.* New York: New Point Press.

Nakad, S.Z. 2003. *Architectures for e-Textiles.* Blacksburg: Virginia Polytechnic Institute.

Rusu, A. 2012a. Weave an augmented reality. Algorithms and fiber art. *Journal of applied mathematics* 5(1): 215–222.

Rusu, A. 2012b. Revitalizing ancient technologies and advancing an ethical design in textile art education. *Procedia-Social and Behavioral Sciences* 51(1): 1061–1065.

UNESCO, About ESD. Available at: http://educationforsustainabledevelopment.com/blog/?page_id = 27, Accessed February 27 2012.

Green Design, Materials and Manufacturing Processes – Bártolo et al. (eds)
© 2013 Taylor & Francis Group, London, ISBN 978-1-138-00046-9

A visual impaired children photography's catalogue as an inclusive tool

A.P.P. Demarchi, B.M. Rizardi, G. Pires & C.B.R. Fornasier
Universidade Estadual de Londrina, Londrina, Paraná, Brasil

ABSTRACT: This article describes the study of the mechanisms of perception of children not seers aiming to understand how the visually impaired understand and make up the space through the tactile perception. It was held first a bibliographical research, to understand the processes of cognition of non-seers presented in this article, as well as the technique of photographs that make possible this practice. After this, an experimental research was carried out, by means of photography workshops with the disabled, when it was applied the ethnographic method, through participative observation to see how the non-seers understand the spatiality in theirs photographs. After the experimental research was carried out the third stage of the project, a case study, through interviews with the disabled using the VPA (verbal protocol Analyses) to describe how they understand the area photographed. The fourth step was the searching for material, which aimed to optimize the way of conversion of the photographs in tactile images.

1 INTRODUCTION

The article aims to understand how the visually impaired understand and make up the space through the tactile perception. The study of the mechanisms of perception of non-seer will be used as the basis for conducting photography workshops with blind children to later convert them into tactile pieces.

This conversion is important to enlarge the inclusion of non-seers in the world of the arts, especially photography, field that is often neglected; both by teachers and by the visually impaired, through the spread of the belief of the inability of the visually impaired fulfill certain tasks. In addition to inclusion, the elements and syntax used in this conversion can be applied later in teaching materials to facilitate children's learning through the study of inclusive design using tactile stimulation.

The inclusive design is one that cares about not only accept the differences, but also try to understand how to eliminate barriers that arise between people and a life of quality. Thus, there is a need for a study on the limitations of the given target audience to develop solutions so that it has access to services and products of equal, inclusive and independent way.

Using preliminary studies on the haptic perception, this work seeks to create perceptions of the moment of the photograph in the pieces that represent, through the application of different textured materials. In this way, the visually impaired can make associations that can perceive and understand their own photographs, as well as provide the same experience to other non-seers.

To this end, the research method used was qualitative of nature applied, with the initial delineation of bibliographical research, which seeks to understand the cognitive processes of children not seers, followed by experimental research, which is addressed through the workshops of photographs with the blind children. Once this is done, we conducted a case study with interviews about the workshops with students, registering their considerations and perceptions about their own photos. Finally, a materials research was realised, seeking to convert the perceptions of non-seers in tactile pieces to create a catalogue of original photographs, reports and record of tactile photos.

Whereas the inclusive design aims to serve the largest possible audience, the article through the reports of children, defines some materials and elements that represent perceptions clearly built at the time of the photo, making it so everyone can "see" what children "not seers "saw", including children seers, inserting the first in the world of the visual arts without excluding the children seers.

2 BLINDNESS AND PHOTOGRAPHY

Evgen Bavcar is a photographer and blind. This statement may cause an initial estrangement to those accustomed to using the oculocenterist vision to perceive the world around them. Photography is light and Evgen uses it at most literal way possible; his photos involve games of light, light-paintings and other unconventional techniques. To photograph he relies on his other senses that progressed to the passage of time, since the total loss of her vision next to 12 years of age. How did happen his construction of mental images from then? How created concepts of spatiality if could not use the vision to realize if the objects were near or far, or whether there were obstacles ahead?

According to Dorina Nowill Foundation (2012) for the blind, "the visual impairment is defined as the total or partial loss, congenital or acquired, of vision. The level of visual acuity may vary, what determines two groups of disability: blindness-for total loss of vision or very little ability to see, which leads the person to need Braille System as a means of reading and writing; low vision or subnormal-vision is characterized by impairment of eye, visual functioning even after treatment or correction. People with low vision can read printed texts stretched or with use of special optical features."

Unsurprisingly the visually impaired, when properly targeted, can meet almost all the day to day tasks. However, this targeting is not always done properly and the learning process of people with disabilities is not satisfactorily by restricting it to specific work fields, either by lack of suitable material, either by the lack of trained teachers or by the bias of the community as a whole. So, is taken by the society that the learnability of the seers is extremely limited and that not allows to interact so full with the world. However, this learning is different-only due to his condition; the perception in some ways is more difficult, while in others it is more developed in relationship to the seers.

All these prejudices lead to visually handicapped person to internalize the speech that is not able to perform tasks involving the use of vision. In this context, there are few people with any disability of the gender in the visual arts field, especially in photography.

However, it is possible that the deficient not only understand what was photographed as its can shoot from specific techniques related to location, adjust light, focus and framing, resulting in works that differ little from photographs of seers?

3 PERCEPTION

3.1 *Tactile perception*

To start the analysis of how the disabled person interacts with the world, we must first understand how works the perception in one general aspect. "According to M. Reuchlin (*apud* Fialho, 2011) perception is a construct, a set of selected information and structured on the basis of previous experience, needs and intentions of the organism involved actively in a given situation."

The perception is related to the cerebral lobes, but distributed in different ways through them.

The somatosensorial processing occurs in the parietal lobe, where the somatosensorial primary cortex receives tactile information, i.e. linked to the pressure, texture, temperature and pain. However, this processing does not occur homogeneously and therefore the area of the cortex dedicated to every part of the body is variable according to the need of use. The perception also follows this logic; the senses are more or less sharp depending on the need to use.

The occipital lobe is responsible for processing vision, coordinating specific areas for different aspects, such as color, shape, movement and location.

The hearing is in charge of temporal lobe.

All this system corresponds with the motor, through activities in the areas of association, joining the intentional behavior with the sensations and body movement.

All this processing of sensory information is regulated by mechanisms of attention, since not all information that we pass to the level of perception, and some of these discarded depending on the attention functions.

"The psychological phenomena of attention enable us to use our limited mental resources sensibly. To lessen the attention on many external stimuli (perceptions) and interiors (thoughts and memories), we can focus on the stimuli that interest us." (Fialho, 2011)

From there, our brain starts the formation of mental images Piaget (apud Cardeal, 2009) "employs the term 'representation' in two very different directions [...]. In the strictest sense, it reduces the mental image or memory-image, i.e. the symbolic evocation of the realities absent. In this sense, one can translate by mental picture by representation imagery".

Building mental images also covers narrations, when through a more general text the individual constructs a minimally detailed situation, with spatial dimensions and particular characteristics. This is the image situaded, whose contents are figured, with individual objects and spatial components.

However the representation by analogy happens when a problem arises. This refers to the memory and is driven by knowledge. Thus, the representation is built in order to understand the situation, while the information retrieved from the memory builds relationships between the objects of the situation (Fialho, 2011).

When it retrieves the image of a dog, not only its form comes to mind, as well as the texture of its fur and even the sound that it emits. All such information is stored in memory, ready to be activated at any time. Thus, the formation of the mental image does not involve only the vision, but all sensations and even cultural processes and individual experiences.

Soon, the person with visual impairment does not face problems for the formation of these images, just goes a different path from that of seers.

3.2 Tactile syntax

It is evidenced that statement when it scans the drawings made by blind children. They present the same difficulties as the younger children with vision. However, they are also able to understand concepts such as perspective and two-dimensional nature of the objects represented, but they face difficulty in positioning the figures on the sheet, due to the absence of "visual feedback" at the time of execution of the design (Gardner, 1994).

People with visual impairments, as well as normal subjects blindfolded, are capable of understanding into simple geometric shapes in high relief by the number of movements performed by hand, as the size difference between the forms— the higher the movement, the time, the larger the object represented.

Therefore, the understanding of forms gives temporal sequential at non-seers, differing from the seers that are characterized by the visual-spatial image record. According to Duarte (2004, p. 8) "[...]a temporal sequence, such as the time taken to traverse with the palm of the hand the edges of a table, for example, does not define the "visual" shape of the object. For the blind, go with the palm of the hand the edges of a rectangular table-top, makes it possible to realize, by feel, the time of a particular sequence until such time that the movement wins necessarily another direction caused by the corner (vertex of the rectangle) or corner of the table. It can realize if tactile sequence in continuity is longer or shorter than previous one, depending on which side of the rectangular table by which started the exercise. In the end, would have covered tactilely consciousness two long and two shorter sequences, alternate, and have found four changes of direction, four corners. But, only with outstretched arms feeling the table in its entirety (if the object's dimensions permitting) will be able to recognize the board as a whole [...]".

So, for a child with impaired vision is possible the notion totalizer of the objects, through appropriate methods and materials, and may even understand contour lines and edges in two-dimensional representations.

4 INCLUSIVE DESIGN

The design is, above all, a tool to meet human needs. When it says this, it is implied that one of the main functions of the design is to facilitate people's lives. However, as far as is this possible?

It is very common for a project to take into account the average population data only, i.e., it is considered that all individuals of a population share the same physical and cognitive abilities. However, the average man does not represent the majority of the population, since each individual is different. Significant amounts of individuals are looked down upon by the statistics and their needs are not considered in the preparation of traditional design projects. A example of this are the numerous products that elderly and disabled cannot use.

The solution to this problem does not lie in the development of specific designs, but on a universal design that can be enjoyed by more people. It is utopian to think of a design that is accessible to all, because there are still major technological limitations. The universal design aims to allow maximum utilization by the largest number of users possible, i.e. promote equity among users, therefore, it is necessary to design specific to minorities that cannot be included in universal design. This approach, which takes place only in functional, without sacrificing aesthetic, provides a listing and a respect for human rights and eliminates barriers that stand between the individual and his freedom to live as they want, improving the quality of life of many users with different needs and abilities.

Thus, this article seeks to include children seers and non-seers in the world of photography, providing both groups the opportunity to realize a photo through the touch.

5 METHODOLOGY

It is an applied research, seeking to apply the knowledge acquired during the course of this in publications and workshops.

It will be a qualitative research, trying to understand the way that disabled people represent the spatiality and composition of these.

As research delimitation, will be held the bibliographical research first in order to understand the processes of cognition of non-seers and photography techniques that make possible this practice to them.

After it will then be made an experimental research, through workshops of photographs with the handicapped, watching as they understand the spatiality in your photographs.

Later, it will be a case study, interviewing these disabled people in search for describe how they understand the photographed space.

The last step will be the research for materials that aims to optimize the ways to translate photographs into tactile pieces with reliefs and textures, to represent with greater fidelity to what was described in the interviews of the non-seers, and finally, assemble the catalog with interviews, photos and tactile parts developed in the project.

6 RESULTS

At the beginning of the technical addressed to students had some difficulties in the seizure of information, which were overcome as it created intimacy with the photographic equipment. The techniques used for guidance of visually impaired had the objective of promoting the autonomy of these at the time of photography. Its main tool is the body itself, used to determine the space in which was situated, the reason for the photo, while the brightness from the warm feeling obtained from light. The hearing also was an indispensable ally, because through it the disabled might create notions of distance necessary.

These techniques, combined with the guidance of supervisors, resulted in photographs that were memorable by the handicapped, which described with reasonable detail, proving its ability to form mental images of the space in which they are located.

7 CONCLUSION

It is possible to affirm that children non-seers have large space capacity and are able to record moments in photography with awareness of what the camera captures. The two-dimensional representation should follow different rules to be understood in relation to perspective, closing, textures and reliefs. However, it is still not possible to say that the representations refer faithfully to photography or to the memory of what was photographed, requiring larger studies in order to develop this technique.

BIBLIOGRAPHY

Cardeal, M. (2009). Imagem e Invisualidade: A leitura tátil de ilustrações em relevo. *Transversalidades nas Artes Visuais Anais do 18º Encontro Nacional da ANPAP/Associação Nacional de Pesquisadores em Artes Plásticas* (18), 3562–3571.

Cardeal, M. (2009). Ver com as mãos: A ilustração tátil em livros para crianças cegas. Florianópolis: *Universidade do Estado de Santa Catarina, UDESC.*

Duarte, M.L.B. (2006). Imagens mentais e esquemas gráficos: ensinando desenho a uma criança cega. [On-line]. Available: <http://www.ceart.udesc.br/posgraduacao/mestradoartesvisuais/batezat_duarte_maria_lucia/blind_aveugles_cegos.swf>

Duarte, M.L.B. (2006). O desenho como elemento de cognição e comunicação ensinando crianças cegas. [On-line]. Available: <http://www.ceart.udesc.br/posgraduacao/mestradoartesvisuais/batezat_duarte_maria_lucia/blind_aveugles_cegos.swf>

Fialho, F. (2011). *Psicologia das atividades mentais: introdução às ciências da cognição.* Editora Insular, Florianópolis.

Freitas, R.O.T. (2009). Os processos geradores das ações comunicacionais táteis no design de superfície. *Anais do X Congresso da APCG,* Porto Alegre.

Fundação Dorina Nowill para Cegos. (2012). Deficiência Visual. [On-line]. Available: <http://www.fundacaodorina.org.br/deficiencia-visual>

Gardner, H. (2004). *Estruturas da Mente: A Teoria das Inteligências Múltiplas* (1st ed.). Artmed, Porto Alegre.

Kennedy, J.M. (1993). *Drawing & the blind: Pictures to touch* (1st ed.). New Haven, CT: Yale University Press.

Morais, D.F.P. (2006). O ensino do desenho para crianças cegas: uma pesquisa-ação junto à Escola de Educação Especial Osny Macedo Saldanha. *VI EDUCERE—Congresso Nacional de Educação—PUC Práxis,* Curitiba.

Morais, D.F.P. (2007). Acessibilidade da Arte ao Público Deficiente Visual: ação educativa inclusiva do Museu de Arte da Universidade Federal do Paraná. *Congresso de Inclusão da Pessoa com Deficiência Visual—Comunicação e Participação Ativa,* Curitiba.

Morais, D.F.P. (2010). A aquisição de conceitos, a formação da imagem mental e a representação gráfica de cegos precoces e tardios: relato de um percurso. *IV Ciclo de Investigações—Deslocamentos Reflexivos,* Florianópolis.

Green Design, Materials and Manufacturing Processes – Bártolo et al. (eds)
© 2013 Taylor & Francis Group, London, ISBN 978-1-138-00046-9

Back to craft

E. Karanastasi
Technical University of Crete, Department of Architecture, Chania, Greece

M. Moelee
Ex.S Architects, Crete, Greece

S. Alexopoulou, M. Kardarakou, M. Nikolakaki, A. Papadopoulou,
X. Papatriantafyllou & A. Terezaki
Technical University of Crete, Department of Architecture, Chania, Greece

ABSTRACT: The paper reports on the interaction between a tutor, six students, the industry of cement blocks production and diverse consultants in the process of manufacturing multiple prototype molds for the eco-stones Wallpot and Triko. Wallpot is a concrete block element for articulated facades of vertical horticulture. Triko is a concrete block for turf-stone-alike green pavements. Throughout the process of mold generation and the research on contemporary modes of production, knowledge was gained on how flexible molding can be through the combination of digital fabrication *and* crafting. The paper presents and concludes on how the design helps in the creation of a *type* and a *system*, the digital fabrication adds a *generic* attribute to the type and crafting enriches it with a *genetic* attribute. Consequently fabrication has triggered crafting more than the design alone would have done and crafting enhanced the engagement with the design for the students.

Keywords: molds, fabrication, vertical green, ecostones, crafting

1 INTRODUCTION

The design process is approached and taught using diverse methods in architectural university programs. Design abilities are considered to be enriched through 1:1 prototype manufacturing and construction of details and objects by the students. 'Fabrication' has displaced the notion of manufacturing and construction, but, at the same time, helped traditional crafting techniques (like molding) flourish and deepen by active application. This paper reports on the process of manufacturing multiple prototype molds for the eco-stones Wallpot and Triko.

The first, Wallpot (Fig. 1), is a beton-block element for articulated facades of vertical horticulture.

Its five variations have an industrial design prototypical licence (design patent) in Greece and its production has been awarded a subsidy from the European Union, the National Strategic Reference Framework (NSRF) programme. All the variations have peculiar shapes. They combine a usual beton block with the shape of a flower pot. It was a privately initiated project by the first author based on an idea of Matthijs Moelee, agricultural

Figure 1. The five variations of Wallpot. Initial 3D representations. Sketch design, 2009.

consultant. It involved students from the Technical University of Crete, where the first author was previously appointed. The students were involved in the design, as well as in a series of prototype production workshops at the professional practice of the author. The prototype production is ongoing through a workshop and a summer school. The second prototype, Triko (Fig. 2), is a concrete tile that was designed as part of the research project for regenerating the coast of Kissamos bay on the north coast of West Crete, in a project commissioned by the Organization of Development of Western Crete to the Technical University of Crete (TUC). The need for extended paved parking areas, together with the ecological framework of the research, lead to the design of a tile-stone that is rainfall permeable while more interesting in shape than the usual turf stones or cement grids. The Triko, a triangular formed tile with three different 'cut-outs' on its edges, based on stable points, produces variable size and shape of holes, depending on how it is combined with the detached tiles (Fig. 3).

For both stones, the process included an educational process, action research, and prototypes' production. Research on how to produce a generic type of mold was done in practice and by combining traditional crafting techniques (in our own practice) and the fabrication laboratory of Technical University of Crete. The research question of this paper focuses on the interaction of digital fabrication and crafting. Is crafting adding a value to parametric design as defined through digital fabrication? In this sense is it adding a value, or is it perhaps generating values for a university level education?

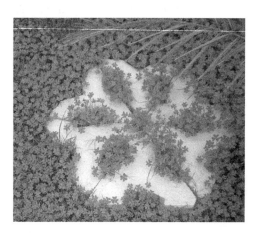

Figure 2. The Triko stone, pilot 2011.

Figure 3. Triko stone: The variations of different shapes of holes, depending on the detached tiles.

2 DIGITAL FABRICATION AND CASTING

2.1 Casting and molding

Casting is most often used for making complex shapes that would be otherwise difficult or uneconomical to make by other methods. It is a manufacturing process that dates back 6000 years. There is a distinction in terms between casting and molding. Often mistaken, casting refers to the metal process, while molding refers to a plastic process. In another distinction, in molding, the liquefied material is poured under pressure, while in casting no external force is required for pouring the material, which flows only with its gravitational force. The distinction in terms used in this paper is defining casting as an action with a 'single-time' or 'one-use' mold, while molding as a process of multiple (prototype) production using the same (generic) mold. We consent additionally to the differentiation in terms according to the output: in molding the outcome is the final finished surface, while in most of the casting processes, the output is unfinished and requires final finishing.

2.2 Digital modeling, fabrication and casting in design education

Digital Modeling and Fabrication is a process that joins architecture, the construction industry and product design through the use of 3D modeling software and CNC machines. In a shifting profession, computer milling and fabrication produces what the designer has envisioned and developed using computer aided design (Fig. 4). In this process, from conception to design to fabrication, the sequence of operations becomes critical. Architects can propose complex surfaces, where the properties of materials are what drive the design. Furthermore, this process allows the architect to examine, first hand, the materiality of their design, an experience beyond that offered by a two-dimensional computer screen and adapt the design accordingly. Digital modeling also assists the fabrication of (usually) more organic forms through the fabrication of actual molds. Whereas traditionally, the making of a mold is an involved process, basing molds off of Digitally Modeled and Fabricated designs possibly involves not only a complex 3D surface, but

Figure 4. Shifting profession: Architects redefine their design while in the Fab-Lab rather than in the office.

one that is closer to complete having already gone through several iterations of design molds.

In Architectural educational processes, that involve molding, students usually design and construct an object where the mold can be used only one time. In many casesthe, hard polyurethane foam is used and is destroyed afterwards. In digital fabrication pedagogy (Hemsath, 2009, p 22) the main research issue is the high level experimentation in form, advances in materials and innovation of articulating techniques but usually not the mass production (Fig. 5).

Pioneering courses offered in universities entitled 'File-to Factory' are a high level of explorative and innovating performative geometries while the interaction with the industry production (factory) and restrictions is not a priority. Although contemporary design and construction processes have been heavily influenced by the system of mass production developed at the end of the 19th century (Ficca, 2013), exploring type as a universal endlessly repeated component has become a fascination for the students. The more and more costless prototyping prevail the production of a system.

2.3 *Crafting and casting*

A craft is a pastime or a profession that requires some particular kind of skilled work. In a historical sense, particularly as pertinent to the Middle Ages and earlier, the term is usually applied to people occupied in small-scale production of goods. In the older days the skills were usually communicated inside the small communities, from father-to-son or from books and manuals. Today, due to the internet, more and more businesses operate from home, from their garage or their garden, offering their crafted products online. Small manufacturers, with crafting techniques that go back many centuries, upload their knowledge on the Internet, in the form of videos of crafting in action, where everyone can view and share it. Our knowledge on molding was enriched through online posts in blogs and videos of crafting in action in online communica-

tion platforms. People shared their local crafting techniques so that an eclecticism of techniques is being produced, as opposed to the old-fashioned 'father-to-son' craftsmanship.

3 CASTING AND MOLDING THE STONES: SEARCHING OR A SYSTEM RATHER THAN A TYPE

3.1 *Casting and molding the Wallpot*

In manufacturing the Wallpot, initially multiple models out of paper were made in order to test the joint possibilities and the irrigation system flow (Fig. 6). Digital fabrication and the industry were catalysts for the design. We collaborated with industry, specifically, with two companies in Crete that are producing concrete blocks. Through our collaboration we had to re-design the Wallpot and re-adjust its height so that it would fit to their machines. We also talked and collaborated with two mold making companies, one in Italy and one in Germany, that are producing molds especially for the machines of the industries and we exchanged drawings and knowledge. We also discussed the mold making with smaller metal workshops and mechanics. All the solutions were very 'fixed', while we needed a mold that can be both re-useable and at the same time adaptable to our mistakes, the demands of the industry and to changes based on functional or aesthetic reasons. Normally, a test-mold for these shapes would have been made from latex, gypsum or from iron by an experienced technician. Combining the input of the industry and our eclectic knowledge gained from numerous short movies we watched online in one-way communication platforms (movies that varied from commercial instructions of molding materials to 'homemade' videos of small artisans making molds in their own workshop) we decided to make multiple molds from wood. This would offer the advantage of producing inexpensive molds that can test both the variations in shape and form of the stone and the variations in chemical consistency. A wooden mold can be re-used and at the same time it is more flexible to crafting alternations than an iron mold. The molds were designed and manufactured with multi-layered triplex wood using the CNC router in the Fabrication Lab (Fab-Lab) of Technical University of Crete (Fig. 7), were the previous appointment of the first author

Figure 5.   Institute for advanced architecture of Catalonia, Fabrication Class, Prof. Edouard Cabay & Tomas Diez, Asissted by Alexandre Dubor. The production of articulated components which are slightly different from one another demands different prototype molds. The alterations of the molds are parametrical but are they corresponding to a system of industrial repetitive production.

Figure 6.   Paper models of Wallpot.

Figure 7. CNC milling of Wallpot molds in Fab-Lab of architecture Dept, TU Crete.

Figure 8. Some of the innumerous 'layered' molds of Wallpot.

Figure 9. Layered mold with conical cavity for the immediate extraction of the prototype.

was. The horizontal layers (Fig. 8). were designed to allow for a slightly conical hollow cavity in the mold, so that the prototype comes out of the mold easily and the mold can be re-used with minimum or no damage. Also considered in the design was the limits of our CNC machine which had a maximum limit of 5 cm thick material and operated in three axes (x,y,z). In order to achieve this conical cavity, every layer of plywood was slightly smaller (0.5 mm) or different than its neighboring one (Fig. 9).

According to the interpretation of workmanship of certainty and risk (Pye, 1971), which "relies on a personal creative knowledge of the tools, materials and techniques" (Boza 2006, Hemsath 2009), we started with the restrictions of the machine and our interpretations of them to end up to the creation of a manufacturing system for a type of flexible molds. This system was also used by other students afterwards for other courses. The students had witnessed part of the process in the Fab-Lab. The first prototypes produced were from a mixture of concrete, lime, pearlite and powder colors (Fig. 10). We realized that the smooth surface we were striving for was disadvantageous compared to the coarse one. Firstly because, for a smooth surface we would need a thinner material consistency to keep the material in the mold until it dries, while for a thicker consistecy we could de-mold immediately, simulating the industry process that we had witnessed (Fig. 11). This would mean a production of 9000 pieces per day.

Secondly, and most important, the layers of the mold would leave a trace, while for the coarse surface they would not. The traces were aesthetically not acceptable. During the process, several changes, mistakes and consults from colleagues affected the mold with smaller or bigger adaptations: one or

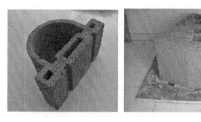

Figure 10. First prototypes.

Figure 11. Industrial production in Crete.

more of the layers would be 're-milled' in a different shape and/or size and subsequently the design file would be revised. In this sense the layered mold became more a 'system' that could generate varieties, rather than a prototype or a 'type' of mold.

3.2 Casting and molding the Triko

The design of the Triko stone followed that of Wall-pot. The experience previously gained was already an attribute in Y. Initially, the design with the 3D modeling and physical modeling (Fig. 12) with cardboard was performed. The design was part of a research project for a public area, while the pilot 'fabrication' and tile placement was part of a private small landscape project of the first author's own practice. Due to the ongoing construction of the latter we decided to use the work force on site for both the molding of the prototypes and their placement on the pavement. At the same time we would make the molds ourselves using the Fab-Lab of the architecture school of TUC, where the research project was initiated. Initially, two molds made out of polyurethane foam were used to make the prototypes for the crash-testing (Fig. 13). Then, thirty two wooden molds were used twice a day (Fig. 14). For the manufacturing of the molds, we used 32×6 wooden plates/layers (Fig. 15), all CNC milled. For the optimal use of material (both wood and concrete mixture) we used cutting optimizer and molding simulation software. Each mold has 6 different layers that have an identical outline, while their inner line is each time set off 0.5 mm, so that the prototype can be extruded without destroying the mold.

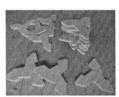 

Figures 12 and 13.   Cardboard models & Crash test of Triko.

Figure 14.   Re-useable molds.

Figure 15.   'Layered' molds.

4 A GENERIC TYPE WITH A GENETIC ATTRIBUTE: CRAFTING AND FABRICATION

Throughout the process of molds' generation and the research on contemporary modes of production, a knowledge base of molding techniques for contemporary product design was built. Above all, the flexibility of molding through the combination of fabrication and crafting was determined. While the Wallpot molding process simulated more of a factory process with the flexibility of craftsmanship, the Triko process was more of an artisanal process, although enriched with the knowledge gained earlier. In other words, we reverted more to craftsmanship while proceeding. In both parts, CNC milling played a crucial role in the genetics of a 'generic' type: in Triko we knew from the beginning that we will work with layers and the CNC router and we designed something that can be generic in its placement. Wallpot was more a project-based learning process: The digital fabrication triggered the production of a systematic mold, which was not the initial intention but influenced other projects in the same university. Concluding the process, we could say that the design helped in the creation of a type and a system. The digital fabrication added a generic attribute to the type and crafting enriched it with a genetic attribute. As a result, Fabrication has triggered Crafting more than the Design alone would have done and Crafting enhanced the engagement with the Design for the students.

CONCLUSIONS

In concluding on the interaction between crafting and fabrication based on these two case studies, we can distinguish three levels; On the level of an educational process: Re-designing while in the Fab-Lab, as well as bringing knowledge from the industry, as mentioned before, can stimulate related activity in the university. Students that took part in this initiative were more informed by the industry than in a limited time of an file-to-factory course. The learning goals of such courses are usually not the objectives of the industry, or in any case they do not share the same priorities in the process of innovation. Based on this observation, we think that university Fab-Labs should be more open to external parties, people with ideas and initiatives, like it happens in private Fab-Labs (Fig. 16).

On the level of the student-designer: There was a much faster generation of the final object and a deeper understanding of the process. The machine 'democratized' the processes that may be too technical for an average student (or a teacher) when done in the traditional way: it has never happened

Figure 16.   Fab Lab Waag Society, Amsterdam.

that there are equally skilled students in one course. There can be persons with talent and motivation but with difficulty in manual skills. On the level of the importance of the workshop: There can be more iterations of an idea, as there is the luxury of the machines "inside the classroom". Endless variations of prototypes are integrated in the design studio, while the proximity of the lab to the studio is crucial for the interaction with a lot of students.

REFERENCES

Anzalone, P., Vidich, J. & Draper, J. *Non-Uniform Assemblage: Mass Customization in Digital Fabrication.* Available:http://scholarworks.umass.edu/cgi/viewcontent.cgi?article = 1041&context = wood, accessed on 2013, 10 February.

Boza, LE. 2006. *(Un) Intended Discoveries Crafting the Design Process.* In Proceedings of the 25th Annual Conference of the Association for Computer-Aided Design in Architecture: 150–157.

Breen, J. & Stellingwerff, M. 2007. *The DigiTile Project, Conceiving, Computing and Creating Contemporary Tiling Prototypes Using Computer Aided Modelling Techniques.* In Proceedings of the eCAADe Conference, Frankfurt am Main: 59–66.

Cheng, NY. & Hegre, E. 2009. *Serendipity and Discovery in a Machine Age: Craft and a CNC Router.* In Proceedings of the 29th Annual Conference of the Association for Computer Aided Design in Architecture, Chicago, Illinois: 284–286.

Hemsath, T.L. 2010. *Searching for Innovation Through Teaching Digital.* eCAADe 28. Available: http://digitalcommons.unl.edu/cgi/viewcontent.cgi?article = 1020&context = arch_facultyschol, accessed on 2013, 10 February.

Kolarevic, B., (ed.). 2003. *Architecture in the Digital Age: Design and Manufacturing.* New York, Spon Press.

Leach, N., Turnbull, D. & Williams, C. (eds.). 2004. *Digital Tectonics.* United Kingdom, John Wiley & Sons Ld.

Pye, D. 1971. *The Nature and Art of Workmanship.* New York, Van Nostrand Reinhold.

www.wikipedia.org, definitions: http://en.wikipedia.org/wiki/Casting & http://en.wikipedia.org/wiki/Digital_modeling_and_fabrication & http://en.wikipedia.org/wiki/Craft, accessed on 2013, 17 February.

Green Design, Materials and Manufacturing Processes – Bártolo et al. (eds)
© 2013 Taylor & Francis Group, London, ISBN 978-1-138-00046-9

Urban Nomads—smart and flexible design strategies for a sustainable future

Nicoline Loeper & Matthias Ott
School of Urban Design, Saxion University of Applied Sciences, Deventer, The Netherlands

ABSTRACT: How could we make the cities more resilient for a unforeseen future? How could we design sustainable and intelligent? The profession reacts with interesting approaches to this fact. Nevertheless education in architecture and urban design seems to be captured in inflexible and static curricula: far removed from the rapid changes in the field. Urban design in The Netherlands has a high level and a rich tradition. That allows us to experiment with new education methods. Four years ago Saxion University of Applied Sciences therefore has started an urban design school. This school works with a new educational approach: "Urban Nomads". As reference projects we use, for example the design methods which are published in "a+t, strategy public" (1) and "Composing Landscapes—Analysis, Typology and Experiments for Design" (2). Urban Nomads is experimenting with 3 keywords of sustainable designing: Spontaneously, independent and aware. We reflect in this abstract about the possibilities and the eligibility to integrate this keywords in education forms.

1 URBAN NOMADS: INTENSIVE TRAVELLING WORSHOPS

In these turbulent times, the context of the design task is very diverse and complex. This makes it essential that the education form can be easily adapted to different circumstances and participants. Therefore the study consist, after the first regular year, of short and intense workshops, with clear results: future proof, flexible and smart design strategies.

These workshops travel like nomads, according to partners, participants and assignments, through the field of activity. Spontaneously, independent and aware.

2 FLEXIBLE DESIGN STRATEGIES INSTEAD OF STATIC MASTERPLANS

The goal of a sustainable urban development plan should be to anticipate, and to respond to unforeseen developments adequately.

Classic urban master plans assume an ideal end state. In practice it is almost always necessary to adapt these master plans at some point. A fixed master plan as an end result can easily lead to 'trouble areas'. To react on unpredictable developments you have to design flexible. Being able to design for flexibility and impermanence is one of the hardest—and yet most important—challenges for an urban designer.

Urban Nomads is exploring the potentials of the dynamics of change, in order to create flexible and smart design solutions: strategies for dealing with changing circumstances. Just in time, together with the field of activity.

3 CASE: "THE IZOLA WATERFRONT" (2012/2013)

3.1 *The assignment*

"The north-eastern part of the old town of Izola in Slovenia, which contains areas protected as natural and cultural heritage, is experiencing significant capital pressure. Today the main characteristics of the area are fenced-in industrial activities and a lack of green programs and systems. Slovenian and foreign investors are seeking permission to build three islands in Viližan Bay and start the construction of large-scale tourist resorts. If the investors can realize these plans, this would completely change the image of Izola and turn it into the largest seaside tourist resort in Slovenia (Šuligoj, 2009/workshop goal University of Ljubljana Faculty of Architecture)."

The assignment of Urban Nomads: develop a sustainable design strategy for the transformation of a harbour area and the waterfront that is related to it. Apply this strategy: make an urban design for Izola.

3.2 *Intensive workshop 1: Izola (location: Izola, Slovenia)*

Urban Nomads was invited for an International Achitectural and Urban Design Workshop

(Waterfront Redevelopment: Izola East, organized by the University of Ljubljana Faculty of Architecture), devoted to a potential transformation of industrial zone into a tourist resort and recreation area.

Among the participants were: urban design students of the Saxion University of Applied Sciences, architecture students of the University of Delft (The Netherlands), architecture students of the University of Ljubljana (Slovenia) and professionals.

The workshop consisted of excursions, lectures by professors of different universities and experts from the field, followed by interdisciplinary workshops on site, coached by professors and experts from the field.

Results: the students developed a toolbox for a design strategy and presented it at the workshop.

3.3 Exploring the potentials of change

During the whole semester the students continued working on the assignment. As already mentioned,

it is necessary to explore the potentials of the dynamics of change, in order to create flexible and smart design solutions. That is why the students were asked to compare the situation in Izola to Dutch harbour areas, that were already transformed, or about to transform. After analyzing these areas, the students would be able to test their strategies in different circumstances, to see if they were indeed as flexible and smart as they expected them to be.

3.4 Intensive workshop 2: Scheveningen (location Scheveningen, The Netherlands)

This workshop was organized by Urban Nomads, with the aim of making the students really

Figure 3.   Jordy Wegman: "Design Principles Izola East".

Figure 1.   Students visit the existing Izola waterfront.

Figure 2.   Interdisciplinary workshops on site, coached by professors and experts from the field.

Figure 4.   Analysis of the transformation processes in the harbour area of Scheveningen and the design of the new waterfront (Manuel De Sola-Morales).

Figure 5. Workshops in the harbour area of Scheveningen, coached by local professionals.

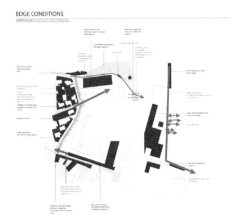

Figure 6. Davy van de Brink: "Edge Conditions".

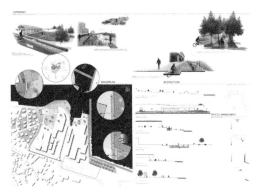

Figure 7. Davy van den Brink: "Experience the Industrial Edges".

Figure 8. Arthur Kramer: "Boulevard, designed for different levels of the sea".

understand a significant reference project. The beautiful and well designed new boulevard of Scheveningen (design: Manuel De Sola-Morales) seems at first sight a static master plan. This is not the case: it is a complex and strategic design, as it includes an ingenious dike that protects the hinterland, that is below sea level, for the coming centuries against the rising sea level. Among the participants were: urban design students of Saxion University of Applied Sciences (The Netherlands) and professionals. The workshop consisted of excursions, lectures by the involved urban designers, municipalities and users of the area and discussion, followed by workshops in the harbour area, coached by local professionals.

Results: the students analyzed the design of the new boulevard, and tested their design own strategies with the new things they learned in mind.

3.5 *The final urban design plan*

Finally the students applied their design strategy on the project location in Izola.

They made an urban design for the harbour area and the waterfront of Izola. This urban design plan consisted of floor plans, sections, details, models and scenarios, that had to prove that the final design was indeed smart, flexible and contributing to a sustainable future.

The results of the workshops were presented to the professionals that were involved and will be published (expected July 2013).

4 CONCLUSION

Spontaneously: The results of the workshop give insight that the students are able to design spontaneously in various circumstances. The quality of the presented results for example Figure 7 are conclusive. The quality of the presented *"Experience the Industrial Edges"* gives clearly insight in the flexibility of the design proposal and the spatial quality of area.

Aware: The results, for example Figure 8 *"Boulevard, designed for different levels of the sea"* gives clearly insight in the awareness of the design approach and focus on the strategic potentials of the design. The design applies the specific local circumstances and strengthens the 'genius loci' of the area.

REFERENCES

a+t ediciones, density series 35/36—strategy public 2003.
Steenbergen Clemens, Composing Landscapes, Birkhäuser Verlag AG, Basel 2008.

Energy efficiency

Green Design, Materials and Manufacturing Processes – Bártolo et al. (eds)
© *2013 Taylor & Francis Group, London, ISBN 978-1-138-00046-9*

Daylighting and ventilation in energy efficient factories

K. Klimke, T. Rössel, P. Vohlidka & H. Riemer
Technische Universität München, Department of Building Climatology and Building Service, Munich, Germany

ABSTRACT: Planners of industrial buildings not only have to pay attention to energy efficiency due to increasing life cycle costs and a growing number of political regulations, they are also required to plan for a high level of user comfort. The fact that companies are building factories all over the world doesn´t make it easier for the architect to react appropriately to the conditions of the different locations. A research project of TU München in cooperation with Siemens Real Estate investigated this issue for two different climate zones. A sensitivity analysis that determines the parameters that have a high influence on the factory's energy demand is combined with a range of energy-input for different industrial processes. The study presents conceptual solution approaches to energy efficient factories to support the planning process.

1 INTRODUCTION

The topics of energy efficiency and sustainability are well established in Germany's building sector. Since the first announcement of the German regulation for energy saving in buildings in 2002 (EnEV 2002), the level of performance has increased steadily and the energy demand of new buildings has reduced gradually over time. Concerning energy efficiency and sustainability in residential and administrative buildings, many comprehensive research reports have been published by different nationwide research institutes and today the so called "zero-energy-standard" can easily be achieved in new construction (Hausladen et al., 2009). However, industrial buildings were mostly excluded from this development. The industry sector in Germany is responsible for more than 35% of the greenhouse gas emissions. In addition to the production-related processes, these emissions are caused by the energy demand for heating, cooling, and ventilation of the building (McKinsey & Company 2007).

To achieve a high level of comfort and energy efficiency for industrial buildings, planners all over the world are facing different challenges. Within a cooperative research project carried out by the TU München and Siemens Real Estate, different locations were investigated. Two different climate zones represented by exemplary cities in Germany and India were analyzed. In further studies, the locations Russia and China were added.

The approaches are based on a sensitivity analysis. The scope of the survey was not restricted to design parameters, like window to wall ratio, air change, and U-values, but also took into account external factors like opening cycles of doors or shift operation. The results were combined in order to provide recommendations for optimizations and are summarized in an Excel-Tool.

From the resulting parameter optimization recommendations are derived, which take energy efficiency, CO_2-emissions, and energy consumption into account. This paper presents excerpts from the study for the locations in Germany (Munich) and India (Mumbai).

2 INVESTIGATION

A typical Siemens industrial building (Fig. 1) is chosen as a "base model" for the following investigations, which are carried out using dynamic building and daylight simulations. The industrial facility has a floor area of 4,355 m^2, a height of 12 m and the entire building volume amounts to 52,260 m^3. Depending on the type of use, internal heat loads of 10, 20, 30, 40 and 60 W/m^2 account for production processes and their heat output. The number of skylights depends on the required smoke ventilation area of minimum of 2% relative to the floor area (ARGEBAU 2000). The ratio of window area to the facade is only 3%. The facility is connected to an office wing consisting of 3 identical floors. Further, the simulation includes a vertical temperature gradient that depends of the internal heat load and the thermal buoyancy. This vertical temperature gradient is within the range mentioned in (FVLR).

Figure 1. Base model with the production facility and the office section.

At the location Germany the requirements for component constructions correspond to benchmarks of Siemens Real Estate representing higher standards than required by the EnEV 2009 standard. At the location India the recommendations of (ASHRAE 90.1 2010) for structural components, thermal insulation, glazing, etc. are used. Furthermore, a 3-shift operation—from Monday to Friday—is assumed in the simulation.

3 SENSITIVITY ANALYSIS

3.1 Daylighting concept (example Germany)

Since natural lighting (daylight) is superior to any artificial lighting in a building (Hausladen & Tichelmann 2009), multiple model variations will examine different methods to increase the comfort for people in the building.

Figure 2 shows the daylight factor per square meter floor area of the basic model. The brighter the area, the higher is the daylight factor.

With an increased daylight factor less artificial lighting is required and the electricity costs can be reduced. Especially for production facilities it is difficult to ensure a consistent distribution of daylight and an adequate daylight supply in the centre of the facility due to the great building depth.

In the following investigations different variations of optimizing daylight supply are analyzed.

In the first step variations with an increased percentage of glazed area on the roof are examined. The glazed area accounts for 4% relative to the floor area. It becomes clear that increasing the glazed area causes a significant improvement of daylight supply. A mere doubling of the number skylights leads to a considerably more evenly illuminated surface: an increase in the area with a daylight factor of >2% from 7% to 41% can be achieved (Fig. 3).

In the next step the windows are arranged vertically instead of horizontally. The proportion of window area is doubled (compared to the basic model) and the percentage of skylights increased

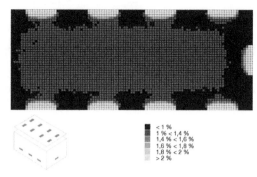

Figure 2. Daylight factor of the base model, location Germany.

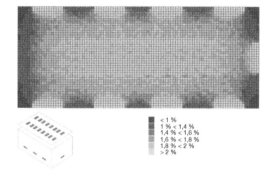

Figure 3. Daylight factor for an increased area of skylight (4% relative to the floor area), location Germany.

from 2% to 3% relative to the floor area. As a result, a considerable increase in the area with a daylight factor of >2% from 7% to 80% can be achieved (Fig. 4).

3.2 Ventilation concept/air flow rate (example Germany and India)

Industrial buildings are usually ventilated mechanically. Particularly because of the considerable heat release and emission of pollutants a low-level displacement ventilation is chosen. Supply air is blown in at the floor level and the exhaust air is extracted in the ceiling area.

For this examination, empirical and design values from different sources are considered. Apart from the German workplace regulation "Arbeitsstätten-richtlinie"—(ASR 5 1979), reference values for the air flow rate in conformity with (ASHRAE 62.1 2010) and (DIN EN 15251 2007) are taken into account. The occurring air flow rates differ significantly from each other due to the different ways of its dimensioning. Defined in ASR 5, these flow rates are determined by the number of persons as well as their activity. Both DIN EN 15251 and ASHRAE 62.1

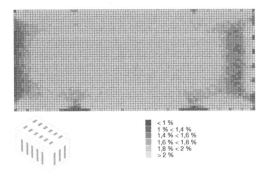

Figure 4. Daylight factor for an increased area of sky-light (3% relative to the floor area) as well as a vertical window arrangement, location Germany.

Table 1. Summary of different examined air flow rates.

Norm/ Directive	Air flow rate [m³/h]	Comments
ASR 5	5,900	Heavy physical work 65 m³/(h*pers)
DIN EN 15251	7,800	Chapter B.1.2, category II, expected percentage dis-satisfied 20%, low-emission building
ASHRAE Standard 62.1	15,700	Flow Rate Procedure— "general manufacturing"
ASHRAE Standard 62.1 + 30%	20,500	Flow Rate Procedure—"general manufacturing"—30% higher, extra LEED point

consider the floor area and the number of persons to calculate the air flow rate (Tab. 1).

3.3 Mechanical air flow rate at the location Germany

Generally speaking—and regardless of the sched-uled internal loads—the heating demand and energy demand for mechanical ventilation increase with an increased air flow rate. The cooling demand, however, decreases in case of an increased air flow rate. The reason is the difference between inside and outside temperature. Especially at high internal loads, the inside temperature is very often higher than the outside temperature. Thus, the fresh air flow rate has a cooling effect on the build-ing interior. This effect of free cooling is enhanced by the three-shift operation and its associated inter-nal heat loads which cause higher inside tempera-tures throughout the day. Furthermore, it appears that the cooling demand is negligibly small at low

internal load (10 W/m²) and the heating demand is close to zero at high internal load, regardless of the chosen air flow.

Figure 5 shows the annual CO_2 emissions for various internal loads in relation to the air flow rate. In this annual balance, the heating and cool-ing demand as well as lighting and electricity demand for mechanical ventilation are considered. Besides, the efficiency of building services and the CO_2 emission factors of individual energy sources are taken into account.

With regard to the CO_2 emissions, it becomes clear that a small air flow rate (e.g. according to ASR 5) at low internal loads (10 W/m²) should be chosen. At high internal loads (60 W/m²) a greater air flow rate (e.g. according to ASHRAE 62.1) is recommendable.

3.4 Mechanical air flow rate at the location India

At the location India, there is no heating demand due to the internal loads, but the higher the investigated internal loads, the higher the cool-ing demand. With a higher flow rate, the cooling demand decreases slightly. Particularly during the night hours the outdoor temperature is colder than the internal temperature which leads to a reduction of the cooling demand in case of higher flow rates. On the other hand, the electricity demand for mechanical ventilation increases at a higher air flow rate. A review of the CO_2 emissions shows that these emissions are largely independent from the selected air flow rate (Fig. 6). This is caused by contrary energy consumption for cooling and ventilation.

In order to obtain a certification in compliance with LEED (USGBC 2011) the air flow rate should be chosen according to ASHRAE 62.1.

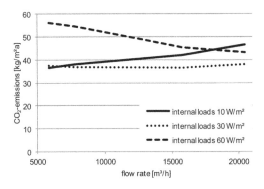

Figure 5. Annual CO_2 emissions for different air flow rates and for internal loads of 10, 30 and 60 W/m², loca-tion Germany. Heating Efficiency: 0.9; Cooling Effi-ciency: 3.5; CO_2-coefficient for heating: 200 g/kWh; CO_2-coefficient for cooling: 563 g/kWh.

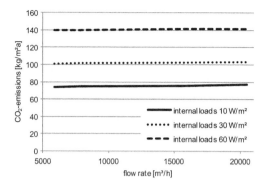

Figure 6. Annual CO_2 emissions for different air flow rates and for internal loads of 10, 30 and 60 W/m², location India. Cooling efficiency: 3.5; CO_2-coefficient for cooling: 800 g/kWh.

4 CONCLUSION

The study shows that the production process should be considered at all locations in the very beginning of the planning phase. An integrated system technology for heating, cooling, and fresh air supply must respond to the industrial production process of the building system services. Two major factors in energy saving can be identified: process heat can be used directly in the building or on site; and an industrial building must be designed differently depending on whether there are large internal loads or whether there is nearly no production-related waste heat. The focus for industrial buildings located in Germany with low internal loads of 10–20 W/m² should lie on the heating demand of the building. If the internal loads are higher, e.g. 30 W/m² or more, the cooling of the building is of higher priority for the interior climate concept. Accordingly, air tightness and thermal insulation are meaningful only at low internal loads.

The importance of the production process is also shown by the analysis of the fresh air supply in industrial buildings. At high internal loads large fresh air flow rates effectively provide a means for passive cooling, thus reducing the need for active cooling. In warm, humid climates with a year-round high solar irradiance such as in Mumbai, India, there is no heating needed in order to comply with the thermal comfort requirements. However, concepts for reducing the cooling demand are necessary for these locations. To lower the effect of the solar irradiation and to improve the energy performance of these buildings highly reflective roof surfaces should be implemented.

Another important finding is that a doubling of the skylight area increases the daylight factor and the visual comfort level significantly. Energy-efficient industrial buildings therefore do not only have a low energy consumption, but also a reduced greenhouse gas emission. The industrial sector thus plays an essential role in fulfilling the goals targeted by the German government as well as the world wide climate change mitigation goals.

REFERENCES

ANSI/ASHRAE 62.1. 2010. *Ventilation for Acceptable Indoor Air Quality.*
ANSI/ASHRAE 90.1. 2010. *Energy Standard for Buildings Except Low-Rise Residential, SI-Edition.*
ARGEBAU. 2000. *Richtlinie über den baulichen Brandschutz im Industriebau (Industriebaurichtlinie-IndBauRL).*
ASR 5. 1979. *Arbeitsstättenrichtlinie 5, Lüftung.*
DIN 15251. 2007. *Eingangsparameter für das Raumklima zur Auslegung und Bewertung der Energieeffizienz von Gebäuden—Raumluftqualität, Temperatur, Licht und Akustik; Deutsche Fassung EN 15251.*
EnEV. 29 April 2009. *Verordnung über energiesparenden Wärmeschutz und energiesparende Anlagentechnik bei Gebäuden—Energieeinsparverordnung.*
FVLR Fachverband Tageslicht und Rauchschutz e.V. *Lichtkuppeln und Lichtbänder.* Heft 10. Detmold.
Hausladen, G. et al. 2009. *Null-Energie-Bürogebäude, Guidebook.* München: TU München.
Hausladen, G. & Tichelmann, K. 2009. *Ausbau Atlas.* München.
McKinsey & Company. 2007. *Kosten und Potenziale der Vermeidung von Treibhausgasemissionen in Deutschland—Sektorperspektive Industrie, BDI initiativ—Wirtschaft für Klimaschutz.*
USGBC. 2011. *LEED 2009 for New Construction and Major Renovations Rating System.*

Green Design, Materials and Manufacturing Processes – Bártolo et al. (eds)
© *2013 Taylor & Francis Group, London, ISBN 978-1-138-00046-9*

Improving sustainability in healthcare with better space design quality

M.F. Castro, R. Mateus & L. Bragança
C-TAC, Department of Civil Engineering, University of Minho, Guimarães, Portugal

ABSTRACT: The hospital project contains different aspects from the most common projects of residential, offices or services buildings. In common buildings, sometimes the user and the client are the same and when they are not, setting the requirements is not difficult since they are common to most inhabitants. In the case of hospital buildings this is not the reality and the project team is usually hired for the purpose of designing a building that includes different spaces and different users, such as doctors, nurses, patients, visitors, cleaning staff, administrators, and others. In this sense it is important to combine different spatial needs, which are always subject to constant changes throughout the period of building operation due to: new features; innovations; enlargement needs; and new healthcare treatment methods. This paper discusses the importance of the design and organization of space quality in the overall sustainability of a hospital building and how this aspect is evaluated in BSA tools.

1 PATH TO SUSTAINABILITY

1.1 *Hospital buildings*

The hospital buildings, not because they are more abundant in the territory, but because they are large consumers of natural resources and energy, should be a major focus of study in the evaluation process of the buildings life cycle (Guenther & Vittori, 2008). The activities implied to the healthcare industry require a lot of energy for heating, refrigeration, etc. On the other hand it is necessary to take into account the use of renewable and non-renewable resources, disposable products, toxic substances and the production of a large quantity of waste (Short & Al-Maiyah, 2009).

The health sector has a strong influence on the economy of nations and their policies, incorporating a group of buildings where the quality of the indoor environment is quite significant. The impacts of this type of buildings are more significant than any other because they are directly related to human health (Guenther & Vittori, 2008). The operation of these equipment for 24 intensive hours, the high number of movement of people, the existence of distinct work zones with different energy needs, the existence of different functions such as treatment, research, rehabilitation, health promotion and disease prevention, the need for the existence of systems strategic reserve of equipment for constant supply of energy, and size of facilities, are key points that differentiate these from other types of buildings and make it a specific case study (Johnson, 2010).

Healthcare providers are not serving patients but serving people. They should design and deliver services to meet the needs of normal people at the most difficult times in their lives (Clark & Malone, 2006).

The hospital project, more than any other, requires a number of concerns with the satisfaction and well being of working teams, patient, administrative staff and other officials. This is a project where all basic design principles (rather generalized and taken into account in the act of designing common buildings) should be considered with the increased responsibilities, since the users' satisfaction and well being demands are more sensitive. The basic design concerns usually considered are: the climate where the building is built; access to solar radiation; the local topography; the program of the building and the interaction between the various elements of the design team; the necessary flexibility and enlargement capacities; the security; the efficiency in the development of activities; the adaptability to new I & D (Dias, 2004).

In this context its possible to say that the design phase is the most comprehensively addressed part of the life cycle in most sustainable building guidelines and evaluation methods (Dias, 2004). The design and the space organization are very important for different areas and can be decisive in environmental, economic and social development of the whole building. Therefore this study is about hospital architecture and how the project design quality can be fundamental for the well being of people and for the sustainability of construction.

2 SPACE DESIGN QUALITY

2.1 The contribution of space design to the sustainable hospital buildings

Healthcare is one of the most complex and rapidly changing industries. It is continually transformed by new technologies, technique, pharmaceuticals and delivery systems (Boone, 2012). In this concern, it is a fact that the hospital architecture incorporates a development project that has as main concerns the adequacy of technological advances in medicine, compliance with rules and regulations (that seek to ensure the quality of designed environments), the complexity and flexibility required for the project and the high cost of premises. This means that the designer often forgets or not gives the adequate importance to sustainable principles that this type of project should follow (Shaw et al, 2010). Consequently the construction of this type of buildings needs to incorporate this evolution and the spaces design can be the way to improve healthcare. The architectural design of the space, its organization, operation and configuration, allows these buildings to respond and adapt positively to the needs for which they are designed. At an early stage, a good investment in their flexible design reduces the need for further improvements (Johnson, 2010).

Analysing the indicators and parameters of the Building Sustainability Assessment (BSA) tools, specifically oriented to hospital buildings, it is possible to assess how important is the use of these methodologies in the architecture design phase to promote the existence of more sustainable buildings in the future. Many of these parameters are easily answered through the spatial and volumetric organization of indoor and outdoor spaces. Therefore it is important to encourage the architects to incorporate these concerns in their projects, avoiding solving future problems resulting from the addition of equipment or other solutions that increase energy consumption, water or other resources, even human. Most times, sustainability assessments are used to comparatively classify the buildings. Nevertheless it is of increasingly importance that such methods are regarded as ordinary work tools in all project phases.

The design phase incorporates many decisions, such as the use of materials, choice of equipment, networks, infrastructure, among others, Nevertheless this paper is focused in the design of effective space, comprising options of building implantation, composition and spatial organization of buildings. Table 1 show some examples of spatial and volumetric organization taken in buildings of recognized quality in terms of sustainability.

Table 1. Example of design options in case studies.

Case study	Description
	**Providence Newberg Medical Centre** • *Recognition*: US Green Building Council LEED (Gold level); Hospitals for a Healthy Environment (H2E) - Environmental Leadership Award (2007). • *Design option*: to implant the building in two volumes in "V" shape. It favors the green spaces, energy efficiency, visual comfort and rainwater use.
	Evelina Children Hospital • *Recognition*: NHS Building Better Health Care Award for Hospital Design, Nov. 2006; Royal Institute of British Architects design competition. • *Design options*: to remove the partition walls between the rooms and the circulation areas. It decreases the footprint, increases the net floor area and allows greater comfort in use by users.
	Kaleidoscope, Lewisham Children and Young People's Center • *Recognition*: Firm won initial CABE design competition. • *Design options*: the plan facilitates deep penetration of daylight and moves the operable windows away from the heavily trafficked urban streetscape.
	BC Cancer Agency Research Centre • *Recognition*: Canada Gren Building Council LEED (Gold level). • *Design options*: the central atrium spine separates different areas from inpatient units. The atrium provides daylight to occupied workspaces, which line the upper floor.
	Boulder Community Foothills Hospital • *Recognition*: First hospital certified by US Green Building Council LEED, in the E.U.A. (Silver level); distinction in 2006 Hospitals for a Healthy Environment (H2E) - Environmental Leadership Award. • *Design options*: to reduce the car parking area and to create adequate parking areas and paths for bicycles, in order to encourage the use of alternative transportation.
	Meyer Children's Hospital • *Recognition*: EU Hospitals Project. • *Design options*: have features vegetated roofs with skylights and light tubes bring daylight deep into the interior. The building barely disrupts the landscape it appears to emerge from.

2.2 Building sustainability assessment tools for healthcare

All over the world there is a growing number of sustainability assessment tools developed efor the building sector and oriented for new constructions, existing buildings and refurbishment/rehabilitation operations. Inside these three groups, most assessment tools are specifically oriented for different type of buildings. In the context of hospital buildings the most well known tools are: BREEEM Healthcare, LEED for Healthcare and Green Star—Healthcare (BREEAM, 2013; LEED, 2013; GBCA, 2013). In addition to these, DGNB is developing a specific methodology for hospitals that is not finished yet, and CASBEE have a system for *new construction* that includes the hospital buildings in the category of residential buildings. Nevertheless the CASBEE tool does not specifically address this type of buildings, but is one tool with different specifications for residential and no-residential buildings. For this reason, this study is focused on BREEAM Healthcare, LEED for Healthcare and Green Star—Healthcare.

The three abovementioned tools have a system of evaluation based in points that are divided over different categories, each of which is based in a series of evaluation parameters (Sauders, 2008). Although there are some differences between these tools, they share the main areas of assessment. Analysing the indicators of each tool it is possible to conclude that there is no sustainability categories directly related with space design quality. Nevertheless there are some sustainability parameters that are indirectly related with that principle.

It should also be noted that an exceptional answer to the category *Innovation in Design* (that allows getting an extra score in all tools) allows correcting a worst performance in other sustainability categories. Credits for innovative performance are awarded for comprehensive strategies, which demonstrate quantifiable sustainability benefits not specifically addressed by other sustainability categories. Table 2 presents the sustainability parameters of the abovementioned

Table 2. Sustainability parameters that are directly influenced by the indoor and outdoor spaces' design quality.

Category	Parameters	B*	L**	G***
Sustainable Sites	Light Pollution Reduction	x		
	Connection to the Natural World—Places of Respite	x		
	Connection to the Natural World—Direct Exterior Access for Patients	x		
Health & Wellbeing	Day lighting	x	x	x
	View Out	x	x	x
	Potential for Natural Ventilation	x		x
	Outdoor Space	x		x
	Arts in Health	x		
	Minimum Indoor Air Quality Performance		x	
Energy	Optimize Energy Performance		x	
	Lighting zoning			x
	Car park ventilation			x
Transport	Proximity to amenities	x		
	Pedestrian and Cyclist Facilities	x	x	x
	Maximum Car Parking Capacity	x		x
	Deliveries and Manoeuvring	x		
	Community Mass-transports			x
	Transport design and planning			x
Materials and Resources	Compactor/Baler	x		
	Storage and Collection of Recyclables		x	
	Resource Use—Design for Flexibility		x	
Land Use & Ecology	Reuse of Land	x		x
	Contaminated Land	x		x
	Mitigating ecological impact	x		x
Innovation in Design	Innovation	x		x
	Integrative Project Planning and Design		x	
	Innovation in Design: Specific Title		x	
	Integrative Project Planning and Design		x	

* BREEAM Healthcare.
** LEED for Healthcare.
*** Green Star—Healthcare.

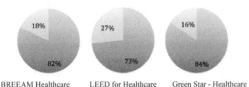

BREEAM Healthcare   LEED for Healthcare   Green Star - Healthcare

Figure 1. The relevance of outdoor and indoor spaces' design quality (in percentage) in the overall sustainable score (represented in light grey).

tools that are directly influenced by the indoor and outdoor spaces design quality. Figure 1 shows with light grey the relevance (in percentage) of these parameters in the overall sustainable score.

3 THE ENVOLVING ROLE OF ARCHITECTURE

3.1 Eco-humanism in hospitals

Analysing the tools presented in the previous section, it is possible to see how important is the careful spaces design to the positive evaluation of several sustainability parameters. Although there are some indirect relations, it is still imperative to retain among the analysis of the categories that there are no categories dealing directly with the sociocultural dimension and functional quality. The social dimension of sustainable development is even more present in the case of hospital buildings due to the importance of the wellbeing of the patient in this kind of projects. On this context, it is in this category that stands out even more the importance of the spatial and volumetric organization of indoor and outdoor spaces, because its quality can give immediate and effective responses to almost all the concerns of this area of interest.

Thus, the category "Sociocultural and functional quality" that DGNB considered in its assessment tool (Table 3), and the similar one "Social, cultural and perceptual aspects" that the International SBTool considers, positively influence the concerns of spaces design quality in the architectural design phase (DGNB, 2013; iiSBE, 2013). This fact promotes the consideration of the patients and users' welfare in this type of buildings

It is within this context that one can speak about Eco-humanism. Eco-humanism in architecture is about having an equal concern for human and ecological wellbeing, and by its nature it touches on many uncomfortable truths (Verderber, 2010). The challenge now is to translate this unprecedented opportunity into action.

The main concerns of the space design are the humans' needs. So, the use of rating systems specifically for the hospital buildings becomes essential to include in the design phase, beyond the importance of historic preservation, and systems of interrelated hierarchies comprised of personal, institutional, and societal constructs (Rokeach, 1979).

3.2 The architectural process

Early environmental design initiatives were focusing only on the reduction of energy demands. Different institutes and governmental initiatives developed tools and policies to address this problem.

In 1980s and 1990s some of the initiatives began to reflect concerns about the sustainability of the construction industry c, and in 1993 the UIA/AIA Word Congress of Architects concluded that it was a bold challenge to the profession of an architect

Table 3. Criteria of the core catalogue of the DGNB tool and International SBTool (DGNB, 2013; iiSBE, 2013).

DGNB tool	International SBTool
Sociocultural and functional quality	Social, cultural and perceptual aspects
• Thermal comfort • Indoor air quality • Acoustic comfort • Visual comfort • User influence on building operation • Quality of outdoor spaces • Safety and security • Handicapped accessibility • Efficient use of floor area • Suitability for conversion • Public access • Cycling convenience • Design and urban planning quality through competition • Integration of public art • Site features	• Access for mobility-impaired persons on site and with the building • Access to direct sunlight from living areas of dwelling units • Visual privacy areas of dwelling units • Access to private open space from dwelling units • Involvement of residents in project management • Compatibility of urban design with local cultural values • Impact of the design on existing streetscapes • Impact of tall structure(s) on existing view corridors • Quality of views from tall structures • Sway of tall buildings in high wind conditions • Perceptual quality of site development • Aesthetic quality of facility exterior • Aesthetic quality of facility interior • Access to exterior views from interior

to put a broader sustainability agenda into practice (Guenther & Vittori, 2008).

In 2000 many of these initiatives turned to incorporate sustainable design strategies as basic and fundamental in standard practice. In 2005, the American Institute of Architects (AIA) established a more aggressive position on the responsibility of design professionals, defending the position that the architects must change the professional actions and work together with the clients changing the actual paradigm of designing and operating a building (AIA, 2005).

The sustainable project requires a revolution in the way of thinking the building design. So it is important that this transformation, that across all phases of the life cycle building, will be reflected in the early stage of architectural design and in the essence of it: the design and organization of space.

If the architectural design should contain the entire patient and users' needs, environmental concerns and generate synergies among all actors of the design team, then this should directly addressed in building sustainability assessment tools. This is essential in order to support architects during the early phases of design and to recognize the efforts of an architect in designing a truly sustainable building.

Michael Lerner (2000) formulated the following question: *"The question is whether healthcare professionals can begin to recognize the environmental consequences of our operations and put our own house in order"* (Roberts & Guenther, 2006). This is not a trivial question, but the foundation of all other issues that may arise around this same concern (Roberts & Guenther, 2006). Based on this principle, Figure 2 illustrates the relationship between human health, medical treatment and environmental pollution that directly affects the mission of the health care industry.

3.3 *Discussion*

It is relevant to promote and discuss the importance of the space organization to the sustainable construction and the influence of the architecture (and not only the building systems) in the Building Sustainable Assessment tools (BSA tools). It is also important that each designer involved in the development and construction of hospital buildings is able to quickly identify a set of parameters that can interfere directly, therefore that later can be considered globally to intervene in each and every one of them. All in all these tools must be bivalent, they must impose the concerns with sustainable construction but also integrate the requirements of each building and each project area, linking priorities and facilitating the integration widespread of

more this concern in the different design projects. This is one aspect that can promote integration and knowledge of these tools in all project teams involved in the construction of this building typology, as well as their use in different phases of buildings life cycle.

4 CONCLUSIONS

The Hospital architecture has a strong social responsibility and impact on the city. Mostly due to various design requirements, these buildings are not designed and operated in a sustainable way. Based on this context it is important to include in BSA tools the best practices in architecture that should be taken into account in the design phase (to support the decisions that contribute to the building sustainability).

Although the design and organization of space encompassing always a great social responsibility, this concern is transverse to the three pillars of sustainable development (economic, social and environmental), since it allows the resolution and fit of many solutions environmentally efficient and economically viable. Based on the conclusions, future developments on BSA tools should give more weight to these aspects of major influence in the building life cycle performance.

ACKNOWLEDGEMENT

The authors acknowledge the Portuguese Foundation for Science and Technology for the financial support under Reference SFRH/BD/77959/2011.

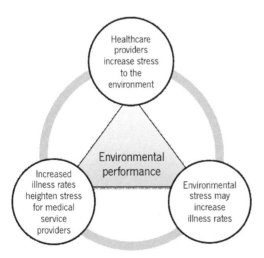

Figure 2. Relationship between environmental performance and health care (Roberts & Guenther, 2006).

REFERENCES

AIA. 2005. *High-Performance Building Position Statements*. Washington: AIA. Retrieved May 18, 2011, from http://www.aia.org/SiteObjects/files/HPB_position_satatements.pdf
Boone, T. 2012, March 29. Creating a Culture of Sustainability. Leadership, Coordination and Performance.
BREEAM. 2013. *Home page of BREEAM*. Retrieved February 20, 2013, from http://www.breeam.org
Clark, P. & Malone, M. 2006. What Patients Want: Designing and Delivering Health Services that Respect Personhood. In S. Marberry (Ed.), *Improving Healthcare with Better Building Design* (1st ed., pp. 15–35). Chicago: Health Administration Press.
DGNB. 2013, January 24. DGNB System—Criteria. *Www.Dgnb-System.De*. Retrieved January 24, 2013, from http://www.dgnb-system.de
Dias, M. 2004, January 25. Resíduos dos serviços de saúde e a contribuição do hospital para a preservação do meio ambiente. *Academia De Enfermagem* 2(2): 21–29.

GBCA, 2013. *Home page of GBCA*. Retrieved February 20, 2013, from http://www.gbca.org.au

Guenther, R. & Vittori, G. 2008. *Sustainable healthcare architecture*. New Jersey: John Wiley & Sons, Inc.

iiSBE. 2013. *Home page of iiSBE*. Retrieved February 20, 2013, from http://iisbe.org/sbtool-2012

Johnson, S. W. 2010. Summarizing Green Practices in U.S. Hospitals. *Hospital Topics 88*(3): 75–81.

LEED, 2013. *Home page of LEED*. Retrieved February 20, 2013, from https://www.leedonline.com

Measurement Decisions in Healthcare. *Health Care Research Collaborative*. 1–32.

Roberts, G. & Guenther, R. 2006. Environmental Responsible hospitals. In S. Marberry (Ed.), *Improving Healthcare with Better Building Design* (1st ed., pp. 81–107). Chicago: Health Administration Press.

Rokeach, Milton. 1979. *The Nature of Human Values*. New York: McGrawHill.

Sauders, T. 2008. *A discussion document comparing international environmental assessment methods for buildings*. breeam. Retrieved from www.dgbc.nl/images/uploads/rapport_vergelijking.pdf

Shaw, C.D., Kutryba, B., Braithwaite, J., Bedlicki, M. & Warunek, A. 2010. Sustainable healthcare accreditation: messages from Europe in 2009. *International Journal for Quality in Health Care 22*(5): 341–350.

Short, C.A. & Al-Maiyah, S. 2009. Design strategy for low-energy ventilation and cooling of hospitals. *Building Research & Information 37*(3): 264–292.

Verderber, S. 2010. *Innovation in Hospital Architecture*. New York: Routledge.

Green Design, Materials and Manufacturing Processes – Bártolo et al. (eds)
© *2013 Taylor & Francis Group, London, ISBN 978-1-138-00046-9*

A linear reciprocating thermomagnetic motor powered by water heated using solar energy

L.D.R. Ferreira, C.V.X. Bessa, I. Silva & S. Gama
Universidade Federal de São Paulo—UNIFESP, Diadema, Brazil

ABSTRACT: The most common use of solar energy is for household solar water heating mainly because the cost and the efficiency of these heaters are attractive notably in those countries with large solar radiation rate per year. In this context this paper presents the development stages followed by experimental results that demonstrate the viability of a thermomagnetic motor powered by water heated by solar energy. The motor main part is formed by one arrangement of permanent magnets and by two equally spaced magnetocaloric compound plates having a Curie temperature (Tc) around 50°C and that remain stationary relatively to the moving magnets. As hot water (70°C) crosses one plate its magnetic state change, releasing the magnets toward the other magnetocaloric plate. In order to make the permanent magnets movement continuous, hydraulic valves, a sensor and one board control are used to toggle the water flow through the plates. Water at room temperature is used as refrigerant fluid.

1 INTRODUCTION

Nowadays, the search for new energy supplies, renewable or finite, is indispensable for all modern and industrialized economies. Also the rational and economic use of all kinds of energy sources is mandatory and the energy itself or its converted forms should be secure, have both relative low production and distribution costs and also be environmentally clean, i.e. greenhouse effect must be attenuated or even avoided. According to (Statistical, 1990) in 1987, 37% of the world energy production was provided by petroleum fuels and projections shows that possibly 15 million barrels per day by 2025 (Avery and Wu, 1994). It is recognized that the petroleum resources of the earth are finite and that world demands for liquid fuels are depleting the reserves at a rapid rate. This could eventually make petroleum fuels so costly that all world economies would be critically affected. Major national security issues are also involved because 80% of the world oil reserves are in the Middle East. It is imperative that cost-effective alternatives to petroleum be developed in time (Avery and Wu, 1994). Among many types of sustainable and clean energies, the solar one is the most important and powerful (Duffie and Beckman, 1980; Iqbal 1983). The main challenge is how to use solar energy efficiently and cheap. The most common use of solar energy all over the world is for house water heating by solar collectors used for washing and other domestic purposes and also for electricity production by photovoltaic cells

(Twidell and Weir, 2006; Boyle, 2004). The later has a higher production cost per square meter of photovoltaic cells. On the other hand, the cost and also the efficiency of solar water heaters are attractive, notably in those countries with large solar radiation rate per year. Besides using solar energy only for water heating e.g. for dishes washing and so on, why not also use the solar heated water to power motors, refrigerators, compressors and etc.

1.1 *The thermomagnetic motor*

In this sense the Thermomagnetic Motors and other Magnetic power conversion systems could represent an alternative to produce energy in a cleaner and sustainable way. This technology can be traced back to the 19th century, when a number of scientist deposited patents on the so called "pyromagnetic generators". At that time permanent magnets were not strong enough nor were the appropriate magnetic materials available in order to build economical and practical devices.

These motors work by the reversible change on the magnetic state presented by some materials when heated above its Curie temperature (Tc).

1.2 *The linear reciprocating thermomagnetic motor*

A simplified motor scheme is shown in Figure 1. The motor core is composed by an arrangement of permanent magnets and by two equally spaced plates made of magnetic material with internal channels

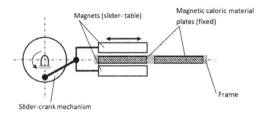

Figure 1. Simplified working scheme of the linear reciprocating thermomagnetic motor.

for the flow of the working fluid. At the first moment hot water (around 70°C), begins to flow across the left plate, as the plate heats it changes its magnetic state, becoming paramagnetic, so releasing the magnets toward the plate placed at the right side and which is magnetized. Thus, in order to make the permanent magnets movement continuous, hydraulic valves, a non-contacting gap sensor and one electronic board control are used to toggle the water flow through the plate arrangements. Water at approximately room temperature (25°C) is used as refrigerant fluid, so bringing the left plate back to its magnetic state. The permanent magnets linear and oscillatory movement is transformed in to revolving movement by means of a slider-crank mechanism.

2 COMPUTATIONAL SIMULATION

In order to optimize the power and the efficiency of the motor it's necessary to make computational simulations allowing for the better understanding of the main working parameters of the motor as well as for the ways in which the design can be improved.

The power produced by a motor is a direct result of the product between its torque and its angular velocity, in thermomagnetic motors the torque produced depends on the magnetic force resultant between the magnetic material and the permanent magnet arrangements, and the velocity produced by the motor is a direct result of the heat transfer speed between the magnetocaloric material and the working fluid.

2.1 Magnetic force

As previously mentioned, the force produced by a thermomagnetic motor is a function of the magnetic field produced by the permanent magnet arrangements, the plate's geometry and also by the magnetization of the magnetocaloric material.

The material used in the prototype here depicted was a compound of GdNdSi that was agglomerated with industrial epoxy to form parallelepipedal plates with internal channels for the flow of the

working fluid. The magnetization of the material in a constant magnetic field of 15,912.78 A/m varies with temperature as shown in Figure 2.

According to the graph shown in Figure 2 the material changes it magnetic state around 330 K, even though this material presents a second order behavior, and therefore slow magnetization decay, it will be used in the plates because of its availability and high magnetization factor.

To produce strong magnetic fields on the motor plates NdFeB permanent magnets were used in a Halbach arrangement (Halbach, 1980). The highest field intensity was around 2 T.

In order to determine the force obtained by the motor a 2D magnetostatic simulation using finite element method (FEM) was carried out. For this simulation the real properties of the plate's material were used as well as the geometry of the plate and its internal channels (see Fig. 4).

The traction force produced by the interaction among one plate and the magnetic field generated by the permanent magnet arrangements is shown in Figure 3. The plate is initially 5 mm away from the magnets and moves horizontally in relation to the magnet arrangements, in the same as it does in the motor (see Fig. 1).

In Figure 3 one can see that the force intensity vary reaching up to 86.229 N, and below zero value when plate is almost entirely under the magnets (52 mm), which means that there is a force with

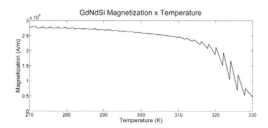

Figure 2. Magnetization versus temperature for the GdNdSi compound.

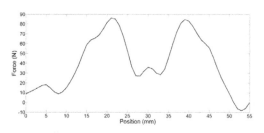

Figure 3. Magnetic traction force on GdNdSi plate along linear motor work direction obtained using finite element simulation (the air gap between the magnet arrangements and the plates is 1 mm).

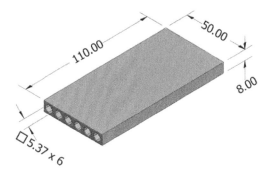

Figure 4. Magnetocaloric plates geometry and dimensions (mm).

a direction opposite to the movement, working to decelerate the plate at the final part of the movement. These variations can be expected due to the nom uniform characteristic of the magnetic field produced by a Halbach arrangement.

2.2 Heat exchange between the plates and the working fluid

Another very important working parameter of the thermomagnetic motors is the heat exchange rate between the magnetocaloric plates and the working fluid. In the linear reciprocating motor water heated by solar heaters was used as the hot source, reaching up to 80°C, and water at room temperature (around 25°C) is used as the refrigerant fluid, and the water flow through the channels with a mass flow rate of 0.01 kg/s.

As previously mentioned the plates have internal channels where the working fluid flows across. The motor constructed plate having six square channels is shown in Figure 4. The plate has a total mass of 104.28 g, the contact area of internal channel is 118.68 cm^2 and the hydraulic diameter of the plate can be calculated to be 33.21 mm.

Although the plate internal square channels configuration adopted in this work isn't ideal, as proved in a previous work carried by the authors where a carefully finite element study was done to optimize the heat transfer rate (Ferreira et al. 2012), these kind of configuration present some advantages to the fabrication of the magnetocaloric plates and therefore will be used in this prototype.

3 EXPERIMENTAL ANALYSIS

In order to demonstrate the viability of the proposed design a prototype of the linear reciprocating thermomagnetic motor was built and some experimental analysis done to identify the working parameters of the motor.

3.1 Magnetic traction force

The magnetic traction force was measured using a load cell placed in the motor main movement direction, the plates were fixed to a screw and had their position measured by a dial indicator. Both plates were placed at the motor in order to identify the working force through the left plate (Fig. 1) a hot water flow was passed, and a cold water flow (room temperature) passed through the right plate.

The experiment was realized a couple of times, and although some points presented considerable variation, especially at counter force part of the movement, the results can be considered very close and are shown with its error bar in Figure 5.

The magnetic traction force measured can reach up to 65.6 N, the difference between the experimental magnetic traction and the simulated magnetic traction can be attributed to the fact that there were some remaining magnetization on the material even above its Curie Temperature (see Figure. 2), and also because of the friction in the slider table that wasn't accounted for in the simulation.

3.2 Operational frequency

In order to determine the operational frequency of the thermomagnetic motor one thermocouple was placed in the water inlet of one of the plates, as the motor changes its position the control electronic board toggle the water flow from the hot to the cold source, generating a step like variation on the water temperature of the inlet.

Another thermocouple was placed in the outlet of the plate, this way it's possible to determine the heating energy that flows to the plate, and also to better identify how long it takes for the plate to reach its Tc.

The measurements were made using a data acquisition board and the water temperature (inlet and outlet) plotted as a function of the experimental time, as shown in Figure 6.

The obtained operational period of the motor was around 12 s, although there were some variations

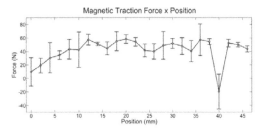

Figure 5. Experimental magnetic traction force on the plate along linear motor work direction.

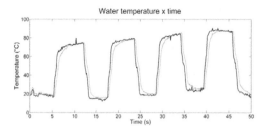

Figure 6. Water temperature in the inlet and outlet of one plate cycling with the transition of the magnetic state of the plates.

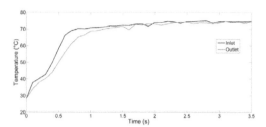

Figure 7. Temperature of the water in the inlet and outlet of one plate during a single heating cycle.

in the period the values were consistent between various experiments.

As previously shown the motor uses a slider-crank mechanism to convert the linear reciprocating movement into a rotary type movement. In order to keep a constant movement a flywheel with a mass of 3.8 kg and a moment of inertia (I_{fl}) of 0.022 kg · m² is used, see Figure 1.

Also a ratchet mechanism is used to prevent the flywheel from stopping during the transition time of the magnetic plates.

The flywheel angular velocity isn't constant because the force produced vary along the movement (see Fig. 3 and Fig. 5) and although the permanent magnets movement can be considered very fast, the plates move from one side to the other on no longer than 0.4 s, the heat exchange is a slow process, and during this interval there is no moving torque acting upon the flywheel, causing it to decelerate and then stop during one cycle of the movement.

The angular velocity was measured using a laser tachometer obtaining the peak value of 7.989 rad/s and mean value (w_m) of 2.458 rad/s.

3.3 Heat transfer analysis

In order to analyze the energy conversion efficiency of the motor it's necessary to determine the total energy that is transferred from the working fluids to the motor, i.e. the total heat exchanged from the fluid to the plate being heated in order for it to change its magnetic state.

The temperature of the water in the inlet and in the outlet of one plate during a single heating cycle, correcting the outlet curve in one second to correspond to the effect of the mixing of the hot water with the cold water still inside the plate, is presented in Figure 7. In this figure one can see that temperatures in the inlet and outlet don't match perfectly at first, as the heat from the water flow is conducted to the magnetocaloric plate.

Figure 7. Temperature of the water in the inlet and outlet of one plate during a single heating cycle.

The heat conducted from the water to the plate at a given moment ($Q(t)$) is defined by equation 1.

$$Q(t)=m_w \cdot c_w \cdot \Delta T(t) \tag{1}$$

Where m_w is the mass of water that flows through the plate, c_w is the specific heat of the water and $\Delta T(t)$ is the difference between the water inlet temperature and the water outlet temperature at a given time.

Integrating both sides of equation 1 in function of time, results:

$$\int_{t_0}^{t_f} Q(t) \cdot dt = m_w \cdot c_w \cdot \int_{t_0}^{t_f} \Delta T(t) \cdot dt \tag{2}$$

In equation 2 its left side corresponds to the heat transferred from the water in time interval ($t_f - t_o$) and the integral of $\Delta T(t)$ in function of time can be obtained from the difference between the area formed by the curves of the water inlet temperature and the water outlet temperature in Figure 7.

Solving equation 2 using the water specific heat capacity, the total time of 3.5 s and the water flow of 0.1 kg/s the total heat transferred obtained (Q_{tot}) is:

$$Q_{tot} = (0.1 \times 3.5) \times 4181.3 \times 7.1397 \tag{3}$$

$$Q_{tot} = 10.449 \text{ kJ} \tag{4}$$

This value represents the total heat loss from the hot water to the heating of the plate, although an important part of the heat loss can be attributed to the mixing of hot and cold fluids inside the magnetocaloric plate and also to the heating of the valves and hydraulic connections necessary for the motor to work.

3.4 Power produced by the motor

The power produced by the motor can be determined by the relation between the energy of the flywheel. (E_{rot}) and the operational period of the motor (t_{rot}), which as measured to be around 12 s, see Fig. 6. So, the energy of the flywheel is determined by equation 4:

$$E_{rot} = I_{f1} \cdot \omega_m^2 / 2 \qquad (5)$$

Solving equation 4 the value of the kinetic energy of the flywheel is 0.066 J, and consequently the total power produced by the thermomagnetic motor prototype is around 5.53 mW.

4 CONCLUSIONS

Although the total power produced by the built linear thermomagnetic motor prototype was considered low, there are a number of improvements to be made in such kind of motors, e.g. the magnets arrangements, the magnetocaloric materials and also de plates configurations and its internal channel design must be optimized, others motors architectures must be constructed, e.g. using multiple plates and magnets arranged in rows or in columns.

The theoretical and experimental results shown comprise the main data of the research and demonstrate the potential for the development of this field as an alternative form of conversion heat energy from many sources, such as solar water heaters, geothermal and industrial waste heat energy.

ACKNOWLEDGEMENTS

This research is sponsored by "Fundação de Amparo à Pesquisa do Estado de São Paulo—FAPESP", grant number: 2009/00013-0 and 2012/09486-0.

The authors would also like to thank the "Coordenação de Aperfeiçoamento de Pessoal de Nível Superior—CAPES", who provided master's level research grant to L.D.R. Ferreira and C.V.X. Bessa.

REFERENCES

Avery, W.H. and Wu, C. 1994, *Renewable Energy from the Ocean—A Guide to OTEC*, Oxford: Oxford University Press, John Hopkins University series.

Boyle, G. 2004, 2nd ed., *Renewable Energy*, Oxford: Oxford University Press.

Duffie, J.A. and Beckman, W.A. 1980, 1st ed.; 1991, 2nd ed. *Solar Engineering of Thermal Processes*, New York: Wiley.

Ferreira, L.D.R., Bessa, C.V.X., Silva, I., Gama, S. 2012, *A Heat Transfer Study Aiming Optimization Of Magnetic Heat Exchangers Of Thermomagnetic Motors*, Grenoble: Fifth IIF-IIR International Conference on Magnetic Refrigeration at Room Temperature, Thermag V.

Fujieda, S., Hasegawa, Y., Fujita, A., Fukamichi, K. 2004, Thermal transport properties of magnetic refrigerants La(FeXSi1-X) and their hydrides, and Gd5Si2Ge2 and MnAs. *Journal of Applied Physicas*: Volume 95, Number 5. March.

Gschneidner, K.A. Jr., Pecharsky, V.K., Tsokol, A.O. 2005, *Recent development in magnetocaloric materials.* Institute of Physics Publishing: Reports on Progress in Physics. Volume: 68. Pag. 1479–1539. Available in: <stacks.iop.org/RoPP/68/1479>.

Halbach, K. 1980, *Design of permanent magnets with oriented rare earth cobalt material, Nucl. Instum. Method* 169: 1.

Iqbal, M. 1983, *An Introduction to Solar Radiation*, New York: Academic Press, reprinted 2004 by Toronto University Press.

Statistical Abstract of the United States, 1990. *U.S. Dept. of Commerce Bureau of the Census,* Table 957.

Twidell, J. and Weir, A. 2006, 2nd ed., *Renewable Energy Resources*, Taylor & Francis.

Green Design, Materials and Manufacturing Processes – Bártolo et al. (eds)
© 2013 Taylor & Francis Group, London, ISBN 978-1-138-00046-9

Environmental quickscans as a decision supporting tool: Scanning the embodied energy of different fibre treatments in the development of biocomposite building products

E.E. Keijzer
TNO, Utrecht, The Netherlands

ABSTRACT: In this article the method of environmental quickscans is introduced. This method is developed in the EU-funded project 'BioBuild'. The goal of BioBuild is to develop biocomposite building products with 50% reduction of embodied energy and no increase in costs, compared to current alternatives. To ensure that the developed products have this improved performance, environmental quickscans are executed for important design decisions related to fibre selection, fibre treatment, matrix selection, painting and coating of the building products. In this article the benefits and lessons learned of the environmental quickscan approach will be explained by using the quickscan on fibre treatment as an example. The quickscan indicates that the embodied energy of fibre treatment processes differs with a factor 6. This shows that investigating the environmental impact of production options is of importance as the options can differ substantially in environmental performance.

1 INTRODUCTION

Environmental assessments become increasingly important to understand the environmental performance of products and services (e.g. Duflou et al., 2009 and Kim, 2011). Often the method of Life Cycle Assessment (LCA) is used for this. LCAs provide insight in all processes and materials which contribute to the environmental impact of the product, taking into account the whole life cycle. In many research & development trajectories, environmental assessments are performed at the end of the line: to see how the newly developed product is performing.

Unfortunately, at the final stage an assessment cannot contribute much in product optimization nor inform the design process; fundamental choices have already been made and assessments can be used for minor adjustments only. The added value of environmental assessments at the end of the product development process can be considered low, as the impact of the assessment is limited to communication of the achieved results.

In BioBuild, a project supported by the European Union's Seventh Framework Programme for Research (FP7; see European Commission, 2012), a different approach is followed in order to support product development. Starting from the beginning of the project and continuing through the whole development process, environmental quickscans are performed in order to support the decision making. The gained insights in environmental impacts of options are applied directly as one of the arguments in decision making, together with information on costs and technical performance.

The objective of the BioBuild project is to develop sustainable biobased building products. The building systems that will be developed contain biobased resins and natural fibres.

As a full Life Cycle Assessment is too data (and labour) intensive to perform many times troughout the product development process, a quickscan approach is developed. Environmental quickscans will be performed to inform a wide variety of design decisions, for example on sort of fibre, matrix, fibre treatment, coating and shaping. The benefits and lessons learned of the environmental quickscan approach will be presented in this article. This will be done by using the example of quickscans performed for the decision on fibre treatment. Fibre treatment is only one of many design decisions for the BioBuild building products for which the quickscan method is used.

Below, the objective of the BioBuild project will be introduced. Thereafter, the goal and scope of this article and the quickscan methodology will be explained. Afterwards, the building product and the different fibre treatments will be introduced. Subsequently, the results of the embodied energy quickscan of the fibre treatments will be discussed and with these results we will indicate the advantages of the quickscan method.

2 ENVIRONMNETAL ASSESSMENT IN BIOBUILD

The goal of the BioBuild project is to develop four biobased building systems with 50% less embodied energy compared to conventional building products (without an increase in costs). The project intends to develop four new building products. For the project a wide range of partners was brought together in a consortium. The consortium contains expertise along the entire supply chain of the specific building products.

In order to reach the goal of 50% embodied energy reduction, it is important that the embodied energy performance of the products is assessed during product development and not only at the final stage of the development process.

The developed sustainability assessment methodology incorporates three stages: 1) benchmarking; 2) quickscans, and; 3) a final assessment. In the benchmarking stage the reference for the product assessment is calculated on the basis of currently used building products. The reference is needed in order to compare the environmental performance of the BioBuild products with potential material alternatives.

The second stage entails a number of quickscans in which the environmental performance of different variants of BioBuild products and production methods are assessed.

The last stage involves a final assessment. This will be done after the product is developed and hence resembles a normal environmental assessment that is done shortly before the product is placed on the market.

3 ARTICLE GOAL AND SCOPE

The focus of this article is on the second stage of the assessment method: the quickscan approach. The aim of this article is to demonstrate this method which incorporates assessment of embodied energy in the whole product development process. The method will be illustrated in this article by assessing the embodied energy of different possible treatments of natural fibres. Information on energy use of the different fibre treatments feeds into the decision making process during the BioBuild product development in order to create a product that is optimized on environmental and energy performance.

4 QUICKSCAN METHODOLOGY

A quickscan gives insight at three levels. First, it identifies the factors that influence the energy impact. Second, differences in impacts between the options can be compared, and third, the effect on the final product is made clear. The quickscan model is structured in a life cycle manner to ensure that the final assessment and the quick scan assessments point in the same direction. The environmental performance is assessed in three indicators: embodied enery, water- and land-use. The results in this article are limited to embodied energy.

5 SYSTEM DESCRIPTION

Of the four construction products under development in the BioBuild project, only one of them is analyzed in this article: the rain screen cladding system (see figure 1).

The function of the rain screen cladding is to protect the insulation and structure behind it from direct external influences such as water and wind. The cladding creates an air cavity directly behind it where an air flow is generated to evaporate any residual moisture in the cavity. Water resistance is a pivotal characteristic of rain screen cladding panels as they are exposed to all climate elements, including rain. Mould growth or swelling and shrinking of the panels is a threat for the durability of the product. The lifetime of the panel is a crucial factor for the energy performance of the product; panels with a short lifespan have shorter replacement intervals.

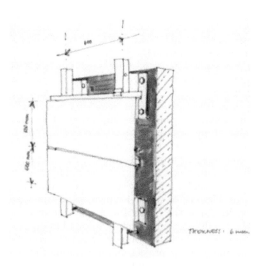

Figure 1. Rain screen cladding system.

6 NATURAL FIBRES

In the last several years, natural fibres have become very popular in the development and production of environmentally friendly fibre reinforced composite materials (Faruk et al., 2012). Replacement of common synthetic fibre reinforcements (often glass fibres) by natural ones and the development of alternative biobased polymer matrices are very promising developments in this regard. Not the least for the development of new composite materials for the building industry.

However, the high flammability, low water resistivity, and restricted mechanical properties of natural fibres constrain their wide application. For outdoor application for example, low water resistivity threatens the durability of the product as a result of risk on swelling and mould growth. To improving the performance of the natural fibres, many treatment methods are being developed.

To enhance the durability of a biocomposite rain screen cladding panels, the fibres should undergo treatments in order to improve the bond between fibres and matrix of the biocomposite. The decision for a specific sort of fibre treatment can be guided by the environmental performance of the different treatments—besides costs and technical performance.

For BioBuild a pre-selection was made of treatments that result in improved durability, water-resistance, co-reactivity and/or better interaction with several polymer matrices (such as (bio)epoxy, (bio)vinyl ester and furan resins). As most of the available fibre treatments are toxic, very complex or energy demanding in their manufacturing, the selected treatment methods are innovative and are supposed to be nontoxic and more environmentally friendly. An environmental quickscan is executed to give insight in the environmental performance of these five fibre treatment methods:

- Water glass (WG) treatment
- Acetylation treatment
- DMDHEU treatment
- Silanization without pre-treatment
- Silanization with pre-treatment

Below a description is given on the expected results of each of these treatments and the fibre treatment process steps.

6.1 WG treatment

Natural fibres undergoing a WG treatment are expected to have improved fire resistance, improved mechanical properties and improved hydrophobicity.

The process contains several steps. Immersion of flax fabric in a bath with diluted WG (Inosil Na-5120 WG diluted with water in the weight ratio of $WG/H_2O = 1:3$ g/g was used in this case) for 20 minutes is followed by hand lay-up pressing-out of non-soaked WG. In the final stage the fibres are dried. This can be done in temperature ranges from RT up to 180°C (2 h at 120°C was used in this case).

During the treatment energy is used for the oven (2.2 kW) and the lab mixer (72 W).

6.2 Acetylation treatment

From all chemical modification reactions of wood and natural fibres, acetylation has the longest history. Reaction of the of acetic anhydride takes place with the cell wall polymeric hydroxyl groups to form an ester bond, with acetic acid produced as a by-product.

Natural fibres undergoing an acetylation treatment are expected to have several improved characteristics. The expected results are reduced water absorption and an increase in durability.

As an example: water absorption of treated wood can be reduced by more than 60%, shrinking and swelling can be reduced by approximately 80% and fungal decay can be reduced to a minimum (highest durability class).

The acetylation treatment takes place at a temperature of approximately 130°C. Catalysts can be applied to initiate or increase the reaction rate. After the treatment the residual acids will have to be removed in a post treatment process.

Acetic acid, the by-product of the acetylation treatment, can be sold on the market and contributes to a lower environmental impact of the treatment: no energy is needed to produce acetic acid since it is the by-product of the fibre treatment.

During the acetylation treatment process energy is used for heating (temperature of 130°C) and the post treatment process.

6.3 Resin treatment using DMDHEU

DMDHEU treatment has been industrially applied for several decades because of its lasting effect on fabrics: it increases its durability, reduces water absorption and increases dimensional stability. Dimethylol dihydroxyethyleneurea (DMDHEU) is an effective agent to cross link cell wall polymers. Cross linking the cell wall polymers is an effective method to prevent water adsorption induced by the polymeric hydroxyl groups and to substantially increase the dimensional stability of the natural fibres.

For wood effective treatment levels of 15 to 20% WPG are recommended. For fibres like flax or jute, treatment levels of 5 to 10% are expected to be effective, most likely resulting in 50% reduction of water absorption and increase of dimensional stability.

The treatment process contains several steps. First the fibres need to be dipped/impregnated. The DMDHEU is applied in an aqueous solution. After this impregnation/dipping step, the fibres need to be dried and cured to finalise the treatment. Depending on the applied catalyst curing temperatures range from 90 to 140 °C. Known catalysts are metal salts (a.o. AlCl3) and ascorbic acid.

6.4 Silanization with and without pre-treatment

Natural fibres undergoing a silanization treatment are expected to perform better: the mechanical characteristics are expected to improve and hence the durability of the flax fibres will increase.

The silanization treatment of flax is executed in several process steps. First the fibres are dried. Afterwards, they need to be immersed in an APS solution for 10 or 30 minutes. Then the drying step is repeated.

The silanization treatment can be expanded with a NaOH pre-treatment. In that case, the fibres need to be immersed in the NaOH solution for 10 or 30 minutes first. After immersion, the fibre packages need to be rinsed excessively with water until they are neutral. Then the first step of the Silanization treatment begins.

7 RESULTS

The results of the quickscans are expressed in embodied energy, as one of the indicators for environmental performance. The results include (1) an identification which factors contribute to the embodied energy of every single treatment, (2) the comparison of the total embodied energy of the different fibre treatments and (3) the contribution of fibre treatment to the overall embodied energy of the final product. Below the results on these three levels are discussed shortly for the example of fibre treatments for rain screen cladding panels.

7.1 Factors influencing embodied energy of fibre treatments

In the figure below the factors influencing the embodied energy of the five different fibre treatments are shown.

The figure shows the main processes and materials which influence the embodied energy results of the five treatments. By application of a cut-off of 1%, in each column, only 2 to 5 processes and materials remain visible as significant factors contributing to the total amount of embodied energy.

For most analysed treatments, except the waterglass treatment, the production of the used chemicals (materials) contributes most to the embodied energy. Energy (electricity) contributes significantly to the energy required for treatment with waterglass and DMDHEU.

Acetylation shows also a negative impact as a result of the generation of useful by-products— acetic acid in this case.

For all five variants it is clear that waste water treatment, packaging and transport, only have a minor contribution to the total embodied energy of the treatments. There are also some processes which do not even reach the 1% cut-off threshold: gas use (in acetylation), the production of demi water and infrastructure for the treatment, like machineries.

The conclusion we can draw from the above is that the key parameter for optimization of embodied energy in each treatment is the amount of chemicals used.

7.2 Total embodied energy of fibre treatments

Figure 2 also makes it possible to compare the embodied energy of the five fibre treatments per kilogram of treated fibres. The difference between the highest and the lowest bar is almost a factor 6. DMDHEU treatment using not even 20 MJ/kg and salinization treatment with pretreatment using more than 120 MJ/kg.

The difference between the silanization with and without pretreatment is 21 MJ or 19%. However the difference compared to the other treatments is much larger, differing with a range from 82 to 113 MJ/kg.

The choice for the specific type of treatment has thus a larger influence on the embodied energy results than the choice for a specific variant (e.g. with or without pretreatment). The choice of

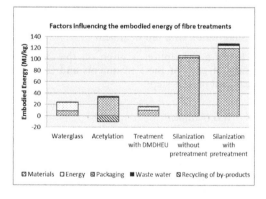

Figure 2. Factors influencing the embodied energy of fibre treatments.

treatment not only depends on the energy needed for the treatment itself, but also on the resulting effectiveness of the fibre in the final product. If the mechanical properties are improved, there is potentially less fibre needed in the final product to reach the product demands. If the moist uptake is reduced, the lifetime of products exposed to rain and other water sources may be elevated, compared to products with untreated fibres. In order to have a final say about which treatment score the best on embodied energy, this information needs also to be added in the equation. Unfortunately, at this stage of the research project this is not possible yet.

In the next paragraph it will be discussed how large these impacts of the different treatments are in its ultimate product application.

7.3 Embodied energy of fibre treatment in relation to total embodied energy for rain screen cladding

The quickscan analysis is performed for the functional unit of 1 m² of rain screen cladding. In Figure 3 the amount of embodied energy needed for fibre treatment is depicted (black bar) as part of the total amount of embodied energy of the production of rainscreen cladding. This figure shows that fibre treatment is not the most important factor contributing to the total embodied energy of rainscreen cladding. Instead, the production of natural fibres accounts for the biggest part of the total embodied energy. Therefore, if fibre treatment contributes to a reduction of the natural fibre amount needed for rain screen claddings, either by improved mechanical properties or enhanced lifetime, the treatment is effective in the reduction of the embodied energy content of the final product.

Waterglass, Acytylation and DMDHEU treatment all have comparable embodied energy impact per rainscreen panel, with a contribution of about

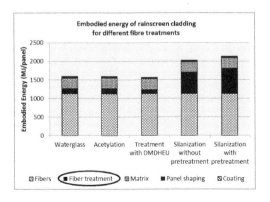

Figure 3. Embodied energy of rain screen cladding for different fibre treatments.

10% to the overall embodied energy of the rain-screen cladding product. Silanization treatment is strongly deviating from this group and contributes approximately 25% to the embodied energy of the rain screen cladding.

The effectiveness of the different treatments (on lifespan of the product, water resistivity etc.) is not yet taken into account in these results. To draw conclusions, estimates have to be made on the expected influence on lifetime and fibre amounts for each of the treatments. Silanization needs to be more than twice as effective compared to the other treatments in reduction of fibre amounts or enhancement of the durability of the final product to become attractive.

For the difference between silanization with and without pretreatment the same argument counts: it is only useful to apply pretreatment if this leads to increased effectiveness of the treatment resulting in reduced amounts of natural fibres needed.

8 DISCUSSION AND CONCLUSION

Often environmental assessments come too late to improve the environmental impact of a product. However, by applying the quickscan method during the product development process, design decisions can be guided by valuable information on the environmental impact of the alternatives, resulting in products with less environmental impact.

The quickscan approach explained in this article enables performance driven support of key decisions in product development. By means of a pre-constructed quickscan model, basic information of experts and a few rounds of iterative communication, important insights can be gained in a quick and comprehensive manner. A critical factor for success is the incorporation of technical preconditions.

The quickscan approach shows that early assessments help to point out key parameters worth emphasizing and items of minor importance which do not require profound effort. For fibre treatment, the energy needed for the production of the chemicals is most relevant, followed by energy use in the treatment process. As most embodied energy of the rain screen cladding can be attributed to the production of the natural fibres, fibre treatment can be effective in reduction of embodied energy content when the amount of natural fibres can be decreased as a result of the treatments (either by improved mechanical properties, or enhanced lifetime). This information is very valuable since it indicates in which part of the production process changes can lead to possible big improvements.

The quickscan results in this article show that the fiber treatments differ considerably in the amount of embodied energy needed. The Silanization treatment contributes approximately 25% to the total embodied energy of the rain screen cladding system, while Waterglass, Acetylation and DMDHEU treatment all have an embodied energy contribution of about 10% to the total. Before discarding Silanization treatment however, the effectiveness of each of the treatments needs to be included, as well as the costs of the treatments.

ACKNOWLEDGEMENTS

This research work was co funded by the European Commission under the call—FP7–2011-NMP-ENV-ENERGY-ICT-EeB—Proposal No: 285689 BioBuild CP-IP.

Project partners include NetComposites; 3XN A/S; ACCIONA; Holland Composites Industrials BV; TransFurans Chemicals; Arup GmbH; DSM Resins B.V.; Cimteclab; KUL; Amorim Cork Composites; Institut fuer Verbundwerkstoffe GmbH; TNO; LNEC; SHR; and FibreForce.

Special thanks goes to the following persons: Edwin Stokes, Dieter Perremans, Sergiy Grishchuk, Bôke Tjeerdsma, Charlotte Heesbeen, John Hartley, Morten Norman Lund, Suzanne de Vos and Jonna Gjaltema. Their contributions to this article were highly valuable.

REFERENCES

CEN January 2012. EN 15804. *Sustainability of construction works—Environmental product declarations—Product category rules*. Brussels: CEN European Committee for Standardization.

Duflou, J.R., J. De Moor, I. Verpoest & W. Dewulf 2009. Environmental impact analysis of composite use in car manufacturing. *CIRP Annals—Manufacturing Technology* 58(1): 9–12.

European Commission 2012. *FP7: the future of European Union research policy*. Retrieved from: http://ec.europa.eu/research/fp7/index_en.cfm. Last visited: 27–01–2013.

Faruk, O., A.K. Bledzki, H-P. Fink & M. Sain 2012. Biocomposites reinforced with natural fibres: 2000–2010. *Progress in Polymer Science* 37:1552–1596.

Kim, K-H. 2011. A comparative life cycle assessment of a transparent composite facade system and a glass curtain wall system. *Energy and Buildings* 43: 3436–3445.

Green Design, Materials and Manufacturing Processes – Bártolo et al. (eds)
© *2013 Taylor & Francis Group, London, ISBN 978-1-138-00046-9*

A bottom-up perspective upon climate change—approaches towards the local scale and microclimatic assessment

A. Santos Nouri
Faculty of Architecture, Technical University of Lisbon, Portugal
Research Scholarship holder from the Portuguese Foundation for Science and Technology—Urbanised Estuaries
and Deltas. In search for a comprehensive planning and governance.

ABSTRACT: Up to now, the issue of 'locality' has mostly been approached through a solely top-down planning perspective. This concentrates on methods of impact prediction by relying on global models as a starting point to anticipate climate change scenarios. Nevertheless, it is becoming consensually established that these models have little specificity and lack information regarding local assessment and adaptation. Although these more encompassing analytical top-down models are fundamental in global adaptation, significant meteorological variables are being frequently overlooked. Having the possibility to contribute to the quality of life and identity within cities, there is an unprecedented interest in the quality/resiliency of urban open spaces due to their role in establishing microclimatic thermal comfort levels. Moreover, this interest shall grow exponentially along with the impending climatic effects within urban outdoor environments. This article discusses how bottom-up orientated planning assessments can contour issues of climatic uncertainty by methodologically approaching these cumulative microclimatic manifestations.

To date, a considerable amount of research relating to 'locality' has been carried out through a top-down approach. This concentrates on methods of impact prediction through global models as a starting point to anticipate climate change scenarios. As these global models have little local specificity, there *"has been a growing interest, however, in considering a bottom-up approach, asking such questions as (1) how local places contribute to global climate change, (2) how those contributions change over time, (3) what drives such changes, (4) what controls local interests exercise over such forces, and (5) how efforts at mitigation and adaptation can be locally initiated and adopted."* (Wilbanks and Kates, 1999, p.601). Consequently, although climate change and uncertainty go hand in hand, climate change adaptation is not a 'vague concept' (Bourdin, 2010). Inversely, it is the concrete bond within specific localities that substantiates its preciseness (Costa, 2011); nevertheless, this means that climatic *"effects cannot be downscaled from a regional weather model, they are complex and require local observation and understanding."* (Hebbert and Webb, 2007, p.125). This article shall discuss how analytical models that are bottom-up orientated can tackle uncertainty by focusing on climate impacts upon the local scale as the *"cities' sustainable development mainly depends on the capacity of the town planners to offer outdoor urban spaces with high environmental qualities (…)*

designing and modifying urban forms induce major and long-term transformations to the environment. The microclimate is one of the fundamental aspects of this process." (Reiter and Herde, 2011, p.1).

1 THE PARADOX OF SCALE

In a world of global environmental change, the variances between 'micro' and 'macro' scale perspectives are currently originating a considerable paradox between top-down and bottom-up approaches. The interchangeable scale at which environmental change has been assessed has been mostly top-down, where the methods of climatic impact analysis are derived from global models. Respectively, Global Climate Models (GCMs) are the primary tools for evaluating global change that provide 'reasonable' simulation accuracy of present climate when viewed from global and hemispheric scales; yet the data presented from these models at more micro scales are often considered erroneous (Hewitson and Crane, 1996). With notable regional and local specificity, there is now a growing interest in considering a reversal of this 'down-scaling' approach within the planning and scientific communities (Wilbanks and Kates, 1999). This elevated weight attributed to bottom-up approaches further underpin the micro-environmental process, socio-economic activities, resource management that arise within the local scale.

1.1 Dominant top-down approaches and relationships with local contexts

In accordance with GCMs, it has been established that the global temperature is to continually rise throughout the 21st century, and that there shall be significant changes in air humidity, wind speed and cloud cover around the globe. However, the Intergovernmental Panel on Climate Change (IPCC) report of 2001 describes the effect of weather and climate on humans with a limitative index that is based on a combination of air temperature and relative humidity. This originates a lack of significant information regarding: (1) meteorological factors including—wind speed and radiation fluxes; and (2) thermo-physiological factors including—activity of humans and clothing. Respectively, this suggests that although these top-down approaches are vital in establishing consequences of global climatic change, they have the tendency to overlook imperative thermo-physiological parameters in local contexts (Matzarakis and Amelung, 2008).

1.2 The downscale of climate change

The GCM's seasonal mean temperature, precipitation and wind speeds within Europe are analysed by sixteen down-scaled Regional Climate Models (RCMs) simulations until the end of the century. These simulations are used to: *"(i) evaluate the simulated climate for 1961–1990, (ii) assess future climate change and (iii) illustrate uncertainties in future climate change related to natural variability, boundary conditions and emissions."* (Kjellstorm, Nikulin, Hansson et al., 2011, p.24).

In the process of downscaling climate change examinations, the establishment of the third IPCC Assessment Report (IPCC, 2001), resulted in new data that scrutinise the possible national territorial impacts of climate change. This resulted in an array of countries establishing their own national studies, and in the case of Portugal, a new SIAM Project Report (Santos, Forbes and Moita, 2002) was thus established. This new outlook endorsed the amalgamation of local, regional, national and international adaptation agendas (Swart, Biesbroek, Binnerup et al., 2009).

2 THE POTENTIALITY OF LOCAL SCALES

When addressing the potentialities of local scales through a bottom-up approach, it is suggested that local dynamics are an important consideration in the climate change agenda (Hebbert and Webb, 2007; Costa, 2011). Although *"it is clear that some of the driving forces for global change operate at a global scale, such as greenhouse gas composition of the atmosphere and the reach of global financial systems. But it seems just as clear that many of the individual phenomena that underlie micro-environmental processes, economic activities, resource use, and population dynamics arise at the local scale."* (Wilbanks and Kates, 1999, p.602). This approach by Thomas Wilbanks and Robert Kates presents various fundamental arguments that express not only why local scales are significant, but also how and where they respectively matter.

Firstly, when considering the domain (i.e. scale) of environmental change, one can reflect upon the global 'snowballing' changes that are resultant of the accumulation of widespread localised change. This presents the opportunity to investigate smaller-scale connotations in order to better comprehend the causes and driving forces of the global universal phenomenon (Root and Schneider, 2003).

Secondly, the previous consideration is strengthened when scrutinising the roles of 'agency' that are intentional human action; and where 'structure' is constituted by formal social affiliations or organisations. This differentiation enforces the prominence of 'agency' having a direct interest regarding local scales and their respective adaptation measures. This depicts upon the *"growing concern among local leaders about the long-term human health or social and environmental effects of inaction as well as the possibility to piggy-back climate change into more urgent local agendas such as improved local environments and liveability of cites ..."* (Corfee-Morlot, Kamal-Chaoui, Donovan et al., 2009, p.33).

Lastly, it is suggested by Wilbanks and Kates that the interactionism between global structure and local agency through varying domains has raised significant interest in consolidating relevant climatic predictions and models. As a result, this suggests a need for both, a multi-scale climatic analysis and a multi-level governance in order to avoid gaps between local action plans and national and global policy frameworks. This approach *"allows two-way benefits: locally-led or bottom-up where local initiatives influence national action and national-led or top-down where enabling frameworks empower local players. The most promising frameworks combine the two into hybrid models of policy dialogue where the lessons learnt are used to modify and fine-tune enabling frameworks and disseminated horizontally, achieving more efficient local implementation of climate strategies."* (Corfee-Morlot, Kamal-Chaoui, Donovan et al., 2009, p.3).

2.1 Locally initiating analytical and adaption measures

Escalating now from the local scale to the global scale, one can note a significant difference in climate change awareness and regulatory practice.

Nowadays, carbon mitigation and/or adaptation to global warming are part of many international agendas with monthly initiatives being disseminated throughout the global scientific community. Although this global dissemination is imperative, it is argued that it is frequently "*focussed*[on] *the exposure of cities to hazards that have a huge impact but low frequency. It has little to say about the high-frequency and micro-scale climatic phenomena created within the anthropogenic environment of the city.*" (Hebbert and Webb, 2007, p.126). Accordingly, this has diminished the comprehension of mitigation and adaptation within the local scale. Local factors such as microclimates are being considerably overlooked, where inclusively "*landscape architects and urban designers strive to design places that encourage* [urban] *activities, places where people will want to spend their time (…) however unless people are thermally comfortable in the space, they simply won't use it. Although few people are even aware of the effects that design can have on the sun, wind, humidity, and air temperature in a space, a thermally comfortable microclimate is the very foundation of well-loved and well-used outdoor places.*" (Brown, 2010, p.2).

Invariably, the considerations upon local climatic variables relays directly to the effects that the climate can have upon the liveability of cities (Hebbert and Webb, 2007). This suggests that urban form can enhance or reduce the quality of urban life, thus signifying the prominence between the spheres of local urban design and climatic adaptation. Issues such as street orientation, street width-to-height ratios, building spacing, architectural detailing in streetscapes, heat-reflectiveness from materials, the location of street trees, parks and water spaces have become a prominent niche for urban climatology and design (Givoni, 1998; Erell, Pearlmutter and Williamson, 2011). Where, with the aid of continual scientific outputs, respective investigations are integrating design concepts and climatic effects with 'biometerological' variables so human comfort levels can be estimated under different climate scenarios and design settings (Nikolopoulou, 2004; Schiller and Evans, 2006; Matzarakis, Rutz and Mayer, 2007; Grifoni, Latini and Tascini, 2010; Wong, Jusuf and Tan, 2011).

Having the possibility to contribute to the quality of life within cities, there is a strong interest in the quality of urban open spaces due to their role in establishing microclimatic thermal comfort levels (Katzschner, 2006). Moreover, this interest shall increase along with the progression of climatic effects upon urban outdoor environments.

In order to approach microclimatic comfort conditions (Fig. 1), open spaces require a precise microclimatic analysis in order to evaluate people's reactions within a given thermal stimuli (Olgyay, 1963).

More specifically, the thermal environmental effects upon humans can be determined with the aid of thermal indices based on the energy balance of the human body. One of the most common and used thermal indices is the Physiologically Equivalent Temperature (PET) (Höppe, 1999). This thermal index includes considerations upon meteorological factors such as air temperature, air humidity, wind velocity and short/long wave radiation that affects humans thermo-physiologically in outdoor environments (Table 1).

Nevertheless, it is often that climatic assessments and thermal comfort studies have resorted to more simplistic and limitative analysis tools. As an example of this discrepancy, the IPCC Report of 2001/7 "*describes the effect of weather and climate on humans with a simple index based on a combination of air temperature and relative humidity. The exclusion of important meteorological (wind speed and radiation fluxes) and thermo-physiological (activity of humans and clothing) variables seriously diminishes the significance of the results.*" (Matzarakis and Amelung, 2008, p.162). Conse-

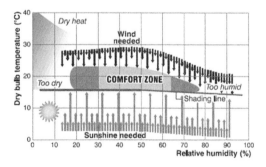

Figure 1. Olgyay's interpretation of moderate European climates—microclimatic requirements through PET values to determine thermal comfort in outdoor environments (Olgyay, 1963).

Table 1. PET Ranges within different grades of thermal perception and resulting physiological/thermal stress on human beings (Matzarakis and Mayer, 1996).

PET	Thermal perception	Grade of physiological stress
≤ 4°C	Very Cold	Extreme Cold Stress
4°C	Cold	Strong Cold Stress
8°C	Cool	Moderate Cold Stress
13°C	Slightly Cool	Slight Cold Stress
18°C	Comfortable	No Thermal Stress
23°C	Slightly Warm	Slight Heat Stress
29°C	Warm	Moderate Heat Stress
35°C	Hot	Strong Heat Stress
41°C	Very Hot	Extreme Heat Stress

quently, examples of such discrepancies can hinder fairly 'obvious' and important microclimatic considerations for a multitude of professionals such as urban designers and architects.

As an example, irrespective of air temperature, there is a constant variance of 5°C between PET values in areas exposed to the sun and those in the shadow (Fig. 2). Interestingly, and comparing this to the fascinating results from Whyte's time-lapse photography (Fig. 3), it is suggested that *"by asking the right questions in sun and wind studies, by experimentation, we can find better ways to board the sun, to double its light, or to obscure it, or to cut down breezes in winter and induce them in summer. We can learn lessons in semiopen niches and crannies that people often seek."* (Whyte, 1980, p.45).

Pioneers of open space design and/or maintenance all indicate the inarguable relationship between microclimates and the vitality of urban spaces (Whyte, 1980; Carmona, 2003; Gehl, 2010; Erell, Pearlmutter and Williamson, 2011). Per se, public spaces can significantly benefit from microclimatic assessments that enable more efficient implementations of climatic strategies (both in analysis and policy criterion) in local contexts.

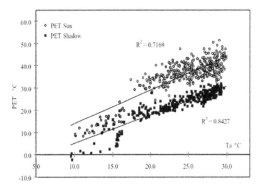

Figure 2. Relation between air temperature and thermal comfort index PET in sun and shadow (Katzschner, 2006).

Figure 3. William Whyte's time-lapse photography of sunlight patterns in New York (Whyte, 1980).

Although more encompassing analytical models are fundamental in the global adaptation to climate change, the local scale is: (1) where the effects of climate change will have their fundamental effects; and (2) directly where adaptation measures will find their niche to regulate these impending effects.

This being said, it is impossible to globally downscale climate change directly into local scales due to high computational demands and costs (Rummukainen, 2010). This 'limitation' is far from restrictive since those that are involved in urban planning and design can start with common meteorological data like air temperature, air humidity and wind speed in order to make significant improvements to the anthropogenic spaces of the city (Matzarakis, Rutz and Mayer, 2007). In terms of methodologically approaching these meteorological factors, one of the most accurate assessment models will be discussed below.

2.2 *RayMan assessment model*

One of the most comprehensive approaches in this analytical arena is the RayMan model, which calculates short and long wave radiation fluxes upon the human body. RayMan is able to deal with complex urban structures/compositions and is an effective tool for urban planners/designers when consulting the outdoor environments of the city.

Unlike other approaches and models, RayMan uses thermal indices such as PET to concretely address the urban bioclimatic and thermal comfort levels within local scales. This method permits *"the human-biometerological evaluation of the atmospheric environment"* and *"to clarify whether or not planning instruments are available for maintaining and improving the human-biometerological situation."* (Matzarakis, Rutz and Mayer, 2007, p.1).

Beyond calculating mean radiant temperature (T_{mrt}), RayMan is able to calculate radiation fluxes due to the input of the following: (1) topography and environmental morphology (Fig. 4a); and (2) fish eye lens photography to aid Sky View Factor (SVF) calculations (Fig. 4b).

These input options allow the user to specify the topography of the area allowing the model to consider the geometric descriptors of the space/canyon. This is further heightened by the possibility to input fish eye lens photography in order to assess the local SVF. This calculation is a vital parameter in order to investigate the: (1) Urban Heat Island (UHI) effect; and (2) exposure of space to diffused solar radiation.

From this information and with the input of commonly accessible meteorological data such as air temperature, air humidity and wind speed—the T_{mrt} and PET can be respectively calculated. This is

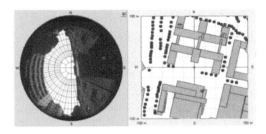

Figure 4. (a) Input window of urban structures and environmental morphology (b) Input window of fish eye lens photographs/drawings. Source: (RayMan Software by Matzarakis, 2000).

a significant step in linking urban climatology and urban design; enabling human comfort levels to be accurately calculated when assessing local climatic implications.

2.3 Thinking "what if?"

Although models such as RayMan only calculate existing microclimatic factors within the local scale, they are evidence that urban design and planning do not have to be hindered by climatic uncertainty. Approaches such as strategic 'what if?' scenarios (Costa, 2011), allow the anticipation of possible impacts within future horizons. Correspondingly, the data input in models such as RayMan can be based on actual figures in order to understand whether a given space offers efficient comfortable PET values. Or, and more interestingly, investigate how these existing biometeorological variables will be affected, by asking questions such as 'what if Europe's land temperature rises by 4.0°C by 2100?' This flexible approach is a pragmatic approach to uncertain climatic events in future horizons. Since these questions can be based on easily accessible meteorological information, local agents can start to comprehend how global climate change will affect the local and anthropogenic environments of the city through a bottom up approach. This endorses a new creative laboratory that fully exploits the potentialities and adaptability of local contexts.

3 CONCLUSIONS

Resultantly, it is becoming consensually recognized that top-down approaches are limitative for 'agencies' whose contributions to climate change lie fundamentally within the local 'domain'. This however does not mean that climatic assessments through GCM's or RCM's are not fundamental to Mankind's challenge in adapting to climate change. Yet it is suggested that these models often overlook

fundamental climatic issues that are more applicable to microscale phenomenon. The often singular reflection and preoccupation with extreme weather effects greatly overlook the less severe yet constant meteorological repercussions such as wind patterns, solar radiation, and air temperature.

This article suggests that given climatic uncertainty, urban design and planning need to find new ways to analyse climatic phenomenon that take place within the local scale. This implies a change in the way in which climate change is to be approached for local entities such as architects and urban designers/planners. This article disputes an innovative reversal of the modern day 'top-down' approach, where microclimatic investigations can be used to combat climatic uncertainty. Implied by this approach, are new forms of thinking that consider both extreme weather effects, and the less severe but more constant effects of meteorological parameters on the public realm. It is consensus that air temperature is rising, yet there is little to be said regarding other meteorological parameters such as wind patterns and solar radiation. This lapse in the climate change adaptation arena therefore requires the understanding of the thermal environmental effects upon humans. This article suggests that this comprehension however can already be achieved, through local meteorological data, and can also be reflected into future contexts through flexible 'what if?' approaches.

Although the approach suggested by this article is still in its initial phases, the benefits are clear for those who are tasked with maintaining dynamic and healthy public realm. The theoretical discussion suggests that this approach raises the opportunity to reflect upon making dominant top-down approaches more useful for local contexts. For example, RCM assessments can be made more relevant for local agents by establishing the correct parameters and analytical scopes for local contexts. Furthermore, the differences in perspective between global structure and local agency can be reduced through multi-scale climatic analysis and multi-level governance in order to avoid gaps between local action plans/interventions and national and global policies. Yet in order to fine tune local implementation of climatic adaptation strategies and measures, microclimatic assessments need to play a more significant role in conjoining the spheres of urban climatology and urban planning/design.

ACKNOWLEDGMENTS

The author would like to acknowledge the research project "Urbanised Estuaries and Deltas. In search for a comprehensive planning and governance.

The Lisbon case."(PTDC/AUR-URB/100309/2008) funded by the Portuguese Foundation for Science and Technology and the European Social Found, 3rd Community Support Framework.

REFERENCES

Bourdin, A. (2010). O Urbanismo depois da Crise. Lisbon, Livros Horizonte.

Brown, R. (2010). DESIGN WITH MICROCLIMATE— The Secret to Comfortable Outdoor Space. Washington Island Press.

Carmona, M. (2003). Public Places, Urban Spaces: The Dimensions of Urban Design UK, Architectural Press.

Corfee-Morlot, J., L. Kamal-Chaoui, M.G. Donovan, I. Cochran, A. Robert and P.-J. Teasdale (2009). Cities, Climate Change and Multilevel Governance OECD Environmental Working Papers Nº14, OECD: pp. 1–126.

Costa, J.P. (2011). 'Climate Proof Cities'. Urbanismo e a Adaptação às Alterações Climáticas. As frentes de água. Lição de Agregação em Urbanismo apresentada na FA-UTL. Lisboa, Universidade Técnica de Lisboa. 2.

Erell, E., D. Pearlmutter and T. Williamson (2011). Urban Microclimate—Designing the Spaces Between Buildings. United Kingdom, Earthscan.

Gehl, J. (2010). Cities for People. Washington DC, Island Press.

Givoni, B. (1998). Man Climate and Architecture Amsterdam, Elsevier.

Grifoni, R.C., G. Latini and S. Tascini (2010). Thermal comfort and microclimates in open spaces. Italy, ASHRAE.

Hebbert, M. and B. Webb (2007). Towards a Liveable Urban Climate: Lessons from Stuttgart. Liveable Cities: Urbanising World: Isocarp 07. Manchester, Routledge.

Hewitson, B. and R. Crane (1996). "Climate downscaling: techniques and application " Climate Research: Clim Res 7: 85–95.

Höppe, P. (1999). "The physiological equivalent temperature—a universal index for the biometeorological assessment of the thermal environment." International Journal of Biometeorology 43: pp. 71–75.

IPCC. (2001). "Climate Change 2001: Impacts, Adaptation and Vulnerability." from http://www.ipcc.ch/ publications_and_data/publications_and_data_reports. htm.

Katzschner, L. (2006). Microclimatic thermal comfort analysis in cities for urban planning and open space design London, Network for Comfort and Energy Use in Buildings NCUB.

Kjellstorm, E., G. Nikulin, U. Hansson, G. Strandberg and A. Ullerstig (2011). "21st century changes in the European climate: uncertainties derived from an ensemble of regional climate model simulations" Tellus Tellus (2011)(63A): 24–40.

Matzarakis, A. (2000). RayMan. M. Institute. Germany, University of Freiburg.

Matzarakis, A. and B. Amelung (2008). Physiological Equivalent Temperature as Indicator for Impacts of Climate Change on Thermal Comfort of Humans Seasonal forecasts, climatic change and human health. Advances in global research 30. T.M. e. al. Berlin, Springer: pp. 161–172.

Matzarakis, A. and H. Mayer (1996). "Another Kind of Environmental Stress: Thermal Stress. WHO Colloborating Centre for Air Quality Management and Air Pollution Control." NEWSLETTERS 18: 7–10.

Matzarakis, A., F. Rutz and H. Mayer (2007). "Modelling radiation fluxes in simple and complex environments— application of the RayMan model." International Journal of Biometeorology(51): pp. 323–334.

Nikolopoulou, M. (2004). Designing Open Spaces in the Urban Environment: a biocliamtic approach. CRES—Centre for Renewable Energy Sources, RUROS—Rediscovering the Urban Realm and Open Spaces.

Olgyay, V. (1963). Design with climate, bioclimatic approach to architectural regionalism. New Jersey, Princeton university press.

Reiter, S. and A.d. Herde (2011). Qualitative and quantitative criteria for comfortable urban public spaces. Proceedings of the 2nd International Conference on Building Physics, ORBi.

Root, T. and S. Schneider (2003). Strategic Cyclical Scaling: Bridging Five Orders of Magnitude Scale Gaps in Climatic and Ecological Studies. Scaling Issues in Integrated Assessment J. Rotmans and D. Rothman. The Netherlands: pp. 179–204.

Rummukainen, M. (2010). State-of-the-art with regional climate models Sweden Swedish Meterological and Hydrological Institute (SMHI). 1: pp. 82–96.

Santos, F.D., K. Forbes and R. Moita (2002). Climate Change in Portugal Scenarios, Impacts and Adaptation Measures—SIAM Project. Lisboa, Gradiva— Publicações, L.da

Schiller, S.d. and J.M. Evans (2006). Assessing urban sustainability: microclimate and design qualities of a new development. PLEA2006-The 23rd Conference on Passive and Low Energy Architecture. Geneva— Switzerland, PLEA.

Swart, R., R. Biesbroek, S. Binnerup, T.R. Carter, C. Cowan, T. Henrichs, S. Loquen, H. Mela, M. Morecroft, M. Reese and D. Rey (2009). Europe Adapts to Climate Change. Comparing National Adaptation Strategies. Helsinki, Finnish Environment Institute (SYKE).

Whyte, W.H. (1980). The Social Life Of Small Urban Spaces. USA, Project for Public Spaces Inc.

Wilbanks, T.J. and R.W. Kates (1999). Global Change in Local Places: How Scale Matters Climatic Change. The Netherlands. 43: 601–628.

Wong, N.H., s. K. Jusuf and C.L. Tan (2011). "Integrated urban microcliate assessment method as a sustainable urban development and urban design tool." Landscape and Urban Planning(100): 386–389.

Green Design, Materials and Manufacturing Processes – Bártolo et al. (eds)
© 2013 Taylor & Francis Group, London, ISBN 978-1-138-00046-9

Energy price differential and industrial production growth in Mexico: A wavelet approach

Gazi Salah Uddin, Sanjib Chakraborty, Rubayet Hossian & Bo Sjö
Department of Management and Engineering, Linköping University, Sweden

ABSTRACT: The aim of this paper is to examine the measure of co-movement of the energy price differential and the industrial production growth in the time–frequency space by resorting to wavelet analysis in Mexico using monthly data from $1978M_5$ to $2011M_{10}$. The co-movement can be studied in either the time domain or in the frequency domain. The results can be sensitive to the frequency of observations. In this paper, we use a wavelet-based measure of co-movement which makes it possible to find a balance between the time and frequency domain features of the data and, which constitutes a refinement to previous approaches. We find that the strength of co-movement of industrial growth and energy price growth difference and changes over the time horizon.

1 INTRODUCTION

Energy is considered as a main source of primary input for industrial growth. This issue is fundamentally important for industrialized economics particularly, for Mexico after the occurrence of the oil shocks in the 1970s. In a study on US, Hamilton (1983) finds that recessions since Second World War II have been proceeded by a dramatic increase in the price of crude petroleum. According to the study by Mork (1989) there exists a statistically significant negative correlation between Gross Domestic Product (GDP) and oil price increases. For US, Japan, Germany, Canada, France, UK and Norway, Mork, Olsen and Mysen (1994) claims that oil shocks have significant influence on income. The inverse relations between these variables are common in oil importing countries. Hooker (1996) investigates the impact of oil price on the macro economy in view of the collapse of the oil market in the mid-1980s. Applying the Markov switching model, Raymond and Rich (1997) analyze the relationship between oil price shocks and the GDP growth and find that the oil prices shocks associated with stock returns. Hamilton (2010) claims that nonlinear transformations of oil price are useful in forecasting GDP growth. On the other hand, Bachmeier, Li and Liu (2008) claim that oil prices have no predictive power for measures of economic activity.

The most common line of research method concentrates on time domain analysis. Croux et al. (2001) have suggested a spectral-based measure, the dynamic correlation which allows one to measure the synchronization between two series at each individual frequency. This concept is similar with the earlier contemporaneous correlation between two series in the time domain. The synchronization in the time domain is not similar with the synchronization in the frequency and time domain. A large number of papers based on empirical research has been found in recent literature on this aspect (see, Crone (2005), Rua and Nunes (2005), Camacho et al. (2006), Eickmeier and Breitung (2006) and Lemmens et al. (2008)). The limitation of the methodology and the procedures to accommodate the analysis of time frequency dependencies between two time series is the focus of pioneering research by Hudgins et al. (1993) and Torrence and Compo (1998) who developed the approaches of the cross-wavelet power, the cross-wavelet coherency, and the phase difference.

In 2010, the share of oil in the Mexico's total consumption is 37 percent and industrial production growth was the 58th highest in the world at 6 percent (EIA, 2010). The aim of this paper is to examine the measure of co-movement of the energy price differential and the industrial production growth in the time–frequency space by resorting to wavelet analysis in Mexico using monthly data from $1978M_5$ to $2011M_{10}$. In this present study, crude petroleum costs are considered as a proxy for energy price and Industrial production index attached as a proxy for GDP due to unavailability of monthly data for GDP. Data are collected from an International Monetary Fund (IMF) CD ROM (2012) of IFS (International Financial Statistics). The study addressed the following questions: (i) to demonstrate if there is anti-cyclical relationship, and (ii) to show in which year these cyclical and anti-cyclical relationships are observed. The empirical application of the proposed wavelet-based measure of co-movement is highlighted as it allows unveiling both the time and frequency varying features.

The paper is organized as follows. In section 2 the wavelet-based measure of co-movement is adopted, while in Section 3 the empirical application is discussed at the different time scales. Finally, the paper concludes in Section 4.

2 A WAVELET-BASED MEASURE OF SYNCHRONIZATION

2.1 The Continuous Wavelet Transform (CWT)

A wavelet is a function with zero mean in both time and frequency scale. A wavelet can be characterized byt its localization in time (dt) and frequency (dω). There is always a tradeoff between localization in time and frequency that was explained by the conventional approach of the Heisenberg uncertainty principle and without the proper definition of dt and dω, it can be noted that there is a limit to how small the uncertainty product (dt · dω) can be. The Morlet wavelet (a particular wavelet) is defined as

$$\psi_\sigma(\beta) = \pi^{-1/4} e^{i\omega_d\beta} e^{-\frac{1}{2}\beta^2} \tag{1}$$

where ω_d and β are dimensionless frequency and dimensionless time. The Morlet wavelet with $\omega_d = 6$ is a best choice to use for feature extraction purpose by wavelet as it gives a good stability between time and frequency localization. To restrict the frequencies of the time series within a certain range the CWT is used to apply the wavelet as a band pass filter. The wavelet is expanded in time by changing its scale (r), so that $\beta = r \cdot t$ and normalizing it to have unit energy. For the Morlet wavelet, with $\omega_d = 6$, the Fourier period (λ_{wt}) is almost the same, i.e. $\lambda_{wt} = 1.03s$. The CWT of a time series x_m, $m = 1, \ldots, M - 1$, M with uniform time steps δt is defined as the convolution of x_m with the scaled and normalized wavelet. The equation can be written as

$$W_t^X(r) = \sqrt{\frac{\delta t}{r}} \sum_{t=1}^M X_{m'} \psi_\sigma\left[(m' - m)\frac{\delta t}{r}\right] \tag{2}$$

The wavelet power is defined as $|W_t^X(r)|^2$. The complex argument of $W_t^X(r)$ can be interpreted as the local phase. To take into account the edge effect, the Cone of Influence (COI) can be introduced, as the CWT has edge artifacts because the wavelet is not completely localized in time. Here we take the COI as the area in which the wavelet power

caused by a incoherence at the edge has dropped to e^{-2} of the value at the edge. By comparing with null hypotheses that the signal is generated by a stationary process with a given background power spectrum (P_k), the statistical consequence of wavelet power can be evaluated.

Data can be generated by the auto regression (AR$_0$ or AR$_1$) stationary process with a certain background power spectrum (P_k). Torrence and Compo (1998) have shown how the statistical consequence of wavelet power can be evaluated against the null hypothesis. The Monte-Carlo simulation process can be used as a more general process for data generation. Torrence and Compo (1998) computed the white noise and red noise wavelet power spectra, from which they derived, under the null hypothesis, the corresponding distribution for the local wavelet power spectrum at each time t and scale r as follows:

$$D\left(\frac{|W_t^X(r)|^2}{\sigma_X^2} < p\right) = \frac{1}{2} P_k \chi_v^2(p) \tag{3}$$

v represents the value of real and complex wavelets and is equal to 1 for real and 2 for complex wavelets. The description of CWT, XWT and WTC used in this paper is heavily drawn from Grinsted et al. (2004).

2.2 The Cross Wavelet Transform (XWT)

The two time series x_t and y_t of the Cross Wavelet Transform (XWT) are defined as $W^{XY} = W^X W^{Y*}$, where W^X and W^Y are the wavelet transforms of a and b, respectively, * denotes complex conjugation. Following from this, the cross wavelet power is defined as $|W^{XY}|$. The explanation of the complex argument W^{XY} can be defined as the local relative phase between x_t and y_t in time frequency space. According to Torrence and Compo (1998), theoretical distribution of the cross wavelet power of two time series P_k^X and P_k^Y with background power spectra can be defined as,

$$D\left(\frac{|W_t^X(s)W_t^{Y*}(s)|}{\sigma_X\sigma_Y} < p\right) = \frac{Z_v(p)}{v}\sqrt{P_k^X P_k^Y} \tag{4}$$

In the equation, $Z_v(p)$ is the confidence level attached with the probability p for a pdf defined by the square root of the product of two χ^2 distributions.

2.3 Wavelet Coherence (WTC)

Fourier spectral approaches define Wavelet Coherency (WTC) as the ratio of the cross-spectrum to the product of each series' spectrum, essentially as the local correlation between two time series, both in time and frequency. The Wavelet power spectrum illustrates the variance of time-series, with times of large power shown by large variance. Following Torrence and Webster (1999), the Wavelet Coherence of two time series can be defined as,

$$R_t^2(r) = \frac{|Q(r^{-1}W_t^{XY}(r))|^2}{Q|(r^{-1}|W_t^X(r)|^2)| \cdot Q|(r^{-1}|W_t^Y(r)|^2)|} \quad (5)$$

where, Q is considered as a smoothing operator. It seems that this definition closely resembles a traditional correlation coefficient and it is significant to consider of the Wavelet Coherence as a localized correlation coefficient in time frequency space. This study will focus on the Wavelet Coherency, instead of the Wavelet Cross Spectrum by following the study by Aguiar-Conraria and Soares (2011).

2.4 Cross wavelet phase angle

In order to calculate the phase difference between two time series it is necessary to calculate the mean and confidence interval of the phase difference. The phase relationships of the circular mean of the phase over regions with higher than 5% statistical significance will be used to quantify these relations. This is a convenient and universal method for calculating the mean phase. The circular mean of a set of angles (x_i, $i = 1, ..., m$) is defined as

$$a_m = \arg(X, Y) \text{ with } X = \sum_{i=1}^{m}\cos(x_i) \quad \text{and}$$

$$X = \sum_{i=1}^{m}\sin(y_i) \quad (6)$$

As the phase angle is not independent it is difficult to calculate the confidence interval of the mean angle reliably. By increasing the scale resolution the number of the angles used in the calculation can be set arbitrarily high. Moreover, it is interesting to know the scatter of angles around the mean. The circular standard deviation can be defined as

$$r = \sqrt{-2\text{In}(R/t)} \quad (7)$$

where $R = \sqrt{X^2 + Y^2}$.

The circular standard deviation varies from zero to infinity and is analogous to the linear standard deviation. The Monte Carlo simulation methods are used for the significance of wavelet coherence.

3 EMPIRICAL FINDINGS OF THE WAVELET

In this section, we report the results of the dynamics of the change in Industrial Growth (DlnIP) co-movement with oil price differential (DlnENG) obtained by applying wavelet analysis approach. The growth rates are calculated as the first difference of the logarithmic transformation of the concerned variables. The result is presented in the following three graphs.

Figure 1 displayed the CWT power spectra of the industrial growth and the return on energy price series for monthly data. The two sets of data display the highest power at different frequencies and also different time scales. Industrial growth has the higher power in the 1–4 month of scale (frequency) which corresponds to the 1990s whereas the change in the energy prices has the higher power in the 1–4, 18–20 month of scale (frequency) corresponding to the 1986s, 5–10 month of scale (frequency) corresponding to the 1990s and 8–20 month of scale (frequency) corresponding to the 2006s. It is evident that for both the data series the high power region is above the 5% significant level. In order to better understand the relations between the industrial growth and the change in energy prices series cross-wavelet transform can be helpful while from the continuous wavelet transform it is very difficult to explain the coincidence of the variables.

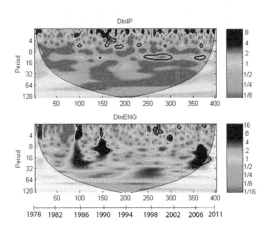

Figure 1. Continuous wavelet power spectra of DlnIP and DlnENG monthly series data.
Note: The continuous wavelet power spectrum of both DlnIP and DlnENG series are shown in the figure. 5% significant level against the red noise is shown by the thick black contour and lighter shade represents the Cone of Influence (COI) where edge effect might distort the picture. The color code for power ranges from black (high power) to white (low power). X-axis represents the time period and Y-axis represents the frequency or scale (in month).

Figure 2 represents the XWT of the industrial growth and the change in the energy price series for the monthly data and the variables are mostly in phase in the significant region. It can be deduced that there is a linear relationship between the signals as the signals have close to zero phase difference for most of the time across the scale. Arrows in the frequency 2–3 month of scale that corresponds to 1986s are pointing right down; this indicates that industrial growth is leading

Interestingly, arrows in the frequency 10–15 month of scale that corresponds to 2006s are pointing right up; indicates that the energy growth is leading. In addition, there are out of phase conditions in the frequency 12–14 month of scale that corresponds to 1990s and arrows are left-down indicated ENG is leading when the variables are out of phase. In addition, outside the area with significant power, the variables are in the phase most of the time. Considering the industrial growth as a proxy for economic growth from the cross-wavelet power it can be implied that there is a strong relationship between industrial growth and the return on energy.

Figure 3 represents the squared WTC of industrial growth and the change in energy series for the monthly data and by comparing with XWT more portions of the data stands out as being significant, also the phase relationship between industrial growth and return on energy is present in all the areas. The area of the time frequency plot below 95% confidence level is not a reliable indication of causality. It is possible for two time series to be perfectly correlated at one specific scale if the scales are approximately weighted for the averaging, but the area of significant correlation might be less than 5% significant level. The result from WTC gives us a very interesting scenario of the causality between industrial growth and energy increase. During 1984–1988, in the frequency band of 25–30 month of scale, arrows are pointing right-up, indicating that the variables are in phase and return on energy variable is leading.

During 2004–2006, in 10–32 month of scale, arrows are pointing right-up, indicating that the variables are in phase and the return on energy price variable is leading.

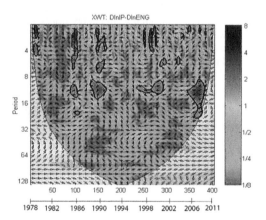

Figure 2. Cross-wavelet transforms of DlnIP and DlnENG monthly series data.
Note: Cross-wavelet power—power on the 95% confidence level compared to red noise which is estimated by using phase-randomized surrogate data from Monte Carlo simulation is represented by the thick black contour and shade represents the Cone of Influence (COI) where edge effect might distort the picture. The color code for power ranges from black (high power) to white (low power). Arrows are used to indicate the phase difference the two series. The variables are in phase indicated by the arrows that are pointing right and out of phase are indicate by pointing left. Arrows pointing right-down indicates that DlnIP is leading and right-up indicates DlnENG leading. Again arrows pointing left-down indicates that DlnENG is leading and left-up indicates DlnIP leading. X-axis represents the time period and Y-axis represents the frequency or scale (in month).

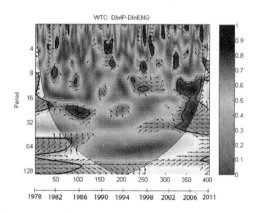

Figure 3. Cross-wavelet coherency of DlnIP and DlnENG monthly series data.
Note: Cross-wavelet coherency—power on the 95% confidence level compared to red noise which is estimated by using phase-randomized surrogate data from Monte Carlo simulation is represented by the thick black contour and shade represents the Cone of Influence (COI) where edge effect might distort the picture. The color code for power ranges from black (high coherency-close to one) to white (low coherency-close to zero). Arrows are used to indicate the phase difference the two series. The variables are in phase indicated by the arrows that are pointing right and out of phase are indicate by pointing left. Arrows pointing right-down indicates that DlnIP is leading and right-up indicates DlnENG leading. Again arrows pointing left-down indicates that DlnENG is leading and left-up indicates DlnIP leading. X-axis represents the time period and Y-axis represents the frequency or scale (in month).

4 CONCLUSION

This paper has described a study of the co-movement of the energy price differential and the industrial production growth in Mexico in the time–frequency space by resorting to wavelet analysis using monthly data from $1978M_5$ to $2011M_{10}$. In this paper, we use a wavelet-based measure of co-movement which makes it possible to find a balance between the time and the frequency domain features of the data and, which constitutes a refinement to previous approaches. We find that the strength of co-movement of industrial growth and energy price growth differ and changes over the time horizon. From the results it is clear that there is a linear relationship between the variables as the waves are in phase maximum of the time across the studied period. It implies that a Government should stimulate the industrial growth through the implementation of energy policy in the medium term.

REFERENCES

Aguiar-Conraria, L.S.M. 2., 2011. Oil and the macroeconomy: using wavelets to analyze old issues. *Empirical Economics*, Volume 40, pp. 645–655.

Bachmeier, L.Q.L. a. D.L., 2008. Should Oil Prices Receive so Much Attention? An Evaluation of the Predictive Power of Oil Prices for the U.S. Economy. *Economic Inquiry*, Volume 46 (4), pp. 528–539.

Camacho, M.P.-Q.G.S.L., 2006. Are European business cycles close enough to be just one? *Journal of Economic Dynamics and Control*, Volume 30, p. 1687–1706.

Crone, T., 2005. An alternative definition of economic regions in the United States based on similarities in state business cycles. *Review of Economics and Statistics.*, Volume 87, p. 617–626.

Croux, C.F.M.R.L., 2001. A measure of comovement for economic variables: theory and empirics. *Review of Economics and Statistics*, Volume 83, p. 232–241.

Eickmeier, S.B.J., 2006. How synchronized are new EU member states with the euro area? Evidence from a structural factor model. *Journal of Comparative Economics*, Volume 34, p. 538–563.

EIA 2010. International Energy Outlook, U.S. Energy Information Administration.

Grinsted, A.M.J.J.S., 2004. Application of the cross wavelet transform and wavelet coherence to geophysical time series. *Nonlinear Processes Geophysics*, Volume 11, p. 561–566.

Hamilton, J., 1983. Oil and the Macroeconomy since World War II. *Journal of Political Economy*, Issue April, pp. 228–248.

Hamilton, J.D., 2010. *Nonlinearities and the Macroeconomic Effects of Oil Prices*, s.l.: mimeo, University of California, San Diego.

Hooker, M.A., 1996. What happened to the oil price-macroeconomy relationship? *Journal of Monetary Economics*, Volume 38(2), pp. 195–213.

Hudgins, L.F.C.M.M., 1993. Wavelet transforms and atmospheric turbulence. *Physical Review Letters*, Volume 71 (20), p. 3279–3282.

Lemmens, A.C.C.D.M., 2008. Measuring and testing Granger causality over the spectrum: An application to European production expectation surveys. *International Journal of Forecasting*, Volume 24 (3), p. 414–431.

Mork, K., 1983. Oil and the macro economy when prices go up and down: An extension of Hamilton's results. *Journal of Political Economy*, Volume 97, pp. 740–744.

Mork, K.A.Ø.O. a. H.T.M., 1994. Mork, K.A.Ø. OlMacroeconomic Responses to Oil Price Increases and Decreases in Seven OECD Countries. *Energy Journal*, Volume 15(4), pp. 19–35.

Raymond, J.E. a. R.W.R., 1997. Oil and the macroeconomy: a Markov state-switching approach. *Journal of Money, Credit and Banking*, Volume 29, pp. 193–213.

Rua, A.N.L., 2005. Coincident and leading indicators for the euro area: A frequency band approach. *International Journal of Forecasting*, Volume 21, p. 503–523.

Torrence, C.C.G., 1998. A practical guide to wavelet analysis. *Bulletin of the American Meteorological Society*, Volume 79, p. 605–618.

Green Design, Materials and Manufacturing Processes – Bártolo et al. (eds)
© *2013 Taylor & Francis Group, London, ISBN 978-1-138-00046-9*

Electrical energy analysis and potential environmental improvements of sheet metal punching processes

G. Ingarao
Department of Chemistry, Management, Computer Science, Mechanical Engineering, University of Palermo, Italy

K. Kellens, R. Renaldi, W. Dewulf & J.R. Duflou
Department of Mechanical Engineering, KU Leuven, Belgium

ABSTRACT: Discrete part manufacturing processes, in particular non-conventional production processes, are still poorly documented in terms of environmental footprint. Within the separating processes, sheet metal punching processes have not been analyzed yet from environmental point of view. The present paper aims to contribute to filling this knowledge gap. In particular, two different punching machine tool architectures were analyzed. Following the previously developed CO_2PE methodological approach, power studies and a preliminary time study have been performed in order to understand the contribution of each sub-unit towards the total energy demand. The influence of the most relevant process parameters (e.g. sheet thickness and punch perimeter) are analyzed regarding the required punching energy. Finally, several potential improvement strategies to reduce the punching energy consumption are reported.

1 INTRODUCTION

Manufacturing processes, as used for discrete part manufacturing, are responsible for a substantial part of the environmental impact of products. Nevertheless such processes, in particular non-conventional production processes, are still poorly documented in terms of environmental footprint. Thus, a thorough analysis on the causes affecting the environmental impact of these processes is necessary.

Duflou et al. (Duflou et al. 2012) provide a comprehensive overview of the state of the art in energy and resource efficiency improvement methods and techniques in the domain of discrete part manufacturing, with attention to the effectiveness of the available measures.

As of this writing, the reported studies predominantly focus on machining processes such as turning, milling and grinding, dealing with the influence of material removal and cutting fluids, in parallel with the electricity consumption (e.g. Gutowski et al. 2006, Diaz et al. 2011, Kara & Li 2011). Despite some exceptions (e.g. Thiriez & Gutowski 2006, Oliveira et al. 2011, Baumers et al. 2012), many non-machining technologies are still not well documented in terms of their energy and resource consumption and related environmental impact. In this respect, the $CO_2PE!$-Initiative ($CO_2PE!2013$) has the objective to coordinate international efforts aiming to document and

analyze the overall environmental impact for a wide range of available and emerging manufacturing processes and to provide guidelines to improve these. In accordance with this initiative, a methodology for systematic analysis and improvement of manufacturing Unit Process Life Cycle Inventory (UPLCI) is provided by Kellens et al. (2012). It employs the DIN 8580(2003) taxonomy to categorise discrete part manufacturing processes, namely: Primary shaping, Forming, Separating, Joining, Coating/Finishing and Processes which change the material properties.

The present paper deals with the energy measurement of sheet metal punching machine tools, which belong to the separating category. Within this cluster, other than the widely analyzed chip forming processes, the environmental impact of CO_2 laser cutting processes (Duflou et al. 2010 and Oliveira et al. 2011) and electrical discharge machining (Dhanik et al. 2011, Kellens et al. 2011) have been thoroughly studied. Since sheet metal punching processes have not been analyzed yet from environmental point of view; the present paper aims to contribute to filling this knowledge gap. In particular, two different machine tools characterized by different architectures were analyzed. Following the methodological approach presented by Kellens et al. (2012), time as well as power studies were performed in order to understand the contribution of each sub-unit towards the total energy demand. The influence of the most relevant process parameters (e.g. sheet thickness and punch perimeter) are analysed regarding the required punching energy. Finally, this paper also reports some potential improvement measures to reduce the punching energy consumption.

2 CASE STUDY

Two different machine tool architectures were analyzed: machine tool A is characterized by a fixed hydraulic pump and a maximum load capacity of 220 kN, while machine tool B has a variable flow pump with frequency convertor and a maximum load capacity equal to 200 kN. Despite both machine tools are equipped with a revolver tool station, the revolver station of machine tool A has one tool holder unit while each tool has its own holder within machine tool B. In consequence, a tool change takes slightly longer for machine tool A.

For both machine tools, six tests were developed with varying process parameters (sheet thickness and punch perimeter). Each test was characterized by a series of 20 similar square holes. In particular, the analyzed process is characterized by two consecutives sequences of 10 punch operations with a tool change simulation in between. As far as the material is regarded, an AISI 304 L steel sheet was used during all tests. Three different sheet thicknesses were considered, namely: 0.8, 2, and 4 mm. For each sheet thickness, two square tools with different perimeter were used: a small tool (10 mm side length) and a large tool (50 mm side length for the 1 and 2 mm thick sheets, 20 mm side length for the 4 mm thick sheet).

The above mentioned choices were made based on the intention to analyze the influence of some process parameters on power/energy demand. The details of the performed tests are reported in Table 1.

The big tool size reduction for Test 4 (20 × 20 instead of 50 × 50) was due to the maximum machine tool load constraint. For each test, the total power demand as well as the power

demand of all relevant sub-units were monitored by using electrical power meters with a sampling rate of 12.8 kHz (results logged and shown are for averaged values over 1 second intervals).

3 POWER STUDY

Sheet metal punching processes typically have no significant direct process emissions or consumables consumption. Therefore, the electrical energy consumption is the main cause of the environmental impact of this type of processes. To quantify the electrical energy usage of the two considered machine tools, the power profiles for all the performed tests were obtained.

3.1 Machine tool A

The power profiles of the total machine tool as well as the hydraulic unit for Test 2 are reported in Figure 1.

It can be seen that the standby mode requires an average power level equal to 3100 watts. As far as the actual punching operations are concerned, the power profile is characterized by 4 peaks. Peak 1 stands for the sheet referencing and displacement. Furthermore, Peak 1 corresponds to a fast displacement of the sheet from the clamping area to the punching zone. During such displacement, the referencing of the sheet is also developed. Peaks 2 and 4 correspond to the two sets of 10 similar punches, Peak 3 represents the tool change simulation. As shown in Figure 1, it is evident that Peaks 2 and 4 are mainly caused by the hydraulic unit power demand. Since the referencing operation corresponds to a fast sheet displacement, Peak 1 can be allocated to a power peak of the drives.

The power demand of relevant sub-units was also measured for all production modes. This helps to understand the cause of the energy consumption and facilitates the identification of strategies to reduce the total energy demand and related environmental impact (e.g. by selectively switching

Table 1. The developed test.

Test id	Sheet thickness (mm)	Tool (mm × mm)
1	0.8	10 × 10
2	0.8	50 × 50
3	2.0	10 × 10
4	2.0	50 × 50
5	4.0	10 × 10
6	4.0	20 × 20

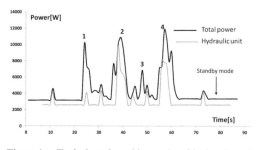

Figure 1. Typical total machine tool and hydraulic unit power profiles for machine tool A (Test 2).

off non-required sub-units). Figure 2 illustrates the sub-unit breakdown analysis determined both for the productive mode (punching operation) and for the standby mode. As can be observed in Figure 2, the hydraulic unit plays a dominant role in the energy consumption. The hydraulic pump accounts for respectively 67% and 83% of the total power consumption during productive and standby mode.

It is worth pointing out that the vacuum system (used to clamp the sheet) accounts for the main part of the remaining 17% of the total energy demand during the productive mode.

3.2 Machine tool B

The power profile of machine tool B (Test 2), shown in Figure 3, differs considerably from the one analyzed for machine tool A.

First of all, due to the adjustable flow pump and frequency convertor, the average power level (1100 Watt) during the standby mode is much lower than what was registered for machine tool A (3100 Watt). The power profile during the punching operations is characterized by two starting peaks and a subsequent high constant value of 4000 Watt. The peaks correspond to the actual punching actions (the two sets of 10 similar punch strokes) and the subsequent constant value corresponds to the buffer refilling. In consequence, from the power profile it is possible to see three production modes: the actual punching mode, standby mode and buffer refilling mode. While the hydraulic unit accounts for 87% of the total power share during the punching and buffer refilling modes, its share during the standby mode drops to 55%.

4 INFLUENCE OF PROCESS PARAMETERS

In this section, the electric energy consumption for varying process parameters (sheet thickness and punch perimeter) is analyzed with particular attention to machine tool A. The energy analysis for sheet metal punching machine tools will not take into account the Energy related to peak 1 since such operation in an industrial production regime would account for a limited time share and moreover such choice allows also to properly isolate the effect of the process parameters on the Power/Energy demand.

The results for all the developed tests are reported in Table 2. Furthermore, the influence of the tool perimeter on both the punching energy is reported.

For 0.8 and 2 mm thick sheets, the tool change (from 10 × 10 to 50 × 50) leads to an increase of the energy demand equal to 10% and 7%, respectively. For the 4 mm thick sheet (Test 5 and 6), where the difference between both tools is smaller (from 10 × 10 to 20 × 20), the increase of the power demand is limited to 4%. These energy increases are exclusively due to an increase of the necessary load to punch the sheet, causing a higher power demand level. Figure 4 shows the comparison of the power profiles obtained for Tests 1 and 2. The increase of the power demand for the test with the bigger tool is evident. Furthermore, sheet thickness proportionally influences the load level of the punching processes. By comparing the punching energy values of Tests 1, 3 and 5, it is possible to understand the influence of the sheet thickness on the energy demand. In particular the punching operation with a thicker sheet requires a higher energy demand of approximately 18%

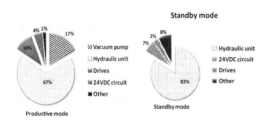

Figure 2. Sub-unit breakdown analysis.

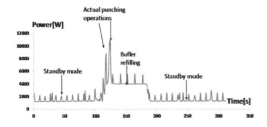

Figure 3. Typical total machine tool power profile for machine tool B (Test 2).

Table 2. Energy usage results for machine tool A.

Test	Punching energy (kJ)	Tool size influence on punching energy (%)
1	158.6	
2	174.1	10%
3	186.5	
4	199.9	7%
5	205.0	
6	212.6	4%

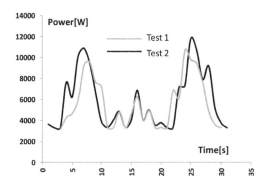

Figure 4. Tool size effect on the power profile.

Table 3. Energy usage comparison.

Test	Machine tool A (kJ)	Machine tool B (kJ)	Additional energy (%)
1	158.6	287.3	81%
2	174.1	365.4	110%
3	186.5	304.3	63%
4	199.9	322.2	61%
5	205.0	286.7	40%
6	212.6	318.6	50%

and 29% for the 2 mm and 4 mm thick sheet, respectively.

When calculating the energy demand for tests on machine tool B, it is necessary to take into account the buffer refilling phase. As a consequence, the energy demand for each test is higher than the energy calculated for machine tool A. Moreover, due to a different load level (and related energy demand), the buffer refilling time is not always the same for each test. Therefore the tests on both machine tools are difficult to compare. Nevertheless, the energy demands for each test for both analyzed machine tools are reported in Table 3. The additional energy required by machine tool B relatively to A is also highlighted.

From Table 3 it can be observed that at the increasing of the energy demand the difference between the two different machine tools substantially decreases both in absolute and relative quantities.

5 INDUSTRIAL PRODUCTION MODES ANALYSIS

In the previous section the energy performance of two different machine tools were analyzed for a selected punching cycle (10 punches, tool change simulation, 10 punches). We can state that machine tool A is more efficient in the productive mode, while the power level of the standby mode of machine tool B is one third of the power level of machine tool A.

It is clear that to evaluate the environmental performance of both machine tools, it is necessary to contextualize their performance in an industrial production regime. Moreover, in order to recommend design modifications or to modify control strategies for non-required subunits, it is necessary to analyze which production mode is dominant during a typical production day of a sheet metal punching machine tool. To do that, a preliminary time study was performed: three punching machine tools (type B) were monitored during fully production activity and the time share of the two main operational modes (standby mode and productive time) was analyzed. Table 4 reports the monitoring time for each machine tool together with the time share for the two different operational modes. Moreover, the overall average values are also reported.

The standby time plays a relevant role in the total monitoring time, indeed for two machine tools the standby time share is even slightly higher than the productive time share. The total average value of the standby mode accounts for 45% of the total time.

By analyzing such industrial measurements, it is straightforward to understand that sheet metal punching machine tools should be characterized by a standby power level as low as possible.

To test such assumption, it is possible to compare the energy consumption of the two machine tools considered in the present study. By considering the production conditions of Test 1, it is possible to calculate the total energy requirement for each machine tool at the varying of the standby time share. In Figure 5, the energy consumption for one hour of production activity is illustrated for both machine tools. Despite the analysed

Table 4. Production modes time shares.

Machine tool	Monitoring time (h)	Standby time share (%)	Productive time share (%)
Punching machine 1	66	52%	48%
Punching machine 2	73	53%	47%
Punching machine 3	72	31%	69%
Total average values		45%	55%

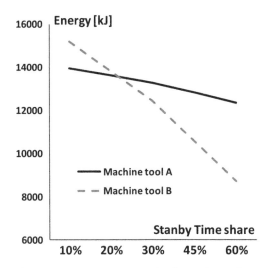

Figure 5. The energy consumption for one hour of production activity in function of the standby time share.

machine tools are characterized by different productivity, the number of punching operations was kept constant and a no switching off strategy was considered. It is clear that machine tool B with a lower standby power level is more energy efficient.

Only in the case of a very limited standby time share (20%), which is rather unlikely, machine tool A is preferable. However, it is worth citing that by applying a switching off procedure, machine tool A (the fastest one), could become the more energy efficient machine tool for all scenarios.

6 CONCLUSIONS

This paper presents an energy consumption analysis of sheet metal punching processes. In particular, two different punching machine tools, characterized by two different machine tool architectures have been analyzed. Both, a power study as well as a partial time study are reported. Furthermore, the influence of the tool size and sheet thickness on the power/energy demand have been analyzed and quantified.

The difference of the power profile of the two studied machine tools has been highlighted and analyzed. These analyses led to the conclusion that machine tool A is more efficient in the productive mode, while the power level of the standby mode of machine tool B is only one third of the power level of machine tool A.

As far as the subunit breakdown analysis is concerned, it is possible to conclude that the hydraulic unit plays a dominant role in the whole energy consumption of both machine tools and for both principal operational modes (standby and productive modes). As a consequence, machine tool builders should focus on the development of strategies facilitating a reduction of the power demand of the hydraulic system, both by changing the machine tool architecture, and by applying a switching off strategy instead of long standby periods. Moreover, the process energy demand is very sensitive to workpiece specific characteristics. Indeed, as for the parameter settings analyzed in the study, on the one hand, the tool size could affect the power demand up to 19%. On the other hand, change of the sheet thickness causes an energy demand increase even up to 25%. Moreover, a partial time study is reported highlighting the importance of the standby mode.

ACKNOWLEDGEMENTS

The authors acknowledge the support of the Institute for the Promotion of Innovation through Science and Technology in Flanders (IWT-Vlaanderen) through its PhD grant N°091232 well as the O&O EnHiPro project (Project No. 090999).

REFERENCES

Baumers, M., Tuck, C., Wildman, R., Ashcroft, I., Rosamond E., Hague R. 2012. Transparency Built-in Energy Consumption and Cost Estimation for Additive Manufacturing *Journal of Industrial Ecology*. DOI: 10.1111/j.1530-9290.2012.00512.x.

CO$_2$PE!—Cooperative Effort on Process Emissions in Manufacturing, last visited 06/03/13, http://www.mech.kuleuven.be/co2pe.

Dhanik, S., Xirouchakis, P., Perez, R. 2011. A System for Resource Efficient Process Planning for Wire EDM; *Proc. 18th CIRP LCE Conference*, Braunschweig, Germany.

Diaz, N., Redelsheimer, E., Dornfeld D. 2011. Energy Consumption Characterization and Reduction Strategies for Milling Machine Tool Use; *Proc. 18th CIRP International Conference on Life Cycle Engineering*, Braunschweig, Germany.

Duflou, J.R., Kellens, K., Devoldere, T., Deprez, W., Dewulf W. 2010. Energy Related Environmental Impact Reduction Opportunities in Machine Design: Case Study of a Laser Cutting Machine, Int. J. of Sustainable Manufacturing 2 (1):80–98.

Duflou, J.R., Sutherland J., Dornfeld D., Herrmann, C. Jeswiet J., Kara S., Hauschild M., Kellens K. 2012. Towards energy and resource efficient manufacturing: A processes and systems approach. *CIRP Annals—Manufacturing Technology* 61: 587–609.

Gibovic, D. & Ciurana De. 2008. Presentation of DIN8580. Girona: Documenta Universitaria.

Gutowski T., Dahmus J., Thiriez, A. 2006. Electrical Energy Requirements for Manufacturing Processes; *Proc. 13th CIRP International Conference on Life Cycle Engineering*, Leuven, Belgium.

Kara, S. & Li, W. 2011. Unit process energy consumption models for material removal processes. *CIRP Annals—Manufacturing Technology* 60: 37–40.

Kellens, K., Dewulf, W., Overcash, M., Hauschild, M.Z., Duflou J.R. 2012. Methodology for systematic analysis and improvementof manufacturing unit process life-cycle inventory (UPLCI)—CO$_2$PE! initiative (cooperative efforton process emissions in manufacturing). Part 1: Methodology description. *Int. J. Life Cycle Assessment* 17: 69–78.

Kellens, K., Renaldi, Dewulf, W., Duflou, J.R. 2011. Preliminary Environmental Assessment of Electrical Discharge Machining. *18th CIRP LCE Conference,* Braunschweig, Germany.

Oliveira, M., Santos, J., Almeida, F., Reis, A., Pereira J.P., Rocha, A. 2011. Impact of Laser-based Technologies in the Energy-Consumption of Metal Cutters: Comparison between commercially available System. *Key Engineering Materials* 473: 809–815.

Thiriez, A. & Gutowski T. 2006. An Environmental Analysis of Injection Molding: *Proc. IEEE Symp. on Electronics and the Environment*, Scottsdale.

Green Design, Materials and Manufacturing Processes – Bártolo et al. (eds)
© *2013 Taylor & Francis Group, London, ISBN 978-1-138-00046-9*

Prototypelab 2012: Mass-customization of energy self-sufficient prototypes and less environmental impact through improved parametric tools, sensors and CNC technology

J. Ballesteros
Universidad Politécnica de Madrid, Madrid, Spain

P. Ferreiro
CIS-Madeira, Ourense, Spain

ABSTRACT: Prototypelab is a project that was born in the University on the bosom of an Educational Innovation Project of the Department of Architectural Design of the ETSAM (Superior Technical School of Architecture of Madrid). It was designed with the aim of exploring the possibilities the parametric architecture provides as a new way of thinking, designing and building architecture, but with a clear vocation of coming close to the industry, technology and the world of innovation. This article presents one of the activities developed in collaboration with CIS Madeira and companies specialized in the sector of timber, on which the research is settled on a prototype which condenses services destined to the public space, energy self-sufficient, a low environmental impact and with the capacity of adapting to different contexts.

1 INTRODUCTION

The prototype presented in this article belongs to the setting of the Master course in Advanced Architecture of the Department of Architectural Design of the ETSAM and on the Educational Innovation Project Prototypelab UPM. It is the aim of it to combine the goals of each participant on an integrated action.

The general aims of this activity are to disseminate, transfer and explore the ways of a new technological context (ICT, the new computer tools for parametric design, nowadays' manufacturing CNC Technologies) as well as the spirit of achieving a desirable environmental balance opens up the spreading of expressive possibilities, the creation and research on the architecture and design in the XXI century, leading to a new culture of design (Willmann 2011).

This project has from its early beginning the main support of the timber sector, to which it is initially addressed. This is so due to the combination of technology and timber which make possible an answer to the request of sustainability and research of expressive possibilities of contemporary architecture. Timber is a material of low environmental impact, and in this sector the implementation of manufacture technology by the use of CNC is wide.

This project also wishes to cooperate in the opening of a debate who might lead to the creation of a reflexion and action pole on these subjects in Galicia.

The team who develops this activity is composed of the members of the Master course in Advanced Architecture of the ETSAM (UPM), the CIS Madeira (Innovation and Technological Services Center for the Sector of Wood in Galicia) and an essential group of companies, producers and traders of timber goods, manufacturers or integrators of technologies devoted to an energy efficiency and companies specialized on the manufacturing and assembling of timber projects.

The project counts too on its first two workshops with the support of the Technology Park of Galicia—Tecnópole, an environment of referent in Galicia for corporate innovative initiatives in technology, research and development. The resultant prototypes of these workshops are initially placed in the Tecnópole's facilities.

This educational and research project tries to connect, through a line of integrated and cross actions, the participant agents following the model of innovation development earlier conceptualised by Etkovitz and Leydesdorf and known as the Triple Helix. In this model there is a combination of efforts from the University as a source of knowledge, the companies as builders of new opportunities of business and the public sector as a provider and source of a favourable environment for innovation (González 2009).

In this sense, the Prototypelab sets as a fundamental and indispensable aim on its development: the collaboration among Universities, innovation agents such as CIS Madeira and the industrial and business world.

Prototypelab seeks for interdisciplinarity and it is addressed to the participation of professionals and students formerly specialised both on tools and on the essential knowledge of parametric architecture (town planners and landscape architects, architects, industrial and engineer designers) as well as those who want to find new tools and means of expression, creatives and artists, with the spirit to promote the exchange of knowledge and experience.

From the academic environment participates the team of the Master course in Advanced Architecture in the Superior Technical School of Architecture of Madrid, where de idea of the project emerged. This group is bounded also to the doctorate course *Automatic, Robotic, Encoded*, that provides theoretical support to the project and constitute a kind of incubator of doctoral thesis which feeds it back.

From the University it is drawn up the thematic content as well as the suggestion of the programs to deal with in order to design the prototype. Additionally, takes the main responsibility for the academic coordination of the teachers.

From the innovation's centre for the timber sector point of view, CIS Madeira, the global project aligns with the purposes defined by the Royal Decree that rules the Technology Centres and the Innovation Support Centres (Ministerio de Ciencia e Innovación 2008).

Among the main purposes of the technological innovation support centres are the contribution to the strengthening of the relationship between the organizations who generate knowledge and the companies, and provide support services to business innovation.

So that, this project wants to offer a context to stimulate innovation, an opportunity to explore the implicit value in the combination of new computer tools and the manufacture CNC technologies, and so that having into consideration the possibilities they offer to create new lines of product or business. This pilot experience could draw attention of other companies of the sector and encourage them to find new paths.

As a centre connected to Galician Public Administration CIS Madeira, Innovation and Technological Services Centre for the Sector of Timber in Galicia, develops a role on making things easier, and it is in charge of providing sup-

port to develop Prototypelab. It coordinates the organization of workshops within its facilities, carries out the main administrative processing and builds bridges and channels the relationships with the companies and collaboration entities. Likewise, technicians from the CIS Madeira provide specialized knowledge on timber in all phases of the project: architectural conception, construction design, structural strategy, manufacturing and assembly.

The third blade of the helix is taken up by the participant companies. They are connected to the workshop by different kind of relationships. On the one hand they can collaborate as sponsorships of the activity, both with economic support and by bringing materials. On the other hand there is the group of companies who cooperate on the process of manufacturing and assembly of the project, putting at the workshop disposal their facilities and the technologies for the cutting and manufacturing of the pieces of the project or providing their technical knowledge and their specialized staff for the assemble of the resultant prototypes.

The participant companies on the activities of this project get a profit from the direct contact with the professionals and the other participant companies, and they know first-hand the keys of the methodologies that might lead to future ways of generating and building in design and architecture.

The project offers the participant companies the opportunity to test their products or processes. The manufacturing companies have the chance to experiment their materials subject to unusual or extreme performance. Other companies, specialized on mechanical cut, can experiment unusual process of mechanization of complex geometries and verify the performing of new tools on a CNC technology basis.

The building of relationships among the participants is lead to create an innovation network to provide future knowledge flow and innovation process. This aim comes from the consideration of networks as valuable tools to face organizational and social innovation and the development of collective talent, as well as the idea of network innovation as an answer to a faster pace of change and a higher complexity. (Cabanelas & Cabanelas 2011).

Moreover, Prototypelab can become a communication tool and provide a broadcasting channel of the use of timber in building, on which this material and its numerous products are linked to contemporary technology showing an image of contemporary and innovation. Represents an opportunity to approach architecture and design on timber to the methodologies of the architectural cutting-edge design, reaffirming the perception of timber as a material for the future.

Such as Pérez de Lama suggests, this kind of projects constitute pilot actions, attempts, as long as it is discovered the role fab lab's are going to play in society, while the role of technology is defined in the future of architecture (2011).

2 ORGANIZATION

During the development of the course the attendees are both invited and guided to go over a complete path from the knowledge of the parametric design tools, going through their put into practice of the development of a project on a team basis, until the manufacturing of one of the selected prototypes.

The workshop is organized in two stages. The first stage, which lasts approximately one week, is based on generating among the participants the competences to use the parametric design software and stimulate on them a reflection on the potential these tools have through the development of the project of a prototype. The participants are given classes to make known the software and tools (Grasshopper, Processing, CNC fabrication, etc.) and in order to generate a dynamic and exchange environment, work teams of about 6 people are built and they are challenged into a project contest. They must develop a common theme for all teams with the restrictions of programme, time, range of materials and technologies available for the manufacturing and assembly of the prototype. On the stage of the project there is constant advice by the technicians of CIS Madeira as well as by the companies specialized on the construction of projects on timber and responsible of the final assembly of the prototype.

During the second stage of the workshop all the teams work on the integral development and the manufacturing of the constructive details of the selected project. It is also a key aspect in this stage the support of the technical knowledge of the specialists of the innovation centre as well as the companies, in order to flesh out the details for the final solution from all point of view: construction, structure, manufacture and assembly. Once the prototype is defined all the members have a personal participation, with the support of the companies, on the process of materialization and manufacturing of the prototype.

3 DEVELOPMENT AND RESULTS

Up to now, Prototypelab has organised two workshops. From the same general scheme it has been opened an evolutionary process of evaluation and continuing education on which gradual changes are implemented on the strategy, organization and contents of the course.

Following there is a description of the concepts who inform the processes of creation and manufacturing of the two prototypes. Since the first prototype has already been objet of analysis in a former article, the explanation focuses specially in the second prototype, which provides title to this document.

3.1 *Parametric strategies for energetic, programmatic and material optimization*

There is an unresolved opportunity presented in the public space of virtually every city in the world: the possibility or rearranging and regrouping all street furniture and infrastructural items into a single organized environment. Traffic posts, mailboxes, street signs, phone booths, benches, bus stops, newsstands, all constitute vital parts of our public space, but are rarely distributed in a coordinated fashion, and thus generate a cluttered environment that makes movement for disabled citizens, elderly people, baby strollers or grocery carts extremely difficult.

This paper attempts to describe our research on generic programmatic solutions for urban public spaces, capable of housing and grouping a significant amount of the elements mentioned above, while simultaneously being energetically self-sufficient, and thus being able to provide additional public services such as cell phone chargers, Wi-Fi access points and certain capabilities of climatic conditioning.

Using the current capabilities of Grasshopper for Rhinoceros to establish a live data link with Autodesk's Ecotec package (using the Geco component set) our paper will describe a methodological approach for progressive energetic optimization towards the most efficient model that simultaneously retains the maximum programmatic capabilities within a generic urban environment.

3.2 *Prototype 1*

As a case study, our research will incorporate the knowledge base accumulated during the development of a full-scale W.E.P.S. mock-up built in 2011, which was based on a fully parametric digital model that offered both a wide range of possible formal and functional variations and a very streamlined digital fabrication logic. This parametric model also generated large surfaces of vertical panels with varying orientations and the structural capabilities to be populated with photovoltaic panels, thus constituting a potential source of solar energy that is directly integrated in the formal qualities of the system.

The proposed research will cover the following points:

– Measurements of the peak direct solar irradiation of the digital parametric model in the form of its built instance.
– On-site measurement of the peak direct solar irradiation on the existing built mock-up. This will lead to a comparative evaluation of the accuracy of the digital Ecotect simulation in terms of irradiation data, and thus enable the possibility of extrapolating this information to predict the real physical behavior of further parametric models.
– Further development of the existing parametric model, formally addressing the simultaneous optimization of its energetic behavior (in terms of maximum solar irradiation) and its functional capabilities (in terms of integration of street furniture and adaptability to site constrains).

This use of parametric modeling linked to live energetic data analysis—and contrasted with real-world physical measurements—will allow us to introduce performance-based feedback into the design process of optimal Wrapped Energetic Public Spaces.

3.3 *Prototype 2*

3.3.1 *Reflection on the urban services space: The convex option*

The detailed review of possible public space occupation, through the elements already available, provides preliminary data that facilitate the first decisions on the model. The concave elements (providers small containers spaces, as, for example, environmental monitoring stations) consume a large amount of public space, with reduced

Figure 1. Fotografía del estado final del Prototipo 1, instalado en el Parque Tecnológico de Galicia—Tecnópole.

usability. At the other extreme, a bin or a traffic signal, achieve efficiency and visibility with an insignificant use of public space. It is what we will call "the convex occupation tendency". While concave elements tend to form an enclosure, and hold public uses inside, the convex projecting its utility to the entirety dimensions of public space. They are, therefore, essentially properly public space, but also more efficient in space consumption ratio/utility program.

3.3.2 *Topological failure generating an ambiguous parametric model*

Given the naturally parametric models effectiveness to fit almost any geometrical behavior, we proceeded to develop a model that, provided the premises of grouping utility, could behave both concave and convex, assuming sometimes specific climatic conditions of a site should require greater protection.

After making invertible parametric models able to behave concave or convex as function of a conditional parameter, we have concluded that these solutions are incompatible with a viable geometrical model in the scenario of CNC fabric, and incompatible in most of the functions that we program for this urban element, which were seriously affected well in one of its forms or another.

3.3.3 *Models and possibilities of the parametric process: The energy-dependent form*

In the constructive resolution of this parametric model have been added qualities and characteristics subsidiaries that have taken shape depending on certain parameters and conditions.

The essential determinant function is self-sufficiency, as this will give viability to another large amount of functions and possibilities that will make sense the model. Thus, there is a lot of orientation surface intended for the collection of solar energy, parametric response gives its characteristic shape, which in turn implies some protection degree for its users. Its indifferent orientation geometry allow anywhere location with the most effective guidance.

The necessary stability with low cost efforts in structural wood, involves the replacement of a single stem by the deployment of the efforts in different brackets and supports as separate as possible from its center of gravity, generating highly effective resistant moments to overturning and wind, which effectively exploit the wood properties as a building material in compression and tension in its different sections, broken down into ribs.

Still, the model shape possibilities remain innumerable, so have adjusted the model variables (parameters) to respond effectively enough to the other required characteristics.

Figure 2.   Imagen digital de un modelo del Prototipo 2.

4   DISSCUSSION OF RESULTS AND CONCLUSION

As research strategy the project has enabled to test the processes and tools of design and manufacturing to bring to reality a prototype generated with parametric design possibilities.

As a collaborative project the development of each prototype serves as a hub of knowledge and generating multiple synergies. Reported to all participating institutions and companies the benefit of teamwork, mutual learning and exchange of perspectives on a common objective. Additionally, this process has generated valuable relationships for future projects.

After the first two models, the evolution of work objectives and the concepts embodied in the project leave a positive balance.

We have succeeded in developing a parametric model adapted to environmental conditions for the obtaining of energy enough to self-provision.

We achieved a manufacturing process directly transferable to industry through CNC cutting board as generated files directly from the parametric model, without the need to draw a single plane. A manufacturing process that allows different models mechanization. This process allows no serialize manufacturing inside the normal industrial procedures, an offer diversity from a single parametric model. Those different results should be produced at the same cost that the usual serialized production but specializing our product in any suitable location and function.

We managed the integrated participation of University, Technology Centres and specialized industrial companies, transmitting knowledge from the areas of research to companies and manufacturers, making possible, communicable also, design factors such as reducing environmental impact, energy self-sufficiency, the sensorized public space and the reactivity to the boundary conditions of place, environmental, usage, etc.

The research opens the door to the fact of exploring mass customization and adaptation to different contexts, from the point of view of the services offered to the public space and in relation to the energy capture.

We also believe that it is essential to guarantee a high quality level in the outcome of the investigations and prototypes built so that they meet real needs detected; they can consolidate the efforts of all participants in the project and scope also the largest possible repercussion in society.

A key part of these projects depends on the ability to communicate their objectives and results to reach the industrial and business world, professionals and technicians who could be recipients of these technologies and society. The necessary improvements for the dissemination of the project must go through a strategic review of the location of the prototype to increase its visibility and a professionalization of the communication strategy.

After the experience of these two prototypes we reaffirm our conviction to continue working in the direction of the stated objectives: to deep in the exploration of the open questions by combining parametric design tools, technology CNC manufacturing, ICT and the principles of sustainability and energy efficiency; progress in stimulating business innovation modeled on the Triple Helix; intensify the exchange of experiences and knowledge transfer with industry to consolidate the third mission of the university (Morales 2008); collaborate in the growth of a network of exchange guided to technological innovation and make our contribution so that Galicia flourish in a center of reflection and creation on this field of knowledge.

REFERENCES

Cabanelas, J. & Cabanelas, P. 2011. *Innovación en redes. Creación de capacidades y control de la innovación.* Vigo: Vigo University. Master Universitario en Innovación Industrial y Optimizacíon de Procesos.

González, T. 2009. El modelo de Triple Hélice de relaciones Universidad, Industria y Gobierno: Un análisis Crítico. *ARBOR Ciencia, Pensamiento y Cultura.* 185 (738): 739–755.

Ministerio de Ciencia e Innovación. Spain. 2008. Real Decreto 2093/2008, de 16 de septiembre, por el que se regulan los Centros Tecnológicos y los Centros de Apoyo a la Innovación Tecnológica en el ámbito estatal y se crea el Registro de tales Centros. *Boletín Oficial del Estado.* 23 July 2009. 7872–7880.

Morales, S.T. 2008. El emprendedor académico y la decisión de crear spin-off: Un análisis del caso español. Economics Faculty. Valencia: Valencia University.

Pérez de Lama, J. 2011. Fab Lab Network y tercera revolución digital. En: Gutiérrez, M. [et al.] (eds.) *FabWorks: Diseño y fabriación digital para la arquitectura. Docencia, Investigación y Transferencia.* Sevilla: Escuela Técnica Superior de Arquitectura. Universidad de Sevilla.

Willmann, J. 2011. El ebanista digital: Hacia la fabricación aditiva robotizada. *Arquitectura Viva* (137): 28–29.

Green Design, Materials and Manufacturing Processes – Bártolo et al. (eds)
© *2013 Taylor & Francis Group, London, ISBN 978-1-138-00046-9*

Energy utilization and output dynamics in Bangladesh

Gazi Salah Uddin & Shair Razin
Department of Management and Engineering, Linköping University, Sweden

Ahmed Taneem Muzaffar
School of Business, University of Western Sydney, Australia

Phouphet Kyophilavong
Faculty of Economics and Business Management, National University of Laos, Laos

ABSTRACT: The aim of this paper is to investigate the relationship amongst energy consumption, carbon emission, industrialization, urbanization and income in Bangladesh within ARDL bounds testing approach using annual data for the period 1972–2009. The results show existence of a long run equilibrium relationship amongst the variables. In the long-run, both the energy and urban intensity drive the national income in Bangladesh. The policy implications are discussed in the text.

1 INTRODUCTION

The last two decades have witnessed a significant increase in the demand for energy in Bangladesh causing concerns for global warming and climate change. Rapid increase in industrialization and urbanization is believed to be a major contributing factor behind this. Emission of Carbon dioxide (CO_2), accountable for at least 60 percent of the total global warming in the world, has a close association with energy consumption. Efficient use of energy is expected to reduce the level of emission. This requires an understanding of the dynamics amongst emission, energy, and income which is influenced by several factors such as composition of economic growth, types of economic activities, intensity of industrialization, and urbanization.

Since the liberalization of financial sector and trade in the 1990s, there has been a rise in the number of industries in Bangladesh. However, the issues relating to the environment were not addressed seriously at that time. This has led to continuous degradation of natural environment and in 2006 the country emitted about one tenth of the world's carbon emission, despite the fact that its population of 160 million represents about 2.4 percent of the world's population (see World Bank 2007). A study by Rahman (1992) reveals that resultant industrial pollution affects a lot of people in Bangladesh. The national pollution profile reveals that different sectors of the industry also contribute to a large percentage of pollution in the country (Islam and Mia 2003). The unsustainable industrial practice has raised concerns and appropriate measures should be taken to reduce pollution (see Bala and Yusuf 2003).

This paper investigates the causal relationship amongst income, industrialization, energy consumption, and carbon emission in the context of Bangladesh using annual data over the period 1972–2009. In particular it seeks to answer three questions: (1) is there a long-run relationship amongst these variables? (2) what is the nature of short-run relationship? and (3) what are the policy implications of the findings?

The remainder of this paper is structured as follows: Section-II explains literature review. In section-III we outline the econometric specification and estimation methodology and discuss how the hypotheses are tested. Section-IV provides a discussion of the findings from the empirical analyses. Finally, section-V summarizes the discussion and provides a concluding remark.

2 REVIEW OF LITERATURE FOR BANGLADESH

A study on Bangladesh's energy and output by Mozumder and Marathe (2007), over the period 1971–1999, finds that per capita GDP Granger causes per capita electricity consumption, but the reverse is not true. However, small sample size is a major limitation for drawing such conclusion. In another study using annual data from 1971 to 2007, Uddin et al. (2011) investigate the causal relationship between energy consumption and economic growth in Bangladesh. Their results show that a unidirectional causality runs from energy consumption to economic growth and the restriction on the use of energy could lead to a reduction in economic growth. There are, however, limitations of these studies. While the first one suffers from small sample size the latter follows a bivariate framework inefficient in determining the causal relationship due to omitted variable bias.

The energy-output dynamics is also investigated by Paul and Uddin (2011) which find that while fluctuations in energy consumption do not affect output fluctuations, movements in output inversely affect movements in energy use. The results of Granger causality tests in this respect are consistent with those of innovative accounting that includes variance decompositions and impulse responses. Their findings from Autoregressive Distributed Lag (ARDL) Models also suggest a role of output in Bangladesh's energy use. However, this study does not consider effect of emission, urbanization and industrialization in the model specification. Our study makes an attempt to fill this gap.

Using annual data between 1971 and 2008, Ahamad and Islam (2011) examine the causal relationship between per capita electricity consumption and per capita GDP of Bangladesh within an error correction framework. Empirical findings reveal that there is a short-run unidirectional causality running from per capita electricity consumption to per capita GDP. One limitation of the study is the failure to incorporate the structural break, valid for Bangladesh (see Paul and Uddin 2011) in the estimation process.

Alam et al. (2012) investigate the relationship amongst energy consumption, CO_2 emissions and economic growth by applying Johansen and Juselius (1990) cointegration framework and the VECM Granger causality approaches. Their results note that energy consumption Granger causes economic growth and CO_2 emissions. Furthermore, economic growth Granger causes CO_2 emissions. However, Amin et al. (2012) found neutral effect between CO_2 emissions and economic growth. Both these earlier papers have shortcomings in that the structural break is not considered in the dynamics estimation which is we try to address in our study. Recently, Hoque and Clark (2012), the most recent study, identify the top ten polluting industries in Bangladesh which are tannery, pulp and paper, fertilizer, textile and cement industries. The authors conclude that compared to leading firms in developed countries, pollution prevention initiatives in Bangladesh are considerably underutilized.

3 DATA AND METHODS

Annual data form 1972 to 2009 on Bangladesh for CO_2 emissions (c) (metric tons) per capita, energy use (e) (kg of oil equivalent) per capita, real GDP (y) (constant 2000 US$) per capita, Industrial value add per capita (i) which is used as a proxy for industrialization and urban population (u) as a proxy for urbanization, was collected from World Development Indicators (WDI-2013), published by the World Bank. The data used in the study was converted to logarithmic form.

The relationship amongst energy consumption, carbon emission, industrialization, urbanization and income in Bangladesh can be expressed in the following basic multivariate model:

$$Ly_t = \alpha_1 + \Omega_1 le_t + \Omega_2 lc_t + \Omega_3 li_t + \Omega_4 lu_t + \varepsilon_t \quad (1)$$

where t is the trend variable, α and Ω are model parameters, and ε_t is the white noise, le is the log of energy consumption, lc is the log of carbon emissions, li is the log of the industrial value add per capita, lu is the log of urban population as a proxy for urbanization and ly is the log of real GDP per capita, The ARDL modeling approach was originally introduced by Pesaran and Shin (1999) and later extended by Pesaran et al. (2001). The ARDL model used in this study can be expressed as follows:

$$\Delta Ly_t = \alpha_2 + \Omega_1 le_{t-1} + \Omega_2 lc_{t-1} + \Omega_3 li_{t-1}$$
$$+ \Omega_4 lu_{t-1} + \sum_{i=1}^{p} \theta_i \Delta le_{t-i} + \sum_{i=1}^{p} \Phi_i \Delta lc_{t-i}$$
$$+ \sum_{i=1}^{p} \Theta_i \Delta li_{t-i} + \sum_{i=1}^{p} \Psi_i \Delta lu_{t-i} + u_{1t} \quad (2)$$

where p signifies the maximum lag length which is decided by the user. This value usually depends on the literature and convention to determine the maximum lag length. The ARDL bounds test approach is to estimate equation (2) using the Ordinary Least Squares (OLS) method. The F-test is used in a bounds test for the existence of the long-run relationship (Pesaran et al. 2001), and it tests for the joint significance of lagged level variables involved. The null hypothesis of the non-existence of a long-run relationship for the equation of $(F_{Ly}, Ly \mid Le, Lc, Li, Lu)$ is $(H_0: \Omega_1 = \Omega_2 = \Omega_3 = \Omega_4 = 0)$ against the alternative hypothesis $(H_1: \Omega_1 \neq \Omega_2 \neq \Omega_3 \neq \Omega_4 \neq 0)$. The Error Correction Model (ECM) is presented using equations (2). To ensure the convergence of the dynamics to the long-run equilibrium, the sign of the lagged error correction (ECM) coefficient must be negative and significant. A general correction model is formulated as follows:

$$\Delta Ly_t = \alpha_3 + \sum_{i=1}^{p} \theta_i \Delta le_{t-i} + \sum_{i=1}^{p} \Phi_i \Delta lc_{t-i}$$
$$+ \sum_{i=1}^{p} \Theta_i \Delta li_{t-i} + \sum_{i=1}^{p} \Psi_i \Delta lu_{t-i} + \lambda_g ECM_{t-1} + u_{2t}$$

(3)

where λ is the speed of adjustment parameter and ECM_{t-1} is the residual obtained from the estimation of equation (3). In order to ensure that the correct statistical methods are applied to the model, diagnostic and stability tests are conducted. The diagnostic tests include testing for serial correlation, function form, normality and heteroscedasticity (Pesaran and Pesaran, 1997). In addition, the stability tests of Brown et al. (1975), which are also known as the Cumulative Sum (CUSUM) and Cumulative Sum of Squares (CUSUMSQ) tests based on the recursive regression residuals, were employed to check the stability of the parameter.

4 EMPIRICAL FINDINGS

We present the empirical findings based on our methodology discussed in the last section.[1] The estimation results for cointegration are presented in Table 1. In order to account for the fact that we have a relatively small sample size, we have produced new Critical Values (CVs) of the *F-test*, computed by stochastic simulations using 20000 replications. If the calculated F-statistics exceeds the upper bound, the null hypothesis of no cointegration among variables in LCO_2 can be rejected. If the calculated F statistics falls below the lower bound, the null hypothesis of no long-run relationship cannot be rejected.[2] The results reported in Table 1 show that there is evidence of cointegration at 10% level of significance when the variables Ly are taken as dependent variables in the model in the case of Bangladesh. This shows that there is one cointegration vector validating the existence of long run relationship between the variables.

---

[1] ADF and PP Unit root tests indicate that each variable has one unit root. The results are not presented but are available upon request.
[2] If the calculated F-statistics falls within the lower and upper bounds, it is inclusive. In addition, Pesaran et al. (2001) cautions that the critical values for the bound test are sensitive to the number of regressors (k) in the model, and Narayan (2004) argues that the critical values of the F-test depend on the sample size.

Table 1. Bounds F-test for cointegration.

Dependent variable	Forcing variables	F-stat	Cointegration
Δly	$\Delta Le, \Delta Lc,$ $\Delta Li, \Delta Lu$	4.482*	Present
Asymptotic critical value	95% critical bounds		90% critical bounds
	I (0)	I (1)	I (0) I (1)
	3.288	4.610	2.719 3.900

Notes: * and ** indicate a rejection of the null hypothesis of no cointegration at the 5% and 10% level of significance, respectively. Δ denotes the first order difference operator.

Table 2. Estimated long-run coefficients.

Regressor	Coefficient	Stan. error	T-ratio
ARDL_SBC ((1,4,0,2,0), Dependent variable: Ly_t			
Le_t	1.77*	0.13	13.43
Lc_t	0.13	0.10	1.39
Li_t	0.02	0.11	0.23
lp_t	−0.67*	0.17	−3.85
Constant	−0.15	0.48	−0.32

Notes: * and ** indicate at the 1% and 5% level of significance, respectively.

Given our sample size, the SBC is preferred to the AIC. The long-run coefficients of the selected ARDL models are presented in Table 2. According to the SBC model specification, the coefficient of energy intensity and urbanization is significant when income is the dependent variable in the model specification. In this situation both energy utilization and intensity of urbanization are the long run forcing variable to explain the national income for Bangladesh. This implies that energy usage leads to economic growth and so limitations on the amount of energy that can be used could result in a drop in economic growth. The results suggest that the estimated log-run coefficients pass standard diagnostic tests, meaning that the underlying assumptions of the statistical model are fulfilled. The long-run model is free of serial correlation, heteroscedasticity, is linear, and has normally distributed residuals at 5% level.

Table 3 provides the error correction representation of the selected ARDL models, when the first difference of real income per capita or economic growth is the dependent variable.

The growth of energy utilization, the change in the industrialization and urban growth is significant when economic growth is the dependent variable. The sign of the error correction term is

Table 3. Error correction representation.

Regressor	Coefficient	Stan. error	T-ratio
Δle_t	0.03	0.08	0.44
$\Delta le1_t$	−0.48*	0.11	−4.10
$\Delta le2_t$	−0.38*	0.10	−3.59
$\Delta le3_t$	−0.27*	0.08	−3.38
Δlc_t	−0.04	0.02	1.56
Δli_t	−0.09*	0.03	−3.12
$\Delta li1_t$	−0.06	0.02	−2.63
dlp_t	−0.22	0.05	−4.09
ecm (−1)	−0.33	0.07	−4.20

Notes: * and ** indicate a rejection of the null hypothesis of no cointegration at the 1% and 5% level of significance, respectively.

Table 4. Diagnostics tests.

Serial correlation	$F(1,6) = 1.34$
Functional form	$F(1,6) = 0.314$
Normality	$\chi^2(2) = 0.03$
Heteroscedasticity	$F(1,31) = 1.72$

significant and negative. The results of diagnostic tests suggest that short run model passes all tests successfully are presented in Table 4. The long-run model is free of serial correlation, heteroscedasticity, is linear and has normally distributed residuals. The value of R^2 is greater than 0.50. It reflects the goodness of the fit of the model. The plots of the CUSUM and CUSUMSQ statistics are well within the critical bounds, implying that all coefficients in the ECM model are stable over the sample period 1971–2009.[3]

5 CONCLUSION

This paper investigates the relations between energy consumption, CO_2 emissions, industrialization, urbanization and income in case of Bangladesh using annual data over the period of 1972–2009. We apply an ARDL co integration approach to explore this relationship and find a long-run steady state among these variables. The result show that there is one co integrating vector validating the existence of long run relationship amongst the variables. The energy utilization and intensity of urbanization are the long run

---

[3] The figures are not presented but are available upon request.

forcing variables to explain the national income for Bangladesh in the long-run, whereas in the short-run the growth of energy utilization, the change in the industrialization and urban growth is significant when economic growth is the dependent variable. The sign of the error correction term is significant and negative. The results of diagnostic tests suggest that short run model passes all tests successfully. The results, in short, are consistent in the context of a developing country. Bangladesh is going through structural changes such as industrialization, urbanization, and greater use of energy which in turn facilitating its economic growth. The challenge for the policy makers is to make sure that this conducive environment remains in order to sustain economic growth.

REFERENCES

Ahamad, M.G. and Islam, A.K.M.N. 2011. Electricity consumption and economic growth nexus in Bangladesh: Revisited evidences. *Energy Policy*, 39: 6145–6150.

Alam, M.J., Begum, I.A., Buysse, J., Huylenbroeck, GV., 2012. Energy consumption, carbon emissions and economic growth nexus in Bangladesh: Cointegration and dynamic causality analysis. *Energy Policy* 45, 217–225.

Bala, S.K. and Yusuf, M.A. 2003. Corporate Environmental Reporting in Bangladesh: A Study of Listed Public Limited Companies. *Dhaka University Journal of Business Studies*, 24(1): 31–45.

Brown. R.L., Durbin, J., Evans, M., 1975. Techniques for testing the constancy of regression relations over time. *Journal of the Royal Statistical Society*, 37:149–163.

Hoque, Asadul and Clarke, Amelia. 2012. Greening of industries in Bangladesh: pollution prevention practices, *Journal of Cleaner Production* 30: 1–10.

Islam, S., Miah, S., 2003. Banglapedia (National Encyclopedia of Bangladesh). Asiatic Society of Bangladesh, Dhaka.

Mozumder, P., Marathe, A., 2007. Causality relationship between electricity consumption and GDP in Bangladesh. *Energy Policy* 35:395–402.

Narayan, P.K., 2004. New Zealand's trade balance: Evidence from the J-curve and Granger causality. Applied Economics Letters 11: 351–354.

Paul, Biru Paksha & Uddin, Gazi Salah, 2011. Energy and output dynamics in Bangladesh. *Energy Economics* 33(3), pages 480–487.

Pesaran, M.H. 1997. The role of economic theory in modeling the long-run. *The Economics Journal*, 107, 178–91.

Pesaran, M.H., Shin, Y., 1999. An autoregressive distributed-led modeling approach to cointegration analysis. *In Econometrics and Economic Theory in the 20th Century. The Ragnar Frisch Centennial Symposium*, ed. Steinar Strom. Cambridge: Cambridge University Press 1999.

Pesaran, M.H., Shin, Y., Smith, R.J., 2001. Bounds testing approaches to the analysis of level relationships. *Journal of Applied Econometrics* 16, 289–326.

Rahman, K., 1992. Industrial Pollution and Control for Sustainable Development. Training Manual on Environmental Management in Bangladesh. Department of Environment (DoE), Dhaka.

Uddin, Gazi Salah and Alam, Md. Mahmudul and Murad, Wahid. 2011. An empirical study on income and energy consumption in Bangladesh, *Energy Studies Review* 18(1): 4.

World Bank 2007. World Bank Development Report: Development and the next Generation, World Bank Press, Washington DC, USA.

Green Design, Materials and Manufacturing Processes – Bártolo et al. (eds)
© 2013 Taylor & Francis Group, London, ISBN 978-1-138-00046-9

Spectral shifters for an enhanced indoor environment

Geoffrey R. Mitchell
Centre for Rapid and Sustainable Product Development, Institute Polytechnic of Leiria, Marinha Grande, Portugal

Fred J. Davis
Department of Chemistry, University of Reading, UK

ABSTRACT: Sufficient light levels are essential for effective working in indoor environments. It is well established that the nature of that light has a significant impact on the health of the occupants. In this work we explore the potential for spectral shifters to provide an enhanced working environment. Spectral shifters absorb incident light and re-emit at different wavelengths thus modifying the spectral distribution of the light. Such shifters would be most effective in buildings which have been designed without sufficient care being given to the level of natural lighting within the building.

1 INTRODUCTION

Sufficient light levels are essential for effective working in indoor environments. It is well established that the nature of that light has a significant impact on the health of the occupants and in particular their working efficiency. Current office building design has focused on the provision of high levels of natural lighting through the inclusion of features such as an atrium. In this work we focus on older buildings with need to retrofit. We explore the potential for spectral shifters to provide an enhanced working environment. Spectral shifters absorb incident light and re-emit at different wavelengths thus modifying the spectral distribution of the light. As a consequence the ambient light can be tailored to suit the needs of the work related activities with a subsequent increase of efficiency through reduced staff-time loss.

2 DAYLIGHT

2.1 Preparing the new file with the correct template

Light is a part of the broad spectrum of electromagnetic radiation. The planet earth is powered by the radiation produced by the Sun, especially the infrared, visible, and ultraviolet portion of the light. On Earth, the sunlight which we receive has been filtered by the atmosphere surrounding the plane, and is obvious as daylight when the Sun is above the horizon.

In situations when the direct solar-produced radiation is not blocked by clouds, it is experienced as sunshine, a combination of bright light and radiant heat. At other times, it is experienced as diffuse light due to the scattering by clouds or other objects.

Human life was transformed in the last 200 years by the development of artificial lighting which enables work and leisure activities to take place inside buildings during the daytime which by their nature greatly reduce the level of sunlight. Artificial lighting also allows activities proceed safely at night-time.

Direct sunlight has a luminous efficacy of about 93 lumens per watt of radiant flux. Luminous flux is the measure of the perceived power of light. It differs from radiant flux, which is the measure of the total power of light emitted by a source, in that luminous flux takes account of the varying sensitivity of the human eye to different wavelengths of light. Icandescent light bulbs exhibit a luminous efficacy of 10–20 lm/W, whilst the recently developed white light LED systems may reach over 100 lm/W. Bright sunlight provides illuminance of approximately 100,000 lux or lumens per square meter at the Earth's surface. The maximum level of energy received at ground level from the sun is 1004 watts per square meter, which is made up of 527 watts of infrared radiation, 445 watts of visible light, and 32 watts of ultraviolet radiation. At the top of the atmosphere sun-light is about 30% more intense, with more than three times the fraction of Ultra-violet (UV), with most of the extra UV consisting of biologically-damaging shortwave ultraviolet.

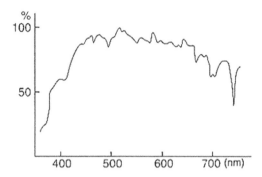

Figure 1.   Spectra of sunlight.

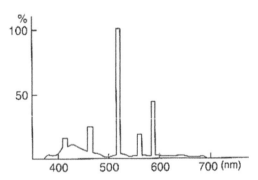

Figure 2.   Spectra of a fluorescent light.

The spectral characteristics of artificial light differ from sunlight as a consequence of the physical processes involved in the light production. Figure 2 shows the spectral distribution of light from a fluorescence tube. The sharp spectral bands correspond to the emissions from the glow discharge which depend on the specific gas involved. It is clearly very different to that for the sunlight shown in Figure 1. Similar comparisons may be made with other sources of artificial light such as halogen bulbs and LED lights. For a LED source, the spectral characteristics are determined by the electronic properties of the semi-conductor employed in the device. Many lights sources such as LEDs and Fluoresence tubes use spectral shifters to provide a more uniform spectral distribution of light from the device. It is possible to purchase what are described as full spectrum lamps.

3   LIGHT CONTROL

The use of materials to change the nature and quality of light which enters in to a building is well established practice. The use of blinds, photochromic windows and other devices is common place

especially where the use of natural lighting is key part of the building design Rosemann (2008). The particulars of the spectral distribution of this light define the impact on the temperature of the building. but also on the building occupants in terms of their individual experience as is discussed in the next section.

A second approach is the use of photo defined refractive index changes in multiphase polymer films, especially using added particles which result in controlled changes in the scattering properties of the plastic film such that the light intensity is reduced and/or the spectral characteristics are modified. This is the basis of polymer dispersed liquid crystal windows Cupellia et al. (2009), Baetens et al. (2010). There are non-liquid crystal technologies available to achieve the same goa for example the gaschromic windows developed by Baetens et al. (2010). The use of plastic films brings flexibility and an ease of integration with other components.

The role of window design in terms of spectral characteristics has been explored by Ye et al. (2012).

Other approaches have used spectral filters on external cladding, for example Schuler et al. (2004), Mertin et al. (2013) used interference filters on external cladding to facilitate the architectural integration of solar collectors in to buildings.

4   LIGHT AND HUMAN BEHAVIOUR

We all have everyday experience on how our working environment affects our efficiency and our health. Surveys show that indoor workers have a strong preference for daylight in offices, Galasiu et al. (2006). It is now ell established that the light which is incident on our eyes has important non-visual biological effects on the human body van Bommel (2006). The most well know of these the winter light syndrone Lewy (2009). More recent studies have shown that the absence of specific spectral components in the ambient light effects the circadian timings within the body, Reppert et al. (2002). Circadian rhythms are internally generated ~24 hour rhythms. Zaidi et al. (2007) have shown that non-visual inputs via the outer retina can disrupt these patterns even in the case of people who are visually impaired. It has been reported that this can lead to depression Yannielli et al. (2004), Stephenson et al. (2012). Appropriate design of offices and their alignment with the sun can minimize the impact of these factors Li and Lam (2001), however many of these studies have focused on the energy minimisation consumption of a building rather than its effective and efficient use, for example Bodart et al. (2001). We propose that more detailed studies are required to develop an active control system which matches the needs of the building occupants in terms of light. We envisage a programme of light changes in which the use of daylight is maximized and supplements by spectrally shifted artificial light.

Begemann et al. (1997) have emphasized the role of poor indoor lighting is the underlying cause of many health and performance problems.

5 SPECTRAL SHIFTERS

The interaction of light with any material may change the nature of light in a variety of ways commonly when light is transmitted through glass it is refracted, but unless the surface is curved (as in a lens) the result is not apparent. However other types of interaction are more obvious; for example scattering of light may result in opacity; and pigments used in for example oil paints rely on the selective reflection of light to provide colour. In the case of a glass containing a material that absorbs light (for the sake of this account visible light say) or for a solution containing a dyestuff, the possibility significant shifts to the spectrum exists The process is as described in the Jablonski Diagram, Gilbert (1991) shown below (Fig. 3).

When a molecule absorbs light from a singlet ground-state (the most likely possibility as it implies the electrons in the sample are all paired) it is excited to a similar singlet state of higher energy. Typically this may lose some energy through vibration and then relax back to the original singlet state (S_o in Fig. XX). This process results in the emission of light known as fluorescence. It is to be expected with such a process that the energy emitted will be lower than that absorbed and as a consequence will be at a longer wavelength. The difference in wavelength between emission and absorption is referred to as the Stoke's Shift.

In some cases the excited singlet state transfers energy through a process known as intersystem crossing to a triplet state. The light emitted here is referred to as phosphorescence. Phosphorescence occurs at longer wavelength and is a time-delayed process since it is quantum mechanically forbidden; thus most spectral shifting is concerned with fluorescence.

A representation of a fluorescence spectrum is shown in Figure 4, as can be seen the distance between the emission maxima and the absorption maxima and thus the Stoke's shift is relatively small, and this is typical for many simple organic chromophores. That being said there are many uses for fluorescent materials including optical brighteners which are used in both paper manufacturing and in washing powders. Such materials typically absorb UV light and re-emit in the visible thus, for example obscuring any yellowing resulting from the presence of chromophores absorbing in the high-energy tail of the visible spectrum, through the addition of emitted light to this region of the spectrum.

In terms of dyes there are two effects—we can remove light of a particular wavelength, or we can shift (albeit with some inevitable energy loss) the a fraction of the light to a longer wavelength. Thus with suitable dyes the light environment inside a structure can be controlled. We have applied this idea to the growth of plants under horticultural plastics. Thus the use of a horticultural structure based on polyethylene containing a phthalocyanine dye can promote the development of plants with reduced height, but no loss of biomass, van Haeringen (1998) by removing the light at 730 nm. Alternatively it is possible to restrict the development of botrytis through the use of a horticultural tunnels which block a fraction of UV light and fluoresce in the visible, West et al. (2000). Some examples of trial tunnels are shown in Figure 5.

There are many advantages to having a large Stokes shifts in applications (notwithstanding the

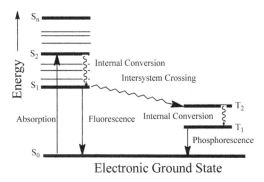

Figure 3. Jablonski diagram showing the origin of fluorescence and phosphorescence.

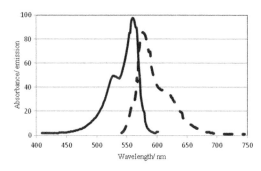

Figure 4. Absorbance (solid line) and emission (broken line) from a typical organic fluorescent compound.

Figure 5. Trial horticultural tunnels with 730 nm absorbing dye (green) and fluorescent UV absorbing materials (purple hue).

huge market for optical brighteners) and there is much activity in extending the values. Lanthanides typically have relatively large Stoke's shifts, but their use might be restricted by availability and cost. Thus a range of approaches have been used to the development of organic materials (largely aromatic) with extended stokes shifts, Ziessel and Harriman (2011), Rihn et al. (2012) and these materials offer many opportunities for new applications.

6 APPLICATIONS TO BUILDINGS

The section relating human behavior to indoor lighting has shown that there is a clear need for a substantial revision of our design of indoor lighting. We have emphasized the need for daylight or daylight equivalence in lighting. The latter may be achieved by the use of LEDs with the inherent directionality of their operation and the possibility of limited spectral control by electronic means. The intergration of light control in windows with the indoor lighting and studies have highlighted the need for this to be personalized. Moreover, we believe that there is considerable potential for the use of spectral shifters to adapt the light pool for an individual worker which could change with time by providing a sequence of spectral shifters under programme control. Current thinking is rather too focus on reducing the energy consumption and not sufficient attention is paid to the building occupants.

7 SUMMARY

We need to rethink the specification of indoor lighting to define both illumination levels but also spectral content and replicate the characteristics of sunlight and provide opportunities for control by indiviuduals of the illumination of their workspace. Such requirements will change throughout the year and the time of day.

REFERENCES

Baetens R., B.P. Jelle and A. Gustavsen 2010 Solar Energy Materials and Solar Cells 94 87–105.
Begemannm S.H.A., G.J. Van den Beld, and A.D. Tenner 1997 International Journal of Industrial Ergonomics 20 231–239.
Bodart M. and A. De Herde 2002 Energy and Buildings 34 421–429.
Cupelli D., F. Nicoletta, S. Manfredi, M. Vivacqua, P. formoso, G. De Flipo and G. Chidichimo 2009 Solar Energy Materials and Cells 93 2008–2012.
Gilbert A. and J Baggott (1991), Essentials of Molecular Photochemistry, Ellis Horwood.
Graw P., S. Recker, L. Sand, K. Krauchi and A. Wirz-Justice 1999 Journal of Affective Disorders 56 163–169.
Li D.H.W. and J.C. Lam 2001 Energy and Buildings 33 793–803.
Mertin S., V. Hody-Le Cer, M. Joly, I. Mack, P. Oelhafen, J.-L. Scartezzini and A. Schuler 2013 Energy and Buildings in press.
Reppert S.M. and D.R. Weaver 2002 Nature 418 935–941.
Rihn S., P Retailleau, A De Nicola, G Ulrich, and R Ziessel 2012 J. Org. Chem, 77, 8851–8863.
Schuler A., C. Roecker, J-L Sacrtezzini, J. Boudaden, I.R. Videnovic, R.S.C. Ho and P. Oelhafen 2004 Solar Energy Materials and Solar Cells 84 241–254.
Stephenson K.R., C.M. Schroder, G Bertschy and P. Bourgin 2012 Sleep Medicine Reviews 16 445–454.
Van Bommel W.J.M. 2006 Applied Ergonomics 37 461–466.
van Haeringen C.J., J.S West, F.J Davis, A. Gilbert, P. Hadley, S. Pearson, A.E Wheldon and R.G.C Henbest, 1998 Photochem. And Photobiol., 67(4), 407–413.
Warman V.L., D.-J. Dijk, G.R. Warman, J. Arendt and D.J. Skene 2003 Neutroscience Letters 342 37–40.
West JS, S Pearson, P Hadley, AE Wheldon, FJ Davis, FJA Gilbert, RGC Henbest, 2000 Annals of Applied Biology Volume 136, Issue 2, pages 115–120, April 2000.
Yanenielli P. and M.E. Harrington 2004 Progress in Neurobiology 74 59–76.
Ye H., X. Meng and B. Xu 2012 Energy and Buildings 49 164–172.
Zaidi F.H., J.T. Hull et al. 2007 Current Biology 17 2122–2128.
Ziessel R. and A Harriman 2011 Chem. Commun., 2011, 47, 611–631.

Green Design, Materials and Manufacturing Processes – Bártolo et al. (eds)
© *2013 Taylor & Francis Group, London, ISBN 978-1-138-00046-9*

Optimalmould: Energy consumption in injection moulding optimization

Carina Ramos, Pedro Carreira, Paulo Bártolo, Igor Reis, Lina Durão & Nuno Alves
Centre for Rapid and Sustainable Product Development, Polytechnic Institute of Leiria, Leiria, Portugal

ABSTRACT: Plastics are popular engineering materials because of their versatility, durability, and relatively low cost. Injection moulding is one of the most used manufacturing processes to produce plastic products, consuming a significant amount of energy. Energy consumption and emissions have been seen as an important problem that needs to be minimized. It is essential to reduce energy consumption and environmental impact of both: injection moulding process and plastic products. Therefore, a methodology is proposed to estimate the energy consumption based on a multi-objective optimization platform that computes optimal parameters in mould design and injection moulding process. A case study was considered to validate the proposed approach and the obtained results show energy savings and environmental impact reduction through the cycle time and wasted plastic material minimization.

1 INTRODUCTION

Injection moulding process enables to replicate complex plastic parts with detailed geometries. However, it processing needs a great deal of energy and produces a waste plastic material resulting from the feeding channels and produced parts that do not meet the quality standards.

Figure 1 shows what are the important points that needs to be considered when are analysing the process energy consumption.

The answers to the questions where, when, why, how much allows defining what is the energy amount and where can be consumption optimized. To know how much electric energy that will be needed, have to define where this energy that will be used and what is the production process. After, is necessary to understand the process behaviour, how many works hours and if needs unproductive times. The next step is define the amount of the plastic parts needed and what is the power required.

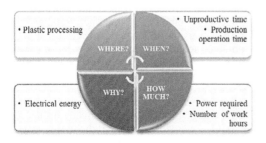

Figure 1. Understanding energy consumption.

In this paper, a methodology is proposed to estimate the energy consumption based on a multi-objective optimization platform that computes optimal parameters in mould design and injection moulding process. Results show energy savings and environmental impact reduction through the cycle time and wasted plastic material minimization.

2 METHODOLOGY TO ESTIMATE THE ENERGY CONSUMPTION IN INJECTION MOULDING

2.1 *Injection mounding process*

The injection moulding process works with an injection machine, where is fixed the mould, which will shape the plastic part.

Currently, the design of an injection mould is a highly interactive and experimental process, involving substantial knowledge from multiple areas such as mould design features, mould making processes, injection moulding equipment and plastic part design, which are highly related to each other (Menges, et al., 2000). Therefore, it is critical to design and produce a mould that is straight forward to manufacture, while providing uniform filling and cooling of plastic parts. At the same time, the injection mould has to be strong enough to resist millions of cyclic internal loads from injection pressures and external clamp pressures, in order to assure the target part's reproducibility (Kazmer, 2007).

In industry, the injection mould is designed according to the best practices that were been considered along of the years, needing plastic part geometry, plastic material and injection machine as inputs specifications (Centimfe, 2003). In complex projects, mould flow simulations are performed to predict some problems that can appear in injection moulding process. After manufacturing the mould, the plastic material processing is developed also according to the best practices, named conventional approach.

A non-conventional approach, named Optimalmould Platform, was developed to be the input database to injection mould design and injection moulding process. Through this platform the optimal mould design and processing parameters are obtained in order to produce quality plastic parts, where the mould is considered as a systems group (Ferreira, et al., 2009) that was transformed in a multi-objective problem. The optimal parameters results from the interaction between optimization tool using genetic algorithms, CAD Software and input database (plastic part geometry, plastic material properties and injection machine specifications) (Ramos, et al., 2012a).

A comparison between optimal and conventional approaches leads to an energetic efficiency and more sustainable production process as shown in Figure 2.

In this research work, the cycle time of injection moulding process is defined as follows (Fig. 3).

Where the filling stage is the needed time to fill the cavity of injection mould, the packing stage is the needed time to packing melt material inside

Table 1. Injection moulding process costs (Bolur, 2013).

Injection moulding	Main costs (%)	Secondary costs (%)
Plastic material	50–75	–
Injection machine	30–15	–
Fixed costs	–	60–45
Electricity	–	30–38
Water	–	7–12
Maintenance	–	3–5
Injection mould	10–5	–
Personnel	10–5	–

cavity, the cooling stage is the needed time to solidification the plastic part, the release stage is the needed time to eject the plastic part from the cavity and open/close stage is the needed time to open or close the mould (Ramos, et al., 2012b). As injection moulding stages are sequentially running, is needed electric energy to melt de plastic material, to reduce the coolant temperature, to move electric systems of injection machine, which depends of plastic material properties (Godec, et al., 2012) and injection machine specifications (Weissman, et al., 2010).

According to Table 1, the main costs that are involved in injection moulding process are plastic material, injection machine, mould and personnel.

The electric energy/electricity, in the injection moulding process costs, represents 5–9% of the global costs, as showed in Table 1, but these percentages can change, depending of the plastic material, part geometry, injection machine.

This table also shows that the plastic material is the more significant cost factor of injection moulding process, due plastic material price. In this sense, it is important analyse injection moulding process, to identify where it is possible to make change in order to reduce their costs and consequently ecological footprint.

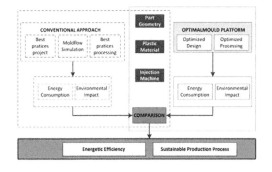

Figure 2. Information flow for energy consumption estimation in injection moulding. Conventional approach vs. Optimalmould approach.

Figure 3. Injection moulding cycle time stages (Ramos, et al., 2012b).

2.2 Electric energy

Nowadays, it is a more used energy form by humanity, due its portability and low rate of energy loss during conversion. In Portugal, the electric energy is obtained mainly through thermoelectric power plants, hydroelectric power plants, wind power plants and nuclear power plants as follows: (EDP, 2009).

These different resources make up of global electric energy production that directly influences the CO_2 emissions and radioactive waste production. However, the electric energy is one of greenest energy (EDP, 2009).

According to Figure 4, the energy origin, during year 2012, the average of CO_2 emissions and radioactive waste are: 228.61 g/kWh and 24.69 µg/kWh, respectively.

In addition, energy consumption must be quantified. Electricity tariffs, in Portugal, are distributed in four time steps along the day, where the hours that have more electricity consumption are also more expensive, as follows (Fig. 5).

The electric energy consumption of injection moulding process is related with injection machine

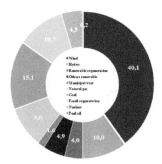

Figure 4. Electric energy origin, in average, during the year 2012, in Portugal (EDP, 2009).

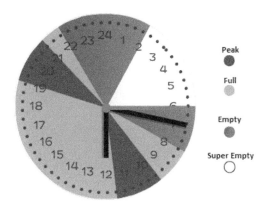

Figure 5. Electric energy time steps (EDP, 2009).

Table 2. Electric energy costs, high voltage 2013 (EDP, 2009).

Tariffs	Active energy (€/kWh)
Peak	0.1109
Full	0.0880
Empty	0.0650
Super empty	0.0605

power in order to correspond with their energy requirements. Therefore, it is important to estimate which is the energy needed in each time step involved in injection moulding cycle time. It can be computed according to the following equation:

$$E_{tc}[J] = E_{fill} + E_{pack} + E_{cool} + E_e + E_{open} + E_{close}$$
(1)

where E_{tc} is the energy consumption in the cycle time; E_{fill} is the energy consumption during filling time; E_{pack} is the energy consumption during packing time; E_{cool} is the energy consumption during cooling time; E_e is the energy consumption during ejection time; $E_{open\ and\ close}$ the energy consumption during open and closing time (both are assumed equals).

As a first energy consumption estimation, the following simplified equation is considered:

$$P_{tc}[kWh] = P_{im} \times t_c$$
(2)

where P_{tc} is the electric energy needed to injection moulding process; P_{im} is the power of injection machine; t_c is the cycle time of injection moulding process.

As future research work, energy consumption in each time step will be defined and computed, representing a new objective function that will be integrated into the Optimalmlould Platform.

3 CASE STUDY

To validate the proposed approach, it was used a case study supplied by an industrial company. The injection mould provided had 4 cavities, cold runners and the parts were produced with Polymethyl methacrylate (Polystyrol 143E BASF) in

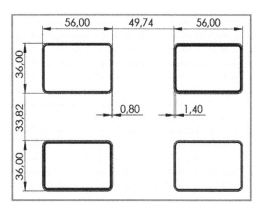

Figure 6. Case study.

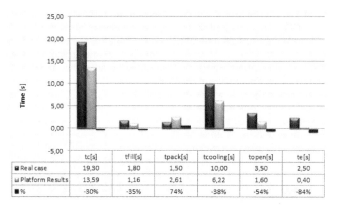

Figure 7. Optimized results from integrated platform (Ramos, et al., 2013).

	tc[s]	tfill[s]	tpack[s]	tcooling[s]	topen[s]	te[s]
Real case	19,30	1,80	1,50	10,00	3,50	2,50
Platform Results	13,59	1,16	2,61	6,22	1,60	0,40
%	-30%	-35%	74%	-38%	-54%	-84%

an injection moulding machine with 130 tonnes of clamping force.

The optimization platform was implemented using optimization software, iSIGHT 5.6, where the optimal parameters were selected according to the Pareto technique. The parameters obtained from this optimization platform were compared with real case parameters (based on best practices). Figure 7 shows that with this new approach it is possible to reduce 30% of the cycle time.

The estimated working cost is 37€/h for the used injection machine. Assuming that this mould was designed to produce 1.500.000 parts, there was a reduction of 2380 h in processing time at the end of the production, representing a reduction of processing costs in approximately 89.000€. By reducing the processing costs with this optimization, the injection machine can more be profitable, which will also result in a production that will finish approximately 397 days before the considered lead-time (considering a work-day of 8 h).

When considering the injection machine costs, it is necessary to consider the following issues: the equipment maintenance that represents 5%, water that represents 12%, power or electric energy that represents 38%, and, finally, fixed costs that represents 45% of the global machine cost.

This presented distribution of the costs is related with this case study. If it were to be applied to another situation, the percentages may need adjustments, depending on the plastic material, part geometry, injection machine, etc.

Considering equation 2, based on the optimal parameters, and considering that the injection machine needs 31 kW of total power, the electric energy costs may be reduced approximately in 30%, corresponding to the reduction of 30% in the cycle time.

This cost reduction can be quantified, considering that the injection moulding company works

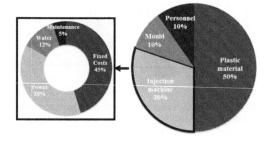

Figure 8. Industrial costs to produce plastic parts (Rosato, et al., 2000).

Table 3. Reductions in injection moulding process.

Description	Minimization
Cycle time [h]	2380
Material waste [kg]	1990
Electric energy [kWh]	77780

during power peak and full time hours, a reduction of 77 780 kWh of consumed power at the end of production is verified. The consumption reduction, according the electricity tariffs, can be quantified in approximately 6490€.

In other hand, with this optimization platform, presented in this case study, also verified a feeding volume reduction in approximately 47%, it is possible to determine a reduction of waste material, that corresponds to a significant cost of injection moulding process, as shown in Figure 8.

With the reduction of power consumption, the production of radioactive waste will also decrease, in a proportional amount.

Through this case study analysis and according to Table 3, it is possible to conclude that the optimization platform presents a significant contribution to the energetic efficiency of the injection moulding industry, reducing the environmental impact.

4 CONCLUSIONS

The problem of optimizing the injection mould's design is a central theme for this industry, due their complexity and diversity.

Based on the results obtained in practical cases, it can be concluded that the optimization platform, Optimalmould, performs well in the case studies where it was tested. The multidisciplinary optimization platform enables to identify the most significant parameters allowing the reduction of the cycle time, without compromising the injection moulding process.

In this case study, a reduction of 30% of the cycle time and 47% of the feeding volume was verified, that are considered significant benefits in cost reductions and consequently in reduction of the ecological footprint enabling a more efficient process.

The results enable to conclude that it is important to add the energy consumption module into the Optimalmould platform, in order to output optimal parameters.

As future work, with the energy consumption module implemented in the platform, when the design mould process begins, the parameters inputs will also consider the minimum energy that is necessary for production and selecting the best injection processing machine.

The presented approach is a tool that enables to have an efficient injection moulding process, reducing the waste of plastic material, making the production more sustainable, where the wastes is minimized and the energy efficiency is optimized.

ACKNOWLEDGMENTS

This work has been supported by Portuguese Foundation for Science and Technology (FCT) through the projects PTDC/EME-PME/108188/2008 and Pest-OE/EME/UI/4044/2011.

REFERENCES

Bolur, P.C., 2013. Understanding energy consumption in injection moulding machine. [Online] Available at: http://www.pitfallsinmolding.com/energyeffic1.html [Accessed 2013].

Centimfe, 2003. Manual do Projectista para Moldes de Injecção de Plásticos.

EDP, E. d. P., 2009. EDP Serviço Universal. [Online] Available at: http://www.edpsu.pt/pt/origemdaenergia/Pages/OrigensdaEnergia.aspx [Accessed 2013].

Ferreira, I., Weck, O. d., Saraiva, P. & Cabral, J., 2009. Multidisciplinary optimization of injection molding systems. Structural and Multidisciplinary Optimization, Volume 41: 621–635.

Godec, D., Rujnic-Sokele, M. & Šercer. M., 2012. Energy efficient injection moulding of polymers. Acta Tecnica Corviniensis—Bulletin of Engineering Tome V, January–March, pp. 103–108.

Kazmer, D.O., 2007. Injection Mold Design Engineering. Hanser, Volume ISBN 978-1-56990-417-6.

Menges, G., Michaeli, W. & Mohren, P., 2000. How to Make Injection Molds. 3ª edição ed. s.l.:Hanser Gardner Publications, Inc., Cincinnati.

Ramos, C., Carreira, P., Bártolo, P. & Alves, N., 2012a. Optimalmould-Part I: Multi-objective Optimization to Moulds Design for Injection of Polymers. Proceedings of the 10th International Conference of Numerical Analysis and Applied Mathematics, Volume 1479, pp. 1569–1573.

Ramos, C., Carreira, P., Bártolo, P. & Alves, N., 2012b. Optimalmould—Part II: Global Optimization of the Injection Moulding Cycle Time. Proceedings of the 10th International Conference of Numerical Analysis and Applied Mathematics, Volume 1479, pp. 1574–1578.

Ramos, C., Carreira, P., Bártolo, P. & Alves, N., 2013. Optimalmould | Cooling System influence in Injection Moulding. Advanced Materials Research, Volume 683, pp. 544–547.

Rosato, D., Rosato, D. & Rosato, M., 2000. Injection Molding Handbook.

Weissman, A., Ananthanarayanan, A., Gupta, S.K. & Sriram, R.D., 2010. A systematic methodology for accurate design-stage estimation of energy consumption for injection moulded parts. Proceedings of the ASME 2010 International Design Engineering Technical Conferences & Computers and Information in Engineering Conference.

Green and smart manufacturing

Green Design, Materials and Manufacturing Processes – Bártolo et al. (eds)
© 2013 Taylor & Francis Group, London, ISBN 978-1-138-00046-9

Sustainable work for human centred manufacturing

M. Santochi & F. Failli
Department of Civil and Industrial Engineering, University of Pisa, Italy

ABSTRACT: Sustainable manufacturing is a term mainly used with reference to processes which eliminate or reduce waste, chemical substances or physical agents hazardous to human health and environment, spare energy and materials as much as possible. Less attention is generally given to social problems and working conditions: but sustainable manufacturing means also sustainable work, seen as a mix of physical and mental safety, satisfaction, acceptable working times, dignitous salary, opportunity of learning, to improve one's professional knowledge and competence. After an introduction dealing with the theories on work organization and related industrial experiences, the paper outlines the main problems of the assembly work and proposes some solutions for a future sustainable workplace in assembly processes.

1 INTRODUCTION

1.1 The "4S principle": Sustainability by safety, satisfaction and salary

Sustainable development is a broad and overused term. The World Commission on Environment and Development (WCED 1987) gave the following definition of sustainability: *sustainable development is a process of change in which the exploitation of resources, the direction of investments, the orientation of technological development, and institutional change are made consistent with the future as well as present needs.*

Manufacturing (Jovane et al. 2009) is doubtlessly a key factor of the economical development: a green or a sustainable process is generally required to eliminate or at least reduce waste, chemical or physical agents hazardous to human health and environment, spare energy and materials as much as possible. However other issues should be considered with the same or even deeper attention since manufacturing is generally based on human work.

Safety of workers, usually seen as absence of disease and injuries risks, should be considered also a mix of physical and mental safety, satisfaction, acceptable working times, dignitous salary, good relationship with colleagues and company management, pride to be a member of the company, opportunity of learning and improving knowledge and competence: in other terms what has been defined a good or sustainable work (Joahnsson et al. 2009).

The increase of the potentialities of the human capital and the attention to working conditions, often raised by workers unions protests, have been recently focused by scientific research programs,

industrial efforts, initiatives of international associations.

An example is the European legislative action focused on the risk of stress caused by some working activities such as assembly. *Work-related stress is experienced when the demands of the work environment exceed the workers' ability to cope with (or control) them* is the definition given in (OSHA 2002). Workers' stress may be a consequence of the shorter and shorter times requested for each operation, reduced rest time in the name of productivity and competition and also of more general factors like increasing use of temporary contracts, increased job insecurity and poor work-life balance (Takala 2009).

According to the Framework agreement dated October 2004 all European companies are now forced to evaluate the risk of stress of employees.

Several messages and proposals have been launched in conferences and workshops. An example is the conference *Sustainable work: a challenge in times of economic crisis* held in 2009 (Gronkvist et al. 2009). Some speakers stressed that a sustainable and socially inclusive economy depends on what happens in the wokplace. A sustainable work is intended as a work crafted to fit the worker's capabilities and interest rather than fitting workers to the job. The work should also promote thinking, reflecting and acting and should allow a high level of participation.

The recent European survey on working conditions (EWCS 2010) underlines the importance of work and working conditions in a sustainable and cohesive growth at individual level but also at societal level, since its quality influences the quality of our lives. Survey data show that the following working conditions are considered important for the wellbeing among workers: giving a say to workers, enabling them to be heard and to make improvements that clarify their roles and tasks, good job design, creating a safe working environment, encouraging collaborative work, addressing job in security, facilitating a good work–life balance. In other terms high motivation, engagement and willingness to remain in the labour market.

The World Economic Forum in the last report (WEF 2012) takes into account the social sustainability among the factors of the competitiveness in the global economy. In particular in a context of social sustainability it is desiderable to meet a certain level of quality of working conditions such as salaries, safety and participation. However a complete analysis of working conditions is not yet available from the International Labour Organization (ILO) since data deal only with a limited number of countries.

1.2 *General proposals to ensure work sustainability*

In the global context a lot of activity towards sustainable work has been performed by the Fair Labor Association (FLA 2013) an international organization of universities, civil society organizations and socially responsible companies dedicated to protecting workers' rights around the world. Typically FLA's work is based on transparent and independent assessments published online and suggestions of innovative and sustainable strategies to help companies to improve their work organization. A recent survey used by FLA in an electronic industry included questions such as: working hours, rest days, social protection, satisfaction with factory facilities, health and safety, worker integration, relationships at work, employees' sense of belonging. It is also worthwhile to outline the FLA's multi stakeholder approach: *The products we buy should not come at the cost of workers' rights. The FLA believes that all goods should be produced fairly and ethically, and brings together three key constituencies—universities, civil society organizations and companies—to find sustainable solutions to systemic labor issues.*

This last sentence underlines the increasing importance of research directed to sustainable work. From this point of view a meaningful and relevant to the European culture example of research project is the project *Factories of the future* of the European Commission (EFFRA 2012). The main ambitious objectives of the section "Social wellbeing" are to provide a stimulating environment for employees, to provide satisfactory workplaces as adaptive human machine interface, to develop human centered production sites by maintaining a level of employment with highly satisfied and skilled workers, by enhancing ergonomics and by integrating factories in their social and urban context.

1.3 *Before and after lean production*

The history of theories on work organization started in 1911 when F.W. Taylor published his *Principles of scientific management* where he proposed a new approach to the management of manufacturing plants especially assembly line. Its most known application was the Ford assembly lines. Taylorism and Fordism may have today a negative meaning even if they represented in those years the transformation from the craft-based to the mass production with recognized advantages for the whole society.

An alternative view, the *humanistic* approach, was developed later on the basis of the study of human behaviour through experimental psychology and psychoanalysis. According to A. Maslow's pyramidal structure of human motivation, other reasons, in addition to money reward, exist to meet satisfaction or a good quality of work: training for a personal grown, sense of belonging to the firm, having a pleasant job and good human relations. Later F. Herzberg, in his book *The motivation to work*, outlined that the satisfaction of basic needs is only a prerequisite and does not mean motivation. Effective motivators are factors like recognition, responsibility, achievement, personal growth.

The humanistic management theory was experienced in Italy by Olivetti in the fifties. Adriano Olivetti was convinced that the aim of an enterprise cannot be only profit but a wide social project based on new relationships between managers and workers: the increase of productivity has to be achieved through workers' motivation and their involvement in the life of the company. A similar experience was made by Volvo in 1989–1993: the assembly line concept was reviewed and organized in parallel by independent teams of skilled workers responsible also for quality control, planning and corrections.

In 1980 a new management concept was born in Japan at Toyota. The MIT in the USA coined the term Lean production to indicate a production system based on aspects like: effective use of resources, co-operative product development, work standardization, process control, preventive maintenance, total quality, clean and well organized workstations, diffuse communication, flexible labour, demand pulled production. The lean methodology seems today to be the dominating and driving approach under the new name of World Class Manufacturing (WCM) developed in USA in '90. However the point of view of employees is not so positive since many Tayloristic aspects still exist. Examples reported typically refer to the assembly process. The main issues are: the positive features of team working are missing since much of the assembly work on the lines is left to individuals, the pace of the lines is increasing, work assignment is standardized, low space remains for autonomy, repetitive job still kills any possibility of job enrichment especially learning, ergonomic problems may arise from one sided work movements.

What are the viewpoints of the main stakeholders, companies and unions?

Most companies obviously think that plants which have requested huge investments must operate without interruptions, quickly, dynamically, with low absenteeism and without strikes: in other words management wants the full governance of plants. Employers' opinion is that competition is hard and the risk of closure of plants and consequent un-employment is true. The consequence is that full usage of available 24 hours operating time from Monday till Friday, working on Saturday, increased usage of permanent night shifts, increased usage of flexible work hours have been recently the used solutions.

On the other side workers' position differs from one country to another as a function of traditions, ideologies and work organization of local companies. In some countries unions have accepted sacrifices: the main reason was to defend job places. A meaningful example is the UAW position (King 2011): *A Fair Deal means sharing in the good times as well as sacrificing in the bad times*. In other words, a pragmatic view free from any ideology. But in other countries the opinions are more conflictual.

1.4 *Recent appreciable initiatives*

It is interesting and appreciable to discover that, in spite of the pressure of globalization and economic crisis, some companies have been starting to give again value to their employees through initiatives in the direction of the "good work", of course asking them for more flexibility. Examples are (Di Vico 2010): welfare actions such as free health care service, organization of summer holidays for children and financial help for their university tuition fees, organizational innovation such as the job sharing among the members of the same family, tele-working, possibility of putting extra working time in time banks in view of future needs, for instance a child, money awards in case of energy saving behavior. Another recent example (De Pommerau 2013) is the attention to elderly workers, problem caused by the increased retirement age and deserving quick solutions to avoid

loss of competitiveness. In assembly lines were introduced adaptable seats, wooden floors softer for the knees, special shoes, easier to read computer screens, more use of daylight, sit positions instead of stand, job rotation during a shift to avoid repetition of the same task, posters along the assembly line to remind workers the importance of stretching exercises and nutrition, recreation areas to rest, more ergonomic workplaces.

The concept of sustainability can be applied to every kind of work and, with reference to manufacturing, to every process where workers are widely involved and cannot be replaced by robots or other automated device. This paper is focused on assembly since it is definitely one of the most meaningful manufacturing activities from the point of view of human involvement for various reasons: every product must be assembled, men cannot be easily or conveniently replaced by robots, therefore human workers are almost always present in assembly lines even if automated. Traditionally the job on assembly lines is a source of diseases, absenteeism, stress, conflicts with unions especially in the automotive industry.

Therefore assembly is an interesting field in the production framework where to analyze the sustainability of work and where to propose new solutions (Santochi 2011).

2 ASSEMBLY WITHOUT STRESS

2.1 *Minimizing the stress, maximizing the efficiency*

One of the most important problems concerning the working conditions on assembly lines is the stress. Everybody remembers the famous movie "Modern Times", produced in 1936: Charlie Chaplin focused the basic issues of an assembly line. Nowadays lot of things are changed, but the stress of a worker forced to occupy a single position and often to keep the same posture for a whole shift remains probably almost the same.

In the past workers' stress was not too important for the company, whose main focus was the avoidance of injuries of operators. Anyway, where more sophisticated production paradigms have been adopted (e.g. Total Quality), this basic approach to manpower safeguard is no more sufficient: actually the existence of relationships between productivity and general work conditions has become evident. Many factors determining undesirable working conditions have been gathered under the name of "stress factors".

Although the classical intervention on ergonomic aspects is clearly very important to minimize the stress, it is not the only applicable improvement to the workspace of an assembly line.

There are three main fields of intervention to minimize the stress of a worker on an assembly line:

– knowledge and involvement;
– confidence;
– ergonomics (social, mental and physical).

2.2 *Providing the necessary knowledge*

A lack of knowledge is often one of the main reasons generating stress. An appropriate knowledge may avoid a series of uncertainty and problems that have been identified as powerful sources of stress in assembly (Kvarnstrom 1997). For instance unpredicted defects in the products or in the process can stress the operator: although he/she probably can identify a reasonable solution (e.g. repair), he/she cannot be sure it is the right action to do. In fact in many cases "right" is a synonymous of "shared", and, in particular, shared with the management. Thus, in some cases a "knowledge injection" (i.e. a quick response of the shift supervisor to an information request, for the mentioned example) can increase the self confidence and the involvement of the operator and decrease his stress.

Receiving information and communication directly at the workstation (Feldmann et al. 2006) can also fill the gaps occurred during the training phase of the operator.

The knowledge related to the operations on an assembly line should be exchanged in two directions: from the line management toward the worker and vice versa. In the former case the information concern:

– product: e.g.: type, function and importance of the assembled part in the final product, where the subassembly will be positioned in the final product (in case of uncertainty concerning the interfaces), etc.;
– process: e.g.: required tools, assembly parameters to use (force, torques, etc.), time available to perform the assembly operation, assembly sequence, accessories required (oil, glue, paint, etc.). etc.;
– production & company: e.g.: pace of the line, planned and current throughput, shift production rate, new company policies, generic internal information, etc.;

In the latter case the information concern:

– difficulties: e.g.: tiredness, maintenance requests, tool substitutions, lack of components, dangerous situations, etc.
– nonconformities: e.g.: nonconformity identification, nonconformity type (concerning the product or the process);
– suggestions on possible improvements of the process: e.g.: possible causes of nonconformity

generation, (useful to improve PFMEA and/ or DFMEA), new tools/machines to introduce in the line, necessity of new or different human resources or infrastructures, etc.

The problems to face are:

– easiness/speed necessary to communicate information;
– completeness of information;
– pertinence of information.

All these needs can be satisfied using multimedia supports. In the paragraph 2.5 an example of workstation embedding multimedia support is shown.

This highly integrated information exchange entails an unavoidable problem: information allow control. In fact the information exchange supplies both technical support to the operator and information on his behaviour to the management.

Thus, all the system has to be based on a strong confidence to be effective. Both workers and management must agree on the fact that the supplied information will be used to remove problems rather than to control people. Each kind of misuse of the collectable information would breakdown the system. A typical example of information misuse is counting the help requests of the operators to detect (and also penalize) the less productive one.

In the next paragraph some methods to establish such confidence are discussed.

2.3 *Enhancing the necessary confidence*

According to Maslow's principles the very basic level of confidence can be established when the salary is correct in worker's opinion and his work is adequate in terms of quantity and quality in employer's opinion. Although the achievement of such condition is still missing in many cases, the most advanced industries consider other methods more interesting. One of the most effective ways to enhance confidence in industry is related to the training activity. By supporting a suitable training activity the management invests in increasing the knowledge and the skill of his workers. In this way the personnel can verify the increase of his value for the management and the management can request more complex (i.e. more added value) tasks, stimulating personnel's empowerment, in a continuous virtuous circle.

Another possibility to increase confidence is the availability of the management to accept suggestions coming from the workshop (see previous paragraph). This is a typical factor considered in Total Quality Management models. In this way also the tendency of the operators to take pride in their own work can be reinforced, creating a virtuous circle of trust.

Remaining into the scope of standard models of quality control, another important method to promote the confidence comes from the well known ISO 9001. One of the most important issues considered in ISO 9001 is actually the necessity for the management to show a clear commitment. There are many ways to demonstrate such commitment, i.e. the perseverance in developing corrective and preventive actions to improve the effectiveness of the processes or ensuring the availability of resources. All these actions demonstrate the will of the management to operate for a continuous improvement of the productive processes and, consequently, for a continuous focus on the working conditions of the operators who are fundamental actors in the processes.

Last but not least, the management should avoid the use of collected information to make the operators' working condition worse. A typical example is the immediate increase of the assembly line pace as a consequence of an operator's suggestion aiming to make the process more efficient. In this case the suggestion has to be considered but also deeply discussed to analyze the effect of the modification on the whole process and on the work of all the other operators in the process: this means involvement.

2.4 Introducing advanced ergonomics

The basic idea of each ergonomic workstation is: the workstation has to be as comfortable as possible within the constraints of each particular production process.

But when can a workstation be defined "comfortable"? It is only a physical issue? Is it determined only by a correct postural set? Currently this is simply the basic goal to achieve, also safeguarded by specific legal standards. Anyway, to overcome the limits of this standard approach to ergonomics, a wider analysis, including all the needs of a generic operator of an assembly line, is probably necessary.

A wider ergonomic concept might include:

– health;
– serenity;
– human relationships.

The first category includes the usual issues on ergonomics. But health means more than avoiding physical injuries. Precautions similar to prescriptions concerning the maximum time for PC monitor usage in offices could be adopted. The information system installed in the workplace could produce a scheduled warning to suggest a pause allowing muscle stretching or other exercises. Of course the compatibility of the pauses with the production pace should be planned and agreed between workers and management. Another method to protect the health of workers and particularly elderly ones could be adopted,

a sort of "continuous check-up". Many stress or fatigue symptoms can be monitored by suitable sensors. Blood pressure and heartbeat are commonly monitored in sport activities by simple and portable devices. The blood oxygen level also can be monitored (often together with heart rate) by portable sensors, named oxymeters.

A workstation equipped with a set of such monitoring devices could be a powerful instrument to face physical stress. But a relaxing condition for the worker does not depend only on physical conditions. The stress is also generated by lack of psychological input like social interactions, group identity or family relations. For instance, parents could need to receive reassuring information concerning their family (children or elderly grandparents).

Thus, an ideal workstation could supply a series of facilities able to fill this kind of gaps.

2.5 Proposal of workstation based on advanced ergonomics

Starting from a normal structure of a typical workstation included in an assembly line, an innovative workstation based on advanced ergonomics can be defined.

In Figure 1 a typical assembly line is represented. Adjusting of classical parameters (height, seat back position, ecc.) could be embedded in the chair of the workplace. A useful feature could be the recording of a specific setting, suitable for each worker. At shift change the adjustement of the chair for the new operator could be very fast and accurate, enhancing its ergonomics.

Around the chair a pair of mobile and adjustable storage shelvings are present. They contain all

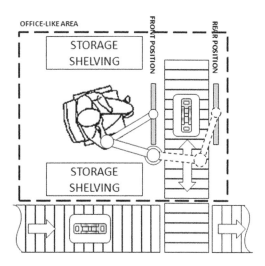

Figure 1.   Top view of the office-like workstation.

165

the tools and components necessary for the assembly process.

In front of the chair the product to be assembled is placed on the conveyor. In this way the workstation structure tends to reproduce an office-like environment.

The heart of the workstation is an articulated arm supporting a special monitor. The monitor is connected to the information system of the plant and is able to show a series of information concerning the product and its assembly process. A keyboard could be available but its functionality can be even included in the monitor by using a touch screen. By this multimedia support the operator can perform all the operation mentioned in par. 2.2 concerning the exchange of useful information. The operator can receive such information using standard multimedia support (text, images and videos) but also using Augmented Reality (AR) features. By AR technique, using a transparent screen instead of a standard screen, the operator can see the requested assembly instruction/information directly related to the real product, independently from the orientation of the screen (see Fig. 2). Through the articulated arm the screen can be moved and superimposed to each element of the workstation. The visualization of information can be managed mainly by moving the screen towards the interesting element in the workstation thanks to screen position sensors. Other standard solution used in AR field (e.g. head mounted display) are not advisable since they are too invasive to be used for the whole shift.

The screen can be used also as a standard monitor. The operator can use the screen to contact colleagues or managers to communicate problems or suggestions or to exchange ideas by an internal chatline (see par. 2.2) or read internal news. The screen also allows the communication with external persons (relatives, children, friends) to reach the goal of a complete advanced ergonomics (see par. 2.4).

3 CONCLUSIONS

The challenging way to sustainable manufacturing made harder and harder by the global competition must not disregard the issue of working conditions. Sustainable work must be one of the key points of sustainable production. A lot of research work can and must be done toward the target of the "good work". Interesting results can today be obtained, as proposed in this paper, through information technology if investments in human resources are considered fundamental as they should be.

REFERENCES

De Pommerau, I. 2013. How BMW reinvents the factory for older workers, http://www.csmonitor.com.

Di Vico, D. 2010 Welfare in cambio di flessibilità, Il Corriere della Sera, 13 Ottobre 2010.

Effra 2012, http://www.effra.eu.

Ewcs 2010, http://www.eurofond.europa.eu/surveys/ewcs/2010.

Feldmann, K.& Lang, S. 2006. Increasing the efficiency of manual assembly structures by innovative information systems, 1st CIRP Int. Sem. on Assembly systems, Stuttgart, November 15–17.

FLA 2013, http://www.fairlabor.org.

Gronkvist, L.& Lagerlof, E. 2009. Report of the conference on Sustainable work: a challenge in times of economic crisis, Stockholm October 2009.

Kvarnstrom, S. 1997. Stress prevention for blue collar workers in assembly line production, ILO Working paper, CONDI/T/WP.1/1997, Geneva.

King, B. 2011. The role of the UAW in the 21st century, http://www.uaw.org/articles/role-uaw-21st-century.

Jovane, F. & Westkamper, E. & Williams, D. 2009. The Manufuture Road, Springer.

Johansson, J. & Abrahamsson, L. 2009. The good work—A Swedish trade union vision in the shadow of lean production, Applied ergonomics, 40, pp 775–780.

OSHA 2002. Work-related stress, osha. europa.eu/en/publications/factsheets/22/view.

Santochi, M. 2011. Sustainable manufacturing: the open issue of human factors in assembly, Proc. of AMST 11, ISBN 978-953-6326-64-8, pp. 35–49.

Takala, J. 2009. OSH in figures: stress at work—facts and figures, European Agency for Safety and Health at Work, European Communities, ISBN 978-92-9191-224-7.

WCED 1987. Report of the world commission on Environment and development. Our common future, Annex to general Assembly document A/42/427.

WEF 2012, World Economic Forum, The global competitiveness report 2012–2013.

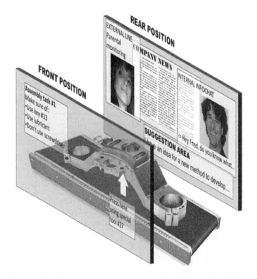

Figure 2. Transparent screen used in AR mode (front position) and in standard mode (rear position).

Green Design, Materials and Manufacturing Processes – Bártolo et al. (eds)
© *2013 Taylor & Francis Group, London, ISBN 978-1-138-00046-9*

Influencing variables on sustainability in additive manufacturing

S. Junk & S. Côté

Department of Business Administration and Engineering, University of Applied Sciences Offenburg, Germany

ABSTRACT: A variety of different additive manufacturing processes have been available for the last three decades. Some of these technologies are very energy-intensive, e.g. laser technology and the manufacture of metal powder. In many areas, the detailed investigation of the energy and material consumption of these new manufacturing methods is still in the beginning. This paper investigates energy and material consumption using 3D colour printing (3DP) as an example. The specific energy required for the layering can be determined from this. This then forms the basis for a comparison of the specific energy consumption with other generative (e.g. Fused Layer Modelling—FLM) and also conventional production processes (e.g. milling and grinding). Thus process selection is facilitated by introducing the specific energy for layering. In addition several variables, in which resource consumption can be reduced are also investigated and compared. For example the influence of the geometry or the positioning of the 3D-printed part in the design space on the consumption are investigated. But also the measuring of different batch sizes is compared. Using the results found, the use of 3D printing can initially be optimized so that less energy, resources and manufacturing time are required.

1 INTRODUCTION

The energy requirements both of a growing world population and of the industrial sector are continually on the rise. This puts a strain on the environment and leads to increased emissions, climate change and a scarcity of fossil resources (Pehnt, 2010). For companies this not only means a rise in energy costs and a tightening of environmental regulations, as for example laid down in the Kyoto Protocol on international climate protection, but also a direct responsibility to increase sustainability by cutting back the negative effects of energy consumption.

In the European Union, a majority of consumed resources are used to generate power, which at the same time is responsible for a considerable amount of the CO_2 emissions produced in the EU (Eurostat, 2011). If companies take measures to save energy and apply it more efficiently, that in turn means a decrease in emissions and in the depletion of raw materials and puts less pressure on the environment. Energy used efficiently leads to a minimization of energy costs, so that the competitiveness of the company and its products can be enhanced (Pehnt, 2010).

Apart from these challenges in the field of energy consumption, companies are also being confronted with changes in market requirements. With many of today's products there is a demand for more variety and smaller quantities, as end consumers prefer individual rather than mass products.

Technical progress means that the life-cycle of a product is considerably shorter, depending on the manufacturing sector. In order to satisfy the demands of the market, companies must customize their product development and product development time (Gebhardt, 2011). One way of doing this is by using Rapid Prototyping technologies.

2 PRESENTATION OF BOTH RAPID PROTOYPING TECHNOLOGIES EXAMINED

Rapid Prototyping (RP) or Additive Manufacturing (AM) is the term used to describe the quick manufacture of samples or prototype construction parts in successive additive layers. The "original material is strengthened or solidified layer by layer by applying energy to it". The separate layers are thereby bonded or fused. The original material may be powdery, fluid or solid. This depends on the chosen procedure methods (Zäh, 2006).

So far only a few AM technologies have been investigated intensely with regard to their energy effectiveness (Kellens et al., 2011). For this reason two widely used methods were examined as examples: on the one hand "Fused Layer Modelling" (FLM) and on the other "3-Dimensional Printing" (3DP) (Junk & Coté 2012a, b).

The materials which were used and the energy required for each of the steps in production with 3DP and also with FLM are shown in Figure 1.

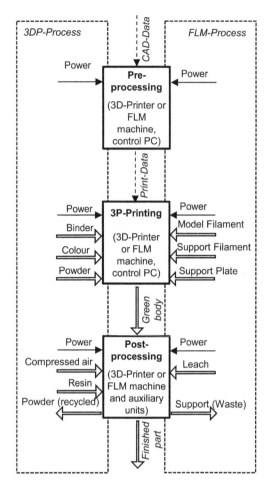

Figure 1. Diagram to illustrate the processes in 3D-Printing (3DP) and Fused Layering Modelling (FLM) with the applied materials and energy.

In this procedure the CAD-model is first preprocessed by reading the data into the control computer, in which printing is prepared with the help of the print software which controls the hardware. Then the virtual model which has been sliced into separate layers is entered into the printer. This normally takes place in a file format which is dependent on the 3DP printer software. In the printer hardware a real 3D model is formed by joining and fusing layers of powdery material (plaster and synthetic materials) with a liquid binder.

With coloured models the colour is added to the surface layer by an additional printing head. Powder serves as a supporting agent for the construction part which at this point is described as a "green body" due to its unstable form. In a subsequent postprocessing procedure the superfluous powder, which can be used again untreated, is blown off

by compressed air. Furthermore the construction part is infiltrated with a two-component synthetic resin, in order to increase its strength and the brilliance of its colour. Now the finished part can be put to use.

In FLM, the second method of technology which was studied, plastic filaments were melted and ejected in a viscous state onto a support plate by means of a heated extrusion nozzle. Additionally a second plastic material was used in the FLM machine for a supporting structure which was removed during the postprocessing in an alkaline bath.

In both technologies the 3D-part is created generatively, which means layer by layer. But the materials used are quite different: in the 3DP-process only one type of plaster powder with a certain amount of polymer powder is used. This powder could be recycled without any kind of regeneration. In contrast the FLM-process uses two different types of filament for the model itself and the support. The support filament has to be removed afterwards as waste by the use of an alkaline bath (leach).

3 SPECIFIC ENERGY FOR LAYER CONSTRUCTION

In the initial approach the energy consumption is examined by taking random measurements on a sample component. This sample component is a throttling valve. Using both RP-Technologies two throttling valves were printed, based on CAD-model generated by the use of the CAD-System CATIA V5 (one for each method with a volume of approx. 124 cm^3). At the same time the electrical power consumption of the machines during preprocessing, manufacturing and the subsequent postprocessing was measured.

Measurements were carried out with the help of a Standby-Energy-Monitor. Therefore all electrical consumers were brought together and connected to the monitor. When 3DP technology is used energy consumption is between 1860 Wh and 2161 Wh, depending on the positioning of the construction part in the construction chamber. This is equivalent to a "specific energy for layer construction" of approx. 55 kJ/cm^3 or respectively 65 kJ/cm^3. In comparison 7791 Wh was required in the manufacture of the same component with FLM. This is equivalent to a specific energy for layer construction of 180 kJ/cm^3 or respectively 230 kJ/cm^3 (depending on the proportional volume and filling level of the supporting material).

Differences can also be seen in the greatly varying lengths of processing time—5h 38 min with 3DP compared with 11h 56 min with FLM.

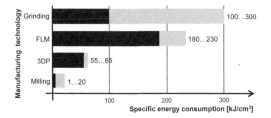

Figure 2. Comparison of specific energy for layer construction and machining methods.

Figure 2 illustrates the required specific energy (proportionate to volume) whereby two traditional subtractive production methods were compared. The values quoted in literature for specific energy consumption in machining production methods (also named as "stock removal energy") vary considerably as they are strongly influenced by the setting of the machine (Gentzen, 2004; Fraunhofer IWU, 2013). It now becomes evident that FLM technology consumes considerably more energy than 3DP technology. That can be explained by considering the operational procedure. With FLM the printer uses a great amount of energy in the preparation phase in order to reach an operating temperature of 270°C for the extrusion nozzle and jets as well as 70°C in the construction chamber, whereas the 3D-printer in 3DP technology merely needs a temperature of 38°C.

In particular the length of time required for FLM is highly dependent on the geometrical shape of the construction part. In this procedure the jet must cover every single point on the layer, first printing the outer contours and then the interior surface. Increasingly complex construction parts lead to a rise in energy consumption and duration of the production process. In addition FLM-technology requires supportive structures, which must be removed after production in an alkaline bath heated to 70°C. In the postprocessing phase of the 3D-printer the construction parts dry within 90 minutes when warmed, independent of the geometry of their structure.

4 EXAMINATION OF THE INFLUENCING VARIABLES

In the following chapter three essential variables are examined which influence the power and material consumption during 3D printing.

4.1 Influence of the component volume and component surface on the consumption

A number of measurements were made with simple test geometry to examine the influence of the

geometric variables (volume and surface) on the consumption of material and power. This geometry is a hollow cylinder of which the wall thickness was gradually increased until a solid cylinder was achieved (see Table 1). This also changes the surface of the component. A comparison of the consumption of power, binding agent and resin, as illustrated in Figure 3, shows that there is obviously a direct connection between the consumption of powder and the volume of the component. This is based on the fact that the component volume also corresponds to the volume of the powder required and, furthermore, that no powder is required for support. Excess powder is blown off and reused completely. Since the resin penetrates mainly the surface, it could be assumed that there is a connection between the size of the surface and the consumption of resin. However in the experiments the consumption of resin showed quite different results. The consumption of resin has proven to be mainly proportional to the component volume. For this example about 0.4 g/cm³ of resin is required for each component volume.

The electrical consumption for the layer build-up rather correlates mainly with the build-up time, i.e. the time required for the actual building process. This value obviously mainly depends on the envelope volume, which means reducing the volume, as with the hollow cylinder, has barely any effect on

Table 1. Variation of the volume and surface of a sample component.

Type	12.5%	25%	50%	75%	100%
Volume V [cm³]	2.47	4.94	9.83	14.82	19.63
Surface A [cm³]	62.02	61.05	58.27	54.40	41.23
Ratio A/V [–]	25.1	12.4	5.9	3.6	2.1
Construction time [min]	57	57	58	59	59

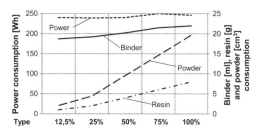

Figure 3. Consumption of power, powder, binding agent and resin for different types.

the electrical consumption. In this case which represents a very small specimen the consumption of electrical energy is approx. 12 Wh/cm³.

The reason is that the area which is "run over" when printing a component layer is considerable for this consumption. And it barely makes any difference whether this surface subsequently belongs to the component or not (cavity). After all, the consumption of binding agents only indicates a low increase in relation to the volume of the component. The binding agent consumption curve is similar to the energy consumption curve. As a result, there is also a similar connection between the envelope volume of the printed component and the consumption. In this example, the consumption per envelope volume is about 0.95 to 1.1 ml/cm³.

An explanation for this is that the printing jets for the binding agents are cleaned regularly after printing several layers (see also Section 4.2). The resulting binding agent consumption therefore mainly depends on the number of printed layers. However, whether the component is a solid or hollow cylinder only has little influence on the results. Another reason is that the saturation with binding agent on the surface with a penetration depth of approx. 1 to 2 mm is higher than in the core of the component (Gatto & Harris, 2011). That is why components with a high surface-volume A/V ratio tend to consume a greater amount of binding agent respectively to their volume.

4.2 Influence of the position on the consumption of electric power

The consumption of the power required to drive the motors during the printing process is influenced mainly by the position of the component in the installation space.

That is why the specimen "Type 100%" from Section 4.1 was set down at five different positions in the installation space (size in x-direction: 253 mm, y-direction: 200 mm) during the preprocessing before measuring the energy consumption. The printing heads for the binding agent and ink move on the gantry in x direction, whereas they are moved together with the gantry in y direction (see Fig. 4). The construction platform moves downwards in vertical z direction (size in z-direction: 200 mm, not illustrated) as soon as a layer has been printed completely.

The power consumption measurement results for 3D printing (manufacturing phase only) in different positions is illustrated in Figure 5. It is clearly revealed here that the lowest consumption of electric power is in the position #3 nearby the position of the cleaning device. This device is regularly run into by the gantry after printing

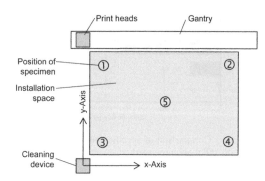

Figure 4. Positioning of specimen in the installation space.

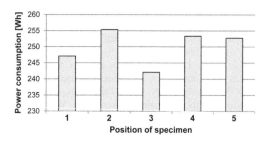

Figure 5. Power consumption (during the manufacturing phase) in different positions in the installation space.

some layers to clean the print heads. If, however, the print heads have to cover long distances to the component, as for example in position #2, the consumption increases, since the motors require more power to move the gantry. Overall not a great deal is saved due to the improved positioning, however, since the difference between the minimum and maximum value is merely 6%.

4.3 Influence of the batch size on the consumption

In each of the previous tests only one component was produced in order to determine the direct influence of the geometry. In the following tests only the batch size was changed in order to determine the effect on the consumption values. The degree of utilisation η of the installation space increases together with the batch size.

A moveable deep groove ball bearing with an outer diameter of approx. 83 mm was used as the sample component (see Fig. 6, left). It was produced in batch sizes of one, three or six items. The arrangement of the components in the installation space of the printer is illustrated in Figure 6, right. However, this increased the degree of utilisation in three stages from η = 5% to 17% and on to 35% in the tests.

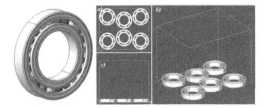

Figure 6. Deep groove ball bearing (CAD, left) and positioning in the installation space (Screenshot, right, η = 35%): a) top view, b) isometric view, c) side view.

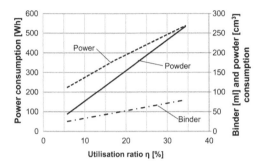

Figure 7. Consumption of energy and material at different batch sizes respectively utilization ratio η.

In this evaluation the degree of utilisation of η = 100% corresponds to the volume defined by the surface area of the installation space and the height of the component (approx. 15 mm). The consumption of powder, binding agent and energy was measured during the production phase. The evaluation of the results also shows here that there is a direct connection between the powder consumption and the volume of the components and that therefore the powder consumption cannot be reduced by increasing the degree of utilisation.

This is different with the consumption of energy and binding agent in the manufacturing phase. As shown in Figure 7 these consumptions do not increase to the same degree as the powder consumption. In other words, considerably lower consumptions can be expected by increasing the batch size and hence the utilisation of the installation space in relation to the component volume. The comparison between batch sizes one and six therefore reveals that the volume-based consumption of binding agent is lowered by 48% and that of energy even by 60%.

The reduction in energy consumption is explained by the fact that the travel distance to the individual parts and to the cleaning station is divided by several components. The considerable reduction of binding agent consumption is based on the distribution of the spray losses when cleaning the printer heads among several components, resulting in a reduction of the losses per component.

5 CONCLUSION AND OUTLOOK

The examinations have shown that the power consumption with FLM is clearly higher than with 3D printing. The introduction of the specific power for the layer build-up allows the power costs per installation volume to be compared for different methods. The result of an extensive comparison with "stock removal energy" in conventional manufacturing technologies is that the two examined RP methods are more favourable than grinding in terms of the consumption but less favourable than milling.

The geometry of the component to be printed also has a considerable influence on the consumption. For example, hollow bodies consume more binding agent than solid bodies in relation to the volume. But the power consumption depends mainly on the printing time, which depends on the envelope volume of the component. Ideal positioning of the components in the installation space also allows the power consumption to be reduced.

The batch size also has a great influence on the consumption. The consumptions per volume unit can be clearly reduced by a high degree of utilisation of the installation space of the 3D printer. Here are the greatest saving options in relation to the two other examined influencing factors. It is therefore of interest for the further examinations to process the results so far in order to allow a general connection to be established between the influencing factors. They could then be used as basis for a prediction of the consumptions in future projects. This would also enable components to be assessed with regard to their energy and material consumption. And, in turn, it would allow components to be assessed with regard to their sustainability prior to their production.

Further examinations are intended, for example, to clarify the effect the orientation in the installation space or the use of colour on the power costs. Further optimisation potentials are assumed here. Furthermore, the examination should be extended also to other RP methods, such as laser sintering or stereo lithography, to enable a comprehensive comparison of the power and material costs.

REFERENCES

European Commission 2011. *Europe in figures—Eurostat yearbook 2011*. Luxembourg: Publications Office of the European Union.

Fraunhofer-Gesellschaft 2013. Fraunhofer-Institut für Werkzeugmaschinen und Umformtechnik—Projekte—Produkte; www.iwu.fraunhofer.de/de/geschaeftsfelder/ressourceneffiente_produktion/energieeffiziente_feinbearbeitung.html [28.01.2013].

Gatto, M., Harris, R. 2011. Non-destructive analysis (NDA) of external and internal structures in 3DP. *Rapid Prototyping Journal* 17/2. pp. 128–137.

Gebhardt, A. 2011. Understanding Additive Manufacturing: Rapid Prototyping—Rapid Tooling—Rapid Manufacturing. Munich: Carl Hanser.

Gentzen, J. 2004. Anwendungspotentiale der Verfahren Schleifen, Superfinischen und Glattwalzen. In Hoffmeister, H.-W. & Tönshoff, H.K. (ed): *Jahrbuch Schleifen, Honen, Läppen und Polieren*. Essen: Vulkan.

Junk, S., Côté, S. 2012a. A practical approach to comparing energy effectiveness of rapid prototyping technologies, *Proceedings of AEPR'12, 17th European Forum on Rapid Prototyping and Manufacturing. Paris. France. 2012.*

Junk, S., Côté, S. 2012b. Energy and material efficiency in additive manufacturing using 3D printing. *Proceedings of the 10th Global Conference on Sustainable Manufacturing. Istanbul*. Turkey.

Pehnt, M. 2010. *Energieeffizienz*. Ein Lehr- und Handbuch. Heidelberg: Springer.

Kellens, K., Yasa, E., Renaldi, R., Dewulf, W., Kruth, J.-P., Duflou, J. 2011. Energy and Resource Efficiency of SLS/SLM Processes. *International Solid Freeform Fabrication Symposium. Austin. Texas*. USA.

Zäh, M.F. 2006. Wirtschaftliche Fertigung mit Rapid-Technologien: Anwender-Leitfaden zur Auswahl geeigneter Verfahren. Munich: Carl Hanser.

Green Design, Materials and Manufacturing Processes – Bártolo et al. (eds)
© *2013 Taylor & Francis Group, London, ISBN 978-1-138-00046-9*

Compressed air system assessment for machine tool monitoring

A. Gontarz, P. Bosshard & K. Wegener
Institute of Machine Tools and Manufacturing (IWF), Swiss Federal Institute of Technology, Switzerland

L. Weiss
Inspire AG Zurich, Switzerland

ABSTRACT: Compressed air is the most common energy supply alongside electrical energy in manufacturing. In addition to the positive properties such as clean application possibilities, safe usage or storability, the compressed air is an underestimated energetic factor. Compressed air supply on machine tools therefore requires particular quantification in machine tool energy assessment and further energy monitoring applications. These energy monitoring applications, which are able to quantify different and combined energy forms, e.g. electricity and compressed air, are not yet available or often based on assumptions.

The following paper evaluates the physical behavior of compressed air and defines the key efficiency parameters within a compressed air system. This knowledge enables the approximation of compressed air in relation to electrical energy and extended assessment possibilities for further measurement, monitoring and optimization purposes.

1 INTRODUCTION

The evaluation and assessment of energy and resources used in manufacturing is addressed in research, legislation and industry. In this regard energy assessments are currently performed by assumptions, modeling or measurement. For compressed air Saidur et al. [2] indicate that reliable information for energy efficiency is not yet present. In accordance with the importance [1, 3] and costs [4] of compressed air, consciously or unconsciously careless use of compressed air for multiple applications can still be observed.

In standardization the ISO/DIS 14955-1 [5] defines the energetic machine tool system boundary for the assessment of resources, including electricity, compressed air and other energy forms, which are used to perform manufacturing processes. A compressed air system is an assemblage of different components such as the compressor, filters, coolers, branched pipes, valves and nozzles. Each of these components represents an energy loss in the form of flow or pressure loss in the system, and these losses must be assessed and taken into consideration when comparing and using energy equivalents.

In machine tools energy in the form of compressed air is generated by integrated equipment or- more often- supplied by the shopfloor. In correspondence with the system boundary definition, not only measurement applications according to Avram et al. and Gontarz et al. [6, 7] but also monitoring applications [8] have to combine various energy forms and resources in order to assess, analyze, compare and optimize the total machine tool energy efficiency within the machine tool system boundary.

This paper reviews potential approaches towards energy assessment through a sensible combination of measurements with simulations based on detailed knowledge of the physical behavior of the components of compressed air systems.

The target of the assessment is the design and quantification of energy saving measures, e.g. leakage prevention, the selection of optimal compressor type, the dimensioning and layout of the distribution network and the use of valves and nozzles. The added value of the approach is a deeper analysis of available or not yet available systems without excessive efforts for measuring on the component level, a step which is time consuming, costly and at times difficult to implement. Key findings are validated punctually against measurements in order to determine the correctness of the approach.

2 STATE OF THE ART

In literature different energy equivalents can be found depending on the compressor type, its capacity, pressure and temperature level. Hinsenkamp et al. [9] indicate the specific power consumption of a screw compressor for 6 bar supply of 0,1006 kWh/m^3 at a capacity of 90%; this value

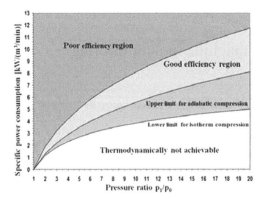

Figure 1. Specific power consumption for compressed air generation [1].

rises to 0,1286 kWh/m³ at 50% capacity. Figure 1 [1] shows an overview leading to <0,1 kWh/m³ for the most efficient compressors.

The given disparity depends on the physical behavior of the compressed air system as well as on cooling, heating and other energetic effects. In the following section narrower focus is given on compressed air subsystems and current assessment tools.

2.1 Assignment of inefficiencies

One of the main subsystems is represented by the compressor. Akbaba [10] states that savings of 8% are possible with high-efficiency motors made up of low-loss materials and constructive improvements in comparison to standard motors. Brunner et al. [11] from the Swiss initiative Topmotors indicate improvement potentials of 20–30%. The improvement of the compressor geometry, in example on a new profile of a screw rotary compressor, leads to 15% as shown by KAESER [12]. An important method for energy saving to be achieved is seen in the heat recovery; with a heat exchanger, the hot compressed gas is cooled by a cooling fluid, e.g. oil. The use of this resulting thermal energy is also discussed by Saidur et al. [2].

An essential element within the compressed air system is given by the filter, which, dependent on the application, results in a pressure drop. Wang et al. [13] and Del Fabbro et al. [14] introduced a model representing the pressure drop at filters; a simplified approach from pressure drop estimation is given by the Darcy equation [15]. Pressure drops with local compressed air supply are possible due to corresponding distribution, pipes and fittings. Coelho et al. [16] focused on energy dissipation by viscous flows. Brkic [17] analyzed friction factors for different flow regimes with approximation

to the Colebrook relation. For pipe fittings, e.g. T-pieces or elbows, the empirical derivated values of Atlas Copco [18] are commonly used and represent an equivalent length of a straight pipe, for which the calculations explained above can be used. Major energy losses are caused by leakage in lengthy and widespread industrial distribution pipe systems for compressed air. An overview of the impact of leaks as well as the industrial procedure to evaluate the volumetric leakage flow are represented by Fraunhofer ISI [1]. Siekmann [19] introduces a model approach for the evaluation of the volumetric leakage flow.

2.2 Simulation of compressed air

Energy oriented simulations on the manufacturing level including the simulation of compressed air are introduced by Herrmann [20] and Thiede et al. [21] while simulation approaches on the machine tool level are given by Avram [6] and Gontarz et al. [22]. For the analysis and evaluation of compressed air systems the software tools AirMaster+ [23] and AirSim [24] are available. These approaches are either too detailed and thus require multiple parameters and detailed measurement values or valid for a particular system layout only, generally limited to the compressor and related components.

2.3 Compressed air measurement

The quantity of air is measured by the measurement of flow $\dot{V}$ at the given pressure or mass flow $\dot{m}$ with built in measurement equipment, e.g. calorimetric or coriolis sensor. Required measurement parameters are flow $\dot{V}$, pressure p and media temperature T. With the measurement based on reference conditions according to DIN ISO 1343 [25] and a standard 6 bar air supply pressure, media temperature can be neglected. Turbulences in the flow influence the measurement in calorimetric measurements, therefore an upstream pipe length of 20 times the pipe diameter, and a respective downstream length of five times is required. For the quantification of the required energy for compressed air generation and distribution the electric power consumption of the compressor in relation to the measured media flow has to be considered. If this measurement is not applicable, e.g. due to complex distribution systems, given power equivalent should be used.

3 METHODOLOGICAL MODELING APPROACH

For the evaluation of compressed air and the quantification of inefficiencies in the form of flow and pressure losses, the compressed air system is classified into six subsystems with respective influencing factors: generation, filtration, dehydration, distribution including leakage, valves and nozzles, and end use on a machine tool or for other purposes.

In a second step the thermodynamic and fluid dynamic behavior for each subsystem is assessed followed by the transcription into a modular modeling, based on known system parameters. Due to the specific system configuration, the design of infrastructure and therefore individual system parameters, inefficiencies cannot easily be identified by a rule-based procedure. In the following sections special focus and modeling examples are given for flow loss, caused by leakage, and pressure drops due to flow friction within the system distribution.

Based on this modeling approach and its validation, options for compressed air energy quantifications for monitoring applications are evaluated.

3.1 Example: Modeling of leakage

Leakage is a direct flow loss without any use. It is dependent on the system design, the choice of components, their actual state of wear or aging, and the usage; therefore special importance is given to leakage air over time, for instance, due to corrosion and aging of hoses and fittings, and through unsealed filters. Leakages often represent a significant cost factor, as shown with some sample values on the basis of a 6 bar supply with an electrical equivalent power of $P_{eq} = 7$ kW/(m³/min), 365 d/24h operation and electricity cost of 0.15€/kWh.

There are two possible ways for leak evaluation:

A. Determination of the leakage volumetric flow V_L. According to Fraunhofer [1] the time is measured while the air tank empties from a start pressure p_s to an end pressure p_e. This measurement can be used for the quanitfication of leakages.

$$\dot{V}_L = \left(\frac{\dot{V}_c \cdot (p_s - p_e)}{t} \right) \tag{1}$$

B. Evaluation of the leakage with an ultrasonic gauge or with infrared thermography. This measurement can be used for the detection of leakages.

Neither evaluation method allows a predictive estimation of leakage of a system before installation; therefore the physics of leakage are further evaluated. Figure 2 shows an idealized leak model. State 1 represents a fully developed flow without any influences of the leakage while state 2 represents the orifice point by the flow downstream the leakage. Ambient conditions are assumed outside of the pipe.

For the discharge behavior the Bernoulli equation for compressible flow is used:

$$\frac{\gamma}{\gamma - 1} \cdot \frac{p_1}{\rho_1} + \frac{\omega^2_1}{2} = \frac{\gamma}{\gamma - 1} \cdot \frac{p_2}{\rho_2} + \frac{\omega^2_2}{2} \tag{2}$$

with the adiabatic coefficient γ, pressure p, media velocity w and density ρ.

This equation can be simplified according to Siekmann et al. [19] by assuming a free jet at orifice with $p_2 = p_{amb}$ and $\rho_2 = \rho_{amb}$, negligible velocity in the pipe with $\omega_1 = 0$ m/s and an adiabatic process p/ρ^γ the velocity at the exit ω_2 to:

$$\omega_2 = \frac{\sqrt{2 \rho_1 p_1}}{\rho_{amb}} \sqrt{\frac{\gamma}{\gamma - 1} \left[\left(\frac{p_{amb}}{p_1} \right)^{\frac{2}{\gamma}} - \left(\frac{p_{amb}}{p_1} \right)^{\frac{\gamma+1}{\gamma}} \right]} \tag{3}$$

The discharge function remains constant for the critical pressure relation $(p_{amb}/p_1)_{crit} \leq 0.528$. This phenomena is called choked flow. The desired leakage volumetric flow is ultimately obtained by multiplying the outflow velocity w_2 with the orifice cross section A_2:

$$\dot{V}_L = \omega_2 \cdot A_2 \tag{4}$$

This model is used for the modeling of leakages and will be validated in Chapter 4.

Table 1. Examples of the costs for an air leak through a hole with different parameters [12].

Hole diameter (mm)	Air consumption (m³/min)	Leakage loss kW	Leakage loss €/year
1	0.065	0.46	604
2	0.257	1.80	2364
4	1.03	7.21	9474
6	2.31	16.17	21247

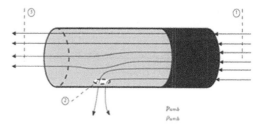

Figure 2. Model of an idealized discharge through a leak orifice.

3.2 *Example: Modeling of the distribution system*

In industrial environments the compressed air distribution represents a major pressure loss within a compressed air system; including all pipes and fittings in between the compressor and the end use. A widely used possibility to calculate the pressure drop between two points is given with:

$$\Delta p_{friction} = \frac{\rho}{2} \bar{w}^2 \cdot f \left(\underbrace{\frac{L}{D}}_{Friction\,part} + \underbrace{\sum \varsigma}_{Frictings\,part} \right). \quad (5)$$

The friction is represented by the Darcy-Weisbach equation. L is the characteristic length of the straight pipe, D the inner pipe diameter and f represents the Darcy friction factor.

For the calculation of this dimensionless factor the flow behavior must be considered. Figure 3 shows the velocity distribution in a straight pipe for laminar and turbulent flow. The friction factor is determined by the Reynolds number which leads to different pipe surface friction.

$$Re_D = \frac{\bar{w} \cdot D}{v} \quad (6)$$

is $\bar{w}$ the mean flow velocity and v is the kinematic viscosity. Up to a critical Reynolds number Re = 2300, a laminar flow is given and the friction factor results in:

$$f = \frac{64}{Re_D} \quad (7)$$

For turbulent flow, Re > 2300, f depends not only on the Reynolds number but also on the pipe surface roughness k_s. For smaller Reynolds numbers and smooth surface roughness $\omega * k_s / v < 5$, where ω describes the shear stress velocity, the following formula is used:

$$\frac{1}{\sqrt{f}} = 2.0 \cdot \log(Re_D \cdot \sqrt{f}) - 0.8 \quad (8)$$

For the model implementation, various supporting points were calculated numerically and implemented into a lookup table. Based on this table, linear interpolation is used in the simulation. To obtain f for very rough pipes with $\omega * k_s / v > 70$, von Karman's explicit formula is used:

$$\frac{1}{\sqrt{f}} = 1.14 - 2 \cdot \log \left(\frac{k_s}{D} \right) \quad (9)$$

As in the entire distribution system transitional roughness is often observed. Colebrook and White [26] developed a curve fit. A simplified alternative for the modeling is used with the explicit Chen approximation according to Brkic [17]. Together with the known pipe length and the gas properties, the pressure loss can be calculated for constant pipe diameter.

In order to quantify the pressure loss at fittings in this simulation a widely used empirical method is chosen. Depending on the fitting geometry an equivalent straight pipe length $L+$ can be obtained which represents the same pressure loss. Atlas Copco [18] and Miller [27] tabulated the values for different pipe diameters.

4 MEASUREMENT AND VALIDATION

For the validation of the findings and implemented models a defined leakage test was performed. As shown in Figure 4a defined 2 mm (8) and 4 mm leakage was measured and compared against the defined model. The entire electrical power input on the compressor (1) against the flow (7) was measured. The tank (2) was filled up to a system pressure at the manometer (3) to 10.6 bar. A ball valve (4) was opened and released the stored compressed air through a high pressure hose (5) and filter (6) to a test pipe with the defined leakage (8).

The leakage model reflects correctly the volumetric flow over time (Fig. 5). As friction and the dynamic accumulated pressure behavior are not modeled the model does not reflect an initial transient behavior and produces an offset in the long run. The offset is caused by the neglected friction and the measurement accuracy on the flow sensor.

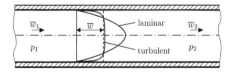

Figure 3. Model of a straight pipe with a laminar and turbulent velocity profile.

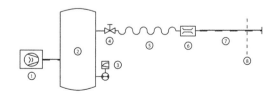

Figure 4. Piping and instrumentation diagram of the leakage measurement set-up.

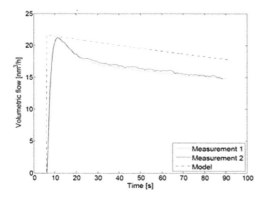

Figure 5. Comparison of measurement and model.

5 CONCLUSION

The given approach and developed model is based on the fundamental physical behavior of the subsystems of the compressed air system and has illustrated that it is possible to estimate the needed compressor dimension and predict its power consumption according to defined parameters. Compared to available software tools it is modular and based on available unmeasured or assumed parameters. In relation to flow measurements, where only the actual flow use of the machine tool including leakage can be quantified, this approach enables the user to quantify energetic weaknesses. In the long term this set of subsystem models included into a software tool is seen as an extension to current measurement and monitoring approaches which enriches the evaluation of compressor efficiency. Depending on given monitoring and evaluation purposes this approach is suitable for compressed air system quantification and optimization.

REFERENCES

[1] Fraunhofer Institut für System und Innovationsforschung (ISI), "Kampagne "Druckluft effizient"," pp. www.druckluft-effizient.de, 2013.
[2] Saidur, R., Rahim, N.A., Hasanuzzaman, M., "A review on compressed-air energy use and energy savings," *Renewable and Sustainable Energy Reviews* vol. 14, pp. 1135–1153, 2009.
[3] Radgen, P., "Efficiency through compressed air energy audits," presented at the Energy Audir Conference, 2006.
[4] Yuan, C.Y., Zhang, T., "A Decision-Based Analysis of Compressed Air Usage Patterns in Automotive Manufacturing," *Journal of Manufacturing Systems,* vol. 25, 2006 2007.
[5] ISO/TC 39/WG12 ISO/DIS 14955, "Environmental Evaluation of Machine Tools," ed, 2012.
[6] Avram, O., Xirouchakis, P., "Evaluating the use phase energy requirements of a machine tool system," *Journal of Cleaner Production,* vol. 19, pp. 699–711, 2010.
[7] Gontarz, A., Weiss, L., Wegener, K., "Energy Consumption Measurement with a Multichannel Measurement System on a machine tool," presented at the International Conference on Innovative Technologies IN-TECH Prague, Czech Republic, 2010.
[8] Hu, S., Liu, F., He, Y., Hu, T., "An on-line approach for energy efficiency monitoring of machine tools," *Journal of Cleaner Production,* vol. 27, pp. 133–140, 2012.
[9] Hinsenkamp, G., Reinhardt, J., Hager, M., "Druckluft. Strömungsfreie, kostengünstige und energieeffiziente Bereitstellung" Energieagentur NRW, Wuppertal2012.
[10] Akbaba, A., "Energy conservation by using energy efficient electric motors," *Applied Energy* vol. 64, pp. 149–158, 1999.
[11] Brunner, C., Nipkow, J., Heldstab, T., Sidler, C., Humm, O., Edel, C., Schlacher, M..(2012, 15.02.2013). *topmotors.ch—Effizienz im Antrieb.*
[12] Ruppelt, E., Hobusch, G., Piendl, S., Bahr, M., "Druckluftseminar.," KAESER Kompressoren GmbH2011.
[13] Wang, F., Yoshida, H., Kitagawa, H., Matsumoto, K., Goto, K., "Model-based commissioning for filters in room air-conditioners," *Energy and Buildings* vol. 37, pp. 1225–1233, 2005.
[14] Del Fabbro, L., Laborde, J.C., Merlin, P., Ricciardi, L., "Air flows and pressure drop modelling for different pleated industrial filters" *Filtration Separation,* vol. 39, pp. 34–40, 2002.
[15] Bear, J., Dynamics of fluids in porous media Dover Publ., 1972.
[16] Coelho, P.M., Pinho, C., "Considerations about equations for steady state flow in natural gas pipelines," *Journal of Brasilian Society of Mechanical Sciencs and Engineering* vol. 29, pp. 262–273, 2007.
[17] Brkic, D., "Review of explicit apporoximation to the colebrook relation for flow friction" *Journal of Petroleum Science and Engineering* vol. 77, pp. 34–48, 2011.
[18] Fordel, P., (2010) Copmressed Air Manual. *Atlas Copco Airpower NV.*
[19] Siekmann, H.E., Thamsen, P.U., *Strömungslehre für den Maschinenbau* Springer 2009.
[20] Herrmann, C., Thiede, S., Kara, S., Hesselbach, J., "Energy oriented simulation of manufacturing systems—Concept and application," *CIRP Annals— Manufacturing Technology,* vol. 60, pp. 45–48, 2011.
[21] Thiede, S., Seow, Y., Andersson, J., Johansson, B., "Environmental aspects in manufacturing system modelling and simualtion—State of the art and research perspectives," *CIRP Journal of Manufacturing Science and Technology,* vol. 6, pp. 78–87, 2012.
[22] Gontarz, A., Züst, S., Weiss, L., Wegener, K., "Energetic machine tool modeling approach for energy consumption prediction," presented at the 10th Global Conference on Sustainable Manufacturing 2012, Istanbul, Turkey, 2012.

[23] U.S.D. o. Energy, "AirMaster+ Software Tool Brochure," U.S.D. o. Energy, Ed., ed. EERE Information Center 2010.

[24] Kissock, K., "AirSim compressed air simulation software," Department of Mechanical Engineering, University of Dayton 2003.

[25] D.F.T.S.C. (NATG), "Reference conditions, normal conditions, normal volume; concepts and values," ed. DIN German Institute for Standardization: Beuth Verlag GmbH, 1990.

[26] Colebrook, C.F., White, C.M., "Experiments with fluid friction in roughened pipes," *Proceedings of the Royal Society of London,* vol. 161, pp. 367–381, 1937.

[27] Miller, D.S., *Internal flow systems,* 1978.

Green Design, Materials and Manufacturing Processes – Bártolo et al. (eds)
© *2013 Taylor & Francis Group, London, ISBN 978-1-138-00046-9*

Crop rotation and association design for N budgeting in organic dairy farms

G. Bukvić, R. Gantner, Z. Steiner & K. Karalić
University J.J. Strossmayer in Osijek, Faculty of Agriculture in Osijek, Osijek, Croatia

ABSTRACT: Aim of the research was to present the design of crop rotation and association by inclusion of nitrogen (N) contributing crops to meet the N needs of simulated organic dairy farm. Proposed crop rotation design with interpolation of legume intercrops and association crops provides the required N for the whole organic dairy farm N needs.

1 INTRODUCTION

One of the fastest growing segments of organic food production is organic dairy farming (McBride & Greene, 2007) where the mineral nitrogen (N) use is prohibited for the required forage production (European Union Council Regulation No 834/2007). However, N is crucially important yield contributing element for majority of crop plants (Graham & Vance, 2000) including forages. Aim of the research is to present the design of crop rotation and association by inclusion of N contributing crops to meet the N needs of simulated organic dairy farm.

2 MATERIAL AND METHODS

2.1 *Prediction of forage and bedding needs and required land area for simulated farm*

For the prediction of required forages there was simulated a small dairy and beef production farm comprising 20 lactating cows with moderate average milk yield of 15 l/day for 305 days of lactation, 3 dried cows, 12 beef and 11 heifers with average Live Weight Gain (LWG) 1 kg/day, of which 3 heifers are intended for a herd renovation. Calving is planned constant through a year with no intentional concentration of feed requirements. Daily diets were diversified among a.m. groups of animals according to their respective needs (Domaćinović, 1999) and feed values of forages (DLG, 1997). Total annual needs for forages were predicted by summation of daily needs across all the animal groups and along a whole year. Requirement for straw as bedding material was calculated upon the need of 3 kg/day/LU (LU = equivalent of 500 kg livestock live weight) during a cold half of year and 2 kg/day/LU for warm half of year

because than animals stay in pastures during a day and require the bedding during a night only. Forage Dry Matter Intake (DMI) during April and May is intended mainly from pasture whilst during June to September pasture is planned at low level. Such pasture partition is set up for the purpose of greater reliance on higher yielding arable forages in summer while compromising the need of pasture utilization for animal health benefits. Predictions of required area for arable forage crops, straw crop and permanent grassland area for rotational grazing were done by dividing the annual forage and straw needs with expected yields per unit area based on the authors' experience.

2.2 *Prediction of livestock N excretion and N available from FYM soil application*

Livestock annual excretion of N through feces and urine is predicted by summation of daily excretion predictions across all the animal groups considering the dietary N intake through feed, and milk production level, live weight gain and other related parameters by using ASAE regression equations (ASAE, 2005). The ratio of urinary/total N is predicted 37% according to Broderick (2003) for moderate dietary protein level. During a grazing season (April–September) only the fecal N and half of urinary is considered for collection to FYM heap since the urine excreted on pasture is not feasible to collect. Gaseous losses of N for a period from excretion till collection to FYM heap are estimated 40% during a warm half of year and 16% during a cold half of year (Moreira & Satter, 2006). Since the bedding straw constitutes FYM, there was accounted 0.59% N in its dry matter (DLG, 1997). N comprised in FYM after 6-months storage in the FYM heap is assumed 80% of the collected N due to gaseous loses (volatization and denitrification, Petersen et al, 1998). The further N loss is accounted 3% for gaseous loses during several hours from FYM spreading over soil till its incorporation by plowing (Chambers et al, 1999). Splitting the prediction of FYM N release over 3–4 year dynamics was not accounted for since the simulation is intended for a dairy farm in a long run, where N is constantly being released from previous FYM applications giving the residual effects of N addition to the partial N release from the last FYM application.

2.3 Assumptions of factors contributing to soil N balance apart from FYM application and plant uptake

Initial mineral N content in arable soils at the beginning of spring vegetation usually ranges from 7 to 21 kg/ha, but the most often found level is at about 15 kg/ha (dr. Krunoslav Karalić, based on the author's experience). N leaching from soil is estimated to contribute −33 kg N/ha/year in average conditions within range of −19 to −65 kg/ha/year (Hansen et al, 2000) depending mainly on soil texture, annual atmospheric water precipitation (rain, snow, etc.), N load of soil and practice of growing catch crops during the absence of main crops in a field. Contribution of gaseous N loses from soil is estimated −24 kg/ha/year in average conditions within range of −13 to −34 kg/ha/year (Kristensen et al, 2005). N release from the humus decomposition contributes 45 kg/ha/year in average conditions (at 2% hummus in soil and 1% annual mineralization, dr. Krunoslav Karalić, based on the author's experience) within the range 11 to 89 kg/ha/year depending mainly on soil organic matter content, C/N ratio of organic matter and microbial activity in soil (Snyder, 2011). Atmospheric precipitation of N is assumed to contribute 10 kg/ha/year in average conditions within wide range of 1 kg/ha/year in uninhabited areas to 7 kg/ha/year around dense populated areas (Kingston et al, 2000) to 16 kg/ha/year in dense populated areas of with intensive livestock farming (Kristensen et al, 2005). N fixation into a new-build soil organic matter is assumed to have negligible contribution since almost all aboveground plant mass is being harvested with no harvest residues incorporation to soil.

2.4 Prediction of crop plants N uptake and the needs for addition of N

For the purpose of this research only the above ground N is taken into account since the N comprised in plant roots remains in soil and does not affect the N balance of a soil. N removal by harvested yield is calculated using expected plant crude protein concentrations in dry matter according to DLG (1997) and dividing it by 6.25 what refers to

the average N concentration of 16% in the plant Crude Protein (CP). Exception was N concentration in wheat grain protein, which was calculated by dividing CP with 5.7. Such N concentration in harvested dry matter was multiplied by the expected DM yield of harvested forages, straw and grain to estimate the N removal from soil. The lack of soil N supply to the required crops is predicted by balance sheet and shown against available N from FYM. Association crops of legumes and grasses are excluded from the balance since they are assumed to comprise sufficient legume partition in a DM yield to be self-sufficient on N (Zemenchik et al, 2001; Boller & Nosberger, 1987) for achieving the predicted DM yields and quality. Therefore, the areas of permanent grassland (grass/legume association crop) for grazing, lucerne/grass sward for cutting and hay drying and winter pea/wheat mixture for haylage production are assumed self-sufficient on nitrogen and consequently excluded from the need for N addition. Lucerne/grass sward is predicted for 4-year exploitation span, therefore releasing 1/4 of its area for succeeding silage maize crop each year. The area of maize 1st year after lucerne/grass sward is assumed too to be sufficiently supplied with N released from lucerne to obtain the desired maize silage yield (equivalent of 13 t/ha grain, Balesta & Lloveras, 2010) and consequently excluded from the balance sheet for calculating the N addition needs.

2.5 Proposing the crop rotation and association design

Proposals for crop rotation and association design were made upon identifying the needs for N addition over FYM available N to required crops, and using the collected data from published researches on amount of N that legumes contribute to the main crop nutrition either as intercrops or companion crops. Implications on interference with main crops seeding terms, competition and eventual yield reductions were discussed.

3 RESULTS AND DISCUSSION

3.1 Needs for forages, bedding, land area and N removal from soil

On a round-year basis, for the simulated organic dairy farm there were required forages and bedding material grown at 34.21 ha (Table 1). Annual N removal from soil by harvested crops was predicted 5008 kg (Table 2), of what 199 kg in straw is assumed to circulate field-stall-field with negligible retention in farm animals and loses.

The portion of 2239 kg of N is harvested from the N-self-sufficient crops (legume/grass companion crops, first 3 crops in Table 2).

Table 1. Annual needs for forage and straw (t) and land area requirements (ha) for the simulated farm.

	Needs (t)		Yield and area	
	Fresh	DM %	DM t/ha	Ha
Pasture herbage	153.2	18	4.6	6.00
Lucerne/grass hay	33.8	85	10.0	2.87
Winter pea/wheat	37.3	40	6.4	2.33
Silage maize	270.0	33	18.0	4.95
Sorghum × sudangr.	153.6	19	7.6	3.84
Fodder beet	28.3	14	8.4	0.47
Barley grain	0.5	85	4.3	0.10
Forage kale	3.1	12	7.2	0.05
Wheat straw	40.3	85	5.0	6.85
(DM = dry matter)			Total	34.21

Table 2. Annual N (kg) removal from soil by harvested crops.

	Harvested crop			N removed	
	DM (t)	CP %	N %	kg.	kg/ha
Pasture herbage	27.6	21.3	3.41	940	157
Lucerne/grass hay	28.7	18.9	3.02	869	302
Winter pea/wheat	14.9	18.0	2.88	430	184
Silage maize	89.1	8.2	1.70	1169	236
Sorghum × sudangr.	29.2	14.8	2.37	691	180
Fodder beet	4.0	8.9	1.42	56	120
Barley grain	0.4	15.3	2.45	11	104
Forage kale	0.4	17.1	2.74	10	197
	Subponder	13.4	Subtotal	4176	
Barley straw	0.6	4.0	0.64	4	38
Wheat straw	34.2	3.7	0.59	195	29
Wheat grain	28.0	13.8	2.42	633	94
			Total	5008	

3.2 N excretion by farm animals, N loses and N from FYM application to soil

Predicted annual N excretion by farm animals rated 79% of the total ingested N through forage consumption (Table 3), what was relatively high excretion of dietary N.

This level of excretion is in line with Broderick's (2003) findings for dairy diets rich in fibers and low in concentrated feeds (even 90% excretion at Broderick's research) and for diets moderate in dietary protein which tend to increase N efficiency and decrease N excretion (73% excretion, Broderick, 2003).

3.3 Spontaneous soil N supply balance

The spontaneous net N supply to crop plants stems from the balance of several contributing factors and is assumed to amount of 13 kg/ha/year in the average arable farming conditions (Table 4).

3.4 Balancing crop N needs, spontaneous soil N supply and FYM N supply

Only the non-N-sufficient crops (all except pasture, lucerne/grass sward and winter pea/wheat) were considered for the balance (Table 5) since grass/legume companion crop associations are assumed with sufficient legume partition to provide N for predicted yields (Zemenchik et al, 2001; Boller & Nosberger, 1987). The small partition of maize area (0.72 ha) is also excluded from balance since it succeeds the 1/4 of lucerne area that is being plowed each year. Such area is considered to provide sufficient N for maize silage yields of 53 to 77 t/ha (Yost et al, 2012).

181

Table 3. Prediction of annual N (kg) available from FYM application (see Material and Methods for explanation).

	N (kg/year)
N consumed by farm animals as dietary protein	4176
N excreted in feces	2090
N excreted in urine (50% collectable, in-stall excr.)	1228
Total excreted N	3318
Gaseous loses of collectable N prior to collection	−757
N in bedding material (what straw)	199
N collected to FYM heap	2146
N gaseous loses from FYM heap during 6 months	−429
N saved in FYM heap after 6 months storage	1717
N gaseous loses from spreading to soil till plow-in	−52
N from FYM incorporated into soil	1665
N released to permanent pasture (50% of urinary N)	614

Table 5. Annual N balance for crops requiring N addition.

	Area requiring N	N contribution	
	ha	kg.	kg/ha
Silage maize	4.23	−998	−236
Sorghum × sudangr.	3.84	−691	−180
Fodder beet	0.47	−56	−120
Barley grain and straw	0.10	−14	−142
Forage kale	0.05	−10	−197
Wheat straw and grain	6.85	−828	−121
Total	15.54	−2597	−167
Net spontaneous soil supply	15.54	+202	+13
Total needs—soil supply	15.54	−2395	−154
FYM available N to soil	15.54	+1665	+107
Lack of N	15.54	−730	−47

Table 4. Assumption of spontaneous annual N (kg/ha) supply from soil (see Material and Methods for explanation).

	N (kg/ha/year)
Initial mineral N content in soil	15
N leaching from soil	−33
Gaseous N loses from soil	−24
N release from soil organic matter decomposition	45
N precipitation with rain and snow	10
Net spontaneous N supply	13

For the predicted crops' yields there was estimated annual lack of 730 kg of N, or 47 kg/ha in average. The predicted lack of N could be decreased by greater reliance on permanent grassland which comprises sufficient clovers partition but, typically, grassland yields less forage DM compared to arable crops, especially in drought summer conditions, so the proposed crops structure can be considered as satisfactory.

Other researchers (Kristensen et al, 2005) have found the surpluses of N per ha in dairy farms that were predicted for leaching to groundwater and gaseous loses. Among the causes for N surplus, the import of supplement feeds to farms was found to contribute between 42 and 51 kg/ha/year.

3.5 *Obtaining the deficient N by crop rotation and association design*

The predicted lack of N for arable crops (Table 5) can be satisfied either by soil incorporation of legume green manure crops (as temporal intercrops for sideration), preceding the crops lacking N or growing the demanding crops in association with N fixing and releasing crops like understory clovers.

Partition of individual annual crops in the total arable area is set to equalize autumn- and spring-seeded area (wheat, pea and barley in total 9.3 ha against maize, sorghum × sudangrass, fodder beet, and forage kale in total 9.3 ha, Table 1) for the purpose of feasible crop rotation design avoiding repeated crops year-by-year. When scheduled over individual crops, the lacking N (730 kg) can virtually appear either on wheat for 6.74 ha (108 kg/ha) or maize for 3.26 ha (223 kg/ha), or in many combinations of other crops. Since the a.m. proposed inter- and understory crops can provide the whole N needs for even the most demanding crop (maize), it is reasonable to apply them just for maize, thus minimizing the area requiring N supply build-up to just 13% of the total arable area (total area excluding pasture and lucerne/grass ley).

However, in the presented pattern of arable crops, there prevails the every-year succession of cool-season cereals (wheat + barley)/warm-season cereals (maize + sorghum × sudangrass). It is worthy to point out that a time-window between harvest of warm-season cereals and seeding cool-season cereals is too short to interpolate an N-fixing intercrop. The only feasible time-window for interpolation remains the period after the harvest of cool-season cereals prior to seeding warm-season cereals.

Among the winter temporal intercrops, winter vetch is probably the most N-yielding. Winter vetch when incorporated into a crop rotation, without N fertilization, allows N for succeeding maize grain yield from 5.8 to 8.8 t/ha (Cook et al, 2010) in Wyoming (USA), or 23 to 30 t/ha of herbage DM in Italy (Caporali et al, 2004), what in average refers to the simulation-required maize silage yields. Incorporation of vetch for building-up the N supply requires a postponed seeding of maize till the vetch flowering, which can reduce maize yields compared to main seeding term, but this reduction is already predicted in the simulation yield. For building-up N supply during summer period there were investigated some warm-season legumes seeded in mid-June in Oklahoma (USA) with total accumulated N of 100 kg/ha in 120 days for soybean, 93 kg/ha for guar and 77 kg/ha for pigeon pea (Rao & Northup, 2009), which were considered unsatisfactory for the simulated farm needs. Among summer N-fixing intercrops, the appropriate solution can be red clover. Agronomy of red clover summer-intercrop comprises the frost seeding of red clover into winter cereal crop at the end of winter, thus enabling the development of clover ground cover after cereal harvest (Singer & Pedersen, 2005). Such cover crop provides a good source of forage during summer and amount of 90–125 kg/ha of fixed N in Ontario (Bruulsema & Christie, 1987) or up to 371 kgN/ha/year in UK (Stopes et al, 1996) which is to be released to subsequent crops. The Kura clover as understory in association with maize (Fig. 1), without N fertilization, allowed maize grain yields from 7.2 to 10.9 t/ha in Wisconsin (USA) in 1996 and 1997 year (Zemenchik et al, 2000) what responds to somewhat lesser silage yields than required by simulation. Thus the application would require certainly

greater maize area and total farm land area. However, the N fixing inter- and understory crops are contributing to soil N supply not only through atmospheric N fixation but also through the catchment of soil nitrate, thus reducing N leaching losses (Ochsner et al, 2010; Rinnofner et al, 2008).

Considering the crops at area requiring N addition (Table 5), a.m. temporal intercrops of winter vetch or red clover can be efficiently built-in crop rotation. Therefore the winter vetch should be seeded in autumn on parcels after wheat or barley harvest, and destroyed in the next spring, prior to seeding of maize or sorghum-sudangrass, thus providing sufficient N to the required area of spring-seeded crops. If red clover is used for N supply build-up, it should be frost seeded into winter cereals and destroyed the next spring prior to the maize seeding. If association cropping of maize with kura clover is to be used for filling the N shortage, there should be established perennial kura-clover living mulch prior to seeding maize. Implications of kura clover living mulch for organic systems are not yet known since the proposed system (Zemenchik et al, 2000) includes herbicide suppression of kura clover during the maize crop establishment. It could be likely to achieve sufficient suppression of kura by low mowing at the maize emergence moment but we cannot advice it now since it is not investigated yet. Furthermore, the compatibility of such living mulch with winter cereals for grain production is not assessed yet but is proved for forage wheat (Contreras-Govea et al, 2006).

4 CONCLUSIONS

The research has shown that N needs can easily be met by growing forage crops as crop-associations of grasses with legumes and/or interpolating legumes into crop rotation as N-fixing intercrops. In the simulated farm conditions there was required relatively small partition of arable area (13%) for N addition through modification of crop rotation or crop association. This was mainly due to the considerable reliance on FYM N and perennial association crops with legumes (pasture and hay sward) and winter pea/wheat arable crop, which together consisted 33% of the whole farm land.

Figure 1. Maize grown in the kura clover understory as a nitrogen self-sufficient crop (provided by the courtesy of dr. Ken Albrecht).

REFERENCES

ASAE 2005. *Manure Production and Characteristics.* St. Joseph: American Society of Agricultural Engineers, USA.

Boller, B.C. & Nosberger, J. 1987. Symbiotically fixed nitrogen from field—grown white and red clover mixed with ryegrass at low levels of ^{15}N-fertilizstion. Plant and Soil 104:219–226.

Bruulsema, T.W. & Christie, B.R. 1987. Nitrogen Contribution to Succeeding Corn from Alfalfa and Red Clover. *Agronomy Journal* 79(1):96–100.

Caporali, F., Campiglia, E., Mancinelli, R., Paolini, R. 2004. Maize Performances as Influenced by Winter Cover Crop Green Manuring. *Italian Journal of Agronomy* 79:96–100.

Chambers, B.J., Lord, E.I., Nicholson, F.A., Smith, K.A. 1999. Predicting nitrogen availability and losses following application of organic manures to arable land: MANNER. *Soil Use and Management* 15: 137–143.

Contreras-Govea, F.E., Albrecht, K.A., Muck, R. 2006. Spring Yield and Silage Characteristics of Kura Clover, Winter Wheat, and in Mixtures. Agronomy Journal 98:781–787.

Cook, J.C., Gallagher, R.S., Kaye, J.P., Lynch, J., Bradley, B. 2010. Optimizing Vetch Nitrogen Production and Corn Nitrogen Accumulation under No-Till Management. Agronom Journal 102(5):1491–1499.

DLG 1997. *Futterwerttabellen Wiederkauer*. Universitat Hohenheim Dokumentationsstelle. Frankfurt am Main: DLG—Verlags GmbH.

Domaćinović, M. 1999. *Livestock Feeding Manual* (in Croatian). University manual. Osijek: University J.J. Strossmayer in Osijek, Faculty of Agriculture in Osijek.

Graham, P.H. & Vance, C.P. 2000. Nitrogen fixation in perspective: an overview of research and extension needs. *Field Crops Research* 65: 93–106.

Hansen, B., Kristensen, E.S., Grant, R., Høgh-Jensen, H., Simmelsgaard, S.E., Olesen, J.E. 2000. Nitrogen leaching from conventional versus organic farming systems—a systems modelling approach. *European Journal of Agronomy* 13(1):65–82.

Kingston, E., Bowersox, V., Zorilla, G. 2000. *Nitrogen in the Nation's Rain*. Champaign: National Atmospheric Deposition Program Office. Illinois State Water Survey, Illinois, USA. NADP Brochure 2000–01c (revised).

Kristensen, I.S., Halberg, N., Nielsen, A.H., Dalgaard, R.L. 2005. N turnover on Danish mixed dairy farms. In: Bos, J., Pflimlin, A., Aarts, F., and Vertes, F. (Eds.). *Nutrient management on farm scale. How to attain policy objectives in regions with intensive dairy farming*. Report of the first workshop of the EGF Working group "Dairy Farming System and environment". Quimper, France, 23–25. June 2005.

McBride, W.D. & Greene, C. 2007. A Comparison of Conventional and Organic Milk Production Systems in the U.S. *American Agricultural Economics Association Annual Meeting*, Portland, Oregon, July 29–August 1, 2007.

Moreira, V.R. & Satter, L.D. 2006. Effect of Scraping Frequency in a Freestall Barn on Volatile Nitrogen-Loss from Dairy Manure. *Journal of Dairy Science* 89:2579–2587.

Ochner, T.E., Albrecht, K.A., Schumacher, T.W., Baker, J.M., Berkevich, R.J. 2010. Water Balance and Nitrate Leaching under Corn in Kura Clover Living Mulch. *Agronomy Journal* 102(4):1169–1178.

Petersen, S.O., Lind, A.M., Sommer, S.G. 1998. Nitrogen and organic matter losses during storage of cattle and pig manure. *Journal of Agricultural Science* 130: 69–79.

Rao, S.C. & Northup, B.K. 2005. Capabilities of Four Novel Warm-Season Legumes in the Southern Great Plains: Biomass and Forage Quality. *Crop Science* 49(3):1096–1102.

Rinnofner, T., Friedel, J.K., Kruijff, R., Pietch, G., Freyer, B. 2008. Effect of catch crops on N dynamics and following crops in organic farming. *Agronomy for Sustainable Development* 28(4):551–558.

Singer, J. & Pedersen, P. 2005. *Legume Living Mulches in Corn and Soybean*. Ames: Iowa State University, University Extension.

Snyder, C.S. 2011. Nitrogen in soil organic matter—How much is released in your field? *Plant Nutrition Today* 6. Norcross: International Plant Nutrition Institute (Georgia, USA).

Stopes, C., Millington, S., Woodward, L. 1996. Dry matter and nitrogen accumulation by three leguminous green manure species and the yield of a following winter wheat crop in an organic production system. *Agriculture, Ecosystems and Environment* 57: 189–196.

Yost, M.A., Coulter, J.A., Russelle, M.P., Sheaffer, C.C., Kaiser, D.E. 2012. Alfalfa Nitrogen Credit to First-Year Corn: Potassium, Regrowth, and Tillage Timing Effects. Agronomy Journal 104:953–962.

Zemenchik, R.A., Albrecht, K.A., Boerboom, C.M., Lauer, J. 2000. Corn Production with Kura Clover as a Living Mulch. Agronomy Journal 92:698–705.

Zemenchik, R.A., Albrecht, K.A., Schultz, M.K. 2001. Nitrogen Replacement Values of Kura Clover and Birdsfoot Trefoil in Mixtures with Cool-Season Grasses. *Agronomy Journal* 93(2):451–458.

Green Design, Materials and Manufacturing Processes – Bártolo et al. (eds)
© 2013 Taylor & Francis Group, London, ISBN 978-1-138-00046-9

Environmental aspects of lightweight construction in mobility and manufacturing

S. Albrecht, M. Baumann, C.P. Brandstetter, R. Horn, H. Krieg, M. Fischer & R. Ilg
Department Life Cycle Engineering (GaBi), Fraunhofer Institute for Building Physics, Germany

ABSTRACT: In automotive, mechanical, medical and plant engineering, and the construction industry, lightweight construction is considered one of the most important future technologies, providing good market opportunities for small and medium-sized companies. Due to the increasing relevance of energy and resource efficiency, not only technical and economic, but also environmental aspects have to be considered when assessing lightweight construction. Only with the consideration of environmental and economic aspects modern technologies can succeed in different fields of application and gain societies acceptance. In the study commissioned by the *e-mobil BW GmbH* in Germany, the topic of sustainability in lightweight construction is analyzed on basis of environment and health (Rommel et al. 2012). This article gives an overview on the most important findings and results of this study with special focus on the Life Cycle Assessment (LCA) of lightweight construction considering the lightweight materials steel, high performance steel, aluminum and carbon fiber composites.

1 BACKGROUND

Lightweight construction is more relevant than ever. Due to the scarcity of resources, these have to be used more efficient. One possibility to increase resource efficiency is lightweight construction—on the one hand by direct material savings, on the other hand by energy savings in the use phase of the products. Lightweight construction is often stated as a key technology and can be understood as "technology instead of renunciation". The aim of lightweight construction is mostly the reduction of moved masses. Hence technical, economic and environmental opportunities arise. This is valid for diverse applications. The reduction of the vehicle weight induces, beside lower fuel consumption, higher acceleration and more transportation charges also more driving pleasure, shorter breaking distances and thus an increase in safety.

In order to apply lightweight technologies environmentally reasonable, one has to assess the whole life cycle. The obtained energy savings during the use phase of the lightweight product for instance, could be offset by an especially energy intensive production phase. Only by considering the full life cycle of products, this shift of burdens from one life cycle stage to another can be avoided. The aim of the study is to provide an independent overview of environmental aspects of lightweight construction. It is analyzed which material is environmentally advisable for which application. At the same time the impacts of lightweight construction

on human health and the environment are considered. Hence, beside the energy and material supplies, also health aspects deriving from the use of lightweight materials are taken into consideration. Thus the study is to provide an orientation, especially for small and medium-sized companies, on the opportunities of lightweight construction and requirements with regard to workplace design and sustainability. It is to facilitate the entry to the new lightweight technologies, especially for small and medium-sized companies, by outlining the complexity of Life Cycle Assessment (LCA) for lightweight construction and by indicating potential challenges in industrial safety and health protection.

This article will give an overview on the most important findings and results of the study named above with special focus on the LCA of lightweight construction.

2 LIFE CYCLE STUDY

2.1 *Goal and scope*

The Life Cycle Assessment study is conducted in accordance with the international standards of environmental Life Cycle Assessment ISO 14040:2006 and ISO 14044:2006. The scope includes the relevant fore- and background systems assuming that they are located in Germany. Within the system boundaries, production, use phase and end of life of the foreground system are modeled—based on the data sets of the database and software GaBi

(2012), which consider the upstream chains such as energy supply in the background system.

The software and database system GaBi (2012) is an engineering tool for Life Cycle Assessment of products and processes along their life cycle used and valued by both industry and science. The reliability of the background data and their industrial relevance has to be emphasized as well. With the operational support of the GaBi software and database, the life cycles of the material types can be environmentally mapped close to reality. As only GaBi was used for modeling, transparency, consistency and timeliness of the utilized data is ensured. A virtual car body part made of different materials and comparable technical properties and function is defined as functional unit, which quantifies the benefit of the investigated product and allows comparing it to other products. Environmental Impacts are analyzed based on taking into account the full life cycle of the car part made of steel, aluminum and Carbon Fiber Reinforced Plastics (CFRP) respectively. In this study, the focus is on the primary energy demand and Global Warming Potential (GWP).

The utilization of steel and aluminum in the mobility sector has been established for decades; hence the availability of data for life cycle analysis is good. Large-scale series of CFRP are currently introduced to the sector. The different requirements for the quality of the carbon fibers due to different technical applications are reflected in the environmental profiles of the production of CFRP. This situation is considered by a specification of bandwidths in the results of the respective impact categories.

2.2 Life cycle inventory—production phase

In automotive applications, steel sheets meeting the requirements of the car industry such as corrosion resistant galvanized steel sheets are used. Preferably used aluminum sheets are of type EN AW-60XX. The carbon fibers are produced on the basis of Polyacrylonitrile (PAN), the matrix is made of epoxy resin. The life cycle modeling is based on data sets that fit these specifications. Depending on the material, different weight saving potentials are possible. Based on a steel structure part (galvanized steel for automotive application) of 10 kg, which is used as reference for the definition of the lightweight potentials, the other material types are dimensioned. Through the use of high performance steel, a weight reduction by 10–20% is achievable. In this study, a weight reduction of 15% is assumed. The assumed weight reduction potential of aluminum is 40% compared to the reference part. Quasi-isotropic CFRP that shows similar mechanical properties for all load

directions is assumed to achieve a weight reduction of 52%. For unidirectional CFRP with optimized mechanical properties for one load direction, a weight reduction potential of 79% is assumed. Based on the used material selection the following weights of the component types are defined as shown in Figure 1.

2.3 Life cycle inventory—use phase

In the study the lightweight parts built of high performance steel, aluminum, quasi-isotropic CFRP and unidirectional CFRP are compared to the reference part built of steel. The differences of fuel consumption are compared to the reference type by the calculation of fuel reduction values according to Koffler & Rohde-Brandenburger (2010). The savings due to secondary effects are considered in these values. These are e.g. the reduction of required driving power to achieve a defined mileage while reducing the vehicle weight. This drive power reduction can be achieved by the adaption of the gear ratio or by the displacement of the engine, as listed in Table 1. The fuel reduction values are calculated based on the New European Driving Cycle (NEDC). The resulting weight-based energy demand is based on calculations of driving physics such as rolling friction and air resistance. To determine the fuel consumption based of occurring energy demands, knowledge of the efficiency of combustion engines is necessary. On the basis of average efficiencies of gasoline-driven naturally aspirated engines and gasoline- and diesel-driven turbocharged engines and with the use of the values of energy demand due to the NEDC, the fuel reduction values are calculated. According to Koffler & Rohde-Brandenburger (2010) the following fuel reduction values (arithmetic medium values) result.

Table 1 shows that the fuel reduction values of gasoline-driven engines are bigger than the values of diesel-driven ones. This can be explained by the higher efficiency of diesel engines.

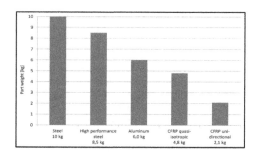

Figure 1. Part weights for reference and lightweight parts.

Table 1. Fuel reduction values according to Koffler & Rohde-Brandenburger (2010).

Fuel type	Adaption	Fuel reduction value FRV
Gasoline	Adapted gear ratio	0.32 ltr/(100 km*100 kg)
Gasoline	Engine displacement	0.39 ltr/(100 km*100 kg)
Diesel	Adapted gear ratio	0.29 ltr/(100 km*100 kg)
Diesel	Engine displacement	0.26 ltr/(100 km*100 kg)

As the values of 0.35 ltr/(100 km*100 kg) for gasoline and 0.28 ltr/(100 km*100 kg) for diesel have been proven as common fuel reduction values, they are also used for this study.

The weight specific reduction of fuel consumption R_w is calculated from the *weight difference* Δw multiplied with the *fuel reduction value FRV*. The *weight difference* is the respective difference between the weight of the lightweight part and the weight of the reference part.

The aluminum part in a gasoline-driven car shows e.g. the following reduction of fuel consumption per 100 km:

$$R_w = \Delta w * FRV$$
$$R_{w,alu} = (6\ \text{kg} - 10\ \text{kg}) * \frac{0.35\ \text{ltr}}{100\ \text{km} * 100\ \text{kg}} * 0.01$$
$$= -\frac{0.014\text{l tr}}{100\ \text{km}} \tag{1}$$

The resulting savings of carbon dioxide result from the chemical composition of gasoline and diesel. The carbon dioxide emissions increase or decrease to the same extends as the fuel consumption.

Based on the consumption values of the lightweight parts resulting from the application of the fuel reduction values, both primary energy demand and global warming potential are determined. Besides the impacts caused by the use of the combustion engines, the results also consider the impacts from the fuel supply. Two mileage scenarios over 150,000 km and 250,000 km are assessed. Both scenarios are based on common assumptions used in the automobile industry. For example, Volkswagen AG (2010) is using a 150,000 km scenario for Life Cycle Assessments while Daimler AG (2012) is expecting a mileage of 250,000 km.

2.4 Life cycle inventory—end of life phase

The production and use phase are followed by the End of Life (EoL), which is the last phase in the life cycle of the components. For the metallic parts, the end of life is represented through so-called recycling potentials. For CFRP thermal recycling is assumed. Recycling potentials have negative environmental impact values, as they reduce the amount of primary materials required, which generally have a higher environmental impact than secondary materials. By conducting thermal recycling, the produced energy substitutes energy generated from other primary resources. The different recycling methods lead to a reduction of environmental impacts, the study focuses on the impacts for global warming potential and primary energy demand.

2.5 Considered life cycle impact categories

The Life Cycle Impact Assessment (LCIA) carried out within the scope of this study focuses on the analysis of the primary energy demand (primary energy demand of lower calorific value, hereinafter primary energy demand), which occurs during the components life cycles, and the environmental impact category "Global Warming Potential" (GWP).

3 LIFE CYCLE ASSESSMENT RESULTS

3.1 Text and indenting

By adding up all life cycle phases, the life cycle contributions of the different structural components for the whole life cycle are assessed. The results of the analysis are shown in Figures 2 and 3 for primary energy demand and global warming potential respectively. The major differences of the life cycle phase's shares are visualized.

The impacts of the conventional steel part are dominated by its use phase due to the higher specific weight and the resulting higher specific fuel consumption. The production phase only plays a minor role. Through the use of the high performance steel component and the associated weight reduction, the impacts during use phase can be

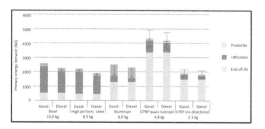

Figure 2. Primary energy demand of entire life cycle (150,000 km).

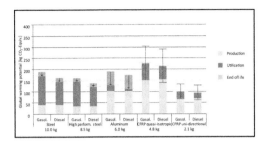

Figure 3. Global warming potential of entire life cycle (150,000 km).

contributions for reduction of the environmental impacts at the end of life, positive contributions for increase of the environmental impacts at the end of life). The presented examples clearly illustrate that the underlying assumptions of the study are of significant importance. If the mileage is increased from 150,000 km to 250,000 km, the relevance of the use phase is increasing, which might change the environmental balance of the assessed parts. In this case, the utilization of the aluminum component compared to the steel component would be already advantageous without consideration of the recycling. This advantage becomes even bigger when taking into account the recycling potential.

For the environmental comparison of components made of different materials, a break-even analysis and depiction of the parts in relation to the mileage, as can be seen in Figures 4 and 5, is useful. In this case, the break-even point describes the mileage at which the higher impact of a components production phase is amortized compared to the reference part of conventional steel.

In this example the break-even point for aluminum is reached at about 137,000 km compared to the conventional component. High performance steel parts have always a lower impact than conventional steel and are therefore preferable.

reduced significantly. The additional impacts during the construction phase are only of small importance. The credits resulting from the recycling have no significant effect on the complete life cycle of both the conventional steel and high performance steel component.

For the aluminum part, a significant shift between the several life cycle phases is already visible. The weight savings lower the contribution of the use phase compared to the steel parts. At the same time an increase of the contribution of the construction phase is stated, so the relevance of this phase increases. When taking into account the recycling potential, the use of the aluminum part can lead to a decreased primary energy demand and global warming potential.

For the CFRP components significantly higher impacts of the production phase are visible. As described in chapter 2.2, the contributions of the production phase of CFRP components show a significant bandwidth, which is shown in the figure through the error bars. Concurrently, the utilization of CFRP promises the biggest impact reduction potentials. This reduction has a beneficial effect on the fuel consumption and therefore on the impacts of the use phase. Thus, the CFRP component types offer the highest fuel saving potentials. Furthermore, the example of CFRP components clearly shows that not only the choice of the material, but also the production technology (quasi-isotropic or unidirectional) and thereby the utilization of the type of material is of great importance for the environmental impacts of products and processes.

Since the advantages of CFRP components are the highest weight reduction and the associated reduction of fuel consumption compared to the other assessed parts, the saving potential is further increased by higher mileages. The recycling potentials imply potential credits at the end of life of the product and are therefore shown separately from the impacts of production and use phase (negative

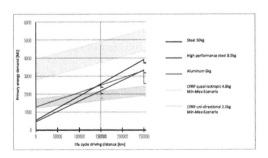

Figure 4. Primary energy demand related to life cycle driving distance (250,000 km).

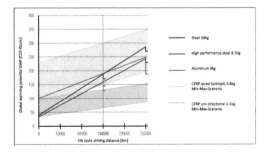

Figure 5. Global warming potential related to life cycle driving distance (250,000 km).

188

When comparing high performance steel parts with aluminum parts, the primary energy demand breaks even after around 240,000 km. However, when taking into account the recycling potential of both materials, aluminum has a better overall performance. The CFRP components are visualized by using bandwidths. These bandwidths include both a low case and a high case scenario resulting from the fiber production and the occurring specific material losses. Regarding the primary energy demands the relatively small bandwidth compared to the one of the global warming potential is notable. This difference results from the energy demand within the production process of the fibers. Concerning power generation the primary energy demand is only influenced little due to the fact that it includes both regenerative and non-regenerative primary energy. At an increasing percentage of renewable energies in power generation, for instance, the primary energy demand is decreasing less than the global warming potential, as non-regenerative primary energy is replaced by regenerative primary energy. Thus the relative bandwidth of the primary energy demand is smaller than the one of the global warming potential. Based on the described bandwidth differences the break-even point also depends on the considered environmental impact (primary energy demand or global warming potential).

4 SUMMARY AND CONCLUSIONS

A general statement about the advantages of lightweight materials compared to conventional components is not reasonable. The environmental profile of the application of lightweight materials in vehicles is strongly dependent on the prevailing circumstances such as field of application, vehicle category, specific fuel consumption and mileage. As all materials have advantages and drawbacks, concrete statements can only be made for specific applications. However, by analyzing the design options and material types in an early stage of production, suitable solutions can be developed and additional value and the full potential of lightweight technologies can be exploited.

Lightweight construction will make major contribution to the future of mobility through new materials, new production technologies and new products. Environmental aspects will play an important role; even more important than they already do today.

Using LCA from "cradle to grave", the optimal processes, materials and methods can be determined. Moreover, the variety of different parameters plays a decisive role for the selection of materials.

In the production phase a major part of the applied primary energy is used for the production of raw materials. Carbon fibers herein are the most energy intensive materials. Big saving potentials consist of the utilization of less energy intensive precursors, such as plastics based on renewable raw materials. Additionally subsequent improved production processes promise optimization potentials.

Regarding metallic lightweight materials, the production of primary aluminum needs more energy than the production of steel. All basic production processes are already technologically advanced and only promise a small environmental benefit. Big improvements will occur in the further optimization of lightweight construction quality through further materials research and improved material properties.

CFRP offers the highest weight reduction, followed by aluminum and high performance steel. The amortization of the higher energy demand of the production phase during the use phase depends on several factors such as component function or useful life.

The recycling processes of steel and aluminum are technically advanced and offer a functioning infrastructure—at least in developed countries. Steel can be recycled without quality decreasing effects, and about 75% of all aluminum ever produced is still in use. CFRP is currently mainly thermally recycled. There are several approaches for the recycling of CFRP and even first facilities on industrial scale, but the recycling involves down cycling, which means a decrease of the quality of the recycled materials. As recycling processes for CFRP are still being developed and improved, the efficiency level is likely to increase in the future, which will then reduce the overall life cycle impacts. The further development of CFRP recycling processes is an important requirement for the utilization on the mass marked and therefore should be intensified.

REFERENCES

Daimler AG 2011. Environmental Certificate for the new B-Class. Life Cycle. Publisher: Daimler AG Mercedes-Benz Cars. Stuttgart, Germany.
GaBi 2012. PE: GaBi5 Software-System and Databases for Life Cycle Engineering. Copyright, TM. Stuttgart, Echterdingen 1992–2012, Germany.
ISO 14040:2006 Environmental Management—Life Cycle Assessment—Principles and Framework, 2006; German and English type DIN EN ISO 14040:2006.
ISO 14044:2006 Environmental management—Life cycle assessment—Requirements and guidelines 2006; German and English type DIN EN ISO 14044:2006.

Koffler, C. & Rohde-Brandenburger, K. 2010. *On the calculation of fuel savings through lightweight design in automotive life cycle assessments.* Int J Life Cycle Assess (2010) 15:128–135. DOI:10.1007/s11367-009-0127-z.

Rommel, S.; Geiger, R.; Schneider, R.; Baumann, M.; Brandstetter, C.P.; Albrecht, S.; Held, M.; Creutzenberg, O.; Dasenbrock, C. 2012. *Leichtbau in Mobilität und Fertigung. Ökologische Aspekte.* e-mobil BW, Landesagentur für Elektromobilität und Brennstoffzellentechnologie Baden Württemberg, Stuttgart, Germany; Fraunhofer Institute for Manufacturing Engineering and Automation IPA, Stuttgart; Fraunhofer Institute for Building Physics IBP, Stuttgart; Fraunhofer Institute for Toxicology and Experimental Medicine, Hannover.

Volkswagen AG 2012. The Golf. Environmental Commendation—Data Sheet. Group Research Environment Affairs Product. Wolfsburg, Germany.

Green Design, Materials and Manufacturing Processes – Bártolo et al. (eds)
© *2013 Taylor & Francis Group, London, ISBN 978-1-138-00046-9*

A step towards sustainable machining through increasing the cutting tool utilization

F. Schultheiss, J.M. Zhou, E. Gröntoft & J.-E. Ståhl
Lund University, Division of Production and Materials Engineering, Lund, Sweden

ABSTRACT: Machining is an important and widely spread manufacturing method which should not be neglected while striving towards sustainable production. Several previous articles have discussed sustainable machining and different approaches towards reaching this goal. However, few have discussed the influence of the cutting tool utilization on the sustainability during machining. This article presents a novel approach of increasing the tool utilization during both turning and milling operations. Experimental results show that by utilizing the cutting tools more efficiently it is possible to increase the tool life by up to 100% depending on machining scenario. Thus, a significant step towards sustainable production could be made.

1 INTRODUCTION

1.1 Sustainable machining

Machining is an important and widely used manufacturing method. It has been estimated that approximately 80% of all products have at one stage or another been machined. Thus reaching sustainable machining is an important step towards achieving sustainable production and in the long run sustainable development. It is however important to stress that sustainable development can only be achieved in combination with technological development in order to become an integral part of the production process.

There exist several different approaches towards increasing the sustainability during machining. For example Pusavec et al. (2010) suggested several different possible approaches such as minimizing waste and improving the use of cutting fluid. Often the use of cutting fluid is seen as the greatest environmental threat during machining, especially since used cutting fluid is considered as hazardous waste. Dahamus & Gutowski (2004) have previously stated that the direct environmental influence of tooling is limited. However, the cutting tool utilization must also be optimized in order to achieve a truly sustainable machining process.

Traditionally machining operations are optimized in order to minimize the manufacturing cost as well as in some cases to comply with technical limitations as previously reported by several different authors (Hägglund, 2002; Kalpakjian & Schimd, 2010; Hinduja & Sandiford, 2004). The obtained results in these cases may in correspond to the most sustainable alternative, however this is far from certain. A possible approach to further consider the environmental aspects were presented by Rajemi et al. (2010). They proposed new model which took the process energy consumption into consideration and thus was better suited to optimize the production process from a sustainable perspective.

1.2 Recycling cemented carbide cutting tools

Coated cemented carbide inserts are manufactured by using powder metallurgy. Tungsten which is an important part of the insert is threated chemically through several different steps before undergoing a process known as carburization and thus forming tungsten carbide (Ståhl, 2012). The energy consumption for producing tungsten carbide powder is approximately 12 kWh/kg if the tungsten carbide is produced from ore concentrates (Bhosale et al. 1990). The tungsten powder is then mixed with other powders and threated through several powder metallurgical steps including pressing sintering and coating to form cemented carbide inserts.

Today the estimated reserves of tungsten are approximately 3 million tones, about half of which may be found in China (Seco Tools, 2010). A previous study has shown that between 1955 and 1991 approximately 60% of the input tungsten was lost (Kieffer & Lassner, 1994). It has been estimated that the tungsten resources will be depleted within 40–100 years and the demands are still rising. Through recycling the cemented carbide scarp it has been estimated that it is possible to prolong the time before the tungsten reserves are depleted with approximately 35% at the same time as the

CO_2 emissions could be reduced by roughly 40% (Seco Tools, 2010).

There exists a wide range of different approaches for recycling cemented carbide inserts (Smith, 1994). However, the two most common are chemical reprocessing which is used for approximately 35% of the cemented carbide scarp and the Zn-method which is used for approximately 25% of the carbide scrap. During chemical reprocessing one of several chemical processes may be used in order to separate the tungsten carbide from the matrix. The Zn-method is instead based on the treatment of the cemented carbide scarp with molten zinc which creates an alloy with the cobalt binder phase resulting in an increase in volume and thus shattering the carbide structure. Depending on the size of the scrap the Zn-method consumes approximately 2–4 kWh/kg of product (Bhosale et al. 1990; Kieffer & Lassner, 1994). For both recycling methods the end result is tungsten carbide powder which may once again be used for manufacturing cemented carbide inserts. However, the cost of the Zn-method is approximately 20–35% lower than for chemical reprocessing depending on circumstances.

2 INCREASING THE TOOL UTILIZATION

During conventional milling- and turning operations it is common practice to use the cutting tools in such a way that the major cutting edge is worn out while the wear on the minor cutting edge is minor. Thus there exist a great potential to increase the cutting tool utilization. By using the minor cutting edge as a "new", slightly worn major cutting edge the tool utilization could be significantly increased. The practical methods of achieving this differs slightly between the milling and turning cases which is further discussed in the following subsections.

Products based on these principles are commercially available (Larssons i Bjärred Mekaniska Verkstad AB, 2009). However, comparatively little scientific research has been published on the effect on the machining process and in particular the sustainability of the machining process.

2.1 Proposed method for milling

Common practice is to only rotate the milling cutter in only one direction. This implies that a majority of the obtained tool wear will appear on the major cutting edge. At the same time the wear on the minor cutting edge is relatively limited, in some cases on the verge of nonexistent. If however the rotational direction is reversed the major and minor cutting edges would swap places. Thus a "new" major cutting edge may be used for

an additional machining operation. A condition is of course that an appropriate milling head is used for both operations. Figure 1 briefly illustrates the proposed method.

It is important to note that use of milling cutters with indexable inserts is essential for implementing the proposed method.

2.2 Proposed method for turning

A similar effect as for the milling case may be obtained by changing the feed direction during longitudinal turning, Figure 2. During the first

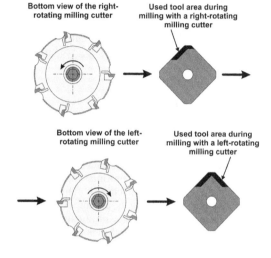

Figure 1. Illustration of the proposed method for milling.

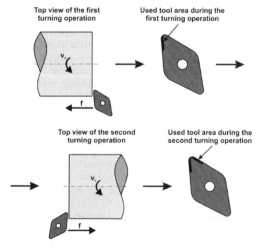

Figure 2. Illustration of the proposed method for turning.

operation the obtained tool wear will primarily appear on the major cutting edge. If the feed direction then is reversed the major and minor cutting edges will swap places and thus the cutting tool utilization could be increased. This is however under the condition that appropriate tool holders are available. In addition it is essential that the workpiece must have a suitable geometry before implementing this method.

3 EXPERIMENTAL VALIDATION

Experimental studies were performed in order to validate the feasibility of the proposed method for both the milling and turning scenario. During all machining experiments the tool wear as well as the obtained surface roughness was measured. All experiments were performed without using any cutting fluid. The reason for this was that used cutting fluid is hazardous for the environment. Thus, if the proposed methods could be performed during dry cutting conditions the advantage in terms of sustainability would be much larger. Only coated cemented carbide cutting tools were used during all experiments primarily since these are generally the preferred choice for the materials being machined according to tool manufacturers.

3.1 Milling experiments

The milling experiments were performed by face milling duplex stainless steel SAF 2304 with coated cemented carbide tools. It was attempted to position the inserts in the best way possible to minimize the variations in relative position to each other which is a common problem during milling operations. However, when analyzing the results this variation should be remembered as it may influence the accuracy of the obtained results.

3.1.1 Experimental setup

During all experiments SEEX09T3AFTN coated cemented carbide inserts were used in an R220.53-0100-09-7A and L220.53-0100-09-7A milling head respectively. Figure 3 shows a view of the rake face of an unworn insert. Note how the major cutting edge changes when varying the rotational direction of the milling cutter. The width of the wiper edge found as illustrated in the figure is approximately 1.5 mm. The milling experiments were performed in a SAJO HMC-40, 4-axis milling machine.

The experiments were performed by initially using the right-rotating milling head with all 7 inserts. Then, when the tool wear approached the flank wear criteria $VB = 300$ μm on the major cutting edge, the inserts where shifted into the

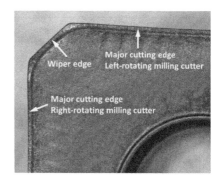

Figure 3. Rake face of the cutting tool during milling.

left-rotating milling head. Then the milling process was repeated. During all experiments the cutting data remained constant with the cutting speed $v_c = 80$ m/min, depth of cut $a_p = 2$ mm and the feed per tooth $f_z = 0.15$ mm/tooth.

3.1.2 Obtained results

Tool wear obtained on the major cutting edge was not only in the form of flank wear during these milling experiments. In addition chipping of the major cutting edge was observed, in some cases significant. In relation to the present study it was however of interest to note that the tool wear on the wiper edge was primarily limited to flank wear. This indicates that any negative effect on the obtained surface roughness should be minimal. It could thus be speculated that the proposed method is better suited for face milling operations than for other milling operations. This since any finished surface machined by the major cutting edge could be thought to have a significantly worse surface roughness.

The average obtained flank wear when machining using a right- and left-rotating milling cutter respectively may be found in Figure 4. As may be seen in Figure 4, the tool wear as a function of machining time on the major cutting edge is almost identical for the right-rotating milling head as for the following left-rotating milling head.

Flank wear on the wiper face was also measured as a function of the machining time, Figure 5. Note that the flank wear on the wiper edge is significantly smaller than on the major cutting edge.

Initially it was thought that the reuse of cutting tools would have a negative influence on the obtained workpiece surface roughness. To investigate this assumption the obtained surface roughness was measured during all milling experiments, Figure 6. Some variations in the surface roughness as a function of the machining time and rotational

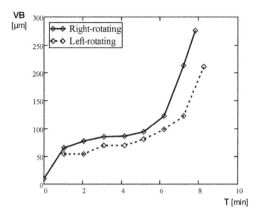

Figure 4. Flank wear VB on the major cutting edge as a function of the machining time T as obtained for both rotational directions.

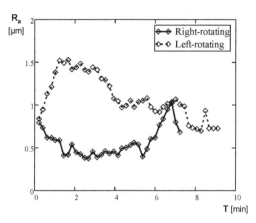

Figure 6. Obtained surface roughness as a function of the machining time T for the right- and left-rotating milling direction.

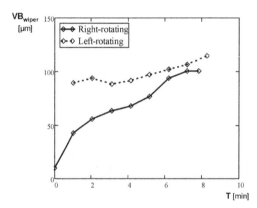

Figure 5. Flank wear on the wiper edge, VB_{wiper}, as a function of the machining time T as obtained for both rotational directions.

direction may be discerned. However, the surface roughness appears to reach its peek value at the end of the right-rotating experiments and at the beginning of the left-rotating experiments. In addition, it of interest to observe that the obtained surface roughness at the end of the left-rotating test cycle is almost identical to that obtained while using a new cutting tool in a right-rotating milling cutter.

3.2 Turning experiments

The turning experiments were performed by longitudinal turning bars of AISI 4340. During the initial operation the feed direction was towards the chuck of the lathe (referred to as "left feed direction"). Later, when the cutting tool had sustained a sufficient amount of flank wear the feed

direction was reversed (referred to as "right feed direction"). To be able to achieve this secondary operation several groves were machined close to the chuck in order to allow space for the cutting tool.

3.2.1 Experimental setup

All turning experiments were performed in a round bar of AISI 4340 with an initial diameter of 168 mm and a length of 960 mm. Due to the length of the workpiece a center hole and tailstock was used during all turning experiments. Throughout the whole set of experiments commercially available CNMG120412 coated cemented carbide inserts were used placed in a DCLNL3225P12 and DCLNR3225P12 tool holder respectively. The turning experiments were performed in a SMT Sajo Swedturn 500 lathe and care was taken to minimize any possible vibrations during the machining process. For all turning experiments the depth of cut a_p remained constant at $a_p = 2.5$ mm while varying the feed f and the cutting speed v_c according to Table 1.

3.2.2 Obtained results

For both feed directions primarily flank wear of the cutting tool was observed. No significant change of the wear characteristics was noted when switching from one to the other feed direction. It was found that the main difference in the tool wear between the two operations is that the secondary operation results in an increased flank wear around the whole tool nose radius. However, no indications were observed that this effect could lead to premature tool failure.

The flank wear VB as a function of the machining time T was measured for each of the 5 turning

Table 1. Cutting data used during the turning experiments.

Test	Cutting speed (m/min)	Feed (mm/rev)	Depth of cut (mm)
1	200	0.25	2.5
2	260	0.25	2.5
3	170	0.30	2.5
4	270	0.30	2.5
5	220	0.40	2.5

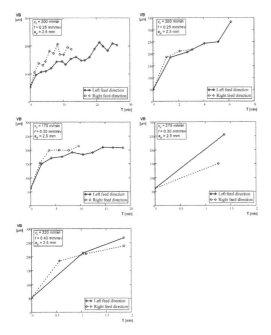

Figure 7. Flank wear VB as a function of the machining time T for each of the 5 turning cases.

cases, Figure 7. As may be seen in the figure when changing from the left to the right feed direction a slightly larger flank wear is obtained for most cases. The opposite is however also true for some cases. In some cases the results could be interpreted as negative tool wear which of course is impossible. Analogous to the milling case this should instead be seen as a result of measuring errors.

Only minor variations were observed between the surface roughness for the left- and right feed direction respectively. Thus, the obtained results indicate that the proposed method does not have any major negative influence on the product quality from this perspective.

4 CONCLUSIONS

Even though the experimental results are limited they appear to strengthen the hypothesis that cutting tools may be reused during both milling and turning operations. Thus a significant increase of the tool life could be obtained as compared to conventional machining operations. The proposed methods have a potential to significantly decreasing the amount of resources and energy needed for producing new cutting tools per product produced. The experimental results prove that it is possible to obtain a tool life up to twice that of an equivalent conventional machining operation depending on circumstances. Further, the influence on the obtained surface roughness was only minor and thus implying that use of the proposed method could be presumed as not having any significant adverse effect on the product quality. However, due to the increased risk of premature tool failure the authors' opinion is that the proposed methods should only be used for rough and semi-finish machining as well as in some cases for finishing of non-critical surfaces.

ACKNOWLEDGMENT

This research is a part of the ShortCut research project financed by the Swedish Foundation for Strategic Research SSF. It is also a part of the strategic research program the Sustainable Production Initiative SPI, a cooperation between Lund University and Chalmers University of Technology. The authors would also like to thank Seco Tools as well as Larssons i Bjärred Mekaniska Verkstad AB for their assistance during this study.

REFERENCES

Bhosale, S.N., Mookherjee, S. & Pardeshi, R.M. 1990. *High Temperature Materials and Processes* 9(2–4): 147–162.

Dahmus, J.B. & Gutowski, T.G. 2004. An environmental analysis of machining. *ASME International Mechanical Engineering Congress and R&D Expo*, Anaheim, USA.

Hägglund, S. 2002. *Global optimization of cutting processes.* Department of Product and Production Development, Chalmers University of Technology, Gothenburg, Sweden.

Hinduja, S. & Sandiford, D. 2004. An optimum two-tool solution for milling 2½D features from technological and geometric viewpoints. *CIRP Annals—Manufacturing Technology* 53(1): 77–80.

Kalpakjian, S. & Schmid, S.R. 2010. Manufacturing Engineering and Technology—Sixth Edition in SI Units. Prentice Hall, Singapore.

Kieffer, B.F. & Lassner, E. 1994. Tungsten Recycling in Todays Environment, *Berg- und Hüttenmännische Monatshefte*, Hefte 9 139:340–345.

Larssons i Bjärred Mekaniska Verkstad AB 2009. *Ecogreenmill™*. Larssons i Bjärred Mekaniska Verkstad AB, Bjärred, Sweden.

Pusavec, F., Krajnik, P. & Kopac, J. 2010. Transitioning to sustainable production—Part I: application on machining technologies. *Journal of Cleaner Production* 18(2): 174–184.

Rajemi, M.F., Mativenga, P.T. & Aramcharoen, A. 2010. Sustainable machining: selection of optimum turning conditions based on minimum energy considerations. *Journal of Cleaner Production* 18(10–11): 1059–1065.

Seco Tools 2010. *Recycling used cemented carbide products*. Seco Tools Inc., Troy, MI, USA.

Smith, G.R. 1994. Materials Flow of Tungsten in the United States. *Bureau of Mines Information Circular 9388*, United States Department of the Interior, USA.

Ståhl, J.-E. 2012. *Metal Cutting—Theories and models*. Division of Production and Materials Engineering, Lund University in cooperation with Seco Tools, Lund/Fagersta, Sweden.

Green Design, Materials and Manufacturing Processes – Bártolo et al. (eds)
© 2013 Taylor & Francis Group, London, ISBN 978-1-138-00046-9

Natural aspect of sustainable food and cosmetics manufacturing

K. Kyriakopoulou, S. Papadaki & M. Krokida
School of Chemical Engineering, National Technical University of Athens, Greece

ABSTRACT: In this study, extraction processes for the acquisition of extracts from natural sources are evaluated. These extracts are considered as ingredients for the production of improved food and cosmetics products. Innovative extraction methods such as ultrasound and microwave assisted liquid extraction using green solvent systems will be evaluated for the isolation of the desired natural compounds. A life cycle assessment was carried out to evaluate the sustainability of these extraction methods. Taking into consideration that the extracts have multifunctional properties due to their richness in both antioxidants and colorants, they are considered to be good replacement for the corresponding synthetic ones. This review targets to suggest effective alternative solutions to food and cosmetics industries, in order to minimize their environmental impact over the product's lifecycle, improving their sustainability and viability, through the use of renewable natural raw materials and non-polluting processes.

1 INTRODUCTION

Nowadays the food and cosmetic industries are using numerous synthetic ingredients of petrochemical origin with potentially dangerous health and environmental implications. These ingredients aim to improve the desired organoleptic characteristics of final products such as color, texture, flavor, taste and shelf life. Some widely used synthetic ingredients in food and cosmetic products are: a) synthetic antioxidants, such as Butylated Hydroxytoluene (BHT) or Butylated Hydroxyanisole (BHA), which are suspected carcinogens (Namiki 1990) and b) artificial pigments such as Brilliant blue FCF (E133), Tartrazine (E102), Green S (E142), Sunset Yellow FCF (E110), Erythrosine (E127) and Allura Red (E129) especially in food products that refer to kids being suspected as potential factors for children behavioral disorder, allergies and cancer (Socaciu 2008, Gultekin & Dofuc 2012, Weiss 2012, Nigg et al. 2012, Randhawa et al. 2009).

The increasing consumers' awareness about the mal effect of the synthetic ingredients on human health has led to augmented demand for natural and safe products. Moreover, the European legislation regarding the product ingredients (especially in food industry Regulation EC 1333/2008) and the environmental aspects of the production process (i.e. Integrated Pollution Prevention and Control Directive 96/61/EEC (IPPC), the regulations on Eco-management and Audit Scheme EMAS 1836/93/EC and on eco-label 880/92/EEC) are getting severer and stricter. This situation affect drastically the profitability of the food and cosmetic industrial sectors, whose products are essential and

used extensively, having direct effects on human health. Therefore, the pre-mentioned industries, in order to ensure their competiveness, viability and sustainability, should turn to natural ingredients and green production processes, minimizing the use of synthetic raw materials, the energy consumption and the environmental pollution.

In this study, multifunctional extracts from carrots have been suggested as ingredients for the production of improved food and cosmetics products. Carrots are a rich source of carotenoids, with most important and valuable the β-carotene. B-carotene is widely used in the food, cosmetics and pharmaceutical industry as coloring, antioxidant and anti-inflammatory agent. With the development of new technologies the range of shades from β-carotene available to food manufacturers has increased and now includes yellow, orange as well as red shades. A very fine dispersion of β-carotene can result in a yellow color and a high coloring strength. As the particle size increases the color solution becomes redder. Therefore, this natural colorant can replace harmful synthetic colorants such as Tartrazine (E102), Sunset Yellow FCF (E110), Erythrosine (E127) and Allura Red (E129) (Food Standards Agency 2011, Mortensen 2006). Moreover, the significant antioxidant activity that beta-carotene presents (range) can sufficiently compete and replace synthetic antioxidants such as BHA, BHT. According to Muller et al. 2011, a weighted average, on the basis of the results obtained in different assays, summarizes the potential of the carotenoids in reducing metal ions (FRAP) or synthetic radical dyes (aTEAC) and peroxyl radicals (LPSC) into a comprehensive value. Most of carotenoids (especially lycopene, α- and β-carotene) showed a greater antioxidant activity than a-tocopherol, BHA and BHT, due to the high peroxyl scavenging activity of carotenoids (Muller et al. 2011).

For effective utilization of β-carotene, the nutrient should first be extracted from the carrots. As with all the carotenoids, β-carotene is oil soluble with water soluble forms being produced by emulsification (Mortensen, 2006). Conventional solvent extractions have been widely used to extract β-carotene from natural sources. These extraction techniques are considered simple and relatively inexpensive but required great extraction time, which may lead to degradation of active components. The excessive use of toxic solvent is also considered as extremely aggravating factors for the human health and the environment. New innovative, fast and greener technique for the extraction carotenoids has been developed using microwaves and ultrasounds. Ultrasound assisted extraction of antioxidants has been widely applied due to its high efficiency and extraction rate, while microwave assisted extraction is considered as an alternative extraction method for releasing the bioactive compounds from the sample matrix into the solvent in significantly short time.

In the frames of this work, a comparative study between conventional and innovative green extraction methods has been conducted. The isolation of the desired natural compounds such as colorants and antioxidants as well as the environmental impact of each extraction method has been the main factors for evaluating the different processes. A comparative Life Cycle Analysis (LCA) was carried out, using proper databases and software, in order to evaluate the selected extraction processes sustainability.

2 MATERIALS AND METHODS

Different extraction methods such as the Conventional Solvent Extraction (CSE), the Soxhlet Extraction (SE), the Microwave Assisted Extraction (MAE) and the Ultrasound Assisted Extraction (UAE) have been considered for the extraction of β-carotene. The concentration of β-carotene in each extract has been evaluated since it is consider a natural colorant and antioxidant ingredient in food and cosmetics products. All the extraction conditions have been estimated according to previous studies (Hiranvarachat et al. 2013; Li et al. 2013, Eskilsson & Bjorklung, 2000).

2.1 Samples and reagents

In the frames of this study, air dried carrots of the genus *Daucus carota* were used for the extraction of β-carotene. Dried fresh carrots were shredded and mechanically ground into fine powder before each experiment.

The used solvents were organic solvents of analytical grade, hexane, acetone and ethanol. The alternative green solvent that was used in UAE extraction was rapeseed oil.

2.2 Extraction experiments

2.2.1 Conventional solvent extraction (CSE)

100 g of dried carrot powder was poured into 0.5 L hexane in a conical flask and mechanically stirred using a Heidolph Mechanical OverHead Stirrer (1000 rpm, 20 W) at room temperature for one hour. The mixture was filtered via vacuum through glass microfiber paper, while the solid residue was collected and re-extracted with fresh hexane under the same conditions.

2.2.2 Soxhlet Extraction (SE)

A laboratory scale soxhlet apparatus was used for the extraction of 100 g of dried carrot powder with 1.5 L of mixed solvent consisted of 50% (v/v) hexane, 25% (v/v) acetone and 25% (v/v) ethanol. The cup was heated to the temperature of 58 °C using a heating mantel (280 W). To maximize the extraction yield the experiments conducted for up to 6 h.

2.2.3 Microwave Assisted Extraction (MAE)

100 g of dried carrot powder was extracted using 250 mL of a mixed solvent (MAE 1), which is consisted of 50% (v/v) hexane, 25% (v/v) acetone and 25% (v/v) ethanol, or hexane only (MAE 2). The mixed solvent system represents a way to prevent degradation of heat-sensitive components and to solubilize non-polar compounds in carrots (Eskilsson and Bjorklund, 2000). A Milestone Start D Microwave Digestion System set at 300 W in order to achieve stable 60°C, was used during the microwave assisted extraction. The extraction time was 5 min.

2.2.4 Ultrasound Assisted Extraction (UAE)

UAE was performed in one-liter toughened glass tank with a high-power ultrasonic processor (Hielscher Ultrasound Technology, Germany), equipped with one powerful ultrasonic transducer (20 kHz, 1000 W) and a versatile power meter which can change all important parameters during ultrasonic processing. The double-layered reaction tank allowed water to circulate controlling the temperature in 40°C. The applied ultrasonic intensity was considered at 22.5 W cm^{-2}. Two extractions have taken place with different solvent systems. In both cases 100 g of dried carrot powder were extracted using 0.5 L of solvent. In the first case the solvent was hexane (UAE 1) while in the second one rapeseed oil was used as a greener replacement of hexane (UAE 2).

2.3 *Life Cycle Assessment (LCA)*

According to International Standards Organization (ISO) 14000series, the technical framework for LCA methodology consists of four phases: (1) goal and scope definition; (2) inventory analysis; (3) impact assessment; and (4) interpretation (ISO, 2006a). Defining the goal and scope involves defining purpose, audiences and system boundaries. The life cycle inventory involves collecting data for each unit process regarding all relevant inputs and outputs of energy and mass flow, as well as data on emissions to air, water and soil. The life cycle impact assessment phase evaluates potential environmental impacts.

2.3.1 *Goal and scope definition*
The goals of this LCA study are: to analyze the different extraction processes that comprise production of multipurpose natural extracts; determine the flow of matter and energy as well as emissions that occur over these processes; and establish and quantify the environmental impacts caused by different emissions throughout the process. In this study we have established that LCA involves the extraction of dried carrots in order to receive natural compounds with great functionality for consumption in the food and cosmetic industries.

2.3.1.1 Functional unit
The functional unit is used to define what the LCA is measuring, and provides a reference to which the inputs and outputs can be related. A direct comparison of single materials is usually not in accordance with ISO/EN14040 as the properties of the materials are varying and therefore an unambiguous definition of their common function might not be possible. To remove performance variation and provide a fair comparison between the extracts, the study defined the functional unit as 100 g of dried carrots extracted to provide the desired natural colorants and antioxidants.

2.3.1.2 Product systems and system boundaries
The systems investigated consist of different production stages, depending on the extraction process used. The impact of drying pretreatment stage is considered to be equal in all studied cases since the sample provided were pretreated under the same conditions. The boundaries of these systems included all major material and energy flows associated with the raw materials. Provision of farm, buildings and pretreatment equipment was excluded from this analysis due to the assumed low attribution of these elements. Provision of the extraction machinery was included however, since this input can contribute substantially to total energy consumption of the extracts. The energy consumption of each component employed in the extraction process (oven, pump, freezer, freeze dryer, rotavapor) was calculated based on their specification (for commercial equipment) and uptime.

2.3.2 *Inventory analysis*
Inventory analysis was carried out according to ISO 14044 (ISO, 2006a,b). The key inventory data along with the database sources Ecoinvent 2.0, ELCD database 2.0 and LCA Food DK. for the three extraction processes at pilot-scale are shown in Table 1.

2.3.3 *Impact assessment*
The life cycle environmental impacts associated with natural extract production were quantified using the problem-oriented approach, CML 2 baseline 2000 v2.04 (Center for Environmental Studies, University of Leiden) method. The impact categories included in this analysis were: Ozone layer depletion (ODP), Human toxicity, Fresh water aquatic ecotoxicity, Marine aquatic ecotoxicity, Terrestrial ecotoxicity, Photochemical oxidation, Global warming (indicator of Carbon footprint), Acidification, Abiotic depletion and Eutrophication. The SimaPro 7.1 software developed by PRé was used to perform LCA calculations.

2.3.4 *Interpretation*
The results of the impact assessment are interpreted based on the need for comparison of the

Table 1. Key inventory data for the extraction of 100 g dried carrots by CSE, SE, MAE and UAE processes.

Data inputs	CSE	SE	MAE 1	MAE 2	UAE 1	UAE 2	Data source
Dried carrots (g)	500	500	500	500	500	500	Ecoinvent & LCA Food DK
Hexane (g)	700	525	140	300	350	–	Ecoinvent
Acetone (g)	–	300	80	–	–	–	Ecoinvent
Ethanol (g)	–	300	80	–	–	–	Ecoinvent
Rapeseed oil (g)	–	–	–	–	–	450	Ecoinvent
Electricity (Wh)	20	360	25	25	350	350	Ecoinvent

different production systems and identification of 'hot spots' and environmentally impacting processes for future improvements; and conclusion and recommendation are drawn from them.

3 RESULTS AND DISCUSSION

In Table 2 and Figure 1 the results of the comparative life cycle analysis of the different carotenoids' extraction techniques are presented. In the frames of this analysis important environmental impact indicators were taken into consideration, with most important ones the Global warming, the Ozone layer depletion and Human toxicity. Regarding these three indicators SE is considered the worst tech nique, since it is a time and energy consuming method, which uses great quantities of organic solvents that may have adverse effects on humans and the environment.

On the other hand, the technique which shows the best carbon footprint is the UAE 2 (Table 2). This innovative green UAE technique has brought

benefits correspond to the principles of green extraction, which summarized as renewable plant resources (carrots), a petroleum-alternative solvent, energy, time and cost savings. Taking under consideration the overall impact of the examined extraction techniques, the most environmental friendly is MAE 1 technique, being the least time consuming as well as economic, effective, simple and consistent, followed by MAE 2. CSE and UAE 1 are two mild techniques with almost equal overall environmental impact, as long as the CSE uses twice the size of solvent amount than UAE 1 and 15 times less energy.

Moreover, according to previous studies it has been observed that UAE 2 of carrots could reach to its highest β-carotene yield (167 mg/100 g d.b.) just in 20 min while CSE using hexane as solvent has shown similar β-carotene concentration of 160 mg/100 g d.b. after one-hour of extraction. These results demonstrated the potential of using seed oils as alternative solvents for the extraction of carotenoids from dried carrots. The MAE at optimum condition can yield up to 24 mg/100 g

Table 2. Comparative life cycle analysis.

Impact category	Unit	CSE	MAE 1	MAE 2	SE	UAE 1	UAE 2
Abiotic depletion	kg Sb eq	0,0205	0,0095	0,0097	0,0358	0,0152	0,0082
Acidification	kg SO_2 eq	0,0064	0,0034	0,0035	0,0165	0,0107	0,0177
Eutrophication	kg PO_4^- eq	0,0008	0,0005	0,0005	0,0015	0,0006	0,0073
Global Warming (GWP100)	kg CO_2 eq	0,8635	0,6320	0,5035	2,1698	0,9653	0,3259
Ozone Layer Depletion (ODP)	kg CFC-11 eq	4,25 E-7	1,49 E-7	2.26 E-7	4.55 E-7	3,53 E-7	2,22 E-7
Human toxicity	kg 1,4-DB eq	0,4657	0,1486	0,2201	0,5267	0,3149	0,3780
Fresh water aquatic ecotox.	kg 1,4-DB eq	0,0629	0,0213	0,0298	0,0735	0,0408	1,2638
Marine aquatic ecotoxicity	kg 1,4-DB eq	251,9871	72,3619	119,3966	265,8253	182,4819	125,5733
Terrestrial ecotoxicity	kg 1,4-DB eq	0,0031	0,0013	0,0017	0,0054	0,0042	0,5606
Photochemical oxidation	kg C_2H_4	0,0016	0,0005	0,0007	0,0021	0,0011	0,0004

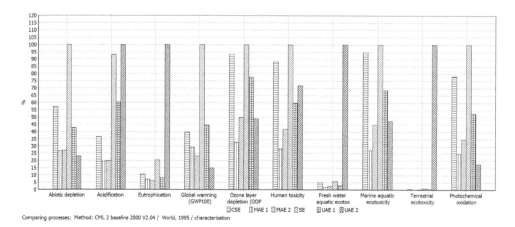

Comparing processes; Method: CML 2 baseline 2000 V2.04 / World, 1995 / characterization

Figure 1. Environmental impact of extraction methods.

d.b. while SE using the same solvent system shows similar extraction yield after 6 h. It was noted that the β-carotene in the case of MAE and SE were significantly lower than those in the case of UAE and CSE extractions, probably due to thermal degradation, leading to lower antioxidant activity.

Finally, regarding the environmental aspect and yielding of the extraction techniques, UAE 2 is considered to be the most sustainable method, providing high yields in sort time using seed oils as solvents minimizing the cost and the carbon footprint of the final product. This green extraction processes also generates a large volume of co-products or by-products which would be used in other applications as well. For example, the oil cakes obtained after the filtration of β-carotenes can be used for animal feed.

4 CONCLUSION

Through the comparison of the different carotenoids extraction techniques that were examined in the frames of this study, it is shown that ultrasound assisted extraction is the sustainable method for carotenoids' extraction, regarding its environmental impact, low cost, high yielding and short time. Microwave assisted extraction is the most rapid and overall eco-friendly technique suffering though from relatively low yielding due to thermal degradation of carotenoids. The conventional solvent extraction, as well as, the soxhlet extraction technique is considered to be time consuming, expensive due to the large amount of solvents used, and therefore potentially hazardous for human health and the ecosystem, affecting the human toxicity and ozone depletion impact categories. The techniques examined were of a laboratory-scale and have shown their potential for scale up, in order to be used in nutraceutical, cosmetic and food industries.

REFERENCES

Dorne J. (2012), Human and animal health risk assessments of chemicals in the food chain: Comparative aspects and future perspectives, *Toxicology and Applied Pharmacology.*

Eskilsson C. and Bjorklung E. (2000), Analytical-scale microwave-assisted extraction, *Journal of Chromatography A*, 902, 227–250.

Food Standards Agency (2011), Guidelines on approaches to the replacement of Tartrazine, Allura Red, Ponceau 4R, Quinoline Yellow, Sunset Yellow and Carmoisine in food and beverages, Report No: FMT/21810/1.

Gultekin F. and Dofuc D. (2012), Allergic and Immunologic Reactions to Food Additives, *Clinic. Rev. Allerg. Immunol.*

Hiranvarachat B., Sakamon D. (2013), Chiwchan N., Raghavan V., Structural modification by different pretreatment methods to enhance microwave-assisted extraction of β-carotene from carrots, *Journal of Food Engineering*, 115, 190–197.

Li Y., Fabiano-Tixier A., Tomao V., Cravotto G. (2013), Chemat F., Green ultrasound-assisted extraction of carotenoids based on the bio-refinery concept using sunflower oil as an alternative solvent, *Ultrasonics Sonochemistry*, 20, 12–18.

Mortensen A (2006), Carotenoids and other pigments as natural colorants, *Pure Appl. Chem.*, 78, 1477–1491.

Muller L., Frohlich K., Bohm V. (2011), Comparative antioxidant activities of carotenoids measured by ferric reducing antioxidant power (FRAP), ABTS bleaching assay (αTEAC), DPPH assay and peroxyl radical scavenging assay, *Food Chemistry,* 129, 139–148.

Namiki, M. (1990). Antioxidants/antimutagens in food. *CRC Critical Reviews in Food Science and Nutrition,* 29, 273–300.

Nigg J.T., Lewis K., Edinger T., Falk M. (2012), Meta-analysis of attention-deficit/hyperactivity disorder or attention-deficit/hyperactivity disorder symptoms, restriction diet, and synthetic food color additives, *Journal of the American Academy of Child and Adolescent Psychiatry*, 51 (1).

Randhawa S. et al. (2009), Hypersensitivity reactions to food additives, *Current Opinion in Allergy & Clinical Immunology.*

Socaciu C. (2008), Food Colorants: Chemical and Functional Properties, 1st edition, 2008, CRC Press, New York.

Weiss B. (2012), Synthetic food colors and neurobehavioral hazards: the view from environmental health research, *Environ Health Perspect* 1201–1205.

Green Design, Materials and Manufacturing Processes – Bártolo et al. (eds)
© 2013 Taylor & Francis Group, London, ISBN 978-1-138-00046-9

Cobalt and manganese recovery from spent industrial catalysts by hydrometallurgy

D. Fontana & F. Forte

ENEA—Italian National Agency for New Technologies, Energy and Sustainable Economic Development, Rome, Italy

ABSTRACT: In the present work, hydrometallurgical techniques were applied in order to recover cobalt and manganese from fly ash coming from incineration of spent industrial catalysts. Washing steps with water were carried out, with the aim of removing the soluble salts. The residue was leached in order to establish the optimal operative conditions. Cobalt/manganese separation was investigated by solvent extraction, using Cyanex 301 in kerosene as an extractant varying pH, extractant concentration and organic/aqueous volume ratio. Stripping tests were performed to re-extract cobalt ions in a new aqueous phase, employing hydrochloric acid as stripping agent.

The proposed process allows to recover 99.7% of cobalt and 86.6% of manganese.

1 INTRODUCTION

Fly ash is considered to be an hazardous waste due to the presence of leachable heavy metals and soluble salts, so it requires adequate treatments before safely disposal.

Separation processes, focused on improving the quality of the residue for further utilization and/or recovering the species of interest (Quina et al., 2008), can be performed through pyrometallurgical and hydrometallurgical techniques. The first ones were largely applied in the past (Jacob et al., 1995, Nowaka et al., 2010), but many of them have high energy consumption and toxic gas emissions. For this reason, in the last decades attention has been moved to hydrometallurgical techniques, that offer some advantages, such as low operating cost, minimization of sludge to be handled and low emissions; moreover they are particularly suitable for material recovery from secondary sources because of their high recovery efficiency.

Valuable material recovery is nowadays gaining great attention among scientific community.

In a recent report, the European Commission made a list of 41 materials defined "critical" at European level due to the economic importance and the high supply risk that they are subjected to. Because of their "criticality", recovery processes from secondary sources are strongly encouraged (European Commission, 2010).

In this paper, hydrometallurgical techniques were applied with the aim of recovering cobalt and manganese from fly ash coming from incineration of spent industrial catalysts. Cobalt is a critical metal mainly produced in DRC (41%), Canada

(11%) and Zambia (9%); its applications are in superalloys, hard metals, permanent magnets, batteries and catalysts (Georgiou et al., 2009). Manganese is largely applied in steel industry (Zhang & Cheng, 2007) and its production is concentrated in China (25%), Australia (17%) and South Africa (14%).

The first step of the hydrometallurgical technique here proposed is a leaching step, aimed at dissolving cobalt and manganese compounds from the ash; then, metal recovery was investigated through solvent extraction.

Cobalt leaching can be performed through ammonia-ammonium carbonate (Katsiapi et al., 2010), chlorine dioxide (Park et al., 2005) and hydrochloride acid (Clark et al., 1996); manganese can be recovered by means of direct reductive leaching processes, including leaching with ferrous iron, sulfur dioxide, hydrogen peroxide, nitrous acid, organic reductants, and bio- and electro-reductions (Zhang & Cheng, 2007). Various organic extractants can be used to complex the metal ions into the organic phase, such as D2EHPA (Hoh et al., 1984), Cyanex 301 in Exxsol D-80 (Tsakiridis & Agatzini, 2004) or in Iberfluid (Ocana & Alguacil, 1998), sodium salts of D2EHPA, PC 88A and Cyanex 272 (Devi et al., 2000), and also binary extractant systems of Cyanex 301 (Jakovljevic et al., 2004).

2 MATERIALS AND METHODS

Fly ash sample was collected from the flue gas treatment section of an Italian incineration plant that treats industrial wastes, such as spent catalysts employed for Trimellitic Anhydride (TMA) production.

All the experiments were performed in duplicate as a check on the experimental technique and precision. Metals concentration was determined by an AAS (Atomic Absorption Spectrophotometer) Shimadzu AA6300.

Cobalt and manganese content was determined by dissolving the ash in concentrated acids. It was found that the ash consist of 19 w/w% of cobalt and 13 w/w% of manganese.

2.1 Leaching tests

Ash samples were leached using deionized water, H_2SO_4, HNO_3, HCl, aqua regia and sulpho-nitric mixed acid, varying concentration, temperature, liquid/solid ratio (L/S) and contact time in order to establish the best operative conditions.

2.2 Solvent extraction tests

Cyanex 301 was employed to extract cobalt ions from the aqueous phase; it is a di-thio phosphinic acid produced by Cytec and it was used as received.

The organic solutions were prepared by dissolving the extractant in kerosene (mainly paraffinic and naphtenic hydrocarbon in C_{10}–C_{14} range). Solvent extraction tests were performed on stock solutions of cobalt and manganese, prepared by dissolving cobalt (II) nitrate and manganese (II) nitrate in hydrochloric acid in a suitable concentration. Equal volumes (10 mL) of aqueous and organic solutions of known concentration were mixed and shaken in stoppered glass tubes at 25 °C, then allowed to settle for at least 1 h. Metal concentration in the aqueous phase was determined by AAS, while ion concentration in the organic phase was calculated using a material balance technique. The distribution of metals between organic and aqueous phases was examined as a function of the hydrogen ion concentration, the extractant concentration and the organic/aqueous volume ratio (O/A).

2.3 Stripping tests

Stripping tests were performed in order to re-extract metal ions into a new aqueous phase.

HCl was employed as stripping agent and different acid concentrations were tested to define the optimal operative conditions. 10 mL of aqueous and organic solutions were mixed and shaken for 30 min in stoppered glass tubes at 25 °C.

3 RESULTS AND DISCUSSION

3.1 Leaching

Water leaching is a necessary fly ash pre-treatment aimed at removing the soluble alkaline salts, so that acid consumption during the following steps can be lowered. Two washing steps were performed at room temperature, for 30 min and employing a L/S of 10 mL/g, resulting in a weight loss of 45.4% (43.5% after a first washing and 1.9% with a second washing). Chemical analysis of the leachates revealed that water washing was ineffective in leaching cobalt and manganese oxides from the ash.

The residue was then leached using different acid solvents. The best leaching efficiencies (w/w%) are reported in Table 1.

The HCl 4M leaching leads to a complete dissolution of cobalt and manganese oxides at low acid concentration and low energy consumption.

3.2 Solvent extraction of Co(II) ions using Cyanex 301 as an extractant

A number of extraction tests were performed to evaluate Cyanex 301 selectivity for cobalt ions and optimize the extraction process. The effect of time on the extraction of cobalt and manganese using Cyanex 301 was studied: it was observed a time of 5 minutes is sufficient to reach the equilibrium. In Figure 1, cobalt and manganese extraction isotherms are reported as a function of the equilibrium pH. E_{org}(%) represents the amount of ions extracted into the organic phase with respect to the initial amount in the aqueous phase.

Table 1. Leaching efficiency.

Leaching agent	Cobalt (w/w%)	Manganese (w/w%)
Water (25°C, 30 min, L/S 10)	0.0	0.0
H_2SO_4 6M (60 °C, 2 h, L/S 50)	63.7	90.2
HNO_3 6M (60 °C, 2 h, L/S 50)	22.6	31.9
HCl 2M (60°C, 2 h, L/S 50)	71.9	90.1
HCl 4M (60°C, 2 h, L/S 50)	100.0	100.0
Aqua regia (60°C, 2 h, L/S 50)	100.0	100.0
Sulpho-nitric mixture acid (60°C, 1 h, L/S 10)	100.0	100.0

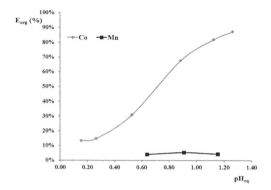

Figure 1. Cobalt and manganese extraction isotherms as a function of pH_{eq} (T = 25°C, O/A = 1:1, [Co] = [Mn] = 0.02M, [Cyanex301] = 0.1 M).

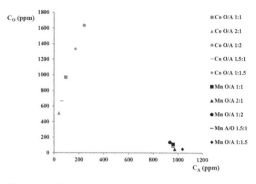

Figure 3. Cobalt and manganese extraction isotherms as a function of organic/aqueous volume ratio (T = 25°C, pH_{eq} = 0.9, MR = 10).

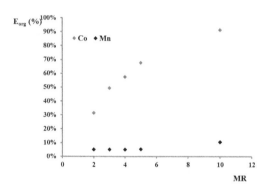

Figure 2. Cobalt and manganese extraction isotherms as a function of extractant concentration (T = 25°C, O/A = 1:1, pH_{eq} = 0.9).

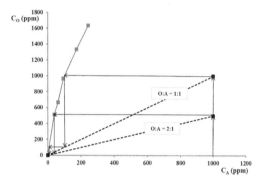

Figure 4. McCabe and Thiele method for determining the number of theoretical extraction stages (T = 25°C, pH_{eq} = 0.9, MR = 10).

The experimental results show that cobalt extraction significantly increases with pH, while manganese extraction is constant and does not exceed 5%.

In the indicated conditions, when pH_{eq} is greater than 1.18, manganese precipitates incorporating cobalt ions.

Extraction tests were carried out varying the extractant concentration, as reported in Figure 2 where MR represents the ratio extractant moles/total cobalt and manganese moles in the aqueous phase.

It is evident that cobalt extraction increases when Cyanex 301 concentration increases, while manganese extraction is almost steady, being the maximum amount extracted equal to 10.4%.

Other extraction tests were carried out varying the organic/aqueous volume ratio O/A, as reported in Figure 3 where C_A and C_O represent respectively the ion concentrations in the aqueous and in the organic phase.

It was found that cobalt extraction significantly increases with the O/A ratio, without relevant effects on manganese extraction Data reported in Figure 3 allow to evaluate the number of theoretical extraction stages required to extract cobalt according to McCabe and Thiele method. Figure 4 shows an example of the graphic constructions with reference to a 1000 ppm cobalt hydrochloric leachate.

Working at O/A = 1:1, two stages are required in order to completely extract cobalt ions from the aqueous phase; increasing the organic/aqueous volume ratio, the number of theoretical stages decreases.

3.2.1 *The stoichiometry of the reaction*
In this section, the stoichiometric of the extraction reaction of Co(II) ions by Cyanex 301 as an extractant is investigated.

We can consider the following equilibrium reaction for an extraction process:

$$M_{aq} + (m+n)HL_{org} \Leftrightarrow (ML_m(HL)_n)_{org} + mH_{aq}^+ \tag{1}$$

where the subscripts org and aq represent respectively the organic and the aqueous phase, M is the metal ion, HL is the extractant and $ML_m(HL)_n$ is the organic-metal chelate.

If HL is a strong acid, it follows that:

$$M_{aq} + mHL_{org} \Leftrightarrow (ML_m)_{org} + mH_{aq}^+ \tag{2}$$

The extraction constant K can be written as follows:

$$K = \frac{[ML_m]_{org} \cdot [H^+]_{aq}^m}{[M]_{aq} \cdot [HL]_{org}^m} \tag{3}$$

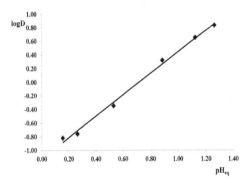

Figure 5. Measured partition coefficient (log D) of cobalt as a function of pH_{eq} (T = 25°C, O/A = 1:1, MR = 5).

and, subsequently, the distribution ratio D of the metal ion between the organic and the aqueous phase becomes:

$$D = \frac{[ML_m]_{org}}{[M]_{aq}} = K \frac{[HL]_{org}^m}{[H^+]_{aq}^m} \tag{4}$$

Then we can write:

$$\log D = \log K + m \cdot \log[HL]_{org} + m \cdot pH \tag{5}$$

In Figure 5 the measured partition coefficient (log D) of cobalt as a function of pH_{eq} is shown:
Experimental data are represented by Equation 6:

$$\log D = 1.50 \cdot pH_{eq} - 1,13 \tag{6}$$

The stoichiometric coefficient m in Equation (1) is equal to 1.50 and then the equilibrium reaction of Co(II) with Cyanex 301 is the following:

$$2Co^{2+} + 3HL_{org} \Leftrightarrow (Co_2L_3)_{org} + 3H_{aq}^+ \tag{7}$$

3.2.2 Cobalt extraction isotherms as a function of pH_{eq} and Cyanex 301 concentration

Cobalt extraction isotherms as a function of pH_{eq} and extractant concentration were obtained as explained below.

In Figure 6 logD as a function of pH_{eq} and MR is reported. The curves were analytically obtained as equations of straight lines passing through the experimental values of logD at MR = 2, MR = 3, MR = 4 and MR = 10, assuming the same slope as that of Equation 6.

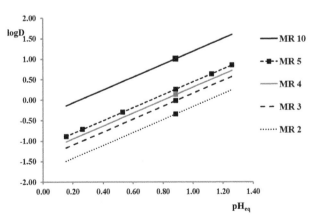

Figure 6. Partition coefficient (log D) of cobalt as a function of pH_{eq} and Cyanex 301 concentration (T = 25°C, O/A = 1:1).

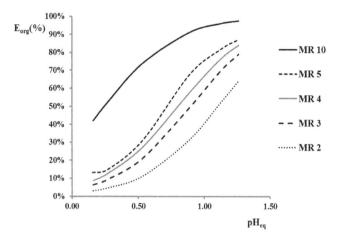

Figure 7. Calculated cobalt extraction isotherms as a function of pH_{eq} and Cyanex 301 concentration (T = 25°C, O/A = 1:1).

By defining y as:

$$y = \log_{10} \frac{[ML_m]_{org}}{[M]_{aq}} \qquad (8)$$

it follows that:

$$10^y = \frac{[ML_m]_{org}}{[M]_o - [ML_m]_{org}} \qquad (9)$$

where $[M]_o$ is the initial cobalt concentration in the aqueous phase.

By substituting the expression of the extraction percentage:

$$E_{org}(\%) = \frac{[ML_m]_{org}}{[M]_o} \cdot 100 \qquad (10)$$

in Eq. (9), it follows that:

$$E_{org}(\%) = \frac{10^y}{1 + 10^y} \cdot 100 \qquad (11)$$

In this way it is possible to represent the extraction isotherms as a function of equilibrium pH and extractant concentration, as reported in Figure 7.

3.3 Stripping

The organic phase coming from the extraction stage was stripped by contacting it with a new aqueous solution. Hydrochloric acid was employed as stripping agent and different concentrations were tested, i.e. HCl 2M, HCl 4M and HCl 6M.

Table 2. Cobalt stripping efficiency.

Stripping agent	Efficiency (%)
HCl 2M	100.0
HCl 4M	100.0
HCl 6M	61.8

Stripping tests were carried out at room temperature, employing a 1:1 organic/aqueous volume ratio and a contact time of 30 min.

Cobalt stripping efficiencies are reported in Table 2.

Experimental data shows that HCl 2M can efficiently re-extract cobalt ions.

4 CONCLUSIONS

In this paper, cobalt and manganese recovery from fly ash by hydrometallurgy was investigated.

It was found that HCl 4M leaching (t = 2h, T = 60°C, L/S = 50 mL/g) leads to a total dissolution of cobalt and manganese oxides.

Cyanex 301 was studied in order to selectively extract cobalt ions from the aqueous phase: the extraction occurs via a proton exchange reaction and its stoichiometry is

$2Co^{2+} + 3HL_{org} \leftrightarrow (Co_2L_3)_{org} + 3H_{aq}^+$; the best operating parameters are: $pH_{eq} = 0.9$, MR = 10 and O/A = 1:1, leading to a cobalt extraction of 94.6%.

Stripping tests show that a complete re-extraction of cobalt ions is possible employing HCl 2M (T = 25°C, O/A = 1:1).

Using this two stages process, it is possible to recover 99.7% of cobalt and 86.6% of manganese.

REFERENCES

Clark, S.J., Donaldson, J.D., Khan, Z.I. 1996. Heavy metals in the environment. Part VI: Recovery of cobalt values from spent cobalt/manganese bromide oxidation catalysts. *Hydrometallurgy* 40: 381–392.

Devi, N.B., Nathsarma, K.C., Chakravorty, V. 2000. Separation of divalent manganese and cobalt ions from sulphate solutions using sodium salts of D2EHPA, PC 88A and Cyanex 272. *Hydrometallurgy* 54: 117–131.

European Commission. 2010. Critical raw materials for the EU.

Georgiou, D., Papangelakis, V.G. 2009. Behaviour of cobalt during sulphuric acid pressure leaching of a limonitic laterite. *Hydrometallurgy* 100: 35–40.

Hoh, Y.C., Chuang, W.D., Lee, B.D., Chang, C.C. 1984. The separation of manganese from cobalt by D2EHPA. *Hydrometallurgy* 12: 375–386.

Jacob, A., Stucki, S., Khun, P. 1995. Evaporation of heavy metals during the heat treatment of municipal solid waste incinerator fly ash. *Environ. Sci. Technol.* 29: 2429–2436.

Jakovljevic, B., Bourget, C., Nucciarone, D. 2004. CYANEX 301 binary extractant systems in cobalt/nickel recovery from acidic chloride solutions. *Hydrometallurgy* 75: 25–36.

Katsiapi, A., Tsakiridis, P.E., Oustadakis, P., Agatzini-Leonardou, S. 2010. Cobalt recovery from mixed Co–Mn hydroxide precipitates by ammonia–ammonium carbonate leaching. *Mineral Engineering* 23: 643–651.

Novaka, B., Pessla, A., Aschenbrennerb, P., Szentannaia, P., Mattenbergerc, H., Rechbergerb, H., Hermannc, L., Wintera, F. 2010. Heavy metal removal from municipal solid waste fly ash by chlorination and thermal treatment. *Journal of Hazardous Materials* 179: 323–331.

Ocana, N. & Alguacil, F.J. 1998. Cobalt/Manganese Separation: the Extraction of Cobalt(II) from Manganese Sulphate Solutions by Cyanex 301. *J. Chem. Technol. Biotechnol.* 73: 211–216.

Park, K.H., Kim, H.I., Das, R.P. 2005. Selective acid leaching of nickel and cobalt from precipitated manganese hydroxide in the presence of chlorine dioxide. *Hydrometallurgy* 78: 271–277.

Quina, M.J., Bordado, J.C., Quinta-Ferreira., R.M. 2008. Treatment and use of air pollution control residues from MSW incineration: An overview. *Waste Management* 28: 2097–2121.

Tsakiridis, P.E., Agatzini, S.L. 2004. Simultaneous solvent extraction of cobalt and nickel in the presence of manganese and magnesium from sulfate solutions by Cyanex 301. *Hydrometallurgy* 72: 269–278.

Zhang, W., Cheng, C.Y. 2007. Manganese metallurgy review. Part I: Leaching of ores/secondary materials and recovery of electrolytic/chemical manganese dioxide. *Hydrometallurgy* 89: 137–159.

Green Design, Materials and Manufacturing Processes – Bártolo et al. (eds)
© 2013 Taylor & Francis Group, London, ISBN 978-1-138-00046-9

Strategic sustainability through a product development tool

Polin Kumar Saha

Unnayan Onneshan, Dhaka, Bangladesh

ABSTRACT: With many other product development tools, Template for Sustainable Product Development (TSPD) has now been introduced in product and service system in many industries. TSPD is now being tested in corporate sectors and several industries in some regions of the world. The generic use of TSPD in an organization is aimed to revise product development and its customers' services through continuous dialogue process. Moreover, in order to fulfill the actual technical and environmental objectives in the company, common sustainability approaches can be derived from the perspectives of integrating TSPD into both the corporate management and environmental services. In following those achievements, the services of TSPD connect to the implications of the whole industrial management services. Study shows that the use of TSPD serves not only for the product development from the business point of view, but also stimulates the existing organizational policies and the whole process recommends some technical and environmental activities to satisfy the overall sustainability needs. Likewise, the study also demonstrates that how this product development tool can be established suitably as an administrative tool for a respective product addressing continuous process of the total sustainability management initiatives.

1 INTRODUCTION

Today, sustainable development is a very challenging term (Jonsson, 2008), since that creates much confusion as surrounded by different meanings of sustainability. As a result of this confusion it is become more confused 'what should be preserved' (Redclift, 1999) in line with various sustainability definitions. So, the misconception, or confusion regarding preservation towards sustainability is another new challenge to the society (Marshall et al. 2005). Considering those challenging issues, the current study realizes to replicate the existing methods and process for sustainable development.

In the recent decades, the concept of product development has been scrutinized significantly as the route of major implementing stage for global sustainability problems (Saha, 2012). As the consequences of, the knowledge about sustainability is found with the main sustainability gap among all industrial activities either in product manufacturing and, or environmental management services (Saha and Seal, 2012). For the human existences on earth and dealing with those un-avoidable circumstances, the study has realized that sustainability reporting and managing of an industry should be framed profoundly with the guidance of a robust tool, particularly with a product development tool (Saha, 2012).

To find the concrete analysis in the study, two supportive product analytical approaches

'Strategic Life Cycle Management (SLCM)' and 'back-casting from basic sustainability principles' have been integrated into the specialized product development tool 'Template for Sustainable Product Development (TSPD)'. The TSPD is first introduced by Henrik Ny and others in 2008 at the case study of a television manufacturing company (Matsushita, one of the largest Japanese multinational electronics companies). Two major steps of TSPD 'current reality' and 'envisioned future' are analyzed with three points of view of a product, e.g. market needs, life-cycle concepts and extended enterprise. On the other hand, SLCM has integrated into TSPD all the ways of life-cycle concepts of a product based on designated basic 'Sustainability Principles' (SPs). SLCM incorporating sustainability principles configures the TSPD for a 'strategic guideline tool' in the product oriented conventional activities toward sustainability.

TSPD is the upgrade version of Method for Sustainable Product Development (MSPD) that introduces sustainability in the early product development phase of a company (Ny et al. 2008). The tool approach is based on the 'framework for strategic sustainable development' and four sustainability principles. This method coordinates creative communication among top management, product developers and stakeholders. On the other hand, SLCM is an advanced technique to make decisions entering the production process of a product (Ny et al. 2006).

In the study, the main process of TSPD is used to seek for the overarching research question for sustainability management. The main objective of the study is to see how the Strategic Sustainable Development (SSD) could be emerged through a product development tool 'Template for Sustainable Product Development (TSPD)'. Based upon the objective of the study the overarching research question is as follows:

What general guiding principles are very useful in integrating strategic sustainability into the company's management process?

2 METHODOLOGY

2.1 *Template for Sustainable Product Development (TSPD)*

Particularly, in TSPD it is aimed to help the organizations in the following ways: 1) shifting the organization by continuous improvements in line with previous environmental performance of a specific product, 2) creating a common under-stating model at different organizational levels, and 3) facilitating a continuous dialogue among sustainability experts at different level (Ny et al. 2008).

The main analytical process of TSPD is designated as ABCD process (see 2.1.1) that is integrated into 6 (six) templates of TSPD. The generic model of the integrated TSPD illustrated in Table 1.

2.1.1 *ABCD analysis*

Backcasting from the SPs (sustainability principles, see 2.1.2) is the practice of interdisciplinary approaches of an organization which is run by the external facilitator through 'ABCD' process (Ny et al. 2008).

A. discuss a system/framework where we have to work for a shared mental model
B. Brainstorming process to assess the current reality of the system in relation to the SPs
C. Brainstorming process to seek for the alternative solution of the present practice of the system in relation to the SPs

D. Stepping prioritized action/investment to minimize the gap between B and C. The actions are taken considering three issues of the system: choosing flexible platform, reducing society's violations of the sustainability principles, and sufficient return on investment.

Figure 1 shows an idea about the ABCD process.

2.1.2 *Sustainability Principles (SPs)*

To analyze the present complexities (through ABCD process) towards sustainability *the natural step international* first introduces 'four system conditions' that focus on socio-ecological perspectives of the capacity to meet the human needs. These basic principles are subsequently used in many analyses as explicit guidance to develop the concept of an organization, community, individual or a product in moving towards sustainability (Robert et al. 2002). The system conditions to success are stated briefly in Figure 2.

2.1.3 *Strategic Life Cycle Management (SLCM)*

SLCM is an advanced technique combined of traditional Life Cycle Analysis (LCA) and backcasting from sustainability principles (Ny et al. 2006). The application of SLCM has been integrated in the second templates of TSPD to identify the 'hot spots' of sustainability based on B and C step of ABCD analysis. This is the process of concept development of a product in life cycle perspectives towards sustainability.

By drawing the conceptual analysis of an organizational product through SLCM tool, the sustainability violations of an organization can be reported. In the process of analysis, the Table 2 is given as guideline.

2.2 *Case study*

Hammarplast AB is a leading plastic manufacturer located in Sweden since 1960. The study considers the company in the analysis. The basic raw material of plastic is used as Polypropylene in the company. The same raw material is used in different assemblies of the product. Hammarplast AB has

Table 1. The generic overview of template for sustainable product development.

B step of ABCD analysis	Template B1: Market desire/needs	Template B2: Concepts	Template B3: Extended enterprise
Current reality	Case study?	Case study?	Case study?
C Step of ABCD analysis	Template C1: Future market desire/needs	Template C2: Future concepts	Template C3: Future extended enterprise
Future opportunities	Case study?	Case study?	Case study?

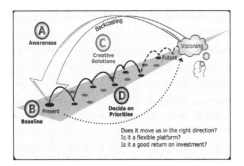

Figure 1. Schematic overview of ABCD process (Source: The natural step international).

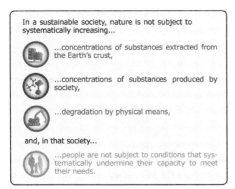

Figure 2. Four sustainability principles with schematic overview (Source: The natural step international).

Figure 3. A Model of the case study product of hammarplast AB storage case.

the EMS ISO 14001 to manage its environmental activities. Among wide range of products of the company, the study selects a popular customer product named 'smart store'. It is used for storing space. Company produces the Smart Store in range of 3–4 Million/year.

Interview is done with the production and environmental managers of the company. Questions are designed to relate the sustainability visions based on study objectives.

3 RESULTS

During the sustainability assessment of the company's activities, the management level of this company usually thinks that sustainability approach is as constrain rather than exploring an opportunity in their product innovation. It is found in the analytical findings of the study that the approach with new product development could be the central platform to work with the challenging issues mostly as surrounded by the sustainability vision of an organization. Literally, on basis of product life cycle, about 80–90% impact of sustainability can be controlled by the decisions during new product development (Gokan et al. 2010). The findings of the TSPD analysis on a product are summarized as follows.

3.1 Current market needs/desires, template B1

The significant issues in this template are found into three categories of product analysis:

Market needs of the product in terms of general customer's demand: efficient space for storing; multipurpose uses; outlook, convenient and portable; and less packaging needs including transportation cost.

Market needs of the product by emerging the issues of sustainability challenges: non-renewable, non-usable and recyclable raw material; heavy metals oriented machinery equipments, non-renewable energy; and disposal system.

Table 2. Strategic Life Cycle Management (SLCM).

Life cycle stages	Sustainability Principles (SPs)			
	SPI: Materials from the earth's crust	SPII: Substances produced by society	SPIII: Physical degradation of earth	SPIV: Human needs
Raw materials				
Production				
Packaging and distribution		Case study?		
Use				
End of life				

211

Market needs of the product in relation with human needs (Max Neef, 1991): leisure needs; freedom (ability to take); idleness with product; good protection; and fostering health practice/subsistence.

3.2 *Future market needs/desires, template C1*

On the basis of sustainability challenges in the template B1, the following significant issues are found for future market needs:

New market needs of the product in terms of general customer's demand and sustainability challenges both: Dematerialization of materials, prevention of wastes, recycling, reusing; increasing eco-friendliness of product; bio degradable raw materials and its hallmark label on product; and less use of heavy metals in machineries

New market needs (related to core business) of the product in terms of satisfying human needs: leisure; understanding of sustainability with product; and subsistence.

New market trends in the competition: customer services; customer spaces; using promising materials (e.g. wood); energy conservation; and use of renewable energy sources.

3.3 *Current product analysis, template B2*

To establish sustainability principles, SLCM is applied here. Current assessment of the product through the process reports not only products related sustainability violations, but also scrutinize the environmental and management activities of the organization (Saha, 2010). The key areas of the product that have been critically violated in sustainability principles, as in the following areas:

Raw Materials: solely use of hydrocarbons; heavy metals use in machineries; mining activities (to physical degradation); transportation; migration of people; and emissions.

Production: consumption of nuclear energy; production equipments; release of health injurious gas; and use of dyes, packaging materials and storage.

End Use: not biodegradable; not continuous recyclable; release of toxic gases at burning; and dumping.

3.4 *Future Product, template C2*

Template "Future Products" categorize the set of goals in lens with "principles of sustainability" and its mechanism i.e. "backcasting from sustainability principles" as for "continuous improvement" defined by environmental management systems of the organization. The key areas for setting goals are as:

Extraction and Use of Materials: dematerialization of all materials and scarce metals; substitution of polypropylene (raw materials) with cotton fibre, food fibre, or bio-plastics; and use of bio-fuels.

Process: energy use from renewable resources; appropriate use of technology; and packaging, labeling and storing.

End Use: recycling and re-use of products.

Competitive Market and Product: efficient marketing with sound knowledge of a product; selling 'storage space' to the customers; and exclusive type of product materials (e.g. bio-plastics, metals, woods etc).

3.5 *Current extended enterprise, template B3*

How product awareness of the customer can pressurize and/or encourage sustainable development and how functioning with the stakeholders (NGO's) can help in safe and sound market image (Bansal & Bonger, 2002)—structured by this template 'extended enterprise'. A mechanism of continuous dialogue with the stakeholders may help the organization in nurturing growth of leadership and sustainability (Robert et al. 2007). Regarding the issue the perception of stakeholders has been extended in some key areas:

Current Preferences of Societal Stakeholders: heavy investment; understanding sustainability; third party R&D inputs; overall quality; alternative commercial materials; certification and law enforcement for sustainability management; enforcing corporate social responsibility; robust alliances among peers; coordination and works of all inter-departments; and capacity building training for the staffs and stakeholders.

Current Value Chain: tied communication model among different stakeholders; cooperation with knowledge based resources for sustainability ventures; and continuous seeking for sustainable raw materials.

3.6 *Future extended enterprise, template C3*

Future extended enterprise aims to help organization to understand sustainable development in relation with natural world surrounded by production, consumption, human intention and ethics. These may suggest appropriate preventive actions for sustainability (Redclift, 1999). The key areas are suggested for such measurements:

Favorable issues in the Future Societal Stakeholders: substituting raw materials; stakeholder engagement; pooling of stakeholders' knowledge and resources; process of gradual shifting and adoption; introducing easy mechanism of sustainability; psychological transformation of societal perception; societal awareness; compensation package for sustainability violations by using of product; and extended producer responsibility.

Favorable issues in the value chain: sustainability awareness in the production and procurement; dematerialization and substitution of raw materials; and selling services.

4 DISCUSSION

Like an organization (e.g. Hammarplast) the same analytical approaches can be executed to find out sustainability challenges as well as sustainability management in other organizations. For sustainability management, the general strategic process in the study can also be applicable in the same directions for a product, organization, or a community within our biosphere. The study observes that problems get more complex in the absence of mass-understanding of sustainability and subsequent missing market and behind the environment to take quantum initiatives.

Methodologically, a backcasting process helps in understanding the organizational nature and system of services, particularly Step 'A of ABCD' process.

By analyzing B and C step of ABCD through TSPD approach, some compelling measures are identified in relation with sustainability principles. Finally, a list of prioritized actions can be transformed into executable sustainability goal ensuring

the justified right direction, flexible platform and good returns on investment—Step D of ABCD.

However, a common sustainability vision has been fostered by the guiding approaches of integrated TSPD model. The dialogue process through this product development tool is the best approach to lead an organization towards sustainable development. The integrated model for strategic sustainability looks like the diagram in Figure 4.

5 CONCLUSION

Finally, by analyzing product development tools on a plastic product of a company (a case study), this study reveals some general recommendations for sustainability management that could be as the integrated policies to the existing management systems of the entire industry. TSPD can be used in different types of manufacturing industries as well as for the service sectors also. The integration process of A-B-C-D methodology and backcasting from sustainability principles into the generic model of TSPD can bring onward a platform which holds on substantial solutions of any product/service in a sustainable manner.

REFERENCES

Bansal, P. & Bogner, W. C. 2002. Deciding on ISO 14001: economics, institutions, and context. *Long Range Planning*, 35(3), 269–290.
Hammarplast AB. *Hammarplast AB*. 2005. Available at: http://www.hammarplast.com/ (accessed August 1, 2012).
Jonsson, M. 2008. Potential Conflict in the relationship between economic growth and environmental sustainabiliy. *Gotenberg: Gotenberg University*.
Marshall, J.D. & Toffel, M.W. 2005. Framing the elusive concept of sustainability: a sustainability hierarchy. *Environmental Science & Technology*, 39(3), 673–682.
Max-Neef M.A., 1991. Human Scale Development: Conception, Application and Further Reflections. *The Apex Press*.
May, G., Taisch, M. & Kerga, E. 2012. Assessment of Sustainable Practices in New Product Development. Advances in Production Management Systems. Value Networks: Innovation, Technologies, and Management, 437–447.
Ny, H., Hallstedt, S., Robèrt, K.-H. and Broman, G. 2008, Introducing Templates for Sustainable Product Development. *Journal of Industrial Ecology*, 12: 600–623.
Ny, H., MacDonald, J.P., Broman, G., Yamamoto, R. & Robèrt, K.H. 2006. Sustainability Constraints as System Boundaries: An Approach to Making Life-Cycle Management Strategic. *Journal of Industrial Ecology*, 10(1–2), 61–77.
Redclift, M. 1999. Sustainability: Life chances and livelihoods. Routledge.

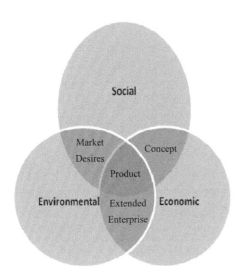

Figure 4. Diagrammatic Model for Strategic Sustainable Development (SSD) through Product Development.

Robèrt, K.H., Basile, G., Broman, G., Byggeth, S., Cook, D., Haraldsson, H. & Waldron, D. (2007). Strategic leadership towards sustainability. *Blekinge Institute of Technology*.

Robèrt, K.H., Schmidt-Bleek, B., Aloisi de Larderel, J., Basile, G., Jansen, J. L., Kuehr, R. & Wackernagel, M. 2002. Strategic sustainable development—selection, design and synergies of applied tools. *Journal of Cleaner production*, *10*(3), 197–214.

Saha, P.K. 2010. Environmental Management Systems and Sustainability: Integrating Sustainability in Environmental Management Systems. *Lambert Academic Publishing.*

Saha P.K. 2012. Incorporating Product Development Tools in Environmental Management Systems (EMS): A new way for sustainability management. Conference Paper of the 2012 Berlin Conference of the Human Dimensions of Global Environmental Change on "Evidence for Sustainable Development".

Saha, P.K., & Seal, L. 2012. A strategic approach to Environmental Management Systems (EMS): An assessment of Sustainability in EMS to move toward Sustainability. *International Journal of Environmental Sciences*, *2*(2), 1093–1102.

The Natural Step International. 2012. Planning for Sustainability: A Starter Guide. *The Natural Step Canada.*

Green Design, Materials and Manufacturing Processes – Bártolo et al. (eds)
© 2013 Taylor & Francis Group, London, ISBN 978-1-138-00046-9

Set up planning for automatic generation of inspection plan

Emad Abouel Nasr
Department of Industrial Engineering, College of Engineering, King Saud University, Riyadh, Saudi Arabia
Mechanical Engineering Department, Faculty of Engineering, Helwan University, Cairo, Egypt

Abdulrahman Al-Ahmari, Osama Abdulhameed & Syed Hammad Mian
Department of Industrial Engineering, College of Engineering, King Saud University, Riyadh, Saudi Arabia

ABSTRACT: Inspection, one of the most important processes in manufacturing industries validates whether manufactured part lies within specified tolerance or not. It is critical to ensure high quality and non-defective products because poor quality might cause serious injury or even fatal accidents which are obviously not good for company's reputation. Automation of inspection activities can meet increased demands of accurate and precise inspection process. However, automated inspection tasks involve number of planning stages such as extraction of design information, probe selection and orientation, feature accessibility, part set up planning etc. Part set up planning is one of most critical steps for effective and efficient automation of CMM inspection. In this paper, two different methods have been introduced to determine part orientation on CMM machine table. The proposed methods have successfully been implemented and tested on prismatic part with many features. Moreover, results have shown that proposed approaches can achieve excellent performance compared to other methods.

1 INTRODUCTION

Dimensional inspection using Coordinate Measurement Machine (CMM) has been one of the most critical steps in manufacturing industries. CMM inspection process usually entails probing different part features using touch trigger probe. The point data thus obtained is used to determine if inspected part meets design specifications or not. In fact, increased demands for better quality have recognized inspection one of the vital processes in production industries. Two different approaches i.e., manual and automated can be used to define different steps for generating inspection plan. When part inspection is performed manually, all decisions regarding part set up, probe accessibility, probe selection and orientation, collision avoidance etc., are taken by CMM operator (Sathi & Rao (2009)). Therefore, each and every step in manual inspection depends on user's expertise and experience. Another approach where inspection plan is completely automated ensures reduced inspection time, better consistency and reliability of measurement results. In automated inspection planning, decisions concerning probing features, probe selection and orientation, collision avoidance, number of set ups etc. are all taken by computer based programs. CMM inspection plan can be divided into two categories: high-level planning and low level

planning (Spitz et al. 1999). High level planning involve set up planning, probe selection, probe orientation etc. whereas low level planning consider strategies to identify coordinates and number of points, probe execution etc. Since, performance of CMM inspection process including measuring time, cost and errors greatly depends on inspection plan therefore inspection planning strategy is an important aspect to achieve accurate and efficient measurement results (Cho et al. 2005). There have been many tasks (shown in Fig. 1) that have to be performed on CMM before inspection process is carried out. In fact, lot of research work is being carried out to investigate each and every stage for generating automated inspection plan.

Corrigall & Bell (1991) successfully minimized inspection time by determining best part set up and probe orientation. Similarly, Ziemian &

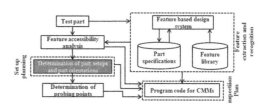

Figure 1. Automated generation of inspection plan.

Medeiros (1998) addressed the issue of inspection planning by introducing a technique for automatic probe selection and part setup. Kweon & Medeiros (1998) utilized Visibility map (VMap) to identify part orientations depending on design and tolerance information. Hwang et al. 2004 have also minimized cycle time for part setups and probe changes. It has been noted that most of previous researches have directed towards overall generation of automated inspection plan rather than focusing on intermediate stages. In fact, improvement in each step can bring lot of positives in form of effective and efficient inspection plan. It has also been found that feature accessibility is greatly affected by inspection feature's orientation, location, dimension and its interactions with other features. Moreover, effective set up planning is very important for generating efficient automated inspection plan. Set up planning determines how part should be oriented on CMM machine table so that maximum number of features can be measured in one set up. Set up planning becomes crucial especially when time needed to change part set up is significant with respect to overall inspection time of part. This paper has focused on determining best part set up for automated inspection plan. Two different approaches have been proposed and implemented for part set up planning.

2 SET UP PLANNING

Main idea for set up planning is to identify part face which has minimum number of inspecting features. This face determines base face for part orientation on machine table. In fact, base face is called as preferential base or primary locating face in set up planning. Any part face that would result in inspection of maximum possible features should define part orientation for best set up plan.

2.1 First rule (numerical method)

This method makes use of Artificial Neural Network (ANN) to predict best set up. Geometric extracting entities and features with same Probe Accessibility Direction (PAD) were identified as input to ANN. In order to implement this approach for given problem, inputs to ANN were determined as follows.

2.1.1 Geometric extracting entities
It included geometric entities such as number of vertices, line edges, circular edges, internal loop, external loop, concave faces and convex faces etc. All these entities were extracted from extraction and recognition file.

2.1.2 Features having same PAD
Features with same PAD were determined as follows:

- Six PADs (+x, −x, +y, −y, +z and −z) and six set ups (S (Right), S (Left), S (Front), S (Rear), S (Top), S (Bottom)) were identified for rectangular block.
- For given set up, number of features that could be accessed for each PAD was determined and presented in form of (m × n) matrix as shown in Figure 2. In this matrix, f11 represents number of features that can be inspected with +x probe direction when right face of part is used as primary locating face. Similarly f54 represents number of features that can be inspected with −y direction when top face act as primary locating face.
- For each PAD, total number of features that could be accessed was calculated. e.g., total number of features for +x PAD was equal to (f11 + f21 + f31 + f41 + f51 + f61). Similarly, total number of features for +z PAD was equal to (f15 + f25 + f35 + f45 + f55 + f65).
- Finally, six inputs in form of summation of PAD$_j$ in each column were obtained.

Therefore, total of 13 inputs (six PADs + seven extracted geometries) were identified for input layer and six nodes including bottom face, top face, front face, rear face, left face, and right face were selected for output layer.

2.1.3 Training and testing of experiments
Training experiments determined number of hidden layers using EasyNN plus to optimize network structure shown in Figure 3. Several training experiments with different number of hidden neurons, learning rate (0.60), and momentum values (0.80) were checked for best training parameters and minimum error.

2.1.4 Result of testing
After several training experiments, network was successfully trained with average validating error of 0.0025 as shown in Figure 4.

Once the training was successfully finished, network was validated for ten examples with validating percentage of 80%.

	+x(PAD)	−x(PAD)	+y(PAD)	−y(PAD)	+z(PAD)	−z(PAD)
S(Right)	f11	f12	f13	f14	f15	f16
S(Left)	f21	f22	f23	f24	f25	f26
S(Front)	f31	f32	f33	f34	f35	f36
S(Rear)	f41	f42	f43	f44	f45	f46
S(Top)	f51	f52	f53	f54	f55	f56
S(Bottom)	f61	f62	f63	f64	f65	f66

Figure 2. Number of features for different PADs and part faces.

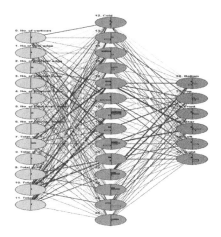

Figure 3. Topological structure of back propagation neural network.

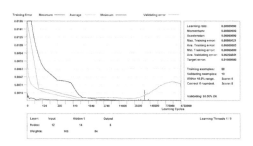

Figure 4. Learning progress.

2.1.5 Implementation

For given prismatic part, following inputs were identified:

- Number of vertices = 52, line edges = 74, circular edges = 4, internal loop = 2, external loop = 28, concave faces = 18, convex faces = 10.
- +x(PAD) = 10; −x(PAD) = 8; +y(PAD) = 27; −y(PAD) = 19; +z(PAD) = 0; −z(PAD) = 16.

After successful training, neural network resulted into best set up with right (or left) face as primary locating face. With this orientation, almost 70% of features could be accessed.

2.2 Second rule (graphical method)

In this method, best set up was determined based on number of interacting features as shown in Figure 5. Face with minimum number of interactions was selected as primary locating face.

Different steps to identify best set up for prismatic part can be defined as follows:

- Divide all faces of prismatic part as primary faces and secondary faces.

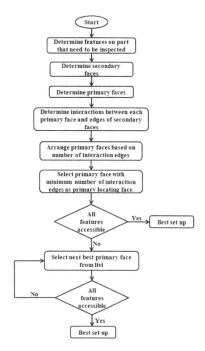

Figure 5. Setup planning algorithm.

Primary faces for rectangular block are faces which determine basic shape of the part. Top, bottom, front, rear, right and left faces fall in to this category.

Secondary faces are faces belong to various features on part such as slot, rib, boss, pocket etc.

- Determine interaction between primary faces and edges of secondary faces. For example, Left face f_{24} (as shown in Fig. 6a) has interaction with only one edge, top faces f_{27} and f_{17} (as shown in Fig. 6b) have interactions with eight edges, right face f_{28} (as shown in Fig. 6c) has interaction with only one edge, bottom faces f_{12} and f_{21} (as shown in Fig. 6d) have interactions with six edges, front face f_7 (as shown in Fig. 6e) have interactions with sixteen edges and rear face f_{16} (as shown in Fig. 6f) have interactions with ten edges.

The primary faces would be arranged in ascending order of number of interactions in order to find best set up.

- Select primary face with minimum number of interactions as primary locating face. For example, Left face f_{24} and right face f_{28} have least interactions (one interaction each). Therefore, either left face or right face can be selected for primary locating face as shown in Figure 7. These faces as primary locating face (left face f_{24} or right face f_{28}) would allow probe to inspect maximum features in one set up. With this orientation, probe

217

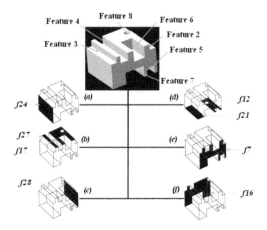

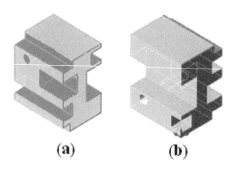

Figure 6. Prismatic part and interactions edges.

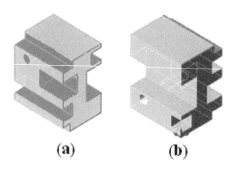

(a)   **(b)**

Figure 7. Setup planning.

can inspect features with feature IDs (2, 4, 5, 6, 7, and 8).

3 CONCLUSION

To generate efficient CMM inspection plan, two approaches for set up planning have been proposed. In minimizing part set up time, numerical method based on ANN and graphical method has been used. These methods can help in great way by minimizing cycle time for part setups and probe changes. The proposed method can reduce inspection time drastically for industrial products. This research has successfully shown the potential of presented techniques in set up planning. Moreover, proposed approaches in this paper can be extended to rotational parts.

ACKNOWLEDGEMENT

This work is funded by National Plan for Science & Technology (NPST), Grant No. 10-INF1280-02, Saudi Arabia.

REFERENCES

Cho, M-W., Lee, H., Yoon, G-S. & Choi, J. 2005. A feature-based inspection planning system for coordinate measuring machines. International Journal of Advanced Manufacturing Technology 26: 1078–1087.

Corrigall, M.J. & Bell, R. 1991. Probe and component set-up planning for coordinate measuring machines. International Journal of Computer Integrated Manufacturing 4 (1): 34–44.

Hwang, C.-Y., Tsai, C-Y. & Chang, C.A. 2004. Efficient inspection planning for coordinate measuring machines. International Journal of Advanced Manufacturing Technology 23: 732–742.

Kweon, S. & Medeiros, D.J. 1998. Part orientations for CMM inspection using dimensioned visibility maps. Computer-Aided Design 30 (9): 741–749.

Sathi, S.V.B. & Rao, P.V.M. 2009. STEP to DMIS: Automated generation of inspection plans from CAD data. 5th Annual IEEE Conference on Automation Science and Engineering, Bangalore, India, August 22–25, 2009.

Spitz, S.N., Spyridi, A.J. & Requicha, A.A.G. 1999. Accessibility analysis for planning of dimensional inspection with coordinate measuring machines. IEEE Conference on Robotics and Automation, Los Angeles, USA, August 22–25, 1999.

Ziemian, C.W. & Medeiros, D.J. 1998. Automating probe selection planning for inspection on measuring machine and part setup a coordinate. International Journal of Computer Integrated Manufacturing 11 (5): 448–460.

Green Design, Materials and Manufacturing Processes – Bártolo et al. (eds)
© *2013 Taylor & Francis Group, London, ISBN 978-1-138-00046-9*

Intelligent fixture planning system for prismatic parts

Abdulrahman Al-Ahmari
Industrial Engineering Department, College of Engineering, King Saud University, Saudi Arabia

Emad Abouel Nasr
Industrial Engineering Department, College of Engineering, King Saud University, Saudi Arabia
Mechanical Engineering Department, Helwan University, Faculty of Engineering, Cairo, Egypt

Awais A. Khan
Industrial Engineering Department, College of Engineering, King Saud University, Saudi Arabia

Ali Kamrani
Industrial Engineering Department, University of Houston, College of Engineering, Houston, TX, USA

ABSTRACT: Fixture Planning (FP) is an important manufacturing activity that has a significant effect on product quality and productivity. FP not only determines the locating scheme to restrict the Degrees Of Freedom (DOF) of work-piece but also locating and clamping datum for final fixture assembly. In this paper, an intelligent fixture planning methodology is proposed. The part geometric data base and process/setup plan is taken as input to an intelligent FP system. At first, a rule based method is developed to determine the correct locating scheme that constraints all DOF of work-piece. Then, the Artificial Neural Network (ANN) is applied to determine feasible locating and clamping faces. The methodology is also validated with an illustrative case study.

1 INTRODUCTION

Substantial research has been conducted to decrease production cost and processing time in manufacturing to meet the competitive market demand. The work-piece handling system, which largely affects the usability of machine tools, is used to clamp, change and store the work-piece in manufacturing (Fleischer et al. 2006).

Fixtures play an important role in work-piece handling. Proper fixture planning can dramatically reduce the manufacturing cost, the lead-time, and labor skill requirements in product manufacturing. Also, the integration of fixture planning with Computer Aided Design and Computer Aided Manufacturing (CAD/CAM) provides an overall optimal solution for product design and manufacture. However, fixture planning is a highly experience-based activity. There are not many fixture planning tools available for industry applications due to the extreme diversity and complexity of work-pieces. New manufacturing technologies, such as Computer Integrated Manufacturing (CIM), Flexible Manufacturing Systems (FMS), Lean Manufacturing (LM), Agile Manufacturing (AM) and internet-based manufacturing, bring new challenges to fixture planning in terms of the

planning theory, planning reliability, and planning efficiency (Kang & Peng 2008).

Babu et al. (2005) developed a software for automatic fixture planning based on 2-D drawing data. The methodology presented for automatic fixture layout planning of machining setups focused on determining the most suitable locating and clamping positions in accordance with the 3-2-1 configuration, considering geometrical and dimensional constraints. Toumi et al. (1989) discussed the planning issues and presented the plans and requirements for automatic setup and reconfiguration of modular fixtures. Wu et al. (1998) developed the geometric analysis technique with modular fixture assembly to present the fundamental study of automated fixture planning. Liu and Wang (2007) presented a hybrid approach in which machining precedence is determined by knowledge based and feature sequencing through geometric reasoning for fixturing setup. Kale and Pande (2000) proposed an intelligent system developed in C++ for automatic generation of fixture layout plan. The designed algorithm is used to determine the most suitable locating and clamping positions by search vector technique. Fuh et al. (1995) developed an integrated logic based CAD-CAPP-CAFP system to allow the generation of manufacturing plans after the part design has been completed.

In this paper, part geometric database created from STEP AP 203 file and Computer Aided Process Plan (CAPP) data are taken as an input to the intelligent FP system. The rule based method is first applied to determine the suitable locating scheme that restricts necessary DOF of work-piece. Then, an ANN is developed to determine the feasible locating and clamping region. The proposed methodology is implemented using Visual C++ that provides the link between the CAD models, the generated process, and fixtures plans.

2 REASEARCH METHODOLGY

The methodology presented in the paper automatically generates modular fixture planning data. The part geometrical database contains geometric information of work-piece in B-rep format. The CAPP database includes technological features data, machining data, and setup plan. The rule based is used to determine the suitable locating scheme for each setup based on work-piece shape, orientation w.r.t the machine tool axis and the cutting tool penetration in primary datum. The primary locating surface is recognized based on the setup orientation. The determination of locating and clamping datum has a lot of variations due to setup plan, orientation of work-piece w.r.t primary reference surface, surface roughness, surface area, etc. Therefore, the ANN is applied to decide the feasible locating and clamping surfaces for each type of fixture setup/orientation. The methodology is presented in the form of flow chart in Figure 1.

2.1 Locating scheme

Location establishes the desired relationship between work-piece and the fixture, which in turn establishes the relationship between work-piece and the cutting tool. The type of location is governed by the type of feature and the number of faces being machined. The purpose of the location is to restrict the degrees of freedom of the work-piece (Chang et al. 1998).

2.2 Degrees of freedom of a work-piece

A work-piece, just like any free solid body, has six degrees of freedom as shown in Figure 2:

– Three linear displacements along the mutually orthogonal co-ordinate axes.
– Three angular displacements with respect to the same axes.

During a set-up, it is necessary to restrict certain degrees of freedom so as to locate and orient the technological feature surfaces with respect to the cutting tools (Nee et al. 2004).

For prismatic shape work-pieces, the most common locating schemes are 3-2-1, base 2-1, and 4-2-1 that restricts all six degrees of freedom.

2.3 Shape, orientation and cutting tool penetration

The work in this paper is restricted to only block shaped prismatic parts. Further classifications of these parts are done according to Optiz GT classification scheme. The Opitz GT scheme classifies the prismatic work-piece into three categories as represented in Figure 3.

To determine the suitable locating scheme, the cube is further classified as small and large cube.

Small cube: A/B <= 3 and A/C < 3
Large cube: A/B <= 3 and 3 < A/C < 4

where A, B, and C are the length, width, and height of the component.

To select the best orientation of work-piece and simplify the geometric reasoning, the primary datum must be selected first. If the feature is present on XY top surface and Tool Access

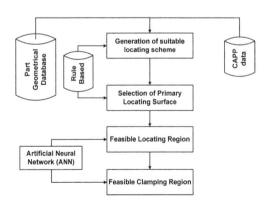

Figure 1.   The proposed methodology.

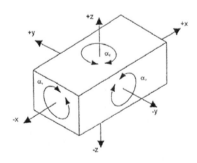

Figure 2.   Six degrees of freedom [10].

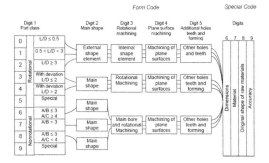

Figure 3. Optiz classification scheme (Girdhar, 2001).

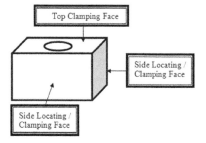

Figure 4. Candidate locating and clamping surfaces.

Direction (TAD) is (0, 0, −1) then work-piece is oriented so that the bottom surface will become the primary datum surface. The cutting tool path in the primary datum face will also determine the correct locating scheme. The feature TAD in the primary datum surface is the indication of cutting tool penetration.

Considering the above conditions, the rules are formulated to determine the correct locating scheme for each type of work-piece orientation.

If work-piece is small cube
And primary datum is XY bottom
And cutting tool is penetrating in the primary datum surface
Then
3-2-1 locating scheme is feasible

2.4 Locating and clamping faces

The candidate locating faces should be along 3 mutually perpendicular planes. The candidate locating faces for secondary and tertiary locations should be perpendicular and adjacent to primary locating/datum face. The surfaces that have larger area will be considered as secondary locating face. The clamping should be positioned to direct the clamping force on the strong, supported part of the work-piece. There are two common clamping types: overhead and side clamps. The overhead clamp applies a force perpendicular to the fixture base plate and the side clamp applies a force parallel to the fixture base plate (Jeng & Gill, 1997) (see Fig. 4).

The primary locating surface is always the primary datum surface of the work-piece. It depends on work-piece orientation with respect to machine tool.

2.5 Artificial Neural Network (ANN)

Artificial Neural Networks (ANNs) are biologically inspired models analogue to the basic functions of

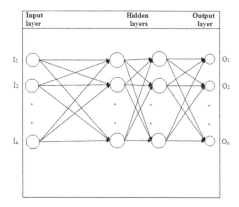

Figure 5. Multilayered neural network.

biological neurons. They have a natural propensity for storing experiential knowledge, and resemble the human brain in the sense that training rather than programming is used to acquire knowledge. A neural network consists of a number of nodes. The nodes are arranged in layers: an input layer, an output layer, and several hidden layers as shown in Figure 5. The nodes of the input layer receive information and transform through the connections to the other connected nodes layer by layer to the output layer nodes. The transformation behavior of the network depends on the structure of the network and the weights of the connections (Amaitik, 2005).

2.6 Data preparation (ANN) for location and clamping faces

There are five input parameters that are taken in the study. These are primary datum face, boundary faces, and orientation with respect to primary datum, surface roughness and surface area.

2.6.1 Primary datum surface
The criteria for determining the primary datum is already described in section 2.3.

2.6.2 Boundary faces

The boundary faces constitutes the boundary of the component as shown in Figure 6. The boundary faces are obtained from geometric information database of the part. The primary datum face is already selected therefore; it will be excluded from the list of boundary faces.

2.6.3 Orietantion w.r.t primary datum

The orientation of each boundary face is calculated by comparing the face normal vector with primary datum face vector. If the direction is same, the face is parallel, otherwise it will be perpendicular.

2.6.4 Roughness value

The surface with smaller roughness value should be selected as locating surfaces for precise location. The roughness value of the surfaces are defined as fine (0–100) or rough value (>100).

2.6.5 Surface area

The larger side locating surface will be considered as secondary surface and the smaller side is tertiary locating surface. If more than 1 face found for

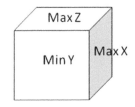

Figure 6. Boundary faces.

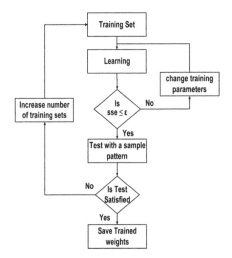

Figure 7. Training and testing procedure of neural network.

secondary or tertiary location then the face with greater area will be selected to ensure the locating accuracy. The formula to find the surface area of face is presented in equation 1.

$$A = 1/2 \sum_{i,j=0}^{n-1} \left((X_j + X_i)(Y_j - Y_i) \right) \qquad (1)$$

2.7 Training and testing experiments

Several training experiments have been carried out to identify the optimal network structure and best training parameters of the neural networks which produce minimum errors during training phase. The adopted procedure is schematically represented in Figure 7. In this way, several training experiments with a different number of hidden neurons, learning rate, and momentum values have been checked.

In particular, a given network structure has been considered able to correctly learn the training set if $sse \leq \varepsilon$ is satisfied. Where ε is the predefined limit value of the error (tolerance value) of the error and sse is the average sum square error resulted at the end of training process. During training the error limit (ε) has been set to 0.00001.

3 CASE STUDY

The block shaped prismatic work-piece is taken as a case study illustrated in Figure 8. The part produced in CATIA contains 7 machining features. The geometric database of the part represents 19 feature faces and 6 boundary faces.

The CAPP data represents one setup. The locating scheme for the setup is determined by formulating the rules based on the work-piece classification, orientation, and checking the tool penetration in selected primary datum surface. Then, the candidate boundary faces, their roughness values, their orientation, and surface areas are determined and coded. The output parameters like top clamping face(s),

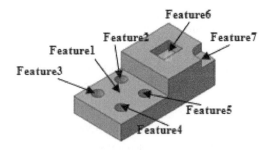

Figure 8. Case study.

Table 1. Network parameters.

Network parameters	Locating and clamping faces
Input neurons	5
Output neurons	3
Hidden layers	2
1st Hidden layer neurons	15
2nd Hidden layer neurons	15
Learning rate	0.1
Momentum	0.95
Number of patterns	54
Number of iterations	270
Average sum squared error	0.0001

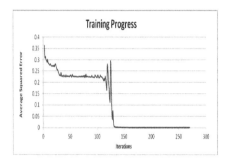

Figure 9. Training progress for locating and clamping.

Figure 10. Fixture planning output.

Table 2. Fixture planning data.

Setup Id	Locating scheme	Primary locating face(s)	Secondary locating face(s)	Tertiary locating face(s)	Top clamping face(s)
1	3-2-1	5	2	15	20

secondary locating faces(s), and tertiary locating face(s) are coded. Each input and output parameter is coded between 0 and 1. A number of training experiments performed to select the optimal structure and training parameters of the neural networks. The results obtained are presented in Table 1.

The graph shown in Figure 9 represents the training set average error on the y-axis against the number of iterations elapsed on the x-axis. The graph shows downward movement of the error rate until the averaged squared error = target error.

The screen shot of fixture planning file is presented in Figure 10. The data is summarized in Table 2.

4 CONCLUSION

The proposed methodology is successfully tested and validated for the selected case study. All the relevant geometrical data is extracted from STEP AP 203 file in the form of faces, edges, and vertices along with their IDs. The process/setup plan generated provides a basis for CAFP. Fixture planning not only decided the feasible locating scheme through a rule based but also the suitable locating and clamping surfaces by using ANN. The significance of generating the suitable locating and clamping surfaces restricts the selection and insertion of modular fixture components for final fixture layout.

ACKNOWLEDGEMENTS

This works is funded by National Plan for Science & Technology (NPST), Grant No. 10-INF1280-02, Saudi Arabia. Moreover, the authors would like to appreciate the Advance Manufacturing Institute (AMI) at King Saud University for their support throughout the research work.

REFERENCES

Amaitik, S., 2005, Development of a step feature-based intelligent process planning system for prismatic parts, *A Phd Thesis*, Middle East Technical University.

Babu, B.S., Valli, P.M., Kumar, A.V.A., & Rao, D.N., 2005, Automatic modular fixture generation in computer-aided process planning systems, *Proc. IMechE Vol. 219 Part C: J. Mechanical Engineering Science.*

Chang, T.C., Wysk, R.A., & Wang, H.P., 1998, *Computer Aided Manufacturing*, 2nd edition, Prentice Hall Inc.

Fleischer, J., Denkena, B., Winfough, B., & Mori, M., 2006, Workpiece and Tool Handling in Metal Cutting Machines, *CIRP Annals—Manufacturing Technology*, 55(2), 817–839.

Fuh, J.Y.H., Chang, C.H., & Melkanoff, M.A., 1995, The development of an integrated and intelligent CAD/CAPP/CAFP environment using logic-based Reasoning, *Computer Aided Design*, vol. 28, No. 3, pp. 217–232.

Girdhar, A., 2001, Expansion of group technology part coding based on functionality, *A Master Thesis*, University of Cincinnati.

Jeng, Y.C., & Gill, K.F., 1997, "A CAD-based approach to the design of fixtures for prismatic parts", *Proceedings of the Institution of Mechanical Engineers, Part B: Journal of engineering Manufacture*, 211–523.

Kale, M.S., & Pande, S.S., 2000, Automatic planning of machine fixture layouts through geometric reasoning of cad part model, *Proc. of 19th all India Mfg. technology design and research conference*, IIT Madras, India, 377–382.

Kang, X., & Peng, Q., 2008, Fixture feasibility: methods and techniques for fixture planning, *Computer-Aided Design & Applications*, 5(1–4), 424–433.

Liu, Z., & Wang, L., 2007, Sequencing of interacting prismatic machining features for process planning, *Computers in Industry* 58, 295–303.

Nee, A.Y.C., Tao, Z.H., & Kumar, A.S., 2004, An Advanced Treatise on Fixture Design and Planning, *World Scientific Publishing Co. Pte. Ltd.*

Toumi, K.Y., Bausch, J.J., & Blacker, S.J., 1989, Automatic setup and Reconfiguration for modular fixturing, *Robotic & Computer-Integrated Manufacturing*, Vol. 5, No. 4, pp. 357–370.

Wu, Y., Rong, Y., Ma, W., & LeClair, S.R., 1998, Automated modular fixture planning: Geometric analysis, *Robotics and Computer-Integrated Manufacturing* 14, 1–15.

Green Design, Materials and Manufacturing Processes – Bártolo et al. (eds)
© *2013 Taylor & Francis Group, London, ISBN 978-1-138-00046-9*

A study of injection moulding with bismuth alloy

A. Kus
Uludag University, Bursa, Turkey

E. Unver & B. Jagger
University of Huddersfield, Huddersfield, UK

I. Durgun
Tofas Automotive Research and Development, Bursa, Turkey

ABSTRACT: There is a demand to produce parts in small quantities, using cost effective injection moulding processes. The emerging economic case suggests there will be advantages in producing injection moulding tools for short duration low volume production runs where the tool is rebuilt for each manufacturing cycle. CAD/CAM, Rapid Prototyping (RP) and digital technologies help to bring new products to market faster. Mould inserts are traditionally produced using expensive and time consuming CNC machining processes. This article presents development of an injection moulding tool where rapid tooling, bismuth-tin mould insert and injection moulding are used to manufacture one hundred plastic parts. It describes the manufacturing methods applied and the use of manual and 3D scanning methods. It also evaluates the surface finishes and dimensions of the parts produced in the tool.

1 BACKGROUND AND INTRODUCTION

Injection moulding is a manufacturing process commonly used for mass produced parts formed by thermoplastic and thermosetting plastic materials. The moulds or die are usually made from aluminum or steel at considerable cost to the manufacturer, which can only be offset by high volume production. Traditional tool manufacturing processes include cutting, abrading, burning, and eroding. Cutting processes involve milling, drilling, turning, planning and reaming to shear material from the workpiece. Abrading processes involve grinding, polishing or sanding. Burning and eroding processes use electricity, chemicals, heat, or hydrodynamics to shape or remove material. Until recently CNC milling machines were the only option to produce moulds. The processes are time consuming, costly and constrained by the machinery used to create the tool.

Rapid prototyping is widely used in the product, automotive, aerospace, and medical industries. RP mostly used for prototyping, rapid tooling and rapid manufacturing. RP is typically used for testing scale, fit and function, as well as providing an effective communication between design, management and marketing teams. RP includes a range of technologies which automatically construct physical prototype models from CAD data using machines recently called 3D printers. In addition to prototypes, RP techniques can be used to make tools (rapid tooling) and even production-quality parts (rapid manufacturing). There are more than six rapid prototyping techniques are commercially available, each with unique strengths.

Recent developments in Rapid Tooling (RT) have changed the design process and economic models for mass manufacture. With the advent of RT, the process can be divided into three segments. The first stage is Prototype or Soft Tooling where a mould is designed to test component functionality, appearance, size and fit. These tools produce only a few hundred parts, whereas in Bridge Tooling tens of thousands of units could be manufactured. These tooling stages facilitate early introduction of new products to the market, prior to or during the fabrication of the Production or Hard Tool. By adopting RT methodologies the manufacturer can gauge the volume of demand for the product before specification and investment in the production tool. RT can meet the needs of a customised market where an inexpensive (softer) tool is sufficiently robust to handle batch outputs. Rapid tooling either directly produces a tool with a rapid prototyping system, or indirectly utilises a rapid prototype as a pattern for the purposes of producing the tool. RT processes complement the RP options by being able to provide higher quantities of parts in a wider variety of materials and can even produce short-run

Figure 1. Initial bismuth casting.

injection moulded parts in the intended production material.

Currently most 3D printing technologies are better suited to the design and testing stages of manufacture, however their material properties are unsuited to RP tool printing for injection moulding where the tool materials must have sufficient hardness to retain good surface quality and accuracy in high volume manufacture. Recent advances in RP 3D metal printing technologies have enabled the production of low volume components for healthcare, aerospace and high-technology engineering and electronics sectors. The initial non-metallic plastics and powder based materials in RP have been extended to include metals such as aluminium and steel. These advances have not only changed how prototype parts are created, but have also facilitated new opportunities to create RP models for custom manufacture at the tool making stages of the manufacturing process. The concept of rapid adaptation in high volume manufacture process has been described as Mass Customization (Atkinson et al. 2008).

In this study a tool insert has been created using RP to create a master for a bismuth alloy insert into a steel injection moulding tool. The process has the advantage of eliminating traditional tool making methods, but its ability to withstand the temperatures and pressures used in an industrial manufacturing setting required testing and analysis. The team chose bismuth alloy for a replaceable insert to produce a RT part as initial testing showed that the material could be cast at temperatures that would not damage the original RP master part. See Figure 1.

2 METHODOLOGY

Injection Moulding is one of the most commonly used manufacturing methods for producing plastics parts where large numbers of identical products are manufactured with relatively low unit costs. Toolmaking is an essential and costly phase of the injection moulding process where tool design, manufacturing and testing are the crucial stages of any plastic injection moulding. The tool configuration, material selection, surface finish and tolerances are important considerations in preparation for injection moulding.

Rapid tooling is the term for either indirectly utilizing a rapid prototype as a tooling pattern for the purposes of moulding production materials, or directly producing a tool with a rapid prototyping system. RT processes complement the RP options by being able to provide higher quantities of parts in a wider variety of materials, even short-run injection moulded parts in the intended production material. Dies and moulds can be produced by using RT and low melting metal alloys. This method has the advantages of reducing the time required to produce the tool, which enables further design iterations and more prototyping cycles with an overall reduction in tool development costs.

In this article, the use of low melting materials for rapid tooling is investigated for low volume production of injection plastic parts. The study describes:

- Mould tool manufacturing methods.
- Use of low melting bismuth alloy for mould iserts.
- Analysis of surface and dimensional changes in the plastic moulded parts (1–100).
- A comparison of traditional 2D and laser assisted 3D inspections techniques.

Bismuth has been known and used throughout recorded history. It has been used in pharmaceuticals as a non-toxic additive in cosmetics and in digestive medicines. Bismuth based non-ferrous alloys are used in the manufacture of solders, semiconductors, batteries, optical, and decorative products because of its distinctive properties of high density, good electrical conductance and low melting point. Typical low melting point alloys are made up from Bismuth, Indium, Lead, Tin and Cadmium. These alloys are produced by mixing two or more of the pure metals in eutectic proportions to achieve combinations that can melt at temperatures as low as hot water and re-solidify consistently even after repeated cycles. Although limited research has been carried out on bismuth alloy injection moulding processes, there are many studies where the use of RP technologies for injection moulding process are discussed and compared with existing methods. Ilyas et al. (2010) reports on the use of indirect Selective Laser Sintering (SLS) and machining processes to create injection mould tools where productivity improvements and energy reduction in injection moulding are discussed. Stucker, and Qu (2003), studied the use of Rapid Tooling directly in conjunction with other processes such as casting, plasma spraying etc. to create metallic or ceramic tooling for limited production and an STL-based finish machining technique for tools and parts made using RP. Pre-process algorithms and machining strategies are explored and software is assessed. Jetley and Low (1998) showed how RP techniques have been adapted to develop a low melting point metal casting alloy tool for injection moulding and compares the results with design and manufacture of mould inserts using traditional methods by examination of the 150 parts produced. Altan et al. (2001), reviewed the variety of technologies available for the manufacture of dies and moulds with particular focus on automotive applications. Dalgarno, and Stewart (2001) reported layer manufacture methods to produce injection mould tools and assesses the economic advantages over existing production methods. Sudershan, Low (2006) studied the use of Low melting alloys for insert design.

3 EXPERIMENTAL STUDY

This research was done in conjunction with Fren Sistemleri Sanayi (FSS) Ltd in Bursa, Turkey to produce a component for a clutch system. The company specialises in the production of Air Brake and Clutch Components for trucks and buses. The research team investigated the use of a recyclable material for low volume injection moulding process and testing. The team used low melting Bismuth 137 Alloy as recyclable material for initial injection tooling before manufacturing of the Steel tool for mass manufacturing.

In this paper the use of low melting materials for rapid tooling is investigated for low volume production of injection plastic parts. The study describes:

- Mould tool manufacturing methods.
- Use of low melting bismuth alloy for mould inserts.
- Analysis of surface and dimensional changes in the plastic moulded parts (1–100).
- A comparison of traditional 2D and laser assisted 3D inspections techniques.

The team used an alloy of Bismuth (58%) and Tin known as Bismuth 137 alloy which has a hardness of 20HB for the tool insert in a steel injection mould. For consistency with the parts already in production the team used PE I20 plastic material on an industrial injection moulding machine. Figure 2 shows the CAD model of the part to be injection moulded.

PE-I20 is a low density polyethylene thermoplastic commonly used for injection moulding of consumer and industrial products. The material is resistant to acids and is commonly used for liquid containers. Material specification is shown on Table 1.

Figure 2.   CAD data of produced part.

Table 1.   PE I20 material parameters.

Property	Value
Tensile strength (kg/cm^2)	100
Coefficient therm. exp. (for −1 °C)	240×10^{-6}
Thermal conductivity (kcal/mh °C)	0.26
Density (gr/cm^3)	0.92
Mould temperature (°C)	30–50–80
Melting point (°C)	165
Shrinkage (%)	2–3

3.1   Mould and insert design

A master male model was created on a CNC turning machine from the Cibatool BM 1051 material shown in Figure 3a. This master model was placed on a metal plate and a frame, shown in Figure 3b. MCP 137 alloy was melted at 150C and poured over the Cibatool shown in Figure 3c&d.

The male and female tools were manufactured using CNC turning and milling machines. As a result of the alloy's density of over 9.78 g·cm^{-3} the overall weight of MCP 137 was 20 kg. The poured alloy was then allowed to air-cool in the die frame for over one hour at room temperature. The low temperature insert tool was removed and finished on a CNC lathe to the required dimensions to fit inside the steel outer mould Figure 3e.

Figure 4 illustrates the 3D design and development of the mould tool using TopSolid CAD software. The external sizes of the tools are 350 × 450 mm. Note for this test, only the male mould insert was made from Bismuth alloy (shown in orange in Fig. 4a). The remaining tool parts are made of German specification Steel 1.2738 with hardness of 32 HRC. The steel is widely used in large moulds for the production of Polystyrenes (PS), PE, and ABS plastic parts. The tool was designed to accommodate a shrinkage allowance

(a)   (b)   (c)

(d)   (e)

Figure 3.   (a) Master model (b) Casting frame (c) Alloy casting (d) Insert models (e) Machined model.

of 2.5% in PE I20. Standard parameters and features in injection mould design were included in the steel components of the mould, such as cooling channels, plastic injection nozzle location and ejector pins.

3.2 *Injection moulding process*

An Arburg Allrounder 420c plastic injection moulding machine used in this experiment is shown in Figure 5a. Figure 5b shows the steel tool

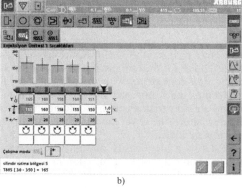

a)   b)

Figure 5.   (a) Steel and bismuth insert. (b) Finished part and male tool.

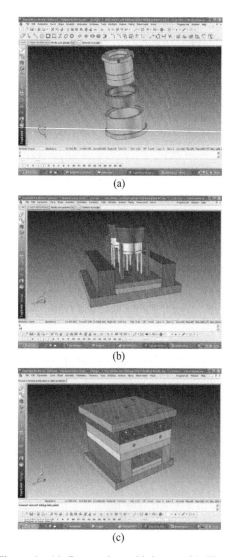

(a)

(b)

(c)

Figure 4.   (a) Part and mould inserts (b) Ejectors (c) mould Set.

a)

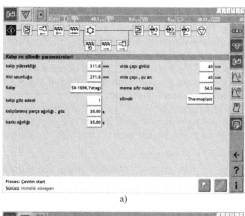

b)

Figure 6.   (a) Injection moulding cycle. (b) Temperature parameters.

and bismuth alloy insert, and Figure 5c shows the white plastic part on the steel male tool.

Figure 6a shows parameters applied to the injection process. The tool closed at 300 mm/s, slowing to 30 mm/s, with a closing force of 40 kN. The pressure on the melted plastic is 1250 bar which delivers the material at 55 cm/s into the mould cavity. A delay of 10.5 sec was allowed to complete the plastic injection into the cavity. The pressure was reduced to 150 bar for 20 sec to allow cooling before the ejector pins parted the component from the mould. This cycle was repeated to produce one hundred components. Figure 6b shows the heater settings used in the injector chamber. During the injection sequence the temperature of the molten plastic fell from an initial temperature of 165C to 150C at the point of contact with the mould. After release from the moulding tool the temperature of the ejected part had fallen to 45C due to automatic cooling of the tool. These settings ensure that the mould insert was kept below the melting temperature of the alloy and also meet the recommended cooling temperature specifications of the plastic manufacturers.

4 INSPECTION AND TESTING

Four samples were selected from the one hundred batch run at numbers 1, 25, 50 and 100. These parts were inspected to measure dimensional changes in diameter and height during the moulding run. Graph 1 shows the changes measured manually with callipers on each sample part. Outside diameter, height, inside diameter, inside thickness and outside thickness were recorded for each of the four selected parts results are shown in Table 2.

For digital comparison, a Breuckmann Opto-TOP HE 3D optical scanning system was used to scan the models. The measurements were then read into Rapidform reverse engineering software for analysis as shown in Figure 7. This method enables not only surface analysis, but also records dimensional changes which are listed below.

1st and 25th parts   : −0.92 to +0.38 mm
1st and 50th parts   : −0.28 to +0.48 mm
1st and 100 parts   : −0.28 to +0.61 mm

The dimensional changes on the sample selection were found to be random with no correlation to increasing production cycle wear in bismuth alloy insert. The team believes this could be as a result of human error during manual measurements or the effect of material properties such as shrinkage during cooling and ambient air temperatures.

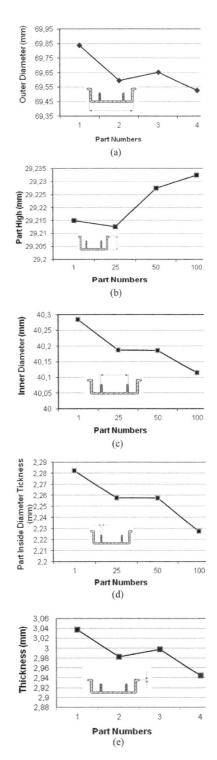

Graph 1. Measurement (a) Outer diameter (b) Height (c) Inner diameter (d) Inner thickness (e) Outher thickness.

Table 2. Measurement analysis.

Measurements	Mean (mm)	Deviation (mm)	
		Min	Max
Outer diameter	69.66	−0.1	0.19
Height	29.21	−0.02	−0.01
Inner diameter	40.2	−0.1	0.1
Inner thickness	2.26	−0.03	0.02
Outer thickness	2.98	−0.05	0.06

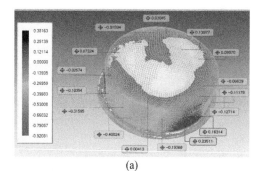

(a)

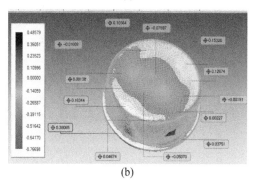

(b)

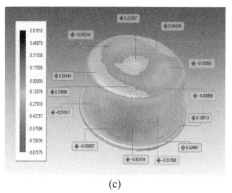

(c)

Figure 7. 3D scanning and automatic analysis of dimensional changes from part 1. (a) part 25, (b) part 50, (c) part 100.

5 FINDINGS

The research showed that bismuth MCP alloys can be used as an insert for injection moulding process for low volume production with the following benefits.

- Bismuth alloys are recyclable materials. Tool manufacturing waste or worn tool inserts can be recycled with very low energy costs.
- Relatively cheap RP models can be used for insert mould making.
- The use of low melting alloy insets to adapt injection tooling will enable globalised production to meet local market demands for customised images, logos or text—Mass Customisation.
- Bismuth insert tooling could be applied to blow, silicon or other moulding processes.
- Bismuth alloy insert tooling has the potential to extend the range of units produced in the Prototype (soft)/Bridge phases of manufacture.
- In hard tooling the bismuth insert enables extended iterations of the product.

There are also disadvantages of using low melting materials in injection moulding processes.

- Low melting temperature and low surface roughness value of Bismuth limits plastic materials selection.
- Without a longer production run it is not possible to determine the maximum number of parts that can be produced with this system but it is probable that it is only applicable for low volume production, it is not a replacement for aluminium or steel mould.
- Low heat transfer co-efficiency results lower cooling time and longer injection moulding time.
- The effects of using bismuth inserts on intricate surfaces and edges will require further testing.

6 CONCLUSION

The experimental process has demonstrated that tooling can be efficiently adapted to meet local consumer needs. Low melting-point alloy inserts have potential as a means of delivering reduced volume Mass Customization with the economic advantages of high volume production. More testing and funding will be required to explore and define the limitations of the process.

Although the melting point of plastic PE I20 material is 165C and exceeds the melting point of bismuth MCP melts at 137C alloy, due to automatic cooling, the mould temperature was kept around 40–45C during the production of one hundred units. The research shows that there were no significant dimensional changes during the hundred-unit production run. Further research is required to determine the upper limits of bismuth alloy insert tooling, which may be very much higher than the total units produced in this experiment. The use of manual and reverse engineering technology, employing 3D scanning of the manufactured parts was to be a more reliable system for monitoring production quality, but also requires a larger volume experiment to determine the tipping points at which either manual or electronic systems begin to measure significant falls in production quality.

At this stage the research team can confidently conclude that it is possible to use low melting-point alloy inserts in short production runs. The technique demonstrates a process that enables insert materials to be recycled efficiently at little cost to the manufacturer. The experiment also found that low cost 3D printing can be used in the manufacture of the bismuth insert tool.

ACKNOWLEDGEMENT

The authors would like to thanks FSS ltd. Bursa, TURKEY for supporting this research.

REFERENCES

Altan, T., Lillg, B., Yen, Y.C. 2001. Manufacturing of Dies and Molds, *CIRP Annals—Manufacturing Technology,* 50(2): 404–563.

Atkinson, P, Unver, E, Marshall, J and Dean, D.T. (2008). *Post Industrial Manufacturing Systems: the undisciplined nature of generative design.* In: Proceedings of the Design Research Society Conference 2008. Sheffield Hallam University, 194/1–194/17.

Dalgarno, K.W., Stewart, T.D., 2001. Manufacture of production injection mould tooling incorporating conformal cooling channels via indirect selective laser sintering, *Proceedings of the Institution of Mechanical Engineers. Part B, Engineering Manufacture,* 215: 1323–1332.

Ilyas, I., Chris Taylor, C., Dalgarno, K., Gosden, J., 2010. Design and manufacture of injection mould tool inserts produced using indirect SLS and machining processes. *Rapid Prototyping Journal,* Vol. 16: 429–440.

Jetley, S., Low, D.K. 2006, A Rapid Tooling Technique Using a Low Melting Point Metal Alloy for Plastic Injection Molding, *Journal of Industrial Technology,* (22)3.

Stucker, B. & Qu, X., (2003), A finish machining strategy for rapid manufactured parts and tools, Rapid Prototyping Journal, Vol.9, No. 4, pp 194–200.

Sudershan J., Low D.K. 2006. A Rapid Tooling Technique Using a Low Melting Point Metal Alloy for Plastic Injection Molding, *Journal of Industrial Technology,* 22(3).

Green Design, Materials and Manufacturing Processes – Bártolo et al. (eds)
© 2013 Taylor & Francis Group, London, ISBN 978-1-138-00046-9

Automated feature extraction from cylindrical parts based on STEP

H.M.A. Hussein
Advanced Manufacturing Institute, King Saud University, Riyadh, Saudi Arabia

Emad Abouel Nasr
Industrial Engineering Department, College of Engineering, King Saud University, Saudi Arabia
Mechanical Engineering Department, Faculty of Engineering, Helwan University, Cairo, Egypt

Awis Khan
Industrial Engineering Department, College of Engineering, King Saud University, Saudi Arabia

ABSTRACT: Automated feature recognition in cylindrical parts is considered a critical parameter in the integration between CAD and CAPP. This integration represents a foundation for many industrial applications. Research work in area is not clear enough. Feature extraction from standard files such as STEP and IGES represents nowadays the main gate to solve many industrial problems. In this paper, the cylindrical parts classified into turning parts and after turning parts. This paper focuses only on turning parts. The turning parts in 3D model are saved as STEP AP 203 and then simplified into 2D drawing. A methodology based on rule based reasoning is used to extract the features from this 2D drawing. The proposed system succeeds to extract the data and features from turning parts. The proposed system constructs as a standalone program based on visual basic 2008 connected with EWDraw module.

1 INTRODUCTION

Computer Aided Process Planning (CAPP) is represented as a bridge which connects between Computer Aided Design (CAD) and Computer Aided Manufacturing (CAM). Feature Recognition (FR) and Feature Based Design (FBD) are also represented as a bridge which connects between the CAD and CAPP.

FR and FBD are two different techniques, which completely different in their concepts. In FR, the proposed program designed to search the part data base to find relations between the data base components. These relations between components combined to each other and contain a "Feature". The rules which control feature extraction are collected from Standard machined rules, such in Shah and Mantyla, 1995, which embedded in the FR program.

The feature extracted from the part represents a kind of surprise for the user. Based on the FR, there is a lot of applications could be used, such as process planning, casting and forging. In the case of FBD, the feature is designed and inserted to the part. The part in this case is constructing feature by feature, so the process plan also build automatically during construction the part.

Both of FR and FBD are important to the industry. FBD is suitable for factories, which have a fixed product. FR is suitable for the factories which have daily variable products.

From the survey, there is a leakage of research work covering the FR in cylindrical parts area. There are different visions to solve this research problem. The first vision was proposed by Amitik, 1994. He defines and classifies his own cylindrical features during the work. For each feature, he suggests an interface to insert the features dimensions manually. He uses a standalone program based on VBasic 4.

Peterka presents a system to work under AutoCAD, in which the system includes customized menu and AutoLISP.

Moustafa, 2010, focus on individual cylindrical features. He compares the new part database in STEP format with stored database. It's not clear if there are features rules stored in his system or not.

Aslan vision depends on solving the problem in 2D only using DXF format. He studies the cylindrical parts in 2D, discuss the data extraction in 1999 and feature recognition in 2005. He estimates a new classification of cylindrical parts as different kinds of created recesses. The recesses include all expected created shapes in cylindrical parts even in turning features or after turning features. There are no features standard names used in his proposed program called "ASALUS". The recesses shapes are coded and impeded in his proposed program.

Another concept depends on simplify the 3D model in DXF format into 2D drawing. Both of Kumari, 2012, and Abdelghafour, 2012, convert their 3D model in 2D drawing, including arcs, (horizontal, inclined and vertical) lines. Abdelghafour focuses only on discussing the upper half above the access line. He takes the vector direction of the line or arc into consideration. Each line or arc direction, get a code. He combines between arcs and lines (including direction) to design features. The code in some cases arrives to 5 characters.

Yousri's methodology is similar to Abdelghafour, but using STEP instead of DXF format. Yousri, 2008 discuss a system includes of CAD/CAPP/CAM in Turning operations based on STEP-NC.

The next vision is depending on FBD. Both of Fidan 2004, and Akkus, 2011, discussed a FBD system includes of about 10 features.

The global vision for the all shapes of machined parts including cylindrical parts is discussed in Amaitik, 2005.

In this paper, the cylindrical parts classified into turning parts and after turning parts. This paper focuses only on turning parts. The turning parts in 3D model is saved as STEP AP 203 and then simplified into 2D drawing. A methodology based on rule based reasoning is used to extract the features from this 2D drawing. The proposed system succeeds to extract the data and features from turning parts. The proposed system constructs as a standalone program based on visual basic 2008 connected with EWDraw module.

2 WORK METHODOLOGY

The proposed methodology successfully extracts the relevant geometric information from STEP AP-203 file. The significance of using STEP AP 203 file is its compatibility with several known CAD systems.

The system deals with two main problems in this work, the first is how to extract and classify the data from the STEP AP 203 file, and the second is how to extract the feature from the simplified 2D drawing. Under those two titles, a lot of steps will be discussed here.

2.1 Data extraction structure

In order to solve the data extraction problem, the STEP AP 203 file structure is discussed and scheduled as in Figure 1. The proposed system read the solid model part data file searching for closed shell term, and then traced the associated data lines as in Figure 1. The system arranges the recorded lines in a table showing all the required information's for feature recognition step.

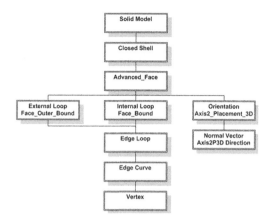

Figure 1. Entity structure of STEP AP-203.

2.2 Feature classification

One of the main problems faced in this work is the absence of the standard features specifications for the cylindrical parts. In many references there are no feature definitions such as Moustafa and Aslan.

Aslan depend in his work on recess coded shapes.

Other researcher's specify their own feature definition which is slightly different from research work to other. The classification made by Abdelghafour, shown in Figure 2 represents the most logical and professional one. In our work, we depend on the classification made by Shah and Mantyla, 1995.

The cylindrical or rotational parts features are mainly could be classified into features done on lathe machine such as (shoulders, grooves, chamfers, tapers, threads, through and blind holes). All those features are axisymmetrical features, and could be named as turning features.

The after turning feature includes features made by milling or drilling operations, such as keyways, and axial holes. In this paper, we focus only on the turning features.

Some common external and internal turning features are represented sequentially in Figures 3 and 4.

2.3 Feature recognition rules

In order to solve the recognition problem, the 3D model is simplified into 2D drafting, by selecting specified lines (edge curves) from the part data base created on the basis of STEP Structure. The basic STEP entities are represented in the form of flowchart as shown in Figure 1.

The part in Figure 5 is representing a 3D model which includes some combined turning features.

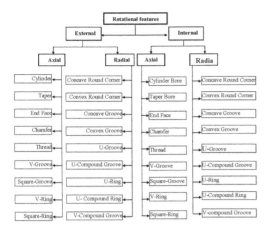

Figure 2.   Classification of rotational parts.

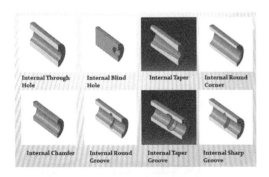

Figure 3.   Sample of external turning features.

Figure 4. Sample of internal turning features.

The features such as External under cut, external square groove, external square shoulder, external taper shoulder, etc, could be easily observed by user eyes, but how can the computer observe it too.

By studying the part data base created from STEP AP 203, there is a lot of observed information

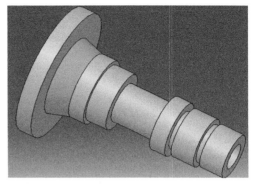

Figure 5.   Combined turning features.

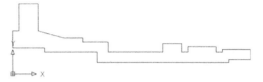

Figure 6.   The associated 2D drawing for the combined turning features as in Figure 3.

could be gained. This information help a lot in constructing the extraction rules, such as: 1. every plane cylinder in 3D contain 4 faces. To represent the cylinder in the form of 2D lines, 12 statements are required. Two cylindrical faces contain 4 line edge curves and 4 circle edge curves and 2 plane faces contain 4 circle edge curves. By selecting the suitable lines from this data base, the 2D drawing such like in Figure 6 could be represented.

In the turning feature, there are three main bodies to deal with, cylinder, cone, and torus. The statements represent the internal features always ended by suffix F as read from STEP AP 203 file, and the statements represent the external features always ended by suffix T.

Based on those observations, the resulted lines and arcs from the part simplify operation must be redraw in 2D and coding. The horizontal line coded as (0), the inclined line coded as (1) the vertical line coded as (2) and finally, the curve coded as (3).

According to the above code technique, the contents of the 2D drawing will be as follow: the external line will be 0T, the external inclined line will be 1T.

The features in 2D drawing is read by combining group of lines and arcs together with the definition if it's external or internal, so, the External square groove will be 0T0T0T and the external taper groove will be 1T0T1T and so on.

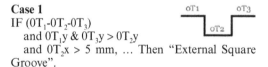

Figure 7. Simulation of the 2D drawing in a table.

To extract the features, the 2D drawing is arranged in a table as shown in Figure 7, in which every relation between 2 items or 3 items is grouped and arranged in a visual shape.

The program automatically searches the machining features contained in the database created from STEP file that satisfy the recognition rules.

The program starts first with the rules which have 3 items such as 0T-0T-0T and check the associated rules, and if it finds this item verify one of the required features, then adds the name of feature, otherwise keeps empty.

After checking all the 3 items, the program moves to check the 2 items, and if it finds any items verify the rules, and then writes the name of feature in the empty field.

The selecting feature from the 3 items has the max priority than the selection in 2 items.

To extract the features, there are some rules and constraints done to control the relations between the 2D drawings items each other. Cases 1 to 7 are selected cases for extracting the features from the 2D drawing, depending on the previous observations.

Case 1

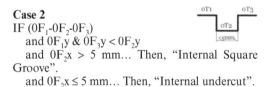

IF $(0T_1\text{-}0T_2\text{-}0T_3)$
  and $0T_1y$ & $0T_3y > 0T_2y$
  and $0T_2x > 5$ mm, ... Then "External Square Groove".
  and $0T_2x \leq 5$ mm, .. Then, "External undercut".

Case 2

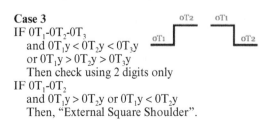

IF $(0F_1\text{-}0F_2\text{-}0F_3)$
  and $0F_1y$ & $0F_3y < 0F_2y$
  and $0F_2x > 5$ mm... Then, "Internal Square Groove".
  and $0F_2x \leq 5$ mm... Then, "Internal undercut".

Case 3

IF $0T_1\text{-}0T_2\text{-}0T_3$
  and $0T_1y < 0T_2y < 0T_3y$
  or $0T_1y > 0T_2y > 0T_3y$
  Then check using 2 digits only
IF $0T_1\text{-}0T_2$
  and $0T_1y > 0T_2y$ or $0T_1y < 0T_2y$
  Then, "External Square Shoulder".

Case 4

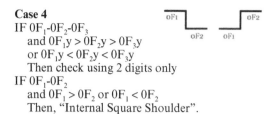

IF $0F_1\text{-}0F_2\text{-}0F_3$
  and $0F_1y > 0F_2y > 0F_3y$
  or $0F_1y < 0F_2y < 0F_3y$
  Then check using 2 digits only
IF $0F_1\text{-}0F_2$
  and $0F_1 > 0F_2$ or $0F_1 < 0F_2$
  Then, "Internal Square Shoulder".

Case 5

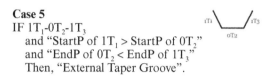

IF $1T_1\text{-}0T_2\text{-}1T_3$
  and "StartP of $1T_1 >$ StartP of $0T_2$,"
  and "EndP of $0T_2 <$ EndP of $1T_3$,"
  Then, "External Taper Groove".

Case 6

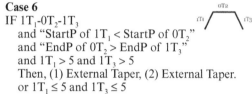

IF $1T_1\text{-}0T_2\text{-}1T_3$
  and "StartP of $1T_1 <$ StartP of $0T_2$,"
  and "EndP of $0T_2 >$ EndP of $1T_3$,"
  and $1T_1 > 5$ and $1T_3 > 5$
  Then, (1) External Taper, (2) External Taper.
  or $1T_1 \leq 5$ and $1T_3 \leq 5$
  Then, (1) External Chamfer, (2) External Chamfer.
  or $1T_1 > 5$ and $1T_3 \leq 5$
  Then, (1) External Taper, (2) External Chamfer.
  or $1T_1 \leq 5$ and $1T_3 > 5$
  Then, (1) External Chamfer, (2) External Taper.

Case 7

IF 1F1-0F2-1F3
  and "StartP of $1F_1 <$ StartP of $0F_2$,"
  and "EndP of $0f_2 >$ EndP of $1F_3$,"
  Then, "Internal Taper Groove".

3 CASE STUDY AND DISCUSSIONS

The proposed system is prepared as a standalone program based on the above rules and connected with EWDRAW module.

As shown in Figure 8, the "import file" button is made to invite the STEP file into the program

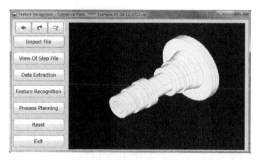

Figure 8. The program main interface.

interface screen. The icons rotate, move, and zoom in the top of the interface screen is made to be user friendly. The "View of STEP File" button, allows the user to go through the original STEP file and direct return back to the 3D model. As shown in Figures 8 and 9.

The "Data Extraction" button, displays the table of extracted data from the STEP file, which made as a report as shown in Figure 10.

The feature recognition button offers 2 shape of reports, one for the individual features such as

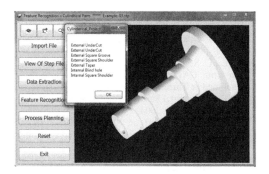

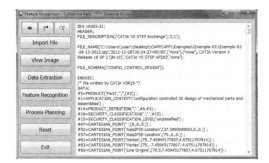

Figure 9. View STEP file module.

King Saud University

Advanced Manufacturing Institute

Data Extraction

Face Number	Face Name	Face Radius
1	Cone	0
2	Cone	0
3	generated cylinder	30
4	generated cylinder	30
5	generated cylinder	50
6	generated cylinder	50
7	Cone	0
8	Cone	0
9	plane	0
10	plane	0
11	plane	0
12	plane	0
13	plane	0

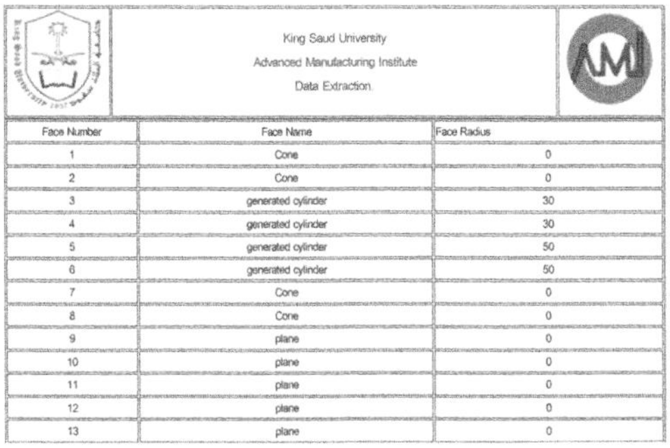

Figure 10. The data extraction module.

Feature Recognition Report.

The Feature Is External Chamfer

Face Number	Face Name	Face Radius
1	Cone	25
2	Cone	25
3	generated cylinder	25
4	generated cylinder	25
5	plane	18
6	plane	25

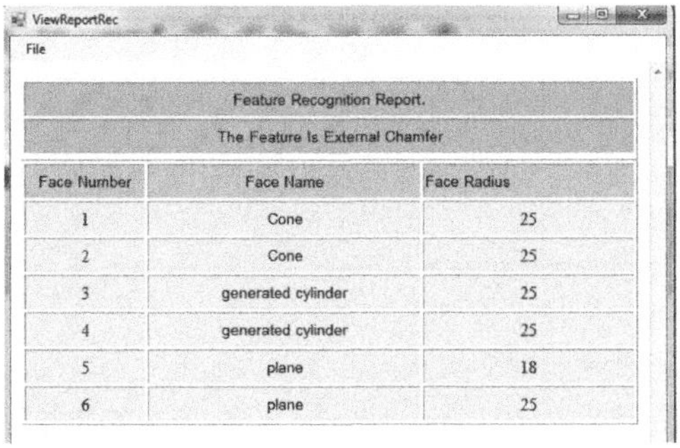

Figure 11. Individual feature recognition module.

Figure 12. The feature extraction module.

"plain cylinder", or "External chamfers" as shown in Figure 11, and the other for the combined feature recognition as shown in Figure 12, which is the main program to discuss in this paper.

4 CONCLUSION

In this paper, a standalone program for extracting the data and features from a solid model part based on STEP AP 203 is prepared. The proposed system succeeds to extract the data and features from turning parts. The proposed system constructs based on visual basic 2008 connected with EWDraw module.

ACKNOWLEDGMENTS

This works is funded by National Plan for Science & Technology (NPST), Grant No. 10-INF1280-02, Saudi Arabia. Moreover, the authors would like to appreciate the Advanced Manufacturing Institute (AMI) at King Saud University for their support throughout the research work.

REFERENCES

Abdulghafour, A.B. and Mithal Ahmed Al-Bassam 2012, "Automatic Features Recognition for Symmetrical Shapes.", Eng. & Tech. Journal, Vol. 30, No. 12.

Akkuş, K. May 2011, "Design of rotational parts using step AP224 features With automatic NC-Code generation.", Master Degree Thesis, Mechanical engineering, The graduate school of natural and applied sciences of middle east technical university.

Amaitik, S.M. 1994, "Computer Aided Interactive Process Planning System For Turning.", Master Degree Thesis, Industrial engineering department Faculty of Engineering, University of Garyounis, Benghazi, Libya.

Amaitik, S.M. 2005, "Development of a step feature-based intelligent process Planning system for prismatic parts.", PhD thesis, Mechanical engineering, Middle east technical university.

Aslan, E. 2005, "process unification and frame preparation of machining parameters for rotational parts.", Journal of engineering sciences, Vol. 11, No. 1, pp. 137–145.

Aslan, E., U. Seker & N. Alpdemir, 1999, "Data Extraction from CAD Model for Rotational Parts to be Machined at Turning Centres.", Tr. J. of Engineering and Environmental Science, Vol. 23, pp 339–347.

Fidan, T. December 2004, "Feature based design of rotational parts Based on STEP.", Master Degree Thesis, Mechanical engineering, Middle east technical university.

Kumari, I. and A.M. Magar, 2012,"DXF file extraction and Feature recognition.", International Journal of Engineering and Technology (IJET), Vol 4 No 2, pp. 93–96.

Mustafa, M.A.M. 2004 "Automatic generation of computer aided process panning and CNC code for rotary parts using geometric and topological similarity.", PhD Thesis, Industrial Engineering Department, Faculty of Engineering, Zagazig University.

Peterka, J., A. Janác & I. Kuric "The configuration utility of AutoCAD software for Computer aided process."

Shah, J.J. and Mantyla, M., 1995, "Parametric and feature-based CAD/CAM", John Wiley & Sons, Inc., New York.

Yusof, Y. and K. Case, 2008, "STEP Compliant CAD/CAPP/CAM System for Turning Operations.", Proceedings of the World Congress on Engineering and Computer Science 2008 WCECS 2008, San Francisco, USA.

Green Design, Materials and Manufacturing Processes – Bártolo et al. (eds)
© *2013 Taylor & Francis Group, London, ISBN 978-1-138-00046-9*

Computer aided feature recognition in free form parts

H.M.A. Hussein
Advanced Manufacturing Institute, King Saud University, Riyadh, Saudi Arabia

H.M. Mousa
Mechanical Engineering Department, Faculty of Engineering, South Valley University, Egypt

ABSTRACT: Automotive exterior shape and body panels are a mostly manufactured using thin formed sheet, which includes many of free form features such as holes and bends. In this paper, generative feature recognition system based on STEP AP203 is discussed briefly. This system construct based on VISUAL BASIC 2008 engaged with EWDraw module. The system methodology able to extract up to 11 different features from the free form shapes. The extracted features include external shapes such as (V-notch, U-Arc notch, U-straight notch, Arc notch), internal shapes such as (Round hole, Square hole, Rectangular hole, Single D-shape, Slots), beside to bends. A typical example is discussed including many internal and external features.

1 INTRODUCTION

The life story of the automotive external shapes and body panel design is passing through different stages. From style stage, to Body-In-White (BIW) design stage to Finite Element Analysis stage, and even in die design procedures, the user deals with automotive body model as a surface not as a sheet metal model.

Only, in some cases of sheet metal dies such as blanking, progressive, bending, and deep drawing dies, the thickness is an important parameter in the design process. In forming die design procedures, the designer works directly on a surface model which represent as a guide for the forming die core and cavity.

Computer Aided Process Planning (CAPP) is a broad term, with many variants covering a wide range of industries. Pairing between CAPP and machined parts leads us to the prismatic, cylindrical, and irregular shape parts. There has been a large numbers of research works in the area of CAPP for machined parts.

Pairing between CAPP and sheet metal parts is dependent on the sheet metal operation main groups, such as shearing, bending and deep drawing or forming. Pairing between CAPP and shearing, especially in blanking operations is called part nesting or sheet metal utilization. Jagirdar, 1995 use a mathematical methodology to extract and recognize sheared features from 2D drawings and automated building of a process planning and sheet metal utilization.

Pairing between CAPP and bending, discusses the arrangement and steps of sheet metal flaps

folding and unfolding. Duflou in 2005 preview the CAPP in bending operation development as a state-of-the-art.

Pairing between CAPP and sheet metal progressive dies is also called strip layout. Kumar in 2005 discuss the knowledge based associated in the design of strip layout.

The pairing between CAPP and deep drawing has its own philosophy, which succeeded till now in axisymmetrical, elliptical and boxed shape parts. Naranje in 2010 discuss the integrated system in CAPP for Axisymmetrical deep drawing parts. Park and Kang, in 1998 discuss a CAPP system in elliptical shape deep drawing parts, and Abdel-Magied in 2006, discuss a CAPP system in boxed shape parts.

From the literature, there are few researchers' attempts at pairing between CAPP and formed parts, such as Sunil and Panda. The reason of this paucity in this branch based on the difficulties in extraction of formed surface features. Feature recognition for free form surfaces is a new area of research work.

The CAPP procedure for automotive free form bodies uses the same stages of other CAPP systems such as data extraction, feature recognition, and finally the process planning. The features of the free form surfaces are close to the features of the sheet metal parts.

There are many sheet metal features in the real industrial world. In the literature, up to the present, there is no comprehensive system available for the recognition of all such features. Liu in 2004, proposed a hierarchical structure system which recognizes many features such as: wall, drawing,

239

bending, cutout, hole, flange, landing, coining, bridging, and slot.

Kannan, 2009 proposed a feature recognition system based on STEP AP-203 format. The system could extract features such as punching, notching, blanking, lancing, embossing, Louvering, V-bend, flanged parts such as (external, internal, collar and simple bends), hem parts such as: open and closed loop and curl parts such as: flat edge and rounded edge.

Zhang, 2006, proposed a staged approach for feature identification and extraction from a sheet metal model. The system is divided into several stages. It is automatically processed the results, and after each stage the results can be interactively corrected and modified for the next stage. The system was prepared mainly to provide for the 3D operation of progressive dies.

Farsi, 2009, proposed a feature recognition module and design advisor system for sheet metal components. Sheet metal features were extracted automatically from a 3D model. These data are modeled based on oriented objects. The system has been implemented in Solid Works 2008 using Visual Basic.

Sunil, 2008, proposed a system for automatic recognition of features from freeform surface of CAD models represented in STL format. The developed methodology has three major steps. STL model preprocessing, region segmentation and automated feature recognition.

In this paper, a short discussion about used free form feature recognition methodology. The proposed system simplifies the part feature in 2D drawing into arcs, vertical, horizontal and inclined lines. The coding for each edge is done automatically via system and a studding of the relation between each edge and the neighbored edges is done in similar to decision table. Finally, the automated feature recognition table is build showing the sequential features discovered in the part. The system is build to serve in the branch of computer aided process planning for the free form parts used in the automotive exterior bodies and BIW design.

2 SHEET METAL FEATURES METHEDOLOGY

Sheet metal is simply described as metal formed into thin and flat pieces. It is one of the fundamental forms used in metalworking industries, which could be cut and bent into a variety of different shapes. The sheet metal features results from a number of part operations, such as shearing or forming to serve in its proposed applications. There is a large number of features use in industry such as holes, slots, bends, etc. Figure 1 shows

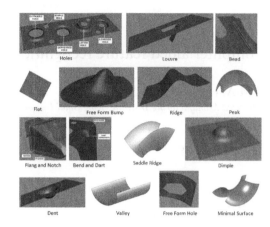

Figure 1. Typical examples of sheet metal features.

some examples of those features. Automated recognition of those features help in the future of the automated die design and selection.

2.1 STEP structure

The Standard for the Exchange of Product Model Data (STEP) is one of the most recent and promising standard, which representing part data in a neutral format. STEP is a family of standards, comprising several classes of documents organized in a database-like architecture. The purpose of STEP is to prescribe a neutral format capable of completely representing product data throughout the life cycle of a product.

2.2 Feature classification

Feature recognition is one of the recent major research issues in the area of automated CAD/CAPP interfaces. Researchers have proposed various approaches and algorithms in this area. Syntactic pattern recognition, volume decomposition, expert systems, graph based approach, and neural-network based approaches are some of those methodologies. Most of the above methods use a solid model as their input, which represents only the purely geometric aspects of the design information.

In this paper, feature recognition represent as a bridge between the CAD based STEP and the process planning. The free form surfaces features which represented in automotive panels could be categorized into stamped features and formed features.

In the current work; feature recognition of stamped features classifies into (internal or inner bound stamped features) and (external or outer bound stamped features) depending on whether they lie on the face outer bound loop of the flattened sheet metal part, or if they form one of the inner bound loops.

Notch features, which cut on the perimeter of the part, are considered as stamped features. Formed features, are bends, draws etc. Only bends and draws have been considered in the current research. Figure 2 shows stamped features with their edges loops.

2.3 Feature design

Automotive panels are manufactured by means of stamping and drawing processes. CATIA V5R19 sheet metal workbench can be used to design a solid sheet metal part with specific features. In case of complex automotive panels all design engineers in big companies use CATIA generative shape design module which deals with surfaces to produce the final CAD panel. Figure 3 shows the proposed feature classification according to STEP file format. More than one stamped feature can exist on one face. Bend features are defined

by a face and a cylindrical surface; however other stamped features are defined by face and a plane. Table 1 illustrates feature definition and design rules for a "*hole*" feature.

2.4 Data extraction stage

Extracting data to recognize features from STEP file with huge number of edge loops and vertex points is not an easy task. Microsoft ACCESS database tables are used to record faces, face outer bound, face bound, edge loops, oriented edges, edge curves, curve type, vertex, and vertex coordinates. All data, extracted from STEP file, and linked together as a tree is used to recognize features. Figure 4 shows a MS-Access table for step

Table 1. Feature definitions and design rules.

Feature: (hole)

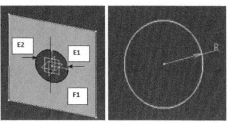

Type:	Face bound feature [shearing]
Feature definition	Advanced face [F1] EDGE_LOOP (two oriented edges) [E1,E2] Edge type [two arcs R1,R2] Feature area = 3.14 R2
Rule	1. Check if edge loop has only two-oriented edges 2. Check if that two-edge curve type is circle 3. Check for D/2 > 0.8 (for automotive panel)
Feature information	1. Hole diameter 2. Center coordinates of hole

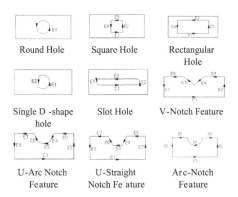

Figure 2. Face bound/internal and external stamped features.

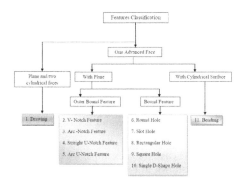

Figure 3. The proposed feature classification.

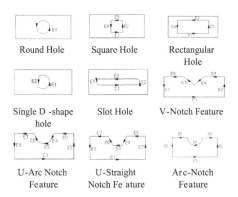

Figure. 4. Data extractions from step file using access tables.

data extraction. Figure 5 shows data extraction from a STEP file.

2.5 Notch—features recognition

It was simpler to recognize inner features than outer notch features, because of the numbers of oriented edges reaching a maximum of four edge curves with specific curve type (line or circle) as described in Table 2. In the case of outer face bound features, these lay on the perimeter of the part, which will increase the number of oriented edges and also increase the number of edge curves. A new technique is used as defined in Table 2, where each feature of these four notches has a rule defined by edge curve type. Table 3 shows geometric data coding for the three cases of line slope and circle.

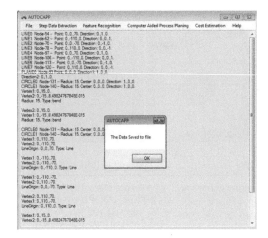

Figure 5. Data extractions from STEP file using access tables.

Table 2. Different codes for proposed notch features.

Feature shape			
Code	(11)	(030)	(131)
Feature name	V-notch	Arc-notch	u-arc-notch
Feature shape			
Code	(101)	(232)	(121)
Feature name	Straight u-notch	Arc-notch	Straight u-notch

Table 3. Geometric data coding.

Case number	Geometric data	Code
1	Line with slope (x = 0)	0
2	Line with slope (0 < x < ∞)	1
3	Line with slope (x = ∞)	2
4	Circle	3

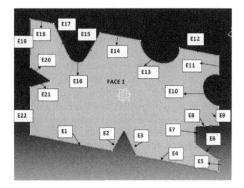

Figure 6. Illustrative example for proposed technique.

Table 4. Proposed features coding.

Feature	Code 1	Alternative code 2
V-notch	11	
Arc-notch	232	030
U-arc-notch	131	
U-straight notch	101	121

Based on Table 3, coding for notch features will be discussed as follows:

V—Notch feature: lies on one edge on the outer perimeter defined by two lines with case number two, so it will not be different if it lies in horizontal or vertical edge.

Arc—Notch feature: can be defined by an arc lies between two lines, hence it will have two codes as shown in Table 2.

U-Arc-Notch feature: can be defined by an arc lies between two included lines as shown in Table 2.

Straight U-Notch feature: This a special case defined by three connected lines two include lines in between horizontal line and it has two cases as shown in Table 2.

A complete case study example for the proposed technique is shown in Figure 6.

Table 4 summarizes feature coding which comes from oriented edges connectivity and

242

Table 5. Illustrative example for proposed technique.

Edge no.	Vertex 1	Vertex 2	Edge. type.	Line. slope.	RO 2E	RO 3E	Feature type
E1	V1	V2	Line	0	01	011	None
E2	V1	V2	Line	1	11	110	V-notch
E3	V1	V2	Line	1	10	102	None
E4	V1	V2	Line	0	02	021	None
E5	V1	V2	Line	2	21	210	None
E6	V1	V2	Line	1	10	101	U-straight notch
E7	V1	V2	Line	0	01	012	None
E8	V1	V2	Line	1	12	123	None
E9	V1	V2	Line	2	23	232	Arc-notch
E10	V1	V2	Cir.	3	32	320	None
E11	V1	V2	Line	2	20	203	None
E12	V1	V2	Line	0	03	030	Arc-notch
E13	V1	V2	Cir.	3	30	301	None
E14	V1	V2	Line	0	01	013	None
E15	V1	V2	Line	1	13	131	U-arc-notch
E16	V1	V2	Cir.	3	31	310	None
E17	V1	V2	Line	1	10	102	None
E18	V1	V2	Line	0	02	021	None
E19	V1	V2	Line	2	21	211	None
E20	V1	V2	Line	1	11	112	V-notch
E21	V1	V2	Line	1	12	120	None
E22	V1	V2	Line	2	20	201	None

relation between each two or three connected edges. Table 5 provides an example to clarify this technique.

3 RESULTS AND DISCUSSION

In this paper, a new standalone system is designed using Visual Basic2008, Ms-Access and EWDraw module. The system extracts the data and features from the free form parts. The system starts by receive the part in its ISO standard format (STEP) in our case, view the part in a standalone interface, which allow the user to rotate, move, zoom in, and out the model part in 3D direction. The system view the STEP file either as a text file or as visual model in its standalone interface. The turns from text to model is done by clicking on the "view the STEP file" button as shown in Figure 7. The system extracts the data and showing in a separate interface by clicking on the "data Extraction" button. The button "Feature Recognition" extracts the features from the free form part, as describe in Table 5. Finally, the reset button is used browse for a new part. Figure 7 shows a flat surface with stamped features. Figure 8 shows a bend part, which include multi surfaces, and multi features.

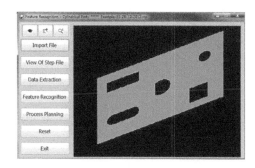

Figure 7. Flat surface with stamped features.

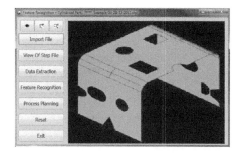

Figure 8. Bend part with multi-surfaces.

4 CONCLUSION

In this paper, a methodology for data extraction and feature recognition from a free form parts is discussed. A new system built using Visual basic 2008, MS. Access, and EWDraw module. The system designed to extract data and features from the free form parts. The system is designed as a CAPP in automotive components.

REFERENCES

Abdel-Magied, R.K.F. 2006, Computer Aided Process Planning for Deep Drawing, PhD. Thesis, Cairo University.

Duflou, J.R., J. Vanxza, R. Aerens, 2005, Computer aided process planning for sheet metal bending: A state of the art, vol. 56, No.7.

Farsi, M.A., B. Arezoo, 2009, Feature Recognition and Design Advisory System for Sheet Metal Components, "5th International Advanced Technologies Symposium (IATS'09)", May 13–15, Karabuk, Turkey.

Jagirdar, R. 1995, Set theoretic and graph based approach for automatic feature recognition of Sheet metal components, PhD Thesis, IIT-Kanpour, 1995.

Kannan T.R. and M.S., Shunmugam, 2009, Processing of 3D sheet metal components in STEP, AP-203 format. Part I: feature recognition system, International Journal of Production Research, vol. 47, No. 4, pp. 941–964.

Kumar S., and R. Singh, 2005, ISSLD: An intelligent system for strip layout design for sheet metal operation on progressive die,14th ISME International Conference on Mechanical Engineering in Knowledge Age, December 12–14, Delhi College of engineering, Delhi-110042, INDIA, Ref. 213.

Naranje, V., S. Kumar, 2010, A Low Cost Knowledge Base System Framework for Design of Deep Drawing Die, World Academy of Science, Engineering and Technology 72.

Park, D.H., W.R. Bae, and S.S. Kang, 1998, Computer Aided Process Planning (CAPP), System for Non-Axisymmetric Deep drawing Process, JMPT, vol. 75, pp. 17–26.

Sunil, V.B., S.S. Pande, 2008, Automatic recognition of features from freeform surface CAD models, CAD, vol.40, pp.502–517.

Zhang, W.Z., G.X. Wang, C. Lu and A.Y.C. Nee, 2007, A staged approach for feature extraction from sheet metal part models, IJPR, vol. 45, No. 15, pp. 3521–3544.

Zhi-jian, L., LI Jian-jun, W. Yi-lin, LI Cai-yuan, X. Xiang-zhi, 2004, Automatically extracting sheet-metal features from solid model. Journal of Zhejiang University Science, vol. 5, No. 11, pp. 1456–1465.

Green Design, Materials and Manufacturing Processes – Bártolo et al. (eds)
© *2013 Taylor & Francis Group, London, ISBN 978-1-138-00046-9*

Package design for frozen foods—principles for sustainability

R.A. Delfino
UTL—Technical University of Lisbon, Lisbon, Portugal

L.C. Paschoarelli
UNESP—State University of São Paulo, Bauru, Brazil

R. Frazão
LNEG—National Laboratory for Energy and Geology, Lisbon, Portugal

ABSTRACT: Packaging for frozen foods is characterized by technological and commercial development, but also leads to negative environmental impact. The present study assessed mock-ups developed by a packaging design method based on principles for sustainability («sustainable») and by a conventional design method («control»). A group of experts analysed eight mock-ups following two criteria: minimization of resources and the application of recycled and renewable materials. The results show that the mock-ups resulting from the «sustainable» process were significantly ($p \leq 0.05$) better assessed. This represents a step forward in methodological studies regarding packaging design that contributes to sustainable development.

1 INTRODUCTION

In recent years, food packaging design has been characterized by the use of new materials and new production technologies, based on new considerations in environmental, social and economic terms. However, food packaging continues to have a high environmental impact, which requires the development of new design methods based on sustainable principles.

Many of those food packages are using materials such as paper and cardboard. With that in mind, and aiming at a greater use of products suitable to the present needs, a method was developed to guide packaging design based on principles of sustainability, through an approach which is closer to production, materials, use and disposal.

The purpose of this study was the assessment of mock-ups designed under this method («sustainable»), comparing them with mock-ups designed using a conventional method («control»). The goal was to determine whether minimization of resources and the application of recycled and renewable materials were highlighted with the developed method.

2 LITERATURE REVIEW

To develop a method for packaging design that meets the principles of sustainability, there should be an analysis and comparison with other tools based on the lifecycle of products.

As a starting point, the definition of the qualitative tool «Design of life cycle» has been considered which, according to Frazão et al. (2006), is intended for the development of sustainable products and has a wide application in many different everyday products. From its principles, it is possible to set a checklist that considers the phases of pre-manufacturing, manufacturing, distribution, use and end of life of products.

In the Global Protocol on Packaging Sustainability 2.0 (2011), a project of The Consumer Goods Forum, The Global Network Serving Shopper & Consumer Needs, an assessment method is proposed that can be qualitative or quantitative, through a set of attributes that provide information about the performance of packaging throughout its life cycle. These attributes are based on standard tools (ISO 14040, 2008 and ISO 14044, 2010) and on European packaging legislation.

The Packaging and Packaging Waste Directive (P&PW) 94/62/EC (1994), currently in use in the European Union, as amended by the Directive 2004/12/EC (2004), provides the general requirements regarding the P&PW management, with emphasis on the control of heavy metals concentration, as well as in the essential requirements regarding the production and composition of packaging and its possibilities of reuse or recovery.

On essential requirements, the European Committee for Standardisation (CEN) published six European Standards (EN) and two Technical Reports (TR), in 2000 (revised in 2004), with the goal of providing practical guidelines on how to comply and demonstrate such requirements. The EN 13430 (2005) particularly specifies the requirements and establishes conformity assessment procedures, especially in its Annex «A» criteria identification for recyclable packaging: manufacturing control/composition and production process, suitability for recycling technologies and environmental emissions caused by the recycling of after use packages.

The Green Paper, Packaging and Sustainability. edited by the EUROPEN—The European Organization for Packaging and the Environment (2011), considers seven areas from the perspective of achieving an «ideal package»: material selection, design, consumer's choice, production, use, end of life for packaging and an innovative business model. These are relevant parameters that match previous methodologies and, to some extent, push for the application of European standards.

The COMPASS (2009) is a web-based application for packaging designers and engineers, which compares the human and environmental impact of packaging. It was developed by the Sustainable Packaging Coalition (SPC), a project by GreenBlue, a non-profit making institution founded by architect William McDonough and chemist Michael Braungart. Its initial goal was to promote the implementation of the Cradle to Cradle project. The metrics of information within the COMPASS include the life cycle and the attributes of packages. The set of metrics was designed under a process guided by the SPC definition of sustainable packaging and ISO 14044.

The Good Practice Guide, produced by Envirowise (2008), considers the following parameters: the use of hazardous substances; minimization of resources; packaging design with recycled and renewable materials; packaging design for reuse; packaging design for recycling and possible composting; and packaging design to final disposal

Using the qualitative tools of Lifecycle Design and adopting some of the questions of the Envirowise guide, a particular design method was developed, aiming at paper and cardboard packaging and the food products category. This consisted of a checklist that considered four major topics: Minimization of Resources, Use of Renewable and Recycled Materials, Use and Disposal, and Recycling and Possible Composting. The method further comprised an analysis matrix on the characteristics of the graphic production processes to be used, as well as a matrix of the symbols about materials and their disposal.

Following this method, packages were designed for frozen foods, thus characterizing a lot of mock-ups that might be assessed.

3 OBJECTIVE

The aim of this study was to compare a group of projects for frozen foods packaging, developed with a design method based on sustainability principles («sustainable»), with another group of packaging designs developed with traditional methods («control»), considering minimization of resources and the use of renewable and recycled materials.

4 MATERIALS AND METHODS

4.1 Object of analysis

Eight mock-ups of frozen foods packaging were divided into two groups of four and analysed. In the first group, «sustainable», the participants adopted the aforementioned method based on sustainability parameters. The second group, «control», designed the packages using a conventional design method. The analysis criteria used were minimization of resources and use of renewable and recycled materials.

4.2 Participants of the panel of experts

The assessment of the packaging mock-ups was performed by a panel of experts. Twelve experts participated in this panel, with broad and diverse academic and professional backgrounds, covering the different stages of packaging production, from design to the end of life: packaging design and design for sustainability, management in the printing and converting industry, graphic technology,

Figure 1.   Packages of the «sustainable» group.

Figure 2. Packages of the «control» group.

marketing in food industry, environmental engineering, chemical engineering, food engineering, paper engineering and urban solid waste.

4.3 Materials

We conducted a survey, concerning each of the eight packages. The survey was available digitally at the Google Drive system. An adaptation of the Likert scale (Tullis & Albert, 2008) was employed here. There are six levels, previously determined to make the process of assigning perceptive values the same for all participants, as well as to allow for an assessment and quantitative analysis of the assigned values. The levels were: «DN/NA» («don't know» or «no answer»—zero); «Very Inadequate» (one); «Inadequate» (two); «Basic or Normal» (three); «Satisfactory» (four) and «Very satisfactory» (five).

For each mock-up a fact sheet was given with detailed information, photographs, technical drawings and a video demonstrating the use and disposal, presenting the package in detail. The identification of the mock-ups of the eight packages was done by numbers and was arbitrary.

4.4 Analysed parameters

In this study, two parameters of the packaging design method for sustainable development were analysed, namely: Minimization of Resources and Renewable and Recycled Materials.

The Minimization of Resources parameter focuses on the issues: production of the package in order to reduce waste; reduction of material and/or accessories, such as layers, loose labels, tapes or adhesives; reduction of internal spaces and elimination of other materials, such as plastic windows. Another important topic is related to the

dimensions and shapes of the primary packaging in order to optimize the carriage.

The second parameter, the use of Renewable and Recycled Materials, considers the main raw material, paper and cardboard, originating in a renewable source; certifications of source, such as FSC and PEFC (certifications of sustainable forests); types of glues, inks and varnishes; use of recycled materials for food products; the barrier to moisture and fat; and the guidelines of the Portuguese Sociedade Ponto Verde (Green Dot Society).

4.5 Data analysis

The data analysis was based on the numerical results of the scales, which were statistically analysed.

The values assigned to the projects «sustainable» and projects «control» were gathered separately, resulting in $N = 48$ for each of the analysed variables. In order to verify the occurrence of significant differences ($p \leq 0.05$) when comparing the two groups of projects, we applied the nonparametric 'Wilcoxon' test in all interactions, since the assumptions of normality of the sample groups were not met ($p > 0.05$) (Shapiro-Wilk's test).

5 RESULTS

5.1 Minimization of resources

Comparing the «sustainable» mock-ups with the «control» mock-ups, the results related to the parameter «Minimization of Resources» demonstrate that the «sustainable» mock-ups (mean 3.58,

Figure 3. Values (and standard deviation) of the «sustainable» and «control» mock-ups, in the parameter «Minimization of Resources».

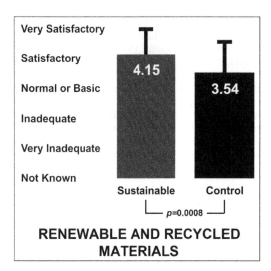

Figure 4. Values (and standard deviation) of the «sustainable» and «control» mock-ups, in the parameter «Renewable and Recycled Materials».

s.d. 1.40) were significantly better rated ($p \leq 0.05$) than the «control» mock-ups (mean 3.08, s.d. 1.40), as shown in Figure 3.

5.2 Recycled and renewable materials

Comparing the «sustainable» with the «control» mock-ups, the results related to the parameter «Renewable and Recycled Materials» demonstrate that the «sustainable» mock-ups (mean 4.15, s.d. 0.82) were significantly better rated ($p \leq 0.05$) than the «control» mock-ups (mean 3.54, s.d. 1.03), as shown in Figure 4.

6 DISCUSSION AND CONCLUSIONS

The results of the comparison between mock-ups of different packaging groups show that the new design method based on principles of sustainability positively influenced the evaluated parameters, Minimization of Resources and Renewable and Recycled Materials.

The presented results, however, are partial and do not fully confirm the proposed design method. Nevertheless, one cannot deny that this method provides a favourable path to a sustainable development, which contributes significantly to studies and to the productive sector of this area.

The parameters related to the graphic technologies of production and the use of symbols regarding to materials and their disposal are still awaiting validation.

ACKNOWLEDGMENT

This study was partially supported by the FCT—Fundação para a Ciência e Tecnologia—Portugal. (Process J044234F4375).

REFERENCES

Comité Europeu de Normalização 2005. NP EN 13427—Embalagem—Requisitos para a utilização de normas Europeias na área de embalagem e resíduos de embalagem, Lisboa: Agência Portuguesa do Ambiente.

Comité Europeu de Normalização 2008. *EN ISO 14040:2006*, 2nd ed. Lisboa: Agência Portuguesa do Ambiente.

Comité Europeu de Normalização 2010. *EN ISO 14044:2006*. Agência Portuguesa do Ambiente.

Directiva 94/62/CE do Parlamento Europeu do Conselho de 20 de Dezembro de 1994, relativa a embalagens e resíduos de embalagens (JO L 365 de 31.12.1994, p. 10) in 1994 L0062—PT—20.04.2009–004.001–1.

Envirowise 2008. Packaging design for the environment: Reducing costs and quantities. Harwell: Envirowise.

Europen 2011. Green Paper, Packaging and Sustainability. An open dialogue between stakeholders. Brussels: Europen.

Frazão, R. & Peneda, C. & Fernandes, R. 2006. *Adoptar a Perspectiva de Ciclo de Vida*. Lisboa: Ineti (Instituto Nacional de Engenharia e Inovação, I.P.) Cendes (Centro para o desenvolvimento Empresarial Sustentável).

Sustainable Packaging Coalition 2009. *Sustainable Packaging Indicators and Metrics Framework*. Charlottesville: GreenBlue.

The Consumer Goods Forum 2011. *Global Protocol on Packaging Sustainability 2.0*. Issy-les-Moulineaux: The Global Network Serving Shopper & Consumer Needs.

Tullis, Tom & Albert, Bill 2008. Measuring the user experience: collecting, analyzing and presenting usability metrics. Burlington: Elsevier Inc.

Green Design, Materials and Manufacturing Processes – Bártolo et al. (eds)
© 2013 Taylor & Francis Group, London, ISBN 978-1-138-00046-9

Synthetic manufacturing—Resilient Modular Systems (RMS)

Wendy W. Fok

Digital Media & Design Program, Gerald D Hines College of Architecture, University of Houston,
atelier//studio WF, New York, USA

ABSTRACT: Resilient Modular System (RMS) (temporal + structural) is a multi-disciplinary research proposal to forge the synergy and efforts between three different colleges/departments within the University of Houston: College of Architecture, Department of Industrial Engineering, and Department of Material Studies and Engineering, designed into a modular system that could be applied as urban interventions within the context of temporary and permanent settings.

The topics and fields of research will include, but will not be limited to: Architecture/Design, Industrial Engineering and Prototyping (Digital/Analogue), Patents, and the Material Sciences. Within the larger understanding of the design-research, all conducted research will require a high level of computational science and bio-engineering support. Each collaborator/Faculty member is a key asset to the development of this project and is experts within their respective fields.

1 RESILIENT MODULAR SYSTEM (RMS)

RMS (temporal + structural) as indicated on the Figure 1 diagram illustrates a multi-disciplinary research project paper that demonstrates the forged synergy and efforts between three different colleges/departments within the University of Houston: College of Architecture, Department of Industrial Engineering, and Department of Material Studies and Engineering, designed into a modular system that could be applied as urban interventions within the context of temporary and permanent settings.

The topics and fields of research will include, but will not be limited to: Architecture/Design, Industrial Engineering and Prototyping (Digital/Analogue), Patents, and the Material Sciences. Within the larger understanding of the design-research, all conducted research will require a high level of computational science and bio-engineering support. Each collaborator/Faculty member is a key asset to the development of this project and is experts within their respective fields. The division of research and development will be as follows: Prof Wendy W Fok (Architecture/Design/Prototyping—Harvard University GSD Doctor of Design doctoral candidate), Generic graduate student (Architecture/Design/Prototyping), Prof Ali Kamrani (Patent/Industrial Engineering/Modular design aspects for form and fit analysis), and Prof Ramanan Krishnamoorti (Material Sciences/Bio-related engineering).

This paper outlines the thorough development and testing required, ensuring that the design development of the temporary structures could

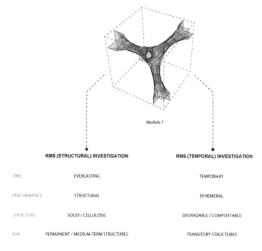

Figure 1. Dualistic research approach of RMS.

be maintained and withstand structural integrity for use within extreme environments, or more significantly, the constructability of these temporary structures by unskilled laborers in variable scenarios.

These applications could include but not limited to:

– Applications as temporary retaining wall within landscape architecture for construction of green public spaces
– During the need of temporary walls to allow for the plant growth to mature

– Along areas like a bayou to clasp structural bracing of falling rock piles, in or from behind
– For domestic use, for urbanites to grow large bushes for dividing spaces with living walls, and as a structure to support plant growth. I.e.: how tomatoes require wire and vine support.

Using both eco-intelligent architectural design objectives, the knowledge and technique of manipulating sustainable materials ultimately pursues a positive impact on the planet as a growth opportunity and engenders a focus on enhancing benefits (not only reducing costs) through its decision-making and actions—taking an approach of optimization rather than minimization. This paper is to outline how the project can understand the perspective of "people, planet and profit," as expansionist and enabling leadership through the achievement of advanced success metrics. For example, the concept of effective design of products and services should move beyond typical measures of quality—cost, performance and aesthetics—to integrate and apply additional objectives addressing the environment and social responsibility.

Through both digital and analogue (physical) prototyping in both architectural and design scales and migrating the opportunity of a full-cycle cradle-to-cradle design process into a Design-Fabrication project—with real-world contextual testing, and use of both repurposed construction waste and biodegradable materials (specifically, biodegradable soy-based polyurethanes, ceramic fillers, and composite plastics)—RMS (temporal + structural) is to find a dualistic opportunity into sourcing ecological solutions of constructing temporary structures within the built environment in locations of need.

2 PROJECT BREAKDOWN

The RMS (temporal + structural) proposal is two-fold. The purpose of the proposal is to A) investigate the opportunities that allow for temporary and biodegradable structures, which could be utilized for temporary construction sites within the built environment, and to B) explore the structural significance and integrity of a modular structure utilising repurposed construction waste.

A) The idea of the RMS (temporal) is the ability of it to become an ecological and resilient modular construct for the built environment that could be subsequently dissolved, yet, in an effort of full-cycle design, also contribute to nourishing the natural landscape.

The temporary proposal is that one of these structures could be possibly constructed as a retaining wall system—similar to the ones that are seen along the side of the highway or a landslide retention wall. The composite within the mixture of this will consist of ceramic filler, broken down glass, and biodegradable plastic as the main composite material. The process of this works as follows: 1) a landslide retaining wall is constructed with the RMS module, 2) due to exposure and UV tested breakdown, when the biodegradable plastic comes to the end-life, 3) the plastic will degrade and dissolve. 4) Since the plastic is made with a mixture of ceramic filler, 5) when the plastic dissolves, 6) the ceramic filler will be left, and since the ceramic filler itself retains moister, 7) when the ceramic filler is deposited into the soil, it would provide itself as a form of nourishment for plantation and development for agricultural growth.

The primary material research for the RMS (temporal), ephemeral structure as illustrated on Figure 2, will be based on agricultural or soy-based biodegradable polymers have been in research since the late 90s and have been improved, bought out and carried forward, by some of the world's largest companies, like food and agricultural giant Cargill, who in 2008, spent over 22 million USD on developing a method to research and use polyols that can replace petroleum-based chemicals. The most effective method is to blend soy protein plastic with biodegradable polymer to form soy protein based biodegradable plastic, and forming the material with the method of extrusion and injection-molding to form useable pieces of plastic. Therefore, using the same traditional methods of constructing plastics, the same design fabricated parts would be used for applying similar 'thermoforming' or 'vacuum' forming techniques into constructing the prototypes.

B) While the secondary research for the RMS (structural) will be research for repurposing construction waste, as a mixture for the remediation of the structural testing and joint detailing, the same modular structure will be utilized to further

Figure 2. Scale, iteration, and tessellation of the RMS modular.

the innovate on studying the structural form/fix/ analysis of the RMS (structural) modular.

Project managers and construction contractors have long recognized the importance of reducing waste and salvaging high value construction and demolition materials such as copper and other metals. Contractors are usually careful about the quantity of materials ordered, how materials are used and how to carefully de-construct valuable materials. In most cases however, materials that are more difficult to separate and that are worth less per unit weight are still going to landfill, even when they are present in large quantities. This represents an inefficient use of natural resources and uses up landfill capacity unnecessarily.

Unfortunately, some contractors do not realize that there are new opportunities for waste minimization, while others are reluctant to implement environmental practices because they believe these practices will increase their project costs. Most contractors are concerned about the cost of the labor that is needed to deconstruct materials for reuse or recycling. However, it has been shown that effective waste management during Construction, Renovation and Demolition (CRD) projects not only help protect the environment, but can also generate significant economic savings. Demonstration projects have shown that the diversion of waste from landfill can reduce waste disposal costs by up to 30%. This is accomplished through reduced tipping and haulage fees and the sale of reusable and recyclable materials.

The diversion of Construction, Renovation and Demolition (CRD) waste from landfill sites is an issue that has been gaining attention within both the public and private sectors. Surveys have indicated that as much as one third of the 20 million tonnes of solid waste of municipal waste streams is generated by construction, renovation and demolition activities. Many of our landfill sites are reaching capacity. In addition, CRD waste is sometimes illegally dumped or burned, causing land, air and water pollution. The increasing costs of disposal are ultimately reflected in project costs, as contractors must incorporate anticipated disposal costs in their bid costing. Realities such as these emphasize the need for initiatives that focus on reducing and diverting as much waste as possible from CRD activities.

Incorporating the 3Rs (reduce, reuse and recycle) into construction, renovation and demolition waste management creates a closed-loop manufacturing and purchasing cycle. This significantly reduces the need to extract raw materials, reduces the amount of materials going to landfill sites and reduces the life-cycle costs of buildings and building materials.

3 OBJECTIVES

With the rise of computer aided technology, the vast amount of rapid prototyping tools prompts designers to question how our visions of objectivity diverge into the tendency to push and understand the limits of different material properties to further the development of architectural design. The premise of this research proposal is to achieve speculative studies within a project framework, which will be presented through a quad-fold process of: design-research, fabrication-construction, exhibition-publication, and international distribution (including patents).

The development of the paper is to demonstrate and will contribute to the design-development and phase I prototype of an exhibition, which will be completed within the first twelve months of the commencement of the project; whereby the phase II processes will include the development of the commercial production for the piece to travel within the region and be displayed within Houston as a case-study. The ambition is also potentially exhibiting the piece at ACADIA 2014 and potentially disseminate to various conferences that include the community, industry partners, academics, designers, architects, urban planners, and policy makers to continue the discussion of novel production and fabrication methods, and the utility of innovative materials within the processes of resilient design.

Design—as illustrated on Figure 3, demonstrates the larger function of the term inclusive of Research and Development of Applied Sciences, Engineering, Technology and Architecture—today could perhaps be described as the relational equations mediated by digital techniques assisted with production and knowledge of fabrication. Like many fields in the modern culture, it strives to be truly integrated wherein the designer can move

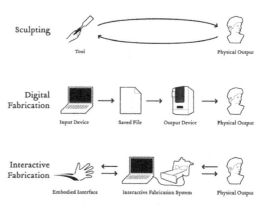

Figure 3.   Research, development, and design process.

seamlessly from concept to production in a single, contained process.

A full-scale (1:1) prototype would be built in Houston, Texas—with space generously offered by the Gerald D Hines College of Architecture, the Burdette Keeland Jr. Design Exploration Center, and the Design and Free Form Fabrication Laboratory at the University of Houston—which would then be transported to different locations for onsite eco-systemic testing.

4 PERFORMATIVE CRITERIA

The current environmental crisis and diminished natural resources has challenged the practice of Architecture to re-think its outdated processes of design and construction (Fig. 5). New processes that act as full regenerative cycle systems are replacing existing wasteful construction models. The scope of this work focuses on the understanding and development of minimal surfaces specifically of those that are Triply periodic (i.e., Periodic in three direction) as an efficient modular building component fabricated out of high content recycle/salvaged construction solid waste. Each building component will be designed utilizing computational generative strategies to find the most optimal performance. Rapid prototyping and digital fabrication methods will be utilized in order to find efficient and economical modular structure systems that perform at three levels: *structurally, environmentally and socio-economically.*

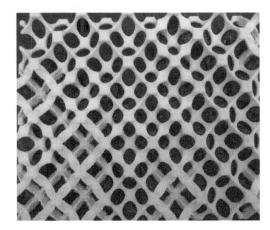

Figure 5.   3D printed ABS prototype.

Figure 6.   Fabrication testing of prototype and mold.

4.1   Structural performance

Figure 4 above illustrates the tetra-pod module chosen as a point of departure as it offers high structure efficiency and construction flexibility. When compared to traditional building material such as a modular brick and concrete masonry block, the tetra-pod outperforms its predecessor in terms of low surface area/weight ratio, performing the same compressive pressures and yield stress ratios.

4.2   Environmental performance

1. Utilization of a mixture of repurposed salvaged/ recycled material, specifically construction debris: Brick, mortar and cement as the prime material for the construction of each module (Figs. 7 and 9).

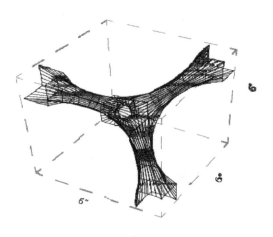

Module 1

Figure 4. Basic modular systematic set-up and constraint.

2. Climate adaptation of the modules. According to the climate conditions of the site to maximize natural ventilation systems and onsite energy generation (Fig. 8).

4.3 *Socio-economic performance*

Fabrication: Figure 6 demonstrates the process to fabricate full-scale modules utilizing rapid prototyping technologies to create molds that could utilize in the mass production of different variations of modules. To develop connection and joint systems that allow for flexible modular configurations.

4.4 *Structural applications/interventions*

Figure 7.   Landscape systems.

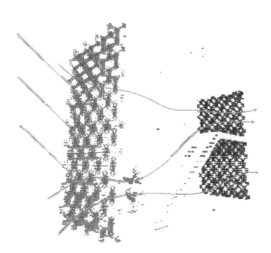

Figure 8.   Residential applications.

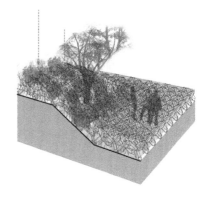

Figure 9.   Pervious surfaces.

5   PROJECT SCHEDULE

Activities: Stage I: Research & Development/ Schematic Design

- Design-Build Architectural Research Studios (Graduate/Intermediate University Level Courses)
- Material and fabrication R&D
- Monthly E-Newsletters and Blog to recognise innovative thinking and give individuals the means to carry out forward-looking opinions that benefit the endeavour of sustainability and cradle-to-cradle design thought within the communities and the wider world
- Design-Fabricate 1:200 scaled prototypes of architectural models
- Design-Test 1:50 scaled prototypes and joint-system of temporary structures

Stage II: Design Development/Construction Development

- Design-Fabricate 1:10 scaled prototypes of architectural structure
- Conferences and Panel discussions inviting industry specialists within academia and professional fields for exchange of ideas and further critique of proposals for group
- Further development into Production of 1:10 & 1:2 scaled prototypes of architectural structure

Stage III: Construction Administration

- Construction of 1:1 scaled prototype and execution
- Official publication of a detailed documented research and conclusive book, which outlines the different stages of the project, developmental research, the physical implementation and construction of the project into the finalized stages, as well as the conferences and exhibitions of that are situated for the project. The publication of the book would be an opportunity to fostered infor-

mation from harvest to growth as a chance for the general public to learn about the project and the life-cycle of cradle-to-cradle design-fabrication systems that contribute to the built environment.

- Exhibitions with established international (Asia (China and India) and Europe) and domestic (North America) publication, conferences, museums and galleries to promote and expand knowledge of our world, improve the quality of life on the planet or contribute to the betterment of humankind.

6 DELIVERABLES

All Design Development and respective stages will be clearly documented through a physical report, and online website documentation, for the opportunity to share the concept with the available consultants, mediators, and trustees of the Digital Media & Design Program, and the College of Architecture. Figure 10 demonstrates one of the few prototypes that are produced through the research.

Schematic Design: (15%)

- Full report on Preliminary Structural analysis drawings
- Conceptual Renderings and diagrammatic information.

Design Development: (15%)

- Environmental Analysis results
- Computer Fluid Dynamics results
- Preliminary Joint and Connection details drawings
- Fabricated joint connection details and material research results.

Construction Documents: (30%)

- Physical Testing
- Detailed Modular Unit Component

Figure 10. RMS module unit.

- Series of variable fabricated modules
- Architectural set of drawings containing Modular Components, Joint Details, Materials and Manufacturer Specifications.

Fabrication: (30%)

- Full Scale modular building component mock up approx. 96″ × 96″ × 24″
  Business and Marketing Plan: (10%)
- Business and Market feasibility plan.

7 STUDENT INVOLVEMENT

The conceptual idea of RMS was developed through the research interests of Professor Wendy W Fok, on the repurposing of construction waste and full-cycle design utilizing digital fabrication methods, and her advising of a Level III graduate studio thesis project on a modularity that investigated into performative structures. The collaboration with Professor Ali Kamrani and Professor Ramanan Krishnamoorti is to further hybridize the effective nature of the research to the involvement of students through the schematic and construction process of the *College of Architecture*, the *Industrial Engineering*, and the *Material Science* departments.

8 MARKET RESEARCH

RMS Modular Units have been designed to work in different scales that can serve as façade screening systems, landscape systems such as ecological freeway shoulder pervious surfaces, ecological fencing systems, and as temporary structures within the built environment. The RMS structural modular units could also serve as porous building envelopes or temporary housing units during times of chaos and disaster.

The primary goal of the RMS (temporal) structure is the ability to market itself to the public and urban markets within the public domain, such as the Texas Department of Transportation, as a structure that is seen as a structural component that could be potentially dissolved and degraded, and would contribute back (full-cycled) into nature as nourishment for soil, due to its composite manufactured. Else, the secondary goal would be for the RMS (structural) to be further marketed within areas, which require a more permanent permeated structure.

Potential Markets

- US Green Build Government Agency
- Texas Department of Transportation
- Federal Emergency Management Agency
- Houston Housing and Community Development
- United States Environmental Protection Agency

9 SAMPLE—PROTOTYPE MODULE
  BUDGET

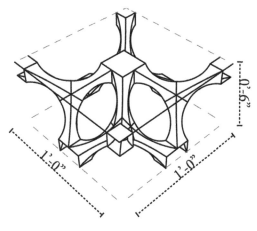

Figure 11.   Unreinforced module 1 (4 single pieces).

48″ × 16 × 4″ Eps foam mold*	$ 25
CNC Router time 1 hour	$ 10 (Keeland exploration center)
Rockite anchoring cement 35% (500 Ml)	$ 5
Aggregate	$ 0 (Repurposed construction waste)
Plywood forming	$ 2
8-clamps	$ 0 (Keeland exploration center)
Tape	$ 3
Mold release	$ 4
Shelac mold protector	$ 4
Total module	**$ 53** (Including mold*)
	$ 5 (Module not including mold and tools)

*Each mold has the ability to produce approximately 5–6 modules.

Green Design, Materials and Manufacturing Processes – Bártolo et al. (eds)
© 2013 Taylor & Francis Group, London, ISBN 978-1-138-00046-9

System-based analysis of environmental effects of liquid food packaging

Zs. Bogóné-Tóth & Z. Lakner
Corvinus University Budapest, Hungary

ABSTRACT: During the last decades there has been a rapid increasing of liquid food (e.g. soft and energy drinks, different types of juices, mineral water, tee, etc.) consumption. Based on development in food technology as well as introduction of new packaging materials, the structure of packaging has been changed too. These processes raise a number of environmental challenges. Based on the comparative analysis of literature the article presents the logic and process of a life-cycle analysis, highlighting the importance of system-based approach. The effects of packaging on greenhouse-effect gases, fossil resources, and the acidification have been analysed. It can be proven, that it is not enough to pick up one or another part of the lifecycle of product, but an integrated approach is necessary, which takes into consideration the environmental burden of raw material production, transportation, recycling, re-using and deposition.

1 INTRODUCTION

In the last three decades we could see a considerable development of food consumption patterns and preservation technologies. This fact can be explained by the changing consumers demands (e.g. switch over from traditional tap water to the mineral waters), increasing demand for convenience, and higher demands for the quality of products (e.g. decreasing consumption of CO_2 containing, carbonated soft drinks to fruit and vegetable drinks). A very important technological breakthrough has been the introduction and rapid proliferation of aseptic technologies.

The basic approach of this process in to achieve a sterile (aseptic) product in a sterile container in a way that maintains sterility. The first aseptic filling plant for milk had been presented in 1961, but from 1970s the aseptic techniques became common practically all over the world. The traditional (one or more way) bottles are replaced by Polyethylene Terephthalate (PET) bottles in an increasing way in fruit juice and milk markets.

The changing consumption patterns and modifications in packaging technology involve an increasing environmental burden, too. The management of ever-increasing quantity of packaging materials means a considerable challenge from point of view of environmental management system. A striking example of this is the fact, that worldwide, approximately 2,65 million tons of PET were have been used juts in the US.

Under these conditions the different governments try to apply different approaches for the waste material disposal and management systems. In Hungary, the government has introduced the product fee system in 1995. This was a green tax on certain products, assumed to create waste in bulk. Among other products these "environmentally sensitive" products include the combined food packaging materials and PET packages. In 2008 the customs authorities took over the bureaucratic administration of the green tax. In 2011 the administrative burden has been diminished by a considerable simplification procedure, but-at the same time-the tax has been increased considerably. A common feature of the tax system modifications has been the fact, that there was not the slightest attempt to determine the real environmental burden of the packaging system. Under these conditions the 18-years long history of the green tax of packaging materials in the long debate of different lobby-groups with antagonistic interests (e.g. one-way glass bottle vs. PET bottle, PET bottle-vs. combined packages, etc ...). The aim of the current article is to determine the methodological base of a more well-founded environmental tax system, which would be able

1. to offer a more realistic picture on environmental burden of different packaging materials;
2. determine the actual environmental burden;
3. base of this knowledge model the optimal monetary flow between different economic entities.

2 METHODOLOGY

The different methods of Life Cycle Analysis are standardised:

The international standard ISO for Life Cycle Assessment are ISO 14040: 2006 and ISO 14044: 2006. They replace the previous standards for Life Cycle Assessment ISO 14040: 1997, ISO 14041: 1999 about Goal and scope definition and inventory analysis, ISO 14042:2000 on Life cycle impact assessment and ISO 14043:2000 on Life cycle interpretation.

The scope of an LCA study based on ISO 14040 includes the following items:

1. The product system to be studied
2. The functions of the product system, or in the case of comparative studies, the systems
3. The functional unit
4. The system boundary
5. Allocation procedures
6. Impact categories selected and methodology of impact assessment, and subsequent interpretation to be used
7. Data requirements
8. Assumptions
9. Limitations
10. Initial data quality requirements
11. Type of critical review (if any)
12. Type and format of the report required for the study.

3 RESULTS AND DISCUSSION

The scope of the studies should be as wide as possible, but there exists a considerable risk of determining the system-boundaries as a too large one. In our opinion the best way is to include into the analysis the material production, but leave out the energy and another inputs, necessary for the production of machines and apparatus, necessary for the production of these packaging materials. Of course, the technological energy and environmental burden of production of the food materials should not be taken into consideration, too.

The next step in LCA is the Life Cycle Impact Assessment.

According to the current standards, this process contains the steps as follows:

1. Selection and Definition of Impact Categories
2. Classification
3. Characterization
4. Normalization
5. Grouping (optional)
6. Weighting (optional)
7. Evaluation and report of LCIA results in order to gain a better understanding of the reliability of the LCIA results.

One of the most difficult part of LCA is the weighting of different factors. The simplest way could be the monetary weighting, but in our opinion there is a considerable danger, that the weighting will be undermined by the distorted by the current European monetary crisis. There seems to be two solutions: the paired preference test and the future scenario planning. The pairwise comparison, introduced by psychometrician L.L. Thurstone is a scientific approach to using pairwise comparisons for measurement. The test is based on the low of comparative judgement.

Another possible method is the future analysis, based on which the environmental consequences of increasing of different environmental burden consequences could be quantified.

The life cycle analysis can be carried over on base of different databases. There are numerous high-quality collection of environmental impact databases, some of them are free to use. The most important are as follows:

- *The European Reference Life Cycle Data System (ELCD core database)* http://lca.jrc.ec.europa.eu/lcainfohub/datasetArea.vm
- *CPM LCA database* http://cpmdatabase.cpm.chalmers.se/
- *US LCI database* http://www.nrel.gov/lci/
- *PlasticsEurope* http://www.plasticseurope.org/plastics-sustainability/life-cycle-thinking.aspx
- *The World Steel Association* http://www.worldsteel.org/index.php?action = programs&id = 62
- *FEFCO (European Federation of Corrugated Board Manufacturers)* http://www.fefco.org/publications/other-publications/european-database.html
- *Ecoinvent* http://www.ecoinvent.ch/
- GaBi (http://www.gabi-software.com/databases/)

Based on these information resources there is a possibility to carry over the environmental analysis itself.

There are two possibilities of LCA:

1. application on specific LCA software;
2. or the application of general-purpose simulation software, using the data on environmental burden from some electronic database. There is a rather wide choice of different LCA software solutions. A relatively early survey on these tools has been published by Jörnbrink et al. (2000). They have determined, that there are different types of LCA software on the market. A detailed comparison of four types LCA software can be found in working paper of Boureima et al. (2007). Based on a detailed survey, they suggest that for the purposes of the work the RangeLCA seems to be the most adequate software, because RangeLCA includes a

Table 1. From https://www.politesi.polimi.it/bitstream/10589/29701/3/2011_10_Latunussa.pdf 16. p.

Impact categories	Elementary flows	Unit
Fossil resourses	Crude oil, natural gas	kg. crude oil eq.
Climate change	CO_2, CH_4, N_2O, $C_2F_2H_4$, CF_4, CCl_4, C_2F_6, R22	kg CO_2 eq.
Summer smog (POCP)	CH_4, NMVOC, benzene, formaldehyde, ethylacetat, VOC, C-total, ethanol	kg Ethene eq.
Acidification	NO_x, NH_3, SO_2, TRS, HCI, H_2S, HF	kg SO_2 eq.
Terrestial eutrophication	NO_x, NH_3, N_2O	kg PO_4 eq.
Aquatic eutrophication	COD, NH_4, NO_3, P	kg PO_4 eq.
Human toxicity	PM10, SO_2, NO_x, NH_3, NMVOC	kg PM10 eq.
Human toxicity, carcinogenic risk	As, B(a)P, Cd, Cr VI, Ni, Dioxin, Benzene, PCB	kg As eq.
Total primary Energy	Hard coal, brown coal, crude oil, natural gas, uranium ore, hydroenergy, other renewable energy	MJ
Non-renewable primary energy	Hard coal, brown coal, crude oil, natural gas, uranium ore	MJ
Transport intensity	Lorry distance	km

powerful statistical tool which allows taking the delivery of all the analysed situations into consideration. A further favourable aspect of the software is the relatively low cost.

It is an important characteristic feature, that all types of the software were commercial. The cost of one professional licence has been between 9800–112000 €/4 years.

In a 2011survey Verdantix research group found that new product LCA software suppliers have entered a market that was previously dominated by complex solutions. Report recommends firms shortlist product LCA solutions with reference to one of the following usage scenarios: product environmental compliance, supply chain optimization, sustainable product design, and sustainable product marketing.

The product LCA software market is still at an early stage of growth, and its price point reflects this—new online applications are available for as little as $85 a month.

A real innovation was on the LCA market the appearance of openLCA software. The idea for the openLCA project and software emerged in 2006. The aim of the project are as follows:

1. Design and build a fast, reliable, high-performance, modular framework for sustainability assessment and LCA modelling, with visually attractive and flexible components, for sophisticated and simple models, in a standard programming language, using only widely available Open Source software.
2. Create a contributing programming community,
3. Build modules for the framework, and enable users to build their own modules. An important

aspect of the project is the well-documented background, supported by a wiki page.

A specific, important feature of the project, that it is able to manage the problem of uncertainity, too (Cirot et al. 2004).

Another possible solution could be the application of discrete event simulation software. The supply of easy-to-use, simple softwares is rather limited. Majority of solution is based on Java or C++application and recquire some programming background. The most widely used applications are Arena and AnyLogic. The application of these type of software offers a favourable possiblity for modelling of effects of re-cycling.

4. LITERATURE

– *The European Reference Life Cycle Data System (ELCD core database)* http://lca.jrc.ec.europa.eu/lcainfohub/datasetArea.vm.
– *CPM LCA database* http://cpmdatabase.cpm.chalmers.se/.
– *US LCI database* http://www.nrel.gov/lci/.
– *PlasticsEurope* http://www.plasticseurope.org/plastics-sustainability/life-cycle-thinking.aspx.
– *The World Steel Association* http://www.worldsteel.org/index.php?action=programs&id=62.
– *FEFCO (European Federation of Corrugated Board Manufacturers)* http://www.fefco.org/publications/other-publications/european-database.html.
– *Ecoinvent* http://www.ecoinvent.ch/.
– GaBi (http://www.gabi-software.com/databases/).

Green transportation

Green Design, Materials and Manufacturing Processes – Bártolo et al. (eds)
© 2013 Taylor & Francis Group, London, ISBN 978-1-138-00046-9

An innovative six sigma and lean manufacturing approach for environmental friendly shipyard

L. Bilgili, D. Deli & U.B. Celebi
Yildiz Technical University, Istanbul, Turkey

ABSTRACT: Ship building is a very old and complex production method. It has many kinds of inputs and processes that are interconnected and each process has an effect on the next stages of a ship's life cycle. Thus, a good planning ought to be developed to make an efficient producing process. Besides, due to the growing environmental awareness, new and innovative methods must be used for manufacturing ships. Organizations have to interest in the issues that lowers the costs for efficiency and surviving in competitive environment. The companies make investments in shipbuilding industry are taking actions for minimization the costs. In this paper, six sigma and lean manufacturing approach are investigated to reach a more environmental friendly shipyard.

1 INTRODUCTION

For the complexity of the ship building, harmful impacts are occurred as a result of different processes, including all solid, liquid and gaseous forms of emissions and discharges. The effect of shipyard on environment and human health can be reduced with technological improvements on shipbuilding industries. New production and innovative technologies to reduce the pollution may be the key to decrease the material inputs, to improve the engineering processes to reuse the materials, to improve the management practices and to use alternative materials to replace toxic chemicals. The six sigma method is a project-driven management approach to improve the organization's products, services, and processes by continually reducing defects in the organization. It is a business strategy that focuses on improving customer requirements understanding, business systems, productivity, and financial performance. Lean manufacturing is an approach that aims to initialize the unnecessary consumptions. Lean manufacturing combines the labor-intensive production and serial production. It aims to increase the production speed and thus, decrease the production time to improve quality, cost and delivery performance at the same time. Total Quality Method, Lean Organization and Process Improvement, Kaizen, Production-Demand Balancing and Human Resources Systems are the methods of Lean Manufacturing.

Great amount of manpower usage, complex work flows, variety of material usage are the main points to describe shipbuilding industry as heavy industry. Production processes of the shipyards are divided into two main divisions: new building and ship repair industry. Production methods of these two divisions are similar characteristics. New ship construction and ship repairing have many industrial processes in common (Celebi and Turan, 2011). During ship production, solid, liquid and gas forms of wastes are released. Some of these wastes are hazardous for human health and environment and also classified as hazardous wastes. To minimize these hazardous wastes, shipyards should keep their responsibilities in a higher level. One of the solutions is to create environmental sensitive shipyards where the waste management principals are well defined and accepted, reuse, waste reduction at the source methods are used to decrease the hazards to human and environment (Akanlar et al, 2009). The prevention of these waste productions at the beginning of the project may be the best solution at all. Alternatives to reduce the pollution may be to decrease the material inputs, to improve the engineering processes, to reuse the materials, to improve the management practices, to use alternatives to toxic chemicals. (Celebi et al, 2010). Supply chain management model of a shipyard improves production processes. Increase in productivity and quality will be obtained with this model (Celebi and Turan, 2011). Innovative approaches such as Six Sigma and Lean Manufacturing methods are needed to minimize the harmful emissions and develop environmental friendly shipyards.

Over the years, many researchers have studied Six Sigma programs and identified many critical decisions of these programs (Chakravorty, 2009). Shah and Ward (2003) examined the relationship between contextual factors and extent of implementation of a number of manufacturing practices that are key facets of lean systems. Yang et al (2011) studied relationships between lean manufacturing practices, environmental management and business performance outcomes. Parast (2011) focused on the effect of Six Sigma projects on firm innovation and performance. Swink and Jacobs (2012) studied the operating performance impacts of Six Sigma adoptions. Jabbour et al (2012) studied to verify the influence of Environmental Management on Operational Performance in Brazilian automotive companies, analyzing whether Lean Manufacturing and Human Resources interfere in the greening of these companies. Six Sigma is a systematic method for process improvement focused on financial results that uses statistical and quality management tools. The main differentials of the Six Sigma methodology compared to the Total Quality practices are expert's organizational structure, focus on metrics and a structured problem solving method (Calia et al, 2009).

The objective of this paper is to understand if the importance of the Pollution Prevention of a shipyard improved with the implementation of the Six Sigma and lean manufacturing methodologies.

2 SHIPYARD OPERATIONS AND WASTE

Shipbuilding period comprises cutting and marking of the steel plates, steel fabrication, assembly of the sections, erection on slipway, launching of the vessel, sea trials and the delivery of the vessel to the Owner. In general, production type in the shipyards is block fabrication. In this production process, the vessel is composed of several blocks subject to the manufacturing and crane capacities of the shipyard. Each block is assembled in the workshops and after completion of the works and surveys; the blocks are erected on slipway. The vessel is launched after completion of all blocks' erection on slipway. The outfitting works of the vessel are completed in the sea and the vessels start their sea trials in order to test the efficiency and control whether they meet the requirements of the Owner and appropriate to the contract items. Consequently, the vessel is delivered to the Owner if the sea trial results can be accepted by the classification societies and the Owners (Turan and Celebi, 2011). Construction processes in the shipyards are very complicated and there are numerous suppliers and subcontractors. Therefore order follow-ups, transportation and storage are difficult factors to manage for all shipyards. In this context, an efficient supply chain management is essential. Thus, they can decrease the costs, delivery times and material handlings/transportations as well as increase the customer satisfaction and their

reputability in the sector (Celebi and Turan, 2011). The management of the wastes mentioned above is extremely important for environmental safety and human health. Waste management is the prevention, minimization, reuse, recycling, energy recovery or disposal of waste materials Source reduction aims to reduce the hazardous materials, pollutants or contaminants to release to environment, prior to recycling, treatment or disposal and minimize the hazards to human health and the environment (Turan and Celebi, 2011).

The effect of shipyard on environment and human health can be reduced with technological improvements on shipbuilding industries (Celebi and Turan, 2011). Most of the processes such as welding, painting, blasting, fiberglass production that have a direct effect on workers' health, i.e. exposure to VOCs, fumes resulting from burning through base metal and from burning the interior and exterior coatings that are often left in place can cause acute and chronic health problems (Celebi et al, 2010). Also these processes produce different types of contaminants harmful to human and environment. Worker exposure reduction with new designs can begrouped under engineering, administrative and Personal Protective Equipment (PPE), as well as work practice controls. These groups have unique legal terms but also more general understanding within the occupational and environmental hygiene community (Celebi et al, 2009a). The safety of workers related to the preventive ways and their consistency by the management. Personal Protective Equipment (PPE) should be defined and classified according to the work and material. The covered areas should be well ventilated, the pollutant air should be monitored and the protective equipment usage such as gas masks should be followed. The welding fume outcome from the process should be removed via the portable fume box (Celebi et al, 2009b). The prevention of these waste productions at the beginning of the project may be the best solution at all. Alternatives to reduce the pollution may be to decrease the material inputs, to improve the engineering processes, to reuse the materials, to improve the management practices, to use alternatives to toxic chemicals. Most of the processes such as welding, painting, blasting, fiberglass production that have a direct effect on workers health, i.e. exposure to VOCs, fumes resulting from burning through base metal and from burning the interior and exterior coatings that are often left in place can cause acute and chronic health problems (Turan and Celebi, 2011). To minimize the hazardous waste materials, to protect the workers health in a long term, these traditional production methods should be replaced with alternative new production technologies. These new alternative methods should be defined and shared with top and medium management levels of facilities (Celebi et al, 2009b). The principles and premises of project management have been evolved over time. Traditionally, project management has been conceived as an organized plan to achieve pre-determined goals within a specified timeline (Parast, 2011). New production technologies to reduce the pollution may be the key to decrease the material inputs, to improve the engineering processes to reuse the materials, to improve the management practices and to use alternative materials to replace toxic chemicals (Turan and Celebi, 2011). Innovative and environmental concepts must be used starting from the concept design process. Also the shipyard must be changed into an industrial place sensitive to environment, occupational health and safety and human health. In this concept, "Green-shipyard" must be developed. It is essential to gather contributors of all fields related to ship industry for consensus. Rules, regulations and applications guide effects of ships on the environment. Special and urgent measure for marine fuels to be considered is reduction of carcinogenic effects (Safa and Celebi, 2011).

3 SIX SIGMA AND LEAN MANUFACTURING APPROACHES

Motorola created the Six Sigma methodology in 1986, in order to increase its competitiveness against Japanese companies in the electronics industry by improving the quality levels. The name of the Six Sigma methodology is derived from the Greek alphabet symbol utilized in statistics for standard deviation, a measurement to quantify variation and process inconsistency (Calia et al, 2009). While Six Sigma has made a big impact on industry, the academic community lags behind in its understanding of Six Sigma. Six Sigma is a concept that was originated by Motorola Inc. in the USA in about 1985. At the time, they were facing the threat of Japanese competition in the electronics industry and needed to make drastic improvements in their quality levels. Six Sigma is an organized and systematic method for strategic process improvement and new product and service development that relies on statistical methods and the scientific method to make dramatic reductions in customer defined defect rates (Linderman et al, 2003).

The principal basis of the Six Sigma methodology is that if you can measure how many defects or failures any business or process has, you can find ways to systematically eliminate them. In statistical terms, the variability (or sigma) of a nearly-perfect process is so small there are six standard deviations between the mean quality of the process and the quality level most customers expect from it.

In practical terms, sigma is a measure of the number of times a process is defective, or fails, per million iterations or units of output. The greater the sigma number, the fewer the defects (Manuel, 2006).

Six Sigma approach analyzes the customer and their expectations better. Customer can see all of the processes and possible losses, holistically. Six Sigma is a method based on the assumption which accepts that zero-error could be reachable as the variables are under control. Thus, Six Sigma is a quality management systems that aims to enhance customer satisfaction and technical outputs and reduce the errors and costs for environment.

Six Sigma method, that completes the Total Quality Management, can be implemented each of the production fields (manufacturing, design, sale, office, health, finance). In Six Sigma method, the amount of total error is divided to the total produced outputs, regardless of the output. The technical tools used in Six Sigma method correspond the requirements of many quality assurance systems such as Total Quality Management, ISO 9000 and AQAP.

The Six Sigma method is a project-driven management approach to improve the organization's products, services, and processes by continually reducing defects in the organization. It is a businsss trategy that focuses on improving customer requirements understanding, business systems, productivity, and financial performance. Six Sigma is a systematic, data-driven approach using the Define, Measure, Analysis, Improve, and Control (DMAIC) process and utilizing Design For Six Sigma method (DFSS). The fundamental principle of six sigma is to 'take an organization to an improved level of sigma capability through the rigorous application of statistical tools and techniques'. The statistical aspects of six sigma must complement business perspectives and challenges to the organization to implement six sigma projects successfully. Various approaches to six sigma have been applied to increase the overall performance of different business sectors (Mehrabi, 2012). From the statistical point of view, the term six sigma is defined as having less than 3.4 defects per million opportunities or a success rate of 99.9997% where sigma is a term used to represent the variation about the process average. Understanding the key features, obstacles, and shortcomings of six sigma provides opportunities to practitioners for better implement six sigma projects. It allows them to better support their organizations' strategic direction, and increasing needs for coaching, mentoring, and training (Kwak et al, 2006). In particular, Six Sigma potentially creates a trade-off between gains in efficiency versus growth. Several important studies suggest that process improvement regimes can stifle innovative exploration in favor of exploitation, thus impeding sales growth. Corporate-wide adoption of Six Sigma often involves considerable investments in consulting support, training, organizational restructurings, and associated information and reporting systems. For example, over a four year period (1996–1999) General Electric reportedly spent more than $1.6 billion on Six Sigma investments. Researchers report that training costs are typically as much as $50,000 per trained worker. Overall, the results indicate that the benefits of Six Sigma adoption tend to more than compensate for associated costs and required investments (Swink and Jacobs, 2012). There are multiple ways to combine the individual practices to represent the multi-dimensional nature of Lean Manufacturing. In combining these practices, the researcher has to contend with the method used to combine and the actual content of the combinations. The dominant method in operations management literature has been to use exploratory or confirmatory factor analysis to combine individual practices in a multiplicative function to form orthogonal and unidimensional factors. Many researchers argue that a lean production system is an integrated manufacturing system requiring implementation of a diverse set of manufacturing practices (Shah and Ward, 2003). The focus of lean manufacturing on internal and process waste reduction to increase efficiency should be extended to a focus on environmental waste reduction increasing environmental efficiency by implementing environmental management practices. Environmental management practices do require additional resources investments. With an increasing social demand of environmental sustainability; firms embrace the strategic importance of environmental management practices for competitive advantage. From a managerial perspective, lean manufacturing and environmental management practices are synergistic in terms of their focus on reducing waste and inefficiency (Yang et al, 2011). Lean Six Sigma has become a business model, a symbol of excellence, with the goal of eliminating waste and reducing the defects and variations in organization's processes. Over the years, the excellence model has also demonstrated his significant importance and power in changing organization culture and employee vision over the changes which occur within organization. Lean Six Sigma refers to a more intelligent management of an organization, which first takes into account customer requirements and his satisfaction by using data and facts for elaborating medium and long term strategies. One of the most important aspects of applying this methodology is to involve all employees in its implementation. Through their involvement, employees are encouraged to contribute to the change which will take place and which will bring all benefits for all of them, thus

feeling more confident both in their own abilities and their work capacity and also in the organization in which theyoperate, developing their creativity and innovation (Pamfilie et al, 2012). Two management areas have gained prominence as targets of effective environmental management. The first is operations/manufacturing management, which, because it processes resources, has significant environmental effects. The second area is human resources, which may influence the performance of new organizational objectives, such as those related to environmental performance (Jabbour et al, 2012). Pollution Prevention is an environmental management approach that reduces the source of pollution by improving the efficiency in the use of resources as energy, materials and water. Thus, most frequently, Pollution Prevention is used not for environmental legislation compliance, but for cost reduction to improve the business' financial performance (Calia et al, 2009).

4 INNOVATIVE SHIPYARD APPROACH

Six Sigma and Lean Manufacturing application in shipyards is presented in Figure 1. In Figure, processes on which Six Sigma and Lean Manufacturing can be implemented are presented such as Small Parts Production, Block Premanufacturing, Block Outfitting, Block Manufacturing, Fabrica/Harbour/Sea Trials and Tests. Error rates are tried to be reduced for each plate, pipe and profile production flow with the enhancement methods during Small Parts Production, Block Manufacturing and Block Outfitting processes. Waste management plan of the shipyard was formed with the former studies.

Best Management Practices are needed to reduce and prevent contamination of the aquatic environment and to improve air quality. Some BMP examples in shipbuilding industry are the treatment of paint washing water containing toxicmetallic components; cleaning of drydock and vessel before floated; usage of alternative raw material such as tin-free, TBT-free, or low-VOC paint; and filtration of volatile organic pollutants of main processes such as blasting, welding, and painting (Celebi et al, 2010). Environmental friendly shipyard concept studies are in practice phase. By this study, which is the first for Turkish shipyards, time, cost and energy savings will be provided. Besides, it is provided for shipyards to consider environment and occupational health and safety into account in all of the production phases from design to launching. Thus, the manufacturing processes will be optimized to operations that have minimized harmful effects to the environment.

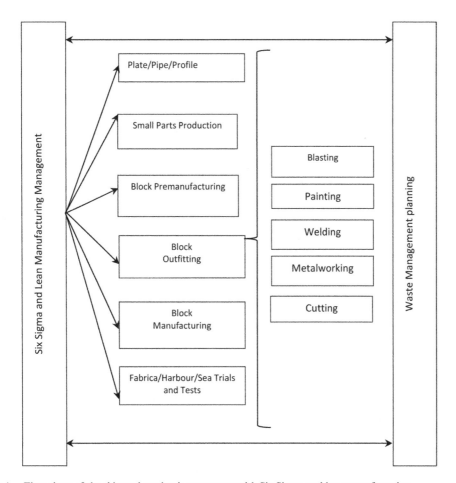

Figure 1. Flow chart of the shipyard production processes with Six Sigma and lean manufacturing.

5 CONCLUSIONS

The management of the wastes mentioned above is extremely important for environmental safety and human health. Waste management is the prevention, minimization, reuse, recycling, energy recovery or disposal of waste materials. The main aim of recycling is to collect the wastes and clean through a process to reuse. The main objective in waste management is, generating the Life Cycle Assessment (LCA). LCA is one of the common implementation for the consideration of environmental impression.

By using Six Sigma method, global world companies have considerable amounts of acquisition. Six Sigma and Lean Manufacturing methodologies, which aim process perfection, provide the companies that use them, profitability, prolificacy and growth in market share. The first Six Sigma and Lean Manufacturing concept implementation

has been begun in Turkish shipyards. By the leading studies integrated with environmental management systems, it has been observed decreases in waste costs. By developing this system, it should be implemented the other shipyards.

Innovative and environmental concepts must be used starting from the concept design process. Also the shipyard must be changed into an industrial place sensitive to environment, occupational health and safety and human health. In this concept, "Green Shipyard" must be developed.

REFERENCES

Akanlar F.T., Celebi U.B. and Vardar N., 2009. Alternative Production Processes and New Technologies for Human and Environmentally Responsive Shipyards, Proceedings of 13th Congress of International Maritime Association of the Mediterranean-IMAM, Istanbul, Turkey.

Calia R.C., Guerrini F.M., Castro M., 2009. The impact of Six Sigma in the performance of a Pollution Prevention program, Journal of Cleaner Production, Vol. 17 pp. 1303–1310.

Celebi U.B., Akanlar F.T. and Vardar N., 2009a. Chemicals and Hazardous Wastes Generated by Shipyard Production and Their Effects on Human Health at Workplace, Fresenius Environmental Bulletin, Vol. 18—No 10. pp. 1901–1908.

Celebi U.B., Akanlar F.T. and Vardar N., 2010. Multimedia Pollutant Sources and Their Effects on the Environment and Waste Management Practice in Turkish Shipyards, Global Warming, Green Energy and Technology, pp. 579–590, DOI: 10.1007/978-1-4419-1017-2_39.

Çelebi U.B., Turan E., 2011. A Supply Chain Management Model for Shipyards in Turkey, Proceedings of 14th Congress of International Maritime Association of the Mediterranean-IMAM, Genova, Italy.

Celebi U.B., Akanlar F.T. and Vardar N., 2009b. Personal protective equipment to minimize the shipyard production processes health effects on shipyard workers, Proceedings of the 2nd International CEMEPE & SECOTOX Conference, Mykonos, June 21–26, ISBN 978-960-6865-09-1.

Chakravorty S.S., 2009. Six Sigma Programs: An implementation model, International Journal of Production Economics, Vol. 119 pp. 1–16.

Jabbour C.J.C., Jabbour A.B.L.S., Govindan K., Teixeira A.A., Freitas W.R.S., 2012. Environmental management and operational performance in aut02motive companies in Brazil: the role of human resource management and lean manufacturing, Journal of Cleaner Production, pp. 1–12 http://dx.doi.org/10.1016/j.jclepro.2012.07.010.

Kwak Y.H, Anbari F.T., 2006. Benefits, obstacles, and future of six sigma approach, Technovation, Vol. 26 pp. 708–715.

Linderman K., Schroeder, R.G., Zaheer, S., Choo, A.S., 2003. Six Sigma: a goal-theoretic perspective, Journal of Operations Management, Vol. 21 pp. 193–203.

Manuel D., 2006. Six Sigma methodology: reducing defects in business processes, Filtration & Separation, Vol. 43, Issue 1, January–February 2006, pp. 34–36.

Mehrabi, J., 2012. Application of six-sigma in educational quality management, Procedia—Social and Behavioral Sciences, Vol. 47 pp. 1358–1362.

Pamfilie R., Petcu A.J., Draghici M., 2012. The importance of leadership in driving a strategic Lean Six Sigma management, 8th International Strategic Management Conference, Procedia—Social and Behavioral Sciences, Vol. 58 pp. 187–196.

Parast M.M., 2011. The effect of Six Sigma projects on innovation and firm performance, International Journal of Project Management, Vol. 29 pp. 45–55.

Safa A., Celebi U.B., 2011. Green Ship: New Ideas for Environmentally Friendly Ships, 1st International Conference on Maritime Technology and Engineering-MARTECH, Lisbon, Portugal.

Shah R., Ward P., 2003. Lean manufacturing: context, practice bundles, and performance, Journal of Operations Management, Vol. 21 pp. 129–149.

Swink, M., Jacobs, B.W., 2012. Six Sigma adoption: Operating performance impacts and contextual drivers of success, Journal of Operations Management, Vol. 30 pp. 437–453.

Turan, E., Celebi U.B., 2011. Advanced Manufacturing Techniques in Shipyards, IMAM 2011, Sustainable Maritime Transportation and Exploitation of Sea Resources CRC Press 2011, pp. 545–550 Print ISBN: 978-0-415-62081-9.

Yang M.G, Hong P, Modi S.B., 2011. Impact of lean manufacturing and environmental management on business performance: An empirical study of manufacturing firms, International Jorunal of Production Economics, Vol.129 pp. 251–261.

Green Design, Materials and Manufacturing Processes – Bártolo et al. (eds)
© 2013 Taylor & Francis Group, London, ISBN 978-1-138-00046-9

Life cycle assessment approach of waste management for ship operation

L. Bilgili & U.B. Celebi

Yildiz Technical University, Istanbul, Turkey

ABSTRACT: Ships have an indispensable role in trading and transportation. Due to the growth of population and the new demands of the growing economies, the need for shipping is increasing. Besides shipping causes a considerable problem for human and environmental issues, increasing amount of shipping activity will constitute a greater danger by more energy consumption and harmful emissions to air and water. Ship operation is the most emission-produced phase of a ship's life cycle. Due to the spillages, accidents, gaseous form of emissions as a result of main and auxiliary engines' activities, and discharges to water (solid and liquid wastes), ship operation phase has a considerable role of the environmental impact of a ship. Life Cycle Assessment (LCA) is a totally new method that is used by universities, companies and even local and international legislators. LCA aims to reduce, minimize and if possible, initialize the consumed energy, harmful emissions and discharges in a whole life cycle of a product. LCA consists all of the phases from the raw material to disposal/recycling. In this paper, LCA method is presented and the method's effects on reducing the emissions and discharges during ship operation. Waste management modeling is highlighted.

1 INTRODUCTION

Shipping runs the major part of the world's total trading operation. For ships can carry big amounts of goods, they are preferred as main carriers. Besides trading capacity, other shipping activities such as fishery, cruise lines and naval forces have significant effects on economy and other issues such as environment and global warming.

In a recent study, it was estimated that there are 90.363 registered ships all over the world at the end of 2001 (Eyring et al, 2005a). Eyring et al have also developed a calculative estimation which projects that there will be 100.400 and 126.800 registered ships in 2020 and 2050, respectively (Eyring et al, 2005b). The calculation was made using the most optimist scenario.

Ship emissions to air and water are already a great problem all around the world, especially coastal zones and port cities. There are two main categories that ship emissions can divide into: (1) emissions to air; (2) emissions to water.

Emissions to air include carbon dioxide (CO_2), nitrogen oxides (NO_x), sulphur oxides (SO_x), Black Carbon (BC) and Particulate Matter (PM), concisely. Emissions to water include ballast water, bilge water, grey and black water, solid wastes and toxic wastes.

Both emissions to air and to water are well documented by recent studies. It has been estimated that the CO_2 emissions from global shipping in 2007 as 943.5 million tons, annually (Psaraftis

and Kontovas, 2009). It has been estimated that shipping is responsible for 2%–4% of global CO_2 emissions (Tzannatos, 2010). It's also predicted the ratio will increase to 12%–18% in 2050 (Heitmann and Khalilian, 2010). Some recent estimations show that shipping is responsible for 6.87 teragram (Tg) of NO_x emissions all over the world, annually (Corbett and Kohler, 2003). It has been estimated that 10%–15% of NO_x emissions are caused by shipping (Lawrance and Krutzen, 1999; Eyring et al, 2005a). Sulphur oxides emissions from international shipping are 6.49 Tg in terms of sulphur (annually) which corresponds 5%–8% of global SO_x emissions (Corbett and Kohler, 2003). Black carbon from shipping has a relatively small portion (1,7% all over the world) but it has considerable effects on Arctic warming (Eyring et al, 2005a). According to a recent estimation, the amount of PM emissions caused by shipping is between 0,9–1,7 million tons, annually (Endresen et al, 2003). It was calculated that a cruise ship with a capacity of 2.000–3.000 passengers can generate some 1.000 tons of waste per day, which can be broken down in 550.000–800.000 liters of grey water, 100.000–115.000 liters of black water, 13.200–26.000 liters of bilge water, 7.000–10.500 kg of solid waste and 60–100 kg of toxic waste during operation. The total emissions of world's cruise lines are more than 182.000 tons per year (Zuin et al, 2009).

The total amount of ballast water unloaded by ships is estimated as 2.200 Mton in 2000. Untreated ballast water discharged amount is 3.500 Mton worldwide (Endresen et al, 2004). It is estimated that over 10 billion metric tons of ballast water is transferred annually across the world (Perrins et al, 2006). 3.000–4.000 species are moved by ballast water into another ecosystem, annually (Göktürk, 2005). Invasive species that are moved by ballast water cause $ 137 billion economic loss in the USA (Wright et al, 2010).

It is obvious that shipping emissions will cause greater problems in the future unless the necessary measures are taken.

2 LIFE CYCLE ASSESSMENT IN SHIP OPERATION

Life Cycle Assessment (LCA) of a product is used to identify, evaluate and minimize energy consumption and environmental impacts holistically, across the entire life of the product. LCA is, therefore, a systematic way of examining the environmental impacts of a product throughout its lifecycle, from raw materials extraction through the processing, transport, use and finally product disposal. LCA could be used also to assist companies to identify and assess opportunities to realize cost savings by making better design and more environmentally friendly products, more effective use of available resources and improving waste management systems. LCA adopts a holistic approach by analyzing the entire life cycle of a product from raw materials extraction and acquisition, materials processing and manufacture, material transportation, product fabrication, transportation, distribution, operation, consumption, maintenance, repair and finally product disposal/scrapping. LCA has three main objectives: (1) minimization of energy consumption, (2) minimization of environmental impacts, (3) rationalization of material used. (Shama, 2005).

LCA is a technique for assessing the environmental aspects associated with a product, system or process over its life cycle, i.e. from cradle to grave. In particular, the interest of using LCA in the waste management sector is increasing. LCA methodology has been utilized to assess the environmental performance of different waste treatment technologies, e.g. incineration, landfill, etc. and to compare the environmental sustainability of waste management scenarios of solid waste. It has also been successfully utilized to assess potential environmental impacts of waste management systems in urban district or region (Zuin et al, 2009).

LCA of ships has three main phases: Manufacturing, operation, disposal/recycling. Operation phase is the most emission-producing phase due to the long life-time as 20–25 years. In Figure 1, the LCA of a ship is presented.

Figure 1. Life cycle assessment of a ship.

LCA for ships is a totally new concept that includes new production methods, innovative technologies and stricter regulations. Waste management is a stage of LCA to reduce, minimize or prevent the harmful wastes.

For up to 80% of GreenHouse Gas (GHG) emissions from conventional diesel fuel occurs during the vessel operation (Winebrake et al, 2006), LCA approach especially must be focused on operation. On the other hand, it must be remembered that LCA is a holistic approach in which the phases cannot be divided into different parts. All phases have effects on the other phases.

Using new technologies during manufacturing may reduce the emissions. Scrubbers can work with or without help of seawater by the addition of chemicals. Scrubbers have been proven to reduce SO_2 emissions up to 85% (EMEC, 2010). Internal modifications to existing engines may reduce 20%–30% of NO_x (Clean Shipping Criteria, 2007). Waste Heat Recovery (WHR) allows up to 12% savings of CO_2 emissions (EMEC, 2010). There are some simpler ways to reduce the emissions such as reducing vessel speed (for CO_2) and using low-sulphur content fuels (for SO_2 and PM). Advanced filtration systems may help to minimize the harmful emissions to air.

New and ballast free designs may initialize the effects of ballast water. Some systems such as heat treatment, ozone based treatment systems, deoxygenation, sonication and innovative filtration may reduce the invasive species up to 99% in ballast water (Badia et al, 2008; Perrins et al, 2006; Wright et al, 2010; McCollin et al, 2007; Holm et al, 2008; Mesbahi, 2004; Tang et al, 2009).

3 CONCLUSION

LCA is an innovative, sustainable and holistic approach for any product to clarify the consumptions, increase the efficiency and reduce the costs, emissions and time. LCA consists all of the phases from raw material through the disposal of the product.

For shipping, a good design brings a good and efficient operation phase and a disposal/recycling phase at an adequate level.

Ship operation phase is the main aim of producing ship. Thus, it must be planned carefully. Besides economic benefits, environmental performance must be in focus. New technologies should be used and if necessary, newer ones should be developed. Educating the workers may have a considerable positive effect on reducing the solid wastes such as rubbish, packages, electronic and medical wastes. Developing stricter rules and regulations will result more control and improve the public awareness. Intense relationship between the phases will correspond the necessities of LCA approach.

For LCA consists all of the phases of a ship, it provides to see the whole picture and relationships. LCA is a developing and totally new approach for ship building industry and it will have more usage and importance in the future of sustainable researches for environmental performance.

REFERENCES

Badia G.Q., McCollin T., Kjell D.J., Vourdachas A., Gill M.E., Mesbahi E., Frid C.L.J., 2008. On board short time high-temperature heat treatment of ballast water: A field trial under operational conditions, Marine Pollution Bulletin, Vol. 56, pp. 127–135.

Clean Shipping Criteria, Guidance Document for Shipping Customers, September 2007. Developed by the Clean Shipping Project, Göteborg, Sweden.

CO_2 Emissions from International Shipping: Burden Sharing under Different UNFCCC Allocation Options and Regime Scenarios. Kiel Working Paper No. 1665.

Corbett J.J., Kohler H.W., 2003. Updated emission from oceanshipping. Journal of Geophysical Research 108 (D20), 4650.

EMEC, 2010. Green Ship Technology Book.

Endresen Ø., Behrens H.L., Brynestad S., Andersen A.B., Skjong R., 2004. Challenges in global ballast water management, Marine Pollution Bulletin, Vol. 48, pp. 615–623.

Endresen Ø., Sørgard E., Sundet J.K., Dalsøren S.B., Isaksen I.S.A., Berglen T.F., et al., 2003. Emission from international sea transportation and environmental impact. Journal of Geophysics Research, 108: D17.

Eyring, V., Köhler, H.W., Lauer, A., Lemper, B., 2005b. "Emissions from international shipping: 2. Impact of future technologies on scenarios until 2050", Journal of Geophysical Research, Vol. 110, D17306, doi: 10.1029/2004JD005620.

Eyring, V., Köhler, H.W., van Aardenne, J., Lauer, A., 2005a. "Emissions from international shipping: 1. The last 50 years", Journal of Geophysical Research, Vol. 110, D17305, doi: 10.1029/2004JD005619.

Göktürk D., 2005. İstanbul Limanlarında Balast Suyu Örneklemeleri, M.Sc. Thesis, in Turkish.

Greece, Atmospheric Environment, Vol. 44, pp. 2194–2202.

Heitmann N., and Khalilian S., October 2010. Accounting for.

Holm E.R., Stamper D.M., Brizzolara R.A., Barnes L., Deamer N., Burkholder J.M., 2008. Sonication of bacteria, phytoplankton and zooplankton: Application to treatment of ballast water, Marine Pollution Bulletin, Vol. 56, pp. 1201–1208.

Lawrance M.G., Crutzen P.J., 1999. Influence of NO_x emissions from ships on tropospheric photochemistry and climate. Nature 402, 167–170.

McCollin T., Badia G.Q., Josefsen K.D., Gill M.E., Mesbahi E., Frid C.L.J., 2007. Ship board testing of a deoxygenation ballast water treatment, Marine Pollution Bulletin, Vol. 54, pp. 1170–1178.

Mesbahi, E., 2004. Latest results from testing seven different Technologies under the EU MARTOB project—where do we stand now? In: Matheickal, J.T., Raaymakers, S. (Eds.), Proceedings of 2nd International Ballast Water Treatment R&D Symposium, IMO London, 21–23 July 2003. GloBallast Monograph Series No. 15. International Maritime Organization, London, pp. 210–230.

Perrins J.C., Cooper W.J., Leeuwen H.J., Herwig R.P., 2006. Ozonation of seawater from different locations: Formations and decay of total residual oxidant-implications for ballast water treatment, Marine Pollution Bulletin, Vol. 52, pp. 1023–1033.

Psaraftis H.N., Kontovas, C.A., 2009. CO_2 emission Statistics for the world Commercial fleet. WMU Journal of Maritime Affairs 8, 1–25.

Shama M.A., 2005. Life cycle assessment of ships, Maritime Transportation and Exploitation of Oceanand Coastal Resources, Taylor & Francis Group, London, ISBN 0 415 39036 2.

Tang Z., Butkus M.A., Xie Y.F., 2009. Enhanced performance of crumb rubber filtration for ballast water treatment, Chemosphere, Vol. 74, pp. 1396–1399.

Tzannatos E., 2010. Ship emissions and their externalities for.

Winebrake J.J., Corbett J.J., Meyer P.E., 2006. Total fuel cycle emissions for marine vessels: A wellto-hull analysis with case study, 13th CIRP International Conference on Life Cycle Engineering, pp. 125–129.

Wright D.A., Gensemer R.W., Mitchelmore C.L., Stubblefield W.A., Genderen E., Dawson R., Dawson C.E.O., Bearr J.S., Mueller R.A., Cooper W.J., 2010. Shipboard trials of an ozone-based ballast water treatment system, Marine Pollution Bulletin, Vol. 60, pp. 1571–1583.

Zuin S., Belac E., Marzi B., 2009. Life cycle assessment of ship-generated waste management of Luka Koper, Waste Management, Vol. 29, pp. 3039–3046.

Green Design, Materials and Manufacturing Processes – Bártolo et al. (eds)
© 2013 Taylor & Francis Group, London, ISBN 978-1-138-00046-9

An innovative method establishment for a green shipyard concept

L. Bilgili & U.B. Celebi
Yildiz Technical University, Istanbul, Turkey

ABSTRACT: Shipbuilding industry is known as one of the hardest metal industry with several chemicals and hazardous material. Most of the production processes such as welding, painting, blasting and fiberglass production have direct impact on workers' health. There are several wastes and pollutants being released during shipbuilding and ship repairing processes. The volume of these wastes and pollutants create a huge amount with major risk on environmental and ecological point of interest. The effect of shipyard on environment and human health can be reduced with technological improvements on shipbuilding industries. The inputs of these processes are various types of products and raw material such as primarily steel and other metals, paints and solvents, blasting abrasives, and machine and cutting oils with an outcome of different forms (solid, liquid, and gaseous) of wastes and pollutants as Volatile Organic Compounds (VOCs), Particulates (PM), waste solvents, oils and resins, metal bearing sludge and wastewater, waste paint, waste paint chips and sent abrasives. Life Cycle Assessment (LCA) is one of the most innovative and effective methods to minimize and initialize the harmful emissions caused by manufacturing and operation. LCA is also aims to minimize the energy consumption. For ship production phase is a long-term and emission-produced stage in a ship's life cycle, LCA method is used to manage the logistics and production process to provide time efficiency and besides, the method is used to reduce the emissions and discharges. In this paper, LCA method is explained and a new LCA method is established for a green shipyard.

1 INTRODUCTION

Ships are built and repaired in shipyards. Ship building is a very complex process and it is known as one of the oldest and heaviest industries. Due to the rapid and positive change of economic rate, ship building industry has been growing for the last decades. Thus, the work load of shipyards are has been increasing, too.

Shipyards are categorized into two main types: Shipbuilding and ship repairing. New ship construction and ship repairing have many industrial processes such as cutting, surface treatment, blasting, painting and coating, solvent cleaning and degreasing, welding and fiberglass manufacturing. There are several wastes and pollutants being released during shipbuilding and ship repairing processes. The amount of wastes and pollutants are a major risk from environmental and ecological point of view. These wastes can be grouped as particulates, Volatile Organic Compounds (VOC's) (styrene, acetone, methylene chloride), metal mists and fumes as pollutants, paint chips, cleaning and paint stripping solvents, oil residues from bilge and cargo tanks, emulsified lubricating and cutting oils and coolants as wastewater and paint chips (containing metals, tributyl-tin), spent abrasives, surface contaminations and cargo tank

residues, spent plating solutions and cyanide solutions, waste cutting oil and lube oils as the residual wastes (Akanlar et al, 2009). During the shipyard processes, various types of products and wastes in solid, liquid and gaseous forms are produced.

A shipyard is a workplace contaminated with spilled petroleum, paints, solvents, Polycyclic Aromatic Hydrocarbons (PAHs), and processed metal slag in relation to shipbuilding and repair activities. Some PAHs, which are usually generated as by-products of incomplete combustion of organic matters, are genotoxic and carcinogenic (Chiu et al, 2006).

Surface preparation is one of the most important steps in the shipbuilding industry. Without proper surface preparation, subsequent surface coatings will prematurely fail due to poor adhesion. Surface preparation is also typically one of the most significant sources of shipyard wastes and pollutant outputs (Celebi and Vardar, 2006). Material inputs used for preparing surfaces include: abrasive materials such as steel shot or grit, garnet, and copper or coal slag; and cleaning water, detergents, and chemical paint strippers (e.g., methylene chloride-based solutions, caustic solutions, and solvents). In the case of hydroblasting, only water and occasionally rust inhibitor are required (USEPA, 1997). Surface preparation process generates air emissions both in solid and gaseous forms. Particulate emissions of blasting abrasives and paint chips are the solid wastes. Some particulates consist toxic metals. Emissions in gaseous form are VOC's and Hazardous Air Pollutants (HAP's) generated by the use of solvent cleaners, paint strippers and degreasers.

The primary residual waste generated is a mixture of paint chips and used abrasives. Paint chips containing lead or antifouling agents may be hazardous, but often in practice the concentration of toxic compounds is reduced due to the presence of considerable amounts of spent blasting medium. Particular attention should be paid to the cleanup of paint chips containing the antifouling TriButyl-Tin (TBT) compounds, which have been shown to be highly toxic to oysters and other marine life. Wastewater contaminated with paint chips and surface contaminants is generated when hydroblasting and wet abrasive blasting methods are used (USEPA, 1997).

Painting is another important step in shipbuilding process. An efficient and adequate painting results with a good hydrodynamics and thus, less fuel consumption. Painting also protects the ship against oxidation and other harmful effects of environment. Thus, the maintenance cost decreases and the life span of ship increases.

Painting can produce significant emissions of VOC's and HAP's when the solvents in the paint volatilize as the paint dries. The solvents in the overspray rapidly volatilize and the remaining dry paint particles can drift off-site or into nearby surface waters. Solid wastes associated with painting are believed to be the largest category of hazardous waste produced in shipyards. Wastewater is also generated when water curtains (water wash spray booths) are used during painting. Wastewater from painting water curtains commonly contains organic pollutants as well as certain metals (USEPA, 1997). VOC's belong to a special category of air pollutants that can adversely affect human health. Industrial operations are important sources of VOC's and there are various technologies that can reduce VOC emissions from these industrial processes (Celebi and Vardar, 2008). Examples of pigments include: zinc oxide, talc, carbon, coal tar, lead, mica, aluminum, and zinc dust some typical solvents include acetone, mineral spirits, xylene, methyl ethyl ketone, and water (USEPA, 2001).

Welding process is the main way to build a ship. Nearly all of the ship building processes includes the welding process. Welding, such as painting, is a source of various types of emissions.

Welding is a common and a highly skilled occupational specialty. Welding processes, involve inhalation exposures, which may lead to acute or chronic respiratory disease. Welding has been associated with many respiratory problems, which vary from acute or chronic respiratory symptoms such as, malaise, cough, dyspnea, chronic bronchitis, interstitial lung disease, pneumonitis, asthma, pneumoconiosis, and lung cancer (Meo, 2003). Hexavalent Chromium has been designated as a priority, pollutant due to its ability to cause genetic mutations and cancer (Kura and Mookoni, 1998).

2 LIFE CYCLE ASSESSMENT IN SHIPYARD OPERATIONS

For shipbuilding is one the heaviest industries, the work in shipyards are very hazardous for human health and environment. Some innovative and new technologies should be applied by the owners to reduce the risks of the production process.

The most beneficial way to minimize timing, costs and risks is to use automated technologies. Instead of manual systems, the automated technologies for transportation, lifting, moving, welding, blasting and painting may maximize the efficiency.

In Figure 1, all of the processes are presented in a shipyard during the shipbuilding. A shipbuilding process begins with preparing plate/pipe/profile, goes on blasting, block manufacturing, block joining and sea trials.

For Life Cycle Assessment (LCA) is a holistic approach, a good co-operation should be done between the processes to achieve an adequate success. All phases are independent from the other but also they are all bounded with.

Shipbuilding is a traditional manufacturing industry for large and heavy machines and devices, which is usually operated under build-to-order strategy. To improve efficiency, shorten delivery time, and reduce cost, module shipbuilding mode has come into wide application in recent years based on grouping technology. In this mode, a ship is decomposed into modules which are usually several tens of tons in weight and several tens of meters in size. Modules are produced in shop floor and then are assembled on shipyard into ship. Accordingly, the change in shipbuilding mode makes module assembly more complicated and important to achieve high productivity (Zheng and Chen, 2008). The fabrication of a curved part starts with forming a flat plate to a desired shape. Since the plate is very thick and heavy, producing a desired shape is not an easy job. Among various techniques for manufacturing a curved plate, the line-heating method is generally used by most shipbuilding companies these days. The fabrication of those plates is a time consuming and labor-intensive process and they work in harsh conditions, threatening workers' safety. So, the

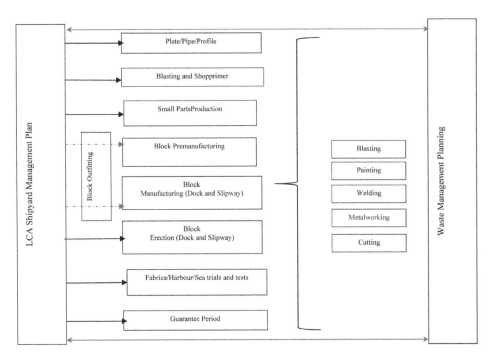

Figure 1. LCA management plan of a shipyard.

necessity of automating the line-heating process has been well appreciated in shipbuilding industries and research on the related topics is being performed actively these days (Park et al, 2007). Recently, the use of robots for painting operation has received great attention as it can save human from a hostile environment and improve coating quality. With the advancement of robotic technology, many sophisticated painting robots are available in the contemporary market. For quality (uniform) coating, the painting applications and the "off-line teaching" method is not promise painting mechanics and parameters (such as spray distance, spray angle, gun orientation, etc.) should be taken into consideration in determining the robot trajectory (Suh et al, 1993). The use of robots for painting operations is a powerful alternative as a means for automation and quality improvement. A typical method being used for motion planning of the painting robot is to guide the robot along the desired path the "lead-through" method (Suh et al, 1991). Automation of painting is a hopeful technology in terms of cost saving in shipbuilding. Painting robot technology is expanding in its importance as a part of new generation shipbuilding research, and it will be essential for modernization of shipyards which utilize to a maximum extent of highly sophisticated production system (Miyazaki et al, 1999).

In general, dry abrasive blasting processes can be divided into three categories based upon how the abrasive media is managed within the process. These categories are: 1) Once-Through Abrasive Blasting (OTAB); 2) Open-Loop Abrasive Blasting (OLAB); and 3) Closed Loop Abrasive Blasting (CLAB). Each of these process types can be characterized by specific sets of associated blasting media, equipment type, surface finish quality, production rates, environmental impacts, safety issues, and others parameters. While it has long been recognized that CLAB has the potential for alleviating the production, environmental, and safety problems associated with OTAB and LAB, the development of successful CLAB equipment has encountered significant engineering obstacles. However, recent developments in turbine blade blast media accelerators, in conjunction with robotic controlled blast head platforms, have now eliminated these obstacles thereby allowing the practical application of CLAB technology (Austin et al, 2002).

Arc welding is considered to be one of the most promising applications of intelligent robots. This situation first stems from a low manual productivity due to the severe environmental conditions resulting from the intense heat and fumes generated by the welding process (Sicard and Levine, 1988). Lead to the construction of numerous offshore steel structures, in which underwater welding is frequently employed. Traditional underwater Manual Metal Arc Welding (MMAW) with covered electrodes is the most common process of underwater engineering. But it is time-consuming because of using divers and cannot guarantee the weld quality because of the rigorous underwater condition. Underwater Flux-Cored Arc Welding (FCAW) can improve the quality of welds and has been used in the construction of offshore structures. Improved automation of underwater welding processes has become increasingly important in the push for higher weld quality and reduced manufacturing cost. To increase the construction efficiency and improve the weld quality, automatic underwater welding device needs to be developed (Shi et al, 2007).

3 CONCLUSION

Ship production and repair in shipyards are very complex processes. Beside complexity, many kinds of harmful or toxic wastes—in solid, liquid or gaseous forms—are produced. Thus, an efficient and effective LCA method must be implemented to the shipyards. For waste management is a key point of LCA, a complex and comprehensive waste management system should be developed.

Waste management systems are used for prevention, minimization, reusing, recycling, energy recovery and disposal of the materials used during the shipyard processes. LCA is a method that compromises all of the phases and determines the most suitable, economic and efficient way to minimize or—if possible—initialize the waste materials.

A green shipyard is also a field in which occupational health is concerned. Processes done by workers must be inspected frequently and the required measurements must be taken. Innovative technologies and methods should be used beginning in design process and throughout the manufacturing operation.

Green shipyard is a subsidiary part of Green Ship concept. Environmental awareness, concern about occupational health and efficient production process are the main goals of green shipyard concept. By using LCA, these goals can be professionally reachable.

REFERENCES

Akanlar F.T., Celebi U.B., Vardar N., 2009. New Automated Technologies in Environmentally Sensitive Shipyards, Second International Conference on Environmental Management, Engineering, Planning and Economics, CEMEPE & SECOTOX Conference, Greece, 21–26 June 2009.

Austin, D., Benze, R. and Kura, B., 2002. Recent Advances in Closed Loop Abrasive Blasting, Ship Production Symposium, Boston, MA, September, 2002.

Celebi, U.B., and Vardar N., 2008. Investigation of VOC Emissions from Indoor and Outdoor Painting Processes in Shipyards', Atmospheric Environment, Volume 42, Issue 22, 5685–5695.

Celebi, U.B., Vardar N., 2006. Wastes and Pollutant Sources Resulted From Shipbuilding Industry in Turkey, Ovidius University Annual Scientific Journal, Mechanical Engineering Series, Vol. 8, No. 1, pp. 24–30.

Chiu, S.W., Ho K.M., Chan S.S., So O.M., Lai K.H., 2006. Characterization of contamination and toxicities of a shipyard area in Hong Kong. Environmental Pollution 142, 512–520.

Kura, B. and Mookoni, P., 1998. Hexavalent Chromium Exposure Levels Resulting from Shipyard Welding. Journal of Ship Production, 14, 4, 246–254.

Meo, S.A., 2003. Spirometric Evaluation of Lung Function (Maximal Voluntary Ventilation) in Welding Workers. Saudi Med J, 24, 656–659.

Miyazaki, T., Nakashima, Y., Ookubo, H., Hebaru, K., Noborikawa, Y., Ootsuka, K., Miyawaki, K., Mori, T., Shinohara, T., Saito, Y., Matsumoto, H., 1999. NC Painting Robot for Shipbuilding. Proc. of ICCAS'99, Boston, USA, No. 2, 1–14.

Park, J.S., Shi, J.G., Ko K.H., 2007. Geometric Assessment for Fabrication of Large Hull Pieces in Shipbuilding. Computer-Aided Design 39 (2007) 870–881.

Shi, Y.H., Wang, G.R., Li, G.J., 2007. Adaptive Robotic Welding System Using Laser Vision Sensing for Underwater Engineering, IEEE International Conference on Control and Automation ThB3-5 Guangzhou, China—May 30 to June 1, 2007.

Sicard, P. and Levine, M.D., 1988. An Approach to an Expert Robot Welding System, IEEE Transactions on Systems, Man, and Cybernetics, Vol. 18, No. 2.

Suh, S.H., Woo, I.K., Noh, S.K., 1991. Development of An Automatic Trajectory Planning System (ATPs) for Spray Painting Robots Proceedings of the 1991 IEEE Intematid conference on Robotics and Autom.

Suh, S.S., Lee, J.J, Choi, Y.J., Lee, S.K., 1993. A Prototype Integrated Robotic Painting System: Software and Hardware Development proceedings of the 1993 IEE/RSJ International Conference on Intelligent Robots and Systems Yokohama, Japan.

USEPA, 1997. Profile of the Shipbuilding and Repair Industry, Office of Compliance Sector Notebook Project, Washington, USA. EPA/310-R-97-008.

USEPA, AP-42, 2001. Preferred and Alternative Methods for Estimating Air Emissions from Surface Coating Operations. Volume II: Chapter 7 July 2001.

Zheng, J., Chen, Z.J.Q., 2008. Minimizing Makespan at Module Assembly Shop in Shipbuilding Service Operations and Logistics, and Informatics, 2008. IEEE/SOLI 2008. IEEE International Conference 12–15 Oct. 2008 1794–1799.

Green Design, Materials and Manufacturing Processes – Bártolo et al. (eds)
© *2013 Taylor & Francis Group, London, ISBN 978-1-138-00046-9*

Social sustainability: A key concern for the recovery of urban roads

A. Annunziata
Department of Civil and Environmental Engineering and Architecture, Cagliari, Italy

ABSTRACT: The nodal concept of efficient, sustainable and equitable city requires a radical reversal in the mode of living and conceiving the urban realm and its spaces. Modern urban landscapes appear to suffer negative outcomes produced by an idea of order tending to segregate urban functions. Urban roads and spaces must be conceived as a medium of a vast information exchange and as a pleasant place designed for people.

1 INTRODUCTION

1.1 *Modern urban landscapes*

The nodal concept of efficient and sustainable city requires a radical reversal in the mode of living and conceiving the urban realm and its spaces. Modern urban landscapes suffer negative outcomes produced by a strict idea of order conceived as a mere visual concept, aimed to evocate a pure and neat structure. The outcome is a perception of order, but not an actual condition for a vivace urban fabric, i.e. able to promote a fecund urban life. This misconception leads to a radical removal of modest scales and to a structure conceived only on a larger scale. Ornament and most tenuous notes are erased and it is produced a net separation of urban functions, as modules composed by nodes of the same content are created. As a result moments of daily life are spread out. Users are prevented from managing a vast range of needs near urban nodes in which most of their time is spent. This results in a significant increase in car use, and in an even more pervasive condition of social segregation.

1.2 *A sensitive urban structure*

In order to coin a more sensitive urban shape a different concept of module is needed, conceived as a set of episodes, as a net of users and urban nodes, tiled by a copious exchange of inputs. Information as a basis for a more effective urban design, is a concept carefully argued by Alexander and Salingaros. A relevant consequence is that urban fabric is supposed to be a means to promote an exchange of inputs of various tone and value, on different levels. A transfer of inputs, of information is described by a cost, by the effort required to grasp the true sense contained in it, content, a scale, i.e. its relevance and its value, due to its

emotive notes. Moreover, an episode is even more frequent or intense if its relevance is ample and its cost is modest. It is also important to observe that the force of a connection results from frequent and intense exchanges of input, source of a specific intimate emotive tone. It is proper to note that a use, an episode, emanates an intimate sense, if it is a usual, routine act. It means that it can be reiterated regularly. In order to be replicated, to become a routine, an episode should be a spontaneous act. As a consequence, its cost needs to be reduced.

1.3 *Information*

Information exchanges include costs related to distance, time, effort required, stress, inconvenience, and externalities (soil consumption and occupation, noise, gases, dust, accidents, caused nuisance or fear, etc.). It is proper to note that costs refer to the means used to transfer data, and its sense, its essence, its value, and mainly its emotive value, all refers to the gain or profit provided by information. Transfers of inputs are various in terms of range, nature, tone, and value. Each episode should be considered as a sign pervaded of a specific sense, i.e. as information to be perceived. Cars, goods or people moving from A to B represent of large- or medium-size exchanges, while conversations or people observing a facade can be seen as small-scale contacts.

1.4 *Episodes and activities*

A definition of episodes sensitive to their nature and aims, is required in order to ponder role and structure of an urban road. Car traffic is to be taken into account considering Access, i.e. small-size motion, penetration, medium-size motion entering a specific urban portion, and Distribution and Transit, to be considered as a large-size relation among distant points of an urban area. Brief car stops can be seen as not onerous uses, while longer car stops, i.e. parking, are medium-size uses which are onerous for the urban fabric in terms of soil consumption. Transfer from pedestrian realm to cars and vice versa are a consequence to be pondered of cars stops. Furthermore, transit and stop of public transport means is to be considered in order to enhance pedestrian routes. An obvious consequence is the need for people to wait at bus stops, and to get on and off the means of transport.

An urban road should also provide spaces to allow people to move by bicycle, which can be considered as a medium size and scale episode. Walking for basic tasks is the most frequent episode, its scale and size is small but its emotive outcome is significant, like paths traveled for leisure whose impact is small in scale and size but based on the prospect of a large emotive gain. An urban road is also to be conceived to persuade people to linger there, for a rest, to find orientation, for leisure, or to observe a detail: above all, an user must be able to sit, so as to recover from fatigue, or to evade in a serene pause, to perform leisure activities, i.e. reading, observing a site, meditating, etc. or being part of broader events, spending time among people. Moreover, it is proper to point out activities, of modest scope and size, but, if frequent, source of an intense emotive gain and of clues about tone and essence of a site.

An urban space is presumed to invite people to listen, to sense scents and sounds, to observe, to meditate, to read, to converse, to perform sport activities and games, to rest to have a snack. It is also important to consider uses aimed at connecting private and enclosed ambits and services of a broader scale, i.e. loading and unloading, storage and waste collection. Moreover, a main concern is to allow access and penetration or transit of emergency vehicles.

1.5 Optimization of information exchange

An urban fabric is a mean for favoring and optimizing information exchange. Several strategies can be adopted. On the one side, an urban realm could be conceived to reduce the cost needed to gain a certain amount of contents. On the reverse side, an urban realm can be intended so as to increase the gain of data earned during an act of given cost. As a consequence, it is not mistaken to conceive a vast exchange of information as a fractal structure in order to reduce an onerous need for long routes and to promote contacts of modest size.

Ensuring that the size of a contact is adequate to its scale and relevance should be a rule. A consequently important principle is to make sure that basic needs are met at a small distance from places where people spend most of their time. A minor cost is supposed to allow people to perform a specific action more often, i.e. it is supposed to be a basis for the rise of frequent episodes and routines.

A positive outcome is a decrease of car use and a significant increase of pedestrian routes; considering pedestrians as main users of an urban realm is thus a natural consequence.

It is also important to reduce costs caused by traveling inconvenience and negative and tedious emotive notes caused by defective media as well as to orient the residual demand for long routes to a more intense use of mass transport. An urban area is also presumed to be designed to increase the gain of notes and inputs: a large-size transfer of information, e.g. a going from A to B is to be considered as a hint for a broader range of modest scope contacts but various in tone and nature, experienced during the motion. Users walking on a road for a certain purpose should be able to observe a façade, gestures of people, to sense a scent or a voice, to notice a vast amount of episodes, floating among urban squares and roads and confined spaces. An urban realm is successful if even a modest act is source of a fecund and profound experience. Not less proper is to assume that a broad range of modest-scope contacts sorted by notes stored in surfaces of an urban space can occur even if the frame of a larger scale episode is removed, even if a cogent reason to move is missing. The vast gain of clues and notes is per se a source of a pleasant episode and a good reason to roam.

Conversely, a space deprived of notes and modest scale data, proves to be a poor and arid space. If modest scope contacts (sitting, talking, lingering, observing, etc) are deleted, episodes caused by strong needs are also emptied and deprived of a fecund part of their sense; basic duties are no longer made less tedious by a vast range of notes sorted by a various and varied urban realm. A poor urban scenario produces as negative outcome, that the need to perform basic duties, or to be part in urban life results in a more pervasive use of car.

2 INFORMATION FIELD

2.1 Relevance of information

Roads and open spaces are not to be reputed as a mono-functional space, conceived for an idea of motion as a stand-alone fact. It is proper to conceive an urban space not as a mere void, but as a piece of a broader realm, pervaded of a vast range of notes and voices that connect it to its users. Data emanates from borders and boundaries, from surfaces and nodes, or events contained in an urban space, i.e. furniture, plants, etc. Data are the content people perceive and the mean allowing them to relate to a space. People need input so as to read the tone and state of a place, to come to an opinion, a concept of it and to set a proper behavior: people act in response to data received from a site and pondered. Data are lost if people do not receive, ponder and confers a sense and a purpose to forms and notes contained on a surface. Users ascribe a broad range of senses and nuances to received data; their reaction adorns spaces of a broader range of features to be noticed and pondered. Users increase the sense and relevance of a space more acutely as the use and fruition of a space becomes more conspicuous and frequent.

It can thus be assumed that the true nature of an urban space, its voice, derives from data and notes pervading it and relating it to people. In order to produce a broader and more fecund sense, an urban space should promote a dense exchange of data, i.e. ensure people perceive a vast amount of inputs. Moreover, a fecund urban life, i.e. more abundant social content, is produced when an urban scenario is designed to foster a fusion of episodes across its space or across its boarder. Episodes promoted by an urban space and nurtured by its information should couple with events prompted by notes and forms conserved in boundaries and with episodes concealed in conterminous nodes.

It is required a sum and a true consonance of data and inputs sorted by an open space and the broader and more fecund range of notes emanated by boundaries. The range of inputs emanated by boarders of a space is therefore to be assumed as a main concern in order to conceive a prosperous urban scenario. The site, and not the a priori idea of it, defines the essence of a road, and requires to set modes of use consonant and sensitive to its true nature. If users base acts and gestures on the inputs received from a site, inputs not only defines modes of use of a space, but also its pace and distribution of episodes and responses within a space. If a space is to be conceived in order to lead users to receive data from surfaces, it can be assumed that its structure derives from or is even concealed in the vast amount of that pervading it. A deviation from the trend of information produces a fracture; uses allowed into an open space are not nurtured by episodes promoted by boundaries and do not give rise to potential uses proposed or implied by surfaces. As a result potential uses do not occur, or take place in different periods and points, so that they are prevented to fuse and to produce a more fecund event. A mere juxtaposition produces a weak ensemble. An open space, considered as a mere void, needs to be distinguished from an urban space as ensemble produced by a profound fusion of void and nodes and boarders. People need a

space meant to preserve and protect episodes and emanating a broad range of cues and data. An urban space, conversely, captures its content from notes and inputs produced by its surfaces, and its value and purpose from people perceiving it and conferring a sense to the receipted inputs. A sensitive design is required to deduce from boundaries, to trace in trend and pace of the input field form and ratio of a space, of its nodes and promenades, in order to lead people closer to sources of notes and features, and to persuade a large number of users to roam, to stop, to linger. The more numerous people the broader the increase in sense and relevance of a space.

2.2 Regions of urban space

Spaces are not to be considered as a static, permanent scenario, composed of a repetitive and periodic sequence of spots or modules. Nodes are an outcome of data field: one is persuaded to stop at points where content is focused. Nodes so induced are transient but frequent, and result in a vast portion of the sense of an urban space. The presence of numerous people increase content and scope of nodes. People cluster and give rise to a vast trade of data, of opinions, of cues. Spontaneous stop may gather to produce broader events or result in more permanent episodes. Urban scenarios are to be designed so as to foster and permit an evolution or a concatenation of events; for instance, urban spaces could propose spots where people are induced to sit and rest.

A pause needs to be filled of a conspicuous gain of inputs, absorbed from people, from boundaries, from pages of a novel etc. Moreover nodes act as steps or stages and alter pedestrian routes: people do not move along an ideal clear line, but move to a node from a previous one. As a result inside an urban space, nodes define regions of different size and tone. An union among clusters can't be ensured by a mere space, since it presumes a mutual dense trade of data: as a result it is needed a scenario conceived so as to sustain pedestrian routes, i.e. a space safe and generous of information. Voids produce a fracture, a caesura, that prevents modest transfer of inputs. A even more net and acute caesura is produced by cars spaces and routes: Cars conquer the entire space, erode road edges, produce noise, gases and dusts, cause a deep sense of nuisance. As a consequence People abandon urban spaces. Policies aimed to reduce costs of usual duties, so as to reduce the presence of cars, are required in order to recover a dense net of spaces to be returned to people, as venue for a broad range of precious episodes of modest scale; besides it is required to protect pedestrian spaces so as to prevent a sense of fear, a nervous emotive status, due to a pervasive car presence. A not vain consideration emerges: pedestrian areas are the core of an urban space and cars space and routes are to be designed so as not to violate them; yet it is proper to permit an osmosis among car spaces and promenades and nodes for people. Proper cautions are to be provided in order to preserve episodes fostered by urban spaces, and to set and to ensure a safe and fruitful exchange among diverse uses. At a broader level, an enormous amount of transfers condensed on a same medium, produces congestion, and censures uses and transfer of input of minor size and scale (Salingaros 2005). A fundamental enunciate arises. An urban realm is to be moulded in order to couple congener modes of use, and to separate, contrasting ones. A consonance relies on, scale, size, and nature or tone or purpose of uses.

3 RECOVER OF URBAN SPACES

3.1 *Essence of an urban space*

A fecund urban life is proved to arise from a copious transfer of content, from a fusion of episodes; nodes and pieces of space are required to reinforce and to be reinforced by contiguous ones; a true consonance in terms of space scansion, visual features, and in terms of modes of use promoted, is needed. As a result, in order to recover an urban space it is proper to:

- Consider a void as a part of a broader unit. Its essence derives from its frontiers, and from nodes related to it via its boarder;
- deduce modes of use of a space from its tone and essence, i.e. from scope, scale, value, tone and sense of the broad range of data that pervade it;
- consider as main uses to be promoted ones performed by pedestrians.

A concern in designing new urban pieces is to conceive nodes as a mean to form, to mould an urban space, and to set and to increase the amount of notes of the potential space (Alexander et al. 1987). The first step of a sensitive recover is to recognise regions of space, pervaded of a specific tone, defined by data emanated by boundaries o via boundaries by nodes. It is proper to discern regions defined by nodes of net cultural value, residential zones, regions connoted by boutiques or stores, or by recreational functions, regions defined by areas of great natural value, spaces connoted by presence of educational nodes, zones defined by spaces for major events (opera houses, sport or concert arenas), spaces defined by presence of poles hosting services of large scale (premises of banks, government offices, universities, hospitals), and roads located in areas signified by production plants.

3.2 *Modes of use*

Contents of an urban space produce episodes of diverse scale, tone, purpose, color: a residential area is a space to be conceived as a place for major social episodes. Spaces are to be provided to allow people to meet, converse, perform games or sports. Spaces connoted by acute cultural or natural value are primarily to be conceived to allow people to sense the site, observe, and notice its content of notes, scents, sounds, forms. Spots conceived to invite people to stop and linger so as to observe are needed in stores and boutiques areas, as spaces designed for people to sit or to stand, so as to rest, are to be provided in areas where major services, educational poles, or event venues are located. A second step consists in defining cars space, ascertaining that these spaces, unsafe and unable to foster and sustain pedestrian routes, do not cut or ruin the core of the space or separate its regions. Carload, size and purpose of car motion are to be deduced in primis from measures, form, and geometry of a road, i.e. from size and scope of the tie it provides inside an urban area. A road not large nor long can sustain sporadic local moves, or can permit access to a node; more frequent or onerous uses are to be prevented. Moreover, carload, and size and nature of car motions define the grade of separation among car space and promenades or spots conceived for people. Local roads do not require or, even more, claim a tenuous distinction of car spaces and pedestrian realms. The road space is conceived for people and cars are mere guests. A tenuous boarder could be useful so as to protect spots or nodes conceived to invite people to sit, or to stand. For roads aimed to sustain a more intense cars presence, i.e. motion of small or medium size, aimed at entering a district, a more marked border is needed, so as to protect and define promenades and spots for people in a clearer manner. Line of bollards, or arboreal curtains, could be useful so as to define a boarder net and porous.

In case of an arterial road overloaded by a copious car traffic, definite borders are required. Roads of this sort are to be designed to promote rapid and intense transit. Separate spaces or side roads are required in case it serves access to contiguous nodes or stops. It is also useful to recover residual voids generated along the road or along its service roads and areas as serene and safe scenarios. A continuous border is needed to separate spaces. Car space should be placed at a different level, in order to create a network of promenades and spaces aimed at suturing the urban realm and providing a spontaneous, safe, and fecund interaction among its parts.

3.3 *Composition of urban spaces*

Once defined modes of uses and main concerns, it is required to define a safe, pleasant and varied scenario. A varied space presumes openness to a vast number of features of urban life, but also means a space dense of tones, a seductive scenario. Taking the proposed uses into account is needed in order to consider an episode as a sum of acts, and more, as part of a series of episodes. An urban space should be aimed at promoting a fluid sequence of actions. It must ensure reception of notes, of cues, in order to favor rise of broader episodes from previous ones, or to foster a sequence or a concurrence of episodes. Positive data to be absorbed and propitious conditions are needed to foster an act and to ensure that a step results in a consequent one. On the other side, it is proper to define adverse conditions to be avoided and prevented. Positive and adverse conditions are to be condensed in a list of pressing needs, and these are to be converted in a precise series of clear requisites. Nodes or spots must be considered as a sum of modules, composed in a severe and clear ratio, and meant as a proper response to a group of basic requisites. In conceiving a piece of an urban space, a proper composition of boundaries is nodal. Boundaries are the main sign to be used so as to structure a space. A border deduces defined spaces from a void and it forges a precise form. Moreover, the vast ranges of notes, of motifs adorning a surface are a piece of sense of a node. It is a nodal concept to opt for varied surfaces, porous, moved by coves and curves and punctuated by pauses. A boarder is, in primis, a mean aimed to protect a space and to set and to foster mergers of congener events. Moreover it is correct to recognise, a greater value to road surfaces and furniture. It is useful to improve content and appearance of pavings, i.e. modulating colors and grain of surfaces so as to produce a more varied sensation, and to permit users to read a place, its spatial scansion. Furniture is to be used so as to evoke a border, to define a spot, a point, or to mean a specific act or part of an episode. A seat, a column, a bollard can be seen as a source of a broad range of inputs, primarily of tactile ones, in order to produce a fecund bond among users and spots or nodes of an urban space.

4 CONCLUSION

Positive outcomes produced by design strategies based on information exchange are proved by the case of Melbourne. A vast operation, led by the Gehl Architects, produced a vast network of public spaces, conceived as a scenario for special events and a venue for spontaneous episodes. A vast piece of the Central City area of Melbourne was considered. This area is bounded by Spencer Street, LaTrobe and William Streets, Victoria Street, Spring Street and the Yarra north bank to the north, and the area of Southbank bounded by

Clarendon Street, Whiteman and Power Streets, Grant Street and St Kilda Road.

The main concern was to transform roads from an indistinct space for cars, to a vivid and vital space. Vast parts of the urban fabric were retrieved, so as to develop a broad range of ties and lanes, suturing spaces and fostering pedestrian routes. For instance, almost 3 km of lanes and promenades were recovered as safe places open to pedestrians, 500 m of which were new lanes or arcades. A vast recovery of spaces oriented to favor episodes required to mould scenarios generous of notes. A key concept was to consider an urban scenario as a sum or a profound union of boundaries and open spaces: the information field produced by boundaries was thus considered as the true and vital content of an urban space, and as a stimulus to increase pedestrian activities. As a result, façades were redesigned so as to offer a vast range of contents of interest to pedestrians: large façades were divided into parts so as to increase articulation. Large windows, as display of information, were provided, to connect open spaces and clusters of social episodes (boutiques, cafes etc); details, scale, colours, and rhythm were modulated so as to emanate a sense of visual opulence. Promenades and pieces of spaces conceived to invite people to stop and rest are consistent with course and trend of information field emanating from boundaries, so as to lead people to perceive notes and motifs proposed by the system of smart façades. Besides, open spaces were designed so as to increase information and to protect paths and nodes: Stone pavements adorn roads surface of seductive textures, lines of trees protect and reinforce tone, enclosure and amenity of urban spaces and nodes, as containers of modest-scale functions (e.g. micro-scale' retail such as fruit stalls, newsstands, information pillars) were provided so as to favour rise of episodes at various levels of scale. As a result a net increase of urban life is obtained (Gehl et al. 2005); The amount of seats in dehòrs of cafes increased up by 177% since 1994 and the number of cafes, restaurants and bars increased from 95 in 1994 to 356 in 2004. Even more notable is the increase in number of episodes fostered by urban spaces. Pedestrian routes recorded during weekdays increased by 39% during the day and by 98% in the evening. On Saturday it is recorded an increase ranging from 9% up to 13%. As regards static episodes, a net increase of 11% is noticed during weekdays, and a positive variation of about 5% is recorded on Saturday. As a result it is not arduous to argue that a sustainable urban area must be conceived so as to reduce consumption of resources, and also in order to promote processes able to reinforce a fecund urban life. A vast net of consonant spaces, conceived to promote a broad range of activities, lends a positive sense of varied and seductive landscape and encourages people to be part of a varied range of social episodes, so as to prevent segregation and to promote participation, to favour trades and contacts; as a result it produces a conspicuous social and economic profit.

REFERENCES

Alexander, C., Neis, H., Anninou, A., King, I. 1987. *A new theory for urban design*. New York: Oxford Press.

Gehl, J., Mortensen, H., Adams, R., Rymer, R., Rayment, J., Campbell, A., Ducourtial, P., Duckett, I.S. 2005. *Places for people—Melbourne 2004*. Melbourne: City of Melbourne.

Salingaros, N. 2005. *Principles of urban structure*. Amsterdam: Techne.

Green Design, Materials and Manufacturing Processes – Bártolo et al. (eds)
© 2013 Taylor & Francis Group, London, ISBN 978-1-138-00046-9

Recovery of roads in urban areas: From an indistinct feature to a specific function

A. Annunziata & F. Annunziata
Department of Civil and Environmental Engineering and Architecture, Cagliari, Italy

ABSTRACT: The management of the existing road system will be the core of this paper: In the future, recovery of existing road system is going to be a strategic issue for the socio-economic development. This is particularly important for the historical centers of urban areas, for the suburbs to be renovated, for railway lines to be partly reused as surface light train systems at different territorial levels, and for the roads which can be assigned, for instance, a role of inter-district streets and equipped axes.

1 INTRODUCTION

1.1 *General considerations*

A road is most commonly defined as an area for public use on which pedestrians, vehicles and animals travel. Regulations establish the standards with which the study of the functional and geometric features of roads need to comply, depending on the type of road. Defining the type of road means identifying uses and users allowed on it, taking into account not only the kind of connection a given route serves, but also the environment it belongs to. However, regulations limit the definition of a road to its functions as a space for motion: the concept of movement of people, vehicles, or goods should be replaced by a more fecund idea of exchange, as a wider phenomenon nourishing the life of an urban area.

A road is the set of a wide osmosis of notes, of content, at various levels, among nodes of a site, among people, and between people and site. These are a significant part of the real essence of a road, evoking a range of activities varying in tone and breath, for which that road should be designed. The importance of the site in sanctioning the role of a road also needs to be taken into account. The result of the design process, both planning new roads and recovering existent ones, is that the work is a sign reinforcing the site it is immersed in, the connection networks pervading it. This assumption emphasizes on a new meaning of the design process as an iterative path in which each choice undergoes continuous checking of outcomes and impacts it causes, in order to achieve a fair balance of functional, economic and environmental needs. As is known, the required result is creating safe infrastructures, sustainable for the surrounding territory and ensuring adequate levels of service throughout their useful life. Sustainability also implies that the infrastructure foster a more equitable structure of a region, and safety also ensures movement: extraordinary events, such as floods, cannot isolate built-up areas. Therefore, a study or an adjustment of road networks according to purposes of civil protection is needed in order to ensure permanent connections to and within a given region.

1.2 *Road networks and context*

Building respecting the site, the context, is thus essential: the road should not be the cause of depletion of flora and fauna, but it needs to maintain and preserve ecological networks without causing fractures. Transport networks need to be conceived as a starting point for creating a fairer distribution of nodes and function poles (Annunziata et al. 2004). An urban realm or a network of smaller urban areas, responding to various levels of diversity, density, and mixture of functions designed to make it easier to access its major nodes, need to be designed considering containment of costs, for both users and the site, deriving from daily tasks or participating to episodes of community life. Moreover, it is considered essential to ensure improvement of accessibility conditions and connectivity within the transport system to which the network belongs.

2 RECOVERING EXISTING ROADS

2.1 *General consideration*

Another important topic is recovery, even to new features, of the existing roads. Adjustment of a road or road network should aim at the same goals mentioned above: it needs to support the primary network, to ensure safe movement for different users, to give the network marked features of connectivity within the transport system to which it belongs. Moreover, it should allow faster and easier access to isolated or depopulating areas, and also contribute to a rebalancing of the system of settlements and services. This also implies considering the relationship between the road and the environmental, geological, hydrological and territorial characteristics of the area, and particularly its relationship with the existing balances. The leitmotif of a general adjustment project should be environmental sustainability and the intrinsic safety of the road. Existing networks can be implemented by building new segments.

This may be the case when the existing roads cannot be adjusted, when its overall functionality requires new infrastructure branches, or when recovery and adjustment require an excessive environmental cost.

2.2 *Tenets for a proper recovery*

It can also be argued that a policy aimed at the recovery of existing networks and channels should maintain the system and each of its elements appropriate to their intended function within its given useful life, while assuring a cost-effective use of resources. This implies the sine qua non of a wider management of roads and transport systems, extended to existing networks, at different levels, as elements of a whole structure. In the first place, role and function of each element are to be defined. This requires a new way of conceiving the role of a road and, thus, specific and rigorous planning tools. The function which the single infrastructure fulfills within a network is to be taken into account as well as that role in its relation within the existing transport networks: how they relate to the road network, providing an alternative link between areas of the same region, in order to assign each element functions as a component of an integrated and intermodal network. Furthermore, a road should preserve and sharpen the forces sustaining life in an area; the meaning and the role of road, both in urban and rural areas, cannot disregard the vast range of connections which pervade the site it is inserted in. This becomes particularly meaningful when a new route allows planners to rethink the function of existing roads and branches which, due to deficiencies in the road network, have been overloaded with unsuitable uses making circulation unsafe and causing or exacerbating the degradation of the site.

2.3 *New functions of restored routes*

It is therefore important to implement a reclassification and subsequent adjustment of the road network so changed, assigning certain elements functions of urban streets serving the exchange mobility of the different sectors and, if the context requires it, giving them functions of environmental-tourist routes. In maintaining the horizontal and vertical alignment, a priority objective of adjustment should be the recovery of empty spaces along the edges of the path, as a scenario and container of a large set of uses and episodes, in order to support and give fresh impetus to the uses of the site, and give a new meaning, new content to a site, promoting a more intense and fruitful union with people. The expected outcomes are therefore to be specified while defining the functional class after adjustment.

2.4 *Evaluation of priorities*

The concept of system and the extent to be given the link between road and site requires that each proposed project is considered in its impact and outcomes on the networks and infrastructures belonging to the same context, as well as on the site, on the phenomena and relations permeating it, according to an idea of system as serving a territory as a whole, although managed by different authorities. Regional or local authorities, if any, are in charge of this coordinated management in order to enable an estimate of the requirements and therefore the identification of the actions needed (Annunziata et al. 2006a,b). The priority among the various actions aimed at defining all or part of the goals considered should take the following into account:

– The role of each route in the context of the network and its efficiency in terms of service provided, safety and environmental critical issues;
– The type of adjustment works needed, in order to evaluate overall programs managing systematic actions which can ensure strong economies of scale in respect of homogeneous performance standards;
– The dynamics of the ongoing processes, on which the standards of functional obsolescence depend, in order to identify the time-frame for action ensuring containment of costs through adjustment of the existing infrastructures.

3 INTEGRATED DESIGN OF ROADS

3.1 *Essence of roads*

In its capacity of involvement of local communities, integrated design, as it unfolds along the planning process shaping the different levels of detail, even in the case of adjustment of existing infrastructures, seems to be the right method to reach a serene union of infrastructures and environment (intended as the site, as set of uses, senses, given by a perpetual human poiesis), of mechanics and safety, of costs and benefits, of aesthetics and statics. The aesthetic essence of a road infrastructure is also an important issue. A road has to give new content to the context, both by promoting consonant uses, and by fostering a positive visual outcome, becoming a new positive content of the site. Whether this is intended as a conceptual goal of the design process or just as a condition, what needs to be considered is not appearance in plant, but rather what the users see.

3.2 *Perception of roads environment*

If the project is based on how users perceive and use the road, in an urban area we need to point out how it is experienced differently by drivers, by pedestrians and/or by cyclists. The way in which the site is exploited varies primarily depending on how fast users travel. Faster users distinguish vaguer content they: i.e. features are not noticed, the weakest signs are lost or glimpsed as a blurred background, and therefore provides no fruitful union, or data exchange. A person on a car tends to notice only the major notes and phonemes. Conversely, a pedestrian notes more subtle tones and nuances (Annunziata et al. 2007). However, we need separate the data whose meaning is unique from the unclear data, whose meaning or the weight varies depending on users, on the purpose for which they move, on their familiarity with the site, on age, gender, duration of their journey, vehicle type, and also on the period in which they move and whether they are drivers or passengers. Users should be able to borrow a large group of consistent data from the site since people respond to data they receive and ponder. Inputs received from environment are the basis on which people come to an opinion on a site and they modulate their behavior accordingly. Various studies point out that sites influence the way of driving: more, the road scenario may also influence the choice of a route depending on the reasons for which a person travels. Among the available routes and paths drivers and pedestrians tend to have preferences leading to opt for a route over another. This usually depends on shorter travel times, comfort or more enjoyable routes; this choice affects the territory crossed, both in rural and urban areas. The road scenario should emanate a wide range of inputs so as to allow users to understand the site, but also to make the journey pleasant.

3.3 *A pleasant environment for pedestrians*

Therefore, what people moving on cars perceive needs to be taken into careful account, providing a series of notes, of object, whose content and position are coherent with the spirit of the journey users can sense fully and comprehensively. But a scenario planned only pondering the larger scale is poor and boring for pedestrians: road scenario thus need to emanate a vast range of notes and phonemes of smaller scale, so that content pedestrians can experience is varied and tinged with a sense of discovery both during travel and stops. This link between the repertoire of notes, forms, and phonemes pervading a site and the ways people use it, that is content, scope and number of episodes it houses, highlights as the edges, which are both source and medium of those notes, and the way by which they spread through and permeate the urban space, are a major issue in creating a pleasant and lively road scenario.

A road should contain conditions, paths, nodes or scenarios, sheltered, pleasant areas in which to sit and linger, favoring uses of different nature and a fine sense of the vast ranks of inputs emanating from the site, from its edges or borders. The aim is thus to give breath to the various events which occur there, allowing as many people as possible to participate and enjoy them, giving an audience to its contents, its views, its nuances. In this sense, a road needs to be conceived not only as a space for the traveling, but in a broader sense, as a suture, as a sign intended to make it a coherent area and favor an exchange intended as movement of people, cars, buses, and goods, but also as content shared by people, and between people and the site. The road should be the set for more than one type of osmosis, a scenario where many kinds of uses rise and are shared by several people, as a means to experience the site where it belongs. The road is not just a way to travel from A to B, but it should be a way to know what lies between A and B.

Thinking of a road as an empty space crowned by full areas creates the desired scenario. This union urges to ponder the link between the road and its edges, the large range of notes and content pervading it, uses and events it houses which, through the openings, the breaks in the walled sides, reach out to the open space. There are therefore many issues to be considered, such as the type of user, the kind of site the road lies on, its contents, the type of uses it causes. The edges give shape to a space, modulate its extension, radiate the range of uses and statutes which pervade it. The edge is the medium with which an empty space is composed and given an essence: the nature of the urban fabric in which a road is inserted defines shape and extension, that is the content of the link between the parts of an urban area it creates, as well as gives a sense, a spirit to the empty space. A study of the full crowning the empty is necessary in order to draw the uses of an urban scenario and the type of users allowed to it.

3.4 *Town center and residential areas*

In this sense, the old town center is the most critical area due to its historical and aesthetic value, and for the density of both of residents and of cultural and economic activities. The identification of the inhabitants with this part of town and the predominantly residential areas implies a remarkable care for values, such as the right to use common areas.

However, regulations are deficient in considering urban roads mainly serving residential areas: issues relating to the needs and wishes of the vulnerable road users, the study of pleasant visual sequences, and street furniture are not addressed. If we accept the definition of quality as "degree of compliance of performance with requirements" as cause and effect of the planning path, and we recognize the wide range of uses, from traffic of cars and pedestrians, up to lingering, sitting, conversing, and enjoying a site, its notes, its tone, and its sounds, it is clear that regulations do not fully allow to design quality roads.

Urban roads, if inserted in historical or highly populated areas, need in the first place to be conceived as scenarios where people gather, talk, take a break: a scenario aimed at giving new breath, new sense, new reach to the living theme promoting uses and episodes based on it. This implies that nodes and scenarios of a road are to be designed to promote those uses, firstly making sure pedestrian realm is safe and persuasive in evoking this feature; creating porous protections between the areas on which vehicles travel and pedestrians areas: hedges, shrubs and furniture can be used to punctuate the space and mark its nodes as well as to evoke varied landscapes full of ideas. Roads should no less serve the urban fabric they belong to, and mass transport should be implemented without neglecting controlled private traffic. Traffic as the following should be avoided:

– Traffic using the roads as shortcuts and deviations from congested routes;
– Excessively fast traffic;
– Parasite traffic, as search of and exit from parking spaces, and parking on sidewalks of vehicles whose destinations are outside the district.

The creation of parking areas for residents, of pleasant and safe areas for pedestrians, a drastic reduction of noise and dust and gases caused by heavy vehicle traffic are ways to promote residence as an element of balance and of city life. Whether city streets emerge into squares or shrink into alleys, whether they are famous or known only by residents and usual users, spaces in residential areas are primarily an intimate setting, a valuable content of the urban fabric: they play a nodal role in promoting productive urban life, in creating a healthy, safe, pleasant urban realm, from which users receive a deep sense of comfort and a calm, emotional state.

4 TRANSPORTATION NETWORKS

4.1 *A tool to design urban fabrics*

Taking the urban context into account, in order to achieve a recovery of its empty spaces aimed at enhancing a larger set of functions more consonant with their nature of opportunity for exchange, the abnormally high number of cars coming daily from neighboring municipalities toward the major cities of our country needs to be reduced. Heavy traffic is a result of the concentration of the "attractors of interest" (workplaces, educational, administrative and commercial services, health services, etc.) in the major cities and it is worsened by a persistent lack of public transport.

Transport system could be considered as a theme related to broader areas: its goal should be integrating the city with its widely inhabited surroundings. It follows that, in planning of urban areas, the relationships between central and external areas are to be preliminarily analyzed as connected functions of territorial rebalancing. It is therefore necessary to build a polycentric network system whose "border" nodes are bridges open to the outside, even redundant, that is, highlighting the need to replicate, even in part, some functions in order to heal a harmful addiction to the center of the suburbs and smaller urban centers. Structure and planning of services are to meet a vision aimed at promoting a real and full union of the living theme with services, mobility and environment, and a strategy to share and locate functions and services in a wider urban area, consistently with the location of demand.

4.2 *Aims of planning transportation systems*

Framework planning of infrastructure works need to be defined and should contain:

– A concentration of new settlements along the public transport paths, enhancing public transport on dedicated tracks;
– Creation of a logistics system by rationalizing the transport nodes in the region;
– Union and cohesion of the different systems of mass transport on dedicated tracks in connection with the surface transport and the parking areas/facilities in order to create a nodal logistics platform meant to successfully reduce vehicular traffic on existing roads. In brief, the framework project goals over time are the following:

– Changing the modal allocation of transport demand, enhancing a mass transport system on dedicated tracks as the backbone of the infrastructure network serving mobility;
– Reducing the number of commuter flows towards the city center, avoiding settling of new attractors of traffic or implementing the existing nodes;
– Reducing the extent of commuter flows toward the city center, locating metropolitan area services outside the central area and placing them along the paths of mass transport as close as possible to users' demand, and therefore more accessible.

A careful transport policy implies making bicycles alternative means of transport; extending pedestrian paths and promenades; starting processes of pedestrianization of the city; returning residential neighborhoods to theirs inhabitants; reorganizing the road network for its specific functions, avoiding important vehicle flows to cross residential areas affecting their continuity and their relationships with population. A transport policy needs to make cars complementary to other transport modes, with people as reference users, especially weak users like children and old people (Annunziata et al. 2004). A transport policy will also reduce traffic pressure on the road network, particularly that accessing the city from the surrounding areas. It means reconsidering social needs besides mobility, recovering the square to its true status of agora, of relational space, and conceiving certain streets in residential areas or historical centers as spaces for people. This means recovering the broader range of urban space meanings which nowadays are sacrificed to a persistent transport model, lacking the features of an integrated and intermodal system. In order to solve mobility problems, a cultured vision of transport policy, honest and prescient of social facts is needed, as essence of a thoughtful strategic idea.

5 CONCLUSIONS

The purpose of this work is therefore thinking roads in urban areas not only as a space for traveling, but as a scenario of a larger range of uses, as a place of cohesion. The status of urban paths, however, is often neglected. Regulation, both the Ministerial Decrees and the National Research Center instructions, consider vehicles as priority users; Pedestrians are ignored, and roads and urban spaces thus become empty scenarios. A new way of thinking the urban area is needed, in order to promote pleasant, fair scenarios whose trait of shared landscape is emphasized. The study of each theme, or

part of it, of a realm urban should aim at this purpose. For example, when planning a network of mass transport on dedicated tracks, mere creation of a coordinated and intermodal system should not be the only priority, but it can be the basis for a rethinking of the ratio of an area: the goal is creating a heterogeneous, dense, mixed realm in which each of its parts, containing nodes of various type, produces a living scenario, a stage for a vast range of episodes. This would allow reducing time and cost of transport, encouraging modes of transport less harmful for the urban landscape, reducing the number of cars overloading it, and their burden. Moreover, it will also encourage a more equitable urban environment, in which people are not forced to use cars to take part even in the most meager urban phenomenon or to meet any modest need. There are several examples, in particular in French cities, of adjustment of existing roads with no clear identity into itineraries aimed at satisfying the prevailing traffic components. Specifically, the proposed innovation, referring to the peculiarities of the Italian context, is transforming existing roads in urban streets as well as in environmental touristic routes, in a logic of enhancement of regions involved. Ongoing discussion on the transformation of the City in the most important research centers in Europe and in the U.S. is in fact focusing on a rethinking of the functions of road networks (Glaeser 2011). The goal, therefore, is creating an urban realm designed for people as a stage for a varied urban life: small-scale urban scenarios, evoking highly emotional uses and episodes, need to be recovered. Is not less important to ensure that the urban landscape, especially if full of poles of fecund urban phenomena, is crossed by scenarios designed only for pedestrians, to promote a strong osmosis between urban nodes and to be set of a vast range of interactions, of modest scale but of acute emotional content, among users and between users and the site. Promenades or dedicated cycle lanes are also an important resource to recover neglected areas, edges, residues, suggesting a new ethic based on thinking even the humblest urban space (the edge of a road, a parking area, the empty spaces along the ramps of a bridge, a river, or a channel) as a starting point to create new scenarios, to be used as sets of fruitful moments. It is important to note that the more the media designed to give rise to a copious flow of people, of goods, of opinions, the more lively the urban area becomes as a stage for a more extensive range of uses. It should be finally noted that a more intense urban life makes the living of each individual more fruitful; this entails the development of extensive people networks, on which a genuine social cohesion can be based, and, not least, a stronger cohesion between civitas and urbs, to promote a more acute sense of belonging and responsibility of people to the context in which they live.

REFERENCES

Annunziata, F. Coni, M. Maltinti, F. Pinna, F. Portas, S. 2004. Progettazione stradale integrate. Bologna: Zanichelli.

Annunziata, F. Porru, R. Maltinti, F. 2006a. Norme funzionali e geometriche per la costruzione delle strade. Part I of La normativa della progettazione stradale. Strumenti didattici n° 15. Cagliari: CUEC.

Annunziata, F. Melis, D. Maltinti, F. Cecere, M. 2006b. Norme per gli interventi di adeguamento delle strade esistenti: alcune riflessioni. Part II of La normativa delle progettazione stradale. Strumenti didattici n° 15. Cagliari: CUEC.

Annunziata, F. Coni, M. Maltinti, F. Pinna, F. Portas, S. 2007. Progettazione stradale. Dalla ricerca al disegno delle strade. Palermo: Dario Flaccovio Editore.

Glaeser, E.L. 2011. The Triumph of the City. New York: The Penguin Press.

Green Design, Materials and Manufacturing Processes – Bártolo et al. (eds)
© *2013 Taylor & Francis Group, London, ISBN 978-1-138-00046-9*

Systematic approach of upscaling aircraft parts and sub-modules to aircraft level

R. Ilg

Department Life Cycle Engineering, Fraunhofer Institute for Building Physics, Stuttgart, Germany

ABSTRACT: Aircrafts are complex product systems containing various supply chains, high grade and specific materials as well as specially customised processing technologies for their part production. Therefore only selective Life Cycle Assessments (LCA) exists containing certain parts or an estimation of an aircraft. The challenge is to model such a complex product with limited information. Therefore a methodology was developed characterising known aircraft parts to use them in order to complement missing information on aircraft parts, where detailed information is not available. Consequently it has to be addressed how parts have to be related to their material composition and the module they belong to.

The presentation deal with the systematic approach by adapting material and process chains related to their functionality in the production phase and quantifies environmental impacts in the aviation sector on a sound basis.

1 INTRODUCTION

LCA offers a methodological framework to quantify the environmental impact reducing potential of such new developments and so provides a solid base for decision making. However, conducting an LCA study can be—up to the amount of data to be collected and analysed—very labour intensive and so easily become economically impracticable. Therefore, simplifications are inevitable and inherent to any LCA study.

Aircrafts are highly complex products made from special materials processed in special processes applying newly developed technologies. The production of aircrafts so majorly differs from the production of products of other industries. Therefore, available LCA data from other industries cannot simply be used as a surrogate for the aviation industry but needs to be upgraded first. This constitutes one aspect of the approach to go beyond LCA state-of-the-art.

Of high importance for the quality of the results of the suggested upscaling and the methodology to be developed is the verification. A first step towards verification is the analysis of various components (with regard to application and function) of different modules and sub-modules and should be discussed with the experts of the respective material supplier.

2 BACKGROUND—LIFE CYCLE ASSESSMENT

The Life Cycle Assessment (LCA) is a suitable tool for analysing and assessing environmental impacts caused through production, use and disposal of products or product systems for specific applications. Depending on the goal and scope of a LCA report, LCA does not produce clear-cut straightforward assertions, but rather it gives diverse and complex results. It supports the process of decision-making by making complex issues transparent. The standard ISO 14044 (EN ISO 14044, 2006) defines an LCA as follows.

LCA is the compilation and evaluation of the inputs, outputs and the potential environmental impacts of a product system throughout its entire life cycle. The concept of an LCA is mainly concerned with the following basic aspects:

- The observation of the whole life cycle of a product—from raw material extraction, processing and production to its use, recycling and disposal.
- The coverage of all those impacts associated with the life cycle on the environment, such as raw material and energy consumption, use of land, emissions to air, water and land, as well as waste.
- Aggregation and assessment of these impacts in view of the possible effects on the environment with the aim of assisting environment-oriented decisions.
- According to the ISO 14044 the application of an LCA can assist in.
- Identifying opportunities to improve the environmental aspects of products at various points in their life cycle.
- Decision-making in industry, governmental or non-governmental organisations (e.g. strategic planning, priority setting, product or process design or redesign).

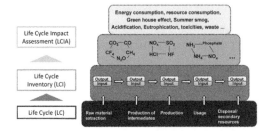

Figure 1. Life cycle thinking as basis for the system model.

– Selecting the relevant indicators of environmental performance, including measurement techniques, and in
– Marketing (e.g. an environmental claim, eco-labelling scheme or environmental product declaration).

An LCA is standardised into the following phases—goal and scope definition, inventory analysis, impact assessment and interpretation (EN ISO 14044, 2006). The figure identifies the reciprocal influences of the individual phases and therefore shows the iterative character of an LCA. The application and the framework of the LCA have been purposely separated to show that an application or a decision is not automatically given through the results of an LCA report. The responsibility for an appropriate application of LCA data remains with the user; it cannot be taken on by the client or the practitioner of an LCA report. The single phases of an LCA are described in the following sections. LCAs cannot cover all interactions with the environment. In this case, other tools—like e.g. Environmental Impact Assessment—must be used to fully observe the environmental impacts.

Figure 1 shows that LCA's are a useful technique for assessing the environmental aspects and potential impacts associated with a product by first compiling an inventory of relevant inputs and outputs of a product system; second to evaluating the potential environmental impacts associated with those inputs and outputs and third to interpret the results of the inventory analysis and impact assessment phases in relation to the objectives of the report.

3 APPROACH

3.1 Fundamentals

Environmental impacts from the production phase of certain reference aircrafts have to be assessed (e.g. airliner, business jet or rotorcraft). As these reference aircrafts are comprised of a huge amount

of different parts manufactured by many different companies a detailed assessment of a whole aircraft will not be possible within a first assessment. Therefore a dummy calculation of those reference aircrafts is to be performed first by extrapolating the environmental impacts of the whole production phase from a limited amount of significant reference parts. The environmental impacts as well as the underlying processes of the manufacturing of these reference parts are known in detail. In order to make reasonable decisions on the applied extrapolation parameters and technique the missing parts have to be compared to the reference part that their environmental impacts are to be extrapolated from. Next to material compositions and structural similarities the underlying manufacturing process chains of the respective parts are important for developing reasonable extrapolation parameters and techniques. However, detailed process chains will not be available for any of the relevant parts of the aircraft. Therefore the missing processes of a certain process chain have to be assumed. This assumption is to be done with the help of a so called "Process-Material-Matrix".

3.2 Systematic categorisation of an aircraft

This Process-Material-Matrix will enable an educated guess of possible processes to fill up the gaps in the process chain. This educated guess is to be made from information on preceding and following processes as well as on the material composition of the respective parts. Therefore, the "Processes-Material-Matrix" will link certain processes technologies to certain groups of materials which can be processed with this technology. Both processes and material groups are to be categorized in a certain hierarchy in order to minimize the amount of possible processes when looking only at processes having a certain function (which can be guessed from preceding and following processes) and a certain material which is to be processed. For the aircraft structure, the categorization of an aircraft was split into main modules (Structure, Propulsion, Cabin and Systems) and various sub-modules, which is given in Table 1.

3.3 Categorisation of the process route

The process route classification covers all the major processing steps to produce, in this as example Aluminium aircraft components from primary and secondary Aluminium. The classification is done with regard to relevance for the function of product. The process steps for Aluminium aircraft component production are shown in Figure 2.

Basically all process groups can be allocated to one of the two basic functions they contribute

Table 1. Modularisation of an aircraft— categorization into modules and sub-modules.

Level	Module	Sub-module
0	Structure	
1		Fuselage
1		Wings
1		Landing gear
1		Fittings elements
1		Tailplane and Fin
1		Pylon
0	Propulsion	
1		Engines
1		Nacelles
0	Cabin	
1		Interior fittings
1		Seats (164)
1		Galley structure
0	Systems	
1		Others
1		Cables
1		Calculators
1		APU
1		Batteries
1		Radar
1		Circuit breaker
1		(T/R)
1		(Convertisseur statique)

Table 2. Process classification by functional contribution.

Process	Explanation
Wrought processing	Processing into basic shapes (plate, sheet, extrusions, ...)
Detailed shaping	Processing into detailed shapes (machining, forming, ...)
Alloy production	Basis for achievable material properties
Microstructural adjustment	Adjusting the final material properties
Finishing	Adjusting the final surface properties
Intermediate conditioning and heat treatment	Provide the necessary shape and material conditions for wrought processing

sub-module and consequently is used within the main modules and finally accounts for the overall aircrafts environmental impacts.

4 EXAMPLE

The fuselage specifications finally used for the upscaling are assumed as given in Table 3 due to some data gaps within the overall sub-module fuselage.

The upscaling takes the specific aviation process routes into account (see chapter 3.3). Each component was modelled respecting underlying production peculiarities. The respective setting of the model parameters is given in Table 4, representing average assumptions given in literature.

The model set up in the LCA software can be found in Figure 3.

5 RESULTS

Results are analysed for the impact categories according to the Institute of Environmental Sciences Leiden (CML) are Global Warming Potential (GWP), Acidification Potential (AP), Eutrophication Potential (EP), Photochemical Ozone Creation Potential (POCP) and the energy consumption "Primary Energy Demand (PED)".

The results of the environmental impact assessment clearly show that the environmental impacts of aircraft component production have been extremely underestimated as the generically quantified impacts are 300 to over 400% of those for standard material and standard process screening (see Table 5).

Moreover, the model still lacks of some data—i.e. for the pretreatment of chemical metal removal—and is based on environmental impacts

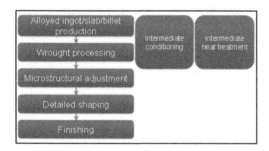

Figure 2. Process classification.

to—shape or material properties. A respective classification is given in Table 2.

Behind this categorisation there is a further split into the different processes, which is important to analyse the overall process chain of the aviation materials. E.g. there are basically three wrought processes relevant for Aluminium aircraft component production today which hence constitute the process alternatives for the process wrought processing, namely rolling (hot, cold), extrusion and forging.

These processes are then used to model the overall process chain of the given parts in a specific

Table 3. Fuselage specifications.

Components	Mass [kg]	Material	Processing
Skin	2900	2024	Sheet
Stringer	1600	2024	Sheet
Frame	2000	2024	Plate

Table 4. Parameter setting for generic Aluminium aircraft fuselage modelling.

Model parameters	Fuselage		
	Skin	Stringer	Frame
Finished component properties			
Mass [kg]	2900	1600	2000
Alloy	2024	2024	2024
Temper	T3	T3	T3
Processing			
Wrought processing	Rolling	Rolling	Rolling
Buy-to-fly ratio	2	1.2	15
Share chem. metal removal [%]	30	30	30
Finishing	CAA	CAA	CAA

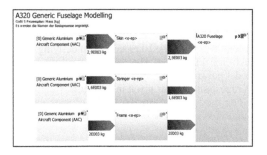

Figure 3. GaBi plan on Aluminium fuselage.

Table 5. Result of the airliner fuselage—standard process vs. aviation specific processes.

Impact category	Standard process screening	Updated process screening	% deviation
A320 fuselage			
GWP [kg CO_2 eq.]	65928	268525	407
AP [kg SO_2 eq.]	342	1122	329
EP [kg PO_4 eq.]	16	66	422
POCP [kg C_2H_4 eq.]	20	71	351
PED [MJ]	1100825	4562992	415

underestimating assumptions—i.e. 100% remelting of mechanically removed metal. Hence the actual environmental impacts of the production of the A320 fuselage can be expected to be even higher than generically assed above.

Additionally, the buy-to-fly ratios of the components of the fuselage are lower than i.e. for the wing where i.e. spars and ribs show the highest buy-to-fly ratios. Hence the average environmental impacts of e.g. Aluminium aircraft components can be expected to be even higher than those of the fuselage components.

6 SUMMARY AND NEXT STEPS

The paper and the upscaling methodology do not represent the final results. This is due to on-going work in cooperation with material suppliers to categorise their materials and processes. The results presented are therefore only an indication for the relevance of analysing aviation specific process routes to get a realistic inventory of aircrafts.

The next steps will be to use the specific aviation process routes e.g. for Aluminium, to analyse the overall impacts of all used Aluminium parts in an aircraft, by considering the specific application and functionality. This also will be applied to further materials, e.g. Titanium and Carbon fibres.

Interpreting the result this does not mean that the environmental impacts of e.g. a specific material and in this case, the Aluminium products, are worse than others, but means that general LCA data in the aviation industry has been underestimated. Therefore this analysis shows the key relevance to quantify the impacts in the aviation industry in order to base further decisions on realistic values for e.g. lightweight design, new process or technology developments.

REFERENCE

EN ISO 14044. 2006. Environmental management—Life cycle assessment—Requirements and guidelines.

Green Design, Materials and Manufacturing Processes – Bártolo et al. (eds)
© *2013 Taylor & Francis Group, London, ISBN 978-1-138-00046-9*

Modeling complex aviation systems—the Eco-Design tool EcoSky

R. Ilg
Department Life Cycle Engineering, Fraunhofer Institute for Building Physics, Germany

ABSTRACT: With the growing environmental awareness in society, the demand for sustainable products and services increases. Hence especially in the expanding aviation sector, a reduction of the ecological impacts is intended to improve the aviation's ecological footprint. Modelling a whole life cycle inventory in a LCA study compliant with the ISO 14040 requires comprehensive expert knowledge. Furthermore aircrafts are complex systems with millions of different parts and various, aviation specific materials. To enable non-experts with little or no LCA knowledge like aircraft designers to assess the environmental impacts of already existing and also of conceptual aircrafts by the method of LCA, the web-based Tool EcoSky was developed with an easy-to-use user interface. In this Tool, the complex LCI modelling is decoupled from the design process. The presentation highlights the characteristics of the aerospace sector and how the EcoSky Tool deals with the specific difficulties.

1 INTRODUCTION

With the growing environmental awareness in society, the demand for sustainable products and services increases. Globalisation is unthinkable without the aviation. The aviation sector so has experienced a rapid growth in its history and is expected to sustainably maintain this growth in upcoming decades. To soundly grow with respect to the environment the aviation sector will have to take responsibility for the environmental impacts it causes. As big parts of today's fleets will have to be replaced in the near future the aviation sector today experiences an unprecedented momentum to rapidly implement new technological developments with potential to reduce environmental impacts.

However, modelling the whole life cycle inventory in a LCA study compliant with the ISO 14040 requires comprehensive expert knowledge. Furthermore aircrafts are complex systems with millions of different parts and various, aviation specific materials. Also no specific tool exist which is applicable for aircraft designers without specific LCA knowledge enabling them to use the benefits of a LCA already in the design process, which is why an Eco-Design tool for the aviation industry called "EcoSky" was developed.

2 BACKGROUND

2.1 Life Cycle Assessment (LCA)

To quantify the environmental aspects of products, processes and services along their whole life cycle,

including production, use and end-of-life phase, the method of Life Cycle Assessment (LCA) is used. In LCA, first an inventory (LCI) of relevant inputs and outputs of a product system is compiled, then the potential environmental impacts associated with those inputs and outputs are evaluated which results will finally be interpreted in relation to the objectives of the study (EN ISO 14040, 2006), (EN ISO 14044, 2006). Detecting strategic risks and identifying the environmentally relevant steps in a product's life cycle, LCAs are already established as a basis for the development and systematic improvement of sustainable products. Next to the competitive advantage by including environmental aspects which improve a product's image, LCAs also support environmental innovations and the decrease of environmental impacts.

2.2 Software for LCA

In order to increase the efficiency in carrying out a LCA, electronic databases and software systems have been created that cover the individual requirements. Many of these databases provide LCI data from various sectors and calculation procedures for the results (e.g. various LCIA methods). Prominent examples of databases, which are embedded in modelling software tools, are such that as of GaBi software. The professional LCA Software-System GaBi (www.gabi-software. com), which is one of the leading tools for LCA and LCE (Life Cycle Engineering) worldwide, is acknowledged for its intuitive software, database quality and volume; hence GaBi (PE International, 2012) is used by companies, industry associations,

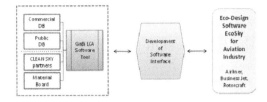

Figure 1. EcoSky tool concept.

government institutions, researchers and consultants. It also contains the datasets provided by the European Commission on their platform (further information can be found: (JRC, 2010)) and was therefore chosen as the best solution.

2.3 Harmonisation of LCA software to enable communication with EcoSky

Based on the state-of-the-art software system for life cycle assessment, GaBi 6, the Eco-Design Software tool EcoSky is an extension of GaBi 6 and supports design for environment. A harmonisation is conducted in order to enable smooth communication and data exchange between the GaBi software and the Eco-Design Software EcoSky. Figure 1 shows the data exchange and how the API (Application Programming Interface) is implemented as an important step towards the development of EcoSky. This Web Service API is a GaBi wrapper. It allows EcoSky to communicate with the server in a seamless way.

By setting up one model of the product under consideration, EcoSky brings the user into the position to solve the above mentioned tasks in one comprehensive tool. Based on the modeled background parts, the designer has the choice of the main parameters of interest, not dealing with the complex LCA system model in the back.

3 ECOSKY—DESIGN TOOL FOR THE AVIATION INDUSTRY

3.1 Approach

To enable non-experts with little or no LCA knowledge like aircraft designers to assess the environmental impacts of already existing and also of conceptual aircrafts by the method of LCA, the web-based Tool EcoSky was developed with an easy-to-use user interface. In this Tool, the complex LCI modelling is decoupled from the design process. The LCI modelling is transferred to a central server, providing aviation specific background data, which is maintained by LCA experts and so ensures a high data quality. The server, a pool of

parameterized LCI models for aircraft parts, relies on the comprehensive GaBi database.

To model the environmental impacts of different aircraft types, designers are able to modify components by varying weights and surfaces, material properties and production processes, partly in predetermined range. Next to a "reference airliner" different scenarios can be modeled to compare the environmental impacts of different design alternatives of aircrafts and their components.

3.2 Requirements

Before setting up the tool, requirements were defined. The requirements are the result of discussions and feedback from the aviation industry and split into requirements on LCA and the Eco-Design tool EcoSky.

Requirements for LCA:

- Organization of specific process data
- No comprehensive expert knowledge in LCA modelling
- Aviation specific background data on processes and materials.

Requirements for the Eco-Design tool EcoSky:

- Easy and intuitive to use
- GaBi as back-end tool
- Compatible with standard LCA Software to ensure a high data quality as background data
- Don't start from the scratch, use the already available LCA databases and software
- Compliancy to standards
- Evaluation of different scenarios
- Variability and flexibility of the interface
- Detailed LCA from Cradle to Grave
- Based on expert background models
- Comprehensive background Database
- Decouple the complex LCI modeling part from the Simplified Tool EcoSky
- Transfer LCI modeling to a central server which is maintained by LCA experts.

3.3 Development of the tool

This tool is developed to model the environmental impacts of different aircraft types to enable the comparison of different design alternatives of aircrafts and their components by selecting parts, materials and process choices.

Ideally all environmental impacts of any component and their design alternatives are made available but this would require an inventory of all material and energy in- and outputs from any process involved under any possible process conditions. Since there is a vast amount of components and component functions as well as various processes, operating conditions and materials, there is a low availability of aviation specific environmental data. To perform LCAs for every single component is just impracticable.

To perform LCAs in aviation, major process routes and their interdependencies with environmental impacts were identified. Those major dependencies were parameterized in generic aircraft LCA models and finally the link between the components and the setting of process parameters was made.

The web-server contains parameterized LCI models for aircraft parts and provides these to the users. With these models the user can evaluate different scenarios. The LCI models can be extended and new aircraft parts can be added. This means a central data storage and maintenance to ensure data consistency.

3.4 *Functionality*

By logging into the EcoSky tool, first an aircraft type has to be selected, namely an airliner, a business jet and a rotorcraft, which is followed by the life cycle phases to be selected production phase, use phase and end-of-life phase (see Fig. 2).

In order to give only a rough overview, we take the example of the aircrafts production phase. All aircraft types are divided into main modules like structure, propulsion, cabin and systems. Those are further split into sub-modules like fuselage (among the main module structure) which are composed of various components like skins, stringers, frames, clips etc. which again are composed of various parts. To select the different modules, sub-modules and parts, a path structure and a tree structure can be used for an easy to use navigation (see Fig. 3).

Furthermore, it is easily possible to analyse the material composition of parts or modules and are shown in pie charts (tool tip function) (see Fig. 4). The environmental impacts, which can be selected from the category (GWP, AP, etc.) or the methodology (CML, EcoIndicators, etc.), are shown in charts but also in tables.

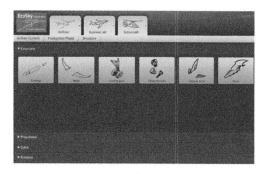

Figure 3. EcoSky tool—Navigation by categorised aircraft modules and sub-modules.

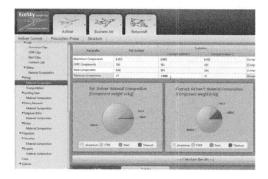

Figure 4. EcoSky tool—Navigation by path/tree structure, aircraft material composition analysis on each module/sub-module/component level, parameter settings for reference airliner and adjustable scenarios.

By varying the parameter settings, it is possible to compare the reference airliner with the adjusted scenarios. The parameters can be set on part level. Changes are highlighted in blue. For some parameters a predefined value range is set as highlighted in Figure 4.

4 SUMMARY

The paper presents a status of on-going work within the aviation industry and is therefore subject to modifications.

To resume Life Cycle Assessment provides the methodology to assess the different environmental impacts of products and systems over their whole life cycle. Also the Eco-Design tool EcoSky makes the link for designers to have an easy access to complex LCA models and don't need specific LCA knowledge to identify weak points and sensitive parameters along their design process.

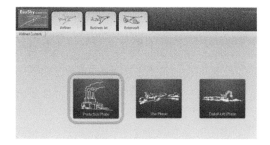

Figure 2. EcoSky tool—Start screen.

A central data storage and maintenance ensure data consistency. Detailed LCA from cradle to grave based on expert background models, relying on comprehensive background data containing specific aviation processes and materials could be used by the designers. This therefore helps to identify the relevant hot spots in the entire life cycle of a single aircraft part up to a complete aircraft and supports the development processes by comparing aviation materials, processes and technologies (e.g. lightweight design).

The results are available as inventories with a detailed list of various, individual selectable emissions or as aggregated environmental impacts and new impact methods can be added as proposed by ISO. This reduces cost and time that would have to be spent on LCA modeling and education.

Eco-Design—integrating LCA based environmental information into decision making as early as in the design of products and systems—constitutes the most viable approach for finding environmentally soundest future air traveling concepts.

REFERENCES

EN ISO 14040. 2006. Environmental management—Life cycle assessment—Principles and framework.

EN ISO 14044. 2006. Environmental management—Life cycle assessment—Requirements and guidelines.

PE International. 2012. GaBi 6 Software System and Databases for Life Cycle Engineering. Copyright, TM. Stuttgart, Echterdingen, Online URL http://www.gabi-software.com/, last access 2013-02-26, 1992–2013.

European Commission—Joint Research Centre (JRC)—Institute for Environment and Sustainability. 2010. Documentation and exchange format for the documentation and publication of Life Cycle Inventory (LCI) and Life Cycle Impact Assessment (LCIA) data sets, Online URL http://lca.jrc.ec.europa.eu/lcainfohub/ datasetArea.vm, last access 2013-02-21, 1995–2010.

Green Design, Materials and Manufacturing Processes – Bártolo et al. (eds)
© 2013 Taylor & Francis Group, London, ISBN 978-1-138-00046-9

New York City blue network: Water as liquid state of earth

M. Louro & F. Oliveira
CIAUD Faculdade de Arquitectura, Universidade Técnica de Lisboa, Lisboa, Portugal

ABSTRACT: Water as a liquid state of earth is assumed as a proposal that integrates the surface of the water that surrounds New York City, as an occupied territory, in extension with the shore. Thus from a flexible, sustainable and self-sufficient structure, which composes the expo program: Clean Tech Expo, that will be held in 2014, five functional clusters are defined linked by a mobility network that is organized in three levels: macro, medium and micro. The macro level includes five mobility units associated with the exhibition and features themes related to each of five boroughs: walking—Manhattan, agriculture—Queens, transformation—Staten Island, energy—Brooklyn and road—Bronx. The medium level integrates collective transport units and the micro level private transport units. The transport network—blue network, links a vast territory that takes land and water as elements of the same urban reality and create the 6th borough of New York.

1 INTRODUCTION

The work presented here (Fig. 1) reflects a research project motivated by the participation in the idea competition of *One Prize 2011: Water as the Sixth Borough*—Open International Design Competition to Envision the Sixth Borough of New York City, promoted by Terreform ONE (www.oneprize.org).

The project carried out in 2011 by a team coordinated by the architects, teachers and researchers of CIAUD—Research Centre for Architecture,

Urbanism and Design of Faculdade de Arquitectura da Universidade Técnica de Lisboa, Margarida Louro and Francisco Oliveira, beyond responding to the request by the assumptions of the competition promoted a reflection on the sustainability of the growth of cities from the paradigm of large metropolis and in particular on the objective intervention in New York. In fact the sharp growth of cities pushes increasingly sustainability of their limits in this paradigm of densification and saturation promoting new logics that would reduce its impact on this small planet. (Louro et al. 2009). If the pretext of great exhibition of 2014, with the thematic of the clean tech, is assumed as an inspiring datum and attractive pole for the consumption of the cities (Louro 2005) it is still necessary to articulate effectively the various boroughs that now constitute the City of New York, in order to promote urban and public qualified spaces (Oliveira 2011).

Thus the great reflection of this research identifies three major themes to which attempts to answer and which are: articulate the issue of densification and saturation of its limits with sustainable forms of formal city; provide the city with attractive clusters that take it as an object of desire and finally the requalification of an efficient transport network that elect public space as an artifice privileged of regeneration.

Figure 1. Water as liquid state of earth—General view of the proposal.

2 THE COMPETITION—MAIN REQUESTS

One Prize is an annual design and science award that aims as objective the promotion of green and sustainable cities. In 2011 the event was mainly focused on New York City and in the intervention on its shores lapped by the different riversides, proposing the regeneration and urban renewal of such spaces. The challenge then was based on the pretext of carrying out the exposure E3NYC Clean Tech World Expo that will be held in New York in 2014, exposition that will celebrate the 50th Anniversary of 1964 New York World's Fair, dedicated to Man's Achievement on a Shrinking Globe in an Expanding Universe and will have as theme, 50 years later, the 21st century promise of clean tech. Thus, the competition through the pretext of harboring this event, proposes to create a network to link the five boroughs that make up the city of New York nowadays, in order to promote linkages that enhance the territory's water as a link, and promoting as well educational, cultural and commercial activities, and demonstrations of clean technology and renewable energy.

The competition destined to architects, landscape architects, urban designers, planners, engineers, scientists, entrepreneurs, economists, artists and students had as major objectives:

– The creation of a transit system: NYC Blue Network that should work with the existing transit system, enhancing the capacity of the utilization during the summer of 2014 of about 10 million visitors within the pretext of Clean Tech World Expo. The core of the transit system should be a series of green transit hubs (with 2500 squares meters), one for each borough: Bronx, Brooklyn, Manhattan, Queens and Staten Island.
– The creation of a spatial and urban concept that will suits the exhibition of 2014 which could host the world largest clean tech event and establish itself as the ultimate green capital on the world. Using the transportation system, the expo is organized into six zones, the five traditional boroughs as well as the water, with the suggestion themes: Bronx: Electric Car City; Brooklyn: Science City; Manhattan: International Pavilions; Queens: Clean Tech Industry City; Staten Island: Hall of Civilization; and East River: Gaia. Gaia is assumed as a floating landmark that will mark the waterway entrance to Clean Tech World Expo, an icon status located in the water.

Each proposal to the competition should include: a plan for the Blue Network with series of green transit hubs incorporating ferries, bike shares, rickshaws, electric car-share and electric shuttle buses and a plan for the Clean Tech World Expo 2014 with the six zones of concentrated activity, one for each borough.

It was strongly encourage the formation of interdisciplinary competition teams, and the invitation to collaborate with professions not usually included in design competitions as engineers, scientists, clean tech entrepreneurs, economists and urban activists along with designers and planners.

The competition was concluded in 2011 and it drew up about 256 people in 86 teams from 18 countries and five continents. It became a big challenge to narrow them down to 30 semi finalists where this project or investigation was included (www.oneprize.org/1semifinalists.html) and which incorporated an exhibition The ONE Prize Award Ceremony and Exhibition Opening that took place on January 18 at the AIANY—Centre for Architecture in New York.

3 THE PROPOSAL—INNOVATIVE IDEIAS

3.1 *Project argument*

Based on the guidelines and requests posted by the competition, it were defined as main objectives of the challenge assume the territory of water as a privileged support of intervention. Thus the motto, water as a liquid state of earth, took over as the project argument that dictated the development of the entire proposal. The idea was to blur the barriers between the water and land assuming stadiums of intervention that circumvent these obstacles by the creation of floating buildings that by their condition of floating transform the configuration of the margin depending on their locations, enhancing the network between districts and enhancing the program of the exhibition.

Accordingly this logic was adopted both in resolution of the transport system as in the creation of a conceptual approach to the spatial resolution of exposure assuming a proposal that integrates the surface of the water that surrounds NYC—New York City, as an occupied territory, in extension with the shore. From the geometry of water crystal a composition derived from a species of new skin, a new epidermal layer that somehow covers the surface of the water harnessing it as a new territory to which new uses, streams, buildings and anchors are add (Fig. 2). The pattern of geometrized hexagons enhances flexibility to the route, tying it with logic and formal urban margins, hitting the metrics of existing geometries, morphologies of diluting urban setting between fixed and mobile, between land and water.

Thus from a flexible, sustainable and self-sufficient structure, which composes the expo program: Clean Tech Expo, five functional clusters are defined linked by a mobility network that is organized in three levels: a macro level, medium level and a micro level. The macro level includes five mobility units associated with the exhibition and features themes related to each of five boroughs:

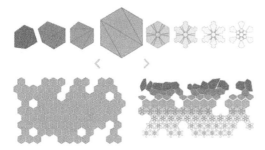

Figure 2. Conceptual composition scheme.

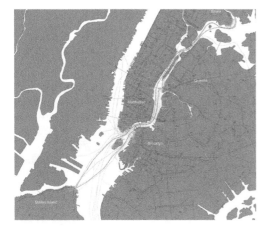

Figure 3. Network of accessibilities.

walking—Manhattan, agriculture—Queens, transformation—Staten Island, energy—Brooklyn and road—Bronx.

The medium level integrates collective transport units and the micro level private transport units. The transport network links the various dimensions of the proposal in a vast territory that takes land and water as common elements in the same urban reality—the 6th borough of New York.

In this way it is formalize a network (Fig. 3) that integrated with the local network (bicycle lanes, paths and routes, air train rail, subway lines and stations, ferry lines terminals) embodies a new network of accessibilities constituting the territory of experiences that will support the exhibition and will formalize the new 6th borough of New York city.

3.2 *Expo and transport*

In terms of the proposal, and emerging from the potential released by hexagonal geometry mesh and from its fractal feature, it were defined different scales of intervention which had consequences

both in the volumetric forms of buildings and urban spaces proposed, as in the proper logics of transport and mobility. The datum of mutability was also assumed as the main vector on the development of the proposal. Associated to the idea of a layer or skin that covers a surface and related with the idea of changing in constant regeneration (like a being where old cells give rise to new cells) a logic of spatial composition of space appears by objects or buildings that somehow can move by this tissue and can occupy multiple positions as a part of a game. To the five districts and following the premises defined by the program of the competition were associated five specific themes for each other: walking to Manhattan associated and incorporated by the thematic of the international pavilions, agriculture to Queens associated to the idea of a clean tech industry, transformation to Staten Island associated with the theme of an hall of civilization, energy to Brooklyn advocating the thematic of the Science City and road to Bronx by the theme of the electric car city. These five proposed themes walking, agriculture, transformation, energy and road, represent activities in order to an ambient of sustainability, and construct by the initials of the concepts, w, a, t, e, r the word "water", the poetical argument of the project and territorial support of all intervention. On which is implemented a new skin or tissue areas that articulate mobile and fixed zones to which are associated modules of exhibition. These modules of exhibition are assumed as the schematic support of the clean tech expo and advocate part of the mutability of the proposal tissue—the macro scale of the mobility. They are buildings of transport that anchored to the various areas of the riverbanks pass through a day of exposure (8 hours daily) in the various districts or ports riverbanks. Thus in addition to promoting a new mode of "transportation" of the visitors throughout the territory claiming the mutability of the landscape as part of a more complex system of accesses composed by other scales of transport (medium and micro) related to smaller groups or individual (Fig. 4).

3.3 *Macro transport, medium transport and micro transport*

From the macro scale of transport comprised on the five units: WATER = Waking + Agriculture + Transformation + Energy + Road, with an ability to travel 5–10 km/hour, powered by electricity from solar panels embedded in each unit and capable of carrying 600 passengers each and thus responding to a displacement of 15000 passengers per day, were associated with other complementary scales embodied in medium scale and micro scale of transport.

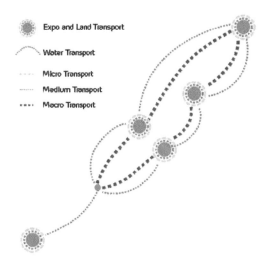

- Expo and Land Transport
- Water Transport
- Micro Transport
- Medium Transport
- Macro Transport

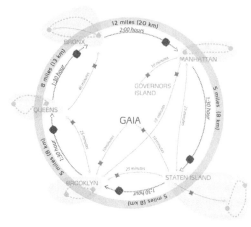

Figure 5.   NYC—blue network system.

Figure 4.   Expo and land transport scheme.

The medium scale is a transport of smaller groups and it is formalized into several types: electric passenger ferries, with a capacity of 18 units travelling 19 km/h with 64000 passengers per day, the water taxis, 30 units with ability to travel 10–30 km/hour with 6 000 passengers per day and the shuttle electric buses, 25 units with a capacity of travel up to 50 km/h with 55000 passengers per day.

The micro scale of transport is intended primarily for personal use and it is proposed in different typologies: the electric car shares, around 340 small electrical units with rechargeable batteries in charging stations and with a capacity to transport up to two passengers with a displacement between 40–50 km/hour, the estimated capacity of this medium of transport is about 13600 passengers per day; the rickshaws, 200 provided in number, they move at about 10–20 km/hour, transport up to three passengers and have ability to respond to a passenger daily displacement of about 8000 passengers and finally the individual displacement systems: bicycles and *segways* proposed in a number of 1600 units with a capacity displacement of 10–20 km/hour and sustain a total number of 16000 passengers per day.

Overall this is the NYC network blue network system (Fig. 5), which in addition to the transport network will respond to the population flow that will visit the exhibition as well as articulate the existing territories of the five districts, their margins and zones of requalification constituting the 6th borough of New York.

The exhibition centers, associated with each of the themes adopted by the tender program will integrate various uses of various playful genres, natu-

Figure 6.   a) Going to the micro transport hub and renting a bike. b) Discovering the city. c) Resting in the garden platforms.

rals, cultural, educational, etc. as: Floating Market, Vertical Parking, Eco Lagoons, Picnic Parks, Solar Power, Electric Car Sales, Swimming Pool, Human Bubble Pool, Multicultural Housing, Restaurants, Hotels, Sightseeing, Water in, Museums, Concerts,

Figure 6. d) Choosing which borough to go next. e) Travelling in the agriculture macro platform. f) Seeing the sunset and swimming in the pool platforms.

Flower plantations, Water collectors, Wind power, Gym, Info, Bicycle routes, Vertical Farms, Shopping, Heated Pool, etc ...

All uses are assumed in order to provide sustainability to all areas, both in economic, social and energy, assuming the motto of clean tech as a comprehensive concept and extended the notions of sustainability that will provide the coexistence of uses as production (agriculture), manufacturing (industry), and recycling the datum of a new paradigm of urban territory.

4 CONCLUSIONS

Through the pretext defined by the international conference on Sustainable Intelligent Manufacturing (SIM 2013), about the reflection and the discussion of innovative ideas in order to promote sustainable development, it seemed to us appropriate the dissemination of this research project. A project where the options reached in several scales of intervention, promote equilibrium and harmony in order to get a sustainable solution and bring the discussion of a creative proposal that is in its essence the conceptual field of the architects.

The solution presented answers to the guidelines laid by the competition ideas, the saturation limits of large cities, the economic viability of their spaces and the promotion of qualified urbanities without disturbing the balance of their environmental impact. Design for efficient cities, challenges and new technologies for the cities of tomorrow, will be the theme of the approach of this proposal in the sense of extensive discussion and in which we thought to contribute to a future consideration that obviously could have new starting points and explorations. An utopian vision that sees itself as a prospecting for new future ways. An open research where different exploration points can be proposed. As the project presented provide different days of an event city (Fig. 6a–6f), in this case from the pretext of New York, but that could be extended to any other city in the world, a day in a clean tech city.

REFERENCES

Louro, M. 2005. www.ciudad.consumo—El impacto de las re-des de consumo en la reorganización del espacio urbano con-temporâneo del área metropolitana de Lisboa. Barcelona: Uni-versitat Politècnica de Catalunya—Base de Datos TDX: http://tdx.cesca.es.

Louro, M., Oliveira, F., Feliciano A.M., Leite, A., Pires A. 2009. Casas para um Planeta Pequeno—Projecto Angola Habitar XXI: Modelos Habitacionais em Territórios de Macro Povoamento Informal. Lisboa: PixelPrint/Pandora.

Oliveira, F. 2011. Chão da cidade: permanência e transformação. De metáfora a impressão digital da cidade/ The City Plan: permanence and transformation. *From the metaphor to the digital imprinting of the city. Revista Proyecto, Progreso, Arquitectura—Permanência y Alteración*, Universidad de Sevilla, n.° 4, Maio 2011 (Ano 2): 138–151.

One Prize Annual Design and Science Award to Promote Green Design in Cities, www.oneprize.org, 2011.

One Prize NYC 6 Boro, www.oneprize.org/1semifinalists. html, 2011.

Life-cycle engineering

Green Design, Materials and Manufacturing Processes – Bártolo et al. (eds)
© 2013 Taylor & Francis Group, London, ISBN 978-1-138-00046-9

Environmental management and use of the water in Food Service Segment: The Life Cycle Assessment as a tool for a sustainable development

M.S. Lourenço
Universidade Federal Fluminense, Faculdade de Nutrição, Niterói, Brasil

S.R.R. Costa
Universidade Federal Rural do Rio de Janeiro, Ciência e Tecnologia de Alimentos, Seropédica, Brasil

M.L. Nunes
Instituto de Investigação das Pescas e do Mar, Lisboa, Portugal

L.S. Xavier & J.A.A. Peixoto
Centro Federal de Educação Tecnológica Celso Suckow da Fonseca, RJ, Brasil

ABSTRACT: The methodology of Life Cycle Assessment (LCA) is a support tool for environmental management that helps to identify and quantify the use of natural resources in production processes of products along their life cycles, allowing identification of potential environmental impacts. The aim of this study was to analyze, applying LCA, the production process of grilled hake steaks in the Food Service Segment in Portugal, focusing the water use and effluents contamination. The results showed the employment of about 1300 liters of drinking water and the production of 1315 liters of effluents per ton. Preeminence is given for the effluent from thawing and preparation stages of hake, which showed high levels of organic matter and solids, as well as appreciable levels of total coliforms and viable microorganisms at 30 °C. Issues of eco-efficiency are commented as contribution to support decision-making for sustainable management of companies of the Food Service Segment.

1 INTRODUCTION

The food industries consume in their production processes, large amounts of natural resources, and this can cause potential impacts to the environment Therefore, if there is no monitoring of the activities performed in production processes and rational use of these resources, such as optimization of water consumption, due to the scarcity of water resources on our planet, they will be contributing to a set of problems that in ultimately, project risks to human survival itself.

The Brundtland report "Our Common Future", published in 1987 is a milestone in the global concern for the preservation of natural resources and the environment. This document highlights the search for balance in the relationship between humanity and nature through Sustainable Development, a process that requires changes, Yet in the end, sustainable development is not a fixed state of harmony, but rather a process of change in which the exploitation of resources, the direction of investments, the orientation of techno-logical development, and institutional change are made consistent with future as well as present needs. We do not pretend that the process is easy or straightforward. Painful choices have to be made. Thus, in the final analysis, sustainable development must rest on political will (WCED, 1987).

The preservation of the environment became one of the most influential factors in shaping the market, spreading very rapidly in the 90s. Thus, companies began presenting solutions to achieve sustainable development while targeting the profitability of their businesses (Andrade et al, 2002), since the economic financial stability must also be maintained.

The present study aimed to examine from the perspective of the methodology of Life Cycle Assessment (LCA) the production process of hake baked in the Food Service Segment in Portugal. This is an important part of the life cycle of hake to the evaluation of eco efficiency associated to the operations at the local and global productive processes involved.

1.1 Environmental management

The strategic role of environmental management for organizations has been evidenced by several findings related to the environment in which firms operate. The social charges on businesses increased in relation to more active posture with respect to liability on its industrial processes, waste and effluents and discarded, as well as to performance of their products and services in relation to the life cycle approach (Seiffert, 2010).

The approach to environmental problems must be multidisciplinary, due to the complexity of the issues involved that require knowledge from many disciplines. The management of the use of natural resources is the factor that can enhance or reduce environmental impacts. This management process is based on three variables, the first would be the diversity of resources extracted from the natural environment, the second would be the speed of extraction of these resources, providing or not to replace them, and the third would be the way to disposal and treatment their waste and effluents. The sum of these variables and how to manage them determine the degree of impact of the urban environment on the natural environment (Phillip Jr et al, 2004).

The European Community in 1994, developed a specific legislation for member countries, establishing standards for the design and implementation of an environmental management system as part of an environmental management system and audit plan, known as Eco Management and Audit Scheme (EMAS). However, the Canadian Standard Association created the standardization of procedures for the implementation the environmental management system and to obtain eco-labeling of products (Nicolella et al, 2004). The ISO 14000 series is one of the most used in the area of corporate responsibility and is recognized internationally as the standard for environmental management. The standards provided by the series are considered effective tools to integrate environmental policy in management practices (Almeida, 2007) in the context of production.

According to Slack et al. (2002), the international standard ISO 14000 allows companies worldwide to evaluate in a systematic and integrated approach how their products, services and processes interact with the environment.

1.2 The Life Cycle Assessment

The Life Cycle Thinking (LCT) means the awareness that not the adequacy of the environmental performance of a single unit of supply chain is not enough, but what matters is that the environmental performance of all stages of the chain is suitable

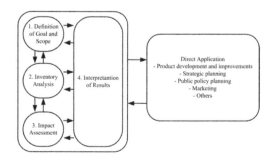

Figure 1. Phases of a Life Cycle Assessment (LCA) Source: ISO, 2006.

for an eco efficient integration. It highlights that environmental performance not only considers the disposal of tailings, but also the consumption of natural resources over all interactions of supply chains (Vilela Jr; Demajorovic, 2006).

According to Coltro (2007), with the use of LCA is possible to evaluate the implementation or alternative processes, products or services. Environmental statements about the product can be based on LCA studies, as well as the integration of environmental aspects into product design and development.

Guinée et al. (2011) pointed out that the second decade of this century will be the decade of sustainability analysis lifecycle covering issues of various products, sectors and mechanisms. LCA is useful, being an information tool to examine alternative routes for future strategic planning (Lundie et al, 2004).

Thabrew et al. (2009) called the attention that LCA is not only a way to examine the environmental impacts of activities, but also a way to understand and visualize a broader set of consequences before and after the decisions in planning, being included mapping each activity, presenting a holistic view of stakeholders.

According to ISO 14040:2006, the inventory is a set of information about the amount of energy and materials critical to carrying out environmental impact studies, used throughout the production chain, and how much of this material was discarded in the middle environment (ISO 2006). The phases of LCA are indicated in Figure 1 (ISO, 2006).

2 FOOD SERVICE SEGMENT

Due to increasing urbanization and industrialization occurring since the 50s, which includes increasing professionalization of women, raising the standard of living and education among other factors, the rising population's access to leisure travel the time it was for feed has changed. With the decrease of this time for the preparation and consumption of meals at home, there was an increase in the consumption of industrialized meals and "outside the home". Consumers seeking an increase in the available time began to realize their meals outside the home, causing an increase in worldwide Food Service Segment (Leal, 2010).

According to Matos (2003), the productive process of meals involves factors such as the number of operators, the type of feed used, the preparation techniques and infrastructure, requiring equipment and tools designed to optimize operations, making them faster and more reliable viewpoint of conformity of the final product. The production system of meals is complex and according to the position of the American Dietetic Association (ADA), to encourage environmentally responsible practices becomes necessary conserving natural resources, reducing the amount of waste generated in the processes of food production, processing, distribution, access and consumption (Harmon; Gerald, 2007).

According to the Fédération Européenne de la Restauration Collective Concédée (FERCO) in the European Union, the market segment Food Collective represents 600,000 jobs across Europe and provides approximately 6 billion meals/year, representing an average of 67 million consumers daily and an annual turnover of 24 billion Euros (FERCO, 2010).

The Food Service Segment in Portugal supplies approximately 148 million meals/year and an annual turnover of 485 million Euros (FERCO, 2010). In hosting companies and foraging sites are approximately 277,645 jobs (PORDATA, 2009).

3 WATER RESOURCES

The various approaches to environmental management stress that water resources are finite and some are already scarce on the planet. However, one of the possible strategies to minimize this framework would be not only the proper management of natural resources, but also the processes of transformation of raw materials that can harm the environment and the welfare and human health. Water is considered the most important element for human survival and all life on Earth (Phillip Jr et al, 2004). With this perspective, the use must be rational, because the progress is in increasing acceleration, and there is a dwindling supply of water under ideal conditions for consumption due to the increasing demand for various activities such as public supply, industrial use, but also for agriculture.

Water management in the industrial sector is typically considered in terms of industrial consumer

surveys, and the total industrial water can be calculated according to the intake and discharge of effluents (UNESCO, 2012). Water pollution has many sources as the origin, among which stand out the domestic sewage, industrial effluents and urban and agricultural load. According to the European Commission (2002), a drop of a hazardous substance can pollute thousands of gallons of water and pollution today may remain for generations in groundwater intended for human consumption.

4 RESEARCH METHODOLOGY

The research was conducted in a Food Service Company (CFS), located in Lisbon/Portugal. The Company has implemented and certified its production of meals, meeting the requirements of Quality of the NP EN ISO 9001:2008 and Food Safety according to NP EN ISO 22000:2005 and NP EN 14001:2004 so it presents an Integrated Management System (IMS).

The CFS works in catering to various segments such as schools and catering. Currently, this supply is about 6,000 meals a day, with various types of menus, and composed salads, soups, meats (fish, poultry, pork and beef), trimmings and desserts. The distribution is cold or hot and the meals are packaged and shipped in accordance with the standards of hygiene and food safety. The approach to the Food Service Segment in Portugal included the monitoring of the various stages of preparation of 100 kg of hake, with emphasis on the consumption of water and generation of waste and effluents. It was carried out the monitoring of the production process of steaks fished from receipt of fished until the cleaning of trays used in cooking (Fig. 2).

The received frozen and headed hake were from the Falkland Islands caught in the Southwest Atlantic. Raw material were weighed, then removed from the packaging and stored at −5 °C over two

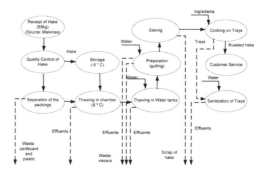

Figure 2. Flowchart of the production process of roasted hake in Food Service Segment in Portugal. Source: Authors, 2011.

days. Further were stored under refrigeration (8 °C) for thawing. The next day, due to food safety procedures, were thawed in tanks with water at room temperature, then were sent to the pre-preparation step (gutting). At this stage the fish were washed and all waste (viscera) removed and further hake were weighed. Later hake were sawed into portions of approximately 150 g. They were then seasoned with salt, garlic oil and soybean oil and baked in combination oven (steam and gas) for 1 hour. After cooking, the product was carried out from the cleaning of the 25 hygiene gastronorms (standard trays of stainless steel) used in the preparation.

This research includes physical, chemical and microbiological analyzes of the wastewater generated. The collection of effluent samples was performed in the following steps: Unfreezing Hake, Hake preparation (Hake washing and evisceration), Washing of the trays after hake roasting, for identifying and detailing the types of environmental impacts, i.e. organic or inorganic pollution. All samples were chilled transported in suitable containers and kept at 4 °C for further analysis.

In physical and chemical analyses Organic Matter and Suspended Solids were considered by gravimetric methods. Regarding microbiological analyses fecal coliform, total coliform and viable microorganisms at 30 °C counts were considered. The method used for fecal coliforms and total coliforms was based on inoculation of specific volumes of the sample (in ice form) culture medium that contains two chromogenic substrates (Salmon-Gal and X-Glu). This medium allows separate coliforms and E. presumptive coli-based coloration of colonies. The following is a confirmation of E. coli Kovac's reagent to produce indole. The greenhouse is adjustable incubation at 37 + 1 °C. The technique was performed according to ISO 16649-2 (ISO, 2001). For the count of total coliform colonies were considered reddish-pink and dark blue or violet. And for E.coli were considered Dark violet. The results were shown in the form of number of colony forming units per gram of sample (N/g), or for total coliforms (total number of colonies pink, dark blue and/or violet) and E.coli (total number of colonies violets showed a positive reaction in contact with the reagent Kovac's. method used was sowing by incorporating different sample volumes in culture, by incubation of a set of plates at 22 ± 1.0 °C for 72 hours and another set at 37 ± 0.5 °C for 24 to 48 hours. Calculation of the number of viable microorganisms per milliliter of sample from the number of colonies obtained.

The collected data on water consumption and waste generation and waste throughout the production process studied was entered into the software UMBERTO ®. This software was used to represent and model the production process of steaks grilled hake, which features graphic representation to describe the steps of this process and the performance of the production, being a tool for guiding decision making and environmental management.

5 RESULTS AND DISCUSSION

The application of LCA methodology, which is limited to water use and wastewater generation and waste in the production process of hake in CFS provides a breakdown of process management. Then it was described the design of this application.

5.1 Production process assessment and life cycle

5.1.1 Defining objectives and scope
The purpose of the application of LCA methodology was to perform Life Cycle Inventory (LCI) of the production process of the CFS fish through the identification and quantification of inputs and outputs to assess the potential environmental impacts related to the activities performed. The target-people for the development of this study are the managers and employees of the Food Service Segment. The research includes monitoring and traceability of the steps of receiving, pre-preparation and cooking of hake in the production process of meals. The functional unit is 100 kg of hake. Therefore, the boundary determined in this study comprises the following process steps: reception, pre-preparation and cooking. This boundary determines the influence of flows in and out of the process and the product that were described in the LCI.

5.1.2 Analysis of the life cycle inventory
At this stage of the LCA, the identification of inputs and outputs of each of the transitions in the production process of roasted hake portions was performed. The quantification of each input and output was carried out by calculating the average monthly quantity of inputs used and the waste generated. Subsequently, the data were entered into the software UMBERTO ® to determine the environmental balance and a graphical representation was produced based on Petri's net conception. In this paper some results form this inventory are summarized next.

The production process of 100 kg hake in the CFS for the preparation of grilled hake steaks correspondeds to the use of 370 liters of water in thawing and washing in water step and fish produced 380 liters of effluent, effluent from this total, approximately 10 liters were partial thawing in the refrigerator (8 °C). In pre-preparation (gutting) step, the fish were washed in running water being consumed approximately 520 liters of water and generated of approximately 520 liters of effluent. Then hake were cut into portions, seasoned and baked in combination oven. Subsequently, at step of cleaning of the gastronorms (trays) used in cooking hake, were consumed approximately 415 liters of water was consumed and more or less 415 liters of effluent generated. The inventory was estimated for monthly processing of 6,000 kg fished in CFS. According to the data obtained, the water consumption would be 78,460 liters. The volumes of wastewater generated in the respective production steps are: thawing 646 liters, water in thawing tanks 22,150 liters, gutting, and washing 31,380 liters and trays washing 24,920 liters.

5.2 Physical-chemical and microbiological analyzes of the production process of hake

In the physical and chemical analyses, the results indicated that the wastewater generated in thawing hake contained high content of organic matter and total solids. The effluent from the fish preparation (gutting) contained appreciable level of organic matter. The effluents from trays washing showed high levels of total solids and organic matter. In the microbiological analysis, the effluents from thawing hake showed appreciable level of total coliforms and high levels of viable microorganisms at 30 °C, the effluent from the preparation of pre-fished and effluent from washing the trays used in the cooking of steaks fished indicated significant levels of viable microorganisms at 30 °C.

Although ISO 14001:2004 have been implemented in CFS, the control of water consumption was not established as a regular activity management system. With the measurement of water held in this research, the interest in this type of study increased greatly, because it clarified the importance of water, not only because it is a strategic natural resource, but also for its contribution to the cost of meals produced. In CFS water meters connected exclusively to the production does not exist, but the need for such monitoring is already a fact of conviction for improvement of the production system.

Due to the consumption of fish constitute a fact very present in dietary habits of the Portuguese, the monthly amount present in the CFS menu is quite relevant, therefore justifying the adoption of environmental protection practices, not only with respect to the quantitative use of water, but as the quality aspects related to previous treatment and effluent generated in processing activities.

Mierzwa & Hespanhol (2005), analyzing processes in the food sector, highlight that it is important to measure, effectively and consolidated company's water consumption, whatever the context, especially in parts of the process where their use is significant. According to Hoekstra (2010), referring to the industrial use water, water use efficiency is expressed as volume of water required to make a product unit. With this view, adopted as a starting point, water users can be encouraged to adopt technology to save water consumption and create awareness to save water, but without losing sight of the relationship of this indicator with broader issues related the echo quality and efficiency, for example, as presented in the calculation of estimates of footprints water (Water Footprint, 2011). This is a way to ensure that local information relate to global information showing scenarios more complete than is good for the environment.

Thrane et al (2009), presenting results on processing of mackerel and herring in two companies in Denmark, found significant savings in water consumption after the monitoring and recording through daily consumption of water meters and changes in work routines, which took to an annual reduction of 120,000 Euros. This study demonstrates that environmental management and employee participation are essential to improving the production process.

Ervim et al. (2009), emphasizes that the main source of organic pollutants in a water body is the urban sewage from households, commercial and industrial on the food and agribusiness connected to the sewer raising and slaughtering of animals in production and storage of agricultural products.

Therefore, sustainability must be intrinsic to the planning and procedures used in the researched scenario, with the view of rational use of water being applied at local and global level, and companies of the Food Service Segment should become aware of the potential impacts associated with their activities, so that they can expand their production capacity and sustainable practices in the production process of Meals.

6 CONCLUSIONS

The application of LCA provided a critical analysis of the production process Meal Food Service Segment in Portugal and provided information for decision making of managers, stakeholders and/ or interested parties involved. For some of these actors, there was opportunity to comprehensively rethink the participation of water in the local and global scenario, in which they operate. At the CFS studied, despite the ISO 14001:2004 certification has been achieved, the control of water consumption and wastewater treatment process, were not systematized. The physicochemical and microbiological analyzes determined the estimated level of contamination that may cause to the environment, and found to be relevant, to assist in identifying the type of contamination of effluents in the stages of production of grilled hake steaks. It should be emphasized that it is evident the importance of a quantitative and qualitative diagnosis, the natural resource is used, in this case water, considering the water footprint in the process, is the influence of the activities performed in the environment, highlighting the effluents and impacts from the same processes studied. With the LCA done, the data indicate that management and procedures adopted must include the optimization of water use in the production steps of meals through improvements that need to be adopted. This will allow a review of environmental management of the productive process in other preparations with respect to water consumption and effluent generation.

The industries/companies can do much by society however the visibility of their environmental and social action is limited. It is necessary to spread the sustainable practices of companies that consumers valorize for their actions and a commitment to the environment.

ACKNOWLEDGEMENTS

The authors acknowledge the support of The Coordination of Improvement of Higher Education Personnel (CAPES) and the Portugueses Institute of Ocean and Atmosphere (IPMA, I.P.).

REFERENCES

Almeida, F. 2007. Os desafios da sustentabilidade. Rio de Janeiro: Elsevier.

Andrade, R.O.B; Tachizawa, T; Carvalho, A.B. 2002. Gestão Ambiental:enfoque estratégico aplicado ao desenvolvimento sustentável. 2nd Ed. São Paulo: Markron Brooks.

Coltro, L. Mourad, A.L.; Garcia, E.E.C.; Queiroz, G.C; Gatti, J.B.; Jaime, S.B.M. Mourad et al. 2007. Avaliação do ciclo de vida como instrumento de gestão. Campinas: CETEA/ITAL, 75 p.

Ervim, L.; Favero, O.B.; Luchese, E.B. 2009. Introdução à Química da água: ciência, vida e sobrevivência. Rio de Janeiro: LTC.

European Comission. A Directiva-quadro da água: Algumas informações. Comissão Européia. Luxemburgo: Serviço das Publicações Oficiais das Comunidades Européias, 2002. in:<http://ec.europa.eu/environment/water/water-framework/pdf/tapintoit_pt.pdf>. (visited: 27, jul., 2011).

FERCO. Fédèration Europèenne de la Restauration Collective Concédée. Situation du secteur en Europe. 2010. in: <http://www.ferco-catering.org/fr/resume.html>. (visited, 8, dec., 2011).

Guinée, J.B.; Heijungs, R.; Huppes, G.; Zamagni, A.; Masoni, P.; Buonamici, R.; Ekvall, T.; Rydberg, T. 2011. Life cycle Assessment: Past, Present and Future. Environmental Science & Technology, v. 45, n. 1, p. 90–96.

Harmon, A.H. & Gerald, B.L. 2007. Position of the American Dietetic Association: Food and Nutrition Professionals can implement practices to conserve natural resources and support ecological sustainability. Journal of the American Dietetic Association, v. 107, n.6, p. 1033–1043.

Hoekstra, A.Y. 2010. The relation between international trade and freshwater scarcity. World Trade Organization. Economic Research and Statistics Division. University of Twente, Netherland, January.

ISO. International Organization for Standardization. 2001. Microbiology of food and animal feeding stuffs—Horizontal method for enumeration of presumptive Escherichia Coli—Part 2: Colony count technique at 44°C using 5-bromo-4-chloro-3-indolyl-β-D-glucoronic acid.

ISO. International Organization for Standardization. ISO 14040:2006. 2006. Environmental management. Life cycle assessment. Principles and framework.

Leal, D. Crescimento da Alimentação Fora do Lar. 2010. Segurança Alimentar e Nutricional, Campinas, São Paulo, v. 17 n. 1, p. 123–132.

Lundie, S.; Peters, G.M.; Beavis, P. 2004. Life Cycle Assessment for Sustainable Metropolitan Water. Journal of Environmental Science and Technology, n. 38, p. 3465–3473.

Matos, C.H. & Proenca, R.P.C. 2003. Work conditions and nutritional status of workers from the food service sector: a case study. Rev Nutr, v.16, n° 4.

Mierzwa, J.C. & Hespanhol, I. 2005. Água na Indústria: Uso racional e reuso. São Paulo: Editora Oficina de Textos., p. 144.

Nicolella, G.; Marques, J.F.; Skorupa, L.A. 2004. Sistema de Gestão Ambiental: aspectos teóricos e análises de um conjunto de empresas da região de Campinas. Jaguariúna: Embrapa Meio Ambiente.

Phillip Jr, A.; Roméro, M.A.; Bruna, G.C. 2004. Curso de Gestão Ambiental. Barueri: Manole.

PORDATA. Base de dados Portugal contemporâneo. 2011. Pessoal ao serviço nas empresas não financeiras: total e por sector de actividade económica. In: <http://www.pordata.pt/Europa/Pessoal+ao+servico+nas+em presasas+nao+financeiras+total+e+por+sector+de+ac tividade+economica-163>. (visited, 8, dec., 2011).

Seiffert, M.E.B. (3rd ed.) 2010. ISO 14001 sistemas de gestão ambiental: implantação objetiva e econômica. São Paulo: Atlas.

Slack, N.; Chambers, S.; Johnston, R. (2nd Ed.) 2009. Administração da produção. São Paulo: Atlas.

Thabrew, L.; Wiek, A.; Reis, R. 2009. Environmental decision making in multi-stakeholder contexts: applicability if life cycle thinking in development planning and implementation. Journal of Cleaner Production. v. 17, n. 1, p. 67–76, January.

UNESCO. United Nations Education, Scientific and Cultural Organization. 2012. The 4th edition of the World Water Development Report. Volume 1: Managing water under uncertainty and risk. World Water Assessment Programme (WWAP). 2012. In: <http://www.unesco.org/new/en/natural-sciences/environment/water/wwap/wwdr/wwdr4-2012>. (visited, 27, mar., 2012)

Vilela Jr, A. & Demajorovic, J. 2011. Modelos e ferramentas de gestão ambiental: desafios e perspectivas para as organizações. São Paulo: Editora Senac.

WaterFootprint. 2011. In:<http://www.waterfootprint.org/?page=files/WaterFootprintAssessmentManual>. (visited, 26, fev., 2013).

WCED, World Comission on Environment and Development. 1987. Our Common Future. Oxford: Oxford University Press.

Green Design, Materials and Manufacturing Processes – Bártolo et al. (eds)
© 2013 Taylor & Francis Group, London, ISBN 978-1-138-00046-9

Life-cycle management of high-performance and sustainable buildings based on enterprise modeling method

Y.M. Shao

Institute of Energy Efficient and Sustainable Design and Building, Technical University of Munich, Munich, Germany

ABSTRACT: In contrast to most enterprise engineering activities, the design, construction and operation of buildings is still highly fragmented and comparatively unsophisticated. In general, silo mentalities and cultures prevail in this domain. Even as new buildings are being designed and constructed to more exacting functional standards, such as LEED, cost effectiveness and high performance cannot be ensured, presumably because they are not designed optimally or operated correctly. This paper argues that to achieve high performance and sustainability, a tight, overarching life-cycle management of all processes and procedures from design to construction on through to operation should be conducted for buildings. Buildings should be planned, designed, built, operated and retrofitted at a level more like that used to design, build, operate and improve enterprises such as refineries and chemical plants. Enterprise modeling, which are widely used in the research of enterprise organization and operation, will be employed in the life-cycle management of buildings in order to achieve high performance and sustainability. The philosophy of enterprise modeling is introduced first. Then the current problems in the field are identified and initiatives are presented, followed by the discussion of the feasibility of translating enterprise modeling into building life-cycle management. A pathway to the new methodology is proposed accordingly.

1 INTRODUCTION

Integration is a magic word in the research of complex systems. In general, the basic goal of integration is to improve the overall system efficiency by managing and controlling it systematically and obtaining a higher effectiveness of the whole system compared with the isolated operation of its elements. Buildings as well as enterprises are rather complex systems with a large number of elements, but in contrast to most enterprise engineering activities, the design, construction and operation of buildings is still highly fragmented and comparatively unsophisticated. The building industry accounts for almost 40% of the global energy use (Hill & Bowen, 1997) which results in improving energy efficiency in buildings a major priority worldwide. But silo mentalities and cultures prevail in this domain and the idea of integration by taking buildings as complex systems are often lacking. Even as new buildings are being designed and constructed to more exacting functional standards, such as LEED, cost effectiveness and high performance cannot be ensured, presumably because they are not designed optimally or operated correctly.

The overarching aim of this study is to bring up with a new perspective of building life-cycle management for high performance and sustainability, leading to a more tight and vigorous management of all life-cycle phases. Enterprise modeling methods, which are widespread in the research of enterprise engineering actives, will be employed in building life-cycle management.

2 BACKGROUND

2.1 *Enterprise modeling methodologies*

Models are abstractions of a given reality and are often used to capture the process functionality and process behavior. Enterprise modeling is the abstract representation, description and definition of the structure, processes, information and resources of an identifiable business (Leondes & Jackson, 1992). An enterprise is a complicated social, economic and physical system. Carrying out enterprise activities expediently and professionally necessitates enterprise integration, which requires understanding, partitioning and simplification of the complexity. Enterprise modeling technology has been considered to play a very important role in dealing with the complexity of designing an enterprise and supporting enterprise management and operation. It is widely used in process industries such as refineries and chemical plants. During the past several decades, a number of enterprise modeling methods have been put forward to describe enterprises. The widely used enterprise modeling methods include CIMOSA, GRAI/GIM, ARIS, and PERA (Kosanke et al. 1999). The IFAC/IFIP Task Force on Architectures for Enterprise Integration has developed an overall definition of a generalized architecture, GERAM (Generalized Enterprise Reference Architecture and Methodology), to organize existing enterprise modeling knowledge, with an integrated modeling method GERA (Generalized Enterprise Reference Architecture). The life-cycle concept of GERA provides for the identification of the life-cycle phases from the start of a project to its final end (Bernus & Nemes, 1996). ISO 19439 defines an international standard for enterprise modeling and enterprise integration based on CIMOSA and GERAM.

Enterprises such as refineries and chemical plants are mostly very complex systems and it is difficult to describe them using a simple and unified model. A typical method used by almost all enterprise modeling methods is to describe the enterprise using several view models such as function view, information view, organization view, etc. Each view defines one aspect from a specific point of view, and then the integration method between the different view models is defined. Here gives a brief introduction of the GERA modeling framework as it will be employed as the main reference architecture in the following study.

GERA defines the general process oriented concepts of enterprise operations and covers various activates in all life-cycle phases from the needs identification to its final ends. Enterprise engineering and integration projects based on GERA principles are closely connected in an enterprise's entire life span. Figure 1 illustrates the GERA enterprise modeling framework and the relationship between the concept of life-cycle phases and the concept of whole life span.

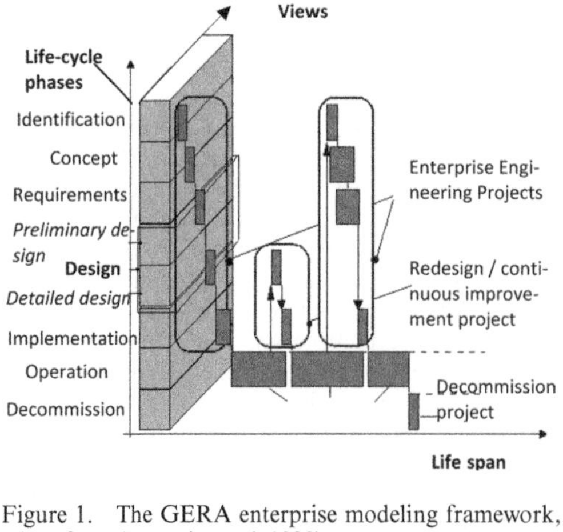

In Figure 1, the vertical axis represents different life-cycle phases of an enterprise, while the horizontal axis represents the enterprise' whole life span. The enterprise's engineering activities among different phases are executed accordingly, and different view models are employed to describe the enterprise.

2.2 Civil engineering applications of enterprise modeling

A lot of efforts have been done to introduce the concept and methodologies of enterprise modeling into the building design, construction and operation. Most of the studies were focused on the manufactured construction. Among them, some scientists and practitioners tried to introduce the methodology of mass manufacturing into construction industry (Mike, 1998; Gann, 1996; Mullens & Arif, 2006) so as to achieve the benefits such as economies of scale, better quality and better control and management which have been quite attractive for this industry. Arif et al. (2011) proposed an enterprise-wide information system that has been implemented in a modular home manufacturing plant.

Murray et al. (2003) pointed out that the use of computer aided design and manufacturing can easily be implemented in offsite construction, thus improving the efficiency of the whole construction process. Pan et al. (2011) suggested that new technologies such as BIM can serve as an effective platform for exchanging information between design, manufacturing and any other department involved in a project, and performing impact analysis on decisions in the system.

3 SUPPORTING BUILDING LIFE-CYCLE MANAGEMENT FOR HIGH PERFORMANCE AND SUSTAINABILITY

3.1 Current problems

The physical life of a building can be subdivided into different key phases from the building's needs identification, concept development, requirement analysis, design, construction, utilization and its final decommission. During the utilization period, the building may go through some refurbishment and retrofit projects. In Germany, HOAI (the German official scale of fees for services by architects and engineers) serves as a guide of classifying the phases. The state-of-art functional standards, such as LEED (Leadership in Energy and Environmental Design), are established to help to design

Figure 1. The GERA enterprise modeling framework, adapt from (Kosanke et al. 1999).

and construct high-performance green buildings. But these meant-to-be high-performance building are underperformed in many cases. On the other hand, Buildings usually consist of many subsystems and components, while achieving high performance and sustainability relates to almost every subsystem of the buildings, so designing and operating a building can be treated as the design and operation of a complex system. Enterprises about process industry are also very complex systems and they adapt enterprise modeling methods to achieve system integration. In contrast to most enterprise engineering activities, the design, construction and operation of buildings is still highly fragmented and comparatively unsophisticated.

3.2 *Enterprise modeling in building life-cycle management*

The general comparisons with other artifacts produced by modern industrial society (e.g., automobiles, aircrafts) which move through the life-cycle are misplaced. A building is not likely going to be an end product of mass production due to the uniqueness of each building. However, buildings should be treated in a higher level. Enterprises are usually very complex systems and they adapt the idea of enterprise modeling to achieve enterprise integration.

To achieve high performance and sustainability, a tight, overarching life-cycle management of all processes and procedures from design to construction on through to operation should be conducted for buildings. In other words, buildings should be planned, designed, built, operated and retrofitted at a level more like that used to design, build, operate and improve enterprises such as refineries and chemical plants. The idea of enterprise modeling is adopted to achieve integrated solutions of buildings. The modeling framework of high-performance and sustainable buildings is shown in Figure 2.

In this framework, different life-cycle phases from the identification of the needs of a building to its demolition or recycling in buildings' whole life span are systemically integrated. Different views models, such as function view, organization view, information view etc. could be used to describe different facets of building life-cycle management accordingly. Building information is stored in the information view model from the very beginning and goes through the buildings' whole life span. Building related activities are executed according to the function view model. High numbers of participants in different life-cycle phases are organized based on the organization view model.

The period of buildings' utilization accounts for the longest time, and it is buildings' function to be

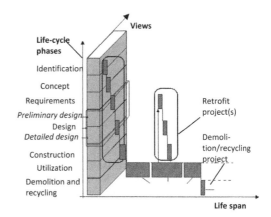

Figure 2. The modeling framework of high-performance and sustainable buildings.

utilized. In the proposed framework, the operation of building systems should be handled synthetically according to the dynamic variation of building energy needs and indoor environment to minimize the energy consumption and to keep high performance. Building retrofit projects are executed according to buildings' historical data, and are performed based on the framework's principle to improve buildings' performance systemically.

3.3 *Pathway to the implementation of the new methodology*

Enterprise modeling plays an important role in dealing with the complexity of engineering activities in enterprise life-cycle management and there are abundant research achievements. These studies could provide valuable resources of translating enterprise modeling into building life-cycle management. Also, the implementation of the new methodology needs state-of-art tools and whole new business models. As a common platform in building industry, Building Information Modeling (BIM) can be a very useful tool to record buildings' information. The Integrated Project Delivery (IPD) based on the development of BIM, can be a promising project delivery approach in the proposed framework. IPD "integrates people, systems, business structures and practices into a process that collaboratively harnesses the talents and insights of all participants to optimize project results, increase value to the owner, reduce waste, and maximize efficiency through all phases of design, fabrication, and construction" (AIA, 2007). It is the goal of BIM implementation—to combine domain technologies, process and policies into one organization.

In order to implement the new methodology, we need a whole new understanding of the building industry and wholly new business models are needed, which may change the roles of every participant. Achieving systematic integration in the building life-cycle management is not an easy task. To move in this direction requires the industry-wide participation of enterprises and related organizations.

4 CONCLUSION

This paper shows that enterprise modeling methods can be used reasonably for life-cycle management of high-performance and sustainable buildings. The proposed framework, based on the GERA enterprise modeling framework, aims at achieving systemic integration of building life-cycle management.

In addition to better design tools, more intelligent building control systems and advanced energy efficiency technologies, we need a whole new understanding of the building industry. Buildings should be planned, designed, built, operated and retrofitted at a level more like that used to design, build, operate and improve enterprises such as refineries and chemical plants. The philosophy of enterprise modeling provides a new perspective of building life-cycle management for high performance and sustainability. But achieving systematic integration in the building life-cycle management is not an easy task. It relates to every corner of this industry, and wholly new business models are needed. To move in this direction requires the industry-wide participation of enterprises and related organizations.

ACKNOWLEDGEMENT

This research was partly supported by the China Scholarship Council (CSC) funded by the State Scholarship Fund of the Education Ministry of the P.R. China Government (File No. 201206210033).

REFERENCES

AIA. 2007. *Integrated project delivery*: a guide. California: AIA California Council.

Arif, M., Kulonda, D., Egbu, C., Goulding, J.S., & Toma, T. 2011. Enterprise-wide information system for construction: A document based approach. *KSCE Journal of Civil Engineering* 15(2): 271–280.

Bernus, P., & Nemes, L. 1996. A framework to define a generic enterprise reference architecture and methodology. *Computer Integrated Manufacturing Systems* 9(3): 179–191.

Gann, D.M. (1996). Construction as a manufacturing process? Similarities and differences between industrialized housing and car production in Japan. *Construction Management & Economics* 14(5): 437–450.

Hill, R.C., & Bowen, P.A. 1997. Sustainable construction: principles and a framework for attainment. *Construction Management & Economics* 15(3): 223–239.

Kosanke, K., Vernadat, F., & Zelm, M. 1999. CIMOSA: enterprise engineering and integration. *Computers in industry* 40(2) 83–97.

Leondes, C.T., & Jackson R. 1992. Manufacturing and automation systems: techniques and technologies. Academic Press.

Mike M. 2003. Rethinking construction: The Egan report (1998). *Construction Reports* 1944–98: 178–194.

Mullens, M.A., & Arif, M. 2006. Structural insulated panels: Impact on the residential construction process. *Journal of construction engineering and management* 132(7): 786–794.

Murray, N., Fernando, T., & Aouad, G. 2003. A virtual environment for the design and simulated construction of prefabricated buildings. *Virtual Reality* 6(4): 244–256.

Pan, W., & Arif, M. 2011. Manufactured construction: Revisiting the construction-manufacturing relations. *Proceedings of the 27Th Annual Conference* 1: 105–114.

Green Design, Materials and Manufacturing Processes – Bártolo et al. (eds)
© 2013 Taylor & Francis Group, London, ISBN 978-1-138-00046-9

Manufacturing renaissance: Return of manufacturing to western countries

B. Kianian & T.C. Larsson
School of Engineering, Blekinge Institute of Technology, Karlskrona, Sweden

M.H. Tavassoli
School of Management, Blekinge Institute of Technology, Karlskrona, Sweden

ABSTRACT: Manufacturing Renaissance, i.e. return of manufacturing to west, has been recently observed. This paper analyzes the patterns observed within each of the four main drivers behind this new phenomenon and delves more deeply into the driver that centers on the new manufacturing technologies such as Additive Manufacturing (AM) and 3D Printing. Next, this paper will make the case that the location of manufacturing will be in west, relying on the established theory that has been able to explain the location of manufacturing, i.e. Product Life Cycle Model (PLC).

1 INTRODUCTION

Twenty-one percent of North American manufacturers reported bringing production back into or closer to North America in the past three months. Surveyed by manufacturing sourcing Web site MFG.com (June 2011). Thirty-eight percent planned to research such a move in the next three months. This new trend is a reverse of what had taken place in the late 60s and early 70s, i.e. western manufacturing has vastly moved to Less Developed Countries (LDCs). The Atlantic (2012). As Norton & Rees (1979) said, the main reasons were "the low labor cost and favorable business climates of such LDC's as South Korea and Taiwan". Vernon (1979) made a similar statement: "Although income, market size, and factor cost patterns have converged among the more industrialized countries, a wide gap still separates such countries from many developing [LDCs] areas". However, much has changed in recent years and the two reasons described above are less significant today than in the late 60s and early 70s. Concerning labor cost, in a recent report Boston Consultancy Group anticipates that the net manufacturing cost in China and US will converge in 2015. (Sirkin et al, 2011). Concerning the business milieu, there has been recent and recurrent complains about IPR problems in China and other Asian emerging economies. Indeed a new trend has been observed which indicates the 'return' of manufacturing to western countries, especially to US (Sirkin et al, 2011); (Economics, 2012); (The Atlantic, 2012).

The aim of this paper is to shed some light on the pattern observed with respect to the locational shift in western manufacturing, i.e. so-called in this paper-manufacturing renaissance. This will be performed by developing arguments within the context of PLC model, while borrowing arguments from transaction cost theory and new economic geography.

The rest of the paper is organized as follows; Section 2 presents the established PLC model, section 3 demonstrates the newly observed trend in location of manufacturing —this is done by adding the additional phase to the established PLC model, and section 4 discusses the factors driving the new pattern(s). Section 5 summarizes and concludes.

2 PRODUCT LIFE CYCLE MODEL (PLC)

The product life cycle approach to international trade and investment provides a systematic explanation of how the location of manufacturing, exporting, and importing of a product changes over time. Such locational shift studied initially at international level (Vernon, 1966, 1979); (Hirsch, 1967). This was followed by studies of PLC model at the interregional level (Rees, 1979); (Norton & Rees, 1979). Vernon (1966)'s original model is presented in Figure 1.

The model proposes that location of the production (and subsequently the export and import patterns) varies based on the maturity level of the product. More specifically, Vernon (1966) argued that the production of a product in its first phase

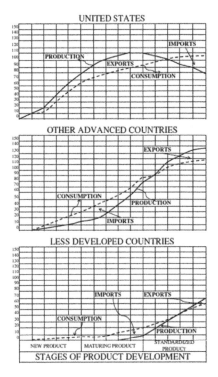

UNITED STATES

OTHER ADVANCED COUNTRIES

LESS DEVELOPED COUNTRIES

STAGES OF PRODUCT DEVELOPMENT

Figure 1. Original Product Life Cycle model (60s and 70s) *Source*: Vernon (1966).

of development (i.e. new product phase) would be located in US. Firstly, this is a result of a higher level of demand for a new product in US market, among other things, due to higher average income in US in comparison with other countries. Vernon (1979). Secondly, there is a larger supply of high skilled labor. Hirsch (1967). Thirdly, swift and effective communication between the agents in the supply/value chain (producer, customers, suppliers, competitors, etc.,) exists in US. Vernon (1966). These three supply factors are essential for overcoming the uncertainty in product specification and market, which exist during the early phase of product development. Utterback & Abernathy, (1975). This early phase of product development is accompanied with higher US export.

The production of the second phase of product development (i.e. maturing product) would be located in other advanced countries. Vernon (1966) argued that as the demand for a product expands, a certain degree of standardization usually takes place; however, there are still efforts for product differentiations. Since there are some degrees of standardization, there would be relatively less need for externalities. Instead, there would be more orientation toward economies of scale and more concerns about production cost rather than prod-

uct characteristics. This is why the manufacturing location of a product would presumably move to other developed countries. Hence, the US-made production would stagnate and the import from other developed country would start, however, US export would still be dominant on US import.

Finally, the production of third phase of product development (i.e. standardized product) would probably move to Less Developed Countries (LDCs), since they can provide competitive advantageous for production location in this phase. Vernon (1996) provides several reasons for such claim. First, the standardized products tend to have lower uncertainty in terms of their specification (unlike new products). Hence, the need for skilled labor and externalities (such as local knowledge) is remarkably reduced, which reduces the dependency of their location on US or other advanced countries. Second, standardized products tend to have lower uncertainty in terms of market, i.e. they have a well-articulated and easily accessible international market, so the marketing cost (from distance) is low. Third, these products are assumed to have high price elasticity of demand (unlike new products) and they are assumed to be mostly sold based on price. This would act as a motivation to take the risk of moving the production to a new location. Fourth, these products require significant labor input for their production, which is (again) an incentive for moving the production to low-cost labor countries, i.e. LDCs. Consequently, it may be wise for the international firm to shift the location of their standardized products into the LDCs, based on the notion that labor costs differences are large enough to offset transportation costs. This would be accompanied with higher import and lower export costs for US.

3 MANUFACTRUING RENAISSNACE: A NEW PATTERN

This part adds an additional phase to the established model of PLC developed by Vernon (1966, 1979) and Hirsch (1967). This is not the first time a study tries to modify the original PLC model based on the observed trend. (Vernon, 1979); (Giddy, 1978). The main motives for introducing the 4th phase here is the recent changes in LDCs (emerging economies). Moreover, the impact of such changes on the behavior of US, and other advanced economies are difficult to distinguish, since they have become homogeneous in terms of various externalities over time. Vernon (1979). The new pattern of production location, import and export for three classic categories of countries is depicted in Figure 2.

Additional 4th phase is the "Renaissance".

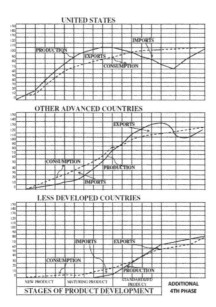

UNITED STATES

OTHER ADVANCED COUNTRIES

LESS DEVELOPED COUNTRIES

| STAGES OF PRODUCT DEVELOPMENT | ADDITIONAL 4TH PHASE |

Figure 2.   Extended product life cycle model.

Assuming the PLC model is valid, the first three phases of Figure 2 are identical to the original PLC model (Fig. 1), while the 4th phase is an add-on. The main argument for adding the 4th phase is that some part of manufacturing production is coming back to western world, especially to US. The reason for proposing the addition of this new phase is based on newly observed pattern, briefly reviewed in introduction section. This can be explained by several driving factors, which are discussed in section 4.

4   FACTORS EXPLANING THE "MANUFACTURING RENAISSANCE"

There are several factors driving the new pattern in location of manufacturing, i.e. manufacturing renaissance. These factors are; raising wage-levels in emerging economies, lower quality of business milieu in emerging economies (LDCs), lower importance of economies of scale for production, and motives for interacting better with customers among US companies. The discussion of each factor is presented in the following subsections.

4.1   Rising wage-levels in emerging economies

Wage-level has always been an important motive for outsourcing manufacturing to LDCs (Norton & Rees, 1979); (Vernon, 1979), especially if economies of scale are already, being fully exploited (Vernon, 1966). Recent evidences also suggest that the wage differential is still one of the most important drivers of outsourcing to LDCs. In examining the motives for outsourcing and offshoring, in a recent survey 50 percent of firms in Denmark, Sweden and The Netherlands state that labor cost savings is the primary reason for sourcing their business functions abroad. Statistic Denmark, (2008). Labor-intensive industries have been among the first to move to LDCs because they are most affected by increases in industrialized countries' wages relative to the rest of the manufacturing sectors. Puga & Venables (1996). In addition, weakly linked industries are also the ones who moved faster to LDCs, because they benefit less from being close to other industries in western world (they neither sell a large fraction of their output to other industries, nor spend a large share of their costs on intermediates produced by them). They are therefore the first to relocate in response to labor cost differentials, being gradually followed by more strongly linked industries. Puga & Venables (1996).

However, new reports point that the labor differential is not in place to the extent that enables the companies to move to LDCs since the 70s up to now. For instance, Boston Consultancy Group argues that wage-level in China is increasing by an average of 20 percent annually and productivity improvement is not enough to offset the labor cost. Hence, it is anticipated that the net manufacturing cost in US and China will converge in 2015. (Sirkin et al, 2011). Such a new situation clearly violates the traditional main driver of moving the manufacturing to LDCs, i.e. labor cost differentials. Recent evidence indeed suggests that increased wages in some LDCs has reduced the US outsourcing to those countries. Swenson (2005).

One can ask why the wage-level in emerging economies has actually increased dramatically in recent years. There are at least two reasons for this. First, as (Puga & Venables, 1996) argued, outsourcing of manufacturing to a country will eventually lead to growth of related industry in that country. This implies the growth for demand in manufacturing within that country. Finally, this leads to bidding up wages in that industry and country and there will eventually be a critical mass. At this point, it is no longer profitable to stay in the home country, hence the manufacturing will move to another country. This is what has actually happened in LDCs, particularly in China. Second, there has been a new trend of what is called "brain circulation", i.e. returning the highly educated Chinese (and some other LDCs) from US back to their home country. Saxenian, (2006). These people usually have higher salaries than ordinary employees do in LDCs. Therefore, by returning to china, they have raised the average wage.

4.2 Lower quality of 'business milieu' in emerging economies

It is shown that entry into new market inherently involves transaction costs and such transaction costs are reduced via proper institutional setting of the host country. Meyer, (2001). Proper institutional setting (business milieu) was indeed one of the reasons that manufacturing has vastly moved to Less Developed Countries (LDCs) in late 60s and early 70s. (Norton & Rees, 1979); (Vernon, 1979). However, there have been recently recurrent complaints about IPR problems in China and other Asian emerging economies, violating the previous image about proper business milieu in these countries. It is argued that China's enforcement of its IP laws has been inadequate (e.g. lack of action against counterfeiting and piracy), although the framework of IP protection has been well established. Wang, (2004). Such lower quality of business milieu can be understood via the concept of opportunism, described by (Williamson, 1981) as dishonest behavior by competing firms. According to Transaction Cost Theory, opportunism represents a source of transaction costs. It is one of the determinants of whether firms will choose outsourcing or vertical integration. Williamson (1981) argued that vertical integration arises out of the need to safeguard against opportunism and contractual hazards.

Such contravention of IPR in (for example) China is combined with the imitation, which has been argued to be one of the factors that can explain the boosting innovation in China. Needless to say, this is a threat to western innovation-based competitiveness Therefore, not only lower quality of business milieu in China in recent years has blurred one of the traditional motivations to move the manufacturing to China, i.e. proper business milieu, but also their imitation skills what it is argued are a thread for innovation-based competitiveness of western companies.

4.3 Lower importance of economies of scale (due to new process innovations)

Economies of scale can reduce the total production cost. It can be achieved through the presence of a large number of suppliers in a particular region (or country). Teece (1986). Such economies of scale have actually been one of the driving factors for western companies to move their manufacturing to China and other LDCs, especially for those western companies who were followers (not first movers) in terms of outsourcing their manufacturing to China and other LDCs.

However, recent process innovations, e.g. 3D printing, degrade the importance of economies of scale. As the magazine Economics wrote recently:

> "It [recent process innovations] will allow things to be made economically in much smaller numbers, more flexibly and with a much lower input of labor, thanks to new materials, completely new processes such as 3D printing, easy-to-use robots and new collaborative manufacturing services available online. And that in turn could bring some of the jobs back to rich countries that long ago lost them to the emerging world." Economics, (April 2012).

Additive Manufacturing (AM) is a relatively new manufacturing method that first came into use in late 1980's. In general, it forms 3D physical objects by solidifying the raw material layer upon layer. Depending on the technology, the solidification mechanism ranges from spraying of a liquid chemical binder, to exposure of the material to various light sources, to electron beam bombardment. The materials used are also very diverse, including various types of polymers, metals and ceramics materials, providing they be in power or liquid form, and again depending on the technologies being used. Originally, due to its limited capacity and low resolution, the method had been used for prototyping and model making, thus the term rapid prototyping. It has since been gradually developed towards providing end-use parts or direct part production, referred to as rapid manufacturing. (Tuck et al, 2008).

As the quality of the AM fabricated products improve, the labor cost related to those products will decrease. This will create a scenario where manufacturers in regions with relatively higher labor costs are able to compete with those that have lower wages in LDS countries. In addition, combining this competitive pricing with the concept of quicker delivery will provide local suppliers with an advantage over their foreign competitors highly competitive in their markets. Wohlers (2011).

Moreover, the rising cost of energy and its efficacy and independency are the major barriers for the future of manufacturing and play a significant role in shaping the geopolitical landscape. Taking in the consideration how AM processes are capable of producing significant waste reduction compared to conventional methods, some concepts like buy-to-fly ratios indicates how wasteful the conventional methods are. Wohlers (2011).

One major source of overall energy costs is the cost of transportation. Much more energy is needed to ship and deliver parts from a long distance than to ship them from a local or regional retailer and supplier. Studies indicates that due to problems such as communication and tool rework and transportation costs, the actual costs

Table 1. Characteristics of the product cycles.

Characteristics	Cycle phases			
	New product	Maturing product	Standardized production	Renaissance
Technology	Short run and rapidly changing	Mass-production and importance of economies of scale	Long run and stable process	Mass customization
Physical capital	Low	High, due to high obsolete rate	High, due to large quantity of specialized equipment	Low, due to new process innovations
Industry structure	Entry is know-how, many firms	Growing number of firms	Stagnation in number of firms	Growing number of spin offs
Human capital	Scientific and engineering	Management	Unskilled	Scientific, engineering, and unskilled
Demand structure	Seller's market, low price elasticity of demand	Growing price-elasticity of demand	Buyer's market, High price elasticity of demand	Closer to customer, shorter technology cycles

Source: New, maturing, and standardized product characteristics are based on Hirsch (1967).

of offshore manufacturing is higher than is anticipated and believed in many cases. Wohlers (2011).

A key point is that the cost of producing much smaller batches of a wider variety (with each product tailored precisely to each customer's need) is falling. The factory of the future seems to have a focus on mass-customization, rather than traditional mass-production. This allows for lesser reliance on economies of scale (available through extensive availability of cheap suppliers in China), which could eventually lead to the return of manufacturing of some parts back to western countries. This is indeed what (Grossman & Helpman, 2005) argued: "disproportionate improvements in the technology for customization in a region can shift the manufacturing toward that region (here referring to US)". Considering the trend toward production on demand, low volume, and need to increase domestic manufacturing and employment, the offshore and overseas production may not be the best choice. Wohlers (2011).

4.4 Interact better with home customers

The PLC model argues that manufacturing production will be outsourced to LDCs in the 3rd phase, assuming the strict standardization of product and lack of the need for customer interaction. This is, however, a strong assumption, especially if one considers the faster life cycle of a product in recent years. This would require close interactions with the home customer, even though a product reaches some degree of standardization. By being close to the home customer, "market-determined inducement" would ease the incremental innovation

for even standardized product. Dosi (1988). In fact, the manufacturing has been outsourced to LDCs because of saving-costs forces. Vernon (1979). Now that those forces are not at play as strong as before, it is reasonable to assume that some part of manufacturing will come back to US and thus have a better interaction with the home customer inter alia.

Following the above reasoning (4.1 to 4.4) for occurrence of manufacturing renaissance, i.e. the 4th phase in PLC model, it is possible to illustrate the characteristics of each stages of PLC. This is illustrated in Table 1. This table is based on Hirsch (1967) and adds the 4th phase.

5 CONCLUSION

Manufacturing Renaissance, i.e. return of manufacturing to west, has been recently observed as a new pattern emerging in western countries, especially in US. This paper identified main drivers of this new phenomenon: (i) rising wage-levels in emerging economies (ii) lowered quality of 'business milieu' in emerging economies (iii) lower importance of economies of scale, due to new process innovations (iv) better interaction with home customers. In doing so, the paper contextualized itself within a well-established theory that explains the locational shift of manufacturing, i.e. Product Life Cycle model (PLC). The paper delves more deeply into one of the drivers that centers on the new manufacturing technologies such as Additive Manufacturing (AM) and 3D Printing.

REFERENCES

Dosi, G. 1988. Sources, Procedures, and Microeconomic Effects of Innovation. *Journal of Economic Literature, 26(3): 1120–1171.*

Giddy, I. H. 1978. The Demise of the Product Cycle Model in International Business Theory. *Columbia Journal of World Business, 13(1): 90–97.*

Grossman, G. M. & Helpman, E. 2005. Outsourcing in a Global Economy. *Review of Economic Studies, 75: 135–159.*

Hirsch, S. 1967. Location of Industry and International Competitiveness. *Oxford: Oxford University Press.*

Meyer, K. E. 2001. Institutions, transaction costs, and entry mode choice in Eastern Europe. *Journal of International Business Studies, 32(2): 357–367.*

Norton, R. D. & Rees, J. 1979. The product cycle and the spatial decentralization of American manufacturing. *Regional Studies, 13(2): 141–151.*

Puga, D. & Venables, A. J. 1996. The Spread of Industry Spatial Agglomeration in Economic Development. CEPDP, 279. *Centre for Economic Performance, London School of Economics and Political Science.*

Rees, J. 1979. Technological change and regional shifts in American manufacturing. *The Professional Geographers, 31(1): 45–54.*

Saxenian, A. 2006. The New Argonauts: Regional Advantage in a Global Economy. *Harvard University Press.*

Sirkin, H. L. Zinser, M. & Hohner, D. 2011. Made in America, Again. *Boston: Boston Consulting Group.*

Statistic Denmark. 2008. International Sourcing, Moving business functions abroad. *Copenhagen: Statistic Denmark.*

Swenson, D. L. 2005. Overseas assembly and country sourcing choices. *Journal of International Economics, 66: 107–130.*

Teece, D. 1986. Transactions cost economics and the multinational enterprise-An Assessment. *Journal of Economic Behavior & Organization, 7(1): 21–45.*

The Atlantic. 2012. The Insourcing Boom. Retrieved 12 04, 2012, from theatlantic.com.

Tuck, C. Hague, R. Ruffo, M. Ransley, M. & Adams, P. 2008. Rapid manufacturing facilitated customization. *International Journal of Computer Integrated Manufacturing 21, no. 3: 245–258.*

Utterback, J. M. & Abernathy, W. J. 1975. A dynamic model of process and product innovation. *Omega, 3(6): 639–656.*

Vernon, R. 1966. International Investment and International Trade in the Product Cycle. *The Quarterly Journal of Economics, 2:190–207.*

Vernon, R. 1979. The product cycle hypothesis in a new international environment. *xford bulletin of economics and statistics, 41(4):255–267.*

Wang, L. 2004. Intellectual property protection in China. *The International Information & Library Review, 36: 253–261.*

Williamson, O. E. 1981. The Economics of Organization: The Transaction Cost Approach. *American Journal of Sociology, 87(3): 548–577.*

Green Design, Materials and Manufacturing Processes – Bártolo et al. (eds)
© 2013 Taylor & Francis Group, London, ISBN 978-1-138-00046-9

Process chain analysis of lightweight metal components—a case study

R. Ilg & D. Wehner
Department Life Cycle Engineering, Fraunhofer Institute for Building Physics, Stuttgart, Germany

ABSTRACT: This document builds on an environmental relevance study of metal processing for the Eco-Design of lightweight components in the aviation industry and highlights the relevance of metal processing for the Life Cycle Assessment (LCA) based Eco-Design of lightweight applications.

The environmental relevance of metal processing is often low when compared to the environmental impacts of the preceding mining and refining of ores as well as their processing into metal ingots, so that the negligibility of metal processing has become a commonly used axiom of many LCA studies. Even for the production of lightweight components in the aviation industry, this credo is commonly recited while referring to the good recyclability of metal.

This paper analyses whether this credo of the negligibility of metal processing in LCA can really hold true for the production of lightweight components and crystallizes its major environmental hotspots.

1 LIFE CYCLE ASSESSMENT FOR ECO-DESIGN OF LIGHTWEIGHT COMPONENTS

1.1 Eco-Design

Evolving conventional design into Eco-Design means to make the minimisation of environmental impacts over the whole life cycle of a product or service the major design objective (see green framed box, Fig. 1). How this objective can be addressed and integrated into existing design structures is illustrated in Figure 1.

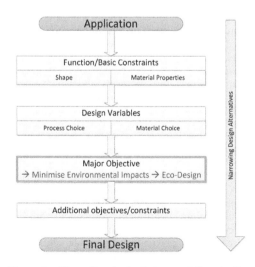

Figure 1. Upgrading basic design workflow to Eco-Design.

The central task of a lightweight component designer is to define shapes of certain material properties to address the necessity of certain functions for a particular application of a designed component. The number of possible design variations the designer could come up with are primarily limited by producibility—and so process and material choice. Within an Eco-Design approach the most environmentally sound design-choice had to be found among all these producible designs. Therefore, the designer needs a direct feedback on the change of environmental impacts coming along with design changes over the whole life cycle of the designed component—from the first mining of materials to the end-of-life of the component. Generating such feedbacks is of tremendous complexity (see chapter 1.2) and so usually cannot go without simplifying assumptions (see chapter 1.3).

1.2 Life Cycle Assessment

According to the ILCD Handbook (European Comission, 2010) Life Cycle Assessment (LCA) is "a structured, comprehensive and internationally standardised method" which is defined in the ISO 14040 and 14044 standard. LCA "quantifies all relevant emissions and resources consumed and the related environmental and health impacts and resource depletion issues that are associated with any goods or services" (European Comission, 2010).

To really account for all relevant emissions and resources the whole life cycle of the assessed goods or services has to be considered. Therefore, all in- and outputs to the different life cycle stages have to be compiled in a so called Life Cycle Inventory. This approach is illustrated in Figure 2.

Only with taking into account all in- and outputs of any involved process all environmental and health impacts can be quantified within a so called Life Cycle Impact Assessment where consumed resources and caused emissions are classified in Environmental Impact Categories. To get a holistic understanding of impacts on environment and health all impact categories needed to be taken into account given that every possible impact was known and comprised within existing impact categories.

As neither are all possible environmental impact types known nor are all known impact types relevant for each analysis, the mandatory steps of the Life Cycle Impact Assessment according to EN ISO 14040 (DIN EN ISO 14040, 2006) starts with the selection of relevant impact categories. An impact category is a "class representing environmental issues of concern to which life cycle Inventory Analysis results may be assigned" (DIN EN ISO 14040, 2006). The assignment of the results of the Inventory Analysis to an impact category is called "classification" which constitutes the second step of the LCIA. Subsequently, the impacts of the different considered categories are quantified by characterization—where the assigned Inventory Analysis results are converted to the same units and then aggregated (European Comission, 2010). The unit, inventory results are aggregated in, is called category indicator (European Comission, 2010).

Commonly considered impact categories are listed in Table 1 along with a short explanation and examples for possibly chosen indicators and LCI items (classification).

The Primary Energy Demand—however—is not an impact category as it does not quantify an impact but energy consumption. Still it is often used to evaluate the energy efficiency of products (DIN EN ISO 14040, 2006).

Selecting a defined number of environmental impact categories is already a simplification of an LCA according to the very basic definition stated above. However, the complexity of LCA does not only derive from the great number of impacts certain emissions create but greatly also from actually assessing all the relevant emissions of the involved products and processes necessary for the production of a component. Therefore LCA can usually not go without further simplification.

1.3 Life Cycle Assessment simplifications

The fact that LCA cannot go without simplification had already been realized within the early work groups for finding standardized LCA methodology which therefore dealt with simplifying LCA. Already in 1995 on the "EPA Conference on Streamlining LCA" (Curran, 1996) in Ohio it was agreed on that a "Full LCA" had actually never been carried out up to the date of the conference as any existing LCA study had been based on simplifications. Only two years later the Society of Environmental Toxicology and Chemistry (SETAC) EUROPE LCA Screening and Streamlining Working Group published its final report on "Simplifying LCA" (Christiansen, 1997). There the problem of the non-existence of Full-LCA was taken into account by, instead of Full LCAs, using the term Detailed LCAs which are a result of an iterative procedure which necessarily involves simplifications along its iteration steps. This procedure is illustrated in Figure 3.

The simplifying options for such models suggested by SETAC Europe—as well as the ones later also by other parties—can all be categorized in one of two categories as they are either about narrowing system boundaries or the use of surrogate data.

The computation of LCA results is commonly carried out by software tools where the analysed system is modelled by a series of unit processes which are described with inventory data. Inventory items which are not taken from or released to the environment but from or to another product system are commonly described with background datasets i.e. provided by the applied software.

Narrowing system boundaries therefore means defining which unit processes and which inventory data are to be taken into account for the study. Using surrogate data means using inventory data which is not optimal for describing a respective inventory item of a product system but good enough with regard to the intended goals of the LCA study.

With applying a background data set out of a LCA software tool commonly both these LCA simplifications are used at the same time. I.e. an aluminium lightweight component producing company buys aluminium sheets for further processing. A particular aluminium sheet may therefore constitute an input item of an inventory of one of the processes planed to model. However, as no dataset for the particular alloy of this sheet

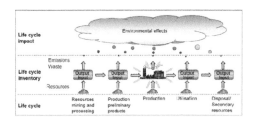

Figure 2.   Life Cycle Assessment (Ilg, 2013).

Table 1. Commonly used environmental impact categories.

Impact category	Explanation	Indicator	Classification
Global Warming Potential (GWP)	Emissions to air which influence the temperature of the atmosphere	kg CO2 eq.	CO2 CH4 ...
Acidification Potential (AP)	Emissions to air which lead to acidification of rainwater	kg SO2 eq.	SO2 CH4 ...
Eutrophication Potential (EP)	Emissions to water and soil which lead to eutrophication	kg PO43- eq.	PO43- NH3 ...
Photochemical Ozone Creation Potential (POCP)	Emissions to air which act as ground-level ozone formation agents	kg C2H4 eq.	C2H4 CO ...
Primary Energy Demand (PE-tot)	Consumption of renewable (PE-r) and fossil energy carriers (PE-nr)	MJ	Crude oil Hard coal ...

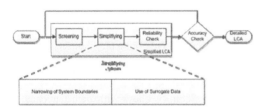

Figure 3. Detailed LCA procedure according to SETAC Europe.

may be available, another dataset of a similar sheet may be applied—surrogate data. In the documentation of the applied data set one is likely to find that the system boundaries have been narrowed for its assessment. Probably i.e. the environmental impacts of the production of the machines for processing aluminium ingots into sheet were not taken into account. The error introduced by these simplifications may in many cases likely be far from significant for the overall results of a particular study so that such simplifications are commonly found throughout the LCA world.

1.4 Commonly applied simplification for the LCA of lightweight components

Lightweight components integrating product systems are often found of tremendous complexity—in particular in the case of aircrafts. As additionally detailed specific data sets for respective finished lightweight components are hardly available, above described simplification methods (see chapter 1.3) are often drawn on and in many cases appear to be stretching the justifiable range of application to its limits.

Particular finished lightweight components are not only modelled by averaged datasets of semi-finished components but in some cases even as ingots of a different type of metal.

The line of argumentation for justification of these assumptions often builds on the publically available environmental data of the aluminium industry which appear to suggest that metal processing is of minor importance for the production of semi-finished components and so may also be of minor relevance for finished components.

This axiom is commonly also transferred from aluminium to other lightweight materials—i.e. Titanium.

Whether the credo of insignificance of metal processing for the environmental impacts of the production of metallic lightweight components can hold true is analysed in the following.

2 ENVIRONMENTAL RELEVANCE OF METAL PROCESSING

2.1 Environmental impacts of semi-finished aluminium products

In the following the relevance of metal processing for the production of the semi-finished aluminium sheets and extrusion profiles is analysed. The analysis makes use of readily available data in the GaBi 5 (PE International, 2012) database and greatly relies on the data from the European Aluminium Industry (EAA) (European Aluminium Association, 2008).

Figure 4 compares the relative environmental impacts for the production of 1 kg of aluminium ingot (reference value), 1 kg of aluminium sheet and 1 kg aluminium extrusion profile. All products

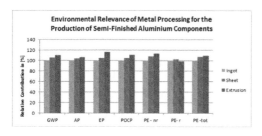

Figure 4. Environmental relevance of metal processing for the production of semi-finished aluminium components.

represent a European product mix. Metal processing includes heat treatment and mechanical metal removal (scrap rates assumed about 30% at 100% re-melting of scrap) as well as application of ancillary material (i.e. lubricants).

The results illustrated in Figure 4: Environmental relevance of metal processing for the production of semi-finished aluminium components how a clear dominance of metal ingot production over metal ingot processing into semi-finished products. Differences in all the different impact categories are found around 10% and may be neglected in studies where such an error is acceptable with regard to its intended results and where the underlying assumptions for metal processing are justifiable.

In the following the results of a detailed process chain analysis of the production of aluminium and titanium lightweight aircraft components are introduced while the potential environmental differences to the metal processing for the production of semi-finished aluminium components are highlighted.

2.2 Process chain analysis of aviation lightweight components

Detailed process chain analysis of the production of lightweight aluminium and titanium aircraft components shows that the production of most of these components can be described by five major processing steps:

- the ingot production
- the wrought processing
- the microstructural adjustment
- the detailed shaping and
- the finishing.

Each of these major processing steps is furthermore comprised of several different processing alternatives.

Therefore, when estimating the environmental impacts of metallic lightweight components with datasets for respective semi-finished components,

the processing steps detailed shaping and finishing are per se neglected.

The major detailed shaping process alternatives can be categorised as either mechanical or chemical metal removal processes. The type of applied mechanical metal removal processes are widely similar or identical with those applied for semi-finished component production. Chemical metal removal for shaping is exclusively found in the detailed shaping. The extent of metal removal for detailed shaping is often found much higher than for the production of semi-finished components. While usually around 1.3 to 1.4 kg of aluminium ingot is processed into 1 kg of semi-finished component, up to more than 20 kg of semi-finished component may be necessary to produce 1 kg of finished lightweight component.

Chemical metal removal does not produce re-meltable scrap but metal sludge which (if at all) can only be recycled in the early stages of metal ingot production.

The applied alloys as well as the required material properties for lightweight aircraft component production may also differ from those majorly comprised in averaged environmental datasets of semi-finished components and so also lead to potential differences in processing steps alloyed ingot production, wrought processing and microstructural adjustment. Scrap of alloys with high tendency to oxidation may not only complicate the recycling of scrap but also i.e. lead to higher metal removal rates in the processing of semi-finished components. Not only, but in particular titanium can be extremely difficult to machine which may lead to significantly increased environmental impacts even in the production of semi-finished components. Moreover, certain alloys may require increased stress relief heat treatment during semi-finished component production.

This great number of qualitative differences between the production of semi-finished standard metal components and finished lightweight metal components clearly call into question that the environmental impacts of semi-finished standard metal components and finished lightweight components are really in a similar range.

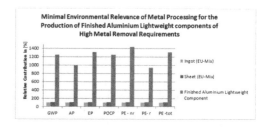

Figure 5. Minimal environmental relevance of metal processing for the production of finished aluminium components of high metal removal requirements.

Table 2. Assumptions for the environmental assessment of the production of a finished aluminium lightweight component.

Process step	Assumption	Comment
Alloyed ingot production	Pure aluminium ingot (EU-Mix)	As for reference ingot
Wrought processing	Scrap rate: 28% Re-melting rate: 100% No reheating for stress relief necessary	As for reference sheet (EU-Mix)
Microstructural adjustment	Only energy screening of solution heat treatment	Impact underestimating by neglecting possible impacts from ancillary material use as well as other possible process steps as cooling, ageing, etc.
Detailed shaping	Scrap rate: 95%	High
	100% re-melting of mechanically removed metal	Underestimating impacts due possible low quality scrap
	No pre-treatment for chemical milling	Underestimation by impact negligence
	30% of removed metal removed by chemical means	Standard value for certain aircraft components
Finishing	Sulphuric acid anodizing (1 m² per component)	Standard settings

2.3 Potential environmental impacts of finished lightweight aluminium components

In this chapter the minimal environmental relevance of metal processing for the production of finished lightweight components of high metal removal requirements are analysed. As no inventory data has been available for several possibly relevant unit processes the so computed environmental impacts have to be considered as minimal impacts.

Figure 5: Minimal environmental relevance of metal processing for the production of finished aluminium components of high metal removal requirements shows the comparison of the environmental impacts of the production of 1 kg aluminium ingot (EU-Mix), 1 kg aluminium sheet (EU-Mix) as well as 1 kg of a finished lightweight aluminium component. The exact assumptions for the analysis of the production of this finished lightweight component are given in Table 2. The environmental impacts of the production of 1 kg aluminium ingot (EU-Mix) constitute the reference for the comparison.

As can clearly be seen from the results illustrated in Figure 6 the increased environmental impacts due to processing of aluminium ingot into finished lightweight components are far from negligible. Even though several impact underestimating assumptions were taken the impacts of the production of high metal removal requiring finished lightweight components are found more than ten times higher than for the production of EU-Mix metal ingots and sheets.

Figure 6 illustrates the distribution of the environmental impacts of the production of the finished lightweight component as described

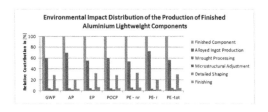

Figure 6. Environmental impact distribution of the production of finished aluminium lightweight components.

in Table 2 over its major processing steps. Results are normalised to the environmental impacts of the production of the whole finished component.

Figure 6 shows that the environmental impacts of the production of the finished aluminium lightweight component are still dominated by the impacts of the alloyed ingot production. As 20 kg of ingot had to be casted for the production 1 kg of finished component this result is not surprising. That the overall contribution to the environmental impacts of production—however—only constitutes about 60% in each of the analysed impact categories shows that also the direct impacts of further processing have tremendously increased. In particular the impacts of detailed shaping are of high significance due to high energy and chemical demand for metal removal. Considering the high absolute impacts of the finished aluminium lightweight component production compared to standard semi-finished component production as well as all the impact underestimating assumptions taken for these processing steps also the direct impacts of the other processing steps are far from negligible.

3 CONCLUSION

The results of this paper clearly show that LCA for the Eco-Design of metallic lightweight components cannot generally rely on the commonly referred to credo of the negligibility of metal processing. Especially where high metal removal rates occur during processing the error of such assumptions may underestimate the actual environmental impacts of production by a factor of higher than ten. If materials are additionally difficult to machine and recycle the error can be expected to be even far beyond factor ten.

Therefore it is most essential for the Eco-Design of lightweight components to correctly assess the quantity and quality of scrap generated during production as well as the recycling routes followed. Moreover, the exact locations of metal removal in the process chain have to be identified as the amount of material processed also triggers the impacts of many other processing steps.

Further research on the yet unknown process steps in lightweight metal component production commends itself in order to gain knowledge of the full dimensions of the environmental impacts of metallic lightweight material production.

The development of generic computer models for the computation of the environmental impacts of specific metallic lightweight components according to their impact triggering production conditions offers to make complex lightweight components containing product systems correctly and economically assessable.

NOMENCLATURE

GWP	Global Warming Potential
AP	Acidification Potential
EP	Eutrophication Potential
POCP	Photochemical Ozone Creation Potential
PE-tot	Primary Energy Demand (total)
PE-r	Primary Energy Demand (renewable resources)
PE-nr	Primary Energy Demand (non-renewable resources).

REFERENCES

Christiansen, K. 1997. Simplifying LCA: Just a Cut? Final report from the SETAC-Europe Screening and Streamlined Working Group. Brussels, Belgium 1997.

Curran, M., Young, S. 1996. Report from the EPA conference on streamlining LCA. The International Journal of Life Cycle Assessment (Int. J. LCA) 1, S. 57–60.

DIN EN ISO 14040. 2006. Umweltmanagement—Ökobilanz—Grundsätze und Rahmenbedingungen. DIN Deutsches Institut für Normung e.V

European Aluminium Association. 2008. Environmental Profile Report for the European Aluminium Industry. Life Cycle Inventory Data for Aluminium Production and Transformation Processes in Europe.

European Comission (Hg.). 2010. International Reference Life Cycle Data System (ILCD) Handbook. Luxembourg, Luxembourg.

Ilg, R. 2013. Shaping the future of aviation. The path towards low-carbon aviation 2011. Available at: http://www.lbp-gabi.de/refbase/files/190_Ilg2011.pdf [accessed 17.02.2013].

PE International. 2012. GaBi. Software-System and Databases for Life Cycle Engineering. Stuttgart, Germany 2012.

Green Design, Materials and Manufacturing Processes – Bártolo et al. (eds)
© 2013 Taylor & Francis Group, London, ISBN 978-1-138-00046-9

Objective monetization of environmental impacts

H. Krieg, S. Albrecht, M. Jäger & J. Gantner
Department Life Cycle Engineering (GaBi), Chair of Building Physics, University of Stuttgart, Germany

ABSTRACT: Many organizations are increasingly aware of the relevance of the environmental impacts of their products and want to measure, and systematically reduce them. Recognized methods such as Life Cycle Assessment (LCA) provide a central basis to do so. LCA takes into account the entire life cycle of products and provides information on their environmental impacts, based on inputs and outputs of materials, resources and energy. However, many companies have difficulties integrating environmental impacts in strategic and operational planning. One way of doing this is the monetization of environmental impacts. The approach presented in this paper allows for a systematical and objective monetization of environmental impacts. It does so by using the simplex algorithm, an established method for production planning. This allows for an economic optimized portfolio planning that takes into account environmental restrictions. This approach represents an innovative way of an systematic integrated economic and environmental portfolio planning.

1 BACKGROUND

1.1 *Type area*

The awareness of environmental impacts of organizations and society has been rising steadily over the last years. While some years ago discussion on global warming and other environmental impacts were only followed by a small share of organizations, sustainability and preservation of the environment is now a core issue for many organizations. This can for example be seen in the rising number of environmental studies conducted and the increased relevance of environmental aspects in legislation and communication.

One method for the quantification is Life Cycle Assessment (LCA). It is a method that allows quantifying environmental impacts of products or processes. This is the basis for a systematical improvement process—only when the source of environmental impacts is known they can be systematically reduced. Thereby, LCA contributes to sustainability efforts of organizations on a product or process level (ISO 2006a & ISO 2006b).

Nevertheless, covering the bridge between sustainability efforts on a product level and sustainability on an organization level is still challenging. The implementation of environmental product footprints into management processes and strategic decision making is still in its development.

One way of supporting the integration of LCA results in management processes and decision making can be the monetization of environmental impacts. This describes the expression of emissions

and other environmental impacts in monetary units, which then again can be basis for incentives like taxes, fees or reduction bonuses (Beckenbach et al. 1998).

Current approaches for monetization are mainly following two approaches; willingness to pay and environmental damage cost.

Willingness to pay describes the amount of money an organization or individual would be willing to pay in order to avoid an environmental impact. As this is a subjective decision which is based on subjective values and believes as well as level of education, region and monetary means, the actual willingness for the same impact can vary widely (Reap et al. 2008).

The idea behind environmental damage cost is to analyze the actual damage caused to the environment and to express this damage in monetary units. There is a lot of uncertainty in this approach. Environment is a very complex system with many interactions between different parts of the system. Therefore it is very difficult to take all chains of effects into account, and then after doing so, to monetize the costs of the environmental impacts. This can lead to widely varying values for the same impact. For example, a meta-study of the German Federal Environmental Agency (UBA) showed for the same systems in different studies a deviation by a factor of 40,000 (UBA 2007).

The approach presented in this paper is therefore following a different approach, which is more objective. It is based on environmental target values that have to be defined, e.g. through legal limitations or reduction goals of companies. It is based on both product specific economic indicators as well environmental impacts and thereby identifies the internal value of environmental impacts for a specific organization, the so-called shadow price while also determining an optimized product portfolio by taking into account the environmental limitations. The approach is based on the simplex algorithm, an established method often used in production planning and operational management to support the optimal allocation of scarce resources (Geiger & Kanzow 2002).

2 METHODOLOGICAL APPROACH

2.1 Life cycle assessment

Life Cycle Assessment is a method widely accepted and applied in both industry and science. It allows to systematically quantifying the environmental impacts of products, processes or services. It is standardized in ISO 14040 (2006a) and ISO 14044 (2006b), which defines the requirement for LCA studies to ensure transparency and reproducibility of results. LCA determines the environmental impacts along the entire value chain, from raw material extraction over production to operation and finally the end of life stage. It identifies the relevance of each life cycle stage and also the contributing processes or materials. Thereby, it offers the basis for a systematic optimization processes for products and services, as it allows to specifically targeting relevant sources of environmental impacts.

LCA results are expressed in impact categories. These are grouping environmental impacts of different emissions according to their contribution to different impacts. Such an impact category is for example Global Warming Potential (GWP). Contributing emissions are in a next step characterized by their significance in relation to a reference unit. For GWP, this is kg CO_2-equivalents. For example, methane contributes to global warming and has a characterization factor of 25. This means that the emission of 1 kg of methane has the same impact on global warming as the emission of 25 kg of CO_2 (IPCC 2009). The approach presented here is using such LCA results as input factors for the monetization.

2.2 The simplex algorithm

As soon as resources are limited, their use has to be planned in order to get the most benefits from the consumption of those resources. This optimal allocation of scarce resources is one of the core tasks in organizational planning. A method to support this allocation is the simplex algorithm, which is often used in production planning to allocate scarce production factors (Geiger & Kanzow 2002).

First, a size to be optimized has to be chosen. The Contribution to Margin (CtM) is best suited here. It is generally described as the difference between selling price and variable costs of a product, thereby representing both costs and revenues of a product or service (Wöhe & Döhring 2005).

In a next step, the restrictions have to be defined, as well as the resource consumption of each product in the portfolio. The simplex algorithm then goes through several iterative steps in order to identify the optimal allocations of the scarce resources in order to identify a portfolio that generates the highest possible CtM (Geiger & Kanzow 2002).

2.3 Using the simplex algorithm to allocate scarce resources

The approach presented in this paper is a combination of environmental impacts based on LCA studies, economic product aspects represented through the CtM and limiting factors such as legal limits or organizational reduction targets.

This information is used as inputs for the simplex algorithm. Through several iterative steps, the algorithm identifies an optimal allocation of scarce resources, in this case which products to produce in order to have the highest possible CtM within the predefined environmental restrictions. Figure 1 illustrates the methodological approach.

The result of this analysis delivers two results that are relevant for decision makers; an optimized portfolio and shadow prices of scarce resources. The portfolio gives information which products and how many to produce or operate under optimized boundaries. The shadow prices determine the internal value of scarce emissions such as CO_2,

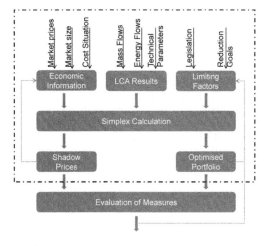

Figure 1. Methodological approach for portfolio planning and monetization.

based on objective factors such as LCA results and the CtM.

3 CASE STUDY

The application of the method is illustrated through a case study. It is based on a virtual logistics company that aims at optimizing their CtM while reducing their emissions. For the sake of simplicity, this case study only takes into account two product variants and two environmental indicators. The environmental indicators are GWP in kg CO_2-eq. and Acidification Potential (AP) in kg SO_2-eq. The products that are being compared are two multi-use plastic crates for the transport of vegetables and fruits. They have a weight of 2 kg and a capacity of 15 kg. The underlying LCA models results for this case study have been taken from a study conducted on behalf of Stiftung Initiative Mehrweg and slightly adapted. More information on the study can be found in Albrecht et al. (2009).

There are two kinds of crates. One is made of conventional crude oil based Polyethylene (PE), the other is made of bio-based PE from sugar cane. For both types of boxes a full LCA is conducted, covering raw material extraction, production, logistics during the life time, washing and end of life processes. In addition to the calculated environmental impacts, the economic aspects are estimated. They are 0.50 € for the conventional PE crate and 0.45 € for the bio-based PE crate. Both types of crates have a technical lifespan of 50 circulations.

It is further assumed that the company operates currently a portfolio consisting to 50% of each type of crate, and that the total market size is 1,000,000 circulations, which corresponds to a total transport of 15,000,000 t of vegetables and fruits.

Based on these assumptions, the organization generates a total CtM of 475,000 €, while emitting 215,000 kg of CO_2-eq. and 2,000 kg of SO_2-eq. For the next planning period, the organization aims at reducing their CO_2-emissions by 10% to 193,500 kg of CO_2-eq. while aiming at maximizing the CtM. The overall market size is limited to 15,000,000 t of vegetables and fruits, corresponding to 1,000,000

circulations of the crates. This can then also be expressed as

$$\text{MAX } 0.50x + 0.45y \quad [\text{CtM}] \qquad (1)$$

Subject to

$0.33x + 0.10y$	$<=$	193,500	[GWP]
$0.0015x + 0.0025y$	$<=$	2,000	[AP]
$x + y$	$<=$	1,000,000	[Market]

The application of the method is not described in detail here, for more detail on the method see Geiger & Kanzow (2002).

After several iteration of the simplex algorithm, an optimal portfolio within the boundary conditions is identified. In total, there are 420,380 rotations carried out with the conventional PE crates and 547,780 rotations with the bio-based PE crates. This results in a reduction of CO_2-emissions by 10% or 21,500 kg, while reducing the overall CtM by 3,8%. A maximum CtM under the new boundary conditions was identified. In addition to this portfolio optimization, shadow prices for the environmental restrictions are calculated. These are 0.85 €/kg CO_2-eq. and 145 €/kg SO_2-eq.

Shadow prices represent the internal value of scarce factors. In this case it means that by emitting an additional kg of CO_2-eq., the overall CtM can be increased by 0.85 €. By emitting another kg of SO_2-eq. the overall CtM can be increased by 145 €. On the other hand, cutting emissions by 1 kg decreases the CtM accordingly. Those values are based on the environmental profiles of each product, their CtM and the emission restrictions. Thereby, they allow an organization specific determination of the monetary value of environmental impacts. For a more detailed case study see Krieg et al. (2013).

4 DISCUSSION

The approach presented here allows for a portfolio planning that takes into account environmental restrictions. It can be adapted to take into account more products or different environmental impact categories. Thereby, it supports both economic planning as well as reduction of environmental impacts of organizations.

Furthermore it allows the objective and organization specific monetization of environmental impacts through shadow prices. These values can then in be used as basis for incentives like an internal environmental tax or to calculate bonuses for environmental improvements. Furthermore, it provides an objective basis for the comparison of internal values and external prices, as it allows for example to compare the shadow price

Table 1. Product properties per crate circulation.

	GWP [kg CO_2-eq.]	AP [kg SO_2-eq.]	CtM [€]
Conventional PE crate (X)	0.33	0.0015	0.50
Bio-based PE crate (Y)	0.10	0.0025	0.45

of CO_2 with the market price for CO_2-emission certificates.

It allows comparing the impact of different reduction goals on the economic success of the company, and can also be used to compare different product variants or sites in different countries.

Thereby, the approach increases the relevance of LCA studies for organizational decision making while also supporting planning and management processes.

REFERENCES

Albrecht, S., Beck, T., Barthel, L., Fischer, M. 2009. The Sustainability of Packaging Systems for Fruit and Vegetable Transport in Europe based on a Life-Cycle-Analysis—Update 2009. On behalf of Stiftung Initiative Mehrweg SIM (Foundation for Reusable Systems under German Civil Law). Stuttgart/Michendorf.

Beckenbach, F., Hampicke, U. & Schulz, W. 1998. Möglichkeiten und Grenzen der Monetarisierung von Natur und Umwelt, *Schriftenreihe des IÖW 20/88*, Berlin, pp. 3–18.

Geiger, C. & Kanzow, C. 2002. *Theorie und Numerik restringierter Optimierungsaufgaben.* Springer-Verlag: Berlin, Heidelberg and New York.

German Federal Environment Agency UBA (2007): Ökonomische Bewertung von Umweltschäden—Methodenkonvention zur Schätzung externer Umweltkosten. http://www.umweltdaten.de/publikationen/fpdf-l/3193.pdf.

Intergovernmental Panel on Climate Change (IPCC) (2009). IPCC Guidelines for National Greenhouse Gas Inventories. Task Force on National Greenhouse Gas Inventories (TFI) of the IPCC. Washington D.C, USA.

ISO 14040-2006a. Environmental management—Life cycle assessment—Principles and framework.

ISO 14044-2006b. Environmental management—Life cycle assessment—Requirements and guidelines.

Krieg, H., Albrecht, S., Jäger, M. 2013. Systematic Monetization of Environmental Impacts. *Sustainable Development and Planning 2013* (accepted for publication).

Reap, J., Roman, F., Duncan, S., Bras, B. 2008: A survey of unresolved problems in Life Cycle Assessment—part 2: impact assessment and interpretation, *Int J LCA,* 2008, pp. 374–388.

Wöhe, G. & Döring, U. 2005. *Einführung in die Allgemeine Betriebswirtschaftslehre*, Vahlen-Verlag: München.

Green Design, Materials and Manufacturing Processes – Bártolo et al. (eds)
© *2013 Taylor & Francis Group, London, ISBN 978-1-138-00046-9*

Assessment tool for building materials

S. Vasconcelos & C. Alho
FA—U.T.L., Lisbon, Portugal

B. Müller
HTW, Berlin, Germany

ABSTRACT: There are some tools regarding building materials sustainability assessment, such as Environmental Declarations (EPDs) and Eco-labels. In addition to these methods, there are software evaluation tools such as: Gabi Software, Sima Pro, AUDIT Solutions, Umberto, etc. They all evaluate the complete life cycle of materials, but only on its ecological aspect, which is not enough from the sustainability point of view. There are no tools that respond holistically and comprehensively to the growing need of making sustainable decisions during the project design. Since this is a more comprehensive tool, MARS_SC will be adjusted for construction materials in this study. The advantage of this type of assessment tool is its potential practical application during the project design, providing easy-to-interpret results so that architects can make informed choices. Sustainability assessment it is a very complex process, and it is a contemporary responsibility of utmost importance to make it more transparent and user-friendly.

1 INTRODUCTION

Architecture has always tried to find answers to contemporary problems and needs. Today the concept of sustainability emerges, since the international crisis with ecological, economic and social issues causes a turning point for architecture and its designers. This is a paradigm shift that requires a system based on preventive actions. Tools that evaluate the performance and damage exert by a construction, a building or even a city, are increasingly necessary in order to assess the planning complexity and multi-disciplinary contexts.

The materials choice is one of the challenges to be taken by designers who aim to achieve sustainable projects. These choices are particularly important because the environmental issues depend on it. The development of assessment tools that measure the sustainability is relevant both to academic studies and also for the industry. There are currently several tools on the market that are used for new projects planning and for the existing buildings assessment (Mateus, 2004). Sustainability considerations are characterized by an analysis of the entire life cycle of a building and a comprehensive inclusion of environmental, economic and socio-cultural aspects (Hegner, 2011). There are several concepts and tools in the product evaluation area, such as the environmental product declarations and label (eg. Blue Angel). The BNB (Bewertungssystem Nachhaltiges Bauen für Bundesgebäude) rating system and life cycle analyzes

evaluate the whole buildings or systems, but aren't specifically related to the building materials. The existing assessment tools for building materials are valuable, but either too complex or to simplified. Environmental declarations create a comprehensive Life Cycle Analysis (LCA), but they are too complex for a quick decision. Labels are lacking of information and they only consider the first phase of building material's life cycle. In both cases, only the ecological aspects are considered, which from a sustainable point of view is not enough. Multiple dimensions and foresight characterize sustainability. Focusing on a single development aspect leads to suboptimal solutions.

Therefore, all efforts should be focused on equally weighing environmental, economic and socio-cultural issues. An important optimization approach to holistic planning, advising and building is the use of modern planning and simulation tools. They could act in the early stages of planning, by evaluating and comparing different solution and variants to demonstrate the feasibility and functionality of solutions before they are built (Herkommer, 2004). Instruments are constantly updated and improved, so that their weaknesses could be corrected. What's required is to design an assessment tool that: acts in the decision making process; is user-friendly; has a transparent calculation method; is flexible enough to be able to analyze a variety of materials; and which attends the constant technological evolution (Mateus, 2004). Certification systems for sustainable buildings play an important role in the international market, but they shouldn't be the only purpose of policy approaches. Developers and construction companies also need fundamentals and tools to achieve the required standards of the certification systems.

There aren't many assessment tools that calculate the impact of building materials on practical, fast and extensive form. This investigation intends to fill this need. The best-known assessment tools are: GaBi Software, Sima Pro, AUDIT Solutions, Umberto, etc. They analyze the whole life cycle of building materials but only from an ecological point of view. MARS-SC (Metodologia de Avaliação relativa de Sustentabilidade—Sistemas Construtivos/methodology for relative assessment of sustainability—Building Systems) is an assessment tool that was developed at the Universidade Minho by engineers Ricardo Mateus and Luís Bragança in 2006. This tool also integrates social and economic aspects in the building systems assessment. This assessment tool has proven to be very user-friendly, in opposition to the previously enumerated instruments, although it is not directly related to building materials but to the entire building systems. An entire building system is, for example, a wall that consists of different building materials, or a roof with different building levels. The structure of the MARS-SC will be explained in the main part of this paper and some improvements will be suggested, so the results will correspond to the primary goal of this study: The evaluation of building materials sustainability by using the right tool.

The initial questions for the development of such tool are: What is a sustainable building material? What qualities are essential? Are the results of the assessment tool clear and simple? Can such an assessment tool actually lead to the sustainable development of projects?

2 MAIN PART

2.1 Conceptual structure of the assessment tool

The MARS SC is based on the generally accepted three-pillar model of sustainability: this model of sustainable development is based on the idea that it can only be achieved by the simultaneous and equal relations of environmental, economic and social objectives. These three forces influence each other (Mateus, 2006).

According to engineer L. Bragança the sustainability assessment should be analyzed in a relative context, which must be related to a particular country/location and time reference solution. Each parameter can be examined to perceive whether it is better or worse in comparison to a standard solution. The evaluation minimum level is the most commonly existing solutions in the market, and it is adapted to the technological development. The best level depends on the existing technology development (BVBS, 2012). From this perspective, the MARS-SC always works with two or more building systems in comparison to each other. This study assumes this perspective and applies it to the building materials, meaning that the materials are analyzed and compared to the same parameters.

2.2 Parameters and data collection

The reliability of such an instrument may be impaired by the use of data from dubious sources, so it is of paramount importance and interest that the sources are free and neutral and that they are updated regularly. In the next paragraph the valuation parameters and their sources are explained.

2.3 Environmental parameters

These parameters evaluate the interaction of building materials with the environment. The following have been selected:

– Global Warming Potential (TP),
– Eutrophication Potential (EP),
– Acidification Potential (VP),
– Ozone depletion Potential (OP),
– Embodied Energy (GE),
– Abiotic stress Resource (AR),
– Water use (WN),
– Solid waste (FA),
– Recyclable/flexibility (RF),
– Type of transportation/distance to the project (TT).

The data for the environmental parameters are predominantly obtained in the database Ökobau. dat. This German database is public and free of charge; it is the basis for the LCA and other types of analyzes to assess the buildings and building materials' environmental performance. It consists of about 950 data sheets, where the following building materials and construction and transport processes categories are described in terms of their environmental effects: Mineral building materials, insulation materials, wood products, metals, coatings and sealants, windows plastic construction components, doors and curtain walls, building and others (BVBS, 2011). The Recycled components parameter contents is only possible if the manufacturer publishes their data. If this is not the case, this parameter is omitted. The parameter Recyclable/flexibility uses a semaphore developed by the author, and divides the materials into groups: non-recyclable, energy (incineration), Recyclable and Reusable. To obtain the parameter transport/

distance to project data it a table from the author Berger (2009) is used, which indicates the values of the ton/km for different transport and fuel types.

2.4 Comfort level parameters

Comfort—consciously or unconsciously, our senses are the means by which we perceive space and measure our level of satisfaction. Our senses are stimulated by the space properties and we interact with the environment depending on the stimulus. The following parameters were chosen:

- Acoustic comfort,
- Thermal comfort (WLF—thermal conductivity) (Santos, 2006),
- Durability (HK) (BVBS, 2011),
- Safety (SH) (DIN, 1998).

The parameters comfort data are purchased from various sources.

For acoustics, a semaphore was developed by the author, which divides the materials into groups: Good absorption/poor sound insulation, absorption less Good/Good soundproofing agent, poor absorption/Good sound insulation.

The parameter for thermal insulation is obtained in ITE50 (LNEC), where the thermal conductivity of the materials is listed (DAA, 2012).

Durability, another comfort parameter is obtained, "according to useful life of components for life-cycle analysis BNB", from the BBSR-table (The Federal Institute for Building Urban and Regional Research/Federal Agency for Construction, Urban Development and Territorial Development) (BVBS, 2011).

Comfort level security is the last parameter; Data is available in DIN 4102-1, where materials are divided into groups regarding their fire behaviour: highly flammable and non-combustible materials (DIN, 1998).

2.5 Economic parameters

Economic parameters are related to the costs for a particular building material to be implemented. The entire life cycle is also considered:

- Construction costs (KK—including manufacturing, transportation and installation) (DAA, 2012) Cost of dismantling and ISPOSAL or reintegration into another product (EK).

Utilization costs are not relevant concerning building materials, since they are only important for the whole building. Maintenance costs could occur, but, since in the study case of this investigation this is not the case, they were not incorporated.

The use of more ecological and comfort level parameters result from their higher complexity.

The global environmental system and the ecological balance can't be evaluated in three or four parameters. In the overall result, all these sustainability parameters are given the same weighting.

2.6 Normalization

After the parameters data has been collected, normalization is carried out in order to turn the data unitless. The normalization of the parameters is important because the units and values of the data are very different. With this procedure it is possible to calculate an overall result. Balteiro (2004) explains this calculation as follows:

When an indicator is of the "more is better" type, one should proceed as follows:

$$\overline{R}ij = 1 - \frac{Rj^* - Rij}{Rj^* - R*j} = \frac{Rij - R*j}{Rj^* - R*j} \forall ij \tag{1}$$

If the value is "less is better", the following formula is valid:

$$\overline{R}ij = 1 - \frac{Rij - Rj^*}{R*j - Rj^*} = \frac{R*j - Rij}{R*j - Rj^*} \forall ij \tag{2}$$

$\overline{R}ij$ = normalized value; Rj^* = ideal value; $R*j$ = worst value (anti-value); Rij = real value.

Using this normalization, the indicators became unitless and the values are converted to a scale, between 0 and 1. An index of sustainability degree is therefore generated, regarding the cases in analysis, 1 being the greater impact value and 0 the least impact value (Balteiro, 2004).

2.7 Aggregation

For better reading and comprehension, an aggregation of the various parameter values for each indicator is made. This aggregation can be done in a measured way to give a relative importance to the parameter on the overall picture, which is assumed when meeting the requirements of the project. However, because the decision-making control should remain with the user, this will not be implemented. The planner, architect etc. should retain the ability to decide whether one aspect is more important than the other, therefore, which parameters have greater importance is not determined in advance by the assessment tool.

The partial performance of the solution at each Indicator (IS) is calculated according to the aggregation method presented in equation:

$$ISi = \sum_{j=1}^{m} P_j \overline{R}_{ij}^P \forall i \tag{3}$$

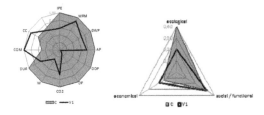

Figure 1. Example of the radar graph: the first with all the parameters and the second with summarized data.

IS is the average without weighting all parameters that have been previously normalized.

2.8 *Weighting*

As described earlier, in this assessment tool there are no weightings made, because there are no official recognized weighting systems. The user is given the liberty and responsibility to make the decisions on its own.

2.9 *Results*

One final summary is made before the results are displayed graphically. The three data, previously aggregated, are added resulting in the level of sustainability [5]:

$$ISi = \overline{R}_{i1} + \cdots + \overline{R}_{in} \qquad (4)$$

After obtaining the final value, it is possible to compare the materials (A and B), concluding one of three possible situations: A > B, A = B, A < B. If an analyzed material can be considered as a reference, more than the comparative result, one gets the notion of how much it is contributing to a choice of impact in terms of what is the national common practice.

2.10 *Graphical representation*

Reporting the results is extremely important to bridge the last objective of this type of analysis: simple and clear understanding of the results in order to ensure its practical applicability in the act of design. Due to the fact that these numbers are not intuitively interpreted, a data graphic representation is proposed as "radar". The results graphic communication illustrates the relationship between the two materials.

3 CASE STUDY

The need for thermal insulation of buildings has been growing to improve the thermal comfort and energy efficiency. To meet the new requirements, the materials used must improve their performance. This need is obvious, since the requirements are becoming increasingly more demanding. The requirements relate not only to thermal conductivity, but also to indoor air quality and to the impact on the environment. The quality of an insulation material depends on its ability to adapt to the type of construction and the national, regional or even local traditions. A material often used in certain areas, could not fit in others places, although they could scientifically replace each other (Labrincha, 2006).

This example is selected because the facade insulation has a great influence on energy efficiency, which is suitable for new construction and existing buildings. A building envelope consists mainly of exterior walls and their insulation and provides a very efficient energy saving measure.

EIFS system (Exterior insulation finishing system) is distinguished from the following points:

- The thermal bridges are reduced, thus thinner insulation can be used and resources saved.
- Thermal inertia of the interior space is maintained.
- In new buildings, it is possible to use thinner walls, which brings more space and less weight, thus saving the structure.
- For existing buildings, the renovation can be carried out without disturbing the residents.
- A bigger diversity to build the facade, for example various grain sizes, colours and textures of plaster (normal, lime or clay plaster), as well as wood, metal, or other cladding combinations.

For this case study three different insulation materials were selected and compared: an organic material, natural cork, and two synthetic materials, polystyrene foam (Expanded polystyrene—EPS) and extruded polystyrene foam (XPS). All three insulation materials can be placed in the air space of a double wall, as an external insulation, or used in suspended ceilings, raised floors and interior walls.

For this investigation, a fictitious example is carried out: the selection of insulation material for an external wall, using an EIFS system.

3.1 *EPS (A)*

The expanded polystyrene (EPS), which is generally known as polystyrene, is produced of expandable styrene polymers by the polymerization of styrene and the incorporation of blowing agents (in general pentanes), and sometimes, certain products can be added to improve the properties of polystyrene's combustion behaviour (Labrincha, 2006).

Polystyrene (PS) is one of the major thermoplastics. Polystyrene was first described in 1839, and the production of plastics began in 1930 at IG Colours in Germany. The major sources of raw materials are petroleum and coal. The polystyrene is supplied as granules to processors that continue to process these products. This process takes place by insulating materials during the extrusion to XPS, or by thermoforming or by blow moulding to EPS. In the preparation of intermediates hazardous materials with a considerable potential risk are involved: benzene and ethylene. Producing styrene is a hazardous activity that is suspected of being carcinogenic. Polystyrene itself is not toxic. Studies have shown that immediately after the preparation of ethyl benzene and styrene from polystyrene, products may outgas in small amounts. These emissions, however, sharply decrease within a few days. Building materials made of polystyrene are considered very stable; long-term stabilizers must be added to prevent degradation from UV light. Products made from polystyrene are not resistant to solvents and petroleum. This must necessarily be considered when applying a coating or in bonding (BVBS, 2012).

In construction EPS is mainly used as thermal and sound insulation.

3.2 Cork (B)

Cork is a material that has accompanied mankind since immemorial time. Its main application in construction is in the Mediterranean countries where it grows naturally. Historically, the main applications were in shipping and construction of seals. The market has expanded greatly since the beginning of the twentieth century, especially through the agglomerate products that can be created from cork. Cork is now used in many different areas, from the stopper up to the aircraft industry, and it is known for its high thermal and acoustic insulation (Gil, 2005).

Cork is a resource that covers the trunks and branches of the cork oak (Quercus suber L.), a related species of oak and the only tree that is able to regenerate its bark, after it was removed. Experienced professionals harvest Cork at a distance of about nine years without harming the tree. Cork is considered an eco-efficient material, because during its lifecycle any residual material is reused in the manufacturing process as biomass to produce new products. There is no waste during the production process, for example of pure expanded cork, in which are needed only hot water vapour and no chemical additives in order to combine the grains. Even the cork powder, obtained during the processing, is used as a biomass for the production of steam or electric power. Virgin cork is ground into gran-

ules, at 350 °C to 370 °C hot water vapour, then it is stretched and compressed. The granules leave a resin (suberic), which allows a compound of the granular particles without the use of foreign additives. The result is a block of expanded cork, after a cooling and stabilizing phase, which it is cut into plates of different thickness. Expanded cork is ideal for heat, sound and vibration insulation (Chiebao, 2011).

3.3 XPS (C)

Extruded polystyrene foam (XPS) is a hard material with a closed cell structure, which is prepared by swelling and extrusion of polystyrene or its copolymers. XPS is part of the group of synthetic organic insulation foam plastics.

The insulation effect is generated by the inclusion of still air or propellant gases in the foamed cells. XPS is produced in a continuous extrusion process. After passing through a cooling zone with downstream equipment, the strand is cut into sheets and the edging is made. The foam membrane is obtained on the top surfaces of the plates. For special applications (eg. to improve the adhesive bonding to concrete, mortar, adhesives) the foam skin is removed. The plate will then have a rough surface. [15]

3.4 Review and results

The three materials were examined with the assessment tool with the same parameters. To calculate the TT (transport type/distance to the project) parameters, Berlin was chosen as the project site. The insulation boards have standard sizes and a density of 30 mm.

3.5 Interpretation of results

The total scores of the two simulations demonstrate that both cork and XPS insulation have a

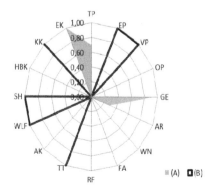

Figure 2. Comparison between (A) (B) with the individual parameters.

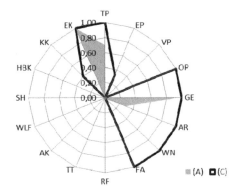

Figure 3. Comparison between (A) (C) with the individual parameters.

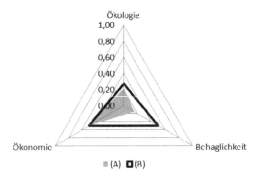

Figure 4. Comparison between (A) (B) after the combining of the results.

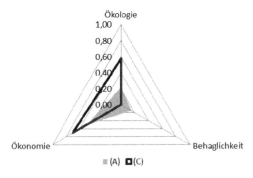

Figure 5. Comparison between (A) (C) after the combining of the results.

worse result compared to EPS, but both for different reasons.

Although cork is generally considered a sustainable building material, this comparative assessment demonstrates otherwise. Cork would be environmentally better than the EPS if the transport/distance parameter was not considered. In addition,

since sustainable rating does not take into account the ecological side only, cork was worst in the economy and comfort evaluation. Its thermal conductivity is greater than the other two materials. To improve this deficit, one has to increase the thickness of insulation boards, therefore consuming more raw materials, which in a renewable resource would cause no major concerns. Nevertheless, this is a decision that belongs to the designer. Even when it comes to security parameters Cork has a lower rating.

XPS is weaker in the ecological and economic parameters, but it has a very good comfort rating.

In the final result the reference material—EPS obtained the best evaluation, since it has a more balanced ratio between the three areas.

The rating system has proved that partial and overall results have different advantages. The partial results can tell the user how the final grade is made, which is very important in this particular case. The overall grade leads to a global sustainability degree, which compared to the reference, shows to the user if the materials have better or worse performance.

4 CONCLUSIONS

The integration of such an instrument, acting as a sustainability assessment tool for building materials, is useful and necessary for a holistic building planning. This rating system has proved to be functional and practical. Building materials known as organic are therefore not always the most sustainable. This versatile approach is very significant because the choices turn out to be more extensive.

The practical implementation of this rating system is simple, since the data needed are obtained in public databases, or completed by semaphores developed specifically for the tool.

The innovation of such an assessment tool is its potential for implementation in planning practice. The assessment of sustainability is a very complex process, but this system simplifies fundamental issues for the user. This is particularly important because the user can freely carry out all comparisons that interest him and can independently make decisions about building materials' selection.

It offers the possibility to make an informed choice. Currently, the user is responsible for deciding what is more important for his case/project. Different choices could be made depending on the building type. On a temporary building it is probably more important the environmental and economic aspect, but in a building which should stay at least 50 years, the comfort is of utmost importance. Even so these decisions are not made by the assessment tool, but taken by the designer.

To make such a decision, the user can read the partial results and decide which property is more important to his case. This responsibility always lies with the user. The assessment tool creates only comparative results, so there is no general assessment or classification of the building materials.

Future expansion could be achieved by creating simple and free Software so that the use of the tool may be even easier.

REFERENCES

Balteiro, L. (2004); In search of a natural systems sustainability index, Ecological Economics 49, 401–405, URL: http://www.is.cnpm.embrapa.br/bibliografia/2004_In_search_of_a_natural_systems_sustainability_index.pdf (05.08.2012).

Bauer, M. (2001); Nachhaltiges Bauen—Zukunftsfähige Konzepte für Planer und Entscheider, in: Deutsches Institut für Normungen e. V. (Hrsg.), Ganzheitliches Planen, Beraten und Bauen, Chapt. 6, Berlin—Wien—Zürich.

Bundesministerium für Verkehr, Bau und Stadtentwicklung, Baustoffdatenbank Ökobau.dat; URL: http://www.nachhaltigesbauen.de/oekobaudat/(05.08.2012).

Bundesministerium für Verkehr, Bau und Stadtentwicklung (2011); Methodische Grundlagen—Ökobilanzbasierte Umweltindikatoren im Bauwesen, PE International, Deutschland, URL: http://www.nachhaltigesbauen.de/fileadmin/oekobaudat/pdf/Methodische-Grundlagen__Version_2-2011_.pdf (05.08.2012).

Bundesministerium für Verkehr, Bau und Stadtentwicklung, Tabelle—Nutzungsdauern von Bauteilen zur Lebenszyklusanalyse nach BNB; URL: http://www.nachhaltigesbauen.de/fileadmin/pdf/baustoff_gebauededaten/BNB_Nutzungsdauern_von_Bauteilen__2011-11-03.pdf (05.08.2012).

Bundesministerium für Verkehr, Bau und Stadtentwicklung, Ökologisches Baustoffinformationssystem—WECOBIS; URL: http://www.wecobis.de/jahia/Jahia/Home/Grundstoffe/Kunststoffe_GS/Polystyrol_GS (05.08.2012).

Berger, B. (2009); Ecology of Building Materials; 2° Edition, Architectual Press Oxford; UK.

Chiebao, F., Technischer Leitfaden—Für Kork als Bau- und Dekorationsmaterial, APCOR—Associação Portuguesa de Cortiça, Santa Maria de Lamas, Portugal, 2011, URL: http://apcor.pt/userfiles/File/Publicacoes/Manual_MCD_DE.pdf (05.08.2012).

DAA—Deutsche Auftragsagentur, Dämmen uns Sanieren; URL: http://www.daemmen-und-sanieren.de/daemmung/daemmstoffe/xps-daemmung (05.08.2012).

DIN 4102:1998-05, Brandverhalten von Baustoffen und Bauteilen—Teil 1: Baustoffe; Begriffe, Anforderungen und Prüfungen.

Gil, L. (2005); A cortiça como material de construção, APCOR—Associação Portuguesa de Cortiça, Santa Maria de Lamas, Portugal, URL: http://www.apcor.pt/userfiles/File/Caderno%20Tecnico%20F%20PT.pdf (05.08.2012).

Hegner, H. (2011), Nachhaltiges Bauen—Zukunftsfähige Konzepte für Planer und Entscheider, in: Deutsches Institut für Normungen e. V. (Hrsg.), Nachhaltiges Bauen in Deutschland—Instrumente und Projekte des Bundes, Chapt. 3, Beuth Verlag, Berlin—Wien—Zürich.

Herkommer, E. (2004); Der Aktuelle Begriff, in: Wissenschaftliche Dienste des Bundestages (Hrsg.), Nachhaltigkeit, Nr.06/2004, Bundestag, Berlin.

Labrincha, J. (2006); Sub Projecto de Isolamento Térmico, 1° Relatório de Progresso, Universidade de Aveiro, Portugal, URL: http://www.aveirodomus.pt/resources/xFiles/scContentDeployer/docs/Doc266.pdf (05.08.2012).

Mateus, R. (2004); Novas tecnologias construtivas com vista à sustentabilidade da construção, Dissertation, Universidade do Minho.

Mateus, R. (2006); Tecnologias construtivas para a sustentabilidade da construção, 1. Auflage, Edições Ecopy, Ermesinde, Portugal.

Santos, C. (2006); ITE50, LNEC, Portugal.

Green Design, Materials and Manufacturing Processes – Bártolo et al. (eds)
© *2013 Taylor & Francis Group, London, ISBN 978-1-138-00046-9*

Set up of a European LCA building rating methodology within the open house project

J. Gantner, K. Lenz & H. Krieg
Fraunhofer IBP, Stuttgart, Baden-Württemberg, Germany

ABSTRACT: Currently, there is no common methodology for a Life Cycle Assessment (LCA) rating within a common sustainability scheme of buildings in a European context. Due to local and regional differences, comparisons between different buildings in different countries are so far not yet possible. An increasing need for a transparent sustainability classification method of buildings on a European level induced the European Commission to fund the project "OPEN HOUSE". The local and regional differences in legal framework, energy demand calculation methodology, weather conditions, etc. made it a prerequisite that the developed LCA methodology enables comparisons of European buildings on the basis of national ratings. Special focus was paid to the development of national benchmarks for the construction and use phase for both a "quick and basic assessment" as well as a "complete assessment". Summarizing, a LCA methodology was created that serves for comparisons and ratings of different buildings within different local contexts.

1 INTRODUCTION

In the last few years Life Cycle Assessments (LCA) (ISO14040 (2006) and ISO14044 (2006)) in the building sector has increasingly gained importance. LCA is applied in the certification of sustainable construction, it provides basis for environmental declaration of building products and progressively serves as well for decision support. In European research projects focusing on building energy efficiency, Life Cycle Assessment is increasingly used as a measure for environmental evaluation and comparison of research results. Unfortunately, there is currently no common methodology for an LCA rating within a common sustainability scheme for buildings in a European context. Due to different local conditions like weather, legal frameworks, energy demand calculation methodologies, etc. it is so far not possible to compare different buildings in different local context with each other. To resolve this problem the European Commission funded the project "OPEN HOUSE" which aims at setting up a transparent and common European building assessment methodology based on existing building certification schemes. This methodology will be peer reviewed by sustainability assessment experts. LCA is here an important part of an "OPEN HOUSE" compliant assessment.

2 LCA METHODOLOGY

The development of a transparent methodology for assessing LCA results of buildings in a European

context was realized based on decisions agreed in workshops and on results from questionnaires (Open House, 2012). This enabled to get the full picture of people's opinions with different knowledge or scientific background and gave people the possibility to take part within the decision process.

First of all, several workshops were held to gather different perspectives on building sustainability or sustainability criteria (e.g. LCA) and how they should be included and weighted for OPEN HOUSE.

Furthermore, questionnaires were developed and sent to the 76 participating case study partners in various countries all over Europe like France, Sweden, Finland, Slovenia, Greek, Cyprus, Czech Republic, Poland, Austria, Denmark, Germany and Italy in order to get a broad picture of local conditions, legal framework, etc. which are important for the evaluation and for the comparison of buildings all over Europe. Within the questionnaire the main topics of interests such as implementation of EU standards, directives or regulations, national standards and regulations, national basic and best practices and building specifics were queried and assessed.

All in all it could be found a multitude of different national particularities, which made it impossible to compare buildings of different countries with each other. Often, crucial LCA input data like energetic calculations (based on national Energy Performance of Buildings Directive (EPBD) implementation) where not established or varying from country to country. Also the calculation methodologies for building areas were different all over Europe. On the other side there are national or even regional legal requirements regarding aspects like fire protection, earth quake protection or others that can have a major impact on the LCA results of a building. For example: it is necessary to protect buildings in Greece from earth quakes by using a very high amount of concrete and reinforced steel in the foundation, leading to higher environmental impacts if compared to what is considered a state-of-the-art foundation in other countries.

Due to the mentioned different local circumstances in every country, especially with regard to the different national EPBD methodologies and building regulations, a classification of buildings within different national contexts is a pre-requisite not only for an environmental but a consistent building assessment scheme. The LCA methodology developed in open house allows the comparison of buildings all over Europe on the basis of national ratings (e.g. with benchmarks of existing building certification schemes).

Starting from the building construction and the building operation the OPEN HOUSE LCA methodology developed, defines two different reference benchmarks (see Fig. 1). The first one (Cref), serves as reference for benchmarking the production of materials, the refurbishment of the built-in materials and the specific End-of-Life. With the second one (Oref), the operational energy use of the building is assessed. The "Construction reference benchmark" (Cref) is based on an average European value derived from the evaluation of all case study buildings. The "Operational phase benchmark" (Oref) can be selected in two different ways. If national benchmarks or national defined limited end energy values of the national EPBD are available, these values should be used. If such values are not available, an average European value shall be used as default value, derived out of the assessment of the OPEN HOUSE case studies. The "constructional reference value" and the

"operational reference value" are then summed up to a "total reference value".

According to the total reference values for each environmental indicator the corresponding target and limit values are calculated by an indicator-specific factor on the reference value (see Fig. 2). For benchmarking of a specific building, the specific LCA results for the building construction (production, refurbishment, End-of-Life) and the building operation have to be defined and summed up. This "building specific value" is compared against the "total reference value" and ranked within the boundaries of "limit reference value" and "target reference value". The individual building is rated based on this limit and target value (see Fig. 3). Following evaluation credits or points are assigned:

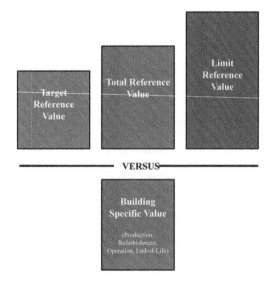

Figure 2. Development of target and limit values based on the reference value.

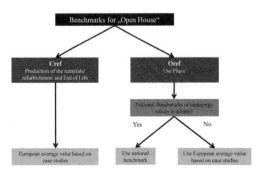

Figure 1. Setting of benchmarks in OPEN HOUSE.

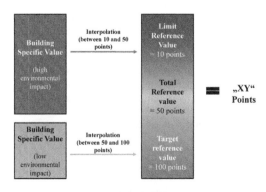

Figure 3. Evaluation of the building.

100 points for reaching the target reference value, 50 points for reaching the reference value and 10 points for reaching the limit reference value. If the building value lies in between, the evaluation points are determined via linear interpolation. This approach is used for defining and calculating evaluation points for each defined single environmental criteria (e.g. global warming, acidification, eutrophication etc.) for which a respective environmental indicator is set up within the OPEN HOUSE sustainability assessment scheme.

In the end, the results for all environmental indicators are weighted using pre-defined weighting factors defined in the OPEN HOUSE assessment methodology and summarized in a global score. This creates a single score value that rates the overall sustainability of a building.

3 RESULTS FROM THE CASE STUDIES

The case studies helped to spot the main differences between the countries and the most important parameters for building LCA. Special focus was paid to the development of national benchmarks for the construction and use phase for both a "quick and basic assessment" as well as a "complete assessment". In order to understand the relevance of certain constructional elements and building materials, sensitivity analysis was performed to verify the necessity of inclusion or exclusion of these elements.

As an outcome of these sensitivity analyses the following lifecycle stages and building elements were assessed.

Table 1. Included/excluded lifecycle stage according to EN15978 (2011).

Lifecycle stage	Included	Excluded
A1-A3: Product stage	X	
A4-A5: Construction process stage		X
B1: Use		X
B2: Maintenance	X	
B3: Repair	X	
B4: Replacement	X	
B5: Refurbishment	X	
B6: Operational energy use	X	
B7: Operational water use		X
C1: De-construction/ demolition		X
C2: Transport		X
C3: Waste processing	X	
C4: Disposal	X	
D: Benefits and loads beyond the system boundaries	X	

Table 2. Included/excluded materials and building elements (open house 2010).

Materials/building elements	Included	Excluded
Exterior walls	X	
Roof	X	
Ceilings (incl. floor coverings/coatings)	X	
Floor slab incl. flooring, floor coverings; floor slab above air	X	
Foundations	X	
Interior walls incl. coatings and supports	X	
Heat generation units	X	
HVAC infrastructure (e.g. cable, tubes, etc.)		X
Internal transport (e.g. lifts)		X
Water and sewage systems		X
Electrical distribution system		X

Some of the building components and their lifecycle stages are currently still excluded due to missing LCA data or expected minor relevance. So far some of the information—needed for the inclusion of certain lifecycle stages (e.g. transport to construction side, energy consumption on site, etc.)—is asked for in the LCA questionnaires, but during the project only few case studies were able to provide this information. For consistency reasons this information was not considered in the LCA assessment, but is nevertheless included in the documentation.

One major outcome on benchmarking was the determination of reference values for each environmental indicator and each level of completeness for the LCA (Quick and Basic, Complete) based on the results of the case study buildings. An LCA software tool supported both "Quick and Basic" as well as "Complete" assessment with easy functionality even for non-LCA experts.

For a "Quick and Basic" assessment a construction catalogue with typical European constructions was furthermore developed and implemented in the LCA software tool, offering case study for users to choose a construction that fitted their building best and to define their surface areas. For the operation phase of the building, the energy sources, the energy demand and potential renewable generated energy had to be defined. For a "Complete" assessment, the pre-defined constructions within the LCA tool were not sufficient. Therefore, the case study users had to model specifically their constructions by detailed defining single layers (e.g. concrete, mineral wool etc.) or service life of constructional elements. For the building operation, the same information as for the "quick and basic" assessment was collected.

Tables 3 and 4 show the detected European average reference values for the case studies.

The results of the case studies show a slight increase of impact from "Quick and Basic" to "Complete" assessment due to a more building specific and therefore more detailed configuration.

Due to the used European LCA database (ESUCO) not only European averages could be assessed, but also country-specific averages could be extracted from the case studies. It has to be noted that the country-specific average values for the

Table 3. European average reference values for "Quick and Basic".

"Quick and Basic"—Europe	Cref	Oref	Total
GWP			
[kg CO$_2$E/(m²*y)]	6.5	33.2	39.7
ODP			
[kg R11E/(m²*y)]	3.1E-07	5.8E-06	6.1E-06
AP			
[kg SO$_2$E/(m²*a)]	2.4E-02	1.8E-01	2.0E-01
EP			
[kg PO$_4^{3-}$E/(m²*y)]	2.7E-03	7.4E-03	1.0E-02
POCP			
[kg C$_2$H$_4$E/(m²*y)]	2.9E-03	1.1E-02	1.3E-02
Penr			
[kWh/(m²*y)]	24.2	168.1	192.3
Pere			
[kWh/(m²*y)]	4.8	15.3	20.1
PEtot			
[kWh/(m²*y)]	29	183.4	212.4

Table 4. European average reference values for "Complete".

"Complete"—Europe	Cref	Oref	Total
GWP			
[kg CO$_2$E/(m²*y)]	10.6	35.2	45.8
ODP			
[kg R11E/(m²*a)]	3.8E-07	6.7E-06	7.1E-06
AP			
[kg SO$_2$E/(m²*a)]	3.3E-02	2.1E-01	2.4E-01
EP			
[kg PO$_4^{3-}$E/(m²*a)]	3,30E-03	8.3E-03	1.1E-02
POCP			
[kg C$_2$H$_4$E/(m²*a)]	3.4E-03	1.2E-02	1.5E-02
Penr			
[kWh/(m²*a)]	33.1	196.5	229.6
PEre			
[kWh/(m²*a)]	4.9	21	25.9
PEtot			
[kWh/(m²*a)]	38	217.5	255.5

different study types are based on a very low number of buildings and also different types of buildings were not considered. Therefore, the results mentioned in Tables 5 and 6 are not to be considered as precise values but as first approach in the direction of developing country-specific reference and benchmark values.

The results for both—the European and the national benchmarks respectively reference values—should be seen as a starting point for ongoing improvements. In some cases, national LCA benchmarks for some environmental indicators are already existing (e.g. DGNB (2013), HQE (2013), LEED (2013), BREEAM (2013), VERDE (2013)) but for many European countries this national reference values are still missing. This challenge forms the bases for harmonization tendencies and

Table 5. Excerpt of country-specific average reference values for "Quick and Basic".

	Cref	Oref	Total
Greece			
GWP			
[kg CO$_2$E/(m²*y)]	8	32.5	40.5
ODP			
[kg R11E/(m²*a)]	3.2E-07	6.2E-06	6.5E-06
AP			
[kg SO$_2$E/(m²*a)]	2.6E-02	1.9E-01	2.2E-01
EP			
[kg PO$_4$3-E/(m²*a)]	3.0E-03	7.4E-03	1.0E-02
POCP			
[kg C$_2$H$_4$E/(m²*a)]	3.6E-03	1.1E-02	1.5E-02
Penr			
[kWh/(m²*a)]	28.2	169.4	197.6
PEre			
[kWh/(m²*a)]	4.2	10.7	15
PEtot			
[kWh/(m²*a)]	32.4	180.1	212.5
Poland			
GWP			
[kg CO$_2$E/(m²*y)	7.3	53.9	61.2
ODP			
[kg R11E/(m²*a)]	3.5E-07	1.1E-05	1.1E-05
AP			
[kg SO$_2$E/(m²*a)]	2.7E-02	3.4E-01	3.6E-01
EP			
[kg PO$_4^{3-}$E/(m²*a)]	2.9E-03	1.3E-02	1.6E-02
POCP			
[kg C$_2$H$_4$E/(m²*a)]	3.3E-03	1.9E-02	2.2E-02
Penr			
[kWh/(m²*a)]	25.3	281.2	306.5
PEre			
[kWh/(m²*a)]	5.9	18.8	24.7
PEtot			
[kWh/(m²*a)]	31.2	300.0	331.2

Table 6. Excerpt of country-specific average reference values for "Complete".

	Cref	Oref	Total
Greece			
GWP			
[kg $CO_2E/(m^2*y)$]	12.1	32.5	44.6
ODP			
[kg $R11E/(m^2*a)$]	4.1E-07	6.2E-06	0.0
AP			
[kg $SO_2E/(m^2*a)$]	4.9E-02	1.9E-01	0.2
EP			
[kg $PO_4^3\text{-}E/(m^2*a)$]	3.3E-03	7.4E-03	0.0
POCP			
[kg $C_2H_4E/(m^2*a)$]	4.4E-03	1.1E-02	0.0
Penr			
[kWh/(m^2*a)]	31.7	169.4	201.0
PEre			
[kWh/(m^2*a)]	2.9	10.7	13.6
PEtot			
[kWh/(m^2*a)]	34.6	180.1	214.6
Poland			
GWP			
[kg $CO_2E/(m^2*y)$]	14.3	53.9	68.2
ODP			
[kg $R11E/(m^2*a)$]	4.0E-07	1.1E-05	0.0
AP			
[kg $SO_2E/(m^2*a)$]	4.2E-02	3.4E-01	0.4
EP			
[kg $PO_4^3\text{-}E/(m^2*a)$]	4.2E-03	1.3E-02	0.0
POCP			
[kg $C_2H_4E/(m^2*a)$]	4.3E-03	1.9E-02	0.0
Penr			
[kWh/(m^2*a)]	34.6	281.2	315.8
PEre			
[kWh/(m^2*a)]	1.5	18.8	20.3
PEtot			
[kWh/(m^2*a)]	36.0	300.0	336.0

research projects like EeBGuide and SBA common metrics (SBA, 2012), which try to establish a commonly agreed LCA methodology and a understanding for LCA data sets. Especially a standardized approach regarding included lifecycle stage and materials is described in (EeBGuide, 2012) and followed up on in Open House to give other research projects the possibility to pick up the results and develop them even further in a consistent and transparent way.

4 OUTLOOK

The OPEN HOUSE LCA methodology proofed to provide a consistent methodology that focus on a regional respectively national scale, but at the same time gives a possibility to compare buildings on European level based on a best-in-class approach. One outcome of the case studies is a first step in the direction of European LCA benchmarks. The small number of case studies does not yet allow statistically significant statements, but the average European benchmarks developed within OPEN HOUSE gives a first range. Further research should be taken in this field in order to increase the number of case studies and therefore develop better benchmarks, but also improve the number and quality of available LCA data sets both on a European and on a national level. Furthermore, especially the calculation of the building surfaces and national EPBD versions should be also taken into account in later research projects.

Both the methodology developed as well as the reference values are milestones on the way to a European Construction Assessment scheme.

REFERENCES

ASIEPI 2010. ASsessment and Improvement of the EPBD Impact (ASIEPI), Comparison of Energy Performance Requirement Levels: Possibilities and Impossibilities. http://www.buildup.eu/publications/9099.

BREEAM 2013. http://www.breeam.org/.

Casals, X.G. 2006. Analysis of Building Energy Regulation and Certification in Europe: Their Role, Limitations and Differences. Energy and Buildings. Vol. 138. No. 5. 2006. pp. 381–392. doi:10.1016/j.enbuild.2005.05.004.

DGNB 2013. http://www.dgnb.de/de/.

EeBGuide 2012. Guidance Document, Part B: Buildings. www.eebguide.eu.

EN15804 2011. Sustainability of construction works—Environmental product declarations—Core rules for the product category of building products. CEN—European Committee for Standardization. Brussels: CEN—CENELEC 2011.

EN15978 2010. Sustainability of construction works—Sustainability assessment of buildings—calculation method. CEN—European Committee for Standardization. Brussels: CEN—CENELEC 2010.

HQE 2013. http://assohqe.org/hqe/.

ILCD 2010. European Commission—Joint Research Centre—Institute for Environment and Sustainability (Ed.): ILCD Handbook. General guide for life cycle assessment: detailed guidance. First edition. Luxembourg: Publications Office of the European Union 2010. ISBN: 978-92-79-19092-6.

ISO14040 2006. Environmental management—Life cycle assessment—Principles and framework (ISO 14040:2006), German and English version EN ISO 14040:2006.

ISO14044 2006. Environmental management—Life cycle assessment—Requirements and guidelines (ISO 14044:2006), German and English version EN ISO 14044:2006.

LEED 2013. http://new.usgbc.org/leed.

Open House 2010. D1.2.1 Assessment of methodologies, normative, standards and guidelines for sustainability of buildings at national, European and International level.

Open House 2012. Complete Assessment Report Environmental Quality—LCA Indicators.

Open House 2013. http://www.openhouse-fp7.eu/.

SBA 2012. Sustainable Building Alliance. Sustainable Building Alliance Research Project: "Piloting SBA Common Metrics". Technical and operational feasibility of the SBA common metrics, 2012.

VERDE 2013. http://www.gbce.es/en/pagina/verde-certificate.

Green Design, Materials and Manufacturing Processes – Bártolo et al. (eds)
© 2013 Taylor & Francis Group, London, ISBN 978-1-138-00046-9

Evaluating the impact of glass and PET packaging for bottled water

H.A. Almeida, C.A. Ramos, H. Bártolo & P. Bártolo
Centre for Rapid and Sustainable Product Development, Polytechnic Institute of Leiria,
Marinha Grande, Portugal

ABSTRACT: The increasing growth of human population and the global development of products and services, with no concern for its environmental impacts, are causing the degradation of the planet. The implementation of sustainable methodologies can contribute to more ecological industrial practices. A computational tool was developed to support sustainable decisions at the conceptual design phase, based on eco-design principles, to get a better understanding of the environmental impact of product manufacturing. This tool enables product designers and clients to investigate and compare different solutions for each product considering the entire life cycle of a product. This work investigates the bottled water sector in Portugal, which has grown dramatically in the last decade. Last year, this sector alone bottled 831.08 million litres of sparkling and natural water. This research work compares two types of water bottle packages, namely glass and PET bottles, regarding all possible end-of-life scenarios, evaluating its environmental impact.

1 INTRODUCTION

An increasing growth rate of human population and the global development of products and services, with no concern for its environmental impacts, are causing the degradation of the planet. The implementation of sustainable methodologies can contribute to the implementation of more ecological practices (Berry, 2004). A computational tool was developed to support sustainable decisions at the conceptual product design phase, based on eco-design principles, to get a better understanding of the environmental impact of current product manufacturing. This tool allows designers and clients to investigate and compare different solutions for each product, taking into account its entire life cycle, either at a conceptual phase or on-going production one.

The bottled water sector in Portugal has grown dramatically in the last decade. In 2012, this sector alone, bottled 831.08 million litres of sparkling and natural water. Niccolucci et al. (2011) stated that water is not only essential for the life of humans and ecosystems, but also a strategic economical resource. It is of particular interest for both human health and the economic and political management of resources (Sen and Altunkaynak, 2009; Zhao et al., 2009; Pfister et al., 2009).

Drinking water is available to consumers in two alternative ways: from the tap or in a bottle. The latter is the most commonly used in developed countries (IBWA, 2010). A positive correlation can be observed between economic growth and bottled water consumption, explained by two main factors:

population increase and economic development (Ferrier, 2001). The last one in particular can have three effects: raising the per capita income, opening up consumption markets to external operators and increasing usage of marketing strategies (IBWA, 2008; Doria, 2006).

This high global growth should be monitored from an environmental and social perspective, as the impact of bottled water consumption can be expected to increase in the future (Gollier et al., 2000). Bottling, trading and transporting water all over the world produces a considerable environmental impact, including pollution, contamination, climate change and depletion of resources (Gleick, 2006; Freire et al., 2001).

This industrial sector frequently overlooks certain aspects, such as the global contributions of water towards the consumption of freshwater resources. Water is both the raw material and the produced good within the entire production and supply chains (Niccolucci et al., 2011). Another aspect relates to the packaging of bottled water, in this case can be either Polyethylene Terephthalate plastic (PET) or glass bottle. This research paper compares between two types of water bottle packages, regarding all possible end-of-life scenarios in terms of its environmental impact evaluation.

2 BOTTLED WATER INDUSTRY

The bottled water industry started in early 1980s, shown a significant increase since then. The global bottled water industry became highly profitable over the past decade and its growth took place worldwide, particularly in Europe and North America.

The bottled water industry is very dynamic and numerous bottled water companies compete in this market. A number of factors contributed to the popularity of bottled water. Many people drink bottled water today, simply because they prefer its taste to tap water (e.g., taste, odor, color) or its higher purity is assured. Consumer focus on healthy-eating habits, with a significant emphasis on sufficient hydration, bottled water is increasingly being considered as a natural product and a vital part of a healthy lifestyle.

Demand is also rising as a result of greater portability and accessibility, via convenience stores, gas stations, supermarkets, foodservice outlets like restaurants and vending machines. The consumption of bottled water has moved the product beyond the niche market into the mainstream, as bottled water has become a basic staple for the society.

In the context of the European economy, the consumption of bottled water (mineral and spring water) represents 45% of the beverage sector, as a result of consumer's choice, intensified by the relevant concerns related to a healthier and balanced diet. Consumers recognize the importance of water consumption, as it a 100% natural product with exceptional value and quality. Figure 1 shows the evolution of the most recent bottled water consumption in Portugal, according to the Portuguese Agency of Statistics.

In 2011, the turnover of the Portuguese sector of mineral waters was approximately 223 million euros. Currently, this sector accounts for 2 percent of the total Portuguese food industry, ensuring a high number of employment and contributing to reduce regional disparities in some regions of Portugal, since the bottling plants cannot be relocated, they must be placed near the springs.

3 BOTTLED WATER PACKAGING

In the context of sustainable development, since the 80's, a set of measures were implemented aiming at minimizing the environmental impact of the water sector and its packaging. The choice of bottled water packaging material is concerned with environmental issues. This industry is highly sensitive to the need of preserving nature, while conciliating the interests of the environment, consumers and the industry itself.

Bottle packaging enables to preserve the quality of the product for the final consumer, ensuring food security, as well maintaining its original properties. In the water sector, the materials used for the primary packaging are glass and PET. Figure 2 illustrates the production of glass and PET packages in Portugal. It is possible to observe that, over the years, the number of bottles between glass and PET is balanced.

The quality of bottled water packages is essential to ensure the quality and the purity of the water. Glass packages can be prepared for a single use or either designed for reuse. Glass is especially used in the channel HORECA, especially in food retail outlets associated with highly prestige brands. Companies and brands offer attractive packaging designs, modern and functional, associating the product with high quality water. Bottled water is sold in a variety of packages: 25 cl, 50 cl and 1 litre. The PET package is the preferred material of the consumer which supplies the traditional food shops and supermarkets. The PET package is lighter making it easier to transport. Bottles are usually of 33 cl, 50 cl, 1 litre, 1.5 litre, 2 litres or 5 litres l of water, and its sale is done individually or in packs.

The challenge for this industrial sector is to reduce the weight of packaging, while preserving the quality and characteristics of the water. This challenge can bring significant environmental benefits in terms of production and transportation. Increasingly, consumers and manufacturers want to know how the various types of containers

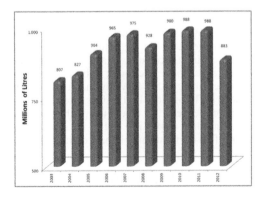

Figure 1. Evolution of consumption of bottled mineral water in Portugal (source: APIAM).

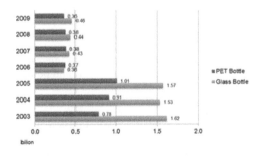

Figure 2. Manufacture of glass and PET packages in Portugal.

actually stack up against one another, in order to minimize storage space, which also decreases its environmental impacts. For instance, more water can be transported in the same vehicle.

Regarding the production of glass or PET bottles, the local industry at Marinha Grande, a Portuguese industrial town, is very significant. Marinha Grande is recognized worldwide for its important industries regarding the manufacturing of Glass and Plastic Injection Moulding. These two different industries provides two different packaging solutions for a very highly demanding industry, in terms of both quality and quantity.

4 D4E COMPUTATIONAL TOOL

The computational tool called Design for Environment (D4E) is based on eco-design principles and it can be easily adapted to all product design practices within an industrial environment. The D4E system uses a Life Cycle Assessment (LCA) approach to evaluate the environmental impact of a product and it enables to quantify all inputs and outputs in a qualitative and quantitative way at all life cycle stages (Santos et al., 2012). It is important establish the idea of a product life cycle. This is generally conceived of as a materials flow process that starts with extraction of raw materials from the earth and ends with the disposal of the waste products back to the earth. The general stages are: material extraction, primary processing and refining, manufacturing, product distribution, use, and final disposition (Gutowski, 2004). This tool uses eco-indicators to quantify the environmental impact for each LCA product. These eco-indicators allow measuring the environmental impact of a material or process throughout its life cycle, considering the materials used along the manufacturing process, its energy consumption, transport and final destination (Finnvedena, 2009).

The first window entitled Life Cycle identifies the project, i.e., the general information on the project is filled out; including name, date, reference, author, company and a project brief summary, as shown in Figure 3a.

The second window entitled Production, refers to the data from each product component in the Production stage including a general description of each component production, the number of components for each type, material composition, mass, manufacturing process and the quantity of processed components, as illustrated in Figure 3b. At this stage, it is also possible to introduce the volume data for each product component, which can be determined by either the 3D product analysis, or the measure of the mass of each component using an analytical balance. These data will be added to the respective table in this window.

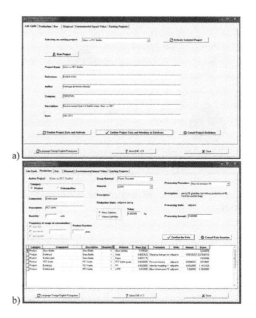

Figure 3. a) Life cycle window; b) Production window.

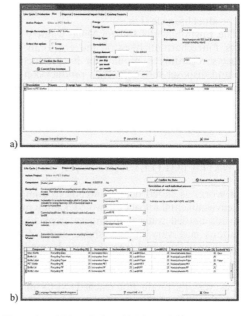

Figure 4. a) Use window; b) Disposal window.

The Use window, refers to the use stage and including the consumed energy or/and the use transport in its distribution phase. Transport data refers to the transport type and distance, as shown in Figure 4a.

347

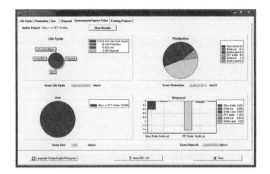

a) b)

Figure 6. a) PET bottle (330 ml) and b) Glass bottle (250 ml).

Figure 5. Example of environmental impact value window.

The Disposal window refers to the final destination of each product component (Fig. 4b). In this step, users can simulate the final destination each product component including recycling, incineration, landfill and waste. The sum of the percentage for each component must be equal to one hundred per cent ($\Sigma = 100\%$), no matter the values obtained for each disposal proportion.

Subsequently, it was possible to visualize charts and environmental impact results for each step of the product lifecycle (Fig. 5).

The Environmental Impact Value window displays four chart analyses of the environmental impacts regarding a product in its production, use and disposal stages for each product component. The final value of the product's environmental impact will be the sum of the environmental impact values obtained for the stages of production, use and disposal.

5 WATER BOTTLE CASE STUDY

This work, as stated before, intends to compare the environmental impact of two packaging materials for bottled water. Current water packaging solutions may vary in size, shape and colours. Regarding the material type, they can be made of either glass or PET. Figure 6 illustrates the two possible packaging options and its volume.

Last year, the production reached 831.08 million litres of sparkling and natural water, as abovementioned. Considering the amount of bottled water and the volume of bottles, in a 100% glass packaging solution, 3324320000 glass bottles could be produced. Conversely, for PET packaging, it would be enough to produce 2518424243 PET bottles.

This research work covered every step from extraction of raw materials, processing, to the manufacturing of the bottles. This data does not include the distribution and transport environmental impacts.

The first step involved the mass measurement of each bottle, using an analytical weighing scale. In this research work, five possible disposable scenarios were considered: recycling, waste treatment, landfill, municipal waste and household waste. The data was then inserted in the D4E tool, and a comparison of the five solutions regarding the environmental impact values was subsequently performed. This tool gives the value of the environmental impact for the complete life cycle of each packaging system, from raw material through finished product, recycling or disposal.

Figure 7 refers to the environmental impact values (in miliPoints) for both packaging solutions, regarding the materials and processing, and the five different disposal scenarios considered. It is possible to observe that the PET packaging solution has a lower material & processing environmental impact value, which is influenced by the high difference in the amount of bottled water between the glass and PET packaging solution. Regarding disposal scenarios, considering only the household waste, the glass packaging solution have a better environmental impact performance.

Figure 8 refers to the global environmental impact values, the total sum of materials and processing with the possible disposal scenarios. Considering the global scenarios, the PET packaging solution has the best performance, except for the household waste scenario which presents similar environmental impact values for both cases. In either packaging solutions, the recycling scenario is by far the best environmental impact scenario.

Another issue frequently overlooked is the production of moulds necessary for the production of both the glass and PET bottles. Regarding the glass or PET processing, in either packing solutions, only one mould (PET) or a set of moulds (glass) is needed to produce the given amount of bottles. For the production of the PET bottles, it was estimated that the steel blow mould would weigh about 50 kg. Regarding the production of the glass bottles, a set of moulds are needed, due

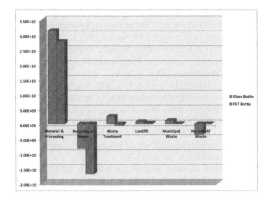

Figure 7.   Environmental impact value of bottles.

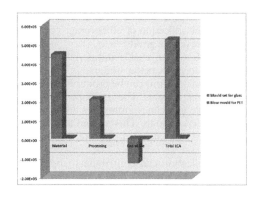

Figure 9.   Environmental impact value of the moulds for the production of either glass or PET bottles.

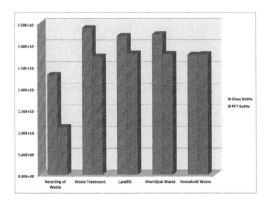

Figure 8.   Environmental impact value of bottles.

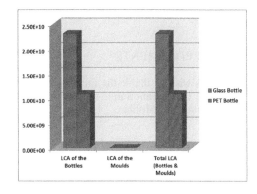

Figure 10.   Global environmental impact value in Portugal (bottles and moulds).

to production requirements, which totals 1850 kg of cast iron in this case. Figure 9 illustrates the environmental impact value for both mould productions. It is possible to observe that the mould set, for the production of the glass bottles, presents by far the highest environmental impact value.

After running the D4E tool, when comparing the environmental impact values of the two different packaging solutions, it is possible to verify that the PET solution presents the lowest environmental impact value, which is valid for both bottles, and the moulds are necessary for the bottle production. The recycle option is the best solution in terms of environmental impact for both products. The design of the both bottles has a strong influence on the consumer regarding its recyclability. Figure 10 illustrates the global environmental impact value of the bottles and the moulds. It is also possible to conclude that the environmental impact of moulds is insignificant when compared to the bottles.

6   CONCLUSIONS

Drinking bottled water has become a trivial habit in people's everyday lives. The tap water taste, high quality water, fitness objectives, safety purposes, amongst other reasons, can lead consumers to buy bottled water. Bottled water may be mandatory in the case of temporary tap water contamination. The trend towards the consuming of more and more bottled water will rise in the coming years. This flourishing market is very profitable for a great number of companies, employing thousands of people worldwide.

The choice of packaging materials increasingly includes environmental concerns. The negative environmental impacts of bottled water can be reduced, just by implementing simple solutions, e.g. re-using bottles of water in adequate sanitary conditions on a local basis, rather than just recycling or re-manufacturing them into new products. When it comes to environmental actions, consumers need to be further educated about recycling, it is requisite to provide a better access to recycling points for the collection of all beverage containers and other recyclable goods.

This D4E system enable industrial companies and clients to get a better understanding on how products and decisions can be produced with lower environmental impacts. The developed tool has a set of simple guidelines aiding designers to meet particular design goals. This system allows a fast and efficient evaluation of the life cycle of real products.

This application can contribute to optimize a more sustainable technological development in the industrial area, establishing itself as an important and user-friendly system for the analysis, optimization and comparison of the life cycle of products. It enables to identify opportunities for the improvement of product design, functionality, applicability, improving its environmental impact.

ACKNOWLEDGEMENTS

The authors wish to thank the following companies for their collaboration in the present research: JRMoldes and VidriMolde.

REFERENCES

Berry, M. 2004. The Importance of Sustainable Development, Columbia Spectator, Canada.

Doria, M.F., 2006. Bottled water versus tap water: understanding consumers' preference. Journal of Water and Health 4 (2), 271–276.

Ferrier, C., 2001. Bottled water: understanding a social phenomenon. Ambio 30 (1), 118–119.

Finnvedena, G., Hauschildb, M., T. Ekvallc et al., 2009. Recent developments en Life Cycle, Journal of Environmental Management, 92 (1), 1–21.

Freire, F., Thore, S., Ferrao, P., 2001. Life cycle activity analysis: logistics and environmental policies for bottled water in Portugal. OR Spektrum 23, 159–182.

Gleick, P. (Ed.), 2006. The World's Water 2006–2007: The Biennal Report on Freshwater Resources. Island Press, Washington D.C, p. 128.

Gollier, C., Jullien, B., Treich, N., 2000. Scientific progress and irreversibility: na economic interpretation of the 'precautionary principle'. Journal of Public Economics 75 (2), 229–253. doi:10.1016/S0047-2727(99)00052-3.

Gutowski, T.G., 2004. Design and Manufacturing for the Environment, Handbook of Mechanical Engineering, Springer-Verlag, pp. 1–25.

International Bottled Water Association IBWA, 2008. Beverage Marketing's 2008 Market Report Findings. Available at: http://www.bottledwater.org/ Last access: February, 17, 2011.

International Bottled Water Association IBWA, 2010. Bottled Water Reporter. Available at: http://www. bottledwater.org/Last access: February, 17, 2011.

Niccolucci, V., Botto, S., Rugani, B., Nicolardi, V., Bastianoni, S., Gaggi, C. (2011) The real water consumption behind drinking water: The case of Italy, Journal of Environmental Management 92:2611–2618.

Pfister, S., Koehler, A., Hellweg, S., 2009. Assessing the environmental impacts of freshwater consumption in LCA. Environmental Science and Technology 43 (11), 4098–4104.

Santos, A.L., Almeida, H.A., Bártolo, H., Bártolo, P.J., 2012, A decision tool for green manufacturing, 2012 Proceedings of the ASME 11th Biennial Conference on Engineering Systems Design and Analysis (ESDA 2012), A. Bernard & F. Chinesta (Eds.), (ISBN 978-0-7918-4487-8):155–162.

Sen, Z., Altunkaynak, A., 2009. Fuzzy system modelling of drinking water consumption prediction. Expert System with Applications 36, 11745–11752.

Zhao, X., Chen, B., Yang, Z.F., 2009. National water footprint in an inputeoutput frameworkda case study of China 2002. Ecological Modelling 220, 245–253.

Renewable energy technologies

Green Design, Materials and Manufacturing Processes – Bártolo et al. (eds)
© *2013 Taylor & Francis Group, London, ISBN 978-1-138-00046-9*

Analysis of the ethanol production chain in the State of Rio Grande do Sul, Brazil: A study based on system's dynamics with a view for exploring scenarios

A. Longhi, G.L.R. Vaccaro, T. Fleck, K. Roos & D.C. Azevedo
UNISINOS—University of Sinos Valley, São Leopoldo City, Rio Grande do Sul State, Brazil

M.H.C. Moutinho
Porto Alegre City, Rio Grande do Sul State, Brazil

ABSTRACT: Brazil plays a leading role in biofuels production, specially Ethanol from sugarcane. With a consolidated production chain, based on large-scale structured farms dedicated to sugarcane monoculture in large tracts of land, and with decades of technological development, the country was responsible for 35% of world production of fuel ethanol. The southernmost state in Brazil, Rio Grande do Sul, participates of this chain producing less than 1% of its local demand, presenting an undeveloped Ethanol production chain based on family farming, even having agricultural zoning that ensures the production of sugarcane and other crops and structured programs of the state government for fostering biofuels production. Through a systems' dynamics approach, highlighting social, technological, economic, environmental, political, and legal dimensions, this study aims at identifying how the actors, and endogenous and exogenous variables, exert influence on this complex system. The research developed was based on a compilation of technical and academic references, and of interviews with experts on the subject. The set of information has given rise to a map of systems' dynamics, presenting the main actors of this peculiar chain, their relationships, and the key variables that influence the system. The map served as basis for future scenario's analysis looking for the key elements for economic sustainability and competitiveness of such chain. The authors hope to contribute to the literature on the subject, presenting relevant for the understanding of the need for a differentiated model for developing a sustainable and intelligent productive chain, when in the presence of small producers based on family farming.

1 INTRODUCTION

Sustainable development involves the comprehension of several aspects of a production chain. It includes a systemic view of the Social, Technological, Economical, Environmental, Political and Legal (STEEPL) aspects, which contribute to constitute the production chain as it is. In order to promote improvements towards a more sustainable and intelligent production chain, it is also needed to understand the foundational drivers, theirs impacts and risks associated to the driving forces that compound the contextual and transactional environments in which such chain evolves (Van Der Heidjen, 2005; Vaccaro et al. 2010). In the pathway for sustainable and intelligent manufacturing, biofuels can be regarded as an alternative for supplying the need for clean power generation, aligned to the emergence of an awareness that the prevailing economic model still leads to the collapse of the economic and environmental structure (MAPA, 2011, MDIC, 2012a).

This paper aims at contributing with the discussion on the subject of sustainable production by presenting a systems' dynamics analysis based on the STEEPL aspects of the Ethanol production chain in the southernmost state of Brazil, Rio Grande do Sul (RS). This chain was selected due to its peculiar characteristics: predominance of family farms, increasing demand and low production of ethanol in the state, despite environmental and political favorable conditions. This paper presents on Section 2 a brief description of the context of the ethanol production chain in the state and then analyses the main driving forces related to the STEEPL dimensions, seeking for elements that contribute towards a sustainable and competitive production chain. Section 3 presents an analysis of the actors and the STEEPL aspects compiled from the literature review and from field interviews. The qualitative and quantitative data were obtained from academic and technical references (indexed publications, technical reports from federal and state government and from civil associations in Brazil), and from interviews with seven experts on the object of study (representatives of sugarcane farmers, rice farmers, ethanol producers, state government agencies and researchers). In Section 4 we draw an analysis based on the processes of systemic mapping and scenarios' planning proposed by Sterman (2000) and Van Der Heijden (2005), we briefly present elements obtained from the systems' dynamics approach and the key aspects that emerged from the scenario's analysis. Finally, some conclusions and suggestions are presented.

The research was conducted combining periods of data collection, data compilation, and modeling and analysis, totaling three cycles of development. Semistructured interviews were conducted with the same specialists in each cycle, seeking for a continued evaluation of the research findings. Modeling and analysis sessions were performed through systematic meetings of the research team. The overall lead-time of the study was of 20 months.

2   THE ETHANOL PRODUCTION CHAIN IN THE STATE OF RIO GRANDE DO SUL

The Brazilian experience with ethanol had prominence in the 70's, with the creation of the National Alcohol Program (PROALCOOL), as an economic response for the Global Petroleum Crisis, after successive increases in the price of oil in the international market, and as a strategic response for the nationalization of power supply sources. After some turbulent times during the 90's, currently Brazil is a world reference in Ethanol produced from sugar cane. With a consolidated production chain with decades of technological development, the country was responsible for a production of 23.7 billion liters of ethanol in 2011, which represented 35% of world production of fuel ethanol (CONAB, 2011). This production was obtained from 588.9 million tons of sugarcane grown on 8,443,430 acres distributed in various regions of Brazil, and it was processed through a network of over 400 mills and distilleries across the country (IEA, 2010). Brazil also exports about 10% of its yearly production to United States, South Korea, Japan, and other countries in South America, Central America and Europe (MDIC, 2012b, ICONE, 2009). The agricultural production profile in the main areas is based on large-scale structured farms dedicated to sugarcane monoculture, a model responsible for 99.5% of the ethanol production in Brazil.

In contrast, RS state has an estimated planted area of 1700 hectares exclusively for ethanol production, producing, in 2011, only 6.58 million liters of ethanol, representing less than 1% of its regional demand (ANP, 2010), making the state one of the largest importers of ethanol in Brazil (ANP, 2010). RS state also has a peculiar profile of sugarcane producers, composed by small farms (the mean size of the property which is destined to sugarcane production is 8 hectares (IEL et al. 2011a, b) and family farming, mostly located in the northwestern region. Another characteristic, due to historical reasons, is that one small ethanol plant is responsible for most of the production. This ethanol plant has an operation model differentiated from the rest of the country by acting as a cooperative. So, in this state's production chain, the sugarcane producers are currently the ethanol plants owners. Other initiatives of ethanol production plants are currently in development, but of no operational relevance yet (IEL et al. 2011a).

Despite the favorable and increasing demand, ethanol production chain in RS state practically did not change in the last decade (ANP, 2010). But, in the same decade, the State Government developed actions seeking for the development of this biofuel chain, mostly focusing on R&D programs, sectorial studies and political acts (ALRS, 2007). In 2009, a law setting the agricultural zoning of sugarcane crops in the state was published, also indicating soil types, seasons and recommended cultivars (MAPA, 2009). As a result, a total of 182 cities (36% of the state) were included in the official map of sugarcane producers, allowing possibility of agricultural insurance, and the disposing of special credit lines for farmers. Furthermore, in 2012, the RS state government launched the Sectorial Program for Biofuels, establishing suggested actions, ranging from qualification of manpower, sharing of R&D, tax incentives for the installation of plants for Ethanol Production and support for family farming (SDPI, 2012). According to Kuiawinski (2008) and Rambo (2006), there is consensus on the need to build a sustainable agribusiness chain that considers the social and environmental aspects, beyond the economic. Herein lies an important condition in terms of chain impacts, the need to consider the family farming in the development of this chain. The main objective of this program was promoting the ethanol chain development in the state by 2016.

In this process, the research centers should support innovation, through research and development aimed at improving cultivars, management technologies, enzymes for ethanol production, new routes and technological advancements to improve the efficiency and reduce the environmental impacts of the cycles production.

Finally sector entities, such as farmers unions, and labor cooperatives, have at their core the association for study, advocacy and coordination of the economic or professional aspects related to a given economic activity. These actors have the responsibility to organize, bringing reality to the discussion of each link in the production chain in order to promote a discussion of individual interests of their constituents (UNICA, 2012).

These elements combined provide a set of subsystems which interact and exchange value, information, goods, money and influence, as presented in Figure 1. Figure 1 also represents the key variables identified during the research. Endogenous variables are presented at each actor while exogenous variables are presented in the relationships between actors.

The production chain in Figure 1 is presented under a context regulated by Federal and State policies. In this context, two different markets need to be recognized:

– The market for ethanol, focused on the use of such biofuel for: (i) automobiles—a strictly regulated market in terms of distribution, allowing sales from ethanol plants only to distributors accredited by the National Agency of Petroleum, mostly Petrobras; and (ii) chemical transformations in industries, allowing direct sales to companies and refineries (ANP, 2011a, b, c); and
– The market for energy and co-products originated form the ethanol production process.

Considering the profile of the ethanol chain in the state, this market represents a crucial role for the sustainability of the chain given the maintenance of the social aspects currently present. The possibility of selling the surplus of energy generated by the ethanol plant by the burn of bagasse, as well as the transformation of other by-products, such as vinasse and fusel oil, into co-products represents an important gain both for the ethanol plant to maintain its operation and for the market to keep acquiring sugarcane production of the family farms (Kieling et al. 2011a, b, c, d, Schneider et al. 2011, Silveira et al. 2011).

On the viewpoint of the sugarcane producers, interaction also happens with other production chains of foods, leading to a constant analysis for the optimal configuration of the crops in order to achieve best financial results. In this sense, variations on the prices of different crops such as corn and soy can lead to a change in the willingness of producing sugarcane, since there is no relevant sugar production in the state. The associated risks to any of the crop choices are expected to be minimized through rural assistance obtained from technical assistance centers and from R&D developments from research centers located in the state. Also, credit lines and state programs benefiting the

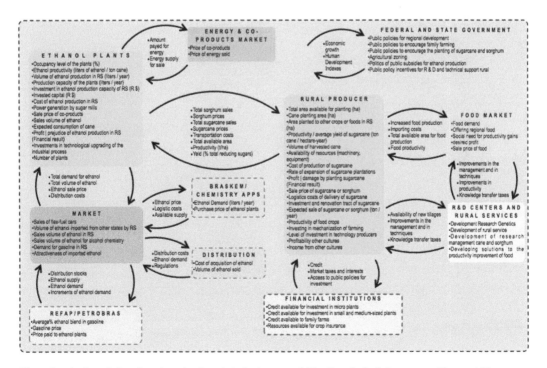

Figure 1. A view of the ethanol production chain in the state of Rio Grande do Sul: actors and key variables.

rural producer are of relevance, as well as the possibility of contracting insurances related to agricultural zoning.

3 ANALYSIS AND DISCUSSION

From the list of elements and variables, and based on the perceptions of specialists, a systems' dynamics map was constructed (Fig. 2), representing influences among the actors described in the previous section. The goal was to enable the complexity of the interrelationships to be seen, so that one could better understand the current and future states of this chain.

In the central part of the map is the farmer. Generally, the farmer directs his choice of planting by two elements: risk mitigation and financial results. From the standpoint of risk, other crops such as soybeans have assured marketing, a production process dominated by farmers, and present profitability, but frequent crop failures caused by droughts in the last years have led the producers to financial losses. In such context, sugarcane and ethanol production appear as an alternative financial income for the farmer, if it is guaranteed a buyer for the production and reduced risk of losses from crop failure (IEL et al. 2011a, b).

Nevertheless, the reduction of area to other crops gives rise to some conflict between the production of foods and raw materials for the production of ethanol. A certain demand for food and other cultivars generates an expectation on the volume of food production, which influences the regional supply of food in the RS. The difference between demand and supply generates a gap, which, after some time influence food prices, increasing the attractiveness (profit) of the sale of food. The presence of the systemic linkages (R1, R2, B3, B4, R3, B7) produces a dynamic equilibrium in the system, often leading to an oscillation among different planting over time. The gap in supply-demand relationship of food generates social necessity of seeking alternative productivity gain. The support of R&D appears as a facilitator of this process through the development of alternatives and genetic improvements and crop management. The expected tendency is that these researches, after some time, generate investments, which in turn increase productivity of food, influencing the sale of food in the same direction, which balances the system again. Gains from the additional sale of food are then reinvested as further improvements by farmers, generating the link R8. On the other hand, Federal and State Governments seek to act as enablers of economic growth and social development. Their policies are aligned in order

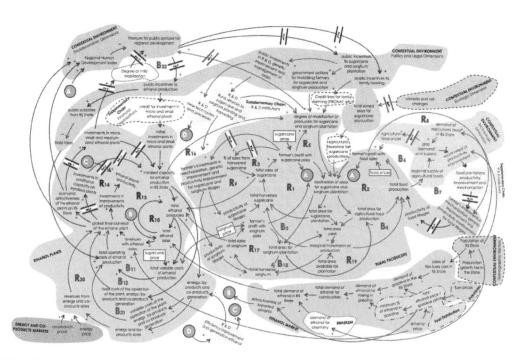

Figure 2. A systems' dynamics map for the ethanol production chain in the Rio Grande do Sul state. R denotes a reinforcement loop while B denotes a balance loop for the involved variables.

to seek for diversified sources of energy and rural setting. The model fostered by the State Government predicts that development of the chain will start from initial investments in micro plants. These micro plants, financed by the government, have the premise that a farmer or a group of farmers intercropped in its agricultural unit can produce ethanol for personal use and can sell the surplus directly to other farmers in the region. Under this view, lies the mindset that these micro plants would promote the chain as nucleating agents for production, generating learning about aspects related to the management of the sugarcane crop. This increased attractiveness would induce the interest of private investors, forming the basis for the development of a supply chain with increased acreage and production capacity through the installation of small and medium-sized plants (enlaces R10 and R14). This view is reinforced by the industrial policy through the biofuel industry program launched in 2012 (SDPI, 2012), containing actions planned to promote the installation of new ethanol plants, such as the fruition of presumed credit of 75% of the taxes generated over the first 4 years of a new venture, and 50% in the next 4 years.

At the same subsystem, the balancer loops B11, B12 and B21 represent the relationship between capacity and increase operating costs and production, impacting the immobilization of assets that tends to negatively impact the overall result of the plants. The balance of power is derived from the relationship between the volume increase of ethanol production, resulting in increased production of by-products and co-products, such as thermoelectric power generated from the burning of bagasse, represented by the link R20.

Some key points emerge from this analysis. The first concerns the importance of treating the chain in an integrated manner. To be economic sustainable the ethanol chain needs to increase the overall revenue of the plants and producers by adding value to by-products and co-products such as bagasse, vinasse and fusel oil, as already discussed earlier (Kieling et al. 2011a, b, c, d, Schneider et al. 2011, Silveira et al. 2011). Second, public policies to encourage the production of ethanol is also a decisive factor. These policies can appear both as reduction of tax rates and investments in installation of micro, small and medium-sized plants. The third aspect is related to the occupancy rate of the plant: in the state, the harvest of sugar cane occupies about 6 months of the year, usually from August to March, so in the rest of the year plants are idled by starvation. Only critical activities are maintained in operation, which increases the fixed cost of the plants and prevents them from joining the constant power supply at the regional system (IEL et al. 2011a, b). R&D and rural services

efforts, such as the use of sweet sorghum as raw material, can help to maintain operation throughout the year, as the harvest of sorghum occurs during the offseason.

Also, initially the specialists understood that the public policies were directly related to resources available to the chain links. However, it could be observed throughout the study, other relevant factors. First, there are the incentives for ethanol production through special credit lines. These incentives are represented on the map indicating the desire of fostering growth by the investment in micro plants and, in turn, generating capacity. Added to these investments, there are public policies to encourage family farming through special credit lines. Furthermore, the study identified the need to promote public incentives for ethanol production through subsidies in tax rates reducing the total cost of production. RS State has a tax rate of 25%, the largest among brazilian states (Demczuk, 2013).

The increase in the need for raw materials such as sugarcane and sorghum will demand a need for investments in R&D focused on genetic improvement and crop management, in order to achieve greater productivity. Is also needed research on new sources of raw materials, such as rice, cassava and maize, new processes like ethanol 2nd generation and development of enzymes for production of ethanol.

Another type of improvement is related to the mechanization of the process, as shown in the link R9. Investments in mechanization tend to generate higher productivity, which in turn generates higher volumes harvested, increasing profit by selling sugarcane and, after certain time, the possibility to reinvest. This same incentive for the production of sugarcane and sorghum tends to require actions to sensitize farmers and communities to adhere to the chain, creating the link R5: the higher the degree of mobilization of farmers, more land is allocated for planting sugarcane and sorghum, a higher volume of sugarcane is produced and sold, generating more profit and consequently more mobilization of farmers, as well as more pressure on the reduction of food crops in the region.

Nevertheless, the low competitiveness and the low scale of the ethanol produced in the state, which results in higher imports from other states, is an aspect pointed out by the experts, and this can compromise the economic sustainability of the local chain. There is a common understanding that large plants will be controlled not by farmers or isolated mill owners, but by big players. In the field of biofuels this trend can be confirmed by the actions of Petrobras that, through its subsidiary Petrobras Biofuels, will invest about

USD 1.2 billion in research and production of ethanol by 2016 (PETROBRAS, 2012, UDOP, 2013). With the entry of big players in the chain of national production of ethanol, there is a tendency of productive clustering, thus reducing production and logistics costs, what will reduce the cost of the ethanol imported by the state, reducing the attractiveness of the ethanol produced in the state. This can discourage the local chain and, after a certain time, would reduce the production of ethanol in the state. In effect, the link R16 negatively influences the mobilization of producers, adversely affecting the allocation of area to cane production (through the link R5) and the profit from sugarcane and sorghum (through links R1 and R17). Moreover, with the negative influence of R1, there would be a positive reinforcement for R2, increasing the allocation of land for food production. These factors may bring losses to the local chain as a whole. The producers and mills mobilized through rural unions and farmers unions, after some time, may generate political pressure on the government to create more public policies for regional development, which can be measured by the Human Development Index (HDI). It should be noted that reducing the HDI would generate political pressure on the state government, which, in turn, should decide on its course of action as a promoter of regional development between: (i) promoting new investments and public policies focused on the chain, thus creating a cycle of reinforcement positive to the chain; or (ii) act to convert the socioeconomic basis of the region to other ways of generating employment and income, forsaking the development of this chain.

The interconnections among the STEEPL dimensions in this chain help to better comprehend the key elements and risks for action scenarios by the involved actors. From the economic standpoint, there is a growing demand for this biofuel. However, the matter is not only economical, but involves regional development issues, family agriculture, sustainability and technological efforts, that should aim at more systemic benefits for the state.

4 FINAL REMARKS

From the technical viewpoint, an Ethanol production chain is a complex agro-energetic system. Thinking this chain in terms of systems, means to seek answers to questions that exhibit characteristics resulting from the interdependence of various factors. In this research, a system's dynamics approach was used seeking to identify the structures of cause-effect-cause relationship among endogenous and exogenous variables, based on events, patterns of behavior and mental models of the actors. This approach also allows enhancing the learning process about a complex system, generating simulation models that help managers to learn about the dynamic complexity of that system, understanding the origins of organizational resistance and design more effective policies (Sterman, 2000; Van der Heidjen, 2005).

Initially farmers, mostly family farmers, face the decision to choose what to plant, seeking to balance higher economic results with lower risks. This decision has a direct impact on other production lines, mainly food, creating potential gaps and social needs of food increments, since the planting area becomes a limiter system, affecting the marginal gains from other cultures.

The government, in turn, believes that encouraging the installation of micro plants will foster the development of the local chain, making it attractive to private investors, that supported by political incentives will expand the installed capacity of the chain. This view is questionable because important aspects need to be sorted out, such as reducing tax burdens and investing in the development of new technologies for ethanol production, genetic research and raw materials.

On the other hand some weaknesses generated by existing mental models can be observed. The chain is strongly based on the effectiveness of public policies. Thus variations in the system can lead actors to seek more support from government public policies. The outcome may be either: a more significant supply of resources by the government, in an attempt to keep the chain in operation, even without fully achieving economic sustainability; or disinterest from the government, leaving the chain to its own sort.

Finally it is important to understand this supply chain not only from an economic point of view: the state most certainly will not become self-sufficient in the production of this biofuel, nor will produce surplus to export it. This chain can be seen as an opportunity for regional development and social inclusion of family farmers, avoiding rural exodus and other social problems related to a lowering HDI, looking towards a more sustainable and intelligent relation of the involved actors in a systemic way.

ACKNOWLEDGEMENTS

The authors wish to thank to the National Council of Research of Brazil (CNPq) for the support provided for this research.

REFERENCES

ANP—Agência Nacional do Petróleo. 2010. *Anuário estatístico 2010*. Available at: <http://www.anp.gov.br/conheca/anuario_2010>.

ANP—Agência Nacional do Petróleo. 2011a. *Resolução nº7*. Available at <http://nxt.anp.gov.br/nxt/gateway.dll/leg/resolucoes_anp/2011/fevereiro/ranp%207%20-%202011.xml>.

ANP—Agência Nacional do Petróleo. 2011b. *Resolução nº66*. Available at <http://nxt.anp.gov.br/nxt/gateway.dll/leg/resolucoes_anp/2011/dezembro/ranp%2066%20-%202011.xml?fn=document-frameset.htm$f=templates$3.0>.

ANP—Agência Nacional do Petróleo. 2011c. *Resolução nº67*. Available at <http://nxt.anp.gov.br/nxt/gateway.dll/leg/resolucoes_anp/2011/dezembro/ranp%2067%20-%202011.xml?fn=document-frameset.htm$f=templates$3.0>.

ALRS—Assembleia Legislativa do Estado do Rio Grande do Sul. 2007. *Relatório da Subcomissão da Cana-de-açúcar, do Álcool e do Etanol*. Agosto de 2007. Available at <http://www.al.rs.gov/>.

CONAB—Companhia Nacional de Abastecimento. 2011. *Acompanhamento da Safra Brasileira Cana-de-Açúcar Safra 2011/2012*. Available at: <http://www.agricultura.gov.br>.

DEMCZUK, A. 2013. Produção de cana-de-açúcar para obtenção de Etanol hidratado no Rio Grande do Sul: Uma análise utilizando dinâmica de sistemas. Masters' Dissertation. Porto Alegre: graduate Program in Management. Federal University of Rio Grande do Sul.

EMBRAPA—Empresa Brasileira de Pesquisa Agropecuária. 2007. In: Assembléia Legislativa do Estado do Rio Grande do Sul. Relatório da Subcomissão da Cana-de-açúcar, do Álcool e do Etanol. Agosto de 2007. Available at <http://www.al.rs.gov/>.

IBGE—Instituto Brasileiro de Geografia e Estatística. 2010a. *Departamento da Cana-de-Açúcar. Evolução da produtividade da cana-de-açúcar no Brasil—2010*. Available at: <http://www.ibge.gov.br>.

IBGE—Instituto Brasileiro de Geografia e Estatística. 2010b. *Produção agrícola 2010*. Available at: <http://www.ibge.gov.br>.

ICONE—Instituto do Comércio e Negociações Internacionais. 2009. *Overview of the biofuels sectors in selected Asian and Latin América Countries*. Available at: <www.iconebrasil.org.br/>.

IEA—International Energy Agency. 2010. *Sustainable Production of Second-Generation Biofuels*. Available at: <http://www.iea.org/>.

IEL—Instituto Euvaldo Lodi, Vaccaro, G.L.R., Moraes, C.A.M., Silveira, C.F.B.; Kieling, A.G., Demartini, F.J., Cabrera, R.S., FLECK, T., Fernandes, I.J. 2011a. Produto 4—Relatório Técnico com Levantamento, Coleta e Análise de Dados Técnicos, Econômicos e Financeiros das Atividades na Fase Industrial de Moagem e Destilação de Álcool para a Cana-de-Açúcar e Sorgo Sacarino. In: *Estudo sobre Produção de Etanol em Indústria de Pequena Escala a Partir de Sistema Integrado de Produção, em Regime Associativo, para a Agricultura Familiar*.

IEL—Instituto Euvaldo Lodi, Vaccaro, G.L.R., Moraes, C.A.M., Silveira, C.F.B., Schneider, J.B.; Kieling, A.G. 2011b. Produto 5—Relatório Técnico com Análise dos Impactos Sociais, Econômicos e Ambientais da Produção de Álcool com Cana-de-Açúcar e Sorgo Sacarino. In: *Estudo sobre Produção de Etanol em Indústria de Pequena Escala a Partir de Sistema Integrado de Produção, em Regime Associativo, para a Agricultura Familiar*.

Kieling, A.G., Moraes, C.A.M., Vaccaro, G.L.R., Fernandes, I.J., Cabrera, R.S. 2011a. Opportunities for the Development of By-Products from Industrial Stage of Ethanol Production from Sugar Cane. In: *ICIEOM 2011—XVII International Conference on Industrial Engineering and Operations Management*. Belo Horizonte: ABEPRO 1:1–13.

Kieling, A.G., Moraes, C.A.M., Fernandes, I.J., Vaccaro, G.L.R. 2011b. Survey Opportunities for Recovery of Sugar Cane Bagasse. In: *5th Deutsch-Brasilianisches Symposium 2011 Nachhaltige Entwicklung—5 Simpósio Brasil-Alemanha 2011 Desenvolvimento Sustentável, 2011, Stuttgart*. 1: 51–51.

Kieling, A.G., Fernandes, I.J., Moraes, C.A.M., Silveira, C.F.B., Vaccaro, G.L.R., Demartini, F.J., Cabrera, R.S. 2011c. Critical Analysis of By-Products Applications from Ethanol Production. In: *ISAF 2011—XIX International Symposium on Alchool Fuels*. Verona: CREAR, 2011. 1: 1–6.

Kieling, A.G., Fernandes, I.J., Agosti, A., Moraes, C.A.M., Vaccaro, G.L.R. 2011d. Mitigação do Impacto Ambiental da Produção de Etanol Através da Utilização dos seus Resíduos para Produção de Alimento Animal. In: *6o Congresso Internacional de Bioenergia*. Curitiba: FIEP-SENAI, 2011. 1: 1–9.

Kuiawinski, D.L. 2008. *Limites e Possibilidades de Desenvolvimento da Cadeia Produtiva do Álcool: Um estudo de Caso no Rio Grande do Sul*. Masters' Dissertation. São Leopoldo: Graduate Program in Production Engineering and Systems. University of Sinos Valley.

MAPA—Ministério da Agricultura, Pecuária e Abastecimento. 2009. *Portaria Nº 332*. Available at: <http://extranet.agricultura.gov.br/sislegis/action/detalhaAto.do?method=consultarLegislacaoFederal>. Accessed in Jan 2nd 2012.

MAPA—Ministério da Agricultura, Pecuária e Abastecimento. 2011. *Cana-de-açúcar e Agroenergia*. Avaliable at: <http:// www.agricultura.gov.br>. Accessed in Mar 2nd 2012.

MDIC—Ministério do Desenvolvimento, Indústria e Comércio Exterior. 2012a. *Álcool Combustível*. Available at: <http://www.desenvolvimento.gov.br>.

MDIC—Ministério do Desenvolvimento, Indústria e Comércio Exterior. 2012b. *ALICEWEB—Sistema de Análise das Informações de Comércio Exterior via Internet. Secretaria de Comércio Exterior (SECEX)*. Available at: <http://aliceweb2.mdic.gov.br/>.

PETROBRAS. 2012. *Etanol: Um Salto para o Futuro*. Accessible by: <http://www.petrobras.com/pt/magazine/post/detalhe-18.htm>.

RAMBO, A.G. 2006. A contribuição Territorial Coletiva e da Densidade Institucional nos Processos de Desenvolvimento Territorial Local/Regional: A Experiência da Coopercana—Porto Xavier/RS. Masters' Dissertation. Porto Alegre: Graduate Program in Geography. Federal University of Rio Grande do Sul.

SDPI—Secretaria de Desenvolvimento e Promoção do Investimento do Rio Grande do Sul. 2012. *Política Industrial—Modelo de Desenvolvimento Industrial do Estado do Rio Grande do Sul 2012–2014*. Available at: <http://www.sdpi.rs.gov.br>.

Schneider, J.B., Silveira, C.F.B., Moraes, C.A.M., Vaccaro, G.L.R., Kieling, A.G. 2011. Environmental Aspects and Impacts Assessment of Ethanol Production Chain. In: *ISAF 2011—XIX International Symposium on Alchool Fuels*. Verona: CREAR. 1: 1–6.

Silveira, C.F.B., Schneider, J.B., Moraes, C.A.M., Vaccaro, G.L.R., Kieling, A.G. 2011. Economical Evaluation of Ethanol Production by a Family Farms Cooperative. In: *ISAF 2011—XIX International Symposium on Alchool Fuels*. Verona: CREAR. 1: 1–6.

Sterman, J.D. 2000. Business dynamics: systems thinking and modeling for a complex world. Boston: McGraw-Hill.

UNICA—União da Agroindústria Canavieira do Estado de São Paulo. 2012. *Estatísticas e projeções*. Available at: <http://www. portalunica.com.br>.

USI—Usinas Sociais Inteligentes. 2012. *Prefeituras, Associações e Instituições conhecem projeto do Bioetanol Social*. Available at: <http://usibiorefinarias.com/view/665/prefeituras-associacoes-e-instituicoes-conhecem-projeto-do-bioetanol-social/>.

UDOP—União dos Produtores de Bioenergia. 2013. *Os investimentos da Petrobras para os próximos dois anos*. Accessible by: <http://www.udop.com.br/index.php?item=noticias&cod=1097655#nc>.

Vaccaro, G.L.R., Pohlmann, C., Lima, A.C., Santos, M.S., Souza, C.B. Azevedo, D.C. Prospective scenarios for the biodiesel chain of a Brazilian state. *Renewable & Sustainable Energy Reviews* 14(1): 1263–1272.

Van Der Heidjen, K. 2005. *The Art of Strategic Conversation*. John Wiley & Sons, Ltd. 2nd ed.

Green Design, Materials and Manufacturing Processes – Bártolo et al. (eds)
© *2013 Taylor & Francis Group, London, ISBN 978-1-138-00046-9*

Use of Jathropha biomass for adsorption of glyphosate in water

H. Nacke
Dynamic Union of Colleges Cataracts, Foz do Iguaçu, Brazil

A.C. Gonçalves Jr., D. Schwantes & G.F. Coelho
Western Parana State University, Marechal Cândido Rondon, Brazil

M.R. Silva & A. Pinheiro
Regional University of Blumenau

ABSTRACT: This study evaluated the use of *Jatropha curcas* L. fruit waste from biodiesel industry for removal of glyphosate in aqueous solutions through adsorption process. For this purpose, two adsorbents materials were prepared: EOJ (endosperm of jatropha fruits after extraction of its vegetable oil) and BOJ (bark of jatropha fruits after extraction of its vegetable oil). The adsorption experiments was studied using different concentrations of glyphosate in solution (0.1 to 1.0 mg L^{-1}) and linearization by Freundlich and Langmuir models. The coefficients obtained from linearizations showed that for the EOJ, Freundlich model is the best fit, suggesting a multilayer adsorption, and for BOJ, Langmuir model have the best fit, indicating a monolayer adsorption. For both materials, the results have revealed that the jatropha fruit waste obtained after the extraction of its vegetable oil can be used for removal of glyphosate in water contaminated with this herbicide.

1 INTRODUCTION

Water resources are used around the world with different purposes, especially water supply, power generation, irrigation, navigation, scenic aquiicultura and harmony, moreover, water is the main constituent of all living organisms. However, this precious natural resource suffers increasingly from pollution generated by human activities, resulting in injury to any environment (Moraes & Jordão 2002).

The contamination of natural resources by pesticides occurs mainly by agriculture, since that are used various applications of these products in crops around the globe. Studies show that these contaminants are found with increasing frequency in different water resources, rivers and soils in recent years (Ayranci & Hoda 2005).

Glyphosate (N-(phosphonomethyl)glycine or 2-[(hydroxy-oxido-phosphoryl)methylamino] acetic acid) is one of the most agrochemical used in the world, and its market continues to grow in line with the increase in the cultivation of glyphosate-tolerant transgenic crops (Duke & Powles 2008).

Although glyphosate present itself less toxic than other herbicides, studies report that this herbicide can cause genotoxicity and toxicity to aquatic organisms and amphibians, teratogenicity in birds and amphibians and endocrine disrupting effects in animals (Mörtl et al. 2013).

The main methods of removing pesticides from the environment are the Fenton coagulation, electro and photo Fenton, ultrasound combined with Fenton, photo oxidation, adsorption and biological degradation (Chang et al. 2011).

The adsorption process can be defined as the selective retention of molecules by chelating, ion exchange or microprecipitation in the surface of an adsorbent, which may be a mineral material, synthetic or natural (Brasil et al. 2007). It should be noted that in the adsorption process, should be sought materials that present low cost, mechanical, chemical and thermal stability and minimal waste generation, characteristics displayed by most of the natural adsorbents (Psareva et al. 2005).

Jatropha curcas L. can produce 2000 L^{-1} of oil annually, therefore, suitable for the manufacture of biodiesel (Silitonga et al. 2011). In this context, the biomass generated in the extraction of oil from jatropha has the potential to be used as natural adsorbent. According to Openshaw (2000), the two main objectives of the incentives for cultivation of jatropha are the use of these plants and their secondary products for rural economic development and environmentally sustainable and to make their cultivation areas self-sufficient in energy, especially in relation to liquid fuels.

The value aggregation in agricultural crops and less waste generation are factors closely linked to the sustainability of agroecosystems. Thus, this study evaluated the use of jatropha (*Jatropha curcas* L.) fruit waste from biodiesel industry for removal of glyphosate in aqueous solutions through adsorption process

2 MATERIALS AND METHODS

This work was conducted at the Western Parana State University (Unioeste) in the municipality of Marechal Cândido Rondon, Paraná, Brazil.

2.1 *Materials*

To obtain the two biosorbents used in this study, fruits of *Jatropha curcas* L. were collected from plants with four years grown at the Unioeste experimental farm. After harvesting the fruits were separated manually in endosperm (EOJ) and bark (BOJ), crushed and then dried in an oven at 60 °C for 36 h. Then, the materials had their oil extracted by a Soxhlet system using the solvent n-hexane for 4 h (International Union of Pure and Applied Chemistry 1988).

After oil extraction, the materials were re-routed to the oven, being kept for 36 h at 80 °C, aiming the complete evaporation of the solvent used in the oil extraction. In order to standardize the particle size of the biosorbents, the granules were sieved, yielding size fractions between 14 and 60 mesh.

2.2 *Materials characterization*

To assess the functional and structural groups of biosorbents, it was performed an Infrared Spectroscopy (IR) in the region between 400 and 4000 cm⁻¹ with a resolution of 4 cm⁻¹, therefore, it was used a Shimadzu® FTIR-8300 Spectrophotometer Infrared Fourier Transform. The transmittance spectra were obtained by using pellets of Potassium Bromide (KBr).

The surface morphology of the biosorbents was evaluated by Scanning Electron Microscopy (SEM) on a microscope FEI® Quanta 200, operating at voltage of 30 kV. The samples were deposited on a double-sided tape attached to a carbon sample holder and then were coated with gold to a thickness of approximately 30 nm by using a sputter coater Baltec Scutter SCD 050.

To study the surface charges of the materials, the point of zero charge (pH$_{PZC}$) was determined. For this purpose, 50 mg of biosorbents mass were added in 50 mL of Potassium Chloride (KCl) solutions at 0.5 mol L⁻¹, with the initial pH values ranging from 2.0 to 9.0, which was adjusted with solutions of Hydrochloric acid (HCl) and Sodium Hydroxide (NaOH), both in a concentration of

0.1 mol L⁻¹. After 24 h stirring (200 rpm), the final values of pH were obtained, thus resulting in a graph of the final pH variation in function of the initial pH, the point that reached the zero value of pH variation corresponded the pH$_{PZC}$ (Mimura et al. 2010).

2.3 *Adsorption experiments*

For the studies of glyphosate removal by the adsorption process, a stock solution was prepared with a glyphosate standard of 99.7% purity (Sigma-Aldrich®). All tests were performed in triplicate.

To prevent the glyphosate adsorption in the glassware used in the experiments, these were silanized using 5% of trimethylchlorosilane in n-hexane for 10 min, then rinsed twice with n-hexane and then with methanol.

All solutions used in this study, including the stock solution, solutions with different concentrations of glyphosate (used in the experiment for adsorption), the solutions used for chromatography (eluent and regenerant) and the silanization solution, were prepared with ultrapure water (Permution® Puritech) with 18.2 MΩ·cm resistivity (25 °C).

The adsorption experiments were conducted with different initial concentrations of glyphosate in solution (0.1, 0.2, 0.3, 0.4, 0.5, 0.6, 0.8, 0.9 and 1.0 mg L⁻¹). The conditions of the adsorptive process, for both biosorbents (EOJ and BAJ), were: 400 mg of biosobent mass; solution pH of 5.5 (adjusted with HCl and NaOH solutions, both at a concentration of 0.1 mol L⁻¹); 60 min of contact time and batch processes with stirring speed of 200 rpm and solution temperature of 25 °C.

After the contact period, the solutions were filtered with the aid of qualitative filter paper and the concentrations of glyphosate were determined by an ion exchange chromatograph Dionex® ICS-90, equipped with a conductivity detector model DS5. Chromatographic conditions: analytical column IonPac® AS22 4 × 250 mm, guard column IonPac® AG22 4 X 50 mm, anion microweb suppressor AMMS® 300 4 mm, eluent solution of 9.0 mmol L⁻¹ Na₂CO₃ and 2.8 mmol L⁻¹ NaHCO₃ with a flow rate of 1.2 mL min⁻¹, regenerant solution of 50 mmol L⁻¹ H₂SO₄ and loop injection of 250 μL (Queiroz et al. 2011).

The amount of glyphosate adsorbed by the biosorbents (Q_{eq} in mg g⁻¹) was calculated with the following equation:

$$Q_{eq} = \frac{(Co - Ceq)}{m} \cdot V \qquad (1)$$

where *m* is the mass of adsorbent used (g), C_o represents the initial concentration of glyphosate in solution (mg L^{-1}), C_{eq} is the concentration of glyphosate in solution at equilibrium (mg L^{-1}) and V is the volume of solution used (L).

The removal percentage of glyphosate (%R) was calculated by the Equation 2:

$$\%R = 100 - \left(\frac{C_{eq}}{C_o} \cdot 100 \right) \qquad (2)$$

The obtained results were used for the construction of adsorption isotherms, which were linearized by Langmuir and Freundlich models.

3 RESULTS AND DISCUSSION

3.1 *Materials characterization*

The results of IR (Fig. 1) show that for two biosorbents (EOJ and BOJ) are observed significant bands at wavelengths of 3400, 2900, 1650 and 1062 cm^{-1}.

The elongated and strong band observed at 3400 cm^{-1} represents the stretching vibration of O-H bonds, found in carbohydrates, proteins, fatty acids, cellulose, lignin units and absorbed water (Stuart 2004, Garg et al. 2008). The band obtained at 2900 cm^{-1} can be attributable to the presence of a stretching vibration of C-H bond, provided by alkanes groups (Barbosa 2007). The band found at 1650 cm^{-1} can be related to the presence of stretching vibrational provided by C-O bonds, present in amides and carboxylic groups (Garg et al. 2008). In 1062 cm^{-1} can be observed a band related to stretching of CO bonds, related to the group of phenols from lignin (Garg et al. 2008).

Among the structural components found in the IR spectrum, can highlight the presence of proteins, carbohydrates, lignin and carboxyl groups,

since they are responsible for adsorption of metal ions (Sharma et al. 2006, Pehlivan 2009).

The images obtained by SEM (figures not shown) demonstrate that the BOJ and EOJ have a morphology with aspect fibrous and spongy, with irregular and heterogeneous structure. This provides a high surface area to biosorbents and enhances its capacity for adsorption.

The results obtained in the determination of the pH$_{PZC}$ of EOJ and BOJ (Fig. 2) show that both biosorbents present point of zero charge at pH values around 6.0, indicating that in situations with a pH value below 6.0, there will be a positive surface charge, which favors the adsorption of anions, and above 6.0, will be a predominance of negative charges on the surface, allowing the adsorption of cations.

Considering that the glyphosate has a greater amount of positives charges below pH 2.0 and a increasing of negatives charges with pH increasing (Coutinho & Mazo 2005, Toni et al. 2006), the best condition for the adsorption would be on acid pH. However, this study aimed the removal of glyphosate from aqueous solutions in natural conditions, where the water pH is close of neutrality (7.0), this way, the pH 5.5 for the solutions of the adsorption tests was chosen, ensuring the presence of positive charges on the adsorbent material and negative charges on the molecules of glyphosate.

3.2 *Effect of glyphosate concentration*

The removal percentages of glyphosate (Fig. 3) demonstrate that the EOJ has greater adsorption at low concentrations compared to BOJ. However, in high concentration both biosorbents have close removal rates and can be considered similar in the higher concentration of glyphosate used in this study (1.0 mg L^{-1}).

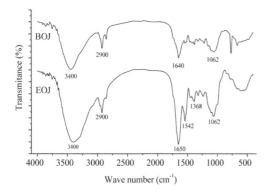

Figure 1. IR spectrum from the endosperm of *Jathropha curcas* (EOJ) and bark of *Jathropha curcas* (BOJ) after extraction of its vegetable oil.

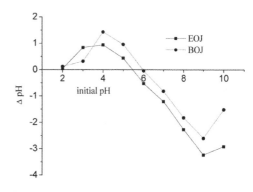

Figure 2. Graphic representation of the point of zero charge (pH$_{PZC}$) from endosperm of *Jathropha curcas* (EOJ) and bark of *Jathropha curcas* (BOJ) after extraction of its vegetable oil.

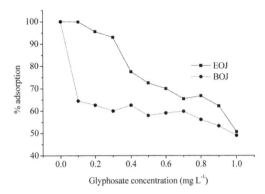

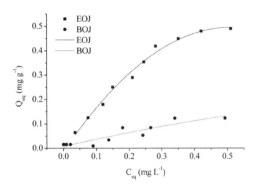

Figure 3. Adsorption porcentages of glyphosate in function of the glyphosate concentration by endosperm of *Jathropha curcas* (EOJ) and bark of *Jathropha curcas* (BOJ) after extraction of its vegetable oil.

Figure 4. Adsorption isotherms obtained in the adsorption of glyphosate by the endosperm of *Jathropha curcas* (EOJ) and bark of *Jathropha curcas* (BOJ) in aqueous solutions.

The study of the adsorption process by isotherms provide information about the relationship between the amount of molecules adsorbed per mass of adsorbent used (Witek-Krowiak et al. 2011). The isotherm linearization further provides coefficients that are extremely important for the interpretation of the data obtained in the process of biosorption.

Figure 4 shows the isotherms obtained in the adsorption of glyphosate by BOJ and EOJ, where it can be observed that for the EOJ, it was obtained an isotherm of L1 type (Giles et al. 1960), which indicates a reduction in active sites of the adsorbent as it increases the concentration of the adsorbate. For BOJ it was obtained isotherm type C1 (Giles et al. 1960), indicating that the number of active sites of the adsorbent is constant and that the adsorbent is porous (corroborating the results of SEM) and has regions of different degrees of solubility for the solute (Falone & Vieira 2004).

The linearization results of the Langmuir and Freundlich models are shown in Figure 5 and Table 1.

For the EOJ, the Freundlich model show the best fit, with a R^2 of 0.994, indicating a multilayer adsorption. The value of the parameter n obtained (1.147) demonstrate the presence of highly energetic sites (Febrianto et al. 2009). The K_f value show a adsorption capacity of 0.15 mg of glyphosate g^{-1} of EOJ.

The results from BOJ show a best fit in the Langmuir model (R^2 = 0.998), which indicates a monolayer adsorption. The adsorption capacity (Q_m) of glyphosate by the BOJ was 0.072 mg g^1 and the adsorption process can be considered favorable (R_L = 0.004), because according to Lin & Juang (2002), for the Langmuir parameter R_L, values

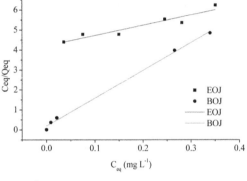

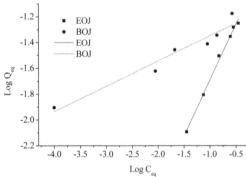

Figure 5. Linearizations obtained by the models of Langmuir (above) and Freundlich (below) for the adsorption of glyphosate in the endosperm of *Jathropha curcas* (EOJ) and bark of *Jathropha curcas* (BOJ) in aqueous solutions.

between 0 and 1 show that the adsorptive process is favorable.

The results obtained in the linearization of Langmuir and Freundlich models, for both biosorbents, corroborates the removal percentages of

Table 1. Parameters of the mathematical models of Langmuir and Freundlich for the adsorption of glyphosate in the endosperm of *Jathropha curcas* (EOJ) and bark of *Jathropha curcas* (BOJ) in aqueous solutions.

	Biosorbent	
	EOJ	BOJ
Langmuir		
Q_m (mg g^{-1})	0.195	0.072
K_L (L mg^{-1})	21.594	2.669
R_L	0.0004	0.004
R^2	0.879	0.998
Freundlich		
K_f (mg g^{-1})	0.152	0.070
n	1.147	5.117
R^2	0.994	0.930

glyphosate (Fig. 3). So it can be stated that the EOJ showed a greater removal capacity of glyphosate in relation to BOJ due to present a multilayer adsorption.

4 CONCLUSIONS

The results obtained in this work show that the endosperm of *Jathropha curcas* L. (EOJ) has a better capacity to removal glyphosate from aqueous solutions in comparison with the bark of *Jathropha curcas* L. (BOJ). However, both biosorbents have potential to remove glyphosate from aqueous solutions by adsorption process, increasing the possibility of utilization of the wastes generated from biodiesel industry.

ACKNOWLEDGMENT

The authors thank the Araucaria Foundation (SETI-PR) and the CNPq-MCTI (REPENSA) for financial support on projects and other aid.

REFERENCES

Ayranci, E. & Hoda, N. 2005. Adsorption kinetics and isotherms of pesticides onto activated carbon-cloth. *Chemosphere* 60(11): 1600–1607.

Barbosa, L.C.A. 2007. Espectroscopia no infravermelho na caracterização de compostos orgânicos. Viçosa: UFV.

Brasil, J.L. Vaghetti, J.C.P. Royer, B. Santos Jr., A.A. Simon, N.M. Pavan, F.A. Dias, S.L.P. & Lima, E.C. 2007. Statistical design of experiments as a tool for optimizing the batch conditions of Cu(II) biosorption using pecan nutshells as biosorbent. *Química Nova* 30(3): 548–553.

Chang, K.L. Lin, J.H. & Chen, S.T. 2011. Adsorption studies on the removal of pesticides (carbofuran) using activated carbon from rice straw agricultural waste. *Engineering and Technology* 76(2): 348–351.

Coutinho, C.F.B. & Mazo, L.H. 2005. Metallic complexes with glyphosate: a review. *Química Nova* 28(6): 1038–1045.

Duke, S.O. & Powles, S.B. 2008. Glyphosate: a once-in-a-century herbicide. *Pest management Science* 64(4): 319–325.

Falone, S.Z. & Vieira, E.M. 2004. Adsorption/desorption of the explosive tetryl in peat and yellow-red argissol. *Química Nova* 27(6): 849–854.

Febrianto, J. Kosasih, A.N. Sunarso, J. Ju, Y.H. Indraswati, N. & Ismadji, S. 2009. Equilibrium and kinetic studies in adsorption of heavy metals using biosorbent: a summary of recent studies. *Journal of Hazardous Materials* 162(2–3): 616–645.

Garg, U. Kaur, M.P. Jawa, G.K. Sud, D. & Garg, V.K. 2008. Removal of cádmium (II) form aqueous solution by adsorption on agricultural waste biomass. *Journal of Hazardous Materials* 154 (1–3): 1149–1157.

Giles, C.H. MacEwan, T.H. Nakhwa, S.N. & Smith, D. 1960. Studies in adsorption. Part XI. A system of classification of solution adsorption isotherms, and its use in diagnosis of adsorption mechanisms and in measurement of specific surface area of solids. *Journal of the Chemical Society* 3973–3993.

International Union Of Pure and Applied Chemistry. 1988. Standard methods for the analysis of oils, fats and derivatives: method 1121. In: C. Paquot & A. Haufenne (eds), *Determination of moisture and volatiles matter content*: 13–16. 7 ed. Oxford: Blackwell.

Lin, S.H. & Juang, R.S. 2009. Adsorption of phenol and its derivatives from water using synthetic resins and low-cost natural adsorbents: A review. *Journal of Environmental Management* 90(3): 1336–1349.

Mimura, A.M.S. Vieira, T.V.A. Martelli, P.B. & Gorgulho, H.F. 2010. Utilization of rice husk to remove Cu^{2+}, Al^{3+}, Ni^{2+} and Zn^{2+} from wastewater. *Química Nova* 33(6): 1279–1284.

Moraes, D.S.L. & Jordão, B.Q. 2002. Water resources deterioration and its impacto on human health. *Revista de Saúde Pública* 36(3): 370–374.

Mörtl, M. Németh, G. Juracsek, J. Darvas, B. Kamp, L. Rubio, F. & Székács, A. 2013. Determination of glyphosate residues in Hungarian water samples by immunoassay. *Microchemical Journal* 107(1): 143–151.

Openshaw, K. 2000. A review of *Jatropha curcas*: an oil plant of unfulfilled promise. *Biomass and Bioenergy* 19(1): 1–15.

Pehlivan, E. Altun, T. Cetin, S. & Bhanger, M.I. 2009. Lead sorption by waste biomass of hazelnut and almond shell. *Journal of Hazardous Materials* 167 (1–3): 1203–1208.

Psareva, T.S. Zakutevskyy, O.I. Chubar, N.I. Strelko, V.V. Shaposhnikova, T.O. Carvalho, J.R. & Correia, M.J.N. 2005. Uranium sorption on cork biomass. *Colloids and Surfaces A: Physicochemical and Engineering Aspects* 252(2): 231–236.

Queiroz, G.M.P. Silva, M.R. Bianco, R.J.F. Pinheiro, A. & Kaufmann V. 2011. Glyphosate transport in runoff and leaching waters in agricultural soil. *Química Nova* 34(2): 190–195.

Sharma, P. Kumari, P. Srivastava, M.M. & Srivastava, S. 2006. Removal of cadmium from aqueous system by shelled *Moringa oleifera* Lam. seed powder. *Bioresource Technology* 97(2): 299–305.

Stuart, B.H. 2004. *Infrared Spectroscopy: Fundamentals and applications*. Chichester: John Wiley and Sons.

Toni, L.R.M. Santana, H. & Zaia, D.A.M. 2006. Adsorption of glyphosate on soils and minerals. *Química Nova* 29(4): 829–833.

Witek-Krowiak, A. Szafran, R.G. & Modelski, S. 2011. Biosorption of heavy metals from aqueous solutions onto peanut shell as a low-cost biosorbent. *Desalination* 265(1–3): 126–134.

Green Design, Materials and Manufacturing Processes – Bártolo et al. (eds)
© *2013 Taylor & Francis Group, London, ISBN 978-1-138-00046-9*

Removal of cadmium from aqueous solutions by adsorption on Jatropha biomass

H. Nacke
Dynamic Union of Colleges Cataracts, Foz do Iguaçu, Brazil

A.C. Gonçalves Jr., G.F. Coelho, L. Strey, D. Schwantes & A. Laufer
Western Parana State University, Marechal Cândido Rondon, Brazil

ABSTRACT: The present study evaluated the use *Jatropha curcas* L. fruit waste from biodiesel industry for removal of Cadmium (Cd) in aqueous solutions through adsorption process. For this purpose, two adsorbents materials were prepared: EOJ (Endosperm of Jatropha fruits) and BOJ (Bark of Jatropha fruits). Based on the optimal conditions, the adsorption process was studied using different concentrations of Cd in solution and linearization by Freundlich and Langmuir models. The two adsorbent materials showed a removal capacity of Cd from solutions of approximately 90%, with a maximum adsorption capacity (Q_m) of 31.69 mg of Cd g^{-1} of EOJ and 29.92 mg of Cd g^{-1} of BOJ. Thus, the results obtained allow to affirm that the jatropha fruit waste obtained after the extraction of its vegetable oil may be used as efficient biosorbent in the removal of Cd from aqueous solutions.

1 INTRODUCTION

The modern industrial and agricultural activities can generate wastes with significant concentrations of heavy metals, contributing significantly to the increase of these contaminants in water bodies (Dal Bosco et al. 2004). Thus, to allow a sustainable development of human activities on our planet, is essential maintaining the natural resources, necessitating the development of technologies that reverse water pollution.

Cadmium is a toxic heavy metal with outstanding importance and may gradually accumulate in the human body, resulting in a number of adverse health effects such as nephrotoxicity and osteotoxicity (World Health Organization 1992).

Actually there are several ways to decontaminate water bodies or wastewater contaminated with heavy metals, highlighting the physical and chemical processes of precipitation, ion exchange, solvent extraction and adsorption (Jimenez et al. 2004).

In the adsorption process, should be sought materials that present low cost, termal, mechanical and chemical stability and minimal generation of waste, characteristics displayed by most natural adsorbents (Psareva et al. 2005).

Abundant waste materials (or products) from industrial and agricultural activities may be potential inexpensive alternatives for heavy metal removal. In recent years, several agricultural wastes have been tested for their heavy metals removal efficiency from simulated wastewaters (Garg et al. 2008).

Jatropha curcas L. can produce 2000 L^{-1} of oil annually, therefore, suitable for the manufacture of biodiesel (Silitonga et al. 2011). In this context, the biomass generated in the extraction of oil from jatropha has the potential to be used as natural adsorbent. According to Openshaw (2000), the two main objectives of the incentives for cultivation of jatropha are the use of these plants and their secondary products for rural economic development and environmentally sustainable and to make their cultivation areas self-sufficient in energy, especially in relation to liquid fuels.

Based on these facts, this study aimed to determine the removal of Cd^{2+} from aqueous solutions under different experimental conditions (mass of adsorbent, contact time, pH and initial concentration of the solutions) using two biosorbents produced from the residue generated in the process of extracting oil from *Jatropha curcas* L. fruits.

2 MATERIALS AND METHODS

This work was conducted at the Western Parana State University (Unioeste) in the municipality of Marechal Cândido Rondon, Paraná, Brazil.

2.1 Materials

To obtain the two biosorbents used in this study, fruits of *Jatropha curcas* L. were collected from plants with four years grown at the Unioeste experimental farm. After harvesting the fruits were separated manually in endosperm (EOJ) and bark (BOJ), crushed and then dried in an oven at 60 °C for 36 h. Then, the materials had their oil extracted by a Soxhlet system using the solvent n-hexane for 4 h (International Union of Pure and Applied Chemistry 1988).

In order to standardize the particle size of the biosorbents, the granules were sieved, yielding size fractions between 14 and 60 mesh.

2.2 Materials characterization

To assess the functional and structural groups of biosorbents, it was performed an Infrared Spectroscopy (IR) with an equipment Shimadzu® FTIR-8300 Spectrophotometer Infrared Fourier Transform.

The surface morphology of the biosorbents was evaluated by Scanning Electron Microscopy (SEM) on a microscope FEI® Quanta 200, operating at voltage of 30 kV.

To study the surface charges of the materials, the point of zero charge (pH_{PZC}) was determined. For this purpose, 50 mg of biosorbents mass were added in 50 mL of potassium chloride (KCl) solutions at 0.5 mol L^{-1}, with the initial pH values ranging from 2.0 to 9.0. After 24 h of stirring (200 rpm), the final values of pH were obtained, thus resulting in a graph of the final pH variation in function of the initial pH, the point that reached the zero value of pH variation corresponded the pH_{PZC} (Mimura et al. 2010).

2.3 Adsorption experiments

For the studies of Cd^{2+} removal under different conditions (mass of adsorbent, contact time, pH and concentration of the initial solutions) a stock solution was prepared with nitrate salt of cadmium tetrahydraten [Cd(NO$_3$)$_2$. 4H$_2$O. AR grade from Sigma-Aldrich®]. All tests were performed in triplicate. In all tests the concentrations of Cd^{2+} were determined by Flame Atomic Absorption Spectrometry (FAAS).

The optimal conditions of the adsorptive process, for each of the biosorbents (EOJ and BAJ), were determined in batch processes with stirring speed of 200 rpm, stirring time of 1.5 h, solution temperature of 25 °C and Cd^{2+} concentration of 10 mg L^{-1}. In this step it was varied the pH of the solutions (4.0, 5.0 and 6.0) and the mass of biosorbents (200, 400, 600, 800, 1000 and 1200 mg).

The amount of Cd^{2+} adsorbed by the biosorbents (Q_{eq} in mg g^{-1}) was calculated with the following equation:

$$Q_{eq} = \frac{(C_o - C_{eq})}{m} \cdot V \qquad (1)$$

where m is the mass of adsorbent used (g), C_o represents the initial concentration of ion in solution (mg L^{-1}), C_{eq} is the concentration of ion in solution at equilibrium (mg L^{-1}) and V is the volume of solution used (L).

The removal percentage of Cd^{2+} (%R) was calculated by the Equation 2:

$$\%R = 100 - \left(\frac{C_{eq}}{C_o} \cdot 100 \right) \qquad (2)$$

To determine the influence of contact time on the adsorptive process, the best conditions obtained in studies of mass and pH were used in different intervals of agitation (5, 10, 20, 40, 60, 80, 100, 120, 140, 160 and 180 min).

In order to evaluate the kinetic mechanism that controls the adsorption process, the models of pseudo-first-order, pseudo-second-order and intraparticle diffusion were applied.

With the best conditions of mass (400 mg), solution pH (5.5) and contact time (60 min) to EOJ and BOJ, tests were performed with different initial concentrations of Cd^{2+} in solution (5, 20, 40, 60, 80, 100, 120, 140, 160, 180 and 200 mg L^{-1}), aiming the construction of adsorption isotherms, which were linearized by Langmuir and Freundlich models.

3 RESULTS AND DISCUSSION

3.1 Materials characterization

The results of IR (Fig. 1) show that for two biosorbents (EOJ and BOJ) are observed significant bands at wavelengths of 3400, 2900, 1650 and 1062 cm^{-1}.

The elongated and strong band observed at 3400 cm^{-1} represents the stretching vibration of O-H bonds, found in carbohydrates, proteins, fatty acids, cellulose, lignin units and absorbed water (Stuart 2004, Garg et al. 2008). The band obtained at 2900 cm^{-1} can be attributable to the presence of a stretching vibration of C-H bond, provided by alkanes groups (Barbosa 2007). The band found at 1650 cm^{-1} can be related to the presence of stretching vibrational provided by C-O bonds, present in amides and carboxylic groups (Garg et al. 2008). In 1062 cm^{-1} can be observed a band related to stretching of CO bonds, related to the group of phenols from lignin (Garg et al. 2008).

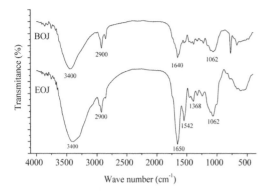

Figure 1. IR spectrum from the endosperm of *Jatropha curcas* (EOJ) and bark of *Jatropha curcas* (BOJ) after extraction of its vegetable oil.

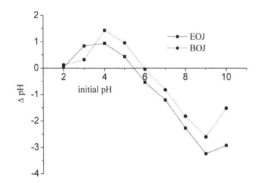

Figure 2. Graphic representation of the point of zero charge (pH$_{PZC}$) from endosperm of *Jatropha curcas* (EOJ) and bark of *Jatropha curcas* (BOJ) after extraction of its vegetable oil.

Among the structural components found in the IR spectrum, can highlight the presence of proteins, carbohydrates, lignin and carboxyl groups, since they are responsible for adsorption of metal ions (Sharma et al. 2006, Pehlivan 2009).

The images obtained by SEM (figures not shown) demonstrate that the BOJ and EOJ have a morphology with aspect fibrous and spongy, with irregular and heterogeneous structure. This provides a high surface area to biosorbents and enhances its capacity for adsorption.

The results obtained in the determination of the pH$_{PZC}$ of EOJ and BOJ (Fig. 2) show that both biosorbents present point of zero charge at pH values around 6.0, indicating that in situations with a pH value below 6.0, there will be a positive surface charge, which favors the adsorption of anions, and above 6.0, will be a predominance of negative charges on the surface, allowing the adsorption of cations.

Once this work aims the adsorption of Cd^{2+} ions, it can be inferred that the adsorption experiments should be conducted in conditions of pH near 6.0.

3.2 *Effect of pH solution and biosorbent mass*

The results of the pH tests (Fig. 3) show that for the BOJ the removal of Cd^{2+} was similar in the four values of pH used. For EOJ the major reductions occurred with pH values of 4 and 5. Considering also, the pH$_{PZC}$ results, the pH values used in the solutions for the experiments of contact time and ion concentration was 5.5.

It should be noted, that these results corroborate those obtained by Garg et al. (2008) when working with the removal of Cd^{2+} from aqueous solutions by *Jatropha curcas* L. cake, who achieved the highest removal efficiency between pH 4 and 5.

In the Figure 3 it is observed that with 200 mg of EOJ and BOJ it was obtained a high removal of Cd^{2+} ion, however, these removals occurred at low concentrations of the ion (10 mg L^{-1}). Thus, to ensure an effective removal of Cd^{2+} were used 400 mg of EOJ and BOJ for the experiments of contact time and ion concentration.

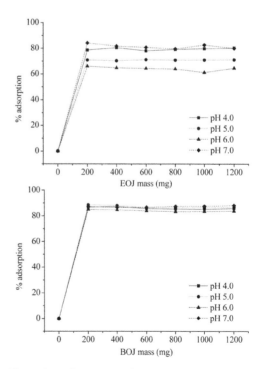

Figure 3. Effect pH solution biosorbent mass in the removal of Cd^{2+} ions by endosperm of *Jatropha curcas* (EOJ) and bark of *Jatropha curcas* (BOJ) in aqueous solutions.

369

3.3 Effect of contact time

The experiments of contact time (Fig. 4) show that after 60 min, to both materials (EOJ and BOJ), occurs the stabilization of the adsorptive process. Thus, the contact time chosen for the isotherm experiments was 60 minutes, confirming the results obtained by Garg et al. (2008).

The contact time of 60 min can be considered fast compared to other studies that used natural biosorbents in the removal of metal ions, which made necessary a contact of 90 min (Goncalves Jr. et al. 2013a, b) and 150 min between metallic ions and the biosorbent (Brazil et al. 2007).

From the results obtained in tests of contact time, the kinetic study of the mechanisms controlling the adsorption process was made by the models of pseudo-first-order, pseudo-second-order and intraparticle diffusion (Table 1).

To evaluate the pseudo-first-order and pseudo-second-order models, Febrianto et al. (2009) report that the R^2 values should be as near as possible to 1 and that the Q_{eq} calculated values must be similar to the Q_{eq} experimental values (average of Q_{eq} values obtained in contact time tests).

Analyzing the results shown in Table 1, it is observed that the pseudo-first-order model does not explain the adsorption process for any of biosorbents used in this work, since the R^2 values are low and Q_{eq} calculated values are not close to Q_{eq} experimental values (1074 mg g^{-1} for EOJ and 0998 mg g^{-1} for BOJ). Therefore, it can be said that the adsorption of Cd^{2+} ions in the sites of BOJ and EOJ does not occur by physisorption, which is indicated by pseudo-first-order model.

The pseudo-second-order model, that indicates the occurrence of chemisorption, is explains satisfactorily the adsorptive process between the biosorbents BOJ and EOJ and Cd^{2+} ions. It is observed in Table 1, that the values of R^2 are very

Table 1. Kinetic parameters obtained in the adsorption of Cd^{2+} by endosperm of *Jatropha curcas* (EOJ) and bark of *Jatropha curcas* (BOJ) for pseudo-first-order, pseudo-second-order and intraparticle diffusion.

	EOJ	BOJ
Pseudo-first-order		
K_1 (min^{-1})	−0.010	−0.012
Q_{eq} (cal.) (mg g^{-1})	0.073	0.079
R^2	0.958	0.757
Pseudo-second-order		
K_2 (g mg^{-1} min^{-1})	0.657	0.674
Q_{eq} (cal.) (mg g^{-1})	1.106	1.033
R^2	0.999	0.999
Intraparticle diffusion	Line A	Line B
EOJ		
K_{id} (g mg^{-1} min$^{-1/2}$)	0.008	0.007
C_i (mg g^{-1})	1.020	1.010
R^2	0.999	0.945
BOJ		
K_{id} (g mg^{-1} min$^{-1/2}$)	0.019	0.007
C_i (mg g^{-1})	0.891	0.932
R^2	0.968	0.919

close to 1 and the values of Q_{eq} calculated are close to Q_{eq} experimental values.

Regarding the intraparticle diffusion model, Gupta & Bhattacharyya (2011) stated that in order to be considered satisfactory, its values of R^2 should be close to 1 and the value of C_i should be close to 0, indicating that the line passes through the origin and that there is an influence of the diffusion process in the adsorption kinetics.

Analyzing the results in Table 1 for the intraparticle diffusion model, can be stated that it does not satisfactorily represents the adsorptive process behavior for the EOJ and BOJ, since the R^2 values are not as close to 1 and the C_i values are different from 0.

3.4 Effect of Cd^{2+} ions concentration

The adsorption isotherms provide information of the relationship between the amount of ions adsorbed per mass of adsorbent used (Witek-Krowiak et al. 2011), the coefficients obtained in their linearizations are extremely important for the interpretation of the data obtained in the process of biosorption.

The linearization results of the Langmuir and Freundlich models are shown in Table 2.

For the EOJ, both the Langmuir model as the Freundlich model, show R^2 values not as close to 1, but may be considered satisfactory, thus indicating that the adsorptive process can occur in both

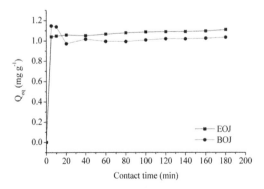

Figure 4. Effect of contact time in the removal of Cd^{2+} ions by endosperm of *Jatropha curcas* (EOJ) and bark of *Jatropha curcas* (BOJ) in aqueous solutions.

Table 2. Parameters of the mathematical models of Langmuir and Freundlich for the adsorption of Cd^{2+} ions in the endosperm of *Jatropha curcas* (EOJ) and bark of *Jatropha curcas* (BOJ) in aqueous solutions.

	Biosorent	
	EOJ	BOJ
Langmuir		
Q_m (mg g^{-1})	31.69	29.92
K_L (L mg^{-1})	0.04	0.01
R_L	0.11	0.28
R^2	0.84	0.97
Freundlich		
K_f (mg g^{-1})	1.53	1.85
n	1.55	1.22
R^2	0.86	0.95

mono and multilayer. The biosorbent BOJ presented a greater R^2 value for the Langmuir model, however, in the Freundlich model, the R^2 value can also be regarded as satisfactory, showing that just as the EOJ, the process occurs in both mono as in multilayer (Montanher et al. 2005).

Regarding the Langmuir parameter Q_m, that indicates the maximum adsorption capacity of the material, EOJ showed ability to remove even 31.69 mg of Cd g^{-1} of biosorbent, and the BOJ a removal capacity of 29.92 mg Cd g^{-1} of biosorbent.

According to Lin & Juang (2002), for the Langmuir parameter R_L, values between 0 and 1 show that the adsorptive process is favorable. In this work, for both biosorbents, the R_L values were in this range, thus, can be stated that the adsorptive process was favorable.

The Freundlich parameter n indicates the reactivity of the active sites from the adsorbent, when these values are higher than 1 there is a strong indication of the presence of highly energetic sites (Febrianto et al. 2009). The two biosorbents studied in this work had n values higher than 1, indicating the presence of high energy sites in its structure, which makes the adsorptive process more efficient by keeping the ions adsorbed on the biosorbent materials for greater periods of time.

4 CONCLUSIONS

The adsorption process occurs by means of chemisorption and is dependent of the adsorbent mass, with low influence of the pH for the two biosorbents tested.

The results obtained in this work allow conclude that the biomass of *Jatropha curcas* L., obtained after the extraction of the vegetable oil from its fruits, have potential to remove Cd^{2+} ions from aqueous solutions by adsorption process.

These results can be helpful for the decontamination of wastewater and hydric resources, as well as promote a major sustainability in the supply chain of *Jatropha curcas* L.

ACKNOWLEDGMENT

The authors thank the Araucaria Foundation (SETI-PR) and the CNPq-MCTI (REPENSA) for financial support on projects and other aid.

REFERENCES

Barbosa, L.C.A. 2007. *Espectroscopia no infravermelho na caracterização de compostos orgânicos*. Viçosa: UFV.

Brasil, J.L. Vaghetti, J.C.P. Royer, B. Santos Jr., A.A. Simon, N.M. Pavan, F.A. Dias, S.L.P. & Lima, E.C. 2007. Statistical design of experiments as a tool for optimizing the batch conditions of Cu(II) biosorption using pecan nutshells as biosorbent. *Química Nova* 30(3): 548–553.

Dal Bosco, S.M. Jimenez, R.S. & Carvalho, W.A. 2004. Application of natural zeolite scolecite in the removal of heavy metals from industrial effluents: competition between cations and desorption process. *Eclética Química* 29(1): 47–56.

Febrianto, J. Kosasih, A.N. Sunarso, J. Ju, Y.H. Indraswati, N. & Ismadji, S. 2009. Equilibrium and kinetic studies in adsorption of heavy metals using biosorbent: a summary of recent studies. *Journal of Hazardous Materials* 162(2–3): 616–645.

Garg, U. Kaur, M.P. Jawa, G.K. Sud, D. & Garg, V.K. 2008. Removal of cádmium (II) form aqueous solution by adsorption on agricultural waste biomass. *Journal of Hazardous Materials* 154 (1–3): 1149–1157.

Gonçalves Jr., A.C. Meneghel, A.P. Rubio, F. Strey, L. Dragunski, D. & Coelho, G.F. 2013a. Applicability of *Moringa oleifera* Lam. Pie as na adsorbent for removal of heavy metals from waters. *Revista Brasileira de Engenharia Agrícola e Ambiental* 17(1): 94–99.

Gonçalves Jr., A.C. Rubio, F. Meneghel, A.P. Coelho, G.F. Dragunski, D. & Strey, L. 2013b. The use of *Crambe abyssinica* seeds as adsorbent in the removal of metals from waters. *Revista Brasileira de Engenharia Agrícola e Ambiental* 17(3): 306–311.

Gupta, S.S. & Bhattacharyya, K.G. 2011. Kinetics of adsorption of metal ions on inorganic materials: A review. *Advances in Colloid and Interface Science* 162(1–2): 39–58.

International Union Of Pure and Applied Chemistry. 1988. Standard methods for the analysis of oils, fats and derivatives: method 1121. In: C. Paquot & A. Haufenne (eds), *Determination of moisture and volatiles matter content*: 13–16. 7 ed. Oxford: Blackwell.

Lin, S.H. & Juang, R.S. 2009. Adsorption of phenol and its derivatives from water using synthetic resins and low-cost natural adsorbents: A review. *Journal of Environmental Management* 90(3): 1336–1349.

Mimura, A.M.S. Vieira, T.V.A. Martelli, P.B. & Gorgulho, H.F. 2010. Utilization of rice husk to remove Cu^{2+}, Al^{3+}, Ni^{2+} and Zn^{2+} from wastewater. *Química Nova* 33(6): 1279–1284.

Montanher, S.F. Oliveira, E.A. & Rollemberg, M.C. 2005. Removal of metal ions from aqueous solutions by sorption onto rice bran. *Journal of Hazardous Materials* 117(2–3): 207–211.

Openshaw, K. 2000. A review of *Jatropha curcas*: an oil plant of unfulfilled promise. *Biomass and Bioenergy* 19(1): 1–15.

Pehlivan, E. Altun, T. Cetin, S. & Bhanger, M.I. 2009. Lead sorption by waste biomass of hazelnut and almond shell. *Journal of Hazardous Materials* 167 (1–3): 1203–1208.

Psareva, T.S. Zakutevskyy, O.I. Chubar, N.I. Strelko, V.V. Shaposhnikova, T.O. Carvalho, J.R. & Correia, M.J.N. 2005. Uranium sorption on cork biomass. *Colloids and Surfaces A: Physicochemical and Engineering Aspects* 252(2): 231–236.

Sharma, P. Kumari, P. Srivastava, M.M. & Srivastava, S. 2006. Removal of cadmium from aqueous system by shelled *Moringa oleifera* Lam. seed powder. *Bioresource Technology* 97(2): 299–305.

Silitonga, A.S. Atabani, A.E. Mahlia, T.M.I. Masjuki, H.H. Badruddin I.A. & Mekhilef, S. 2011. A review on prospecto of *Jatropha curcas* for biodiesel in Indonesia. *Renwable and Sutentainable Energy Reviews* 15(8): 3733–3756.

Stuart, B.H. 2004. *Infrared Spectroscopy: Fundamentals and applications*. Chichester: John Wiley and Sons.

Witek-Krowiak, A. Szafran, R.G. & Modelski, S. 2011. Biosorption of heavy metals from aqueous solutions onto peanut shell as a low-cost biosorbent. *Desalination* 265(1–3): 126–134.

World Health Organization. 1992. *Environmental Health Criteria*. Geneva: WHO.

Green Design, Materials and Manufacturing Processes – Bártolo et al. (eds)
© 2013 Taylor & Francis Group, London, ISBN 978-1-138-00046-9

Guidelines for redesign a commercially available product: The case of a split air conditioning unit

P.N. Botsaris
Democritus University of Thrace, Xanthi, Greece

ABSTRACT: There comes a time during the product life cycle that needs a change or a redesign. The product may have an outdated look and feel, needs additional functionality or the competition has caught up the product, and/or design faults may be found during its lifetime. Redesign improves product quality, maintainability, serviceability, life cycle, and reduces the cost and environmental impact. There are many different guidelines for developing product redesign techniques. This study examines the redesign process of a commercially available product, a split air conditioning unit, based on the assumption of reduced operating cost and an environmentally conscious approach. The study examines the approach to redesign a commercially available split air conditioning unit that operates with a dc current source (photovoltaic panels). The results show a successful and very promising design consideration especially for countries like Greece.

1 INTRODUCTION

There comes a time in any product's lifecycle that it needs a change or a redesign. The product may have an outdated look and feel, needs added functionality or the competition has caught up to the product, and/or design faults may be found during its lifetime. Redesign improves product quality, maintainability, serviceability, life cycle, and reduces costs and environmental impacts. Usually during the redesign phase an available product is modified according to user needs or prior designs. Redesign process focuses on resolving prior design conflicts and current product needs e.g. energy efficiency, climate change e.t.a. Most approaches start by choosing a reference design which reduces conflicts between user needs and product functions, as much as possible. Remaining conflicts, depending upon their degree, are resolved by changing component attributes, replacing components, or changing the structure of the original design (Han et al., 2006, Li et al., 2006). New innovative products are only introduced when major conflicts remain between customer needs and update products. However, business success is also strongly related to product innovation (Howard et al., 2011). New designs, approaches, and products capture customer interest and keep successful companies at the forefront of their industries.

2 REDESIGN APPROACHES

There are many different guidelines for developing techniques for product redesign. These guidelines

and the cohort techniques vary based upon the reasons for redesigning a product (Smith et al., 2012). So there are techniques that detecting design faults, identifying functional dependencies and reusing design information to create new solutions, or reducing e.g. production costs including the environmental impacts. Other redesign techniques have been developed for creating new products based on prior design concepts or reducing costs by combining them into groups or families.

Today the most common reasons for redesigning a product are to rebate design faults, minimize costs (product or operation) and optimize efficiency. Design faults or conflicts occur when a finished product does not meet users' requirements. Similar conflicts can occur when customers change their demands after the introduction of a product into the market. The redesign process generally starts by creating a base product model. The model is generally used to identify design conflicts or functional dependencies. Conflicts and dependencies, in turn, identify the parts or components that must be redesigned to meet requirements. Techniques for detecting design conflicts, identifying functional dependencies, and reusing design information to create solutions are all useful elements for any redesign process. However, the described methods all focus on making modifications to a specific existing product, rather than on developing a brand new product from prior designs.

Another parameter that bounds a product's useful life time is the perceived value, the ratio of product quality to product price. Customers must be willing to purchase a product at a given price. Company success is also related to profit margin, the difference between product price and product cost (Alizon et al., 2007). Quality can be improved by modifying a product to improve performance with respect to requirements. However, different techniques generally needed to reduce costs. In today's market, techniques for reducing costs must generally consider the entire product life cycle, including environmental impacts. Researchers like Bovea and Wang (2007) considered environmental requirements with customer demands and product costs in a holistic redesign approach for developing environmentally conscious products. The approach uses QFD, life cycle assessment and cost, and contingent valuation techniques to weigh increased product costs for environmentally friendly products against increased customer preference and revenue potential. The approach uses a component-level analysis to determine overall product value, individual component values, and components that need to be redesigned to improve environmental impacts or reduce costs.

Other approaches focus on narrower products fields like the electrical and electronic products, describe methods for redesigning into modules improving maintainability, reusability, and recyclability (Yang et al., 2011). The methods use functional and physical risk assessments to ensure the redesigned product functionality and manufacturability. Functional risk assessment is based on principles like the Axiomatic Design Theory or the TRIZ approach. Physical risk assessment is based upon an assembly graph model. The method uses intelligent algorithm to find an optimized modular design that meets functional and physical constraints.

Other redesign methods focus on the cost (production and/or operational) reduction. However, they are also focus on making modifications to a specific existing design. Costs can also be reduced by combining products manufacturing into families. A product family is a group of related products, which shares common features, components, and subsystems, designed to satisfy a variety of market segments. A product family is typically designed around a product platform, which is the set of common elements implemented across the range of products (McGrath, 1996). Redesigning products into a product family promotes component reuse between products, which generally improves product quality, reduces product costs, and simplifies the product design process. Methods for redesigning products into families consider multiple reference products. However, the methods aim to increase component commonality, reduce design effort, and increase profit margin, without changing the overall functionality of any product.

3 TRENDS FOR REDESIGN A RESIDENTIAL SPLIT AIR CONDITIONER UNIT

About half of the cooling needs of buildings in Europe covered by residual split air conditioners as long as the rest is covered by central units (chillers). Cooling demand is growing rapidly in many parts of the world, especially in hot climates, such as Greece. This led to a rapid increase in energy (electricity) demand on hot summer days, which causes an undesirable increase in the use of fossil fuels and also threatens the stability of electricity and charges the environmental impact.

Today's homeowners have several options to supply the cooling needs of their homes. Each solution has its own advantages and disadvantages. Air conditioners can mainly be divided into three main categories:

- central systems which can serve more than one space (chillers)
- fixed single-room units
- moveable single-room units

By far the most common type of room air conditioner is the fixed, single split system. It accounts for over 75% of sales (weighted by cooling capacity). Also, new system designs, based on the no. 2 section referred to redesign guidelines, exceed the conventional air conditioning systems. Over the past years, manufacturers have increased energy efficiency by over 50% through enhanced system and new technologies. Typical measures used to achieve this performance are variable speed compressors, well-managed defrost controls and carefully designed heat exchangers and fans. These technologies also result in the greater improvements of part-load efficiency (which are generally higher than at full load).

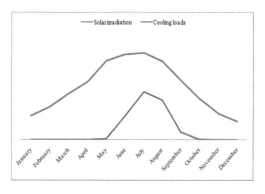

Figure 1. Typical cooling loads variation vs solar irradiation on a city in north eastern Greece (latitude: 41.14°, mean elevation: 40 m).

Table 1. Barriers for solar assisted air conditioning units.

Barriers	Thermal driven cooling units	Electrical driven cooling units	
		Direct	Indirect (with batteries storage)
Lack of units with small capacities	√	–	–
Lack of units for residential and small commercial applications	√	√	√
Low efficiency	√	–	–
Higher initial investment costs compared with conventional units	√	√	√
Lack of awareness of solar cooling	√	√	√

Although, outstanding is that cooling applications, such as air conditioning, have a high dependency with the availability of solar irradiation. The solar cooling obviously has a high potential to replace conventional cooling machines based on the power grid especially in parts of the world with high solar irradiation such as Greece, as. Figure 1 presents. Larger solar cooling systems have been demonstrated successfully and smaller machines, which could be used in a (small) residential and office buildings, entering the market.

In solar cooling systems, the solar energy is used to drive the cooling process (solar assisted air conditioning) others as heat or electricity. Such as heat, solar energy drives cooling machines, such as absorption or adsorption chillers for many years but in recent years, the demonstration projects have shown the ability to use solar energy as electricity (photovoltaic) to drive directly or indirectly split system cooling units.

Direct version means that the unit has the appropriate parts (e.g. current converter DC/DC) to handle the DC current that flows from the photovoltaic panels into the main body of the unit while the indirect version sided by external parts such as batteries and DC/AC current inverters. Because most of the available solar assisted thermal chillers have large cooling capacities (often several hundred kW), the focus of current R&D is largely in developing smaller thermal cooling units and overcome techno-economic barriers to the development of solar energy (direct or indirect) powered units by improving the overall design unit as Table 1 shows.

4 CONCLUSIONS

There are many different guidelines for developing techniques for product redesign. These guidelines and the cohort techniques vary based on the reasons for redesigning a product. So there are techniques that detect design faults, identify functional dependencies and reuse design information to create new solutions, or reducing e.g. production or operation costs including the environmental impacts.

A case of such a redesign is the solar assisted air conditioning split unit. Cooling applications, such as air conditioning, have a high dependency with the availability of solar irradiation. The solar thermal and cooling obviously has a high potential to replace conventional cooling machines based on grid electricity especially in parts of the world with high solar irradiation such as Greece. In solar cooling systems, the solar energy is used to drive the cooling process (solar assisted air conditioning) indirect as heat or direct as electricity.

Although, there are many techno-economical barriers that even the current R&D is trying to overcome as the development of new components, advanced modeling and simulation tools, financial incentives and regulatory measures.

REFERENCES

Alizon, F., Shooter, S.B., & Simpson, T.W. 2007. Improving an existing product family based on commonality/diversity, modularity, and cost. *Design Studies* 28(4): 387–409.

Bovea, M.D., & Wang, B. 2007. Redesign methodology for developing environmentally conscious products. *International Journal of Production Research* 45(18): 4057–4072.

Han, Y.H., & Lee, K. 2006. A case-based framework for reuse of previous design concepts in conceptual synthesis of mechanisms. *Computers in Industry* 57(4): 305–318.

Howard, T.J., Culley, S.J., & Dekoninck, E.A. 2011. Reuse of ideas and concepts for creative stimuli in engineering design. *Journal of Engineering Design* 22(8): 565–581.

Li, Z.S., Kou, F.H., Cheng, X.C., & Wang, T. 2006. Model-based product redesign. *International Journal of Computer Science and Network Security* 6(1): 99–102.

McGrath, M. 1996. *Setting the pace in product development. A guide to product and cycle-time excellence. Revised Edition.* Butterworth-Heinmann, USA.

Smith, S., Smith, G., & Shen, Y.-T. 2012. Redesign for product innovation, *Design Studies* 33(2): 160–184.

Yang, Q., Yu, S., & Sekhari, A. 2011. A modular eco-design method for life cycle engineering based on redesign risk control. *International Journal of Advanced Manufacturing Technology* 56: 1215–1233.

Green Design, Materials and Manufacturing Processes – Bártolo et al. (eds)
© 2013 Taylor & Francis Group, London, ISBN 978-1-138-00046-9

To develop a framework for the implementation of Miscanthus-fuelled ESCOs for non-domestic heat in Ireland—state of the art report

E. Stilwell & D. O'Sullivan

Institute of Technology, Carlow, Co. Carlow, Republic of Ireland

ABSTRACT: The security and sustainability of Irish energy supply is at risk, while renewable energy implementation policies have been disproportionally allocated across energy-usage sectors. As one of the most promising biomass sources in the Irish context, Miscanthus offers a potential solution. This paper specifically examines the multifaceted issue of renewable heat provision, alongside an assessment of potential contribution of Miscanthus. Socio-economic and environmental benefits as well as established Irish agricultural heritage provide an ideal backdrop for the usage of energy crops, yet progress has not been forthcoming. Barriers to widespread adoption are non-technical, largely concentrated at the implementation stage. Education and willingness to address these difficulties are highlighted. Other key factors of success include financial incentives and long-term supply contracts plus management of risk and supply chain by energy service companies.

1 INTRODUCTION

Currently the security and sustainability of Irish (the Republic of Ireland only) is at risk, exacerbated by a peripheral location and small market scale. 94% of energy used in 2011 originated from non-renewable sources, and overall import dependency for 2011 was 88%, poor compared with the EU-15 average of 56% (NCC, 2010; Howley et al., 2012b). Large financial penalties from carbon taxes and Greenhouse Gas (GHG) emissions will result if heavy fossil fuel usage continues. Based on the first four years of the review period, preliminary figures show that Ireland is on course to meet the national Kyoto target of GHG emissions maintained at a maximum 13% above 1990 levels for the years 2008–2012 (EPA, 2012). However, this reduction could be circumstantial rather than intentional, resulting from the decline in energy use since 2008 due to the global recession.

Heat is the primary end-use of energy in Ireland, comprising 41% of Total Final Consumption (TFC) in 2011. Renewable heat (RES-H) provision reached 4.8%, and failed to reach the 2010 target of 5% (Howley et al., 2012b). RES-H provision has been neglected to date, mainly due to its decentralised nature and unsuitability for long-distance distribution. The focus of policy-makers has been disproportionately on the smaller, more-easily addressed electricity sector, creating an imbalance in renewable energy (RES) provision. Where RES-H has been implemented, feedstock has primarily come from forestry, continuing the European trend where 'woody' biomass is the most important single source of renewable energy (UNECE, 2011). Wood-based fuels have significant advantages such as ease of combustion and existing supply infrastructure, however the large area of agricultural land verses forestry (64% and 11% of total land area, respectively; (Teagasc, 2011) offers significant potential if utilized for energy crop production. Ireland cannot sustain the levels of wood-fuel use of many European countries, especially as the adoption of RES energy provides the opportunity to improve energy independence.

The objective of this paper was to examine the multifaceted issue of RES-H provision for non-domestic heat in Ireland. The non-domestic sector consumed the majority of thermal energy in 2011 (53% TFC_{th}; Howley et al., 2012a) and provides economies of scale that suit the adoption of new technologies. *Miscanthus X Giganteus* (herein referred to as 'Miscanthus') was selected as the principal energy crop due to significant advantages over similar biomass fuels, particularly in the Irish context. The methodology consisted of a) comprehensive literature review, b) limited economic analysis, c) review of policy and legislative documents and, d) primary research and case studies. All required information was gathered from peer-reviewed journals and technical reports. In order to benchmark Irish progress, Germany was selected as a peer-review country based on its preeminent position as a world-leader in RES implementation.

2 STATE OF THE ART IN HEAT FROM ENERGY CROPS

Energy crops are agricultural crops that are expressly produced for use as fuel; the high levels of lignin and cellulose increase the embodied energy useful for heat production (Lewandowski et al., 2003). Currently, RES-H from energy crops is almost non-existent in Ireland, having been overlooked in favour of simpler and more abundant wood-based fuels. Progress to date has been dispersed, and collectively has not had a significant impact on the sector. Growth of RES-H in Germany is equally underdeveloped, despite world-leading progress in other RES sectors.

Several factors are evident in successful biomass installations in Ireland and Germany. Efficiency of biomass RES-H systems is maximised by running the boiler at full capacity, using a buffer tank to regulate the heat supply. Boiler technology differs from mainstream wood-based boilers; the number of boiler manufacturers ensuring compatibility with Miscanthus is increasing. Dedicated personnel for maintenance and operation are crucial for larger installations; this may incur additional expense but is generally offset by the energy cost savings achieved. Fuel quality and consistency is also central to efficiency, while long-term fixed price supply contracts for locally sourced fuel reduce uncertainty for the consumer. The management of the supply chain from farm-gate to delivered heat by an Energy Service Company (ESCO) is advantageous, as the separate organisations normally involved provide several points of possible failure. Financial assistance, no matter how limited, is a success factor in overcoming initial investment costs.

3 RES-H FROM MISCANTHUS

3.1 *Fuel production, conversion and logistics*

A C4 perennial, rhizomatous grass with lignified stems, Miscanthus is a sterile hybrid of *M. Saccha-riflorus* and *M. Sinensis* and the only species in the genera with biomass potential. Miscanthus takes 3–4 years on average to become established, with a modest yield of 6–7 tonnes Dry Matter (tDM) ha^{-1} harvestable at the end of year 2. Yields begin to plateau after year 4 when the crop reaches full growth height (approximately 3 m) and provides an annual harvest for the remaining stand lifespan (15 years on average; Teagasc & AFBI, 2010a).

As a relatively new enterprise establishment has often been poor or uneven leading to decreased yields (Clifton-Brown & Lewandowski, 2000). As yields vary within literature, an average yield of 10 tDM ha^{-1} a^{-1} will be assumed for this paper.

A Net Calorific Value (NCV) of 17.7 GJ tDM^{-1} and an input-output energy ratio of 1:32 gives Miscanthus the advantage over similar energy crops (Atkinson, 2009; Teagasc & AFBI, 2010a).

It is vital to consider energy crops as fuel and not simply agricultural produce; careful handling and preparation is required to maintain high quality standards, with preparation beginning at harvest. Miscanthus is usually baled or chipped directly at harvest using conventional farm machinery; baled material is more common, however chipped material involves a more simple process leading to significant financial savings over baled material. Harvest should ideally occur during March when the crop has dried to approximately 30% Moisture Content (MC). Transport is achieved using conventional transportation vehicles, however distances should be minimised in order to reduce GHG emissions (Shifeng Wang et al., 2012).

3.2 *Environmental impacts*

Miscanthus provides several positive impacts for the environment, including reduced soil disturbance, little or no pesticide use, and minimal fertilizer application (Lewandowski et al., 2003). To date, no serious negative impacts have been documented; the sterile nature of the hybrid *Miscanthus X Giganteus* prevents uncontrolled spread, and the crop can be removed by harvesting the rhizomes.

Agriculture and the provision of thermal energy contributed 31% and 22% of national GHG emissions in 2011, respectively. Miscanthus could potentially assist in a net-reduction of total GHG emissions by substituting GHG-emitting fossil fuels in energy provision, displacing GHG-intensive livestock production and increasing soil carbon-sequestration in agriculture (Styles & Jones, 2007). Even a limited conversion of land could significantly reduce regional GHG emissions, however economic factors of existing land-use must be considered.

3.3 *Economics*

In order to ascertain the cost of supplying Miscanthus in Ireland, figures from previous studies were used and adjusted accordingly. Styles et al., (2008) found the discounted, annualised production cost (adjusted for inflation) for chipped and baled Miscanthus to be 47.8 € tDM^{-1} and 53.4 € tDM^{-1}, respectively. Including a profit for the grower and transport costs (converted to € using late-2012 mid market rates; Shifeng Wang et al., 2012) the total delivered cost for chipped Miscanthus, including farm profit, was calculated as 78 € tDM^{-1}, assuming a delivery distance of 50 km. However the authors believe that a value closer to 156 € tDM^{-1}

Table 1. Energy content and delivered energy price comparison of biomass and common fossil fuels.

	Energy (kWh/unit)	Price (€/unit)	Cost (cent/kWh)
Miscanthus* chips, 20% MC	3805 /t	130/t	3.4
SRC willow* chips, 30% MC	3500/t	130/t	3.7
Wood pellets** bulk, 10% MC	4800/t	220/t	4.58
Heating oil**	10.55/l	1.10/l	10.45
Natural gas**	–	–	3.97
Electricity**	–	–	16.1
Hard coal**	7760/t	48/t	0.62
Peat***	2140/t	29.50/t	0.33

*General prices valid as of 2010 (Teagasc & AFBI 2010b). **Average business gas and electricity price bands; valid as of S2-2011 (SEAI 2012). ***Peat included as reference; valid Aug 2010 (Forfás 2010).

(130 € t^{-1} at 20% MC), as outlined in Teagasc & AFBI (2010b), is required for sustainable market growth. Even at this price level, chipped Miscanthus is one of the most competitively priced fuels, as shown in Table 1.

Significant socio-economic benefits from investment in the bioenergy sector were confirmed by IrBEA (2012). In order to meet the 2020 targets for RES-E, RES-H, and RES-T approximately €825 million will be spent in the Irish economy over the period 2012–2020. Permanent employment is anticipated to reach 3600 FTEs by 2020, with the majority located in rural areas, thereby delivering more balanced regional economic development and sustaining rural economies (Segon & Domac, 2012).

4 SALIENT ISSUES FOR EXPANDED UTILISATION

4.1 Barriers to production and utilisation

The term 'barrier' refers to the dynamic and changeable restricting factors that affect the implementation of RES systems.

4.1.1 Land availability
Biomass production will lead to increased competition for land use, most notably with food production (Segon & Domac, 2012). The issue of land distribution is complex and provision for food, fuel, and fibre need to be accounted for equally, taking into consideration their contribution to the national economy. For example, while Gerbens-Leenes & Nonhebel (2002) found that only 0.24 ha

capita^{-1} of agricultural land is required for indigenous food production, the agri-food sector and primary agriculture contributed 7% and 2.5% of GDP in 2011, respectively (Teagasc 2011).

4.1.2 Perception, finance and risk
Biomass is often perceived as a fuel of the past, associated with poverty and low status (Hall & Scrase, 1998). Negative publicity from the failure of badly implemented biomass installations can also influence the perceptions of the public and government. Miscanthus presents greater risk to the grower than conventional agriculture, requiring long-term commitment and delayed economic returns. Crop establishment costs are high, and biomass boilers and feed-in systems are complex, making financial incentives a necessity. Clancy et al., (2011) identified farm size to be a significant contributory factor in the uptake of energy crops, even over farm profitability. Uptake has been slow and less than 2500 ha have been planted since 2006.

4.1.3 Supply chain
Any supply of Miscanthus is dependent on growers' willingness to produce the crop, which will only occur once there is a guaranteed market for their produce. Conversely, widespread adoption by consumers will only occur once a consistent supply of fuel is available; hence, a classic 'chicken-and-egg' situation. Supply chains often require the cooperation of multiple organisations and thus become susceptible to failure (McCormick & Kaberger, 2007).

4.1.4 CAP reforms
In general, Common Agricultural Policy (CAP) has made energy crop production more appealing to growers (Ericsson et al., 2009). However the effects of reforms due in 2013 have yet to be determined, depending on the classification of Miscanthus and a number of 'greening' requirements. Growers may be incentivised if ecological focus areas (7% of eligible farm area) can be utilised for energy crop production. A minimum of three crops is required on all arable areas over three hectares; if included, Miscanthus may be incentivised on arable farms but disincentivised on grassland farms. Maintenance of permanent pasture levels from 2014 may see the exclusion of Miscanthus from grassland, and the elimination of milk quotas may increase the competition for land (Matthews, 2012).

4.2 Incentives to production and utilisation

Little has been done to incentivise the production and utilisation of Miscanthus. The Bioenergy Establishment Scheme (BES), funded by government, has been the sole support mechanism for Miscanthus, designed to lower the initial outlay for growers wishing to adopt Miscanthus and Short Rotation Coppice Willow (SRCW) for the exclusive use as bioenergy for heat and electricity. The scheme for 2013 will provide funding up to 1300 € ha^{-1} for 50% of the establishment costs (DAFM, 2012).

The Sustainable Energy Authority of Ireland (SEAI—formerly SEI) administered the Renewable Heat (ReHeat) Deployment Programme which provided €26 million in funding over 4 years for the deployment of non-domestic RES-H systems; 77.65 MW of biomass boilers were installed, however only wood fuel was supported and therefore did little to incentivise the adoption of energy crops.

4.3 *Peer country review*

Germany currently ranks first and third in the world for installed capacity of solar photovoltaic and wind power, respectively (REN21, 2011), however these world-leading standards do not extend to RES-H; only 3.8% of total investment in construction of renewable installations in 2011 was made in the RES-H sector. Two specific policy measures, the Market Incentive Programme (MAP) and the Renewable Energies Heat Act (EEWärmeG), have attempted to promote RES-H. However Miscanthus does not feature in the data, and considering the small supply area (approximately 3000 ha), cannot be contributing significantly to the sector.

Several site visits in Germany showed positive progress. Combustion issues were solved over time by tweaking the system and its parameters. Finance was secured, despite the level of local government funding being below expectation. Each site had a dedicated 'boiler-man' for maintenance and day-to-day operation, illustrating the delicate nature of biomass systems. Despite budgetary constraints leading to an insufficiently sized boiler, savings of a third were being achieved at Himmerod Abbey. Bio-Energie Hoffenheim GmbH is an ESCO supplying heat through a Miscanthus-fired District Heating (DH) system without any significant problems, suggesting Miscanthus may be more suited to the lower operating temperature of a DH system compared with conventional boilers. Heat is successfully supplied to local premises at 6.5–7 c kWh^{-1}, undercutting oil at 10c kWh^{-1}. The management of the supply chain by one company is credited for the seamless functioning of fuel supply and heat delivery.

4.4 *Alternative markets—electricity from co-firing*

Peat is one of Ireland's only major indigenous sources of fuel, but its use is declining due to environmental pressure and increased emissions control. Coupled with obligations outlined in EU Directive 2009/28/EC, the Irish government set a target of 30% co-firing in the three peat-fired power stations (Edenderry, Lanesboro, and Shannonbridge), leading to the replacement of 0.9 million tonnes of peat with renewable biomass by 2015 (Clancy et al., 2012). To date, Edenderry has been the focus of co-firing trials, successfully testing a range of indigenous and imported biomass, yet experience with Miscanthus has not been successful; boiler corrosion has resulted from the chlorine content of the fuel and as such, Edenderry are no longer seeking Miscanthus supply contracts. Experience in the UK has been more positive; Drax coal-fired power station in North Yorkshire is heavily committed to replacing coal with biomass to become a predominantly biomass fuelled electricity generator. Biomass accounted for 7% (0.5 million tonnes) of the total fuel burned in 2012. Drax utilise Miscanthus pellets extensively without significant issues; pellets provide greater control of particle size and MC than the chipped material used in Edenderry (Wood, 2013). However using pellets in Ireland may prove prohibitively expensive.

The Renewable Energy Feed-In Tariff (REFIT) is the primary monetary support mechanism for co-firing in Ireland; energy crops attract a REFIT rate of 9.5 c kWh^{-1} which the Irish Farmers' Association (IFA) indicate is insufficient to encourage farmers to diversify into growing energy crops at the scale required to meet Ireland's RES targets, instead proposing a rate of 14c kWh^{-1} (IFA, 2011). Miscanthus could not exclusively supply the co-firing requirements as the 3–5 year lead-in period for energy crop establishment is too great to meet the 40,000 ha requirement (Clancy et al., 2012) by 2015, and may be completely omitted based on the combustion issues encountered to date.

5 DISCUSSION

Miscanthus has the potential to make a significant contribution to Irish energy supply concerns, in combination with all other available RES sources. However two main issues need to be tackled. Firstly, challenges in RES-H provision need to be addressed from a government level. The appointment of a 'minister-for-heat' may help redress the current imbalance in energy policy, with funding based on past incentive schemes. Secondly, feedstock supply chains require management on two levels. Nationally, the appointment of a semi-state organisation with existing expertise in biomass supply to coordinate nationwide implementation would encourage greater synchronicity between organisations and individuals. ESCOs (Fig. 1)

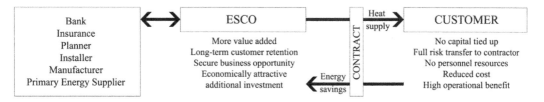

Figure 1. ESCO model as defined by the Swiss Contracting Association. Taken from ENVIROS (2005).

offer potential coordination of local supply and demand, providing all necessary utilities for service delivery through established long-term contracts. A network of biomass trade centres would simplify the supply process, providing a central fuel depot and additional opportunities for rural enterprise (Gavigan, 2012). Existing wood fuel supply infrastructure could provide a working supply chain model.

A significant success factor evident in case studies was a desire for success, with long-term vision of the benefits and determination to overcome short-term issues. McCormick & Kaberger (2007) conclude there are no absolute barriers to the widespread adoption of biomass. Miscanthus (and biomass) implementation presents several challenges, however the time of ignoring these difficulties in favour of dependence on easier fossil-fuelled equivalents has now passed. Education will be required in order to overcome negative public perception.

6 CONCLUSIONS

As this paper is a brief summary of research currently underway by the authors final conclusions cannot be drawn, yet certain realities still become evident. Central management of supply systems, both national and local, reduce risk and ensure consistent service delivery. Government policies need to be extended to include energy crop biomass, while financial incentives must be made available for both grower and consumer.

ACKNOWLEDGEMENTS

This paper forms part of on-going research currently underway at Institute of Technology, Carlow, and has been funded under the ITC Postgraduate Research Scholarship Programme 2011–2012. The authors would like to acknowledge the invaluable contributions of the many industry personnel in Ireland and Germany who contributed to this paper.

REFERENCES

Atkinson, C.J. 2009. Establishing perennial grass energy crops in the UK: A review of current propagation options for Miscanthus. *Biomass and Bioenergy* 33(5): 752–759.

Clancy, D., Breen, J.P., Moran, B., Thorne, F. & Wallace, M. 2011. Examining the socio-economic factors affecting willingness to adopt bioenergy crops. *Journal of International Farm Management* 5(4): 1–16.

Clancy, D., Breen, J.P., Thorne, F. & Wallace, M. 2012. The influence of a Renewable Energy Feed in Tariff on the decision to produce biomass crops in Ireland. *Energy Policy* 41: 412–421.

Clifton-Brown, J.C. & Lewandowski, I. 2000. Overwintering problems of newly established Miscanthus plantations can be overcome by identifying genotypes with improved rhizome cold tolerance. *New Phytologist* 148(2): 287–294.

DAFM 2012. 2013 Bioenergy Scheme. 2013. Available at: http://www.agriculture.gov.ie/ruralenvironment/climatechangebioenergybiodiversity/bioenergy/bioenergyscheme/ [Accessed September 21, 2012].

ENVIROS 2005. *Assessment of the Potential for ESCOs in Ireland* Enviros, ed., Sustainable Energy Ireland.

EPA 2012. Ireland's Greenhouse Gas Emissions in 2011, Wexford.

Ericsson, K., Rosenqvist, H. & Nilsson, L.J. 2009. Energy crop production costs in the EU. *Biomass and Bioenergy* 33: 1577–1586.

Forfás 2010. The Irish Energy Tetralemma: Fuel Reports, Dublin.

Gavigan, N. 2012. Biomass Trade Centre II. In *IrBEA National Conference 2012*.

Gerbens-Leenes, P. & Nonhebel, S. 2002. Consumption patterns and their effects on land required for food. *Ecological Economics* 42(1–2): 185–199.

Hall, D.O. & Scrase, J.I. 1998. Will biomass be the environmentally friendly fuel of the future? *Biomass and Bioenergy* 15: 357–367.

Howley, M., Dennehy, E., Holland, M. & Ó Gallachóir, B. 2012a. *Energy in Ireland—Key Statistics 2012*, Dublin.

Howley, M., Dennehy, E., Holland, M. & Ó Gallachóir, B. 2012b. *Energy in Ireland 1990–2011: 2012 Report*, Dublin.

IFA 2011. Investing in Agriculture to Deliver Economic Growth—The Irish Farmers' Association 2012 Budget Submission, Dublin.

IrBEA 2012. The Economic Benefits from the Development of BioEnergy in Ireland to meet 2020 Targets D. Economic Consultants & R. Consulting Engineers, eds.

Lewandowski, I., Scurlock, J.M.O., Lindvall, E. & Christou, M. 2003. The development and current status of perennial rhizomatous grasses as energy crops in the US and Europe. *Biomass and Bioenergy* 25(4): 335–361.

Matthews, A. 2012. CAP reform and bioenergy. In *IrBEA National Conference 2012*.

McCormick, K. & Kaberger, T. 2007. Key barriers for bioenergy in Europe: Economic conditions, knowhow and institutional capacity, and supply chain co-ordination. *Biomass and Bioenergy* 31(7): 443–452.

NCC 2010. Annual Competitiveness Report 2010—Volume 1: Benchmarking Ireland's Performance, National Competitiveness Council.

REN21 2011. Renewables 2011 Global Status Report, Paris.

SEAI 2012. Commercial & Industrial Comparison of Energy Costs, Dublin.

Segon, V. & Domac, J. 2012. Part IV: Socio-economic impacts of bioenergy production, Biomass Trade Centre II.

Styles, D. & Jones, M.B. 2007. Energy crops in Ireland: Quantifying the potential life-cycle greenhouse gas reductions of energy-crop electricity. *Biomass and Bioenergy* 31: 759–772.

Styles, D., Thorne, Fiona & Jones, M.B. 2008. Energy crops in Ireland: An economic comparison of willow and Miscanthus production with conventional farming systems. *Biomass and Bioenergy* 32: 407–421.

Teagasc 2011. Agriculture in Ireland. *Teagasc Corporate Website*. Available at: http://www.teagasc.ie/agrifood/ [Accessed February 20, 2012].

Teagasc & AFBI 2010a. *Miscanthus Best Practice Guidelines* B. Caslin, J. Finnan, & L. Easson, eds., Teagasc.

Teagasc & AFBI 2010b. *SRC Willow Best Practice Guidelines* B. Caslin, J. Finnan, & A. McCracken, eds.,

UNECE 2011. State of Europe's Forests 2011: Status and Trends in Sustainable Forest Management in Europe, Oslo.

Wang, Shifeng, Wang, Sicong, Hastings, A., Pogson, M. & Smith, P. 2012. Economic and greenhouse gas costs of Miscanthus supply chains in the United Kingdom. *GCB Bioenergy* 4(3): 358–363.

Wood, R. 2013. Email to E. Stilwell, 20 February.

Green Design, Materials and Manufacturing Processes – Bártolo et al. (eds)
© *2013 Taylor & Francis Group, London, ISBN 978-1-138-00046-9*

Modeling of multi-assortment production of CO_2-extracts

E.P. Koshevoy, V.S. Kosachev & V.U. Chundyshko
Kuban State Technological University, Krasnodar, Russia

N.N. Latin
Company Caravan, Krasnodar, Russia

ABSTRACT: Manufacture of valuable high quality products—CO_2-extracts—from various kinds of vegetative raw material of wide assortment is made by small parties on in parallel—working installations of periodic action. The purpose of work was mathematical modelling multi-assortment manufactures of CO_2-extracts in connection with development of information system of decision-making at planning. Use of algorithm based on full sorting variants is offered for short-term planning. Borders of applicability of the given algorithm for search of a subset of optimum schedules are established.

1 A GENERAL CHARACTERISTIC MANUFACTURES OF CO_2-EXTRACTS AND MODELLING OF PROCESS EXTRACTION

CO_2-extracts are producing from various kinds of vegetative raw material with use of solvent—dioxides of carbon [Koshevoy and Bljagoz, 2000; Meretukov Z.A., Koshevoy E.P., 2010]. Extraction—the basic process—realize manufactures in periodically working devices under a high pressure. As an example such manufacture is realized at the enterprise «Company Caravan» (Russia, Krasnodar). The operational circuit manufactures of CO_2-extracts is submitted in Figure 1.

Production with use periodically working installations can be submitted as consecutive operations of transformation of raw material in finished product (1—crushing; 2—loading; 3—extraction; 4—plums of an extract; 5—clearing of an extractor at transition to other raw material). Thus receive

marc—6 and an extract—7. The level of expenses of time at work of group installations for extraction determines their competitiveness and it is necessary to find ways of decrease of expenses. The longest and variable for various kinds and quality of raw material a stage of technology is extraction. Forecasting an output of an extract that is carried out on experience installation is important. Dependences of an output extractable substances from a bed of various materials can be presented as:

$$E(t) = 1 - \exp(-b \cdot t) \tag{1}$$

Dependences of an output extractable substances in an industrial extractor, which is characterized by a high bed (in comparison with a layer in experience an extractor frequency rate of height h), are received as:

$$E_h(t) = 1 - e^{(-b \cdot t)} \cdot \sum_{k=0}^{h-1} \frac{(b \cdot t)^k}{k!} \tag{2}$$

Dependence of profitableness of a stage extraction has received:

$$D(t, b, C_E, G_R, C_O, C_T)$$
$$= C_E \cdot G_R \cdot C_O \cdot \left[1 - \exp(-b \cdot t) \cdot \sum_{k=0}^{n-1} \frac{(t \cdot b)^k}{k!} \right] - C_T \cdot t \tag{3}$$

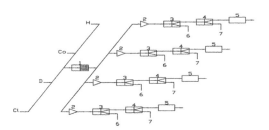

Figure 1. As an example, it is accepted four installations for extraction and it is processed four basic kinds of raw material (H—Hop; Co—Coriander; D—Dill; Cl—Clove).

where C_E—cost of an extract, rub/kg; G_R—weights of raw material loaded in an extractor, kg; C_0—contents of extractable substances in raw material, kg/kg; C_T—cost of operation, rub/h. It is possible to note on dependences for the investigated kinds of raw material presence of a maximum and according to optimum duration of process. Investigating on an extremum the equation (3) concerning time extraction we receive value of criterion (D) in a point of a maximum.

2 MODELLING THE SCHEDULE OF WORK MULTISTAGE AND OF PROCESSING MULTI-ASSORTMENT RAW MATERIAL

For sequence of operations installations the matrix model of the schedule is offered, allowing to determine parameters of schedules on duration of operations and number of parties of processed raw material:

$$w_{i,j} = \left[\Phi(i) - \Phi(i-1) \right] \cdot \sum_{k=0}^{j} \left[d_{\Phi(i-1),k} \right]$$
$$+ \Phi(i-1) \cdot \left\{ d_{\Phi(i),j} + \sum_{k=0}^{j} \left[d_{\Phi(i-2),k} \right] \right\} \qquad (4)$$

where $w_{i,j}$—time of end i-th operations ($i = 0,1 \ldots$) j-th parties of raw material ($j = 0,1 \ldots$); $\Phi(x)$—step function Hevisaid; $d_{i,j}$—duration i-th operations ($i = 0,1 \ldots$) j-th parties of raw material ($j = 0,1 \ldots$).

Applicability of a method of full search for short-term scheduling of group operations installations at processing of parties multi-assortment raw materials checked comparison with Johnson's method for calculation of optimum loading of two mashins which allows to solve a problem for the limited number of works and raw materials parties at observance of following restriction:

$$\min(A[i], B[i+1]) \leq \min(A[i+1], B[i]),$$
$$i = 1, \ldots, n-1 \qquad (5)$$

where $[i]$ designates work which has i-ю sequence of service.

The method of full search is more universal, than Johnson's method. However the decision of a problem for any number of installations and parties a method of search demands computing expenses.

The method of computer modelling investigates borders of applicability of algorithm full sorting which are limited to 15 elements of set (product of number of installations and parties of processed raw material) in the analyzed schedule, that corresponds to short-term planning. The analysis of results of computer modelling of loading of several extractors (n) at change in a set of raw material has allowed to establish the common dependence of duration of a production cycle on number of recurrences of use of each extractor on one kind of raw material (K):

$$T_x(K,n) = n \cdot D_x + (Z_e + E_x + S_e) \cdot K$$
$$+ O_e \cdot sign(n \cdot K) \qquad (6)$$

where $T_x(K)$—duration of performance of works on processing various kinds of raw material (for example, $x = 1, \ldots, 4$) from frequency rate (K) uses of an extractor; D_x—duration of crushing of raw material; Z_e—duration of loading and unloading of an extractor; E_x—duration extraction raw material; S_e—duration a plum of an extractor; O_e—duration of clearing of an extractor; $sign(n \cdot K)$—alarm function equal 0 if the argument is equal 0 and equal 1 if the argument is more 0. More detailed models can be received at use of the matrix approach at calculation duration process. For the considered case the matrix duration (M) can be submitted by the following block structure:

$$M_{2,n\cdot k} = \begin{Vmatrix} d_{1,1} & \cdots & d_{1,k} & \vdots & d_{1,i} & \cdots & d_{1,k\cdot(n-1)+1} & \cdots & d_{1,n\cdot k} \\ s_{2,1} & \cdots & s_{2,k} & \vdots & s_{2,j} & \cdots & o_{2,k\cdot(n-1)+1} & \cdots & o_{2,n\cdot k} \end{Vmatrix}$$
$$\qquad (7)$$

where n—frequency rate of use of installations; k—number of used installations; $d_{1,i}$—crushing of raw material, $i = 1, 2, \ldots, n \cdot k$; $s_{2,j}$—Σ (loading, extraction, plums), $j = 1, 2, \ldots, (n-1) \cdot k$; $o_{2,p}$—Σ (loading, extraction, plums, clearing), $p = (n-1) \cdot k + 1, \ldots, n \cdot k$. In this case the matrix of end of works (Z) can be designed under the following matrix formula:

$$Z_{2,n\cdot k} = \begin{vmatrix} m_{1,1} & \sum_{i=1}^{2} m_{1,i} & \cdots & \sum_{i=1}^{j} m_{1,i} & \cdots & \sum_{i=1}^{n\cdot k} m_{1,i} \\ m_{1,1}+m_{2,1} & \begin{matrix} \sum_{i=1}^{2} m_{1,i}+m_{2,2} \\ +(c_0+c_1\cdot k)\cdot floor\left(\dfrac{2-1}{k}\right) \end{matrix} & \cdots & \begin{matrix} \sum_{i=1}^{j} m_{1,i}+m_{2,j} \\ +(c_0+c_1\cdot k)\cdot floor\left(\dfrac{j-1}{k}\right) \end{matrix} & \cdots & \begin{matrix} \sum_{i=1}^{n\cdot k} m_{1,i}+m_{2,j} \\ +(c_0+c_1\cdot k)\cdot floor\left(\dfrac{n\cdot k-1}{k}\right) \end{matrix} \end{vmatrix}$$

$$(8)$$

where $m_{i,j}$—elements of a matrix duration; *floor* (x)—function of truncation of argument (x) up to the smaller whole. Use of the matrix approach allows to define not only end of all stages of process, but also their beginning—subtraction from a matrix (Z) matrixes (M). Use of elements of the received matrix of the beginning of works $(R = Z - M)$ allows to define duration of idle times of installations pending loadings by their raw material under the formula:

$$S_\Delta = \sum_{j=1}^{n \cdot k} \left(r_{2,j} - z_{1,j} \right) \tag{9}$$

Apparently from the received matrix equations the bottom border of applicability is caused by presence of two frequency rates of use of installation, and the top border is defined by use of factors of the equations of idle times, and also a condition received positive values. In the specified borders full factorial experiment at three levels has been realized. As a result of the carried out calculations it has been established, that these data can be described by the equation:

$$S_\Delta(k,n) = -90 \cdot k \cdot n + 12{,}5 \cdot k^2 \cdot n$$
$$+ 90 \cdot k \cdot n^2 - 12{,}5 \cdot k^2 \cdot n^2 \tag{10}$$

Modelling system of extractors is carried out in view of changes in a set of raw material. At transition from one raw material on another duration of the second stage increases for the period of clearing the extractors. The determining factor of such schedule becomes operation of crushing. Let for performance on one installation set simultaneously acts works $N = \{1, 2 ..., n\}$ and duration of performance of each work t_i, $i \in N$ is known. The method of the decision of a problem essentially depends on criterion of efficiency. We shall enter the following designations: t'_i—time of the beginning of performance of work $i \in N$; t''_i—time of the ending of performance of work $i \in N$; d_i—directive time during which performance of work $i \in N$ should be completed; α_i—the penalty for expectation of work i in unit of time till the moment of the beginning of its processing $i \in N$. Time of the beginning and the ending of performance of works is connected by the following dependence:

$$t''_i = t'_i + t_i \tag{11}$$

Time T necessary for performance of all works of set N, does not depend on the order of performance of works and is equal to the sum of times of performance of all works:

$$T = \sum_{i \in N} t_i \tag{12}$$

Taking into account the accepted designations, it is possible to write down, that a delay z_i (excess of directive term of stay in system) works i will make:

$$z_i = \max_{i \in N}\left(0, t''_i - d_i\right), i \in N \tag{13}$$

Thus, the criterion of efficiency determining size of total costs, connected with delay in performance of works to the set terms, will look like:

$$\Phi_1 = \sum_{i \in N} \alpha_i \cdot z_i \tag{14}$$

The criterion, allowing calculating the maximal penalty connected also with delay in performance of works, looks like:

$$\Phi_2 = \max_{i \in N}\left(\alpha_i \cdot z_i\right) \tag{15}$$

At the decision of various problems criteria (14) and (15), as a rule, are required to be turned into a minimum. Frequently there is a criterion:

$$\Phi_3 = \sum_{i \in N} \alpha_i \cdot t_i \tag{16}$$

which can be used for minimization of the sum of the connected means. Taking into account presence of losses of target components in processed raw material criteria representing the sum of the penalties connected to expectation of works in system are most important. In a result the basic tabulared structures of information system of short-term scheduling of manufacture of CO_2-extracts are developed. The main thing is the directory of characteristics of the raw material, containing the necessary data for the analysis and planning of duration of a production cycle by criterion of maximization of profitableness installations. Optimum control of processing multi-assortment raw material is based on the developed models of schedules.

3 CONCLUSIONS

Thus, in the presented work the technique, mathematical and the software of information system of decision-making is developed, approved in practice at planning multi-assortiment manufactures of CO_2-extrakt's. Following scientific and practical results are received.

1. Problems of ordering of schedules dare the offered algorithm, based on full search of variants. Borders of applicability of the given algorithm for search of a subset of optimum

schedules are established. Use of the combined approach for the decision of difficult multidimensional problems of integer optimisation is offered.

2. The model of process extraction scaled and identified on its key parametres which allows to prove optimum by economic criterion duration of process extraction is developed.

3. The developed matrix models of construction of schedules received on the basis of permutable algorithms allow to calculate time parametres of all stages of processing multi-assortiment raw materials in manufactures of CO_2-extrakt's.

4. The scaled structure of information system of decision-making is developed for construction of suboptimum schedules in manufactures of CO_2-extrakt's which operatively considers changes in volumes and processing terms multi-assortiment raw materials.

5. Borders of applicability of the developed models of schedules that has allowed to solve problems of construction of schedules with use of algorithms of global search are established.

6. Offers on perfection of management of industrial production at the enterprise «Company Caravan» (Russia, Krasnodar) are developed at processing multi-assortiment raw materials.

REFERENCES

Koshevoy E.P., Bljagoz H.R., 2000, Extraction dioxide of carbon in food technology. Maikop, Publishing house MGTI. (in Russian).

Meretukov Z.A., Koshevoy E.P. Preparation of vegetative raw material for extraction by extrusion processing. Materials of the 19th Internatonal Congress of Chemical and Process Engineering, CHISA 2010 and the 7th European Congress of Chemical Engineerring ECCE-7 Praha Czech Republic, 28 August-1 September 2010, Summaries 5 Systems and technology, p. 2110.

Tanaev V.S., Sotskov J.N., Strusevich V.A., 1989, Theory of schedules. Multiphasic systems. M.: Science. (in Russian).

Green Design, Materials and Manufacturing Processes – Bártolo et al. (eds)
© 2013 Taylor & Francis Group, London, ISBN 978-1-138-00046-9

Natural ventilation potential on thermal comfort of a light-steel-framing residential building

A. Craveiro
ISISE, Polytechnic Institute of Leiria, Leiria, Portugal

A. Gameiro Lopes
ADAI, University of Coimbra, Coimbra, Portugal

P. Santos & L. Simões da Silva
ISISE, University of Coimbra, Coimbra, Portugal

ABSTRACT: Nowadays, given the increasing comfort requirements that are imposed on buildings and the overall energy concerns, it is not reasonable dissociate ventilation from the thermal comfort. In fact, ventilation, as passive strategy, when properly studied and implemented on a building, may contribute to the thermal performance increase, and also, to minimize the energy expenditure required to achieve comfortable conditions with consequent improvement of building energy efficiency. In order to predict natural ventilation potential on the comfort increase, as passive strategy in a purely climatic analysis, a simplistic analysis on two different European scenarios is presented using the Autodesk Ecotect weather tool. In addition, some outcomes of numerical simulations, using Design Builder software, aims to highlight the importance of natural ventilation on the thermal behaviour of a light-steel-framing residential building: CoolHaven building.

1 INTRODUCTION

In Europe, buildings account for 30% of the annual emissions of greenhouse gases and consume about 40% of all energy (UNEP 2009). Revealing the importance about this issue, the European Commission adopted in 2002 a Directive about thermal performance of buildings (EPBD 2002), reformulated in 2010 (EPBD 2010), where ventilation of buildings assumes a great importance in this field.

Ventilation is a very important issue in buildings, mainly, due to the role that it assumes in the preservation of indoor air quality, durability of materials, thermal behaviour, and energy efficiency. In fact, ventilation when properly studied and implemented in buildings, assumes relevance as a measure with high potential contribution for the buildings energy efficiency.

Nevertheless, the selection and optimization of ventilation strategies are not easy, being the ventilation strategy choice and implementation very important at design and construction stages, respectively. A major difficulty lies in the dual scenarios which can be established for buildings ventilation: a reduced ventilation leads to problems of indoor air quality and, on the other hand, a larger ventilation can cause discomfort and waste of energy in buildings (PNNL & ORNL 2010). For these reasons and attending to the problem relevance and complexity, carefully attention and a suitable study of ventilation of buildings is justified.

2 ECOTECT WEATHER TOOL ANALYSIS

To get a better understanding of natural ventilation potential in thermal comfort, an analysis will be performed using the Autodesk Ecotect weather tool, on two European climatic scenarios: Coimbra (Portugal, PRT) and Karlstad (Sweden, SWE). This computational tool enables architects and engineers to use a wide range of features that help to get a better and earlier understanding of the effects of climatic factors (sunshine, temperature, shading, lighting and ventilation) on buildings performance (Ecotect 2012). This paper intend to predict natural ventilation potential, as Passive Design Strategy (PDS), on the comfort increase, in an purely climatic analysis applied to two different European climate scenarios (Southern and Northern Europe). Thus, this simplistic approach, does not consider the building geometry (building shape, openings size and position, etc.).

2.1 Climate classification

The Autodesk Ecotect weather tool needs reference weather data files for the cities under study.

The weather file for the city of Coimbra was created based on the information contained in the program SOLTERM-LNEG (2011), recommended by Portuguese legislation, and supplemented with additional data (precipitation, wind direction and intensity and atmospheric pressure), obtained from the International weather for energy calculations IWEC (2011). Nevertheless, the climate data file of Karlstad was fully obtained from the climate database IWEC (2011). Detailed climate information for the two cities under study is presented in Table 1. According Köppen classification, the climate for Coimbra is type "Csb" (warm temperate, warm summer dry) and the climate for Karlstad is type "Dfb" (cold winter (snow), fully humid and warm summer).

Although strongly dependent on weather conditions, natural ventilation as passive comfort measure, is favoured by the location of Portugal (with major influence of Mediterranean and Atlantic climate), earning 6 points as energy saving measure on a scale ranging from 0 to 7 (less to more important) according to UNEP (2007). Analogously, taking into account the previous classification, and considering the climatic zone of Sweden (Continental), natural ventilation as passive comfort measure earns 4 points as energy saving measure. According to the foregoing considerations, the contribution of natural ventilation as a passive design strategy to improve comfort and energy efficiency of buildings is well evident.

2.2 Additional comfort

The Autodesk Ecotect weather tool allows the quantification of comfort percentage achieved, with and without natural ventilation as PDS, according to a psychrometric analysis. Pursuant to this, the additional comfort represents the difference between the comfort percentage evaluated with and without natural ventilation as PDS, respectively. However, results may vary according to the level of activity and the air speed. This computational tool enables the classification of the activity level as: low, sedentary, light, medium and heavy, and the air speed as: still, pleasant, noticeable, draughty, and annoying. It is not desirable that an office goes beyond "noticeable" as this can cause undesirable effects; however, anything too far below "pleasant" can lead to a feeling of stuffiness (Ecotect 2012).

This paper presents the results of the additional comfort percentage, according to the air speed and the activity level, on a monthly (see Fig. 1) and an annual (see Fig. 2) analysis.

Figure 1 illustrates the additional comfort percentage for the two European cities under study:

Table 1. Climate data for Coimbra and Karlstad (SOLTERM-LNEG 2011, IWEC 2011).

	Coimbra	Karlstad
Köppen classification	"Csb"	"Dfb"
Summer duration	Jun–Aug	Jun–Aug
Extreme summer week*	34 ± 10.977°C	26.30 ± 7.471°C
Typical summer week**	20.28 ± 0.372°C	15.67± 0.255°C
Winter duration	Dec–Feb	Dec–Feb
Extreme winter week***	2.20 ± 5.054°C	−19.50 ± 10.294°C
Typical winter week**	9.40 ± 0.448°C	−2.89 ± 0.115°C
Autumn duration	Sep–Nov	Sep–Nov
Typical autumn week**	15.89 ± 0.658°C	5.82 ± 0.313°C
Spring duration	Mar–May	Mar–May
Typical spring week**	13.19 ± 0.114°C	4.90 ± 0.655°C

*Nearest maximum temperature; **Nearest average temperature; ***Nearest minimum temperature.

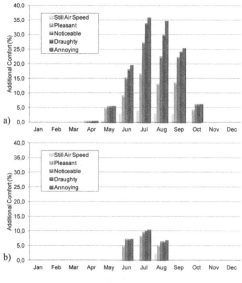

Figure 1. Additional comfort percentage, for different air speeds and sedentary activity (monthly values): a) Coimbra (PRT); b) Karlstad (SWE).

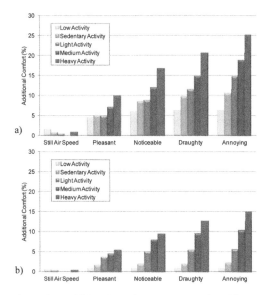

Figure 2. Additional comfort percentage, for different air speeds and activity levels (annual values): a) Coimbra (PRT); b) Karlstad (SWE).

Coimbra (Fig. 1a) and Karlstad (Fig. 1b), in a monthly analysis, for a sedentary activity and different air speeds.

The monthly analysis presented in Figure 1a shows, for Coimbra, a null percentage of additional comfort in the months of November to March, probably because these months are characterized by lower temperatures and, therefore, natural ventilation is not useful to increase the occupants comfort. Hence, the additional comfort becomes gradually relevant only in the months of April to July, induced by the higher temperatures. Inversely, a slight decay on the additional comfort percentage is verified in the months of August to October, compatible with the temperatures decrease. Notwithstanding, the additional comfort percentage is higher in July (higher temperatures), reflecting the importance of natural ventilation for cooling the buildings. Moreover, it is verified that the air speed variation has a greater impact on the warmer months (June to September), due the greater detachment verified in the additional comfort percentage during these months (Fig. 1a).

Similarly, the monthly analysis presented in Figure 1b shows, for Karlstad, a null additional comfort percentage in the months of September to May, compatible with the assumptions referred previously. Moreover, the additional comfort becomes gradually relevant in the months of June and July. However, the relevance of natural ventilation on comfort for Karlstad is smaller than in Coimbra, being higher in the warmer month (July).

Nevertheless, in Karlstad it is verified that the air speed variation has a smaller impact in the provided additional comfort, as illustrated in Figure 1b.

Figure 2 illustrates the additional comfort percentage for the two European cities under study, in an annual analysis, and for different air speeds and activity levels. In general terms, for both cities, one may note a gradual increase on the additional comfort percentage with the increase of air speed velocity and the level of activity (highest values for "annoying" air speed and heavy activity). However, the presented values reveal that, as expected, natural ventilation as PDS is a more efficiency strategy for the city of Coimbra.

3 DsBUILDER NUMERICAL SIMULATION

The numerical simulations are intended to predict and highlight the importance of ventilation on the thermal behaviour of the CoolHaven building.

3.1 *CoolHaven case study*

The CoolHaven building is a residential building of typology T3, located at the village of Antanhol, Portugal, and represents one of the several types of buildings that CoolHaven company offers (CoolHaven 2012).

This innovative solution is on the market since 2009, and it is intended to meet the requirements of (i) aesthetics, (ii) functional and environmental efficiency, (iii) easy transportation and assembly, and (iv) differentiated costs (Gervásio et al. 2011). A major concern of this project is the thermal performance and building energy efficiency, to be optimized through the implementation of strategies that are aimed at the minimization of the energy consumption and, simultaneously, at the reduction of the energy needs through renewable energy sources.

3.2 *Modelling*

The computational model of the Coolhaven building, illustrated on Figure 3, was obtained using Design Builder software (DsBuilder 2011), that uses the Energy Plus as calculation processor (Energy Plus 2011). The model comprises eight thermal zones that correspond to the several divisions of the CoolHaven building (Craveiro et al. 2011).

The U-Factor values of the opaque envelope and glazed areas are indicated in Table 2. The linear thermal bridges were discarded on the modelling options. Additionally, other conditions were assumed for the computational formulation. In this sense, it was considered a building occupation for 4 persons. In addition, the internal gains

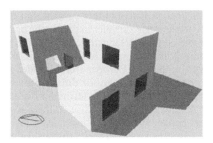

Figure 3. Design builder model of CoolHaven building: Southern perspective.

Table 2. U-Factor values of CoolHaven opaque envelope and glazed areas.

	U [w/m²ºC]
External walls	0.27
External pavement	0.38
Internal pavement	0.36
Covering	0.29
Glazed areas	0.63

were considered according to RCCTE (2006) for a residential use type (4 W/m²), and the shading devices were programmed to be activated whenever the solar radiation exceeds the value of 120 W/m². In order to reproduce the local weather conditions as close to reality in the computational analysis, a single weather file compiling all relevant climate information for the city of Coimbra was used (Section 2.1).

3.3 Approaches and scenarios

The analysis of the thermal behaviour of the Cool-Haven building was performed following two different approaches, which gather various scenarios, as shown in Table 3. The first approach seeks to impose a constant renewal air rate through the building. This approach includes three distinct scenarios with constant values for the renewal (0.0, 0.6 or 6.0 ACH—Air changes per hour).

Notwithstanding, the second approach is defined by a variable renewal air rate. Hence, this ventilation rate is defined in a range of values bounded below by a minimum value of 0.6 ACH, to ensure indoor air quality (RCCTE 2006), and above by a maximum value (6 ACH). In order to avoid overheating, in approach 2 the ventilation is reduced to a minimum (0.6 ACH) whenever the outside temperature is higher than the indoor temperature. Additionally, the control of ventilation is of great importance, and it is foreseen through

Table 3. Description of approaches and ventilation scenarios used for computational analysis.

Approach	Scenarios	ACH [h⁻¹]	SPT [ºC]
Approach 1*	Scenario 1	0.0	–
	Scenario 2	0.6	–
	Scenario 3	6.0	–
Approach 2**	Scenario 4	[0.6; 6.0]	25
	Scenario 5	[0.6; 6.0]	20

*Constant ventilation; **Variable ventilation.

the definition of two ventilation activation temperatures (Set-Point Temperature or SPT): 25ºC (Scen. 4) and 20ºC (Scen. 5). These temperatures were defined based on the reference comfort conditions: 20ºC for the heating season and 25ºC for the cooling season RCCTE (2006).

3.4 Thermal behaviour

The thermal behaviour analysis of the CoolHaven building was done based on the indoor temperature dependence with the ventilation rate, considering a passive analysis (disabled cooling system). Simulations were performed for a typical summer week (10–16 August) in an hourly schedule. The presented results are average values obtained for the entire building, for the various scenarios according to the approaches 1 and 2. Figure 4a illustrates the inner temperatures obtained for constant ventilation rate (Approach 1), for a typical summer week. Analysis of this figure, shows that the indoor temperatures in the absence of ventilation (Scen. 1) reach the highest average values when compared with the other considered scenarios. As expected, the ventilation rate increase (Scen. 2 and Scen. 3) leads to a decrease in the indoor average temperatures, due the effective cooling ventilation capacity. In general terms, the ventilation cooling capacity increases with the renewal air rate, showing better average results for Scen. 3 (–4.7ºC compared to Scen. 1). However, because these scenarios are characterized by constant ventilation, they are not optimized. Thereby, an uncontrolled ventilation during the cooler periods of the day can lead to a decrease of the indoor temperature below 20ºC (winter comfort temperature), as observed on Scen. 3 (Fig. 4a). In fact, a constant air renewal rate can help to cool the spaces in summer. However, it is important to control the ventilation rate during the warmest periods of the day, whenever the outdoor temperature exceeds the indoor temperature, to avoid the risk of overheating the building. A similar reasoning can be applied to avoid an unnecessary cooling on periods of the day with lower temperatures.

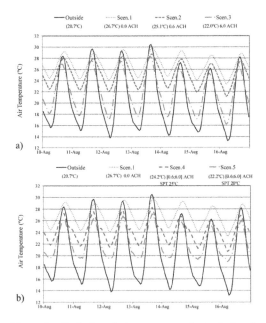

Figure 4. Evolution of air temperature for typical week summer (Coimbra): a) constant, and b) variable renewal air rate.

Figure 4b illustrates the inner temperatures obtained for an imposed variable renewal air rate (Approach 2), as described in Table 3.

A SPT 20°C is characterized by higher thermal amplitudes and thereby by an higher average cooling of 4.5°C for Scen. 5, assuming Scen. 1 as reference. On the other hand, a SPT 25°C is characterized by lower thermal amplitudes and by an average cooling of 2.5°C for Scen. 4, considering Scen. 1 as reference. This situation (Scen. 4) can be justified taking into account that natural ventilation is only triggered when the inner temperature reaches 25°C and, probably, in these circumstances, the outside air temperature will be higher than the indoor temperature. Thereby, ventilation will be reduced to a minimum rate, which leads to a much reduced ventilation period and consequently to a reduced ventilation cooling capacity. Unlike what happens considering a constant renewal air rate, it is verified that when the outside temperature is markedly reduced during the night, the internal temperature remains in an acceptable range of values of approximately: 22°C for STP 25°C and 20°C for STP 20°C.

4 CONCLUSIONS

Through the ECOTECT climatic analysis, it was possible to assess that natural ventilation potential

assumes more importance in summer time to cool buildings. For the two tested locations, Coimbra (Southern Europe) and Karlstad (Northern Europe), the first is the one that benefits most from natural ventilation and shows more sensitivity to airspeed. Therefore, the monthly analysis (sedentary activity and an "annoying" air speed) reveals, for July, an additional comfort percentage for Coimbra about 25% times higher than in Karlstad. Additionally, in the same month it is verified, for Coimbra, an additional comfort variation of about 32.5%, considering the air speed variation, unlike to the smaller impact of 10% observed for Karlstad. The annual analysis, shows an evident increase of additional comfort percentage with the increase of the air speed and the level of activity. This situation, reveals significant results for Coimbra, with an increase of additional comfort percentage of about 10% times higher than in Karlstad, considering a heavy activity and an "annoying" air speed.

Several numerical simulations were also performed in order to relate ventilation and thermal behaviour of a light-steel-framing residential building. The passive analysis allow to concluded that a constant ventilation leads to an effective average cooling of the building in summer. Furthermore, it was verified that the average cooling capacity is higher for the maximum renewal air rate i.e. 6.0ACH (−4.7°C for Scen. 3 considering Scen. 1 as reference). However, controlled ventilation allow to obtain better indoor temperatures conditions, during the outside temperature peaks (more severe conditions). Therefore, the obtained results reveal the importance on a SPT definition. In fact, a SPT of 20°C lead to better results than a SPT of 25°C, allowing an average cooling of 4.5°C and 2.5°C, respectively. It was concluded that ventilation can help to an effective cooling of spaces in the summer station. However, the control of ventilation is essential and should be regarded in terms of two essential parameters: i) relation between the outside temperature and indoor temperature, ii) indoor activation temperature.

ACKNOWLEDGEMENTS

Financial support from QREN 2009 under contract n°. 2009/5527 is gratefully acknowledged.

REFERENCES

CoolHaven 2012, www.coolHaven.pt. Accessed in February, 2012.
Craveiro, A. et al. 2011. Ventilação e eficiência energética em edifícios residenciais com estrutura em aço—caso de estudo. *VIII Congresso de Construção Metálica e Mista*: II-265–274.

DsBuilder 2011, www.designbuilder.co.uk. Accessed in 2011.

Ecotect 2012, www.students.autodesk.com. Accessed in November, 2012.

EnergyPlus 2011, www.energyplus.gov. Accessed in October, 2011.

EPBD 2002, Directive 2002/91/EC. Directive of the European Parliament and of the council of 16 December 2002 on the Energy Performance of Buildings.

EPBD 2010, Directive 2010/31/EU. Directive of the European Parliament and of the council of 19 May 2010 on the Energy Performance of Buildings.

Gervásio, H. et al. 2011. Construção metálica modular e eco-eficiente. *VIII Congresso de Construção Metálica e Mista*: II-255–264.

IWEC 2011, International Weather for Energy Calculation: apps1.eere.energy.gov/buildings/energy-plus/weatherdata_about.cfm. Accessed in October, 2011.

PNNL & ORNL 2010, Pacific Northwest National Laboratory & Oak Ridge National Laboratory. *Building America Best Practices Series. Retrofit Techniques & Technologies: Air Sealing*. Building Technologies Program. Office of Energy Efficiency and Renewable Energy. United States Department of Energy, Volume 10.

RCCTE 2006, Portuguese Law: Decreto-Lei n.º 80/2006. Regulamento das Características de Comportamento Térmico dos Edifícios.

SOLTERM-LNEG 2011, Software for performance analysis of solar systems in Portugal. Accessed in October, 2011.

UNEP 2007, United Nations Environment Programme. *Building and Climate Change—Status, Challenges and Opportunities*. ISBN: 978-92-807-2795-1.

UNEP 2009, United Nations Environment Programme— Sustainable Buildings & Climate Initiative. *Building and Climate Change—Summary for Decision-Makers*. Sustainable Buildings & Climate Initiative. ISBN: 987-92-807-3064-7.

Green Design, Materials and Manufacturing Processes – Bártolo et al. (eds)
© *2013 Taylor & Francis Group, London, ISBN 978-1-138-00046-9*

Investigation of the feasibility of constructing a biofuel plant in the region of North Evros

A.N. Papadopoulos & M. Tsatiris
Democritus University of Thrace, Orestiada, Greece

ABSTRACT: This study investigates the possibility of setting up a plant for biofuel production in the region of North Evros. This is mainly an Ex Ante financial evaluation, in account of a potential investor who would be prepared to invest in a biofuels' plant, in order to assist the decision-making process. At the same time, the impact of an investment of that kind on the rural population of north Evros prefecture and consequently on the local economy, is also examined but it isn't analyzed. The methodology followed is purely financial and the net present value, the internal rate of return and the payback period of the investment period, were estimated. The results in this study are in favor of creating a biodiesel plant in North Evros.

1 INTRODUCTION

1.1 *General*

This thesis investigates the viability of a biofuel plant—e.g. liquid fuel from biomass conversion—in the region of N. Evros and the consequent impacts, the transition from traditional crops to energy crops would cause to the agricultural production and farmers' income of the region.

1.2 *Scope of the thesis*

This is primarily an Ex Ante economic evaluation on behalf of the potential investor, who considers investing in a biofuel production plant, in order to assist the decision making process (Simpson & Walker 1987). This is done by taking into account the capital and operating cost of the plant and the cash income of the business. It is also investigated the optimum type and size of the plant and some aspects of the supply chain.

At the same time, it is also evaluated, the impact that this investment will have in the rural population of N. Evros and consequently to the local economy, the ability of available land to supply the factory with necessary raw material and the possible changes of local agricultural production.

Finally this thesis tries to achieve a coupling of economic and technical data and sustainability aspects of the two poles of bioenergy, i.e. the farmers on one hand, and investors on the other.

2 LITERATURE REVIEW

2.1 *Energy evolution*

Considering all above, it is clear that an energy policy should reflect a balance between state intervention trying to ensure energy security and environmental protection and, on the other hand, a deregulated, competitive market with the aim of energy costs reduction. Although each state or international geopolitical region poses different objectives and priorities for energy policy (Carriquiry 2007), depending on the level of economic development and energy status (importer, producer, exporter of energy), the categorization of these objectives may be based on the following three goals:

– 1st goal: Ensuring energy supply
– 2nd goal: Competitiveness and efficiency of energy sector
– 3rd goal: Environmental protection.

2.2 *Environmental impacts of fossil fuel*

The greater environmental impact of fossil fuel, among many, is the emission of carbon dioxide, which is the main greenhouse gas. The burning of fossil fuels for electricity, heat and transport combined adds about 25 billion tons of carbon dioxide (CO_2) into the atmosphere each year. The increasing concentration of carbon dioxide (CO_2) in the atmosphere, produced by burning fossil fuels, is the main reason of climate change through the greenhouse effect.

2.3 Biofuels

Biofuels are any solid, liquid and gaseous fuel derived from renewable sources (plants, biomass etc). The most common are:

a. biodiesel which is produced by vegetable oils and animal fats through a process called trans-esterification. It can be used in existing diesel engines either solely or in admixture with diesel fuel.
b. bioethanol which is produced by plants containing sugar, cellulose and starch through a process called alcoholic fermentation. It can be used in admixture with gasoline in existing petrol engines after little or no modifications, depending on the content of the mixture.
c. biogas (methane) which is produced by the decomposition of organic waste
d. biomass, i.e. residues of agricultural crops and forest products.

2.4 Advantages of biofuels

Biofuels, being produced by renewable sources, are clean, non-toxic and do not contain dangerous compounds for human health. Probably, their most important advantage is that during combustion they do not affect carbon dioxide (CO_2) balance in the atmosphere, because the plants we use to produce biofuels, had previously absorbed the same amount of the CO_2 during the photosynthesis process.

This fact—i.e. keeping the balance of carbon dioxide in the atmosphere—as well as the reason that the raw materials we use to produce biofuels, can be cultivated in Europe's countries' fields have led the EU to enforce the use of them by the member countries, with a specific timetable (Steenberghen & Lopez 2008).

Moreover, biofuels' emissions of sulfur dioxide (SO_2) is zero or very low compared to conventional fuels, dew to their very low or zero sulfur content. Also, they don't contain aromatic hydrocarbons and in addition produce low emissions of nitrogen oxides (NO_x), carbon monoxide (CO), unburned hydrocarbons and soot (particulate matter) (Kouroussis, Karimi, 2008).

2.5 Disadvantages of biofuels

Apart from the environmental benefits arising from the use of biofuels, there are also some drawbacks.

I we want to have a broader picture of the environmental impact of the usage of biofuels, we must take into consideration their complete life cycle. The general environmental impacts and thus the disadvantages of the usage of biofuels must focus to the following.

The energy that is required for the production, cultivation, harvesting and transportation of raw materials, as well as the production and distribution of biofuels, is usually derived from fossil fuels.

The widespread and intensive cultivation of energy crops leads to monoculture, degradation of land and significant impacts on biodiversity (removal of birds and insects), water supply due to increased demands for irrigation of energy crops and soil quality.

The usage of fertilizers and pesticides based on compounds of nitrogen, sulfur and ammonia increases the acidity of the soil and water, while creating eutrophication conditions.

The engines burning biodiesel emit more nitrogen oxides (NO_x) than the ones burning fossil diesel.

The production of biofuels may today be more expensive than other ways of reducing emissions of carbon dioxide (CO_2).

In a global market of raw materials and fuel and in a global greenhouse gases emissions market, it is very likely that only the developed countries could take advantage of the benefits of the usage of biofuels (Lucia & Nilsson, 2007), by reducing emissions in the transport sector, while the disadvantages of plant cultivation and production of raw materials, will possibly damage the third world countries that devote large areas for energy crops. So while the developed countries seem to achieve their goals based on the Kyoto Protocol, developing countries will be faced with increased emissions, increased fertilizer usage and soil erosion for the production of "clean" biofuels (Smeets et al. 2005).

Finally, the growing demand for fuel can lead the poor, developing tropical and subtropical countries, where the cultivation is highly efficient (up to ten times more than the corresponding crops in temperate regions) to farm energy plants, reducing the land devoted to food production (L. Peskett et al. 2007). Such a practice would have dire consequences for the residents of these areas, since biofuel production will slightly improve their income and living standards, while they could be deprived their subsistence food.

2.6 What are biofuels produced from

Biodiesel is made by vegetable oils by the process of transesterification. These come mainly from rape-seed, sunflower, soybean, cottonseed and palm oil (Ralph et al. 2006). Biodiesel can also be produced by used cooking oils from restaurants and also by animal fats from slaughterhouses etc., which deliver though, fuel of lower quality.

Bioethanol is an alcohol produced by the fermentation of sugars derived by sugar beet, sugar cane, wheat, corn and sorghum sugar, even from wine (Zappi et al. 2003). Today research is conducted for future production of 2nd generation biofuels by materials of lesser value, as forest residues, biomass waste, stems etc., which will reduce the currently high cost of biofuels.

3 METHODOLOGY

The following data are the inputs for the cost benefit analysis (Clinch 2004):

– The study was undertaken on behalf of a potential investor and examines the viability of the investment.
– Base year for the study is 2009.
– The city of Orestiada-N. Evros was chosen as the plant's location, because the surrounding farmland holds traditionally the lead in growing energy crops in our country, with 70,847 thousand acres of sunflower (2008) while the production contracts keeps on growing at a large pace. Additionally raw material could be transported by neighboring area of Rhodope (Panoutsou, 1998; Panoutsou et al. 2000).
– The plant will be a vertically integrated unit, with sunflower seeds as raw material, exporting oil and converting it into biodiesel.
– The investment to be evaluated is a biodiesel plant with a capacity of 40,000 tons/year.
– The level of investment, operating costs, productivity, and selling prices of products are taken from the bibliography. (Boukis et al. 2008), (Edwards et al. 2008).
– Evaluating method is NPV. In addition the Internal Rate of Return (IRR) and payback period is also calculated. We took into account only what can be cash valuated, because this is what the potential investor is concerned about. This method is most appropriate when considering the profitability of an investment on behalf of the investor. Intangible or indirect values are taken into consideration in the discussion.
– For the calculations we used a simple and approachable template by "engineering solutions on-line", which is based on the MS-EXCEL.
– Total investment for the plant, according to the bibliography is 38.000.000 €. For the region of Evros this could be subsidized up to 60% by the Government (Development Act). Also the 20% of the investment will be granted by a bank as a loan at the current interest rate of 4%, with 10 years repayment period. So the directly invested capital is calculated at 20% of the investment, ie 7.600.000 €.

– Operating expenses were calculated according to the bibliography (Boukis, et al. 2008) at 7.806.400 €/year.
– The sunflower was selected as the main raw material, because is a plant traditionally cultivated in the area (Kallivroussis et al. 2002).

4 DISCUSSION OF RESULTS

The results obtained by the previously analyzed methodology are presented below. On the first part

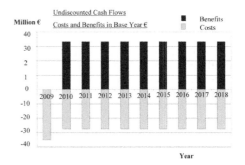

Figure 1. Diagram of undiscounted cash flows.

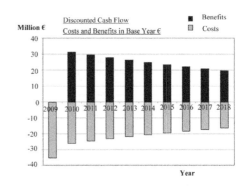

Figure 2. Diagram of discounted cash flows.

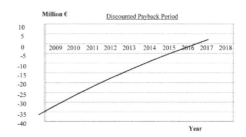

Figure 3. Diagram of discounted payback period.

the cash flows of investment and operational costs were calculated. On the second part the cash flows of benefits were calculated. On the third part are the results showing the NPV and the IRR. Then we present the following charts:

– Chart of undiscounted net cash flow
– Chart of discounted net cash flow
– Chart of payback period.

Finally, in order to examine the sensitivity of the results, a sensitivity analysis was performed. This was done by the following method using MS-Excel:

Regarding the cost, it was chosen to examine the sensitivity depending only by the operating cost, because the cost of raw material we assume is connected to the selling price of biodiesel. Therefore increases in the price of raw materials will subsequently increase the selling price of biodiesel. Furthermore it is also examined through the variance of the selling price of biodiesel, setting the value of raw material as a constant.

Regarding the benefits, it was chosen the sensitivity of the results for the price of biodiesel to be examined, which is the largest percentage of benefit comparing to byproducts.

The fluctuation range was chosen to be from −10% to +10% for each data, in steps of 5% (Mergos, 2007).

For each combination of two input data—costs and benefits—the new resulting IRR was calculated.

Table 1. Historical price progress of sunflower in Orestiada region.

Year	Price (€/t)
2006	200
2007	250
2008	280
2009*	300
2010	370
2011	400
2012	420

*Base year.

Table 2. Product prices for base year 2009.

Year	Price (€/t)
Biodiesel	730
Pie	150
Glycerin	200

As shown in following tables, the results of the estimated net present value of the investment in a ten year horizon is 3.508.601 €, while the internal rate of return is estimated at 8%.

Below are presented the following charts:

– Chart of undiscounted net cash flow
– Chart of discounted net cash flow
– Chart of payback period.

5 CONCLUSION

With an IRR at 8%, the investment in question is an attractive one for the potential investor. But in order to have a wider view of our subject we must discuss some other aspects too. These are:

– The attractiveness of the investment for the entrepreneur
– The impact of the investment on the local economic environment and the rural income of the region
– The impact of investment on the local and general natural habitat

The first aspect was answered by the econometrical results of this work. Regarding the second, the progress of prices of sunflower in Orestiada's region (Table 1), promotes the cultivation of sunflower as particularly advantageous for farmers during recent years, since the framework for introducing biofuels in the Greek energy market was implemented. So while in 2006 farmers received around 200 €/ton, in 2010 the Union of Agricultural Cooperatives of Orestiada bought sunflower crop at 370–390 €/ton and 2012 at 420 €/ton. In addition, cost of production has not increased accordingly. This is a very positive development for the economic environment of the region of N. Evros, during a difficult circumstance for the rural income.

Therefore the introduction of biofuels in the Greek market had a clear positive impact on cultivation of sunflower and on farming income. Thus, the Union of Agricultural Cooperatives of Orestiada, must be the first to consider an investment in the construction of a biodiesel factory, improving the financial outcome of the institution itself and at the same time the financial rewards of local farmers (Iliadis et al. 2004).

As for the third aspect of the discussion, the investment will not have any impact on the local natural habitat, neither good nor bad. Speaking generally though, the dependency of the modern world from fossil fuels and increasing environmental problems of the planet, leads to the need to find alternatives. Such a solution is likely to be biofuels produced from agricultural raw materials. The study of their life cycle shows that their effects on the environment is more favorable compared to fossil fuels, provided, however, that local soil conditions, irrigation and climate are favorable (Edwards et al. 2008).

Moreover, if you can use the byproducts of fuel production appropriately, e.g. heat and electricity coproduction (CHP) (Gustavsson 2011), then the balance in favor of biofuels is definitely positive (Panoutsou 2008). The environmental impact of both biofuels' and fossil fuels' usage, are not expected to change drastically in the future. However, the advantages of biofuels will probably increase, when the new technologies and processes—such as the production of 2nd generation biofuels (Zinoviev et al. 2007), or the use algae (seaweed) grown in controlled tanks—are implemented.

So biofuels may not be a global solution to the environmental problems of fossil fuels. But they can certainly be part of the solution (Edwards et al. 2008), along with other options like regulations for vehicles' emissions reduction, plug-in hybrid and electric cars, electricity production by wind and solar power, alternative use of biofuels, e.g. biomass burning and increased investments in energy efficiency of buildings and industrial processes.

Looking locally, however, biofuels can be an important step in the development of the crisis ridden rural economy of the region.

REFERENCES

Boukis I., Vassilakos N., Kontopoulos G., Karellas S., 2008. Policy plan for the use of biomass and biofuels in Greece. P1. *Renewable Sustainable Energy Review.* doi:10.1016/j.rser.2008.02.007.

Boukis I., Vassilakos N., Kontopoulos G., Karellas S., 2008. Policy plan for the use of biomass and biofuels in Greece. P2. *Renewable Sustainable Energy Review.* doi:10.1016/j.rser.2008.02.008.

Carriquiry, M. 2007. A comparative analysis of the development of the United States and European Union biodiesel industries. *Briefing Paper 07-BP 51.* Center for Agricultural and Rural Development-Iowa State University-Ames, Iowa 50011-1070-www.card.iastate.edu.

Clinch, J. 2004. Cost–Benefit Analysis Applied to Energy. *Encyclopedia of Energy.*, Volume 1, Elsevier.

E.C. DG. AGRI G-2/WM D, 2007. Economic analysis, perspectives and evaluations. Economic analysis of EU agriculture: The impact of a minimum 10% obligation for biofuel use in the EU-27 in 2020 on agricultural markets-Ref.: *Impact assessment of the Renewable Energy Roadmap*, Brussels, 30 April 2007.

Edwards, R., Szekeres, S., Neuwahl F and V. Mathiew. 2008. Biofuels in the European Context: Facts, Uncertainties and Recommendations. *JRC EUR 23260 EN–2008. DOI 10.2788/69274*, Petten (N.-H.) Netherlands.

Gustavsson, L. 2011, Coproduction of district heat and electricity or biomotor fuels. *Energy* Vol 36, No 10: 6263–6277.

Iliadis, L., Koutroumanidis T. and Arabatzis G. 2004. Evaluation and ranking of the financial status of the Greek rural cooperatives union by a Decision Support System. *Agricultural Economics Review.* Jan 2004, Vol 5, No 1.

Kallivroussis, L., Natsis, A. and G. Papadakis. 2002. The energy balance of sunflower production for biodiesel in Greece. *Biosystems Engineering.* 81(3): 347–354.

Kouroussis, D and S. Karimi. 2006. Alternative Fuels in Transportation. *Bulletin of Science Technology Society*, 26; 346, DOI: 10.1177 /0270467606292150.

Lucia, L. and L. Nilsson. 2007. Transport biofuels in the European Union. *Transport Policy.* 14: 533–543.

Mergos, G. 2007. *Socio-economic investment evaluation.* G. Benos Publications, Athens.

Panoutsou, C. 2008. Bioenergy in Greece: Policies, diffusion framework and stakeholder interactions. *Energy Policy* 36: 3674–3685.

Panoutsou, C., Namatov, I., Lychnarasb, V. and A. Nikolaou. 2007. Biodiesel options in Greece. *Biomass and Bioenergy.* 32: 473–481.

Peskett, L., Slater, R., Stevens, C and A. Dufey. 2007. Biofuels, Agriculture and Poverty Reduction. *Natural Resource Perspectives* 107, Overseas Development Institute 2007, ISSN 1356–9228.

Ralph, E., Sims, H., Hastings, A and P. Smith. 2006. Energy crops: current status and future prospects. *Global Change Biology.* 12: 2054–2076.

Simpson, D and J. Walker. 1987. Extending cost-benefit analysis for energy investment choices. *Energy Policy.* (15) 3: 217–227.

Smeets, E., Junginger, M and A. Faaij. 2005. Supportive study for the OECD on alternative developments in biofuel production across the world. *Report NWS-E-2005-141, ISBN 90-8672-002-1*, Copernicus Institute for Sustainable Development, Department of Science, Technology & Society–Utrecht University.

Steenberghen, T and E. Lopez. 2008. Overcoming barriers to the implementation of alternative fuels for road transport in Europe. *Journal of Cleaner Production.* 16: 577–590.

Zappi, M., Hernandez, R., D. Sparks, D., Horne, L. and M. Brough. 2003. A Review of the engineering aspects of the biodiesel industry. Mississippi University Consortium for the Utilization of Biomass. *MSU E-TECH Laboratory Report ET-03-003.* Mississippi Biomass Council, Jackson, MS.

Zinoviev, S., Arumugam, S. and S. Miertus. 2007. Biofuel Production Technologies. *WORKING DOCUMENT prepared by Area of Chemistry*, ICS-UNIDO.

Green Design, Materials and Manufacturing Processes – Bártolo et al. (eds)
© *2013 Taylor & Francis Group, London, ISBN 978-1-138-00046-9*

Comparison of working (operating) efficiency of electric and gas heat pump regarding to renewable heat source

D. Rajković
Faculty of Mining, Geology and Petroleum Engineering, Zagreb, Croatia

M. Sentić
INA, Oil Industry PLC, Zagreb, Croatia

ABSTRACT: Rational use and managing with energy is basic presumption of sustainable development. Economy with energy today is directed to using and promotion of pure technologies, with high energetic efficiency and using of renewable sources, and all in purpose of environmental protection. During the last decade heat pump technology has attracted increasing attention as one of the most promising technologies to save energy. Exactly, heat compressor pumps represent technology with high energetic efficiency. Heat pump is a device for moving heat from a low temperature heat source to a higher temperature heat sink using the power for compressor running. In most cases, electric engine runs the compressor. The efficiency of electric heat pump is defined with Coefficient of Performance (COP—heating factor). It is defined as the ratio of heat delivered by the heat pump and the electric power for compressor running and auxiliaries. But, COP is not observed from primary energy source, it is observed after fuel energy transformation into electric energy (after primary energy transformation). In this article the electric heat pump running will be compared with the gas heat pump running from primary energy source (PER—primary energy ratio). PER is ratio between useful heating energy and primary energy. Gas heat pump uses a primary energy source without energy transformation and has got a lower CO_2 emission. In this article, the major advantage of gas heat pump, the utilization of exhausted heat from gas engine (from cooling water and exhausted gases) will be described. Also, it will be described the two renewable heat sources: air and sea water with advantages and disadvantages.

1 INTRODUCTION

The basic idea of all heat pump concepts is that heat source is absorbed by a medium, which releases the heat at a higher temperature. During the last decade heat pump technology has attracted increasing attention as one of the most promising technologies to save energy.

What is the heat pump?

Heat pump is a device for moving heat from a low temperature to a higher temperature level using power for compressor running.

Useful heat delivered by the heat pump is greater (3 to 4 times) from the power supplied to the compressor and auxiliaries (fans or pumps).

So, with using environmental renewable heat source (air or sea water) heat pump produces valuable useful heat on a higher temperature level.

Table 1. Basic characteristics of renewable heat sources: air and sea water.

	AIR/ZRAK	SEA WATER/MORE
Criterions for utilization (using)	Outside air	Sea water
Temperature or energetic level	−25 °C deg C to +20 deg C	Minimum 0 deg C
Site availability	Everywhere	Somewhere
Time availability	Always	Always
Time compatibility of consumption and available energy	No coherent, at least energy on availability when is neccessary	Coherent
Possibility of independetly utilization	Yes	Yes
Chemichal or physical properties which complicate utilization	Ice forming	Soils, salts, seaweds
Influence on energetic environment balance	No significant influence	Mostly negligible

The text should fit exactly into the type area of 187 × 272 mm (7.36" × 10.71"). For correct settings of margins in the Page Setup dialog box (File menu) see Table 1.

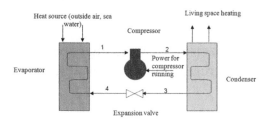

Figure 1. Heat pump scheme.

2 CHARACTERISTICS OF HEAT SOURCES FOR HEAT PUMPS

Without renewable heat source heat pump cannot operate. Renewable heat source must ensure safety, reliable and economical work.

Heat pump demands for heat source are: temperature level, availability on site, time availability, conformity between heat source and heat demands, possibility of independent utilization, energy consumption between energy source and evaporator, influence on the environment balance and environment pollution.

Most important demands for renewable heat source are:

heat source need to ensure always amount of heat and on higher temperature level (if it is possible), energy for transportation from heat source to evaporator needs to be as least as possible, expenses for connection needs to be as least as possible (Pavkovic, 2006).

Main characteristics of air and sea water are shown in Table 1.

Air as the renewable heat source is mostly used, but has got some limitations. Air limitations are: temperature variability, unfavorable thermodynamics properties (low specific heat, low heat conductivity). The advantages of sea water as renewable heat source are: excellent thermodynamics properties and approximately constant temperature through whole year of the heating period.

3 BASIC PRINCIPLE OF COMPRESSOR HEAT PUMP

The majority of heat pumps work on the principle of the vapor compression cycle. An air or sea water source heat pump extract heat from cold outside air or sea water in the winter. Heat pump scheme is shown on Figure 1, while the T-s diagram process scheme is shown on Figure 2. In this cycle, the circulating substance (refrigerant) is physically separated from the heat source (air, with a temperature of T_{in}) and user (heat to be used in the process, T_{out}) streams, and is re-used in a cyclical fashion, therefore called 'closed cycle'. In the heat pump, the following processes take place:

1. In the evaporator the heat is extracted from the heat source (air or sea water) to boil the circulating substance;

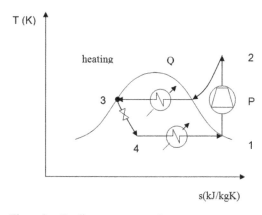

Figure 2. T-s diagram process scheme.

2. The circulating substance (refrigerant) is compressed by the compressor, raising its pressure and temperature;
3. The heat is delivered to the condenser and use for space heating;
4. The pressure of the circulating substance (refrigerant) is reduced back to the evaporator condition in the expansion (throttling) valve.

Useful heating output:

$$Q = m x (h_2 - h_3) (kW) \qquad (1)$$

Power for compressor running:

$$P = m x (h_1 - h_2) (kW) \qquad (2)$$

Heat is extracted (Q_o) from the heat source (air or sea water) and rejected at a higher temperature level at T_{out}. Useful heat Q at higher temperature is shown on Figure 2. The difference between the temperatures where the heat is extracted—the "source", and the temperature where the heat is delivered—the "sink" is called the lift. For a heat pump, the typical sources and sinks available translate into typical pressure differentials (lift) between which compressor and auxiliaries. COP is greater if the difference between heat source and sink is lower (lift). The smaller the lift, the higher efficiency will be, so the COP value could reach even more than 4.

What is difference between electric and gas heat pump?

The basic difference is in the power for compressor running. Electric heat pump uses electric energy for compressor running, while gas heat pump uses gas engine for compressor running. Gas heat pump can utilize exhausted heat from the gas engine (from cooling water and exhaust gases) and operate at low outside air temperatures with the same heat capacity. At electric heat pump heat capacity decreases on lower temperatures.

3.1 Heat pump electric efficiency—COP

The performance of a heat pump is theoretically the product of the heat extracted from the low temperature heat source and the power needed to drive the cycle. The steady state performance of an electric compression heat pump at a given set of temperature conditions is referred to as the Coefficient of Performance (COP). COP is observed after fuel energy transformation into electric energy (after primary energy transformation). It is defined as the ratio of heat delivered by the heat pump and the electric power for compressor running.

$$COP = \frac{\text{Useful heating output (kW)}}{\text{Compressor and auxilliares power output (kW)}}$$

(3)

3.2 Primary energy ratio

PER (Primary Energy Ratio) is observed from primary energy and represents the whole efficiency of the heat pump system. PER is defined as ratio between useful heating and primary energy.

$$PER = \frac{\text{Useful heating energy (kW)}}{\text{Primary energy (kW)}}$$

(4)

4 ELECTRIC HEAT PUMP DISADVANTAGES

Electric heat pump with air as renewable heat source is limited at lower outside temperatures. Today, the electric heat pumps are very popular for heating living space. But, during winter at very low outside temperature the electric heat pump has operating problems and heating capacity decreases. In fact, electric heat pump doesn't run under −5 °C. Another problem is efficiency at lower outside temperature. COP rapidly decreases at temperature of 1 °C approx., even until 30% at temperature below

0 °C. And the cost of heating increases. With temperature drops it is more difficult to get a targeted temperature in the living space.

5 GAS HEAT PUMP ADVANTAGES

Gas engine driven heat pumps are very similar to electric motor driven heat pumps.

The advantages of gas engine driven heat pumps are:

– Use a primary energy without energy transformation;
– Greater PER (primary energy ratio);
– Utilization of exhausted heat from the gas engine (from cooling water and exhaust gases);
– Ability to recover waste heat;
– No defrosting operations;
– A smaller electrical service;
– Longer durability.

The major advantage of a gas heat pump is utilization of exhausted heat from the gas engine (from cooling water and exhaust gases).

In order to recover engine exhaust heat efficiently, the gas heat pump with air as renewable heat source employs a small size, high performance plate type refrigerant heater and a high performance refrigerating circuit unique in which the (refrigerant) heater is arranged in parallel with the air heat exchanger (Kasara et al, 2002).

Refrigerant heater utilizes exhausted heat from the gas engine from the cooling water and exhausted gases. Its purpose is to obtain a stable heating capability, using both the air heat exchanger and the refrigerant heater during heating operation, while on the other hand, switching to the operation of the refrigerant heater only when the evaporation performance of the air heat exchanger becomes low at very low outdoor temperatures.

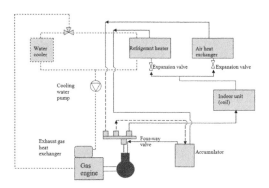

Figure 3. Gas heat pump with air as heat source scheme (Kasara et al, 2002).

Electric heat pump doesn't have refrigerant heater.

Gas heat pump operates in two different ways with different heat sources (air and sea water). On Figure 4 is shown air gas heat pump. Additional heat exchanger is placed before compressor, parallel with air heat exchanger.

At gas heat pump there is no heating capacity reduction with the outside air temperature decreasing.

Gas heat pump with sea water as renewable heat source has got constant temperature, so the exhausted heat (from cooling water and exhausted gases) is utilized after condenser.

Gas heat pump shows better performances. The comparison of energetic transformation PER between electric and gas heat pump is shown on Figure 5.

It is visible that higher heat energy efficiency is with gas heat pump than with electric heat pump with regard to primary energy (ASUE, 2004). PER of gas heat pump is 1,49 and PER of electric heat pump is 1,19.

Acronyms from Figure 5 are explained bellow:

η_e—efficiency of electric power plant	0,36
η_t—efficiency of gas transportation	0,96
η_m—efficiency of gas engine	0,3
ε—COP electric heat pump	3,3
ε—COP gas heat pump	3,5

Also, on Figure 5 it is shown that the part of the environmental heat source is greater on electric than on gas heat pump, and that amount is 70% at electric heat pump and 47% at gas heat pump. We can also compare the effective heat/primary heat ratio between gas and electric heat pump with following expressions:

a) Electric heat pump

$$\frac{\text{Effective heat}}{\text{Primary heat}} = \eta_k \times \text{COP} \qquad (5)$$

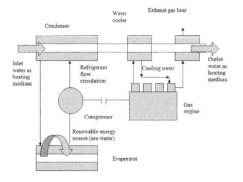

Figure 4. Gas heat pump with sea water as heat source (Samsalovic, 1987).

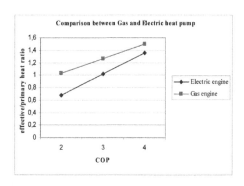

Figure 6. Comparison between gas and electric heat pump (Recknagel, 1984).

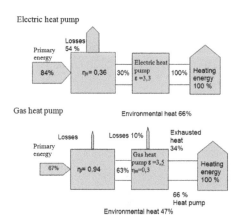

Figure 5. Comparison of energetic transformation PER between electric and gas heat pump.

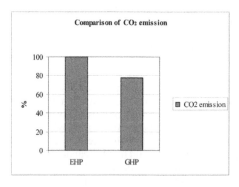

Figure 7. Comparison between gas and electric heat pump CO_2 emission (Japan Gas Association, 2003).

Table 2. CO_2 emission at different types of heating for family house with 150 m^2 (ASUE, 2004).

Heating with/	Primary energy	Primary energy per year (kWh/a) (annual heating energy consumption is 10 500 kWh/a)	S spec. emission (kg CO2/kWh primary energy)/	CO_2 emission (kg CO$_2$/annual)/
Electric energy	278%	29 190	0,20	5 838
Fuel oil	117%	12 285	0,26	3 194
Natural gas	109%	11 445	0,20	2 289
Compressor electric heat pump	84%	8 820	0,20	1 764
Gas heat pump	67%	7 035	0,20	1 407

where is: η_κ electric power plant efficiency
(COP) – Coefficient of Performance
b) Gas heat pump

$$\frac{\text{Effective heat}}{\text{Primary heat}} = \eta_t \times \eta_m \times COP$$
$$+ \text{exh heat from gas eng} \qquad (6)$$

where is:
η_t–transport efficiency
η_m–gas engine efficiency
(COP)—Coefficient of Performance

Gas heat pump has got much greater effective heat/primary heat ratio shown on Figure 6.

Gas heat pump advantage is utilization of exhausted heat from the gas engine (from cooling water and exhaust gases), another advantage is better gas heat pump PER ratio with the same COP shown in Figure 6.

Gas heat pump operates using natural gas—a clean fuel without SO_x emmision or soot.

In comparison with electrical heat pump, the CO_2 emission (a principal cause of global warming) is reduced for about 20% shown on Figure 7.

CO_2 emission is analyzed on family house with 150 m^2 and annual heating energy consumption is 10 500 kWh/a, with specific annual energetic consumption 70 kWh/m^{2}a (ASUE, 2004).

In Table 2 it is shown that gas heat pump has got the lowest CO_2 emission per year in compare with electric heat pump and other types of heating.

6 CONCLUSION

Heat pump uses renewable heat source. Today, air as renewable is mostly used. But, sea water could be used as renewable heat source. Sea water advantages are: excellent thermodynamics properties and approximately constant temperature through whole year of the heating period. Gas engine driven heat pump has got many advantages in comparison to the electric heat pump.

Gas heat pump uses primary energy without transformation into electric energy. Gas heat pump efficiency is greater and could be even 25% higher.

Also, a big advantage is utilization of exhausted heat from the gas engine (from cooling water and exhaust gases).

At lower outside temperatures gas heat pump operates with the same heating performances.

Electric consumption and loading is reduced with gas heat pump running.

Gas heat pump requires no defrosting operation.

Gas heat pump requires a smaller electrical service.

Gas heat pump is environmentally friendly technology with CO_2 emission reduced for about 20%.

REFERENCES

ASUE (2004): "Gaswarmepumpen", 2004.
H. Kasara, Yoshimura, R. Yoshimura, H. Iwata: High-Perfomance, High-Reliability Gas Heat Pump, Mitsubishi Heavy Industries, Ltd. Technical Review Vol. 39 No. 2 (Jun. 2002).
Japan Gas Association, 2003.
Pavković B.: Radni procesi i toplinski izvori za dizalice topline, Tehnički fakultet, Rijeka, 2006.
Recknagel—Sprenger: Grejanje i klimatizacija, Građevinska knjiga, Beograd, 1984.
S. Šamšalović: Toplotne pumpe u primeni, SMEITS, Beograd, 1987.

Green Design, Materials and Manufacturing Processes – Bártolo et al. (eds)
© *2013 Taylor & Francis Group, London, ISBN 978-1-138-00046-9*

System of thermal energy accumulation in a soil layer in a combination with wind energy

A.V. Bunyakin

Institute of Oil, Gas and Energy, KubSTU, Krasnodar, Russia

ABSTRACT: In the work a combination of three saving energy methods is presented, and this variant is represented as optimum for the certain conditions. The combination consists of petrothermal wells with hydro-isolation from a soil; wind-power station of vortical kind for compensation of losses of energy on circulation of the heat-transfer-agent system; and thermo-active protections, that is with a supply of heat inside of a protection between heat-isolation layers.

1 INTRODUCTION

Important form of energy is a heat and ecological energy sources are sunlight, wind, heat of Earth. The system of heat accumulation in soil with saving of its till 50% at conditioning the houses using wind energy is offered. Average power (in time of year) of the system is depend from size of it and from the region conditions. Thus approximately mechanical power of system is 5 kW at speed of wind 3 m/s for following technological parameters of basic elements of system: wind energy station of vortical (cyclone) type of the maximal height near 20 m with diameter of the basis 10 m; wells depth 50 m (diameter 100 mm) with hydro-isolation. The thermal power depend from number of wells and from properties of soil. In the paper consider the basic structural elements of system, including a principle of their action and of regime parameters (entrance data are taken for an example, parameters are estimated approximately).

2 A WELLS AND SYSTEM OF CIRCULATION OF THE HEAT-TRANSFER-AGENT

The petrothermal method is a system of accumulation of heat at its surplus (summer) and heating by this heat the soil near wells with subsequent use of this (accumulated thermal energy) at its lack for heating in the winter. In the summer the heat-transfer-agent selecting warmly from an attic or from within overlapping of the top floor, passed in ring space of a wells. The wells is completely isolated from a soil entrance by a pipes inside of which the column of technological pipes is contained.

In the winter mode heat-transfer-agent is started up in system of the circulation, including pipelines in a cellar or inside of overlapping a floor of a ground floor. The layer of soil near well is cooled so to not admit yet formation of an ice. Thus, the well and soil volume near it work as the accumulator of heat.

Inside of a soil layer around of a well the temperature varies (as a first approximation) under the logarithmic law by radius from axis of a well, that is $T = T_C + T_K - T_C/\ln(R_K/r_C)\ln(r/r_C)$. Here r_C, T_C—radius of a well and temperature outside from it. R_K, T_K—radius of a contour (conditionally cylindrical surface on which the soil has the natural temperature, not dependent on thermal influence from a well) and temperature there. From here $dT/dr = T_K - T_C/\ln(R_K/r_C)1/r$, then thermal power at stationary heat exchange $\dot{q} = \Lambda 2\pi r_C H(dT/dr)|_{r=r_C} = 2\pi\Lambda H(T_K - T_C)/\ln(R_K/r_C)$. Here Λ, H factor of heat conductivity of a soil and height of its layer. Let's for an example of calculation $r_C = 0,05\,m; R_K = 15\,m; T_K - T_C = 5\,K; \Lambda = 0,8\,W/m \cdot K$ (a soil on a clay basis) $H = 25\,m$, we shall receive $\dot{q} = 110\,W$—the thermal power corresponding stationary heat exchange at downturn of temperature near well on 5°C from normal value. In the calculations resulted above necessary, that as a result of circulation of the heat-transfer-agent inside of a well the temperature falls up to 4°C.

Then such thermal power corresponds to mass submission of the heat-transfer-agent $\dot{m}$ and $\dot{q} = c\dot{m}\Delta T$, at $\Delta T = 1\,K$, $c = 4000(W \cdot s/kg \cdot K)$ (it is close to a thermal capacity of water) $\dot{m} \approx 0,0275(kg/s)$. Let's length of all pipeline system $L = 500\,m$, diameter $d = 0,025\,m$ equivalent factor of hydraulic resistance $\varsigma = 0,03$, hydraulic losses $h = 8(\varsigma L/g\,d)(\dot{m}/\rho\pi d^2)^2 \approx 0,1\,m.$

3 WIND-POWER STATION OF VORTICAL TYPE

It has as basis a metal-carcass of the cylindrical part and conic dome [1]. Cylindrical part are made by of some moving doors of the jalousie type. At special installation of jalousies the input of air in internal volume is opens. Jalousies from the counter party for a wind stream (on Fig. 1 are on frontal side) should be established so that air entering inside was more, than let out outside through jalousies alee (back side).

At change of wind direction the jalousies are reinstalls. In the variant with installation of all jalousies on an identical corner then reinstallation it is not required. The question of optimization is possible also.

The difference of input and output of air flows in volume of a conic dome, here its twists, and in process of rise upwards speed of rotation is increased (preservation of the moment of quantity of movement), further the vortex flow goes to the turbine (Fig. 2). The basic idea used in this

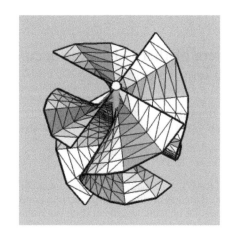

Figure 2.   Rotor of the turbine.

wind-power station, consists that before directing on the turbine, the flow is specially prepared (twists in a whirlwind).

The turbine is the device intended for contribution of energy from the twirled stream (unlike axial wind-power stations where energy is making from stream along an axis). The quantity of blades can be not four, and more. There is a question on what is advantage of such wind-power station in comparison with existing designs. We shall specify only three from them:

– First, the main part of station (everything, except for the turbine) represents a usual building (metals—plastic) construction. It can be executed on the basis of metal frame modules and plates of cellular polycarbonate. Wind-power station of such design can be made without application of heavy techniques, and manufacturing of separate modules possible from the widespread building materials;
– Secondly, a design strong enough from the point of view of static (gravitation) and dynamic (wind) actions, therefore its sizes can be great enough (up to 100 m). Accordingly the flow of air through the turbine can be enough greater, so speed of twisting too will increase at greater concentration in a vortical plait (it is effect of formation of typhoons and tornado). In general it is possible to tell, that instead of blades of the big size for an axial stream (as in traditional wind stations) and small speed of rotation the turbine with a vertical axis, smaller diameter is offered, but speed of rotation is more;
– Thirdly, wind-stations of standard types require almost constant direction and speeds of a wind, differently power becomes much less. The given scheme can work both at various directions of

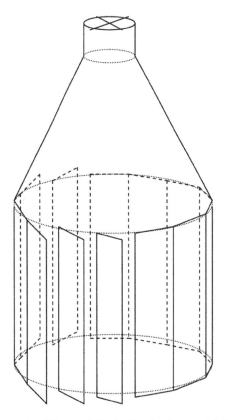

Figure 1.   Schema of the basis—directing part (with special installation of jalousies).

a wind, and at frequent change of its speed. For definiteness it is possible to consider, that all jalousies are established under one corner (without active management), the internal volume of rotating air plays a role of the accumulator of kinetic energy of a stream. Besides the given wind-power station can work near to the large barrier creating an obstacle to a wind (buildings, trees, hillsides). In stated below materials of useful power and speed of rotation of the turbine are approximate estimations, and also (by criteria of physical similarity) rules of scale transition to the any size and wind speed are present.

First we shall make the approached estimation of speeds of flows. Let for example speed of wind $V_* = 3 (m/s)$, stagnation pressure $P_* = \rho(V_*^2/2)$. We shall assume, that in the field of leaving a vortical path (by air transit through a jalousies on the back side, fig. 1) boundary layers obstacle in flow is liquidated by the blowing and the pressure here is equal to half of stagnation pressure (it is the overestimated, there can be and vacuum). That is, movement of a flow from inlet to final section of jalousies along an internal surface of a cylindrical part of wind-power station occurs under difference of pressure $P_*/2$.

This value is accordingly underestimated (possible vacuum is not considered). Accepting total factor of hydraulic resistance at movement of air in internal volume of cylindrical part of wind-station equal 3/2 (a moving between jalousies) we shall write down following equation: $P_*/2 = \rho(V_*^2/4) \approx (3/2)\rho(V_0^2/2)$, that is $V_0 \approx V_*/\sqrt{3}$. It is the approached estimation of air speed on external circle of vortical movement inside volume of a cylindrical part. From here $V_0 = 1,7 m/s$, this value will be used for an estimation of power (a little underestimated), contributing from a flow at its release upwards (through top of a conic dome where the turbine is established).

The volume of air of internal space of station receives a rotary impulse (the moment of quantity of movement), we shall discuss a value of circulation on a contour which is passing near to an internal wall Γ_0, and also surplus of entering air (through jalousies on a cylindrical part) in comparison with air leaving. This difference of streams is created is artificial and defines average value of air speed in an ascending flow inside of a conic dome.

In process of ascending of air volume to narrowed space of a conic dome both values of speeds (vertical and rotary) will increase. The increase in a vertical component speaks reduction of the area of section of a cone (square of orthogonal crossing of vertical axis) at approaching top. The increase in a rotary component speaks the law of preservation

of the moment of an impulse (its hydrodynamic analogue is a theorem of preservations of circulation). We shall stop more in detail on it, and we shall give the explanatory.

Analysis of distribution of speed we shall make, using approach of ideal gas in the follow form: $\partial V/\partial t + grad(V^2/2 + P) + [rotV,V] = 0$. Here V is a vector of speed, $P = \int dP/\rho$ or $P = P/\rho$ (the last at condition of constant density), square brackets designate vector multiplication.

Having taken integral from this equation on the closed contour, we shall receive (triangular brackets designate the mixed multiplication):

$$\frac{d}{dt}\oint V \cdot dl + \oint \langle rotV,V,dl \rangle = 0; \quad \Gamma = \oint V \cdot dl$$

Let's consider as a contour of integration a circle with an axis of the symmetry conterminous with an axis of a conic dome (integration is conducted counter-clockwise at a sight from above). The field of speed in space under a dome, at a qualitative level, can be approximated by functions of a following kind (the axis z is directed vertically upwards, system of coordinates is Cartesian) $V = (V_x, V_y, V_z)$, $V_x = \omega_0 y - ax$; $V_y = -\omega_0 x - ay$; $V_z = bz$. Positive values of parameters a,b,ω_0 correspond to a flow twirled clockwise at a sight from above, with narrowing and ascending spirals trajectories (these parameters in considered approach are considered as constants) $rotV = (0,0,-2\omega_0)$, $dl = (dx,dy,dz)$. Then the mixed multiplication in the left part of integrated equality is:

$$\langle rotV,V,dl \rangle = \begin{vmatrix} 0 & 0 & -2\omega_0 \\ \omega_0 y - ax & -\omega_0 x - ay & bz \\ dx & dy & dz \end{vmatrix}$$

$$= -2\omega_0 \begin{vmatrix} \omega_0 y - ax & -\omega_0 x - ay \\ dx & dy \end{vmatrix}$$

$$= 2\omega_0 a(x\,dy - y\,dx) - 2\omega_0^2(x\,dx + y\,dy).$$

So, $x\,dy - y\,dx = r^2 d\varphi$, where r—radius of a circle of integration, $d\varphi > 0$ at integration counter-clockwise. Then we have $d\Gamma/dt + 4\pi\omega_0 a r^2 = 0$.

Whereas $\Gamma < 0$ for flow under a dome turns out, than at approaching to top of a cone absolute value of circulation has increasing (that is, $a,\omega_0 > 0$ then $d|\Gamma|/dt > 0$). This phenomenon is known under the name «formation of a vortical plait» (meets in nature in the form of whirlpools, typhoons). It is qualitatively characteristic, that the layers nearest to an axis of rotation, are move more quickly, circulation aspires to a constant (independence from r) if in limit ω_0 and a aspires to zero.

The total "integrated" moment of quantity of movement remains almost constant, than flow is closer to an ideal (nonviscous) limit. Pressure decreases at approach to axis of a vortical plait (in a considered case it means, for air will be streamlining moving to the top of a dome)—one more reason testifying to efficiency of the offered aerodynamic scheme.

Proceeding from the law of preservation of circulation, absolute value of twisting speed V on the approach to top of a dome can be estimated approximately (underestimated) from equation $2\pi V_0 r_0 = \Gamma_0 \approx \Gamma = 2\pi V r$, that is $V/V_0 \approx r_0/r$, and the equation the more precisely, than more sharply a corner of a cone. Really, reduction of a corner at top of a cone up to zero corresponds to reduction of parameter a (see above) also up to zero.

The device intended for contribution of mechanical energy from a rotating flow—the turbine (Fig. 2) consists of the rotor with blades of the special form qualitatively similar to the its analogs with supplied directing device. The turbine is established in the top of a cone in cylindrical stator with minimally possible backflows. The quantity of blades is a subject to optimization by criterion of maximization of power. It corresponds to minimization of hydraulic resistance in between blades channels.

The rotor can have various parities of geometrical parameters (external and internal diameters, height, a corner at a "sharp" edge of the blade, etc.). It is a corner of crossing of the blade with a plain, an orthogonal axis of rotation, on input and output between blades channels. The given parameters are subjects of optimizations by criterion of the maximal power at the certain (most probable) working mode of the turbine. Not stopping in detail on various optimizations problems statements for a finding of parameters of the turbine, we shall specify only key parameters and ways of their approached estimation.

Key parameters of the turbine are the corner of a sharp edge of blades α and value of optimum angular speed of rotation ω. All parameters in calculations has positive values. The corner α is connected with components of axial V_z and rotation velocities $V = \Gamma/2\pi r$ by parity $tg\,\alpha = V_z/V$ that corresponds to a condition smooth (almost continuous) input of flow on an edge of blades. The given condition can be precisely executed only on special radius R, according to the accepted rules for calculation of working blades of the turbine, settlement value R undertakes approximately corresponding 2/3 from the radial size of the blade (further from an axis). Optimum speed of rotation is $\omega = V/R$.

Accepting conditionally (for an estimation of the values order) speed of a twisting of a flow in internal space of cylindrical part $V_0 = 1,7\,(m/s)$

(see above) equal speed of flow through entrance section. The square across of entrance section $S_0 = 10\,m^2$ is taken conditionally (height of a cylindrical part 10 m, width of jalousies 1 m). Radius of a basis—cylindrical part $R_0 = 5\,m$, we receive blowing of air $Q_0 = V_0 S_0 = 17\,m^3/s$, circulation $\Gamma = 2\pi V_0 R_0 = 53,4\,m^2/s$. Accepting the square across of flow through passage section of the turbine $S = 1,7\,m^2$ (head diameter of cylindrical dome is near 1,5 m), the settlement radius undertakes $R = 0,5\,m$, then speed of flow rotation on this radius is approximately estimated as $V = V_0(R_0/R) = 17\,m/s$.

Let's conditionally take, that 1/4 part of the blowing air tend to the turbine $Q = Q_0/4$, we shall find $V_z = Q/S = 2,5\,m/s$, then a corner of an edge of blades $\alpha = arctg(V_z/V) = 8,4°$. Optimum speed of rotation $\omega = V/R = 34\,rad/s$ (about 5 turns in second). The power contributing from a flow, thus will have the value $N = 1/2\rho Q\Gamma\omega = 5\,kW$, air density $\rho = 1,3\,kg/m^3$.

Really, specific on volume Ω the moment of quantity of movement K (the kinetic moment) is similar to circulation of a vector of the speed multiplied by density, that is $dK/d\Omega = \rho\Gamma$. Energy of rotation $E = mr^2\omega^2/2$ (for a material point) is connected with the kinetic moment $K = mVr = mr^2\omega$ by the equation $E = (\omega/2)K$ which is true and for a hard body and for our discussed case of flow with ring lines of flow and constant circulation of a vector of the speed. Therefore the equation for such energy $dE/d\Omega = (\omega/2)(dK/d\Omega) = (1/2)\rho\Gamma\omega = N/Q$ on the volume gives expression of power, suitable for an estimation of the order of values. Power increases (proceeding from reasons of physical similarity) is proportionally cube of speed of a wind, and a square of the linear size of wind-station.

There is one more reason on which it is possible to consider the given aerodynamic scheme more effective, than in standard designs. Even at full wind calm if vortical power station is under influence of a sunlight the temperature of air in internal volume will be more external. Warmer air will rise in the top part of a conic dome, suction of air from the outside through jalousies which will give to it rotation. This effect especially essential, than is more sizes of wind-power station.

Besides at a choice of a direction of flow rotation in a cylindrical part it is necessary to note that in northern hemisphere it is necessary to rotate clockwise, and in southern—on the contrary (at a sight from above). It speaks what even at absence of a wind and sunlight the effect of rotation of air will work in northern, a hemisphere clockwise (because of acceleration at rotation of the Earth). Probably described effect it will not be appreciable to be shown on a background of wind action, but with increase in the size of wind-station it will become more and more essential, as the volume of air delivered inside (also due to a thermal lifting) will become the more.

4 ACTIVE THERMAL PROTECTION

Simple and effective method from the point of view of conditioning internal premises is installation of a radiator in a floor of a ground floor for heating and accordingly in a ceiling of the top floor for cooling [2] is represented. There is a problem of optimization of this system—definition of heat-protection layers thickness on both parties from a radiator (according to heat-conducting properties of their material) at the given power of heat-transfer circulation from wells and the square of heat transfer layers.

Let's designate as T_1, T_2 temperatures outside and inside of indoors (on both parties from a protection), λ_1, λ_2 factors of heat conductivity of outside and inside layers of a protection (on both parties from a layer of an active supply or heat removal), δ_1, δ_2 accordingly thickness of these layers, λ factor of heat conductivity inside of active thermal layer and 2δ its thickness. Let's designate through $\dot{q}/s$ a relative heat transfer power of a supply or heat removal. We approximate the change of temperature inside active thermal a layer by parabola (while we consider, for example, the case that is warmly brought) $T = T_0 + (a/2)(x - \delta_1 - \delta/\delta) + b/4(x - \delta_1 - \delta/\delta)^2$, the axis x is a coordinate across a layer, the zero corresponds to an external surface of a protection. Let's consider temperatures T_-, T_+ on the boundaries of a active thermal layer then equality of thermal streams at a heat transfer through these is in the form:

$$\lambda_1 \frac{T_- - T_1}{\delta_1} = \lambda \frac{d}{dx}\left(\frac{a}{2} \frac{x - \delta_1 - \delta}{\delta} + \frac{b}{4}\left(\frac{x - \delta_1 - \delta}{\delta} \right)^2 \right)\Bigg|_{x=\delta_1} \tag{1}$$

$$\lambda \frac{d}{dx}\left(\frac{a}{2} \frac{x - \delta_1 - \delta}{\delta} + \frac{b}{4}\left(\frac{x - \delta_1 - \delta}{\delta} \right)^2 \right)\Bigg|_{x=\delta_1+2\delta}$$
$$= \lambda_2 \frac{T_2 - T_+}{\delta_2} \tag{2}$$

So, $T_+ - T_- = a$, it follows directly from the equation of approximation, and also $\lambda(b/\delta) = \lambda_2(T_2 - T_+/\delta_2) - \lambda_1(T_- - T_1/\delta_1) = -(\dot{q}/s)$.

The last follows from the equations (1, 2), we shall copy them in the form:

$$\lambda_1 \frac{T_- - T_1}{\delta_1} = \lambda\left(\frac{a}{2\delta} - \frac{b}{2\delta} \right) = \lambda \frac{T_+ - T_-}{2\delta} + \frac{\dot{q}}{2s} \tag{1'}$$

$$\lambda_2 \frac{T_2 - T_+}{\delta_2} = \lambda\left(\frac{a}{2\delta} + \frac{b}{2\delta} \right) = \lambda \frac{T_+ - T_-}{2\delta} - \frac{\dot{q}}{2s} \tag{2'}$$

Accepting for example of calculation the values of temperatures $T_1 = -20, T_2 = 20,$ $T_- = 4, T_+ = 5$ in $°C$, $\lambda_1 = \lambda_2 = 0,01 W/m \cdot K$ factors of heat conductivity of external and internal heat-protection layers (a material of glass-wool type), $\dot{q}/s = 100 W/50 m^2 = 2$ relative heat power (the supply of heat near 100 W is coordinated with parameters of one well also temperatures T_-, T_+). The factor of heat conductivity inside active thermal a layer $\lambda = 0,5 W/m \cdot K$ corresponds to the heat-transfer-agent of type of water, is warmly brought by «a water shirt», semi-thickness of it $\delta = 0,125 m$. Then from the equations (1', 2') we shall receive accordingly $0,24/\delta_1 = (0,5/0,25)+1,$ that is $\delta_1 = 0,08 m$ and $0,15/\delta_2 = (0,5/0,25)-1,$ that is $\delta_2 = 0,15 m$.

Instead of «a water shirt» there can be a radiator of type of a flat coil, the prisoner between metal sheets, and it can be simulated (approximately) as proportional reduction of values λ and δ.

Thus, optimum in energy saving for the given parameters of a protection is an asymmetrical heat-protection layers in both parties from active thermal layer (radiator). So heating (the winter—importing of the active thermal layer in a floor of a ground floor) should differ from summer period (in a ceiling of the top floor). Namely, it is necessary to notice, that symmetry (invariant properties) of the equations (1', 2') that is replacing system of inequalities $T_1 < T_- < T_+ < T_2$ on $T_1 > T_- > T_+ > T_2,$ can be only such: mutual substitution T_1 on T_2, T_- on T_+, δ_1 on δ_2, and replacement of sign $\dot{q}$. It corresponds to transition from a winter mode to summer and on the contrary.

Variants of horizontal radiators instead of wells, in particular cases of the water-sated soil—marsh, for example, are possible also.

5 CONCLUSION

Such energy sources as a wind and heat of Earth (having in view of wells with depth less 50 m) are distributed, therefore power of separate station cannot be great. Therefore for mass reception of energy from such sources the network of the stations distributed on the big territory is required. Then there are problems of increasing of reliability of each object and decreasing of price of its. Such requirements are provided by simplification of system and by minimization of risk at executing of it. Risk for systems of the specified type are infrasonic noise from wind-power stations and danger of frosts of a soil. The presented system is an attempt of statements and the solution of a problems of optimization for such systems which are meeting the requirements of simplicity, reliability, safety.

REFERENCES

1. Bunyakin, A.V., Mikhailov, K.M. 2011. Patent for useful model RU106920U1 from 3/30/2011.

2. Gagarin, V.G. 2009. The method of the economic analysis of increase of a level of a heat-shielding of protecting designs of buildings // ABOK (Architectural questions of protecting designs), №1 P. 10–16, №2 P. 14–23, №3 P. 62–66.

Reuse and recycling techniques

Green Design, Materials and Manufacturing Processes – Bártolo et al. (eds)
© *2013 Taylor & Francis Group, London, ISBN 978-1-138-00046-9*

Design of experimental composting of animal carcasses in universitary unit of treatment aimed a correct final disposition and soil improve

J.C.L. Fonseca
Central Office, UNESP—Univ Estadual Paulista, São Paulo, Brazil

M.R.R. Marchi
Institute of Chemistry, UNESP—Univ Estadual Paulista, Araraquara, Brazil

L.T. Braz
School of Agrarian and Veterinarian Sciences, UNESP—Univ Estadual Paulista, Jaboticabal, Brazil

A.A. Cecílio & L.V.S. Sacramento
School of Pharmaceutical Sciences, UNESP—Univ Estadual Paulista, Araraquara, Brazil

ABSTRACT: For many years, composting has been used as a result of the recycling of organic matter. There is significative animal carcasses accumulation from teaching and researching activities of the university veterinary hospital. Every year, Unesp University needs to dispose correctly about 180 tones of this waste and the composting seemed to be the most sustainable alternative. Piles of animal carcasses were prepared using peanut hulls and tree pruning as bulking agent and water to the first phase of this process. The extracts pH values no impediments for offering germination and indicated a good addition to the soil management. The germination index showed no impediment to the seeds germination on any type of compost and the extracts concentrations not influenced this biological process. No parameters studied assigns risks of contamination of carcasses for the compost development in Unesp according to the proposed design.

1 INTRODUCTION

Composting can provide animal producers with a convenient method for disposing of animal mortalities and also provide a valuable soil amendment. In addition, the finished compost can be stockpiled and reused to compost other mortalities. Composting is a natural decomposition process conducted by microorganisms that happen under controlled conditions. It reduces the size of the material by removing organic products, water, and energy as carbon dioxide, steam and heat. Also, the pathogens are destroyed by the high temperatures reached during the composting process, which require time and space, and some specialized equipment may be necessary. If the process is not done correctly, pathogens survive and odors may occur, attracting flies and vermin, as well as vultures that can uncover the carcasses. Composting of organic wastes is a biooxidative process involving the mineralization and partial humification of the organic matter, leading to a stabilized final product, free of phytotoxicity and pathogens and with certain humic properties (Zucconi and de Bertoldi, 1987). During the first phase of the process the simple organic carbon compounds are easily mineralized and metabolized by the microorganisms, producing CO_2, NH_3, H_2O, organic acids and heat. The accumulation of this heat raises the temperature of the pile. Composting is a spontaneous biological decomposition process of organic materials in a predominantly aerobic environment. During the process bacteria, fungi and other microorganisms, including microarthropods, break down organic materials to stable, usually called compost. The composting also implies the volume reduction of the wastes, the destruction of weed seeds and of pathogenic microorganisms. Incineration, burial, and composting are the main methods used for livestock mortality disposal. Due to bio-security concerns and increasing costs associated with buria, livestock producers are showing a growing interest *on-farm* disposal methods. Incineration is commonly used for small carcasses, but is less practical for larger carcasses as it requires substantial amounts of fuel and can cause air pollution. Although burial is the most common *on-farm* mortality disposal method, its use is declining due to groundwater and soil contamination concerns. Frequently composting is being adopted as a viable method for *in situ* treatment of livestock carcasses. Because the successful use in the poultry and swine industry, it is perceived to be an economical and environmentally friendly process. The high temperature acquired through microbial metabolism has potential to inactivate pathogens. Therefore it has been used for pathogen related carcasses disposal in USA and Canada. Glanville et al. (2006) reported excellent viral pathogen inactivation and animal tissue decomposition using a windrow type cattle composting system. Previous researchers reported animal tissues (internal organ and soft tissues) were fully decomposed in 4–10 months in windrow type composting systems (Glanville et al., 2006; Sander et al., 2002). They observed and determined animal tissues decay extent during excavating composting piles. Properly estimated carcass decomposition rate is valuable for designing and controlling animal mortality composting systems. However, it is still difficult to evaluate the decomposition rate inside of the field scale compost piles. Every year, Unesp (São Paulo State University) needs to dispose about 540 tones of this waste and the composting seemed to be the most sustainable alternative. In this study, 750 kg of carcasses were composted in two compost boxes with damp peanut hulls and tree pruning as cover materials. These carcasses are from teaching and researching activities of the university veterinary hospital.

1.1 *The composting process*

The necessary ingredients to compost carcasses are a bulking agent and water. The bulking agent soaks up the leachate produced by the decomposing carcass, provides aeration, and increases the carbon-to-nitrogen (C:N) ratio. The carcass contains a high concentration of nitrogen and water, and the wood product, high in carbon, wicks up the moisture. Traditional bulking agents include: sawdust, wood shavings, and wood-based bedding and manure. Other materials besides sawdust can be used for composting, such as straw or corn stover (Elwell et al.,1998), but require additional management for water control. The process is also recommended to poultry mortality, poultry litter and straw (Murphy and Carr, 1991). This work shows an efficient alternative to be used by zoo, small farmers and Veterinary Schools, because it presents simple managed and an investment of low cost. It is possible to use as bulking agent, local tree pruning as well peanut hull, like it has been utilized in North Carolina (cited in NPPC, 1997).

2 MATERIAL AND METHODS

2.1 *Design composting process*

In this study, animal carcasses piles in the three different cells (8 m³) were prepared using peanut hulls and tree pruning. Initially, rendering of animal mortalities was made to increase the contact surface and to get the time of composting shorten. The design was arranged in layers using bulking agent (peanut hulls and tree pruning) and 50 kilos of carcasses from equines, chickens and cattle (Fig. 1), in ratio 2:1. Small animals as poultry or lab animals, until 2 kilos were disposed in one peace. Enough water was added to the bulking agent for the first phase of this process using a ratio 1:1 for moisture content. The peanut hulls were used to allow the action of air and microbial underneath the carcass. The material was disposed carefully, aiming to maintain space between them, avoiding superposition. The last layer was 30 cm (1 ft) of the damp peanut hulls and tree pruning and works like the biofilter for odor control around the pile and insulates the pile to retain heat. Odors may be released when an inadequate cover is used or when it is too dry. The released odors may also attract scavenging animals and pets to the pile.

2.2 *Data collection*

Monitoring temperature is necessary to measure progress of the composting process. The temperature of the composting piles was measured using a thermocouple, Hanna Instruments, model HI93530 of 1 meter long to reach several points in piles (three points at least). Temperatures should rise to 60–65 °C and remain there for several weeks (whereas this temperature ensure the pathogen

Figure 1. Pieces of carcasses arrangement in the box disposed over bulking agent. The new amount of bulking agent was disposed over.

ensure destrucion). The compost of carcasses after trituration was subjected to extractive process in water at 60 °C, using a ratio of 10% (w/v) for 30 minutes. After cooling for 2 hours, the mixture was subjected to 15 minutes of stirring (120 rpm) and stand 30 minutes, then proceeding to a filtration and subsequent storage of the filtrate (extract) in a refrigerator. After the extraction, it was made an application of the crude extract (100%) and diluted to 50%, 25%, 12.5%, 6.25% in petri dishes containing 20 seeds of tomato (*Lycopersicon esculetum* Mill.). Then it was subjected to germination conditions in the BOD incubator at 30 °C. The counting of the germinated seeds were daily for determining the germination rate, and at the end of 7 days was estimated the percentage of germination and root length. The pH values of compost and vegetable matter extracts were determinate using indicator papers for rapid qualitative determination.

3 RESULTS AND DISCUSSION

The permanence time was defined by the temperature stabilization, what happened in about 120 days. The partial quality and maturity of the obtained composting was monitored by organic C, total N content, C/N ratio (Table 1).

Nutritional balance is mainly defined by the C/N ratio. Microorganisms require an energy source (degradable organic-C) and N for their development and activity. The adequate C/N ratio for composting is in the range 25–35, because it is considered that the microorganisms require 30 parts of C per unit of N (Bishop and Godfrey, 1983). High C/N ratios make the process very slow as there is an excess of degradable substrate for the microorganisms. But with a low C/N ratio there is an excess of N per degradable C and inorganic N is produced in excess that can be lost by ammonia volatility or by leaching from the composting mass. In this study C/N ratio was about 19, it was considered appropriate, since it was not perceived release of ammonia. The germination index and germination rate of tomato seeds treated with compost and vegetable matter extracts has been utilized

as an indirect parameter to evaluate the compost maturity. The addition of the carcass to vegetable matter reduced C/N ratio of 2.4 times shavings demonstrating the N contribution to the compost. The pH of the extracts ranged between 6.0 and 7.0 no impediments for offering germination and indicated fine conditions to the disponibilization nutrients of the compost. The germination rate for seeds treated with the extracts differ from values assigned to seeds treated with water (Fig. 2), in greater proportion considering the 100% extracts for the extracts prepared with equine and poultry carcasses. Compost derived from miscellaneous

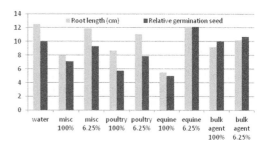

Figure 2. Root length (cm) and relative germination seed of tomato (*L. esculetum*) germinated with crude extract (100%) and diluted extract (6.25%) of carcasses composting in comparison with water and bulking agent.

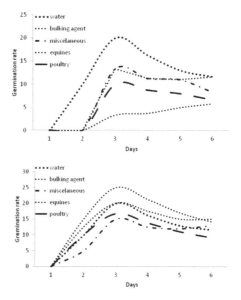

Figure 3. Germination rate of the seeds of tomato (*L. esculetum*) with crude extract 100% (A) and diluted extract 6.25% (B) of carcasses composting in comparison with water and bulking agent.

Table 1. Average carbon and nutrients values for compost obtained at the piles.

	C/N	%N	%C	%H	%S
Bulking agent	33	1.39	46.68	5.78	0.33
Equine	19	2.36	44.96	5.35	0.57
Poultry	18	2.45	43.94	5.24	0.51
Miscelaneous	19	2.44	45.59	5.41	0.49

meat decreased the germination of tomato seeds in relation to water in 30%. Because the bulking agent presents germination rates like the control treatment (water), it was considered an inert material to the used evaluation criterion. To the root length they were observed responses very similar to the process of germination, however, for the bulking agent there was a small reduction in root development in relation to the water. A search in the environment area publications about this subject did not find similar results. Figure 3 shows the germination rate over the days of evaluation. In (A) it is possible to see that the lowest rate of germination occurred for carcasses compost of equines during the six days of monitoring. Between the second and third days, to the other materials (poultry, miscellaneous and bulking agent) the germination rate showed the same values of speed that water, remaining similar to control (water) until the end of the process, however showing lower rates. In (B) the curves profiles the answers get closer to water and show that the substances present on the 100% extract demonstrate action dependent on concentration, for the 6,25% extracts represent 1/16 of crude extracts.

4 CONCLUSION

Carcasses composting, when developed correctly with proper attention to the design, layout, monitoring, maintenance, and environmental impacts of the system used, may be considered an efficient and safe method of disposing of animal carcasses.

The germination index did not show impediment to seed development on different types of tested composts and the extracts concentrations did not influence in the biological process. Any of the studies parameters assigns risk to initial plant development of carcasses for the compost produced in the UNESP. According to the proposed design may be used in other Universities.

REFERENCES

Bishop, P.L., Godfrey, C., 1983. Nitrogen transformation during sewage composting. *Biocycle* 24, pp. 34–39.

Elwell, D.L., Moller, S.J., Keener, H.M. 1998a. *Composting large swine carcasses in three amendment materials.* In: Proceedings of Animal Production Systems And The Environment. Des Moines, Iowa. pp. 15–20.

Glanville, T.D., Richard, T.L., Harmon, J.D., Reynolds, D.L., Ahn, H.K., Akinc, S. 2006. Final project report: *Environmental impacts and biosecurity of composting for emergency disposal of livestock mortalities.* Iowa Department of Natural Resources (IDNR project # 03-7141-08).

Murphy, D.W., and Carr, L.E. 1991. *Composting dead birds.* Fact Sheet 537. Cooperative Extension Service, University of Maryland System.

NPPC. 1997. *Swine mortality composting module.* National Pork Producers Council, Clive, IA.

Sander, J.E., Warbington, M.C., Myers, L.M. 2002. Selected methods of animal carcass disposal. *Journal of the American Veterinary Medical Association* 220:1003–1005.

White, D.G., Regenstein, J.M., Richard, T., Goldhor, S. 1989. *Composting Salmonid fish waste: A waste disposal alternative.* Sea Grant publication. Cooperative Extension NY State, Cornell University, Ithaca, N.Y.

Zucconi, F., de Bertoldi, M., 1987. *Compost specifications for the production and characterization of compost from municipal solid waste.* In: de Bertoldi, M., Ferranti, M.P., L'Hermite, P., Zucconi, F. (Eds.), Compost: Production, Quality and Use. Elsevier, Barking, pp. 30–50.

Green Design, Materials and Manufacturing Processes – Bártolo et al. (eds)
© 2013 Taylor & Francis Group, London, ISBN 978-1-138-00046-9

Kinetics, equilibrium and thermodynamics of the adsorption process of lead using cassava industry wastes

D. Schwantes, A.C. Gonçalves Jr., L. Strey & V. Schwantes
Western Parana State University, Marechal Cândido Rondon, Brazil

H. Nacke
Dynamic College of Cataracts, Foz do Iguaçu, Brazil

ABSTRACT: The present work aimed to evaluate the use of solid cassava wastes as natural adsorbents for the removal of waters polluted with Pb^{+2}. In order to evaluate the factors that influence the adsorption process, the following tests were conducted: pH range influence (4.0 to 6.0), adsorbent mass, contact time of adsorbent/adsorbate, adsorption isotherms and temperature influence. The adsorption kinetics was evaluated by pseudo-first order, pseudo-second order, Elovich and Intraparticle Diffusion. The isotherms were linearized by Langmuir, Freundlich and Dubinin-Radushkevich. The mass tests demonstrate that the proportion 8 mg mL^{-1} of Pb^{2+} is enough for an efficient removal, the more adequate pH range is at 5.5, with an ideal contact time at 60 min, being the process endothermic. Pseudo-second order and Dubinin-Radushkevich had the best fitting. It can be concluded that the use of the solid cassava wastes are potential biosorbents for the decontamination of Pb^{2+} polluted waters.

1 INTRODUCTION

The hydric contamination with heavy metal ions is a great concern (Oliveira et al. 2001), because these cause severe toxic effects to the human beings, animals and the environment (Li & Bai, 2005) and because they have been excessively released to the environment (Wan Ngah & Hanafiah, 2008).

Between these metals it can be emphasized, for it toxicity, the cadmium (Cd^{2+}), chromium (Cr^{6+}), mercury (Hg^{2+}), nickel (Ni^{2+}), lead (Pb^{2+}) and, in minor degree (Cu^{2+}) and zinc (Zn^{2+}) (Pereira, 2004).

One of the alternatives for removal of contaminants in waters is the adsorption process, being this technology extensively used for the removal of organic and inorganic pollutants from aqueous solutions. The activated coal is one of the most used adsorbent for a great variety of pollutants in waters, however, it high cost is one of its primary disadvantages (Lin & Juang, 2009).

The adsorption of metallic ions is more viable when it is used natural adsorbents like for example wastes from the industry or agriculture. Those adsorbents constitute themselves as an excellent alternative for the chemical remediation for its high adsorption capacity, low cost and large availability (Demirbas, 2008).

Many authors have researching alternative biosorbents for the removal of metallic ions from contaminated solutions, e.g. banana and orange peels (Annadurai et al. 2002), cocoa peel (Meunier et al. 2003), rice peel (Montanher et al. 2005), dry mass of *Eichhorniacrassipes* (Gonçalves Jr. et al. 2009); mussel shells (Peña-Rodríguez et al. 2010), sugar cane bagasse in natura and modified (Dos Santos et al. 2011), and others.

However, there are not many studies relating the adsorptive capacity of the solid fraction of the wastes from cassava root processing industry (peel, bagasse and peel + bagasse) for its utilization as biosorbents of contaminated hydric bodies.

The industrial processing of cassava generates large amounts of solid wastes, which can cause severe environmental problems when destined incorrectly.

In this way, this work aimed to evaluate the viability of the use of the cassava solid wastes (peel, bagasse and the mix peel + bagasse) as potential biosorbents in the remediation of hydric bodies contaminated with Pb^{2+}.

2 MATERIAL AND METHODS

2.1 *Biosorbents obtaining and characterization*

The biosorbents: Peel (P), Bagasse (B) and the mix Peel + Bagasse (P + B), were obtained in an agroindustry located at the city of Toledo—Brazil. The materials were dried at 60°C for 48 h, milled and sieved, collecting the portion between 14 to 60 mesh. It was not performed any treatment that could modify chemically the biosorbents structures.

The solution contaminated with the metallic ion Pb2+ was prepared from lead nitrate [Pb(NO$_3$)$_2$ content > 99% Sigma-Aldrich].

To study the surface charges of the materials, the point of zero charge (pH$_{PZC}$) was determined. For this purpose, 50 mg of biosorbents mass were added in 50 mL of Potassium Chloride (KCl) solutions at 0.5 mol L^{-1}, with the initial pH values ranging from 2.0 to 9.0, which was adjusted with solutions of Hydrochloric acid (HCl) and Sodium Hydroxide (NaOH), both in a concentration of 0.1 mol L^{-1}. After 24 h stirring (200 rpm), the final values of pH were obtained, thus resulting in a graph of the final pH variation in function of the initial pH, the point that reached the zero value of pH variation corresponded the pH$_{PZC}$ (Mimura et al. 2010).

2.2 Adsorption studies (pH x adsorbent mass)

In this first step were maintained constant the stirring speed (200 rpm), stirring time (1.5 h), temperature (25 °C) and the concentration of the Pb^{2+} solution (10 mg L^{-1}), varying only the pH of the solution and the adsorbent masses.

The Pb^{2+} solution was adjusted in 3 conditions of pH (4.0, 5.0 and 6.0) and the masses of the adsorbents were 200, 400, 600, 800, 1,000 and 1,200 mg.

The determination of the Pb in the equilibrium was determined by FAAS (Welz & Sperling, 1999). By the obtained value for equilibrium concentration the adsorbed quantity was calculated (Equation 1).

$$Q_{eq} = \frac{\left(C_o - C_{eq}\right)}{m} \cdot V$$ (1)

in which: Q_{eq} is the quantity of ions adsorbed by unit of adsorbent in equilibrium (mg g^{-1}), m is the used adsorbent mass (g), C_o correspond to the initial concentration of the ion in solution (mg L^{-1}), C_{eq} is the concentration of the ion in equilibrium solution (mg L^{-1}) and V is the volume of solution used.

The % of removal is calculated by Equation 2:

$$\%R = 100 - \left(\frac{C_{eq}}{C_o} \cdot 100\right)$$ (2)

in which: $\%R$ is the percentage of removal of ion by the adsorbent, C_{eq} is the ion concentration in equilibrium solution (mg L^{-1}) and C_o correspond to the initial concentration of the ion (mg L^{-1}).

2.3 Adsorption studies (contact time)

In order to determine the influence that the time of contact have in the adsorptive process, the finest results obtained in the mass and pH studies were used in different periods of stirring.

The periods of agitation were 5, 10, 20, 40, 60, 80, 100, 120, 140 e 160 min, when the period was finished the samples were filtered and the concentration determined by FASS (Welz & Sperling, 1999).

In order to evaluate the kinetic mechanism that controls the adsorption process, the mathematical models of pseudo-first-order, pseudo-second-order, Elovich and intraparticle diffusion were used.

2.4 Adsorption isotherms

By the results of the previously studies were performed tests with crescent concentrations of Pb^{2+} (5, 20, 40, 60, 80, 100, 120, 140, 160 and 200 mg L^{-1}), for the building of adsorption isotherms, which were linearized by the mathematical models of Langmuir, Freundlich and Dubinin-Radushkevich (D-R).

2.5 Influence of temperature

The influence of temperature in the adsorption process was evaluated in five conditions: 25, 35, 45, 55 and 65 °C. For that it was used 50 mL of Pb^{2+} solution at pH 5.5, in initial concentration of 100 mg L^{-1}. The solution was added to erlenmeyers flasks of 125 mL with 400 mg of adsorbent material, they were stirred at 200 rpm during 60 min. After this period of time it was performed the determination of the metal concentration by FAAS (Welz & Sperling, 1999).

By the obtained data were evaluated the thermodynamic parameters and investigated the process nature. For that purpose were calculated the Gibbs free energy (ΔG), the enthalpy (ΔH) and the entropy (ΔS) (Sari et al. 2007).

3 RESULTS AND DISCUSSION

3.1 Characterization of the biosorbents

The obtained results from the pH$_{PZC}$ (Fig. 1) indicate that the pH which corresponds to the equivalence point between positive and negatives charges for the peel is 6.00, for the bagasse is 6.17 and for the mix peel + bagasse is 6.24.

In this way, the cations adsorption, Pb^{2+} in this case, will be favored by pH values superiors to pH$_{PZC}$ (Tagliaferro et al. 2011).

3.2 Influence of pH and adsorbent mass

According Hossain et al. (2012), the pH controls proprieties of the surface of adsorbents, functional

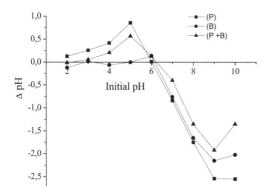

Figure 1. Point of zero charge (pH_{PZC}), KCl 0.5 mol L^{-1} for the adsorbents Peel (P), Bagasse (B) and Peel + Bagasse (P + B).

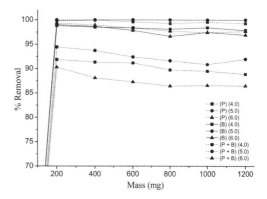

Figure 2. Effect of the adsorbent mass and pH solution in the removal percentage of Pb^{2+} for the biosorbents Peel (P), Bagasse (B) and Peel + Bagasse (P + B).

groups and ionic state of metallic species affecting greatly the adsorption of metallic ions.

According to Mimura et al. (2010), the adsorption of metallic ions in the positive form is favored in pH ranges which the negative species are predominant in the adsorbent surface. However, according to the obtained results in the mass and pH tests (Fig. 2), it can be seen that the removal of Pb^{2+} superior in the pH 5.0.

It can be observed in the Figure 2, adsorbent masses higher than 400 mg did not caused higher ion removal, and in this way in the next tests were used 400 mg of adsorbent with Pb^{2+} solution at pH 5.5.

The pH of the contaminant solution was chosen at 5.5 because despite the best removal was found at pH 5.0, the values of pH_{PZC} indicate that near the pH 6.0 is where is found the neutrality of the adsorbent, so it was chosen to work with a pH close from the pH_{PZC}.

Also is possible to see in the Figure 2, that even in lower concentrations as 10 mg L^{-1}, the materials peel, bagasse and peel + bagasse were capable of removing almost all concentration of Pb^{2+} from the solutions, reaching more than 90% of removal.

Many materials from vegetal origin has it removal influenced by the solution pH, as verified the mesocarp and endocarp of macadamia (Vilas Boas et al. 2012), bark of *Pinuselliottii* (Gonçalves Jr. et al. 2012), peel of rice and rice bran (Mimura et al. 2010), peanut hulls (Runping et al. 2008) and others.

3.3 Influence of the contact time

The influence of contact time between the adsorbent/adsorbate for Peel (P), Bagasse (B) and Peel + Bagasse (P + B) is shown in the Figure 3.

In the Figure 3 it is observed in the course of time, occur an increase of the adsorbed quantity

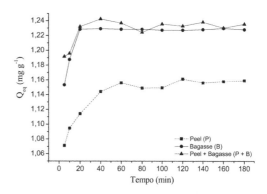

Figure 3. Effect of time (min) in the Pb^{2+} adsorbed quantity by Peel (P), Bagasse (B) and Peel + Bagasse (P + B).

of Pb^{2+}, and, after 60 min of stirring, the system reach the dynamic equilibrium, not occurring higher adsorption in superior periods of time, what indicate fast adsorption.

The obtained results were linearized by the kinetic models of adsorption, the results are presented in the Table 1.

According to the obtained values of the Table 1, the models of pseudo-first-order and Elovich don't fit convincingly to the data, because of the low values of R^2.

However, the obtained R^2 presented by the pseudo-second-order model are satisfactory, suggesting that the process is from chemical nature for all three biosorbents.

According to the values of R^2 for the intraparticle diffusion (Table 1), it can be observed that even splitting the data in small lines (multilinear), the values of R^2 are very low, indicating that this model don't explain the sorption that it was observed.

Table 1. Kinetic parameters for Pb^{2+} biosorption by Peel (P), Bagasse (B) and Peel + Bagasse (P + B).

Pseudo-first-order
(P)

K_1 (min^{-1})	−0.018	−0.013	−0.009
Q_{eq} (cal.) (mg g^{-1})	0.056	0.008	0.025
R^2	0.814	0.122	0.354

Pseudo-second-order
(B)

K_2 (g mg^{-1} min^{-1})	1.322	4.543	6.436
Q_{eq} (cal.) (mg g^{-1})	1.163	1.230	1.235
R^2	0.999	1.000	0.999

Elovich
(P + B)

A (mg g^{-1} h^{-1})	1.040	1.149	1.182
B (g mg^{-1})	0.024	0.017	0.011
R^2	0.919	0.652	0.585

	Line A	Line B	Line C

Intraparticle diffusion
(P)

K_{id} (g mg^{-1}min$^{-1/2}$)	0.015	0.0012	–
C_i (mg g^{-1})	1.042	1.142	–
R^2	0.972	0.141	–

(B)

K_{id} (g mg^{-1}min$^{-1/2}$)	0.033	−0.0002	−0.0004
C_i (mg g^{-1})	1.079	1.229	1.222
R^2	0.995	0.224	0.227

(P + B)

K_{id} (g mg^{-1}min$^{-1/2}$)	0.013	−0.007	0.002
C_i (mg g^{-1})	1.160	1.287	1.215
R^2	0.850	0.856	0.092

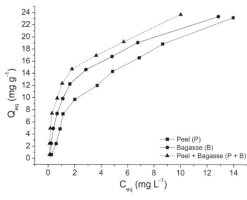

Figure 4. Isotherms of adsorption of Pb^{2+} for the biosorbents Peel (P), Bagasse (B) and Peel + Bagasse (P + B).

Table 2. Parameters of Langmuir, Freundlich and Dubinin-Radushkevich (D-R) for Pb^{2+} biosorption.

	P	B	(P + B)
Langmuir			
Q_m (mg g^{-1})	29.26	25.16	24.81
K_L (L mg^{-1})	0.052	0.003	0.002
R_L	0.489	0.628	0.746
R^2	0.961	0.989	0.983
Freundlich			
K_f (L mg^{-1})	3.814	6.18	8.643
n (mol g^{-1})	1.224	1.449	1.704
R_2	0.87	0.772	0.712
D-R			
Q_d (mol g^{-1})	0.0015	0.0006	0.0004
E (kj mol^{-1})	10.733	13.245	14.712
R^2	0.997	0.984	0.991

Still in Table 1, we observe the adsorbents (P) and (B), in the Line B, with R^2 of 0.972 and 0.995 respectively, at first sight satisfactory, however in the two cases the values of C_i are higher than '0', indicating that the line don't pass in the origin of the axes, and that the intraparticle diffusion is not the limiting step of the adsorption process (Gupta et al. 2011).

3.4 Equilibrium studies

The isotherms of adsorption obtained for the biosorbents: Peel (P), Bagasse (B) and Peel + Bagasse (P + B) in the adsorption of Pb^{2+} are shown in Figure 4.

According to Giles et al. (1960), the isotherms from the Figure 4 follow the behavior of "*group L*" (from Langmuir) and the "*subgroup 1*" indicating a slow saturation of the biosorbent surface.

The isotherms were linearized according to the models of Langmuir, Freundlich and D-R. The obtained parameters are presented in the Table 2.

Observing the obtained values of R^2 for the mathematical models (Table 2), it is evident that Langmuir presents a good fitting to the adsorption phenomenon, what suggest a monolayer adsorption.

In relation to the Langmuir parameters (Table 2), the Q_m presented expressive results, all above 20 mg g^{-1}, indicating that these materials, although naturals, are potentially biosorbents for Pb^{2+} remediation.

Evaluating the values of R_L from Langmuir (Table 2), it is observed that favorability in all cases, once that the R_L values comprehended between '0' and '1' (Lin & Juang, 2009).

According to Wan Ngah et al. (2008), when the medium energy of sorption (E) shows values above 8 kJ mol^{-1} indicates chemisorption, suggesting that the adsorption of Pb^{2+} by the biosorbents is chemical.

Table 3. Values of $Q_{eq\,(experimental)}$ and thermodynamic parameters for Pb^{2+} adsorption by Peel (P), Bagasse (B) and Peel + Bagasse (P + B).

| Biosorbents | TEMP °C | Q_{eq} | Thermodynamic parameters | | | |
			ΔG	ΔH	ΔS	R^2
(P)	25	5.681	−0.646	7.601	27.675	0.991
	35	5.747	−0.923			
	45	5.793	−1.200			
	55	5.824	−1.477			
	65	5.841	−1.753			
(B)	25	6.050	−3.412	16.28	66.081	0.976
	35	6.056	−4.073			
	45	6.072	−4.734			
	55	6.148	−5.394			
	65	6.159	−6.055			
(P + B)	25	6.125	−3.776	33.243	124.226	1.000
	35	6.133	−5.018			
	45	6.174	−6.261			
	55	6.182	−7.503			
	65	6.200	−8.745			

3.5 Thermodynamic of adsorption

In order to better comprehend the effect of temperature in the sorption process of the metallic ions by the biosorbents and analyses the nature that rules the process, were analyzed some thermodynamic parameters, which are showed in Table 3.

It can be observed in Table 3, that the increase of temperature causes a gradual increase of adsorption.

Also, the values of ΔH are positives, indicating an endothermic system (Wan Ngah & Fatinathan, 2010).

According to Wan Ngah & Hanafiah (2008), negative values of ΔG indicate spontaneous nature, whereas positive values for ΔS indicate increase of the disorder and randomness of the system (Table 3).

The obtained results show that higher values of temperature caused higher rates of adsorption, however, the adsorption process is not operated in higher temperatures, because that increase to much the operational costs.

4 CONCLUSION

By the obtained results it can be concluded that the use of the cassava wastes are potential biosorbents in the decontamination of hydric bodies polluted with Pb^{2+}.

It is verified that this practice complements the final steps of the productive chain of cassava, providing a new destiny to the solid wastes from this agroindustrial activity.

ACKNOWLEDGMENTS

The authors thank the Araucaria Foundation (SETI-PR) and the CNPq-MCTI (REPENSA) for financial support on projects and other aid.

REFERENCES

Annadurai, G.; Juang, R.S.; Lee, D.J. 2002. Adsorption of heavy metals from water using banana and orange peels. *Water Science & Technology* 47(1): 185–190.

Demirbas, A. 2008. Heavy metals adsorption onto agro-based waste materials: A review, *Journal of Hazardus Materials* 157: 220–229.

Dos Santos, V.C.G.; Souza, J.V.T.M.; Tarley, C.R.T.; Caetano, J.; Dragunski, D.C. 2011. Copper ions adsorption from aqueous medium using the biosorbent sugarcane bagasse *in natura* and chemically modified. *Water, Air and Soil Pollution* 216: 351–359.

Giles, C.H.; Macewan, T.H.; Nakhwa; S.N.; Smith, D. 1960. Studies in adsorption. Part XI. A system of classification of solution adsorption isotherms, and its use in diagnosis of adsorption mechanisms and in measurement of specific surface areas of solids. *Journal Chemical Society*: 3973–3993.

Gonçalves Jr., A.C.; Selzlein, C.; Nacke, H. 2009. Use of water hyacinth (Eichornia crassipes) dry biomass for removing heavy metals from contaminated solutions. *Acta Scientiarum.Technology* 31(1): 103–108.

Gonçalves Jr., A.C.; Strey, L.; Lindino, C.A.; Nacke, H.; Schwantes, D.; Seidel, E.P. 2012. Applicability of the Pinus bark (*Pinuselliottii*) for the adsorption of toxic heavy metals from aqueous solutions. *Acta Scientiarum. Technology* 34(1): 79–87.

Gupta, S.S.; Bhattacharyya K.G. 2011. Kinetics of adsorption of metal ions on inorganic materials: A review. *Advances in Colloid and Interface Science* 162(1–2): 39–58.

Hossain, M.A.; Ngo, H.H.; Guo, W.S.; Nguyen, T.V. 2012. Biosorption of Cu(II) from water by banana peel based biosorbent: Experiments and models of adsorption and desorption. *Journal of Water sustainability* 2(1): 87–104.

Li, N.; Bai, R. 2005. Copper adsorption on chitosan-cellulose hydrogel beads: behaviours and mechanisms. *Separation and Purification Technology* 42(3): 237–247.

Lin, S.; Juang, R. 2009. Adsorption of phenol and its derivatives from water using synthetic resins and low-cost natural adsorbents: A review. *Journal of Environmental Management* 90(3): 1336–1349.

Meunier, N.; Laroulandie, J.; Blais, J.F.; Tyagi, R.D. 2003. Cocoa shells for heavy metal removal from acidic solutions. *Bioresource Technology* 90(3): 255–263.

Mimura, A.M.S.; Vieira, T.V.A.; Martelli, P.B.; Gorgulho, H.F. 2010. Utilization of rice husk to remove Cu^{2+}, Al^{3+}, Ni^{2+} and Zn^{2+} from wastewater. *Química Nova* 33(6): 1279–1284.

Montanher, S.F.; Oliveira, E.A.; Rollemberg, M.C. 2005. Removal of metal ions from aqueous solutions by sorption onto rice bran. *Journal of Hazardous Materials* 117: 207–211.

Oliveira, J.A.; Cambraia, J.; Cano, M.A. 2001. Cadmium absorption and accumulation and its effects on the relative growth of water hyacinths and salvinia. *Revista Brasileira de Fisiologia Vegetal*, 13(3): 329–341.

Peña-Rodríguez, S.; Fernández-Calviño, D.; Nóvoa-Muñoz, J.C.; Arias-Estévez, M.; Núñez-Delgado, A.; Fernández-Sanjurjo, M.J.; Álvarez-Rodríguez, E. 2010. Kinetics of Hg (II) adsorption and desorption in calcined mussel shells. *Journal of Hazardous Materials* 180: 622–627.

Pereira, S.R. 2004. Identificação e caracterização das fontes de poluição em sistemas hídricos. *Revista Eletrônica de Recursos Hídricos* 1(1) 20–38.

Runping, H.; Han, P.; Cai, Z.; Zhao, Z.; Tang, M. 2008. Kinetics and isotherms of neutral red adsorption on peanut husk. *Journal of Environmental Sciences* 20: 1035–1041.

Sari, A.; Tuzen, M.; Citak, D.; Soylak, M. 2007. Equilibrium, kinetic and thermodynamic studies of adsorption of Pb (II) from aqueous solution onto Turkish kaolinite clay. *Journal of Hazardous Materials* 149(2): 283–291.

Tagliaferro, G.V.; Pereira, P.H.F.; Rodrigues, L.A.; Da Silva, M.L.C.P. 2011. Cadmium, lead and silver adsorption in hydrous niobium oxide prepared by homogeneous solution method. *Química Nova* 34(1): 101–105.

Vilas Boas, N.; Casarin, J.; Caetano, J.; Gonçalves Jr, A.C.G.; Tarley, C.R.T.; Dragunski, D. 2012. Biosorption of copper using the mesocarp and endocarp of natural and chemically treated macadamia. *Revista Brasileira de Engenharia Agrícola e Ambiental*. 16(12): 1359–1366.

Wan Ngah, W.S.; Fatinathan, S. 2010. Adsorption characterization of Pb(II) and Cu(II) ions onto chitosan-tripolyphosphate beads: Kinetic, equilibrium and thermodynamic studies. *Journal of Environmental Management* 91(4): 958–969.

Wan Ngah, W.S.; Hanafiah, M.A.K.M. 2008. Biosorpiton of copper ions from dilute aqueous solutions on base treated rubber (*Hevea brasiliensis*) leaves powder: kinetics, isotherm, and biosorption mechanisms. *Journal of Environmental Sciences* 20(10): 1168–1176.

Welz, B. & Sperling, M. 1999, Atomic absorption spectrometry. 2 ed. Weinheim: Wiley-VCH, 941 p.

Green Design, Materials and Manufacturing Processes – Bártolo et al. (eds)
© *2013 Taylor & Francis Group, London, ISBN 978-1-138-00046-9*

Equilibrium of the adsorption process of glyphosate using wastes from the cassava industry

D. Schwantes, A.C. Gonçalves Jr., G.F. Coelho, J. Casarin & J.R. Stangarlin
Western Parana State University, Marechal Cândido Rondon, Brazil

A. Pinheiro
Regional University of Blumenau, Brazil

ABSTRACT: The present work aimed to evaluate the use of solid cassava wastes as natural adsorbents for the removal of waters polluted with glyphosate. The biosorbents peel, bagasse and the mix peel + bagasse of cassava roots were characterized by infrared spectroscopy and pH_{PZC}. In order to evaluate the adsorption of this organic molecule, tests with glyphosate gradual concentrations were performed at constant temperature of 25°C, with the results used for the built of adsorption isotherms which were linearized by Langmuir, Freundlich and Dubinin-Radushkevich. The IR suggests the presence of functional groups like starch, amides, carboxyl groups, lignin, protein and carbohydrates in the biosorbents constitution, what indicates that these adsorbent materials are potential biosorbents. The obtained isotherms showed best fitting by the Dubinin, suggesting chemisorption of glyphosate. It can be concluded that the use of the cassava wastes are potential biosorbents for the decontamination of waters polluted with glyphosate.

1 INTRODUCTION

Since its commercial introduction in 1974, glyphosate [N-(phosphonomethyl)glycine] has become the dominant herbicide worldwide (Duke & Powles, 2008).

This herbicide is used on many food and non-food crops as well as non-crop areas such as roadsides and due to the different application methods and weather conditions, a significant amount of herbicide can contaminate the soil (Khoury et al., 2010) and groundwater.

The removal of such compounds at low levels from water always constitutes a problem (Konstantinou et al., 2000).

One of the alternatives for the removal of contaminants from waters is the adsorption process, being this technology extensively used for the removal of organic and non-organic pollutants from aqueous solutions (Lin & Juang, 2009).

Adsorption enables the separation of selected compounds from dilute solutions. Compared to alternative technologies, adsorption is attractive for its relative simplicity of design, high capacity and favorable rate, insensitivity to toxic substances, ease of regeneration and low cost (Soto et al., 2011).

The adsorption of ions is more viable when are used natural adsorbents as for example, wastes from industry and agriculture. These adsorbents are one excellent alternative for chemical remediation by its great capacity of adsorption, low cost and high availability (Demirbas, 2008).

Many authors have researching alternative biosorbents for the removal of ions from contaminated solutions, e.g. banana and orange peels (Annadurai et al. 2002), cocoa peel (Meunier et al. 2003), rice peel (Montanher et al. 2005), dry mass of *Eichhornia crassipes* (Gonçalves Jr. et al. 2009); mussel shells (Peña-Rodríguez et al. 2010), sugar cane bagasse (Dos Santos et al. 2011), pinus bark (Gonçalves Jr. et al. 2012), and others.

However, there are not many studies relating the adsorptive capacity of the solid fraction of the wastes from cassava root processing industry (peel, bagasse and peel + bagasse) for its utilization as biosorbents of contaminated hydric bodies.

The industrial processing of cassava generates large amounts of solid wastes, which can cause severe environmental problems when destined incorrectly.

In this way, this work aimed to evaluate the viability of the use of the cassava solid wastes (peel, bagasse and the mix peel + bagasse) as potential biosorbents in the remediation of hydric bodies contaminated with glyphosate.

2 MATERIAL AND METHODS

2.1 *Biosorbents obtaining and characterization*

The biosorbents: Peel (P), Bagasse (B) and the mix Peel + Bagasse (P + B), were obtained in an agroindustry located at the city of Toledo—Brazil. The materials were dried at 60°C for 48 h, milled and sieved, collecting the portion between 14 to 60 mesh. It was not performed any treatment that could modify chemically the biosorbents structures.

The contaminated solution was prepared with N-(Phosphonomethyl)glycine 99.2% (Sigma-Aldrich®), using ultrapure water a from reverse osmosis (Permution®) and a ultra-water purifier microprocessed (Permution®), i.e. nanopure water at 25°C (18 MΩ cm⁻¹) was used for all procedures.

All the laboratory glassworks were silanized with 5% trimethylchlorosilane in n-hexane solution, and after 10 min of contact were washed twice with n-hexane and methanol, preventing the adsorption of glyphosate by the glassworks.

To study the surface charges of the materials, the point of zero charge (pH$_{PZC}$) was determined according to Mimura et al. 2010.

For the evaluation of the possible functional groups that are responsible for the ligation with glyphosate molecules was performed an infrared spectrum characterization of the biosorbents, using a spectrophotometer Shimadzu Infrared Spectrophotometer FTIR-8300 Fourier Transform, in the region between 400 and 4,000 cm⁻¹ with resolution of 4 cm⁻¹. The spectrum was obtained by transmittance, using tablets of KBr.

The glyphosate concentrations were determined by an ion exchange chromatograph Dionex® ICS-90, equipped with a conductivity detector model DS5. Chromatographic conditions: analytical column Ion-Pac® AS22 4 × 250 mm, guard column IonPac® AG22 4 × 50 mm, anion micromembrane suppressor AMMS® 300 4 mm, eluent solution of 9.0 mmol L⁻¹ Na$_2$CO$_3$ and 2.8 mmol L⁻¹ NaHCO$_3$ with a flow rate of 1.2 mL min⁻¹, regenerant solution of 50 mmol L⁻¹ H$_2$SO$_4$ and loop injection of 250 µL (Queiroz et al. 2011).

By the obtained values for equilibrium levels the quantity adsorbed was calculated (Equation 1).

$$Q_{eq} = \frac{\left(C_o - C_{eq}\right)}{m} \cdot V \qquad (1)$$

in which: Q_{eq} is the quantity of ions adsorbed by unit of adsorbent in equilibrium (mg g⁻¹), m is the used adsorbent mass (g), C_0 correspond to the initial concentration of the ion in solution (mg L⁻¹), C_{eq} is the concentration of the ion in equilibrium solution (mg L⁻¹) and V is the volume of solution used.

The % of removal is calculated by Equation 2:

$$\%R = 100 - \left(\frac{C_{eq}}{C_o} \cdot 100\right) \qquad (2)$$

in which: %R is the percentage of removal of ion by the adsorbent, C_{eq} is the ion concentration in equilibrium solution (mg L⁻¹) and C_0 correspond to the initial concentration of the ion (mg L⁻¹).

2.2 Adsorption isotherms

The isotherms of adsorption were performed by tests with crescent concentrations of glyphosate (0.1, 0.2, 0.3, 0.4, 0.5, 0.6, 0.7, 0.8, 0.9 and 1.0 mg L⁻¹) with 400 mg of adsorbent mass, solution pH at 5.5 and 60 min of stirring at 200 rpm and 25 °C. The results were linearized by the models of Langmuir, Freundlich and Dubinin-Radushkevich (D-R).

3 RESULTS AND DISCUSSION

3.1 Characterization of the biosorbents

The obtained results from the pH$_{PZC}$ (Fig. 1) indicate that the pH which corresponds to the equivalence point between positive and negatives charges for the peel is 6.00, for the bagasse is 6.17 and for the mix peel + bagasse is 6.24.

In this way, the glyphosate adsorption will be favored by pH values inferiors to pH$_{PZC}$ (Tagliaferro et al., 2011).

For the infrared analysis of the biosorbents (Fig. 2) were observed the following bands: 3440, 2920, 1730, 1650, 1420 e 1030.

According to the Figure 2, it can be observed the presence of a strong and wide band at 3440 and 3330 cm⁻¹, what suggest one stretching vibration of O–H. This band is characterized by the vibrational stretching of hydroxyl groups presents in carbohydrates, fatty acids, proteins, units of lignin, celluloses and absorbed water (Stuart, 2004; Feng et al., 2011; Han et al., 2010).

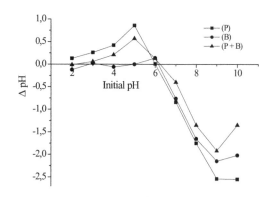

Figure 1. Point of zero charge (pH$_{PZC}$), KCl 0.5 mol L⁻¹ for the adsorbents Peel (P), Bagasse (B) and Peel + Bagasse (P + B).

424

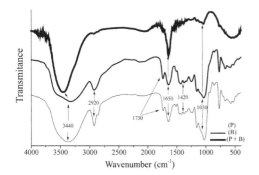

Figure 2. Infrared spectrum for Peel (P), Bagasse (B) and Peel + Bagasse (P + B).

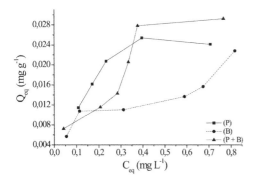

Figure 3. Glyphosate isotherms of adsorption for the biosorbents Peel (P), Bagasse (B) and Peel + Bagasse (P + B).

The band at 2920 cm^{-1} suggests a stretching vibration of C–H in alkanes (Barbosa, 2007).

According to Horn et al. (2011), the bands at 1730 cm^{-1} suggest the presence of starch, lignin and hemicelluloses.

The bands at 1420 and 1650 cm^{-1} suggests the stretching vibration of the bonds of C–O from amides and carboxylic groups (Han et al., 2010). The bands at 1030 cm^{-1} can indicate the presence of C–O bonds, also suggesting the presence of lignin (Pascoal Neto et al., 1995), which according to Guo et al. (2008), presents compounds like carboxylic groups which provide adsorption sites.

These results indicate that the cassava solid wastes are potential biosorbents for the adsorption of ions in solutions.

3.4 Equilibrium studies

The isotherms of adsorption obtained for the biosorbents: Peel (P), Bagasse (B) and Peel + Bagasse (P + B) in the adsorption of glyphosate are shown in Figure 3.

The isotherms were linearized according to the models of Langmuir, Freundlich and D-R. The obtained parameters are presented in the Table 1.

The sorbent peel presented best fitting for D-R ($R^2 = 0.964$), with E values higher than 8 kJ mol^{-1}, suggesting chemisorption (Wan Ngah et al., 2008).

According to Table 1, the biosorbent bagasse presented excellent fitting for Freundlich and D-R models ($R^2 = 0.990$ and 0.989 respectively), suggesting a chemisorption multilayer, since the values of E are higher than 8 kJ mol^{-1}.

According to Sodré et al. (2001), the parameter n of Freundlich indicates the reactivity of the active sites of the adsorbent, being that when the values of n are higher than '1' that indicates the presence of highly energetic sites in the adsorbent surface, what can be observed clearly in Table 1.

The adsorbent peel + bagasse presented best fitting by D-R, what, according to the obtained values of E, suggest a chemisorption of glyphosate by this biosorbent (Table 1).

The values of R^2 obtained by the Langmuir model are lower when compared to the values obtained by the models of Freundlich and D-R, indicating that this model does not explain satisfactorily the biosorption process of glyphosate.

Figures 4–6 present the linearizations by Langmuir, Freundlich and D-R in the adsorption of glyphosate by the biosorbents Peel (P), Bagasse (B) and Peel + Bagasse (P + B).

Table 1. Parameters of Langmuir, Freundlich and Dubinin-Radushkevich (D-R) for glyphosate biosorption.

	P	B	(P + B)
Langmuir			
Q_m (mg g^{-1})	0.0297	0.0170	0.0359
K_L (L mg^{-1})	146.9052	376.8300	135.7890
R_L	3.403 10^{-5}	1.327 10^{-5}	3.682 10^{-5}
R^2	0.956	0.961	0.946
Freundlich			
K_f (L mg^{-1})	0.0474	0.0175	0.0283
n (mol g^{-1})	1.614	2.581	2.156
R_2	0.957	0.990	0.893
D-R			
Q_d (mol g^{-1})	0.996	0.997	0.997
E (kJ mol^{-1})	11.352	14.434	12.953
R^2	0.964	0.989	0.999

Q_m (mg g^{-1}): max adsorption capacity, K_L ou b (L mg^{-1}): constant related to the forces of interaction adsorbent/adsorbate; R_L: Langmuir constant; R^2: determination coefficient; K_f (L mg^{-1}): related to the adsorption capacity; n: related to the heterogeneity of the solid; Q_d (mol g^{-1}): max capacity of adsorption; E (kj mol^{-1}): sorption medium energy.

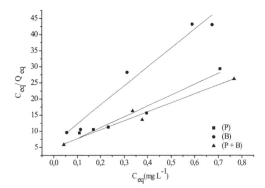

Figure 4. Linearizations obtained by the model of Langmuir for the adsorption of glyphosate for Peel (P), Bagasse (B) and Peel + Bagasse (P + B).

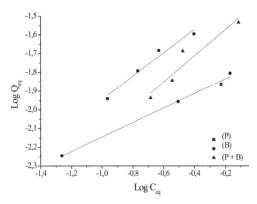

Figure 5. Linearizations obtained by the model of Freundlich for the adsorption of glyphosate for Peel (P), Bagasse (B) and Peel + Bagasse (P + B).

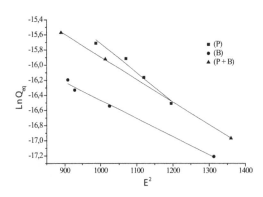

Figure 6. Linearizations obtained by the model of D-R for the adsorption of glyphosate for Peel (P), Bagasse (B) and Peel + Bagasse (P + B).

4 CONCLUSION

By the obtained results it can be concluded that the use of cassava wastes are potential adsorbents in the decontamination of waters polluted with glyphosate.

It is verified that this practice complements the final steps of the productive chain of cassava, providing a new destiny to the solid wastes from this agroindustrial activity.

ACKNOWLEDGMENTS

The authors thank the Araucaria Foundation (SETI-PR) and the CNPq-MCTI (REPENSA) for financial support on projects and other aid.

REFERENCES

Annadurai, G.; Juang, R.S.; Lee, D.J. 2002. Adsorption of heavy metals from water using banana and orange peels. *Water Science & Technology* 47(1):185–190.

AOAC. 2005. Official methods of analysis of the Association Analytical Chemists. Maryland: AOAC, 3000p.

Barbosa, L.C.A. 2007. Espectroscopia no infravermelho na caracterização de compostos orgânicos. Viçosa, 189 p.

Demirbas, A. 2008. Heavy metals adsorption onto agrobased waste materials: A review, *Journal of Hazardus Materials* 157: 220–229.

Dos Santos, V.C.G.; Souza, J.V.T.M.; Tarley, C.R.T.; Caetano, J.; Dragunski, D.C. 2011. Copper ions adsorption from aqueous medium using the biosorbent sugarcane bagasse *in natura* and chemically modified. *Water, Air and Soil Pollution* 216: 351–359.

Duke, S.O.; Powles, S.B. 2008. Glyphosate: a once-in-a-century herbicide. Pest *Management Science* 64(4): 319–325.

Feng, N.; Guo, X.; Liang, S.; Zhu, Y. Liu, J. 2011. Biosorption of heavy metals from aqueous solutions by chemically modified orange peel. *Journal of Hazardous Materials* 185(1): 49–54.

Gonçalves Jr., A.C.; Selzlein, C.; Nacke, H. 2009. Use of water hyacinth (Eichornia crassipes) dry biomass for removing heavy metals from contaminated solutions. Acta Scientiarum.Technology 31(1): 103–108.

Gonçalves Jr., A.C.; Strey, L.; Lindino, C.A.; Nacke, H.; Schwantes, D.; Seidel, E. P. 2012. Applicability of the Pinus bark (*Pinus elliottii*) for the adsorption of toxic heavy metals from aqueous solutions. *Acta Scientiarum. Technology* 34(1): 79–87.

Guo, X.; Zhang, S.; Shan, X. 2008. Adsorption of metal ions on lignin. *Journal of Hazardous Materials* 151(1): 134–142.

Han, R.; Zhang, L.; Song, C.; Zhang, M.; Zhu, H.; Zhang, L. 2010. Characterization of modified wheat straw, kinetic and equilibrium study about copper ion and methylene blue adsorption in batch mode. *Carbohydrate Polymers* 79(4): 1140–1149.

Horn, M.M.; Martins, V.C.A.; Plepis, A.M.G. 2011. Effects of starch gelatinization and oxidation on the rheological behavior of chitosan/starch blends. *Polymer international* 60(6): 920–923.

Khoury, G.A.; Gehris, T.C.; Tribe, L.; Sánchez, R.M.T.; Afonso, M.S. 2010. Glyphosate adsorption on montmorillonite: An experimental and theoretical study of surface complexes. *Applied Clay Science* 50(2): 167–175.

Konstantinou, I.K.; Albanis, T.A.; Petrakis, D.E.; Pomonis, P.J. 2000. Removal of herbicides from aqueous solutions by adsorption on Al-pillared clays, Fe–Al pillared clays and mesoporous alumina aluminum phosphates. *Water Research* 34(12): 3123–3136.

Lin, S.H..; Juang, R.S. 2009. Adsorption of phenol and its derivatives from water using synthetic resins and low-cost natural adsorbents: A review. *Journal of Environmental Management*, 90(3): 1336–1349.

Meunier, N.; Laroulandie, J.; Blais, J.F.; Tyagi, R.D. 2003. Cocoa shells for heavy metal removal from acidic solutions. *Bioresource Technology* 90(3): 255–263.

Mimura, A.M.S.; Vieira, T.V.A.; Martelli, P.B.; Gorgulho, H. F. 2010. Utilization of rice husk to remove Cu^{2+}, Al^{3+}, Ni^{2+} and Zn^{2+} from wastewater. *Química Nova* 33(6): 1279–1284.

Montanher, S.F.; Oliveira, E.A.; Rollemberg, M.C. 2005. Removal of metal ions from aqueous solutions by sorption onto rice bran. *Journal of Hazardous Materials* 117: 207–211.

Pascoal Neto, C.; Rocha, J.; Gil, A.; Cordeiro, N. Esculcas, A. P.; Rocha, S.; Delgadillo, I.; De Jesus, J.D.; Correia, A.J. 1995. 13C solid-state nuclear magnetic resonance and Fourier transform infrared studies of the thermal decomposition of cork. *Solid State Nuclear Magnetic Resonance* 4(3): 143–151.

Queiroz, G.M.P.; Rivail, M.; Bianco, R.J.R.; Pinheiro, A.; Kaufmann, V. 2011. Glyphosate transport in runoff and leaching waters in agricultural soil. *Química Nova* 34(2): 190–195.

Sodré, F.F.; Lenzi, E.; Costa, A.C. 2001. Utilização de modelos físico-químicos de adsorção no estudo do Comportamento do cobre em solos argilosos. *Química Nova* 24(3): 324–330.

Soto, M.L.; Moure, A.; Domínguez, H.; Parajó, J.C. 2011. Recovery, concentration and purification of phenolic compounds by adsorption: A review. *Journal of Food Engineering* 105(1): 1–27.

Stuart, B.H. 2004. *Infrared Spectroscopy: Fundamentals and applications*. 1ed. John Wiley and Sons, 224 p.

Tagliaferro, G.V.; Pereira, P.H.F.; Rodrigues, L.A.; Da Silva, M.L.C.P. 2011. Cadmium, lead and silver adsorption in hydrous niobium oxide prepared by homogeneous solution method. *Química Nova* 34(1): 101–105.

Wan Ngah, W.S.; Hanafiah, M.A.K.M. 2008. Biosorpiton of copper ions from dilute aqueous solutions on base treated rubber (*Hevea brasiliensis*) leaves powder: kinetics, isotherm, and biosorption mechanisms. *Journal of Environmental Sciences* 20(10): 1168–1176.

Welz, B. & Sperling, M. 1999, Atomic absorption spectrometry. 2 ed. Weinheim: Wiley-VCH, 941p.

Green Design, Materials and Manufacturing Processes – Bártolo et al. (eds)
© 2013 Taylor & Francis Group, London, ISBN 978-1-138-00046-9

Use of bark of *Pinus elliottii* as a biosorbent in the removal of glyphosate from aqueous solutions

L. Strey, A.C. Gonçalves Jr., G.F. Coelho & D. Schwantes
Western Parana State University, Marechal Cândido Rondon, Brazil

H. Nacke
Dynamic College of Cataracts, Foz do Iguaçu, Brazil

C.R.T. Tarley
Londrina State University, Londrina, Brazil

ABSTRACT: The objective of this work is the use of bark of *Pinus elliottii*, a residue generated in agroforestry industry, as a biosorbent on adsorption of glyphosate in aqueous solutions as an alternative to conventional methods of treating contaminated water bodies. After the characterization of the biosorbent, was verified the adsorption capacity of glyphosate by pine bark. Adsorption tests were performed in order to obtain the adsorption isotherms, which were linearized by the mathematical models of Langmuir and Freundlich. From the results, it was concluded that the bark *of Pinus elliottii* is a viable and promising alternative in the adsorption and removal of Cd from contaminated water bodies, complementing the final stages of the production chain of this kind, where the use of byproducts and waste of the industry agriculture and forestry is an ideal inserted into the question of sustainability of agroecosystems.

1 INTRODUCTION

With the rapid development of agriculture and the need to supply a growing population expansion, there has been an increased need related to the use of herbicides in crop management. In this context, herbicides formulated by glyphosate have gained expression and importance because of the growing areas sown with genetically modified crops (Rodrigues & Almeida 2005). Glyphosate, although it is the most widely used in agriculture, it is also used for controlling aquatic weeds emerging in surface waters or shores of water bodies (Salomon & Thompson 2003). Thus, in recent years, it has been observed that both world records surface water and groundwater is contaminated by pesticides are used in agriculture.

Thus, various studies and techniques aimed a decontamination of these water bodies are being developed, aiming at minimizing the impacts of these contaminants. The main methods of decontamination of water bodies are contaminated with glyphosate coagulation with Fenton, photo and electro Fenton, ultrasound combined with Fenton, photo oxidation and biological degradation (Chang et al. 2011). However, these methods often become limited to be technically or economically unviable, making it difficult to apply (Ferreira et al. 2007).

As a result, the adsorption becomes an alternative method of treatment is effective for removing heavy metals and pesticides, where, during this process, due to the accumulation of a certain element or substance at the interface of the surface of adsorbent material (Kanitz Jr. et al. 2009).

The adsorption process involves the use of various adsorbent materials. However, when considering the need to reduce costs involved in this process should be sought materials that present low cost, thermal stability, mechanical and chemical, minimal waste generation and high availability features provided by most natural adsorbents (Psareva et al. 2005).

Byproducts and waste from agriculture and forestry industries, like the bark of *Pinus elliottii*, because they require little processing and are abundant in nature, are considered low cost materials. This material contains various organic compounds such as lignin, cellulose and hemicellulose, which have polyphenolic groups which may be useful for the binding of heavy metal ions (Gonçalves Jr. et al. 2012).

The objective with this work is use the bark of *Pinus elliottii*, a residue generated in agroforestry industry, as a biosorbent on adsorption of glyphosate in aqueous solutions as an alternative to conventional methods of treating contaminated water bodies.

2 METHOD AND MATERIALS

2.1 Obtaining and characterization of biosorbent

For the preparation of biosorbent, was used the bark of *Pinus elliottii*. After collection, the material was oven-dried, homogenised and standardized crushed in 35 mesh sieves. Then the biosorbent was characterized by scanning electron microscopy (SEM), infrared spectroscopy (FT-IR) and point of zero charge (pH$_{PZC}$).

2.2 Equilibrium adsorption

To verify the adsorption capacity of glyphosate by the bark of *Pinus elliottii*, adsorption tests were performed in order to obtain the adsorption isotherms. For this, solutions containing glyphosate were prepared from a standard certificate with concentrations of 0.10 to 1.00 mg mL⁻¹. These solutions, adjusted and buffered at pH condition 5.5, were added in Erlenmeyer flasks with the biosorbent. The flasks were shaken for 90 min at 200 rpm and 25 ° C in a water bath thermostated. After the stirring period, the final concentration of glyphosate in solution was determined by high performance liquid chromatography (HPLC) (Amarante et al. 2002).

The glyphosate amount adsorbed was determined using the Equation 1:

$$Q_{eq} = \frac{\left(C_0 - C_{eq}\right)}{m} \cdot V \quad (1)$$

which Q_{eq} = amount of glyphosate adsorbed by adsorbent mass, C_0 = initial glyphosate concentration, C_{eq} = equilibrium concentration, V = volume, m = mass.

2.3 Adsorption isotherms

From the results obtained, the adsorption isotherms were constructed and linearized according to mathematical models of Langmuir and Freundlich.

The Langmuir model (Equation 2) suggests that the adsorption on a uniform surface composed of a finite number of sites assumes a monolayer adsorption (Witek-Krowiak et al., 2011).

$$\frac{C_{eq}}{Q_{eq}} = \frac{1}{q_m \times b} + \frac{C_{eq}}{q_m} \quad (2)$$

in which q_m = maximum capacity of adsorption, b or K_L = adsorbent-adsorbate interaction forces.

Unlike the Langmuir model, the Freundlich model (Equation 3) describes the multilayer

adsorption and it is applicable to heterogeneous surfaces (Witek-Krowiak et al., 2011).

$$\log Q_{eq} = \log K_f + \left(\frac{1}{n}\right) \times \log C_{eq} \quad (3)$$

in which K_f = adsorption capacity, n = intensity of adsorption and the adsorbent-adsorbate interaction.

3 RESULTS AND DISCUSSION

3.1 Characterization of biosorbent

Figure 1 shows the infrared spectra obtained for the biosorbent from the bark of *Pinus elliottii*. The broadband and strong 3334 cm⁻¹ may be attributed to stretching vibrational O-H connections of primary amines/amides and secondary, and O-H bonds to hydroxyl groups present on cellulose, lignin, and water adsorbed on the surface of the adsorbent (Goncalves Jr. et al., 2010). The band at 2925 cm⁻¹ is related to the stretching vibration of CH bonds of the groups of alkanes and aliphatic acids (Iqbal et al. 2009). The presence of bands at 1612, 1571 and 1446 cm⁻¹ may be attributed to stretching vibration of C-O amide bonds and carboxyl groups (Pavan et al. 2008). The band at 1056 cm⁻¹ suggests the presence of lignin (Pascoal Neto et al. 1995), which has carboxyl groups as compounds which provide adsorption sites for the metal ions (Guo et al. 2008). The observed structures react with metal ions by forming coordination complexes with the same solution, facilitating the adsorption process (Gonçalves Jr. et al 2013).

In Figure 2 are shown the images obtained by SEM of the biosorbent with magnification of 800 and 20000 times. The surface of the adsorbent appears to be quite irregular and with major and

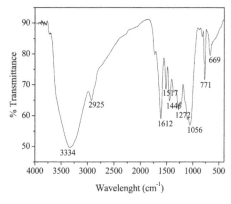

Figure 1.   Spectra in the infrared region to the bark of Pinus elliottii.

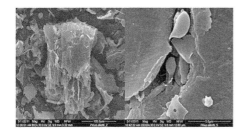

Figure 2. Image by scanning electron microscopy of the adsorbent material in magnification 800 and 20000 times.2010).

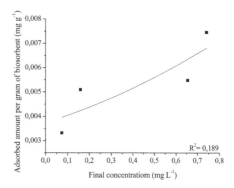

Figure 3. Adsorption isotherm for glyphosate on the bark of *Pinus elliottii*.

significant voids. It's also possible to identify the structure in the form of chips, introducing on the whole surface of the material, cracks and pores, thus providing a large surface area the same.

The value of pH_{PZC} founded is 3.5. The pH_{PZC} value is the point at which the functional groups of the surface of the adsorbent does not contribute to the pH of the solution. Above this pH the surface charge of the adsorbent becomes negatively charged, causing the cations occupying the adsorption sites with greater interaction (Marcilla et al. 2009), so the adsorption of cations, such as Cd^{2+} is favored at pH values higher than the pH_{PZC}.

3.2 *Equilibrium adsorption*

From the results of adsorption obtained, it was possible to construct the adsorption isotherm for glyphosate on *Pinus elliottii*, which can be seen in Figure 3.

As can be seen, the isotherm behave favorably, where highest rates of removal occurring at the lowest concentrations of glyphosate.

In Figure 4 presents the linearization of Langmuir and Freundlich adsorption isotherm for glyphosate. The values of the parameters of the mathematical models are shown in Table 1, where we can observe a better fit by Langmuir model ($R^2 = 0.990$), which indicates that the adsorption of glyphosate on the bark of *Pinus elliottii* occurs in monolayers

In view of this, one can calculate the maximum amount of adsorption of glyphosate per gram of biosorbent (Q_m), which showed a value of 0.0056 mg g⁻¹. Furthermore, we obtained a high bonding energy (K_L) between the glyphosate and the pine bark value of 26.4065 mg L⁻¹.

The mathematical model of Freundlich, despite not fit satisfactorily ($R^2 = 0.620$), had a high amount of reactivity of the active sites ($n = 5.487$), indicating the presence of high-energy sites on the surface of the adsorbent material.

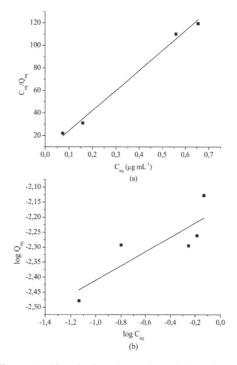

Figure 4. Linearizations by mathematical models of Langmuir (a) and Freundlich (b) for the adsorption of glyphosate by *Pinus elliottii*.

Table 1. Parameters of adsorption of glyphosate by *Pinus elliottii*.

Langmuir constants			Freundlich constants		
Q_m (µg g⁻¹)	K_L (L mg⁻¹)	R^2	K_f (mg g⁻¹)	n	R^2
0.0056	26.4065	0.990	1.4348	5.487	0.620

431

The best fit to the Langmuir model suggests that the adsorption sites are all the same energy and that adsorption occurs in monolayers (Gonçalves Jr. et al. Despite the low adsorption capacity of the biosorbent adsorption sites showed highly energetic and therefore a high binding energy between adsorbent/adsorbate (Gonçalves Jr. et al. 2012). Thus, the molecules of glyphosate are more strongly adsorbed on the surface of the biosorbent and unlikely to be desorbed back into solution.

4 CONCLUSIONS

From the characteristics obtained in the surface of the biosorbent study of its composition and structure, as well as the results obtained by applying the mathematical models of Langmuir and Freundlich, it can be concluded that the bark of *Pinus elliottii* is a biosorbent with excellent sorption characteristics, making it a viable and promising alternative to glyphosate adsorption and removal of contaminated water bodies. Moreover, this practice complements the final stages of the production chain of this kind, to give a proper and sustainable destination for the waste generated in their activities.

ACKNOWLEDGEMENTS

The authors thank the Araucaria Foundation (SETI-PR) and the CNPq-MCTI (REPENSA) for financial support on projects and other aid.

REFERENCES

Aksu, Z.; Çalik, A.; Dursun, Y. & Demircan, Z. 1999. Biosorption of iron (III)-cyanide complex anions to *Rhizopus arrhizus*: application of adsorption isotherms. *Process Biochemistry* 34 (5): 483–491.

Amarante, J.O.P.; Santos, T.C.R.; Brito, N.M. & Ribeiro, M.L. Métodos de extração e determinação do herbicida glifosato: breve revisão. *Química Nova* 25 (1): 420–428.

Chang, K.L.; Lin, J.H. & Chen, S.T. 2011. Adsorption studies on the removal of pesticides (carbofuran) using activated carbon from rice straw agricultural waste. *Engineering and Technology* 76 (2): 348–351.

Ferreira, J.M.; Silva, F.L.H.; Alsina, O.L.S.; Oliveira, L.S.C.; Cavalcanti, E.B. & Gomes, W.C. 2007. Equilibrium and kinetic study of Pb^{2+} biosorption by *Saccharomyces cerevisiae*. *Química Nova* 30 (5): 1188–1193.

Gonçalves Jr., A.C.; Nacke, H.; Fávere, V.T.; Gomes, G.D. 2010. Comparação entre um trocador aniônica de sal de amônio quaternário de quitosana e um trocador comercial na extração de fósforo disponível em solos. *Química Nova* 33 (5): 1047–1052.

Gonçalves Jr., A.C.; Strey, L.; Lindino, C.A.; Nacke, H.; Schwantes, D.; Seidel, E.P. 2012. Applicability of the Pinus bark (Pinus elliottii) for the adsorption of toxic heavy metals from aqueous solutions. *Acta Scientiarum. Technology* 34 (1): 79–87.

Gonçalves Jr., A.C.; Meneghel, A.P.; Rubio, F.; Strey, L; Dragunski, D.C.; Coelho, G.F. 2013. Applicability of Moringa oleifera Lam. pie as an adsorbent for removal of heavy metals from waters. *Revista Brasileira de Engenharia Agrícola e Ambiental* 17 (1): 94–99.

Guo, X.; Zhang, S. & Shan, X. Adsorption of metal ions on lignin. *Journal of Hazardous Materials* 151 (1): 134–142.

Iqbal, M.; Saeed, A. & Zafar, S.I. 2009. FTIR spectrophotometry, kinetics and adsorption isotherms modeling, ion exchange, and EDX analysis for understanding the mechanism of Cd2+$^{2+}$ and Pb^{2+} removal by mango peel waste. *Journal of Hazardous Materials* 164 (1): 161–171.

Kanitiz Júnior, O.; Gurgel, L.V.A.; De Freitas, R.P. & Gil, L.F. 2009. Adsorption of Cu(II), Cd(II) and Pb(II) from aqueous single metal solutions by mercerized cellulose and mercerized sugarcane bagasse chemically modified with EDTA dianhydride (EDTAD). *Carbohydrate Polymers* 77 (3): 643–650.

Marcilla, A.; Gomez-Siurana, A.; Muõz, M.J. & Valdés, F.J. 2009. Comments on the Methods of Characterization of Textural Properties of Solids from Gas Adsorption Data. *Adsorption Science & Technology* 27 (1): 69–84.

Pascoal Neto, C.; Rocha, J.; Gil, A.; Cordeiro, N.; Esculcas, A.P.; Delgaldillo, I.; Jesus, J.D. & Correia, A.J. 1995. ^{13}C solid-state nuclear magnetic resonance and Fourier transform infrared studies of the thermal decomposition of cork. *Solid State Nuclear Magnetic Resonance* 4 (3): 143–151.

Pavan, F.A.; Lima, E.C.; Dias, S.L.P. & Mazzocato, A.C. 2008 Methylene blue biosorption from aqueous solutions by yellow passion fruit waste. *Journal of Hazardous Materials* 150 (3): 703–712.

Psareva, T.S.; Zakutevskyy, O.I.; Chubar, N.I.; Strelko, V.V.; Shaposhnikova, T.O.; Carvalho, J.R. & Correia, M.J.N. 2005. Uranium sorption on cork biomass. *Colloids and Surfaces A: Physicochemical and Engineering Aspects* 252 (2): 231–236.

Rodrigues, B.N. & Almeida, F.S. 1995. *Guia de herbicidas*. Londrina: IAPAR.

Solomon, K.R. & Thompson, D.G. 2003. Ecological risk assessment for aquatic organisms from over-water uses of glyphosate. *Journal of Toxicology and Environmental Health B* 6 (3): 211–246.

Walker, G.M. & Weatherley, L.R. 2001. Adsorption of dyes from aqueous solution — the effect of adsorbent pore size distribution and dye aggregation. *Chemical Engineering Journal* 83 (3): 201–206.

Wang, S. & Zhu, Z.H. 2006. Effects of acidic treatment of activated carbons on dye adsorption. *Dyes and Pigments* 1 (1): 1–9.

Welz, B. & Sperling, M. 1999. *Atomic Absorption Spectrometry*. Weinheim: Wiley-VCH.

Witek-Krowiak A., Szafran R.G. & Modelski S. 2011. Biosorption of heavy metals from aqueous solutions onto peanut shell as a low-cost biosorbent, *Desalination* 265(1–3): 126–134.

Green Design, Materials and Manufacturing Processes – Bártolo et al. (eds)
© 2013 Taylor & Francis Group, London, ISBN 978-1-138-00046-9

Kinetics, equilibrium and thermodynamics of cadmium adsorption by a biosorbent from the bark of *Pinus elliottii*

L. Strey, A.C. Gonçalves Jr., D. Schwantes & G.F. Coelho
Western Parana State University, Marechal Cândido Rondon, Brazil

H. Nacke
Dynamic College of Cataracts, Foz do Iguaçu, Brazil

D.C. Dragunski
Paranaense University, Umuarama, Brazil

ABSTRACT: This paper proposed the use of the bark of *Pinus elliottii*, a residue from the timber industry, as biosorbent on adsorption and removal of Cd from contaminated water. After characterization of the biosorbent were determined the optimal conditions of pH, adsorbent mass and contact time for adsorption and, from these data, obtained the adsorption isotherms. Was also evaluated, the desorption ability of the material and the temperature influence on the adsorption process. From the results, it was concluded that the bark *of Pinus elliottii* is a viable and promising alternative in the adsorption and removal of Cd from contaminated water bodies, complementing the final stages of the production chain of this kind, where the use of byproducts and waste of the industry agriculture and forestry is an ideal inserted into the question of sustainability of agroecosystems.

1 INTRODUCTION

With the reduction in the availability of raw material supply for the timber industry, reforestation programs have been implemented to meet the needs of consumption and supply in this sector, where species of the genus Pinus have been widely used (Machado et al. 2002). However, the generation of large amounts of forest residues, coupled with the improper disposal of these materials, has caused numerous environmental problems such as soil and water resources, and air pollution (Budziak et al. 2004).

One of the ways, in order to minimize these environmental impacts, is the use of these materials as biosorbents in processes of adsorption and removal of metal ions from polluted effluents (Idris & Saed 2003), an alternative to conventional treatment of contaminated water bodies, such as precipitation, ion exchange, flocculation and filtration, often limited because they are technically or economically unfeasible (Ferreira et al. 2007).

The contamination of effluents and hence the environment for metal ions is one of the issues that have brought greater concern to researchers and government agencies. In this context, Cadmium (Cd) is one of the most toxic metal ions to humans and the environment, being widely used for coating materials, paints and pigment in the plastic industry, may be added to the ground by means of urban waste or industrial sewage sludge and phosphate fertilizers.

Therefore, this work proposes the use of the bark of *Pinus elliottii*, a residue from wood processing, as biosorbent in the adsorption of metal ion Cd from contaminated water, being an alternative to conventional methods of treating contaminated water bodies.

2 MATERIAL AND METHODS

2.1 *Obtaining and characterization of biosorbent*

For the preparation of biosorbent, was used the bark of *Pinus elliottii*. After collection, the material was oven-dried, homogenised and standardized crushed in 35 mesh sieves. Then the biosorbent was characterized by Scanning Electron Microscopy (SEM), infrared spectroscopy (FT-IR) and point of zero charge (pH_{PZC}).

2.2 *Tests of mass biosorbent and pH solutions*

Ideal conditions were determined from the adsorption tests with different masses biosorbent (0–1200 mg) in contact with 50 mL of Cd solution

(10 mg L^{-1}) adjusted and buffered at three pH conditions: 5.0, 6.0 and 7.0. After 90 min of shaking at 200 rpm and 25 °C, there was the determination of Cd in solution by atomic absorption spectrometry, flame mode (FAAS) (Welz & Sperling 1999).

2.3 Kinetic adsorption

From the ideal conditions for mass and pH, tests were conducted varying the contact time: 20 to 180 min. At each time, we performed a determination of Cd by FAAS. The adsorption kinetics was evaluated according to the kinetic models of pseudofirst and pseudosecond orders, Elovich and intraparticle diffusion.

2.4 Equilibrium adsorption

With the results of mass, pH and contact time adsorbent/adsorbate, adsorption tests were performed in order to obtain the adsorption isotherms, which were linearized by the mathematical models of Langmuir, Freundlich and Dubinin-Raduskevich (DR). Therefore, the biosorbent was put in contact with Cd solutions with concentrations ranging from 10 to 90 mg L^{-1}. The system was stirred for 40 min at 200 rpm and 25 °C and then held determining the content of Cd in solution by FAAS.

2.5 Thermodynamic

The influence of temperature on the process of adsorption was evaluated in five conditions: 25, 35, 45, 55 and 65 °C. From the data obtained the thermodynamic parameters were evaluated and investigated the nature of the process, through the Gibbs free energy (ΔG), enthalpy (ΔH) and entropy (ΔS).

2.6 Desorption

For the desorption process, the biosorbent used in adsorption processes, was placed in contact with a solution of hydrochloric acid at 0.1 mol L^{-1}, being stirred at 200 rpm for 40 min at 25 °C. The final concentration of Cd in solution was determined by FAAS.

3 RESULTS AND DISCUSSION

3.1 Characterization of biosorbent

Figure 1 shows the infrared spectra obtained for the biosorbent from the bark of *Pinus elliottii*. The broadband and strong 3334 cm^{-1} may be attributed to stretching vibrational OH connections of primary amines/amides and secondary, and OH bonds

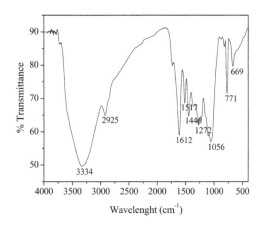

Figure 1. Spectra in the infrared region to the bark of *Pinus elliottii*.

to hydroxyl groups present on cellulose, lignin, and water adsorbed on the surface of the adsorbent (Goncalves Jr. et al. 2010). The band at 2925 cm^{-1} is related to the stretching vibration of CH bonds of the groups of alkanes and aliphatic acids (Iqbal et al. 2009). The presence of bands at 1612, 1571 and 1446 cm^{-1} may be attributed to stretching vibration of CO amide bonds and carboxyl groups (Pavan et al. 2008). The band at 1056 cm^{-1} suggests the presence of lignin (Pascoal Neto et al. 1995), which has carboxyl groups as compounds which provide adsorption sites for the metal ions (Guo et al. 2008). The observed structures react with metal ions by forming coordination complexes with the same solution, facilitating the adsorption process (Gonçalves Jr. et al. 2013).

In Figure 2 are shown the images obtained by SEM of the biosorbent with magnification of 800 and 20000 times. The surface of the adsorbent appears to be quite irregular and with major and significant voids. It's also possible to identify the structure in the form of chips, introducing on the whole surface of the material, cracks and pores, thus providing a large surface area the same.

The value of pH$_{PZC}$ founded is 3.5. The pH$_{PZC}$ value is the point at which the functional groups of the surface of the adsorbent does not contribute to the pH of the solution. Above this pH the surface charge of the adsorbent becomes negatively charged, causing the cations occupying the adsorption sites with greater interaction (Marcilla et al. 2009), so the adsorption of cations, such as Cd^{2+} is favored at pH values higher than the pH$_{PZC}$.

3.2 Tests of mass biosorbent and pH solutions

The tests showed that mass 8 g L^{-1} (400 mg) of biosorbent are sufficient for efficient removal of the

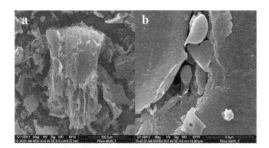

Figure 2. Image by scanning electron microscopy of the adsorbent material in magnification 800 (a) and 20000 (b) times.

metal in solution. From the tests of pH, the optimal condition obtained was 7.0.

3.3 Kinetic and equilibrium adsorption

The ideal time for the adsorption of Cd by the bark of *Pinus elliottii* was 40 min. In Table 1 can be seen the values of the Kinetic and equilibrium parameters adsorption of metal ion Cd^{2+} by *Pinus elliottii*.

The R^2 values (Table 1) demonstrate that the models that best explain the adsorption kinetics of Cd^{2+} are the pseudo-second order ($R^2 = 0.999$) and intraparticle diffusion ($R^2_1 = 0.941$ and $R^2_2 = 0.845$). Confirming these results, calculated Q_{eq} (cal.) to pseudo-second order for all cases is very close to experiment Q_{eq} (exp.). The best fit model by pseudosecond order, is an indication that the limiting step is the chemisorption process, where there is the occurrence of chemical bonds involving valence forces and sharing of electrons between the adsorbate and adsorbent (Feng et al. 2011).

Regarding Cd adsorption on pine bark, the best fit model was the DR ($R^2 = 0.999$), however, for the Langmuir model and Freundlich adjustment proved satisfactory with values of R^2 and very close together. Thus, relying on the DR model, the vicinity of the surface of the solid is characterized by a series of equipotential surface having the same sorption potential. Values of free energy of sorption (E) above 8 kJ mol^{-1} indicate a chemical nature of the adsorptive process (Fávere et al. 2010).

The mathematical models by adjusting the Langmuir and Freundlich adsorption indicates that the Cd adsorption occurs in a heterogeneous manner in mono or multilayer respectively, due to binding sites having different adsorption energies, showing variation as a function of surface coverage (Walker & Weatherley 2001). Although the binding energy (K_L) is relatively low, the parameter n, with values between 1 and 10, indicated the presence of high energy sites, which favors the adsorption process (Gonçalves Jr. et al. 2012).

Table 1. Kinetic and equilibrium parameters adsorption of metal ion Cd^{2+} by *Pinus elliottii*.

Pseudo-first order	
K_1 (min^{-1})	−0.011
Q_{eq} (cal.) (mg g^{-1})	0.105
Q_{eq}(exp.) (mg g^{-1})	1.181
R^2	0.715
Pseudo-second order	
K_2 (g mg^{-1} min^{-1})	0.294
Q_{eq} (cal.) (mg g^{-1})	1.188
Q_{eq}(exp.) (mg g^{-1})	1.181
R^2	0.999
Intraparticle diffusion	
K_{id1} (mg g^{-1} min$^{-1/2}$)	0.025
C_{i1} (mg g^{-1})	0.957
R^2_1	0.941
K_{id2} (mg g^{-1} min$^{-1/2}$)	0.006
C_{i2} (mg g^{-1})	1.100
R^2_2	0.845
Elovich	
A	0.951
B	0.043
R^2	0.838
Langmuir	
Q_m (mg g^{-1})	6.301
K_L (L mg^{-1})	0.089
R_L	0.053
R^2	0.985
Freundlich	
K_f (mg g^{-1})	1.250
n	3.035
R^2	0.984
D-R	
Q_d (mol g^{-1})	1.13 e^{-4}
E (kJ mol^{-1})	13.558
R^2	0.999

Q_{eq} = amount adsorbed at equilibrium. K_1 = constant of pseudofirst order. K_2 = constant of pseudosecond order. A = rate constant initial chemisorption. B = constant extension of surface coverage and activation energy of chemisorption. K_{id} = intraparticle diffusion constant. C_i = the thickness of the boundary layer effect. Q_m = maximum amount adsorbed. K_L = binding energy. K_f = adsorption capacity. n = heterogeneity solid. R^2 = coefficient of determination.

3.4 Thermodynamic and desorption

Table 2 presents the thermodynamic parameters obtained for the adsorption of Cd as a function of temperature.

The negative values of ΔG (Table 2) indicate that the adsorption process is spontaneous and positive, where the more these values are negative,

Table 2. Thermodynamic parameters of Cd adsorption on the bark of *Pinus elliottii*.

Temperature (°C)	ΔG (kJ mol^{-1})	ΔH (kJ mol^{-1})	ΔS (J mol^{-1})
25	−76.816	40.684	121.249
35	−78.029		
45	−79.241		
55	−80.454		
65	−81.666		

is the more energetically favorable process. The positive value of entropy (ΔS) suggests an increase in disorder in the solid-solution interface and ΔH values above 40 kJ mol^{-1} confirm the endothermic nature of the process, characterizing a process of chemisorption (Zhu & Wang 2006).

Regarding Cd desorption of adsorbed amount (50.03% on average), 65.08% was desorbed into the solution, this being an important characteristic to consider the reuse of these materials to new adsorption processes. Thus, based on these results, it can be stated that the process of adsorption of Cd by the bark of *Pinus elliottii* is mainly controlled by the chemisorption.

4 CONCLUSIONS

From the results obtained in the characterization, kinetic and equilibrium adsorption, it is concluded that the bark of *Pinus elliottii* is a biosorbent with excellent adsorption characteristics and is therefore a promising and viable Cd adsorption of contaminated water bodies. Moreover, this practice complements the final stages of the production chain of this species through the use of byproducts and residues from agriculture and forestry industry in which it operates, and this is an ideal inserted into the question of sustainability of agroecosystems.

ACKNOWLEDGEMENTS

The authors thank the Araucaria Foundation (SETI-PR) and the CNPq-MCTI (REPENSA) for financial support on projects and other aid.

REFERENCES

Budziak, C.R.; Maia, C.M.B.F. & Mangrich, A.S. 2004. Transformações químicas da matéria orgânica durante a compostagem de resíduos da indústria madeireira. *Química Nova* 27 (1): 399–403.

Fávere, V.T.; Riella, H.G. & Rosa, S. 2010. Cloreto de n-(2-hidroxil) propil-3-trimetil amônio quitosana como adsorvente de corantes reativos em solução aquosa. *Química Nova* 33 (7): 1476–1481.

Feng, N.; Guo, X.; Liang, S.; Zhu, Y. & Liu, J. 2011. Biosorption of heavy metals from aqueous solutions by chemically modified Orange peel. *Journal of Hazardous Materials* 185 (1): 49–54.

Ferreira, J.M.; Silva, F.L.H.; Alsina, O.L.S.; Oliveira, L.S.C.; Cavalcanti, E.B. & Gomes, W.C. 2007. Equilibrium and kinetic study of Pb^{2+} biosorption by *Saccharomyces cerevisiae*. *Química Nova* 30 (5): 1188–1193.

Gonçalves Junior, A.C.; Nacke, H.; Fávere, V.T.; Gomes, G.D. 2010. Comparação entre um trocador aniônica de sal de amônio quaternário de quitosana e um trocador comercial na extração de fósforo disponível em solos. *Química Nova* 33 (5): 1047–1052.

Gonçalves Jr., A.C.; Strey, L.; Lindino, C.A.; Nacke, H.; Schwantes, D.; Seidel, E.P. 2012. Applicability of the Pinus bark (Pinus elliottii) for the adsorption of toxic heavy metals from aqueous solutions. *Acta Scientiarum. Technology* 34 (1): 79–87.

Gonçalves Jr., A.C.; Rubio, F.; Meneghel, A.P.; Coelho, G.F.; Dragunski, D.C.; Strey, L. 2013. The use of Crambe abyssinica seeds as adsorbent in the removal of heavy metals from waters. *Revista Brasileira de Engenharia Agrícola e Ambiental* 17 (1): 306–311.

Guo, X.; Zhang, S. & Shan, X. Adsorption of metal ions on lignin. *Journal of Hazardous Materials* 151 (1): 134–142.

Idris, A. & Saed, K. 2003. Possible utilization of silica gel sludge for the removal of phenol from aqueous solutions: laboratory studies. *The Environmentalist* 23 (1): 329–334.

Iqbal, M.; Saeed, A. & Zafar, S.I. 2009. FTIR spectrophotometry, kinetics and adsorption isotherms modeling, ion exchange, and EDX analysis for understanding the mechanism of Cd2+$^{2+}$ and Pb^{2+} removal by mango peel waste. *Journal of Hazardous Materials* 164 (1): 161–171.

Machado, S.A.; Conceição, M.B. & Figueiredo, D.J. 2002. Modelagem do volume individual para diferentes idadese regimes de desbaste em plantações de Pinus oocarpa. *Revista Ciências Exatas e Naturais* 4 (2): 185–197.

Marcilla, A.; Gomez-Siurana, A.; Muõz, M.J. & Valdés, F.J. 2009. Comments on the Methods of Characterization of Textural Properties of Solids from Gas Adsorption Data. *Adsorption Science & Technology* 27 (1): 69–84.

Pascoal Neto, C.; Rocha, J.; Gil, A.; Cordeiro, N.; Esculcas, A.P.; Delgaldillo, I.; Jesus, J.D. & Correia, A.J. 1995. ^{13}C solid-state nuclear magnetic resonance and Fourier transform infrared studies of the thermal decomposition of cork. *Solid State Nuclear Magnetic Resonance* 4 (3): 143–151.

Pavan, F.A.; Lima, E.C.; Dias, S.L.P. & Mazzocato, A.C. 2008 Methylene blue biosorption from aqueous solutions by yellow passion fruit waste. *Journal of Hazardous Materials* 150 (3): 703–712.

Walker, G.M. & Weatherley, L.R. 2001. Adsorption of dyes from aqueous solution—the effect of adsorbent pore size distribution and dye aggregation. *Chemical Engineering Journal* 83 (3): 201–206.

Wang, S. & Zhu, Z.H. 2006. Effects of acidic treatment of activated carbons on dye adsorption. *Dyes and Pigments* 1 (1): 1–9.

Welz, B. & Sperling, M. 1999. *Atomic Absorption Spectrometry*. Weinheim: Wiley-VCH.

Green Design, Materials and Manufacturing Processes – Bártolo et al. (eds)
© 2013 Taylor & Francis Group, London, ISBN 978-1-138-00046-9

Adaptation of the creativity tool ASIT to support eco-ideation phases

B. Tyl
APESA, Innovation Department, Bidart, France

J. Legardeur
ESTIA, Bidart, France
IMS, UMR 5218, Talence, France

D. Millet
SUPMECA, Toulon, France

F. Vallet
UTC Roberval Laboratory, Compiègne, France

ABSTRACT: Today the challenges of sustainable development require new offers and new uses to be developed within the framework of an eco-innovation process integrating environmental and societal approaches. The aim of this paper is to present the final result of a three year research about the advantages of using creativity tools to foster new concept and idea generation phases during eco-innovation processes. To do so, this paper describes an eco-innovation tool based on the modification of the creativity tool ASIT in the perspective to answer to the challenges and issues raised by sustainable development. Also, by using state of the art in eco-innovation tools, we propose a strategy for adapting the ASIT tool for eco-innovation. This adaptation concerns the preparatory stage, when the group needs to formulate the initial problem, and the eco-ideation stage, when ideas and concepts have to be generated in a sustainable way. This research confirms the need to better support the ideation phase of eco-innovation processes. More generally this paper shows that the eco-innovation processes require stimulation mechanisms in order to lead to more efficient eco-innovations in all the aspects of sustainable development.

1 INTRODUCTION

The environmental consequences of means of mass manufacturing and consumption today require us to completely rethink our way of designing, manufacturing and consuming by implementing a responsible innovation strategy. In the 1980's, quality, and then innovation approaches were put forward, to deal with global competition.

Today it is essential for us to work on new approaches (methods, tools, and organisations ...) which tend to a more responsible form of innovation to reduce the environmental impact of anthropic activities while enabling useful products and services with meaning for society to be developed.

We need to encourage the emergence of new, more innovative and creative practices as well as new fields to be explored. Erhenfeld (2008) states that "reducing unsustainability is not the same as creating sustainability". Eco-innovation must be this new way of integrating the environment in a "positive and creative" manner in the product and service development process. It is defined by Fussler & James (1996) as "the process of developing new products, processes or services which provide customer and business value but significantly decrease environmental impact". Eco-innovation as a departure point for environmental criteria seems to be a more motivating approach, notably due to the fact that it does not only focus on reducing the impact but aims to generate new products and services. Thus, the innovation process is a positive value creation process (Beard & Hartmann, 1999).

The eco-innovation approach opens up some advantages: increasing the competitiveness of businesses by creating new markets for desirable goods (Baroulaki & Veshagh, 2007), creating opportunities to review their skills in the context of sustainable development (Dewberry & de Barros Monteiro, 2009). It involves considering new opportunities by studying the system and its components, and therefore requires methods to be implemented which offer new values to create new solutions (Dewberry and de Barros Monteiro, 2009).

This paper presents an eco-innovation tool, developed within the framework of a PhD thesis (Tyl, 2011).

To do so, our paper is divided in 5 sections as follows: first a presentation of the eco-innovation concept and eco-innovation tools. Then we will expose the creativity tool ASIT and its adaptation in an eco-innovation tool EcoASIT and we will conclude with some perspectives.

2 ECO-INNOVATION CONCEPTS AND TOOLS

2.1 Eco-innovation tools

Numerous tools (Table 1) have been developed to support the eco-innovation process, enabling the approach to be systematized and generalised as well as to direct users/designers towards well defined objectives. We can note that these tools can be either totally original such as the Information-Inspiration tool (Lofthouse, 2004) or adapted from a tool which already exists, such

Table 1. List of some eco-innovation tools.

Eco-innovation tool	Tool's main contribution	
	Evaluation	Ideation
PIT Diagram (Jones et al., 2003)		X (weak)
Eco-Compass (Fussler & James, 1996)	X	
LiDS Wheel (Brezet, 1997)	X	
Ten golden rules (Luttrop & Lagerstedt, 2006)		X (weak)
Eco functional matrix (Lagerstedt, 2003)	X	
ecoQFD (Rahimi & Weidner, 2002; Wolniak & Sedek, 2009)	X	
Information-Inspiration (Lofthouse, 2004)		X (weak)
TRIZ—Contradiction matrix (Chen & Liu, 2003)		X (strong)
TRIZ—Final Ideal Result (Jones, 2003)	X	
TRIZ—Nine screen (O'Hare, 2010)	X	
TRIZ—CBR (Yang & Chen, 2011)		X (strong)
TRIZ—Eco guidelines (Russo & Regazzoni, 2008)		X (strong)
TRIZ—LCP Planner (Kobayashi, 2006)		X (strong)
TRIZ—Eco-MAL'IN (Samet, 2010)	X	X (strong)

as ecoQFD (Rahimi & Weidner, 2002; Wolniak & Sedek, 2009).

By studying these tools, we can identify two main families. The first family concerns eco-innovation tools based above all on the evaluation of concepts. These include Eco Compass (Fussler & James, 1996) or the Brezet Wheel and Lids Wheel (Brezet, 1997). The second family concerns tools developed to support more specifically the ideation process. These include the PIT diagram tool which is based on the principles of mind mapping to structure the eco-innovation idea generation phase (Jones et al., 2001), and the TRIZ-based method, which was the most studied for its relevance in the eco-innovation approach.

2.2 Need for creativity in eco-innovation

Table 1 shows that tools currently developed have shortcomings with regard to eco-innovation issues. The various existing methods focus more on implementing a strategy by assessing concepts rather than generating ideas (Jones, 2003; Bocken et al., 2011).

This limited development of the ideation phase in éco-innovation can be explained by the fact that very few creativity tools have been adapted for eco-innovation. Only the TRIZ theory was widely studied for the eco-innovation process went, but the different tools resulting from this theory have heterogeneous performance with regard to eco-innovation issues (Jones, 2003). Nevertheless, the ideas generation phase, or ideation, is in fact the central part of a creativity session, although it is often neglected in eco-innovation (Bocken et al., 2011).

Moreover, we can see that the majority of eco-innovation tools follow an approach which is above all technical and focuses on products (Table 1). However, eco-innovation needs to be detached from this technical approach of the environment in order to call into question the function itself of the system studied, its social and cultural context, and to go beyond eco-design (Sherwin, 2004).

To answer these shortcomings, the aim of your research was to implement a tool that focuses on the idea generation phase by proposing mechanisms which stimulate reflection and creativity with regard to the different aspects of sustainable development.

To do so, following the example of other works (O'Hare, 2010), the second stage in our research consists in identifying an existing creativity tool to test it in an eco-innovation process. This process enables the existing performance of this tool to be capitalised on and thus to benefit from the trust that users place in this tool and its abilities (Lindahl, 2006).

Lastly, the final stage consisted in identifying possible adaptations of the creativity tool with considering the strategies and constraints required by eco-innovation so as to implement a prototype eco-innovation tool usable by companies. This proposed eco-innovation tool was then tested in order to analyse how it worked and also to clarify its deficiencies and weaknesses (not presented in this paper).

3 IDENTIFYING A PRE EXISTING TOOL

In the literature, there is a significant number of creativity tools: intuitive methods such as brainstorming or more converging methods such as TRIZ (Altshuller, 1988). TRIZ was a widely studied tool for eco-innovation in ideation phases. Yet as we have seen earlier, the theory disadvantage of this theory is that it is quite complex and seems to be "cumbersome" and "time consuming" when it is used. The TRIZ method is efficient, but only under certain conditions and within the restricted perimeter of solving well-identified technical issues.

Thus the aim was to find a tool that would be close enough to this method but less complex and with more global stimulation mechanisms so as to better meet eco-innovation processes. The SIT method (Horowitz, 1999) and the related ASIT tool (Horowitz, 2001), resulting from the simplification of TRIZ, were chosen to implement an eco-innovation tool focussing on the ideation phase.

3.1 *SIT AND ASIT*

The SIT method (Structured Inventive Thinking), subsequently renamed ASIT (Advanced Systematic Inventive Thinking) was imagined by Roni Horowitz based on work from the early 1990's which aimed to simplify use of the TRIZ method (Horowitz, 1999; Horowitz, 2001).

It is a problem-solving tool based on a postulate which considers that the most innovative ideas are always close to the initial problem.

Consequently, instead of thinking "out of the box" as creativity would lead us to believe, ASIT attempts to think "inside the box" and condition the search for solutions within the problem's universe.

To do this, the tool offers two possibilities to condition the search for solutions:

– The Closed World (CW) which considers that "the world of the solution" does not introduce a new type of object with regard to the "world of the problem": creative solutions are very close in their "genealogy" to conventional solutions,

– and the Qualitative Change (QC) which removes the conflict between the cause of the problem and its effect or exceeds it (Maimon & Horowitz, 1999).

These conditions thus enable the problem to be reformulated. Research carried out shows that when these two criteria are met, solutions are more innovative (Maimon & Horowitz., 1999).

The method proposes 5 tools, or "breaking operators" to generate simple phrases which can be used during a group session to provoke ideas.

Thus, the SIT/ASIT process, consists in:

1. Formulating the problem using the closed world condition and the qualitative change condition. This stage consists in formulating objects from the "world of the problem" and then formulating the problem.
2. An idea generation phase using 2 strategies: Extension/restructuring:
   – The extension strategy attempts to solve the problem by assigning a new use to an existing object (Unification tool) or an object of the same type (Multiplication tool);
   – The restructuring strategy attempts to solve the problem by removing an object from the system (Object Removal tool), by dividing an object and reorganising its parts (Division tool) or by transforming a symmetrical situation into an asymmetrical situation (Symmetry breaking tool).

Empirical studies have shown that the SIT method improves the problem solving rate (Horowitz, 1999) to such an extent that we can consider that the tool is academically validated (Reich et al., 2010). To conclude, Horowitz's work was followed by various other works which proposed improvements to the ASIT tool (Takahara, 2009) or an evolution of ASIT based on the C-K method (Reich et al., 2010).

3.2 *ASIT and eco-innovation*

By studying the ASIT tool with a view to transforming it into an eco-innovation tool, we here wish to present its advantages for eco-innovation. Initially, one of the conditions to obtain innovative solutions is the "closed world" which consists in achieving the desired objective solely using the system's objects. Contrary to the vision of creativity tools which attempt to go beyond their framework, this condition seems to suit eco-innovation issues because it refers to a finite and limited world of resources in which systems must be designed without additional resources (Turner, 2003).

Furthermore, in order to illustrate the Closed World condition, Maimon and Horowitz use the example of the electric car. Car pollution applied to the condition of the closed world would namely prevent using the electric motor to solve the problem because it is an element outside the system (Maimon & Horowitz, 1999). In eco-innovation logic, this approach is therefore extremely relevant. In fact, it invites us to initially think about energies available in the vehicle or reduce the energies needed to make the vehicle work, and not to attempt to think about a purely technical and immediate approach of energy substitution.

Subsequently, ASIT proposes stimulation mechanisms within the logic of sustainable development. ASIT proposes wide enough stimulation mechanisms to meet global issues. Furthermore, the tool proposes mechanisms, such as "Object removal" which aims to remove objects and thus gain simplicity, in line with eco-innovation.

Finally, eco-innovation issues must be positioned upstream in the project, during the design process' conceptual phases. However, one of the features of conceptual phase issues is that their defining phase is not static and evolves with interactions between the issue and the solution (Mulet & Vidal, 2008). This feature therefore needs a flexible and quickly usable tool so as to encourage this iteration. The ASIT tool seems to be highly relevant for this because it is the result of a major simplification of TRIZ.

4 ADAPTATION OF THE ASIT TOOL AS AN ECO-INNOVATION TOOL

This section presents our adaptation of the existing ASIT creativity tool so as to optimize it according to environmental issues. We modified ASIT in order to come close to the features of an efficient eco-innovation tool. To do this, we developed the EcoASIT tool with main improvements concerning:

– Its ability to deal with eco-innovation issues on higher systems levels
– The way it takes into account a life cycle approach
– The way it takes into account the environmental and societal dimensions of sustainable development.

The EcoASIT process (Fig. 1) uses the bases of a creativity tool and adds to it a sustainable dimension. The tool's architecture therefore consists in a problem formalisation phase, which enables the group to define the boundaries for the system which is being studied and to draw a reflection strategy. Then, during the idea generation phase, the group tries to distort the system so as to achieve the desired objective using the appropriate stimulation mechanisms.

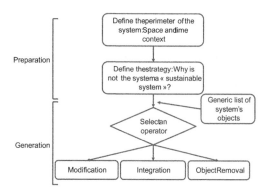

Figure 1. EcoASIT process.

4.1 Preparation

4.1.1 9 screens (tool not specific to EcoASIT)
The eco-innovation approach requires the problem to be positioned at high systems levels. For this, not only the systems studied, but also its context of use and the boundaries of the study need to be clearly defined.

The "9 screen" study is relevant when it comes to encouraging this positioning. This tool is the result of Altshuller's work and enables the product to be recontextualised with a space and time framework (Altshuller, 1988). It makes the group position the system in time (past-present-future) and also in the environment in which it is inserted (sub-system/system/super-system).

Therefore the tool makes the group define the boundaries of the system it wants to study in addition to the level of questioning of the system.

This provides a global vision of the problem with a space and time framework, taking into account the desired elements.

4.1.2 Defining an objective: Ideal final product
After defining the problem's framework, EcoASIT proposes to assess the system using a 5-axis diagram which enables the group to identify the why the current system is not an ideal and sustainable system. These five axes correspond to 5 major problems:

1. The system consumes natural resources (water, energy, materials)
2. The system generates waste and/or pollution
3. The system does not participate in local dynamism
4. The system is not perceived as a sustainable system
5. The system does not correspond to uses.

Assessing the system according to these 5 axes (Fig. 2) helps to quickly formalize the problem and to identify a goal allowing the group to spend more

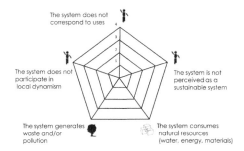

Figure 2. EcoASIT diagram.

Table 2. The world of the problem in EcoASIT.

Life cycle objects	Impact type objects
Raw materials, production, distribution-sale, use, end of life	Natural resources, waste, use, perception of the product, local activity

effort generating ideas instead of formalising the problem.

4.1.3 *Closed world principle*

The ASIT method requires the initial problem to be formulated by describing the "world of the problem", listing the "problem's objects" and the "environment's objects". This condition therefore proposes a framework for thought within which solutions can be generated.

Unlike ASIT, where objects are formulated by the group, in EcoASIT these objects are pre-defined to instantly direct solutions towards eco-innovative concepts. The objects of the problem have been formulated according to two principles of eco-innovation: the product life cycle and the description of the environmental, social and economic impacts of the system. Table 2 describes the objects in the "world of the problem".

Thus, by grouping together certain "similar" objects, EcoASIT proposes a generic "world of the problem" made up of the following objects: natural resources, production, sales, waste, perception, use and local activity (Fig. 9).

This involves a major difference with SIT. The initial method requires the group to formalise objects according to the problem, whereas the EcoASIT method proposes generic objects. These objects are then used during the idea generation phase to generate innovative concepts.

4.2 *Generation of ideas*

The EcoASIT tool also offers to use two ASIT strategies to stimulate the group by proposing standard phrases to unstructure the problem. Nevertheless, the operators of these two strategies have been adapted to eco-innovation:

1. The extension strategy which attempts to solve the problem by modifying the use of an existing object.
   - Modification operator: Modify {the raw materials, production, sale, waste, perception, use and local activity} will enable to achieve my objective.
2. The restructuring strategy which aims to solve the problem by removing an object from the system or by linking two objects.
   - Object Removal operator: Modify {the raw materials, production, sale, waste, perception, use and local activity} will enable to achieve my objective.
   - Integration operator: Link {the raw materials, production, sale, waste, perception, use and local activity} and {the raw materials, production, sale, waste, perception, use and local activity} will enable to achieve my objective.

5 DISCUSSIONS AND PERSPECTIVES

Through a critical analysis of the scientific literature and several experiments, this research confirms the need to better support the ideation phase of eco-innovation processes. Thus, we have shown that eco-innovation processes require stimulation mechanisms in order to lead to more efficient eco-innovations in all the aspects of sustainable development.

For this, we have shown the relevance of using creativity tools and adapting them to eco-innovation issues. Although the validation of a new design tool is extremely complex, we tried to validate the EcoASIT tool according to two perspectives. The first validation, is theoretical. It consisted in studying comparatively the conceptual heritages and the foundations of the EcoASIT tool with the ASIT tool. The second validation, not presented in this paper, is an experimental validation which first demonstrated the advantage of the ASIT creativity tool with regard to environmental issues. In a second test, we confirmed that the EcoASIT tool preserved the positive characteristics of the ASIT tool while making reflection focus on eco-innovation issues.

Future work must analyse more specifically the relevance of the ideas generated. In fact, a study carried out during the first test showed that if we consider all the ideas provided, there is a very weak correlation between the different experts' assessments with regard to the environmental relevance of ideas. Therefore research remains to be done on the environmental evaluation of concepts in the upstream phases of projects.

Finally, this research work was carried out within a structure whose aim is to support SMEs in implementing eco-innovation. Thus, our future research aims to see how to make implementation and deployment of this eco-innovation tool within partner companies more independent. This research will thus help to analyse the tool's long-term validation in order to integrate sustainable development criteria into the early phases of design processes in these companies.

REFERENCES

Altshuller, G.S. 1988. Creativity as an Exact Science Gordon and Breach, ISSN 0275-5807, New York.

Baroulaki, E., Veshagh, A. 2007. Eco-Innovation: Product Design and Innovation for the Environment. In *Proceedings of the 14th CIRP Conference on Life Cycle Engineering*, Waseda University, Tokyo, Japan.

Bocken, N.M.P., Allwood J.M, Willey A.R., King J.M.H. 2011. Development of an eco-ideation tool to identify stepwise greenhouse gas emissions reduction options for consumer goods, *Journal of Cleaner Production* 19(2): 1279–1287.

Brezet, J.C. 1997. "Dynamics in ecodesign practice", *UNEP Industry and Environment*, vol 20 (1–2): 21–24.

Chen J.L., Liu C.-C. 2003. An eco-innovative design approach incorporating the TRIZ method without contradiction analysis, *the Journal of Sustainable Product Design* 1: 263–272, 2001.

Dewberry, E.L., de Barros Monteiro, M. 2009. Exploring the need for more radical sustainable innovation: what does it look like and why?, *International Journal of Sustainable Engineering*, 2(1): 28–39.

Ehrenfeld J.R. 2008. Sustainability by Design: A Subversive Strategy for Transforming Our Consumer Culture. *Yale University Press*, New Haven, CT.

Fussler C., James P., 1996. Driving eco-innovation, 1996, London: Pitman.

Horowitz R 1999. Creative problem solving in engineering design, PhD Thesis Tel-Aviv University.

Horowitz, R 2001. From TRIZ to ASIT in 4 steps. *The TRIZ Journa*

Jones, E., Stanton N.A., Harrison D. 2001. Applying structured methods to Eco-innovation. An evaluation of the Product Ideas Tree diagram, *Design Studies* 22(6): 519–542.

Jones, E. 2003. Eco-innovation: tools to facilitate early-stage workshop, PhD Thesis Department of Design, Brunel University.

Kobayashi, H. 2006. A systematic approach to eco-innovative product design based on life cycle planning, *Advanced Engineering Informatics* 20: 113–125.

Lagartsted J. 2003. Functional and Environmental Factors in early phases of product development-Eco-Functional Matrix, PhD Thesis, Royal Institute of Technology-KTH, Stockholm.

Lindahl, M. 2006. Engineering designers' experience of design for environment methods and tools e Requirement definitions from an interview study, Journal of Cleaner Production 14: 487–496.

Lofthouse, V. 2004. Final report of results for 'Information/Inspiration' the DTI/EPSRC Sustainable Technologies Initiative funded project, Loughborough University, Loughborough, May 2004. www.informationinspiration.org.uk. Retrieved 20/08/2011.

Luttropp, C., Lagerstedt, J. 2006. EcoDesign and The Ten Golden Rules: generic advice for merging environmental aspects into product development, *Journal of Cleaner Production* 14: 1396–1408.

Maimon OZ, Horowitz R. 1999. Sufficient Conditions for Inventive Solutions. IEEE transactions on systems, man, and cybernetics—Part C: application and reviews, vol 29, no 3

Mulet E., Vidal R., 2008. Heuristic guidelines to support conceptual design, *Research in Engineering Design* 19:101–112.

O'Hare, J.A. 2010. Eco-innovation tools for the early stages: an industry-based investigation of tool customisation and introduction, PhD Thesis Department of Mechanical Engineering, University of Bath.

Rahimi, M., Weidner, M. 2002. Integrating Design for Environment Impact Matrix into Quality Function Deployment Process, *The Journal of Sustainable Product Design* 2: 29–41.

Reich, Y, Hatchuel, A, Shai, O, Subrahmanian, E. 2010. A theoretical analysis of creativity methods in engineering design: casting and improving ASIT within C-K theory', *Journal of Engineering Design*. First published on: 16 July 2010.

Russo D., Regazzoni, D., (2008), TRIZ Law of evolution as eco-innovative method. In *Proceedings of IDMME—Virtual concept 2008*, Beijing China.

Samet W. 2010 Samet Kallel, W., Développement d'une méthode d'éco-innovation: Eco-MAL'IN, Thèse de Doctorat ParisTech.

Sherwin C. 2004. The Journal of Sustainable Product Design 4:21–31.

Takahara T. 2009. Logical Enhancement of ASIT. *The TRIZ journal.*

Turner, S., 2009. ASIT—a problem solving strategy for education and eco-friendly sustainable design, *International Journal of Technology and Design Education* 19 (2).

Tyl B. 2011. L'apport de la créativité dans les processus d'éco-innovation—Proposition d'un outil pour favoriser l'éco-idéation de systèmes durables, PhD thesis, Université Bordeaux 1.

Wolniak, E.R., Sedek, A. 2009 Using QFD method for the ecological designing of products and services, Quality Management of Process and Product Department, Organization and Management Faculty, Silesian Technical University, Roosvelta 24–26, 41–250, Zabrze, Poland.

Yang, C.J., Chen, J.L. 2011. Accelerating preliminary eco-innovation design for products that integrates case-based reasoning and TRIZ method, *Journal of Cleaner Production*, doi:10.1016/j.jclepro.2011.01.014.

Green Design, Materials and Manufacturing Processes – Bártolo et al. (eds)
© *2013 Taylor & Francis Group, London, ISBN 978-1-138-00046-9*

Reducing packaging waste by GIS applications

Adrián Horváth & Ákos Mojzes
Szechenyi Istvan University, Department of Logistics and Forwarding, Győr, Hungary

Gábor Takács
Szechenyi Istvan University, Infrastructural Systems Multidisciplinary Doctoral School of Engineering, Győr, Hungary

ABSTRACT: In the 21st century, the role of electronic gadgets has reached an unexpected limit where they can literally support every aspect of our lives. Not only have the variety of their knowledge-base and the services they provide reached surprising high levels which are now able to support mobility around the globe, but also these gadgets have created new possibilities to measure, collect and save data.

The purpose of this article is to present current packaging methods, but also to investigate and specify the actual stress on our products so that we can quantify the real exposure we want to avoid. The importance of this investigation is that if we could find a method to measure and quantify the correct data of the actual stress, it would provide a effective base for reducing the quantity of the packaging materials used. This is especially the case of those products transported on pallets and therefore we could significantly reduce not only the quantity of the base material used for packaging but also the amount of the left over waste resulting from packaging material globally.

As an addition to all the above mentioned factors, the topic also gives an interesting perspective on how the savings on the packaging materials could reduce the cost of these products and so provide companies a way to serve their clients in a more cost effective way and an option to offer their products for a much cheaper price and hence gain a larger market share.

1 INTRODUCTION

Our current society is all about mobility and information technology that provides the user the much needed data and support quickly and immediately wherever whenever and whatever they need. Most of the required information is available via different information communication networks on gadgets with different sizes, colors, software and software environments.

If we think it through, these smart gadgets, the basics of which we have inherited from the past,— are currently effectively little personal computers, which are suitable not only for providing different communication channels, but more importantly offering the advantage of network data communication, recording and saving data immediately into their memories. More and more different sensors have been built into these gadgets while their size has been decreasing continuously, so now these devices are suitable for a lot of different functionalities.

Only the software environment could limit the use of these new functionalities with the existing sensors for whatever purpose. In this article the purpose is find methods to measure, collect and save real data. We are trying to look into solutions on how these gadgets could serve to measure the required data in a more cost efficient way based on different software environments.

It is essential to find a supportive application. In particular, a cost effective way of using that method which could help us to measure the actual stress on packaged object as discussed above, so that we could realize the savings on the packaging materials and preserve waste consuming packaging in global sizes. "From a corporation's perspective, the integration of new information technology is most commonly understood as innovation." (Erdős, 2008).

2 IMPORTANCE OF SUSTAINABILITY

During the examination process as well as during the usage of the equipment it is an essential consideration point that it has to be sustainable, so it should be worth the investment.

Developing a software application, if it supports real requirements from the market in a profit-oriented field, is quite sustainable. The costs of development are one off i.e. defining the problem and our requirements, resolution analysis, requirement specification, planning of optimized model, physical realization, implementation, although the cost of the equipment procurement could be higher. However once this investment is in place and it is well used it will be sustainable and will show results.

The main advantage of developing such a system is that packaging tests could be performed immediately and locally wherever packaging/stress data is required.

Considering that the currently existing methods require the products to be evaluated (e.g. two pallets of goods) to be transported to a laboratory environment where the equipment used for examination are available, it would provide a huge cost saving for companies if this extra examination, transportation, waiting period and the harmful consequences on those examined products could be avoided.

3 CONSIDERING LOGISTIC STRESS TYPES DURING THE PLANNING OF PACKAGING

During a continental road transportation, which also includes the handling and storage processes, it is quite easy to define the qualitative characteristics of the logistical stress. These kinds of stresses, which include both mechanical and environmental stress, have high importance in the field of product—packaging developments (Soroka, 2002).

Before designing or developing any process or equipment we have to be aware of all kinds of logistic stress. The stresses have to be well defined by values, indicators or indexes.

In the following paragraph we briefly describe those affects, which are of high importance during road transportation.

– Vibrational stresses are considered to be the most frequently occurring stress. During the complete transportation process, the product and its packaging system, like all transport loads, are under a continuous vibration affect. These vibrations nay arise from the unevenness or roughness of the road (like railway crossings, pot-holes, etc.), and/or the combination of the vehicle's spring characteristics and/or the force of inertia with any unbalanced loads. The vibrations on the load platform are known as stochastic type, which means that the combination, frequency and the strength of the vibrations and their amplitudes are strongly fluctuating. There are frequencies, where the amplitude exceeds the acceleration.

In these moments, the load becomes detached from the platform, and as a result of the continuously platform vibration the load when it returns to the platform could be displaced. (Mojzes, 2008, Panczel, 2006) In those cases, where both vertical and horizontal load position displacement appear, we often observe packaging damages. In extreme cases, these affects appear as a shock stress.

Figure 1 describes the simplified mass—spring model of a stacked load on a truck platform. The $k_{1..n}$ means the spring stiffness, the $c_{1..n}$ mean the spring stiffness.

– The shock and impact stresses are also commonly observed. But if we compare two different transports on the same route, the results will show some differences both in the number and the intensity of the impacts.

Those derivers who are highly influenced by for example personal negligence and technical parameters. i.e. an improperly passed railway crossing or road section or passing through with a high vehicle speed might generate an extremely high number of impacts with extreme g values. The shocks and impacts also can appear in both vertical and horizontal axes. The horizontal impacts can cause slipping of the loads. (Mojzes, 2007). For stacked loads the vertical impacts also can cause toppling or tip over. If we know the exact transportation route, the vertical impacts may be defined quite accurately, and critical areas and locations could also be predicted. The parameters and stress effects of handling and warehousing are also predictable. (Burgess, 1990). Figure 2 maps the type and correlation of possible shock parameters, and also describes where the packaged product will most probably be damaged. The threshold values of the non–damage areas can be defined only by valid and appropriate measurements.

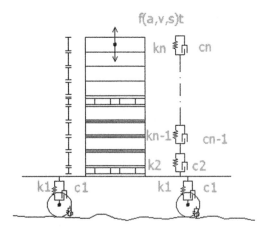

Figure 1. Mass spring damper model.

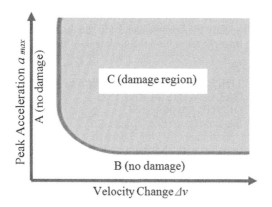

Figure 2. Damage boundary curve of a packaged product.

- Piling up stock during transportation or storage is a popular method, as this process quickly and effectively improves logistical indicators. This parameter is essential during the product packaging system development, but when we also count the above mentioned stress types, we will get a very complex task, which may conceal more design problems. Stacking stresses can appear as both a dynamic stress and as a static stress. Both types are influenced by both technical and personal parameters over a wide range. The distributed stress can often turn into a local stress during an improper transportation and/or handling of the load, which can also cause product damage (Böröcz et al, 2008).
- During the well defined logistical procedure, which includes the transportation, warehousing and handling of the stock, there is one more stress which is continuously present and always appear linked to another mechanical stress type. These are the different types and forms of environmental and/or climate stress. During continental road transportation, the form and the exact value or parameters of the climate affects can be well described and accounted for. Regarding transport packaging which is specifically developed for road transportation we have to consider two major effects: the relative humidity and the temperature of the air, and its changes/variations. The parameters which relate to appearance can highly influence the product protection function of the packaging system. There are many existing transportation models or techniques, where the developed packaging system provides good logistic and economic indicators and values, but the unknown or not considered environmental stress types can reduce the efficiency and cause possible product damages (Böröcz, 2007).

4 THE SUMMARY OF THE DATA COLLECTION FOR THE PLANNING OF THE CORRECT PACKAGING METHOD

The measuring device has to be strongly fixed to the load itself so that they will move together when an impact happens on the road. The device has to be able to capture the timing, the latitude, the longitude and the measured values of the displacement and be able to save these data into its memory. After forwarding the collected data this is copied into a computer. With the collected data, the system is not only able to bulid a database and analyze the data but also will be able to identify the root-cause of damage. Our primary task is to specify the time and the location of the damage and once this is done, we will be able to identify if the damage was caused by the driving method, or road conditions, or weather conditions or other external factors. As a second step, we should use our collected database to create a map with all the damages reflected on specific locations plus adding the value of the damage itself. This map would provide us sufficient information to decide which route to use for transportation and to evaluate the value of any damage caused and whether a change of route with extra transortation costs to go around specific places to avoid the damage. This is why when damage happens during transportation it is essential to write down the values of the damage (Hirkó, 1997).

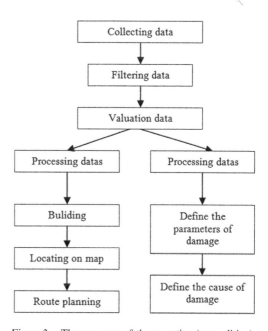

Figure 3. The summary of the operation (own editing).

445

Figure 4.   Steps of filtering.

The most difficult task is to evaluate the collected data and identify those which are significantly higher than the standard values.

So the task is to analyze the data of one route. The scale and the type of the value can explain the type of damage. When these data are paired with latitudes and longitudes, the place of the events can be exactly located. With the help of the map it is easy to decide whether the quality of the road or the driving method or other different external event were causing the damage.

A parameterised map helps to choose the best or the "safest" routes between two points prior to the actual transport date. Different routes have different distances, different road quality, different weather conditions, etc. So we can decide, that our packaging is strong enough for a shorter but harder route or we must choose a longer and more expensive road because the type of the packaging system and the used packaging materials are too sensitive against the different extreme conditions. Or if we have the possibility, we have to choose our packaging methind and materials according to the choosen road (Yao et al, 2008).

The model must point out the different types of logistic stress. (Hirkó et al, 2005). When most of the collected data show the same values on the same latitudes and longitudes then the possible stress can easily be determined. The latitude, longitude and the effect of the damage can be saved in the map database which is useful for the route planning.

So with the help of the map and the database we can specify the suitable route and the exact cost of the transportation and the packaging itself.

5   THE ROLE OF THE COLLECTED DATA IN THE PLANNING OF THE PACKAGING

The parameters of the packaging are in close relationship with the characteristics of the transportation route. The packaging must protect the product from the effects of the route. The protecting factor of the packaging is dependent on the quantity and quality of the used materials. These values influence the costs of the packaging itself. (Pánczél, 2008).

As we mentioned the most important stress types are the mechanical stress and the environmental stress. The environmental stress is a fixed stress type, these are considered to be fairly constant over large geographical areas. However the values of the mechanical stress can vary over a wide scale. Two different routes between the same two cities could result in very different stress values.

Sustainability in packaging not only means lower use of materials and as a result reducing

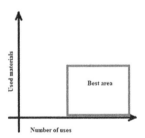

Figure 5.   Goal area in planning of packaging.

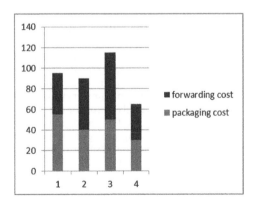

Figure 6.   Possible combinations of costs.

waste but also making an effort to create a packaging method that might be used several times.

The goal is to minimize the usage of packaging materials and to maximize the number of the times it can be used while still effectively protecting the product. (Horváth, 2009).

All in all if we consider all the above goals the best way to achieve them if we use a route which creates the lowest stress on the packaging itself (and hence on our product).

On the other hand it is also important to consider those costs that can be foreseen. The most important component is the distance and the gasoline consumption. Generally speaking we can state that the shortest route between two points are not always the fastest way. for example if the used roads are mainly minor roads, where the speed limit is low, the consumption and the impacts will be higher. The fastest route between two points generally highways where the speed and so the consumption is optimal and the impacts are lower. (Taylor, 1991).

If we analyze the connection between above mentioned factors, we will most probably realize that the fastest way has the most minimal impact on our packaging. So in general terms the optimum of the two sides will be equal with the global optimum.

However we have to consider the possibility that due to somespecial reasons our goal is not necessary to aim for the optimum.

The system presented here helps to collect data and with our saved database it will be easier to design a packaging method that is strong enough to protect our product against the different stress types and also helps in reducing the wasted material resulting from packaging after the transport has been delivered. (Bajor et al, 2008).

6 CONCLUSIONS

Waste reduction and cost efficiency are significant factors to be considered in this topic. It is clearly visible, that the correct and realistic value identification, measurements making and accurate data collection about the chosen route has an essential role in cost effective and globally sustainable transportation and packaging.

This paper has presented a system, which can be useful for creating the correct data collection methods to monitor the stress types and values during transportation. The described methodology should be a good start to investigate the scales of the expected stress and should be able to provide a good support to develop more cost efficient product packaging systems with lower packaging waste combined with cost efficient transporting.

REFERENCES

Burgess, G. 1990: Consolidation of Cushion Curves, Packaging Technology and Science Vol, 3, 189–194.

Böröcz P., Mojzes Á. 2008: A csomagolás jelentősége a logisztikában Transpack –szakmai folyóirat. VIII./2.

Böröcz P. 2007.: Játékelmélet alkalmazási lehetőségei a logisztikai rendszerekben—az egy- és többutas szállítási csomagolási eszközök közötti döntéselméleti probléma elemzése, I. Logisztikai Rendszerek és Elméletek Tudományos Konferencia, Győr.

Erdős F. 2008.: The Innovation Through IT Investments at SMEs in Hungary. Schriftenreihe Informatik 25. Trauner Verlag, Linz. pp. 311–318.

Hirkó, Bálint, Logisztikai folyamatok számítógépes szimulációja, Közlekedési Nyári Egyetem, Győr, pages 65–69, 1997.

Hirkó, Bálint and Németh, Péter, Áruelosztási feladatok stratégiai tervezése, Magyar Logisztikai Egyesület, pages 82–95, 2005, ISBN 1218-3849.

Horváth Adrián: Elosztási rendszerek kialakítását és működését befolyásoló tényezők, 33–40 o. Műszaki és informatikai rendszerek és modellek III. 2009 ISBN: 978-963-7175-54-1.

Horváth, A, Mojzes, Á., Takács G.: Csomagolt termékek szállításának és nyomonkövetésének gazdasági haszon elemzése a térinformatika támogatásával. Logisztikai évkönyv2013: pp. 177–186. (2013).

Keith Taylor: Computer systems in logistics and distribution, Kogan page, London, 1991.

Mojzes, Á. 2007: Fejlesztési—tervezési irányzatok a csomagolástechnika műszaki, gazdasági és ökológiai egyensúlyban betöltött szerepének optimalizálására, I. Logisztikai Rendszerek és Elméletek Konferencia tud. konferencia kiadványa, Győr.

Mojzes, Á 2008.: Theories and Methods to Develop the Systematic Approach for Package Design Technologies, Acta Technica Jaurinensis, volume 1, number 2, pages 397–408, ISSN 1789-6932.

Pánczél, Z. 2008: The Significance of Logistic Package System Design, Acta Technica Jaurinensis, volume 1, number 2, pages 247–258, ISSN 1789-6932.

Pánczél, Z., Mojzes, Á 2006: Importance of package planning and laboratory testing from the aspect of the logistic stresses, during transportation and warehousing, Management of Manufacturing Systems, Presov—Slovakia pp.: 64–69.

Peter Bajor, Adrian Horvath: The role of decision-making parameters in constructing and re-engineering of distribution networks, 55–63 o. FIKUSZ 2008, Fiatal Kutatók Szimpóziuma 2008.11.07., BMF, Budapest ISBN: 978-963-7154-78-2.

Soroka, W 2002: Fundamentals of Packaging Technology. Inst. of Packaging Professionals; 3rd edition.

Yao, Y., Dresner, M.: The inventory value of information sharing, continuous replenishment, and vendor-managed inventory, in Transportation Research Part E 44, 2008, pp. 361–37.

Green Design, Materials and Manufacturing Processes – Bártolo et al. (eds)
© 2013 Taylor & Francis Group, London, ISBN 978-1-138-00046-9

EIA implementation and follow up: A case study of Koga irrigation project—Ethiopia

Wubneh B. Abebe
ADSWE, Bahir dar, Ethiopia

M. McCartney
IWMI, Colombo, Sri Lanka

ABSTRACT: In Ethiopia, the importance of follow-up in the Environmental Impact Assessment (EIA) process is clearly recognized. Follow-up involves the implementation of measures taken to mitigate the adverse environmental impacts of a project and monitoring to determine their effectiveness. This paper reports on a study of the follow-up of EIA-recommended mitigation measures in the Koga irrigation and watershed management project. This research found that the EIA documents, which were prepared during the feasibility study, were generally satisfactory. But, monitoring of impacts and the implementation of mitigation measures are currently very poor. Public participation in the project is also very limited. Moreover, one weakness in the EIA was the poor estimation of downstream flow requirements, which did not take into account the natural variability of flow nor the livelihoods of people dependent on downstream fisheries. Hence, unless improvements are made it is likely the sustainability of the project may be severely compromised.

1 INTRODUCTION

Environmental Impact Assessment (EIA) is a process which attempts to identify, predict and mitigate the ecological and social impacts of development proposals and activities. It also helps to assist decision-making and to attain sustainable development. The effectiveness of EIA depends on several factors, among which the quality of EIA guidelines, EIA reports and implementation and follow-up of EIA recommendations is of particular importance (Arebo, 2005). According to EPA Australia (1995), EIA follow-up is needed because relatively little attention is paid to the actual effects arising from project construction and operation. Without some form of systematic follow-up to decision making, EIA may become just a paper chase to secure a development permit, rather than a meaningful exercise in environmental management to bring about real environmental benefits.

The aim of this study was to determine the critical factors affecting the successful implementation of EIA mitigation measures, developed to minimize environmental and social impacts of the Koga irrigation and watershed management project in the district of Mecha, Amhara National Regional State—Ethiopia. The research questions addressed were:

- To what extent are EIA-recommended mitigation measures implemented by the project proponent?

- How do regulatory bodies ensure implementation of EIA-recommended mitigation measures?
- How and to what extent did the public participate in the EIA process?
- What are the likely downstream impacts of the project and to what extent where they considered?

2 METHODS

The research method comprised both a literature review and field work. The literature review centered on issues of sustainability and links to EIA and the MDGs as well as EIA experiences in Ethiopia and other countries. Both semi-structured and structured questionnaires were used. This enabled the perceptions and opinions of specialists (from the project and EPA), the community (upstream/downstream) and management bodies (from the project, EPA and other organs) to be gathered. The extent of public participation in the project was assessed using "the Aarhus practice evaluation criteria for public participation"; adopted from European convention on public participation (Hartley and Wood, 2004). Besides the Environmental Management Plan, accomplishment reports, monitoring reports and permit conditions of the project were reviewed. Finally, field observations were undertaken to independently assess the accomplishments of EIA-recommendations. Analysis of results was done by comparing the perceptions of different stakeholders on the accomplishment of the project and by comparing the accomplishment reports with the Environmental Management Plan.

3 KOGA PROJECT: A REVIEW OF EIA RELATED DOCUMENTS

Most of the documents (Acres and Shawel, 1995; WAPCO and WWDSE, 2005; KIWMaP, 2006; EPLAUA, 2006; MacDonald, 2004a & 2004b; ADF, 2000; ADF, 2001; and McDonald, 2006a & 2006b) fulfill requirements and give satisfactory information on impacts and mitigation measures to be undertaken to minimize environmental problems. The document for Environmental Flow Assessment (EFA) of the project tells us that it used Q95 method. This method could not address variable nature of the hydrological regime. Environmental Management Plan (EMP) document has some draw backs—consultation process not mentioned, it is without project scenarios/alternatives, no monitoring plan for erosion and siltation, and public participation did not get clear emphasis. Summary for some of implementations taken for three years accomplishment report in comparison with the plan shown on the environmental management plan is tabulated below (KIWMaP, 2006).

4 KOGA PROJECT: OPINIONS, PERCEPTIONS AND FIELD OBSERVATIONS

4.1 Interviews

4.1.1 The community (Farmers)
The interview focuses on public participation and implementation of mitigation measures. Among the ten criterion used for this assessment 'communication' is one. It is used to determine whether the project material is presented in a non-technical format and is understandable to lay people. The overall interpreted result for this criterion based on the percentage of interviewee is: 19% of the interviewees agreed that *communication* criterion for the project is completely fulfilled, 14% nearly fulfilled, 26% partially fulfilled and 41% of the community said that the project did not provide project materials in a clear format, implies that the criterion not fulfilled (Fig. 1). This realizes that nearly half of the interviewed people living in Koga catchment have more or less no clear understanding of the project document or the project itself based on the materials provided through training and meetings undertaken concerning the project.

Table 1. Summary table of three years implementation of mitigation measures.

Environmental Mgt plan/EMP/	Accomplishment report
1. Watershed mgt plan	COD = 23%, terrace = 13.5%, Gully treatment = 14.6%
2. Resettlement plan	Resettled = 9%, compensation = 11%
3. Control of pollution from labour camp	No report/no activity
4. Public health	Farmers training = 30%, No other activity
5. Control of air pollution	No report/no activity
6. Maintenance of riverine fishery	No report/no activity

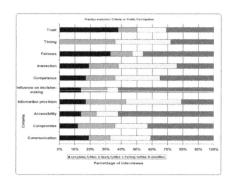

Figure 1. Aarhus Practice Evaluation Criteria for Public Participation, community's opinion.

4.1.2 Specialists
The interview with specialists focuses on implementation, permit condition and downstream impacts. The results obtained from the interview conducted for specialists indicated that more than 70% of the specialists said that the project has less satisfactory condition on the appropriateness and feasibility of mitigation measures designed for the environmental impacts the project is imposing. In addition, the sufficiency of the Environmental Management Plan (EMP) in terms of institutional arrangement, time schedule, cost, integration of EMP with the project schedule and fulfillment of expert staff is judged to be below satisfactory by more than 90% of the specialists.

4.1.3 Management bodies
The interview of the management bodies focuses on follow-up activities and permit condition. The different components of EIA follow-up activities were tested by management bodies' interview. Koga irrigation and watershed management has no official

permit certificate based on the EIA legislation as the interview result indicated. Rather the lender organisation has set an obligation to undertake EIA and because of this obligation the project has prepared its own EIA document and a group of AfDB experts has gone to the site and approved the EIA document after a certain study and preparing its own EIA summary (ADF, 2000). Most of the management bodies from stakeholder organisations said they know nothing (no opinion) about the permit condition, monitoring activity and others at all.

5 CONCLUSIONS AND RECOMMENDATIONS

5.1 *Conclusions*

- The result based on reports showed that among the 20 major plans indicated on the environmental management plan for implementing EIA-recommended mitigation measures two activities, namely: forest seedling plantation and livestock development are found to be satisfactory; other three activities, namely: watershed management measures (conservation and production measures), public health and resettlement/compensation payment are unsatisfactory; and the remaining plans left untouched or not reported.
- Interview and documents review results showed that either the project proponent or consultants have no specialists in this regard i.e. environmentalist. So, the proponent is not undertaking formal monitoring. Besides the regulatory agency, EPLAUA, has undertaken surveillance/monitoring only once after four years of the commencement of the project.
- A review of the monitoring report proved that the report was very weak in the number and contents of the activities monitored and recommendations given as compared to the number of activities in the monitoring plan which need to be monitored.
- Results showed that the performance of the project on public participation is poor.
- Potential downstream impacts due to change in flow regime is not mentioned. Q95 method is used in deciding the monthly flow releases to maintain the downstream river ecology considering only maintenance requirements for normal year but there is no consideration for drought period requirements. There is also no mention of higher flow requirements.

5.2 *Recommendations*

- EIA should consider all feasible alternatives which may include different methods of undertaking a development; and impacts indicators should be quantitative in order to make easy the monitoring work, which are not the cases for koga project EIA.
- Experiences showed that preparing and undertaking *community workshops* and then organizing strong "farmers' development teams" all over the Koga river catchment can improve public participation in the project.
- The drawbacks of the Q95 method should be considered again to ascertain whether the right decisions are taken to flow releases. Although koga project has failed to do so, as other activities of the EIA, environmental flow assessment needs consensus of the affected and interested parties on the desired state of the river.

REFERENCES

Acres, Shawel, 1995. Feasibility study of the Birr and Koga irrigation project, Koga catchment and irrigation studies, annexes Q-T.

ADF, 2001. Koga Irrigation and Watershed Management Project: Appraisal Report. (Available on http://www.afdb.org).

ADF, 2000. Summary Environmental Impact Assessment, Koga Irrigation and Watershed Management Project.

Arebo, S.G. (2005), Improving Environmental Impact Assessment for Enhancing Sustainable Water Resources Developments in Ethiopia, MSC thesis, UNESCO-IHE Institute of Water Education, Delft.

EPA, 2000. Environmental impact assessment guideline document; Federal democratic republic of Ethiopia.

EPA Australia, 1995. International Study on the Effectiveness of Environmental Assessment. Canberra, Australia: Environment Protection Agency.

EPLAUA, 2006. EPLAUA, Environmental protection study policy and control department; Koga Irrigation and Watershed Management Project field monitoring report, Bahir Dar, Ethiopia.

Hartley, N., Wood, C. (2004), Public participation in environmental impact assessment-implementing the Aarhus Convention, *Environmental Impact Assessment Review*, 25, 319–340.

KIWMaP1, (2006) Koga Irrigation and Watershed Management Project; Progress of Watershed management project component implementation, Preliminary draft report.

MacDonald, M. 2004a. Koga irrigation and watershed management project, Interim Report.

MacDonald, M. 2004b. Koga irrigation and watershed management project, hydrology factual report.

MacDonald, M. 2006a. Koga irrigation and watershed management project, Dam Design final report.

MacDonald, M. 2006b. Koga Dam and Irrigation Project; Design final Report; Part 2: Irrigation and Drainage.

WAPCOS, WWDSE, 2005. Environmental monitoring and resettlement plan for Koga irrigation and watershed management project, draft, Volume I and II.

Green Design, Materials and Manufacturing Processes – Bártolo et al. (eds)
© *2013 Taylor & Francis Group, London, ISBN 978-1-138-00046-9*

Waste prevention and reuse of synthetic textiles: A case study in a Brazilian garment industry

I.S.B. Martins, C.P. Sampaio & U. Perez
Universidade Estadual de Londrina, Paraná, Brazil

ABSTRACT: This article presents the results of a research project on design and sustainability developed with a group of fashion design and graphic design students of a Brazilian university in 2012. The study aimed to find solutions, through design, to environmental problems caused by synthetic textile waste generated by Brazilian clothing industries. The object of study was a women's clothing firm in Londrina, located in the northern region of the state of Paraná. The research was started with a literature review on the environmental impacts of the apparel and fashion industry, especially in the state of Paraná, which concentrates much of the clothing producers in Brazil. This review focused on the main issue of textile waste, the environmental problems caused and existing proposals for its reduction. Then, the project was carried out considering two approaches: a preventive, in which solutions were sought to avoid or minimize the generation of waste, through the *zero waste* methodology, and other corrective seeking new applications for waste through the creation of new products. The preventive approach was divided into three stages: diagnosis of waste generation, identifying opportunities for prevention and implementation of the proposed actions. The corrective approach included the collection and sorting of waste samples, the generation and selection of ideas through brainstorming, testing using waste with various types of glues, the development of the most promising ideas, test and manufacture of functional prototypes. As a final result, were fabricated two prototypes of luminaires, which were presented in an exhibition dedicated to products made with various textile wastes, from the studied company. The paper also presents a discussion on the challenges and lessons learned by the students with the project.

1 INTRODUCTION

1.1 Context of research

This article presents and discusses the results of the research project "Innovation and sustainability in the use of textile waste from clothing sector of Londrina and region", held at the Design department of the State University of Londrina, a medium sized city of Paraná, Brazil.

The research began with an analysis of proposals for waste prevention using a specific "zero waste" methodology for the industry. A case study was performed in a partner company in which, through the insertion of a team member, was made the diagnosis and accounting of waste.

They were then generated solutions that were viable for the reality of the company, and developed solutions for both greater eco-efficiency of processes and for the reuse of waste as raw material for new products in order to increase the life cycle of the material that would be discarded, and also educate the customer about more sustainable consumption through information contained in the new product line developed.

1.2 Current scenario of apparel and fashion industry and its environmental impacts

The textile and clothing industry is one of the largest industrial sectors in Brazil, representing, according to data from ABIT (2011), 16.4% of jobs in industry transformation and 5.5% of revenues, with $ 60.5 billion, comprising 3.5% of total GDP in Brazil. With an average production in the clothing sector of 9.8 billion pieces, represents the fourth largest producer of clothing in the world. Although the economic contribution is relevant, the sector is also responsible for significant environmental impacts, in which solid waste have significant importance (Guimarães & Martins, 2010).

1.3 The issue of textile waste and environmental problems caused in the national and local levels

The generation of waste occurs daily in business clothing and fashion garments and at various times of the production process, becoming a serious environmental problem, since such waste is produced at a different speed than the environment can absorb. The textile and clothing industry of Paraná state generate solid waste in large volume.

The toxicity of the waste is not significant, but the high volume affects other environmental variables, such as CO_2 emissions and depletion of non-renewable natural resources (Martins et al, 2011). The generation of waste causes a critical impact, representing the main direct environmental impact of companies in this sector. Although McQuillan & Rissanen (2011) and SEBRAE (2004) point out that the rate of wastage in the sector of production of clothing and fashion is about 15%, to Conrad (2010) this waste can reach up to 20%.

This represents, according SEBRAE (2004), 1397.2 kg of waste generated annually by each firm and average annual wastage of R $ 32,783.00. Thus, the generation of solid waste is characterized as a problem not only for the environment but also causes economic impact to industries sector, as it represents a huge waste of raw materials and investments.

1.4 Preventive and corrective approaches to the issue of textile waste

The waste from the production process of the textile and clothing sector, until recently regarded as an intrinsic part of the production, is now treated as waste and is essential to study and redesign production processes to reduce their impacts (Guimarães & Martins, 2010).

According to these authors, not only companies currently seek to resolve outstanding problems concerning the generation of waste but also begin to prevent them and avoid them, for the prevention reduces costs in the production process and increases the profitability of the company (SEBRAE, 2004).

For a preventive approach is needed, according to Perez & Martins (2012), to survey data regarding the consumption of raw materials and diagnose waste by mapping the production process in order to understand the process in which moments are generated waste and then propose solutions to minimize them.

In corrective approach, the starting point was the reuse of large amount of synthetic waste generated by the partner company. Then the research team developed a line of products with value-added design, made from such waste, while it saw a way to correct, in part, the waste of raw material due to its production process.

2 METHODOLOGY

2.1 Case study: Women's clothing firm in Londrina, Paraná

For the research, a partnership was established with a small business of fashion apparel and fitness day-by-day, located in Londrina (PR), by inserting a member of the research team in the production process.

The research was started with analysis of existing proposals for the reduction of solid waste, such as manuals and methods of prevention, especially SEBRAE (2004), INETI (2007) and Guimarães & Martins (2010). From this study, we defined the methodology used in this work, which includes the diagnosis of the production process, analysis of waste generation, waste accounting and mapping of the main stages of waste generation, allowing the identification of opportunities for prevention or reduction waste generation.

2.2 Intervention project

For preventive approach, was used the "zero waste" methodology (Guimarães & Martins, 2010) which proposes, among other ideas, to reduce waste by prioritizing the use of measures to avoid waste still in the stage of product development. In fashion, it consists of modeling techniques that aim to reduce or even eliminate the waste of tissue resulting from the fit and cut.

Following this logic, any remaining residues are targeted for other applications, especially the creation of new products, featuring a corrective approach. The reuse of waste textiles is one way to reduce environmental impacts and add value to waste, which become input for a new production (Kazazian, 2009). These approaches are used in complementary fashion following the logic of the three R's: first reduced, then reuse, and then recycle, since the first two alternatives are exhausted.

3 RESULTS

3.1 Diagnosis of the situation found in the company

For more than 10 years in the market, the company studied is a small company specializing in sportswear whose differential is the use of technological fabrics and design with anatomical cutouts that protect the joints and value the body shapes.

With production capacity of four thousand pieces per month, the company has developed two collections per year and is located in the city of Londrina, concentrating the product development sector in the city of São Paulo. The structure involves a total of 44 direct employees and associates indirectly more than 15, and most of the production done in the factory, although some pieces are sent to factions outsourced.

Most tissues used by the company has in its composition elastane, polyamide or polyester, which are not biodegradable chemical fibers. However, according to Teixeira & Caesar (2005, p. 56), "it is biodegradable does not make a material completely eco-efficient, since this feature does not guarantee the reduction of waste."

Furthermore, the use of technological fabrics is a trademark of the company differential studied and, therefore, not possible to replace them. However, is difficult for the company to find individuals or institutions interested in collecting their waste, so it is essential to find solutions to prevent their generation.

While the company considers low wastage rate of approximately 20%, when comparing the accounting of its waste with data SEBRAE (2004) perceives that the waste in the company is very high when compared to other clothing industries, as can be seen in Table 1.

The analysis of the production process, in turn, shows the step where more waste is generated is cut, but the source of waste occurs in previous steps, because the fitting and use of tissue are planned in modeling stage.

However, such software is used for docking and there is concern in modeling to improve the use of the material, it is clear that the most critical step for the generation of waste is product development, where models are created that do not provide good fit in modeling and use of materials is not intended to prioritize the best possible use. As stated by Seiffert (2011), the generation of waste can be indicative of flaws in the design of product.

3.2 *Proposal for preventive intervention: Solutions to avoid or minimize the generation of waste through zero waste approach*

In the method of zero waste design, fitting deserves special attention, still being planned at the time of creation and modeling, being able to change the format of the molds during conception in search of the best fit possible. In this method, "rather than imposing a design to fabric and a preconceived model, the designer becomes a facilitator, allowing the form to emerge and guiding its evolution," reshaping the form and manner so that all parties fit into one another, resulting in different forms and innovative integration by the clothing

fabric that would be wasted in the cut (Fletcher & Grose 2011, p. 48th).

Although there are some software to optimize the use of the fabric in the groove and cut, their effectiveness is limited because "they are not able to adapt to completely new concepts for making clothes, and therefore may slow the emergence of innovations related to waste reduction and the new aesthetic that they can reveal" (Fletcher & Grose 2011, p. 48).

The creative process is different from the traditional approach, because the parts are not created through sketches, but from the study of modeling, docking, and as recommended by McQuillan (2010), manufacture of prototypes in 1:2 scale modeling to test and make the necessary changes. Before you begin building, however, it is necessary to choose the fabric, because, as stated & Souza Queiroz (2010, p. 6), "the study of the width of the fabric is of great importance since the molds and fit depend on this measure".

3.3 *Proposal for corrective intervention: Creation of new products from textile waste*

The creation of products from synthetic fabric waste included the following activities:

– Survey and analysis of similar products (luminaires);
– Survey and analysis of manufacturing techniques that could be used in the project;
– Experimentation with practical techniques surveyed;
– Choose techniques more viable and with more aesthetic potential for creating different forms;
– Generate product ideas through brainstorming, from the results of practical experimentation;
– Manufacture of prototypes of the ideas chosen;
– Testing of prototypes, and adjustments.

After research and analysis of similar products and manufacturing techniques, the research team conducted a practical experiment, which sought alternatives of how to join pieces of fabric and shape it. For this, tests were made using maize starch gum and white glue PVA and mixed with pure water. The different blends were applied on the fabric pieces supported by flat surfaces of objects used as template to see if the tissue conforms to those surfaces.

Finally, the starchy tissues were taken for drying in the sun, a natural way, and in oven at high temperature to determine what form of drying was more efficient and adequate for creating shapes.

After drying it was found that, among the glue methods tested, only the pure glue dried naturally in the sunlight united well the tissue and conformed over the object without losing stability (Fig. 1).

Table 1. Comparative data on waste in the company studied with data of SEBRAE.

	Average wastage	Waste/year (kg)	Average annual waste
Studied company	24,26%	1.601,52	U$ 28.754,83
SEBRAE	12,5%	1.397,2	U$ 16.608,24

Figure 1. Test of textile material bonded with white glue (PVA) and naturally dried in the sun.

In experiments with other adhesives, tissue was too soft and not fully assumed the shape of the object.

Thus, the method using pure white paste was chosen for being the most effective, easier to perform, especially by using PVA only white glue which is washable and biodegradable. Furthermore, the use of few materials in the composition of the product is more environmentally desirable as it facilitates separation and recycling at the end of product life (Manzini & Vezzoli, 2002).

From these testing processes, the team discussed possible products which could be developed and defined the design theme for creating domes for luminaires. For this, a test was made of light on the sample that performed the best, and it was noted that he had great ability to pass light.

Then we conducted a brainstorming to generate ideas luminaires that could be produced with the molding technique chosen. As a result, two techniques have been proposed for conformation on surfaces: one using various types of rigid packaging (PET, metal, glass) as template, and another glued pieces of fabric about inflatable shapes such as balls and air balloons. The goal of these two variants was investigated which of them would produce more technical viable and aesthetic interesting result.

After drying it was observed that in the applications on rigid packages, the flaps of tissue were took off to withdraw the container from the inside of the fabric layer bonded. The same occurred in the applications made on rubber balls, in which pieces of fabric were glued, making it difficult to demolding.

In tests with balloons, however, it was found that it was possible to remove the tissue after molding by simply empty them. Thus, it was possible to obtain the original shape of the molded bladders, resulting in dome shapes with very interesting

visual result (Fig. 2). Tests were also made of waste composition on bladders, to check the visual possibilities in terms of colors and forms of waste.

From these tests, four luminaires were produced: two for use on surfaces such as tables and shelves (Fig. 3), and two for use in a ceiling or attached to some support (Fig. 4). As light source were used compact fluorescent lamps and LED lamps, and it was found that fluorescent light showed more attractive effect.

Furthermore, compositions were tested with the use of other types of waste material, such as MDF and granite, with satisfactory results, which indicates a path rather interesting in terms of reusing other types of waste.

The results of these experiments were displayed in the exhibition held in April, 2012 at the State University of Londrina, along with other products made by the students of the 3rd year class of Fashion Design Course, in partnership with the garment industry studied.

Figure 2. Application of synthetic fabric waste with white glue on air balloon (composition process of the dome).

Figure 3. Luminaire for use on tables or shelves.

Figure 4.   Cable suspended luminaire.

4   CONCLUSIONS

Solid waste represent much of the environmental impacts of industry making clothing and fashion in Brazil, and we need to evaluate the entire production process to identify its cause and origin. As demonstrated in the case studies, although the waste is generated in greater quantities when cutting the pieces, the origin of this waste occurs in product development.

In this sense, we propose a proactive approach in which, for the effective reduction of waste we need to modify not only the product but also its development process and modeling, avoiding waste generation from the beginning through the methodology zero waste, and planning it is yet when the engagement of creating and modeling.

On the corrective approach, through the use of waste products for the creation of products, the experiments demonstrated the feasibility of the materials and processes used, although its use on an industrial scale has not been tested. This will be one of the next steps of the research project described in order to implement the experiments for an industrial process of molding on complex surfaces. In this sense, the hot-pressing and vacuum-forming can be considered viable alternatives to enable the solutions found, and that should be explored in depth.

REFERENCES

Associação Brasileira da Indústria Têxtil e de Confecção—ABIT. 2012, *Perfil do Setor: Dados gerais do setor atualizados em 2012, referentes ao ano de* 2011. [Online] Available at: http://www.abit.org.br/site/navegacao.asp?id_menu=1&id_sub=4&idioma=PT [Acessed March 12, 2012].

Conrad, F.R. 2010. *Fashion Tries on Zero Waste Design.* 2010. The New York Times, August 15, 2010: ST1. [Online] Available at: http://www.nytimes.com/2010/08/15/fashion/15waste.html?partner=rss&emc=rss [Acessed September 20, 2012].

Fletcher, K. & Grose, L. 2011. *Moda & Sustentabilidade: design para a mudança.* São Paulo: Senac.

Guimarães, B.A. & Martins, S.B. 2010. Proposta de metodologia de prevenção de resíduos e otimização de produção aplicada à indústria de confecção de pequeno e médio porte, *Projética* 1(1): 184–200.

INETI. 2007. *Manual para a prevenção de resíduos: estudo de caso para o sector têxtil.* [Online] Available at: http://preresi.ineti.pt/actividades/demonstracao/Manual_EC_T_Confeccao.pdf [Acessed February 12, 2012].

Kazazian, T. 2005. *Haverá a Idade das Coisas Leves.* São Paulo: Senac.

Manzini, E. & Vezzoli, C. 2002. *O Desenvolvimento de Produtos Sustentáveis.* São Paulo: Edusp.

Martins, S.B., Sampaio, C.P. & Castilho, N.M. 2011. Moda e Sustentabilidade: proposta de sistema produto-serviço para Setor de Vestuário, *Projética* 2(2): 126–139.

McQuillan, H. 2010. *Zero-waste pattern cutting process.* [Online] Available at: http://centerforpatterndesign.com/content/Zerowaste.pdf [Acessed September 20, 2011].

McQuillan, H. & Rissanen, T. 2011. *Yield: making fashion without making waste.* [Online] Available at: http://yieldexhibition.com/yieldexhibition-catalogue.pdf [Acessed April 17, 2012].

Perez, I.U. & Martins, S.B. 2012. Metodologia de prevenção de resíduos sólidos para o setor de vestuário e moda. In: *International fashion and design congress,* CIMODE 1, 2012. Proceedings, Guimarães, 2012.

Souza, P.M. & Queiroz, J.C. 2010 O desenvolvimento sustentável de produtos de moda para a viagem de mochileiros. In: *Congresso brasileiro de pesquisa e desenvolvimento em design, n. 9,* 2010, São Paulo.

Teixeira, M.G. & César, S.F. 2005. Ecologia industrial e eco-design: requisitos para a determinação de materiais ecologicamente corretos. *Revista Design em Foco* 2 (1): 51–60. Salvador: EDUNEB.

Green Design, Materials and Manufacturing Processes – Bártolo et al. (eds)
© 2013 Taylor & Francis Group, London, ISBN 978-1-138-00046-9

Controlling the properties of materials manufactured using recycled PVC

C.M. Pratt & F.J. Davis
The Department of Chemistry, The University of Reading, Reading, UK

G.R. Mitchell
Centro de Desenvolvimento Rápido e Sustentado de Produto, Instituto Politécnico de Leiria, Portugal

ABSTRACT: PolyVinyl Chloride (PVC) is a readily available material which has found use in a range of applications. Such applications are enhanced through the presence of property modifiers added to the PVC. These additives include processing aids, stabilizers, lubricants and plasticers. While these additives provide excellent control of the properties in virgin PVC their present in recycled PVC may be a disadvantage depending on the application. In this presentation we discuss how PVC recycled PVC can be modified to produce materials with desirable properties.

1 INTRODUCTION

There is a growing awareness of the role of recycling in reducing energy consumption and the stocks of pertrochemical based materials. There are considerable societal pressure for industry to increase the proportion of materials which are recycled. Recycling of plastics is unlikely to be as universal as is the case with glass, paper and steel. However, in situations where large volumes of a single polymer are available without the complications of sorting and collation, recycling can make a valuable contribution.

PVC is rarely used as a neat polymer but with the right additives, thousands of diverse products can be made, such as those listed in Table 1 [Carroll 1992]. For example, water pipes, garden furniture, window frames and clothing can made from PVC but rely on the presence of additives for their functionality. Such additives include heat stabilizers which prevent dehydrohalogenation and subsequent colour formation; other additives used include processing aids lubricants, impact modifiers, and plasticizers. Though recently health concerns surrounding some of the phthalates used as plasticizers has resulted in restrictions in the use of PVC for example in toys [Borrell Fontelles 2005] and there is continuing concern about the use of such material in medical devices [Messoria 2004].

The data in Table 1 shows that there is clearly a huge amount of PVC available and this may come from a variety of sources but the source available may vary substantially in properties. For example

Table 1. US PVC sales volume, 1990, by application.

Application	Sales volume (1000 tonnes)
Pipe, tube and fittings	1700
Siding and accessories	387
Packaging film and sheet	216
Wire and cable	185
Flooring	111
Window profile/Gutters/Skirts	180
Bottles	100
Other calendering/Extrusion	555
Other coating/Moulding/Sealants	328
Export	421
Total	4183

material recovered from cable may be relatively hard, while material recovered from flooring may contain large amounts of plasticizer and be extremely flexible but may be reinforced with particulates such as glass fibre. In this presentation we shall look at the way the material properties of PVC can be altered focusing particularly on material obtained from cable.

Approximately 227000 tonnes of PVC enter the wire and cable market annually. Because of the maturity of this product, millions of tonnes of decommissioned wire and cable scrap are available for the recycling market. Recycling PVC cable coatings is particularly economical as it is a secondary recyclate as a byproduct of the extraction of highly valuable copper wire. For these reasons, cable coatings are one of the main PVC products recycled by reprocessing at present [Brebu 2000]. Typically, the copper cable is stripped out leaving by a process which involves chopping and grinding the cable, liberating the conducting metal [Scheirs 1998]. The metal is then separated from the PVC by a gravity separator. Further separation can be done using hydrocyclones. This leaves the coatings cut up into small chippings, which can be fed through an extruder and moulded into various products by injection or compression moulding [Wikes 2005].

2 CONTENT OF PVC

A comprehensive understanding of how to utilize recycled PVC requires knowledge of the materials available. As stated in Section, this may vary depending upon the source of the recycled materials, thus soft PVC such as often found for flooring, will contain a higher level of plasticizer than the more rigid PVC found in cable waste, and indeed the properties can be altered by blending the two materials. uPVC, will not contain plasticizer, but is likely to contain a range of other additives. Thus understanding the opportunities for recycling requires knowledge of the main constituents of the material. Unfortunately this may present significant challenges. We have applied a full range of analytical techniques to a sample of PVC obtained from stripped cable. The techniques include NMR infrared and EDX, from electron microscopy; some soluble components had to be further isolated using column chromatography. Despite the use of this range of top-end analytical techniques a proportion of the material remained unidentified though this also may reflect the overall accuracy of the techniques.

The need for high-end analytical equipment seems to present an insurmountable problem to the control of recycling conditions, however, it is possible to develop simpler tests, based on solubilities and densities *etc* that can be used to identify the important groups of components, namely inorganic fillers, polymer, and plasticizer, although this will not allow for complete identifications which might be required under safety regulations. Thus for example PVC may be soluble in tetrahydrofuran, but not in ethanol; while plasticizer is generally soluble in this latter solvent. A particularly important component which is not an intrinsic part of PVC is water.

3 WATER CONTENT

Notwithstanding the suitability of the PVC material for any specific application a major consideration is the presence of water in the sample; for large scale production water may accumulate in considerable amounts during both storage and transportation. This is a particular problem if (as is likely) the manufacturing process involves treating molten polymer, as in a process such as injection moulding. Water contamination in the PVC causes problems with the extrusion process both by virtue of the energy required to boil of the water and the volume of steam produced. Figure 2 shows the decline in rate of extrusion of PVC (cable) from an extruder following the addition of different amounts of water, this is because steam being generated interferes with the extrusion of polymer. Thus in addition to energy problems, there is a cost in terms of the reduced productivity by virtue of the increased time required to complete the moulding. In addition to this, where the production facility is less automated, a substantial risk to the work-force.

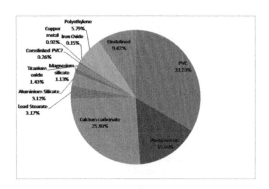

Figure 1. Summary of the composition of a sample of waste obtained from cables.

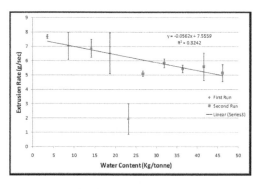

Figure 2. Decline in extruder performance with water contamination.

4 PLASTICIZER CONTENT

Plasticizer content is important in terms of the final uses of the materials, since it will have a considerable effect on particularly the mechanical properties of the polymer. Different sources of PVC will have different plasticizer content and this need to be considered prior to manufacture. Thus we have found that cable waste contains typically 33% PVC, 17% plasticizer and 50% particulates; in contrast flooring contains 35% PVC, 39% plasticizers and only 25% particulates. This clearly makes the latter a softer material, but this may not be appropriate for certain applications, the particulate quantities are low, but in some applications the nature of the particulates (for example glass fibre) may present an issue.

A solution to the inconsistent plasticizer content may be simply to mix the two materials, but these may have significantly different costs. A further option is to add smaller amounts of a material known as plastisol. Plastisol is a term commonly used to describe a suspension of PVC in plasticizer, it has many commercial applications for example in producing PVC coatings or dip moulding. As implied above this material countains much higher amounts of plasticizer and can be used to supplement plasticizer content. Fortunately there are waste streams that produce plastisol, so this can be obtained at relatively low cost.

Clearly plasticizer content largely is controlled by the material property requirements of the product, although inorganic fillers may also be used to fine tune such properties. One interesting aspect of adding plasticizer is that it can act as an aid to processing by lowering the temperature at which the polymer can be extruded. Figure 3 shows the fusion temperature for PVC containing varying amounts of plastisol (as a plasticizer) and calcium carbonate (as a filler). As can be seen where there are significant quantities of filler (which serve to raise the apparent working temperature), the presence of plasticizer servers to reduce the fusion temperature.

The use of plastisol as a means of increasing the plasticizer content obviously has significant potential, but since the material is a (viscous) liquid, this represents challenges to simple injection moulding systems to ensure the correct dose is applied.

5 FILLERS

A range of fillers can be used in recycled materials, the advantage of this approach is firstly, the control of mechanical properties in particular, but also other properties such as the density.

Previously studies have been carried out on the effects of fillers on the properties of a range of recycled polymers. For example, Sole and Ball [1996] studied blends of polypropylene and the fillers talc, barium sulfate, calcium carbonate and fly ash. Tests of the yield strength showed the latter two fillers reduced the yield tensile strength with increasing filler content in a more or less linear fashion; in contrast the talc had substantially less influence of the mechanical properties.

5.1 Calcium carbonate as a filler

Calcium carbonate is an attractive option for use as a filler for PVC [Wypych 2008]. It is relatively cheap and thus reduces he overall cost per unit mass and it can add stability during processing, since the carbonate will neutralize any HCl liberated by thermal degradation. A further advantage (in some applications at least) is that its presence will allow the manipulation of the density. Its presence as an additive will also alter stiffness and hardness; Figure 4 shows the Young's modulus for a range of samples of the waste cable to which calcium carbonate had been added.

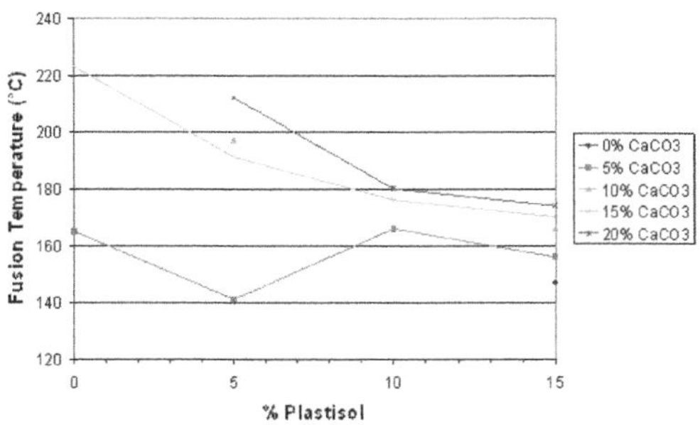

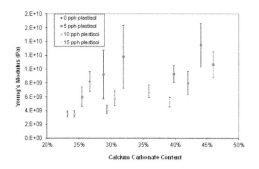

Figure 3.   Effect of plastisol on fusion temperature of mixtures.

Figure 4.   Plot of Young's Modulus for PVC vs. calcium carbonate content.

Broadly this increases with increasing quantities of the carbonate. However the ultimate stress possible for the sample starts to decline above about 25% chalk addition (Fig. 5) and the rupture point shows a broadly linear decline with calcium carbonate content.

5.2 *Fly ash*

The ease of using chalk as a filler and the advantages that clearly accrue lead us to explore the possibility of using an even cheaper filler, namely fly ash; an abundant and problematic waste source from coal-fired power stations.

We have attempted measuring the effectiveness of a using fly ash on PVC mixes with a range of samples Scrap cable coatings (Be-Ha-Rec, Einbeck, Netherlands), Plastisol (Vinyl Recycling Ltd., Mossley, UK) and Fly ash (Didcot Power Station, Didcot, UK and Tilbury Power Station, Tilbury, UK) were used mixed together in varying proportions as described in Table 2 below. These mixes were then passed through an extruder and moulded into constant shapes (square-faced blocks for bases). The resulting materials were then subjected to a simple drop test at varying temperatures to check the stability of the moulded product.

Production of the moulded items was done in three steps: Weighing and blending of the materials, heating and mixing and compression moulding. The required weights of cable and plastisol were placed into a box (mixes 1–3), onto the floor (mixes 4–6) and mixed with a spade (mixes 1–6), or were placed into a portable cement mixer and mixed for about 5–6 minutes (mixes 7–36). The required weight of fly ash was then added and mixed for about another 5–6 minutes. The mixes were then placed in a box and either extruded straight away (mixes 1–7) or left overnight (mixes 11–36). Each mix was

Table 2. Mixes used for trials with fly-ash (samples in bold italics are those passed a stability drop test).

Mix	Cable coatings		Plastisol		Fly ash	
	Kg	%	Kg	%	Kg	%
1	22.50	90.0	2.50	10.0	0.00	0.0
2	*23.75*	*95.0*	*1.25*	*5.0*	*0.00*	*0.0*
3	*22.50*	*90.0*	*1.25*	*5.0*	*1.25*	*5.0*
4	*21.25*	*85.0*	*1.25*	*5.0*	*2.50*	*10.0*
5	*20.00*	*80.0*	*2.50*	*10.0*	*2.50*	*10.0*
6	*20.00*	*80.0*	*1.25*	*5.0*	*3.75*	*15.0*
7	*18.25*	*75.0*	*2.50*	*10.0*	*3.75*	*15.0*
11	21.25	84.5	2.66	10.6	1.25	5.0
12	21.25	85.0	1.25	5.0	2.50	10.0
13	*18.75*	*75.0*	*1.25*	*5.0*	*5.00*	*20.0*
14	17.50	70.0	2.50	10.0	5.00	20.0
15	17.50	70.0	1.25	5.0	6.25	25.0
16	*16.25*	*65.0*	*2.50*	*10.0*	*6.25*	*25.0*
18	15.00	60.0	2.50	10.0	7.50	30.0
22	12.50	50.0	2.50	10.0	10.00	40
35	24.38	97.48	0.63	2.52	0.00	0.0
36	25.00	100.0	0.00	0.0	0.00	0.0

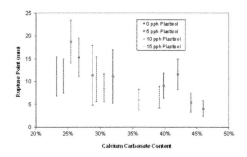

Figure 6. Plot of rupture point for PVC vs. calcium carbonate content.

then extruded sequentially through a Leistritz Doppelsckenextruder LSP125/1 twin screw extruder. The extruded mix was then compression moulded into square faced blocks.

In Table 2 the samples shown in bold italics are those which passed the simple drop test. As can be seen the blocks made with smaller amounts of fly ash were less likely to break when dropped at −20°C. Below 10% the blocks all passed. Above 25% the blocks all failed. This would suggest blocks bases could be manufactured with between 10 and 25%. The small number of blocks produced precludes calculating the probability of a block failing the tests or narrowing down the range at which fly ash could be added. The failure of block bases made with larger amounts of fly ash indicates that fly ash causes the material to become more

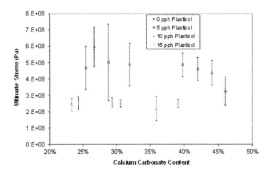

Figure 5. Plot of ultimate tensile stress for PVC vs. calcium carbonate content.

brittle and weaker. In this sense it is not dissimilar to the calcium carbonate.

Scanning electron microscope (FEI Quanta FEG 600 Environmental Scanning Electron Microscope) images were taken of the fracture breakages from the samples. These Fracture breaks were prepared by cooling sample using liquid nitrogen and then snapping them. Figure 7 shows the sample prepared with no fly ash; Figure 8 shows Sample 5 with 10% fly ash/10% plastisol and Figure 9 shows Sample 7 with 15% fly ash/10% plastisol.

In Figures 8 and 9 spheres from the fly ash were clearly visible and the distribution of particles appeared to be largely uniform; as expected the fly ash particles were more numerous with the samples prepared with larger amounts of fly ash. In general it is feasible to utilize this material as a filler in these systems, although we note that there might

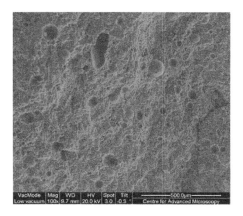

Figure 9.   SEM micrograph of Sample 5 with 15% fly ash.

be some abrasive properties associated with silica and other materials present in the fly ash. This may prove problematic in some moulding processes.

6   CONCLUSIONS

Recycled PVC is generally considered to be of poorer quality than virgin PVC; however, specific properties of materials can be controlled by varying the additives. A knowledge of the make-up of recycled feedstocks is crucial in the design of any process involving recycled materials and careful storage and transportation to eliminate water contamination is important. Even in recycled materials additional additives and fillers to control mechanical and other properties is important but this can often be achieved at a relatively low cost.

Figure 7.   SEM micrograph of Sample 1 (Table 2) with no fly ash.

REFERENCES

Borrell Fontelles, J. 2005. Directive 2005/84/EC of the European Parliament and of the Council.

Brebu M., *et al.*, 2000. *Polymer Degradation and Stability*, 67(2): p. 209–221.

Carroll W.F. Jr.,, Elcik, R. G., Goodman, D., *Polyvinyl Chloride*, in *Plastics Recycling*, Ehrig R.J., Editor. 1992, Carl Hanser Vrlag: München. p. 133.

Messoria, M. Toselli, M. Pilati F., E. Fabbri E., Fabbri P., Pasquali L., & Nannarone S. 2004. *Polymer* 45 () 805–813.

Scheirs J., 1998, *Polymer Recycling*. Polymer Science, ed. J. Scheirs., Chichester: Wiley.

Sole, B.M. and A. Ball, *On the abrasive wear behaviour of mineral filled polypropylene,* 1996. Tribology International,. 29(6): p. 457–465.

Wikes C.E., Summers, J.W., Daniels, C.A., 2005. *PVC Handbook*.: Hanser Gardner Publications, Cincinnati.

Wypych G., 2008 in PVC Degradation and Stabilization, Chemtec Publishing, Toronto, Canada.

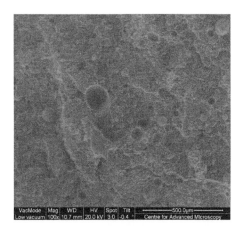

Figure 8.   SEM micrograph of Sample 5 with 10% fly ash.

Smart design

Green Design, Materials and Manufacturing Processes – Bártolo et al. (eds)
© *2013 Taylor & Francis Group, London, ISBN 978-1-138-00046-9*

Mass customization of ceramic tableware through digital technology

Eduardo Castro e Costa & José Pinto Duarte
CIAUD, Faculdade de Arquitetura, Universidade Técnica de Lisboa, Portugal

ABSTRACT: Research on mass customization of ceramic tableware is presented. We document a first attempt to implement a system of mass customization, through the development of a design system into a shape grammar that automatically generates tableware elements, its computational implementation through parametric modelling, and subsequent production through digital fabrication technology, namely 3d printing. This first experience acts as a mockup for intended subsequent research involving local ceramic industry.

1 INTRODUCTION

Mass customization is pointed as a direction towards competitiveness (Pine, 1993), which is essential to guarantee the sustainability of companies. This paradigm allows companies to offer innovative and differentiated products to ever more demanding consumers. Research is presented on the application of the mass customization paradigm to the production of ceramic tableware. The aim of this research is to develop a system that allows the end users to customize their own tableware set. Traditionally established in Portugal, companies in the ceramic industry can benefit from the competitiveness granted by mass customization.

According to Duarte (2008), the implementation of a mass customization system implies development on three fronts: a design system, which encapsulates the rules that govern the shape and decoration of the tableware elements, generating the corresponding digital models; a production system, which allows to automatically materialize these digital models into usable tableware elements; and a computational system, which implements the design system and articulates it with the production system.

This paper presents the development of a first mockup, which poses as a proof of concept for implementing a mass customization system. In this mockup, the Shape Grammar apparatus, invented by Stiny and Gips (1972) and further developed by Stiny (1980), was used for developing the design system, encoding the rules for the shape generation of the tableware elements into a parametric shape grammar. The grammar was then computationally implemented using the visual programming interface Grasshopper. The implemented system allowed using the shape grammar rules and

manipulating the corresponding parameters in order to generate instances of tableware elements and the equivalent digital models. These models could then be automatically produced using 3D Printing technology.

2 DEVELOPMENT OF THE DESIGN SYSTEM

The development of the design system is the first step into the implementation of a mass customization system (Duarte, 2008). The objective of the design system is to generate sets of tableware elements according to a certain style, defined by the user. The versatility of such a system will be determined by the number of styles that it can consistently generate.

However, as a starting point, the design system was developed for generating one single existing collection, although keeping in mind its future extension into other styles. As a case study, in order both to guide and test the design system, a collection of tableware elements was selected, comprising basic types of plates and cups. The design system should be able to generate all the elements of the selected collection, in terms of both shape and decoration.

The selected tool for encoding the design system was a shape grammar. As defined by Knight (2000), "a shape grammar is a set of shape rules that apply in a step-by-step way to generate a set, or language, of designs". The developed grammar is further explained.

2.1 *Functional analysis*

The development of the shape grammar was preceded by a functional analysis. From observing the different types of elements within the selected collection (Fig. 1), and from trying to identify what is common among them and what distinguishes them from one another, it was considered that all the elements are, to a certain extent, topologically similar to each other.

For example, a charger plate is very similar to the dinner plate, only wider. And the soup plate is also similar to the dinner plate, only taller, so it can hold liquids in it. These dimensional differences can be described by parametric variations, justifying the use of a parametric shape grammar, and the subsequent use of parametric modelling for the implementation of the design system.

Conclusions from the functional analysis are expected to be verified for other collections. However, what distinguishes collections from one another is their style. Hence, style will be the primary subject of the shape grammar rules.

2.2 Shape grammar rules

Rules were developed to generate both the base shape and the decoration of the elements in the selected collection (Fig. 2).

The base shape can be described as a solid of revolution, since the selected collection features circular elements. Therefore, rules and derivation steps regarding base shape generation are represented by a bi-dimensional profile (Fig. 2, rules 1 to 7).

The definition of the profile is thus the result of a series of operations on the profile—such as subdivision, skewing and substitution—(Fig. 2, rule 1 to 7). The base shape corresponds to the surface generated by revolving the resulting profile.

In the selected collection, decoration is based on relief rather than on painting (Fig. 3). This was a specific criterion while selecting the collection to work with, so that focus would maintain on three-dimensional form, rather than to disperse over themes like color or texture.

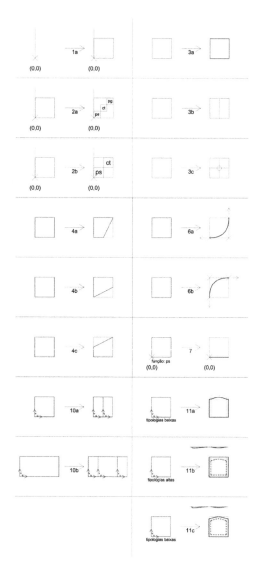

Figure 2. Rules of the shape grammar.

Figure 1. Original collection as the shape grammar corpus.

Figure 3. Decoration detail of the original collection.

468

Decoration is achieved through the application of rules to the resulting base shape surface, based on operations such as surface subdivision (Fig. 2, rules 10), and application of a formal motif onto the resulting subsurfaces (Fig. 2, rules 11).

2.3 *Derivation*

As mentioned before, to generate a result using a shape grammar, rules must be applied step-by-step. This sequential combination of rules is commonly referred to as a derivation, which can be considered to be the algorithm that generates a particular design within the language defined by the grammar.

Specific combinations of the rules generate the elements of the original collection, the corpus. In Figure 4 and Figure 5 a derivation is presented for the soup plate belonging to the selected collection.

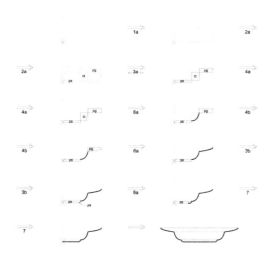

Figure 4.   Derivation of the base shape of a soup plate.

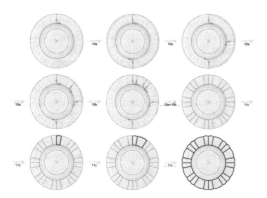

Figure 5.   Derivation of the decoration of a soup plate.

However, different combinations can generate solutions that are different from the original corpus, but that are still apparently part of the same language. Such an experience will be addressed further in this paper, when we present the generated digital models.

3   IMPLEMENTATION OF THE DESIGN SYSTEM

As mentioned before, the differences among tableware types can be described as parametric variations. Therefore, it made sense to implement the design system through parametric modelling. Therefore, the system was implemented in Grasshopper, a visual programming interface that interacts with modelling software Rhinoceros (Fig. 6).

A parametric model built in Grasshopper generates digital models, according to input parameters. It should be noted that the Grasshopper model is not considered to be an implementation of the shape grammar. However, if we consider that the result of the derivation of a parametric shape grammar is a parametric model, than we can argue that we are implementing a derivation.

Using the parametric model, two types of variations can be considered: parameters and rules. While, because of the very nature of the parametric model, parameter changes are dealt with automatically, rule changes have to be made manually. However, this manual manipulation of rules is manageable, since for now the number of rules is still small.

For purposes of illustrating the parametric model, and for later production, four instances of the soup plate were generated (Fig. 7). In all four instances parameters have been changed, concerning both base shape and decoration, and in two of the instances we applied rules different from the other two. Observing the results, we might say that the purely parametric variations appear to be more true to the selected collection, but while

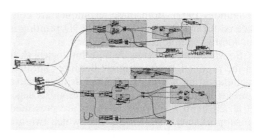

Figure 6.   Parametric model as partial implementation of the shape grammar.

Figure 7. Digital models of variations of the soup plate.

Figure 8. 3D printing of the soup plate variations.

variations in the shape grammar rules seem to generate solutions further apart from the original collection, they still look similar, hence belong to the same language.

4 TESTING WITH A PRODUCTION SYSTEM

From the digital models generated by the Grasshopper program, the corresponding physical model were produced using digital fabrication technology, namely 3D Printing (Fig. 8). 3D Printing is an additive manufacturing technique by which models are produced through addition of layers of material. One of the advantages of additive manufacturing is the ability to accurately produce complex shapes, such as curved surfaces, like the ones usually found in tableware. 3D Printing features especially high resolution, approximately 0.1 mm (3D Systems 2013).

However, 3D printing is very different from the processes traditionally used in manufacturing ceramic tableware, either handcrafted, or in an industrial context. While handcrafted, tableware elements are usually modelled in a potter's wheel. In fact, this can account for the traditionally circular form of tableware elements, such as plates or cups. In an industrial context, these elements are fabricated through the use of molds, adequate for the production of large numbers of objects (Fagundes 1997). Both these techniques are considered formative techniques, while 3D printing is an additive technique.

So why should we consider a technological shift in the production of ceramics? Because "additive manufacturing has great potential as an effective tool for more sustainable product design" (Diegel et al. 2010, 70). Several arguments are presented to support such a statement, from which two are noted. First, additive manufacturing is an adequate technique for producing mass-customized products, which by being suited to the customer's

needs, have their longevity extended, and thus their environmental impact is reduced. Also, when compared to other manufacturing techniques, namely subtractive techniques such as CNC milling, additive manufacturing produces very little waste (Diegel et al. 2010).

The application of 3D printing to ceramic production has already given its first steps. Although it has been recently published some experiences in this field (Huang and Hudson 2013), only a few have been applied to an industrial context (Hoskins 2012). We intend to explore this form of manufacturing, assessing its suitability as a production system for the mass customization of ceramic tableware.

5 RESULTS AND FURTHER DEVELOPMENTS

The results of this mockup indicate a good probability of implementing mass customization at a larger scale.

The next step of the investigation should be to actually apply the system to handcrafted ceramics. Using the 3D printed models, molds can be cast, and later be used to produce corresponding ceramic replicas. This approach will be useful to assess mistakes or frailties in the shapes generated by the design system. These frailties should be analyzed in order to correct the design system, and therefore eliminating them. This feedback approach is intended to optimize the whole mass customization system.

We also intend to involve a local company in order to develop the mass customization system at an industrial level. This collaboration with the industry should be useful for gathering information necessary for the development of the system, be it technical data, best practices, or even information regarding organizational processes. On the other hand, the company can benefit first hand from the development and implementation of the system itself.

REFERENCES

3D Systems. 2013. "ZPrinter® 350." Accessed February 27. http://www.zcorp.com/en/Products/3D-Printers/ZPrinter-350/spage.aspx.

Diegel, Olaf, Sarat Singamneni, Stephen Reay, and Andrew Withell. 2010. "Tools for Sustainable Product Design: Additive Manufacturing." *Journal of Sustainable Development* 3 (3) (August 19): P68. doi:10.5539/jsd.v3n3P68.

Duarte, José Pinto. 2008. "Synthesis Lesson—Mass Customization: Models and Algorithms—Aggregation Exams". Agregação, Lisboa: Faculdade de Arquitectura, Universidade Técnica de Lisboa.

Fagundes, Arlindo. 1997. Manual Prático De Introdução à Cerâmica. Caminho.

Hoskins, Stephen. 2012. "Towards a New Ceramic Future." *Ceramic Review* 255: 64–65.

Huang, Mary, and Alan Hudson. 2013. "Sake Set Creator." *Shapeways.com.* February 14. www.shapeways.com/creator/sake-set/.

Knight, Terry W. 2000. "Shape Grammars in Education and Practice: History and Prospects." *Shape Grammars in Education and Practice: History and Prospects.* http://www.mit.edu/~tknight/IJDC/.

Pine, B. Joseph. 1993. Mass Customization: The New Frontier in Business Competition. Harvard Business Press.

Stiny, G. 1980. "Introduction to Shape and Shape Grammars." *Environment and Planning B: Planning and Design* 7 (3): 343–351. doi:10.1068/b070343.

Stiny, George, and James Gips. 1972. "Shape Grammars and the Generative Specification of Painting and Sculpture." In *Information Processing 71*, edited by C.V. Freiman, 1460–1465. Amsterdam: North Holland.

Green Design, Materials and Manufacturing Processes – Bártolo et al. (eds)
© 2013 Taylor & Francis Group, London, ISBN 978-1-138-00046-9

Robotically controlled fiber-based manufacturing as case study for biomimetic digital fabrication

N. Oxman, M. Kayser, J. Laucks & M. Firstenberg
Massachusetts Institute of Technology, Cambridge, Massachusetts, USA

ABSTRACT: The research explores the process of silk deposition generated by the silkworm *Bombyx mori* and proposes a novel fiber-based digital fabrication approach inspired by its biological counterpart. We review a suite of analytical methods used to observe and describe fiber-based constructions across multiple length-scales. Translational research from biology to digital fabrication is implemented by emulation in the design of fiber-based digital fabrication techniques utilizing a KUKA robotic arm as a material deposition platform. We discuss the ways in which the silkworm *Bombyx mori* constructs its cocoon and scaffolding structure and speculate regarding the possible applications and advantages of fiber-based digital fabrication in the construction of an architectural pavilion as case study.

1 INTRODUCTION

1.1 *Background*

Additive manufacturing and digital fabrication processes such as 3D-Printing typically involve the layered deposition of materials with constant and homogeneous physical properties (Gibson *et al.*). Yet all natural materials and biological systems are made of fibrous structures that are locally aligned and spatially organized to optimize structural and environmental performance (Mitchell and Oxman 2010). Furthermore, construction processes found in the animal kingdom such as woven spider nets or aggregate bird's nests are characterized by the animal's ability to generate, distribute, orient, dandify and assemble fiber-based materials (Benyus 2002; Hansell 2005). As a result biological structures (including animal architectures) are considered highly sustainable natural constructions (Benyus 2002; Robbins 2002; Hansell 2005). Many of these constructions are "designed" by insects well known for their ability to construct highly sustainable structures made of fiber composite materials such as silk (Sutherland and Young *et al.* 2010).

The paper reviews a suit of analytical protocols designed to examine the process of constructing a silk-cocoon by the silkworm. Following we demonstrate a set of design tools created to reconstruct the cocoon in various length-scales using a 6-axes KUKA robotic arm.

1.2 *Problem definition*

Fiber-based 3D constructions with spatially varying composition, microstructure and fiber-orientation are omnipresent in Nature (Seidel and Gourrier *et al.* 2008). In contrast to natural materials and biological structures, industrially fabricated constructions, such as concrete pillars and façade panels are typically volumetrically homogenous (Oxman 2011). Additive manufacturing platforms such as 3-D printing provide for the generation of highly complex geometrical forms. However, despite their formal complexity, these products and building components are still typically manufactured from materials with homogeneous properties. Compared with biologically constructed fiber-based materials, homogenous constructions fabricated using additive manufacturing technologies are much less sustainable: from a material perspective—homogeneous materials offer less potential for structural optimization; and from a fabrication perspective—additive manufactured components are constructed in layers, relying on the deposition of significant amounts of wasted support material (Oxman and Tsai *et al.* 2012).

2 FIBER-BASED CONSTRUCTION IN NATURE

2.1 *Introducing the silkworm bombyx mori*

Silk is one of the most ancient, expensive, and highly valued materials in the world (Omenetto and Kaplan 2010). It has many applications in textile, medicine, and industry (Frings 1987). The silk produced by the domesticated silkworm *Bombyx mori*. It constructs its cocoon using composite fibrous material made of fiber (fibroin) and binder (sericin) in order to provide shelter during its transitional stage of pupation (Zhao and Feng *et al.* 2005; Rockwood and Preda *et al.* 2011). A single fiber is used to construct the cocoon, which is approximately one kilometer in length. The silkworm starts by spinning a scaffolding structure in any three-dimensional space given it can triangulate and attach its fibers parasitically to its immediate environment. While spinning this scaffolding it will close in onto itself to begin to construct its cocoon within the scaffolding structure. The cocoon itself can be characterized by changes in fiber quality transitioning from the inner layers to the outer ones (Zhao and Feng *et al.* 2005).

2.2 Silkworm motion tracking

Various methods for motion tracking data were considered. Popular methods include visual routines using cameras and/or sensor-based systems (Black and Yacoob 1995). The fact that the silkworm cocoons itself *within* its structure eliminated the use of video-based techniques unable to capture construction processes internal to the cocoon. The challenge was to create a motion-tracking rig on a very small scale that could capture motion data of the silkworm from *inside* the cocoon as well.

An experimental sensor rig 40 mm × 40 mm × 40 mm in dimension was developed using magnetometer sensors placed on 3 planes of the cube. This allowed for data capturing from a 1 mm × 2 mm magnet attached to the silkworms' head (Fig. 1). After attaching the magnet to the silkworms' head, the silkworm was placed within the described space. As expected the silkworm attached its fiber scaffolding structure to the walls of the described

rig and constructed its cocoon within this defined space.

From the collected data set of Cartesian x, y and z points, a point cloud was visualized (Generative Components Software) as a path, sequenced in time as seen in Figure 2.

2.3 Motion tracking data evaluation and speculation for robotic emulation on larger scale

The captured data demonstrates a clear overall cocoon shape constructed from over 1,000,000 points. The detailed motion path is slightly disrupted by the polar positioning of the magnet as the silkworm spins its cocoon.

This experiment establishes the possibility to convert biological data into robotic motion. The silkworms' actual motion path can be translated into a readable language (Cartesian x, y, z points) and passed on to a robotic arm or any multi-axis material deposition system. This in turn can inform the robotic arm movement in terms of distribution of fiber structures as well as precise fiber placement as this work-in-progress path simulation demonstrates.

2.4 SEM imaging across multiple scales

In order to investigate local fiber placement of the cocoon as well as to gain a better understanding of the scaffolding structure, SEM images were taken of the outer layer of the cocoon.

Figure 3 shows the overall all cocoon form. Its shape and curvature radii largely depend on

Figure 1. Motion tracking of the silkworm *Bombyx mori* using magnetometer sensors and a 1 mm × 2 mm magnet.

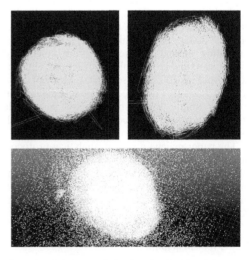

Figure 2. Motion path (top) and point cloud (bottom) in Generative Components (GC) software.

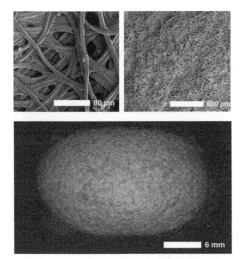

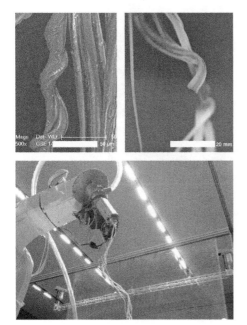

Figure 3. SEM images of the silkworm *Bombyx mori* cocoon across multiple scales (Photo credit: James Weaver, 2013).

Figure 4. SEM image of silk fibers (top left), HDPE extrusion (top right), and thermoplastic extruder on KUKA robotic arm (bottom).

the silkworm's own body structure (i.e. its bending radii) as well as its overall length. This could also inform robotic principles in terms of using the robotic arm's reach envelope as a limiting factor or constraint. Furthermore, fiber self-alignment

shown in Figure 4 constitutes a significant aspect of fiber-based construction, as the stresses seem to be locally equalized across varying fiber distribution on global scale. This aspect is further discussed in section 3 below.

3 FIBER-BASED DIGITAL FABRICATION

3.1 Strategies for robotic fiber-based construction on larger scales

Based on the analytical protocols developed and reviewed above, a synthetic approach for translating the biological process into a digital fabrication protocol was developed. Several synthesis methods were developed each mimicking a distinct aspect of the silkworm's fiber placement process and its material organization strategies across scales. Three robotic-end-arm-tools were developed to test and analyze novel avenues for fiber-based robotic construction inspired by the silkworm's construction methods.

The first approach explores 3D digital construction using a single fiber or a combination of several composite fibers forming a single structural element. A thermoplastic extruder was developed in order to accomplish fiber or multi-strand continuity.

The second approach explores the dual stages in the silkworm's cocoon construction process: (a) parasitic construction and (b) cocoon spinning.

3.2 Synthesis 1: Multiple strand thermoplastic extrusion

A 'Free-Form-Printing' tool—inspired by concepts of fiber self-alignment—was developed and built. A specially designed nozzle for a custom-built High-Density-Polyethylene (HDPE) thermoplastic extruder was built to allow for local self-alignment of individual strands (Fig. 4). Self-alignment of fibroin and sericin as observed in the silk fiber inspired the design of an extruder nozzle, which combines fiber and binder as a single material system.

The extruder nozzle contains multiple outlets laid out in a circular configuration around a single central and larger opening. In this way the HDPE polymer can flow through, before being rapidly solidified by active air-cooling. In this method, the central strand is stabilized by the surrounding thinner strands as well as the outer strands reconnecting to previously extruded strands in close proximity to the overall structure (Rauwendaal 2001). Figure 4 compares between the biological extrusion process using silk and its digital-fabrication counterpart using composite HDPE.

Based on the mono-material synthesis approach using thermoplastic further experimental synthesis approaches were developed and simulated.

3.3 *Synthesis 2: Fiber placement tools*

In this approach the silkworm cocoon construction is divided into two stages: the first being the parasitic scaffolding and the other being the cocoon construction process itself, as the enclosure within the scaffolding. Fiber-placement in these two phases of the silkworm cocoon construction differs greatly in material quality, organization and function. As the silkworm constructs the scaffolding it "parasites" to its environment, attaching its fibers and pulling it across, connecting to another part of the space repeatedly, building up a three-dimensional web. In the second stage it builds its cocoon in figure-8 pattern, building up wall thickness for the cocoon over

time by constantly reconnecting the fibers locally inside the previously built scaffolding.

3.4 *Synthesis 3: Parasitical attachment and fiber pulling*

Two robotic end arm tools were developed. The first pulls a continuous 2 mm polypropylene fiber through epoxy resin and is used in combination with a robotic rig describing the reach envelope of the robotic arm. This 'scaffolding tool' is designed to attach the resin-soaked fiber from point to point on the provided external rig. This system relies on a modular hook system on which the robotic arm can attach the fiber 'parasitically' to the hooks.

As seen in Figure 5 the scaffolding of the silkworm *Bombyx mori* consists of a loose-networked structure, which relies on an external three-dimensional space to which it

Figure 5. Fiber pulling tool (top), fiber pulling tool on KUKA robotic arm with external hook structure (middle), comparison of epoxy resin-soaked polypropylene twine (bottom right) and SEM image of scaffolding silk fiber (bottom left).

attaches itself to. For the synthesis of this process we developed a rig to which the robot can attach fiber whilst pulling the fiber through a resin bath right at the tool head not unlike the biological process.

3.5 *Synthesis 4: Fast deposition tool*

A secondary end-arm tooling was developed, which is a combination of depositing fiber in controlled speeds while spraying binder onto the fiber. This method also requires a robotic rig and is used in accordance with a previously made scaffolding structure to adhere to, which is described above. The scaffolding structure would act as a mold for the 'cocoon' shell to be placed upon, and can vary in density according to previously mentioned distribution maps acquired through motion tracking. This tool places a fiber on top of this scaffolding structure by pushing 1 mm polypropylene string by means of two motorized rollers (Fig. 6) whilst spraying them with contact adhesive. The speed of the deposition and the robotic movement must be synced in order to achieve varying densities. These fibers build up a layer of fiber at a loose configuration based on the 8-figure patterns. Depending on the robotic movement and speed, varying densities and gradients can be achieved.

The experiments demonstrate that an external structure (equipped here with hooks) is required in order for the robotic tool to "print" with fibers. These works-in-progress demonstrate that it is possible to create fiber-based rigid structures, which may be use for the manufacturing of products such as lightweight furniture and building components. The secondary process of fast fiber deposition demonstrates the possibilities for creating

Figure 6. Fast deposition tool on KUKA robotic arm, SEM Image of outer silkworm cocoon surface, composite material from polypropylene twine and contact adhesive.

additional structural integrity in a component as well as varying properties across its inner wall. The combination of these two processes could lead to a novel and customizable robotic construction process for large-scale fiber-based composite parts.

4 DIGITAL FABRICATION OF FIBER-BASED CONSTRUCTIONS AS A CASE FOR SUSTAINABLE MANUFACTURING: RESULTS, CONCLUSION AND OUTLOOK

The research demonstrates the need for sophisticated analytical tools in translational research of fiber-based systems across scales. Such analytical protocols are required for the synthesis of robotic fabrication processes via the development of robotic-end-arm tooling to facilitate experiments in the field of sustainable digital fabrication. Two synthetic approaches in digital fabrication were presented using three distinctive custom end arm tools.

Further research into the topic will include the combination of the described robotic-end-arm-tools with motion tracking data, enabling direct comparisons between micro scale structures and their robotic macro-scale counterparts. The process of data collected from the biological world combined with experiments into novel fiber placement methods will lead to integrative and sustainable fiber-based manufacturing using Nature as inspiration and technological advances as facilitators.

ACKNOWLEDGEMENTS

The research was carried out by the Mediated Matter group at the MIT Media Lab. Additional Research Assistants who have contributed to this work include Carlos Gonzales Uribe and Jorge Duro-Royo.

This work was supported in part by NSF EAGER Grant Award No. #1152550 "Bio-Beams: Functionally Graded Rapid Design & Fabrication". We gratefully acknowledge the support of the Institute for Collaborative Biotechnologies through Grant No. W911NF-09-0001 from the U.S. Army Research Office and the support of MIT Institute for Soldier Nanotechnologies (Contract No. DAAD-19-02-D0002). The content of the information does not necessarily reflect the position or the policy of the Government, and no official endorsement should be inferred.

We would like to acknowledge the assistance of James C. Weaver for SEM & CT Imaging; Prof. Fiorenzo Omenetto, Nereus Patel, Leslie Brunetta, and Catherine Craig for their consultation; Steven

Keating and Dave Pigram for Robotic Control and Ben Peters for mechanical support. In addition we would like to thank Undergraduate Research Assistant Ayantu Regassa. The authors also wish to thank our network of colleagues and advisors of MIT including Prof. W. Craig Carter, Prof. Christine Ortiz, Prof. Mary C. Boyce, Prof. Lorna Gibson and Prof. David Wallace.

REFERENCES

Benyus, J.M. 2002. *Biomimicry: Innovation inspired by nature*, New York: William Morrow Paperbacks.

Black, M.J. and Yacoob, Y. 1995. Tracking and recognizing rigid and non-rigid facial motions using local parametric models of image motion. *Computer Vision, 1995. Proceedings., Fifth International Conference on,* Ieee Computer Society Washington, DC, USA

Frings, G.S. 1987. *Fashion: from concept to consumer,* London, Prentice-Hall.

Gibson, I. Rosen, D.W, and Stucker, B. 2010. *Additive manufacturing technologies.* New York, Springer US.

Hansell, M.H. 2005. *Animal architecture*, Oxford University Press Oxford, UK.

Mitchell, W.J. and Oxman, N. 2010. *Material-based design computation*, Massachusetts Institute of Technology. Munich: Carl Hanser Verlag.

Omenetto, F.G. and Kaplan, D.L. 2010. New opportunities for an ancient material. *Science* 329: 528–531.

Oxman, N. 2011. Variable property rapid prototyping. *Virtual and Physical Prototyping* 6(1): 3–31.

Oxman, N. Tsai, E. *et al.* 2012. Digital anisotropy: A variable elasticity rapid prototyping platform: This paper proposes and demonstrates a digital anisotropic fabrication approach by employing a multi-material printing platform to fabricate materials with controlled gradient properties. *Virtual and Physical Prototyping* 7(4): 261–274.

Rauwendaal, C. 4th edition (2001). *Polymer extrusion*: 57–63.

Robbins, J. 2002. Second Nature. *Smithsonian* 33(4): 78–84.

Rockwood, D.N. Preda, R.C. *et al.* 2011. Materials fabrication from *Bombyx mori* silk fibroin. *Nature protocols* 6(10): 1612–1631.

Seidel, R. Gourrier, A. l. *et al.* 2008. Mapping fibre orientation in complex-shaped biological systems with micrometre resolution by scanning X-ray microdiffraction. *Micron* 39(2): 198–205.

Sutherland, T.D. Young, J.H. *et al.* 2010. Insect silk: one name, many materials. *Annual review of entomology* 55: 171–188.

Zhang, Y.-Q. 2002. Applications of natural silk protein sericin in biomaterials. *Biotechnology advances* 20(2): 91–100.

Zhao, H.-P. Feng, X.-Q. *et al.* 2005. Mechanical properties of silkworm cocoons. *Polymer* 46(21): 9192–9201.

Green Design, Materials and Manufacturing Processes – Bártolo et al. (eds)
© 2013 Taylor & Francis Group, London, ISBN 978-1-138-00046-9

Freeform 3D printing: Towards a sustainable approach to additive manufacturing

N. Oxman, J. Laucks, M. Kayser, E. Tsai & M. Firstenberg
Mediated Matter Group, MIT Media Lab, Cambridge, Massachusetts, USA

ABSTRACT: Most additive manufacturing technologies, such as 3D printing, utilize support materials in the fabrication process. Beyond the technical challenges of support removal, these materials are wasteful—increasing fabrication and processing time while impacting quality. This paper presents "Freeform Printing", a novel design approach for 3D printing without additional auxiliary structures. A 6-axis KUKA robotic arm is repurposed as a 3D printing platform onto which custom-designed thermoplastic extruders are attached. We demonstrate freeform extrusion using a round nozzle attached to an active air-cooling unit, which solidifies the material upon extrusion. In addition, we present a method for printing geometrically complex structures using a multi-strand extrusion nozzle. The experiments presented in this paper, combined with their evaluation and analysis, provide proof-of-concept for Freeform Printing without support materials. They represent a sustainable approach to additive manufacturing and digital fabrication at large, and point towards new possible directions in sustainable manufacturing.

1 INTRODUCTION

1.1 *Biological systems*

Material systems in Nature are typically composed of graded composites grown and adapted from a single material system rather than an assembly of parts (Oxman et al. 2011). More so, growth in the plant and animal kingdoms rarely follows rectilinear paths confined to single planes but instead spreads through space in response to various factors and stimuli (Braam 2004). Identical systems and matching fabrication processes can result in substantially different structures depending on external environmental constraints. Such is the case with *Cecropia* silkworms, which can produce silk cylinders and sheets in addition to the canonical silk cocoon (Van der Kloot & Williams 1953).

In Nature, form typically follows function such that the composition and properties of a material system vary locally as part of the fabrication process. *Bombyx mori* silkworms vary the porosity and amount of sericin—a bonding agent between fibers found throughout the layers of their cocoons producing a highly bonded network in inner layers compared with outer layers (Chen et al. 2012). Orb-weaving *Araneus diadematus* spiders spin a wide variety of different silks, ranging from the incredibly stiff and strong dragline-silk to the glue-coated and highly extensible viscid silks (Guerette et al. 1996).

In all of Nature's systems, the optimization of material usage and hence metabolic cost plays a necessary role. In tensile systems such as spider webs, there is little material waste as structural support is an integral part of the web design (Gosline et al. 2004).

2 BACKGROUND

Additive manufacturing technologies emerged in the 1980s as a promising method for fabrication and construction automation (Jacobs 1992). Today, these additive fabrication technologies operate across a wide variety of materials and are used in applications ranging from medical implants to large-scale prototyping. Fused Deposition Modeling (FDM) systems in particular are found in both hobbyist and professional 3d printing platforms such as the MakerBot Stepstruder, and more professional grade systems such as the Stratasys Dimension 3d printer. Consistent to all of these systems is the need to use support materials to fabricate certain thin-walled or particularly complex geometries (Levy et al. 2003).

3 GOALS AND OBJECTIVES

Interest in Freeform Printing was inspired by the concept of self-supporting fiber-based construction in contrast to the scaffold-based robotic weaving and wrapping explored previously (Tsai et al. 2012). Here, various natural examples were examined, including silk producing bi-valves,

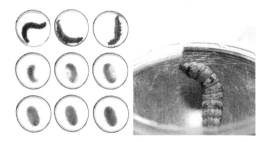

Figure 1. **A)** (L.) *Bombyx Mori* silkworm cocoon sequence. **B)** (R.) *B. Mori.* Silkworm spinning in half-sphere.

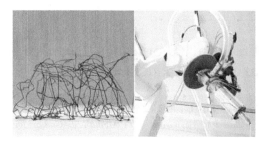

Figure 3. **A)** (L.) ABS Filament freeform extrusion. **B)** (R.) Custom HDPE extruder mounted on KUKA.

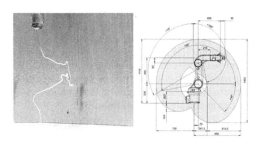

Figure 2. **A)** (L.) Single strand of High-Density Polyethylene (HDPE) freeform print. **B)** (R.) KR5 sixx R850 robotic work envelope.

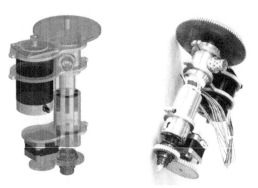

Figure 4. **A)** (L.) Digital model of extrusion tool. **B)** (R.) Extrusion tool with variable tip.

spiders, and silkworms both wild and domesticated (Fig. 1A). One of the guiding principles observed among the different natural systems was the concept of producing a continuous fiber based construction method. (Fig. 1B).

Initial extrusion of a mono-material was pursued through the employment of a 6-axis KUKA KR5 sixx R850 robotic arm as a means to increase the scale capacity (build volume) of the final result (Fig. 2). The ability to create knitted or woven structures requires high levels of robotic dexterity or multiple agents, otherwise resulting in the process being rate limited by fiber component length and splicing or tangling of the material.

4 METHODOLOGY

4.1 *Initial exploration*

Preliminary tests explored the use of a Stepstruder tool-head with MakerBot Acrylonitrile Butadiene Styrene (ABS) filaments to test the concept of drawing a fluid plastic material through space (Fig. 3A). As a departure from small-scale ABS tests, a custom extrusion tool for attachment to

a robotic arm was developed. The design of the tool was based on research into current extrusion devices in industrial applications. The core of the tool is a large 20.6 mm diameter auger-type masonry drill bit cut to a length of 184.5 mm. The goal was to make the housing as compact as possible in order to achieve a high degree of control over the maneuverability of the tool in the robot workspace. (Fig. 3B).

4.2 *Tool development*

A 3d model of the tool was developed in Rhinoceros 5 for design development, visualization and fabrication (Fig. 4A). The housing for the extrusion chamber was constructed out of 50 mm diameter aluminum round stock.

The body and other cylindrical parts of the extruder were turned from round stock on a CNC lathe. The housing was bored to accept the auger bit and then machined from the other end to allow the extruder to accept various interchangeable extrusion tips via three setscrews. The output end of the tool also retained a substantial wall thickness between the bore and the exterior to allow for the placement of up to twelve cartridge-heating elements. Near the top of the tool, the auger bit was machined with an indexed shank to accept a series of water jet cut aluminum spur gears. Near the top of the housing at the furthest point from the heater elements an opening was created to feed plastic pellets. Aluminum motor mounts were created using a flexural design to carry the NEMA 23 stepper motor. The motor is controlled by a Gecko G201X Digital step driver to drive the gears turning the auger bit. When the auger bit is turned by the gears, a steady supply of plastic pellets is fed from a hopper through flexible tubing via a venturi for material advancement (Fig. 4B). As the pellets are transferred down through the housing, the heater cartridges heat the pellets to about 130° C as regulated by an Arduino-controlled thermistor, while the downward pressure advances the molten material out through the selected tip.

4.3 Material tests

For the proof-of-principle experiments, we chose High-Density Polyethylene (HDPE), commonly used today for a variety of applications, ranging from storage containers and furniture products to professional lenses and pipes. In contrast to Low-Density Polyethylene (LDPE), the HDPE polymer backbone has no branches, yielding stronger intermolecular forces and denser packing. It is therefore more crystalline and exhibits a higher ratio of tensile strength to density—a property crucial to its ability to support itself during printing. In addition, its relatively low melting temperature of 130° C allowed us to melt, extrude, and harden it in the air using a compact setup that is easily mountable on the robotic arm.

4.4 Tip development

A variety of tips were explored and developed based on material properties and deposition processing constraints. The initial tip was developed as a variable diameter and cross section tip (Fig. 5A). With an additional stepper motor mounted near the bottom of the extruder this tip is able to vary between a 10 mm round extrusion to an 8 mm triangulated extrusion profile.

Following initial experiments with the variable tip (Fig. 5B), a series of interchangeable tips were developed, including two tapered single diameter extrusion tips of different length. The diameter of these single extrusion tips consisted of a 3 mm extrusion hole resulting in a 3.5 mm final extrusion.

Additional tips were developed to enable more complex extrusion profiles. For example, one of the tips was designed with a flat 'ribbon-type'

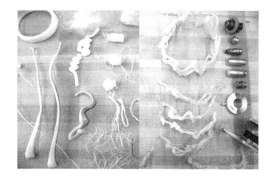

Figure 5. **A)** (L.) Tubular and triangular cross-section extrusions. **B)** (R.) Custom extrusion tips & freeform tests.

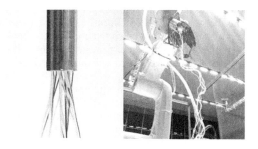

Figure 6. **A)** (L.) Multi-strand HDPE extrusion close up. **B)** (R.) Multi-strand HDPE extrusion in space.

extrusion cross-section. The extrusion clearance measurements were 3 mm by 16 mm and resulted in a ribbon extrusion of 3.5 mm by 16.25 mm. Another tip enabled the generation of a hollow tube-like extrusion with a series of internal fins allowing the molten plastic to flow around and reconnect between the interior walls of the tip and a cylinder shaped interior wall. Advanced versions of this tip incorporated a multi-strand approach. Two multi-strand tips were developed, one with a variety of self-similar holes and another with varying holes. The holes of the second tip contained larger diameter strands on the interior retaining heat for reconnection; and thinner strands to cool more quickly to support the printing in 3D space (Fig. 6A).

5 RESULTS AND DISCUSSION

5.1 Testing

The initial variable extrusion was found to be promising in modulating the extrusion profile from a complete round strand to a triangulated tapered design. The single strand extrusion profile proved to be the best balance of both heat and rapid cooling for initial print in-space experiments. The first of the multi-strand extrusion experiments proved to be a success and allowed for a quicker vertical extrusion test with the fibers cooling in air. The multiple-strands have the potential for multiple-strand bundling as a way of providing additional support (as the structure progresses in vertical space) and self-alignment due to the forces of gravity.

The final multi-strand printing nozzle was modified for the original design to be both longer and thinner for increased agility when printing more complex structures. One of the challenges in many of the freeform printing tests was to provide for material connectivity to plastic parts previously cooled and hardened. The revised multi-strand tip utilizes five thicker diameter holes at the center and along the outer perimeter, allowing for a balance between quickly cooling strands for structural support as the path is extruded and thicker slower cooling stands, which retain more heat and allow for better reconnection to existing cooled extrusions (Fig. 7A).

It was also found upon attempting more complex path planning and part printing exercises that a longer extrusion tip length allowed for much greater flexibility in the maneuverability of the extruder while attached to the six axis robotic arm (Fig. 7B).

5.2 Future development

While initial tests were highly dependent on developing custom tooling, future work may explore the further development of active heating and cooling at the tooltip, allowing for greater freedom of possible print geometries. Larger printed systems may be explored through leveraging the strand-like nature of larger printed 'cells' that utilize fibrous interfaces at their edges to assemble a larger fibrous aggregate system.

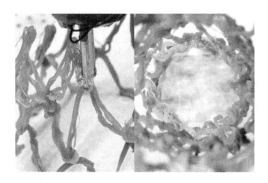

Figure 7.   **A)** (L.) Close-up of HDPE freeform prototype. **B)** (R.) Detail of finished HDPE freeform prototype.

Future tests could also benefit from substantially expanding the working envelope of the machine. This could, for instance, be a larger scale industrial robotic arm or an autonomous robotic system capable of transporting deposition material to the final desired location(s).

6   CONCLUSION

As of yet additive manufacturing methods typically rely on the use of support materials in the fabrication of certain geometries. These materials are used to support overhanging features and undercuts during the construction process, and are typically removed or dissolved upon completion of the print. In powder-based Selective Laser Sintering (SLS) processes the excess material acts as the support of the printed structure, whereas thermoplastic deposition and resin curing processes require additional support structures that are themselves printed.

The research and experiments presented in this paper focus on the intersection of biologically inspired design, fibrous construction, mono-material construction and the development of free-form printing. Applications for this novel process are varied and range from product fabrication to furniture and architectural scale construction. With the elimination of support material in the printing process, printing speeds are increased, and waste is eliminated. The experiments presented in this paper provide proof-of-concept for Freeform Printing without support materials. They represent a sustainable approach to additive manufacturing and digital fabrication at large, and point towards new possible directions in sustainable manufacturing.

ACKNOWLEDGEMENTS

The research was carried out by the Mediated Matter group at the MIT Media Lab. Additional Research Assistants who have contributed to this work include Carlos Gonzales Uribe and Jorge Duro-Royo.

This work was supported in part by NSF EAGER Grant Award No. #1152550 "Bio-Beams: Functionally Graded Rapid Design & Fabrication". We gratefully acknowledge the support of the Institute for Collaborative Biotechnologies through Grant No. W911NF-09-0001 from the U.S. Army Research Office and the support of MIT Institute for Soldier Nanotechnologies (Contract No. DAAD-19-02-D0002). The content of the information does not necessarily reflect the position or the policy of the Government, and no official endorsement should be inferred.

We would like to acknowledge the assistance of James C. Weaver for SEM & CT Imaging; Prof. Fiorenzo Omenetto, Nereus Patel, Leslie Brunetta, and Catherine Craig for their consultation; Steven Keating and Dave Pigram for Robotic Control, and Ben Peters for mechanical support. In addition we would like to thank Undergraduate Research Assistant Ayantu Regassa. The authors also wish to thank our network of colleagues and advisors of MIT including Prof. W. Craig Carter, Prof. Christine Ortiz, Prof. Mary C. Boyce, Prof. Lorna Gibson and Prof. David Wallace.

REFERENCES

Braam, J. 2004. In touch: plant responses to mechanical stimuli. *New Phytologist* 165(2): 373–389.

Chen, F., Porter, D. & Vollrath, F. 2012. Silk cocoon *Bombyx mori* Multi-layer structure and mechanical properties. *Acta Biomaterialia* 8(7): 2620–2627.

Gosline, J.M., DeMonet, M.E. & Denny, M.W. 2004. The structure and properties of spider silk. *Endeavour* 10(1): 37–43.

Guerette, P.A., Ginzinger, D.G., Weber, B.H.F. & Gosline, J.M. 1996. Silk properties determined by gland-specific expression of a spider fibroin gene family. *Science* 272(5258): 112–115.

Jacobs, P. 1992. Rapid prototyping & manufacturing: funda mentals of stereolithography, Sme.

Levy, G.N., Schindel, R. & Kruth, J.P. 2003. Rapid manufacturing and rapid tooling with layer manufacturing (LM) technologies, state of the art and future perspectives. *CIRP Annals—Manufacturing Technology* 52(2):589–609.

Oxman, N., Keating, S. & Tsai, E., 2011. Functionally Graded Rapid Prototyping, Innovative Developments in Virtual and Physical Prototyping: Proceedings of the 5th International Conference on Advanced Research in Virtual and Rapid Prototyping, Leiria, Portugal, pp. 483–490.

Tsai, E., Firstenberg. M., Laucks, J., Sterman, Y., Lehnert B., & Oxman N. 2012. CNSILK: Spider-Silk Inspired Robotic Fabrication of woven Habitats in *RobArc 2012 Robotic Fabrication in Architecture, Art and Design*. Vienna, Austria.

Van der Kloot, W. G. & Williams, C.M. 1953. Cocoon Construction by the Cecropia Silkworm II. The Role of the Internal Environment. *Behaviour* 5(3): 157–174.

Green Design, Materials and Manufacturing Processes – Bártolo et al. (eds)
© 2013 Taylor & Francis Group, London, ISBN 978-1-138-00046-9

How to design smart-buildings without smart-people?

Marie-Christine Zélem
University of Toulouse II, France

ABSTRACT: Despite of a sensitivity increasing to the environmental problems, despite of rising energy costs, despite of increasingly powerful of smart-equipments, despite of smart-zero-energy-buildings, house-holds don't really change their energy consumption. Intelligent buildings imagine that people will develop virtuous behaviors, and fully compatible with the objectives of innovative technologies: automation, communication systems, computers, controllers, cells, etc. Why the occupants did not become "smart-people"? From a sociological point of view, it is clear. The occupants are very difficult to comply with the instructions for use of technical systems, or they develop a range of behaviors cons-performance (rebound effects for example), that invalidates the project to reduce the consumption of which new buildings are invested in.

1 INTRODUCTION

To reduce climate change, states are engaged in significant reductions of greenhouse gas emissions (factor 4 in 2050). By 2020, Europe should reduce final energy consumption and emissions of greenhouse gases by 20%. It must increase the proportion of renewable energy to 20%. In France, the Grenelle gives a target of 38% reduction of energy consumption in buildings by 2020. The objective is to rehabilitate 400,000 units per year on energy. These commitments help to fight against the depletion of fossil resources and the growing importance of energy imports in the trade balance.

In France, cities participate in this energy transition through Territorial Climate Plan. Some of them are committed to meet these goals by 2020. They undertake projects that experiment a new approach of energy (eco-district). Energy transition is based on several elements: reducing the percentage of fossil fuels and producing renewable energy (solar, biomass, wind ...); regional planning and development of public transport and soft modes of circulation ... allowing lifestyles more energy efficient, programs of energy rehabilitation of buildings, on the principle of energy efficiency; energy sobriety.

Cities have discuss about and set up the devices to go towards what we name the energy efficiency. If the first three points have found their rhythm of cruise, we notice a real difficulty to incite the inhabitants to reduce their energy consumptions.

We can develop alternative systems of energy production. We know how to improve the energy performance of buildings. We know how to make electric equipments more successful. On the other hand, we don't still know how to make the practices of the consumers sober. This situation leads to an unexpected phenomenon: The arrival of the inhabitants in a house leads to a loss of energy performance. While the engineers thought that users would immediately become allies of technical devices, we observe that users don't develop behaviors consistent with the expected objectives: they are relatively incompetent to manage their energy consumption. What has happened?

1.1 *Smart-devices to support the energy transition?*

In France, the building sector is responsible for 18% of greenhouse gas (23.1% of national CO_2 emissions) and it consumes nearly 46% of primary energy, 70% of which are attributable to residential sub-sector (against 30% for the tertiary sector). According to the Grenelle Environment and in the perspective to support the transition energy project, planners, architects, engineers and technicians are invited to build neighborhoods or cities low-consumption: Eco-districts and eco-cities with buildings BBC[1], BEPOS[2] or zero-energy[3], which together contribute to the development of smart-cities.

These technical translations of the objectives of the Grenelle prefigure the the energy-efficient city of tomorrow. These projects are based in particular on the increase of smart equipments

---

[1]Low Consumption Building, ideally designed to consume around 50 kWh/m²/year.
[2]Energy POSitive building that produces more primary energy than it consumes.
[3]Buildings which respect the standard Minergie.

(thermostats, regulators, time switches and other technologies, including smart-meters ...) in residential and tertiary sectors. These technologies, ideally in network, are then supposed to allow buildings to use less energy, or to produce more than they consume. The challenge is to manage so-called "smart" energy. The term "intelligent" means that the expected goal is to achieve energy savings by making consistent sources of power generation equipment and consumers behavior. Besides these smart equipments, devices of centralized management ensuring the piloting of all the technical and electronic equipments of the building, are designed to realize until to 30% of energy savings, by optimizing the comfort of the occupants[4]. But if the efforts are successful from the standpoint of technology, both in the tertiary or residential sectors, occupants' behaviors don't really follow.

In fact, smart buildings, and smart-devices are based on the assumption that inhabitants are collaborating on the project and implement virtuous behaviors, consistent with the aims of innovative technologies in the housing: automation, communication systems, technical management and network security, smart meters, etc ... But are people really willing to become "smart people", to live in hyper-technified homes? This is the condition for the new generation of buildings, or renovated buildings on principles of efficiency, are eco-efficient and effectively contribute to the goal of reducing the energy consumption of a "factor 4" by 2020.

1.2 Smart-meters and smart grids to support policies

Control of Energy Demand involves being able to control its electricity needs. This means both reduce its overall consumption (playing on the energy efficiency of equipment and the sobriety of its behavior) and deport them off-peak hours. This is where the smarts meters may contribute to the regulation of energy consumption. They give to their users a dynamic representation of their consumption (display and recording of consumption, item by item, if it's possible) to enable them to adjust their ways to use their devices. Then smarts-meters invite consumers to think about ways to use their equipment. They are supposed to influence behavior. They function like a support system for energy management. Then, smart-equipment can take many forms: communicating appliances, interfaces to control equipment such as wall units and smart temperature sensors (electric heating, water heaters), intelligent plugs for dishwasher or

washing machine, specific interfaces for air conditioning, energy-box, CO_2 sensors, temperature sensors, humidity sensors, opening sensors, light sensors, motion detectors, wall outlets controlled, drive for electric meters (pulse, ICT) ... The overall challenge is to provide factual information in real time, or information about a given period, to enable consumers to monitor their ways of using energy, then change its when possible (by providing better equipment energy class, for example).

The implicit in this scenario is that people adhere to the project, they comply with the specific features and equipment available to them, they understand how to interact with the smart-meters and then, become smart-people. The underlying assumption is that the information thus provided can be sufficient in itself to induce behavioral changes more energy efficient.

1.3 Why consumers struggle to become sober?

Buildings are more energy efficient than expected. The results of the first office buildings High Environmental Quality show that the actual performance is much lower than expected (Carassus, 2011). This is confirmed in the residential housing (Sidler, 2011). Occupants do not comply with the instructions for use of technical systems. Sometimes they have cons-performance behavior. The "demonstrators" show energy consumption significantly larger than the values calculated in the laboratory at the design stage. This contradicts the project to reduce consumption to be achieved by new buildings. Occupiers' Liability is immediately pointed: they make resistance to change (Zélem, 2010).

Smart-technologies involve learning that can not be reduced to reading manuals. They are based on standardized designs of use, comfort, which refer to technical standards (19 ° in buildings) which tend to establish themselves as social norms. They often differ humans-being by increased automation, programming, intelligent devices ... People are considered like "external variables", as well as climatic data or energy prices. Their commitment to reduce their consumption issues, or to learn technical performance, are little considered: Either the integration of these two aspects of human behavior is determined by the distribution of manuals or, the user is left to his intuition and intelligence to adapt himself to the technical aspects of the building, or, it's necessary to think about devices to support change.

In any case, the world of technical thinks in terms of social acceptability. Engineers and technicians study how users use "correctly" or not technical devices designed to reduce the energy consumption of buildings. However, it is clear that even the simplest devices are struggling to generate

---

[4]www.actu-environnement, 31 oct 2011.

interest, either because they are not explicit in their handling, either because they don't provide value to their users (Van Dam *et al*, 2010), or because the issue of energy savings is less important than other concerns of life (employment, health, family ...). The debate should be refocused on the conditions of social integration of technologies. And scientists should think more in terms of socio-technical feasibility of low-energy housing project.

1.4 *The socio-technical feasibility at the heart of the energy transition*

In reality, energy consumption, so energy performance, refer to the combination of factors, both technological and cultural. Energy consumption results from the convergence of standards, practices and technological developments, that contribute to build a socially acceptable definition of normality in terms of comfort at home or at work (Bartiaux, 2011). Differences between the level of theoretical consumption of buildings and the real level of consumption of the users could be explained by failing to take into account the interconnection between technology and infrastructures, social norms and conventions about comfort, convenience and social practices.

These are not technical equipment consume, or even people. Energy expenditure is the product of the interplay of interactions between men and their devices, considering changing settings (configuration household, report to work, lifestyle, distinctive multi-equipment ...) or events (climate, local policies ...). This way of thinking about energy follows the theoretical vision of Bruno Latour (1993): « *No one has ever seen techniques and nobody has ever seen a human. We see only assemblies, crises, disputes, inventions, trade-offs, substitutions, translations, arrangements that involve increasingly complicated still more elements* ». However, in the project of smart-cities, housing is seen as a technical object, which, at best, involves intuition and, at worst, requires a simple instruction manual for its occupants. Then the relationship is asymmetrical. Apart from the technical couriers, it's organized around instructions, labels and energy meters against which users most often remain ignorant, so much so that Christian Morel (2007) speaks about "hell of buttons" and about the tragedy of manuals.

The "energy practices" of occupants of eco-efficient building must be considered in their sociotechnical context. According to Elizabeth Shove (2003), the technical context, social standards and practices refer to the 3Cs: comfort, cleanliness and convenience that are in constant co-evolution in the sense of increased energy consumption. However, the use of a low-energy building redefines the way to use appliances and well-being. It means there must be an experimental phase to test objective and subjective comfort in these new living spaces, which is rarely the case.

2 CONCLUSION: A SOCIO-TECHNICAL APPROACH TO REDUCE TECHNOLOGICAL UTOPIAS

The shared challenge of energy transition project is to provide more effective responses to contemporary energy issues. Building technologies, alone, will not be enough to achieve the goal of sustainable and sober society. They must integrate the social dimension of technology in a sociotechnical approach. This approach leaves its place to the technique as part of regulatory practices and consumption. It recognizes skills of human actors and their mastery of technology. It recognizes their intelligence and common sense. It also calls for developing a culture of energy conservation compatible with the culture of consumption. It is therefore to introduce a sociology of lifestyles and energy use at the heart of building engineering. At the same time, it's necessary to oblige technical and architectural objects to meet human actors, and to build better conditions for sociotechnical energy performance. Without this, sustainable society would be a technician utopia.

REFERENCES

Bartiaux, F., et al., 2006, Socio-technical factors influencing residential energy consumption, SEREC, Bruxelles.

Beslay, C., Gournet, R., Zélem, MC., 2013, Le bâtiment économe: une utopie technicienne ? in: J. Boissonnade (dir), "Sociologie des approches critiques du développement et de la ville durables", Paris, Ed. Petra, col. Pragmatismes.

Carassus, J., 2011, Les immeubles de bureaux « verts » tiennent-ils leurs promesses ? Performances réelles, valeur immobilières et certification « HQE Exploitation », Paris, CSTB.

Cayre, E., Allibe, B., Laurent, MH., Osso, D., 2001, There are people in the house ! How the results of purely technical analysis of residential energy consumption are misleading for energy policies, ECEE 2011, Sumer study, Energy efficiency first: the foundation of a low-carbon society.

Gras, A., Joerges, B., Scardigli, V., 1992, *Sociologie des techniques de la vie quotidienne*, Paris, L'Harmattan.

Latour, B., 1993, *Petites leçons de sociologie des sciences*. Paris, La Découverte.

Leysen, E., 2010, Retour d'expérience: la tour Elithis est-ce vraiment un « bâtiment à énergie positive » ? *Le moniteur*, 20 avril.

Morel, C., 2007, *L'enfer de l'information ordinaire*. Paris, Gallimard.

Scardigli, V., 1996, *Les sens de la technique*, Paris, PUF.

Sidler, O., 2011, De la conception à la mesure, comment expliquer les écarts ? Colloque *Évaluer les performances des bâtiments basse consommation*, CSTB/CETE de l'Ouest, Angers.

Shove, E., 2003, Comfort, cleanliness and convenience: the social organisation of normality, Berg, Oxford and Nework.

Van Dam, SS., Bahher, CA., Van Hal, JDM., 2010, Home energy monitors: impact over the medium-term, *Building Research and information*, n°38, (5), pp. 458–469.

Zélem, MC., 2010, Politiques de maîtrise de la demande d'énergie et résistances au changement. Une approche socio-anthropologique, Paris, L'Harmattan.

Green Design, Materials and Manufacturing Processes – Bártolo et al. (eds)
© 2013 Taylor & Francis Group, London, ISBN 978-1-138-00046-9

An irregular discretization process for climate-responsive façades

E. Vermisso, M. Thitisawat & M. Feldsberg
Florida Atlantic University, Fort Lauderdale, USA

ABSTRACT: The project focuses on developing façade assemblies—which may include moving or static components—that are inherently responsive to climate (i.e. sunlight and daylight). The design constraints encourage "synthetic" workflows that combine various software and hardware interfaces to reach optimized solutions for building components that reduce heat gain in South Florida and promote daylight utilization. The paper discusses regular as well as non-homogeneous geometries (voronoi) to create light apertures of variable scale and density on a façade surface. The authors would like to assess the work with regards to both performance and process, evaluating the tools and identifying shortcomings in the parametric/simulation workflow dynamics.

1 INTRODUCTION

1.1 *Project scope*

This research encourages "synthetic" workflows that combine software and hardware interfaces to design building components for reduced energy consumption. The authors are interested in configuring facade assemblies that are responsive to minimize solar gain and optimize day-lighting conditions. The research explores the combination of parametric modeling and simulation to create performative or performance-based design, which allows for generative or synthesis act (Oxman, 2009). While exploring workflows and design possibilities, form finding process is still not fully automated. Benefits of the combination include an ability to quantitatively assess building performance using generative model, and ability to perform parallel computing for the assessment on different alternatives (Greenberg, et al. 2013).

1.2 *Precedence: Natural geometrical patterns*

The project began by breaking down a facade to repetitive modular components that open and close variably according to solar insolation values calculated on the facade surface. The current state of design examines the possibility of using "voronoi" geometries-often encountered as a naturally occurring pattern (Fig. 1)—to create light apertures of varying size and density. Voronoi are encountered in various forms in living organisms and can manifest as structural logic (i.e. dragonfly wing), visual surface configuration (i.e. sea turtle skin) or volumetric behavior (i.e. soap bubble aggregation).

Figure 1. Naturally occurring "voronoi" patterns: dragonfly wing; sea turtle skin and shell; hole formation in baked dairy product.

2 DISCUSSION: DESIGN PROCESS

2.1 *v.01: Parametric model scaled by distance*

Initially, the voronoi cells were created in Rhinoceros®/Grasshopper® and scaled parametrically according to their distance from the solar path using attractor points in Grasshopper® ("attractor points" alter the properties of a geometry depending on its distance from a changing sun position in scaled sky dome as a reference). The resulting geometry (v.01), although exhibiting some sort of inhomogeneity that is visually attractive, does not realistically reflect the distance from the solar path diagram to the panel (at such a large distance, the difference between one panel and another would be negligible).

2.2 *v.02: Curved façade scaled by sun angle*

A second iteration was therefore modeled, where the aperture of the panels is controlled by the angle of the sun on the solar path diagram with reference to the point generating the initial (voronoi) geometry (Fig. 2). The resulting meshes are linked through the GECO® plug-in to Autodesk® Ecotect®

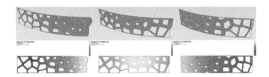

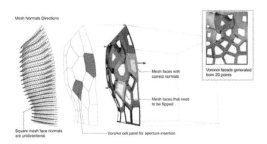

Figure 2. Simulations of voronoi generated apertures based on sun angle on a curved surface at 6 am, 12 pm and 7 pm respectively.

Figure 3. Face normal on a square 20′ × 20′ grid mesh and a curved voronoi surface with 20 panels. Faces in blue indicate wrong normal direction, those with grey are correct.

Analysis to calculate their insolation (radiation over time) levels.

This process is more accurate and realistic than the simple distance definition.

2.2.1 *Environmental simulation: Solar analysis*

The curved surface analysis in Ecotect® did not yield results that seemed fitting to our anticipation of solar gain due to inconsistencies in the mesh geometry: the mesh surface normal generated in Grasshoper® did not all face to the right direction, and so the faces were not recognized as solids in Ecotect®. As a result, we had to distinguish between those faces to keep and the ones to flip (Fig. 3). This issue revealed an obstacle towards achieving a fully automated process since we had to manually adjust the surface normals prior to simulation.

2.3 *Process evaluation/problem solving*

The voronoi parametric models evidenced a certain unsuitability to prepare information fast for simulations. This prompted us to look back at a more regular geometry in order to solve the problems at a simpler level and then adapt the solution to the more complex pattern.

2.3.1 *v.03: Flat Façade scaled by solar values*

A flat façade was modeled using a 16′ × 20′ dimension, and was simulated for peak incident radiation

levels at different times of the day, over two different seasons: Summer and Winter. The incident radiation values are uniform on the flat surface, except between the seasons but vary greatly on the curved surface (Fig. 4).

2.3.2 *v.04: Curved façade scaled by solar values*

A surface that displayed certain curvature variation was modeled with the intent to generate varying solar radiation simulation results. Opening sizes vary inversely to the amount of incident radiation on their locations. Glass windows can allow radiation to transmit through it. Hence, the opening sizes respond to the amount of solar heat gain (Fig. 6).

The range of incident radiation in winter is higher than that of summer due to lower winter profile/ altitude angle that allows the sun to strike the façade more directly, and more intense direct radiation due to lower cloud cover. This simplified panel design offers more practical solution in terms of autonomous opening control responsive to the sun.

The modeling process brought up a few issues that needed addressing, primarily regarding the parametric definitions to generate the scaling factors for the panels. In one of the simulations (June 21@12pm), the panelization process left three faces void. This error occurred because the respective scaling factors for these panels were

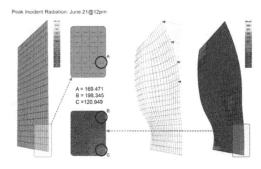

Figure 4. Comparison of solar gain on flat and curved façades based on a 20 × 20 grid mesh.

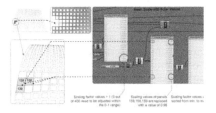

Figure 5. Missing panels in solar value-generated mesh; Parametric definition for adjusting over-scaled panels for June 21st at 12 pm (see Fig. 5 for solar diagram).

greater than 1. The definition had to be adjusted in order to replace these values (the appropriate indices of a list) with ones that were lesser than one, and close to the ones around them (Fig. 5).

2.3.2.1 Environmental simulation: Sunlight analysis

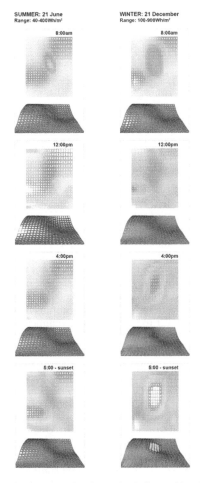

Figure 6. Summer & Winter simulations of insolation levels on a curved surface with 400 rectangular subdivisions that have been scaled based on the insolation levels. Simulations executed at 8 am, 12 pm, 4 pm and 5 pm (front and bird's eye views). The range of value intensity varies from 40–900 Wh/m² (summer min. 40, max. 400 and winter min. 100, max. 900 Wh/m²).

2.3.3 v.05: Voronoi façade scaled by solar values

We would like to apply the scaling logic of the curved façade panel aperture to a curved voronoi façade. This proved to be fairly complex due to the mesh properties within the modeling

software, and therefore our digital and parametric models required some data management in Grasshopper®. Figure 7 shows a voronoi façade with 20 panels whose apertures are scaled based on the peak incident solar radiation levels on the solid voronoi mesh. A problem that we encountered is the number of faces in each voronoi cell was more than one, resulting in multiple apertures as opposed to only one. We would therefore need to reduce the faces within each cell to make the scaling work.

2.3.3.1 Environmental simulation: Lighting simulations

An imported model underwent natural light simulation in Ecotect. The simulation allowed us to examine characteristics of an attached room in terms of daylighting performance (Fig. 8). Observation points were placed in the room to produce equidistant sky dome projections that revealed how much the façade affected daylight conditions in different locations. The façade is tall in relation to the depth of the room. As a result, the difference in terms of daylight factor between area close to the façade and the back of the room is small.

2.4 Workflow assessment

The project has an overall flexible character that allows integrating data towards an associative conceptual model, by combining modeling, virtual programming and simulation to both generate and evaluate design solutions. The schema for utilizing these tools is repetitive and iterative:

Rhinoceros® > Grasshopper® > GECO® > Ecotect® > Grasshopper® > GECO® > Ecotect® > Grasshopper®

Starting within the modeling platform the parametric model is then exported through the Environmental engine for analysis, its results are reimported into the parametric model for second analysis and the final results are imported back to the modeling platform. The recursive nature of the

Figure 7. Panelization for aperture scaling based on solar analysis: each cell in the voronoi mesh includes several faces.

491

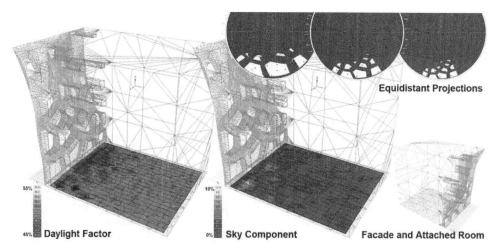

Equidistant Projections

55%

45% **Daylight Factor**

10%

0% **Sky Component**

Facade and Attached Room

Figure 8. Screen developed from 20 voronoi cells for simulation; Light levels within a room of 16′ × 16′ × 20′; Equidistant sky dome projection from the back, middle and near the façade with dots showing equal daylight factor sections.

data flow ensures that the results obtained are relatively reliable.

The greater the above cycle, the more optimized the results. A possible drawback of this process is the need to manually set up the cycle; other ways to look for optimization may be worth pursuing, like applying genetic algorithms to expedite the recursive process.

3 CONCLUSION

In conclusion, we have been able to derive scaling of regular panelization patterns on doubled-curved surfaces from localized conditions. We have further identified the constraints for applying the same design logic as irregular voronoi geometries due to the complexity induced by the adjacent voronoi cells. In the process we have addressed issues of modeling, model optimization for data transfer, simulation parameters and relation of geometry to environmental performance.

Our current model tries to configure the exact settings to produce variations of a voronoi-grid facade where the density of the panels as well as their sizes can reflect the heat gain from solar exposure and consider daylighting. Currently, this strategy could be applied to a fixed number of facades, after a study to reflect of local climate conditions. As the project evolves we intend to optimize the design of this panel as well as its operation strategy, to an adjustable one that would adapt real-time to the sun movement during the course of the day. We are in the process of designing a physical prototype that incorporates sensors,

to perform live, full scale testing, first with the regular panels and subsequently using irregular grids.

In the case of irregular panels, we can either configure static facades that reflect an optimized condition of i.e. certain time of the day or use the average insolation levels from a season that demonstrates the 'worst' case scenario; alternatively, a second layer of voronoi may be connected to the outer one with some flexible membrane that allows the inner layer to track the sun dynamically.

A possible issue we need to address at a later stage of the design is the customized panel resulting from the uniqueness of each voronoi cell; we believe that the randomness of this geometry is an interesting aesthetic driver for creating a facade whose appearance reflects the climate of its respective context. We are interested in subdividing each voronoi cell to a subset of smaller triangular panels that may be standardized. The final goal is creating a palette of a minimal number of panels that can be combined to form any voronoi configuration.

In retrospect, the authors believe the importance of this project lies within its integrative nature, assembling various digital and physical tools in order to optimize a design condition and would like to explore further a process of multi-objective optimization using genetic algorithms to generate a fully automated workflow.

ACKNOWLEDGMENTS

The authors would like to thank Henry Marroquin for doing early ground work on which the later experiments were based. Furthermore, we thank

FAU Broward Campuses for funding this project with an undergraduate research award. Finally, we would like to extend our gratitude to Nathan Hoofnagle (Handel Architects LLP) and Keith van de Riet (FAU) for their advice in matters of parametric design.

REFERENCES

Greenberg, D., K. Pratt, B. Hencey, N. Jones, L. Schumann, J. Dobbs, Z. Dong, D. Bosworth and B. Walter "Sustain: An Experimental Test Bed for Building Energy Simulation" Energy and Buildings, 58 March 2013, 44–57.

Oxman, Rivka. "Performative Design: A Performance-based Model of Architectural Design." Environment and Planning B: Planning and Design 36 (2009): 1026–1037.

http://genetichouse.blogspot.com/2010/08/voronoi-like-shapes-in-nature.html.

http://danimu.ch/libelle/.

Green Design, Materials and Manufacturing Processes – Bártolo et al. (eds)
© 2013 Taylor & Francis Group, London, ISBN 978-1-138-00046-9

Designing a sustainable and healthy living environment with smart technology

J.H. Lan

National Taichung University of Science and Technology, Taichung, Taiwan, P.R. China

ABSTRACT: The research presents a design project to illustrate how to design and develop a sustainable and healthy living environment based on smart technologies. The ILE project integrated network infrastructure, wireless sensor technologies and smart devices with physical space components for sustainable and healthy system performances. The design issues regarding with integrating space components with smart technologies and devices were studied. A system prototype to demonstrate the overall system performance was developed in the research.

1 INTRODUCTION

The recent developments of Information and Communication Technologies (ICT) have not only enabled better performance of equipments, but also provide innovative applications of new technologies. These technologies provide convenient, comfort and efficient performances according to various human living needs. Based on the integrations of ICT technologies, many researchers have explored the visions of future lives by applying smart technologies. (Aldrich 2003, Chiu & Chiang 2006, Cook & Das 2005, Chen & Jeng 2005, Jeng 2005, Mahdavi 2001a, OH et al. 2006, Trulove 2002).

With the trend of information technologies, how to design an intelligent living space to face the challenging issues regarding with security control, living convenience, energy saving, healthy environment and living comfort have become important research issues today.

2 THE RESEARCH GOALS

This paper presents an intelligent space design project developed with sustainability and healthy living considerations. The Intelligent Living Environment (ILE) project was conducted and the construction work was done. The ILE project provided an intelligent solution for designing a sustainable and healthy space based on smart technologies. The important issues regarding with sustainable and healthy living, such as intelligent energy saving, sustainable material utilizing, air quality control, thermal quality control, and lighting quality control, were studied in the research.

3 THE ILE PROJECT

3.1 *Overview*

In this study, the ILE project was designed and constructed to illustrate how to integrate smart technologies with physical space to provide a sustainable and healthy living environment. A typical classroom with 50 m^2 was re-designed to various space areas, including entrance area, living experience area, lighting experience area, dining area, kitchen area, and so on, Figure 1.

The ILE project was developed based on modular point of view. The design integration issues regarding with the application of smart technologies into space components were studied. In addition to general design issues, how to coordinate

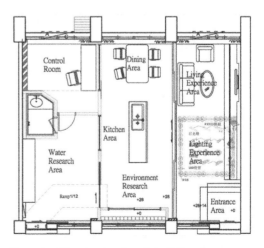

Figure 1. The plan drawing of the ILE project.

and cooperate with different system consultants to conduct the ILE project became a complicate problem in the study. The ILE project integrated network infrastructure, wireless sensor technologies and smart devices with physical space components to provide various sustainable and healthy system performances.

3.2 *The design concept*

As the design concept, first of all, the building materials used in the ILE project all complied with current green building rules. Secondly, an intelligent security control system with motion detecting technology was developed in the entrance area. Figure 2 shows the design concept. Once a motion event is detected, a web camera will be triggered to record video information into database. An authorized user then can get into the space through a RFID tag.

Besides, in the living space area, the design concept of an intelligent environmental control system is shown as Figure 3. The environmental control system can monitor the air quality, thermal quality and lighting quality of interior space, which provides specific healthy living situation for the occupants in the space. Moreover, an air conditioning control system with wireless sensor technology was developed to provide the intelligent energy saving function in the ILE project.

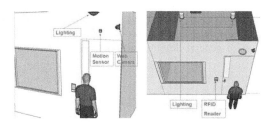

Figure 2. The design concept of security control system.

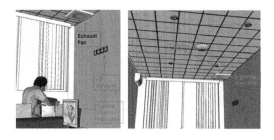

Figure 3. The design concept of environmental control system.

4 SYSTEM PROTOTYPE

Based on the design concepts mentioned above, the ILE project developed a system prototype with domain knowledge to monitor and control the overall system performance. The system prototype was developed with six modular functions, including Living Space Central Control System, Indoor Air Quality Control System, Air Conditioning Control System, Lighting Control System, Security Control System and Network Gateway Control System. A variety of sensors were installed with physical space components to detect the environmental situations from various perspectives. Once a typical environmental situation is detected, a corresponding event is triggered based on domain knowledge. The ubiquitous computing devices, such as remote sensors, RFID reader, touch screen, tablet PC and web camera were integrated with the ILE project to demonstrate the entire system performances. Figure 4 shows the spatial prototype of the ILE project after construction.

4.1 *Living Space Central Control System*

The Living Space Central Control System allowed system manager to monitor and control smart devices installed within the space components by incorporating various network technologies. The system provided a server database with designed user interface to monitor the overall indoor environmental situations detecting by smart devices. The domain knowledge regarding to controlling indoor environmental quality was programmed into the system function. The developed system then can provide knowledge-based intelligent interaction to specific environmental situation automatically. The user can also access the system functions by using a tablet PC through wireless network. Figure 5 shows the interface of Living Space Central Control System developed in the ILE project.

Figure 4. The spatial prototype of the ILE project.

Figure 5.   The Living Space Central Control System.

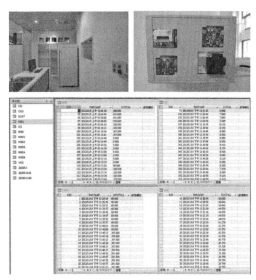

Figure 6.   The Indoor Air Quality Control System.

Figure 7.   The Air Conditioning Control System.

Figure 8.   The performance of Lighting Control System.

4.2   Indoor Air Quality Control System

The Indoor Air Quality Control System is the information center and control center of air quality in the ILE project. The system can monitor and record air quality information, including CO, CO_2, VOC and DUST, into server database at fixed period. For each air sensing module developed, it provides specific kind of detecting function and can transmit data collected into server database through wireless technology. Once detecting a specific extraordinary air situation, the exhaust fan will turn on to improve the air quality situation automatically. Figure 6 shows the related devices and interface of Indoor Air Quality Control System.

4.3   Air Conditioning Control System

The Air Conditioning Control System provided intelligent energy saving and control function to air conditioner in the ILE project. The system can monitor the thermal information and control the air conditioner based on the thermal situation detected by wireless sensors. Figure 7 shows the related devices of the Air Conditioning Control System.

4.4   Lighting Control System

The Lighting Control System provided functions to monitor lighting situation and control lighting devices. The system can detect the lighting quality information and adjust lighting illumination automatically based on domain knowledge. Besides, a variety of lighting scenarios can be programmed by users to experience lighting effects from various perspectives. Most lighting devices in the ILE

project were using LED technology for energy saving purpose. The user can also use a tablet PC or a touch screen interface to access and control the lighting devices through a web interface. Figure 8 shows the spatial performance of Lighting Control System.

In addition to the lighting functions mentioned above, a RGB-mixing lighting system was developed to provide fantastic lighting experience in the ILE project. The RGB-mixing lighting system allows user to program various lighting scenarios through the system interface. The user then can directly experience the lighting effects performed by the programmed scenarios. Figure 9 shows the interface and performance of RGB-mixing lighting system in the ILE project.

4.5 *Security Control System*

The Security Control System provided functions to assure door entrance security. Figure 10 shows the related hardware devices, including a web camera, a RFID reader, and an infrared ray detecting device.

In the Security Control System, a motion detecting module was developed by adopting an infrared ray technology. Once a motion event is detected, a specific light in the entrance area will turn on automatically to improve the lighting quality, which is important for taking clear video information next. And then a web camera is triggered to record video information in the entrance area into server database. Moreover, the system used the RFID technology to authorize specific user to get into the space. Figure 11 shows the interface design of the Security Control System.

Figure 10. The hardware devices of Security Control System.

Figure 11. The interface design of Security Control System.

4.6 *Network Gateway Control System*

The Network Gateway Control System provided the common network communication protocols for the ILE project. The Gateway Control System integrated Ethernet, Internet, Local Area Network (LAN), Wireless Local Area Network (WLAN) and Wireless Sensor Network (WSN) to provide web server functionality for developing client-server applications in the ILE project.

5 CONCLUSIONS

With the developments of smart technologies, how to design an intelligent space to provide sustainable and healthy living environment is an important issue today. To face the challenge, the research developed an intelligent design project by adopting various smart technologies. The developed ILE project demonstrated the intelligent solution for the challenging issue. Although there were still some limitations according to the system performances, the ILE project did provide valuable developing experiences for further references.

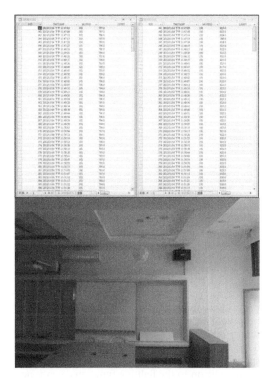

Figure 9. The interface and performance of RGB-mixing lighting system.

5.1 Results

In this research, the system prototype of the ILE project was developed with six modular functions. The Living Space Central Control System integrated and managed the overall environmental information detected by smart sensors and devices in the ILE project. The Indoor Air Quality Control System monitored the air quality information and provided intelligent air quality control mechanism. The Air Conditioning Control System monitored the thermal quality information and provided intelligent energy saving mechanism. The Lighting Control System monitored lighting quality situation and provided intelligent lighting control mechanism. The Security Control System monitored and controlled the door entrance security. The Network Gateway Control System provided the wired and wireless network communication protocols for the ILE project.

Based on the research findings, the overall system performance provided great potentials in delivering sustainable and healthy living services with smart technologies. The system prototype provides valuable experiences to develop an intelligent space design project for future references.

5.2 Further studies

With the limitation of research progress, the design issues regarding to coordinating various system consultants to conduct the ILE design project didn't discuss in this paper. However, during construction progress, we did find it was quite a complicated problem to coordinate various system consultants to cooperate with each other across multiple domains. To establish an effective communication platform among designer, contractor, ICT consultant, lighting consultant and various smart system consultants is a key to ensure the success of an intelligent space design project. Since the ILE project should be done within a tight schedule, the effective discussion among various system consultants became impossible. Therefore, some design issues regarding to integrating smart devices with spatial components did not study very well. For further study, the design integration issues and coordinating issues across multiple consultants are worth studying further.

ACKNOWLEDGEMENTS

The research is partial supported by National Science Council in Taiwan, NSC101-2221-E-025-015. The author is grateful to this support.

REFERENCES

Aldrich, F.K. 2003, Smart Homes: Past, Present, and Future, in R. Harper (ed.), *Inside the smart home*, Springer, pp. 17–39.

Chiu, M.L. & Chiang, B. 2006, Communicating with Space and People: Smart Interface Design for Enhancing User Awareness and Interactions, *proceedings of eCAADe*, Volos, Greece.

Cook, D.J. & Das, S.K. (eds.) 2005, *Smart Environment— Technology, Protocol and Applications*, Wiley Interscience.

Chen, J. & Jeng, T. 2005, A Context-Aware Home for Child-minding, *Proceedings of CAADRIA2005*, New Delhi, India, pp. 403–412.

Harper, R. (ed.) 2003, *Inside the smart home*, Springer, London.

Jeng, T. 2005, Advanced Ubiquitous Media for Interactive Space: A Framework, *Proceedings of CAADFutures*, Vienna, Austria.

Mahdavi, A. 2001a, Aspects of self-aware buildings, *International Journal of Design Sciences and Technology*, Europia: Paris, France, Volume 9, Number 1, pp. 35–52.

Mitchell, W. 2000, *e-topia*, The MIT Press.

OH S. et al. 2006, A Prototype System for Use in Designing Ubiquitous Environmental Spaces, *Proceedings of CAADRIA2006*, Kumamoto, Japan, pp. 355–362.

Trulove, J.G. 2002, *The Smart House*, HarperCollins Publishers, New York.

Weiser, M. 1991, *The Computer for the 21st Century*, Scientific America, Vol. 265, No. 3.

Green Design, Materials and Manufacturing Processes – Bártolo et al. (eds)
© *2013 Taylor & Francis Group, London, ISBN 978-1-138-00046-9*

Climate change adaptation and strategies: An overview

A. Santos Nouri
Faculty of Architecture, Technical University of Lisbon, Portugal
Research Scholarship holder from the Portuguese Foundation for Science and Technology, Urbanised Estuaries and Deltas. In search for a comprehensive planning and governance

M. Matos Silva
Faculty of Fine Arts, University of Barcelona, Spain
Ph.D. Scholarship holder from the Portuguese Foundation for Science and Technology, Urbanised Estuaries and Deltas. In search for a comprehensive planning and governance

ABSTRACT: Anthropogenic activity has influenced global climate change since the industrial era and will continue to affect societies in the long term. Attached to this certainty, the adaptation agenda straightens its pertinence which, on the other hand, is followed by the importance of acknowledging not only its meaning but also the range of possible alternatives. Through a bibliographical review, this paper will analyse the concept of adaptation and discuss some of the most relevant suggested strategies. The goal is to contribute to diminish the gap between theory and action, being a step forward and closer to deciding on the best measures to advocate in light of climate change.

1 RELEVANCE OF CLIMATE CHANGE ADAPTATION

When considering future horizons, the subject of climate change will always have to cope with factors of uncertainty and the consequential disagreement among scientists (particularly climate scientists). Nevertheless there is strong evidence in a key issue: that anthropogenic activity has influenced global climate change since the industrial era and will continue to affect societies in the long term (Min, Zhang et al. 2011; AMS 2012).

As present times prove, mitigation strategies have not been as successful as previously predicted. Every year Mankind is progressively more distant of the goals initially set by the Kyoto protocol (that through the reduction of emissions, global temperatures should not rise more than two degrees from current levels). Looking at today's registered emissions, even if mitigation policies presented immediate results, several authors are of the opinion that there already are certain unavoidable impacts (IPCC 2007; Schaeffer, Hare et al. 2012). Underlying this frustration is the adaptation agenda that sees its relevance enhanced as a fundamental complementary strategy (Costa, Matos Silva et al. 2012).

1.1 *An overview of concepts*

Although adaptation to anthropogenic climate change may be a new phenomenon, this is not the case for human adaptation to environmental change. Though not always successful, climate adaptation followed our evolution as a species (Burton 2004).

'Adapt' means to make suitable to requirements or conditions through the processes of adjustments or modifications (Online Oxford Dictionary). However, the terms have more specific interpretations in particular disciplines; for instance, biological adaptation is not the same as cultural adaptation (Smit, Burton et al. 2000).

In climate change literature, definitions of adaptation are also numerous and, still, there is no standard terminology. In order to simplify, Table 1 lists some of the most relevant definitions since 1992.

It is commonly known that the United Nations Framework Convention on Climate Change (UNFCCC) definition of climate change adaptation was the first to be inaugurated (UNFCCC 1992). On the other hand, the classical definition, or the most quoted, belongs to McCarthy et al. (2001) in IPCC's Third Assessment Report. A recent forthright definition is presented by UK's Adaptation Sub-Committee (ASC), summarizing it as *"any adjustment of behaviour to limit harm, or exploit beneficial opportunities, arising from climate change"* (ASC 2011, p.92).

Through a quick analysis of the assembled definitions, one can note that while authors vary in the emphasis given to particular sub-subjects

Table 1. Climate change adaptation—summary of descriptors.

Source	Description
(UNFCCC 1992)	Adaptation to climate change is adjustment to natural or human systems in response to actual or expected climate *stimuli* or their effects, which moderates harm or exploits its beneficial opportunities.
(Burton 1993)	Adaptation to climate change is the process through which people reduce the adverse effects of climate on their health and well-being, and take advantage of the opportunities that their climatic environment provides.
(Smit 1993)	Adaptation involves adjustments to enhance the viability of social and economic activities and to reduce their vulnerability to climate, including its current variability and extreme events as well as longer term climate change.
(Stakhiv 1993)	The term adaptation means any adjustment, whether passive, reactive or anticipatory, that is proposed as a means for ameliorating the anticipated adverse consequences associated with climate change.
(Smith et al.1996)	Adaptation to climate change includes all adjustments in behaviour or economic structure that reduce the vulnerability of society to changes in the climate system.
(Watson et al, 1996)	Adaptability refers to the degree to which adjustments are possible in practices, processes, or structures of systems to projected or actual changes of climate. Adaptation can be spontaneous or planned, and can be carried out in response to or in anticipation of change in conditions.
(IPCC 1996) Second assessment report	Adaptation to climate change includes measures that serve the dual purpose of (a) reducing the damages from climate change; and (b) increasing the resilience of societies and eco-systems to the impacts of the climate change.
(Downing et al. 1997)	Adaptation is synonymous with "downstream coping".
(Burton et al. 1998)	Adaptation refers to all those responses to climate change that may be used to reduce vulnerability.
(Smit et al. 2000)	Adaptation as adjustment in ecological-social-economic systems in response to actual or expected climatic stimuli, their effects or impacts.
(McCarthy et al. 2001) IPCC—Third assessment report	Adaptation is the adjustment in ecological, social and economic systems in response to actual or expected climatic stimuli and their effects or impacts. This term refers to changes in processes, practices, or structures to moderate or offset potential damages or take advantage of opportunities associated with changes in climate. It involves adjustments to reduce the vulnerability of communities, regions, or activities to climatic change and variability.
(Pielke 1998)	Adaptation refers to adjustments in individual, group and institutional behaviour in order to reduce society's vulnerabilities to climate.
(Scheraga and Grambsch 1998)	Adaptive actions are those responses or actions taken to enhance resilience of vulnerable systems, thereby reducing damages to human and natural systems from climate change and variability.
(UNDP 2002)	Adaptation is a process by which strategies to moderate, cope with and take advantage of the consequences of climatic events are enhanced, developed and implemented.
(Easterling, Hurd et al. 2004)	A successful adaptation is defined as one that follows a climate change causing adverse impacts and maintains a system at approximately the same level of welfare or services as was provided before the change in climate.
(Adger et al. 2007) IPCC—Fourth assessment report	Adaptation as adjustments to reduce vulnerability and enhance resilience in response to observed or expected changes in climate and associated extreme weather events.
(McGray, Hammill et al. 2007)	Adaptation is a process, not an outcome.
(Alexandre Magnan, Benjamin Garnaud et al. 2009)	Adaptation is a continuous learning process.
(ASC 2011)	Adaptation is any adjustment of behaviour to limit harm, or exploit beneficial opportunities, arising from climate change.

Source: Adapted from (Smit, Burton et al. 2000; Schipper 2007; Bahinipati 2011).

(natural or human systems; health and well-being; social/behaviour, ecological and economical activities; individual, group and institutional behaviour), there is a general accordance of what is the core interpretation. On the other hand, and more importantly, it proves the that adaptation can no longer be considered as an exclusive process of the evolution of plants and animals and should be promoted as a concept for guiding policy and actions to confront climate change. According to Schipper, "*the main difference in biological adaptation and climate change adaptation is the level of planning and consciousness by which adjustments are carried out*" (Schipper 2007, p. 4) and, in this sense, climate change adaptation encompasses different strategies.

Choosing the optimal measure for a particular situation is neither easy nor simple. Sometimes it is even dangerous, namely when adaptation does not fulfill its objectives and ultimately raises vulnerability. Taking this into consideration, and in order to diminish the cases of 'maladaptation' (Barnett and O'Neill 2010), specific strategies and consequent actions must be carefully comprehended. These shall be considered in the next section.

2 CLIMATIC ADAPTATION STRATEGIES

2.1 *The terminology of adaptation strategies*

Although to date a considerable amount of authors and institutions have written about climatic adaptation strategies, there is no standard terminology. As such, there is a significant amount of discrepancy and interchangeability of terminology usage in existing literature.

2.2 *Resemblances and dissimilarities of adaptation strategies*

In order to understand how climatic strategies are approached by different authors and entities (such as Frankhauser & Smith, Hallegatte, Smit & Burton, Rijke & Veerbeek, Peel, Bruij & Klijn, International Panel on Climate Change, United Nations Development Program, United Kingdom Climate Impacts Programme, Plan BLEU and The Nature Conservancy), Table 2 can be broken down into specific adaptation strategies, as shown in Tables 3–6.

When considering anticipatory adaptation strategies, Fankhauser et al. IPCC, and Smit et al. define the strategy very similarly (Table 3). Nevertheless, if one considers different terminologies from Table 2, similar anticipatory objectives are discussed. Respectively: (1) the TNC-UA's Response strategy (*14c*), 'assists the unit or system to follow changing climates' (TNC-UA 2007);

(2) the IPCC's Prevention of loss strategy (*2a*), involves 'actions to reduce the susceptibility to impacts of climate change' (IPCC 1994); and (3) Bruij et al.'s Resistance strategy (*13b*), enforces the anticipatory 'threshold of significant negative flooding consequences' (Karin de Bruij, Klijn et al. 2009). Consequently, this comparison illustrates that the use of terminology can be diverse when they have similar or identical approaches.

Unlike anticipatory strategies, reactive strategies are those that take place after impacts of climate change have been observed (Smit, Burton et al. 2000; IPCC 2001). This is distinguishably different to autonomous adaptation strategies as they are 'spontaneous' and are not a result of a 'conscious' responses to climatic stimuli (Fankhauser, Smith et al. 1999; IPCC 2001). Furthermore, both reactive and autonomous adaptations are straightforward in that there is little confusion with other adaptation strategies between authors.

In contrast, planned adaptation strategies fall victim to different perceptions of the various entities as discussed in Table 2.

Although Fankhauser et al. and the IPCC (2001/7) have similar descriptions of planned adaptation (Table 4), if one is to again consider other terminologies from different authors, this circumstance changes. When addressing Hallegatte's adaptation strategies, it is clear that they are also grounded upon planned approaches: (1) reversible strategies (*5b*), that are 'reversible and flexible i.e. easy-to-retrofit' (Hallegatte 2009); (2) safety margin strategies (*5c*), that 'reduce vulnerability at null or low costs (usually requiring over pessimistic estimations in design phases' (*ibid.*); and (3) synergy with mitigation strategies (*5f*), that 'consider future environmental costs such as energy and fossil fuel consumptions of delineated adaptation measures' (*ibid.*).

Similarly, the UNDP also present adaptation strategies that require comprehensive planning when applying their approaches: (1) no-regrets adaptation strategy (*7a*), that 'are justified by current climate conditions, and further justified when climate change is considered' (UNDP 2004); and (2) low-regrets strategy (*7b*), that are 'those made in light of climate change but at minimal cost' (*ibid.*).

Although there are many other examples of strategies (such as, *8b, 9b, 12a/b/c, 14a/b/e*) that also are based on 'planned adaptation', the last example will be consider PLAN BLEU's approach to climatic adaptation: (1) prevent effects (hard solutions) strategy (*10c*), that 'involves preventing the effects and impact of climate change through structural measures, i.e. through technology and infrastructure' (BLEU 2012); (2) prevent effects (soft solutions) strategy (*10d*), that 'involves preventing

Table 2. Comparative analysis of adaptation strategy terminology used by different authors and institutions.

Source	N°	Climatic adaptation strategies	Adapt. sector
(Fankhauser, Smith et al. 1999)	*1a)*	Reactive adaptation	General
	1b)	Anticipatory adaptation	
	1c)	Autonomous adaptation	
	1d)	Planned adaptation	
(IPCC 1994)	*2a)*	Prevention of loss	General
	2b)	Tolerating loss	
	2c)	Spreading or sharing loss	
	2d)	Changing use or activity	
	2e)	Changing location	
	2f)	Restoration	
(IPCC 2001)	*3a)*	Anticipatory adaptation	General
	3b)	Autonomous adaptation	
	3c)	Planned adaptation	
	3d)	Private adaptation	
	3e)	Public adaptation	
	3f)	Reactive adaptation	
(IPCC 2007)	*4a)*	Anticipatory adaptation	General
	4b)	Autonomous adaptation	
	4c)	Planned adaptation	
(Hallegatte 2009)	*5a)*	No-regret	General
	5b)	Reversible	
	5c)	Safety margin	
	5d)	Soft	
	5e)	Reduce decision-making horizon	
	5f)	Synergy with mitigation	
(Smit, Burton et al. 2000)	*6a)*	Passive adaptation	General
	6b)	Reactive adaptation	
	6c)	Anticipatory adaptation	
(UNDP 2004)	*7a)*	No-regrets adaptation	General
	7b)	Low-regrets adaptation	
(UKCIP 2005)	*8a)*	Living with risks/bearing loses	Coastal
	8b)	Preventing effects/reducing exposure	
	8c)	Sharing responsibility	
	8d)	Exploiting opportunities	
(CERCCS 2011)	*9a)*	Do nothing	Coastal
	9b)	Managed realignment	
	9c)	Hold the line	
	9d)	Limited intervention	
(BLEU 2012)	*10a)*	Bear risks/losses "do nothing"	Water
	10b)	Share risks/losses	
	10c)	Prevent effects (hard solutions)	
	10d)	Prevent effects (soft solutions)	
	10e)	Change/reorganise uses and activities	
	10f)	Improve climatic knowledge	
	10g)	Strengthen capacities/raise awareness	
(Rijke, Veerbeek et al. 2010)	*11a)*	Opportunistic	Water
	11b)	Active	
	11c)	Business as usual	
(Peel 2009)	*12a)*	Retreat	Coastal/
	12b)	Attack	water
	12c)	Defend	

(Continued)

Table 2. (*Continued*)

Source	Nº	Climatic adaptation strategies	Adapt. sector
(Karin de Bruij, Klijn et al. 2009)	*13a)*	Resilience	Water
	13b)	Resistance	
	13c)	Do nothing	
(TNC-UA 2007)	*14a)*	Resistance (reactive adaptation)	Ecological
	14b)	Resilience (facilitative adaptation)	
	14c)	Response (proactive adaptation)	
	14d)	Realign	
	14e)	Reduce	
	14f)	Triage	

Table 3. Description of anticipatory adaptation strategies.

Anticipatory adaptation strategies	Description
(Fankhauser, Smith et al. 1999)	Strategies that are deliberate decisions to prepare for potential effects of climate change
(IPCC 2001)	Strategies take place before impacts of climate change are observed
(IPCC 2007)	Same as previous
(Smit, Burton et al. 2000)	Strategies that act before climatic phenomenon

Table 4. Description of planned adaptation strategies.

Planned adaptation strategy	Description
(Fankhauser, Smith et al. 1999)	Strategies that require conscious intervention
(IPCC 2001)	Strategies that are resultant of a deliberate policy decision, based on the need for action
(IPCC 2007)	Same as previous

Table 5. Description of 'do nothing' strategies.

'Do nothing' strategy	Description
(CERCCS 2011)	Strategies that follow a 'wait and see' approach
(Karin de Bruij, Klijn et al. 2009)	Strategies that maintain existing practices

Table 6. Description of tolerating loss and living with risks/bearing loses strategies.

Tolerating loss strategy	Description
(IPCC 1994)	Accept adverse impacts as they can be sustained in the long term
Strategies with similar terminology	
(UKCIP 2005)	Living with risks/bearing loses
	Accepting loss of assets as they are no longer worth sustaining

the effects and impact of climate change through non-structural measures, i.e. through town planning standards' (*ibid.*); and (3) strengthen capacities/raise awareness strategy (10 g), that 'lengthen planning time frames and increase public awareness and acceptability of public policy alterations' (*ibid.*).

When considering strategies that emphasise the 'do nothing' approach, one can note significant dissimilarities between the authors (Table 5). Although based on the same terminology, CERCCS's approach (*9a*), is described as 'strategies that follow a wait and see approach permitting the destruction of coastal infrastructure'

(CERCCS 2011). This demonstrates that although the strategy enforces the 'do nothing' approach, it still enforces mitigating climatic hazards and considers future climatic scenarios. On the other hand, Bruij et al.'s approach (*13c*), are 'strategies that maintain existing practices' (Karin de Bruij, Klijn et al. 2009). This means little or no attention is paid to climatic changes, and existing attitudes are continued regardless of future horizons.

Although significantly interlaced to the 'do nothing' approach due to inaction attitudes, toleration of loss will very likely be required in any of Mankind's adaptation approaches to climate change. Nevertheless, there is still significant disparity between different authors and institutions regarding the acceptance of loss (Table 6).

As an example, although the IPCC (1994) and the UKCIP (2005) have similar approaches to loss toleration, their attitudes are considerably different. The tolerating loss strategy (*2b*), are those 'that accept adverse impacts in the short term because they can be sustained without long-term damage' (IPCC 1994). This implies that there can be an 'acceptance' of loss when not substantial in the temporal perspective (i.e. in the long-term). On the other hand, the living with risks/bearing loses (*8a*), are those 'that accept pre-impact systems, behaviours are no longer sufficient—accepting loss of assets as they are no longer worth sustaining' (UKCIP 2005). This shows a more functional perspective, in that the losses are accepted because they are not worth sustaining in the long run. In this comparative analysis of loss toleration, others strategies can also be addressed such as (*10a/b* and *12a*).

2.3 *Coalescing strategies for climatic adaption*

As shown in previous tables, it is clear that different authors and institutions describe their strategies differently, even if they use the same terminology. This demonstrates a lack of consensus in existing literature that can hence confuse public decision makers.

Additionally, approaches such as 'planned adaptation strategies' can also represent a vast arena of possible meanings, attitudes and concepts. Defined as strategies that require conscious intervention or policy decision, it can be argued that this description contributes very little in guiding agents who are to apply particular and specific adaptation strategies.

Independent of generic approaches being important, relying solely on one strategy would be to reduce a fertile scope in which to deal with climate change. As such, a range of examples must be fully comprehended in order to form a particular, original, opportune and most fitted strategy.

Conversely, it is also important to consider the general nature of the analysed strategies, be it 'anticipatory', 'planned', or 'business as usual/bear losses'. Yet this does not imply that their overall nature cannot sustain other strategies. As an example, Table 1 intentionally breaks down different strategies into 'adaptation sectors', illustrating that general adaptation can be strengthened by authors that exclusively focus on 'water', 'coastal', or 'ecological' adaptation.

3 CONCLUSIONS

Existing literature is a starting point for both the private and public sector to initiate and pursue their respective adaptation strategies. This article considers that adaptation is an ongoing process that can exploit the opportunities that are presented alongside the effects climate change. Nevertheless, important decisions need to be made,

and routes need to be chosen from a vast array of alternatives.

Having discussed the relevance and definitions of climate change adaptation, it is clear that uncertainty plays a fundamental role and cannot but be included in strategies and consequent measures. Through the comparative analysis of the literature review regarding different strategic approaches, those that incorporate this uncertainty and that furthermore take advantage of its opportunities, are considered as the most relevant.

Having compared the resemblances and dissimilarities of different strategies, with a focus on anticipatory, reactive, autonomous, planned, 'do nothing', and tolerating losses; further conclusions can be extrapolated from this article.

Firstly, although authors may use the same terminology in their strategies, their concepts can be considerably different. This is illustrated by the presented disparity between 'do nothing' strategies and between the 'tolerating of loss' strategies.

Secondly, the broadness and interchangeability between strategies can also be highly beneficial when considering uncertain horizons, as: (1) anticipatory adaptation strategies can incorporate other approaches such as 'response', 'prevention of loss', and 'resistance' strategies; and (2) planned adaptation strategies can incorporate approaches such as 'no-regrets', 'low-regrets', and 'reversible' strategies.

Thirdly, and accepting the broadness and dissimilarities of climatic adaptation strategies, the breakdown of strategic 'adaptation sectors' facilitates agents to consider both general and sector specific adaptation.

Lastly, it is the initial comparative analysis of climatic adaptation strategies that launches climatic adaptation into a broad, fertile, and maturing arena of existing knowledge for climate change—where agents that are to propose adaptation measures can focalise their own response to address uncertainty and propose both flexible and resilient measures for eventful horizons.

ACKNOWLEDGMENTS

The authors would like to acknowledge the research project "Urbanised Estuaries and Deltas. In search for a comprehensive planning and governance. The Lisbon case." (PTDC/AUR-URB/100309/2008) funded by the Portuguese Foundation for Science and Technology and the European Social Found, 3rd Community Support Framework; and the Portuguese Foundation for Science and Technology for the individual doctoral grant (SFRH/BD/76010/2011) funded by POPH—QREN.

REFERENCES

AMS (2012). An Information Statement of the American Meteorological Society. Climate Change. Boston, American Meteorological Society: 7 pp.

ASC (2011). Adapting to climate change in the UK—measuring progress. Adaptation Sub-Committee of the Committee on Climate Change.

Bahinipati, C.S. (2011). Economics of Adaptation to Climate Change: Learning from Impact and Vulnerability Literature. Working paper no. 213, Madras Institute of Development Studies.

Barnett, J., et al. (2010). "Maladaptation." Global Environmental Change 20(2): 211–213.

Bleu, P. (2012). Water Resources and Natural Environment. Building the Mediterranean Together. France, PLAN BLEU.

Burton, I. (2004). Climate Change and the Adaptation Deficit. Adaptation and Impacts Research Group (AIRG), Meteorological Service of Canada.

CERCCS (2011). Information Sheet: Managed adaptation options. Australia, Coastal Ecosystems Responses to Climate Change Synthesis Project: pp. 1–4.

Costa, J.P., et al. (2012). A adaptação às alterações climáticas, os processos ecológicos e o desenho da infraestrutura de gestão das inundações urbanas. Palcos da Arquitectura. AEAULP. Lisboa, Academia de Escolas de Arquitectura e Urbanismo de Língua Portuguesa. **Vol. II:** 506–515p.

Fankhauser, S., et al. (1999). "Weathering climate change: some simple rules to guide adaptation decisions." Ecological Economics 20(1): pp. 67–78.

Hallegatte, S. (2009). "Strategies to adapt to an uncertain climate change" Global Environmental Change 19(1): pp. 240–247.

IPCC (1994). Technical Guidelines for Assessing Climate Change Impacts and Adaptations Prepared by IPCC Working Group II [Carter, T.R., M.L. Parry, H. Harasawa, and S. Nishioka (eds.) and WMO/UNEP. CGE-IO15-'94. Univerdity College-London. UK, and Centre for Global Environmental Research, Tsukuba, Japan, 59 pp.

IPCC (2001). Summary for Policy Makers, A Report of Working Group II of the Intergovernmental Panel of Climate Change, Prepared by IPCC Working Group II.

IPCC (2007). Climate Change 2007: Impacts, Adaptation and Vulnerability, Intergovernmental Panel on Climate Change.

IPCC (2007). Climate Change 2007: The Physical Science Basis. Contribution of Working Group I to the Fourth Assessment Report of the Intergovernmental Panel on Climate Change. S. Solomon, D. Qin, M. Manning et al. Cambridge, United Kingdom and New York, USA, Intergovernmental Panel on Climate Change: p. 996.

Karin de Bruij, et al. (2009). Flood risk assessment and flood risk management—An introduction and guidance based on experiences and findings of FLOODsite (an EU-funded Integrated Project), FLOODsite Consortium.

Min, S.-K., et al. (2011). "Human contribution to more-intense precipitation extremes." Nature 470(7334): 378–381.

Peel, C. (2009) "Facing up to Rising Sea-Levels: RETREAT? DEFEND? ATTACK?", 15.

Rijke, J., et al. (2010). Adapting where we can. Rotterdam, Paper presented at the Deltas in Times of Climate Change International Conference.

Schaeffer, M., et al. (2012) "Long-term sea-level rise implied by 1,5°C and 2°C warming levels." Nature Climate Change, 4 DOI: 10.1038/NCLIMATE1584.

Schipper, E.L.F. (2007). Climate Change Adaptation and Development: Exploring the Linkages, Tyndall Centre for Climate Change Research.

Smit, B., et al. (2000). "An anatomy of adaptation to climate change and vulnerability." Climatic Change 45(1): pp. 223–251.

TNC-UA (2007). Conceptual Management Approaches to Climate Change Adaptation. C. Enquist. New Mexico, TNC-UA Workshop for managers on adaptation.

UKCIP (2005). Measuring Progress: Preparing for Climate Change through the UK Climate Impacts Programme (Technical Report). United Kingdom, UK Climate Impacts Programme (UKCIP).

UNDP (2004). Adaptation Policy Frameworks for Climate Change: Developing Strategies, Policies and Measures. United Kingdom, Cambridge University Press.

UNFCCC (1992). United Nations Frame Convention on Climate Change, United Nations: 24.

Green Design, Materials and Manufacturing Processes – Bártolo et al. (eds)
© *2013 Taylor & Francis Group, London, ISBN 978-1-138-00046-9*

Digital fabrication and rapid prototyping as a generative process

Filipe Coutinho & Mário Kruger
Departamento de Arquitectura, Universidade de Coimbra, Coimbra, Portugal

José P. Duarte
Faculdade de Arquitectura da Universidade Técnica, Lisboa, Portugal

ABSTRACT: This paper shows a construction of a shape grammar integrating data gain from Digital Fabrication (DF) and Rapid Prototyping (RP). The goal of this experiment is to see the impact it may have in the process of both evaluating and validating the grammar that uses Leon Battista Alberti *De Re Aedificatoria* treatise rules. A part of the Column System shape grammar from Alberti's treatise will be used as an analysis object. A generation of a column systematization of a specific building and a capital from this column will be the subject of this paper. This grammar was constructed directly from the treatise rules. The grammar generates detailed elements and it has the capacity to optimize such elements in an accurate and better informed way. Its derivations have produced buildings elements that were built as artifacts. The goal of this experiment is to better understand the methodology used and to improve the Column systematization shape grammar.

1 INTRODUCTION

This paper shows a shape grammar construction integrating data gain from Digital Fabrication (DF) and Rapid Prototyping (RP). The goal of this experiment is to see the impact it may have in the process of validating the grammar, ultimately teaching and communicating subjects related with Alberti's theory using built artifacts.

A shape grammar can be seen as a generative process to design buildings in a specific style. Traditionally it is presented as computation formalism with shapes through its recognition and its replacement controlled by rules. A set of rules of transformation applied recursively to an initial shape generates new or existing designs.

A Column Systematization shape grammar from Alberti's treatise, *De Re Aedificatoria*, will be used as an analysis object. This grammar was generated directly from the treatise. This shape grammar is a grammar of detail and it may allow a better and more intelligent control of the column elements to be fabricated. Its derivations have produced a range of buildings constructed by Alberti, as well as others from the Portuguese heritage. This grammar is built encompassing 4 views (plan, elevation, section) and a solid 3D model. This model is consistent with most DF tools, particularly 3D printing and CNC Milling. This article will gain insights from a construction grammar (Sass, 2007), particularly to show the use of RP in the construction of the Column Systematization shape grammar.

The research described in this abstract is part of a wider project aimed at decoding the treatise by inferring the corresponding shape grammar using the computational framework provided by description grammars (Stiny, 1981) and shape grammars (Stiny and Gips, 1972) to determine the extension of such an influence in the counter-reform period in Portugal (Kruger, 2011). Based on document sources that reveal links between Portuguese architects and Alberti's work, some historians have pointed that such an influence was real, but none was able to determine its extent so far. The idea is to shed some light on this discussion by translating the treatise into a shape grammar (Stiny and Gips, 1972) and then trace the influence of Alberti's work, determining to which extent the grammar can account for the generation of Portuguese classical buildings (Kruger, 2011). This approach follows the transformations in the design framework proposed by Knight (1983) according to which the alteration of one style into another can be explained by changes of the grammar underlying the first style into the grammar of the second. The project foresees the development of grammars as showed in this article, encoding the rules for designing the column system as well as the rules for designing buildings, namely churches and in this case a *Loggia*.

2 METHODOLOGY

Alberti's treatise, *De Re Aedificatoria*, can be seen as a set of algorithms that explain how to design

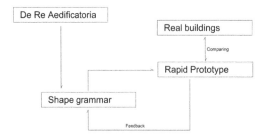

Figure 1. Integrating the data gain from RP and real building.

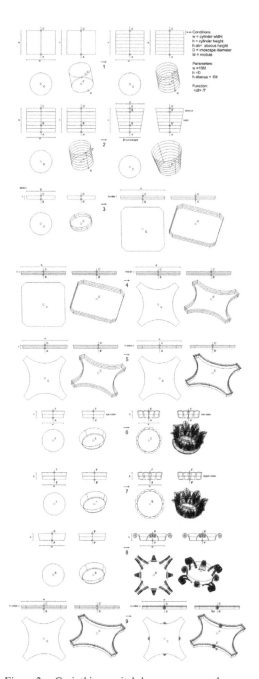

Figure 2. Corinthian capital shape grammar rules.

segments or overall buildings in such detail that is possible to decompose those descriptions in a shape grammar and a description grammar.

The parametric design system gain from this approach is then used in a computer program to automate the generation of building elements to be fabricated using a Zcorp 3d printer, as is the goal of this experiment. The fabricated 3d artifact analysis will help to clarify some formal aspects of the column system elements. When comparing these fabricated elements with real built elements, some elements will change and new geometries will be applied to the shape grammar rule. (Fig. 1).

3 DESCRIPTION OF APPLICATION

The grammars were developed as parallel grammars encompassing four views: plan, section, elevation, and axonometric.

The first three of these views are developed in the Cartesian product of the algebras U12 and V12, and the fourth in the Cartesian product of the algebras U13, U33 and V13. Each rule has a section containing both parameters, descriptions and, when needed, a set of functions to be coordinated with other grammars (just seen in rule 1). In this article we reveal the rules of a Corinthian Capital: It is given an insertion setting of subelements and then proceeds to apply various rules recursively to generate the capital using elements of ornament described by Alberti. The Corinthian Capital grammar starts with a primitive object in rule 1, a cylindrical element "the vase" that is divided in 7 parts. Rule 2 transforms this cylinder in a truncated cone. In Rule 3, 1/7 part of the cone is taken and used as the height for the Abacus. Rule 4 gives the geometry of the abacus. In rule 5 the height of the abacus is divided in 3 parts, one for the collar, the remaining for the operculum. Rule 6 inserts 8 generic leafs between labels A and E. Rule 7 inserts 8 generic leafs between labels F and E. Rule 8 inserts volutes. Rule 9 inserts 4 flowers in the Abacus as seen in Figure 2. Rules derivation seen in Figure 3.

A column system is used and consists of a base, a shaft, the Corinthian capital and an arch composed by a Doric architrave mould. This column and arch is a portion of the Loggia Rucellai grammar presented (Coutinho, 2013). The column and the Loggia are not presented due to space restrictions.

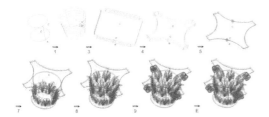

Figure 3. Corinthian capital shape grammar derivation.

Figure 4. Corinthian capital and column with arch RP model.

Both elements were fabricated at different stages of the grammar construction. First the Corinthian capital, then the column with arch. Finally the Loggia.

4 EVALUATION AND VALIDATION USING RP

Two elements, among others, were fabricated.

The Corinthian Capital has different levels of complexity. The leaf contains geometry with double curvature and it has several brunches that are faced with its surface. Other complex elements are the volutes and stalks. The volutes were built, as a two centered point spiral, as described by Alberti in the treatise section explaining how to generate the Ionic Capital. The stalks were generated using a Nurbs curve, achieved by observation of real capitals and then tested in the prototype produced.

The second element fabricated was the column with an arch. This column is a sum of a Doric base, a plain shaft, the Corinthian capital, referred to above, and an arch composed by a Doric architrave mould. The column and the arch have another element called latastrum that is applied to make the structural load transition from an arch to a Corinthian or Composite capital. This element connects the abacus to the starting section of an arch. Several experiments were done to see the best way to connect this element with the arch in order to better produce them. A Doric mould of 3 different fillets generated the geometry of the arch. This prototype helped us to detect some mistakes on the positioning of the arches of the roof and how they unload in latastrum of the Corinthian capital. There were also other elements digitally manufactured using different techniques throughout the development of the main grammars (Coutinho et al, 2012). Aiming the accuracy verification of the generation, the digital model obtained was printed on a 3D printer of Zcorp, a model of Corinthian capital and its integration of the column with arch is demonstrated in Figure 4.

All these new geometries were used in specific rules and in some aspects new rules were structured.

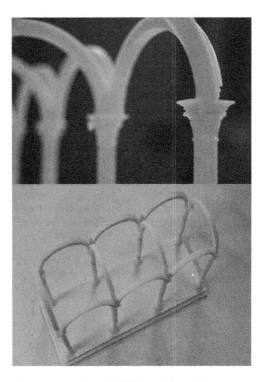

Figure 5. Loggia rucellai RP parcial model.

As in the case of the Loggia Rucellai shape grammar, here the treatise and the visual surveying were not enough to structure all the rules required for such generation as seen in Figure 5.

5 CONCLUSIONS

Traditionally there is a historical separation between digital fabrication and generative design. This article attempts to show that such a division is not entirely accurate. If we consider the derivation of the rules of the shape grammar a generative process (generative design) and if this grammar absorbs for its construction aspects arising from artifacts fabricated, than we can consider rapid prototyping as an active part of the construction of the grammar and its use.

The grammar presented in this paper is grammar of detail and the control of the column elements may facilitate a more intelligent use of those elements in terms of constructability.

The RP revealed to be of great employment for:

- Demonstrate aspects relative to the geometry of fittings between different portions of the buildings;
- Verify aspects of the geometry of different elements and how they could improve the shape grammar rules;
- Communicate with people without much knowledge of the Alberti's theory, how the column system elements relate to each other. Show that a complete element of this system has proven useful and has captured the interest of students as well as a wider public.
- Some less positive aspects relate to the time consumption taken to manufacture the elements and the price that these elements often cost.

Fabricating the elements studied, may give a better understanding of the construction processes and may clarify the best way to use those elements.

This would be a complete and effective approach in understanding the theory of Alberti.

REFERENCES

Duarte, J.; Figueiredo, B; Costa, E; Coutinho, F; Kruger, M. 2011. Alberti Digital: investigando a influência de Alberti na arquitetura portuguesa da contra-reforma. In: BRANDÃO, Carlos Antônio Leite; FURLAN, Francesco; CAYE, Pierre; LOUREIRO, Maurício. Na gênese das racionalidades modernas: em torno de Leon Batista Alberti. Belo Horizonte: Editora UFMG, 2013.

Coutinho, F.; Castro e C.E, Duarte. J.P.; Kruger, M. 2011. Interpreting De re edification—a shape grammars of the orders system. Proceedings of International Conference Eccade 2011, Ljubliana, Eslovenia.

Knight, TW. 1983. Transformations of language of design, Environment and Planning B: Planning and Design 10. 125–177.

Kruger, M.; Duarte, J.P.; Coutinho, F. 2011. Decoding De Re Aedificatoria: using grammars to trace Alberti's influence on Portuguese classical architecture, in Nexus Network Journal Vol. 13, No. 1.

Sass, L. 2007. A Palladian construction grammar-design reasoning with shape grammars and rapid prototyping. Environment and Planning B: Planning and Design 34. 87–106.

Stiny, G. 1980. Introduction to Shape and Shape Grammars. Environment and Planning B: Planning and Design 7. 343–352.

Oxman, N. 2007. Digital Craft: Fabrication Based Design in the Age of Digital Production, in Workshop Proceedings for Ubicomp 2007: International Conference on Ubiquitous Computing. September; Innsbruck, Austria; 534–538.

Green Design, Materials and Manufacturing Processes – Bártolo et al. (eds)
© 2013 Taylor & Francis Group, London, ISBN 978-1-138-00046-9

Architectural fabrication of tensile structures with flying machines

A. Mirjan, F. Gramazio & M. Kohler
Architecture and Digital Fabrication, ETH, Zurich, Switzerland

F. Augugliaro & R. D'Andrea
Institute for Dynamic Systems and Control, ETH, Zurich, Switzerland

ABSTRACT: The paper presents a manufacturing process for the erection of tensile structures with flying robots. Analogies are drawn to existing methods of aerial construction and robotic fabrication in architecture. Firstly, we describe a set of aerial building instructions for the vehicles, such as an assembly of a node and the erection of a link. Secondly, we investigate combinations of these instructions as prototypical structural arrangements and identify distinct characteristics for architectural production.

1 INTRODUCTION

The manoeuvre of aerial machines in construction is difficult and dangerous (Smith 2010). Today, manually operated flying vehicles are only used in challenging situations where no traditional construction method can be applied. However, new developments in sensing, computation and control allow to create autonomous flying machines that are able to perform complicated maneuvers in unstructured environments. The use of aerial robots opens up new possibilities in architectural production. This paper presents construction methods suitable for erecting tensile structures with flying machines (Fig. 1). It demonstrates the ability of a flying robot in construction to reach distant points in space and the unique skill to fly through and around existing objects while performing building tasks.

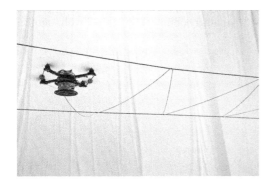

Figure 1. Autonomous erection of a tensile structure by a flying robot.

2 AERIAL ROBOTIC CONSTRUCTION OF TENSILE STRUCTURES

2.1 *Aerial construction*

Speculative architecture with flying machines exists for centuries. As a fantastic novel (Swift 1726), an architectural utopia (Krutikov 1928) or a hypothetical concept (Cook 1969), aerial architecture has always been a research interest.

In architectural fabrication, flying machines are applied on construction sites since the 1950's (Carter et al. 1963). Helicopters in construction are most commonly used to lift building materials and to transport it to remote locations with no access to streets. In highline construction such aerial cranes carry power poles to designated locations where they are assembled by workers on the ground. Helicopters are also applied to lift rotors for the erection of wind wheels or to string cables for the construction of suspension bridges (Cooper 1998). Another method of construction with aerial vehicles is the use of balloons (Gablenz & Spaltmann 2011) using lighter than air technology to generate lift. The balloons filled with helium hoist construction elements similarly to a crane. The position of the balloon and its cargo is controlled by individually adjusting the length of three cables connecting the balloon to the ground. A third technique of aerial construction was presented for the assembly of the Siduhe River Bridge (Wang et al. 2009). The installation of the suspension cables across the 500 m deep valley began with the placement of two pilot cables. The 1300 m long ropes were attached to two rockets and fired over the canyon to erect a link between the two sides.

2.2 Flying robots in architectural research

Research on robotic construction in architecture dates back to the early 1990s (Andres et al. 1994). Although highly advanced, these developments did not find access to the market since they were not flexible enough to adapt and react in different design situations (Gramazio & Kohler 2008). In the course of the recent shift towards digital technologies in architecture, universities have set up research facilities for construction with industrial robots resulting in adaptable (Helm et al. 2012) and sustainable (Oesterle et al. 2012) construction methods. Such novel technologies motivate new approaches to the design of architectural structures and advanced constructive systems. However, conventional robotic systems have predefined working areas. Stationary robotic arms or CNC-machines have a limited scale of action, constraining the size of the work-piece they act upon. These machines are usually smaller than buildings. This limits their use in architecture to the scale of a small artifact or building component (Kolarevic 2003). Mobile robots like dimRob (Helm et al. 2012), extend the working range of the machine in two dimension but are still constrained in elevation. Flying machines, however, do not have such tight boundaries of movement. The space they act upon is substantially larger than they are themselves making them apt to work at the full scale of architecture. The vehicles are not fixed to a base. This allows them not only to reach points in space otherwise not accessible by conventional machines but also to fly through and around existing objects while performing construction tasks. This unique feature has no other computer controlled construction machine today.

Research in aerial construction with flying robots is a recent topic. The Flight Assembled Architecture installation (Willmann et al. 2012) demonstrated the ability of quadrocopters to autonomously erect a highly differentiated structure by assembling a 6 meter tall tower out of 1500 foam elements. First steps into aerial construction with quadrocopters were also demonstrated by building cubic structures consisting of bars containing magnets (Lindsey et al. 2013). The ARCAS project (ARCAS 2011) focuses on aerial assembly by helicopters equipped with robotic arms. In parallel, hover-capable Unmanned Aerial Vehicles (UAVs) such as quadrocopters (Michael et al. 2010) and duct-edfan vehicles (Marconi at al. 2012), and their interaction with the environment are nowadays a research topic in many groups.

Today, quadrocopters offer an excellent compromise between payload capabilities, agility and robustness (Mahony et al. 2012).

2.3 Tensile elements as constructive material

Research into construction with flying machines requires on the one hand the development of adequate methods for hovercapable UAVs to physically interact with the environment and on the other hand, the investigation of light material systems and new constructive processes that are both robotically transportable and configurable at heights (Kohler 2012). In this context, this paper explores, with a series of experiments, the building of lightweight tensile structures by quadrocopters. Tensile elements, such as cables, are relatively lightweight, have a high structural strength and can span large distances. Objective of the investigation was the development of a set of building instructions for the vehicles to erect and manipulate tensile building elements. Different combinations of these instructions result in the realisation of different tensile structures. Therefore, we first defined characteristic instructions for the construction and translated them later into trajectories for the quadrocopters. Various configurations of these instructions were then tested to erect distinct prototypical formations of tensile structures. Experimental results validate the feasibility of the approach.

3 PARAMETRIC BUILDING INSTRUCTIONS

The flying vehicles lift, place and connect the linear building material to existing objects or already built structural components. The flexible tension elements dynamically react to the behaviour of the quadrocopters they are connected to. The sequencing and trajectory of the vehicle therefore directly influences the construction. In this section we present the basic building primitives we have used during the assembly processes.

3.1 Node

A node is a point of intersection of the cable with another object (Fig. 2) or with another

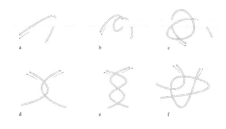

Figure 2. Node building instructions. a) Single turn hitch, b) (Multi-) round turn hitch, c) Knob, d) Elbow, e) Round turn, f) Multiple ropes knob.

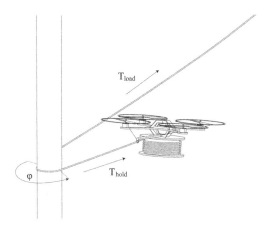

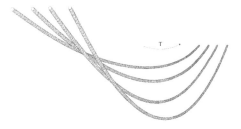

Figure 4. Control algorithms allow to apply a desired force (T) on the cable when spanning a link.

Figure 3. A small holding force on one side can carry a much larger loading force on the other side.

cable element. The material characteristics of the construction elements are used to connect to the support point by tying or weaving around it (Fig. 3). Following the capstan equation (Morten & Hearle 1993), depending on the coefficient of friction (μ) between the cable and a support point, and the amount of turns around the object (angle in radians φ), the loading force (T_{load}) can be calculated from the holding force (T_{hold}).

$$T_{load} = T_{hold}\, e^{\mu\varphi} \qquad (1)$$

This allows to parametrically design the nodes. It can be specified whether it is a gliding connection (Fig. 2a) or a fix node with a large holding force. Because of its exponential nature, a few rotations around an object already prevent the unreeving of the cable and hence generate a knot (Ashley 1944).

3.2 Link

A cable spanned between two structural support points generates a link. During the fabrication process we distinguish between static and dynamic supports. Already existing structural elements are static supports. The flying vehicles manipulating and guiding the cables from one static support to another are dynamic supports. The flying vehicles are controlled by appropriate methods that enables them to track desired trajectories and apply a desired force to the cable (Augugliaro & D'Andrea 2013). Furthermore, the vehicles automatically orient themselves along the cable direction, allowing for a smooth cable deployment. The tension of each link can be defined parametrically (Fig. 4).

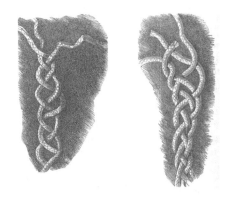

Figure 5. A braiding (Semper 1860) could not be constructed by a single UAV. However, the cooperative sequencing of three vehicles would allow to assembly the braid on the left. Four machines could produce the braid on the right.

3.3 Multivehicle cooperation

Digital control of the robots enables the vehicles to communicate and synchronize their actions among themselves. The machines can collaborate to lift particularly heavy loads (Micheal 2010). In addition, cooperation can be exploited during the assembly process. The vehicles don't merely distribute the workload among themselves but perform building tasks an individual machine could not accomplish alone, independently of the payload capacity. The flying robots can have complementary abilities with different skills. During the manufacturing of a node, for example, where the working end of a cable has to be carried through a loop (Fig. 2f), one vehicle could guide the working end while another forms the loop for the first one to fly through. However, working with multiple vehicles poses additional challenges. For example, the wash generated by the propellers of one vehicle could affect the performance of the others. Despite these difficulties, the cooperative performance of multiple flying machines widens the spectrum of possibilities in architectural production (Fig. 5).

3.4 *Experimental setup*

The experiments are performed in the Flying Machine Arena (www.FlyingMachineArena.org), a 10 × 10 × 10 meter indoor space for aerial robotic research. The space is equipped with a motion capture system that provide vehicle position and attitude measurements. This information is sent to a PC, which runs algorithms and control strategies and sends commands to the quadrocopters. The vehicle of choice are quadrocopters. These flying robots have demonstrated their dynamic capabilities performing flips (Lupashin et al. 2012), balancing poles (Hehn et al. 2011), learning fast maneuvers (Schoellig et al. 2012) and juggling balls (Muller et al. 2011).

The vehicles are equipped with a cable dispenser and a roller on which the tension elements are wound up. The friction of the roller can be adjusted and thus influencing the tension of the cable during its deployment. For the experiments described in this paper we worked with Ultrahigh-molecular-weight polyethylene rope (Dyneema). The material stands out due to a low weight-to-strength ratio, making it suitable for aerial manipulation. A 100 m long rope with a diameter of 4 mm weighs 1.1 kg and supports 1400 kg. The method allows to install different kinds of tensile elements such as cables, ropes or wires. The further description refers to the tensile elements as cables.

4 STRUCTURAL TYPOLOGY

Following the definition of a set of building instructions, we present in this section distinctive combinations of them, forming characteristic structural elements.

4.1 *Linear structure*

The linear structure is a tensile element spanning between two support points. It arranges a node, fastening the cable to an existing structural element and establishes a link, arranging an additional node at a second support point (Fig. 6).

With this basic building element, we demonstrate the ability of a flying machine to reach any point in space in order to erect a structure. Whether it is a link between two skyscrapers or a connection over 500m deep valley, the vehicle performs this task independently of the conditions on the ground. The tensile strength of the structure can be increased by adding additional links between the two supports.

4.2 *Surface structure*

The two-dimensional intersection of linear structures constitute a surface structure (Fig. 7). The vehicles establish nodes and link the tensile

Figure 6. Linear structure. The vehicle autonomously fastens the cable at the existing structure before spanning the link.

Figure 7. Surface structure. The vehicle constructs a planar structure by flying through and around built components.

elements to a structural entity. The loads and stresses of the intersecting members interact to find a structural equilibrium. The form of the structure adapts to the loads applied to it and hence dynamically changes during the deployment of additional nodes.

The surface structure experiment demonstrates the ability of the flying machines to fly through and around already constructed members of the structure while manipulating it. This is a unique feature of aerial machines. Conventional robotic arms in contrast would intersect with the structure while performing this manoeuvre.

4.3 *Volumetric structure*

The three-dimensional intersection of linear structures establishes a volumetric structure. Similarly to the surface structure described above, the volumetric structure members seek a tension equilibrium. Varying the tension on individual links

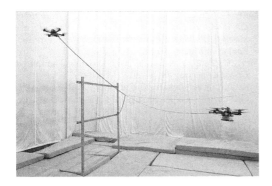

Figure 8. Volumetric structure. Two vehicles cooperate in order to accurately position a node in three-dimensional space.

allows to shift the position of a single turn node in space. Additional degrees of freedom can be added by having two vehicles cooperating. The vehicles can freely place a connection in space by performing a synchronised node manoeuvre at the designated position (Fig. 8). The controlled situating of a node within the design space could not have been done manually and demonstrates the possibility for new forms of architectural materialisation.

5 CONCLUSION

Architectural fabrication with digitally controlled flying machines is a new research direction, still in its fledgling stages. The construction method introduced in this paper, addresses the ability of flying machines in architectural production. It takes advantage of the capability of the vehicles to reach any point in space, allowing robots to erect suspending structure at locations otherwise not accessible by conventional construction machines. Further on, it makes use of the unique skill of aerial machines to manoeuvre in and around existing objects to fasten construction elements, and to fly in and around already built structural elements to manipulate them. Finally, it shows the knack to add degrees of freedom by multivehicle cooperation during the fabrication process, allowing to position nodes freely within the three-dimensional space.

From the architectural as well as from the robotic perspective, various aspects of the approach seek further exploration. The fabrication process is less constrained by traditional assembly and build up parameters, such as the need to build from the ground up. This profound difference calls for new design strategies incorporating nonlinear fabrication sequencing to materialise architecture that could not have been built before.

REFERENCES

Andres, J. Bock, T. & Gebhart, F. 1994. First results of the development of the masonry robot system ROCCO. *Proceedings of the 11th ISARC*: 87–93.

ARCAS. 2011. *www.arcas-project.eu.* accessed 12.03.2013.

Ashley, C.W. 1944. *The Ashley Book of Knots.* London: Faber and Faber Limited.

Augugliaro, F. D'Andrea, R. 2013. Admittance control for physical human-quadrocopter interaction. *European Control Conference ECC13.*

Carter, E.S., Decker, R.S. & Cooper, D.E. 1963. Handling Qualities Considerations of Large Crane Helicopters. *Annals of the New York Academy of Sciences* 107: 5–18.

Cook, P. 1969. *Instant City.* Courtesy of Collection FRAC Centre. Orleans. France.

Cooper, J.D. 1998. World's Longest Suspension Bridge Opens in Japan. *Public Roads* 62(1).

Gablenz, C.H. & Spaltmann, D. 2011. Das CargoLifter Ballonkransystem—der 'CargoLifter Lufthaken' ist da!. *19. Internationale Kranfachtagung 2011.* Magedeburg. Germany.

Gramazio, F. & Kohler, M. 2008. Towards a Digital Materiality. Kolarevic, B & Klinger, K. (eds.). *Manufacturing Material Effects: Rethinking Design and Making in Architecture.* New York: 103–118.

Hehn, M. & D'Andrea, R. 2011. A Flying Inverted Pendulum. *IEEE International Conference on Robotics and Automation*: 763–770.

Helm, V. Ercan, S. Gramazio, F. & Kohler, M. 2012. Mobile robotic fabrication on construction sites: DimRob. *International Conference on Intelligent Robots and Systems (IROS). IEEE/RSJ*: 4335–4341.

Kohler, M. 2012. Aerial Architecture. *LOG* (25). New York: 23–30.

Kolarevic, B. 2003. *Architecture in the Digital Age: Design and Manufacturing.* New York: Spon Press.

Kruitkov, G. 1928. *The City of the Future.* Courtesy of A.V. Schusev State Museum of Architecture. Moscow.

Lindsey, Q. Kumar, V. 2013. Distributed Construction of Truss Structures. *Springer Tracts in Advanced Robotics (86)*: 209–225.

Lupashin, S. & D'Andrea, R. 2012. Adaptive Fast Open-Loop Maneuvers for Quadrocopter. *Autonomous Robots* 33(1–2): 89–102.

Mahony, R. Kumar, V. & Corke, P. 2012. Multirotor Aerial Vehicles: Modelling, Estimation, and Control of Quadrotor. *IEEE Robotics & Automation Magazine* 19(3): 20–32.

Marconi, L. & Naldi, R. 2012. Control of Aerial Robots. Hybrid force/position feedback for a ducted-fan. *IEEE Control System Magazine* 32(4): 43–65.

Michael, N., Fink, J. & Kumar,V. 2010. Cooperative manipulation and transportation with aerial robots. *Autonomous Robots* 30(1): 73–86.

Morton, W.E. & Hearle, J.W. 1993. *Physical properties of textile fibres.* Textile institute.

Muller, M. Lupashin, S. & D'Andrea, R. Quadrocopter Ball Juggling. In IEEE/RSJ International Conference on Intelligent Robots and Systems: 5113–5120.

Oesterle, S. Vansteenkiste, A. & Mirjan, A. 2012. Zero Waste Free-Form Formwork. *2nd International Conference on Flexible Formwork (ICFF)*.

Schoellig, A.P. Muller, F.L. & Raffaello D'Andrea. 2012. Optimization-Based Iterative Learning for Precise Quadrocopter Trajectory Tracking. *Autonomous Robots* 33(1–2): 103–127.

Semper, G. 1860. *Der Stil in den technischen und tektonischen Künsten, oder Praktische Aesthetik*. Frankfurt a. M.: Verlag für Kunst und Wissenschaft.

Smith, M. 2010. *Two injured in Vermont helicopter crash*. http://www.wcax.com/story/12348670/2-injured-in-vermont-helicopter-crash. Accessed 12.03.2013.

Swift, J. 1726. *Travels into Several Remote Nations of the World, in Four Parts. By Lemuel Gulliver, First a Surgeon, and then a Captain of several Ships*. London: Benjamin Motte.

Wang, C., Peng, Y. & Liu, Y. 2009. Crossing The Limits. *Civil Engineering* 79: 64–80.

Willmann, J. Augugliaro, F. Cadalbert, T D'Andrea, R. Gramazio, F. Kohler, M. 2012. Aerial Robotic Construction Towards a New Field of Architectural Research. *International Journal of Architectural Computing* 10(3): 439–459.

Green Design, Materials and Manufacturing Processes – Bártolo et al. (eds)
© *2013 Taylor & Francis Group, London, ISBN 978-1-138-00046-9*

Space time information on display: Smart devices and spatial cognition

D.P. Henriques
Faculty of Architecture, Technical University of Lisbon, Portugal

ABSTRACT: Smart devices play an essential role in our lives, aiding navigation and solving other problems in both urban and non urban environments. In this paper, we analyze how these devices, such as smartphones, tablets and other interactive screens, are changing human spatial cognition of our surrounding environment. We examine recent research and applications for spatial cognition, and we establish a relation between space time information display on smart devices, as external imagery, and an ancient philosophical tradition that combines mental imagery with thought processes. We also suggest clues regarding possibilities of innovation. We conclude that the use of these devices can play a significant role for a sustainable development through energy and resource efficiency in a mediated reality. In this ubiquitous presence of smart devices, we can also consider software as a material that combines design innovation with social, environmental and economical issues, still relevant in mobile information societies.

1 INTRODUCTION

In this last decade, we have witnessed an exponential development of space time technologies that play an essential role in our daily lives, aiding navigation and solving other problems in both urban and non urban environments, through digital media such as smart devices connected to internet (e.g. via WiFi, 3G, 4G), global positioning systems (e.g. GPS, GLONASS, Galileo, Compass) and Location-Based Services (LBS).

Lynch (1960) argued that the cities' *imageability* could guide the spatial behavior and experiences of its inhabitants. The ease with which the interrelation of the city's elements (paths, edges, districts, nodes and landmarks) could be recognized and organized into a coherent pattern, was not only the result of several external characteristics, but also a product of its inhabitants, the memory of their past experiences, and the symbolic devices by which they oriented themselves in the city's environment. For example, Lynch recognized that the diagrammatic map of an underground transportation system could structure and build the identity of a metropolitan city like London (see also Monteiro 2008), and this can be seen today in both western and eastern metropolises.

Nowadays, smart devices with interactive applications support the performance of several categories of tasks in distinct situations (e.g. cities, natural parks, neighborhoods, airports, hospitals, museums, schools, and retail), adjusting to external change and producing, through digital display, both external and internal imagery of our surrounding environment. Through interactive representations of complex and temporally extended patterns of data to amplify not only human perception but also cognition, these smart devices are opening new possibilities for design innovation and business opportunities.

Can the images captured, viewed and saved in smartphones, tablets and other interactive touch screens enhance our thought processes? It is possible to develop human spatial cognition with these smart devices, combining design innovation with social, environmental and economical issues? We examine recent research and applications for spatial cognition and digital media, and we try to establish a relation between space time information display on smart devices, as external imagery, and a philosophical tradition that combines mental imagery with thought processes. We also suggest some clues regarding possibilities of innovation in the use of these smart devices. Finally we draw some conclusions on how the use of smart devices in our daily lives can play a significant role for a sustainable development.

2 APPLICATIONS FOR SPATIAL COGNITION

Montello & Raubal (2012) define human spatial cognition as the area of research that analyses activities that involve explicit or potentially explicit mental representations of space. While some functions of spatial cognition are used to solve primarily problems that involve spatial properties (e.g. location, size, shape, connectivity, direction, overlap, hierarchy, distance, dimensionality, pattern) as their core component, human spatial cognition plays a significant role in the performance of several categories of tasks such as: (1) location allocation, (2) using spatial language, (3) acquiring and using spatial knowledge from direct experience, (4) imagining places/reasoning with mental models (5) using spatially symbolic iconic representations and also (6) wayfinding as part of navigation. They also suggest some important application areas for spatial cognition research such as: (a) Location-Based Services (LBS), (b) geographic and other information systems, (c) information display, (d) architecture and planning, (e) personnel selection and (f) spatial education (Montello & Raubal, 2012, Montello, 2001; see also Dolins & Mitchell, 2010, for a comparative view between human and non-human animal spatial cognition; Ishikawa et al. 2008 and Wessel et al. 2010, for comparative studies between navigation with GPS-based mobile devices, direct experience of routes and paper maps; and Henriques, 2013, for a view on how GPS-based mobile devices change both external and internal imagery).

We can see today digital display apps running on smart devices (whether commercial or non-commercial) for all the application areas suggested above. For example, almost every area of the planet is covered by digital and interactive maps accessible via internet and it is possible to say that, in recent years, this application area of spatial cognition through information systems apps plays a substantial role in the smart devices market: Apple decided to compete with Google, by developing their system of interactive maps to be installed on their own devices. Also in architecture and planning, several institutions and companies are developing databases of future projects that can be experienced on site (e.g. SARA Nai in the Netherlands), and also of existing urban areas to be promoted (e.g. Dashilar app for Beijing Dashilar District in China) through digital display on smart devices. These applications combine design innovation with social, environmental, and economical issues, while developing human spatial cognition through the mediation between digital space time information on display, in smart devices, and existing physical reality.

3 MEMORY, IMAGERY AND MEDIA

About this ubiquitous presence of smart devices, Burnett (2012) suggests that 'the interface between reality and human perception has shifted from our eyes to eyes/screens/events/experiences/screens' in a 'vast and expanding process of annotation and visualization' that involves multiple layers of new *interface realities*. He argues that this process 'now defines experiences according to the strength with which they have been recorded', and that the image files captured, viewed, and saved in smart devices influence the formation of the environmental image, not only for the individual but also for the collective. Can the external imagery produced by smart devices, be called an artificial memory? Can it enhance our thought processes?

3.1 *Recalling the art of memory*

Although is not easy to define the meaning of mental imagery, Thomas (2010) suggests that 'imagery experiences are understood by their subjects as echoes, copies, or reconstructions of actual perceptual experiences' and 'at other times they may seem to anticipate possible, often desired or feared, future experiences.' Thomas also refers that 'according to a long philosophical tradition' mental imagery 'plays a crucial role in all thought processes,' like memory, visuo-spatial reasoning and creative thought. And he then recalls Aristotle, who saw images 'as playing an essential and central role in human cognition, one closely akin to that played by the more generic notion of mental representation in contemporary cognitive science' (Thomas, 2010).

The ancient Greeks also developed a technique of impressing 'places and images' on memory known as *method of loci*, that could be described as follows: '[the persons] must select places and form mental images of the things they wish to remember and store those images in the places, so that the order of places will preserve the order of things, and the images of the things will denote the things themselves, and we shall employ the places and images respectively as a wax writing-tablet and the letters written on it' Cicero, *De Oratore*, II, lxxxvi (Thomas, 2010, Yates, 1966). This 'artificial memory is established from places and images' *Ad Herrennium*, IV, and it was widely used, at that time, to organize memory (e.g. for speech) through the mental association and later *promenade* of real (or imaginary) architecture or urban spaces (Yates, 1966).

Also referred above in the quote from Cicero was the wax writing-tablet employed by ancient Greeks and Romans (see Fig. 1), which consisted essentially in a tablet of wood covered with a layer of wax where it was possible to write using a stylus, and that could be erased and reused by warming its surface.

Cicero was describing an internal imagery process through an external annotation device that is not very far, at least in its essential form, from recent smart digital devices like laptops or tablets.

Figure 1. Ancient greek man with a wax tablet. Excerpt from Painting by Douris (about 500 BC). Berlin, Staatliche Museen. Author: Pottery Fan. CC-BY.

In ancient Greece, this technique of the *method of loci*, also described like an artificial memory, was based on internal visual imagery, or visualization, to organize, store and recall information into coherent patterns. Contemporary research 'has shown that even a simple and easily learned form of the *method of loci* can be highly effective,' as well 'as other imagery based mnemonic techniques' (Thomas, 2010).

3.2 *Visuo-spatial representation and memory*

As we have seen, this new process of annotation and visualization is vast and in expansion. Ware (2004) argues that even the word *visualization* changed from 'being an internal construct of the mind' to 'an external artifact supporting decision making.' He goes on suggesting that in the 1970s this word was used as a construction of 'a visual image in the mind', while today visualization is more related to 'a graphical representation of data or concepts' (Ware, 2004). The use of the term visuo-spatial sketchpad, coined by Baddeley & Hitch's model of working memory in 1974 (Baddeley, 2000, see also Klauer & Zhao, 2004), and the sketchpad device, developed in 1963 by Ivan Sutherland, also show how internal processes and external digital technology have been influencing each other for the last decades. Research is being carried to unify both. The development of human spatial cognition, through research and also through its

multiple applications, is deeply related to the development of digital technologies and smart devices.

Shelton & Yamamoto (2009) argue that although there is a need for a unifying theory in the spatial cognition literature, recent data shows that 'individuals rely heavily on visual information for spatial learning.' They also conclude that 'although we can clearly establish aspects of spatial representation that are not strictly visually dependent, it is clear that vision and visual memory play a significant role in many different aspects of spatial cognition.' Jiang et al. (2009) argue that 'visual memory allows us to maintain spatiotemporal continuity in this constantly changing environment. It enables us to visualize without actually seeing, and it helps us see things we already experienced more efficiently.' Couldn't we say the same thing about smartphones, phablets, tablets, laptops and other smart devices?

4 SKETCHES ON A MEDIATED REALITY

In the last decades, information visualization has been an extremely successful area in both research and application domains, using several technologies.

We highlight the personal art project, by Fernanda Viégas & Martin Wattenberg, Wind

Figure 2. Mediated reality running on Apple iPhone.

Map: accurate surface wind data, revised once per hour via NDFD (National Digital Forecast Database in the United States of America), is visualized online through a delicate tracery of visual motion. Although it was conceived as an artistic exploration, implemented in HTML and JavaScript, this recent *data viz* project represents a beautiful example of the potential of software to externalize invisible dynamics.

Most of the researches in data viz are focused on enhancing data analysis, through visualization, to clarify intuitive though processes. However, other features of information visualization, such as the concretization of mental models through visual display to verify ideas/hypotheses, or the organization and sharing of data, can also be developed in applications that aid users to perceive and understand space time information (Lau et al. 2009, Tory & Moller, 2004).

Software implementation for digital display apps, running on smart devices, is evolving very fast and the same can be said for hardware. Micro-Electro-Mechanical Systems (MEMS) and sensor technology are also being developed very fast (e.g. ambient light and proximity sensors, gyroscope, accelerometer, compass), broadening the sensing and tracking capabilities of smart devices to captured and measure spatial properties (e.g. location, size, shape, connectivity, direction, overlap, hierarchy, distance, dimensionality, pattern).

We are particularly interested in spatial cognition application areas concerned with spatial education, architecture and planning: visualizations of invisible dynamics and other space time data, sensed, implemented and displayed on smart devices for terrain or building analysis, and virtual model visualization juxtaposed on site. The use of these devices, as sketchbooks, can fuel innovation on early stages of spatial media design process, but it can also enhance rigorous control over later stages of construction processes. We also foresee more possibilities to study, enhance, and externalize cognitive maps with smart devices, since they can support different views (in both allocentric and egocentered spatial systems).

5 CONCLUSION

In a mediated reality, space time information displayed with smart devices can play a significant role for a sustainable development. Both software and hardware research and applications are enhancing energy and resource efficiency: their autonomy, mobility, connectivity, geo location and multitask capabilities, if combined with a low cost to their users, can have an positive impact in low income countries and rapid growing economies,

through a more open access to information, services and social media.

We see such smart devices not only as generators of external imagery but also as an external artificial memory of mobile information societies—remember Facebook's timeline or LinkedIn home?—that can amplify human perception and cognition, opening new possibilities for design innovation and business opportunities. They can enhance thought processes, but also the performance of several tasks that involve spatial cognition, such as rigorous navigation or *flâneur* exploration.

Smart devices can also be a valuable tool to foster innovation on spatial media design, for terrain or building analysis, and virtual model visualization juxtaposed on site.

In this ubiquitous presence of smart devices, we can also consider software as a material that combines design innovation with social, environmental and economical issues (particularly if there would a unification of operating systems, but maybe this would forget the economical issues of commercial business, and diversification strategies), still relevant in mobile information societies.

Future work will be focused on spatial education, architecture and planning, both on research and apps of invisible dynamics and other space time data, sensed, implemented and displayed on such devices.

ACKNOWLEDGMENTS

We deeply thank our master dissertation supervisors Professor Filipa Roseta Vaz Monteiro and Professor Francisco Santos Agostinho, Faculty of Architecture, Technical University of Lisbon. We also thank, for their time and advices, Professor Zhu Li, Confucius Institute, University of Lisbon and Professor Jorge Fava Spencer, Faculty of Architecture, Technical University of Lisbon.

REFERENCES

Baddeley, A. 2000. The episodic buffer: a new component of working memory? In *Trends in cognitive sciences* 4 (11): 417–423.

Baer, K. 2008. *Information design workbook: graphic approaches, solutions, and inspiration + 30 case studies.* Beverly: Rockport Publishers.

Burnett, R. 2012. Interface realities. In *Critical approaches to culture + media* (weblog) November 17, 2012 <http://rburnett.ecuad.ca/ronburnett/2012/11/17/interface-realities> [accessed in 2013–4–15].

Card, S.K. et al. 1999. *Readings in information visualization: using vision to think.* San Diego: Academic Press.

Carlson, L.A. et al. 2010. Getting lost in buildings. In *Current directions in psychological science* 19 (5): 284–289.

Csikszentmihalyi, M. 1996. *Creativity: flow and the psychology of discovery and invention*. New York: HarperCollins.

Damásio, A. 2010. *O Livro da consciência: a construção do cérebro consciente*. Maia: Círculo de Leitores.

Dolins, F. & Mitchell, R. 2010. *Spatial cognition, spatial perception—mapping the self and space*. Cambridge: Cambridge University Press.

Gibson, D. 2009. *The wayfinding handbook: information design for public spaces*. New York: Princeton Architectural Press.

Golledge, R.G. (ed.) 1999. *Wayfinding behavior: cognitive mapping and other spatial processes*. Baltimore, Maryland: The John Hopkins University Press.

Henriques, D.P. 2013. O (re)verso da paisagem: percepção e cognição espacial auxiliada por sistemas de posicionamento global (GPS) [The (re)verse of landscape: spatial perception and cognition aided by GPS]. In *O (re)verso da paisagem—filosofias da pobreza e da riqueza* (CD-ROM): 8 pp. Lisboa: Centro Editorial e de Comunicação da Faculdade de Arquitectura da Universidade Técnica de Lisboa.

Ishikawa, T. et al. 2008. Wayfinding with a GPS-based mobile navigation system: A comparison with maps and direct experience. In *Journal of environmental psychology* 28 (3): 74–82.

Jiang, Y.V. et al. 2009. Visual memory for features, conjunctions, objects, and locations. In James R. Brockmole (ed.), *The visual world in memory*: 33–65. Hove: Psychology Press.

Klauer, K.C. & Zhao, Z. 2004. Double dissociations in visual and spatial short-term memory. In *Journal of Experimental Psychology: General* 133 (3): 355–381.

Lau, H.Y.K. et al. 2009. A VR-based visualization framework for effective information perception and cognition. In *Advances in intelligent and soft computing* 60: 313–332.

Lynch, K. 1960 [1994]. *The image of the city*. Cambridge: The MIT Press (original work published in 1960).

Monteiro, F. 2008. *Underground: from zapping to musical cityscape*. Presentation at the IXth International Conference on History on Comparative History of European Cities, Lyon.

Montello, D. 2001. Spatial cognition. In N. Smelser & P. Bates (eds.); *International encyclopedia of the social & behavioral sciences*: 14771–14775. Oxford: Pergamon Press.

Montello, D. Raubal, M. 2012. Functions and applications of spatial cognition. In D. Waller & L. Nadel (eds.), *The APA handbook of spatial cognition*: 555–591. Washington DC: American Psychological Association.

Reas, C. et al. 2010. *Form+code in design, art, and architecture*. New York: Princeton Architectural Press.

Shelton, A.L. & Yamamoto, N. 2009. Visual memory, spatial representation, and navigation. In James R. Brockmole (ed.), *The visual world in memory*: 140–177. Hove: Psychology Press.

Thomas, N. 2010. Mental imagery. In Edward N. Zalta (ed.), *The stanford encyclopedia of philosophy*. <http://plato.stanford.edu/archives/win2011/entries/mental-imagery/> [accessed in 2013-4–15].

Tory M. & Moller, T. 2004. Human factors in visualization research. In *IEEE transactions on visualization and computer graphics* 10 (1): 72–84.

Valhouli, C.A. 2010. *The internet of things: networked objects and smart devices* (research report). New York /Bradford, MA: The Hammersmith Group.

Ware, C. 2004. *Information visualization: perception for design*. San Francisco: Elsevier.

Wessel, G. et al. 2010. GPS and road map navigation: the case for a spatial framework for semantic information. In *Proceedings of the international conference on advanced visual interfaces AVI'10*: 207–2014. New York: ACM.

Yates, F.A. 1966 [2001]. *The art of memory*. Kent: Pimlico (original work published in 1966).

Green Design, Materials and Manufacturing Processes – Bártolo et al. (eds)
© *2013 Taylor & Francis Group, London, ISBN 978-1-138-00046-9*

Fabricating selective elasticity

Maria Paz Gutierrez

Department of Architecture, University of California, Berkeley, USA
Energy Climate Partnership of the Americas, US Dept. of State, USA

ABSTRACT: Bioinspiration can fuel conceptual leaps for the creation of membranes with selective elasticity capable of responding synergistically to multiple environmental conditions. Recent advances in new methods designed to produce elastomeric substrates with variable densities, geometries, and flexibilities through digital fabrication can be pivotal for technological shifts in this frontier. However, pioneering opportunities of integrating nano and microengineering design and fabrication to advance the field of self-regulated membranes with multifunctional capabilities remains largely unexplored. Researchers face crucial challenges in this endeavor. While material science moves swiftly into establishing new elastomeric composites, often these advances are aimed towards applications under the millimeter scale. As such, innovations frequently prove inadequate for building applications primarily due to insufficient climatic and structural endurance. Another major barrier in the development of multifunctional elastomeric membranes in architecture pertains to fabrication. Scalability obstacles derived from the gap between 3d printing technologies and microscale manufacturing are significant. *SABERs* research provides a framework to discuss an interdisciplinary and multiscale design approach to challenge scalability barriers. The research discusses key potential opportunities in digital fabrication of lightweight building envelopes with selective elasticity. Programming selective matter can prove significant for future building resilience.

1 INTRODUCTION

Physical elasticity has been a vehicle in architecture as much as for technological leaps as for supporting sociocultural transformations. Air-based dirigibles, atmospheric railways, zeppelins, and Joseph Plateau's mid-nineteenth century paradigmatic research on minimal surfaces, are some of the inventions and discoveries which led architects to consider pneumatics as a material worth experimenting with. Advances in air physics and elastic materials developed after WWII, supported a plethora of architectural trials and developments in lightweight, portable, and rapidly deployable pneumatic systems. On the other hand, elastomer based inflatables, are widely acknowledged to have been an essential means to support cultural transformations in the 1960's (Pidgeon 1968, Banham 1966, Dent 1972). Inflatables provided alternatives to conventional construction systems. They streamlined opportunities of change in architecture, from adjustments to environmental conditions to ease of transportability (Steiner 2005).

From the latter quarter of the twentieth century, advances in membrane technologies have streamlined experimentations with minimal surfaces and later with structural pneumatic advances from rigidified elastomeric composites to Tenseairity systems (Campbell 1963, Luchsinger et al. 2008).

The growing interest in advancing elastomer membranes encourages researchers to turn towards nature in search of models of structural efficiency for lightweight and flexibility. Not only does nature design membranes with variable elasticity, it does so seamlessly. Variable functions are calibrated through geometries and biochemical operations that result in structural differentiations across scales (Ball 1997). Frequently, these scale-based variations are consequences of variable rigidity and elasticity created by permutations of both organic or mineral structures and their infinite combinations (Huebsch & Mooney 2009).

Aware of the advantages provided by nature's design of variable elasticity, science has advanced particularly in the last decade our capacity to create smart materials that can benefit from these advantages. By emulating an abalone shell's structural strength through compressive mineral "columns" and flexible organic "slabs" to Gecko-inspired adhesive structures, we are constantly drawn to nature as a way to progress in the field of elastomers and flexible systems (Meyers et al. 2008, Mengüç et al. 2012). This interest has sparked great attention towards understanding that organisms implement variable elasticity in order to respond to environmental conditions. Elasticity is deployed in search of integrated functional optimization based on selective responsiveness (Behl et al. 2010).

Researchers strive to advance functional capabilities of matter based on nano and microengineering (Ratner & Bryant 2004). It is becoming progressively clearer that new interdisciplinary collaborations and funding models established to advance sustainable building technologies is leading researchers to evaluate the potential of multifunctional materials in construction. In part, this field is concentrating specifically in the potential of multifunctional materials for membranes or "living walls" characterized by self-regulated capabilities and integrated functions (http://www.nsf.gov/eng/efri/fy10awards_SEED.jsp). Variable rigidity can play a critical role in this frontie, since it supports resilient adaptability.

This paper presents SABERs research as a framework for discussing the potential of new 3d printing technologies for manufacturing elastomeric architectural membranes designed for environmental control actuation. By combining microphotonic lenses to reactive polymers, SABERs tests the boundaries of fabrication scalability (Jeong et al. 2006, Gutierrez 2011a).

2 ELASTOMERS AND ARCHITECTURE

The incorporation of rubbers into architecture has been anchored in two intertwined foundations. For one part, it has been rooted in the continuing advances in elastomer substrates and composites carried since early twentieth century evolving from the introduction of the gutta percha (J. Tressendant, 1956), the vulcanization of rubber (T. Hancock—patent 1844), and the first synthetic elastomer synthesized from butadiene (S. Lebedev, 1910) to recent advances in high performance elastomeric composites (Barlow et al. 1994, Wood 1940, Ha et al. 2006). For one part the vulcanization of rubber with sulphur provided a technological leap in structural resistance and resilience in moldings and products revolutionizing industries (Loadman 2005). Comparative progress was established through the later introduction of synthetic rubber with exponential growth during the WWII period. The incorporation of elastomers across industries ranging from aerospace to architecture fuelled technological and socioeconomic paradigm shifts (Tully 2011).

Significant transformations in architectural theory and technology stemmed from the convergence of progressive technological advances in rubbers and the conceptual shifts triggered by a growing understanding of elastomers' unique capacity to house maximum volumes with minimum surface (Dent 1972, Ball 2011). From the early air experiments of Boyle and Hobbes, the microphysical investigations of Maxwell and Faraday, to Plateau's largely known studies on soap bubbles, the singular advantages of the soap bubble geometries fueled new concepts and productions in architecture, particularly in the 1960's (Shapin & Shaffer 2011, Shirley 1951, Herzog 1976) (Thompson & Bonner 1961). Material elasticity endowed structural experimentation with organic forms, temporal and flexible constructions, streamlining new concepts of adaptability.

During the decades to follow, elastomers permeated architectural experimentation, primarily as a substrate for deployable and lightweight structures (Otto 1973, Forster 1994, LeCuyer et al. 2008, Fernandez 2012). More recently, architecture research has begun to experiment with coating technologies and hybrid structural systems. High-performance materials for pneumatics applications include *Rigidified Pneumatic Composites* (RPC), where a chemical (surface application) hardens the elastomer to become a rigid and permanent structure (Van Dessel & Chini 2001).

Innovation in air-based structure is moving into new systems. A key example pertains to the coupling of tensegrity to air pressure modules: Tensairity®. This hybrid structural triples the compressive strength of typical pneumatic systems (Luchsinger & Crettol 2006). Structural and material innovations of pneumatic systems, such as, RPC and Tensairity®, are providing unforeseen opportunities for lightweight and large span applications (Bechthold 2008).

To a large extent, the fields of robotics and aeronautics are the key drivers of present innovation in potential applications of elastomeric structures. Innovation ranges from experimentation in artificial muscles to intelligent flexible air structures (Zhang et al. 2010, Speck et al. 2006). An example is NASA's *InFlex* technology development program initiated in 2005. The program was created to study the potential incorporation of recent advancements in material science and electrical engineering in inflatable structures' membranes. The primary method developed in the studies used a simple breakpoint detection system of conductive traces or fiber optics embedded in the elastomer. InFlex employed conformal coated conductive ink applied to polyimide, polyester, and polyurethane. This was done through standard ink-jet printing techniques allowing for finest pitch resolution of the traces and particle size detection (Cadogan et al. 2006). Polyimide-based capacitance sensing has been successfully developed in other research studies, by exploiting modifications in capacitance (Brandon et al. 2011). While the intersection of these advances remains undoubtedly distant to architecture, in the years these advances will permeate research and construction. In this quest, it is anticipated that current fabrication technologies of elastomers for architectural applications will require transformations of current digital fabrication technologies.

3 PROGRAMMING ELASTICITY IN DIGITAL FABRICATION

3.1 *Current digital fabrication and elastomers*

Architects are recently developing innovative approaches in flexibility by probing variable rigidity through digital fabrication processes. Composites made with particles of traditional construction materials such as wood and cement, as well as polymers, are examples of new materials being developed with "phased elasticity" (Oxman 2012). These attempts mark new approaches towards material flexibility. Architects continue to pioneer into advancing material systems particularly through 3d printing technologies with variable polymeric combinations which open new opportunities for programming elasticity (Oxman 2011). 3D printing technologies with elastomers have also been researched through informal experimentations i.e. do-it-yourself in the past years (www.youtube.com/watch?v=DzT0LJWOoBA).

Both formal and informal 3d printing investigations are paralleled by explorations in air responsive structures, mechatronics, and uses of rubbers as active molds both in practice and academia (Hensel & Menges 2007, Khan 2008, Fox & Kemp 2009). Elastomeric studies in academia are testing traditional boundaries of pneumatics. Figure 1 presents an inflated shape formed a doubly curved parabolic configuration known for strength in tensile structures, but unseen in pneumatics. Decreasing the width of the perforated areas caused an increasing in the curvature at a rate proportional to the width reduction. The structural implications of such an inflatable unit remain yet untested. However, it provides an interesting paradox of tensile curvature achieved through inflation.

New frontiers frequently bring unanticipated challenges. The 3D printing of elastomeric

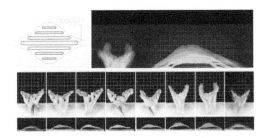

Figure 1. Bioinspired air structure unit by K. Gravier (Studio F2012, director: MP Gutierrez, UC Berkeley).

substrates does not escape from this trend. From restricted fine grain resolution, required for higher actuation sensitivity, to limited elastomeric combinations, the exploration of new material composites and variable structural hierarchies is yet to be solidified (Wosnick & Shoichet 2008, Ghajar et al. 2008). In response, architects, engineers, and scientists are establishing new collaborations to challenge existing restrictions in fabrication that supports broad elastomeric uses. Moreover, advances to support multiscale design where nano and micro scale fabrication can be connected to architectural scales of production will likely become increasingly sought after (Gutierrez 2011b).

3.2 *Scalability barriers in digital fabrication*

Recreating the capacity to selectively exchange functions with environments is known to bear strong potential for architecture. Pioneering studies in this, such as Lang's pneumatic multi-control membrane, have sought this aim (Laing 1967). To establish materials with programmed selective reactivity is, nevertheless, challenging. The transfer of behavioral principles from nature into architecture requires structural definitions across variable scales. Hence, fabrication processes in multiscale design present some of the obstacles faced in pursuit of multifunctional matter (Meyers et al. 2008a, Srinivasan et al. 1991).

A fundamental component of the selective reactive capabilities of natural organisms is the capacity to respond and balance multiple functions through an integrated material network. It is this multifunctional capacity that endows a synergistic and optimized balance with environmental dynamics. With the aim to advance multifunctional materials, SABERs research, strives to establish a new building membrane that can synergistically and selectively react through adaptive and integrated material responses to environmental dynamics (National Science Foundation award #CMMI-103002).

Fueled by models from nature (insect's sencilla and stoma) characterized by high calibration sensibility to hygrothermal and light conditions, SABERs research aims to establish a building envelope that can self-regulate ("The Bio-Engineered Façade—Architect Magazine." 2013). Multifunctional and self-regulation are designed to balance in an integrated fashion light, humidity, and temperature demands for human comfort. SABERs establishes a combination of reactive polymers and microlens arrays (Fig. 2).

SABERs incorporates the intersection of microphotonics and reactive biopolymers to create an envelope composed of microventuri tubes array designed to control light transmission and air rate

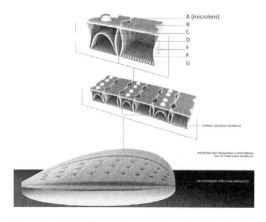

Figure 2. SABERs membrane schematic diagram (PI: Maria—Paz Gutierrez and Luke Lee, UC Berkeley).

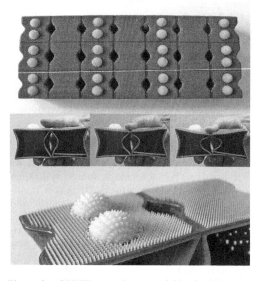

Figure 3. SABERs membrane variable elasticity—3D prints—top (scale 4:1; middle and bottom 20:1 scale) (Maria—Paz Gutierrez and Charles Irby, UC Berkeley).

increase as required by lightweight architecture situated in hot and humid environments. Advancing our understanding and capacity to code multifunctional behavior in matter can yield unforeseen opportunities to interface with the environment.

SABERs early experiments concentrated in establishing a new photoresponsive biopolymer membrane (Fig. 4). Since the elastomeric substrate was produced in the lab, a major research aim has been to test its potential scalability for the architectural scale. Consequently, early studies are being developed to explore multiscale elastomeric fabrication through existing 3d printing technologies (Fig. 3).

The initial phase has encompassed the custom design and sourcing of a 3d printer created to support elastomeric printing without any limitations of material selection (Fig. 5). One and two syringes prototypes are testing the resolution, speed, overhang angles, and volume of elastomeric printing with variable rigidity (Figs. 6 and 7).

Early tests are presenting a comparative speed to standard polymer 3d printing (PLA). A fundamental advantage of SABERs early prototypes is the implementation of one step printing process and unlimited elastomeric uses. Our digital fabrication prototypes have the potential of decreasing material losses and energy (no need of supporting material), as well as, of supporting the development of natural elastomeric fabrication (Fig. 8). The next phase of research will continue to test methods to improve resolution and scalability where polymerization can be integrated to one step 3d printing.

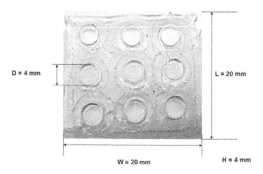

Figure 4. SABERs photoreactive biopolymer: early experiment model (Younggeon Park, Maria—Paz Gutierrez and Luke Lee, UC Berkeley).

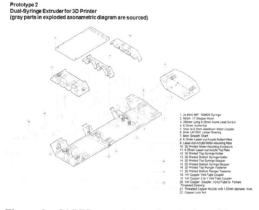

Figure 5. SABERs—prototype 2 extruder with two syringes.

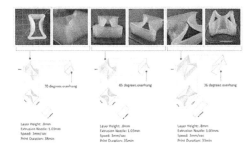

Figure 6. SABERs: SABERs: silicone rubber overhang angles tests with 1.003 nozzle and 8 mm layer height at 3 mm/s. Probe angles vary from 70 to 35 degrees (Maria-Paz Gutierrez and Pablo Hernandez, UC Berkeley).

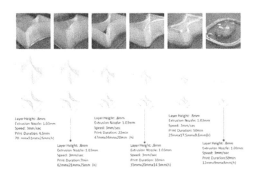

Figure 7. SABERs: silicone rubber scalability tests with 1.003 nozzle and 8 mm layer height at 3 mm/s. Probes footprint range from 70.5 mm × 54.2 mm × 29 mm (height) to 11.7 mm × 8.7mm × 4.8 mm (height) (Maria-Paz Gutierrez and Pablo Hernandez, UC Berkeley).

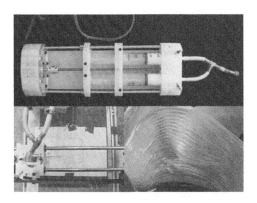

Figure 8. SABERs: photos of dual syringe frame (top), nozzle, latex 3D print test, and variable elasticity 3D print test (right bottom) (Maria-Paz Gutierrez and Pablo Hernandez, UC Berkeley).
(UC Berkeley BIOMS Research Group: M.P. Gutierrez (founder/lead), C. Irby, P. Hernandez, P. Suen, J. Lee, C. Lee and D. Campbell).

4 CONCLUSION

The power of integrating nano and microengineering to advance smart materials architecture, particularly for multifunctional and self-regulation capabilities is beyond dispute. A growing awareness is rising around this frontier in architecture (Ashby et al. 2009). Thus, it is expected that architects will be progressively more drawn towards advances in elastomers produced through nano and microengineering fabrication systems, i.e. electroactive polymers, self-healing elastomers, and rigidizable coatings. The need to advance sustainable multiscale design and fabrication to generate multifunctional and reactive elastomers is strong. Fabricating selective elasticity can render architectural resilience possible.

Support from the National Science Foundation is acknowledged.

REFERENCES

Ashby, Michael F., Paulo JSG Ferreira, and Daniel L. Schodek. 2009. *Nanomaterials, Nanotechnologies and Design: An Introduction for Engineers and Architects.* Elsevier.
Ball, Philip. 1999. *Made to Measure: New Materials for the 21st Century.* Princeton University Press.
2011. *Shapes: Nature's Patterns: a Tapestry in Three Parts.* Vol. 1. OUP Oxford.
Banham, Reyner. 1966. Zoom Wave Hits Architecture. *New Society*, March 3.
Barlow, Colin, Sisira Jayasuriya, and C. Suan Tan. 1994. *The World Rubber Industry.* Routledge.
Bechthold, Martin. 2008. *Innovative Surface Structures: Technology and Applications.* Taylor & Francis Group.
Behl, Marc, Muhammad Yasar Razzaq, and Andreas Lendlein. 2010. Multifunctional Shape-Memory Polymers. *Advanced Materials* 22 (31): 3388–3410.
Brandon, Erik J., Max Vozoff, Elizabeth A. Kolawa, George F. Studor, Frankel Lyons, Michael W. Keller, Brett Beiermann, Scott R. White, Nancy R. Sottos, and Mark A. Curry. 2011. Structural Health Management Technologies for Inflatable/deployable Structures: Integrating Sensing and Self-healing. *Acta Astronautica* 68 (7): 883–903.
Cadogan, David, Craig Scheir, Anshu Dixit, Jody Ware, Janet Ferl, Dr Emily Cooper, and Dr Peter Kopf. 2006. Intelligent Flexible Materials for Deployable Space Structures (InFlex). *In SAE International Conference on Environmental Systems, 06ICES-91*, Virginia, Beach, VA.
Campbell, J.B. 1963. Synthetic Elastomers. *Science* 141 (3578) (July 26): 329–334.
Dent, Roger Nicholas. 1972. *Principles of Pneumatic Architecture.* Halsted Press Division, Wiley.
Fernandez, John. 2012. *Material Architecture.* Routledge.
Forster, Brian. 1994. Cable and Membrane Roofs—a Historical Survey. *Structural Engineering Review* 6 (3): 145–174.

Fox, Michael, and Miles Kemp. 2009. *Interactive Architecture*. Princeton Architectural Press.

Ghajar, Cyrus M, Xiaofang Chen, Joseph W Harris, Vinod Suresh, Christopher CW Hughes, Noo Li Jeon, Andrew J Putnam, and Steven C George. 2008. The Effect of Matrix Density on the Regulation of 3-D Capillary Morphogenesis. *Biophysical Journal* 94 (5) (March 1): 1930–1941.

Gideon, Siegfried. 1948. Mechanization Takes Command. *Nova Iorque*: Oxford University Press.

Gutierrez, Maria Paz. 2011a. Matter: Sense and Actuation-Self-Activated Building Envelope Regulation. *In Project Catalogue of the 31st Annual Conference of the Association for Computer Aided Design in Architecture (ACADIA)*, 114–120. Banff (Alberta).

2011b. Innovative Puzzles. *In Project Catalogue of the 31st Annual Conference of the Association for Computer Aided Design in Architecture (ACADIA)*, 70–71. Banff (Alberta).

Ha, Soon M., Wei Yuan, Qibing Pei, Ron Pelrine, and Scott Stanford. 2006. Interpenetrating Polymer Networks for High-Performance Electroelastomer Artificial Muscles. *Advanced Materials* 18 (7): 887–891.

Hensel, Michael, and Achim Menges. 2007. Morpho-Ecologies: Towards Heterogeneous Space In Architecture Design. London: AA Publications.

Herzog, Thomas, Gernot Minke, and Hans Eggers. 1976. *Pneumatic Structures: a Handbook of Inflatable Architecture*. Oxford University Press.

Huebsch, Nathaniel, and David J. Mooney. 2009. *Inspiration and Application in the Evolution of Biomaterials*. Nature 462 (7272): 426–432.

Jeong, Ki-Hun, Jaeyoun Kim, and Luke P. Lee. 2006. Biologically Inspired Artificial Compound Eyes. *Science* 312 (5773): 557–561.

Khan, Omar. 2008. Reconfigurable Molds as Architecture Machines. *In Silicon + Skin: Biological Processes and Computation." In Proceedings of the 28th Annual Conference of the Association for Computer Aided Design in Architecture (ACADIA)*, 286–291. Minneapolis, Minnesota.

Laing, N. 1967. The Use of Solar and Sky Radiation for Air Conditioning of Pneumatic Structures. *In IASS, Proceedings of the 1st International Colloquium on Pneumatic Structures*, University of Stuttgart, Stuttgart, Germany, 11–12.

LeCuyer, Annette W., Ian Liddell, Stefan Lehnert, and Ben Morris. 2008. *ETFE: Technology and Design*. Birkhäuser.

Lee, Haeshin, Bruce P. Lee, and Phillip B. Messersmith. 2007. A Reversible Wet/dry Adhesive Inspired by Mussels and Geckos. *Nature* 448 (7151): 338–341.

Loadman, John. 2005. *Tears of the Tree: The Story of Rubber-a Modern Marvel*. Oxford University Press.

Luchsinger, R.H., and R. Crettol. 2006. Adaptable Tensairity. *Adaptables*, Eindhoven, The Netherlands.

Luchsinger, R.H., R. Crettol, and T. S. Plagianakos. 2008. Temporary Structures with Tensairity. *In Proceedings of 3rd Latin American Symposium on Tensile-Structures*. Acapulco: International Symposium IASS-SLTE.

Mengüç, Yiğit, Sang Yoon Yang, Seok Kim, John A. Rogers, and Metin Sitti. 2012. Gecko-Inspired Controllable Adhesive Structures Applied to Micromanipulation. *Advanced Functional Materials*.

Meyers, Marc André, Po-Yu Chen, Albert Yu-Min Lin, and Yasuaki Seki. 2008a. Biological Materials: Structure and Mechanical Properties. *Progress in Materials Science* 53 (1): 1–206.

———. 2008b. Biological Materials: Structure and Mechanical Properties. *Progress in Materials Science* 53 (1): 1–206.

Otto, Frei. 1973. *Tensile Structures, Vol. 1: Pneumatic Structures and Vol. 2: Cables, Nets and Membranes*. The MIT Press, Cambridge, Massachusetts.

Oxman, Neri. 2011. Variable Property Rapid Prototyping. *Virtual and Physical Prototyping* 6 (1): 3–31.

———. 2012. Programming Matter. *Architectural Design* 82 (2): 88–95. doi:10.1002/ad.1384.

Pedretti, Dr Mauro. 2004. TENSAIRITY®. *In European Congress on Computational Methods in Applied Sciences and Engineering* (ECCOMAS), 1–9.

Pidgeon, Monica. 1968. Pneu World. *Architectural Design*, June.

Ratner, Buddy D., and Stephanie J. Bryant. 2004. Biomaterials: Where We Have Been and Where We Are Going. *Annu. Rev. Biomed. Eng.* 6: 41–75.

Shapin, Steven, and Simon Schaffer. 2011. *Leviathan and the Air-Pump: Hobbes, Boyle, and the Experimental Life*. Princeton University Press.

Shirley, John W. 1951. The Harvard Case Histories in Experimental Science: The Evolution of an Idea. *American Journal of Physics* 19: 419.

Srinivasan, A.V., G.K. Haritos, and F.L. Hedberg. 1991. Biomimetics: Advancing Man-made Materials Through Guidance from Nature. *Appl. Mech. Rev* 44 (11): 463–482.

Speck, T., R. Luchsinger, S. Busch, M. Rüggeberg, and O. Speck. 2006. Self-healing Processes in Nature and Engineering: Self-repairing Biomimetic Membranes for Pneumatic Structures. *Design and Nature* 3: 105–114.

Stalder, Laurent, and Jill Denton. 2010. Air, Light, and Air-Conditioning. *Grey Room*: 84–99.

Steiner, Hadas. 2005. The Forces of Matter. *The Journal of Architecture* 10 (1): 91–109.

"The Bio-Engineered Façade—Architect Magazine." 2013. Accessed March 18. http://www.architectmagazine.com/blogs/postdetails.aspx?BlogId=mindmatterblog&postId=94834.

Thompson, D.W., & Bonner, J.T. (1961). *On growth and form*. Cambridge: Cambridge University Press

Tully, John. 2011. *The Devil's Milk: A Social History of Rubber*. NYU Press.

Van Dessel, Steven, and Abdol R. Chini. 2001. Rigidified Pneumatic Composites: Use of Space Technologies to Build the Next Generation of American Homes. *National Science Foundation* Award Abstract #0122022.

Wood, Lawrence A. 1940. Synthetic Rubbers: a Review of Their Compositions, Properties, and Uses. *Rubber Chemistry and Technology* 13 (4): 861–885.

Wosnick, Jordan H., and Molly S. Shoichet. 2007. Three-dimensional Chemical Patterning of Transparent Hydrogels. *Chemistry of Materials* 20 (1): 55–60.

Zhang, Zhiye, Michael Philen, and Wayne Neu. 2010. A Biologically Inspired Artificial Fish Using Flexible Matrix Composite Actuators: Analysis and Experiment. *Smart Materials and Structures* 19 (9): 094017.

Green Design, Materials and Manufacturing Processes – Bártolo et al. (eds)
© 2013 Taylor & Francis Group, London, ISBN 978-1-138-00046-9

Digital manufacture: Robotic CAD/CAM protocol for low cost housing

Vasco Portugal
MIT PP—IST, Lisbon, Portugal

ABSTRACT: This paper proposes a new housing fabrication protocol; the aim is to translate basic digital geometry directly into robotic instructions, such that a simple architectural model quasi-automatically generates actual fabrication instructions. This should allow anyone to rapidly design panelized buildings and fabricate them directly as high-quality structure that can be basically self-erected. The project further expanded to evaluate the technical validity of using panelized system logic as a low-cost unitary structure/enclosure methodology, which would offer a highly streamlined fabrication process, benefitting from CAD/CAM methods to radically challenge current building-industry methods. The ultimate goal of this research is to offer a scalable solution to increase access to low-cost housing, while stressing the need to clearly recognize the role of fabrication as an alternative building practice.

1 INTRODUCTION

1.1 *CAD/CAM and architecture*

Architecture has turned to Computer Aided Design (CAD) and Computer Aided Manufacturing (CAM) systems to help develop and produce complex geometry. More recently digital fabrication or CAD/CAM starts showing potential to emerge as a real value proposition for full scale projects with some built examples[1] and experimental structures (Sass & Botha, 2006, Menges, 2011). The integration between the design and manufacturing stages of CAD/CAM-based production processes promises to overcame traditional construction processes in expense, ease of implementation, and efficiency by enabling the evaluation of the design and the manufacture to be undertaken using the same system of encoding geometrical data. Able to generate mass customized structures on budget while maintaining a high quality of manufacture and assembly. Just as important, CAD/CAM can give architecture a closer control over the construction/fabrication development, allowing a bottom-up and integrated approach. This suggests that in the coming decades the ideas of construction and fabrication of structures could extend from the builders to the common user as a do-it-yourself project because of the availability of these fabrication tools.

1.2 *Paper scope*

Through the deployment of a particular fabrication protocol, this study intends to look into the relation between design and manufacturing parameters to implement a methodology to facilitate the streamline of the housing fabrication process. The above stated is enabled by a proposed methodology that combines fabrication equipment with computer aided design tools and procedures. Therefore, the question pursued is: How to develop an integrated method that combines design, evaluation and fabrication within the same digital work environment and how to simplify the design to construction process.

The results of the research and the discussion about the proposed fabrication protocol and the possible future implications on the architectural design process are presented in this paper.

2 METHOD

2.1 *Construction system*

The housing construction system proposed in this paper is a variant of a Structural Insulated Panel (SIP), a sandwich panelized system that compresses a highly insulating polymer such as expanded Polystyrene between two skins (interior, exterior) of Oriented Strand Board (OSB). The change from standard panels is that instead of using aluminum beams or wood profiles to attach the panels to each other, it has an interlocking, self-aligning joint logic built in the panel OSB skins, which allows fastening the panels without additional elements. There are two procedures for this joint logic a) for "L" joints, where the panels join at a ninety degree angle, and b) coplanar "I" joints used to connect the panels in the same plane. Both of them follow

[1]http://www.facit-homes.com/.

an arrangement of interlocking tabs and slots symmetrically disperse along the panel edges.

2.2 Translate digital geometry directly into a housing construction system

The novelty is on the methodology to translate digital geometry into a full-scale architectural space and not in the building structure. So for the purpose of the study, this method was executed in simple geometry, reserving the chance of exploring more complex forms in further research.

The base geometry should be a parallelepiped that could be modeled in any standard 3D software. This will be the first input for a component built in C# to implement the formerly outlined construction system to the input geometry.

The component is divided in two main sections, the input parameters that are essentially the information/geometry required to perform the embedded algorithm, and the output parameters that is the geometry generated from the input data. The first and main input is the base geometry; from this model the component will breakdown and tag the geometry into its elements (surfaces, edges and vertices). The inputs—length—and—width—establish the maximum dimensions of a single panel.

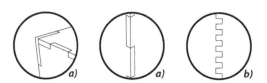

Figure 1. joint logic a) for "L" joints and b) for coplanar "I" joints.

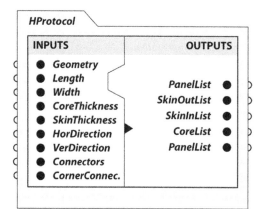

Figure 2. HProtocol component to translate 3D geometry into a construction system, identification of the necessary inputs and outputs.

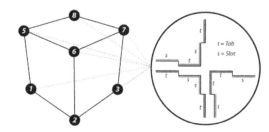

Figure 3. Edge joint logic for the full geometry and detail of a three edge angle.

This implies that each time there is a surface larger that this maximum the component splits the surface in two and generates a second panel; if it is three times bigger it splits the surface into three panels and so on. One of the first issues was understanding if when generating the interlocking system, it should be arrayed computing each single panel or instead run through an holistic approach where it determines the connections for the complete geometry and then subdivide into panels and solve the coplanar joints. Both solutions were tested, but the reductionist solution revealed several difficulties of execution, particularly when faced three edge vertices of the parallelepiped. The 'panel per panel' method worked perfectly to attach two panels that are next to each other, where the edge joint on the next panel is an inverse of the previous, but when three or four panels meet, this same logic stops working and requires different generative rules. The solution was to just generate initially the edges of the entire geometry and then split that geometry into panels and corresponding connections. The slots and tabs length in the corners is defined by the input—CornerConnections—.

After calculating the edge angles it generates the joints for the surfaces edges, the number of tabs and slots is created based in the—connectors—input, where the user introduces the length of each slot/tab then the component takes that number, divides it by the total length of the edge and rounds it into an odd integer that characterizes the number of divisions to the edge. This division generates a list of points along the edge; this list then will be copied in the vector direction of the following panel, the distance for the copy is led by the—SkinThickness—input. The two point lists are then connected according to Figure 4 scheme.

This scheme automatically sets the interlocking system in each edge. The input—HorDirection—sets the vector direction of the panels that are in the horizontal, i.e. roof or floor, the—VerDirection—defines the direction of the wall panels. The options in these inputs are numerical values that represent the direction vectors of the panels

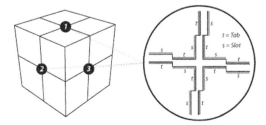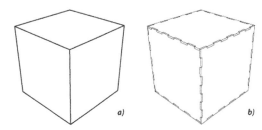

Figure 4. Edge joint logic for the full geometry and detail of a three edge angle.

Figure 6. 1 × 1 m box—a) original geometry b) generated geometry.

Figure 5. Point List to generate interlocking system.

[0 = {x;z} or 1 = {z;x}] for VerDirection and [0 = {y;z} or 1 = {y;x}] to HorDirection. The result is a collection of tabs around the edges of the original geometry, which with the inputs Skin-Thickness provides the depth of the exterior and interior skin of the geometry resulting in Figure 5 instance.

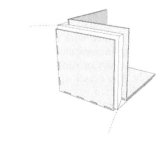

Figure 7. Section of geometry output. When panels join at 90° the core is cut with a 45° plan.

The CoreThickness input selects the core thickness, whenever panels are coplanar there is straight cut between the cores of two panels, if there is a connection of any angle, the cut is always bisectional, e.g. if we have a 90° connection the cut between panels should have 45°.

The interior skin follows exactly the same principles of the exterior skin, starting from an offset geometry within the base geometry. The offset distance of this geometry is set by the sum of Skin-Thickness + CoreThickness.

Figure 8. a) 3.6 × 7.2 m geometry translated into 1.2 × 2.4 m panels b) Integration of standard or customized fabricated windows.

2.3 Housing details

Windows and doors sizing can be generated following a previously explored method (Portugal, 2011) or using a BIM database adapted to the same construction method. using a component to identify edges of the opening in the geometry and generate joints for standard or customized windows. These component should classify the window dimensions and generate studs, plates and shoes to receive the windows.

The base geometry can be more complex and further construction details can be added to the construction system logic. e.g. in Figure 9 there is a design with some variations, presenting acute and obtuse edge angles in the roof to wall connections, interior walls and a window crossing different surfaces in the geometry.

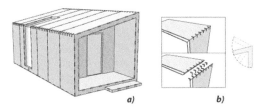

Figure 9. a) Design variant b) Obtuse and acute edge angles system.

2.4 From geometry to robotic instructions

A second component is created to digitally unfold the HProtocol component output lists of geometry, tag it and nest the geometry within a pre-set bed size preparing it for CNC milling. This operation is led on a piece by piece basis. Each panel is shaped by three pieces and two different materials, two

OSB skins and one Polystyrene core. For the OSB skins the unfolding and toolpath making is very simple, because all the robotic instructions can be flattened to 2d polyline and unfolded and nested into world UCS to be cut in a CNC machine. For the core there is a need for three-dimensional cutting instructions since there are cuts with various angles.

3 CONCLUSIONS

3.1 *Alternative building practice*

The goal was to present a method to accelerate the housing production through the creation of a panelized system compatible with digital design and precision cutting techniques, enabling delivery of a prefabricated wall, floor or roof element for rapid on-site assembly. The presented process is a two-fold: first, pays attention to the assembly boundaries and material properties when translating digital geometry into an actual building system. Second, directly arranges the geometry for manufacture, rushing the process of transforming a simple solid into an actual physical space.

3.2 *Faster, cheaper and more efficient*

Prefabricated panels offer a number of advantages to delivering more sustainable buildings including modular, rapid on-site assembly (substantially faster and safer) which reduces cost, construction activity impacts and waste. This method contrasts from conventional pre-fabrication methods by diminishing the requirement of external services and by simplifying the all process. Avoiding patents, expensive equipment, long transportations and eases the all process of conceiving a household. It might potentially by a worthy solution for low-income housing because the levels of skills required are lower than in standard buildings, takes the most of local resources and participation, presenting itself as an employment opportunity for unskilled labor and community involvement.

REFERENCES

The Instant House: Design and digital fabrication of housing for developing environments. Botha, Marcel and Sass, Lawrence D. 4, Cambridge, Massachusetts: International Journal of Architectural Computing, 2006, Vol. 4.

Integrative design computation: Integrating material behavior and robotic manufacturing processes in computational design for performative wood constructions. Menges, Achim. Calgary: ACADIA Proceedings, 2011.

Informed Parameterization: Optimization of building openings generation. Portugal, Vasco and Guedes, Manuel Correia. Lima, Peru: PLEA2012, 2012.

Green Design, Materials and Manufacturing Processes – Bártolo et al. (eds)
© 2013 Taylor & Francis Group, London, ISBN 978-1-138-00046-9

The haptic/visual image as an inclusive tool

A.P.P. Demarchi, B.S. Pozzi & C.B.R. Fornasier
Universidade Estadual de Londrina, Londrina, Paraná, Brazil

ABSTRACT: Considering the varied needs that emerge in contemporary times and the search for design to address them in order to accept the differences and eliminate barriers between individuals, this work aims to study how tactile images must be developed so that they are understood both by seers as not seers, producing the same meaning, and by the junction of the visual image with the tactile image, convey the same information during the learning process. It will deal with concepts such as disability, inability and disadvantage; construction processes of perception in sighted and non-sighted and the image's role in the learning process. The study of the perception of people seers will be used as a reference for comparison of the process in the blind. Records and reports of individuals sighted and sighted not about his feelings and perceptions will be used, and will be discussed the role of perception of individuals through the use of universal illustrated books, aimed at children.

1 INTRODUCTION

According to Dorina Nowill Foundation, an impairment is any loss or abnormality of psychological, physiological function or structure or anatomical. A disability is any restriction or lack of ability (resulting from an impairment) to perform an activity in the manner considered normal for a human being. A downside is derived from an incapacity or disability, which limits or prevents the fulfillment of a person's normal role. The downside is the socialization of the incapacity or disability and as such, reflects the consequences on the individual—cultural, economic and environmental—arising from the presence of the incapacity or disability.

In a world governed by the cultural strength of film, photography and television, is necessary visual literacy teaching for communicators and to those to whom the communication is addressed (Dondis, 2000). In addition to the visual images, there are other modes of access to basic information, such as the auditory mode (voice, tonality, acoustic signals) and haptics (Braille, embossed symbols, buttons, movements, positions, vibration), being often required the conversion between modes in order to be able to share the information that must be handled in different ways (Gandhi, 2001). Therefore, the visual images need to be properly adapted to the haptic system, to be understood by the visually impaired, because as says Sen, (2001) to have a society of active individuals, participatory and with quality of life, it is necessary to adjust the physical environment to the capacities of the people rather than require them to adapt to the already designed.

This article uses ethnographic research and as a strategy the VPA (Verbal Protocol Analysis) applied in children congenital blind to understand how the construction of perception happens. So, it will be necessary first a research on secondary data, to conduct a field research that will aid in the understanding of how the construction of perception in blind children, congenital.

2 PERCEPTION

The perceptual process involves three basic steps: selection, organization, and interpretation. During the selection, by means of selective attention the brain considers some stimuli and not others, influenced by physiological, psychological factors, and by the habituation of stimuli. The sensory data are organized according to the shape-explained through laws of Gestalt organization—the principle of figure and background, continuation/closure, closeness, contiguity and similarity—and through the perceptual stability which enable the identification of a standard even in different conditions of distance, lighting, color, etc. In interpreting the brain uses information organized in the explanation and construction of judgments of the world. This process is influenced by factors such as: preliminary experiences, expectations, cultural factors, motivations, needs and referrals (Huffmann et al. 2003) acquired by means of touch.

The active touch, called haptics system, consisting of cutaneous and more physical components, constitutes the information forms through which the impressions, sensations and detected vibrations are interpreted by the brain. So, straight shapes, curves, volumes, textures, density, thermal oscillations, among others, "are properties that generate tactile sensations and mental images important for communication, the aesthetic, the formation of mental representations and concepts." (Domingues et al. 2010) explain that the tact in motion can be directed and guided willingly, assisting in the detection of stimulus and information about a given object.

The concepts formed by congenital blind differ from those formed through visual experiences (Domingues et al. 2010). It is not known with certainty how are mental representations of a blind person constructed from haptic perception, but there are indications that are representations of mental images of seers (Lima & Silva, 2000). The mental image of something is a concept and in the blinds the formation of congenital concepts occurs through remembering acquired and by remaining senses (Novi, 1996). It is important that the visually impaired develop these senses since birth, and learn to organize his perception so as to receive and sort relevant information. Thus, the blind children with more opportunity for learning and varied experiences acquire a good cognitive development. Domingues et al. (2010) explain that the congenital blind constructs images and mental representations of the world not only through the senses, but also by the activation of higher psychological functions, in which lie the memory, attention, imagination, thought and language, which are systems that contribute to the Organization of all aspects of life. The sense and meaning of things is mediated by language, when knowledge must be associated with the content and life experiences.

Analyzing shooting of the hands from participants while verifying different attributes of objects, Lederman & Klatzky (*apud* Ballesteros & Heller in: Grumwald, 2008) showed that the manipulation occurs according to the information you want to extract, called Eps—stereotyped hand movements exploratory procedures, or stereotyped exploration procedures of movements of hands. The authors propose six varieties of movements that take place systematically according to the attribute you want to understand of the object. When the observer wants to assess the weight of an object, it raises the surface (unsupported holding movement). If the attribute being evaluated is the texture, the hand is moved back and forth along the surface, with lateral movements. To check the firmness of a surface or object, the motion is of the pressure, and when it intends to perceive the temperature, is the static contact. Finally, there is the movement of enclosure, related the global shape and volume, with the movement to follow the contour used to realize the exact shape of an object. These

movements are considered windows to exploration of objects.

Tact and vision have similar performance on a range of textural stimuli, but the vision seems more suitable to interpret large configurations, and has advantages over the tact in perception of large-scale spaces, and large objects. This does not happen when the irregularities of the surfaces are very small. For more delicate textures, the touch has more advantages. (Ballesteros & Heller in: Grumwald, 2008).

Regardless of the presence or absence of vision, knowledge and understanding of the concept of colors is of great importance for all children, because the colors are present in all spheres of life, such as leisure, at work, at school, entertainment content, symbols, art, literature, among others (Domingues et al. 2010). When asked about the possibility of working the colors from a pattern to be taught, a professor at the Institute Londrina of Instruction and Work for the Blind (ILITC) and congenital blind says that wouldn't be possible, because "who saw will already be registered in his memory, now for those who never saw is something very blurred".

A study by Bustos et al. (2004), sought to relate color and texture through the supply of existing textures in nature and designed by man-questioning blind participants individually, what color they came to mind when realized certain material. Some conclusions were highlighted, such as the fact that both for users who are blind as acquired blindness; it's easier to associate textures with colors than the reverse. In addition, sharp and rough textures were associated with dark colors like brown and black; smooth textures with bright colors like yellow, pink, blue; soft textures and satin with white and the direct association of tree leaves with green and orange color, with orange fruit. The authors also realized that the blind differentiate shades in colors, and associate with temperature, whereas the yellow as hot and the blue as cold.

3 THE TEXTBOOKS

The textbook adapted is transcribed for the Braille system and has standardized specifications of size, page layout, graphics, maps and illustrations, in accordance with the standards and criteria established by the Brazilian Commission of Braille. Already the book accessible is designed under the universal model, and aims to reach all readers, being conceived from a matrix that allows the production of the material in different ways—Braille, audio, digital and extended typography. This model is not yet available in bookstores and libraries, being an object of debate regulatory and dependent negotiation between the government and the parties of the productive chain of the book. Meanwhile, the first books of children's literature printed in ink and in Braille, embossed designs, punctual and isolated initiatives. (Sá et al. 2007).

The books adapted for the blind children are a topic much discussed yet. To Nuernberg (2010), the illustrations in children's books hold the role of facilitating understanding and involvement in the narrative, and therefore, tactile adaptations should represent the elements and characters that are part of the story. However, only transform images with perspective, representatives of three-dimensional objects in relief points, ultimately limiting access by the visually impaired. In general, the book is limited to two dimensions, but the concrete experience of blind child with the objects of the world is three-dimensional, based in kinesthetic, olfactory information, auditory and tactile.

Nuernberg (2010) reports that despite performing more details and dimensions, the tactile illustrations of various materials (EVA, felt, buttons, toothpicks, etc.) reproduce the same problems of tactile illustrations based on points raised, because the reference in its construction is still the printed image in paint. According to the author, different from what happens with the children seers, these tactile two-dimensional illustrations, not reach the statusof representation accessible to the blind child. In this case, tactile illustration will only have meaning to the blind child if there is a previous knowledge of the object to which it refers. Following this thought, Halliday (1975) noted that for a miniature or a replica object can have some meaning for the child, it is necessary that before she is familiar with the actual object. Thus, the child must have multiple experiences with a ball, or a car, that can play before with clay and give a form to a ball or to a car.

Grijp et al. (2010) state that the disbelief on the capacity of the visually impaired in perceiving tactile images lies in the misunderstanding of the failure of the people when they can't recognize an embossed pattern. Access to two-dimensional patterns by visually impaired does not occur with the same frequency as for the visionaries, and when offered, are not properly taught. Also, when doing the transcription of the visual image to the two-dimensional, it is not consider specific modalities of haptic system (Lima, 2001 apud Lima et al. 2010). Part of the difficulty in recognizing high-relief figures can be caused by difficulty in locating categories or names of the figures, and not the perception of patterns. In addition, congenital blind are not familiar with the conventions of pictorial language, which does not mean that the haptic system is not able to recognize images in relief. (Lima, 2011).

Recent studies demonstrate the ability of the blind in two-dimensional images recognition. Keneddy (Lima & Silva, 2005) showed that individuals understand drawings in relief and reached a basic understanding of space in their designs. With enough time, congenital blind are capable of producing representations in perspective in your drawings, as well as interpret them in an illustration (Heller, 1990 apud Lima & Silva, 2005), and given the right conditions, can recognize embossed designs with the same success than the others (2011).

4 METHODOLOGY

Of all methods of empirical research and observational study, for the examination of the activities of individuals, the Protocol Analysis is the one that has received attention and contributed to more applications (Cross, 2011). Considered to be the method that best explains, at least to some extent, the mysteries of how the individual constructs knowledge. Considering the above, this article used the Protocol Analysis to illustrate the theoretical constructs. There were analyzed the attitudes and the skills used by a blind person when are building their perception by means of touch, during the process of learning.

The research is oriented to the content, try to reveal the thoughts of individual, by Verbal Protocol Analysis-VPA, when the Protocol (the blind congenital individual) first performs the action, and then outsource what was thinking and doing. There were used for the research a camera to film and Photograph camera, as the knowledge of the protocols are difficult to be verbalized.

With this, it was possible to observe the skills and attitudes used by the protocol during the action, so you can understand how the knowledge construction was and how the skills inherent in the individual assisted in this process.

The survey was conducted with teachers and a student of ILITC, where the teacher is congenital, blind, the teacher is seer, and the student has acquired blindness.

5 RESULTS

The student read three didactics books, two with images in reliefs of points and a third produced by the Institute itself, with different textures like wool and suede paper in the figures. In all of them she performs exploratory movement, surface with your fingertips to recognize patterns and textures. In the books with relief points was noted greater difficulty to understand the figures, even when the teacher helped to locate the eyes and mouth of the face of a boy, within the relief of points. The enlarged images (like the detail trouser bar without the whole pants to be on the page), made it even more difficult its interpretation.

It offers greater ease of reading of images with reliefs of textures, but still has difficulties and constantly question to the teacher if she are identifying correctly ("this is the King? This is a boy? Here is the mouth?). The student can easily associate some textures such as cotton to a character's beard. However, she is confused, because in the same figure there are other soft surfaces, which leads to identify how the coat collar as a beard of the same character.

According to professor, material considered universal book, with 100% of inclusion, it still leaves much to be desired:

"you have the format of the drawings here, but sometimes they don't give the correct dimension of the drawing, whether it is a duck, if it's a cock ...and then you would have to do this book, but in a slightly larger binder, for you make the design the way it is. So to you to know that here was a sun, where would that have a texture, an alternative material here. Because, in fact, what we have is only the outline. It is a thing which leaves a lot to be desired. All material that is adapted to the visually impaired have to have a concrete thing".

This opinion is shared by the professor seer, which states

"If you catch like a stopper, makes the texture of the trunk (the tree), alike, as close as possible to the real, because ... What is this here? (showing the book with relief of points) for her (the student), she sees well, because she has seen it all, have this recall ... and yet still have enough difficulty. Need to always have someone helping the reading, or it don't matters, not exploits. Often does not need to be so worked, have so much, but be real's as close as possible".

6 CONCLUSIONS

The blind child should have with the representations the same relationship the seer child has with the visual images, and for this, the book affordable, adjusted for the universal model must be produced in order to meet this need. Currently, the representations are just as a literal translation of the image for relief of points, which is not efficient. Whereas it is possible that there is a standardized interface in the way visually impaired perceives color and texture, these must be used so that, through perception, can help in his learning and apprehension of the world.

REFERENCES

Cross, Nigel. 2011. *Design Thinking*. Oxford: Berg.

Dondis, Donis A. 1997. *Sintaxe da linguagem visual*. 2. ed. São Paulo: Martins Fontes.

Domingues, Celma dos Anjos; Sá, Elizabet Dias de; Carvalho, Silvia Helena Rodrigues de; Arruda, Sônia Maria Chadi de Paula; Simão Valdirene Stiegles, 2010. *A Educação Especial na Perspectiva da Inclusão Escolar: os alunos com deficiência visual: baixa visão e cegueira*. Brasília: Ministério da Educação, Secretaria da Educação Especial.

Grijp, Ana Carolina; LIMA Francisco José de; Guedes Lívia Couto; OLIVEIRA, Leny Ferreira de, 2010. A produção de desenho em relevo: da imagem visual para a representação tátil. *Revista Brasileira de Tradução Visual*. vol. 04, n. 04.

Godinho, Francisco. Design Universal nas Tecnologias da Informação e comunicação. *Anuário 2001 Design Inclusive*. Centro Português de Design. Ano nove. P. 88–89.

Grunwald, Martin, 2008. *Human Haptic Perception Basics and Applications*. Basel-Boston-Berlin: Birkhauser.

Halliday, Carol,1975. Crescimento, aprendizagem e desenvolvimento da criança visualmente incapacitada, do nascimento à idade escolar. Fundação para o livro do cego no Brasil.

Huffman, Karen; Vernoy, Mark; Vernoy, Judith. 2003. *Psicologia*. São Paulo: Editora Atlas. 5a edição.

Lima, Francisco José de; Silva, José Aparecido, 2000. Algumas considerações a respeito do sistema tátil de crianças cegas ou de visão subnormal. *Revista Benjamin Constant*, n. 7.

Lima, Francisco José de; Silva, José Aparecido, 2005. *O desenho em relevo: uma caneta que faz pontos*. Disponível em:<http://www.lerparaver.com/lpv/desenho-relevo-caneta-que-faz-pontos> Acesso em 01/12/2012.

Lima, Franciso José de, 2011. Breve revisão no campo de pesquisa sobre a capacidade de a pessoa com deficiência visual reconhecer desenhos hapticamente. *Revista Brasileira de Tradução Visual*, vol. 06, n. 06.

Novi, Rosa Maria., 1996. *Orientação e mobilidade para deficientes visuais*. São Paulo: Cotação da Construção.

Nuernberg, Adriano Henrique, 2010. Ilustrações táteis bidimensionais em livros infantis: considerações acerca de sua construção no contexto da educação de crianças com deficiência visual. *Revista Educação Especial*, v. 23, n. 36.

Salmaz, Carla e Netto, Carlos Alexandre, 2004. A memória. *Ciência e Cultura* vol. 56 n. 01.

Sá. Elizabet Dias de; Campos, Izilda Maria de; Silva, Myriam Beatriz Campolina, 2007. *Atendimento Educacional Especializado*. São Paulo: MEC/SEESP.

Simões, Jorge Falcato, 2001. Design Universal Porquê? *Anuário 2001 Design Inclusive*. Centro português de design. Ano nove. P. 82–83.

Green Design, Materials and Manufacturing Processes – Bártolo et al. (eds)
© 2013 Taylor & Francis Group, London, ISBN 978-1-138-00046-9

The limits of inclusive design in the current design practice

E. Zitkus, P. Langdon & P.J. Clarkson
University of Cambridge, Cambridge, UK

ABSTRACT: The adoption of inclusive design principles and methods in the design practice is meant to support the equity of use of everyday products by as many people as possible independently of their age, physical, sensorial and cognitive capabilities. Although the intention is highly valuable, inclusive design approaches have not been widely applied in industrial context. This paper analyses the findings of an empirical research conducted with industrial designers and product managers. The research indicates some of the hindrances to the adoption of inclusive design, such as the current way the market is considered and targeted, and; the way the designers are driven by the project's brief and budget to orient their research strategy and activities. The paper proposes a way to improve the current industrial mode by strategically supplying clients, designers or both together with information about inclusivity.

1 INTRODUCTION

Many countries have faced the challenges related to the growing proportion of the elderly population. One of the challenges is the need to address problems that naturally results from the ageing process, such as the loss of physical, sensorial and cognitive capabilities. In this case, the best practice in new product development would be to consider a wide range of user capabilities while creating new design features to promote independent living among the elderly population. This is exactly the design principle advocated by inclusive design theory and practice. Consequently, the adoption of inclusive design in new product development seems to be appropriate in the current scenario.

However, differently of what could be expected in an ageing society, inclusive design approach has not been widely used in industry (Goodman-Deane et al, 2010; Vanderheiden & Tobias, 2000; Sanford et al, 1998). In fact, the available tools and methods created along the last years to evaluate accessibility of new design concepts have been scarcely used. This may be a result of incompatibility issues of the available techniques with the design practice (Zitkus et al, 2011); or the deficit of incentive to companies to adopt inclusive design (Dong et al, 2004: p 13); or both.

1.1 The compatibility of the tools

Previous studies reviewed current tools of inclusive design and compared to their adaptability of the design process in industrial contexts (Zitkus et al, 2011, 2012b; Cardoso et al, 2004). Two major issues were found among the techniques:

1. they are time consuming, and;
2. the stages in the process where they are applied are either too earlier or too late to change the design under development.

To better integrate the techniques to the design practice it is necessary to understand the design process and the differences among design domains (Zitkus et al, 2012a). However, it is also necessary to understand the motivations behind the non-adoption of more user-centered process.

1.2 The adoption of inclusive design

According to Goodman-Deane et al (2010) the adoption or non-adoption of inclusive design in the process is greatly influenced by the client—the company that is requesting the product or design. The authors state the need of providing information about inclusive design to clients, as well as improving current ways to inform designers.

The need of recognizing clients and designers drivers in order to enhance the practice was the reason of the study presented in this paper. The methods and findings are briefly presented in the following sections.

2 METHODS

2.1 Sample size

A total of 22 industrial designers and 7 project stakeholders participated in the study. The sample of industrial designers was formed through six design agencies based in the United Kingdom and one multinational company; the stakeholders were from the two large multinational companies and one small enterprise. The stakeholders interviewed are often the people responsible for commissioning the design to external design agencies. In this paper they are called clients, who represent the interests of the company that owns the final product.

2.2 *Data collection*

Data was collected through unstructured interviews and observation of the designers at work. The interviews supported in-depth investigation of the design activity through opinions, knowledge, behaviour and experience of the participants while the observations contextualised what was mentioned in the interviews or they brought new insights to the research (Patton, 1987).The participants were encouraged to talk about their background and experience in the field, as well as to give a broad picture of their role in the consultancy or in the company. The major aim of the interviews was to understand new product development and user requirements from distinct point of views—clients and designers. Hence, the spine of the interview was always the description of the design process, with examples. This enabled the researcher to recognise some drivers that guide designers and clients decisions. The interviews were audio recorded and transcribed afterwards.

After analysing the data, the findings were represented in a framework format to be presented to designers and stakeholders in order to receive their feedback regarding the understanding of the design activity and how user requirements were dealt alongside the process. The feedback sessions supported the study by correcting misunderstandings and confirming some of the results.

2.3 *Data analysis*

The transcripts were coded and categorised using Atlas.ti, software developed to support qualitative research analysis. It enables the user to cross-compare the reoccurrence and co-occurrence of codes, to divide them into code families and to map their connections. The categorisation was based on utterances related to the same idea among different participants. Every time a relevant fact was recognised in a transcript, old transcripts were re-analysed to find out the views of past participants related to that aspect (Corbin and Strauss, 1990). For example, as the research evolved, the role that clients play in the design process had to be clarified, as a result of which other questions were raised and past transcripts were re-analysed.

The codes were mapped according to their importance (reoccurrence) and their connection to other codes or family codes (co-occurrence).

Care was taken to ensure that the same code was not duplicated for a single participant under the same interview topic. This procedure prevented the reoccurrence of codes only based on single views.

3 FINDINGS

3.1 *The brief driving the design activity*

The participants mentioned a similar process that happens at the initial stages of the design process: usually designers are guided by the 'design brief'. The brief is the initial source of information about the new product's functionality, components, manufacturing, disposal and also the potential user's information. Regarding the latter, it was highlighted that user's data are restricted to market views, which means target market and commercial requirements. User's information is normally general demographic information like age or social class.

Regarding accessibility and usability in the brief, it was mentioned that it "*is not something that always got designated time within the process*" (Designer 1). In fact, according to the designers, sometimes the brief is focused on a main issue or a key requirement that drives the design activity, compromising other requirements. For instance, some designers mentioned 'design for manufacturing' or 'emotional design'—focused on the appearance of the product.

Nevertheless, according to the designers if it is part of the project requirement to consider accessibility, then they usually seek for data on books, tables, internet, or they would look for specifications in guidelines.

3.2 *The research conducted by the designers*

As user's data is quite limited on the brief, designers have to manage their time and budget to get user's information from other sources. The designers mentioned that often some research start taking place earlier in the conceptual phase. The research can happen in different ways to provide different information, such as competitors' data, technical specifications and also user's information.

It was observed that the internet was very useful to supply one of the designers with technical data of components; another with ergonomic data, and; two other designers with technical data about materials. The interviewees confirmed that the internet is used to find out more about end-users, but designers outlined that they also follow guidelines while designing.

3.2.1 *How designers use guidelines at this stage*
The responses indicate that designers mainly rely on guidelines, though their comments also

highlighted that they find the information on these sources deficient and sometimes incompatible to their needs. They mentioned that they balance the deficiency of the guidelines by including some live-assessments, such as self-evaluations and user-trials.

3.2.2 *How designers run user trials at this stage*

Although it was mentioned the possibility of incorporating users in accessibility tests, all the designers interviewed stressed that user observation or user trials only take place if the research allowance considers that, which rarely happen. As a result, the users are hardly ever involved in the process.

It is important to underline that the designers highlighted that the user's needs, such as those related to accessibility and usability are only one part of the requirements that the designer has to deal with. They emphasized that design is a compromise activity, where decisions are made all the time and costs are involved in every option taken.

The observation did not highlight how rigorous the process of searching for users' data is.

3.2.3 *The designers views about inclusive design*

It seems to us that among the designers interviewed there were two groups:

1. Those who are more proactive in terms of user requirements research, who would look for other means to understand the end-users. They responded positively to inclusive design tools and methods as a way to add value to the design.
2. Those who are used to and satisfied with self-evaluation of concept designs. They seem to think that their experience and knowledge about users are enough to cope with accessibility and usability of the new designs, given the usual constraints of project's resources. Hence, for them the need of inclusive design techniques to be applied to the design process is not in their top priorities.

Both groups however highlighted that they would rely on clients' requirements. Therefore, according to all designers interviewed, the views of the client regarding the end-user are mandatory to implement a more user-centered design process.

3.3 *Company requirements that drive client's decisions*

The interviews with clients highlighted the aspects they have to consider while planning a new project and before commissioning the project to designers. The responses indicate the drivers of clients' decisions and outline how they understand the end-users.

3.3.1 *Target market, market response and final price*

According to the clients, the user requirements are consumer requirements based on market research. Market research normally covers the market needs, market share and consumer expectations. It does not cover user requirements related to accessibility and usability. The comment below highlights the characteristics of market research:

"We have a research agency, which is specialized in market research. And the way we typically do market research, on top line level, we do a lot of online [research]. And the reason we do online is because it is very easy to get a big number of respondents which give us a very robust data. We also did some store research, which we literally send people to hang around in Tesco and Dixons and Argos and watch people buying the [product]. Once they bought it, they go and ask them questions: 'why did you buy the [product]?'; 'what is that? Because of the features, the color, is it the brand?' ... 'And then, from that we bring the research into our design agency and brief them around the types of [products] we wanted.'

"The other piece of research we did, was very much market focused, looking at what price point is important, how much people would spend, where they bought the [products]." (03:15—Client 2).

According to the participants, the tendency is to divide the market into groups of consumers and then to target at those specific markets.

3.3.2 *Volume of sale, market share and competitors*

Changes on design features, including changes towards improving accessibility or usability would happen if they make a difference on sales or bring advantages over competitors, which would increase the market share. The clients mentioned that *"there is a kind of trade off"* within the company's requirements before they make decisions.

A common comment among the interviewees is that they make decisions based on the top priorities of the company. The number one priority according to the majority (6 of 7 clients) is the volume of sale and the delivery of profit to the business. One participant highlighted that they prioritize the projects based on *"the volume value that the product lands into the business"*. Thus, in this case, they start by prioritizing the development of the two products that represent the best sales, and after developing those, they would look to new products.

3.3.3 *The clients views about user-centered design*

When asked about the research conducted with end-users, the clients normally talked about market research. One of the interviewees explained the attention the company has with end-user and described a recent project in which they conducted research with the conceptual design proposals, very early in the process:

'[The design agency] did design 3 or 4 different products and we test very heavily in research.'

When asked about the way the tests were carried out, whether they used rapid prototypes, the client explained the following:

'We do with images actually, because lot of brand is bought on line or catalogues. So, actually, it is quite a good way to test them, to test them online, because it quite simulates the buying experience. So, if you convey the aspects into an image in that environment, then you are likely to be able to do it in a catalogue, which gives us an opportunity to get quite a number of people, which you wouldn't do if you have prototypes. If you have prototypes, then you might go, do 50 people. Going to 8000 gives you a rich and robust quantitative data.' (19:30—Client 2).

Among all the clients interviewed improving inclusive design would mean delaying the design process and adding extra costs to the process. For some of them inclusive design seems to be associated to technology or concepts that are specific targeted at elderly or disabled people.

The clients seem to be satisfied with the way new products are developed. They assume the designers will cope with accessibility and usability issues, as underlined in the comment below:

"Good designs always take into account people's needs anyway. There will always be extremes of uses that you might decide that you can't afford, because it would make the product too expensive to deal with [ ] So, you might decide that it makes more sense to design products in a first place to meet the mass market." (02:09—Client 4).

4 DISCUSSION

According to the findings, clients and designers, make decisions related to the design. This may happen separately while clients are defining the project, even before briefing the designers; or when designers are creating new concepts, before presenting new ideas to the client. Or design decisions are made by both together, when clients and designers meet to discuss the project. However, for each of them the reasoning process is based on different requirements and drivers.

4.1 The designers decisions

The designers described the design process as a trade-off activity that has to consider the functionality, aesthetics, manufacturing process, materials, components, usability, disposal and other requirements.

According to them, the requirements are prioritized normally based on the brief, which often contains the key purpose of the project and the desirable characteristics of the design. The decisions made by the designers consider not only the brief, but also the results of research, tests, evaluations and their own knowledge and experience. Figure 1 frames the design decision process based on design requirements and drivers.

The influence that clients exert in the design activity is indeed a fact that has to be considered. In fact, the brief, the research and the evaluations carried out by the designers seem to be sturdily dependent on clients' views, procedures and funding for the project. As already mentioned by Gill (2009) Small to Medium-sized design consultancies tend to face the pressure of costs and tight deadlines from the client, which constrain the designer's decisions. The clients' knowledge of end-user influences the emphasis given to research and the way it is conducted.

4.2 The clients decisions

The clients explained new product development as a trade-off activity that have to consider the market, competitors, costs of the project, impact on brand, final price and volume of sale since the very beginning, even before commissioning projects to design agencies. Figure 2 illustrates the design

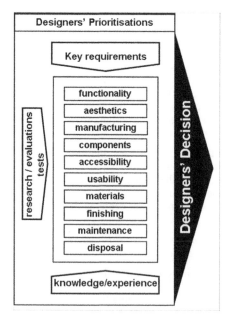

Figure 1.   Decisions based on design requirements and drivers.

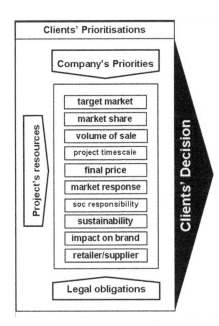

Figure 2. Decisions based on company's requirements and drivers.

decision process based on company's requirements and drivers. The clients consider the company priorities to make design decisions, which may vary from one company to another, but among the participants, a similar number one priority was highlighted: the profitability of the investment.

The business mindset of the clients seems to influence the way the end-user is recognized. The interviews with clients underlined the client mindset focused on developing the right product to the market to improve volume of sales. For them the end-user is a consumer and thus, the research with end-users is a research with consumers, which focus on understanding the reasons behind their purchase.

4.3 *User-centered approach*

One of the interviewees described a recent research conducted with end-users; although the stage in the process where early in the process and thus, the right one to run usability and accessibility tests, the description of the way the research was carried out highlighted the focus on sale, not on usability.

The responses of the designers in this study confirmed that direct involvement with users rarely happens in commercial projects; a problem already underlined in past literature (Dong et al, 2005; Sanford et al, 1998).

Users are consumers and their profiles are brought to designers in the form of target markets,

which generally consist of a demographic view of potential buyers. User's requirements, such as those related to the diverse range of physical, sensorial and cognitive capabilities, are not part of the market information normally provided by the client. Clients' views of the market seem to be limited to mass market, where the elderly and disabled people are not included.

In the current scenario it seems that clients assume that designers will consider usability and accessibility while designing, whilst designers assume the client will request user-centered process or allow more resources (time and budget) to conduct user trials. As a result of their assumptions the inclusivity of the final product is normally very poor.

4.4 *Improving the current design practice*

In industrial contexts, the design process commonly ignores the need of taking into account different individuals capabilities to promote equity of use of everyday products. On the one hand it is a consequence of the way companies run their business and how they view the market. Policies or incentives that benefit the accessibility and usability of new products rarely exist. On the other hand it is a result of the way designers conduct their work, which reflects the way design is taught in the university. Design higher-education has neglected their role in changing the designers' attitudes towards user-centered design or co-design.

This paper however will not discuss the amplitude of problems that result in the current design mode. Nevertheless, it suggests strategic places in the process where clients and designers could be supplied with information about inclusivity.

4.4.1 *Before commissioning the project: informing clients*

Before commissioning the project to designers and even before elaborating a brief, information about inclusive design could raise clients' awareness. If information about accessibility and inclusion were considered by clients when they select a project to commission to designers, they would understand the gap between market data and user's needs. This could encourage them to build a brief focused on user, which would influence the entire process in adopting user-centered approaches.

The priorities established in the brief are powerful influences alongside the project. Therefore, inclusivity should be part of the brief as a key requirement if user needs are expected to be considered in the design process.

4.4.2 *During ideas generation: informing designers*
Tools and methods could support design decision towards more inclusive design. During the time

that designers are making decisions they can be influenced by research, test and evaluation results that promote inclusivity.

4.4.3 *During design meetings: informing clients and designers at the same time*

Design projects are pretty much driven by the decisions made by clients and designers. Strategic stages where these decisions happen are the meetings or 'checkpoints' established by designers and clients. The meetings are used to specify the project, to discuss the project and to present and discuss the ideas.

Therefore, to support the development of more inclusive designs, one way is to inform both—designers and clients—about inclusivity while design decisions are made.

5 CONCLUSION

The paper describes an exploratory study that confirms a problem indicated in the literature: inclusive design has not been part of the current design practice in industry. It acknowledges that normally the designers do not consider the adoption of user-centered design process unless the clients request. The clients however do not consider their consumers as individuals with different physical, sensorial and cognitive capabilities which should be considered carefully. Consequently, they assume that the way designers evaluate usability and accessibility is satisfactory.

Although the current design practice is a result of a broad problem in different sectors (for instance, industry, government and higher-education), here the problem is approached as a result of deficient information alongside the design process. Therefore, it is proposed three ways to promote greater inclusivity into the design practice: 1) by informing clients while elaborating the brief; 2) by informing designers while generating new concepts, and; 3) by informing designers and clients in design meetings. This would enhance the decision-making process with information about inclusivity, which in turn could result in more inclusive designs.

REFERENCES

Cardoso, C., Keates, S. & Clarkson, P.J. 2004. 'Comparing Product Assessment Methods for Inclusive Design', in S. Keates, J. Clarkson, P. Langdon & P. Robinson (eds), *Designing a More Inclusive World.* Springer: London.

Corbin, J.M., & Strauss, A. 1990. Grounded theory research: Procedures, canons, and evaluative criteria. *Qualitative Sociology, 13*(1), 3–21.

Dong, H., Keates, S., Clarkson, P., Stary, C., & Stephanidis, C. 2004. Inclusive Design in Industry: Barriers, Drivers and the Business Case. *In User-Centered Interaction Paradigms for Universal Access in the Information Society.* 3196: 305–319. Springer Berlin: Heidelberg.

Dong, H., Clarkson, P.J., & Cassim, J.a.K.,S. 2005. Critical user forums—an effective user research method for inclusive design. *The Design Journal, 8*(2), 49–59.

Gill, S. 2009. Six Challenges Facing User-oriented Industrial Design. *The Design Journal, 12*(1), 41–67.

Goodman-Deane, J., Langdon, P., & Clarkson, J. 2010. Key influences on the user-centred design process. *Journal of Engineering Design, 21*(2-3), 345–373.

Patton, M.Q. 1987. Program evaluation kit. 4, How to use qualitative methods in evaluation. Sage: Newbury Park, London.

Sanford, J.A., Story, M.F., & Ringholz, D. 1998. Consumer participation to inform universal design. *Technology and Disability, 9*(3), 149–162.

Vanderheiden, G., & Tobias, J. 2000. Universal Design of Consumer Products: Current Industry Practice and Perceptions. *Proceedings of the Human Factors and Ergonomics Society Annual Meeting, 44*(32), 6-19-16-21.

Zitkus, E., Langdon, P., & Clarkson, J. 2011. Accessibility Evaluation: assistive tools for design activity in product development. In H. Bartolo (Ed.). *In International Conference on Sustainable Intelligente Manufacturing. Proceedings.* 659–670. IST Press: Leiria, Portugal.

Zitkus, E., Langdon, P. & Clarkson, J. 2012a. 'Can computer graphic systems be used to inform designers about inclusivity?' *in 12th International Design Conference proceedings.* Dubrovnik, Croatia.

Zitkus, E., Langdon, P. & Clarkson, J. 2012b 'Design Advisor: How to Supply Designers with Knowledge about Inclusion?' in J Clarkson, P Langdon, P Robinson, J Lazar & A Heylighen (eds), *in Designing Inclusive Systems.* Springer: London.

Green Design, Materials and Manufacturing Processes – Bártolo et al. (eds)
© 2013 Taylor & Francis Group, London, ISBN 978-1-138-00046-9

Rapid construction with functionally graded designs

F. Craveiro, H.A. Almeida, L. Durão, H. Bártolo & P. Bártolo

Centre for Rapid and Sustainable Product Development, Polytechnic Institute of Leiria, Marinha Grande, Portugal

ABSTRACT: The construction industry is increasingly optimizing its performance reducing costs and minimizing its environmental impact. New technologies, growing client expectations and a shift in design thinking are motivating major improvements in the construction sector towards more integrated systems using novel computational fabrication processes.

A RapidConstruction System, based on extrusion technologies, was developed to integrate the concept of material space and form to construct eco-efficient buildings with complex forms and geometries. This new 3D digital processing system was used to fabricate functional graded structural components with different material compositions and shape, so its functional requirements can vary with location. The development of this functional design concept, using material gradient and/or geometrical gradient, will enable the fabrication of more efficient structures regarding thermal, acoustic and structural conditions.

1 INTRODUCTION

Design in nature is quite flawless and several engineering design solutions and its applications are frequently inspired in it. Biomimetics investigates the structure and functionality of biological systems as models for the design and engineering of materials and machines. All biological features required for a structure, such as energy savings, beauty, functionality and durability were already created and optimized by nature. In order to replicate nature's design and implement it in an architectural design work, a high level of engineering and biological knowledge is essential. Yet living things in the natural world know nothing about load bearing or architectural principles, in spite of having the capability of adjusting and adapting to new loads. Figures 1 and 2 illustrates two biologically inspired architectural designs, the Crystal Palace in London and the Munich Olympic Stadium.

One of the topics discussed in biomimetics is Functionally Graded Materials (FGM). FGMs can be found in many natural biological structures, for example bones, bamboo, mollusc shell, etc. These structures present compositional or microstructural gradients, such as the gradation in the density of fibers along the bamboo stems, or the density of the trabecular tissue along the femoral bone (Fig. 3). A continuous bulk functionally graded material has the potential to be an ideal orthopedic implant for load bearing applications.

Architecture and Construction are highly interdisciplinary fields, integrating numerous professionals and engineering domains to

Figure 1. The Munich Olympic Stadium is inspired in the Dragonfly Wing design.

Figure 2. The Crystal Palace in London was inspired in the Water Lily design.

Figure 3. Section view of a femoral bone combined with a detailed view of the Eiffel Tower.

produce structures with different levels of scale and complexity. Construction is usually considered very conservative, risk averse and reluctant to invest in new ideas, is facing an increasing pressure to be more efficient to survive in the current economic conditions. Conversely, it needs to address the sustainable and climate change issues, as the environmental impact of the construction sector is huge with buildings accounting for 40% of the European Union energy demand.

In today's global economy, the construction industry is increasingly challenged to optimize its performance reducing costs and minimizing the impact on the environment. New technologies and rising client expectations are motivating radical improvements in the construction sector (P. Barrett, 2008). The industry is currently evolving towards a full digital based design system through a better integration of structure, materials and form through new fabrication technologies, which can create new opportunities and introduce new global challenges to maintain competitiveness.

Additive technologies are a class of manufacturing processes, in which a part is built by adding layers of material upon one another. Additive technologies, usually called rapid prototyping technologies, are one of the most rapidly growing manufacturing technologies in the world (T. Wohlers, 2011), due to their capability of producing highly complex geometric products with gradient functionality in terms of material and geometric forms.

These technologies (F. Craveiro et al., 2011) are currently applied in a great variety of industries, such as the aerospace, architectural, automotive or medical fields, though its application for construction industry practices is less suitable for two main reasons: i) they are only used for medium or small-scale objects, and ii) the majority of these systems cannot easily process more than one material. The uniqueness of the construction sector constitutes a challenge for the direct adaptation of these technologies.

A new system was developed to introduce additive manufacturing technologies in architecture and construction, overcoming the limitations of scale and dimension, a new system for automatic construction was developed at Leiria, by the Centre for Rapid and Sustainable Product Development, based on an extrusion process, aiming at the construction of eco-efficient buildings with complex forms and geometries.

2 A RAPIDCONSTRUCTION SYSTEM

The concept of developing highly automated tools and techniques for application in construction is still embryonic. Improving the quality and reliability of manufactured products has been a relevant issue for quite a while. In 1983, Ayres and Miller (1983) reported several advantages on the use of robots in manufacturing, obtained by a detailed survey of 40 major U.S. manufacturers.

It is fundamental to prevent construction site accidents and promote safety for everyone involved in the construction sites, converting the construction activity into a more sustainable one. New challenges and opportunities are emerging to develop modern and innovative methods, which to be successful need to be integrated in a very demanding human and harsh environment.

In recent years, significant advances in technology have created numerous opportunities for innovation in construction automation. Khoshnevis et al (2006) developed a concept for the automatic fabrication of a house called Contour Crafting, consisting of the automatic fabrication of the building walls layer by layer, until the creation of a formwork filled, after the cure, by mortars mainly composed by cement. Lim (S. Lim et al., 2009) developed a concrete printing strategy to produce 3D customized products. This concrete printing system is also based on the extrusion of cement mortars. Dini (L. Dini, 2011) developed the D-Shape process that uses a powder deposition binder similarly to the FDM process. In this process, each build material layer is laid to the required thickness, compacted and then the nozzles on the gantry frame deposit the binder in a selective way. Since 2009, the Centre for Rapid and Sustainable Product Development is developing a novel 3D additive technology system, using the know-how acquired in the production of bio-structures for cell support (scaffolds), using materials mainly composed by cement, polymers and clay (F. Craveiro et al., 2009).

This fabrication system, called RapidConstruction, comprises a computer controlled mobile crane integrating multi-deposition heads with various degrees of freedom (F. Craveiro et al., 2012). This equipment uses fast curing thixotropic materials with low shrinkage (F. Craveiro et al., 2011). This crane has two parallel rails to enable its movements. This additive manufacturing automation process will allow building house walls layer by layer, in a continuous way. As the head moves along the walls of the structure, the construction material is extruded and troweled using a set of actuated, computer controlled trowels. An extrusion head will continuously deposit material until it approaches a window or door opening space, then it slows down until stopping at these previously selected points (F. Craveiro et al., 2012). The use of computer controlled trowels allows producing smooth and accurate surfaces. A large scale prototype illustrated in Figure 4 is being tested for material deposition strategies. Figure 5 illustrates the proposed system at real scale.

The optimization of this new system aims at integrating the concept of material space, material

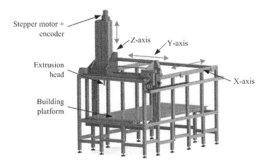

Figure 4. Large scale prototype of the RapidConstruction System.

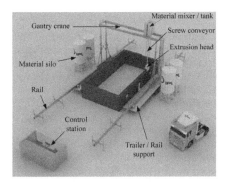

Figure 5. The proposed system for the automatic construction of a building (real scale).

Figure 6. Building concept capable of being produced by the RapidConstruction System.

composition information of building heterogeneous components with geometry information, assigning different spatial features according to requirements. The proposed multi-material deposition system will enable the construction of more efficient buildings regarding thermal, acoustic and structural conditions. Figure 6 illustrates a building concept capable of being produced by the RapidConstruction System.

3 FUNCTIONALLY GRADED DESIGN

Functionally Graded Design can be achieved through two concepts, by either geometric shapes and forms or material density variations. In this research work, both concepts are presented.

3.1 Graded shape functionality

In the last decade, Hyperbolic Surfaces attracted the attention of many researchers from several engineering domains. Hyperbolic geometries commonly exist in natural shapes and structures. Among the several existing Hyperbolic Surfaces, Minimal Surfaces are the most studied. If a Minimal Surface has a space group symmetry, it is periodic in three independent directions, and is often called Triply Periodic Minimal Surfaces (TPMS) (Wang, 2007). TPMS describe several natural shapes, such as lyotropic liquid crystals and colloids, zeolite sodalite crystal structures, diblock polymers, silicates, lipid bilayers bicontinuous composites, detergent films, hyperbolic membranes (found in the prolamellar structure of choloroplasts in plants), echinoderm plates (interface between the inorganic crystalline and organic amorphous matter in the skeleton), cubosomes and certain cell membranes (Larsson et al., 2003; Lord and Mackay, 2003). Two important sub-classes of TPMS are the so-called Schwartz and Schoen primitives, considered in this research work (Fig. 7).

The geometric modelling of the Schwartz and Schoen models were obtained through a commercial CAD software. Figure 8 illustrates Boolean operations by the addition of the basic units into an arbitrary unit, with thickness variation

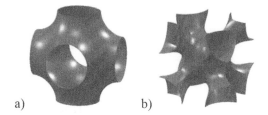

Figure 7. a) Schwartz and b) Schoen TPMS primitives.

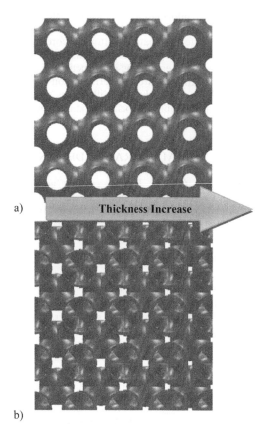

a)

Thickness Increase

b)

Figure 8. CAD models illustrating thickness gradient within the structures for the a) Schwartz and b) Schoen geometries.

resulting in a construction block with a thickness gradient. Figure 9 illustrates the production of both construction blocks through an extrusion-based additive manufacturing system.

To get a better understanding of the structural behavior proposed for the construction blocks, structural simulations were performed to understand the influence of the thickness gradient. A displacement solicitation along the direction of the thickness gradient was defined to undergo

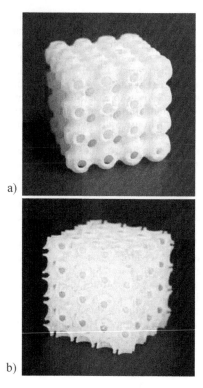

a)

b)

Figure 9. Physical models of a) Schwartz and b) Schoen geometries produced through extrusion-based additive manufacturing.

the simulations. Results show that as the thickness of the elementary units in both construction models increase, the tensile variation tends to lower in value, becoming more homogenous as illustrated in Figure 10.

Figure 11 illustrates a building concept based on TPMS surfaces where these natural surfaces can be well combined with existing cities such as Lisbon.

3.2 Graded material functionality

There is an increasing interest in tailoring building structures so the functional requirements can vary with location (Y. Miyamoto et al., 1999). In a Functionally Graded Material (FGM), both the composition and the structure can gradually change over the volume, resulting in varying material properties.

The processing system (F. Craveiro et al., 2011) combines mixing and extrusion processes to enable the continuous manufacturing of finished components using different materials in a one-step process. The raw materials can be varied continuously (Fig. 12) or by using multiple extruders with different raw materials to build up walls with material variations with the required characteristics.

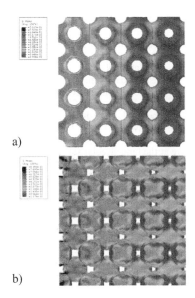

a)

b)

Figure 10. Variation of Tensile Stresses along the thickness gradient for both structures the a) Schwartz and b) Schoen geometries.

Figure 11. Building concept inspired on TPMS primitives.

Figure 12. Representation of a material with a functional gradient.

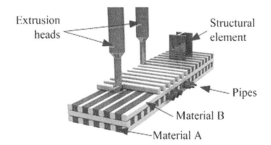

Figure 13. The concept of FGM applied to a building wall.

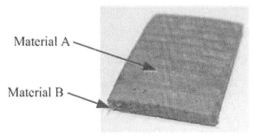

Figure 14. PU polymers with different compositions.

The direction of the gradient can determine its features, allowing predefining a more advantageous direction regarding a better thermal insulation, for instance. The same material can be lighter, with more gaps, in a specific location, so more weak and be denser in another position and stronger. The use of lighter materials can reduce the building weight without compromising its structural safety. The material distribution can be optimized, using the same or different materials according to structural, thermal or acoustic needs (Fig. 13).

A functionally graded component was fabricated using PU polymers with different compositions (Fig. 14). This pilot test was inspired by the biomedical cell supports (scaffolds), though the layer thickness needs to be reduced to allow a better control on the spatial properties of the material.

4 CONCLUSIONS

Modern information technologies and the rapid progress of Computer-Aided Design (CAD), with the advances in digital design and fabrication are motivating a shift in architecture and design towards the production of full integrated buildings. The industry is currently evolving to new processes of design fabrication combining material, form, structure and construction, creating new business opportunities. High public expectations and the dynamic nature of the construction industry are motivating new advances in building technologies, introducing an increased efficiency and competitiveness.

The RapidConstruction system, a multi-material deposition equipment based on extrusion-based technologies, presents several advantages over traditional approaches, namely greater geometrical freedom, structural optimization, multi-material, faster and lower construction costs.

The optimization of this new system enable to design and construct complex and best-adapted buildings, regarding thermal, acoustic and structural conditions.

REFERENCES

B. Khoshnevis, B.D. Hwang, K. Yao, Z. Yeh, Mega-scale Fabrication using Contour Crafting, International Journal of Industrial & Systems Engineering, Vol 1 (No. 3) (2006) 301–320.

F. Craveiro, J.P. Matos, H. Bártolo, P.J. Bártolo, Advanced innovation in building manufacturing, Proc. Fraunhofer Direct Digital Manufacturing Conference, Berlin, Germany, 2012.

F. Craveiro, J.P. Matos, H. Bártolo, P.J. Bártolo, An innovation system for building manufacturing, Proc. ASME 2012 11th Biennial Conference On Engineering Systems Design And Analysis, Nantes, France, 2012.

F. Craveiro, J.P. Matos, H. Bártolo, P.J. Bártolo, Automation for building manufacturing, in: P.J. Bártolo et al. (Eds), Innovative developments in virtual and physical prototyping, Taylor & Francis, 2011.

F. Craveiro, J.P. Matos, H. Bártolo, P.J. Bártolo, Automatisation de la construction de bâtiments, Proc. AEPR'11—16th European Forum on Rapid Prototyping and Manufacturing, Paris, France, 2011.

F. Craveiro, J.P. Matos; N.M. Ferreira; H. Bártolo, P.J. Bártolo, Construção Automática de Edificações, Proc. Engenharia' 2009 Conference, Covilhã, Portugal, 2009.

L. Dini, D-Shape, 2011, www.d-shape.com.

Larsson, M., Terasaki, O. and Larsson, K. 2003 "A solid state transition in the tetragonal lipid bilayer structure at the lung alveolar surface" Solid State Sci, 5(1):109–14.

Lord, E.A. and Mackay, A.L. (2003) "Periodic minimal surfaces of cubic symmetry", Current Science, 85(3):346–362.

P. Barrett, Revaluing Construction, Blackwell, Oxford, 2008.

R.U. Ayres, S.M. Miller, Robotics, applications and social implications, Ballinger Publishing Company, Cambridge, Mass., 1983.

S. Lim, T. Le, J. Webster, R. Buswell, A. Austin, A. Gibb, T. Thorpe, Fabricating construction components using layered manufacturing technology, Proc. Global Innovation in Construction Conference, Loughborough University, Leicestershire, UK, 2009, pp. 13–16.

T. Wohlers, Tooling and Manufacturing: State of the Industry, Rapid Prototyping, Wohlers Associates, USA, 2011.

Wang, Y. 2007 "Periodic surface modeling for computer aided nano design" Computer-Aided Design, 39:179–189.

Y. Miyamoto, W.A. Kayser, B.H. Rabin, A. Kawasaki, R.G. Ford, Functionally gradded materials design, Processig and Applications, Kluwer Academic Publishers, Boston, 1999.

Smart materials

Green Design, Materials and Manufacturing Processes – Bártolo et al. (eds)
© *2013 Taylor & Francis Group, London, ISBN 978-1-138-00046-9*

New patent on nanomaterials for preserving stone and wood structures

Santina Di Salvo
Department of Architecture, University of Palermo, Palermo, Italy

ABSTRACT: Nowadays, we recognize the need for a renewed commitment to the questions posed by the contemporary city to imagine new constructability scenarios paying attention to energy efficiency and cost saving, in order to obtain the recovery of identity of a city, achieving efficiency and effectiveness of the results. This paper focuses on developing effective strategies, based on sustainability and policies applicable to the built environment. Scientific experiences of the Author, demonstrated by scientific and technological patents, regarding the implementation projects on Nanomaterials, nanostructured inorganic oxides, and more particularly titanium sesquioxide and silicon, show that it is possible to obtain materials with high level of biocompatibility that can be used for the consolidation of ancient wood and stones.

1 INTRODUCTION

1.1 Premise

The search for coherence between the need for the recovery of identity and sustainability of interventions in the built heritage, highlights the issue on methods for selecting most appropriate strategies and tools to achieve the purpose, both at national and European level (Brandon & Lombardi 2011). In Italy, the current activities on the preservation and appreciation of the built environment and predictions about its future lead to considerations on the European scenery that can help to better understand the future behaviour in our national context. The theme of recovery of old buildings, applying non-traditional technologies, requires, in particular, a comprehensive overview of strategies for reliable interventions and a methodology to achieve goals that are consistent with the concept of sustainability.

1.2 The sustainable approach

In this perspective, the preservation and appreciation of the built heritage is of particular interest, through experimentation and innovative technologies. As evidenced by the good results obtained in some interventions both at national and European level, activities for the conservation of built heritage must be the result of synergies and collaboration of multiple groups, science and knowledge: architecture, engineering, technology, sociology, economics, urban planning, legislation must be managed and planned to work together. In fact, if sustainability ultimately means learning to think and act in terms of guaranteeing the prosperity of interdependent natural, social, and economic systems, then the built heritage, with its unique values and experiences must be contextualized and integrated with this view (Ayong Le Kama 2001).

Since the 1970s sustainability has evolved as a significant mode of thought in nearly every field of intellectual activity. As we know, decisions concerning conservation of the built environment have in the past been the domain largely of architectural historians, urban planners, conservation specialists, and related professionals. But conservation cannot remain a closed and solely self-referential profession, and indeed it has not (Macinnes 2004). With particular regard to the conservation of historic neighbourhoods and city centres, as well as individual monuments and sites of unique beauty, the challenge facing the conservation community is to develop a set of strategies and priorities that will permit it to focus its efforts on the conservation of those resources where the benefit-cost ratio is most favorable (Matero & Teutonico 2001). In this case sustainability means controlling change and choosing directions that capitalize most effectively on the inheritance from the past. In any decision about change and about the impact of the future on the remains of the past, therefore we should be conscious of two separate questions: the first is how to reconcile minimizing loss with the needs of the present; the second is how to ensure that the balance we strike does not reduce too greatly the options for future generations when they come to understand and enjoy their inheritance. Actually, built environment and heritage conservation should provide a dynamic vehicle by which individuals and communities can explore, reinforce, interpret and share their historical and traditional past and present, through community membership as well as through input as a professional or non-professional affiliate.

As claimed by Marion King Hubbert, *during the last two centuries we have known nothing but exponential growth, and in parallel we have evolved what amounts to an exponential-growth culture, a culture so heavily dependent on the continuance of exponential growth for its stability that it is incapable of reckoning with the problems of non-growth. Since the problems confronting us are not intrinsically insoluble, it behoves us, while there is yet time, to begin a serious examination of the nature of our cultural constraints, and of the cultural adjustments necessary to permit us to deal effectively with the problems rapidly arising* (Hubbert et al. 1949). The reflection of Hubert, highly topical, highlights how little has been recorded on our growth model in the last thirty years, and the need to start, with ambition and care, structural changes in our economies and in urban cultures to face the current crisis. The city we live in, today is incapable of performing its functions of "structure", it is not able to be a "cultural guide" anymore. The use of innovative technologies can be a test and a challenge for the re-construction of the rules to redevelop the built and "buildable" environment, by inserting the ecological variable and the resulting technologies and manufacturing solutions (Fairclough 1999).

2 INNOVATIVE TECHNOLOGY

2.1 *Inventions in progress*

In the building industry, sustainability has become synonymous with "green architecture", or building designed with healthy work environments, energy conserving systems, and environmentally sensitive materials. For historic tangible resources—whether cultural landscape, town, building, or work of art—the aim is notably different, as the physical resource is finite and cannot be easily regenerated (Stubbs & Makaš 2011). Instead, sustainability in the preservation of built environment means ensuring the continuing contribution of heritage to present through the thoughtful management of change responsive to the historic environment and to the social and cultural processes that created it. By shifting the focus to perception and valuation, conservation becomes a dynamic process involving public participation, dialogue and consensus, and understanding of the associated traditions and meanings in the creation, use, and re-creation of heritage. Sustainability emphasizes the need for a long-term view. But in the transformation of our physical environment, what relationships should exist between change and continuity, between the old and the new? Only when history is rightly viewed as continuous change can conservation

affect an integrated and sustainable environment. Conservation, based on the concept of sustainability, helps to extend places and things of the past into the present and establishes a form of mediation critical to the interpretive process that reinforces these important aspects of human existence. The fundamental objectives of conservation concern ways of evaluating and interpreting cultural heritage for its preservation and safeguarding now and for the future (Bennet 1996).

2.2 *The research: A new patent on nanomaterials for preserving stone and wood structures*

Thorough investigations give rise to new studies on the evaluation of characteristics, opportunities and effects of re-involvement of technology in knowledge, enhancement and communication of the built heritage, both ancient and modern. Research papers relating to technological innovation for the preservation and appreciation of the built heritage are many. Scientific experiences of the Author of the present contribution, shown by scientific and technological patents, regarding the implementation projects on nanomaterials, nanostructured inorganic oxides, and more particularly titanium sesquioxide (Ti_2O_3) and silicon (Si_2O_3), have demonstrated that it is possible to obtain materials with a high level of biocompatibility that can be used for the consolidation of archaeological wood and stones.

For example, the invention, entitled *Innovative sonochemical process that employs ultrasonic cavitation for the synthesis of monodispersed amorphous silicon dioxide nanoparticles, and a method for producing high-performance water-soluble lithium silicate compounds, for the application in the consolidation in situ of ancient stone and wood structures* (Di Salvo S., Patent Pending PA2011A000012), represents a powerful breakthrough in the synthesis of new materials for the protection of buildings, therefore for the improvement of the built environment. This patent has been stimulated by the need to obtain nanomaterials to be used for the consolidation and more particularly for the conservation and protection of natural stone and wood of ancient structures. As we know, the deterioration of stone and wood with which the ancient structures and monuments are built is a complex physical-chemical process caused by the interaction of several factors: climate of the locations, urban pollution, and the same material properties. Every direct method to consolidate and, more particularly, to protect and conserve the ancient structures must have the following characteristics: a) to be respectful of the environment; b) to be careful not to damage the wood surfaces or stone material of the ancient structures; c) not to affect the structural characteristics of the material that forms the structure to be preserved; d) to be well absorbed by capillarity; e) not to produce any change of colour of the treated material; f) to have a good penetration and ensure a high degree of consolidation. One of the most promising inorganic compounds for the consolidation of stone materials is lithium silicate. At the state of the art, two patented inventions in the United States, respectively US n. 4.443.496 (Obitsu et al., Application No. 400,820) and US n. 4.521.249 (Obitsu et al., Application No. 567,028), claim the use of lithium silicate to impregnate concrete surfaces. The first invention (1984) is titled *Agent and method for modifying surface layer of cement structures* and the second invention (1985) is titled *Silicate containing agent cement surface modified with this agent*. The descriptions of said patents highlight a method for the formation of a silicate coating on the surface of concrete. However, the inventions claimed by the above-mentioned U.S. patents do not address the problem of consolidation and conservation of stone, of which the pH is much lower than the matrix of cement. The patents provide for the use of a plasticizer which is sodium salt of naphthalene-sulfonate condensed with formaldehyde. This polymer has a typical dark brown colour and its application produces an undesirable colour formation of substrate (brown staining). Furthermore, the introduction of said organic material produces a microbial infestation of the stone.

Currently, the research aims to constantly create new materials capable of consolidating structures of stone and wood of ancient buildings, without generating any of the aforementioned problems. The function of the present invention is to create economically and conveniently nanoparticles of silicon dioxide, through a new process of synthesis which employs ultrasonic cavitation, totally unknown in the state of the art of science and technology. Ultrasonic cavitation is the energetic effect which is basically used by ultrasound. To be more precise, the ultrasonic cavitation is a physical phenomenon consisting in the creation of vacuum "tears" commonly referred to as "bubbles" in a fluid which immediately and violently implode. This compound of silicon dioxide in nanoscale structure can subsequently react with lithium hydroxide and/or carbonate, in water, to form a specific inorganic material of high performance, with new characteristics. The nanoparticles of silicon dioxide have a crucial role in the creation of a new form of water-soluble lithium silicate, to be applied *in situ* to improve the performance of stone and wood materials of ancient structures, essential for many applications in various fields of technology. This water-soluble compound can be used as a hardener, characterized by the unique ability to penetrate by capillary action in the pores, cracks and lesions of structures of stone and wood. This new material of lithium silicate performs the task perfectly respecting the environment, without changing the physico-chemical and mechanical structures of the treated materials, and all with no lasting effects. These nanoparticles, characterized by an average diameter of ~20 nm (Fig. 1),

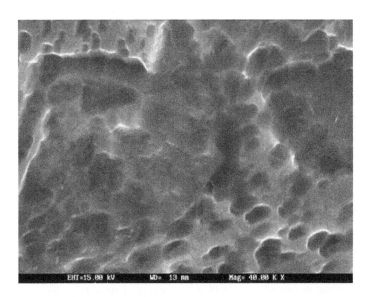

Figure 1. Image SEM (Scanning Electron Microscopy). Morphology of the surface of a porous stone sprayed with a 10% solution of lithium silicate which contains nanoparticles of silicon dioxide (~20 nm).

are synthesized through the use of the ultrasonic cavitation that determines the optimal conditions for the formation of molecules with a well-defined, regular spherical structure. More particularly, the Author of the present invention has found that an adequate ultrasonic frequency is indispensable, in order to avoid the agglomeration of the nanoparticles silicate dioxide. To have an optimum cavitation, capable of generating spherical nanoparticles well-defined of silicate dioxide, this ultrasonic frequency must be between 20 KHz and 60 KHz.

In fact, various experiments conducted by the author of this patent have shown that an ultrasonic frequency too high may generate the formation of nanoparticles of silicate dioxide characterized by irregular shapes, while an ultrasonic frequency too low may delay the formation of nanoparticles. The consolidating power of lithium silicate water-soluble is highly dependent on the size of the amorphous and mono-dispersed nanoparticles of silicon dioxide. The sonochemical technique, which uses ultrasonic cavitation, has the following advantages in the synthesis of nanoparticles of silicon dioxide: versatility and ease of execution, purity, consistency and high performance of the material obtained. In fact, the self-cleaning ability of this new type of silicon dioxide, having a nanoscale structure, applied directly to the stone surfaces, allows us to preserve their condition unchanged, without any alterations to their appearance of technical features, preventing biological pollutants and corrosion impurities, effectively counteracting the deterioration of the surfaces of stone materials and significantly reduces maintenance costs (Di Salvo 2012).

3 CONCLUSIONS

As we imagine, the city of the future will no longer be comparable to today's cities, largely because the technological challenges we face are so immense. Several sentiments reflect respect for antiquity and a distaste for modernity, a common conservative instinct, rather than an active concern for conservation. It is a persistent element in the culture of all modern societies, which eventually leads to a demand for government action to preserve the relics of the past. Cultural heritage is the mirror of society. It constitutes the legacy of tangible artefacts, such as historical buildings and monuments, as well as intangible features, such as traditions, customs and practices. Built environment and cultural heritage operate through a symbiotic relationship, whereby the physical symbols serve as evidence of underlying norms and values of a culture. Taking this into consideration, the importance of protecting tangible cultural heritage is

significant not only in order to reflect on and to better understand the past but also to maintain identification in the future. The cultural heritage of the European Union is crucial for establishing a shared European identification through progressive integration. All the projects, the interventions and recent new patents show the commitment and interest in the experimental research in innovative materials and reliable systems to ensure the preservation, enhancement, appreciation and enjoyment of the built environment. In recent years, several lines of research have developed innovative methods and new production processes that allow nanostructured particles to become an advantageous and indispensable component for the preservation, enhancement and appreciation of the built environment. These new nanomaterials can be successfully tested and verified to help bringing history to life, to protect our built environment, setting the stage for a really accessible and safeguarded city in the future.

Finally, the combined approach of multiple disciplines is a methodological strength of all inventions and projects (Pearce et al. 2012). In shared projects, a large number of transverse phases may well allow research teams to join in the development of common activities, useful to compare experiences and adopt the successful patterns of scientific knowledge. Experience has shown that innovation occurs when a process of change reaches a critical mass able to overcome the inertia of the "traditional system", and that it is only by focusing on new and innovative processes it is possible: 1) to establish groups of interdisciplinary research designed to implement plans and practices much more ambitious than the current ones; 2) to develop a set of reliable strategies which can be relevant both locally and internationally; 3) to improve the responsibility of groups involved; 4) to reinforce the concept of participation on clear objectives. In this scenario, it will be possible to read the signs of a possible different future, the feasibility of a new relationship between technology and the built environment.

REFERENCES

Ayong Le Kama, A.D. 2001. Sustainable growth renewable resources, and pollution, *Journal of Economic Dynamics and Control* 25 (12): 1911–1918.

Bennet, G. 1996. *Cultural Landscape. The conservation challenge in a changing Europe.* London: Institute for European Environmental Policy.

Brandon, P. & Lombardi, P. 2011. *Evaluating Sustainable Development in the Built Environment.* Oxford, United Kingdom: Wiley-Blackwell.

Di Salvo, S. 2012. *Methodological approaches for the enhancement of c.ltural heritage.* Rome: Aracne.

Fairclough, G. 1999. Protecting Time and Space: undestanding historic landscape for conservation in England, in Ucko, P.J. & Layton, R. (eds), 1999. *The Archaeology and Anthropology of Landscape: Shaping your landscape*, One World Archaeology 30, London: Routledge, 119–134.

Hubbert, M.K, Daniels, F. & Wigner, E.P. 1949. Our Energy Resources. *Physics Today* 2 (April 1949), 19–22.

Macinnes, L. 2004. Historic Landscape Characterization, in Bishop and Phillips (eds) *Countryside Planning: New approaches to Management and Conservation*. London: Earthscan. 155–169.

Matero, F. G. & Teutonico, J.M. 2001. *Managing change: sustainable approaches to the conservation of the built environment*. 4th Annual US/ICOMOS International Symposium organized by US/ICOMOS, Program in Historic Preservation of the University of Pennsylvania, and the Getty Conservation Institute 6–8 April 2001, Philadelphia.

Pearce J., Albritton S., Grant G., Steed G., & Zelenika I. 2012. A new model for enabling innovation in appropriate technology for sustainable development, *Sustainability: Science, Practice, & Policy* 8(2), 42–53.

Stubbs J.H & Makaš E.G. 2011. *Architectural conservation in Europe and the Americas*. Hoboken: Wiley.

Green Design, Materials and Manufacturing Processes – Bártolo et al. (eds)
© *2013 Taylor & Francis Group, London, ISBN 978-1-138-00046-9*

Transparent and sustainable materials

A. Lanza Volpe
Department of Architecture, University of Palermo, Palermo, Italy

ABSTRACT: The search for greater transparency in buildings characterizes the current architectural scenario: the use of glass is more and more common in both the design of new buildings and for renovation. However, the design of the transparent part of the building envelope is critical to the achievement of efficiency and sustainability. There are several research companies that aim to improve the energy performance especially of the glass. Today there is a wide range of products capable of providing different solutions, depending on the aesthetic and technical requirements of the structure in question. The aim of this article is to provide a clear overview of the possibilities of the transparent materials, their properties, their performances in relation to different climatic zones and their limits, to prevent negative effects in the design choices both in economic terms and in energy terms.

1 INTRODUCTION

Transparent materials are fundamental in the building envelope: they provide with the ability to both let light into rooms while at the same time keeping inclement weather outside and providing views and connection with outdoor spaces. Glass is the transparent material par excellence. The architecture of glass became a leading player in the course of the 19th century, with the industrial revolution, when the first iron and glass buildings such as arcades, markets, greenhouses and railway station canopies were erected in every major city. Since then the use of glass has evolved towards increasingly sophisticated shapes and technological solutions in various project typologies, some newly built, such as service buildings (Fig. 1) or for shopping, for advertising and image, other for re-qualifying existing buildings, to enhance the relationship between old and new (Paoletti & Romano 2010). The search for greater transparency and de-materialisation in building claddings characterizes the current architectural scenario in very different climatic zones: today fully glazed buildings are common in cold countries, but also in very hot climates, with the risk of energy consumption and discomfort if they were poorly designed (Butera 2005). When windows let heat escape on cold winter nights (causing the heating system to use more energy) and when they admit solar radiant heat on hot summer afternoons (causing the air cooling system to use more energy) they increase the building's energy costs. How can they reduce energy use? (McCluney & Jindra 2000) In 1981 the architect Mike Davies proposed a theoretical but potentially applicable smart glass wall. He entitled it the "polyvalent wall", a wall for all seasons.

The wall would control the flow of energy from the exterior to the interior using extremely thin layers that are multifunctional to minimize the energy consumption. Although this smart glass is still a prototype, the energy needs can decrease through an appropriate choice of glass characteristics associated with microclimatic parameters, such as natural ventilation, night cooling, sunlight, etc. (Schittich et al. 1999). In fact, glass alone will not

Figure 1. Stadttor, Düsseldorf, Germany, 1998 (Petzinka, Overdiek und Partner). The total gross floor area is some 30.000 m².

provide all the answers in the needs of a building. Orientation, overhangs, shading devices and window size all have a bearing on how well a glass product can perform.

2 ENERGY PERFORMANCE OF GLAZING

2.1 *Parameters for control of glazing energy performance*

The great development in technology has provided glass products with those features which were not originally a characteristic of glass material, for example the improvement of energy performance. An energy efficient window should provide good lighting during the day and good thermal comfort both during day and night. This implies that overheating as well as excessive cooling should be minimized. The key concepts for the evaluation of the energy characteristics of the glass surfaces are: thermal insulation, solar control and spectral selectivity. Thermal insulation is the reduction of heat transfer between environments of differing temperature. In buildings the glass is normally used as sheets of very small thickness, typically from 4 to 8 mm, and so the thermal insulation that a single sheet can provide is not very high. A good thermal insulation improves energy efficiency directly by lowering the "U-value" that represents the rate of transfer of heat through conduction, convection and radiation between two separate spaces. A monolithic glass has a transmittance of 5.8 W/m^2 K for a thickness of 4 mm that become 5.7 W/m^2 K for a thickness of 8 mm. Thus it is almost useless to act on the sheet thickness. Generally the U-value is mainly influenced by the quality of the glazing (Daneo & D'Este 2009).

The solar control is the reduction of the sun's direct heat energy through the glass. In a hot environment or in buildings with high internal loads, solar control glass can be used to reduce the effect of the sun's heat and to eliminate glare. The solar control, expressed by the Solar Heat Gain Coefficient "SHGC" or "g-value", depends on the amount of absorbed and reflected radiation by the glass. There is a compromise between solar control and transmission of light because a glass more controls the incoming heat and more it reduces the amount of light inside (Carmody et al. 2004).

Solar and terrestrial radiation is spread across a large spectrum of wavelengths. One of the most powerful facts about transparent glazing materials is that they affect different wavelengths of energy differently. The most important advance in glazing in the last twenty years has been the development of selectively reflective surfaces that can be tuned to block only certain portions of the spectrum. The variables of the glazing material, its thickness, and the angle of incidence of the incoming radiation determine whether energy of a given wavelength will be transmitted, absorbed or reflected. The spectral selectivity is the ability of the glass to allows the components of the solar and terrestrial spectrum to be teased apart and filtered out, depending on the specifics of an application. It can be obtained by reflection or absorption. In a building where heating is the primary need, such as a house in a cold climate, the ideal spectrally selective glazing would admit all of the visible and near-infrared radiation of the sun, and reflect back into the room all of the far-infrared terrestrial radiation. In a building where cooling is the primary concern the ideal glazing would admit only the visible light, while screening out all of the near infrared heat that accompanies it. This would allow daylight to be substituted for electric lighting, lowering the waste heat load in the space as well as saving electricity directly through the reduced use of electric lights. An adequate light will be allowed with high Visible Light Transmission and a low Solar Heat Gain Coefficient, while blocking significant amounts of solar radiation. The ratio of the Visible Light Transmission "VT" to Solar Heat Gain Coefficient "SHGC" or "g-value" is known as the Light-to-Solar Gain ratio, LSG. The recommended ratio is 1.25 or higher. The higher the ratio, the higher the daylighting benefit. It also means that the glass transmits more light than heat to the interior (Wasley & Utzinger 1996).

2.2 *Energy efficient glazing*

In a recent period the research has been carried out with the aim to limit thermal losses, but also to limit overheating, using the glass as solar protection. There are three fundamental approaches to improving the energy performance of glazing products (two or more of these approaches may be combined):

- To assemble various layers of glazing. The standard composition are laminated glass and insulating glass units. The laminated glass is manufactured by bonding two or more layers of glass together with layers of PVB, under heat and pressure, to create a single sheet of glass. When broken, the PVB interlayer keeps the layers of glass bonded and prevents it from breaking apart. From energy point of view, the PVB interlayer blocks 99% of incoming UV radiation. The insulating glass units are double or triple glass window panes separated by an air or other gas (argon, krypton or xenon) filled space to reduce heat transfer across a part of the building envelope. The contribution of the gas introduced in the cavity is important, not only for its low thermal conductivity compared to the glass, but also because the viscous forces in the gas filling prevent the motions of natural convection inside the interspace, reducing the heat transmitted through the glazing system.

– To alter the glazing material. Current examples are the body-tinted glass and the ceramic fritted glass. The body-tinted glass is a normal float glass with special inorganic additive to obtain a high absorption coefficient of radiation. Thus, it is used as solar control glass because it absorbs more heat than clear glass and radiates a majority of the absorbed heat to the outside by means of natural convection. Tinted glazings retain their transparency from the inside, although the brightness of the outward view is reduced and the colour is changed. Every change in colour or combination of different glass types affects visible transmittance, solar heat gain coefficient and reflectivity (Compagno 2002). The most common colours are neutral gray, bronze, and blue-green, which do not greatly alter the perceived colour of the view and tend to blend well with other architectural colours. Tints do absorb solar radiation, but only the green and blue colours selectively absorb more infrared radiation than visible radiation (Wasley & Utzinger 1996). The ceramic fritted glass is made by silk-screening. The process involves screen printing ceramic frit paint onto the glass and fusing it onto the surface during the toughening or heat strengthening process. Frit consists of tiny glass particles, pigments and various chemicals for the curing. The patterns are usually lines or dots not perceived in distant vision applied to the interior side of the glass. The size, density and colour of patterns determine the opacity and shading. The results are an increase in solar reflection and absorption on the interior glass ply and a lower shading coefficient. The innovative use of ceramic fritted glass can offer significant benefits in terms of controlling unwanted solar heat gain or visual glare from solar energy. The silk-screened elements can have a gradient of variation which depends on the angle of solar incidence. This explains how a material that is apparently inappropriate for acting as a shelter and modulating daylight provides a wide range of possible forms of regulation thanks to use of advanced production techniques and calculation systems for correct positioning of panels.

– To apply a coating to the glazing material surface. The glass surface can be coated with microscopically thin deposits, "coating", generally metals, noble metals or oxides layers. Recent technological developments have made available a large group of coating that can greatly improve the glass energy performance (Paolella & Minucci 2004). Reflective coatings such as silver are thin enough

to be transparent, but thick enough that they reflect wavelengths throughout the solar spectrum more or less equally. Therefore, they significantly reduce the Solar Heat Gain Coefficient, but also the light transmission and glare. Advanced low-e coatings use various metals and thicknesses to arrive at a layer that interferes with certain portions of the spectrum while being transparent to others. A great evolution is represented by the application of low-emissivity coating that reflect radiant infrared energy, encouraging radiant heat to remain on the same side of the glass from which it originated, while letting visible light pass. This often results in more efficient windows because radiant heat originating from indoors in winter is reflected back inside, while infrared heat radiation from the sun during summer is reflected away, keeping it cooler inside. The transmittance of the glass and the Solar Heat Gain Coefficient are reduced by lowering the radiative heat flux (Pfrommer et al. 1995). Today there are glasses with emissivity of 0.01 against to the value of 0.89 for standard clear glass. The Low-E glass is primarily designed for use in glazing units. However, solutions with hard coating may be used also in the case of monolithic or laminated glass. The solar reflectance of the low-e coatings can be manipulated to include specific parts of the visible and infrared spectrum. There are low-emissivity coating in the far infrared, for both cold and hot climates, and low emission glass throughout the infrared, for hot climates. Depending on the climate and orientation it is necessary to choose low-e coatings with different characteristics. In a cold climate a higher Solar Heat Gain Coefficient is required, in a hot climate a lower Solar Heat Gain Coefficient, with a consequent reduction in the direct heat gain (Carmody et al. 2007).

3 INNOVATIVE TRANSPARENT MATERIALS

The scenario of transparent systems includes a series of innovative and high-performance materials an advanced stage of development. The main innovations can be summarized in the following points:

– High insulation glass. Technology based on the use of aerogel and geometric media, transparent polycarbonate or polymethylmethacrylate structures, in the cavity of the classic double glazing. These products are characterized by very low values of thermal transmittance, comparable to those of opaque structures, and high light transmission.

– Chromogenic glazing (or smart glass). This technology allows to change the colour of the glass and consequently the light and solar characteristics.

The transition coloured state, including the intermediate transitions, may be activated by: electrical pulses, temperature, solar radiation. The most advanced technology is electrochromic glazing, on the market with many high costs.
- Daylighting systems. These materials and components are able to intercept the solar radiation and to direct it outside or, alternatively, towards the more internal and less bright areas of the building. The main technologies are: prismatic glass, holographic films and high reflective shading devices (Zinzi 2009).
- Solar windows. A company of Maryland, New Energy Technologies, is developing the first-of-its-kind SolarWindow™ technology enables see-through windows to generate electricity by 'spraying' their glass surfaces with New Energy's electricity-generating coatings. This technology utilizes an organic solar array composed of a series of ultra-small solar cells fabricated using environmentally-friendly hydrogen-carbon based materials.

In the field of transparent materials, nowadays there is a wide range of products other than glass capable of providing different solutions, depending on the aesthetic and technical requirements of the structure in question. The beehive polycarbonate is characterized by a structure that gives lightness and provides an excellent insulating effect, but it has lower transparency than glass. The compact polycarbonate or polymethylmethacrylate "PMMA" sheet is transparent to visible light and to infrared radiation but it is sensitive to scratches and abrasions. One of the problems of such plastic materials is the loss of transparency due to the deposition of particulates and other substances present in the atmosphere which degrade their aesthetic appearance and their functionality. A step forward is represented by the co-polymer ethyl tetrafluoroethylene "ETFE" foil. It is essentially a plastic polymer created by taking the polymer resin and extruding it into a thin film. It is used due to its high light transmission properties. Transparent windows are created either by inflating to or more layers of foil to form cushions or tensioning into a single-skin membrane. Weighing approximately 1% the weight of glass, simple-ply ETFE membranes and ETFE cushions are both extremely light-weight (Wilson 2009). However, it can not to replace the glass for domestic applications for many reasons: the material is very transparent to sound meaning not only that sound exits the building but sounds come in from outside too; it requires special installation techniques and heavy costs; it can be easily damaged even with a sharp pencil; the system relies on constant air pressure—that requires special air pumps—which in turn need to be connected and maintained to the electrical supply.

4 CONCLUSIONS

Glass boasts of ancient origins, but at the same time possesses, thanks to the intensive research activity of recent decades, an extraordinary modernity, standing out as a highly technological and functional material, capable of meeting the most diverse of requirements. Depending on climatic conditions, it is possible to combine different types of sheets in order to have thermal insulation, solar control or spectral selectivity. In recent years, several lines of research have also developed innovative materials and new production processes that will allow for the transparent systems to become an increasingly dynamic component of the built environment.

REFERENCES

Butera, F.M. 2005. Glass Architecture: is it sustainable? in Santamouris, M. (ed), *Passive and Low Energy Cooling for the Built Environment, Proc. 1st International Conference on Passive and Low Energy Cooling (palenc 2005)*. Santorini: Heliotopos Conferences.

Carmody, J., Selkowitz, S., Arasteh, D., & Heschong, L. 2007. *Residential Windows: A Guide to New Technology and Energy Performance*. New York: Norton & Company.

Carmody, J., Selkowitz, S., Lee, E., Arasteh, D. & Willmert, T. 2004. *Window Systems for High Performance Buildings*. New York: Norton & Company.

Compagno, A. 2002. *Intelligent Glass Facades*. Berlin: Birkhäuser.

Daneo, A. & D'Este, A. 2009. Considerazioni controintuitive sull'isolamento termico dei vetrocamera con gas, *Rivista della Stazione Sperimentale del Vetro*, 39 (1): 13–17.

McCluney, R. & Jindra, P. 2000. *Industry Guide to Selecting the Best Residential Window Options for the Florida Climate*. Cocoa: Florida Solar Energy Center—University of Central Florida.

Paolella, A., Minucci, R. (ed.) 2004. *L'efficienza energetica degli edifici. L'uso del vetro per la riduzione degli effetti negativi derivanti dai mutamenti climatici*. Roma: Edicomprint.

Paoletti, I. & Romano, M.G. 2010. Transparent roofing. History and technologies, *Frames*, 146: 108–113.

Pfrommer, P., Lomas, K.J., Seale, C., & Kupke, C. 1995. The radiation transfer through coated and tinted glazing, *Solar Energy*, 54(5), 287–299.

Schittich, C., Staib, G., Balkow, D., Schuler, M., Sobek, W., 1999. *Glass Construction Manual*, Basel: Birkhäuser.

Wasley, J.H. & Utzinger, M. 2000. *Vital Signs: Glazing Performance*. Johnson Controls Institute for Environmental Quality in Architecture, School of Architecture and Urban Planning, University of Wisconsin-Milwaukee.

Wilson, A. 2009. ETFE: The new Fabric Roof, *Interface. Extraordinary and Unusual Roofs*: 4–10.

Zinzi, M. 2009. Caratterizzazione e valutazione di materiali trasparenti innovativi e sistemi schermanti, *Report Ricerca Sistema Elettrico*, 9.

Green Design, Materials and Manufacturing Processes – Bártolo et al. (eds)
© 2013 Taylor & Francis Group, London, ISBN 978-1-138-00046-9

Photocatalytic degradation of textile effluent using ZnO/NaX and ZnO/AC under solar radiation

T.M.P. Schimidt, F.R. Soares & V. Slusarski-Santana
Chemical Engineering Department, Universidade Estadual do Oeste do Paraná (UNIOESTE), Toledo, Paraná, Brazil

F.F. Brites-Nóbrega & N.R.C. Fernandes-Machado
Chemical Engineering Department, Universidade Estadual de Maringá (UEM), Maringá, Paraná, Brazil

ABSTRACT: The aim of this study was to evaluate the efficiency of ZnO both in suspension and immobilized on NaX zeolite and Activated Charcoal (AC) on the degradation of dye solutions and industrial effluent of the laundry of jeans under solar radiation. Catalysts containing 5 and 10% (wt%) of ZnO were prepared by a wet impregnation method and characterized by textural analysis, X-ray diffraction, temperature-programmed desorption of ammonia and scanning electron microscopy. The results show that increasing the mass of the catalyst resulted in an increase in the process efficiency. The activity of ZnO in suspension was maintained on the degradation of the laundry effluent, demonstrating the potential application of solar radiation to degrade textile effluents. The results showed that both the zeolite and activated charcoal are good supports. Among the supported catalysts, 5% ZnO/NaX and 5% ZnO/AC catalysts were the most efficient (73% and 69% of degradation, respectively).

1 INTRODUCTION

Many industries generally contribute to the contamination of water bodies. In particular, the textile industries consume large volumes of water and generate large amounts of wastewater which have a toxic organic load, a characteristic color and are resistant to biodegradation (Kritikos et al. 2007).

The application of the heterogeneous photocatalysis in the treatment of industrial waste water has been studied intensively over the past three decades, mainly using catalysts in suspension. However, the practical application of this approach is limited due to the necessity of a post-separation process. Thus, research on the semiconductors immobilizations in an inert support are being performed (Sobana & Swaminathan, 2007, Sobana et al. 2008, Santana et al. 2010). Different types of supports may be used: glass spheres and slides (Santana et al. 2010), activated charcoal (Sobana & Swaminathan, 2007) and zeolites (Petkowicz et al. 2010). Activated charcoal and zeolite are interesting because they exhibit a high surface area and porosity as well as mechanical and thermal resistance.

The aim of this study was to evaluate the efficiency of ZnO suspended and immobilized on NaX zeolite and activated charcoal. The effect of the concentration of the active phase and the influence of the support type on the photocatalytic degradation of the textile effluent under solar radiation were also investigated.

2 EXPERIMENTAL

2.1 Materials and methods

ZnO (Nuclear) mass and immobilized on NaX zeolite (Bayer with 3 mm of diameter) and Activated Charcoal (AC) (BRASILAC with 2 mm of diameter) were used as catalysts. The supported catalysts with a catalyst loading of 5 and 10% (wt%) were prepared by wet impregnation from solutions of zinc nitrate (DYNAMIC PA). All catalysts were calcined at 400 °C for 5 h.

2.2 Characterization of catalysts

The catalysts were characterized by textural analysis from adsorption-desorption isotherms of N_2 at 77 K using Quantachrome Nova 1200 equipment, X-Ray Diffraction (XRD), CuKα source, using Shimadzu D600 equipment, temperature-programmed desorption of ammonia (NH_3-TPD) using CHEMBET 3000 Quantachrome Instruments equipment and Scanning Electron Microscopy (SEM) using SHIMADZU IC-50 equipment.

2.3 *Photocatalytic activity*

The experimental apparatus (glass reactor of 30 cm diameter) were exposed to solar radiation between 10:00 a.m. and 15:00 p.m (5 h) on non-cloudy days, during summer. The mean global solar irradiance was determined using LI-COR Radiation Sensor LI2000AS pyranometer equipment from the Technological Institute of Meteorology of the Brazilian Parana State (SIMEPAR), located in Toledo, Paraná (Brazil). The mean global solar irradiance for these tests with the reactive 5G dye and industrial effluent were 546.0 and 509.4 $W \cdot m^{-2}$, respectively.

The photocatalytic tests consisted of irradiating 500 mL of the reactive blue 5G dye ($10 \, mg \cdot L^{-1}$) or industrial effluent of jeans laundry containing different catalysts which contains some chemicals as softener and surfactant used during the treatment of washing, besides the dye which coming out during of jeans washing. At regular intervals, samples were collected, filtered and analyzed by UV-Vis spectrophotometry (Shimadzu UV-1800) for the wavelength range 200 to 800 nm. The discoloration was determined from the decrease of absorbance at 600 and 750 nm for the dye solution and effluent, respectively.

3 RESULTS AND DISCUSSION

3.1 *Characterization of catalysts*

ZnO showed adsorption isotherms of type II with $5 \, m^2 \cdot g^{-1}$ of specific surface area, $0.064 \, cm^3 \cdot g^{-1}$ of total pore volume and a mean diameter of the pores of 51.5 Å The activated charcoal and zeolite have different pore distributions. Activated charcoal has a large pore size distribution, while the NaX zeolite which is formed by sodalite cages in a tetrahedral arrangement has a narrow regular pore size distribution (Li et al. 2005). The high specific surface area of the supports ($423.6 \, m^2 \cdot g^{-1}$ for NaX and $790.9 \, m^2 \cdot g^{-1}$ for charcoal) was reduced with impregnation of ZnO and particularly so for NaX (60% reduction).

The XRD patterns of the catalysts and supports can be seen in Figure 1 and they showed the characteristic peaks of ZnO in the catalyst mass and catalysts supported only on activated charcoal. The XRD analysis also showed that the zeolite is crystalline, while the activated charcoal is amorphous.

The results of the NH_3-TPD analysis showed that the basic character of NaX, it was unable to neutralize the acidity of the active phase. All catalysts supported on NaX showed acidity, a characteristic of the impregnated oxides, and weak acid sites (desorption temperature of around 530 K). The acidity was 0.831 and 0.902 mmol $NH_3 \cdot g_{cat}^{-1}$ to 5 and 10%

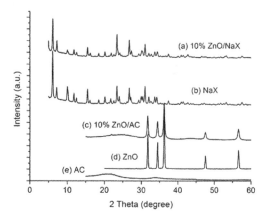

Figure 1. XRD patterns of the supports and catalysts with 10% of ZnO.

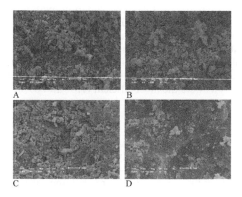

Figure 2. SEM image of supported catalysts: (A) 5%ZnO/NaX, (B) 10%ZnO/NaX, (C) 5%ZnO/AC, (D) 10%ZnO/AC. All pictures have 2000× magnification.

ZnO/NaX, respectively. The increase of the content of active phase caused a slight increase in the acidity of the catalyst. All catalysts supported on activated charcoal showed no acid sites. Probably the activated charcoal annulled the intrinsic acidity of the ZnO by the interaction support-oxide, changing the acidity after adsorption of oxides.

SEM photographs of the supported catalysts are shown in Figure 2. As can be seen, the catalyst 10%ZnO/NaX showed more spherical particle agglomerates compared to 5%ZnO/NaX catalyst. Already the catalysts supported on activated charcoal presented porous surface with some agglomerated particles.

3.2 *Photocatalytic activity*

Initially, the effect of the mass of ZnO in suspension (0.4 and 1.0 $g \cdot L^{-1}$) was evaluated on the reactive 5G dye degradation (10 mg·L⁻¹). When working with 0.4 g·L⁻¹ of ZnO in suspension, discoloration of 69% was obtained with a constant apparent rate (k_{ap}) of a pseudo-first order process of 0.265 h⁻¹, following the Langmuir-Hinselwood model (Brites et al. 2011). The increase of mass to 1.0 g·L⁻¹ caused an increase in process efficiency (80% of discoloration and k_{ap} = 0.373 h⁻¹). UV-Vis absorption spectra of the dye solution before and after the photocatalytic treatment can be observed in Figure 3. We propose that discoloration occurred by breaking of the chromophoric groups (reducing the characteristic peak at 600 nm) and mineralization by breaking of aromatic rings (reduction near 300 nm), especially when it was used 1 g·L⁻¹.

The supported catalysts were not efficient on the degradation of the dye solution; about 2.6% of discoloration was obtained (Fig. 4).

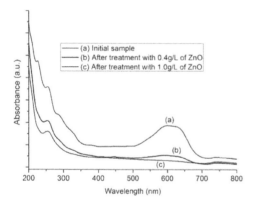

Figure 3. UV-Vis absorption spectra of the dye solution before and after photocatalytic treatment with ZnO in suspension.

In tests of degradation of industrial effluent, the efficiency of ZnO in suspension and supported, the effect of the active phase concentration (5 and 10%) and the influence of support type (NaX and activated charcoal) were evaluated.

ZnO is highly photoactive under solar radiation and the efficiency of the photocatalytic process was maintained when working with industrial waste-water, which makes the practical application of solar radiation promising for the treatment of industrial effluents.

The effluent degradation using supported catalysts was lower than with the catalyst in suspension, such results can be visualized in Figure 4. However, even with a smaller amount of ZnO in the catalyst 5%ZnO/NaX (0.05 g), its activity was slightly lower (decolorization of 73% versus 85%).

This result indicates a synergistic effect between active phase and support that favored the photocatalytic process of industrial wastewater under solar radiation. Another relevant factor relates to the distribution of the active phase in the zeolite when working with small amounts of ZnO. The increase in the ZnO content in the supported catalysts caused a slight decrease in process efficiency, possibly due to a more irregular distribution over the support surface, preventing the activation of all active sites by solar radiation.

The UV-Vis absorption spectra of the supported catalysts are shown in Figure 5. The 10%ZnO/AC catalyst was less efficient compared to other catalysts and as can be seen in Figure 5, it discolored the industrial effluent through the breaking of the chromophore groups of the organic molecules, forming intermediate species that absorb between 250 and 300 nm which were responsible for the increased absorbance in this range of wavelength.

Under solar radiation, the activated charcoal was effective as support, especially when working

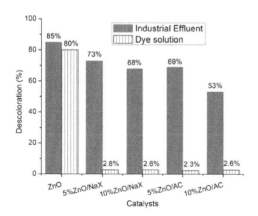

Figure 4. Discoloration of dye solutions and industrial effluent under solar radiation.

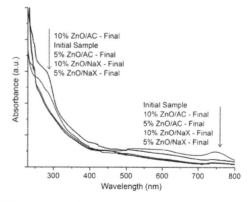

Figure 5. UV-Vis spectra of industrial effluent before and after photocatalytic treatment with supported catalysts.

with 5% of ZnO. This result is contrary to that observed with artificial UV radiation (Brites et al. 2011) and literature which states that the high adsorption capacity of charcoal inhibits the mineralization of the dye (Bhattacharyya et al. 2004). In this case, a large fraction of visible radiation of the solar spectrum favored the interaction between ZnO and the surface of activated charcoal on the degradation of laundry effluent.

4 CONCLUSIONS

The results of dye degradation showed that the ZnO in suspension has a high photoactivity and that the greater the catalyst mass, the greater the number of active sites activated by solar radiation and the greater the degradation.

The effluent degradation was slightly superior to that obtained for the dye solution, demonstrating the potential application of solar radiation to degrade industrial effluents which are generally more difficult to degrade due to their complex composition.

The supported catalysts were not efficient on the dye solution degradation under solar radiation, however they were able to degrade the industrial effluent of jeans laundry (around 70% at 5 h), probably due to the composition of this effluent (presence of softener and surfactant).

The results showed that the zeolite and activated charcoal are good supports. Among the supported catalysts, 5%ZnO/NaX and 5%ZnO/AC catalysts were the most efficient (73% and 69% of degradation, respectively).

ACKNOWLEDGEMENTS

Authors thank CNPq and UNIOESTE/PRPPG for the financial support of this study.

REFERENCES

Bhattacharyya, A., Kawi, S., Ray, M.B. 2004. Photocatalytic degradation of orange II by TiO_2 catalysts supported on adsorbents. *Catalysis Today* 98: 431–439.

Brites, F.F., Santana, V.S., Fernandes-Machado, N.R.C. 2011. Effect of support on the photocatalytic degradation of textile effluents using Nb_2O_5 and ZnO: Photocatalytic degradation of textile dye. *Topics in Catalysis* 54: 265–269.

Kritikos, D.E., Xekoukoulotakis, N.P., Psillakis, E., Mantzavinos, D. 2007. Photocatalytic degradation of reactive Black 5 in aqueous solutions: Effect of operating conditions and coupling with ultrasound irradiation. *Water Research* 41(10): 2236–2246.

Petkowicz, D.I, Pergher, S.B.C., Silva, C.D.S., Rocha, Z.N., Santos, J.H.Z. 2010. Catalytic photodegradation of dyes by in situ zeolite-supported titania. *Chemical Engineering Journal* 158: 505–512.

Santana, V.S., Mitushasi, E.O., Fernandes-Machado, N.R.C. 2010. Avaliação da atividade fotocatalítica de Nb_2O_5. *Acta Scientiarum. Technology* 32(1): 55–61.

Sobana, N., Muruganandam, M., Swaminathan, M. 2008. Characterization of AC-ZnO photocatalytic activity on 4-acetylphenol degradation. *Catalysis Communications* 9: 262–268.

Sobana, N., Swaminathan, M. 2007. Combination effect of ZnO and activated carbon for solar assisted photocatalytic degradation of Direct Blue 53. *Solar Energy Materials & Solar Cells* 91: 727–734.

Green Design, Materials and Manufacturing Processes – Bártolo et al. (eds)
© *2013 Taylor & Francis Group, London, ISBN 978-1-138-00046-9*

Comparison of residue of cassava industry modified and "in natura" for adsorption of methylene blue dye

R.B. Pardinho & J. Caetano
Universidade Estadual do Oeste do Paraná, Toledo, Brazil

J. Casarin & A.C. Gonçalves Jr.
Universidade Estadual do Oeste do Paraná, Marechal Candido Rondon, Brazil

L.F. Bonfim Jr. & D.C. Dragunski
Universidade Paranaense, Umuarama, Brazil

ABSTRACT: This work used the residue of the cassava industry "in natura" and modified to promote the adsorption of Methylene Blue (MB) in aqueous solution. This residue was modified by using sodium hydroxide and citric acid in order to increase adsorptive capacity. The best pH for adsorption of the dye was 7.0 with a time of 180 minutes for both residues. The adsorption mechanism for the residues "in natura" and modified follows the model of Pseudo-second order with r^2 (Linear Correlation) worth 0.99931 and 0.99996, respectively. The model of isotherms with best answer to the residues with r^2 of 0.9912 and 0.9861 respectively was Freundlich.

1 INTRODUCTION

It is estimated that about 15% of the world production of dyes is lost to the environment during the synthesis, processing or applications. This represents a release of 1.20 tons per day of this class of compounds into the environment (Guarani 1999). In order to reduce the environmental impact of the dye industry, several methods have been developed for the removal of synthetic dyes from industrial effluent, that based on the adsorption by actived carbon is the most commonly used (Dural 2011). Techniques used for the removal of dyes include process of membrane, filtration with coagulation, coagulation with ozonation, adsorption, precipitation, chemical degradation, electrochemical and photochemical, biodegradation (Guaratini 2000, Cardoso 2010).

Adsorption is a physical process of great interest with associated low cost with high rates of removal. In some cases, it is not a destructive method and enables adsorption of the dye recovered without loss of its chemical identity. The retention of the dyes by adsorbent materials involves several attractive forces, such as ionic interaction, Van der Waals forces, hydrogen bonds and covalent bonds (Deng et al. 2011, Geada 2006).

An alternative to the adsorption process is a method which uses biomass such as agricultural by products biosorbents as adorbent materials.

The main advantages of using the process of biosorption compared to currently used methods include, low operating costs, minimization of volume of chemical and/or biological materials to be discarded, and high efficiency in detoxifying effluents (Marques 1999, Cho 2003). Among the residues studied, cellulose (Annadurai 2002), rice bran and wheat (Geada 2006), residues pineapple (Weng 2009), peanut (Özer 2007), and others have been studied. This work will use as the adsorbent material the agroindustrial residues of cassava. At present this is a large volume crop grown for the high starch content, in over 80 countries, of which Brazil participates with more than 15% of world production (Embrapa 2011).

Thus, the objective of this work is the use of the natural residues of the cassava industry i as an adsorbent material for removal of Methylene Blue (MB).

2 MATERIAL AND METHODS

2.1 *Preparo do biosorvente*

The adsorbent materials was washed, oven dried (85° C) for approximately 24 hours. After drying, they were ground and sieved (42 Mesch). For modification the residue was treated with NaOH 0.1 mol L^{-1} and citric acid 1.2 mol L^{-1} (Rodrigues 2006).

2.2 Adsorption experiments

The analyzes of adsorption as a function of pH and time for the modified residue, were performed using 50 mL of the dye with a concentration of 100 mg L⁻¹, they were stirred with 0.5 g of residue using an orbital shaker for 24 hours. To evaluate the adsorption of the dye as a function of pH, measurements were performed over a range of pH ranging from 3 to 11. The data obtained were used to calculate the amount of dye adsorbed per gram of residue (Q_{eq}) in function of pH (Equation 1).

$$Q_{eq} = \frac{C_o - C_{eq}}{m} V \qquad (1)$$

where: Q_{eq} is the quantity of heavy metal adsorbed (mg g⁻¹), m is the mass of residue (g), C_0 is the initial solution concentration (mg L⁻¹), C_{eq} is the final average concentration of the metals in balance in the solution (mg L⁻¹) and V is the utilized solution volume (L).

The measures of kinetics adsorption were performed at pH 7 using 8 samples, the time intervals were varied from 10 min. to 24 h. The adsorption isotherm was performed to evaluate the adsorption capacity versus concentration of dye. Measurements were made in pH 7 to a concentration range of methylene blue ranging from 10 mg L⁻¹ to 900 mg⁻¹. The samples were placed under stirring for 4 hours. All measurements (pH, time and concentration) were made in batch and to perform each analysis, the resulting liquid was filtered, diluted and analyzed using a Shimadzu UV-1601PC spectrophotometer.

3 RESULTS AND DISCUSSION

3.1 Influence of pH

The Figure 1 shows the change in Q_{eq} with pH for both the residues. It can be observed that the retention of MB for the residues "in natura" and modified, presented a small change in adsorption capacity as a function of pH of each solution, this fact may be explained by the chemical groups present in the structure of the residue, being of great importance for a wastewater treatment system that can be used in mild acidic to basic. However, it can be seen that the adsorbent modified presented higher degree of recoat, possibly due to the expansion of the adsorption sites.

Thus, we chose to work with 7.0 for both residues, because besides the greater amount of dye adsorbed, this is the neutral pH of this solution, which may be easier for applications with real samples.

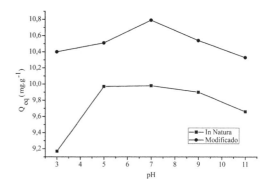

Figure 1. Influence of the pH in the adsorption of AM to the residues of cassava modified in nature.

Measurements were made of the adsorption kinetics, seeking time required for the system to reach equilibrium. In Figure 2 it can be seen that the equilibrium time for the "in natura" and modified residues was reached in about 180 minutes. However, the modified residue submitted greater adsorption capacity (Q_{eq} 10.4 mg·g⁻¹) when compared with "in natura" residue (Q_{eq} 8.7 mg·g⁻¹) at the same time of the analysis, in other words, for the adsorption of the same amount of dyes, the residue modified only required 39 minutes.

The adsorption mechanisms of MB in the cassava industry residues were evaluated using the model of Pseudo 1ª Order, Pseudo 2ª Order, intraparticle diffusion and Elovich. These depend on the characteristics physical and/or chemical adsorbent, as well as of the mass transport process.

In the model of pseudo 1st order, the adsorption rate of the solute in solution is proportional to the number of free sites. To evaluate the kinetic model and determining the constant speed (k_1) and Q_{eq} was used to equation 2.

$$\log(Q_{eq} - q_t) = \log Q_{eq} - \frac{k_1 x t}{2.303} \qquad (2)$$

where, Q_{eq} (m$_{eq}$ g⁻¹) and q_t are the amounts of adsorbed AM by the biosorbent in equilibrium and each time t, and k_1 is the constant of the first-order biosorption speed (min⁻¹) (Özer 2007).

In the model of Pseudo 2st order, the velocity of sorption is proportional to the square of the number of free sites. The values of k_2 and Q_{eq} and modelo kinetics were evaluated by equation 3.

$$\frac{t}{qt} = \frac{1}{k_2 Q eq^2} + \frac{1t}{Qeq} \qquad (3)$$

where, k_2 is a second-order speed constant of biosorption (g mg⁻¹ min⁻¹). Parameters q_e and k_2 are

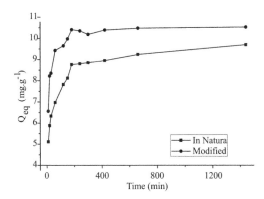

Figure 2. Effect of time (min) on the adsorption of MB to the residues of cassava modified and "in natura" for a solution of 100 mg L^{-1} in pH 7.0.

calculated on the inclination and the interception the t/q_t versus t graphics (Horsfall 2006).

The Elovich kinetic model of the adsorption follows the behavior of chemisorption (Pérez-Marin 2007), and has been successfully appliedto a wide range of processes with a slow rates of adsorption occurring on heterogeneous surfaces, being represented by the equation 4 (Cheung 2000).

$$q_t = A + B \ln t \qquad (4)$$

where A and B are constants Elovich.

The model of intra-particle diffusion was explored using the equation (Cheung 2000):

$$q_t = K_{id}\, t^{\frac{1}{2}} + C_i \qquad (5)$$

where, K_{id} (mg g^{-1} min$^{-1/2}$) is the constant rate of intra-particle diffusion, C (mg g^{-1}) is the constant that suggests the thickness of each limiting layer, that is, the higher the value of C is, the greater the effect on the limiting layer is. The calculated kinetic parameters for the four kinetic models are show on Table 1.

It can be observed that the adsorption mechanism for the residues "in natura" and modified follow the model of Pseudo-Second order with r^2 (Linear Correlation) worth 0.99931 and 0.99996, respectively. The values of Q_{eq} (adsorption capacity of the MB) are close to those obtained experimentally, indicating a strong interaction of the dye.

Another model that fits in the residue "in natura" is the intra-particle diffusion, presenting two straight lines in this way, it can be said that the residue presents porosity difference, being that initially occurs a rapid filling of the macropores and subsequently a fill of micropores, ie two-stage diffusion.

Table 1. Parameters for the kinetics of pseudo-first-order and pseudo-second-order models, Elovich's model and intra-particle diffusion model for the MB adsorption of residues of cassava starch: "in natura" and modified.

	In natura	Modified
Pseudo-first-order		
K_1 (min^{-1})	0.035	0.060
Q_{eq} (cal.) (mg g^{-1})	3.090	1.710
R^2	0.804	0.750
Pseudo-second-order		
K_2 (g mg^{-1} min^{-1})	4.222	1.240
Q_{eq} (cal.) (mg g^{-1})	9.780	10.590
R^2	0.999	0.999
$Q_{eq}exp$ (mgg^{-1})	8.950	10.530
Elovich		
A (mg g^{-1} h^{-1})	0.980	0.758
B (g mg^{-1})	3.073	5.860
R^2	0.958	0.811
Intraparticle diffusion		
Fist line		
K_{id} (g mg^{-1}min$^{-1/2}$)	0.321	0.086
C_i (mg g^{-1})	4.360	8.324
R^2	0.979	0.461
Second line		
K_{id} (g mg^{-1}min$^{-1/2}$)	0.039	–
C_i (mg g^{-1})	8.182	–
R^2	0.989	–

3.2 Adsorption isotherms

The adsorption isotherms show the equilibrium between the solution and the dye retained in the sorbent. In order to investigate the adsorption isotherm, the experimental data were fitted to mathematical models: Langmuir, Freundlich, Temkin and Dubinine Radushkevich (DR), equations 6, 7, 8 and 9 respectively (Özacar 2003, Dogan 2004, Dahiya 2008).

The isotherm linearity, using mathematical models, provides basis for this conclusions through the mathematical parameters provided by these models. These models are important in describing the adsorption behavior, because when the adsorption equilibrium state is reached the isotherm may indicate the distribution of dye molecules between the liquid phase and the solid phase. This is significant for an understanding the adsorption behavior to identify the most appropriate model of adsorption isotherm.

The isotherm Langmuir (equation 6), is based on the fact that adsorption occurs at sites uniformly becoated in a monolayer, in other words, is limited by a single layer and has an affinity-ionic independent of the amount of adsorbed material (Sodré 2001).

$$\frac{C_{eq}}{q} = \frac{1}{q_{max}b} + \frac{C_{eq}}{q_{max}} \qquad (6)$$

where: C_{eq} is the concentration in the balance (mg L^{-1}), q the adsorbed quantity in the balance divided by mass unity of the soil (mg g^{-1}), q_m (mg g^{-1}) is the maximum amount of MB per unit of bio-sorbent mass when all active sites are occupied and b (L mg^{-1}) is the adsorption equilibrium constant that is related to the affinity of the linking sites.

The Freundlich model, equation 7, admits a multilayer adsorption with interaction between the adsorbate molecules, being used for systems with surfaces heterogeneous (Kalavathi 2005).

$$\log q = \log K_f + \left(\frac{1}{n}\right)\log C_{eq} \qquad (7)$$

where: C_{eq} represents the concentration in the balance (mg L^{-1}), q the adsorbed quantity in the balance divided by the mass unity of the soil (mg g^{-1}) and K_f and n are Freundlich constants and are related to the adsorption capacity and intensity of the biosorbent, respectively. (Bulut & Aydin 2006, Han 2011).

The Temkin model, predicts that the heat of all molecules of layer decreases linearly with the coverage due to interaction of the adsorbent-adsorbate. (Temkin & Pyzhev 1940).

$$q = B_1 \ln K + B_1 \ln_C \qquad (8)$$

where, K is the linking equilibrium constant, corresponding to the maximum linking energy and B_1 is related to the adsorption heat.

The Dubinine-Radushkevich isotherm (DER) (equation 7), evaluates the adsorption has characteristics physical or chemical.

$$\ln q_e = \ln q_d - B_d E^2 \qquad (9)$$

where, B_d is a constant related to the average free energy of adsorption per mol of adsorbate (mol^2 J^{-2}), q_d is the theoretical saturation capacity and E is Polanyi's potential which is equal to RT ln $(1 + (1/C_e))$, where R (J mol^{-1} K^{-1}) is the gas constant and T (K), the absolute temperature.

The Table 2 presents the values obtained for the four isotherm models.

It can be observed that the model that best answer to the residue "in natura" and modified, with r^2 of 0.9912 and 0.9861 respectively was Freundlich, showing that adsorption occurs via multilayer indicating a residue which is more heterogeneous, this result confirms the model of Pseudo-Second order indicating strong adsorption between adsorbate and adsorbent. Meanwhile, comparing the two the

Table 2. Parameter comparison of Langmuir, Freundlich, Temkin and Dubinine (DER) models for the MB adsorption of residues of cassava starch: "in natura" and modified.

	In natura	Modified
Langmuir		
Q_m (mg g^{-1})	95.880	112.110
b (L mg^{-1})	0.012	0.0009
R^2	0.980	0.737
Freundlich		
K_f (L mg^{-1})	7.030	7.444
n (mol g^{-1})	1.110	1.098
R^2	0.991	0.986
Temkin		
K_t (K J mg^{-1})	0.216	0.274
B_1 (dm^3 mg^{-1})	29.710	23.790
R_2	0.874	0.745
D-R		
Q_d (mol g^{-1})	20.362	250.400
E (kj mol^{-1})	0.0002	0.008
R_2	0.590	0.605

biosorbents, the modified residue obtained a larger K_f, thus signifying a capacity greater than adsorption compared on the "in natura" biosorbent.

4 CONCLUSIONS

The adsorbent used from residues of the cassava industry ("in natura" and modified) is effective in the adsorption of methylene blue, following the kinetic model of pseudo-second order and isotherm follows the model proposed by Freundlich, making it clear that the modification made with biosorbent not change its molecular structure. However, the residue modified from that obtained from the starch manufacturer has greater efficiency in adsorption of the dye compared to "in natura".

ACKNOWLEDGEMENTS

The authors are grateful to the Fundação Araucária and Brazilian National Counsel of Technological and Scientific Development (CNPq) and by the Brazilian Ministry of Science and Technology (MCTI) for its financial support.

REFERENCES

Annadurai, G.; Juang, R.S.; Lee, D.J. 2002. Use of cellulose-based wastes for adsorption of dyes from aqueous solutions. *Journal of Hazardous Materials* 92(3): 263–274.

Bulut, Y.; Aydin, H. 2006. A kinetics and thermodynamics study of methylene blue adsorption on wheat shells. *Desalination* 194: 259–267.

Cardoso, N.F. 2010. Remoção do corante azul de metileno de efluentes aquosos utilizando casca de pinhão in natura e carbonizada como adsorvente. Universidade Federal do Rio Grande do Sul, Porto Alegre. Mestrado em Química 42p.

Cheung, C.W.; Poter, J.F.; Mckay, G. 2000. Sorption kinetics for the removal of copper and zinc from effluents using bone char. *Separation and Purification Technology* 19: 55–64.

Cho, D.H.; Kim, E.Y. 2003. Characterization of Pb^{2+} from aqueous solution by Rhodoturulaglutinis. *Bioprocess and Biosystems Engineering* 25: 271–277.

Dahiya S.; Tripathi R.M.; Hegde A.G. 2008. Biosorption of heavy metals and radionuclide from aqueous solutions by pre-treated arca shell biomass. *Journal of Hazardous Materials* 150: 376–386.

Deng, H.; Lu, J.; Li, G.; Zhang, G., Wang, X. 2011. Adsorption of methylene blue on adsorbent materials produced from cotton stalk. *Chemical Engineering Journal* 172: 326–334.

Doğan, M.; Alkan M.; Türkyilmaz, A.; Özdemir, Y.; 2004. Kinetics and mechanism of removal of methylene blue by adsorption onto perlite. *Journal of Hazardous Materials* 109: 141–148.

Embrapa—Empresa Brasileira de Pesquisa Agropecuária. Disponível em: http://www.cnpmf.embrapa.br/index.php?p=pesquisa-culturas_pesquisadas-mandioca.php.

Geada, O.M.R.N.D. 2006. Remoção de corantes têxteis utilizado resíduos agrícolas da produção de milho. 128p. Dissertação Faculdade de Engenharia da Universidade do Porto, Portugal. Mestrado em Engenharia do Ambiente. 128p.

Guaratini, C.C.I.; Zanoni, M.V.B. 2000. Corantes têxteis. *Química Nova* 23(1): 71–78.

HAN, X.; Wang, W.; Ma, X. 2011. Adsorption characteristics of methylene blue onto low cost biomass material lotus leaf. *Chemical Engineering Journal* 171: 1–8.

Horsfall Jr. M.; Abia, A.A.; Spiff, A.I. 2006. Kinetic studies on the adsorption of Cd^{2+}, Cu^{2+} and Zn^{2+} ions from aqueous solutions by cassava (Manihot sculenta Cranz) tuber bark waste. *Bioresource Technology* 97: 283–291.

Kalavathi, M.H.; Karthikeyan, T.; Rajgopal, S.; Miranda, L.R. 2005. Kinetc and isotherm studies os Cu (II) adsorption anto H$_3$PO^{-4} activates rubber wood Sawdust. *Journal os Colloid and Interface Scienc* 292: 364–362.

Marques, P.A., Pinheiro, H.M.; Teixeira, J.A.; Rosa, M.F. 1999. Removal efficiency of Cu^{2+}, Cd^{2+} and Pb^{2+} by waste brewery biomass: pH and cation association effects. *Desalination* 16: 36–44.

Özacar, M.; Şengil, İ.A. 2003. Adsorption of reactive dyes on calcinedalunite from aqueous solutions. *Journal of Hazardous Materials* 98: 211–224.

Özer, D.; Dursun, G.E.; Özer, A. 2007. Methylene blue adsorption from aqueous solution by dehydrated peanut hull. *Journal of Hazardous Materials* 144: 171–179.

Pérez-Marín, A.B.; Zapata V.M.; Ortuño, J.F.; Aguilar, M.; Sáez, J.; Lloréns, M. 2007. Removal of cadmium from aqueous solutions by adsorption onto orange waste. *Journal of Hazardous Materials.* 139: 122–131.

Rodrigues, R.F.; Rodrigues, R.F.; Trevenzoli, R.L.; Santos, L.R.G.; Leão, V.A.; Botaro V.R. 2006. Heavy metals sorption on treated wood sawdust. *Eng. Sanitária Ambiental* 11(1): 21–26.

Sodré F.F.; Lenzi E.; Costa A.C. 2001. Applicability of adsorption models to the study of copper behaviour in clayey soils. *Química Nova* 24: 324–330.

Temkin, M.I.; Pyzhev, V. 1940. Kinetic of Ammonia Synthesis on Promoted Iron Catalist. *Acta physiochim* 12: 327–356.

Weng, C.; Lin, Y.T.; Tzeng, T.W. 2009. Removal of methylene blue from aqueous solution by adsorption onto pineapple leaf powder. *Journal of Hazardous Materials* 170: 417–424.

Green Design, Materials and Manufacturing Processes – Bártolo et al. (eds)
© *2013 Taylor & Francis Group, London, ISBN 978-1-138-00046-9*

Use of films based starch and casein for conservation of guavas (*Psidium guajava*)

L.F. Bonfim Jr., A.R. Lopes & D.C. Dragunski
Universidade Paranaense, Umuarama, Brazil

J. Casarin & A.C. Gonçalves Jr.
Universidade Estadual do Oeste do Paraná, Marechal Candido Rondon, Brazil

J. Caetano
Universidade Estadual do Oeste do Paraná, Toledo, Brazil

ABSTRACT: This study aimed to produce biodegradable films from cassava starch, casein and glycerol as plasticizer, in order to increase the shelf life of guavas. It was found that fruit coated with the films showed better results, increasing by three days the storage time for the film containing starch and glycerol. However, it was found that the addition of casein did not significantly alter the results, being impracticable due to their value. Guavas with the films were greener and firmer, moreover, they exhibited less weight loss and less sugar, indicating an increase in the ageing time of the fruit. Thus, the incorporation of biodegradable films is an important alternative for preserving fruit without damaging the environment.

1 INTRODUCTION

Brazil is one of the four largest producers of guava (*Psidium guajava*), with most of the production is destined for industry (Faostat 2005).

According to Campos et al. (2011), guava is a climacteric fruit exhibiting rapid maturation after harvest and its shelf life is 3 to 5 days under ambient conditions. The main factors which indicate a depreciation of the quality, post-harvest of this fruit, are the rapid yellowing, softening and the high incidence of rot and wilt, making it difficult to export or send it to more distant consuming centers, as there will be heavy losses during transport.

However, there are several methods of conservation, highlighting, besides cooling and modified atmosphere, the use of plastic bags, but most of that packaging generates an impact on the environment due to low degradation rate (Vicentino et al. 2011).

Faced with this impasse, in recent years there has been a growing interest in studies aimed at sustainable development. Therefore the application of biodegradable coatings in the surface of products perishable vegetables becomes one of the alternatives. Several tests have been conducted with the starch, proteins and other polymers in order to control perspiration and respiratory metabolism of fruits. (Bolzan 2008, Souza et al. 2009).

Starch films without plasticizers are tough and with little elasticity. Increasing content of plasticizer makes these materials more flexible and deformable (Mali et al. 2010). Thus, further studies are necessary to improve the mechanical properties of starch films, an alternative is to incorporate new substances, such as gelatin, casein and polyvinyl alcohol.

The coatings based on the protein casein, which is isolated from whey, are biodegradable, transparent, mechanically strong and relatively water resistance. This coverage provides mechanical protection to the fruit, which helps to minimize injuries, besides being excellent barrier to O_2 and CO_2, so a good option for fresh fruits coating (Cerqueira et al. 2011).

In order to increase the flexibility of films based on proteins and polysaccharides, plasticizers are employed; glycerol is one of the most used to improve the physical and mechanical properties of these films, and it reduces the interactions arising from hydrogen bonds, increasing the intermolecular spaces (Cerqueira et al. 2011).

Therefore, the objective of this study was to evaluate the effect of coating of guavas (*Psidium guajava*), using films modified cassava starch and casein containing glycerol as plasticizer.

2 MATERIAL AND METHODS

The fruits were purchased from a commercial orchard in the city of Altônia, Paraná, Brazil, in August 2012, located approximately 310 m of height, with geographic coordinates 23° 52′ 28″

south latitude and 53° 54′ 06″ longitude west of Greenwich, produced according to practices in the region. The selection criteria for harvest we regular surface characteristics, color and sizes homogeneous and no physical damage.

Guavas were sent directly to the laboratory for storage. Initially they were immersed in aqueous sodium hypochlorite (0.01%) for 30 minutes, after this operation the fruits were hung to dry for 24 hours at room temperature. Then three films were produced containing Acetylated Starch (AS), Casein (CA) and glycerol, containing the following proportions: 1-(AS) 4 g Starch containing 30% glycerol, relative to the total weight; 2-(AS-CA) 3 g of casein and 1 g of starch containing 30% glycerol in relation to the total mass; 3-(CA) 4 g casein containing 30% glycerol in relation to the total mass. CA film was produced by dissolving casein at room temperature, the AS and AS-CA films were prepared using water bath at approximately 85 °C until complete gelatinization. All films were produced containing 100 mL of water.

After obtaining the film forming solutions, they were cooled to approximately 35 °C, then guavas (*Psidium guajava*) were immersed in these solutions for 1 minute while the control group were immersed for the same period of time in distilled water. Subsequently, the fruits were hung to complete drying. A total of five tests were performed: colorimetric, texture, Total Acidity (TA), Total Soluble Solids (TSS), weight loss and determination of vitamin C, in eight days of shelf.

The fruits used in non-destructive analysis (weight loss and colorimetric) were separated and identified. In the analysis of weight loss four fruits were used in each group. The fruits were weighing every day, with the aid of an analytical balance. The results are given in percentages by equation 1:

$$PM = \left(\frac{(PI \times PF) \times 100}{PI} \right) \quad (1)$$

where PI = initial mass of sample (g), PF = final mass of the sample in the sampling periods and PM = percentage of weight loss during the period.

The colorimetric indexes were obtained using a Minolta CR400 colorimeter model, taking into account the results for L*, a* and b*, where L = variance in luminance 0 (black) to 100 (white), the a* ranges from green (−60.0) to red (+60.0) and b* color ranging from blue (−60.0) to yellow (+60.0) (Reis et al. 2006).

Was calculated total color difference (ΔE) according to the equation 2:

$$\Delta E = \left[\sqrt{\left((\Delta L)^2 + (\Delta a)^2 + (\Delta b)^2 \right)} \right] \quad (2)$$

The Δ = difference between each color parameter of the initial sample (time zero) and stored sample (Moura et al. 2007).

Also calculated was the change in color in plant products (Hue) according to equation 3.

$$Hue = \left(\frac{b}{a} \right) \quad (3)$$

The destructive analyses were conducted in duplicate for each group. The texture analysis of fruit was conducted using a penetrometer (PTR-100—Instrutherm) pressure of 8.0 mm diameter. The fruits were drilled on opposite side, being performed in duplicate with two punctures per fruit, expressing results in Newton. (Lima et al. 2005).

The parameters of Titratable Acidity (TA) were determined in duplicate. A sample of 1 ml of guava juice was titration using sodium hydroxide (NaOH) 0.01 mol L^{-1}, the results were expressed in percentages of citric acid per 100 g of the fruit by neutralization of the solution (Food and Drug Administration Publisher 1992). Equation 4 shows the calculation to determine the values of TA:

$$TA = \left(\frac{V \times M \times PM}{10 \times P \times n} \right) \quad (4)$$

where V = volume of spent sodium hydroxide titration (mL) M = molarity of the sodium hydroxide, P = pipetted volume (mL) M = molecular weight of citric acid corresponding (g) (constant = 192); n = number of ionizable hydrogens (constant = 3).

The Total Soluble Solids (TSS) were observed with the aid of a refractometer Quimis 9002. Readings were taken in duplicate by placing 0.05 mL of guava juice on the prism unit, expressing the results in degrees Brix (° B).

The determination of vitamin C was performed using 5 mL of the guava juice, and 50 mL of distilled water, 10 mL of 20% sulfuric acid, 1 mL solution of potassium iodide to 10% and 1 mL of 1% starch, performing subsequent titration with potassium iodate 0.02 mol L^{-1}, until blue color expressing the result in percentage by equation 5:

$$VitC = \left(\frac{100 \times V \times F}{P} \right) \quad (5)$$

where V = volume of spent iodate titration, F = 0.8806 to 0.002 KIO$_3$ molL^{-1}, and P = volume of sample in mL.

3 RESULTS AND DISCUSSION

3.1 Loss of mass

It was observed during the analysis period, that there was a gradual increase of weight loss in all treatments (Fig. 1). However, these losses are still within the acceptable limit for consumption that is 15% (Silva et al. 2012). The control treatment showed the greatest loss of mass from the 3rd to the 10th day of storage when compared to treatments that received films, losing approximately 14% of the mass at the end of the experiment. The fruits coated with AS-CA (starch + casein) and CA (casein) had a weight loss of about 12%. This loss was observed for the control treatment on the 8th day of storage, increasing 2 days of shelf life guava. Among the coatings studied, the lowest mass loss was observed for guavas covered with treating starch (approximately 11%), while the control treatment showed the same rate during the 7th day of storage, showing an extension in the shelf life of guavas in 3 days.

It can be observed that for films containing casein had not a large difference in the starch containing film. Therefore, for this parameter can be stated that the addition of casein is not recommended, due to the market value of this protein.

3.2 Titratable acidity

The ripening stage corresponds to that in which fruit is fully ripe and becomes more palatable because of specific flavors and odors that develop in conjunction with increased sweetness and acidity (Chitarra & Chitarra 2005).

It was observed an increase in the ATT from the 6th to 7th days in the control treatment (Fig. 2) due

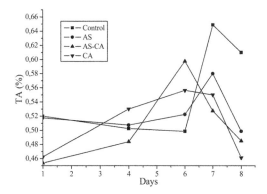

Figure 2. Variation of titratable acidity in for guavas percentage, without cover, and coated films: starch (AS), starch and casein (AS-CA) and casein (CA), according to the storage time.

to hydrolysis of the starch was converted to glucose and subsequently into lactic acid. Treatments for AS, AS-AC and AC, TA rates were constant, possibly a delay of ripening proves that the biodegradable films provided.

From the 7th day there was a decrease in the values of total acidity during storage to control (Fig. 2). These decreases are phase characteristics and senescence of fruit results from the metabolism of organic acids (Botrel et al. 2010).

Again films containing casein did not present major differences regarding the movie composed solely of starch, it is noted that the peak acidity for the film AS-CA occurs before the peak for the control and for the film AS, indicating a possible change in the taste of the fruit.

3.3 Total soluble solids

According to Oliveira & Cereda (2003), is expected that there is an increase in the levels of TSS due to fruit ripening, this increase may be due to hydrolysis of starch, fruit dehydration and degradation of cell wall polysaccharides (Chitarra & Chitarra 2005).

It was found that the control treatment showed an increase in the levels of TSS ratio compared to other treatments. Moreover, the TSS quantity for treatment AS, AS-CA and CA remained practically constant, this fact indicates that the films were effective (Fig. 3), confirming the values found for TA.

The film AS-CA was what enables a lower TSS value, indicating that for this parameter the casein helps in delaying the ripening of the fruit.

3.4 Texture

With the advancement of fruit ripening is expected a decrease in values of texture, because

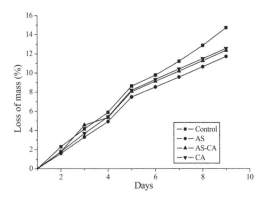

Figure 1. Variation of mass loss in for guavas percentage, without cover (control), and coated films: starch (AS), starch and casein (AS-CA) and casein (CA), according to the storage time.

575

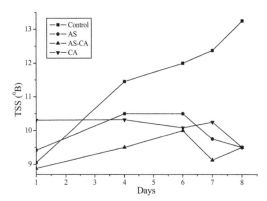

Figure 3. Variation of total soluble solids in for guavas percentage, without cover (control), and coated films: starch (AS), starch and casein (AS-CA) and casein (CA), according to the storage time.

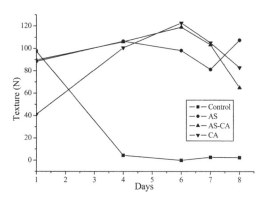

Figure 4. Variation of texture in for guavas percentage, without cover (control), and coated films: starch (AS), starch and casein (AS-CA) and casein (CA), according to the storage time.

despite being a physical parameter is related to the solubilization of pectin, which have carboxylic groups bound to calcium (protopectin), and, fruits with high percentage of soluble pectin have low resistance. Therefore, with the maturation occurs solubilization of pectin, ie the transformation of insoluble pectin with pectin release of soluble calcium pectates and thereby softening of the fruit (Oliveira & Cereda 2003). This behavior was observed for the control treatment during the storage period, but treatment with AS, AS-AC and AC observed an increase in values of texture until the 6th day of storage (Fig. 4), indicating that films delayed ripening and increased the time of senescence of fruit, this fact is of great value, especially for the conservation of the unique features of these fruits.

In this parameter it was found that the fruits coated with casein films were more resistant, between the fourth and seventh days of storage, indicating that casein gives a better mechanical protection to the fruit.

3.5 *Vitamin C*

The nutritional importance of fruits and vegetables is attributed to their content of vitamins and minerals (Yamashita et al. 1999). Vitamin C is highlighted as an important factor in human nutrition. This reductive substance is easily oxidized when exposed to heat, light and oxygen, and may also be lost during handling of the products, being relatively stable in acid medium (Brunini et al. 2004). It can be observed that in all treatments the amount of vitamin was increased on the 6th day of analysis (Fig. 5), primarily due to water loss.

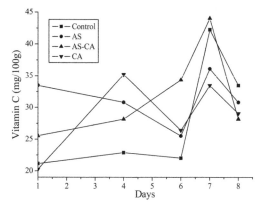

Figure 5. Variation of vitamin C in for guavas percentage, without cover (control), and coated films: starch (AS), starch and casein (AS-CA) and casein (CA), according to the storage time.

3.6 *Colorimetry*

According Hojo et al. (2007), a factor very important for the fruit quality is the colouration, being preferred products of strong and brilliant color, although the color does not necessarily contributing to an effective increase in nutritional value.

For determining the colorimetric indexes L, a and b were determined, and was subsequently calculate the variation of the color delta E (ΔE) and Hue. The control treatment showed a wide variation in parameters ΔE (Fig. 6) and Hue (Fig. 7), when compared to the other treatments, these variations are due to degradative processes (Reis et al. 2006). For treatments containing films were observed little variation in color, showing that the films prevented some guava degradative processes, supporting all data.

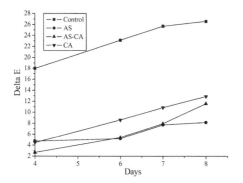

Figure 6. Variation of delta E in for guavas percentage, without cover (control), and coated films: starch (AS), starch and casein (AS-CA) and casein (CA), according to the storage time.

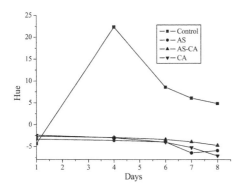

Figure 7. Variation of Hue in for guavas percentage, without cover (control), and coated films: starch (AS), starch and casein (AS-CA) and casein (CA), according to the storage time.

4 CONCLUSIONS

Fruits coated with biodegradable films AS-CA, CA had increased within 2 days of life, while guavas coated with AS showed extension of 3 days compared to the control group without films, indicating that the incorporation of casein did not promote improvements the shelf life of fruits.

The application of the films generated an increase in fruit firmness, moreover, left the fruits greener, more attractive to the consumer, further extending the shelf life, because after 8 days the fruit still retained its original features.

ACKNOWLEDGEMENTS

The authors are grateful to the UNIPAR, the Araucaria Foundation (SETI-PR) and the CNPq-MCTI (REPENSA) for financial support on projects and other aid.

REFERENCES

Bolzan, R.P. 2008. Biofilmes comestíveis para conservação pós-colheita de tomate "dominador". Universidade Estadual do Paraná, Curitiba.

Botrel, D.A.; Soares, N.F.F.; Camilloto, G.P.; Fernandes, R.V.B. 2010. Revestimento ativo de amido na conservação pós-colheita de pêra Williams minimamente processada. *Ciência Rural* 40(8): 1814–1820.

Brunini, M.A.; Macedo, N.B.; Coelho, C.V.; Siqueira, G.F. 2004. Physico chemical characteristics of West Indian cherry provenients of differents regions of cultivation. *Revista Brasileira de Fruticultura* 26(3): 486–489.

Campos, A.J.; Campos, A.J.; Fujita E.; Costa, S.M.; Neves L.C.; Vieites, L.; Chagas, E.A. 2011. Gamma radition and passive modified atmosphere on the quality of guavas 'Pedro Sato' *Revista Agro@mbiente On-line* 5(3): 233–239.

Cerqueira, T.S.; Jacomino, A.P.; Sasaki, F.F.; Alleoni, A.C. 2011. Protein and chitosan coatings on guavas. *Bragantia, Campinas* 70(1): 216–221.

Chitarra, M.I.F. & Chitarra, A.B. 2005. Pós-colheita de frutas e hortaliça: Fisiologia e Manuseio (2.ed.). Lavras: UFLA.

Faostat. 2005. Food and Agriculture Organization of the United Nations, Statistics.

Food and Drug Administration Publisher. 1992. In: *Bacteriological analytical manual.* 7ed. AOAC International, Arlington, VA.

Hojo, E.T.D.; Hojo, E.T.D.; Cardoso A.D.; Hojo R.H.; Vilas Boas, E.V.B.; Alvarenga, M.A.R. 2007. Use cassava starch films and pvc on post-harvest conservation of bell pepper. *Ciência e Agrotecnologia* 31(1): 184–190.

Lima, L.C.; Dias, M.S.C.; Castro, M.V.; Martins, R.N.; Ribeiro Júnior, P.M.; Silva, E.B. 2005. Post-harvest conservation of unripe figs *(Ficus carica L.)* cv. "roxo de Valinhos" treated with sodium hypochlorite and stored under refrigeration in passive modified atmosphere. *Ciência Agrotecnologia* 29(4): 810–816.

Mali, S.; Grossmann, M.V.E.; Yamashita, F. 2010. Starch films: production, properties and potential of utilization. *Semina: Ciências Agrárias* 31(1): 137–156.

Moura, S.C.S.R.; Berbari, S.A.; Germer, S.P.M.; Almeida M.E.M.; Fefim, D.A. 2007. The determination of dehydrated apple shelf-life using accelerated assays. *Ciência e Tecnologia de Alimentos* 27(1): 141–148.

Oliveira, M.A.; Cereda, M.P. 2003. Pós-colheita de pêssegos (Prunus pérsica L. Bastsch) revestidos com filmes a base de amido como alternativa a cera comercial. *Ciências Tecnologia Alimentar* 23: 28–33.

Reis, K.C.; Elias, H.H.S.; Lima, L.C.O.; Silva, J.D.; Pereira, J. 2006. Japonese Cucumber (Cucumis sativus L.) submitted of the treatment with cassava starch film. *Ciência e Agrotecnologia* 30(3): 487–493.

Silva, D.F.P; Salomao, L.C.C.; Zambolim, L.; Rocha, A. 2012. Use of biofilm in the postharvest conservation of 'Pedro Sato' guava. *Revista Ceres* 59(3): 305–312.

Souza, P.A.; Aroucha, E.M.M.; Souza, A.E.D.; Costa, A.R.F.C.; Ferreira, G.S.; Bezerra Neto, F. 2009. Postharvest conservation of eggplant fruits by the application of cassava edible coating or PVC film *Revista Horticultura Brasileira* 27: 235–239.

Vicentino, S.L.; Floriano, P.A.; Caetano, J.; Dragunski, D.C. 2011. Films of starch cassava to coat and conservation of grapes. *Revista Quim. Nova* 34(8): 1309–1314.

Yamashita, F.; Benassi, M.T.; Kieckbusch, T.G. 1999. Effect of modified atmosphere packaging on kinetics of vitamin C degradation in mangos. *Brazilian Journal of Food Technology* 2(1–2): 127–130.

Green Design, Materials and Manufacturing Processes – Bártolo et al. (eds)
© *2013 Taylor & Francis Group, London, ISBN 978-1-138-00046-9*

Biomimetic materials for design

L. Pietroni & J. Mascitti

University of Camerino—School of Architecture and Design, Ascoli Piceno, Italy

ABSTRACT: This paper aims to understand the contribution of bio-inspired materials to the design culture. Taking nature as a reference for solving technological and design problems of humans, the objective of this paper is to show dynamics behind the "resilient behavior" of materials and systems that nature has decided to develop during four billion years of evolution and transfer them into project of sustainable products. The biomimetic materials present many new features but one of the most interesting properties of biological materials is their capacity to self-healing. Today self-healing polymers, metals, paints and concrete mixes that are able to react to damage, are being studied. These represent the most innovative frontier of bio-inspired materials. In the future, these will make possible to create smart and sensitive products able to adapt or react to external stimuli, allowing us to greatly extend their useful lives.

1 LEARNING FROM NATURE

Watching nature to learn to do what it has developed in billions of years is a challenge for the man that has cropped up for centuries.

If the biological metaphor has always been the favourite field for the study of form, of "how it is made", in the last years the coming of Biomimicry has shifted the aim to a more important question: "how does it work"?

Biomimicry is the science that tries to look for technological and design solutions to problems of humans through the study of nature (Benyus 1997) considered as an efficient and fully developed context in which living beings (with their structures, strategies and behaviours), completely developed, can be rightfully considered "winners".

Therefore the aim is not to imitate a shape but to understand which purpose the nature had using it. In other words, nature tells us "this is the answer". We, as scientists, but even simply as human beings, have to work out questions (Santulli 2012).

However, the aim is harder than it seems because nature, on a macro scale we know well, shows only a little portion of its reality. In fact, we are discovering that a lot of dynamics of the exceptional complexity of Mother Nature, take place on a micro and nano-scale and a deep understanding of them can help, in perspective, an effective process of imitation.

This way, an entirely new scenario made of new performances and "non-intuitive behaviours" of materials, different from expectations based on our glaring experience and completely new, opens to the world of design (Asby 2002; Narducci 2008).

And yet we should wonder: why have we started to observe the natural world again today?

Basically the answer consists in two main factors: the first one, already mentioned, is the development of new scientific instruments and nanotechnologies and nano-sciences that nowadays allow a different degree of vision and comprehension of natural dynamics; the second one is the necessity to realize a drastic change of our production and consumption systems through a radical reduction of energy and environmental resources, trying to improve affluence of most of the world population, using less energy and materials (Pietroni 2011).

If materials are the basis for each design project and one of the main focus of designers' work, here we can restart to understand how Biomimicry can give an important contribution to future design.

2 THE GECKO: AN EMBLEMATIC EXAMPLE

An emblematic example of how the interaction of described factors can change the world of design, is given by the gecko's foot and by its extraordinary ability to adhere to almost all surfaces. For a long time we believed that its ability to move on vertical surfaces and upside down was due to a system of pads acting because of a sucker or friction effect.

Because of the impossibility to understand how this little reptile could scoff at force of gravity, designers tried to copy the foot's morphology, hoping to obtain the perfect surface grip. This is shown by Vibram sneakers where the pattern looks like the characteristic pad's shape, and toes, that are free, are used to optimize the grip on the ground and to assure the max traction.

Only in the last years, the vision on a nano metric scale of the inner structure of the animal's leg has enabled us to understand the real dynamic. In fact, we have found out, that the little reptile can walk without being worried about the inclination of the plane on which it lies, thanks to intermolecular forces or forces of van der Waals, that work on pads, covered with micro and nano *setae*, and on the plane (Autumn et al. 2000). On a macroscopic scale, *setae*, with a concentration of 14.000 per millimetre square, seem a tick undercoat that covers the foot. At the end of each *setae*, hundreds of *spatulae* branch out, as branches from a trunk. Adhesion occurs when these join the surface, producing interaction forces able to contrast the more powerful force of gravity. The force operates under 30° *spatula*-surface contact angle, over this point the leg moves away and allows the gecko to move forward.

The real consequence of this discovery has been the creation of a series of nano-structured materials, called "gecko tape", able to adhere one-way to several surfaces without using chemical agents. Experimental uses are numerous: super adherent tyres, high adhesion soled shoes, climbing robots and few centimeters square of adhesive tape able to support the weight of a television.

The time when designers can assembly materials, even the discordant ones, without having to find definite solutions as riveting, gluing or welding, is not far away. Therefore, it will be possible to plan efficiently and simply the disassembling of very complicated objects, adapting design and studying surfaces of adherence and directions of assembling.

3 A WORLD OF MATERIALS TO COPY

If the gecko is a methodological model to refer to, to approach the study, development and use of biomimetic materials, there are other examples that allow to widen the range of peculiar and unexpected features offered to future designers.

Starting from super-hydrophobic lotus leaf of *Nelumbo Nucifera*, Lotusan Paint, a paint able to make water and dirt slide over any surfaces, has been developed.

In a natural model this happens thanks to some micron tick waxy microcrystals that cover the leaf's surface and avoid drops of water adhering.

Upholstered furniture and clothing stain-resistant are the first application of this paint, but also concrete and ceramic outdoor products with self-cleaning ability under the action of rain.

To realize surfaces with anti-adherent capacity superior than the Teflon one, scientists have been inspired by carnivorous plants from *Nepenthes* family. They feed on insects that are attracted by the scent of a liquid on the bottom of a long shaped sac similar to a goblet, where they are melted and absorbed as soon as they fall into it. Once they get into the sac there is no chance to climb up because internal surfaces are so slippery they stop even ants, well-known climbers, from climbing up. Tests done by researchers from Harvard on a derivative material, called SLIPS (Slippery Liquid-Infused Porous Surface), show an extraordinary ability to reject any solid or liquid elements that try to cling to the surface, even below zero ice and bacteria that can't proliferate (Epstein et al. 2012). Its structure is composed by two basic elements: a solid nanostructured surface on which a lubricating liquid is laid. The applications with the most potential are found in the development of products with extreme temperatures (such as the polar ones) in aeronautics and in biomedical fields.

Even the shark or better still its skin, has antibacterial properties and an extraordinary ability to minimize water friction. Thick dermic plates, similar to scales, that cover it, show small grooves put in the same way of the animal's direction. These create micro vertexes around the body that have a double effect: improving hydro dynamism and avoiding seaweeds and cirripeds to cling to it (Ball 1999). The first one is well-known by professional swimmers whose performance noticeably increased wearing Fastskin costumes made by Speedo.

The anti-vegetative action is the goal of "shark coating", a coating made for boat hulls composed by billions of small reliefs, each one is about fifteen nanometre high, that moving and striking one another, prevent seaweed adhesion.

The study of the structural colour of *Morpho* butterfly's wings, of some beetles and mussels' shells led the carrying out of materials able to assume different colourings without using pigments. In fact, the colour is obtained from interaction of light wave and overlapping layer surfaces so small to "filter" light waves depending on their frequency. The result is that the colour of a surface depends only on frequencies we want to absorb. Morphotex fabric, developed by researchers of Tokyo Institute of Technology, is a multilayer structure fibre made of different nanometer size, able to assume various colours when lights reflected from overlapping surfaces, interact. Changing the fibre thickness is even possible to reflect infrared and ultraviolet rays controlling, this way, heat transfer too (Nose 2005).

In 2009, the Australian fashion designer Donna Sgro presented the first dress entirely made with this yarn and characterized by iridescent colors.

The same principle is used by ChromaFlair paint that aims to reproduce opals' iridescent effect. In its interior there are synthetic scales, about one micrometer tick, able to change tone depending on visual angle and light incidence.

Instead, the researchers of University of Southampton are able to change color of metals like gold through the "molecular structuring" of surface. They can determine the visible color of various metals generating nano-reliefs on the surface and modulating their height (Zheludev 2011). Green gold, red silver and blue aluminum will be usual requests in jewelry but more promising prospects are opened to the design world: the staining process of metal surfaces like as painting, anodizing and plating will no longer be used, making metal base products greener.

Vibro harveyi sea bacteria's bioluminescence is the basis for "Bio-light" lighting system proposed by Philips Research and Design Lab, in 2011. The emission source is a liquid with high concentration of these bacteria that emits green light in an environment full of methane. Bioluminescence is a characteristic quality of bacteria, seaweeds, mushrooms, shellfish and, of course, insects as lightning bugs, that takes advantage of a chemical reaction where an enzyme called "luciferase" interacts with a molecule called "luciferin" that emits light.

The examples shown are only the smallest part of what we are rediscovering from nature in the field of materials. Despite all improvements in the research phase, there are several production limits with nowadays technologies. One of these is the impossibility to copy the nature innate propensity to "build from the bottom", through self-assembling processes in low pressure and temperature (Salvia et al. 2009).

4 SELF-HEALING MATERIALS

One of the most interesting features that Biomimicry is investigating is the nature ability to repair its own "products". For example, the fact that bones and tissues are able to heal, gets to two important remarks: the first one is that we can observe a continuous adaptation process, according to changeable external conditions, in the material's structure and density; the second one is that the damaged part can be removed and replaced, preserving, this way, the system structural integrity (Fratzl 2007).

The problem is that, up to now, materials engineers and designers have been worried about empty spaces thinking of them as defects. From nature we are learning, at our cost, that in a material with no defects nothing moves, no liquid can pass through, so development is blocked. The living organism that does not develop is destined

to have a precarious life and to be mechanically fragile (Santulli 2012).

That's why we are studying different typologies of materials whose goal is self-healing. We are optimizing polymers able to heal cracks and lacerations, through microcapsules and a catalyst incorporated in the matrix (Brown et al. 2002; Blaiszik 2010). When a crack occurs, microcapsules get in contact with the air and polymerize, filling it up. Tests show a 90% recovery of primary resistance after restoration.

In the same way, rubbers with very interesting "healing" properties can join the two detached sections together again, in a little more than fifteen minutes.

Attractive power comes from hydrogen bond that the two detached sections are able to recover (Cordier et al. 2008). Application fields are difficult to be circumscribed, because they can potentially cover from toys for children to passive protection systems up to the aerospace field where a material performance security, even if damaged, is vital.

In this way, in design and development process of innovative products, a designer will have the chance to evade the golden rule of oversize the weight and the thickness of used materials, to prevent potentially harmful events.

Self-recovering powers of hydrogen bonds and the electrical conductibility of nickel nanoparticles are the basis for the first promising testing of artificial skin. The material created at Stanford University has shown the capability to get deformed and self-healing of a polymer and, at the same time, to conduct electricity (Lipomi et al. 2011).

On the contrary, researchers of Freiburg University studied a paint that can locally get back to liquid state and fill any scrapes on the surface if it is exposed to ultraviolet light as the lamp's one. The first uses have already started in the motor field and in little electronic devices.

To extend the life of a material, as versatile and hard as concrete and at the same time fragile if it is cracked, a new mixture has been thought. It promises to make outdoor furniture and buildings' life longer, thanks to bacteria put in microcapsules and incorporated in them. Eating acid water from seepages and calcium lactate, bacteria produce calcite (a natural mineral made of calcium carbonate) that fills in cracks and little holes.

5 CONCLUSIONS

In the future a further development of biomimetic materials will allow products to have a complexity of matter similar and similar to the biological organisms' one and permit the artificial world to replicate the natural one.

Super-hydrophobia, anti-adherence, anti-friction, protection of bacteria proliferation, structural colour, bioluminescence, self-cleaning and self-healing will be the characteristics of new materials available for designers and will determine a new scenario of "resilient objects", able to react virtuously to external stimuli potentially traumatic.

The word "resilience" (from the Latin *resalio:* jump, bounce) belongs to engineering and describes a material's ability to support strong stresses roughly given without being structurally damaged. Today this word has several meanings that are used in psychology, economy and computer science too. In these new meanings there are two conditions that are necessary and adequate to identify the resilience process: the presence of a highly stressful and potentially traumatic condition and the consequent evolution that restores and improves the starting conditions. So, the process is characterized, not only by a high degree of resistance, but shows a strong inclination towards transformation and adaptation (Magrin 2008).

Thanks to development and implementation of new biomimetic and resilient materials, in the future designers will have truly promising tools to develop sustainable products. In fact, easy disassembling, self-cleaning, self-repairing and self-regenerating will be very important features of products, developed by bio-inspired design in a sustainability scenario, in terms of energy efficiency, of improvement of their environmental performances and, above all, of lengthening of their useful lives.

REFERENCES

Ashby, M., & Johnson, K. 2002. *Material and design*. Oxford: Elsevier Ltd.

Autumn, K. Liang, Y. Hsieh, T. Zesch, W. Chan, W.-P. Kenny, T. Fearing, R. & Full, R.J. 2000. Adhesive force of a single gecko foot-hair. *Nature* 405: 681–685.

Ball, P. 1999. Shark skin and other solutions. *Nature* 400: 507.

Benyus, J.M. 1997. *Biomimicry: innovation inspired by nature*. New York: Perennial.

Blaiszik, B.J., Kramer, S.L.B., Olugebefola, S.C., Moore, J.S., Sottos, N.R., & White, S.R. 2010. Self-healing polymers and composites. *Annual Review of Materials Research* 40: 179–211.

Brown, E.N., Sottos, N.R., & White, S.R. 2002. Fracture testing of a self-healing polymer composite. *Experimental Mechanics* 42 (4): 372–379.

Cordier, P., Tournilhac, F., Soulié-Ziakovic, C. & Leibler, L. 2008. Self-healing and thermoreversible rubber from supramolecular assembly. *Nature* 451: 977–980.

Epstein, A.K., Wong, T.-S., Belisle, R.A., Boggs, E.M., & Aizenberg, J. 2012. Liquid-infused structured surfaces with exceptional anti-biofouling performance. *Proc Natl Acad Sci USA* 109: 13182–13187.

Fratzl, P. 2007. Biomimetic materials research. What can we really learn from nature's structural materials? *J. R. Soc. Interface* 4: 637–642.

Lipomi, D.J., Vosgueritchian, M., Tee, B.C., Hellstrom, S.L., Lee, J.A., Fox, C. H., & Bao, Z. 2011. Skin-like pressure and strain sensors based on transparent elastic films of carbon nanotubes. *Nature nanotechnology* 6 (12): 788–792.

Magrin, M.E. 2008. Dalla resistenza alla resilienza: promuovere benessere nei luoghi di lavoro. *Giornale Italiano di Medicina del Lavoro ed Ergonomia* 30 (1): A11–A19.

Narducci, D. 2008. *Cosa sono le nanotecnologie. Istruzioni per l'uso della prossima rivoluzione scientifica*. Milano: Sironi Editore.

Nose, K. 2005. Structurally colored fiber "Morphotex". *Annals* 43: 17–21.

Pietroni, L. 2011. Il contributo della Biomimesi per un design sostenibile, bio-ispirato e rigenerativo. *Op-cit* 141: 15–36.

Salvia, G., Rognoli, V. & Levi, M. 2009. *Il progetto della natura. Gli strumenti della biomimesi per il design*. Milano: FrancoAngeli.

Santulli, C. 2012. *Biomimetica: la lezione della natura. Ecosostenibilità, design e cicli produttivi del terzo millennio*. Padova: Ciesse Edizioni.

Zheludev, N.I. 2011 A roadmap for metamaterials. *Optics and Photonics News* 22 (3): 30–35.

Green Design, Materials and Manufacturing Processes – Bártolo et al. (eds)
© *2013 Taylor & Francis Group, London, ISBN 978-1-138-00046-9*

Particle model for orange peel pyrolysis

J.A. Rodríguez, J.A. Loredo & R.C. Miranda
Facultad de Ciencias Químicas, Universidad Autónoma de Nuevo León, San Nicolás de los Garza, Nuevo León, México

ABSTRACT: The present work consists on the establishment of a mathematical model for the pyrolysis process of the orange peel. The main objective is to predict concentration and temperature profiles on radial direction within a particle of biomass. The model is established from partial differential equations, which represent mass and energy conservation. In order to solve the partial differential equations that describe the phenomena of mass and heat from the process, the method of orthogonal collocation was applied. This work is a complement for experimental work to obtain several products from the orange peel components. The model was finally solved by a program developed in FORTRAN 90®.

1   INTRODUCTION

The focus of this work is the modeling. It has become a fundamental tool that allows to describe the system behavior and, on this way, to get a better understanding of it. Other applications are the control, optimization, design and cost evaluation of the process. There are different ways to describe a system, but for these kind of models, partial or ordinary differential equations are obtained and solved by analytical or numerical methods.

The renewable energy depletion due to the burning of fossil fuels is a problem that take place in our days.

These fossil fuels produce high emissions of greenhouse gases, causing damage to the environment and global warming. In order to resolve this problem the biomass is considered such as a new renewable energy source. The biomass can be converted in different forms of energy (electricity, heat as steam) by processes such as pyrolysis, direct combustion, etc. It is believed that fossil fuel can be replaced by biofuels generating a decreasing pollution. The biomass is organic matter obtained from trees, plants, animal waste, agriculture (corn, coffee, rice waste), sawmill, and urban residual. The conversion process that is studied in this work is the pyrolysis, it is define such as the physico-chemical decomposition of organic matter under heat and in absence of oxygen.

The present work consists on the establishment of a mathematical model to study nutshell pyrolysis, considering that it can be useful to reactors design. The model is established from partial differential equations, which represent mass and energy conservation. In order to solve the partial differential equations that describe the phenomena of mass and heat from the process, the method of orthogonal collocation was applied.

2   METHODOLOGY

The methodology includes five stages: a) Solution of a mathematical model for a single biomass pyrolysis to obtain concentration profiles b) Solution of a mathematical model for a single biomass pyrolysis to obtain temperature profiles c) Particle model adjustment with experimental data (TGA) and d) Effective parameters determination.

Kinetic model. There are many different kinetic models that describe the chemical reactions that are carried out in a biomass pyrolysis process. In the present work the kinetic model proposed by Miranda et al. (2004) was considered.

Hemicelullose $\rightarrow V_1 + R$

Cellulose $\rightarrow V_2 + R$

Lignin $\rightarrow V_3 + R$

The model consists in three parallel reactions that represent the meanly components decomposition of biomass (hemicelullose, cellulose and lignin). They decompose independently, considering there isn't interaction between the products of the reaction.

2.1   Particle model

Matter balance: The mass transfer is due to diffusion phenomena, which is the result of a concentration gradient, and it includes the term of reaction. The biomass is constituted by hemicellulose, cellulose and lignin, therefore there were carried out three material balances:

$$\frac{\partial [He]'}{\partial t'} = D_{He} \frac{1}{r'^2} \frac{\partial}{\partial r'}\left[r'^2\left(\frac{\partial [He]'}{\partial r'}\right)\right] \tag{1}$$
$$-A_{He}e^{-(E_{He}/RT')}[He]'$$

$$\frac{\partial [Ce]'}{\partial t'} = D_{Ce} \frac{1}{r'^2} \frac{\partial}{\partial r'}\left[r'^2\left(\frac{\partial [Ce]'}{\partial r'}\right)\right] \tag{2}$$
$$-A_{Ce}e^{-(E_{Ce}/RT')}[Ce]'$$

Lignina:

$$\frac{\partial [Li]'}{\partial t'} = D_{Li} \frac{1}{r'^2} \frac{\partial}{\partial r'}\left[r'^2\left(\frac{\partial [Li]'}{\partial r'}\right)\right] \tag{3}$$
$$-A_{Li}e^{-(E_{Li}/RT')}[Li]'$$

2.2 Particle surface

Energy balance: When heating a particle into an inert atmosphere, the heat is transfered from the atmosphere to the surface by convection and/or radiation and then inside the particle by conduction.

$$\rho Cp\left(\frac{\partial T'}{\partial t'}\right) = k\left[\frac{1}{r'^2}\frac{\partial}{\partial r}\left(r'^2\frac{\partial T'}{\partial r'}\right)\right] \tag{4}$$
$$+(-\Delta H)R(c',T')$$

The dimensionless is carried out in order to achieve a versatile mathematical model, that is more practical and direct to solve, and also allows the understanding of how a parameter will affect a variable.

Therefore the mass and energy balances as follows:

Particle model:

Matter balance

$$\frac{\partial C_i}{\partial t} = \frac{D_e \cdot t_{max}}{R_p^2} \cdot \frac{1}{r^2} \frac{\partial}{\partial r}\left[r^2 \frac{\partial C_i}{\partial r}\right]$$
$$-\frac{A_i \cdot t_{max}}{(C_o - C_\infty)} \cdot e^{-E/R[T(T_o-T_\infty)+T_\infty]}$$
$$\cdot \left[C_i(C_o - C_\infty) + C_\infty\right] \tag{5}$$

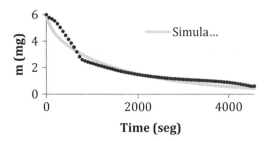

Figure 1.   Mass vs. time graphic (10 °C/min).

Energy balance

$$\frac{\partial T}{\partial t} = \frac{k_e \cdot t_{max}}{Cp \cdot \rho \cdot R_p^2} \cdot \frac{1}{r^2}\frac{\partial}{\partial r}\left[r^2 \frac{\partial T}{\partial r}\right]$$
$$+\sum_{l=1}^{3} \frac{(-\Delta H_i) \cdot t_{max} \cdot A_i}{(T_o - T_\infty)Cp \cdot \rho} \cdot e^{-E_i/R[T(T_o-T_\infty)+T_\infty]}$$
$$\cdot \left[C(C_o - C_\infty) + C_\infty\right] \tag{6}$$

3   RESULTS AND DISCUSSION

Particle model

The experimental data used for the simulation were taken from previous studies obtained at laboratory scale carried out at the Facultad de Ciencias Químicas, UANL.

A spherical particle with 0.005 m of radio was considered with a composition of the biomass of 10% hemicellulose, 30% cellulose and 70% lignin. The initial temperature was 240 °C, this temperature was taken because it is the approximate temperature at which moisture is removed from the sample and when the thermal decomposition starts.

The obtained results from the simulation were compared with experimental data from thermogravimetric analysis for a biomass particle (nutshell) at different heating rates. The heating rate of 10 ° C/min was selected, because it gives us a better approach to the behavior of the system.

4   CONCLUSIONS

o The main conclusion from this study concerns the possibility of modeling the pyrolysis of a biomass particle, coupling equations of heat and mass transfer with chemical kinetics.

o The simulated results obtained using the particle model developed in this work were adjusted

584

to the experimental data (TGA) from previous studies in the Facultad de Ciencias Químicas, UANL.

o The main contribution of this work was primarily the study of the mass transfer obtaining coefficients mass transfer, which vary between $4.81 \times 10^{-7} - 9.05 \times 10^{-5}$ m/s, and the effective diffusivity between $6.98 \times 10^{-10} - 1.06 \times 10^{-9}$ m^2/s, which are similar to those reported in literature.

o A correction factor was used for the Biot number of 0.00453 to include the effect of both convective and radiative transfer, as well as the different stages involved in the process.

o The mechanisms of pyrolysis reactions were simulated by taking three parallel reactions, of which each represents the main components of the biomass, (model proposed by Miranda et al. 2004).

o The mass and energy balances are a great addition as they allow reactor design, calculate system costs, to determine the efficiency of the process and understand the behavior of the system at larger scale.

o The mathematical model of particle developed in this work allows us to find the right parameters for a good adjustment with the experimental data, which will be useful for predicting the behavior of the system on a larger scale or to modify some variables.

o The deviation of the simulated data with experimental attributes to: In the mathematical model considers three simultaneous reactions, whose activation temperatures, it has experimentally detected that are very specific, however, provided kinetic models fail to detect an activation temperature.

REFERENCES

Acevedo, J., Garza V. y López E., *Modeling and Simulation of Coal and Petcoke Gasification in a Co-current flow reactor,* Elsevier B.V., (2005).

Babu, B.V., y Chaurasia A.S., *Modeling for pyrolysis of solid particle: kinetics and heat transfer effects,* Energy Convers. Mgmt., 2251–2275 (2003).

Finlayson, B.A. (1980). *Nonlinear analysis in chemical engineering.* Washington: McGraw-Hill .

Perez Garcia Manuel, The Formation of Polyaromatic Hydrocarbons and Dioxins During Pyrolysis: A Review of the Literature with Descriptions of Biomass Composition, Fast Pyrolysis Technologies and Thermochemical Reactions, Junio 2008, Universidad del Estado de Washington.

Wakao N. y Kaguei S., Heat and Mass Transfer in Packed Beds, Gordon and Breach, Science Publishers, Yokohama National University, Japan, Inc., Capítulos 4, 5, 6 y 8, pp.

Green Design, Materials and Manufacturing Processes – Bártolo et al. (eds)
© 2013 Taylor & Francis Group, London, ISBN 978-1-138-00046-9

Advanced smart polymer/nanographite composites for environmental pollution control

M. Knite, J. Zavickis, G. Sakale, K. Ozols & A. Linarts
Riga Technical University, Riga, Latvia

ABSTRACT: Our recent achievements in development of a new Smart Polymer Nanocomposites (SPNC) for use as sensitive elements in different kinds of transducers are presented. The main task was to elaborate and investigate a sort of materials that could be easily integrated, for example, in construction materials for civil engineering, automotive engineering, biosynthesis equipment etc. All of our elaborated materials are composites made of flexible insulating polymer matrix and electrically conductive carbon nanostructure filler. By choosing appropriate polymer matrix it is possible to create SPNC for humidity sensing, pressure sensing, mechanical vibrations control as well as for Volatile Organic Coumponds (VOCs) sensing. Completely flexible mechanical vibrations sensor element as well as integrated multisensors carpet have been developed. Functioning elements based on SPNC for selective sensing of different VOCs in this paper are demonstrated.

1 INTRODUCTION

VOCs are attributed to dangerous environmental pollution. In its turn, mechanical vibrations are frequently avoidable. Commercial solid state multifunctional sensors or transducers for use in smart systems are available and well known, for example, piezoelectric ceramics pressure or mechanical vibrations sensors, metal oxides gas sensors etc. All the mentioned transducers have high sensitivity and good repeatability of sensing cycles but their drawbacks are brittle structure and higher cost.

The number of publications about elaboration and investigation of Polymer/Conductive Nanostructured filler composites (PCNC) as materials for different sensors applications in last decade have increased. The electrical resistance of PCNC is sensitive to various external influences [1–4], such as pressure, deformation, different gaseous environment and temperature. In certain applications PCNC promise to replace conventional rigid inorganic sensors due to their flexibility, ease of processing and low price. Electrical resistivity (ρ) of polymers may be tuned by the introduction of conductive filler particles. A continuous insulator to conductor transition is observed in two-component systems at gradual increase of the amount of homogeneously dispersed conductive particles in an insulative matrix. Most often such transitions, called percolation transitions, are described by the model of statistical percolation [5]. In the vicinity of the percolation transition, the electrical conductivity σ of the composites changes under the slightest deformation of the matrix. On the microscopic level the increase of the electrical resistance of PCNC under strain can be explained as the result of changes in the percolation structure of conductive particles network. Irreversible change of the electrical resistance at deformation by stretch or pressure has been found in the case of micro-size particles of good conductors as well as Low Structured Carbon Black (LSCB) [2].

New interesting properties have been expected in case the composite contains dispersed nano-sized conducting particles. Polymer/electrically conductive nanostructure composites offer attractive alternatives for developing of new generation of flexible, both—small and large-size sensors because of their superior mechanical and electrical properties.

Recently we reported on our investigations of several distinguished sensing effects inherent to polyisoprene matrix/High Structure Carbon Black (HSCB) composites with filler concentration values in the region of percolation transition [6,7].

In this paper our very recent results on elaboration and investigation of Smart Polymer/Nanographite Filler Composites (SPNC) are presented.

2 EXPERIMENTAL

2.1 Concepts of materials design of SPNC

The SPNC samples were prepared using different polymer matrices and different nanographite fillers. Under the term "nanographite" following fillers are understood: Extra-Conductive Highly Structured Carbon Black (EHSCB), Carbon Nanotubes (CNT), Thermally Exfoliated Graphite (TEG) as well as recently discovered graphene. All of them have a sp^2-hybridized crystal structure like graphite, and at least one dimension is smaller than 100 nm. The selection of appropriate composite matrix depends from a kind of desirable sensing effect. If pressure or deformation sensor is being created, an elastomer should be chosen for composite matrix. In case of VOC sensor the matrix should be able to swell remarkably in presence of chemical vapour. So, the swelling causes the structural change in conductive nanographite channel grid with the subsequent change of ρ. To elaborate composites for thermal resistors the polymer matrix with higher thermal expansion coefficient is preferable. The expansion of matrix under the influence of rising temperature destroys the 3 D nanographite grid, so the ρ of composite raises. The kind of used nanographite filler affects the level of sensitivity as well as sensing speed of acquired SPNC samples.

2.2 Preparing SPNC samples

Four different types of nanographite fillers have been used: 1) 0-dimensional nanostructures—(EHSCB) Printex XE2 obtained from Degussa®—specific surface area 950 m^2/g, average primary particle diameter 30 nm, DBP absorption 380 ml/100 g; 2) 1-dimensional nanostructures—Long Multi-Walled Carbon Nanotubes (LMWCNT) produced by Sigma-Aldrich®—outer diameter 40–60 nm, length 0,5–500 μm; 3) 1-dimensional nanostructures—Short Multi-Walled Carbon Nanotubes (SMWCNT) obtained from CheapTubes®—outer diameter 50–80 nm, length 0,5–2 μm; 4) 2-dimensional nanostructures—TEG obtained from Kyiv National Taras Shevchenko University.

Depending on the type of the filler two different raw SPNC production methods were used for preparing of pressure sensing elements as well as for elaboration of VOC sensors. In the case of EHSCB filler natural Polyisoprene (PI) with necessary vulcanization ingredients (sulphur, stearic acid, zinc oxide and N-Cyclohexyl-2-Benzothiazole Sulfenamide) were mixed with various EHSCB concentrations using cold roll mixing. Further in the text these two-component systems are called Polyisoprene/Carbon Black (PICB) composites.

However for LMWCNT, SMWCNT and TEG fillers a multi step solution mixing method was used: 1) PI with curing ingredients was stirred and dissolved in chloroform at a room temperature for 24 hours; 2) dispersion of nanographite filler in chloroform (dispersed using ultrasonication with specific power 1 W*5 min/1 ml) were added to the PI solution and stirred for 24 hours; 3) obtained mixture was poured into Petri dishes and left for 24 hours in drying chamber for chloroform to evaporate; 4) films were homogenized using cold rolls. To obtain mechanical vibration sensors the homogenized raw material was filled in hot stainless steel mould and vulcanized using Rondol thermostated press under 3 MPa of pressure at 150°C temperature for 15 minutes to enable the cross-linking of raw rubber. The design and the method of creation of completely flexible vibration sensor element as well as multisensor carpet is going to be described further in the text.

To manufacture chemical sensors (sensitive to VOC) the obtained composite mixture (after 2nd step) was dip coated onto an epoxy laminate substrate with integrated brass wire electrodes. The last procedure performed was vulcanization by molding at 3 MPa pressure and 150°C temperature. Composite samples for chemical sensors were prepared in dimensions of $14 \times 10 \times 0.4$ mm.

After curing, the samples were shelf aged at room temperature for at least 24 hours before any measurements were made.

2.3 Experimental setups

The optimal curing conditions were determined during vulcanization phase using Monsanto 100 dynamic rheometer.

The Direct Current (DC) electrical conductivity of each SPNC sample was measured using Keithley 6487 Picoammeter/Voltage source. The piezoresistive effect in DC mode was determined using Zwick/Roell Z2.5 universal material testing machine coupled with Agilent 34970A data acquisition/switch unit. Due to technical limitation of this measuring equipment SPNC samples with conductivity lower than 10^{-8} S/m were not tested for piezoresistivity.

The Alternating Current (AC) measurements were made using Agilent E4980A precision LCR meter. The changes of both AC conductance G and capacitance C of SPNC sample under different values of uniaxial pressure were measured in the frequency range 20 Hz–2 MHz.

Original experimental equipment and method were used for investigation of chemical sensing effect as well as for simultaneous electrical resistance, mass and length measurements of the composite samples during and after exposure to VOC as described elsewhere [8].

3 RESULTS AND DISCUSSION

3.1 Percolation transition

First the electrical resistivity ρ of all elaborated SPNC was measured. Figure 1 represents electrical

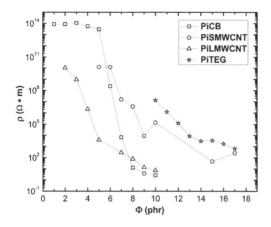

Figure 1. Electrical percolation curves of SPNC with different nanographite fillers.

percolation curves (electrical resistivity versus concentration of conductive filler) for PI matrix and different nanographite filler composites. Filler concentration is defined as mass parts of filler per hundred parts of polyisoprene rubber (phr). The abbreviation phr is often used in rubber industry as a unit of filler concentration.

According to the classical percolation theory, the electrical percolation is obtained when adjacent conductive particles have connected geometrically to form a 3 D conducting channel grid. Bauhofer and Kovacs showed that polymer tunneling barriers exist between MWCNT and they have a dominant effect on the conductivity of polymer/MWCNT composite [9]. Also our previous investigations [6,8] prove the existence of tunneling currents in percolative grids of different nanographite/polymer composites. The various percolation curves in Figure 1 we explain due to diffe-rent shape of nanographite as well as due to different number of tunneling barriers between nanographite. The lowest percolation threshold is observed for PiLMWCNT due to LMWCNT very high aspect ratio and entangled structure. However the steepest percolation transition is for PiCB composites that could be explained in terms of different composite production methods for SPNC manufacture. Generally our composite production methods distinct from each other whether or not ultrasonification where used and one can see that for PCNC with ultrasonification these transitions are more gently sloping. According to this it's possible to conclude that ultrasonification reduces the average size of nanographite agglomerates leading to more uniform PCNC structure.

Research of percolation transition was necessary, because according to literature [6] successful manufacturing of SPNC requires the values of nanographite concentration to be located in the region of percolation transition.

3.2 DC sensors for mechanical vibration control

Based in Figure 1 results certain raw rubber compositions were chosen to elaborate Layered Flexible Pressure Sensor Element (LFPSE) and Layered Flexible Pressure Multi-Sensor Carpet (LFPMC). In Figure 2 the schematic design of LFPS is shown.

LFPSE consist of 1 piezoresistive PIBC layer placed between 2 well conductive PICB layers (flexible electrodes) with incorporated comparatively small brass foil electrodes which could be soldered to wires. Brass foil electrodes have been used because of their good adhesion and electrical connection to flexible PICB electrodes during vulcanization phase. Insolating polyisoprene rubber shell covers all the sensing system.

Figure 3 shows already finished functional LFPSE prototype as well as the cut of the prototype in three parts to show the layered structure. Simultaneous measurements of both

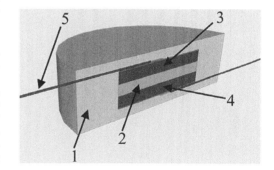

Figure 2. The schematic cross-cut of layered flexible pressure sensor element, consisting of: 1—non conductive outer shell, 2—piezoresistive PIBC, 3—electro-conductive PIBC, 4—brass foil electrode, 5—wires.

Figure 3. The image of finished layered flexible pressure sensor element and its cut in three parts.

cyclic compressive loading and relative DC electrical resistance response of LFPSE are shown in Figure 4.

One can see very accurate response of electrical resistance to periodic changes of pressure in case when PICB composite is incorporated as piezoresistive element. Positive piezoresistive effect assures the rise of resistance with increasing pressure as it is explained in our previous work [6]. The 3 D percolative grid is partially destroyed due to transverse (shear) deformation caused by uniaxial pressure and as a result—the tunneling currents diminishes. Thus, due to reversible piezoresistive effect, the sensing of the mechanical vibrations by means of elaborated LFPSE should be possible.

In Figure 5 finally vulcanized LFPMC is presented, that comprises 6 piezoresistive PICB (8 phr of EHSCB) composite elements connected in series by means of appropriate number of internally vulcanized PICB with good electrical conductance (10 phr of EHSCB) composite layers.

3.3 Pressure sensors working in AC mode

Previously discussed LFPSE (Fig. 3) has also been tested for pressure sensing possibility in AC mode. The changes of conductance depending on frequency and applied fixed values of mechanical pressure are shown in Figure 6. Stable proportion of G change depending on pressure can be seen only at low frequencies although the ability of the sensor itself to sense pressure is maintained up to 2 MHz. The decrease of conductance versus pressure at low frequencies and partially at high frequencies is in good agreement with resistance rise versus pressure (Fig. 4) in DC mode.

Rather large changes of capacitance of LFPSE can be seen in Figure 7. Largest relative capacitance change depending on uniaxial pressure appears around 10 kHz.

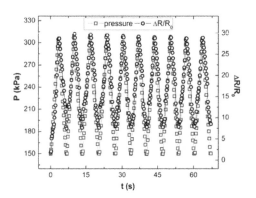

Figure 4. Results of cyclic compressive loading of layered flexible pressure sensor element. Pressure p and relative DC electrical resistance $\Delta R/R_0$ change versus time.

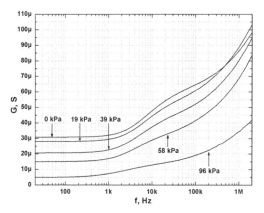

Figure 6. Dependence of conductance G on AC frequency f at different fixed values of applied mechanical pressure.

Figure 5. The image of finally vulcanized layered flexible pressure multi-sensor carpet with 6 PICB composite sensing elements.

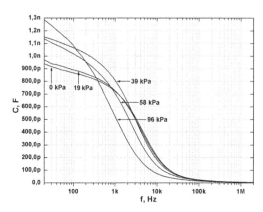

Figure 7. Capacity C dependence on AC frequency f at different fixed values of applied mechanical pressure.

The mechanism of the comparatively large capacitance change depending on pressure is not yet completely clear. In our opinion, the layered structure built from PICB composite layers with different conductivities and capacities as well as different conductive percolative 3 D grid, different degree of separately located clusters of EHSCB are responsible for the observed large capacitance change at certain frequency.

3.4 Chemoresistivity of SPNC

The produced composite samples (Fig. 8) for chemical sensors were exposed to different VOCs. An electrical resistance change was measured to determine PNCC selectivity to particular VOC.

SPNC selectivity is determined by VOC and matrix compatibility, which we evaluated, firstly, by comparing dielectric permittivity values (ε). In Figure 9 the relative electrical resistance changes

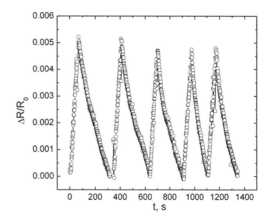

Figure 10. Relative electrical resistance change of PISMWCNT17 sample cyclically kept in toluene vapour (400 ppm). Each cycle is realized as follows: the sample is kept in vapour for 60 seconds and then 200 seconds in air.

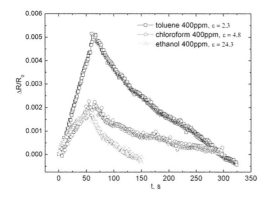

Figure 8. The experimental package of 4 parallel chemical SPNC sensors independently connected to the measurement system.

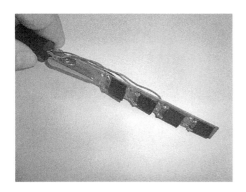

Figure 9. Relative electrical resistance change of PISMWCNT17 sample held in VOCs for 60 seconds with subsequent relaxation in air.

in the presence of VOCs with different polarity are shown for polyisoprene/short multiwalled carbon nanotubes (17 phr) composite sample (PISMWCNT17).

From obtained results one can conclude that response of electrical resistance of SPNC greatly differs if the type of detectable VOC is changed. In addition to that, as ε value of VOC is closer to ε of polyisoprene the observed electrical resistance change of SPNC is higher. It is experimentally been confirmed, that the highest relative resistance change response is observed when a polymer matrix ($\varepsilon \approx 3$) sensor is placed toluene vapour (Fig. 9). The obtained results indicate that the rate of electrical resistance change of SPNC in the presence of VOC is mainly determined by both VOC and composite matrix compatibility. Our experiments also showed that VOC molecule diameter has minor effect on electrical resistance change.

Figure 10 represents good repeatability of chemical sensing effect of sensing element mentioned above.

4 CONCLUSIONS

The finally produced SPNC samples do not contain hazardous elements as, for example, volatile Pb as in case of PZT piezoelectric ceramics.

The SPNC samples based on PICB composite showed the best sensing properties for future elaboration of mechanical vibrations detectors.

For the first time a remarkable piezopermitivity has been realized in layered flexible pressure sensor element on the PICB composite basis.

The SPNC based on PISMWCNT composite has promising chemicals sensing properties and can be used in development of advanced VOC detectors.

REFERENCES

1. Zhang, X.W., Pan, Y., Zheng, Q., Yi, X.S., Time Dependence of Piezoresistance for the Conductor-Filled Polymer Composites. *Journal of Polymer Science Part B: Polymer Physics*, 2000, vol. 38, p. 2739–2749.
2. Das, N.C., Chaki, T.K., Khastgir, D., Effect of Axial Stretching on Electrical Resistivity of Short Carbon Fibre and Carbon Black Filled Conductive Rubber Composites. *Polymer International*, 2002, vol. 51, p. 156–163.
3. Wu, T.M., Cheng, J.C., Morphology and Electrical Properties of Carbon-black-filled Poly(ε-caprolactone)/poly(vinyl butyral) Nanocomposites. *Journal of Applied Polymer Science*, 2003, vol. 88, p. 1022–1033.
4. Ryan, A., Shevade, A.V., Zhoud, H., Homer, M.L., Polymer–Carbon Black Composite Sensors in an Electronic Nose for Air-Quality Monitoring, *MRS bulletin*, 2004, vol. 29, p. 714–719.
5. Staufer, D., Aharony, A. *Introduction in to percolation theory, 4th ed.* London: Taylor Francis, 1985.
6. Knite, M., Teteris, V., Kiploka, A., Kaupuzs, J., Polyisoprene-Carbon Black Nanocomposites as Strain and Pressure Sensor Materials. *Sensors and Actuators A*, 2004, vol. 110, p. 142–149.
7. Knite, M., Ozols, K., Shakale, G., Teteris, V., Polyisoprene and High Structure Carbon Nanoparticle Composite For Sensing Organic Solvent Vapours. *Sensors and Actuators B*, 2007, vol. 126, p. 209–213.
8. G. Sakale, M. Knite, V. Teteris, Polyisoprene-nanostructured carbon composite (PNCC) organic solvent vapour sensitivity and repeatability, Sens. Actuators, A 171 (2011) 19–25.
9. W. Bauhofer, J.Z. Kovacs, A review and analysis of electrical percolation in carbon nanotube polymer composites, Compos. Sci. Technol. 69 (2009) 1486–1498.
10. J. Zavickis, M. Knite, K. Ozols, G. Malefan, Development of percolative electroconductive structure in piezoresistive polyisoprene-nanostructured carbon composite during vulcanisation, Materials Science & Engineering C, 2011, V31, p 472–476.

Green Design, Materials and Manufacturing Processes – Bártolo et al. (eds)
© *2013 Taylor & Francis Group, London, ISBN 978-1-138-00046-9*

Behavior of high strength concrete with raw rice husk exposed to high temperature effect

B. Ucarkosar, N. Yuzer & N. Kabay
Yildiz Technical University, Istanbul, Turkey

ABSTRACT: The physical and mechanical properties of concrete change when concrete is exposed to high temperatures. As a result of these changes, concrete may exhibit damages, such as cracks and spalling. These damages are crucial, especially in High Strength Concretes (HSCs) due to the lower pore ratios in high strength concrete. In this study, Raw Rice Husk (RRH) and Polypropylene (PP) fibers were incorporated in concrete at 0.5–3% and 0.2–0.5% by weight of cement respectively and HSCs were produced. Concrete specimens were exposed to elevated temperatures of 20 (control) and 300°C. Before and after this exposure, compressive and splitting tensile strength was determined. The pore structure of the concretes was characterized by determining ultrasonic pulse velocity. In addition, during the high temperature effect, the amount of the released harmful gases (CO, CO_2) and oxygen was measured by a gas detector.

1 INTRODUCTION

Buildings such as houses, schools, factories, tunnels, bridges or oil platforms might encounter high temperature effects due to their function or fire. High temperature effect on concrete and reinforced concrete structures is still a topic of interest and has been researched since 1920s. The researches focused on the effect of high temperature on normal strength concrete until 1990s (Khoury, 2000; Khoury, 2003). However recently in many modern and industrial structures, widespread use of high performance and high strength concrete can be noticed. These concretes have a lower porosity and a dense structure and thus exhibit a weak performance at high temperatures when compared to normal strength concrete (Scherefler *et al.*, 2003). At high temperatures concrete may exhibit damages such as cracks and spallings due to the increase in thermo-mechanical stresses and vapor pressure (Kanema *et al.*, 2007). It is therefore necessary to use thermal barriers, polypropylene fiber, air entraining additives and aggregates with low coefficient of thermal expansion to decrease the damage. (Khoury, 2003a).

Kalifa *et al.* (2001) reported that an amount of 1 kg/m³ of PP fiber addition would be sufficient to avoid spalling of concrete. Diederichs *et al.* and Nishida *et al.* found that the use of 0.1% of PP fiber provided a decrease in thermal stress induced spalling damage of high strength concretes, compared to those with no fiber. Therefore, in this study, in order to prevent spalling due to high

temperature effect, concretes containing various amounts of PP fiber were cast.

Rice is one of the main nutriments and according to FAOSTAT, 701.1 million tons of rice was produced in 2010 all over the world and RRH is about 20% of the overall production (www.faostat. org). RH is used as a combustible fuel in stoves in the countryside, as a raw material in refractory production in countries such as Egypt and Japan, as a lightweight aggregate in lightweight concrete, as a pozzolan in cement and as an insulator in the steel industry (Mazlum, 1989). Active carbon is obtained by burning RRH in air-free conditions. Due to its high active carbon content, RRH is also used in industry as color and scent compensator. Additionally, the use of RRH in concrete is expected to absorb toxic gases due to its silica content and higher surface area and it is also expected that RRH could facilitate the release of vapor by generating micro pores and channels in concrete (Yüzer *et al.*, 2012). Therefore, the main objective of this study is to assess the feasibility of the use of rice husk in concrete production instead of polypropylene fibers.

Besides their advantages, it is known that PP fiber and RRH at high temperatures induces porosity and thus cause a decrease in compressive strength of concrete.

Yüzer *et al.* (2012) performed a series of research in normal strength concrete containing RRH as 1.5%, 3% and 5% of cement by weight and studied the compressive strength of concrete at elevated temperatures (300, 600 and 900°C). They reported a loss of 4%, 16% and 25% in compressive strength of concrete by the increase in the amount of RRH. They also noted that 1.5% of RRH addition did not significantly affect the strength and that the compressive strength was reduced by 6%, 34% and 60% at 300, 600 and 900°C respectively.

CO levels generated by the combustion of polypropylene were sufficient to produce the lethal effects (Purohit & Orzel, 1988). Every year 53.4 million tons of PP fibers are produced in all over the world and for each tons of production, 9.9 kg (air, water and waste) pollution occurs.

Raw Rice Husk (RRH) is an agricultural co-product that outermost layer of the paddy grain. It is hard to storage due to its low bulk density. Utilization of RRH in concrete saves economy and sustainability. Through its morphology under the high temperature, it can be used in concrete production as an absorber instead of polypropylene fibers.

In this study, usability of RRH instead of PP fibers is investigated. Consequently, HSC mixes were prepared by incorporating different amounts of PP and RRH in concrete. Possible harmful gases (CO and CO_2) release, by the combustion of PP fiber was analyzed to evaluate its influence on human health. In addition, mechanical properties such as compressive and splitting tensile strength were determined before and after the high temperature effect.

2 EXPERIMENTAL PROGRAMME

2.1 Materials and mix design

The materials used in this research include limestone coarse aggregates, natural river sand, crushed limestone sand (Table 1), ordinary portland cement (CEM I 42,5R) and two types of pozzolanas; silica fume and granulated blast furnace. The physical properties of aggregates are presented in Table 1, and chemical composition of cement and pozzolanas are given in (Table 2).

To minimize the spalling problems in concrete, two different dosages of PP fiber and RRH with a density of 0.122 g/cm³ and combustion temperature

Table 1. Physical properties of materials.

Material	Particle density (g/cm³)	Aggregate max size (mm)
Limestone coarse aggregate	2.76	12
Crushed limestone sand	2.72	4
Natural river sand	2.61	4

Table 2. Chemical composition of cement and pozzolanas.

Chemical composition (%)	Cement	Silica fume	Granulated blast furnace slag
CaO	64,31	<1	33,75
SiO2	19,75	>85	40,96
Al2O3	4,28	–	13,22
Fe2O3	3,48	–	1,15
MgO	1,14	–	7,25
SO3	2,7	<2	0,35
Loss on ignition	2,57	<4	0
Specific gravity (g/cm³)	3,15	2,33	2,78
Specific surface (Blaine, cm²/g)	3591	–	5465

Table 3. Physical properties of polypropylene fibers.

Type	Density (g/cm³)	Length (mm)	Melting point (°C)	Tensile strength (MPa)
Homopolymer polypropylene	0.91	19	160	570–660

Table 4. Sieve analysis of raw rice husk.

Sieve size (mm)	1	2	4	8
Passing (%)	0	60	100	100

of 350°C was used. Physical properties of PP fiber are presented in Table 3 and sieve analysis results of RRH are presented in Table 4.

Cement weight and water to binder ratio of the mixes were kept constant 450 kg/m³ and 0.25 respectively. In order to provide a slump of 16 ± 2 cm, a superplasticizer was used at varying contents. In all mixes silica fume and granulated blast furnace slag was added to mixes as 7% and 10% of cement weight respectively. Concrete mix proportions were determined according to the relevant Turkish Standard TS 802 and mix proportions are presented in Table 5.

2.2 Specimen preparation and curing

Five concrete mixes were cast with different percentages of PP fibers and RRHs. The proportions of PP fibers were as 0.2 and 0.5% and RRHs were 0.5 and 3% by weight of cement.

For all tests, cylinder specimens of 100 mm diameter and 200 mm height were prepared and cured in water for 28 days. The specimens were

Table 5. Concrete mix proportions.

Mixture code	N	PP02	PP05	RRH05	RRH3
Cement (kg/m³)	450	450	450	450	450
Water (kg/m³)	127,5	127,2	127	128	127
Coarse aggregate (kg/m³)	1128	1126	1126	1098	943
Crushed sand (kg/m³)	371	371	371	361	310
Natural sand (kg/m³)	355	355	355	345	297
Silica fume (kg/m³)	31,5	31,5	31,5	31,5	31,5
Granulated blast furnace slag (kg/m³)	45	45	45	45	45
PP fiber (% of cement weight)	–	0.2	0.5	–	–
RRH (% of cement weight)	–	–	–	0.5	3

air-dried in the laboratory for another 28 days and oven dried at 105°C for 48 hours to evaporate free water. The specimens were then subjected to elevated temperatures of 20 and 300°C with a heating rate of 4°C/min and kept at the target temperature for 2 hours. The specimens were then allowed to cool to room temperature before testing.

2.3 Testing procedure

For each concrete mix, compressive strength, splitting tensile strength and ultrasonic pulse velocity was determined. In addition, the amount of CO and CO_2 was measured during high temperature exposure.

3 RESULTS AND DISCUSSIONS

3.1 Compressive strength

Compressive strength test was conducted on 6 cylindrical (Φ100/200) specimens for each mix according to EN 12390-3. Average of compressive strength values are presented in Figure 1. It can be noticed that except for the reference concrete (N), there has been a distinct reduction in compressive strengths by the increase in temperature. This reduction is more notable in RRH mixes, compressive strength decreased by 30% at 300°C.

3.2 Splitting tensile strength

Splitting tensile strength test was conducted on 4 cylindrical (Φ100/200) specimens according to EN 12396. The average of tensile strength values are shown in Figure 2.

It can be seen that there occurred a slight reduction in tensile strength of all mixes by the increase in temperature.

3.3 Ultrasonic pulse velocity

Ultrasonic pulse velocity was determined according to BS1881 on 4 cylindrical (Φ100/200) specimens

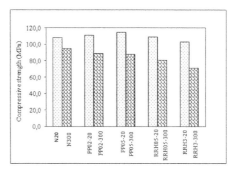

Figure 1. Compressive strength of mixes before and after high temperature exposure.

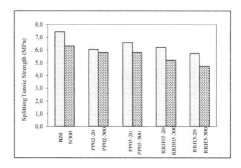

Figure 2. Splitting tensile strength of mixes before and after high temperature exposure.

for all mixes. The average values are shown in Figure 3. It was seen that the pulse velocity values did not differ significantly with the addition of PP fibers and RRH. However ultrasonic pulse velocity of all mixes more or less decreased at 300°C as can be seen in Figure 3.

3.4 Gas measurement

In order to determine the amounts of released CO and CO_2 at elevated temperatures, 10 cylindrical specimens (Φ100/200) were used for each concrete mixture. These samples were placed in the oven and exposed to 300°C. The gas measurements were performed at specified time intervals and the maximum values are determined and shown in Figures 4 and 5.

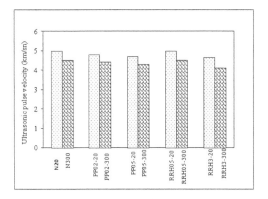

Figure 3.   Ultrasonic pulse velocity of mixes before and after high temperature exposure.

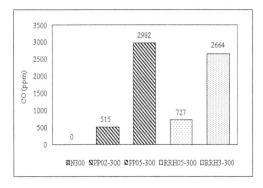

Figure 4.   The amount of CO released during 300°C.

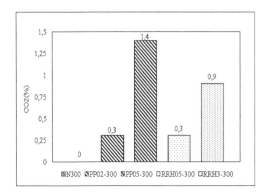

Figure 5.   The amount of CO_2 released during 300°C.

It can be easily noticed that the reference concrete (N) releases no CO and CO_2 gases at 300°C. However the mixes with PP fiber and RRH more or less release these gases. An increase in percentage of PP fibers causes an obvious increase in CO and CO_2 amounts. This increment can also be seen in RRH series but not as much as the concretes with PP fibers. When we compare PP05 and RRH05 series (who have the same amounts in concretes), it can be noticed that the released CO and CO_2 amounts in the concrete mix with PP fiber (PP05) are more than four times greater than the concrete mix with RRH (RRH05).

4   CONCLUSIONS

Through this research, it is possible to conclude:

- Increase in the temperature causes decrease in mechanical properties of all mixes. This decrease was notable especially in series with PP fiber and RRH.
- The reference concrete (N) released no harmful gases at 300°C. When used at the same amounts, it was observed that the harmful gas amount of concrete with PP fiber was more than four times greater than the mixes with RRH.
- Although the use of RRH results in a decrease in mechanical strength at 300°C, it prevents the release of harmful gases when compared to PP fiber, when used at same amounts.
- The results obtained from this study confirm that RRH can be used instead of PP fibers as a protective material against fire. Since RRH release less harmful gases, the use of RRH in concrete will help to decrease casualties in case of fire.
- When we consider that the combustion temperature of RRH is 350°C, its effect on concrete should be further researched for higher temperatures.

REFERENCES

Kalifa, P., Chene, G., Galle, C., 2001. High Temperature Behaviour of Hpc with Polypropylene Fibers from Spalling to Microstructure, Cement and Concrete Research, 31: 1487–1499.
Kanema, M., Morais, M.V.G., Noumowe, A., Gallias, J.L. ve Cabrillac R., 2007. Experimental and Numerical Studies of Thermo-Hydrous Transfers in Concrete Exposed to High Temperature, Heat Mass Transfer, 44: 149–164.
Khoury, G.A., 2000. Effect of Fire on Concrete and Concrete Structures, Progress in Structural Engineering and Materials, 2: 429–447.
Khoury, G.A., 2003. Fire & Assessment, International Centre for Mechanical Sciences, Course on Effect of Heat on Concrete, Udine/Italy.

Khoury, G.A., 2003. Spalling, International Centre for Mechanical Sciences, Course on Effect of Heat on Concrete, Udine/Italy.

Mazlum, F., (1989). Pirinç Kabuğu Külünün Puzzolanik Özellikleri ve Külün Çimento Harcının Dayanıklılığına Etkisi, in Turkish, PhD Thesis, İstanbul Technical University, Institute of Science, İstanbul.

Purohit, V., Orzel, R.A., 1988. Polypropylene: A Literature Review of the Thermal Decomposition Products and Toxicity, International Journal of Toxicology, 2: 221–242.

Rice Production statistics all over the world in 2010, Retrieved April 11, 2013. http://faostat3.fao.org/home/index.html#VISUALIZE_TOP_20.

Scherefler, B.A., Gawin, D., Khoury, G.A., Majorana, C.E., 2003. Physical, Mathematical & Numerical Modeling, International Centre for Mechanical Sciences, Course on Effect of Heat on Concrete, Udine/Italy.

Tanyildizi, H., (2009). Statistical Analysis for Mechanical Properties of Polypropylene Fiber Reinforced Lightweight Concrete Containing Silica Fume Exposed to High Temperature, Materials and Design, 30: 3252–3258.

Yüzer, N., Aköz, F., Çınar, Z., Biricik, H., Kabay, N., Kızılkanat, A., Yalcin, Y., 2012. The Effect of Rice Husk Addition on High Temperature Performance of Concrete, Scientific Research Project Report (2009-05-01-ODAP01), Yıldız Technical University Research Foundation.

Green Design, Materials and Manufacturing Processes – Bártolo et al. (eds)
© *2013 Taylor & Francis Group, London, ISBN 978-1-138-00046-9*

Novel plastics for sustainable building design

Geoffrey R. Mitchell

*Centre for Rapid and Sustainable Product Development, Polytechnic Institute of Leiria,
Marinha Grande, Portugal*

ABSTRACT: The construction and operation of any building has a substantial and wide-ranging impact on the environment in both the short and long term., In the European Union, buildings currently consume 40% of the energy and produce 36% of the CO_2 emissions. The building will generate waste and may emit harmful gases and other emissions directly in to the atmosphere. Designers face considerable and many challenges to meet demands for facilities that are accessible, secure, healthy, and productive whilst minimizing the impact of the building on the environment and the consumption of energy. These challenges will be exacerbated by the anticipated changes in the global climate and the accompanying legislative programmes to reduce energy consumptions and CO_2 emissions. This paper is focused on an examination of new plastic based materials which facilitate novel design approaches especially in terms of the control of the environment of the building. The types of plastics range from energy storage, generation and conversion, light filtering and spectrum conversion, smart materials for micro ventilation and air quality control. A particular of this study is the emphasis on materials which have both smart functionality and a role as an integral structural component.

1 INTRODUCTION

The construction and operation of any building has a substantial and wide-ranging impact on the environment in both the short and long terms. In each of the phases in the life of a building, namely, construction, occupation and demolition, the building uses significant resources including energy and raw materials. In fact, in the European Union, buildings consume 40% of the energy and produce 36% of the CO_2 emissions EU (2010). Moreover the building will generate waste and may emit harmful gases and other emissions directly in to the atmosphere. Designers face considerable and many challenges to meet demands for facilities that are accessible, secure, healthy, and productive whilst minimizing the impact of the building on the environment and the consumption of energy (Fig. 1). Moreover as Alwaer and Clements Croome (2010) have identified, different groups of stakeholders and different individuals will place rather different priorities with regard to the key performance indicators and the issues such as (environmental, social, economic and technological factors) used to provide an assessment of sustainability of buildings. These challenges will be exacerbated by the anticipated changes in the global climate Solomon et al. (2007) and there are now programmes to reduce energy consumptions and CO_2 emissions, for example Alwaer and Clements Croome (2010).

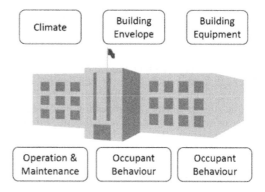

Figure 1. Key factors which define the energy consumption of a building (after Yoshino (2011)).

Plastics are widely used in current building design and construction. There is also a strong programme of building improvements which exploits current plastic materials. It is fair to say that the majority of the existing applications of plastics in buildings are direct replacements for earlier technologies, window frames and waste piping are two such examples. There is considerable attention directed at the use of recycled plastics and the development of plastics from sustainable sources Azapagic et al. (2003). However, there is another important component in the field of sustainable design which is the development and deployment of new materials which allow a more intelligent design of buildings which facilitates the use of active components and functional elements which are incorporated or integral to the building and its operation. In other words, these materials extend the scope of the toolbox for designers which allow reductions in energy use during the operation of the building, allow a greater use of natural energy, light and air movements and new features for aesthetics and the design of building with special needs of individuals at the core of the design.

This paper is focused on an examination of new plastic based materials which facilitate novel design approaches especially in terms of the control of the environment of the building. The types of plastics range from energy storage, generation and conversion, light filtering and spectrum conversion, smart materials for micro ventilation and air quality control. A particular of this study is the emphasis on materials which have both smart functionality and a role as an integral structural component.

2 PLASTIC MATERIALS

Plastics are a class of materials which is a very important component of the materials portfolio used in the building sector. Current estimates give that 20% of all the plastics used in Europe are for products in the building industry (Fig. 2). This means that after packaging, the building industry is the second highest user of plastics. There is substantial use of plastics for pipes for transporting water or sewage, soffits and related items, window and door systems, floor coverings, foam and panels for insulation. Other applications include the foundations of the house and paint and other coatings. These applications arise from the fact that plastics can be strong, resistance to weather and heat as well as and flexible. Plastics are very lightweight and they require very little maintenance. This

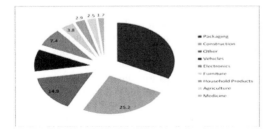

Figure 2. The use of plastics in Germany in 2007 (after Engelsmann (2010)).

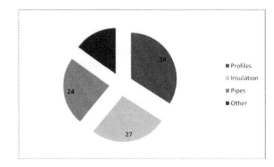

Figure 3. The application of plastics in construction in Germany in 2007 (after Engelsmann (2010)).

makes them ideal for the construction industry. The use of plastics in construction is shown schematically in Figure 3. This based on data for construction in Germany in 2007. It will naturally vary from geographical location but this can be seem as the 'traditional' use.

3 ENERGY HARVESTING

There are many sources of energy but here we focus on solar energy and motion energy as they provide capture systems which can be readily integrated in to building components.

3.1 Solar energy

Solar energy represents an energy source which to a certain extent is already used by all buildings but which can be harnessed further on an individual basis. If a building harnesses solar energy further than the 'natural' absorption of solar radiation by the exterior shell, it has no impact on the climate other than slightly changing the proportion of radiation reflected back in to the atmosphere.

Passive solar architecture has been widely exploited for many years, although on a practical basis implementation depends greatly on geographic location and is often most prominent in minimizing the impact of solar energy Balcomb (1992). Key issues are the orientation with respect to the Sun, a low surface area to volume ratio, the use of selective shading overhangs and thermal mass. There is use of active elements including switchable windows (see light section), pumps and fans to improve system performance.

In terms of more fully harnessing solar power, there is widespread use of solar collectors especially in China and on a population adjusted basis Cyprus. In terms of the contribution that plastics can make to this area, the work of Dorfling et al.

(2010) is a good example. They considered plastic solar collectors fabricated from a novel thermoplastic extrudate known as a micro-capillary film. These films contain many capillaries which run continuously along th length of the film. The diameter of these embedded capillaries can be varied in the range 0.01 to 1.0 mm. Experiments and numerical modelling show that this novel collector performs as well as existing solar collectors. However, it offers considerable potential advantages in the form of low cost and mechanical flexibility. The production process can be used with a variety of materials and there is a greater possibility of incorporating such collectors as an intrinsic component of a building skin.

The use of solar cells which convert solar energy in to electricity is a fast moving area in terms of both implementation and research Krebs (2009), Nielsen et al. (2010). There is the possibility of utilizing polymer films extensively as the substrate to provide a flexible and adaptable device in which the functionality may be more easily incorporated or used as building skin components. There is a strong growth is the development of all organic/polymer solar cells which build on the major advances in plastic electronics in recent years. Both approaches have a greater capacity for intelligent deployment in buildings.

Any solar energy capture has to address the unavoidable fact that the building moves it orientation with respect to the sun throughout the day and a seasonal basis. The development of materials which exhibit a photo-activated mechanical transducer effect, for example most recently with liquid crystal elastomers Sanchez-Ferrer et al. (2011) provides potential for systems which track the sun using functional materials, rather than electromechanical devices.

3.2 Motional energy

The energy associated with ambient vibrations is well established but for most practical systems the energy associated with such vibrations was too small. However, the development of new materials and micro and nanotechnology moves the capture of such energy in to the realm of possibilities. This greatly helped by the reducing power requirements of many electronic circuits, and lighting. The power requirements for heating are largely unchanged.

Seiko commercially introduced the kinetic watch in 1998 in which a minature generator in the watch converts the motion of the watch, essential the motion of the hand into electricity. Since then many researchers have explored the use of induction, capacitance, or electroactive materials to convert vibrations into electricity. The simplest device contains piezoelectric materials such as electro-active polymers. Another possibility perhaps more suited to use in flooring and other building components is capacitive devices. The power densities of the piezoelectric and electrostatic conversion have been reported to be ~250 and ~50 µW/cm^3 Roundy et al. (2003). Clearly these will not heat a building in northern Europe but could power localized building sensors, wireless devices and other electronics which are needed in an intelligent building design.

4 ENERGY STORAGE

Building naturally store energy in terms of their large mass. Utilising this mass for more functional purposes is an area attracting considerable interest. MESSIB (www.messib.eu) is a European project focused on 'the development, evaluation and demonstration of an affordable Multi-Source Energy Storage System (MESS) integrated in building, based on new materials, technologies and control systems, for significant reduction of its energy consumption and active management of the building energy demand'. There is considerable interest in the exploitation of phase change materials Cabeza et al. (2011) which are compounds with a substantial heat of fusion which by, melting and solidifying at a defined temperature, are able to store and release significant levels of energy. Heat is absorbed or released when the compound transforms from solid to liquid and the reverse on cooling. Thus phase change materials can store a larger quantity of energy than simply related to the heat capacity. The challenges are both the amount of energy that can be stored and the integration of these phase change materials in to the building system. Encapsulation is an important part of this process and the transfer of the energy in to the building environment. Park (2010) has reported on the use of electrospun fibres in energy storage through phase change materials. Electrospinning is a process for generating fine fibres (nano or micro scale) by means of applying an electric field to droplets of a polymer solution or melt passed from tip of a fine orifice Greiner and Wendorff (2007). There are advantages in encapusulating the phase change material and Park used the coaxial electrospinning technique to encapsulate polyethylene glycol in polyvinylidene fluoride. Stability is the key to the performance of any phase change material system. In that study the resultant electrospun fibres exhibited no fatigue in the temperature or enthalpy of the phase change after 100 thermal cycles.

5 LIGHT

The use of materials to change the nature and quality of light which enters in to a building is well established practice. The use of blinds, photochromic windows and other devices is common place especially where the use of natural lighting is key part of the building design Rosemann (2008). The particulars of the spectral distribution of this light define the impact on the temperature of the building but also on the building occupants in terms of their individual experience, for example the winter light syndrone Lewy (2009). There is a growing interest in horticulture in the use of plastic films which alter the spectral content either passively or actively to control plant growth and inhibit disease, for example Fletcher (2004). This type of system depends on the precise molecular structure of the additives included in the plastic. Such technology is readily straightforward to transfer to building design.

A second approach is the use of photo defined refractive index changes in multiphase polymer films, especially using added particles which result in controlled changes in the scattering properties of the plastic film such that the light intensity is reduced and/or the spectral characteristics are modified. This is the basis of polymer dispersed liquid crystal windows Cupellia et al. (2009) Baetens et al. (2010). There are non-liquid crystal technologies available to achieve the same goal. The use of plastic films brings flexibility and an ease of integration with other components.

6 VENTILATION AND AIR QUALITY

A key factor is the quality of an individual building occupants experience is the quality of the air and the control of that quality. Electrospinning is a technology which natural generates filtration membranes Griener and Wendorff (2007). The ability to encapsulate active ingredients, antimicrobial, slow release etc in to the nano and micro scale fibres, means that such filters can be much more than simple particulate traps. There is much to learn from the natural world—the so-called biomimetic approach regarding ventilation John et al. (2005). The construction of complex ventilation as found in nature using synthetic materials is now an achievable goal in light of the range of responsive polymers to stimuli such as humidity, temperature and light. The replacement of electromechanical systems with large area membranes which open and close, powered by the external stimuli both reduce power and provide the ability for very localized control.

7 SUMMARY

We have focused on the examination of plastic materials which can facilitate novel design approaches especially in terms of the control of the environment of the building. It is clear that there significant opportunities in energy harvesting and storage, light filtering and spectrum conversion as well as smart materials for micro ventilation and air quality control. From these brief considerations we can see the potential for reducing energy consumption and reducing CO_2 emissions. It is essential to integrate the materials focused research with building design if we are to make significant progress in these areas.

ACKNOWLEDGEMENTS

This work was funded in part by the Fundação para a Ciência e a Tecnologia, Portugal.

REFERENCES

Alwaer H. and Clements Croome D.J. 2010 Building and Environment 45 799–807.

Azapagic A., Emsley A., and Hamerton I. 2003 Polymers: The Environment and Sustainable Development Wiley.

Baetens R., Jellea B.P. and Gustavsen A. 2010 Solar Energy Materials and Solar Cells 94, 87–105.

Balcomb, J.D 1992. Passive Solar Buildings. Massachusetts Institute of Technology. ISBN 0262023415.

Cabeza L.F., Castella F., Barreneche C., de Gracia A. and Fernández A.I. 2011 Renewable and Sustainable Energy Reviews 15, 1675–1695.

Cook-Chennault, K.A., Thambi, N., and Sastry, A.M. 2008 Smart Materials Structures, 17, 1–33.

Cupellia D., Nicoletta F.P., Manfredi S., Vivacqua M., Formoso P., De Filpo G. and Chidichimo G. 2009 Solar Energy Materials and Solar Cells 93, 2008–2012.

Directive 2010/31/EU of 19 May 2010 on the energy performance of buildings.

Dorfling C., Hornung C.H., Hallmark B., Beaumont R.J.J., Fovargue H., and Mackley M.R. 2010 Solar Energy Materials and Solar Cells 94, 1207–1221.

Engelsmann S., Spalding V. and Peters S. Plastics 2010 in Architecture and Construction Birkhauser Greiner A, and Wendorff JH. 2007 Angew. Chem. Int. Ed., 46; 5670–5703.

Fletcher J.M., Tatsiopoulou A., Hadley P., Davis F.J., and Henbest R.G.C., 2004 Acta Horticulturae, 633, 99–106.

John G., Clements-Croome D. and Jeronimidis G. 2005 Building and Environment 40, 319–328.

J-S Park 2010 Adv Nat Sci: Nanoscience and Nanotechnology 1 043002.

Krebs F.C. 2009 Solar Energy Materials and Solar Cells 93 394–412.

Lewy A.J., Emens J.S., Songer J.B., Sims N., Laurie A.L., Fiala S.C. and Buti 2009 Sleep Medicine Clinics 4, 285–299.

Nielsen T.D, Cruickshank C., Foged S., Thorsen J. and Krebs F.C., 2010 Solar Energy Materials and Solar Cells 94, 1553–1571.

Rosemann A., Mossman M. and Whitehead L. 2008 Solar Energy 82 302–310.

Roundy, S., Wright, P.K., and Rabaey, J. 2003 Computer Communications, 26, 1131–1144.

Sánchez-Ferrer A., Merekalov A., Finkelmann H. 2011 Macromol. Rapid Commun., 32, 672–678.

Solomon, S., D. Qin, M. Manning, Z. Chen, M. Marquis, K.B. Averyt, M. Tignor and H.L. Miller (eds.). 2007 Contribution of Working Group I to the Fourth Assessment Report of the Intergovernmental Panel on Climate Change Cambridge University Press, Cambridge, United Kingdom and New York, NY, USA.

Yoshino H. 2011 Energy Conservation in Buildings and Community Systems Project 53.

Green Design, Materials and Manufacturing Processes – Bártolo et al. (eds)
© *2013 Taylor & Francis Group, London, ISBN 978-1-138-00046-9*

Phase change materials as a tool for climate change mitigation

Ana P. Vieira, Helena Bártolo, Geoffrey R. Mitchell & Paulo Bártolo
Centre for Rapid and Sustainable Product Development, Polytechnic Institute of Leiria, Marinha Grande, Portugal

ABSTRACT: In the EU, buildings consume 40% of the energy and produce 36% of the CO_2 emissions. This study focuses on increasing the efficiency of energy use in buildings by exploiting phase change materials as energy storage systems. In particular, we focus on the practical application of phase change materials to buildings and their incorporation in building materials. We propose that the most effective implementation will be achieved by incorporating the phase change materials in to the building blocks used to construct the building envelope.

1 INTRODUCTION

The use of Thermal Energy Storage (TES) systems as received increasing interest, which has been recognized as one of effective approaches to reducing energy consumption of buildings, for heating and cooling. The drive here is to reduce the consumption of energy in buildings. An EU report reports that in the EU, 40% of all energy is consumed in buildings which results in 36% of all CO_2 emissions arising from buildings. The larger part of this consumption arises from residential buildings EU (2010).

In buildings, Sensible Heat Storage (SHS) using thermal mass of the structural elements is the most common approach for the storage of thermal energy. This type of storage depends on the material mass and specific heat as well as the temperature difference. In general there is little opportunity for optimization as the mechanical and other properties of the structural materials is of prime concern. One approach to improving this situation, is the use of Latent Heat Thermal Storage (LHTS) by exploiting Phase Change Materials (PCMs).

PCMs have been incorporated in passive LHTES systems as buildings walls, windows, ceilings or floors. It is well established that the building envelope plays a major role in solar heat gain, since the external surfaces of it are in direct contact with outdoor air. A variety of approaches have been taken to incorporate PCMs in the materials which make up the building envelope materials, as glazing, facades, stone or concrete Soares et al. (2013), Baetens et al. (2010) and Tyagi (2007).

The focus of this study is to find a possible solution for the effective integration of PCMs with building materials, as a solution for the reduction of the energy consumption and the improvement of the energy conservation in buildings as this could lead to enhancing energy efficiency and sustainability of buildings,

For that, is necessary understand the phase change process, the proprieties, classification and types of PCM. The development of a latent heat thermal energy storage system, involves the understanding of three essential subjects: phase change materials, containers materials and heat exchangers, Sharma et al. (2009). Different stages are involved in the development of a latent heat storage system must be considered, Figure 1.

The most important property of phase change materials is the capacity of storage the heat energy in latent form, leading to a greater heat storage capacity per unit volume than that exhibited by conventional building materials. Moreover the advantage of using thermal storage is that it can contribute to match supply and demand energy when they do not coincide in time Saoes et al. (2013).

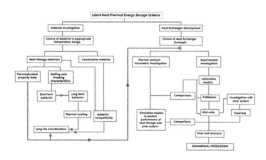

Figure 1. Flow chart showing different stages involved in the development of a latent heat storage system, after Sharma et al. (2009).

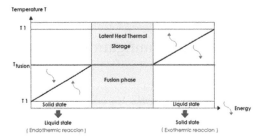

Figure 2. Phase change mechanism.

As the ambient temperature rises, PCMs change phase from a solid to a liquid as a consequence of the melting of the crystalline solid. Since this reaction is an endothermic process, they absorb heat. As the temperature decreases, PCMs will return to the crystalline solid state liquid to solid. On cooling they release the absorbed heat, since this reaction is exothermic process as shown schematically in Figure 2. The principle of PCMs use is very simple, but evaluating the effective contribution of the latent heat loads in the enhancement of the energy performance of the whole building is a challenge Soares et al. (2013).

A large number of PCMs are known to melt with a heat of fusion in the required range, Tyagi and Buddhi (2007). The selection of an appropriate PCM for any application requires the PCM to have a melting temperature or switch temperature within the practical range of application, Zhou et al. (2012).

2 PCM CONCEPT AND PROPERTIES

When phase change materials are include in building constructions, it is necessary that the PCM exhibits specific thermal, physical, kinetic, chemical, economic and environmental properties and the key properties are presented in summary form in Table 1.

Also is necessary for select the appropriate Phase Change Material and its correct switch temperature take into account the local climate, the building characteristics, its orientation and occupation profile, as well as the purpose and type of PCM application.

The switch temperature is one of the most determinant factors in the performance of PCM applications; an inadequate phase change temperature may make the application completely non-functioning.

A large number of phase change materials (organic, inorganic and eutectic) are available in

Table 1. Essential PCM properties.

Thermal and physical properties	• Suitable phase-change temperature in the desired operating range; • High thermal conductivity and good heat transfer; • High latent heat of transition per unit mass; • High specific heat and high density; • Congruent melting and long term thermal stability; • Favourable phase equilibrium and no segregation; • Small volume change on phase-change; • Low vapour pressure at operating temperature;
Kinetic properties	• High nucleation rate to avoid super cooling of the liquid phase; • High rate of crystal growth, so that the system can meet demand of heat recovery from the storage system;
Chemical properties	• Complete reversible freeze/melt cycle; • No degradation after a large number of freeze/melt cycle; • No corrosiveness to the construction materials; • Non-toxic, non-flammable and non-explosive material for safety;
Economic properties	• Abundant and available; • Cost effective;
Environmental properties	• Low embodied energy; • Separation facility from the other materials and recycling potential; • Low environmental impact and non-polluting.

Table 2. Commercial PCM manufacturers worldwide.

Manufacturer	Temperature range	Number of PCMMs listed
Rubitherm (www.rubitherm.de)	−3 °C to 100 °C	29
Cristopia (www.cristopia.com)	−33 °C to 27 °C	12
Teap (www.teappcm.com)	−50 °C to 78 °C	22
Doerken (www.doerken.de)	−22 °C to 28 °C	2
Mitsubishi Chemical (www.mfc.co.jp)	9.5 °C to 118 °C	6
Climator (www.climator.com)	−18 °C to 70 °C	9
EPS Ltd (epsltd.co.uk)	−114 °C to 164 °C	61

any required temperature range. The properties of each subgroup which affects the design of latent heat thermal energy storage systems using PCMs are in shown in Table 2.

For buildings applications the PCM, independent of its classification, should have the phase change temperature restricted around 15–70°C, Cabeza et al. (2011).

3 PCM CLASSIFICATION

A large number of phase change materials (organic, inorganic and eutectic) are available in any required temperature range. These may be grouped in to the categories shown in Figure 3. The objective is to maximise the Latent heat of the transition and to dfine the switch temperature. The properties of each subgroup which affects these parameters are listed in Table 1.

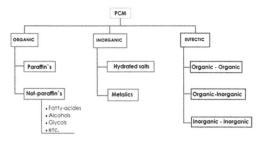

Figure 3. Classification of PCM used in energy storage.

4 COMMERCIAL PCM

Any realistic implementation of PCM in building systems requires an adequate supply of the PCM itself. In table 2 we list the currently available PCMs.

Table 3 summarises the key thermo-physical properties for selected PCMs from those shown in Table 2.

Sharma et al. (2013) have noted that it is important that the data provided by the PCM manufacturers could be erroneous, uncertain and overoptimistic. It is clear that after using such data to select the appropriate PCM that measurements in a realistic environment must be performed to verify the effective thermal properties of the PCM. This is usually achieved using Differential Scanning Calorimetry (DSC) and the Differential Thermal Analysis (DTA).

5 PCM FORMAT/CONTAINMENT

During the heating and cooling of the storage cycle, the PCM transforms from a solid to liquid. As a consequence it is necessary to contain or localise the PCM. This may be achieved by the incorporation of the PCM in to a building material or through the use of a separate containment material. Clearly such containment must not impede thermal transfer. Several studies have been conducted to study heat transfer enhancement techniques in Phase Change Materials (PCMs) and include finned tubes of different configurations, bubble agitation, insertion of a metal matrix into the PCM and, using

Table 3. Thermophysical properties of commercial products for comfort applications in buildings between 22 and 28C After Cabeza et al. (2011).

Ref	Type	Tm (°C)	Heat of fusion (kJ/kg)	Thermal conductivity (W/m K)	Source
RT 20	Paraffin	22	172	0.88	Rubitherm GmbH
Climsel C 23	Salt hydrate	23	148	–	Climator
E23	Salt hydrate	23	155	0.43	EPS Ltd.
Climsel C 24	Salt hydrate	24	108	1.48	Climator
TH 24	Salt hydrate	24	45.5	0.8	TEAP
RT 26	Paraffin	25	131	0.88	Rubitherm GmbH
RT 25	Paraffin	26	232	–	Rubitherm GmbH
STL 27	Salt hydrate	27	213	1.09	Mitsubishi chemical
S27	Salt hydrate	27	207	–	Cristopia
AC 27	Salt hydrate	27	207	1.47	Cristopia
RT 27	Paraffin	28	179	0.87	Rubitherm GmbH
			146	0.2	
RT 30	Paraffin	28	206	–	Rubitherm GmbH
E28	Salt hydrate	28	193	0.21	EPS Ltd.

PCM dispersed with high conductivity particles, micro-encapsulation of the PCM and or shell and tube (multitubes) Rodriguez-Ubinas et al. (2012) and Agyenim (2010).

6 BUILDING APPLICATIONS FOR PHASE CHANGE MATERIALS

There are many reports on the use of PCMs in building which amplify the thermal energy storage benefits to lightweight constructions, decreasing the temperature fluctuation in the interior and reduce their energy consumption, but this depend of the particular PCM materials and applications.

The applications of PCM can be organized in to different levels, the way that PCM are incorporated in buildings and the factors related with their successful usage. The first level is related to the possibility of the PCM be incorporated in a passive or active system Zhu et al. (2009). The second level is related to how PCM are used in the buildings, three different possibilities can be identified: PCM as components, PCM integrated into construction materials and PCM in storage units Tyagi et al. (2007) and Rodriguez-Ubinas (2012).

Table 4. PCM applications.

First level	
Passive system	Passive applications are the ones that do not use mechanical devices or systems. The heat or cold are stored or released automatically when the air temperature rises or falls beyond the PCM melting point. The passive systems can take advantage of the direct and indirect solar gain, as well as the internal thermal gains
Active system	In active application, the PCM thermal energy charging and discharging is achieved with the help of mechanical equipment.
Second level	
PCM as components	When the PCM element is one of the layers or parts of a construction section
PCM integrated into construction materials	When the PCM is mixed with, impregnated or incorporated to a construction material
PCM in storage units	PCM in storage units, can only be applied to active systems, and it is generally thermally separated from the building by insulation.

The placement of PCMs in specialised storage units is most appropriate for retrofit applications—most residential buildings that will be used in the future are already built.

The optimum approach is to incorporate the PCM in to the construction material to facilitate thermal transfer and to provide a low cost approach to the development of a practical system.

7 OVERVIEW OF THE MAIN PCM PASSIVE LHTES SYSTEMS FOR BUILDING APPLICATIONS

Table 5 reviews the key element of passive LHTES systems for building applications Sharma et al. (2013).

Romero-Sanchez et al. (2012) explored the use of natural stone as a matrix for PCMs. The objective was to improve the thermal properties of natural stone by exploiting associated latent heat storage phenomena. As a consequence, natural stone treated with PCMs could be used as a construction material with the ability to store thermal energy leading to reduction of the overall buildings' energy consumption, This work incorporated the PCM in natural stones. In the first stage, the porosity and thermal properties of natural stone treated with PCMs were measured. The effect of PCM integration in the natural stone thermal behaviour was experimentally investigated. Samples of Bateig azul marble have been selected, the stone is blue and composed of calcite and quartz, with an average porosity of 14%. The PCM utilized for the treatment of Bateig azul was the Micronal DS 5000 (provided by BASF), with a melting temperature of approximately 26°C. Bateig azul was impregnated with the PCMs, by immersion in PCM water-based solutions. Although this is a very attractive concept, unfortunately the level of incorporation of the PCM in to the stone was rather small, a few percent and hence any gains were limited. It may be that other stone types might be more effectively used. However, the key to the performance will be most the type of PCM and the level of its incorporation. We propose that this can only be achieved through the use of fabricated building components such as bricks and blocks which are designed specifically to incorporate PCMs. The loading in a building is given by

Volume of PCM in building block × volume of building blocks in building envelope

Now although the former of these two factors will be lower for a filled brick compared to a dedicated storage panel, the latter factor will be substantial higher providing an effective passive thermal storage system. Of course this has to be included at the design and build stages rather than through retrofit.

Table 5. Overview of the main PCM passive LHTES systems for building applications.

Building applications	Study cases
PCM enhanced wallboards	The efficiency of these elements depends on several factors: • how the PCM is incorporated in the wallboard; • the orientation of the wall; • climatic conditions, • direct solar gains; • internal gains; • color of the surface; • ventilation rate; • the PCM chosen and its phase-change temperature; • the temperature range over which phase-change occurs; • the latent heat capacity per unit area of the wall, etc.
Other PCM walls	• enhanced cellulose insulation; • sandwich panels for prefabricated walls; • Structural Insulated Panel (SIP) outfitted with PCMs; • inserting PCM inside dry assembled walls, generally characterized by low thermal inertia.
SSPCM enhanced elements	• SSPCM plates, combined with night ventilation; • Mixed type PCM-gypsum and SSPCM plates; • Structures integrated with SSPCM wall;
PCM bricks	• Two dimensional model for a common building brick with cylindrical holes containing PCMs; • PCM in conventional and alveolar brick construction for passive cooling; • Incorporating macro-encapsulated paraffin into a typical Portuguese clay brick masonry wall; • Incorporation of PCM in hollow thermal-insulation bricks;
PCM enhanced concrete systems and mortars	• Innovative concrete with PCM in order to develop a product which would not affect the mechanical strength of the concrete wall; • PCM enhanced concrete to store solar energy in floors for moderate sea climates; • A concrete roof with cone frustum holes filled with PCM; • thermally enhanced mortar with 25% of microencapsulated PCM on the mass fraction;

(*Continued*)

Table 5. (*Continued*)

Building applications	Study cases
PCM Trombe wall	• The introduction of PCMs in Trombe wall systems could contribute to the development of light, portable, movable and rotating systems fully adapted to the lightweight buildings category;
PCM shutters, window blinds and translucent PCM walls	• Exterior PCM shutters containing PCMs are movable structural shading elements associated to windows facades; • A TIM-PCM external wall system for solar space heating and daylight composed of Transparent Insulation Material (TIM) and translucent PCM was theoretically and experimental investigated; • Wall made of hollow glass bricks filled with PCM for thermal management of an outdoors passive solar test-room; • PCM facade panel for day-lighting and room heating;

8 SUMMARY

Phase change materials provide a route to enhancing the energy storage in a building in order to reduce the energy consumption. The choice of the PCM with its specific switch temperature is critically. The incorporation of the PCM in to blocks used for constructing the building envelope provides a route to a high volume of storage material. This approach needs to be defined at the build stage. Most of the currently available implementations are most appropriate for retrofit applications.

REFERENCES

Agyenim, F. N. Hewitt, P. Eames, M. Smyth—"A review of materials, heat transfer and phase change problem formulation for Latent Heat Thermal Energy Storage Systems (LHTESS)" 2010 Renewable and Sustainable Energy Reviews, volume 14 (2010), pp. 615–628, http://dx.doi.org/10.1016/j.rser.2009.10.015.

Baetens, R. B.P. Jelle, A.Gustavsen 2010 "Phase change materials for building applications: A state-of-the-art review"—Energy and Buildings, volume 42, (2010), pp. 1361–1368, http://dx.doi.org/10.1016/j.enbuild.2010.03.026.

Cabeza, L.F. A. Castell, C. Barreneche, A. de Gracia, A.I. Fernández 2011 "Materials used as PCM in thermal energy storage in buildings: a review"—Renewable and Sustainable Energy Reviews, volume 15, pp. 1675–1695, http://dx.doi.org/10.1016/j.rser.2010.11.018.

Directive 2010/31/EU of 19 May 2010 on the energy performance of buildings.

Edwin Rodriguez-Ubinas, Letzai Ruiz-Valero, Sergio Vega, Javier Neila 2012 "Applications of Phase Change Material in highly energy-efficient houses"—Energy and Buildings, volume 50 (2012), pp. 49–62, http://dx.doi.org/10.1016/j.enbuild.2012.03.018.

Grove, A.T. 1980. Geomorphic evolution of the Sahara and the Nile. In M.A.J. Williams & H. Faure (eds), The Sahara and the Nile: 21–35. Rotterdam: Balkema.

Na Zhu, Zhenjun Ma Shengwei Wang 2009 "Dynamic characteristics and energy performance of buildings using phase change materials: A review"—Energy Conversion and Management, volume 50 3169–3181, http://dx.doi.org/10.1016/j.enconman.2009.08.019.

Osterman, E. V.V. Tyagi, V. Butala, N.A. Rahim, U. Stritih 2012 "Review of PCM based cooling technologies for buildings"—Energy and Buildings, volume 49 (2012), pp. 37–49, http://dx.doi.org/10.1016/j.enbuild.2012.03.022.

Romero-Sánchez, M.D. C. Guillem-López, A.M. López-Buendía, M. Stamatiadou, I. Mandilaras, D. Katsourinis, M. Founti 2012 "Treatment of natural stones with Phase Change Materials: Experiments and computational approaches"—Applied Thermal Engineering, volume 48 pp. 136–143, http://dx.doi.org/10.1016/j.applthermaleng.2012.05.017.

Sharma, A. V.V. Tyagi, C.R. Chen, D. Buddhi 2009—"Review on thermal energy storage with phase change materials and applications"—Renewable and Sustainable Energy Reviews, volume 13, pp. 318–345, http://dx.doi.org/10.1016/j.rser.2007.10.005.

Soares, N. J.J. Costa, A.R. Gaspar, P. Santos—2013 "Review of passive PCM latent heat thermal energy storage systems towards buildings energy efficiency"—Energy and Buildings, volume 59, pp 82–103, http://dx.doi.org/10.1016/j.enbuild.2012.12.042.

Tyagi, V.V. D. Buddhi—"PCM thermal storage in buildings: a state of art"—Renewable and Sustainable Energy Reviews, volume 11, (2007), pp. 1146–1166, http://dx.doi.org/10.1016/j.rser.2005.10.002.

Zhou, D. C.Y. Zhao, Y. Tian 2012 "Review on thermal energy storage with phase change materials (PCMs) in building applications"—Applied Energy, volume 92 (2012), pp. 593–605, http://dx.doi.org/10.1016/j.apenergy.2011.08.025.

Sustainable business models

Green Design, Materials and Manufacturing Processes – Bártolo et al. (eds)
© *2013 Taylor & Francis Group, London, ISBN 978-1-138-00046-9*

Assessment of the costs and benefits of environmental investment

Mohamed E. Hussein & Gim S. Seow
Department of Accounting, University of Connecticut, USA

Kinsun Tam
Department of Accounting and Law, State University of New York at Albany, USA

ABSTRACT: This study develops a methodology to analyze environmental investment based on Ansari et al's (1997) classification of environmental costs into prevention costs, assessment costs, control costs and failure costs. BP's Deepwater Horizon oil spill demonstrated that the costs of failure significantly exceed the costs of prevention, assessment and control. With better preemption, there is less need for remediation. We argue that BP could have averted the Deepwater Horizon disaster if it had applied our proposed methodology of environmental risk analysis.

1 INTRODUCTION

Economic sustainability is the premise of a sustainable society. A sustainable economy requires efficient use of finite resources, and reduction in waste and emissions (Smith and Ball, 2012), which in turn require investment in processes and technologies that yield long-term benefits but lower short-term profits. Since these benefits and costs are not entirely captured in the traditional accounting system, many organizations do not foresee all the environmental costs until problems occur. For example, a manufacturer might choose cheaper materials that enhance short-term profitability but may prove costlier to dispose of because of toxicity. The additional disposal costs might exceed the cost savings of the cheaper materials over more expensive green materials. A good example was reported by Ansari et al. (1997) where Allied Signal's engineers recommended spending $30,000 to replace disposal pipes in its Bendix plant for fear of PCB leaking. Allied Signal's management overruled the engineers and later had to spend $287 million to clean up the contaminated site.

Smith and Ball (2012) argue that the extant literature on sustainable manufacturing is "generic and high level" (see, for example, Abdul Rashid et al. 2008, and Seliger et al. 2008), and therefore, provides little practical guidance on how to attain sustainable manufacturing. To correct this deficiency in the literature, this study provides a specific methodology for assessing the *ex ante* risk and comparing expected costs of preemption to expected costs of remediation and demonstrates with a recent case of an environmental disaster in offshore drilling.

The purpose of this study is to develop a process to analyze environmental investment based on the classification of environmental costs proposed by Ansari et al (1997) who divide environmental costs into prevention costs, assessment costs, control costs and failure costs. Ansari et al's framework is presented in Table 1. Failure costs can be divided into remediation costs, and damages and penalties costs. Prevention, assessment, and control costs adversely impact short-term profit but, *ex post*,

Table 1. Environmental Costs.

Type of costs	Description
Prevention	Actions to prevent or reduce environmental accidents. These actions include employee training, product and process redesign, use of nontoxic materials, etc.
Assessment	Periodic review of potential sources of environmental damages, evaluating policies, procedures and employee awareness.
Control	Extra control at points where risks are high, e.g. reinforced concrete where there is high risk of leaks, special valves that relieve pressure and/or sound early warning.
Failure	Costs of repairs and clean up. Medical and compensation costs for employees and others injured or affected by the failure.
Damages and penalties	Damages awarded to affected parties and penalties imposed by regulatory agencies.

Source: Ansari et al. 1997.

compare favorably to costs of internally discovered failures or failures discovered by external monitoring agencies. The example of Allied Signal mentioned above as well as many other cases, including BP's Deepwater Horizon oil spill to be discussed in details later, have shown that the costs of failure in terms of remediation, damages and penalties significantly exceed the costs of prevention, assessment and control. With better preemption, there is less need for remediation. While environmental investment data is useful on its own, there is a need to assess the economic impact of such investment. When making economic decisions, for instance, managers need to know the impact of investment in emission remission or waste treatment on the overall profitability of the firm. Thus companies must pay attention to all their operations from facilities and processes to design, production, distribution and disposal of products.

2 THE ECONOMIC DIMENSION OF SUSTAINABILITY

Prior research (e.g., Norman and MacDonald 2004 and Carter and Rogers 2008) has viewed sustainability as comprising of three dimensions: environmental, economic, and social. This paper focuses on the first two, namely environmental and economic. This path is chosen because there is little in the accounting literature concerning the economic analysis of environmental costs. Thus this paper proposes a methodology that can be used by managers to assess the costs and benefits of environmental investments. The methodology will provide an enterprise with a systematic method for monitoring its environmental costs, generate performance metrics on which decisions can be made by managers to reduce waste and/or pollution and assess environmental risks associated with processes and outputs.

An example of such assessment is the decision by the Design for Environment team at Herman Miller, a major office furniture manufacturer, to replace Polyvinyl Chloride (PVC) with Thermoplastic Urethane (TPU) for the arm pads of their Mirra chairs (Lee and Bony, 2009). While TPU is more expensive and costlier to fabricate and assemble than PVC, TPU does not generate toxic fumes and therefore is "greener" and cheaper to dispose of at the end of the chair's useful life. This change also reduces the risk to employees' health from working with PVC, and thereby helps Herman Miller avoid medical costs and damages payments. Furthermore, Herman Miller avoids any future liabilities if their customers dispose of chairs with TPU arm pads into landfills that do not accept toxic materials.

3 THE METHODOLOGY

The methodology entails a systematic analysis of the risks and costs of environmental situations (see Fig. 1). The analysis starts with identifying the environmental risks in facilities, processes, designs, production, distribution and disposal at end of product life-cycle. Each environmental risk is evaluated for the worst case scenario, expected case scenario and best case scenario. The probability assessment of environmental risk is based on prior environmental failure incidents.

The second step is to estimate the different types of environmental costs of remediation, damages & penalties. The third step is to identify actions to avert environmental failures. These will include prevention in terms of waste treatment, designs, equipment, use of green materials, training of employees, and assessment and control activities that help detect problems before they turn into disasters. The fourth step is to calculate the net benefit of the environmental investment which is the difference between the investment on prevention, assessment & control and the estimated costs of remediation, damages and penalties.

The next section uses the environmental disaster of BP Deepwater Horizon oil spill in the Gulf of Mexico as an example. We show that this disaster could have been avoided if BP had done the right analysis of prevention, assessment and control costs in comparison to the expected costs of failure. We argue that BP had experienced two disasters (i.e. the Texas City Refinery in 2005 and the Prudhoe Bay oil spill in 2006) that took place not too long before the Deepwater Horizon disaster. Such events should have made BP and the rest of the industry sensitive to risks of environment disasters. We chose the Deepwater Horizon disaster rather than a manufacturing example because it provides a vivid illustration of what could happen when a company ignores environmental risks. In addition, this case had attracted worldwide attention.

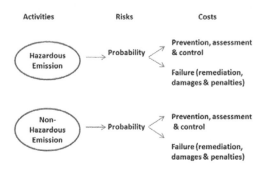

Figure 1.   Environment risks and costs.

Hence, there is enough publicly available data to demonstrate the methodology.

4 THE CASE OF BP DEEPWATER OIL SPILL

On April 20th 2010, the Deepwater Horizon oil rig exploded and sank. The explosion caused the death of 11 workers and spilled 4.9 million barrels of oil before it was capped on July 15, 2010 (Wikipedia, 2013). The spill caused damage to marine life and the eco-system of the Gulf of Mexico. It seriously impacted the livelihood of fishermen and people who depended on tourism. It affected the health of the people of the Gulf and of the cleanup workers. According to Relocate Magazine (February 5th, 2013), BP's estimate of the costs of the disaster reached £42 billion by the end of 2012. There are still cases going through the courts that would add to the £42 billion.

Many experts have argued that BP's emphasis on cutting costs was a major cause to the disaster. BP's own investigation concluded that the accident was caused by "a complex and interlinked series of mechanical failures, human judgments, engineering design, operational implementation and team interfaces." BP's report found that the cement and shoe track barriers failed to contain the hydrocarbons within the reservoir as they were designed. Over a 40-minute period the Transocean (subcontractor of the rig) rig crew failed to recognize and act on the influx of hydrocarbons into the well until the hydrocarbons were in the riser and rapidly flowing to the surface. After the well-flow reached the rig it was routed to a mud-gas separator, causing gas to be vented directly on to the rig and into the engine rooms through the ventilation system. The rig's safety system did not prevent the subsequent ignition. Even after the explosion and the fire had disabled its crew-operated controls, the rig's blow-out preventer on the sea-bed should have been activated automatically to seal the well. It also failed to operate probably because critical components were not functioning.

BP could have better handled the incident if it had followed a methodology of environmental investment assessment similar to what we propose. Considering that BP had two environmental disasters not long before the Deepwater Horizon oil spill, BP would be more aware that the cost of failures far exceeded the costs of prevention, assessment and control. In Table 2, we apply our methodology to the Deepwater Horizon oil spill. Our analysis is based upon the findings detailed in a letter from the Energy & Commerce Committee of the United States House of Representatives to BP CEO Hayward.

The following three statements from the Energy & Commerce Committee's letter provide clear evidence that BP did not have appropriate prevention, assessment and control procedures in the Deepwater Horizon project:

i. "At the time of the blowout, the Macondo well was significantly behind schedule. This appears to have created pressure to take shortcuts to speed finishing the well. In particular, the Committee is focusing on five crucial decisions made by BP: (1) the decision to use a well design with few barriers to gas flow; (2) the failure to use a sufficient number of "centralizers" to prevent channeling during the cement process; (3) the failure to run a cement bond log to evaluate the effectiveness of the cement job; (4) the failure to circulate potentially gas-bearing drilling mud out of the well; and (5) the failure to secure the wellhead with a lockdown sleeve before allowing pressure on the seal from below. The common feature of these five decisions is that they posed a trade-off between cost and well safety."

ii. On April 19, one day before the blowout, BP installed the final section of steel tubing in the well. BP had a choice of two primary options: it could lower a full string of "casing" from the top of the wellhead to the bottom of the well, or it could hang a "liner" from the lower end of the casing already in the well and install a "tieback" on top of the liner. The liner-tieback option would have taken extra time and was more expensive, but it would have been safer because it provided more barriers to the flow of gas up the annular space surrounding these steel tubes. A BP plan review prepared in mid-April recommended against the full string of casing because it would create "an open annulus to the wellhead" and make the seal assembly at the wellhead the "only barrier" to gas flow if the cement job failed. Despite this and other warnings, BP chose the more risky casing option, apparently because the liner option would have cost $7 to $10 million more and taken longer."

iii. "BP's decision not to conduct the cement bond log test may have been driven by concerns about expense and time. The cement bond log would have cost the company over $128,000 to complete. In comparison, the cost of canceling the service was just $10,000.45. Moreover, Mr. Roth of Halliburton estimated that conducting the test would have taken an additional 9 to 12 hours. Remediating any problems found with the cementing job would have taken still more time."

It is evident that BP had overlooked or underestimated the importance of prevention, assessment and control. BP ignored the technical information

Table 2. Analysis of environmental risks and costs of the Deepwater Horizon oil spill.

	BP's Actions
Risks	
The influx of oil or gas is the primary risk. The rig design and choice of materials used should eliminate or reduce the risk of the influx. A related risk is that the workers on the rig do not have the training to be aware of the risk of the oil or gas influx nor the skills to respond when an influx takes place.	BP underestimated the risks and opted for the lower common denominator when deciding design options: 1. Using a well design with few barriers to gas flow, 2. Using insufficient number of "centralizers" to prevent the formation of mud channels in cement columns during the cement process, 3. Failure to run a cement blog, 4. Failure to circulate potentially gas-bearing drilling mud out of the well, 5. Failure to secure the wellhead with a lockdown sleeve. BP also did not ensure that subcontractor employees on the rig were adequately trained. They took 40 minutes to react to the problem.
Prevention	
Because of the high safety risk in oil drilling, especially offshore, the design and choice of materials used should emphasize safety. Additionally, construction and operating crews should be highly trained.	If BP has accepted the better design, the liner option of cement casing which is easier to cement into place properly but costs an additional amount of $7 to $10 million and add more days, the disaster might have been averted. A better training of the crew on the rig would have resulted in a quicker reaction to the problem. The crew training cost could not be comparable to the cost of failure.
Assessment	
Considering the high risk, the monitoring and assessment of the construction are just as important as the design. An acoustic cement bond test should have been conducted.	The cement bond test would have cost over $128,000 and took an additional 9 to 12 hours. We do not have an estimate of what the other assessment and monitor actions would cost but we can assume that they will not be in the hundreds of millions of dollars.
Control	
High quality cement and liners should have been used to protect against the influx of oil and gas. Circulation of gas-bearing mud out of the well must be controlled.	BP's five crucial decisions outlined in the congressmen letter are strong evidence that BP did not establish the necessary controls.
Failure	
Lost 11 lives. Impact on the environment would continue for many years.	The costs so far of remediation, damages and penalties are £42 billion and counting!

that the design of the well had weaknesses. The training of employees working on the rig was inadequate as it took 40 minutes for them to recognize and act on the influx of the hydrocarbons. If BP had done an analysis similar to that outlined in Table 2, it might have made different decisions and avoided the disaster. Although this is an *ex post* analysis, even with significantly lower estimates of the cost of failure would have justified spending money on the risk items identified by the Energy & Commerce Committee and by BP's own investigations.

replete with cases of environmental disasters that have led to bankruptcies of major companies. Emphasis on short-term profits and analysis that understate the risks and costs of environmental disasters can lead companies such as BP and its subcontractors to catastrophic results. Better classifications and measures of environmental risks and costs can lead to more efficient decision making and balancing between the short-term and the long-term. The proposed methodology is a small contribution towards better classifications and measures.

5 CONCLUSION

Commitment to environmentally sound policies is a sound business strategy that reaps benefits and safeguard assets, save lives, and uphold corporate reputations in the long run. The world is

REFERENCES

Abdul Rashid, S.H., S. Evans, and P. Longhurst. 2008. A comparison of four sustainable manufacturing strategies. International Journal of Sustainable Engineering 1 (3), 214–229.

Ansari, Shahid, Jan Bell, Thomas Klammer, and Carol Lawrence. 1997. Measuring and Managing Environmental Costs. Irwin McGraw-Hill.

BP. 2010. Deepwater Horizon Accident Investigation Report. http://www.bp.com/liveassets/bp_internet/globalbp/globalbp_uk_english/incident_response/STAGING/local_assets/downloads_pdfs/Deepwater_Horizon_Accident_Investigation_Report.pdf.

Carter, C.R. and S. Rogers. 2008. A framework of sustainable supply chain management: Moving toward new theory. International Journal of Physical Distribution & Logistics Management 38 (5), 360–387.

Ditz, Daryl, Janet Ranganathan, and R. Darryl Banks. 1995. Green ledgers: Case studies in corporate environmental accounting. Washington, DC: World Resources Institute.

Energy & Commerce Committee of the United States House of Representatives. 2010. BP Deepwater Oil Spill—Energy and Commerce Committee's Letter Outlining Risky Practices in Anticipation of Hayward's Thursday Testimony http://www.theoildrum.com/node/6604.

Gale, Robert. 2006. Environmental management accounting as a reflexive modernization strategy in cleaner production. Journal of Cleaner Production 14, 1228–1236.

Gibson, Kathleen C. and Bruce A. Martin. 2004. Demonstrating value through the use of environmental management accounting. Environmental Quality Management 13 (3), 45–52.

Gopalakrishnan, Kavitha, Yahaya Y. Yusuf, Ahmed Musa, Tijjani Abubakar, and Hafsat M. Ambursa. 2012. Sustainable supply chain management: A case study of British Aerospace (BAe) Systems. International Journal of Production Economics 140, 193–203.

Hoyt, David and Stefan Reichelstein. 2011. Environmental Sustainability at REI. Stanford Graduate School of Business Case # SM-196.

Jasch, Christine. 2003. The use of EMA for identifying environmental costs. Journal of Cleaner Production 11, 667–676.

Lee, Deishin and Lionel Bony. 2009. Cradle-to-Cradle Design at Herman Miller: Moving Toward Environmental Sustainability. Harvard Business School Case # 9-607-003.

Norman, W. and C. MacDonald. 2004. Getting to the bottom of 'triple bottom line.' Business Ethics Quarterly 14 (2): 243–262.

Relocate Magazine. 2013. BP continues to count cost of Deepwater Horizon disaster. http://www.relocatemagazine.com/relocation-news-blog-format/3-general-relocation-news/7146-bp-continues-to-count-cost-of-deepwater-horizon-disaster.

Seliger, G., H.-J. Kim, S. Kernbaum, and M. Zettl. 2008. Approaches to sustainable engineering. International Journal of Sustainable Engineering 1 (1–2), 58–77.

Smith, Leigh and Peter Ball. 2012. Steps towards sustainable manufacturing through modeling material, energy and waste flows. International Journal of Production Economics 140, 227–238.

U.S. Environmental Protection Agency. 1995. An introduction to environmental accounting as a business management tool: Key concepts and terms, EPA 742-R-95-001. Washington, DC. June 1995.

Wikipedia. 2013. Deepwater Horizon oil spill. http://en.wikipedia.org/wiki/Deepwater_Horizon_oil_spill.

Green Design, Materials and Manufacturing Processes – Bártolo et al. (eds)
© 2013 Taylor & Francis Group, London, ISBN 978-1-138-00046-9

Managerial decision support tools for sustainable actions

M. Dolinsky
Pan European University, Bratislava, Slovakia

V. Dolinska
Matej Bel University, Banská Bystrica, Slovakia

ABSTRACT: This paper deals with decision support tools developed on a basis of a Game theory devoted to the group of proactive SMEs already developing their own concepts of CSR policy. Nowadays, new business challenges are being brought into the daylight, and emerging thriving markets valuating triple bottom line philosophy are one of the most promising ones. This paper is aimed at the description of an assistance universities can offer to Small and Middle Sized companies in order to help them to succeed at these markets. This involves creation of Managerial decision support tools based on mathematic models. A decision support tool being described in this paper is based on Cournot game principles and helps to a management to sketch the potential market situation at the market valuating principles of sustainable development. We were using data mined in Italian company PintInox, S. p. A.

1 EUROPEAN COMPANIES ENDANGERED

1.1 Current threats typical for European SMEs

At the beginning of our cooperation with Italian cutlery producer PintInox, S. p. A. in June 2012, we officially labeled this SME as a reactive one, seeing their traditional global beliefs that Corporate Social Responsibility policy is time and money consuming and does not have a real business potential. After the implementation of suggested sets of indicators (sustainability measurement metrics), management of PintInox, S. p. A. has found their usage as one of the main drivers of their differentiation strategy. According to the manager of marketing department, Dr. Mauro Romani, the philosophy of being responsible towards society is one of the remaining tools to be used in a competitive battle against Chinese companies and their cost leadership caused by severe, deplorable working conditions enabling savings in personal costs. Mentioned savings of Chinese companies are forming a steady ground of their success in penetrating of European markets.

PintInox, S. p. A. initiated a competitive-response in lowering the proportion of personal costs (*current social indicators are measuring well-being of employees with valid labor contract, therefore we could not explicitly state that labor force optimization process is in every case in a conflict with Corporate Social Responsibility philosophy*) by massive investments into the new technologies and innovations—some of them already identified as environmental ones in previous papers.

As Dr. Romani points out, the labor costs always formed the significant portion of production costs of PintInox, S. p. A., therefore, they were constantly automating setup of production in order to lower the proportion of personal costs until they reached 17.37% in 2011 (annual reports of PintInox, S. p. A.), which was the ceiling for improving their bottom line by personal costs optimization (an important remark is that they were doing their cost optimization by making employees redundant, not by worsening working conditions!). Applied optimization programmes were a reaction to Chinese cutlery producers and their low-cost leadership. However, as management of PintInox, S. p. A. discovered afterwards, due to disparities between Italian labor standards and Chinese labor standards, PintInox, S. p. A. will never be able to imitate Chinese cost effectiveness.

1.2 Foundations of Chinese cost leadership

To do a verifiable comparison of Italian and Chinese labor standards is a tough issue, due to missing International Labor Organization data about *Chinese rates of occupational injuries* and *workdays lost due to occupational injuries*. This kind of statistics would be helpful to compare the quality of working life in both countries. There are also discrepancies in a collection and reporting of the data to the ILO. In Italy, ILO uses abbreviation *"compensated injuries"* whilst in China—*"reported injuries*". The reason behind is that in Italy: "*Insurance against occupational injury is compulsory for all employees carrying out paid manual work on a permanent or casual basis in specific activities or processes*"(ILO databases). It means that every injury in Italy is automatically being recorded by insurance company whilst in China, the records are being made by Labor inspectorate.

However, the results are not trustable in our opinion, because China still didn't ratify "*C081— Labour Inspection Convention*". C081 convention contains for example the legally binding system of labor inspections, which has an obligation to: "*secure the enforcement of the legal provisions relating to conditions of work, such as wages, safety, health and welfare, the employment of children and young persons*"(ILO databases). The direct consequence of non-ratification of ILO directives are in our opinion inadequate labor standards. We are analyzing trustworthiness of Chinese data in a Table 1, comparing an amount of fatal and nonfatal injuries in 2001, 2002 and 2010 (there is an information gap between 2002 and 2010 regarding China in ILO databases).

In order to analyze the situation in a Table 1, we have used descriptive statistics, percentile rank statistics and correlation analysis of the whole dataset available in order to identify, whether the Chinese results are trustworthy. According to information reported by Italian companies and by www.chinalaborwatch.org, the labor conditions in China are much worse than those in Italy, however, from the ILO database, we are evidencing that only 0.19 employees per 100 000 Chinese workers were experiencing an injury in 2010 (China therefore lies at 0.00 percentile of a dataset)!. We have found this number as suspicious and therefore we have prepared two scenarios in order to identify outlier in the dataset. In the first scenario, we excluded China from dataset which has changed *Skewness* from 1.06 to 1.03. In other words, dataset without China is also right-side skewed (mean > median), but a bit closer to symmetric distribution than before. In

terms of correlation analysis (correlation does not imply causation), in our first scenario, correlation coefficient changed from 0.0004 into 0.7592. In second scenario, where we projected dataset without Italy, correlation coefficient has changed from 0.004 into 0.0036.

1.3 *Possible explanations*

Above all, we identified China as an outlier in the dataset, and therefore the data delivered by Chinese labor inspectorate as not trustworthy. In our opinion, the critically low indices describing China are the results of non-reporting a significant proportion of occupational injuries.

We agree with the opinion of Italian managers from Lombardia that the scarce concern for health and safety rules and precarious employment enables money savings, what then makes Chinese companies more cost effective in comparison with Italian companies. PintInox, S. p. A. is already utilizing its full innovation potential in automation of a production, they are employing the necessary minimum of workers, but altogether with other factors (e.g.: Material effectiveness, pace of production, high innovation potential, etc.), Dr. Mauro Romani still considers their ability to realize a cost leadership strategy as a very limited. After cooperation with our team, he believes that philosophy of being responsible towards the society is one of the few possibilities how to successfully compete with Chinese. Thanks to emerge of thriving markets valuating triple bottom line principles, there is a market space which cannot be entered by companies with low social and environmental performance. In this paper, we are presenting a "*Managerial decision support tool*" enabling to proactive companies like PintInox, S. p. A. to get a picture about their potential to succeed in these new markets. In such markets, the game is changed, which disables Chinese companies to enter (e.g.: Insufficient health and safety standards lead to poor "*social performance*"). In following paragraphs, we are explaining this Managerial decision support tool based on a Cournot game principles.

Table 1. Amounts of injuries (fatal + non fatal) occurring in selected countries.

		2001	2002	2010	Employed population in 2010 (Paid, self-employment in thousands)	No. of workers facing occupational injury per 100 000 workers (2010)
Italy	No. of non-fatal injuries	629 014	598 608	775 339	22 872.3	3393
	No. of fatal injuries	1 077	934	845		
	Total (fatal + non-fatal)	630 091	599 542	776 184		
China	No. of non-fatal injuries	4 141	3755		761 050	!!0.19!!
	No. of fatal injuries	12 554	15 924			
	Total (fatal + non-fatal)	16 695	19 679	1 475		
Slovakia	Total (fatal + non-fatal)	20 989	19 526	9 172	2 317.5	395
Czech Republic	Total (fatal + non-fatal)	93 280	91 073	51 678	4 885.2	1057
Norway	Total (fatal + non-fatal)	28 720	26 989	15 416	2 500	614
Finland	Total (fatal + non-fatal)	58 318	57 804	50 590	2 447.5	2067

Table no. 1, Source: http://laborsta.ilo.org/STP/guest ; www.inail.it – Instituto Nazionale per L'Assicurazione Contro gli Infortuni sul Lavoro (document called: Infortuni sul Lavoro e Malatie Profesionali, 31.10.2012)

2 SUSTAINABLE DEVELOPMENT AS A BUSINESS OPPORTUNITY

2.1 *Thriving markets analyzed by Cournot game*

In our research, we have decided (among other approaches) for sustainability measurement metrics based on game theory. Using a Cournot game, we are capable of thinking in dimension of competitiveness. Using system of equations, Hessian matrix, and regression analysis, we were able to shift from *"cost dimension"* into *"competitiveness dimension"*. As we are persuaded that sustainable development contains lots of behavioural issues (e.g.: Decisions to wisely allocate resources, a need to predict/outguess behaviour of entities, etc.), we are aimed at the use of Game theory as a theoretical base for managerial decision support tools being developed by our team.

2.2 Demand function creation

The main purpose of our Cournot game was to identify, how successful will be PintInox, S. p. A. at the new type of market valuating triple bottom line. This new type of market is accessible only for companies with verifiable economic, social and ecological performance and with trustable Corporate Social Responsibility policy. PintInox, S. p. A. belongs to this category after almost one year of cooperating with our team. *Managerial decision support tool* we are describing in this part is based on a Cournot game. Here, the firm's strategies are quantities. Firms simultaneously choose their respective output levels q_i, from feasible sets $Q_i = [0,x)$ (Fudenberg & Tirole 1991). They sell their output at the market-clearing price $p(q)$, where $q = q_1 + q_2$. Firm i's total profit is then generally expressed by equation (1) and Cournot reaction functions specifying each firm's optimal output for each fixed output level of every opponent in the game (2), (3). There is a PintInox, S. p. A. in our game and its competitors from Italy. Chinese competitors were not able to enter this market due to their inability to come up with functional and verifiable Corporate Social Responsibility policy—which enables to create in Italy a natural oligopoly composed of Italian companies.

$$u_i(q_1, q_2) = q_i p(q) - c_i(q_i) \qquad (1)$$
$$r_i : Q_j \to Q_i \qquad (2)$$
$$r_j : Q_i \to Q_j \qquad (3)$$

For the explanation of natural oligopoly, we are selecting definition provided by M. Horniaček, because this definition of natural oligopoly allows firms in the industry to have different cost functions—which is also our case. The definition of a natural oligopoly in a single product industry is as follows (4), (5) (Horniaček 2011):

Consider $k \in \{1, \ldots, m\}$ and output $Q_k \in (0, \Sigma_{j \in J^{(k)}} \chi_j^{max}]$. A set of firms $J^{(k)*} \subseteq J(k)$ with $\#(J^{(k)*}) \geq 2$ is a natural oligopoly in the industry producing type k of good for output level Q_k if there exists an output vector $y^{(k)*} \in \prod_{j \in J^{(k)*}} Y_j$ such that

$$\sum_{j \in J^{(k)*}} y_j^{(k)*} = Q_k \qquad (4)$$

And

$$\sum_{j \in J^{(k)*}} c_j\left(y_j^{(k)*}\right) \leq \sum_{j \in J^{(k)}:y_j>0} c_j\left(y_j\right), \forall y \in$$
$$\prod_{j \in J^{(k)}} Y_j \text{ with } \sum_{j \in J^{(k)}} y_j = Q_k \qquad (5)$$

A market demand is being represented by Accor hotel chain and is geographically limited to Italy. Hotel chains are typical B2B customer of PintInox, S. p. A. and its competitors.

The demand is being represented by 65 hotels residing in Italy and belonging to the Accor chain (Mercure, Sofitel, Ibis, Novotel). We assume one typical product of PintInox, S. p. A. and its competitors (a fork in this case) per one hotel quest. We presumed the full utilization of hotel capacities—maximum allowed amount of guests in a hotel room (standard or family rooms). The total demand for the Italy is then 8766 forks. We have considered only hotels with obtained internal certification "Planet 21". The reason behind was that this form of standardization is the key element setting up sustainability boundaries—a hotel which possess this internal certification being received from Accor headquarters in Paris is supposed to track the origin of compounds and auxiliary materials being used during previous phases of a life-cycle (this life cycle dimension is being required by ISO 14 040) of products being purchased by the hotel management. As we already remarked, hotel chains are one of the traditional B2B customers of cutlery producers and in this concrete case, the Accor is a direct creator of a market based on sustainability principles. This market is on the other hand closed for competitors with weak environmental performance, because cost effectiveness itself does not represent a competitive advantage here. After selection of a price from the catalogue and the discussion with management about potential demand referring to modelled price, we have set up an inverse demand function (7), from the demand function (6). The roots of the function were taken from regression analysis.

$$Q_d = 8787 - 104_p \qquad (6)$$
$$P = 84.5 - 0.009Q \qquad (7)$$

We have started our game with the total amount of 11 players. We have indentified 10 direct competitors of PintInox, S. p. A. in Italy, however, we didn't go into the further specifications of each, since we didn't have direct access to their facilities. We've only assumed their levels of production costs together with management of PintInox, S. p. A. (this step in not a weak point, since companies are supposed to use this tool in a future by creating their own possible scenarios based on different competitor's cost effectiveness). In this kind of industry, PintInox, S. p. A. possess detailed knowledge about technology variations typical for cutlery production and related costs. In the sample of 10 companies, we are comparing various cost levels of production of an identical product (a model of fork called "*Sirio*" produced by PintInox, S. p. A. and related substitutes being made by different companies). Cost functions we have created for each company in a game are based on 30 observations occurred during one week. In PintInox, we were observing production process of a model *Sirio* during one week. We took into a consideration fixed costs and all types of variable costs.

- Included in fixed costs: Total production costs of a shearing form used in cutting process to cut the exact shape of a fork.
- Included in variable costs: Salaries and wages, material costs and other types of costs varying during the production.

Standards PintInox, S. p. A. has to comply with are described in "*Contratto collective nazionale—Industria metalmeccanica private e della installazione di impianti*". It is a general agreement made 15. October 2009 in Rome between six parties (representatives of employers and unions). There are numerous obligations PintInox, S. p. A. shall be fulfilling on a periodic basis (e.g.: prescribed probation period, equal opportunities norms, compulsory content of labour agreement, classification of work types in metal industry and related remuneration, etc.) (AI Bresciana 2009). Variability of variable costs in our game was influencing also a concrete employee who was operating the machine during our observation. During one week, four employees belonging to different category each (there are 7 wage categories based on character of duties, difficulty and work experience) were rotating at observed workplace. The monthly salary levels in effect from 1st January 2012 are depicted in a Table 2.

Table 2. Prescribed monthly salary levels valid in Italian metal industry.

Category	Monthly salary levels in effect from 1st January 2012
1st	1.206,23 EUR
2nd	1.327,19 EUR
3rd	1.466,17 EUR
4th	1.528,32 EUR
5th	1.634,56 EUR
5th superior	1.748,28 EUR
6th	1.876,27 EUR
7th	2.038,21 EUR

Source: contratto collettivo nazionale, Associazione Industriale Bresciana. Grafica F.G. Brescia 2009. p. 281.

3 RESULTS

3.1 *Identification of a player's market share*

Based on principles described in part 2.2., using regression analysis, we have created a cost functions for PintInox, S. p. A. and for 10 Italian competitors. Our regression is based on a large sample (30 observations). According to some authors, If $|t|>2$, it indicates significant difference between the two groups at the 95% confidence level. $|t|>1.68$ indicates the coefficient is significant at the 90% confidence level. Our datasets meet the 95% confidence threshold when $N = 30$ observations. In terms of PintInox, S. p. A., coefficient for amount of production does not exceed $|t\text{-statistic}|$ of 2. The coefficient for q (amount of produced forks) has estimated standard error of 3.03, t-statistic of -1.17 and p-value of 0.25. It is therefore statistically insignificant at significance level $\alpha = 0.05$ as $p > 0.05$. In this regression, we cannot reject the possibility that differences (various cost levels) were occurring by chance, and that they were not caused by total outputs. It is because we cannot reject H_0 hypothesis. In order to explain this situation, we have to go deeper in our analysis and to explain firstly the use of degrees of freedom, since it is important for t-statistic. The degrees of freedom are given by the number of observations minus the number of coefficients in the fitted regression model. For a quadratic curve model, this gives n-3 since we are fitting two slopes and one intercept. Having 27 degrees of freedom, the critical value for t_0.025 one tailed and t_0.05 two tailed is 2.052. Using the critical value approach, we computed $t = -1.495$ [(coefficient—1)/st. error]. So, we do not reject null hypothesis at level 0.05 since $t = |-1.495| < 2.052$. From the statistical point of view, we are not able to confirm causality between level of output and total costs in PintInox, S. p. A. In case of competitor's data, there are two competitors from 10, where we are not able to reject zero (H_0) hypothesis—on the other hand, it doesn't mean, we have to accept H_0. The data from regression analysis are to be used in the Cournot game, which is our *Sustainability decision support tool*. The data feed will be realized by companies (in our case by PintInox, S. p. A.), and it is upon their decision, what kind of "*typology*" of competitors they will decide for. They may create various scenarios based on different cost levels of production. The data in our case are from the real production plant and were confirmed by the management of PintInox, S. p. A. The competitor's data were also confirmed by management of PintInox, S. p. A. as one of possible scenarios. Therefore, we consider them relevant and trustworthy. The computed cost functions based on regression analysis are as follows (8), (9), (10), (11), (12), (13), (14), (15), (16), (17), (18):

$0.0055q_1^2 - 3.5306q_1 + 5562$ PintInox, S. p. A. (8)

$-0.0408q_2^2 - 19.78q_2 + 4184$ Competitor x_2 (9)

$0.007q_3^2 - 1.08q_3 + 5573$ Competitor x_3 (10)

$-0.0017q_4^2 + 4.02q_4 + 5096$ Competitor x_4 (11)

$-0.0130q_5^2 + 7.43q_5 + 4567$ Competitor x_5 (12)

$-0.0056q_6^2 + 7.43q_5 + 5096$ Competitor x_6 (13)

$-0.0075q_7^2 + 10.49q_7 + 4540$ Competitor x_7 (14)

$-0.0241q_8^2 + 22.7q_8 + 3951$ Competitor x_8 (15)

$-0.0031q_9^2 + 6.55q_9 + 4818$ Competitor x_9 (16)

$0.0021q_{10}^2 - 0.61q_{10} + 5681$ Competitor x_{10} (17)

$0.0088q_{11}^2 + 0.002q_{11} + 4930$ Competitor x_{11} (18)

Using regression analysis, we could have modelled also linear and cubic cost functions, but we were interested in having quadratic ones, since we were supposed to do also second order condition. Only those quadratic cost functions which could have passed first order and second order conditions became a part of a Cournot game (19):

$$\left(84.5 - 0.00961 \sum_{P_{i \neq j=1}}^{8} q_j - 0.00961q_i\right)q_i - a_iq_i^2 - b_iq_i - \gamma_i$$

First derivation: $84.5 - 0.009 \sum_{i \neq j=1}^{8} q_j - 0.01922q_i - 2a_iq_i - b_i$.

Second derivation: $-0.01922 - 2a_i \leq 0 \leftrightarrow a_i \geq -0.01922/2$.

$a_i < 0 \rightarrow |a_i| \leq 0.00961$ (19)

As we can see, competitors x_2, x_5, and x_8 were not able to meet second order condition, and were therefore excluded from the game (absolute value of a coefficient for q squared was higher than 0.00961). Remaining quadratic cost equations were processed together with demand function in a following way (20), (21), (22):

$$Q = q_1 + q_2 + q_3 + q_4 + q_5 + q_6 + q_7 + q_8$$ (20)

$u_i = p * q_i - TC_i$ (21)

$u_i = (84.5 - 0.00961Q)q_i - (a_iq_i^2 + b_iq_i + \gamma_i)$ (22)

Computed equations were then ordered into a system of eight linear equations with 8 unknowns (see Fig. 1).

Solving the system, we have come into results being depicted in a Table 3, x_1 stands for PintInox.

Economic interpretation of negative results is zero level of production. Competitors x_4 and x_5 have completely lost in a competitive battle and have appeared with zero production. Other three players were dividing market among each other, but to calculate the exact amount of forks delivered to the market valuating triple bottom approach, we had to solve the system of equations once again, with competitors having zero production levels excluded. The final results are available in Table 4 (x_1 stands for PintInox, S. p. A.). The suggestion made for PintInox, S. p. A., respecting the market situation—especially the *"strength"* of competitors is to set up production programme for cca. 680 units of products. Or, on the other hand, the result is very useful for conducting a feasibility study (e.g.: Computing break-even point analysis, payback period analysis, etc.).

$0.029x_1 + 0.009 x_2 + 0.009 x_3 + 0.009 x_4 + 0.009 x_5 + 0.009 x_6 + 0.009 x_7 + 0.009 x_8 = 88.03$

$0.009 x_1 + 0.0194 x_2 + 0.009 x_3 + 0.009 x_4 + 0.009 x_5 + 0.009 x_6 + 0.009 x_7 + 0.009 x_8 = 83.42$

$0.009 x_1 + 0.009 x_2 + 0.0146 x_3 + 0.009 x_4 + 0.009 x_5 + 0.009 x_6 + 0.009 x_7 + 0.009 x_8 = 80.48$

$0.009 x_1 + 0.009 x_2 + 0.009 x_3 + 0.0068 x_4 + 0.009 x_5 + 0.009 x_6 + 0.009 x_7 + 0.009 x_8 = 76.08$

$0.009 x_1 + 0.009 x_2 + 0.009 x_3 + 0.009 x_4 + 0.003 x_5 + 0.009 x_6 + 0.009 x_7 + 0.009 x_8 = 74.01$

$0.009 x_1 + 0.009 x_2 + 0.009 x_3 + 0.009 x_4 + 0.009 x_5 + 0.0118 x_6 + 0.009 x_7 + 0.009 x_8 = 77.95$

$0.009 x_1 + 0.009 x_2 + 0.009 x_3 + 0.009 x_4 + 0.009 x_5 + 0.009 x_6 + 0.0222 x_7 + 0.009 x_8 = 85.11$

$0.009 x_1 + 0.009 x_2 + 0.009 x_3 + 0.009 x_4 + 0.009 x_5 + 0.009 x_6 + 0.009 x_7 + 0.0356 x_8 = 84.5$

Figure 1. System of equations.

Table 3. Optimal output levels—first round.

$x_1 = \mathbf{1360.4911}$	$x_2 = 2173.0598$	$x_3 = 3510.6825$
$x_4 = -6936.2828$	$x_5 = -2198.3037$	$x_6 = 6117.7936$
$x_4 = 1840.1380$	$x_8 = 890.2188$	

Table 4. Optimal output levels—second round.

$x_1 = \mathbf{679.2885}$	$x_2 = 863.0549$	$x_3 = 1077.8163$
$x_4 = 1252.0612$	$x_5 = 808.0129$	$x_6 = 3591.3469$

623

4 CONCLUSIONS

We are using Cournot game as a *Managerial decision support tool,* and as a matter of fact, it helps companies to get a picture about their competitiveness and abilities to place certain amount of environmentally friendly production to the market valuating sustainability principles. Based on our experience gained thanks to our cooperation with PintInox, S. p. A., we are dividing traditional wrong managerial global beliefs about sustainable development into two categories—1) beliefs typical for reactive companies and 2) beliefs typical for proactive ones. Attitude of PintInox, S. p. A. towards philosophy of sustainable development was at the beginning of our cooperation typical for reactive companies. Management of PintInox, S. p. A. was persuaded that sustainable *"mindset"* is not feasible since it is only time and money consuming. When we applied our first sustainability decision tool—a model able to identify already realized eco-innovation and its % contribution to overall footprint improvement (Molnar & Dolinsky 2012), first partial wins in a form of results possessing marketing value gained for free, have changed the "mindset" of managers from being reactive, into being proactive in terms of acceptance of sustainable development as an integral part of their differentiation strategy. However, this new proactive "orientation" goes hand in hand with changes in requirements for *Managerial decision support tools.* Management of PintInox, did the step forward, and consequently, required a Managerial decision support tool assisting them in decision making process aimed at utilization of their newly created concept of Corporate Social Responsibility policy in market/business terms, especially when they found a sustainable development philosophy as one of the very few possibilities how to combat Chinese competitors. The main discovery of this paper is the fact, that economic efficiency remains important also in sustainable development. Because, after beating up low-cost competitors with verifiable CSR policy, it is again an economic efficiency which makes the difference between members of natural oligopoly.

REFERENCES

Associazione Industriale Bresciana. 2009 Industria metalmeccanica privata e della installazione di impianti. Brescia: Grafica F.G.

Fudenberg, D. & Tirole, J. 1991 *Game Theory.* Boston: The MIT Press.

Horniaček, M. 2011 Cooperation and Efficiency in Markets. *Lecture Notes in Economics and Mathematical Systems* vol 649. Springer: Heidelberg.

Molnár, P. & Dolinský, M. 2012. Environmental Performance Assessment Model as s Sustainability Decision Tool for SMEs. *International Journal of World Academy of Science, Engineering and Technology* V67: 952–959.

Green Design, Materials and Manufacturing Processes – Bártolo et al. (eds)
© 2013 Taylor & Francis Group, London, ISBN 978-1-138-00046-9

Exploring Green IT and Green IS: Insights from a case study in Brazil

T.A. Viaro & G.L.R. Vaccaro
UNISINOS, São Leopoldo, Brazil

ABSTRACT: The role of Information Technology (IT) has been highlighted in the discussions around environmental sustainability in organizations. As an alternative to tackle the problem, the concepts of Green IT and Green IS has emerged in the current literature and has opened possibilities to guide companies in the direction of sustainable business. We present here a Green IT/IS framework built considering four units of analysis: economy, environment, process and technology. This model seeks to investigate how constructs like energy efficiency, optimized equipment utilization and sustainable information systems are considered by an organization in terms of practices, policies and corporate strategy. In this paper, we argue that organizations which are evolving towards sustainable business go through different maturity stages until contribute effectively to the environment. In this sense, although Green IT, that is more focused in reducing the impact caused by IT, be an important milestone for the company, it has a lesser impact in the long term than Green IS, that has the potential to change and create new business models and processes. Moreover, we present a case study that was conducted in a large multinational company from the software industry in order to verify the depth of Green IT/IS strategies and practices in place. At last, the empirical findings are analyzed and presented in the light of the theoretical model proposed in this study.

1 INTRODUCTION

There are several companies which still may not perceive value in the idea of sustainable development, due to the belief that more environmentally compliant denotes less competitive or less cost effective. This hypothesis comes from the trade-off involving, on one hand, the social and environmental benefits of creating sustainable products and processes, and, on the other hand, the financial cost from those decisions (Nidumolu et al. 2009). Instead, several organizations already perceived that the search for sustainability does not have to be a burden. Making a company environmentally friendly can help reduce costs and increase revenue and profits (Porter; Van der Linde, 1995, Ambec & Lanoie, 2008, Nidumolu et al. 2009). Companies that invest in improving environmental performance can obtain economic and financial results through better access to certain markets, differentiation of products and services in their portfolio, and reduced costs associated with materials, energy and services (Ambec & Lanoie, 2008). In this scenario, the latest manifestations involving sustainable business practices and technology involve the idea of Green Information Technology (Green IT) and Green Information Systems (Green IS).

In this paper, we consider Green IT more focused in the achievement of environmental sustainability goals via technical infrastructure measures, such

as cost center efficiency, as an example (Velte et al. 2009; Murugesan, 2008). On the other hand, Green IS is seen tackling a broader challenge: to turn entire systems more sustainable instead of only reducing energy consumption in the IT infrastructure (Boudreau et al. 2007).

Despite the dimension of the goals, in order to implement Green IT/IS programs, one organization needs to acknowledge which are the strengths and weaknesses in its structure. This perception is relevant to enable the organization going through different maturity stages towards the implementation of sustainable practices. In order to support this task, we propose a preliminary Green IT/IS framework, comprehending four dimensions of analysis: economics, environmental, processual and technological. The framework is based on theoretical and empirical evidences from a case study developed in one World Class Software Company with developing units located in Brazil. Focusing on the empirical findings we aim to contribute to the discussion of the role of Information technologies and Information Services towards a concept of more sustainable and intelligent organizations.

To continue this paper, the concepts of Green IT and Green IS will be detailed in sections 2 and 3, respectively. In section 4, a brief view of the methodological procedures is presented. Then, in section 5, we present a description of the case study conducted, focusing on exploring Green IT/IS constructs. Finally, in Section 6 we draw some final remarks and suggestions of future discussions on the subject of Green IT/IS as support for sustainable, responsible and intelligent organizations.

2 GREEN IT

IT, due its pervasive and constant presence, is considered responsible for costs associated with energy consumption in a great extent. As an example, in Australia, ICTs contribute for over 1.5% of the national CO_2 emissions (Molla, 2008), and it is estimated to contribute with about 2% of global CO_2 emissions (Gartner, 2007, Reuters, 2009), with an estimative of yearly growth of 6 percent (Reuters, 2009) due to the increasing demand of software and computing power. In face of such reality, and considering the potential of costs reduction and processes improvements associated to IT solutions, organizations seek for Green IT actions both as a way to reduce their carbon footprint as well as to reduce their expenses in energy (Molla, 2009a, Molla & Cooper, 2009b). Technological development has the potential to turn non-clean technologies obsolete, like occurred with the introduction of wireless broadband communications. Its advance: (i) reduced the need of physical wire infrastructure; (ii) contributed to reduce raw material consumption with elevate effect over the environment (Hart & Milstein, 2003); (iii) reduced installation and infrastructure maitenance costs; and (iv) aggregated perceived value to consumers by allowing high mobility and connectivity. As another example, South Korea established a five year plan to become the world's seventh most competitive country by 2020 in terms of energy efficiency and ability to adapt to climate change (BizTechReport, 2010, Korea, 2013), by inducing energy efficient transport options, green buildings and new technology, with special attention to LEDs, solar cells, hybrid cars and the efficient use of IT. Moreover, accordingly to Molla (2008), the development of analytic tools and information systems that support decisions with a focus on energy consumption reduction, and the implementation of systems to manage and to neutralize carbon emissions generated by business practices may be actions correlated to Green IT strategy. Therefore, Green IT can be faced as an opportunity to enable the application of Green Management practices, developing sustainable and responsible business in the social and environmental contexts, and adding competitiveness to the organization. For most companies, the major question is not whether adopt Green IT, but rather on how to do it (Marcus & Fremeth, 2009).

In a wider context, Green IT can be found as a milestone for responsibility and sustainability, as proposed by Viaro et al. (2010a) and Viaro & Vaccaro (2010b). Associated to pressure for non-palliative environmental solutions, strategic requirements (Marcus & Fremeth, 2009, Siegel, 2009) and value generation for stakeholders are added to organizational demands. In this sense, organizations need to address green practices strategically, aligning the Green IT adoption to its strategy and in response to the environmental and social pressures (Hart & Milstein, 2003, Viaro et al. 2011), realizing that the benefits for the adoption are associated with how these practices will be adopted (Ambec & Lanoie, 2008).

Green IT can be seen as the study and practice of efficient and effective design, manufacture, use and disposal of computers, servers and associated subsystems, with minimal or no impacts to the environment. It also pursues economic viability, addressing energy efficiency and total cost of ownership, which includes the cost of disposal and recycling (Murugesan, 2008). Lamb (2009) also presents Green IT as the study and practice of using computing resources efficiently, corroborating the definition that many organizations commonly refer to Green IT: a synonym for efficiency in data centers (Molla, 2008). Aligned with this vision, Velte et al. (2008) present the component elements of Green IT as including: virtualization of data centers, e-waste, recycling, and redesign, amongst others. This view maintains the definition of Green IT still closely tied to the physical elements of Information Technology.

On the other hand, the impact of Information and Communication Technologies (ICTs) may be understood as beyond the above definition: it should be expected, from ICTs, to provide means for organizations enabling their core businesses to become 'green' businesses—i.e., a business that is sustainable from an environmental point of view (Marcus & Fremeth, 2009, Nidumolu et al. 2009). Through the use of ICTs (e.g., tools for monitoring emissions of greenhouse gases and water consumption) organizations can reduce their carbon footprint while also establishing strategies for production and service markets. This leads to a deeper and broader understanding of Green IT: the ability of an organization to implement environmental sustainability on different technical elements of the Information Technology (design, production, acquisition, use and disposal), and also for human and managerial resources (Molla, 2009a). In the Information Systems literature (Boudreau et al. 2007, Brooks et al. 2012, Watson, 2010), this understanding of Green IT has been presented also as Green IS.

3 GREEN IS

As stated before, whereas Green IT may be seen more associated with energy efficiency and equipment utilization, Green IS has been considered a broader concept, which supports sustainable business processes (Watson et al. 2010, Brooks et al. 2012). In other words, it refers to design, construction and implementation of information systems that contributes to sustainable business practices.

Watson et al. (2010, p. 2) gave some examples on how Green IS can help an organization:

- Reduce transportation costs with a fleet management system and dynamic routing of vehicles to avoid traffic congestion and minimize energy consumption;
- IS can move remote working beyond telecommuting to include systems that support collaboration, group document management, cooperative knowledge management, and so forth;
- Track environmental information (such as toxicity, energy used, water used, etc.) about the creation of products, their components, and the fulfillment of services;
- Monitor a firm's operational emissions and waste products to manage them more effectively;
- Provides information to consumers so they can make green choices more conveniently and effectively.

In this way, Green IS aids organizations to move towards eco-effectiveness, concept described by Chen et al. (2008) as the ultimate solution for ecological problems, i.e., not only be restricted to the things right (efficiency) but do the right things (Watson et al. 2010). While eco-efficiency pursue consumption reduction of finite resources, eco-effectiveness aims at finding a new solution to switch to renewable resources or even to innovate the whole process in order to change the current solutions minimizing or eliminating ecological impact. The concept of eco-effectiveness is also supported by closed loop initiatives on manufacturing, such as the Cradle-to-Cradle approach (McDonough & Braugart, 2010), which aims to create a positive footprint through eco-effectiveness.

Watson et al. (2010) bring a good example of how Green IS can make a difference when used wisely, in this case with transportation systems. The more successful systems have created information systems that helps the transportations systems to mitigate their physical limitations. Consequently, it minimizes environmental impact, meeting citizens' fundamental needs. Conversely, the research also found that the transportation system that had the physical implementation before the information system has been set up fail in different ways, what

indicates that Green IS must be strategically considered by organizations which aims to put it in place.

With very frequent calls for urgent changes, mainly because global warming requires fast actions to tackle the causes of the environmental problems, Green IS seems to be an alternative to create new business models that supports sustainable practices.

4 METHODOLOGICAL PROCEDURES

This paper is based on part of a research focusing on assessing the impacts and maturity levels of Green IT/IS on companies. The research was conducted as a qualitative study, comprehending the following steps: (i) exploratory literature review and group discussion, aiming to define the scope and main concepts related to the object under analysis; (ii) in-depth literature review and compilation of concepts; (iii) proposal of a preliminary Green IT/IS framework; (iv) empirical case studies, seeking for refining the proposed framework; (v) refinement of the proposed framework; (vi) analytic generalization (Eisenhardt, 1989) and results communication.

The data sources utilized comprehend papers from indexed journals and magazines (EBSCO, SCOPUS, SCIELO, CAPES), as well as books and documental evidences. The empirical study presented in this paper was performed through direct observation, semistructured interviews and documental research.

Data was analyzed by using content analysis and descriptive statistics. Results were compiled and discussed by the research team and, afterwards, presented to the key contacts in the involved company, for refining and accreditation.

5 EMPIRICAL FINDINGS

In this section we present the main results of a case study developed in a large IT organization located in Brazil, in order to evaluate Green IT/IS maturity. The study was conducted during a period of eight months, assessing different parts of the organization.

The preliminary conceptual model supporting the case study is presented in Figure 1, whose definition is described in Viaro et al. (2010) and Viaro et al. (2011). The aim was to propose a conceptual model to evaluate, considering the understanding of Green IT/IS strategies and practices (Viaro et al. 2010; Molla et al. 2010; Chen et al. 2008), how to classify an organization in order to either figure out how mature certain market segment is or to

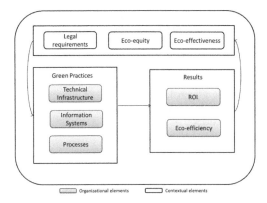

Figure 1. Green IT/IS conceptual model (Viaro et al. 2011).

provide a company an outlook of their maturity. It would aim the organization to improve, adapt, innovate, invest in terms of technical infrastructure and information systems.

In the conceptual model, sustainable practices are divided into two dimensions: Technology and Processes. The technology dimension contemplates the concepts of technical infrastructure and information systems, which has been extensively discussed in the recent literature as exposed in Sections 2 and 3. Viaro et al. (2011) propose the Process dimension in order to identify sustainable processes present in an organization with direct relation with the IT/IS life-cycle, namely: sourcing or acquisition, use or utilization and disposal or recycle. These processes are not separated from the infrastructure technology or information systems present in the organization. They are seen as a flow of assets and information inside the organization that require specific handling from sustainability perspective.

From the collected data, we could infer that company has an advanced maturity level for operations and systems prospects. Over server virtualization, printing optimization, automated cooling and virtual collaboration, environmental sustainability steers IT management, as many authors has defended (Velte et al. 2009, Lamb, 2009, Molla, 2009a, Molla & Cooper, 2009b). Systematically, the company is reducing greenhouse gas emissions through, for instance, efficiency improvements in power supply operations and cooling of corporate IT assets, and by taking advantage of Green IS to bear actions in sustainability. These activities are aligned with company's strategy for environmental sustainability, considering economic, social and environmental aspects, i.e. fostering sustainable development (WCED, 1987).

In terms of processes, we found no institutionalized politics or processes regarded to IT acquisition

and disposal are in place in the organization. This allows us to infer that the company still remains in an early development stage in the Processual dimension.

The company has properly disposed IT assets that reach their end of life in the subsidiaries located in countries with legal regulations that require such measures. Although, in Brazil, the destination of IT assets is more linked to social actions, as there were no legal requirements obliging organizations to dispose according to environmental laws. Notebooks that reach their end of life, for example, are donated to needy schools, but there is no further association from the company with its future destination. However, government initiatives such as the National Policy on Solid Waste (Brasil, 2011) will regulate, when in effective application, environmentally sound disposal of waste, among other things. Based on the evaluation presented in this study, the company started a project to better define processes for adequate IT assets management.

To include strategic product management in organizations requires stakeholders (suppliers, customers, partners, etc.) engaged in IT sourcing and end-of-life management planning (Hart & Milstein, 2003). However, the lifecycle administration crosses company's traditional boundaries, including additional costs related to procurement and assets disposal (Hart & Milstein, 2003, Nidumolu et al. 2009). Nevertheless, at the same time, it may be used to leverage corporate reputation and legitimacy, as evidenced with sustainability reports that large corporations publish annually. These factors, considered in this empirical study, have a strong tendency to contribute to the lack of practice from the organization in this regard.

Finally, the systematic application of Green IT/IS practices within the company is reflected in the economic and environmental outcomes achieved. The company performs actions in Green IT directly tied to cost savings, which can be observed, in particular, when it comes to reducing energy consumption. The sustainability report released by the company clearly relates the efficiency of data centers as a strategic factor for reducing costs and CO_2 emissions, which is supported by several internal initiatives in terms of server virtualization, optimization, IT cooling, to name a few initiatives in these sustainability projects within the IT industry's global enterprise. Additionally, Green IS activities were encountered with the tracking of environmental information, CO_2 emissions monitoring, travel footprint reduction (air and land) through virtual collaboration tools. The utilization and development of such tools denote the company's tendency to not be restricted to eco-efficiency actions, but be biased towards eco-effectiveness proceedings.

These initiatives are linked to Green IT/IS strategy in a wider environmental sustainability perception within the organization, which considers various factors in Green Management, as shown by several authors (Hart, 1997, Hart & Milstein, 2003, Ambec & Lanoie, 2008, Nidumolu et al. 2009, Siegel, 2009, Marcus & Fremeth, 2009).

6 FINAL REMARKS

A high challenge for organizations is to build and utilize physical and informational systems that can turn society life environmentally sustainable. In this paper we presented evidence on how Green IT/IS may be a sustainable driver for improvements in an organization, through the understanding of the dimensions we perceive as most important: processual and technological, under a view of eco-effectiveness. The research findings from the empirical study allowed us to analyze the impact of such dimensions on the decisions taken by the organization regarding a more sustainable, intelligent operation.

Research findings also demonstrated that the company's Green IT/IS maturity level is high, since there is synergy between Green IT/IS practices and the company's strategy. This synergy results not only in cost reduction and eco-efficiency, but in ecological outcomes, which can be reached by successfully eco-effectiveness actions.

Green IT/IS are still being conceptualized by many authors (see Brooks et al. 2012 review), what evidences their importance and, at same time, how fresh these concepts are in existing literature. There are several research possibilities in order to identify current sustainable practices in technical infrastructure and informational systems in organizations. In this study, we attempted to contribute to the existing literature with a case study in a large IT enterprise in Brazil. We hope our research can foster other Green IT/IS discussions and a broader understanding of such concepts, as we believe Green IT/IS practices can highly further foster environmental sustainability.

REFERENCES

Ambec, S. & Lanoie, P. 2008. Does It Pay to Be Green? A Systematic Overview. *Academy of Management Perspectives* 22(4): 45–63.

BiztechReport. Green Technology and Sustainable Development in India and South Korea. Asia-Pacific Business & Technology Report. February 1st, 2010. Available at <http://www.biztechreport.com/story/352-green-technology-and-sustainable-development-india-and-south-korea>. Accessed in May 1st, 2011.

Brasil. Presidência da República. Lei N° 12.305. Available at: <http://www.planalto.gov.br/ccivil_03/_ato2007–2010/2010/lei/l12305.htm>. Acessed: Jan. 2011.

Boudreau, M.-C. Chen, A.J., and Huber, M. (2007). Green IS: Building Sustainable Business Practices. In Watson, R.T. (ed.) *Information Systems*: 1–15. Athens, GA: Global Text Project.

Brooks, S.; Wang, X. & Sarker, S. 2012. Unpacking Green IS: A Review of the Existing Literature and Directions for the Future. In Brocke, J., Seidel, S. & Recker J. (ed.) *Green Business Process Management: Toward Sustainable Enterprise:*15–37. Berlin: Springer Berlin Heidelberg.

Chen, A.J.W.; Boudreau, M. & Watson, R.T. 2008. Information Systems and Ecological Sustainability, Journal of Systems and Information Technology 10(3): 186–201.

Eisenhardt, K.M. 1989. Building Theories from Case Study Research. *Academy of Management Review* 14(4): 532–550.

Gartner Group. Gartner Estimates ICT Industry Accounts for 2 Percent of Global CO2 Emissions. Press Release, 2007. Available at <http://www.gartner.com/it/page.jsp?id=503867>. Accessed in March 1st, 2013.

Hart, S.L. Beyond Greening: Strategies for a sustainable world. Harvard Business Review, Vol. 75 n. 1, p. 67–76. 1997.

Hart, S.L. & Milstein, M.B. 2003. Creating sustainable value, *Academy of Management Executive* 17(2): 56–69.

Korea. Green Growth Korea. Available at: <http://www.greengrowth.go.kr>. Accessed: Mar. 2013.

Lamb, J. 2009. The Greening of IT: How Companies Can Make a Difference for the Environment. Boston: IBM Press.

Marcus, A.A. & Fremeth, A.R. 2009. Green Management Matters Regardless. *Academy of Management Perspectives* 23(3):17–27.

McDonough, W. & Braungart, M. 2002. *Cradle To Cradle: Remaking the Way We Make Things*. New York: Northpoint Press.

Molla, A. 2008a. GITAM: A Model for the Acceptance of Green IT, *19th Australasian Conference on Information Systems,* Christchurch, New Zealand, December 3–5.

Molla, A. 2009a. Organizational Motivations for Green IT: Exploring Green IT Matrix and Motivation Models, *Pacific Asia Conference on Information Systems*, Hyderabad, India, July 10–12.

Molla, A. 2009b & Cooper, C. Green IT Readiness: A Framework and Preliminary Proof of Concept. *Australasian Journal of Information Systems* 16(2).

Murugesan, S. 2008. Harnessing Green IT: Principles and Practices. *IT Professional* 10(1): 24–33.

Nidumolu, R., Prahalad, C.K. & Rangaswami, M.R. 2009. Why Sustainability Is Now the Key Driver of Innovation. *Harvard Business Review* 87(9): 25–34.

Porter, M.E. & van der Linde, C. 1995. Toward a new conception of the environment competitiveness relationship. *Journal of Economic Perspectives* 9 (4): 97–118.

Reuters. 2009. <http://www.reuters.com/article/2009/01/15/us-computing-carbon-emissions-idUSTRE50E5QO20090115>. Accessed: Mar. 2013.

Siegel, D.S. 2009. Green Management Matters Only If It Yields More Green: An Economic/Strategic Perspective. *Academy of Management Perspectives* 23(3): 5–17.

Velte, T.; Velte, A. & Elsenpeter, R. 2008. Green IT: reduce your information system's environmental impact while adding to the bottom line. New York: McGran-Hill.

Viaro, T.A.; Vaccaro, G.L.R.; Azevedo, D.C.; Brito, A.; Tondolo, V. & Bitencourt, C.C. 2010a. A conceptual framework to develop Green IT—going beyond the idea of environmental sustainability. *Proceedings of Conf-IRM 2010. International Conference on Information and Resources Management. Montego Bay, Jamaica, 16–18 May 2010.*

Viaro, T.A. & Vaccaro, G.L.R. 2010b. A conceptual framework to develop Green IT—going beyond the idea of environmental sustainability. *Proceedings of ICEOM 2010, International Conference on Industrial Engineering and Operations Management. 1–12. São Carlos, Brazil.*

Viaro, T.A.; Vaccaro, G.L.R. & Scherrer, T. 2011. Conceptual model to analyze Green Maturity in Organizations: Proposition and Case Study. *Proceedings of XVII ICIEOM, Belo Horizonte, Brazil, 4–7 October 2011.*

Watson, R.T., Boudreau, M.-C., & Chen, A.J. 2010. Information Systems and Environmentally Sustainable Development: Energy Informatics and New Directions for the IS Community. *MIS Quarterly* 34(1): 23–38.

Watson, R.T., Boudreau, M., Chen, A.J.W., & Sepúlveda, H.H. 2011. Green projects: An information drives analysis of four cases. *Journal of Strategic Information Systems* 20: 55–62.

WCED. 1987. *Our common future.* Oxford: Oxford University Press.

Green Design, Materials and Manufacturing Processes – Bártolo et al. (eds)
© *2013 Taylor & Francis Group, London, ISBN 978-1-138-00046-9*

Sustainable supply chain and collaboration: What is the link and what are the benefits?

Y. Dwi, I. Nouaouri, H. Allaoui & G. Goncalves
Laboratoire de Génie Informatique et d'Automatique de l'Artois (LGI2A), Université Lille Nord de France, Béthune, France

ABSTRACT: In this paper, we deal with simultaneously design of a supply chain at strategic level and capacity planning at tactical level to optimize the three bottom lines of sustainability. We show the main benefits of collaboration in sustainable supply chain. Our approach is hierarchical and based on simulation. In order to show the benefits of collaboration we simulate and then we compare three scenarios modeling the spring mix supply chain. The first one is without collaboration; the second is with vertical collaboration and the third is with horizontal collaboration. Numerical results show the impact of collaboration on three bottom lines.

1 INTRODUCTION

Nowadays, productive enterprises are facing increasing challenges to balance business performance and economic gains with environmental and social issues. The concept of sustainability has become increasingly important in supply chain management. Furthermore, companies are undergoing pressure from various stakeholders such as governmental agencies, workers, organisations and some customer segments to deal with social and environmental issues related to their supply chains (Vachon and Klassen 2006; Welford and Frost 2006).

In this context, collaboration is considered as the key requirements of successful sustainable supply chains. The term supply chain collaboration refer to those activities among and between supply chain partners concerned with the cost effective, timely, and reliable creation and movement of materials to satisfy customer requirement (Muckstadt et al. 2001).

Many papers deal with industrial sustainable supply chain (A. Chaabane et al. 2012) (Fan Wanga, et al. 2011). In this paper we concentrate on sustainable collaboration. Collaboration includes practices like sharing distribution, information, transportation, etc. to reduce cost, environmental impact and improve social aspects.

Economic aspects: supply chain collaboration denotes the sharing of materials, information, capital (Kumar and van Dissel, 1996), risks, technology and other resources. Materials can be shared in terms of reserving manufacturing capacities (Xu and Yang, 2004). Supply chain actors share information about inventory levels, production and delivery schedules throughout the supply chain (Webster, 1995; Xu and Yang, 2004). (Kumar and van Dissel, 1996) and (Lee-Mortimer, 1993) discuss the sharing of information technology platforms, to the arrangement of common product specifications in workshops. Environments aspects: stakeholders can work together on environmental purchasing, product and process design, environmental innovation (Geffen and Rothenberg, 2000), etc. Social aspects: it's being socially responsible in terms of equal human development, ensuring health and safety of employees and contributing to humanity and the environment (GülçinBüyüközkan and ÇiğdemBerkol, 2011) (Beske et al. 2008; Spence and Bourlakis, 2009). This aspect can be improved only if there is dialogue and collaboration with several stakeholders including government, trade unions, non-governmental organizations (NGOs) and consumer groups.

Collaborative supply chain is a complex system that has a great influence to viability of an industry. Quantitative tools for analyzing complex systems like collaborative supply chain such as optimization, simulation (Herrmann et al. 2003,), and decision theory, have been used for some time to gather knowledge of system behavior. However, simulation of complex systems such as supply network contains a rich description of the system, which can clarify the interpretation of results and improve understanding or cause and effect relationships in a way that can be hard to obtain in optimization (Barnett and Miller, 2000).

The rest of the paper consists of five sections. After this introduction, a short review of the relevant literature helps to establish a link among collaboration, supply chain management, and sustainable development. In section 3, we describe the case study of spring mix supply chain. Section 4 presents the simulation of three scenarios modeling this supply chain. The first one is without collaboration; the second is with vertical collaboration and the third is with horizontal collaboration. We analyze simulation results according to strategic and tactical level. Finally, section 5 states some general conclusions and further research.

2 LITERATURE REVIEW

A supply chain is a set of firms acting to design, engineer, market, manufacture, and distribute products and service to end customers. Collaboration is about organizations and enterprise working together and can be viewed as a concept going beyond normal commercial relationship (Matopoulos et al. 2007). The term supply chain collaboration to refer to those activities among and between supply chain partners concerned with the cost effective, timely, and reliable creation and movement of materials to satisfy customer requirements (Muckstadt et al. 2001). However, what is sustainable supply chain collaboration and what are the benefits?

Environmental and social issues are becoming increasingly important themes on many businesses agendas. While much of the existing research on sustainability to date has been focusing on the question whether it pays off to be sustainable (e.g. Pagell et al. 2004), there is a growing consensus that this question is becoming irrelevant, because businesses will have to in one or another way deal with environmental and social issues (e.g., Kleindorfer et al. 2005; Corbett and Klassen, 2006).

In this context, companies cannot address sustainability agenda in isolation. They have to do this in conjunction with their external parties—i.e. customers and supplier, and adopt a sustainable supply chain management view (Carter and Rodgers, 2007; Pagell and Wu, 2009). Therefore, companies should collaborate together—i.e. simultaneously addressing economic, environmental and social aspect of sustainability across the whole supply network.

Supply chain collaboration enables firms to achieve above-normal returns via differential performance, as they can access and combine resources and routines that exists among a diverse members of a supply network (Dyer and Singht, 1998, Fawcett et al. 2012). Research by Facett et al. (2008), and Mentzer et al. (2008), shown that it is the identification and combination of complementary capabilities (e.g. knowledge, production and logistics capacities and information sharing

routines) that leads to superior performance and consequently to competitive advantage.

However in practice, Barratt (2001) and Fawcett et al. (2008, 2012) noted that few companies managed to achieve the high-level supply chain collaboration required for achieving differential performance.

In supply chain, there are different concepts of collaboration. (1) Dimensionality that refers to internal collaboration (various departments and functions within a company should work collaboratively) (Kahn and Mentzer (1998) and Pagell (2004)), and external collaboration (the integration between a focal company (e.g. a manufacturer) and its external environment (e.g. suppliers and/or buyers)) (Flynn et al. (2010)). (2) Selectivity that refers to a notion that companies should not collaborate with everyone (Lambert and Burduroglu, 2000). (3) Trust and relationship management means that The existence of internal as well as external trust is one of the key building blocks of supply chain collaboration. (4) Information sharing and IT systems integration (Cachon and Fisher, 2000; Chen et al. 2000; Gilbert and Ballou, 1999; Lee et al. 2000; Yu et al. 2001). (5) Joint planning and decision-making, such a joint forecasting and capacity planning (Lee and Whang 2000; Sabath and Fontanella 2002). (6) Finally, alignment of incentives and performance metrics.

In the literature, the collaboration is also seen as "green" (de Giovanni, 2009; Simpson et al. 2007; Vachon and Klassen, 2006), and as part of the CSR (Corporate Social Responsibility) (Perez-Aleman and Sandilands, 2008; Rasche, 2010; Argenti, 2004). Companies' initiatives are working in a collaborative manner with each other in order to improve the environmental performance and/or their social impacts within the supply chain and on external stakeholders. This often derives from external pressure.

(Matopoulos et al, 2007) detail the benefits of collaboration identified from the literature. Example of benefits: (1) less time searching for new suppliers and tendering, (2) lower stock holdings, increased asset utilization, (3) better quality following from involvement of supplier in design, (4) increasing product quality, minimize supply disruptions, (5) increasing responsiveness, (6) faster and flexible delivery, (7) more accurate forecast, joint resolution of forecast exceptions, and (8) improvements in lead times.

3 DESCRIPTION OF SPRING MIX SUPPLY CHAIN

The following case will compare 3 models of a spring mix supply chain. The first model is simple spring mix supply chain (no collaboration), the second model is vertical collaboration between manufacturer in main spring mix supply chain and retailers, and the last model is horizontal collaboration between grower 1 and grower 2. For each model, we simulate the supply chain at strategic level and capacity planning at tactical level.

3.1 Description

The following case will compare 3 models of a spring mix supply chain. The first model is simple spring mix supply chain (no collaboration), the second model is vertical collaboration between manufacturer in main spring mix supply chain and retailers, and the last model is horizontal collaboration between grower 1 and grower 2.

Simulations of the supply chain that will be created are based on the SCOR model. The simulation is about how to regulate the flow of products in the supply chain start from suppliers (growers) until received by end customer and measure the sustainable development indicators that related to all stage in supply chain.

In this case, first we considered a simplified supply chain with three suppliers, one manufacturer (factory), one distributor, and four retailers. The model of supply chain is based on an original supply chain layout that used in spring mix supply chain. Below is the mainstream supply chain.

The real system that is modeled in Figure 2 represents the spring mix supply chain. Retailer order finished products from the central warehouse (distributor) via order lists (process D1.2). Retailer orders are processed instantly and only once (process D1.3). Picking lists are generated and the stock levels of the corresponding products are updated. According to the order lists and the picking lists, the following processes are carried out: D1.9, D1.10, D1.11, and D1.12. Within the

process D1.12, shipment lists for the retailers are launched.

Parallel with these processes, at the central warehouse, purchasing of the finished product from the manufacturer is performed (S1.1). The process D1.2 decreases inventory of finished products, this process initiate process S1.1. Within the process S1.1 purchase lists are launched which correspond to manufacturer sales orders.

At the manufacturer, these orders are processed instantly. Now, processes D1.2–D1.12 are performed. Process D1.2 is carried out only once and after order receiving. Process D1.12 at the manufacturer, triggers process S1.2 at the central warehouse that will start after arrival of the products. Next, processes S1.2–S1.4 are executed at the central warehouse.

Process S1.2 generates the following document—records list. Update of the inventory level is done within process S1.4 after the product becomes available for sale. After each order processing, at the manufacturer, checking of preconditions for the start of the manufacturing processes is performed. If the condition for manufacturing is fulfilled, a working list is generated. The working list includes the quantity of finished products which should be produced and also required quantities of the components according to the bill of materials.

All these activities are performed within the process M1.1. The following actions are carried out within process M1.2: issuing of components against the working lists, launching of component requisition from the manufacturer storage, and adequate updating of the components inventory levels. Next, processes M1.3–M1.5 are executed, and after that, the products are available for sale (customer order processing). Each issuing of components initiates the process of component purchasing from the supplier (process S1.1). After that, processes S1.2–S1.4 are performed and then components become available for the production processes. The resume of each activity modeled below.

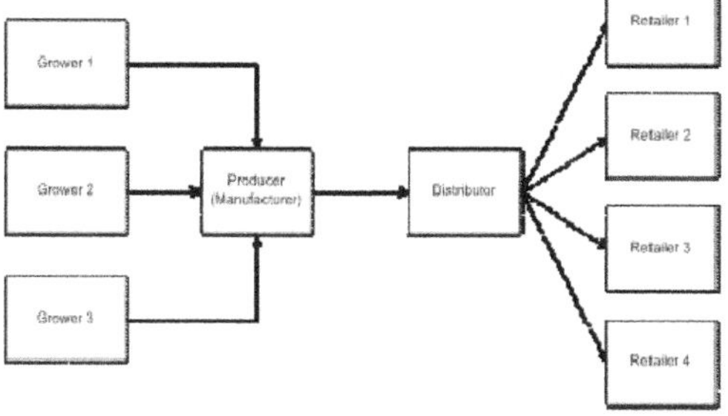

Figure 1.   Mainstream spring mix supply chain.

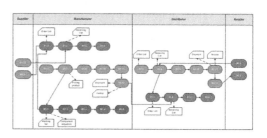

Figure 2.   Supply chain structure—UML activity diagram.

3.2 Assumption

We consider that the capacity of each vehicle that used in the distribution process is unlimited and the demand of each retailer is between 20 and 36 packs per month. Distances (miles) between different stakeholders ∈ [60, 170], velocity (mph) ∈ [40, 45], fuel use by vehicle (gallon) ∈ [10, 32] and Carbon per shipped (pounds) ∈ [200, 700].

Production capacity: 100 packs/day,
Supply capacity of supplier 1: 50 lbs/day,
Supply capacity of supplier 2: 70 lbs/day,
Supply capacity of supplier 3: 80 lbs/day,
Lot size distributor to factory = 50 pack,
   1 pack = 1 lbs.

3.3 Sustainable supply chain indicators

The sustainable indicators used in this case study according to the three bottom lines:

Environment (total carbon emissions from driving), economic (transportation cost, production cost, overtime cost, and revenue), and social (job satisfaction based on the overtime).

4 SIMULATION AND RESULTS

For each model (without collaboration; with vertical collaboration and with horizontal collaboration), we simulate the sustainable supply chain at strategic level and capacity planning at tactical level to optimize the three bottom lines.

4.1 Simulation of scenario

4.1.1 Tactical Level (TL)
Scenario 1: there is no collaboration with other supply chain outside the main supply chain. The model represents the spring mix supply chain without collaboration.

Scenario 2: there is a vertical collaboration between distributor and retailers. The collaboration is happen between members of supply chain whose has different level. The objective of this collaboration is to satisfy the needs and decrease the inventory cost caused by need to fill an order from retailer that usually not in lot size. This scenario implementing the theory of vendor managed inventory. Vendor-Managed Inventory (VMI) is a family of business models in which the buyer of a product provides certain information to a supplier of that product and the supplier takes full responsibility for maintaining an agreed inventory of the material. In this case, the distributor plays role as supplier and retailers as buyer. Retailers inform to distributor about their demand that become purchase order and then the distributor will calculate

the products that will be delivered. The distributor will give discount to their retailers as the reward of implemented the VMI model.

Scenario 3: In this scenario there is a horizontal collaboration between main spring mix supply chain and other spring mix supply chain. The collaboration is happen between members of supply chain whose has same level. The objective of this collaboration is to satisfy the needs and decrease the transportation cost. In this case, there is collaboration between supplier 1 and supplier 2. The case is supplier 1 and supplier 2 want to share transportation resource. The rule for this collaboration is when supplier 1 get an order and supplier 2 too, supplier 1 will send their raw material to supplier 2. Then from supplier 2 will be delivered to producer. The value to be analyzed is the transportation cost. The expectation from sharing transportation resources activity is the reduction of transportation costs, reduced fuel usage, and reduced air pollution from vehicle fuels.

4.1.2 Strategic Level (SL)
The objective here is to analyze the impact of decision in strategic level. The decision is to open a new factory which is located closer to the supplier 3, but have capacity production less than the main factory. There is additional cost for build a new factory that will be adding in initial cost of simulation.

Scenario 4: there is no collaboration.

Scenario 5: this scenario has the same objective with scenario 2. There are collaboration between distributor and retailers to implement the theory of VMI.

Scenario 6: this scenario simulates the horizontal collaboration. The objective is the same with scenario 3.

4.2 Results

We simulated scenarios according to the carbon emissions, the job satisfaction and the profit (euros) for short term (1 month), midterm (1 year), and long term (3 years).

Tables 1 and 2 show the results of total profit.

Table 1 shows the largest profit for supplier when we implement the horizontal collaboration. It's because between supplier 1 and 2 collaborate their utilization of vehicle, so the impact is reduction of the transportation cost and CO_2 emission, and increase the profit. For the other player the profit became high when we implement the vertical collaboration, the management inventory by distributor give sharing inventory cost between the distributor and the retailer. The retailer gets their safety stock and the distributor can reduce their inventory cause by the policy of lot size from factory.

Table 1. Results for total profit for a midterm (1 year).

Player	No collaboration	Horizontal collaboration	Vertical collaboration
Tactical level			
Supplier 1	7441.97	12328.73	7441.97
Supplier 2	9676.07	10199.30	9676.07
Supplier 3	3001.74	3001.74	3001.74
Factory	10567.62	10567.62	10567.62
Distributor	6230.12	6230.12	7277.094
Retailer 1	25889.50	25889.50	26407.29
Retailer 2	25398.34	25398.34	25533.43
Retailer 3	24863.52	24863.52	24931.08
Retailer 4	24066.90	24066.90	24189.43

Table 2. Results for total profit for a long term (3 years).

Player	No collaboration	Horizontal collaboration	Vertical collaboration
Strategic level			
Supplier 1	22325.91	36986.49	37696.49
Supplier 2	29028.21	30297.93	30297.93
Supplier 3	9005.22	9005.22	9005.22
Factory	27312.37	29678.37	35889.56
Distributor	4469.02	4659.45	4293.85
Retailer 1	18728.85	18728.85	93401.45
Retailer 2	77687.28	18728.85	19103.43
Retailer 3	16206.24	16206.24	16292.44
Retailer 4	74620.48	74620.48	74823.25

The impact of build a new factory is improvement in the supply chain's profit but in the long term horizon time. Because there is an initial cost for building the factory and for long term the factory can take more order because they have more production capability.

5 CONCLUSION

This paper Build an integrated understanding of collaboration sustainable supply chain and the main benefits of collaboration.

We modeled and simulated a case study; comparing three scenarios of spring mix supply chain using ARENA simulation software.

The scenarios in (SL) outperforming scenarios in (TL), because the scenarios in (SL) produce a fairly large product and reduce the overtime cost. Based on the result of profit in all players, vertical collaboration gives the largest profit. Implementation of vendor manage inventory provide the improvement of profit because there are sharing inventory cost between distributor and retailer that favorable for both. However, horizontal collaboration is the best way to implement, because sharing

vehicle not only reduces CO2 emissions but also transportation cost.

From an environmental standpoint, scenarios in (TL) generate less CO2 emission to air than scenarios in (SL). The horizontal collaboration gives better results because sharing vehicle not only reduces CO2 emissions.

From social standpoint, the decision to open a new factory reduces the overtime cost and so the overtime work hour. Therefore labors do not need to work over to fill the over order.

ACKNOWLEDGEMENT

This work was supported by INTERREG IVB NWE program: SCALE project. This support is gratefully acknowledged.

REFERENCES

Argenti, P.A. (2004). *Collaborating with activists: how Starbucks works with NGOs*. California Management Review, 47(1), pp. 91–116.
Barnett, M.W., & Miller, C.J. (2000). *Analysis of the Virtual Enterprise Using Distributed Supply Chain*

Modeling and Simulation: An Application of e-SCOR. Paper presented at the Proceedings of the 2000 Winter Simulation Conference.

Barratt, M. and Oliveira, A. (2001). *Exploring the experiences of collaborative planning initiatives.* International Journal of Physical Distribution & Logistics Management, 31 (4), pp. 266–89.

Beske, P., Koplin, J., Seuring, S. (2008). *The use of environmental and social standards by German first-tiersuppliers of the Volkswagen.* AG.Corporate Social Responsibility and Environmental Management 15 (2), pp. 63–75.

Büyüközkan G., Berkol C. (2011). Designing a sustainablesupplychainusing an integratedanalytic network process and goal programmingapproach in quality function deployment. Expert Systemswith Applications, 38, pp. 13731–13748.

Cachon, G.P., Fisher, M. (2000). *Supply chain inventory management and the value of shared information.* Management Science 46 (8), pp. 1032–1048.

Carter C & Rogers D (2007). *A framework of sustainable supply chain management: Moving toward new theory.* International Journal of Physical Distribution & Logistics Management, 38(5), pp. 360–387.

Chaabane A., Ramudhin A., (2012). *Paquet M., Design of sustainablesupplychainunder the emissiontradingscheme.* International Journal of Production Economics, 135 (1), pp. 37–49.

Chen, F., Drezner, Z., Ryan, J.K. and Simchi-Levi, D. (2000). Quantifying the bullwhip effect in a simple supply chain: the impact of forecasting, lead times, and information. Management Science, 46(3), pp. 436–443.

Dyer, J.H & Singh, H. (1998). The relational view: Cooperative strategy and sources of interorganizational competitive advantage. Academy of Management Review, 23, pp. 660–679.

Fawcett, S.E., Fawcett, A.M., Watson, B.J. and Magnan, G. (2012). *Peeking Inside the Black Box: Toward an Understanding of Supply Chain Collaboration Dynamics.* Journal of Supply Chain Management, 48, pp. 44–72.

Gilbert, S.M., & Ballou, R.H. (1999). *Supply Chain Benefits from Advanced Customer Purchase Commitments.* Journal of Operations Management, 18 (1), pp. 61–77.

Giovanni D., (2009). *A differential game of knowledge accumulation and management.* Proceeding of the 10 Knowledge Management, 1.

Herrmann, J.W., Lin, E., & Pundoor, G. (2003). *Supply Chain Simulation Modeling Using the Supply Chain Operations Reference Model.* Paper presented at the Proceedings of DETC'03, Chicago, Illinois.

Kahn, K.B., Mentzer J.T. (1998). *Marketing's integration with other departments.* Journal of Business Research 42, pp. 53–62.

Lambert, D.M., Burduroglu, R. (2000). *Measuring and selling the value of logistics.* The International Journal of Logistics Management, 11 (1).

Lee, H.L. and Whang, S. (2000). *Information sharing in a supply chain.* International Journal of Technology Management, 20 (3), pp. 373–87.

Lee, H.L., So, K.C., Tang, C.S. (2000). *The Value of Information Sharing in a Two-Level Supply Chain.* Management Science, 46(5), pp. 626–643.

Matopoulos, A., Vlachopoulou, M., Manthou, V., & Manos, B. (2007). *A conceptual framework for supply chain collaboration: empirical evidence from the agrifood industry.* Supply Chain Management: An International Journal, 177–186.

Mentzer J.T., T.P. Stank, T.L. Esper, (2008). *Supply Chain Management and its Relationship to Logistics.* Marketing, Production and Operations Management, Journal of Business Logistics, 29(1), pp. 31–46.

Muckstadt, J.A., Murray, D.H., Rappold, J.A., & Collins, D.E. (2001). *Guidelines for Collaborative Supply Chain System Design and Operation.* New York: School of Operations Research And Industrial Engineering College of Engineering Cornel University.

Pagell M, Wu Z (2009). Building a more complete theory of sustainable supply chain management using case studies of 10 exemplars. Journal of supply chain management, 45(2), pp 37–56.

Pagell, M. (2004). Understanding the factors that enable and inhibit the integration of operations, purchasing and logistics. Journal of Operations Management, 22 (5) pp. 459–487.

Perez-Aleman, P., & Sandilands, M. (2008). Building Value at the Top and the Bottom of the Global Supply Chain: MNC-NGO Partnerships. California Management Review, 51(1) 24–49.

Rasche, A. (2010). *Collaborative Governance 2.0, in: Corporate Governance.* The International Journal of Business in Society, 10(4), pp. 500–511.

Sabath, R.E., Fontanella J., (2002). *The unfulfilled promise of supply chain collaboration.* Supply Chain Management Review, 6(4), pp. 24–29.

Senge, P., Lichtenstein, B., Kaeufer, K., Bradbury, H. & Carroll, J., (2007). *Collaborating for Systemic Change.* MIT Sloan Management Review, 48 (2) pp. 44–53.

Simpson D, Power D, Samson D. (2007). *Greening the automotive supply chain: a relationship perspective.* International journal of operations & production management, 27(1), pp 28–48.

Spence, L., Bourlakis, M. (2009). *The evolutionfromcorporate social responsibility to supply chain responsibility: the case of Waitrose.* Supply Chain Management: An International Journal 14 (4), pp. 291–302.

Spence, L., Bourlakis, M., (2009). *The evolutionfromcorporate social responsibility to supplychainresponsibility: the case of Waitrose.* Supply Chain Management: An International Journal 14 (4), pp. 291–302.

Vachon S, Klassen RD (2008). Environmental management and manufacturing performance: the role of collaboration in the supply chain. International journal of production economics, 111, pp 299–315.

Vachon S, Klassen RD, (2006). Extending Green Practices Across the Supply Chain, The impact of upstream and downstream integration, International journal of production economics, 26(7), pp 795–821.

Wanga F., Laib X., Shi N. (2011). *A multi-objective optimization for green supply chain network design*, Decision Support Systems, 51 (2), pp. 262–269.

Welford R., and Frost S. (2006). *Corporate social responsibility in Asian supply chains.* Corporate Social Responsibility and Environmental Management, 13(3), 166–176.

Yu, Z., Yan, H. and Cheng, E. (2001). *Benefits of Information Sharing With Supply Chain Partnership*, Industrial Management & Data Systems, 101 (3), pp. 114–119.

Green Design, Materials and Manufacturing Processes – Bártolo et al. (eds)
© 2013 Taylor & Francis Group, London, ISBN 978-1-138-00046-9

Collaborative actions, technology and smart design

I.S.L. Xavier & P.J. Silva Junior
Universidade Federal Fluminense UFF, Niterói, RJ, Brazil

ABSTRACT: The future of civil construction industry in Brazil is directly connected to the adoption of a new industrial logistic and the enhancement of collaborative actions allied to sustainability. We know that this innovation has a direct relation with the economic, historical, business and cultural moment of the country. In being so, innovation will happen when these variables are capable to produce competitive advantage and increased productivity. We live in a moment of impasse, when there is no longer place to improvisation or "re-work". The low quality of projects and the lack of compatibilization are still the main problems of our civil construction. To work while thinking and to think while working is one of the most relevant aspects in the use of technology and in its application on architecture. The BIM modeling can allow the development of new actions of information on how the elements are constructed (parameterized geometries) and technical-constructive performance, including sustainability. The most targeted desire is that AEC industry in Brazil will change the constructive chain, leaving the low process of conventional production behind, to become agile, efficient, economical, profitable and with quality: make more with less. In our recent market, it is difficult to deal with little quality of projects and their compatibilization and with manpower (disqualified workers and lack of them). The adoption of new technologies necessarily comes from the complete reformulation of the design process conception—the adoption of smart design. Our current model cannot evoke new techniques or technologies. It is expensive, heavy and slow and wastes physical, human and financial resources. Companies must rethink the process of project conception, the use, maintenance and destination of final waste. It is a challenge that goes through planning, technological development and cultural changing.

1 COLLABORATIVE ACTIONS

1.1 Benchmarking and collaborative actions

Nowadays it is a consensus that groups of professionals and companies share knowledge about problematical situations in order to find better solutions. (Boxwell, 1994). This sharing of knowledge through benchmarking is crucial to the development of indicators, allowing performance comparison which will show the best practices of design process, management and sustainability actions.

Benchmarking collaborative groups might happen between companies as well as between individual beings. In the process of designing, by example, through internet groups and social networks, the company or team work may receive suggestions for the enterprise's name and useful ideas for design, sustainability, and technology. Afterwards, these collaborative actions handle information to the creation of the infrastructure: efficient systems to the conception of smart design, like the use of local material, water reuse and construction waste recycling. Therefore, collaborative groups develop knowledge, generate innovation and make the process clear, establishing an open and trustful environment with equality for experiences exchanges. This sort of group allows commitment between its participants.

2 THE AEC INDUSTRY IN BRAZIL

Beside the design technologies, Brazilian constructive system, focused on multifamily housing, experienced great changes over the past 20 years: company's size was reduced, several barriers were established in the entrance of the industry and the period was marked by strong competition, with great reduction of profit margins, difficulties in financing and high interest rate.

Since the establishment of PAC (plano de aceleração do crescimento), a Brazilian governmental plan, the market's behavior was marked by made-to-order buildings and by real state incorporation (pre-sales housing market) with high competition influenced by the plan. The funding offer and the decline of interest rates set a new scenario and a new economic order, with clear definition of market prices for conventional products.

These issues induced developers to adopt a new economic sustainability for the sector with the releasing of products more appropriated to consumers (the market as the great definer of the prices). In order to cooperate with these new demands, companies have had to increase their actions of planning and construction management, besides constructive optimization process and quality programs.

In order to answer these new demands, companies are seeking certification processes and outsourcing of the most necessary activities for production. This search is still based on the adoption of incipient technologies and on the improvement of technology insome areas of the constructive process, still based on the main use of concrete cast in loco. The main goal is to reduce the costs, benefiting the internal customer (IIR).

Some companies have realized that the adoption of a more conscious activity at the stage of design conception (design quality improvement) increases the value of the products from the user point of view, and not only the economic price of the product, privileging the final user and attending the requirements of constructive technical and building performances. (Koskela, 2000; ISO – 6241; ABNT, 2000; Foliente, 2005).

What happens, however, despite the desire of pursuing these new requirements, is that Brazilian construction industry is conservative and slow in innovation processes. The owners aren't willing to change this reality yet. Companies are economically weak, with no political power, low-tech, low competitive intelligence, low power to make profits and short time of performance in market.

These companies haven't much experience in mechanization, so their production is subjected to changes in the weather. They have low productivity and high levels of accidents plus much improvisation on site. There are also performance problems with accuracy, quality and losses (human, material and financial), besides under skilled and temporary manpower that work on dirty, dangerous and tiresome environment work.

Another relevant aspect is the lack of social commitment: the non-preoccupation with social, urban and environmental consequences. Companies are focused only inbuilding new units in short time, sonormally the products are expensive, with low quality and high level of inadequacy.

The desired future through collaborative actions, technology and smart design is innovative, flexible and sustainable construction. For this model it is necessary the adoption of actions that reduce losses and the design focused on industrialization with the use of ICT (information and communication technology). Besides environmental sustainability,

there must be incorporated social and economic sustainability, which generate good work conditions and career plans, capable of attracting more qualified professionals.

This dichotomy between the reality and the desired future has as main cause the lack of dialogue between companies, universities, government and professionals. This conflict is the cause of many problems, as the high consume of material and manpower and the high production of waste along the building's circle of life. These problems could be reduced if greater efforts were dedicated in the beginning of the design process. Experience shows that some failures are more expensive and difficult of being solved after this period of planning and conception is over.

3 ADOPTION OF NEW TECHNOLOGIES

3.1 Misuse of technology

New design technologies based on CAD, help architects to produce more precise designs. However, they still produce problems related to the compatibilization of design (rework and waste at construction period). These problemspoint to the misuse of software in the process of design. As stated by Richard Sennett (2012) "How could such a useful tool be misused? (...) What we conclude from this misapplication of CAD is that what appears on the screen provides an impractical coherence, composed in a way that never occurs in the physical view."

Designing in CAD can reveal some basic problems that result from the misuse of software: difficulties in understanding the different phases of a design, the limited view of the design on the screen and the effects of the zoom that hide mistakes and design failures. Another relevant fact is the disconnection between the proportions that are represented on screen to the designer, because of the use of different scales, and that will never be substituted by someone on site, which means that there are solutions on screen that will never be verified on the sight of someone on site.

The difficulties of civil construction point to mistakes of design as the great villain of incompatibilities between the different sorts of designs (architectural-structural-infrastructure) that produce rework and delay in building chronograms. The origin of this kind of error is due to the lack of knowledge of the potentialities of a tool so useful and yet, so misused. This is a big problem of capacitation that starts at college graduation.

The use and teach of CAD reflects recent university graduation. Few colleges make an effort to pass a deeper knowledge of the program to the students. Many times not even teachers dominate it. The rising of new other 3D designing tools too make it difficult to know which one is the best option in the process of graduating the student into a professional. Not even the professionals have a consensus about this. Most students that dominate any designing software have gained that knowledge by themselves or at stages.

3.2 Change of technology Culture

Civil construction in Brazil must be allied to an industrial logistic. The constructive model "modeled in-loco" must give way to new technologies of production and for setting the building up, with software of management associated to smart design. To simply put the pieces together on site and make a building has got to become a reality, for it is no longer admitted improvisation that leads to a rework which wastes time, material and, therefore, money. Yet, the overheating of the market and the imposition of each time shorter deadlines push the conventional construction, filled with pathologies, into changes. Complains about deficiencies on delivered buildings and on the delay of the deliverance of them point to the frailty of the present model of construction. The actions focused on quality management system, TIC (communication and information technologies), processes of production and adoption of new constructive techniques may add quality to the product, diminishing construction pathologies.

3.3 The manpower problems

The reality of civil construction nowadays reveals an arising on the difficulty to find manpower to work on site, for the same manpower is disputed by different sectors. The existing model isn't capable of providing better payments, besides the inexistence of incentive programs and of guaranty of maintenance of the workers in possible moments of crises. Manpower blackout might be overcome with productivity, training, planning and awareness of businessman and trade associations. Insome building phases, workers are better paid by civil construction than by factory, but this is a very punctual aspect. The dispute for manpower is so that, being qualified or not, there is lack of workers.

4 BIM TECHNOLOGY AND ITS APLICATION IN AEC

4.1 Design process quality

BIM (Building Information Modeling) is a technological innovation that modifies the process of building in civil construction and its product (Andrade & Ruschel 2011). BIM proposes the integration in the development processes of design through the collaboration and the multidisciplinary work of the team (stackholders). By a change in the paradigm of the design process, all people involved in it work in a single virtual model that contains all information (geometrical and non-geometrical) that characterizes the building, with significant reduction of time, costs and rework, and adding quality in the process and in the product presented. (Eastman et al. 2011).

According to Crespo and Ruschel (2007), BIM is a digital model composed by a data bank that allows adding and producing information in the virtual model of the building for various purposes and with a gain of productivity and rationalization in the process. It is the development and use of a computational model to simulate the construction and operation of an enterprise. (Nascimento et al. 2011) The model is a representation of the building with parameterized elements of which it is possible to obtain and manipulate technical and physical information that help in the decisions on the design, on the maintenance and on the operation of the building.

4.2 Difficulties on implementation

As any other technological information, there are difficulties in the implantation of BIM in the whole AEC chain. Some of them are:

a. Costs on implantation—the price of the software and license, besides indirect costs generated by the low initial productivity during the process of training the staff.
b. Resistance by professionals—BIM represents a change in the paradigm of the design process and so finds resistance among professionals already familiarized with the traditional processes (Senior's culture) and that are not willing to learn new design processes. There is also a natural resistance to any kind of innovation. For Junior & Amaral (2008), construction industry has a great resistance by professionals in assuming risks on uncertainties in implanting technological innovation.
c. Training staff—training professionals on the use of BIM means new investments, which might be another barrier on the implantation of this new technology. Companies must invest on the design staff with time, resources, physical space, etc. However, businessman are not always willing to invest in this, even that it means possibilities of future return.

d. Latest technology—for BIRX (2006), the fact that BIM is a new technology implies that softwares are still being improved. The fact that the chain of designs and other consulters and collaborators still haven't adhered to the innovation implies on the need of trained professionals for the market and a long transition period.

4.3 *The software interface and interoperability*

BIM bases its design process on the interoperability, which is the capacity of exchange of information of the model between different softwares and disciplines. Interoperability in BIM proposes sharing information through archives in IFC format. IFC is a model that has been elaborated since 1994 by IAI (International Alliance for Interoperability), now called Building Smart International and approved by ISO (International Standardization Organization) that facilitates the interchange of architectural and construction information between softwares with intelligence based on the object—like BIM. (Oliveira, 2011; BIAGINI, 2007). Interoperability is the ability to exchange data between applications, which smoothes workflows and sometimes facilitates their automation. (Eastman et al. 2011).

Interoperability guaranties that various experts on design in different phases of the process add information independently on the software that is being used or the phase of the enterprise. The virtual model of the design on BIM concentrates various objects as well as parameterized information.

4.4 *Noise about creative process an new technologies*

The implementation of a BIM design process suggests a change on the steps of the traditional process. It is believed that this change is only going to happen when new methodologies of architectural design incorporate the dilution of the rigid division between the phases of the conception and development, seeking to strengthen the architectural conception. (Bulhoes, 2012; Florio, 2007; Eastman, 2011).

Today's paradigm among designing professionals e those that perform the work of building is the distance between the diverse knowledge. While some know about design tools (CAD and BIM), others dominate the knowledge necessary to make the building come true. This feedback of knowledge is a great challenge to the professional reality. It is important to establish an exchange of information between those that dominate tools of design and those who dominate construction techniques.

4.5 *Sustainability actions*

One of the advantages of BIM is the possibility of simulations while designing. BIM enables architects and engineers to use digital information to analyze and understand their projects' performance before they are built, developing and evaluating multiple alternatives. At the same time, it enables easy comparison and informs better sustainable design decisions.

Design decisions made in the beginning of the process can deliver significant results when it comes to the efficient use of vital resources. Employing sustainable analysis tool helps architects and engineers to make more informed decisions earlier in the design process and enable them to have a greater impact on the efficiency and performance of a building design. (Autodesk, 2010).

BIM provides superior ways to measure simulation and analyze performance, with its integrated data of building & service system elements, GIS context and definition of human and related activities (Mitchell, 2011). With this anticipation, designers can make better decisions and use a smart design to build. BIM encourages trying better solutions in design process.

5 CONCLUSION AND RECOMMENDATIONS

The big challenge in our days is the production of sustainable buildings with decreasing costs, reduction of deadlines without losing quality, technological innovations of design and smart design. For these characteristics, there should be expected:

– Opened professional environment, with equality and with collaborative actions;
– Planning attitude. Investments in technology, incorporation of actions related to sustainability and TIC for adapting to digital revolution;
– Smart design through new designing technologies associated to the resolution of faults, imprecision, incompatibility and haggard orientations when compared to the digital model;
– Digital enhancements to minimize and avoid rework in the production caused by the misuse of design technology;
– Improvement and made current the practice of simulation to anticipate solutions for the design and evaluating environmental, technical and constructive performance.

This action starts since the conception of the design, passing through the construction of the building, operation and maintenance of it, till recycling and discarding waste. It doesn't mean just the reuse of the rain's water or implementing solar plates, etc. Today's challenge is wider, involving implantation of systems supported by new technologies to the development of design and the transformation of the model 'modeled in-loco' to 'pre-modeled'.

Therefore, our design, laws, culture and demands should evolve at the same time. The development of new products, technology and capacity is the true technological revolution that must have too a cultural revolution in the development of the country.

BIBLIOGRAPHY

Autodesk, Sustainable Design Analysis and Building Information Modeling. Autodesk Ecotect Analysis 2010. In Autodesk.com, 2010.

Andrade, m.; Ruschel, r. Building Information Modeling (BIM). In. O processo de projeto em arquitetura: da teoria à tecnologia/ Doris C.C.K. Kowaltowski, Daniel de Carvalho Moreira, João R.D. Petreche, Marcio M. Fabricio (orgs.) São Paulo: Oficina de textos, 2011.

Boxwell, Robert (1994), Benchmarking for a Competitive Advantage, McGraw Hill, 1994.

Bulhoes, M.C.S. Impactos do Building Information Modeling (BIM) no Processo de Projeto. Dissertação (Mestrado), Universidade Federal Fluminens FF, Niterói-RJ, Brasil, 2012.

Eastman, Chuck et al. BIM Handbook: a guide to Building Information Modeling for Owners, Managers, Designers, Engineers and Contractors. New Jersey: Jonh Wiley & Sons, 2011.

Koskela, Lauri (2000): An exploration towards a production theory and its application to construction, Espoo, VTT Building Technology. 296 p. VTT Publication, 2000.

Mitchell, John. Bim & Building Simulation. In Proceedings of Building Simulation 2011:12th Conference of International Building Performance Simulation Association, Sydney, 2011.

Sennett, Richard. O Artifice. Ed Record. 2012.

Foliente GC et al. Performance Based Building R&D Roadmap. CIB, Rotterdam, The Netherlands, 2005.

Florio, Wilson. Contribuições do Building Information Modeling no Processo de Projeto em Arquitetura. In: III ENCONTRO DE TECNOLOGIA DE INFORMAÇÃO E COMUNICAÇÃO NA CONSTRUÇÃO CIVIL, 3, 2007, Porto Alegre. Anais ... Porto Alegre—RS, Brasil, 2007.

Green Design, Materials and Manufacturing Processes – Bártolo et al. (eds)
© *2013 Taylor & Francis Group, London, ISBN 978-1-138-00046-9*

Eco-efficiency in six small and medium enterprises in Mexico

L. Iturbide, P. Lozoya & P. Baptista
Universidad Anáhuac México Norte, Mexico

ABSTRACT: This paper documents the process and results of six small manufacturing firms in the Valley of Mexico, which participate in an eco-efficiency program designed by the government agency SEMARNAT (Ministry of the Environment and Natural Resources). The objective of this SEMARNAT program is to become a leader in helping small and medium businesses achieve competitive advantages through mechanisms of environmental management with emphasis on waste control, energy and water efficiency. Each industry prepared its eco-efficiency project, seeking economic gains and ecological efficiency. This study explores the advantages of implementing the SEMARNAT methodology by evaluating the results of the six participating industries.

1 INTRODUCTION

The environmental challenge has acquired greater relevance among the general public, because resources are depleting and the quality of life is dangerously threatened. A vision in which nature is seen as in inexhaustible source of resources and a waste disposal must transit towards corporate responsibility and sustainability. Companies need to understand that *doing business as usual* has no place in today's world, and governments must present an appropriate legislation for this transit to happen. In this context there is a pressing need to integrate an eco-efficient perspective into Small and Medium Enterprises (SMEs). In Mexico, micro, small and medium enterprises constitute more than 90% of all business units, constituting 72% of employment, and contributing 52% of the GNP (INEGI, 2009). SMEs, (small and medium enterprises) represent 33% of business units in México, and tend to ignore the environmental challenge, believing that the adoption of eco-efficiency measures is a costly investment and *out of their league.*

2 THEORETICAL FRAMEWORK

2.1 *Background information*

The current debate on climate change has two main strands: adaptation or mitigation. The first rests on the premise that damages to the ozone layer are irreversible, and that the impact of extreme forces of nature have always had their effect, hence the current challenge is merely an adaptation to new scenarios. The second strand presses for the cleaning and preservation of the environment, claiming that pollution is reversible through voluntary and costly mechanisms that should to be paid by the polluter. It is said that voluntary actions in favor of the preservation of the environment are commendable, but a system of penalties and rewards based on national and multinational environmental policies are fundamental. The present suggestion is not only compliance to the regulatory framework, but the surpassing of it. Thus, this second perspective contemplates the implementation of financially viable mechanisms for all countries, through market-based solutions, as suggested by Stavins (2009).

Much of the accumulation of CO2 in the atmosphere is due to emissions in industrialized countries. However, the emissions attributed to developing countries increase each day, as noted by Fullerton since the nineties (1998). The bottom line is that climate change is a social problem which requires the participation of every individual, business and country. Every year that redirection toward a sustainable path is delayed, will increase by the billions the cost of corrective actions in the future. The way of doing business nowadays has changed and this demands an outlook based on profitability with social responsibility and sustainability.

2.2 *Mexico's position*

Mexico is twelfth in the emission of greenhouse gases (WRI, 2007), (WCSC 2007), which is why in recent years many programs and laws have been created for environmental protection. The Mexican Government has implemented specific policies to achieve sustainable development. In this sense, certain national territories have been declared to be protected biospheres. Furthermore, action has been taken in favor of wastewater treatments, waste management and reforesting for the conservation of ecosystems. Additionally, Mexico has acquired international commitments at summits such as the Kyoto Protocol (Japan, 1997) whose main objective was to reduce carbon emissions by 5%. Also, Mexico is committed with the Special Climate Change Program (Gallegos, 2011) to reduce greenhouse gas contamination by 8% per year. The private sector has expressed discontent with Mexico's international commitments. They argue that more environmental regulations harm their profits. This goal of reducing carbon dioxide emissions by 8% implies the reduction of 640 million tons per year, which is considered too costly for the industries involved. They claim (CCE, 2009) that such drastic reduction in emissions would threaten the country's economic growth. Companies directors avoid committing because they say that, compared to the rest of the world, Mexico's emissions are insignificant. This is an important motive for creation of the Environmental Leadership for Competitiveness Program by SEMARNAT. The following section presents the methodology of this program, followed by the results obtained by 6 small companies in Mexico.

3 METHODOLOGY

3.1 *Environmental leadership for competitiveness program by SEMARNAT*

The methodology of this program was designed to create specific and differentiated solutions for companies, depending on their capabilities. It helps to exploit opportunities and neutralize risks, allowing companies to implement their own project, according to their challenges and possibilities. It is especially aimed at SMEs that want to improve their competitiveness in value chains, through environmental management with emphasis on eco-efficiency, 450 SME's in the country were invited to enter the program.

This study was done by Anahuac University Business Accelerator during 2009–2010 based on the results of six small firms. It provides a documented answer to the research question that enquires about the efficiency and feasibility of this program, exploring the possibility of its adaptation by similar industries.

3.2 *Characteristics of participating industries*

Table 1 shows the characteristics of the 6 participating industries in the study: Type of industry and its size are expressed in number of employees and annual sales in pesos. This sample has 5 manufacturing small industries and one medium company (a gas LPG distributing company). In the last columns *critical points* and *possible competitive factors*, are those identified by the six small company directors in Module 1 of the intervention.

3.3 *The intervention*

Six Mexican SME directors voluntarily participated in the Environmental Leadership for Competitiveness Program which consisted in 2 parts. The first provided the conceptual tools necessary for the development of eco-efficiency projects, such as the eco-map, value stream map, eco-balance, life cycle analysis, matrix MED, risk analysis and environmental auditing. With these conceptual tools the entrepreneurs themselves defined the company's standing faced to this issue, and identified environmental problems in aspects of planning, management and verification, among others. In this 40-hour workshop, participants applied eco-indicators to calculate costs of their inefficiency. After completing this workshop each director produced at least one eco-efficiency project. In the second part, each company implemented their eco-efficiency project. This phase was benefited by the support of consultants, both present and through a website. The objective was to implement the project that had been designed in the first part

Table 1. Characteristics of participating industries.

Industry	Size		Identified factors	
	Staff	Sales*	Critical	Competitive
1. Aluminium railings	5	5,422	Bad lighting	Price, image, raw materials
2. Plastic packaging	344	72,979	Material waste	Tech, designs logistics
3. Cardboard & paper	119	90,327	Energy costs	Innovation, recycling
4. LPG	129	237,427	Fuel costs	Quality service
5. Scafolds & construction	27	10,000	Paint & energy	Competitive prices
6. Light fixtures & bulbs	12	N/A	Energy costs	Customized designs

*Annual sales in million pesos (1US dlls = 13.20 pesos approximately).

of the program and to measure advances with the eco-indicators.

The organizational performance of the companies was measured in economic and environmental dimensions. Water and energy consumption, as well as waste, were the main indicators evaluated before and after the intervention. The economic part was accounted as the total payment in pesos made for the consumption of water and energy. The environmental part for the energy indicator was measured as the number of kilowatts per hour consumed by the companies. The water indicator was accounted for by the number of liters consumed. And waste reduction was measured by the number of kilograms of waste decrease for each company in the study.

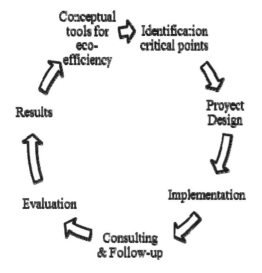

Figure 1. Eco-efficiency Project Cycle (Paola Lozoya (2010).

4 RESULTS

Table 2 summarizes the gains obtained by each company after applying their own eco-efficiency project. It condenses the solutions which were proposed to tackle the critical points previously identified. In the same column, ROI (return of investment) is conveyed in the number of months the company had to wait in order to perceive economic gains after implementing their eco-efficiency projects. The next column represents the investment each director made or spent on the solutions. In the next, the economic gains, expressed in annual savings for the each company, are indicated. The last column indicates the environmental gains.

The six participating industries emit annually 101.77 tons of CO_2 less than before implementing the program. They also saved 28,618 liters of gasoline and 122.95 tons of garbage waste per year. Their energy savings amount to 59,767.48 kWh, which suffices to provide 27 families with electric energy for a whole year.

Table 2. Economic and environmental gains.

Industry	Solutions & ROI**	Investment*	Savings*	Eco-efficiency
1. Aluminium railings	Leds' bulbs 7.02	1754	3,276	1,450 KwH & 0.95 tons of CO_2
2. Plastic packaging	Cutting molds & recycling 29	1000000	814700	122.95 tons of waste reduction
3. Cardboard & paper	Ceiling Domes 9.2	116,256	106236	52,452 Kwh and 34.29 tons of CO_2
4. LPG	Gasoline substitution (LPG) 9	43,400	58,279	25,018 liters of gasoline & 54.79 tons of CO_2
5. Scafolds & construction	Training courses, HVLP painting tools 4.3	10,743	57,200	3,600 liters of gasoline, 7.88 tons of CO_2 and 0.72 tons of paint
6. Light fixtures & bulbs	Leds' bulbs 15.3	7,820	20,599	5,865 KwH & 3.86 tons of CO_2

*In pesos (1US dll = 13.20 pesos aprox.).
**In months.

5 CONCLUSIONS

These economic and environmental gains demonstrate that the Environmental Leadership for Competitiveness Program produced great benefits for the participating industries. All implemented innovative methods which improved in the efficient use of water, heat and electricity, as well as waste disposal.

Industries also developed capabilities for continuous economic and environmental improvements. As they increased their efficiency by reducing costs and emissions of greenhouse gases, their reputation with stakeholders increased as they were recognized to be environmentally responsible companies. The latter promoted their growing and market permanence. The results speak for themselves. They demonstrate the scope and impact of a well-focused public policy, such as the Environmental Leadership Program for Competitiveness. This can also be described as an experience in dialogue between the public and private sectors which increased confidence and enthusiasm in directors of SMEs. It is expected that participating companies will continue to advance in eco-efficient strategies.

REFERENCES

Consejo Coordinador Empresarial (2009) Cambio Climático: Visión del Sector Empresarial. CCE, México, D.F.

Gallegos, R (2011) Evaluación especial del Programa Especial del cambio climático, ECCP. Documento del Instituto Mexicano de la Competitividad 2009–2011e ICI del Ministerio Federal de Medio Ambiente, Protección de la Naturaleza y Seguridad Nuclear del Gobierno Alemán. México.

Fullerton, D. (1998). How Economics See the Environment. *Nature*, 395(2): 433–434.

INEGI (2009) Instituto Nacional de Estadística y Geografía, México. www.inegi.org.mx.

Kyoto Protocol (1997). United Nations Framework Convention on Climate Change. United Nations, 1998. Document available http://unfccc.int/resource/docs/convkp/kpeng.pdf.

Lozoya, P. (2010) Ganancias en eficiencia de PYME's a través del ahorro de agua, energía y reciclaje. Tesis de Licenciatura en Economía. Facultad de Economía y Negocios. Universidad Anáhuac México Norte.

Stavins, N.R. (2009). Can countries cut carbon emissions without serious damage to their economies?, in *World Economic Outlook*, International Conference. HIS Global Insight, Boston, Ma. EE.UU. October 21, 2009.

SEMARNAT (2011). Lineamientos para el Programa de Liderazgo Ambiental para la Competitividad. Secretaria del Medio Ambiente y Recursos Naturales México, 10 de enero 2011. www.semarnat.gob.mx/apoyossubsidios/liderazgo/.

WCSC (2007) Promoting Small and Medium Enterprises for Sustainable Development. Document published in collaboration with The SNV Netherlands Development Organization. World Council for Sustainable Development. Geneve, Switzerland.

Green Design, Materials and Manufacturing Processes – Bártolo et al. (eds)
© 2013 Taylor & Francis Group, London, ISBN 978-1-138-00046-9

A method to select best nuggets from eco-innovation sessions

F. Vallet
UTC Roberval Laboratory UMR CNRS 7337, Compiègne, France

B. Tyl
APESA, Innovation Department, Bidart, France

D. Millet
SUPMECA Toulon LISMAA EA 2336, France

B. Eynard
UTC Roberval Laboratory UMR CNRS 7337, Compiègne, France

ABSTRACT: For the last decades, designers have been facing a new challenge: design products and services with lower environmental impact and greater value to customers and society. To meet these expectations, the environmental assessment of solutions in design teams must be relevant and commonly accepted. This is especially true in the early phases of the Product Development Process when details of the concepts are not yet defined. Literature tends to prove that evaluations of ideas vary substantially, meaning that designers do not have the same understanding of ideas or criteria. To organize a more efficient assessment stage and to obtain a reliable assessment, it is necessary to implement new procedures. The subsequent research question is: what are the relevant factors to assess early design concepts from the environmental perspective and create an adhesion to promising ideas (called 'best nuggets') in a design team? To do so an experiment was conducted with four groups of eco-design experts. They were asked to evaluate the environmental relevance of the design concepts generated during two previous eco-design sessions. Some criteria were common to all groups of designers, such as the potential reduction of environmental impact on each life cycle stage, and the potential reduction of each type of environmental impact (i.e. resource consumption, toxicity, etc …). Besides, specific evaluation criteria and their effect on the group adhesion were analyzed: the risk of an environmental impact transfer and the potential positive influence on user. Some results and recommendations are presented in this paper.

1 INTRODUCTION

For the last decades, designers have faced a new challenge: design products and services with lower environmental impact and greater value to customers and society.

During the design process, designers regularly have to assess the environmental impact of solutions. This environmental assessment in design teams appears to be an important issue. This is especially true in the early phases of the Product Development Process when details of the concepts are not yet defined and always carried out in a context of great uncertainty to select the "right project" (Ullman 2009) with a strong potential of reduction of environmental impact. During those phases designers have more opportunities to significantly reduce the environmental impacts of the product.

In a traditional innovation process, the concept evaluation phase remains a critical stage with major consequences on the later development of the eco-innovation process. This stage consists on evaluating the relevance of generated ideas and on prioritizing ideas that should be further explored in the next steps of the design process. It is always a complex decision making situation where different actors, directly affected by the result of the assessment, can express their point of view.

Nevertheless, the assessments carried out by the evaluators are irregular and often lack rigor, as underlines Ferioli (2010). Amabile (1983) underlines that if some methods help designers to estimate the innovative potential of concepts, most of them are based only on subjective notions.

Thus, many recent research work agree on the lack of reliable and appropriate assessment methods, which leads to an under-estimation of the issues of this stage by the designers. Moreover, from the industrial point of view, it also causes a waste of time and money (Geng et al. 2010).

If we add the environmental aspect to these remarks on 'classic innovation', the assessment phase becomes more critical, subjective and uncertain. Indeed, each eco-innovative concept must be assessed in a multi criteria approach all along its lifecycle (ISO 14062, 2002).

Bocken wonders if it is possible to predict "which concepts give most improvement in emissions performance with least implementation difficulty, before committing organizational resources to detailed evaluation of these options" (Bocken 2012).

In a previous research work, we noted an important dispersion of environmental assessment from experts (Tyl et al. 2010). We assumed that originality and environmental relevance were not reliable enough criteria to allow a neutral comparison of eco-ideation sessions because they involve experts' subjectivity. The results tend to prove that the three experts' assessments of the ideas differ substantially meaning that they do not have the same understanding of (1) the ideas; (2) the criteria (Tyl et al. 2010). Such variability makes it necessary to implement new procedures to obtain a reliable assessment. Our research question is therefore the following.

What are the relevant factors to assess early design concepts from the environmental perspective and create adhesion to promising ideas in a design team?

Key items on assessment of design and eco-design concepts are presented in detail in section 2. The empirical study based on existing eco-design outcomes is proposed in section 3. Results concerning the influence of the choice of environmental assessment factors on potential adhesion in design teams appear in section 4. Finally section 5 introduces conclusion and perspectives for future work.

2 ASSESSMENT OF DESIGN CONCEPTS

2.1 *Assessment of early design concepts*

A significant amount of work on the evaluation of early design creative concepts can be reported in the traditional design field.

Generally speaking, Ferioli et al. (2010) enumerates some constraints for an early concept evaluation tool: a large amount of idea evaluated, undetailed ideas, a relatively quick assessment. Moreover, Ferioli et al. (2010) consider two parts in the evaluation stage: the first one is objective and more oriented on the technical feasibility criteria, whereas the second one is more subjective, based on the 'feeling' of evaluators.

One of the most used evaluation method is the Weighted Objectives Method (WOT) which consists

in establishing a list of criteria to compare different concepts. Within this family of tools, Justel et al. (2007) identified six different methods adapted to assess concept: Pahl & Beitz' method, Pugh's method, Analytic Hierarachy Process (AHP), QFD matrix method, Fuzzy method, Hypothetical Equivalents and Inequivalents Method (HEIM). Nevertheless he considers that no method was developed to assist the evaluation of the innovative potential of conceptual design.

Moreover recent works include discussions on evaluation metrics. Amongst these can be found Van der Lugt's study of the self-evaluation of ideas by participants (Van der Lugt 2003). Shah's work, took up by Nelson proposed two criteria directed towards the assessment of ideas: the *originality* of an idea (i.e. the unexpected and original aspect of an idea compared with another) and the *quality* of the idea generated (measuring the feasibility of an idea and its response to the initial specifications) (Shah & Vargas-Hernandez 2003, Nelson et al. 2009). More recently, some authors have evaluated their early concepts using the branching preference (Lewis et al. 2011) or the Function-Behaviour-Structure (FBS) model (Howard et al. 2010).

2.2 *Environmental assessment of concepts*

It is widely acknowledged that quantitative environmental assessment of products and services can be operated by means of an LCA-based methodology, provided that enough information is available. Such information encompasses production details, scenarios for the use phase and for the end of life for instance.

For designers, the need for environmental assessment is often related to the comparison of performances for different versions of the same product. When comparing products sharing the same Functional Unit and main components, it is allowed to focus only on the differentiating components in a life cycle perspective. This is called 'streamlined LCA' (Hunt et al. 1998). Another way to compare environmental performances with reasonable efforts and resources is to implement the so-called 'Fast-track LCA' method (Vögtlander 2010). In this method, the impact indicators are chosen from the outset, making life cycle inventory useless. For instance, indicators may be cost, Cumulative Energy Demand or carbon footprint. Thus the idea is to "multiply the inputs and outputs directly by eco-burden factors" available online in the Ecocost 2007 databases (Vögtlander 2012). It has to be noted that, with this set of LCA-based methods, it is necessary to define precise inputs and outputs. But this may not be reached with sufficient accuracy in the conceptual phases of product development, and assessment in the early design stages is difficult to support by 'exact' numerical figures.

Some attempts have been reported to account for the environmental performance of very coarsely defined concepts. In the case of eco-innovation, Jones (2003) led an experiment where concepts generated in 6 groups with different tools were eventually assessed by two judges against mixed criteria: originality, appropriateness, environmental relevance, radical nature of concepts and system levels. An environmentally relevant concept shows potential to reduce the environmental impact (of dishwashing in this case) with or without rebound effects. However, no detail is provided on the deep understanding of this criterion by judges.

Another study used the self-evaluation of ideas proposed by Van der Lugt (2003) to measure the creativity of ideas according to environmental information (Collado-Ruiz & Ostad-Ahmad-Ghorabi 2010). Finally, Cluzet et al. (2012) used the criteria proposed by Shah & Vargas-Hernandez (2003) and adapted them to an industrial project. More specifically, they also propose to take into account the study boundaries and temporality (long and short term) to classify ideas.

Finally Bocken developed a simple tool to assess concepts against two criteria: the implementation difficulty and the emissions benefits of the concept. To conclude, we can state that there is very little literature about evaluation of concepts in the early stages of the eco-innovation process and all attempts have shortcomings. On the one hand, most evaluations only use a single, global and subjective environmental criteria (for instance Jones (2003) proposes the "environmental relevance" criteria). On the other hand, the evaluation sometimes consists in a mono-criterion environmental approach, based on the carbon emissions for Bocken (2012).

3 ENVIRONMENTAL ASSESSMENT WOKSHOP

3.1 Hypothesis

The aim of this initial experimental study was to test the effect of some result-oriented criteria on the dispersion of designers' assessments. To do so, we adopt the hypothesis that designers differently assess eco-innovative concepts should they consider:

- the life cycle stage where the main impact reduction appears,
- the category of environmental impact,
- the risk of impact transfer,
- the potential user's behaviour, i.e. if the concept has potential to enhance more environmentally friendly behaviour.

3.2 Background of the empirical study

Two eco-design experiments were previously carried out and provided creative outcomes as basis of this research work (Tyl et al. 2010, Vallet et al. 2011). The first experiment (called Barbaraz) concerned the redesign of a disposable razor, whereas the second one focused on an outdoor lighting system. Ten concepts were extracted from each design session, reframed and presented in an homogeneous manner (Table 1).

3.3 Empirical procedure

It was proposed to investigate the influence of assessment criteria on the potential adhesion to concepts in groups of eco-design experts.

38 participants from industry and academia gathered for a quarterly eco-design seminar were involved in this free research work. The participants were split into 5 groups of 7 to 8 persons. Their experience of eco-design ranged from 1 to more than 15 years, and different experience levels were mixed in each group to ensure homogenous group profiles. One reference group (G_0) only assessed concepts against life cycle stages of main impact and environmental impact category. In groups G_1 and G_2, the risk of impact transfer was added. Lastly in groups G_3 and G_4, the potential influence on user's behaviour was also considered. On each topic of investigation (Razor and Lighting), individual marking and ranking was followed by a group discussion to rank the Top 3 concepts (Fig. 1).

Table 1. List of concepts considered for the assessment session.

Name	Razor	Lighting
A	Half-sized handle	Lamp embedded in a treillis
B	Associate with organic cosmetics	Lamp in urban furniture
C	Blow out hairs	Lamp screwed in ground
D	Spherical handle	Fluorescent painting
E	Multipurpose handle	Lighting on demand
F	Thread razor	Lighting service
G	Bamboo handle	Painted ground to reflect light
H	Change of stiffness	Integrated reflecting system
I	Water jet cleaning	Central lighting system
J	Gauge (water/temperature)	Ball and joint light

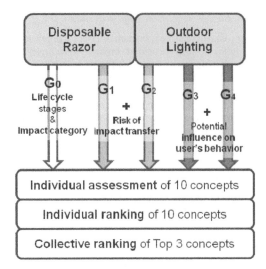

Figure 1.   Agenda of the experiment.

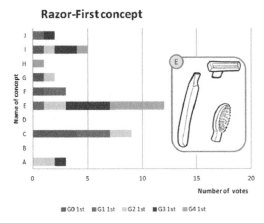

Figure 2.   Concepts ranking first for the Razor case.

3.4 *Environmental assessment criteria*

The criteria common to the five groups (namely life cycle stages of main impacts and environmental impact category) were implemented with a 3-point Likert scale: 0 = NI (No Idea); 1: Low reduction of impact; 2: Medium reduction of impact; 3: High reduction of impact. As for the risk of pollution transfer (the potential influence on user's behaviour respectively), on a 4-point Likert scale, 1 means 'High risk of transfer' ('Very negative influence' for user's behaviour) and 4 means 'No risk of transfer' ('Very positive influence' for user's behaviour).

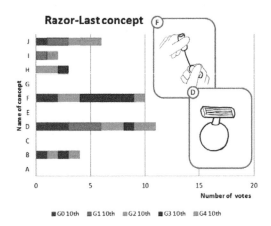

Figure 3.   Concepts ranking last for the Razor case.

4 INFLUENCE OF ENVIRONMENTAL ASSESSMENT FACTORS ON CREATING ADHESION

4.1 *Overview of first and last ranking concepts*

Firstly it is envisaged to examine the concepts that were identified as first (i.e. best) and last (i.e. worst) by individuals in the five groups for each topic: Razor (Figs. 2 and 3) and Lighting (Figs. 4 and 5).

For the Razor case, concept E (multi-purpose handle) raised more than 31% of the top ranking votes (12/38), followed by concept C with 24%. As for the worst concept, opinions across the five groups are balanced between concept F (26%) and concept D (29%). More surprisingly, concepts H, I and J raise votes for first position (1, 5, 2 votes respectively) *and* for last position (3, 2, 6 votes respectively).

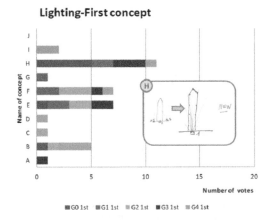

Figure 4.   Concepts ranking first for the Lighting case.

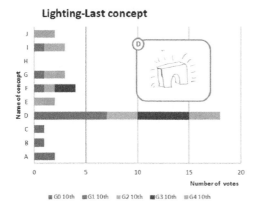

Lighting-Last concept

Number of votes

G0 10th   G1 10th   G2 10th   G3 10th   G4 10th

Figure 5.   Concepts ranking last for the Lighting case.

For the Lighting case, concept H (integrated reflecting system) is far ahead in cumulated votes with 29%, followed by concepts F and E (18%).

As for worst concept, opinions across the five groups seem very consensual concerning concept D (fluorescent painting) with 45% (17/38) of total number of votes. As previously noted, concept F is both appreciated as *best* for 4 participants and *worst* for 7 other participants.

In the next section, a deeper analysis is carried out to identify promising concepts in each group and to study the influence of the assessment criteria on the choices made by groups.

4.2   *Identification of promising concepts*

It is decided to characterize the potential of adhesion created by a concept in a group. Such an idea is named a 'nugget concept' and should equally appear in a significant number of Top3 ranking results across groups (in our case for more than half of the participants in a group) AND stand in first position for at least one participant. In our opinion this indicates a fuzzy agreement of a majority of evaluators combined with a strong conviction of one or few other evaluators.

Our approach develops as follows.

1. Isolate Top3 concepts for each participant in each group.
2. Identify 'nugget concepts' among Top3 concepts by a systematic examination of two indicators:
   • the total number of occurrences (or votes) within a group in first, second or third position;
   • the total number of occurrences (or votes) within a group in *first position* only.
3. Validate the method by comparing the identified 'nugget concepts' with Top3 concepts obtained after group discussion.

4. Study the influence of assessment criteria on the selected 'nugget concepts'.

For an easier understanding the total number of votes is segmented into 3 classes for 7 or 8 evaluators in a group.

• Class 1: 0 to 3 votes (low to medium adhesion)
• Class 2: 4 to 6 votes (high adhesion)
• Class 3: 7 to 8 votes (very high adhesion).

For the Razor case, the identified 'nugget concepts' are visible in Figure 6. For instance in test group G0, concept C raises 5 votes in Top3 and ranks first for four evaluators. Thus we assume that there is a strong adhesion to this concept (grey shadow on Fig. 6). Besides, a potential adhesion is also associated to concepts J and E (G0) with 4 votes in Top3, or concept A (G2) with 3 votes in Top3, two of which are a 'first'. The whole result of this analysis is shown in Figure 7.

	Nb of votes	TOP3			1st
		Class 1 [0;3]	Class 2 [4;6]	Class 3 [7,8]	
G0	C		5		4
	J		4		0
	E		4		0
G1	C		4		3
	E		4		1
	J	3			1
G2	G		6		1
	E		6		2
	C	3			2
	A	3			2
G3	E			8	4
	I		4		2
	A		4		1
G4	E		6		5
	G		4		0

Figure 6.   Identification of 'nugget concepts' (Razor case).

	RAZOR	LIGHTING
G0	C	H
	J	B
	E	E
G1	C	H
	E	F
	J	E
G2	G	B
	E	F
	C	C
	A	
G3	E	H
	I	A
	A	F
G4	E	B
	G	J
		H

Figure 7.   Summary of 'nugget concepts' (Razor and Lighting cases).

651

Table 2. Top 3 concepts from group discussion.

Group	Razor	Lighting
G_0	C E J	H I E
G_1	A C E	H E F
G_2	G/E C A	B C/A E/F
G_3	*NA**	*NA**
G_4	E J G	*NA**

NA: No Answer.

In order to validate our method to characterize the potential of adhesion, a comparison is made between 'nugget concepts' of Figure 7 and Top 3 concepts resulting from groups discussions (Table 2). The scope of comparison is reduced since G3 (and G4 for the Lighting case) did not provide any group ranking. Nevertheless results are very consistent for both the Razor case (G0, G2 and G4) and the Lighting case (G1, G2). Results are partially consistent for G1-Razor and G0-Lighting with two out of three concepts that are common to groups and individuals.

4.3 *Influence of assessment criteria*

In previous section it has been shown that individual ranking of concepts makes it possible to reveal 2 to 4 promising concepts in each group. The next questions are thus:

- Is the set of 'nugget concepts' similar in groups sharing the same assessment criteria?
- Is there a difference in 'nugget concepts' in control group G0 and other groups?

In order to form an opinion on the similarity or difference in concept selection, we propose to implement a new rule: based on Figure 7, a similar way of judging promising concepts leads to a minimum of two common concepts across groups.

According to this rule, it appears that G1 and G2 express a similar confidence in concepts C and E in the Razor case, but it is not true for the Lighting case. In both cases, G3 and G4 do not evaluate 'nugget concepts' similarly. Besides, G0 and G1 seem to assess ideas in a similar way: selection of three common concepts of razor (C, E, J) and two common concepts of lighting (H, E).

Some conclusion may be drawn from this analysis. First, we do not observe similar results between G1 and G2, or G3 and G4, meaning that the use of specific assessment criteria does not seem to clearly influence the type of promising 'nugget' concepts. Moreover, we notice a similar tendency in the selection of concepts between G0 and G1, which should require further investigation.

5 CONCLUSION AND FUTURE WORK

In this article we have emphasized stakes and difficulties linked with the assessment of early design concepts, should it be operated in traditional innovation and in eco-innovation.

In order to reach a reliable assessment in design teams, it is necessary to implement new procedures with the aim to create adhesion of teams to promising ideas or 'nugget concepts'.

Through an experiment on two cases with eco-design experts (Razor and Lighting), the main contribution of this article is (1) to build a new method of selection based on individual and collective rankings of concepts (2) to investigate the influence of assessment criteria (risk of impact transfer, potential influence on user) on the creation of adhesion. Our method is based on the examination of individual ranking of Top3 concepts by individual evaluators. The proposition is the following. 'Nugget concepts' should be ranked in Top3 by more than half of the group AND ranked first by at least one evaluator, demonstrating their potential of adhesion. Some 'nugget concepts' could be identified in the two different cases in consistency with group discussions. In conclusion the proposed assessment criteria are useful for experts to build a more reliable evaluation on environmental performance than observed in previous experiments (Tyl et al. 2010, Vallet et al. 2011). Nevertheless more in-depth investigation is needed to see how assessment criteria influence ranking and adhesion within design teams.

In perspective to this work, it is envisaged to adapt and test the method to larger groups to avoid the 'small group' bias. Another important issue is raised, linked to the question of the technical assessment, and more particularly the functional requirements. In this work the focus has been put on the environmental assessment and performances, but this should not be separated from a functional assessment, and also from the system boundaries to make the evaluation more reliable. For example, Bocken (2012) proposes to evaluate the implementation difficulties of each concept. But another stage of the evaluation should be implemented to compare the environmental reduction potential of each 'nugget concept' with the acceptability from the end-user.

REFERENCES

Amabile, T.M. 1983. Brilliant but Cruel: Perceptions of Negative Evaluators, *Journal of Experimental Social Psychology* 19: 146–156.

Bocken N.M.P, Allwood J.M., Willey A.R.,.King J.M.H. 2012. Development of a tool for rapidly assessing the implementation difficulty and emissions benefits of innovations, *Technovation* 32: 19–31.

Cluzet F, Yannou B, Millet D, Leroy Y (2012) Identification and selection of eco-innovative R&D projects in complex systems industries, *International design conference Design 2012, Dubrovnik.*

Collado-Ruiz D, Ostad-Ahmad-Ghorabi H. 2010. Influence of environmental information on creativity, Design Studies 31(5): 479–498.

Ferioli, M. 2010. Phases amont du processus d'innovation: proposition d'une méthode d'aide à l'évaluation des idées, PhD thesis, INPL.

Ferioli M., Dekoninck E., Culley S., Roussel B., Renaud J. 2010. Understanding the rapid evaluation of innovative ideas in the early stages of design, *International Journal of Product Development* 12(1): 67–83.

Geng, X., Chu X., Zhang Z. 2010. A new integrated design concept evaluation approach based on vague sets, *Expert Systems with Applications* 37: 6629–6638.

Howard T.J., Dekoninck E.A., Culley S.J. 2010. The use of creative stimuli at early stages of industrial product innovation. *Research in Engineering design* (21): 263–274.

Hunt RG, Boguski TK, Weitz K, Sharma A (1998): Case Studies Examining LCA Streamlining Techniques. *International Journal of LCA* 3(1): 36–42.

ISO/TR 14062 2002. Environmental Management, Integrating environmental aspects into product design and development. Technical Report, ISO, Geneva.

Jones, E. 2003. Eco-innovation: tools to facilitate early-stage workshop, PhD Thesis Department of Design, Brunel University.

Justel D., Vidal R., Arriaga E., Franco V., Val-Jauregi E. 2007. Evaluation method for selecting innovative product concept with greater potential marketing success. In:*Proceedings of the international conference on engineering design (ICED'07), 28–31 August, Paris, France.*

Lewis W.P., Field B.W., Weir J.G. 2011 Assessing quality of ideas in conceptual mechanical design. In: *Proceedings of the international conference on engineering design (ICED'11), Copenhagen.*

Nelson, B.A., Wilson, J.O., Rosen, D., Yen, J. 2009 Refined metrics for measuring ideation effectiveness, *Design Studies* 30 (6): 737–743.

Shah, J., Vargas-Hernandez, N., Smith, S. 2003. Metrics for measuring ideation effectiveness. *Design Studies* 24(2): 111–134.

Tyl B., Legardeur J., Millet D., Vallet F. (2010) Stimulatecreative ideas for eco-innovation: an experiment to compare eco-design and creativity tools. In: *Proceedings of IDMME Virtual Concept 2010, 20–22 October. Bordeaux, France.*

Ullman, D.G. 2009. The Mechanical Design Process, Mc Graw Hill(ed.), New York.

Vallet F., Millet D., Eynard B. 2011. Requirements and Features Clarifying for eco-design tools. In A. Bernard (Ed). *Global Product Development.* Springer.

Van der Lugt R. 2003. Relating the quality of the idea generation process to the quality of the resulting design ideas. In: *Proceedings of the international conference on engineering design (ICED'03), Stockholm.*

Vögtlander J. 2012. LCA a practical guide for students, designers and business managers. 2nd Edition. VSSD: Science and Technology.

Green Design, Materials and Manufacturing Processes – Bártolo et al. (eds)
© 2013 Taylor & Francis Group, London, ISBN 978-1-138-00046-9

Implementing concurrent engineering in the construction industry: A different approach for a better consideration of users

X. Latortue & S. Minel
ESTIA, Bidart, France

ABSTRACT: French construction industry is facing radical changes: the new 2012 Thermal Regulations as well as an increasing tendency from decision-makers to select the lowest tender force the constructors to change their practices. Those constraints have a deep impact on the industry. Reconsidering constructors behaviour appears to be essential. Current efforts focus on the adoption of manufacturing concepts including Concurrent Engineering (CE). Implementation of CE fosters a more efficient collaboration and increases the users needs consideration. Using 4D CAD or Building Information Modeling can also be a great contribution. Practices changes should succeed through CE trainings and it is crucial to develop procedures that will be followed by the construction actors. As a conclusion, case studies are rare and studying the implementation conditions of concurrent methodologies in the construction industry appears as a relevant approach to investigate.

1 INTRODUCTION

French construction industry is at a turning point of its history: the new 2012 Thermal Regulations and a negative economical context that increases the tendency from decision-makers to select the "lowest tender" (Katz, 2011) in building projects force the constructors to change their practices. The new Thermal Regulations obligates designers to consider the building as a whole from the design phase with architects but also with multidisciplinary teams of craftsmen, without forgetting the finishing works. The objective of this paper is to present the interest of studying user-centred concurrent methodologies in the construction industry in order to face these new challenges.

These new obligations are time-consuming and require changing part of their working process. Moreover, the additional cost directly related to these thermal regulations is estimated around 10–15% (ADEME, 2010).

However, considering that the lowest bidder usually wins the contacts, construction companies have to improve their productivity and save as much money as possible on their projects to be able to comply with the new regulations without increasing the cost of the construction.

Those antagonistic constraints will soon be the norm and will have a deep impact on the Architect, Engineering and Construction (AEC) industry uses. AEC industry has to find innovative and perennial solutions to overcome those issues without damaging the quality of the constructions.

Considering the whole diversity of users (Denton et al., 1973; Shen et al., 2012) is essential to be able to design an adequate building that takes into account the variability of needs and expectations of the future inhabitants. Therefore being creative will be necessary to elaborate new design and building processes. Moreover it becomes essential to take into account the integration of the building into its whole surroundings: its block, its neighbourhood, and all its interactions (Transports, shops, schools etc...) that will inevitably influence the occupants and impact on their behaviour.

2 A FRAGMENTED PROCESS

2.1 A sequential process

Those new challenges focus on global characteristics of the building and will be difficult to overcome by providing isolated technical solutions that are specific to each profession. It is then essential to adopt a global approach to bring adequate answers that consider constraints from each building actors. Unfortunately it goes against the fragmented nature of the construction industry.

Indeed, construction projects follow a sequential process that generally consists of a series of distinct steps during which the client asks for a feasibility study, the technical specifications, defines a budget and has a tender. The traditional organization of the construction process is very sequential

(Fig. 1), with a clear separation of roles between actors at different phases of the project.

Obtaining the cooperation and a constructive dialogue between the different actors is a complex task, but is necessary to carry the project to completion.

2.2 Fragmented stakeholders

Furthermore, the different stakeholders not only have their own vocabularies and practices, but also have different and sometimes opposite goals and interests to defend.

This fragmentation between different actors (Fig. 2) throughout the different phases, the one shot type of relationships and projects that are

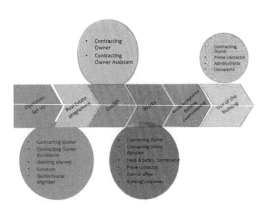

Figure 1. Sequential process of construction.

Figure 2. Construction project stakeholders (Olander, 2003 adapted from Cleland, 1999).

built by them and the constraint of "the prevalence of a strategy of lowest bidder" (Benhaim, 1997; Campagnac & Winch, 1998 cited by Crespin-Mazet & Ghauri, 2007) complicates the management of such a project. Unfortunately, this lowest bidding strategy added to the difficulties of coordination, exposes the project to a greater risk of non-quality (Vivier, 2011). Therefore the savings on the cost of an efficient project management is relativized by the cost of damages caused by non-quality as well as its associated insurances.

One consequence of this attitude is that the construction industry has been largely dominated by "hostile relations" (Crespin-Mazet & Ghauri, 2007) between the different stakeholders involved in the project, where accountability and partners engagement is difficult to establish. Thus, according to Guillou et al. (2003, cited by Crespin-Mazet & Ghauri, 2007) "each project being unique, contractors employ all their efforts to win the bid, even if it pushes to adopt a winner-loser relationship with the other actors."

As a consequence, the use of project management methods employed in traditional manufacturing industries appears difficult to implement as such and require adopting a co-construction approach to adapt those methodologies to the AEC industry.

2.3 Integrating the construction process

Reconsidering the behavior of construction professionals, in the manner what the manufacturing industry has undergone during the second industrial revolution, appears as an essential step to face these new challenges (Gobin, 2001).

To overcome these constraints the industry will have to find durable solutions for the building while also considering the occupant in its neighborhood (transports, shops, schools etc…) in order to prevent a decreasing quality of the buildings or of the welfare of the users.

The integration of the rather fractionated processes of the construction industry is seen as a path to improve not only the efficiency of the construction industry (Howard et al. (1989) Egan (1998) and quoted by Kamara et al., 2000a; House of Commons, 2008) but also the cohesion of actors (Anumba et al., 2007).

Efforts to develop strategies to integrate the construction process focus on the adoption of several manufacturing concepts including Concurrent Engineering (CE) (De la Garza et al., 1994; Love & Gunasekaran, 1997). According to Christophe Gobin (2001) the objective "is to integrate the various stakeholders into a collective to achieve true value for money as well as an effective consultation instead of solving a juxtaposition of local optima."

3 CONCURRENT ENGINEERING, AN INTERESTING APPROACH

3.1 *Benefits of CE implementation*

The disparities among different professions inevitably lead to disagreements or misunderstandings between individuals. In addition to causing delays on the construction sites and thus increase project costs, those difficulties arise as an obstacle to the application of CE, which is by its very nature based on cohesion and shared objectives between the parties (Minel, 2003).

The compromises are therefore necessary between each of the actors, and throughout the project, all along the different phases of the project, and sometimes even within each stakeholder.

The arguments in favor of the adoption of the CE in construction are however numerous and have been featured in an important number of publications:

- Benefits due to the use of the CE in the manufacturing industry with a reduction in product development time are up to 70% (Prasad, 1996);
- Construction can be seen as a manufacturing process, regarding the process involved in the design and production of the building. Moreover if one perceives the building as "a production of works for a single destination (...) (it can be considered that) the specificity of the building is not in its production but in its use." (Gobin, 2001).
- The relevance of the objectives of the CE for the resolution of obstacles and challenges faced by the construction industry (integration process of construction, customer satisfaction with regards to the cost, time and value).

Those common points with the manufacturing industry encourage the use of CE concepts that have had positive results in the manufacturing industry (Mohamad et al., 2008; Egan, 1998; Sanvido & Medeiros, 1990; Crowley, 1996).

Beyond the energy issue, given that one of the objectives of the CE is the final satisfaction of the customer, the implementation of CE practices must foster collaboration of the actors on aspects that affects directly the client: understanding their needs (whether they are expressed or not), and the inclusion of these needs in the design and development of the project. Indeed, their participation in the project is crucial to the success of the latter, and is seen as a key to accelerate progress in industry (Kamara & Anumba, 2007).

Including the customer in the predesign and needs collecting phases provides an environment that is more convenient to the use of CE (Kamara & Anumba, 2007). Indeed, it appears that effective study of customer needs can stimulate the practice of CE because it involves constructive exchanges between the various trades including regarding the constraints and tradeoffs related to design (Fiksel & Hayes-Roth, 1993). Better consideration of the needs of the user goes hand in hand with the implementation of the CE in construction (Kamara et al., 2000b). This phase of user requirements recollections must be established as a phase in itself, and be carried out by qualified players with a methodology and media appropriate to remedy the current deficiencies (Kamara & Anumba, 2001). It is crucial to consider the customer in its diversity: from the investor that conceive it as a profitable investment, to the occupant that is seeking a place to live that will fulfill his needs. However, the implementation of such a methodology requires a profound remodeling of the current processes, either on the entire project between professions, but also within each profession.

While the major corporations in the construction industry in France are already focusing on these issues, and on the new methodologies, smaller size construction companies can't deploy that much effort into redesigning their process. In 2010 (source: INSEE), 88% of the construction companies in France counted six or less employees. This figure may foreshadow difficulties to change methodologies, but may at the same time bring hope for a greater flexibility in business processes.

3.2 *Technical tools as a support for CE*

Even if the implementation of CE in the construction industry is mainly based on the implementation of new collaborative approaches along the different phases of the process, some technical tools can encourage those new practices. Those new technological tools start to be common on big construction project. Using 4D CAD or BIM (Building Information Modelling) laser can be a great contribution to the improvement of the processes (Staub-French & Fischer, 2007; Shen et al., 2012).

4D (also known as 5D) CAD associates time and financial constraints to the 3D CAD representation of the building. It also provides a common basis and unique representation for all the stakeholders that can take into account all the tasks that are programmed to be performed by the different builders: shell, plumbing, wiring etc...

The fact that there is a single representation of the model is a revolution *per se,* as in a typical construction project, a building may be drawn up to a dozen times by various consulting firms that focus on their very own tasks characteristics.

Grouping together the different stakeholders on a unique representation imposes conditions of interoperability among project collaborators. If such interoperability can appear as a constraint in the first place, the benefits associated with the use of a digital model are numerous as explained by Staub-French & Fischer (2007), among others:

- Shorter estimating time,
- Fewer quantity take-off errors,
- Less rework,
- Increased productivity,
- Construction realized on time and within budget...

On the other hand BIM allows construction workers to retrieve all the spatial data from an existing building that requires renovation into a 3D CAD representation by using a laser scanner. It is particularly useful to visualize inaccessible areas of the building that could have been, for example, altered since the edification of the building.

The use of BIM technology saves considerable time compared with manual measurement and will probably become mandatory in the future for UK public projects.

However, it is important to distinguish the tools from the process to not forget that the priority of concurrent engineering is to firstly focus on methodologies related to work then to contribute to these methodologies by implementing specific tools.

4 CONCLUSION

Practices changes will succeed through CE training of the stakeholders. The CE is often described as rather vague guidelines; it seems then necessary to develop real tangible procedures that embody the theory and will be followed by the different actors (Anumba et al., 2007). Nonetheless, these procedures should not be focused on the use of new tools, but rather impact the intrinsic process of each profession.

Beyond the process, in order to improve the efficiency of construction projects, it is necessary to achieve a true reflection on the organizational level, but also on the building site to improve not only the operation between professions but also inside each professions. If the arguments in favor of the CE are numerous, case studies of the implementation on actual projects are rare. Feedbacks and a critical analysis of results would draw interesting conclusions and provide elements to improve the methodology.

Moreover, in order to persuade actors and decision-makers from the construction industry that could be reluctant to the implementation of CE in construction projects, it is essential to quantify the progress that is related to its implementation.

Although it goes without saying that progress can't always be quantifiable. Of course, this quantification requires working on the parameters to be measured and on the measurement protocol. Case studies on implementation of actual projects are rare. Therefore testimonials and a critical analysis of results would allow to draw conclusions and to provide elements to improve the methodology.

Confronted to this situation, building professionals have to find innovative solutions. There will be two types of complementary solutions. The first one appears to be using technologies that are adapted to use changes while reducing the cost of projects and the energy consumption of buildings and their environment. The other one, which is essential to the success of technological solutions, is to integrate the various stakeholders in a collective in order to achieve the optimization of the resources that are mandatory to meet the needs of users.

Studying the implementation conditions of user-centered concurrent methodologies in the construction industry appears therefore as a relevant approach to help the construction industry to solve these new constraints.

REFERENCES

ADEME 2010. Le bâtiment basse consommation, une obligation du grenelle environnement et déjà une réalité. Relations Presse ADEME/TBWA-Corporate (2010).

Anumba, C.J., Cutting-Desselle, A.F. & Kamara, J.M. 2007. Concluding notes. Concurrent Engineering in Construction Projects. Edited by Chimay Anumba, John Kamara and Anne-Françoise Cutting-Decelle. 2007.

Benhaim, M. 1997. Interfirm Relationships within the Construction Industry: Towards the Emergence of Networks? A Comparative Study between France and the UK, DBA Thesis, Henley Management College, Brunel University.

Campagnac, E. & Winch, G. 1998. Civil engineering joint ventures: The British and French models of organization in confrontation. In R.A. Lundin & C.Midler (Eds.), Projects as Arenas for Renewal and Learning Processes (pp. 192–206). Dordrecht Kluwer Academic Publishers.

Cleland, D.I. 1999, Project Management—Strategic Design and Implementation, third edition, McGraw-Hill.

Crespin-Mazet, F. & Ghauri, P. 2007. Co-development as a marketing strategy in the construction industry. *Industrial Marketing Management* 36 (2007) 158–172.

Crowley, A. 1996. Construction as a manufacturing process. In: Kumar B, Retik A, editors. Information representation and delivery in civil and structural engineering design. Edinburgh: Civil-Comp Press, 1996. p. 85 ± 91.

De la Garza, J.M., Alcantara, P., Kapoor, M. & Ramesh, P.S. 1994. Value of concurrent engineering for A/E/C industry. *Journal of Management in Engineering* 1994;10(3):46 ± 55.

Denton, T., Ind, P.D., McCollum, J. & Stutsman R. 1973. Types Of User Building Evaluation. *Environmental Design Research Association* (EDRA).

Egan, J. 1998. Rethinking construction. Report of the Construction Task Force on the Scope for Improving the Quality and Efficiency of UK Construction. Department of the Environment, Transport and the Regions, London, 1998.

Fiksel, J. & Hayes-Roth F. 1993. Computer-aided requirements management. Concurrent Engineering: Research and Applications 1993;1:83 ± 92.

Gobin, C. 2001. L'ingénierie concourante—Un nouveau professionnalisme. *Dossier Techniques de l'Ingénieur.*

Guillou, M., Crespin-Mazet, F. & Salle, R. 2003. La Segmentation dans les Entreprises Travaillant par Affaires: l'Exemple de Spie Batignolles dans le Secteur du BTP. *Décisions Marketing*, 31, 63–71.

House of Commons Business and Enterprise Committee 2008—Construction matters. *Ninth Report of Session* Volume I 2007–08.

Howard, H.C., Lewitt, R.E., Paulson, B.C., Pohl, J.G. & Tatum C.B. 1989. Computer integration: reducing fragmentation in AEC industry. Journal of Computing in Civil Engineering; 3(1):18 ± 32.

Kamara, J.M., Anumba C.J. & Evbuomwan N.F.O. 2000a. Assessing the suitability of current briefing practices in construction within a concurrent engineering framework.

Kamara, J.M., Anumba, C.J. & Evbuomwan N.F.O. 2000b. Establishing and processing client requirements—A key aspect of concurrent engineering in construction. Engineering, *Construction and Architectural Management* 2000;7(1):15 ± 28.

Kamara, J.M. & Anumba, C.J. 2001, 'A Critical Appraisal of the Briefing Process in Construction', *Journal of Construction Research*, Vol. 2. No. 1, pp. 13–24.

Kamara, J.M. & Anumba, C.J. 2007. The voice of the client within a Concurrent Engineering design context. Concurrent Engineering in Construction Projects. Edited by Chimay Anumba, John Kamara and Anne-Françoise Cutting-Decelle. 2007.

Katz, P. 2011. EDITO—La politique du moins disant—ou du plus disant et du moins faisant. Le bulletin de la société française des architectes n°48.

Love, P.E.D. & Gunasekaran, A. 1997. Concurrent engineering in the construction industry. Concurrent Engineering: Research and Applications 1997;5(2).

Minel, S. 2003. Démarche de conception collaborative et proposition d outils de transfert de données métier. Application à un produit mécanique "le siège d'automobile". Thèse EN-SAM 2003.

Mohamad, M.I., Baldwin, A.N. & Yahya, K. 2008. Application Of Concurrent Engineering (CE) For Construction Industry. In Issues In Construction Industry 2008.

Olander, S. 2003. External Stakeholder Management in the Construction Process. Licentiate Dissertation, May 2003. Published by *Division of Construction Management*. Department of Building and Architecture. Division of Construction Management, Lund Institute of Technology, Lund University.

Prasad, B. 1996. Concurrent engineering fundamentals, Vol. 1: Integrated products and process organization. NJ: Prentice Hall.

Sanvido, V.E. & Medeiros, D.J. 1990. Applying computer-integrated manufacturing concepts to construction. *Journal of Construction Engineering and Management* 1990.

Shen, W., Shen, Q. & Sun Q. 2012. Building Information Modeling-based user activity simulation and evaluation method for improving designer user communications. *Automation in Construction* Volume 21, January 2012.

Staub-French, S. & Fischer, M. 2007. Enabling Concurrent Engineering Through 4D CAD. Concurrent Engineering in Construction Projects. Edited by Chimay Anumba, John Kamara and Anne-Françoise Cutting-Decelle. 2007.

Vivier, A. 2011. Le dumping démonétisé la prestation. Le bulletin de la société française des architectes n°48.

Green Design, Materials and Manufacturing Processes – Bártolo et al. (eds)
© *2013 Taylor & Francis Group, London, ISBN 978-1-138-00046-9*

Sustainable supply chain management: A systematic literature review about sustainable practices taken by focal companies

A.M. Nascimento & R.L.C. Alcantara
Federal University of São Carlos, São Carlos, São Paulo, Brazil

ABSTRACT: Business organizations have increasingly been charged to act aiming a sustainable development, assuming economic, environmental and social responsibilities. In this scenario, it is essential to expand the thought about sustainability to beyond the boundaries of firms, considering supply chains, and think about the concept of Sustainable Supply Chain Management. It is noteworthy the important role of focal companies, which represent the most influential organizations in their chains, and have the responsibility to ensure that sustainable actions occur along the supply chains they are part of. Thus, the aim of this study is, from a systematic literature review, the presentation of the sustainable practices undertaken by focal companies and that are expanded upstream and downstream the supply chain.

1 INTRODUCTION

Business organizations have increasingly been charged to act aiming a sustainable development, the one that meets the needs of present generations without compromising future generations to meet their needs (WCED, 1987). In this scenario, one of the modern challenges is to think about business models, products and services that enable citizens to meet their needs at the lowest social and environmental cost (Carvalho & Barbieri, 2012).

In a broader view, sustainability can be seen as consisting of three components: social, environmental and economic responsibilities (Carter & Rogers, 2008). The consideration of these three pillars—environmental, social and economic—is also known as Triple Bottom Line (TBL) concept outlined by Elkington (2004) to describe the organizational sustainability. In this sense, an organization will be sustainable when its activities are governed by these dimensions, acting with ecological caution, economic efficiency, and generating social equity (Barbieri & Cajazeira, 2009).

However, it is crucial to expand the thought about sustainability to beyond the boundaries of firms, considering then supply chains. According to Christopher (2009), supply chains can be defined as networks of organizations that are involved through upstream and downstream linkages in the different processes and activities that produce value in the form of products and services to the final consumer.

This is relevant because of the paradigm shift in modern business management, in which individual businesses no longer compete as autonomous entities, but rather as supply chains (Lambert et al, 1998), which occurs due globalization and increasing outsourcing in various industries (Seuring et al., 2008). With this concept in mind, it is still worth noting the concept of Supply Chain Management, which represents the integration of all activities associated with the processing and flows of goods and services, including the flow of information necessary for its success (Mentzer et al. 2001).

Given the above, it is emphasized that the ideal is the adoption of the Sustainable Supply Chain Management (SSCM), defined by Seuring & Muller (2008) as the "strategic, transparent integration and achievement of an organization's social, environmental, and economic goals in the systemic coordination of key inter organizational business processes for improving the long-term economic performance of the individual company and its supply chains".

The SSCM incorporates a variety of concepts such as the Green Supply Chain Management, which represents the integration of the environmental thinking to the management of the supply chain, in order to minimize the impact generated (Walker & Jones, 2012; Srivastava, 2007). Moreover, the SSCM also includes the consideration of social issues such as the effort to guarantee that suppliers ensure adequate working conditions to their employees (Walker & Jones, 2012). It is also important to note that the environmental and social dimensions of SSCM must be taken from a clear and explicit recognition of the economic goals of the firm and its supply chains (Walker & Jones, 2012; Carter & Rogers, 2008).

In this context, it is noteworthy to highlight the important role of focal companies, the most important organizations along supply chains, which have the responsibility to ensure that sustainable actions occur in supply chains they are part of (Seuring & Muller, 2008). Therefore, it is important that these organizations start to establish sustainable practices in their production processes by considering environmental and social impacts generated both upstream as downstream their supply chains.

This being done, firms and their supply chains can reduce the negative effects generated to the environment and society, gain a better reputation with their consumers, and also improve their relationships with stakeholders such as non-governmental organizations and activists' communities (Zailani et al., 2012).

Therefore, it is important that these organizations start to establish sustainable practices in their production processes by considering the environmental and social impacts generated both upstream as downstream.

Given the above, the aim of this work is to present a framework that relates the strategies that have been incorporated by the focal firm to improve sustainability in their supply chains. More specific, the authors main to answer the following question: "How can sustainable issues be incorporated to supply chains by focal companies?". For this purpose, the methodology used in this article was the systematic literature review.

2 METHOD

Literature reviews can be conducted for a variety of purposes, whether to provide the theoretical framework for future research in a certain subject, learn about the depth of research in a particular topic, or answer practical questions by understanding what the existing research has to say on the matter. (Okoli & Schabramm, 2010). In this context, Okoli & Schabram (2010) highlight the existence of a type of review that differs from the traditional, which is able to create a solid starting point for research on a particular topic: a systematic literature review.

Thus, this article was conducted from a systematic literature review. This type of review is different from traditional reviews for being more systematic and explicit in the selection of the studies involved. It also employs more rigorous and replicable assessment methods (Denyer & Trandfield, 2009).

This kind of review describes the evidence in order to allow that clear conclusions about what is already known and what is unknown about the subject can be drawn (Denyer & Tranfield, 2009).

It also permits the identification and presentation, in a synthetical way, of the relevant knowledge about the topic, highlighting the themes studied, and helping to position future studies.

Thus, from the literature about sustainable practices, a selection and evaluation of relevant contributions was done. An analysis and synthesis of relevant data was also conducted.

The selection of the articles occurred in November of 2012 from queries to the databases SCOPUS and PROQUEST. For the query to SCOPUS two searches were performed, which are described at Table 1.

Considering the terms described at Table 1, at first, to restrict the number of results, the options "Article title, Abstract, Keywords" and "Article and Review" were selected. As another way to limit the results, only articles related to "Physical Sciences" and "Social Sciences & Humanities" were included, considering those areas the more adequate for this study.

It is believed that the terms chosen (Table 1) represent those able to present strategies within the supply chain so that it can become more sustainable. Thus, the search aimed to look for practices such as as environmental programs, certifications, training, collaboration, and assessment of suppliers, which are actions that are stimulated and sometimes implemented by focal companies. The term "operations" was used because of the possibility to aggregate sustainable actions described among traditional operations. However, this study aims to go beyond the operations already described by Srivastava (2007), highlighting those which the focal organizations are leaders of.

There is also the combination of the terms "practices" and "operations" with the words "suppliers" and "consumer OR customer", as a way to incorporate upstream and downstream results that might not have been given in the first search.

After these initial searches, a series of filters were applied, so the pertinent results for the purpose of this study were selected. The protocol used for the selection is described in Figure 1.

After the filtrations, a total of 34 articles were obtained at the search in Scopus database.

Table 1. Terms used in the search at SCOPUS database.

Searches	Terms
1	(sustainable practices OR sustainable operations) AND supply chain
2	("sustainable practices" OR "sustainable operations") AND ((customer OR consumer) OR supplier))

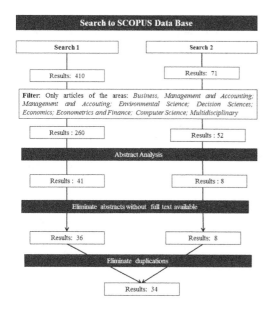

Figure 1. Search protocol at SCOPUS database.

For the query on the **PROQUEST** database, the terms used for the search were "sustainable supply chain AND (practices OR operations)" in the Advanced search of the database. These words were considered in all fields, only for articles, and with no restriction on the language. In this search, two articles were selected.

It is noteworthy that the extent of the search words occurred in a way that the largest number of results involving the supply chain as a whole could be obtained, considering practices that the focal companies extend to suppliers and consumers.

Finally, after the appropriate filtering and analysis of the results, a final number of 36 articles pertinent to answer the research question were selected. These articles were then evaluated for the extraction of the relevant data, which considerations are described in the next section, along with the presentation of the framework.

3 FRAMEWORK

The study of sustainability should be done considering the extended supply chain, being necessary to leave the boundaries of the firm, which represents a challenge to the success of sustainable practices (Gimenez et al, 2012). This success is also anchored in the actions taken by focal companies, which can use their influence to encourage their suppliers to act in more responsible ways, and also work with the consumers on issues facing their awareness,

for example. Thus, sustainability issues must be rooted in the supply chain from the action of all its members, with the focal company acting as leader.

In this context, from the mapping and analysis of the selected articles, it's possible to conclude that the literature shows 13 practices that may be established by the focal organization in their chains. From these practices, ten are related to suppliers, and may take two strategies—assessment and collaboration.

As ways of evaluating suppliers, focal organizations can establish the requirement of codes of conduct, certification, green purchasing and can also monitor working conditions and the impact of the activities in external communities.

Moreover, among the actions that can be established through collaboration are environmental design, development of technologies that generate less environmental impacts, establishment of good working conditions, actions with external communities and information sharing between the focal company and its suppliers.

Gimenez & Tachizawa (2012) point out that both the actions of evaluation and collaboration have a positive impact on environmental and social performance of the chain. However, studies have highlighted that the adoption of only evaluation actions are not enough for a significant improvement, being in collaboration the greater chances of success and increase of social and environmental performance throughout the supply chain (Ageron et al, 2012; Gimenez et al, 2012; Gimenez &Tachizawa; 2012).

With respect to consumers, it is noticed that the number of articles that address this aspect is significantly lower. Accordingly, only three practices taken downstream by the focal firm were found: information sharing, labeling, and use of appropriate technologies.

Finally, the last practice found in the literature is the reverse logistics, which in this study was not classified as neither upstream nor downstream, since this is an operation that promotes integration between these parties. Reverse logistics deals with the management of returns, and plays an important role in the sustainability of the supply chain since, besides creating new market opportunities, has the capacity to significantly mitigate environmental problems (Tang & Zhou, 2012). For the success of this operation, it is essential that occurs cooperation between all members of the chain (Chan, 2007), hence the importance of information sharing and joint planning objectives. To this end, there is again the important role of the focal organization, which should act as a leader in this environment.

Some authors explored, in the context of Reverse Logistics, aspects related to the Closed Loop Supply

Chain. This concept is important because it relates to the integration of the flow of materials, information and finances in all directions of the supply chain (HUANG et al, 2007), "closing" the chain. To illustrate these results, Figure 2 presents a framework with a summary of what was presented. Thus, the focal firm influences the strategies of the supply chain upstream and downstream, acting from several practices between these links. It must be mentioned that, for a successful implementation of sustainability along the supply chain, it is crucial that exists a constant flow of information among members, and the incorporation of CLSC as a way to close the circuit of products, materials, information and finances.

Finally, considering all the studies used in this article, it is important to highlight that environmental practices are more recorded, performed and stimulated than the social ones, existing even fewer works that consider these two issues simultaneously. To illustrate that, Figure 3 represents all the articles used in this paper. Thus, it is observed that, from the 36 papers evaluated in this study,

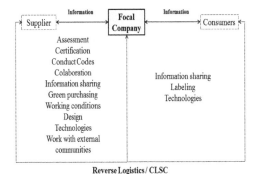

Figure 2. Framework—sustainable practices taken by focal companies along the supply chain.

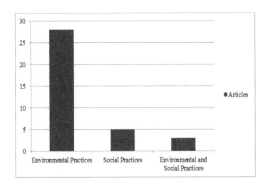

Figure 3. Framework—sustainable practices taken by focal companies along the supply chain.

28 were related to environmental strategies. On the other hand, only five articles referred to social issues, and only three addressed these aspects simultaneously.

4 CONCLUSIONS

The charge for business organizations to act aimed at sustainable development is increasing. In this context, it is necessary for these organizations to consider, besides the economic responsibilities, social and environmental issues in their objectives, taking sustainability as part of the business, and not as isolated actions. Moreover, it is fundamental to the success of sustainable thinking to go beyond the boundaries of firms and consider supply chains, adopting the Sustainable Supply Chain Management. Therefore, it is central to focal organizations, those considered the most influential in the chain, to act stimulating the development of sustainable supply chains, either upstream and downstream.

In this context, from a systematic literature review, this article aimed to highlight the practices that can be taken by the focal firms with their suppliers and customers in order to induce the issues of sustainability in their supply chains.

When considering suppliers, the strategies undertaken by the focal firm can be aimed at both the evaluation of suppliers, through the establishment of rules and standards to be followed by them, as there may be actions of collaboration, in which suppliers and focal company act together in planning activities to pursuit sustainability in the supply chain. Importantly, the success of such collaborative actions is dependent on the exchange of information between the links, ensuring that there is transparency between parties.

Regarding consumers, it was perceived a lower quantity of literature, with labeling, information sharing and use of appropriate technologies among the most common practices. The lack of studies on this topic can be seen as an area for future research, aimed at finding new ways to bring awareness of the importance of sustainable development for consumers.

Finally, the literature also addressed issues related to reverse logistics and closed loop supply chains, fundamental actions for the "closure" of the chain with respect to the flow of materials, information, finance, etc.

It is worth noticing that most part of the literature regards environmental issues, indicating a lack of studies that consider sustainability within the concept of the Triple Bottom Line, considering its three dimensions. In this sense, it is important to highlight the importance of expanding the thought—from Green to Sustainable—so this issue can be studied in an integrated way.

As a perspective for future studies, in addition to more studies about sustainability practices that can be conducted with consumers, more work addressing sustainability within the TBL is needed.

REFERENCES

Ageron, B., Gunasekaran, A., & Spalanzani, A. (2012). Sustainable supply management: An empirical study. International Journal of Production Economics, 140(1), 168–182.

Barbieri, J.C. & Cajazeira, J.E.R. 2009. Responsabilidade Social Empresarial Sustantável. São Paulo: Saraiva.

Carter, C.R., & Rogers, D.S. (2008). A framework of sustainable supply chain management: moving toward new theory. International Journal of Physical Distribution & Logistics Management, 38(5),

Carvalho, A.P. & Barbieri, D.C., (2012). Innovation and Sustainability in the Supply Chain of a Cosmetics Company: a Case Study. Journal of Technology Management and Innovation, 7(2), 144–156.

Chan, H.K. (2007). A pro-active and collaborative approach to reverse logistics—a case study. Production Planning & Control, 18(4), 350–360.

Christopher, M. 2009. Logística e gerenciamento da cadeia de suprimentos: criando redes que agregam valor. São Paulo: Cengage Learning.

Denyer, D., & Tranfield, D. (2009). Producing a Systematic Review. In: Buchanan, D.A., Bryman, A (Eds), The SAGE Handbok of Organizational Research Methods, pp. 671–689. London.

Elkington, J. (2004), Enter the triple bottom line. In Henriques, A. and Richardson, J. (Eds), The Triple Bottom Line: Does It All Add up?: 1–16. Earthscan, London.

Gimenez, C., Sierra, V., & Rodon, J. (2012). Sustainable operations: Their impact on the triple bottom line. International Journal of Production Economics, 140(1), 149–159.

Gimenez, C., & Tachizawa, E.M. (2012). Extending sustainability to suppliers: a systematic literature review. Supply Chain Management: An International Journal, 17(5), 531–543.

Huang, X.-Y., Yan, N.-N., & Qiu, R.-Z. (2009). Dynamic models of closed-loop supply chain and robust H ∞ control strategies. International Journal of Production Research, 47(9), 2279–2300.

Lambert, D.M., Cooper, M.C., & Pagh, J.D. (1998). Supply Chain Management: Implementation Issues and Research Opportunities. The International Journal of Logistics Management, 9(2), 1–20.

Mentzer, J.T., Keebler, J.S., Nix, N.W., Smith, C.D., & Zacharia, Z.G. (2001). Defining Supply Chain Management. Journal of Business Logistics, 22(2), 1–25.

Okoli, C., Schabram, K. (2010). A Guide to Conducting a Systematic Literature Review of Information Systems Research,. Sprouts: Working Papers on Information Systems, 10(26). http://sprouts.aisnet.org/10-26.

Seuring, S., & Müller, M. (2008). From a literature review to a conceptual framework for sustainable supply chain management. Journal of Cleaner Production, 16(15), 1699–1710.

Seuring, S., Sarkis, J., Müller, M., & Rao, P. (2008). Sustainability and Supply Chain Management—An introduction to the special issue. Journal of Cleaner Production, 16, 1545–1551.

Srivastava, S.K. (2007). Green supply-chain management: A state-of-the-art literature review. International Journal of Management Reviews, 9(1), 53–80.

Tachizawa, Elcio M.; Thomsen, C.G.; Montes-Sacho, M.J. (2012). Green Supply Management Strategies in Spanish Firms. IEEE Transactions on Engineering Management, 1–12.

Tang, C.S., & Zhou, S. (2012). Research advances in environmentally and socially sustainable operations. European Journal of Operational Research, 223(3), 585–594.

Walker, H., & Jones, N. (2012). Sustainable supply chain management across the UK private sector. Supply Chain Management: An International Journal, 17(1), 15–28.

World Commission on Environment and Development (1987), Our Common Future, Oxford University Press, New York, NY.

Zailani, S., Jeyaraman, K., Vengadasan, G., & Premkumar, R. (2012). Sustainable supply chain management (SSCM) in Malaysia: A survey. International Journal of Production Economics, 140(1), 330–340.

Green Design, Materials and Manufacturing Processes – Bártolo et al. (eds)
© 2013 Taylor & Francis Group, London, ISBN 978-1-138-00046-9

Multi-criteria evaluation of reaction to the economic recession in modernization of multi-apartment buildings

L. Kelpsiene
Department of Civil Engineering, Siauliai University, Lithuania

M.L. Matuseviciene
JSC "Meba", Siauliai, Lithuania

ABSTRACT: Climate change and limited energy resources stimulate exploitation of energy more efficiently in a residential dwelling. This paper discusses modernization of multi-apartment buildings in Lithuania. Nearly 96 % flat-houses were built before 1993. The author examined two parallel projects in 2007 and 2009 by applying the support system for making decisions. Socio-economic conditions, the macro, meso and micro environment was very different, even though these objects are very similar technically. All houses renovation projects lines decreased in 2009 in comparison to 2007's of values. The project price fell down 18.54%. Construction site costs have changed the least, it's change results only in 9.06%, and the profit has decreased most of all—73.93%. The government approved the proposal to set up a Public Investment Development Agency in 28 November, 2012. The European Union and national resources will be used more efficiently and faster for implementation of renovation of multi-apartment buildings. At the same time it aims to develop the JESSICA financial model in other areas: urban development, environment, sports infrastructure, education, culture, tourism and transport.

1 INTRODUCTION

Any commercial structure is focused on profit seeking under market conditions. To achieve this goal, entities are constantly facing risk and competition in a particular legal and political environment. A burst of the real estate bubble caused very tangible consequences of the crisis for the construction and real estate sector both, in the world and in Lithuania. Enterprises which are outside construction severely reduced investments in their building economy or real estate rent. International organizations and national governments as well as executives of specific businesses are trying to respond and mitigate the effects of the global crisis in different ways. In order to justify decisions, scientists refer to the management experience of the former crises and try to simulate the impact of certain measures.

In investigating sustainable development of construction (Black 2004), three equivalent and closely interconnected components are distinguished: ecological dimension, economic aspect and *social dimension*. It depends on the stability of the social dimension, to what level social values will be preserved for the future, social identity, social relations and social institutions. This social aspect is illustrated by:

– widely recognised and durable norms and values, e.g., bilateral communication, fairness of law-courts and respecting the rights;
– preservation of personal identity and cultural diversity (this would be a social equivalent of biodiversity);
– ability of social institutions to satisfy people's needs;
– ability of social institutions to respond to unexpected changes, e.g., environmental, economical and technological changes.

The construction and real estate market in Europe is non-homogeneous. The current situation and prospects of each Member State depend on its situation considering needs, demographic trends, the basic economic principles and so on. They also depend on the time when corrections in the real estate market took place as well as on economic openness of a specific country. Finally, they depend on the type of long-term recovery measures to be chosen and on the degree of their successful effect on the entire sector (Detemmerman 2009).

The residential sector is the largest final energy consumer and is responsible for 30% of total carbon dioxide emissions in Hungary. In order to address the general poor condition of the building stock and resulting inefficiency in energy use, from 1990 onwards, the government and local authorities initiated energy efficiency support programmes in the residential buildings sector (Czakó 2012).

Climate change and limited energy resources stimulate exploitation of energy more efficiently in a residential dwelling. This paper discusses modernization of multi-apartment buildings in Lithuania. Nearly 96% flat-houses were built before 1993. Heat energy consumption is two times higher in these buildings than in Scandinavian countries, and 1.75 times higher than in newly built apartments in Lithuania. The State promotes the modernization of multi-apartment buildings. The State provided 50% of support for investment projects until 2010.

2 MODERNIZATION OF MULTI-APARTMENT BUILDINGS IN LITHUANIA

2.1 The structure of housing resources in Lithuania and energy needed for heating the dwellings

A dwelling is understood as a part of a dwelling house consisting of one or several rooms for living, a kitchen and some subsidiary rooms separated from spaces of common use, other flats and premises by partition constructions. A dwelling house is a building accommodated for living in, where more than a half of useful space is suitable for living in. Thus, the main feature that distinguishes dwelling from other kinds of real estate is that its useful space should be suitable for people to live in.

A house of three or more apartments is considered to be a multi-apartment house. By their number multi-apartment buildings comprise merely 7.9% of total number in Lithuania (Table 1), nevertheless, when estimating the living space, one-or two—apartment houses occupy nearly the same number of square meters as multi-apartment buildings do (National Land Service State and Enterprise Centre of Registers 2012).

A high percentage of multi-apartment houses is 20 and more years old (Fig. 1). Since 1991 the number of one-or two-apartment houses has increased by 9.9%, while the housing resources in multi-apartment buildings have grown merely by 8.6%.

The European building stock is a major contributor to energy waste and CO_2 emission. Multi-storey buildings form a significant proportion of this building stock—13% in the old EU member states and more than one third in the new Central and Eastern European member states. Here, the prefabricated housing stock is predominant, characterized by an enormous maintenance backlog and very low structural and thermal quality (Zavadskas et al. 2008).

Data of investigation on energy consumption in households show that in 2009 household apartments have consumed about one third (more than 31%) of the total ultimate energy of the country. 81% of the energy has been consumed by households for heating the apartments and preparing hot water, nearly 12% have been consumed for lighting and electrical appliances and more than 7% for cooking food (Department of Statistics Lithuania).

Table 1. Distribution of number and area between multi-apartment and one- or two-apartment houses in the counties of Lithuania.

| County | Multi-apartment | | One or two apartment | |
	Number of buildings	Total area, 100 m²	Number of buildings	Total area, 100 m²
Alytus	1543	2265	33126	3537
Kaunas	7472	10670	77521	10419
Klaipėda	4665	6664	29182	4050
Marijampolė	1930	1765	31659	3496
Panevėžys	2760	3476	47296	5404
Šiauliai	3906	4462	53494	6101
Tauragė	1645	1175	23263	2821
Telšiai	1667	1948	24710	3060
Utena	1975	2289	37536	3909
Vilnius	9773	16961	76125	9602
Total in Lithuania	37336	51675	439767	53482

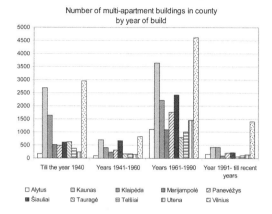

Figure 1. Number of multi-apartment buildings in county by year of build.

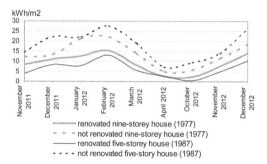

Figure 2. Amount of energy consumed to heat 1 square meter during the heating season.

Thermal properties of partition walls of the houses are built according to obligatory requirements that were valid during their construction. Since 1959 till 1992 heat transfer coefficients fixed in the regulations were 0.9–1.3 W/(m²·K). Comparing modernised and not modernised houses built in the same year and of the same design (Fig. 2), the amount of energy needed to be consumed for heating may differ by 20%–90%.

That buildings refurbishment not is only decreases energy consumption but also improves whole condition of the building: its exploitation, noise insulation conditions, exterior, and comfort; prolongs buildings life cycle, increases value of the buildings, reduces negative impact to environment and guarantees healthy living and working conditions. Satisfaction of these requirements is obligatory in sustainable refurbishment provision (Mickaityte et al. 2008).

Till the end of 2009 a renovation financing model has been applied, according to which the state compensated 50% of renovation expenditure,

however, the interest rate for borrowing money has not been fixed.

Since 2010 financial instrument JESSICA (Joint European Support for Sustainable Investment in City Areas) has been introduced: 1) up to 3% preferential credit interest up to 20 years, 2) 50% support for the project preparation, 3) 15% support for the implementation, 4) 100% support for low-income families. The citizens have chosen the primary model of the housing modernization more actively than that adopted in 2010. Construction prices have fallen on the overall economic downturn. Because the conditions of project sponsorship were reduced it resulted in interruption to for the citizens to have advantage of the low prices.

Up to now, merely two communities of multi-apartment houses have used the recent financing model.

The government approved the proposal to set up a Public Investment Development Agency in 28 November, 2012. The European Union and national resources will be used more efficiently and faster for implementation of renovation of multi-apartment buildings. At the same time it aims to develop the JESSICA financial model in other areas: urban development, environment, sports infrastructure, education, culture, tourism and transport.

2.2 The impact of the global economic crisis on the construction and real estate sector

The state not only regulates the construction process by directly setting the rules of the "game" but it also may affect it by its financial policies. For example, in Sweden an easily obtained credit, government subsidies and strong economy in the second half of the 1980s resulted in the boom of investments in the construction sector, which reached its peak in 1990 when new residential construction amounted to 69 600. During the crisis of the early 1990s, there was a decline in construction mainly due to the reform of 1990–1991 which was aimed at the elimination of state subsidies (Economic sectors: Construction 2008).

Australian scientists McKibbin and Stoeckel (2009), by applying the model of dynamic stochastic general equilibrium (DSGE), have investigated how macroeconomic measures chosen by the governments of one or another country can influence more rapid or slower recovering from the consequences of global financial crisis in separate branches of economy. 6 economic sectors (energetic, mining, agriculture, production of durable goods, production of consumer goods and services) and 15 countries/regions (the USA, Japan, UK, Germany, EU, Canada, Australia, OECD countries China India, the rest of Asia, Latin America, the rest of the countries using languages of Latin origin, East Europe and the former Soviet Union, OPEC). were chosen. Exploded real estate bubble in the USA has changed the attitudes towards risks both of households and of business. Decline in financial activities created conditions for sudden unemployment growth and caused political response. To protect the industry within the country, different measures are applied, beginning with subsidies and ending with guarding the borders. Research results have shown that the effects of the crisis in the world market were not so dramatic if the risks arisen to other countries were not so overestimated. This caused a great shrinkage of production and trade. Households are worried about their future income and limit consumption. Because of changes of expectations investments shrink and go down to the level of previous four years, and return on the investments may last even a decade. Fiscal measures have only a temporary positive effect.

A significant growth in 2004 is noticeable when comparing the range of the construction input price indices (Fig. 3) in Lithuania since 2000. This was mostly determined by Lithuania's accession to the European Union and the increased investments.

A leap of the residential housing curve in 2007 and in the beginning of 2008 expresses the boom of residential loans and the consequent increase in construction prices (especially in Vilnius). 2008 year-end and 2009 year-beginning statistics (Department of Statistics Lithuania) reflects an obvious drop in the construction price index due to the global financial crisis and the attempt of banks to reject high-risk lending. The comparison of the construction sector development in Lithuania with the same in the European Union (Fig. 4) shows an extremely sharp leap but also a greater decline in 2004–2005 (Eurostat).

Due to publicly funded projects, growth was typical of the Central and Eastern European

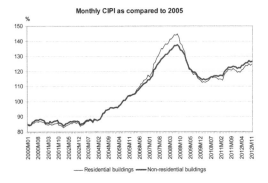

Figure 3. Construction increase with growing amounts of construction work and decrease in the period of crisis.

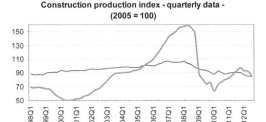

Figure 4. Construction production in European Union and Lithuania.

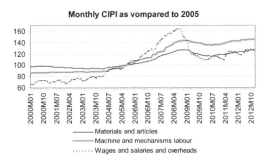

Figure 5. A monthly variation in the main construction inputs grouping.

construction sector until the middle of 2008. This branch held a higher share of GDP in the country compared to Western European countries: the construction sector in the Czech Republic makes up to 15% of GDP while it accounts for only 6.6% in Great Britain's economy. Czech builders receive orders to establish the necessary infrastructure and public buildings (hospitals and courthouses). The growth of the living standard is accompanied by increasing possibilities for the residents to take out residential loans. The Austrian group Strabag embraced the notion of growth in the East for these reasons long ago. One-third of its annual revenue of 9.43 million euros comes from this region (Spink, 2008). Since 2002, wages and overhead costs have been the fastest-growing components of the construction price (Fig. 5).

Since the beginning of economic difficulties, materials and products inputs have shifted slightly while wages and overhead costs have declined from 23.9 percent in January 2008 to −23.7 percent in March 2009 (Department of Statistics Lithuania).

In 2004–2008, the granting of loans by banks obviously intensified in Lithuania. Loans could be very easily obtained by both, construction firms planning to build blocks of flats or houses with one or two apartments and residents willing to buy their own dwelling-house (Fig. 6). The findings produced by the Bank of Lithuania in 2009 show that residential loans have been granted to 11.2% of Lithuanian households. The income of households which obtained residential loans was nearly twice the average Lithuanian household income level. Their family's average monthly income is LTL 4 540 meanwhile the average family income in Lithuania is LTL 2 422 (Bank of Lithuania).

Such a desire to borrow was determined by both, active advertising (the idea of living on credit was hardly characteristic of the post-communist society) and low interest rates on loans (which is the most expensive part of the credit). In 2005, the interest rates on residential loans fell to a record low level at around 3.9% (Fig. 7). This tendency was reinforced by favourable residential property credit terms (Bank of Lithuania).

The banks began to grant residential loans up to 100% of their value, the maximum duration of residential property credit extending to as long as 40 years. The following tax credits contributed to the development of the housing market:

– personal income tax reduction from 33% to 24%;
– the possibility to deduct interest on the residential property credit from taxable income.

Distrust in other market participants with regard to business deals has increased due to the general economic crisis. During the growth period of the construction sector, the company manager used to make verbal arrangements with the customer on the scope of works whereupon the project documentation was prepared and subcontractors hired.

Now contractors, no matter how well acquainted with each other, take no action until signing of the contract.

The crisis leads to social and psychological stress in the society as well as in a particular company as a team. This stress is conditioned by lack of information, insecurity and scarcity of social guarantees and fear of future difficulties. The tendencies of moral decline are dismaying because the crisis will pass but spiritual decline will have lasting consequences.

Heintz et al. (2009) simulated the probable economic impact of the public investment increase in infrastructural areas such as transport, water systems, energy and public school buildings. The model adopted by the US Department of Commerce showed that a billion dollar investment in infrastructure development would create 18 000 jobs; in the meantime, the equal support for households through tax reduction would create 14 000 jobs at the maximum because part of the funds would be allocated to the purchase of imports. The authors calculated the need for indispensable (basic) investments in general infrastructure for five years—87 billion dollars annually, about 54 billion of which should come from the state and 33 billion would be private investments. For comparison, they programmed the five-year demand for the high yield investments providing 148 billion dollars per year. Under the said program, 93 billion should come from the public sector and 55 billion would be investment of the private sector. The basic infrastructure program will create approximately 1.6 million new jobs in the United States of America and the high yield investment program will create about 2.6 million jobs. The construction sector shall gain the most of them—about 40 percent. This would mean about 640 000 jobs as usual scenario and 1 million jobs in accordance with the high yield investment program.

3 THE CONCEPTUAL MODEL OF CONSTRUCTION IN TIMES OF ECONOMIC RECESSION

A Model (A. Kaklauskas et al. 2011) for a construction in times of economic recession was being developed step by step as follows:

– A comprehensive quantitative and conceptual description of the construction in times of economic recession, its stages, interested parties and environment.
– Development of a complex database based on quantitative and conceptual description of the research object.
– Adaptation of methods of multiple criteria analysis developed by E.K. Zavadskas and A. Kaklauskas (2001) to carry out multivariate

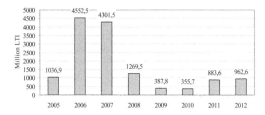

Figure 6. Residential loans.

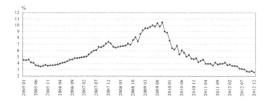

Figure 7. Interest rates on new residential loans.

design of a construction crisis, determine the utility degree of the alternative versions obtained and set the priorities.

– Development of a Multiple criteria decision support system to be used in computer-aided multivariate design of a construction in times of economic recession, determining the utility degree of the alternative versions obtained and setting the priorities.
– Multivariate design of construction in times of economic recession.
– Multiple criteria analysis of construction in times of economic recession.
– Selection of the most rational version of construction in times of economic recession.
– Analysis of micro, meso and macro level environment factors influencing a construction and possibilities to alter them in a desired direction.
– Development of rational micro, meso and macro level environment.
– The above model will be now described in more detail.

Any national economic environment is directly affected by tax, monetary, capital flow, investment, lending and interest rate policies enforced by national authorities. Furthermore, economic environment also depends on demand, supply, competition, pricing and other economic factors (Zavadskas et al. 2004). The main indicators of an economic environment are cyclic development, inflation and unemployment.

The natural environment is the one in which the built and human environment exists. Lithuanian industry of building materials and products (except for the energy industry) consumes the biggest portion of electricity and other types of energy. Lithuanian companies use outdated manufacturing facilities, hence manufacturing of building materials and products is far more energy guzzling than any equivalent in Western Europe.

Technological environment. Companies need continuous analysis of potential new technologies and their possible effect on operation. New technologies open new markets and opportunities. The more sophisticated the technology, the more specialised it is. Often not only companies specialise in a narrow technological field but also only some employees within the company are qualified for certain operations. The indispensability factor makes the labour more expensive thus raising the price of products or services.

Socio-cultural environment includes institutions, traditions and other factors that help to shape and perceive the main social values, views and norms of behaviour. Individuals grow in certain societies that shape their key values and beliefs. Settled world-views latter affect individual's relations with other people (Kaklauskas et al. 1998). Labour relations are one of the more significant types of social and cultural interaction. The level of labour relations, career opportunities and use of state-of-the-art technologies help organisations involved in construction and real estate industry make the best of their potential and improve the performance.

Booming sales of real estate were mostly determined by overestimated creditworthiness of buyers and giving in to massive efforts at improvement of living conditions. People often changed their homes succumbing to the influence of others and wishing to stand comparison with their neighbours. Similar psychological aspects prevent revival of normal market relations in times of a crisis. Someone about to buy a dwelling changes the mind just upon hearing talk about possible reorganisation in the workplace or even news about deteriorating economic indicators in a neighbouring country. People fear for their future and abandon planned investments.

Microenvironment mostly encompasses elements in direct relation to the client, the project promoter and users. The micro-level factors, of course, depend on the macro-level factors. For instance, all activities related to building's lifecycle are regulated by a range of laws and norms adopted at the macro-level.

4 MULTI-VARIANT DESIGN AND MULTIPLE CRITERIA ANALYSIS OF CONSTRUCTION IN TIMES OF ECONOMIC RECESSION

Methods for performing multivariate design and multiple criteria analysis of the alternatives developed by E.K. Zavadskas and A. Kaklauskas have been adapted to carry out multivariate design of a construction in times of economic recession, determine the utility degree of the alternative versions obtained and set the priorities (Zavadskas et al. 2001, Kaklauskas et al. 2010, Kaklauskas et al. 2011):

– A method of complex determination of the weight of the criteria taking into account their quantitative and qualitative characteristics was adapted. This method allows to calculate and coordinate the weights of the quantitative and qualitative criteria according to the above characteristics.
– A method of multiple criteria complex proportional evaluation of the alternatives enabling the user to obtain a reduced criterion determining complex (overall) efficiency, priority and utility degree of the variant was adapted. This generalized criterion is directly proportional to

the relative effect of the values and weights of the criteria considered on the efficiency of the alternative. According to this method the alternatives utility degree are directly proportional to the system of the criteria adequately describing them and the values and weights of these criteria.

– A method of multiple criteria multivariant design of a alternative construction in times of economic recession enabling the user to make computer-aided design of up to 100,000 alternative versions was adapted. Any alternative construction variant obtained in this way is based on quantitative and conceptual information.

Construction opportunities in times of either economic boom or economic recession are compared by solving the problem, which analyses implementation of two equivalent projects–refurbishment of apartment buildings, each with a total floor area of 2,016 m² —in 2007 and in 2009. Taip pat lyginti dviejų devynaukščių namų modernizavimo projektai, vykdyti 2008 ir 2010 metais, tiek įmonės, tiek investuotojo požiūriu.

The information used in the analysis is both quantitative and qualitative. The criteria systems for evaluation of construction opportunities at different economic conditions and criteria weights are determined using the integrated method for weight establishment. Objective weights of quantitative criteria depend on the level of dominance among indicators. The first portion of calculations includes 6 stages:

Stage 1. Calculation of the sum of the values for each quantitative criterion (expressed in LTL).

Stage 2. Calculation of the sum of the values of all quantitative criteria related to the project in question.

Stage 3. Calculation of the weight of each quantitative criterion.

Stage 4. Weights of quantitative and qualitative criteria are matched to come up with the benchmark *E*. Experts chose the price of materials as the benchmark for comparisons of refurbishment projects in apartment buildings, hence for calculations $E_{DNM} = 0.422$.

Stage 5. Expert methods are used to establish the initial weights of qualitative criteria pertaining to projects.

Stage 6. Weights of all qualitative criteria are combined considering the benchmark weight of the quantitative criterion.

When the weights of criteria are already known, multiple criteria analysis methods for projects are employed to estimate the degree of priority and weight of compared options (Zavadskas, 1987; Kaklauskas, 1999; Kaklauskas et al., 2001). The degree of priority and weight is directly and proportionally related to the system of defining criteria for alternatives (both their values and weights). Calculations cover Stages 7–10.

Stage 7. The initial decision-making matrix D is normalised.

Stage 8. The sums of evaluated normalised minimising indicators S_{-j} (lower value is better, e.g. standard VAT rate, interest rates, etc.) and maximising indicators S_{+j} (higher value is better, e.g. GDP growth, tax credits for loan holders, etc.) describing the option j are calculated.

Stage 9. Positives S_{+j} and negatives S_{-j} of all compared projects are considered to establish the relative weight of the options Q_j.

Stage 10. The compared options may vary only slightly, but may also diverge farther apart. The difference is easier to assess when the utility degree of alternatives is known N_j.

The developed systems of environment factors affecting construction projects provide a thorough description of construction alternatives in diverse circumstances. The systems are used for practical problem solving.

Integrated databases of criteria which define the components of the construction project environment were developed; they define the alternatives of such components both from qualitative and quantitative perspectives.

Comparing renovation of five-storied houses, in 2007 initiative of the employees as the most important feature has been distinguished (2,7%), and in 2009—index of confidence in construction (10,7%). Comparing the projects of refurbishing nine-storied houses, in 2008 the criterion of average gross payment for work in construction is notable (possible correction of the value may be 1.9%), while in 2010—the portion of additional expenditure (possible correction of the value could amount to 8.4%).

5 CONCLUSIONS

The given sector is highly influenced by bank lending policy. A marked change of this factor is often an essential condition for the crisis to occur. Faulty and risky lending policy is the underlying cause of the loan crisis in most countries. The given complex crisis forecasting models for construction and real estate allow the assessment of the distinct characteristics of the loan and the borrower as well as of varying macroeconomic conditions.

Research has shown that in 2010 both macro- and micro environment was favourable for developing renovation of multi-apartment houses, nevertheless, the policy of the state (reducing support for the residents) did not create suitable conditions for using the situation.

Even the best crises management solutions applicable in other countries cannot be copied blindly. They should be chosen exclusively depending on the specific economic, political, legal, technological, organizational, institutional, social, cultural, psychological, educational and environmental condition. There is no such thing as an integrated solution to crises management that could be suitable to all countries.

REFERENCES

Bank of Lithuania. [2013-02-03. (Website http://www.lb.lt/).

Black, A. 2004. The quest for sustainable, healthy communities. *Paper presented to Effective Sustainability Education Conference, NSW Council on Environmental Education*. UNSW, Sydney, Feb 18–20. (Website http://www.environment.nsw.gov.au/resources/cee/alanblack.pdf).

Czakó, V. 2012. Evolution of Hungarian residential energy efficiency support programmes: road to and operation under the Green Investment Scheme. *Energy Effeciency*. May 2012, Vol. 5, Issue 2, pp 163–178.

Detemmerman, V. 2009. Impact of the crisis on the construction industry. [2009-12-21]. (Web publication http://eesc.europa.eu/sections/ccmi/Hearingsandconferences/Thepast/Financial_crisis/documents/Detemmerman_Vincent.ppt).

Department of Statistics to the Government of the Republic of Lithuania. [2013-02-01]. (Website http://www.stat.gov.lt/).

Economic sectors: Construction. (Industry overview) (Geographic Overview) / Economist Intelligence Unit: Country Profile: Sweden. Economist Intelligence Unit N.A. Incorporated, 2008. (Website http://find.galegroup.com/itx/start.do?prodId=ITOF).

Eurostat. [2013-02-02]. (Website http://epp.eurostat.ec.europa.eu/).

Heintz, J., Pollin, R., Garrett-Peltier, H. 2009. How Infrastructure Investments Support the U.S. Economy: Employment, Productivity, and Growth, Political Economy Research Institute and Alliance for American Manufacturing, January, 2009. (Website http://www.peri.umass.edu/fileadmin/pdf/other_publication_types/green_economics/PERI_Infrastructure_Investments).

Kaklauskas, A., Zavadskas, E.K., Ambrasas, G. 1998. Increase of student study efficiency through the application of multiple criteria decision support systems, in Monash Engineering Education Series. *Seminar Proceeding* [2nd Baltic Region Seminar on Engineering Education Riga Technical University, Riga, Latvia]. Edited by Zenon J. Pudlowski and John D. Zakis. UNESKO International Centre for Engineering Education (UICEE) Faculty of Engineering Monash University, Clayton, Melbourne, VIC 3168, Australia, 128–134.

Kaklauskas, A. 1999. Multiple criteria decision support of building life cycle: Research report presented for habilitation. Vilnius Gediminas Technical University, Vilnius.

Kaklauskas, A., Zavadskas, E.K., Vainiūnas, P. 2001. Efficiency Increase of real estate e-business systems by applying multiple criteria decision support systems. Eighth European real estate society conference: Alicante, June 26–29, 2001: proceedings [CD]. University of Alicante, 11–24.

Kaklauskas, A., Zavadskas, E. K., Bagdonavičius, A., Kelpšienė, L., Bardauskienė, D., Kutut, V. 2010. Conceptual modelling of construction and real estate crisis with emphasis on comparative qualitative aspects description. *Editorial material. Transformations in business & economics. Vilnius University: Kaunas Faculty of Humanities* 9(1): 42–61.

Kaklauskas, A., Kelpsiene, L., Zavadskas, E.K., Bardauskiene, D., Kaklauskas, G., Urbonas, M., Sorakas, V. 2011. Crisis management in construction and real estate: Conceptual modeling at the micro-, meso- and macro-levels/Original Research Article. *Land Use Policy* 28(1): 280–293.

McKibbin, W. J., Stoeckel. A. 2009. Modelling The Global Financial Crisis. *Centre For Applied Macroeconomic Analysis Working Paper Series*. The Australian National University September, 2009. 44 p. (Website: http://cama.anu.edu.au/Working%20 Papers/Papers/2009/McKibbin_Stoeckel252009.pdf).

Mickaityte, A., Zavadskas, E.K., Kaklauskas, A., Tupenaite, L. 2008. The concept model of sustainable buildings refubishment, *International Journal of Strategic Property Management*, Vol. 12, pp. 53–68.

National Land Service State and Enterprise Centre of Registers. [2013-01-31]. LIETUVOS RESPUBLIKOS NEKILNOJAMOJO TURTO REGISTRE ĮREGISTRUOTŲ STATINIŲ APSKAITOS DUOMENYS 2012 M. SAUSIO 1 D. (Website http://www.nzt.lt).

Saiuliu energija [2013-02-04]. (Website http://www.senergija.lt/; http://mano.senergija.lt/NamoInfo/).

Spink, Ch. Construction: building the new Europe: the construction sector in Central and Eastern Europe continues to grow at above the rate of many of the region's burgeoning economies. This is encouraging acquirers to look at consolidation opportunities. (CENTRAL EASTERN EUROPE REPORT)./ Acquisitions Monthly 283 (May 2008): S18(4).

Zavadskas, E.K. 1987. Multiple criteria evaluation of technologica decisions of construction: Dissertation of Dr Sc (in Russian). Moscow Civil Engineering Institute, Moscow.

Zavadskas, E.K., Kaklauskas, A., Banaitiene, N. 2001. Pastato gyvavimo proceso daugiakriterinė analizė. Vilnius: Technika. 379 p. ISBN 9986-05-441-9.

Zavadskas, E.K., Kaklauskas, A., Viteikienė, M. 2004. Decision support of web-based system for construction innovation, in The international Conference "Reliability and Statistics in Transportation and Communication" (RelStat' 04), 14–15 October 2004, Riga, Latvia: programme and abstracts. Riga: Transport and Telecommunication Institute, 1–8.

Zavadskas, E.K., Kaklauskas, A., Tupenaite, L., Mickaityte, A. 2008. Decision-making model for sustainable buildings refurbishment. Energy efficiency aspect. The 7th International conference Environmental engineering, Vilnius, pp. 894–901.

Integrated environmental and economic assessment in the construction sector

H. Krieg, S. Albrecht & J. Gantner
Department Life Cycle Engineering (GaBi), Chair for Building Physics, University of Stuttgart, Germany

W. Fawcett
Cambridge Architectural Research Ltd., Cambridge, UK

ABSTRACT: The consideration of the full life cycle of products supports to systematically account for and reduce both life cycle costs and environmental impacts. This is especially relevant for products with a long life cycle, as it is in the construction sector. With a longer life cycle, the share of costs and environmental impacts occurring during the operation phase becomes more relevant in comparison to the initial investment costs and environmental impact. In order to facilitate results interpretation, methods for intuitive results presentation have to be applied. One method to present multidimensional economic and environmental results are ecoportfolios. This paper describes the approach that is being applied for results presentation within the CILECCTA project.

1 BACKGROUND

1.1 *Sustainability in construction*

Within the last decades, sustainability has been given increasing attention by people, organizations and states. Taking into account the preservation and protection of resources is required to allow ongoing economic activities for current and future generations.

When assessing the sustainability of products or processes, life cycle thinking should be applied. This means the consideration of all life cycle stages, from raw material extraction to production, operation and end of life. This holistic view avoids what is called a shift of burdens, that is, optimizing one life cycle stage while increasing the impact of another stage, and thereby not reaching an overall improvement.

The CILECCTA project (funded by the European Community's Programme FP7/2007–2013 for Research, Technological Development and Demonstration Activities, under European Commission Grant Agreement No. 229061; www.cileccta.eu) aims at supporting sustainability in the construction sector by providing a software tool that allows the systematic assessment of both environmental and economic aspects of construction projects. Therefore CILLECTA connects environmental Life Cycle Assessment (LCA) and monetary Life Cycle Costing (LCC) in one combined LCC+A approach.

1.2 *Life Cycle Assessment*

Life Cycle Assessment (LCA) is a method for the quantification of the environmental impacts of products or processes. The method is standardized in ISO 14040 (2006a) and ISO 14044 (2006b) and has been applied in industry for more than 20 years. The method is based on life cycle thinking. At each step along the value chain, a Life Cycle Inventory (LCI) is generated. This is an extensive list of all physical inputs and outputs that are required for and caused by each respective material, product or process. LCI data is very detailed and has some thousand inputs and outputs. In order to make the processing and interpretation of this vast amount of information easier, a Life Cycle Impact Assessment (LCIA) is conducted. Emissions are classified according to their contribution to different environmental impacts (such as Global Warming (GWP), Acidification (AP) or Eutrophication (EP)) and through a characterization factor expressed in the reference unit for each impact category (e.g. GWP in kg CO_2-equivalents). The emission of 1 kg of Methane has the same contribution to global warming as the emission of 25 kg of CO_2, therefore the characterization factor is 25 (IPCC 2009).

Through these steps, LCA expresses the environmental impact of products or processes along their entire life cycle with a small number of indicators, typically between four and eight for most studies.

1.3 Life Cycle Costing

Life Cycle Costing (LCC) is, like LCA, based on life cycle thinking. However, it focuses on the economic aspects of products or processes measured in monetary units. It is standardized for buildings and construction in ISO 15686-5 (2008). As in LCA, a detailed list of economic inputs and outputs for all materials, products and processes is created for all life cycle stages. The result of this is a detailed schedule of costs and benefits. Unlike LCA, these costs and benefits cannot be simply added up, because the time when they occur is relevant for economic analysis. This is due to interest, both on deposits and loans, and the risk aversion of investors. Therefore, cash flows are discounted with a discount factor that represents the market interest rate as well as the individual's or organization's risk rate. The sum of all discounted cash flows is called the Net Present Value (NPV). The NPV expresses the present value of long-term projects in monetary terms, and thereby provides a basis on which projects can be compared.

1.4 Life Cycle Costing and Assessment (LCC+A)

While both LCC and LCA are based on life cycle thinking, they cannot be simply combined. This is due to differences in scope, assumptions, data basis and cut-off criteria. Up to now, LCC and LCA have been regarded as separate studies using separate evaluation tools, leading to duplication of effort and problems of inconsistency between the two methods.

To overcome these problems the CILECCTA project has developed an integrated methodology that ensures the compatibility between LCC and LCA. This joint methodology is called Life Cycle Costing and Assessment, or LCC+A. It defines consistent system boundaries for the assessment, allowing LCA and LCC to be carried out in parallel within the CILECCTA software tool.

An LCC+A study on the economic and environmental impacts of the heating systems in a non-domestic residential building is used as a case study for explaining the CILECCTA tool in this paper.

1.5 Life Cycle Costing and Assessment (LCC+A) under future uncertainty

Current methods of life cycle evaluation, both LCC and LCA, assume that exact data about the future is available over the whole building life or study period. This can be called the deterministic approach to LCC and LCA. The precise data inputs produce precise output values for life cycle costs and environmental impacts, but in reality precise data is rarely if ever available, and the apparently precise output value are actually approximations.

CILECCTA moves from a deterministic to a probabilistic approach, taking account of future uncertainty. In CILECCTA it is possible to describe factors like component service lives and future costs by a range of values, not a single predicted value. With inputs of this type, the software runs multiple scenarios over a range of possible futures that are consistent with the data ranges. The findings of this approach to life cycle evaluation take the form of probability distributions, providing a more realistic basis for decision-making when our knowledge of the future is uncertain.

CILECCTA's probabilistic approach allows a further advance, by providing for the evaluation of flexible design alternatives. Flexibility can be an effective strategy when the future is uncertain, because it allows new decisions to be made in the light of unfolding events that are unpredictable at the time of design. By reducing the impact of uncertainty, flexibility increases project value. And the greater the level of uncertainty, the greater the value of flexibility (Fawcett et al, 2012).

2 MULTIDIMENSIONAL RESULT INTERPRETATION

2.1 Multi-dimensional result and the ecoportfolio method

The CILECCTA tool aims at providing a software tool that can be widely applied in the construction sector. It allows calculation of both environmental and economic aspects of construction projects. The environmental and economic results fall into two fundamentally different dimensions that have to be compared. Furthermore, LCA results are expressed in several impact categories. For the underlying study, there were a total of six environmental indicators: Global Warming Potential (GWP), Abiotic Resource Depletion Potential (ADP), Acidification Potential (AP), Eutrophication Potential (EP), Ozone Depletion Potential (ODP) and Photochemical Ozone Creation Potential (POCP). In addition, two energy indicators assessed: renewable and non-renewable Primary Energy Demand (PED). In combination with the monetary NPV, this adds up to a total of nine result indicators. Figure 1 shows the results of comparing five different heating systems.

As it can be seen, there is no clearly dominant scenario that performs best. This is the case for most studies. Therefore, results have to be interpreted in order to rank different alternatives and assess their suitability to fulfill certain targets. However, this task requires detailed knowledge of

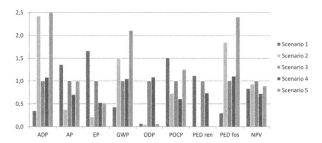

Figure 1. Combined LCA and LCC results.

both environmental impacts as well as of target values.

In order to simplify the decision making process, ecoportfolios have been developed, as defined in the international standard for eco-efficiency assessment ISO 14045 (2012). In those portfolios, environmental impacts are summarized to a single impact factor, based on normalization and weighting. They are then combined with the so-called product system value. This is defined as the "numerical quantity representing the product system value" (ISO 14045-2012). It can be expressed in monetary values, but also other quantified values. This allows the presentation of different alternatives as points within a two-dimensional portfolio.

2.2 The CILECCTA ecoportfolio

The result presentation within CILECCTA is closely based on the definition as found in the standard. However, in some points it goes beyond the specifications of ISO 14045 (2012).

As CILECCTA offers an integrated solution of environmental and economic aspects, the numerical expression of the system value is always expressed in monetary units. The economic assessment is based on the LCC approach and is expressed as the NPV of different alternatives. Therefore, the numerical quantification is here always the net present value; depending on the scope of the study the NPV can include costs, incomes or both.

For the environmental part of the assessment, the current version of CILECCTA uses Ökobau. dat (2009) data, a standardized German construction database. Environmental impacts are reported in six environmental impact indicators and two energy indicators. As noted above, these are the indicators that are taken into account for the environmental assessment.

CILECCTA follows a slightly different approach for the normalization and weighting of environmental indicators when integrating them to a single point indicator.

For the normalization in CILECCTA, objective values are used. This means that the alternatives are not normalized in relation to other alternatives, but on the basis of annual European emissions. When one alternative is used as the basis on which other alternatives are normalized, the ranking of scenarios can change. This can be due to extreme values in some impact categories, which then are transmitted to the overall evaluation. Therefore, different alternatives can appear to have the lowest overall environmental impact, depending on which is used as basis for normalization. This can be prevented by using objective values, which is why CILECCTA is using the European emissions as a base for normalization. This ensures further objectivity of the normalization step.

For the weighting of environmental impacts, two methods are available. One is to rely on expert weighting sets. These are available within dedicated LCA software tools such as the GaBi software (PE 2013), or can be taken from sustainability assessment schemes such as BREAM, or they can be based on judgment of sustainability and LCA experts. Another method that CILECCTA will offer is individual weighting factors, allowing planers or organizations to express their preferences or environmental strategies in a weighting. By offering both expert and individual weighting, the ecoportfolio can be widely applied. It can be used by non-LCA experts to generate easy to understand environmental impact factors that can be the basis for decision making and internal communication. It also allows more experienced users to bring in their experience or to represent their organization's goals. If users are interested in more detailed results, CILECCTA makes the results available on an indicator level in a transparent and consistent way. It also provides documentation of weighting and normalization factors.

While the weighting and normalization of environmental impacts adds subjectivity, it also allows integrating a large number of indicators to a single environmental indicator. This again allows generating a two-dimensional ecoportfolio, that allows to easily compare different scenarios based on their environmental and economic aspects.

2.3 Decision making with the CILECCTA ecoportfolio

The CILECCTA ecoportfolio takes a further normalization step. As this is based on previously weighted and normalized results, it has no impact on the ranking of alternatives. One of the alternatives is used as a base scenario; the others are then normalized to this alternative. The final portfolio presentation is shown in Figure 2.

In this case, alternative 1 was used as basis for normalization, as it is the currently used heating system. Because factors like component service lives and future costs by ranges of values, not single predicted values, the findings of the life cycle evaluation take the form of probability distributions. For this reason the results on the ecoportfolio are not single points but look more like clouds, as can be seen in Figure 2. The probabilistic results provide a more realistic basis for decision-making. The size of the cloud represents the am ount of uncertainty involved in each alternative: a bigger area means more uncertainty, while a smaller area is related to smaller deviations from the mean value. In addition the intensity of different areas in the cloud represents the probability of occurrence. The paler the area is the less is the probability of occurrence.

Derived from the base alternative, there are four different segments:

– Right top: alternatives in this segment have both higher costs and higher environmental impacts. They are therefore inefficient and should not be selected
– Left bottom: alternatives in this segment have both lower costs and lower environmental impact. They are more efficient and allow improvement of both economic and environmental aspects

– Left top: alternatives in this segment have higher costs but lower environmental impacts. It depends on the preferences of the user whether they are preferable or not
– Right bottom: alternatives in this segment have lower costs, but higher environmental impacts. It depends on the preferences of the user whether they are preferable or not.

As it can be seen in Figure 2, there are only two alternatives in the study that are efficient, alternative S1 and alternative S4. Other alternatives can be neglected for further assessment.

The preference between alternatives S1 and S4 depends on the decision maker's relative weighting of economic factors against environmental factors; this is likely to vary between decision makers. The relative weighting for any given decision maker can be represented by contour lines on the ecoportfolio. The decision maker then seeks the alternative that lies on the highest contour. If decision makers have different relative weightings they will also have different contours, and this may lead to the selection of different alternatives.

Thereby, this approach is an easy to use interpretation tool for integrated LCA and LCC studies.

3 SUMMARY

The approach presented in this paper is closely in line with the ISO 14045 (2012) standard and enhances it when necessary to improve the transparence and simplification aspects. It transfers a large number of environmental indicators into a single environmental impact factor, which can be presented in an easily understandable two-dimensional ecoportfolio. This allows the interpretation of both LCA and LCC results to a wider public, increasing the relevance of both methods for organizational decision making. This is an important step to increase support for sustainability efforts.

While being closely in line with the ISO standard there are some significant differences, especially the use of objective values for the normalization of the environmental indicator values, which ensures a stable order of alternatives. If one of the alternatives under assessment is used as basis for normalization, the ranking can change. Therefore, CILECCTA only uses objective values for the calculation of the environmental impact factors.

The second normalization step has another advantage, apart from the creation of four segments. By representing the NPV as a normalized value without a monetary unit, the chance of misinterpretation is reduced. Non-LCC experts sometimes understand the NPV as the actual costs

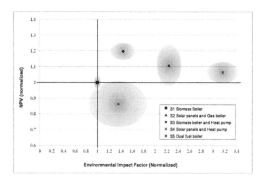

Figure 2. Ecoportfolio.

of a product system. Even though it is expressed in Euro or any other currency, it does not represent the costs, but gives information about the present value of all cash flows occurring in the product system.

Nevertheless, it has to be kept in mind that weighting of impacts brings in subjectivity, so the results are therefore not strictly objective. Also, when creating a single environmental impact factor, different environmental impacts are related and compared to each other. When creating single results, it is important that the weighting and normalization factors are documented. Furthermore, the underlying LCA results have to be available for a more detailed interpretation. When these requirements are fulfilled, ecoportfolios are a suitable tool for the presentation of LCA and LCC results in an integrated way.

REFERENCES

W Fawcett, M Hughes, H Krieg, S Albrecht, A Vennstrom. (2012). Flexible strategies for long-term sustainability under uncertainty Building Research & Information vol. 40, no. 6.

Intergovernmental Panel on Climate Change (IPCC) (2009). IPCC Guidelines for National Greenhouse Gas Inventories. Task Force on National Greenhouse Gas Inventories (TFI) of the IPCC. Washington D.C, USA.

ISO 14040 (2006a). Environmental management—Life cycle assessment—Principles and framework.

ISO 14044 (2006b). Environmental management—Life cycle assessment—Requirements and guidelines.

ISO 14045 (2012). Environmental management—Eco-efficiency assessment of product systemrinciples, requirements and guidelines.

ISO 15686-5 (2008). Building and constructed assets—Service-life planning—Part 5: Life-cycle costing.

Ökobau.dat (2009). LCA database for sustainability assessment in construction. http://www.nachhaltiges-bauen.de/baustoff-und-gebaeudedaten/oekobaudat.html.

PE (2013): GaBi 5. Software-System and Databases for Life Cycle Engineering. Copyright, TM. Stuttgart, Echterdingen 1992–2013.

Green Design, Materials and Manufacturing Processes – Bártolo et al. (eds)
© 2013 Taylor & Francis Group, London, ISBN 978-1-138-00046-9

Sustainable development and business: Past and future

M. Leščevica
Vidzeme University of Applied Sciences, Valmiera, Latvia

ABSTRACT: Sustainable development means balance among social, environmental and economic intentions. This article discusses the past and the future research concerned with sustainable development, business and entrepreneurship. Entrepreneurship has been acknowledged as a major origin for sustainable products and processes, and new ventures are being held up as the only solution for many social and environmental concerns. There remains considerable uncertainty regarding the nature of business's role and how it may develop. There are also suggestions for further development and research.

1 INTRODUCTION

Originally the term 'sustainable development' was first mentioned at the United Nations Conference on the Human Environment in 1972.

Later more described in a report to the United Nations by the World Commission on Environment and Development (WCED, 1987).

Definition coming from this report, emphasizes the dynamic aspect of sustainability (see Fig. 1).

*"**Sustainable Development** is development that meets the needs of the present generation without compromising the ability of future generations to meet their own needs" (WCED, 1987).*

Later, it has been recognised that in economics as in ecology, the rules of interdependence apply and that isolated actions are not recommended or even impossible.

For example, a policy which is not carefully thought through will have various adverse effects, not only on the economy but equally for the environment.

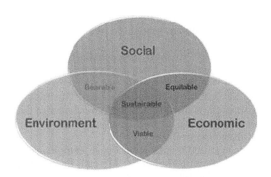

Figure 1. Sustainable development (WCED, 1987).

Also social aspect of development will have affected on environmental and economical issues and vice versa.

Research's also identified that sustainable development can be applied also to the entrepreneurship and entrepreneur, where basic idea may come from three fields of development.

Several researchers in journal articles had advanced the idea that entrepreneurship may be a solution for many social and environmental concerns (Brungmann and Prahalad, 2007; Handy, 2003; Hart and Milstein, 1999; Senge et al., 2007; Wheeler et al., 2005).

Many research books that declared warnings of environmental disaster often end on an optimistic note—salvation of civilization rests upon the shoulders of heroic social and environmental entrepreneurs (Brown, 2006; Homer-Dixon, 2006; Lovins et al., 2004; Vaitheeswaran, 2003).

2 MODERN TYPE OF ENTERPRISE

World globalisation, crises and economy transformation has led to the more challenging division of enterprises (see Fig. 2).

The classical point of view is '**economic enterprises**' which development is driven mainly by profit making—business goal. Prime value of such enterprises is financial returns/gains, where other values are secondary and often forgotten.

In the past a variety of official definitions of social enterprise reflected varying approaches within the European Union and the European Commission and created confusion. Member states national definitions are focusing on legal form or social problems to be solved.

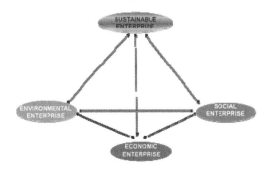

Figure 2. Different types of enterprise (Young & Tilley, 2003).

New EU definition used in Social Business Initiative:

- Social enterprises seek to serve the community's interest (social, societal, environmental objectives) rather than profit maximisation.
- They often have an innovative nature, through the goods or services they offer, and through the organisation or production methods they resort to.
- They often employ society's most fragile members (socially excluded persons).
- They thus contribute to social cohesion, employment and the reduction of inequalities. (www.nesst.org).

EU regional, employment and social policy for 2014–2020 foresees funding for following activities:

- **extend the support** given to microcredit providers under the current European Progress Microfinance Facility (launched in 2010)
- provide funding for **capacity-building** of microfinance institutions
- include investments for **developing and expanding social enterprises**, i.e. businesses whose primary purpose is social, rather than to maximise profit distribution to private owners or shareholders.

The total proposed budget for the microfinance and social entrepreneurship axis is around EUR 192 million for the period 2014–2020. Access to microfinance would receive EUR 87 million that could result in EUR 400 to 450 million of microloans. Institutional capacity building would receive almost EUR 9 million and EUR 95.5 million would be dedicated to support to social enterprise development (www. ec.europa.eu).

Social enterprises in some countries like Latvia are understood (miss-interpreted) as non-profit

organizations, but increasingly there are changes and opinion of for-profit social ventures or hybrid profit/non-profit is starting to prevail.

Prime value for such companies is social values, and usually economic and environmental values play a secondary role (see Fig. 2).

As examples, William Drayton can be mentioned (founder of Ashoka), the first to promote social entrepreneurship in the World. Davis (2002) follows Drayton with inventing the term 'social entrepreneur'. Drayton (2002) and Ulhoi (2004) considers both social and economic entrepreneurs as having the same core temperament. They recognise the need for systematic change in society and are seeking to find the way to shift the balance of society onto a new path.

Social entrepreneurs play the role of change agents in the social sector by:

- Adopting a mission to create and sustain social value,
- Recognising and relentlessly pursuing new opportunities to serve the mission,
- Engaging in a process of continuous innovation, adaptation and learning,
- Acting boldly without being limited by resources current in hand,
- Exhibiting heightened accountability to the constituencies served and for the outcomes created." (Dees, 2001).

Environmental enterprises are those who activities seek to improve the environment.

Usually the target is environmental benefits, but economic and social values play a secondary role.

There are two main types of environmental enterprises—environment-conscious entrepreneurs and green business.

Volery (2002) describes environment-conscious entrepreneurs as being aware of the issues but not operating in the environmental marketplace. They more typically follow a business case for their environmental activities by striving for eco-efficiency in the use of resources. This type of ecopreneurship may be more common but they have had limited success in the move toward sustainability.

Isaak (2002) defines **'green businesses'** as pre-existing organisations that discover the advantages of environmental innovation or marketing. They can be found in all industry sectors. This type of ecopreneur is more radical than the former in that they are seeking to find environmentally-centred business opportunities.

Volery (2002) calls them '**green entrepreneurs**', still seeking a profit. Isaak (2002) calls them '**green-green businesses**', they are the product of the counter culture in that *the entrepreneur seeks not* *just profit, but also to be environmentally responsible whilst making a social statement.*

These entrepreneurs are aware of environmental issues, because their organisations operate in the environmental marketplace:

- recycling and disposal of waste;
- remediation of polluted areas;
- water treatment;
- engineering and consulting services.

The intention of the entrepreneur is to design products and processes that are 'green'.

The **Sustainable enterprise** seeks everywhere **sustainable wealth generation** (economic, social, and environmental wealth). Prime values for these enterprises are Economic, Social, Environmental; it means all values must be considered equally.

3 SUSTAINABLE ENTERPRISE

Crals and Vereeck (2004) described and defined sustainable entrepreneurship as continuing commitment by businesses.

- to behave ethically
- contribute to economic development
- while improving the quality of life for the workforce, their families, the local and global community as well as future generations.

Sustainable business model is different from traditional business model. Sustainable business model has following characteristics:

- Use all three goals with equal importance;
- Uses waste as resource for the system development;
- Uses internal resources more than external;
- Calculates and includes costs of environment.

Sustainable enterprise uniqueness is based mainly on sustainable development appliance to traditional enterprise.

Reasons for a Business to become Sustainable:

- Volatile energy prices;
- Increases in raw material costs;
- Increases in waste and disposal costs;
- Changes in waste legislation;
- Changes in environmental laws;
- Changes in customer demands and expectations;
- Changes in competitive advantages
- Transparency issues;
- The acquisition, retention and motivation of employees;
- The cost of procrastination;
- Additional costs as result from waste (Scott, 2010).

These are the main reasons for companies to change their attitude to other goals for sustainable development.

4 CONCLUSIONS AND FUTURE RESEARCH FIELDS

This article outlined some of the recent contributions exploring the role of entrepreneurship in sustainable development.

For the potential future research directions concerned with sustainable entrepreneurship, transformation is needed to reduce or balance environmental and societal impacts created by currently unsustainable business practices.

Ecopreneurship is described, explained and compared in relation to other forms of environmental management.

Sustainable development still remains fuzzily defined; it has been described as an influential concept for entrepreneurship policy, practice, and theory.

The relationship between sustainable development and business is more prescriptive than descriptive and at this development stage, overly optimistic.

The emergence and legitimization of sustainable development within business and entrepreneurship policy brings changes to the rules of the game. There will be winners and losers with implications for both existing and new companies.

For future research it remains an important question as to whether, and to what extent, entrepreneurs have the potential for creating sustainable economies. Proving this scientific remark will require Author's insights into a number of related researches, some of which are fundamental to entrepreneurship.

There must be done more research on sustainable business to clarify facilitating conditions under which entrepreneurship can simultaneously achieve economic growth, advancing social and environmental goals.

REFERENCES

5.1.1.1 **New EU Programme for Social Change and Innovation, www.ec.europa.eu, accessed February 28, 2013**.

5.1.1.2 **New EU Programme for Social Change and Innovation, www.ec.europa.eu, accessed February 28, 2013**.

Brown, L. 2006. Plan B 2,0: Rescuing a Planet under Stress and a Civilization in Trouble. W.W. Norton, New York.

Brungmann, J., Prahalad, C., 2007. Cocreating business's new social impact. Harvard Business Review 85 (2), p. 80–90.

Crals, E., Vereeck, L., 2004. Sustainable entrepreneurship in SMEs. Theory and Practice, 3rd Global Conference in Environmental Justice and Global Citizenship, February, Copenhagen, p. 2.

Davis, H., 2002. The Howard Davies Review of Enterprise and the Economy in Education, HMSO, Norwich.

Davis, H., 2002. The Howard Davies Review of Enterprise and the Economy in Education, HMSO, Norwich.

Dees, G.J., 2001. The Meaning of Social Entrepreneurship. Stanford University, Graduate School of Business, Stanford.

Dees, G.J., 2001. The Meaning of Social Entrepreneurship. Stanford University, Graduate School of Business, Stanford.

Drayton, W., 2002. The citizen sector: becoming as competitive and entrepreneurial as business, California Management Review 44.3, p.120–132.

Drayton, W., 2002. The citizen sector: becoming as competitive and entrepreneurial as business, California Management Review 44.3, p.120–132.

Handy, C., 2003. Helicoptering up. Harvard Business Review 81 (8), p. 80–93.

Handy, C., 2003. Helicoptering up. Harvard Business Review 81 (8), p. 80–93.

Hart, S., Milstein, M., 1999. Global sustainability and the creative destruction of industries. Sloan Management Review 41 (1), p. 23–33.

Hart, S., Milstein, M., 1999. Global sustainability and the creative destruction of industries. Sloan Management Review 41 (1), p. 23–33.

Homer-Dixon, T., 2006. The Upside of Down: Catastrophe, Creativity, and the Renewal of Civilization. Random House, New York.

Homer-Dixon, T., 2006. The Upside of Down: Catastrophe, Creativity, and the Renewal of Civilization. Random House, New York.

Isaak, R., 1998. Green Logic: Ecopreneurship, Theory and Ethics. Sheffield, UK, Greenland Publishing.

Isaak, R., 1998. Green Logic: Ecopreneurship, Theory and Ethics. Sheffield, UK, Greenland Publishing.

Isaak, R., 2002. The making of the ecopreneur. Greener Management International 38, p. 81–92.

Isaak, R., 2002. The making of the ecopreneur. Greener Management International 38, p. 81–92.

Lovins, A., Datta, E.K., Bustnes, O., Koomey, J.G., Glasgow, N.J., 2004. Winning the Oil Endgame. Rocky Mountain Institute, Snowmass, CO.

Lovins, A., Datta, E.K., Bustnes, O., Koomey, J.G., Glasgow, N.J., 2004. Winning the Oil Endgame. Rocky Mountain Institute, Snowmass, CO.

NESsT develops sustainable social enterprises that solve critical social problems in emerging market economies, www.nesst.org, accessed February 28, 2013.

NESsT develops sustainable social enterprises that solve critical social problems in emerging market economies, www.nesst.org, accessed February 28, 2013.

Senge, P., Lichtenstein, B., Kaeufer, K., Bradbury, H., Carroll, J., 2007. Collaborating for systemic change. MIT Sloan Management Review 48 (2), p. 44–53.

Senge, P., Lichtenstein.

Transformation of Economy, www.naturalstep.org, accessed February 28, 2013.

Ulhoi, J.P., 2004. The social dimension of entrepreneurship. Technovation. Open source development: a hybrid in innovation and management theory. Management Decision, 42, p. 1095–1115.

Vaitheeswaran, V.V., 2003. Power to the People: How the Coming Energy Revolution Will Transform Industry, Change Our Lives, and Maybe Even Save the Planet. Farrar, Straus and Giroux, New York.

Volery, T., 2002. Ecopreneurship: Rationale, current issues and future challenges, in Fugistaller, U., Pleitner, H.J., Volery, T., and Weber, W. (eds.) Radical changes in the world—will SMEs soar or crash? Rencontres Conferences, St. Gallen, Switzerland.

WCED: The World Commission on Environment and Development, 1987. Our Common Future, Oxford University Press, New York.

What is Social Entrepreneur? Ashoka Foundation, 2003, http://edit.ashoka.org/social_entrepreneur, accessed February 28, 2013.

Wheeler, D., McKague, K., Thomson, J., Davies, R., Medalye, M., Prada, M., 2005. Creating sustainable local enterprise networks. MIT Sloan Management Review 47 (1), p. 33–40.

Young, W. and Tilley, F. (2003) Can Businesses Move Beyond Efficiency? The Shift toward Effectiveness and Equity in the Corporate Sustainability Debate, paper presented at the Greening of Industry Conference, San Francisco.

Green Design, Materials and Manufacturing Processes – Bártolo et al. (eds)
© *2013 Taylor & Francis Group, London, ISBN 978-1-138-00046-9*

Corporate sustainability measurement in Portuguese manufacturing organizations

Pedro Mamede
Process Advice, Lda., Portugal
University of Coimbra—School of Economics, Portugal

Carlos F. Gomes
University of Coimbra—School of Economics, Portugal
ISR—Institute of Systems and Robotics, Portugal

ABSTRACT: Given the recent emphasis on sustainable development and its influence on organizational performance measurement and management, the objective of this study is to examine the current views of manufacturing executives on key aspects of performance measuring. Specifically, this research focuses on current practices related to extent of use, predictive value and availability of information for seventy-four (74) performance measures. The results point to the lack of a broad perspective on corporate sustainability performance measurement.

1 BACKGROUND

In the last decade, business organizations have been increasingly under pressure to embrace the sustainable management philosophy and to integrate it into their performance measurement systems (Bonacchi & Rinaldi, 2007; Hubbard, 2009). In this context, business organizations are facing three main challenges:

- The dichotomy between people and their natural environment, as they seek to meet present aspirations (Barkemeyer *et al.*, 2011).
- Inter-generational justice, to promote future development and establish temporal equity (Sikdar, 2003).
- The assumption that all the world is interconnected and responses should be on a global level and as though to a single unit (Blasco, 2006).

To overcome these challenges and incorporate sustainability into their activities and processes, executives of business organizations should cover three important dimensions (Blasco, 2006; Tregidga & Milne, 2006; Labuschagne *et al.*, 2004):

- The economic dimension, which is based on prosperity through value creation, tradable on markets.
- The environmental dimension, which is based on the preservation of biodiversity as a result of the balance between human needs and the regenerative ability of the environment.
- The social dimension which is based on equity through inclusive processes for human rights and freedoms, universally accepted.

The conceptual foundations of sustainability and its applicability to business organizations can focus on these three main approaches:

- Business organizations should only be concerned with the creation of value by focusing on economic sustainability and shareholder satisfaction (Friedman, 1970). In this context, business organizations tend to gradually integrate environmental and social sustainability forced by values, customs, and legislation.
- Business organizations should integrate the impacts of their activities according to critical global issues in terms of ecological and social systems (Robèrt, 2000; Richards & Gladwin, 1999). Therefore, executives of business organizations should avoid planning and controlling their operations based solely on external pressures. They must anticipate the changes ongoing in society and in the environment (Richards & Gladwin, 1999).
- Business organizations should promote their operations within a framework that should meet stakeholder expectations (Epstein & Roy, 2001; Schaltegger & Burritt, 2009; Schaltegger, Herzig, Kleiber, & Müller, 2003; Skouloudis, Evangelinos, & Kourmousis, 2009; Hubbard, 2009). This approach underlines the importance of dialogue with stakeholders as a means of ensuring sustainability (Tregidga & Milne, 2006; Wilson, 2003).

Executives who want to follow a sustainability approach in the design, implementation, and utilization of organizational performance management systems may face several challenges, including:

- the involvement of stakeholders in the analysis and formulation of corporate strategy (Borga *et al.,* 2009);
- the consistent integration of objectives, targets and actions in accordance with the corporate sustainability vision (Schaltegger & Wagner, 2006);
- the promotion of transparency in reporting compliance (Lamberton, 2005).

According to the literature the corporate Sustainability Performance Measurement Systems (SPMS) fall into four distinct groups:

- Global systems—Based on global/world sustainability indicators translated into strategic and process indicators at enterprise level (Richards & Gladwin, 1999; Robèrt, 2000).
- Stakeholder systems—Based on the identification of expectations and critical issues through dialogue with stakeholders, translated in the formulation of indicators associated with the results of the engagement process (Von Geibler *et al.*, 2006; Bonacchi & Rinaldi, 2007).
- Triple Bottom Line (TBL) systems—Based on the methodological structure across the three dimensions of sustainability (economic, environmental, and social), including product life cycle (Sikdar, 2003; Bakshi & Fiksel, 2003; Hubbard, 2009).
- Adapted systems—Based on the traditional methodologies used in strategic and operational contexts originally not sustainable-based (eg. Sustainability Balanced Sscorecard), integrating one or several aspects of corporate sustainability (Schaltegger & Wagner, 2006; Bonacchi & Rinaldi, 2007; Staniškis & Arbačiauskas, 2009).

The existing Performance Measurement Systems (PMS), such as the Balanced Scorecard, have been used as a shortcut to satisfy corporate sustainability performance needs. These PMS have characteristics similar to those required to measure the performance of corporate sustainability, including:

- the involvement of stakeholders (Lo, 2010),
- the linkage between the needs of stakeholders and company operational activities (Blasco, 2006; Hubbard, 2009; Schaltegger & Wagner, 2006),
- the integration of the three sustainability dimensions (economic, environmental and social) (Epstein & Roy, 2001; Hubbard, 2009; Skouloudis *et al.,* 2009; Schaltegger *et al.,* 2003; Schaltegger & Burritt, 2009; Bansal, 2005; Adams & Frost, 2008; Kleine & von Hauff, 2009).

The utilization of sustainability reports has been the most common response to sustainability

performance monitoring (Hubbard, 2009). However this practice has some limitations:

- they are not a component of the conventional economic reports (Schaltegger & Wagner, 2006; Hubbard, 2009);
- they only focus on the positive aspects of performance (Hubbard, 2009);
- they are descriptive and lack measures to be used to benchmark performance (Lamberton, 2005; Cooper & Owen, 2007);
- frameworks used to collect, analyse, report and audit have an internal orientation and do not involve other stakeholders (Perrini & Tencati, 2006; (Hubbard, 2009);
- they are much more focused on the environmental dimension than on the social dimension (Hubbard, 2009).

Despite of the progressive integration of the sustainability approach (Bansal, 2005), business organizations that want to be competitive in the global market still find it hard to change their traditional performance systems. The following are the main barriers to this change:

- Value creation continues to be assessed by the financial performance dimension (Bansal, 2005).
- The perception that measuring financial value is most important (Robinson *et al.,* 2006).
- The lack of experience in measuring the non-financial performance aspects (Perez & Sanchez, 2009).

In order to override these barriers, the sustainability measures selected need to be specific to each business organization and be (Tanzil & Beloff, 2006; Székely & Knirsch, 2005; Staniškis & Arbačiauskas, 2009):

- consistent and reproducible
- complementary to legislation and regulations
- useful in decision-making
- able to systematically consider each stage in the product life.

In this context, the objective of this study is to gain an understanding of practices related to corporate sustainability performance measures and measurement based on twelve major sustainability indicator guidelines. The performance measurement practices in terms of utilization, relevance, and availability of information are studied for a sample of executives in Portuguese manufacturing firms.

2 METHODOLOGY

2.1 *Sample and instrument*

The data for this exploratory investigation were obtained from a convenience sample.

The participants belong to the PROCESS ADVICE marketing database.

Twenty-nine (29) completed responses from executives of Portuguese manufacturing organizations were used in this research.

Almost all of these manufacturing organizations are small-to-medium enterprises, according to European Union classification. Only three of these organizations did not implement any type of certification.

Based on the analysis of twelve of the major sustainable measurement guidelines, seventy-four (74) performance measures were included in the research instrument. According the classification used by these guidelines, measures are organized in three dimensions: economic, environmental, and social. For each of the measures included in the instrument, executives were asked to classify the nature and characteristics of each measure used in the instrument on a 1 to 5 Likert-type scale.

2.2 Models, variables, and data analysis

The data obtained from the participants was analysed using regression analysis and gap analysis. The objective of the data analysis was to obtain a profile of the participating executives in terms of the extent to which they used measures to assess the different aspects of their organizational performance. This profile was obtained using multiple regression analysis in which the frequency of use of a given performance measure (FU) was assumed to be a function of its Predictive Value (PV) and the ease with which information for the measure could be acquired (EA). Thus, the model tested was:

$$FU = f(PV, EA) \tag{1}$$

This model was used to evaluate the profile of the executives in relation to the relative use of economic, social, and environmental measures.

Finally, gap analysis was utilized to gain a better understanding of the relative importance of the particular performance dimension as perceived by the executives. The differences between the predictive value and the ease of information acquisition for each of the 74 measures were examined. These differences were then multiplied by their predictive values to find the GAP indicator, as below:

$$GAP_i = (PV_i - EA_i)PV_i \tag{2}$$

The differences were multiplied by their predictive values to provide scores that reflect the relative importance of the predictive value for the measure utilized (Dempsey et al., 1997; Gomes et al., 2004; Gomes et al., 2011). In this context, the larger the gap indicator the greater the disparity between

the usefulness of the measure and its information availability.

3 RESULTS

3.1 Regression analysis

The model initially proposed in the methodology section was used to evaluate the utilization profile of performance measures. Thus, the linear function to be estimated was:

$$\overline{FU}_i = \alpha_0 + \alpha_1 \overline{PV}_i + \alpha_2 \overline{EA}_i + e_i \tag{3}$$

The observation unit used in this model was the average of the responses of the executives surveyed for each measure. The use of regression analysis in this manner is consistent with (Hair et al., 2009).

After verifying the assumptions relevant to linear regression a stepwise procedure was used to select variables to include in the model. This procedure resulted in only one of the independent variables being included (PV).

The regression results (Table 1) point to a high R^2 of 0.954, thus showing that ninety-five percent of the total variability in the frequency of use has been explained only by the predictive value.

Since a behaviour profile is being analysed deviations from the profile must be evaluated. It is therefore important to evaluate the measures used in this study to assess their departure from the profile. The regression model below was used for this.

$$FU = -0.730 + 1.212 \, VP \tag{4}$$

Table 2 reveals a set of measures, with positive signs representing the measures most used. It also has a set of measures with negative signs representing those least used.

The group of measures most used includes five measures from the economic category and three from the social category. The greater utilization

Table 1. Regression results.

R	R^2	Adjusted R^2	Std. error of the estimate
0.977	0.954	0.952	0.1957

Model	Unstandardized coefficients		T	Sig.
	B	Std. error		
(Constant)	−0.730	0.089	−8.167	0.000
PV	1.212	0.132	9.165	0.000

Table 2. Departure of residual errors from the estimated manufacturing profile.

Measure (category)
Significant positive residuals*
Net sales (EC)
Net profit (EC)
Revenue from sales of assets (EC)
Return on equity (EC)
Non-conformities detected during production cycle (SO)
Value of contributions to public policies (SO)
Product/service risk assessment related to client (SO)
Return on assets (EC)
Significant negative residuals**
Location and size of land owned (EN)
Total number of external training/education actions (SO)
Monetary value of fines related to product utilization by client (SO)
Total number and volume of significant spills (EN)
Lower wage/local minimum wage ratios (EN)
Accidents and/or occupational diseases (SO)
Number of occurrences of child labour (including subcontractors) (SO)

*more use; **less use.

of traditional performance measures (financial and non-financial) in manufacturing environments should be noted.

Within the least used measures group there are three measures from the environment category, and four from the social category. The inclusion of only non-financial measures confirms the difficulty of using this type of performance measure, mentioned in the literature.

3.2 GAP analysis

To understand the reasons behind the apparent lack of relative use of some of the performance measures, by executives, the relationships between the predictive values and the ease of information acquisition values for each of the 74 measures were examined using the GAP indicator.

As mentioned in the methodology section, the larger this indicator (GAP) the greater the disparity between the usefulness of the measure and its availability. Negative or relatively small values for the gap indicator indicate a surfeit of information. Thus the measures studied were divided into two groups.

The first group included forty measures with negative indicators, indicating an excess of information about the usefulness of these measures. Table 3a presents the five measures with the lowest values. All the measures in this group are social in nature. This excess of information seems to be

Table 3a. Measures with a negative GAP indicator.

Measure (category)
Total number of training hours (SO)
Total workforce by employment type (SO)
Total of health and safety training hours (SO)
Value of fines related to product utilization by client (SO)
Non-conformities detected during production cycle (SO)

Table 3b. Measures with a positive GAP indicator.

Measure (category)
Impact of activities, prod., and serv. on biodiversity (EN)
Cost per unit produced (EC)
Environment externality cost (EC)
Total direct and indirect greenhouse gas emissions (EN)
Indirect economic impacts (EC)

associated with organizational requirements, which means that such information must be provided to official entities (e.g. total number of training hours, and total workforce by employment type) and to quality management systems (non-conformities detected during production cycle).

The second group included nine measures with GAP values above the average of positive GAP values. Table 3b shows the five measures with the largest disparity between their usefulness and their information availability reflecting the lowest availability of information. Among these measures, three measures from economic category and two measures from environmental category are found. This group seems to be influenced by two types of difficulty in information acquisition, namely environmental information related to global issues (e.g. impact of activities, products and services on biodiversity, environmental externality costs), and related to control of the production cycle (e.g. cost per unit produced).

4 CONCLUSION

The objective of this study is to gain an understanding of current practices related to corporate sustainability performance measurement and measures in manufacturing organizations. Specifically, the extent of use, importance and availability of information for a group of seventy-four (74) economic, social, and environment measures, are were examined. The results, which were derived from multiple regression analysis and gap analysis, suggest some important conclusions which have both practical and future research implications.

First, the regression analysis results appear to indicate that executives of Portuguese manufacturing organizations are using sustainability performance measures based only on their predictive value.

Second, these results also indicate that these executives are emphasizing the financial and product related performance dimensions. Meanwhile, an under-utilization of social and environmental related measures is noted.

Third, the GAP analysis results indicate a lack of information related to global environmental issues. Maybe manufacturing executives do not sufficiently value performance related measures, and so they are reluctant to pay for this information.

Overall, it seems that Portuguese manufacturing organizations are following a closed-system performance measuring model, with an unbalanced approach to corporate sustainability performance measurement.

Corporate sustainability performance measurement is a recent and evolving process. The results of this study seem to indicate that executives are still trying to escape from the efficiency/internal focus of the closed system orientation. Applied research is therefore called for to facilitate the practical transition to an open system orientation where corporate sustainability performance is viewed and measured as being multi-faceted.

REFERENCES

Adams, C.A. & Frost, G.R., 2008. Integrating sustainability reporting into management practices. Accounting Forum, 32(4), pp. 288–302.

Bakshi, B.R. & Fiksel, J., 2003. The quest for sustainability: Challenges for process systems engineering. AIChE Journal, 49(6), pp. 1350–1358.

Bansal, P., 2005. Evolving sustainably: a longitudinal study of corporate sustainable development. Strategic Management Journal, 26(3), pp. 197–218.

Barkemeyer, R. et al., 2011. What Happened to the "Development" in Sustainable Development? Business Guidelines Two Decades After Brundtland. Sustainable Development.

Blasco, J.L., 2006. Indicadores para la Empresa, Fundación Santander Central Hispano.

Bonacchi, M. & Rinaldi, L., 2007. DartBoards and Clovers as New Tools in Sustainability Planning and Control. Business Strategy and the Environment, 473, pp. 461–473.

Borga, F. et al., 2009. Sustainability Report in Small Enterprises: Case Studies in Italian Furniture Companies. Business Strategy and Environment, 176, pp. 162–176.

Cooper, S.M. & Owen, D.L., 2007. Corporate social reporting and stakeholder accountability: The missing link. Accounting, Organizations and Society, 32(7–8), pp. 649–667.

Dempsey, S.J. et al., 1997. The use of strategic performance variables as leading indicators in financial analysts' forecasts., 2(4), pp. 61–79.

Epstein, M.J. & Roy, M.-J., 2001. Sustainability in Action: Identifying and Measuring the Key Performance Drivers. Long Range Planning, 34(5), pp. 585–604.

Friedman, M., 1970. The Social Responsibility of Business is to Increase its Profits, pp. 1–5.

Gomes, C.F., Yasin, M.M. & Lisboa, J.V., 2004. An examination of manufacturing organizations' performance evaluation: Analysis, implications and a framework for future research. Journal of Operations & Production Management, 24(5), pp. 488–513.

Gomes, C.F., Yasin, M.M. & Lisboa, J.V., 2011. Performance measurement practices in manufacturing firms revisited. International Journal of Operations & Production Management, 31(1), pp. 5–30.

Hair, J.F. et al., 2009. Multivariate Data Analysis 7th ed., New Jersey, USA: Prentice Hall.

Hubbard, G., 2009. Measuring Organizational Performance: Beyond the Triple Bottom Line., 19, pp. 177–191.

Kleine, A. & Von Hauff, M., 2009. Sustainability-Driven Implementation of Corporate Social Responsibility: Application of the Integrative Sustainability Triangle. Journal of Business Ethics, 85(S3), pp. 517–533.

Labuschagne, C., Brent, A.C. & Van Erck, R.P.G., 2004. Assessing the sustainability performances of industries. Journal of Cleaner Production, 13(4), pp. 373–385.

Lamberton, G., 2005. Sustainability accounting—a brief history and conceptual framework. Accounting Forum, 29(1), pp. 7–26.

Lo, S.-F., 2010. Performance Evaluation for Sustainable Business: Corporate Social Responsibility and Environmental Management, 17, pp. 311–319.

Perez, F. & Sanchez, L.E., 2009. Assessing the evolution of sustainability reporting in the mining sector. Environmental management, 43(6), pp. 949–61.

Perrini, F. & Tencati, A., 2006. Sustainability and Stakeholder Management: the Need for New Corporate Performance Evaluation and Reporting Systems. Business Strategy and the Environment, 15, pp. 296–308.

Richards, D.J. & Gladwin, T.N., 1999. Sustainability metrics for the business enterprise. Environmental Quality Management, 8(3), pp. 11–21.

Robinson, H.S. et al., 2006. STEPS: a knowledge management maturity roadmap for corporate sustainability. Business Process Management Journal, 12(6), pp. 793–808.

Robèrt, K.-H., 2000. Tools and concepts for sustainable development, how do they relate to a general framework for sustainable development, and to each other? Journal of Cleaner Production, 8, pp. 243–254.

Schaltegger, S. et al., 2003. Sustainability Management in Business Enterprises—Concepts and Instruments for Sustainable Development Centre for Sustainability Management (CSM), ed., German Federal Ministry for the Environment and Federation of German Industries.

Schaltegger, S. & Burritt, R.L., 2009. Sustainability accounting for companies: Catchphrase or decision support for business leaders? Journal of World Business, 45(4), pp. 375–384.

Schaltegger, S. & Wagner, M., 2006. Managing Sustainability Performance Measurement and Reporting in a Integrated Manner. In S. Schaltegger, M. Bennett, & R. Burritt, eds. Sustainability Accounting and Reporting. Springer, pp. 681–697.

Sikdar, S.K., 2003. Sustainable Development Sustainability Metrics. AIChE Journal, 49(8), pp. 1928–1932.

Skouloudis, A., Evangelinos, K. & Kourmousis, F., 2009. Development of an evaluation methodology for triple bottom line reports using international standards on reporting. Environmental management, 44(2), pp. 298–311.

Staniškis, J.K. & Arbačiauskas, V., 2009. Sustainability Performance Indicators for Industrial Enterprise Management. Environmental Research, Engineering and Management, 2(48), pp. 42–50.

Székely, F. & Knirsch, M., 2005. Leadership and Corporate Responsibility Metrics for Sustainable Corporate Performance., pp. 1–54.

Tanzil, D. & Beloff, B.R., 2006. Assessing Impacts: Overview on Sustainability Indicators. Environmental Quality Management, pp. 41–57.

Tregidga, H. & Milne, M.J., 2006. From Sustainable Management to Sustainable Development: a Longitudinal Analysis of a Leading New Zealand Environmental Reporter. Business Strategy and the Environment, 241, pp. 219–241.

Von Geibler, J. et al., 2006. Accounting for the Social Dimension of Sustainability: Experiences from the Biotechnology Industry. Business Strategy and the Environment, 15, pp. 334–346.

Wilson, M., 2003. Corporate sustainability: What is it and where does it come from? Ivey Business Journal, (March/April), pp. 1–5.

Green Design, Materials and Manufacturing Processes – Bártolo et al. (eds)
© *2013 Taylor & Francis Group, London, ISBN 978-1-138-00046-9*

An economic approach on the urban development of squatter settlements in hillsides high-risk areas

S.R. Soares & S.T. Moraes
Universidade Federal de Santa Catarina, PGAU Cidade, Brazil

ABSTRACT: This study discusses possible uses of urban strategies and instruments provided by the Brazilian urban legislation, as a path for the economic development and improvement of squatters' settlements in hillsides high-risk areas. As case study, we have compared the social and economic structure of the middle-class neighborhood of Saco dos Limões to the adjacent urban squatter settlement of Alto da Caieira, both located in Morro da Cruz, Florianopolis, Southern Brazil, where recent implementation by the Federal Government of actions of the "Acceleration Growth Program" (PAC) occurred.

1 INTRODUCTION

The main target of this study is to highlight the need of better strategies for the insertion of the poor communities in risky areas on the Brazilian urban development policies.

In the 1980s, the city of Florianópolis had a huge population growth. This urban expansion resulted in 16 slums on Morro da Cruz, built on steep slopes, in soils of high risk of landslides. The Morro da Cruz massif (see Fig. 1) is 285 meters high, being the culminating point of the central area of Florianópolis, embracing a population of

Figure 1. Location and overviews: 1.1. Alto da Caieira and the working class community, Base Map from IPUF, 2010; 1.2. Location of Saco dos Limões working class community; 1.3. Overview of Saco dos Limões neighborhood from the top of Caieira; 1.4. Overview of Caieira neighborhood from the bottom of Saco dos Limões, Photos February, 2011.

about 61,899 inhabitants according to the 2010 Census of IBGE[1].

In this massif 40% of the population has a very low income profile (less than US$ 1000,00/month). One of the 16 slums on the slopes is Alto da Caieira, characterized by the lack of urban infrastructure, land tenure undocumented. This community presents poor living conditions and environmental degradation and a half of their inhabitants has a family monthly income between 0 to 3 minimum wages (less than US$ 33 per day).

Besides Alto da Caieira, Saco dos Limões is another neighborhood with a middle-class and low middle-class profile known for a traditional working-class community (see Fig. 1). Approximately 24% of the total inhabitants of Morro da Cruz live in Saco dos Limões. The two neighborhoods are very different in relation to urban infrastructure of services and public equipments. Saco dos Limões also presents much bigger diversity in commercial activities than Alto da Caieira. Data from IBGE show that Saco dos Limões' average monthly income is the double of Alto da Caieira.

In the last decades in Brazil, new possibilities for interventions on cities are available by a new (2001) progressive federal law known as "Statute of the City" (see Box 2). Among others legal urban development tools provided by the law, there is the "Urban Operation Trust" (UOT)[2], that consist in a group of actions and measures on urban environment, coordinated

---

[1]IBGE—Brazilian Institute of Geography and Statistics.
[2]"Urban Operation Trust" is a free translation for "Operações Urbanas Consorciadas".

by the Municipal Government and associated to owners, housekeepers and private investors aiming to implement or complete the infrastructure, social improvements and environmental valorization.

The perspective of a new presence of the State regulating, inducing and controlling the processes of the city production is a bet on managing the land use with long-term effects. (Brasil, Ministério das Cidades, 2005 pg. 126).

Another important intervention is the government policy called "Acceleration Growth Program" (PAC) launched in 2008 as an attempts to minimize the effects of the irregular occupation problem with some technical and social solutions. Aiming to intervene in urban areas of risk this program presented the opportunity to change the inadequate living conditions and poor areas, integrating them in the infrastructure of the formal city.

Landslides contention and the deployment of public roads recently implemented in Morro da Cruz as part of PAC interventions minimized but not eliminated the danger of landslides. Unfortunately, those punctual actions should not have been part of a bigger plan that could preview the inclusion of Morro da Cruz into Florianopolis city dynamics. In spite of this, PAC still is a great opportunity to fight the historic ineffectiveness of low-income housing policies; the lack of adequate legislation for environmentally fragile areas; the inefficacy of control system of urban land use and occupancy; and the lack of technical support for construction of houses for low-income population.

2 METHODOLOGY

In order to draw guidelines for a more rational commercial and institutional occupancy aiming the urban and economic development of Morro da Cruz by attracting private investors, we examined and criticized the intervention of " PAC inside this area. We also have compared the social and economic structure of the middle-class neighborhood of Saco dos Limões to the adjacent urban squatter settlement of Alto da Caieira, examining urban infrastructure of services and public equipments trying to check up possibilities of integration and economic growth of the two neighborhoods. For this scope, we divided the study in three different steps:

1) Studying and mapping the conditions and territory structures existing under the economic approach of analysis;
2) Defining the main sectors and axis of interventions for commercial endings;

3) Identifying the urbanization restrictions in edge of the areas studied and environmental fragility in areas of slopes.

This division allowed us to understand and map the sectors and axis commercially relevant inside the analyzed settlements (Saco dos Limoes and Alto da Caieira), considering also the "risk cartography" as basis to be used together with the urban planning instrument.

3 FORGOTTEN ENVIRONMENTAL ISSUES

By environmental studies we were able to identify the conditions that can restrict the occupancy and evidence the fragile environmental structure (soil and vegetation). Using the risk cartography basis to establish parameters for the urban expansion on the slopes and for defining instruments of urban planning might draw a more balanced environment.

These findings are still shortly defined and used and do not provide parameters for classifying what would be sustainable. This still includes the search for reaffirming the important role of the geo studies as restrictive parameters for the urban planning of land use and occupancy.

The recent public actions did not take this issues into account and the land tenure are in the sense of permitting construction inside irregular areas. The allowance of the land tenure of irregular high risk landslides urban areas can contribute for the growth of natural disasters. On the other hand, reducing the risks of natural disaster is no longer related to the poverty reduction.

The inclusion of environmental studies as parameters to better define and quantify the housing deficit using the available resources of PAC should be an excellent measure to guarantee successful proposals.

4 FORGOTTEN ECONOMIC ISSUES

Besides the environmental approaches, we intend to discuss an approach for understanding economic inclusion allowed by recent applied urban policies. The economic approach in this context has two key issues: a) the use and occupancy of urban land for commercial ends and b) the instruments of induction of urban development.

Basically there are two urban planning tools inside the City Statute we would like to take into consideration when referring to instruments of urban development: 1) UOT and 2) The Additional Building Potential convertible to the right to build. Therefore it will be possible to suggest urban instruments of incitement for the rearrangement of the existing commercial areas, taking in account those findings.

The urban policies in Brazil present some guidelines we consider as opportunities for development such as the UOT and the Additional Building Potential convertible to the building rights. These instruments are consequences of urban deregulation in association with urban public and private partnerships.

"This tool was based on the "solo criado" idea, a Brazilian version of the French "Plafond legal de Densite" much in vogue in the 1970s Urban Reform discussions in Brazil, and originated the concept of granting building rights costly, which would become the basis of urban operations." (Nobre, 2012, pg. 3).

In the UOT there would be a counterpart from the entrepreneur for an affected area for the onuses/ damages that an enterprise could cause.

The central issue for this instrument is related to the possibility to create public-private partnerships. Throughout the private market finances indirectly the requalification of the city. However, the UOT is one of the most polemic instruments of the City Statute due to it can be used to enrich areas with urban infrastructure of services and public equipments, losing its redistributive desirability. (Brasil, Ministério das Cidades, 2005 pg. 130).

One of the ways of avoiding these distortions would be on the possibility of extending the territorial areas meant for the UOT beyond the interest of the market, including areas of squatter settlements.

Figure 2. Interesting urban spots, important nodes and crossings: 2.1 Alto da Caieira—bottom; 2.2 Alto da Caieira—top; 2.3 Grêmio Recreativo Consulado Samba School and AMOCA; 2.4 Saco dos Limoes neighborhood center Abdon Batista square; 2.5 Núcleo Comercial da R. Fco. Elesbão de Oliveira; 6 Núcleo da Escola Básica Getúlio Vargas; 2.7 José Mendes neighborhood; 2.8 Carvoeira neighborhood; 2.9 Armazém Vieira.

So that, it would be possible to create ZEIS (Special Zones of Social Interest) inside the Urban Operation Trust and invest the financial sources from the selling *"building rights"* for those aims. (Brasil, Ministério das Cidades, 2005 pg. 130).

In this sense, studying the results of the recent infrastructural interventions of PAC in Morro da Cruz it has been evident that the government actions lost opportunities to increase improvements to eliminate the environmental risks of degradation of an urban sector. Unfortunately, the interventions did not take care of several environmental and social problems neither reduced the risk of landslides nor even provided better access to city in order to create better integration with adjacent commercial sectors and axis (vectors).

In the Morro da Cruz' communities there is a lack of institutional partnerships related to the urban environment and social urban planning programs and policies that could avoiding conflicts of competence among the different tiers of government.

These improvement and development of political and institutional articulation could reinforce urban policies in order to put under practice the instruments of those policies according to values of justice and social inclusion with principles established in the City Statute (2001).

Moreover, there is a narrow interface between public administration, non-governmental organizations and civil society marked by an inhibited real consciousness due to the strong socioeconomic inequality, despite the mandatory community participation on planning in Brazil.

5 INVESTING IN THE MORRO DA CRUZ

By our analyzes, we have defined the main sectors and axis of possible interventions for commercial ends in Saco dos Limões and Alto da Caieira neighborhoods.

In Alto da Caieira settlement we defined two sectors: one at the bottom of the hill and another at the top, where there are some institutional and commercial buildings such as a church, a soccer game field and two small supermarkets. The commercial axis match to the two buses (see Fig. 3) lines. We observed two net commercial sectors, one near the bus stop of Alto da Caieira from where all areas of the community can be accessed easily. (see Fig. 03 sector 01).

A good possibility to stimulate investments in this area is to work where there is a strong market activity guiding the financial sources of UOT for the provision of urban infrastructure of services and public equipments for the poor surrounding

Figure 3. Location of the two commercial activity centers in Alto da Caieira. Base Map from Google Earth, 2011.

neighborhoods. (Brasil, Ministério das Cidades, 2005 pg. 76).

We understand those instruments of urban development could be properly implemented integrating Saco dos Limoes and Alto da Caieira neighborhoods. The first one might be more attractive due a bigger development of commerce and urban infrastructure and the second presenting more needs and requiring more investments.

6 PROMOTING THE ECONOMIC DEVELOPMENT

Some principles and guidelines for the desired development should start in reviewing the Municipal Master Plan[3].

Actions for demarcation of areas for the urban development planning should take into consideration:

– Identify environmentally fragile areas not suitable for urban occupancy;
– Provide urban equipments (day care centers, schools, health care centers, workplaces) public transportation, with good accessibility among others to support the expansion of housing neighborhoods;
– Adequate the sanitation policy to fit the demands of the estimated urban expansion aiming at the environmental conservation and at the public investments, promoting better infrastructure for the cities;

---

[3]The master plan is a field of the political action and new social and economic development building and where/in which there is a disputation about the notion of the development, from the different points of view of the citizens about the city they desire. (Brasil, Ministério das Cidades, 2005 pg. 42).

– Outfit expansion areas with infrastructure able to receive vertical expansion that generate building rights;
– Implement low income housing projects considering the proximity to the working areas for promoting the economic development, avoiding the construction of outlining areas with lack of adequate infrastructure and urban services;
– Deploy an attractive urban design able to attract private investors and convince them to finance the project.

Aggregating forces on the society basis, political and public concern, educational and scientific institutions struggle to build a new political cycle which integrates economy and ecology searching new urban patterns for the discussions of urban development.

7 CONCLUSIONS—TOWARDS THE NEW CHALLENGES

The intention of this paper is to discuss how a better use of strategic urban instruments can provide a real socioeconomic and urban territorial inclusion.

Moreover, we expect as results from our thoughts some guidelines for occupancy of sectors and axis (vectors) with financial potential including aspects of urban and architectural planning.

This study also criticizes the unfair use of urban land regulation and development instruments emphasizing the need of the poor population living in areas of high landslide risk as Alto da Caieira, being included in the city dynamics.

The turning point is based on the idea that the current model of urban development is not sustainable, once the problems are faced in a fragmented way. Government care a lot of drainage issues, for instance, but do not care about social issues in the same environment.

So, reformulate urban planning methods, approaches and practices to embrace the complexity of the city under sustainable standards might assure the quality of urban space through strategic urban interventions to avoid urban expansion without development.

Some of the urban challenges are the multiple approaches of urban planning and management imagining what should provide more lively neighborhoods. One of the main roles of urban planning and management is towards this new challenge: facing poverty through the right to a qualified city and moreover environmental, the social sustainability.

REFERENCES

BRASIL. 1988 Constituição da República Federativa do Brasil, promulgada em 5 de outubro.
_____. Estatuto da Cidade. Lei Federal 10.257/2001.
_____. Medida Provisória nº. 2220 de 04 de setembro de 2001. Disponível em: http://www.planalto.gov.br/ccivil_03/mpv/2220.htm acesso 17/02/2013.
_____. Ministério das Cidades. 2005. Plano Diretor Partici pativo: guia para elaboração pelos municípios e cidadãos. 2ª. Edição/Coordenação Geral de Raquel Rolnik e Otilie Macedo Pinheiro—Brasília: Ministério das Cidades; Confea.

NOBRE, Eduardo Alberto Cusce. 2012. Who wins and who loses with the great Urban Projects? Operação Urbana Comsorciada Água Espraiada evaluation in Sao Paulo.
United Nations/ISDR. 2008. Linking Disaster Risk Reduction and Poverty Reduction. [online] Available at: http://www.unisdr.org/files/3293_LinkingDisaster-RiskReductionPovertyReduction.pdf.

Green Design, Materials and Manufacturing Processes – Bártolo et al. (eds)
© *2013 Taylor & Francis Group, London, ISBN 978-1-138-00046-9*

Secure e-Manufacturing: Here there and everywhere

D. Meehan
College of Engineering and Built Environment, Dublin Institute of Technology, Dublin, Ireland

ABSTRACT: By 2015 it is predicted that there will be 25 billion connected products and services. This is impressive when set against a projected world population of 7 billion. It illustrates the need for a safe and secure Future Internet capable of allowing these devices and services to communicate without interference from cybercriminals and other intruders. Examples of this are the possible misuse of PLC (Programmable Logic Controlled) strategic equipment by criminals having gained access via the internet. Manufactured equipment capable of operating on the internet would be inherently more secure if Biometric access protocols were adopted. Biometrics I now a well established technology which has many security features that have been tried and tested. Examples of such are Banking, Passports etc. Biometric technologies have proven to be suitable for identity verification. However there are concerns about data protection and storage. Case studies illustrating the potential of this technology to improving the security of internet connected manufactured products and services will be presented. The employment potential of this technology will be outlined.

1 INTRODUCTION

The world is moving towards an information and Knowledge Society (IS), in which technology will play a key role in important areas including social interactions, economics, politics, and industry. This new paradigm is characterized by continuous transformations and globalization, resulting in an increasingly racially and culturally heterogeneous society that is not limited by traditional national boundaries [1]. The need for identification is growing, and the deployment of the European Biometric passport is a good example [2].

By 2015 it is estimated that there will be 25 billion connected products and processes like Smartphones, Tablets, Smart grids as well as in Connected Health and Independent Living systems. This is impressive when compared to a projected world population of 7 billion. This illustrates the need for a safe and secure Future Internet capable of allowing these devices to communicate without interference from cybercriminals and malicious intruders. Examples of this are the possible misuse PLC's (Programmable Logic Controlled) strategic equipment by criminals having access via the Internet.

Manufactured equipment capable of operating on the Internet would be inherently more secure if Biometric access protocols were adopted.

Biometric technologies like Fingerprint, Iris, Voice, Face, & Gait [3] have proven to be suitable for ensuring that the identity of people, but there are still important concerns about their uniqueness as an identifier. Biometrics is now a well-established technology which has many security features that have been tried and tested. Examples of such are Banking and Passports. This technology still has some concerns related to data protection and retention but in general they are being addressed.

The paper wishes to address these questions with the establishment of research in developing a robust identification process. Other important considerations that are addressed are user acceptance and Biometric data protection [4].

In today's world, unimodal Biometric technologies have not become as popular as expected for various reasons:

- Privacy has not been treated carefully: Citizens are aware of the risks and potential abuses of private information. In the case of Biometrics, the problem of privacy becomes more acute—can we cancel our Biometric Data? [5].
- Problems with scalability, adaptability, universality and user-acceptance. Current biometric systems must improve their performance in these areas for better accuracy, lower computational costs, and have wider use under multiple scenarios.
- Multimodality and correlated biometrics have not been well-developed yet [6]. These factors increase user acceptance by providing alternative ways of identification as well enhancing security through additional authentication methods.

- There is no integration between identification frameworks and existing Biometric functionalities. This makes it difficult to easily incorporate Biometrics into existing identification schemes.

The preservation of liberties and fundamental rights for people is a milestone in the world. Concepts such as e-Inclusion [7], privacy, and sensitivity to minorities have a great impact in society nowadays. Furthermore, the economic impact of the resilience, integration, trustworthiness of Biometric systems is vast. The position of European Biometric companies must be strengthened to those from the USA, India, Japan and China [7] [8].

Again Biometrics was identified as a technology a helping to a secure identification system for cloud users. The new Irish governments Department of Social Welfare has a new Biometric public service card This card aims to make €625 million in savings this year [9]. I initiated and advised on this pilot for the Irish Government in 2011.

2 MOTIVATION

It has been established that the (European Union) EU is very good at basic Research and Development but poor at exploiting these activities to establish sustainable industrial growth to enable it to compete with the US and China and more lately the BRIC (Brazil, Russia, India, China) countries who are now developing at a rapid pace. The unemployment levels in the EU now reach 25 million, 30% of the under 25 age group have no employment. These include extremely highly educated people up to PhD level who are now with a poor future. This is a unacceptable scenario both economically and socially [10].

The EU needs to establish a global R&D synergy around the application of Biometrics to address the major issues of industry and job creation to improve all our lives. It would be envisaged that the with the Irish government holding the EU Presidency in 2013 that it would play a pivotal role in establishing the EU as a center for such developments as Europe needs a technology that has a well-established R&D background that is ready to be developed to secure Global growth from spin off activities. The present motto of the Irish Government presidency is stability, jobs and growth.

3 DEVELOPMENTS AND CASE STUDIES TO DATE

The goal would be to leverage EU-wide infrastructures to support advanced experiments demonstrating the versatility of Biometrics across a multiplicity of heterogeneous environments. Several European regions or urban areas are increasingly being equipped with advanced infrastructures (e.g. sensor platforms, advanced broadband wireless networks, energy grids and content delivery networks) Smart Cities [11] that need to have secure and robust security protocols that as stated to prevent terrorist and malicious attacks.

Large industries as well as Small and Medium Enterprises (SME's), which compose about 90% of the industries in the EU, could develop these Next Generation Access services (NGA) in collaboration with operators and infrastructure companies. Common ground is possible in such projects with the use of open source software that is vendor independent. The use of pilots as a platform for learning and developing new solutions to improving urban transport other public services is now possible.

The increased adoption of Cloud based services both in the Public (Smart Cities) [12] and Private (Social Networking Sites) domain increases the need and adoption of personal identity techniques that are secure over computer networks. IDC [13] predicts that more than 80% of new applications will be deployed/distributed via the cloud and that cloud platforms will gradually displace the client/server approach as the dominant model for application and solution delivery. This is to be encouraged as the Green agenda of the EU requires that we use less resources to manage our economies hence using less of the earth's nonrenewable resources.

The use of a robust Biometric Identification (ID) process with a low false error recognition index is critical to the successful use of these NGA services.

A number of centers of excellence in Biometrics in the US and other countries have been identified. It is vital for the EU to establish co-operation agreements with such centers.

A co-operation agreement with the Technical University of Brno, Czech Republic [14] has been established as a result of on a short visit last year by the author. Also established was a unique bilateral agreement under the Erasmus scheme between that University and Dublin Institute of Technology (DIT) [15] where Masters students as well as R&D staff could be exchanged over short periods of time.

Research and development appropriate to the needs of the EU is required to further its goals of developing sustainable growth in this multi-billion euro worldwide industry. Within Ireland we have a number of centers of excellence e.g. CASALA at Dundalk Institute of Technology, [16] and TSSG [17] at Waterford Institute of Technology.

It is anticipated that a center of excellence for Biometrics could be established in Dublin.

This would have a global remit and would be a platform for Irish research and development which would also provide a platform for job creation for the many graduates of Engineering and Computer Science from the various Universities and Institutes of Technology in Ireland. At present we do not have such a center. Activities such as global academic and student exchange could also be established.

It could also establish co-operation agreements with some of the BRIC countries in for example the development of Smart City technologies. An initiative in Smart Cities technologies in China is under investigation at present by the author of this paper.

4 CONCLUSION

There is a quote from William Shakespeare seems apt in the present circumstances of the Irish economy "There is a tide in the affairs of men which taken at the flood, leads on to fortune omitted, all the voyage of their life is bound in the shallows". The tide being Biometric opportunities that can be developed in Ireland as we have the technology graduates who have English as their main tongue.

We are on the periphery of Europe with very good and stable exports at present in Electronics and Software compared to other countries with a similar economic base. In the US we have many Irish who are extremely positive towards Ireland and its future development. Only recently we had Bill Clinton at a Dublin Castle economic forum praising our positive export growth in this area and indicating that the US is willing to assist Ireland in its future economic development.

All the technology indicators are pointing in the direction of a Secure Internet Knowledge Society where privacy of the individual, as stated in the opening paragraph of this submission, will play a key role in important areas such as social interactions, economics, politics and industry. Biometrics is one of the key cornerstones of this development.

REFERENCES

[1] Cassells. M (2005). Alianza Editorial (Ed), The information era, Thenetsociety (5thEd) Madrid: Alianza Ed.

[2] Hornung G' The European Regulation on Biometric Passports': Legislative procedures, Political Interactions, Legal Framework and Technical Safeguards, (2007) 4:3 SCRIPT Ted 246.

[3] Jain, A and Ross A, Multibiometric Systems, Communications of the ACM, Vol 47, No1, Jan 2004.

[4] http://www.riseproject.eu; http://www.hideproject. org.

[5] "The EPDS and EU Research and Technological Development", EDPS Policy Paper, April 2008.

[6] Drygailo, A, "Multimodal Biometrics for Identity documents and Smart Cards: European Challenge", Proceeding of Euripco, 2007.

[7] G. Hornung, "The European Regulation on Biometric Passports: Legislative Procedures, Political Interactions, Legal Framework, and Technical Safeguards", (2007) 4:3 SCRIPT ted 246 http://telecentreeurope. ning.com/profiles/blogs/report-on-the-public.

[8] http://www.actuity-mi.com/FOB Report.php.

[9] The Irish Times, 8 Octoberr 2011, http://www.irish-times.com.

[10] V. Reding, "Social Networking in Europe: Success and challenges", Safer Internet Forum, September 2008.

[11] CIP ICT PSP 2012, CIP ICT PSP 20013, http://ec.europa.eu/ict_psp.

[12] SmartSantander-2-Open-Call, 24th September 20012, http://www.irishtimes.com smartsantanderopencalls@timat.unican.es.

[13] IDC, www.idc.com.

[14] http://www.vutbr.cz/en.

[15] www.dit.ie.

[16] http://casala.ie.

[17] http://www.tssg.org.

Green Design, Materials and Manufacturing Processes – Bártolo et al. (eds)
© 2013 Taylor & Francis Group, London, ISBN 978-1-138-00046-9

Parametric places 22@: Smart urban analysis tools and place branding value

P. Speranza

Bottom Up Urban Design Lab, University of Oregon, Portland, Oregon, USA

ABSTRACT: Parametric relationships between protected cultural buildings and public open space enhance the sustainability and place branding values of existing manufactured built form in city districts. The research investigates the development of smart urban design tools to understand these relationships in the 22@ information activities district of Barcelona. These tools identify design criteria parameters and use computation to create a unit block prototype that is subsequently informed with real world external data across the blocks of the 22@ study area. The outcome of these urban analysis tools is the ability to use computation to understand multiple urban design criteria for qualitative values. The value is sustainable in the reuse of built form, social in the acknowledgement of culture and economic in the place branding value these tools provide to the ongoing evolvable identity of places.

1 INTRODUCTION

435€ billion: the value of the Eiffel Tower to the French Economy. (Monza & Brianza 2012).

Smart urban design tools identify and optimize relationships between *modernisme* historic built fabric manufacturing from previous generations and public open space to enhance the place brand identity of city districts such as Barcelona's 22@ information activities district. Protection and acknowledgement of *patrimonio* or culturally historic buildings, provides not only place branding value but also a sustainable approach toward the built environment as a non-polluting process, energy conservation and the minimizing the use of natural resources. This new bottom up approach to urban design planning builds on open-ended guidelines of the Digital Cities movement of the 1990's and 2000's particularly tested in 22@ Barcelona. Smart and bottom up design differs from traditional top down planning to allow for a more current attachment to place using intelligent/smart technology for planning to adapt to each generation.

2 BACKGROUND

In 2000 the city government of Barcelona conceived of an information activities district, similar to Silicon Valley, located in the post-industrial neighborhood of Poblenou (Barcelo 2001). Its dual purpose was to diversify the city's business activities and to continue urban renewal along the waterfront. Unlike the time constrained *tabula rasa* top-down urban planning for the 1992 Olympic Games that demolished large expanses of the city, the planning of 22@ promoted a plurality of small and medium sized enterprises by protecting historic industrial fabric and newly specified block-by-block guidelines for minimum requirements of 10% open space, 10% protected residential and 10% 7@ social service uses. Rather than adopting traditional axial structure to connect blocks, the *connectivity* of the 22@ blocks is left undefined. Likewise undefined are the relations between the 10% open spaces and the protected *modernisme* built fabric.

The term place brand in the context I will use here explains the identity of a place as originated from over time from use, or what I call bottom up design. In the case of Barcelona and 22@ the place brand is the Poblenou factory village once known as the Manchester of the Iberian Peninsula and the economic engine of Catalunya (22@ 2011). Modernisme, or the Catalan Art Nouveau movement, and particularly Anton Gaudí's work in Barcelona, provides a clear economic place brand value for the city of Barcelona as exemplified with the Sagrada Familia ranking third at 90€ billion according to the Monza & Brianza Chamber of Commerce study of 2012. In the new 22@ information activities zone with greater density including towers up to fifteen stories tall and the removal of most built fabric, the material character of the built fabric provides a unique and value *modernisme* identity that is not optimized for the public experience of the place.

3 APPROACH

Parametric Places describes the research objectives used to create urban analysis tools to draw relationships between otherwise unrelated aspects of 22@ planning guidelines. The research is done primarily in the Parametric Places media course with additional data from the life | city | adaptation: Barcelona Urban Design Program, both within the research lab of the Bottom Up Urban Design Lab, at the University of Oregon, Portland and Eugene. The tools developed in this research were subsequently presented to the 22@ district planning office within the Ajuntament de Barcelona.

Although the research is technology innovation driven, the individual project tools are user driven to solve applications problems to enhance the needs of people. The research operates from two ends, an understanding of urban design principles and an investigation of parametric media. The urban design criteria and the parametric variables become one in the same. The objective is to identify the criteria in the real world problem and building a parametric system that is unit, organization and external force, to simulate that condition.

In the case of 22@ Barcelona we began by understanding the real world culture, history, material built fabric, transportation, urban structure, street sections and use of the 22@ district in Barcelona. In parallel we developed parametric urbanism understandings working from singularities through systems based parametric design using Grasshopper software and in some cases optimization algorithms using Galapagos software (Rutten 2007).

We began with existing design criteria of the 22@ district. To capture new real estate values and change from the existing 22a heavy industrial use to contemporary information technology use property developers needed to provide within each block: 10% public open space, 10% protected housing and 10% public services. Floor area ratios were subsequently increased from 2.0 to 3.0 (22@ 2011). Additionally, and important to our place branding research, selective *patrimonio,* or culturally historic buildings, were protected in the district. A systems based approach was begun with analysis of the underlying Cerdà Eixample plan and patterning exercises explored the analog method to understand unit to whole patterning variations and urban scale. Case studies of parametric urban design work included the following: Stan Allen and James Corner's Freshkills Park; Patrik Schumacher and Zaha Hadid's Kartal-Pendik Masterplan; MVRDV Datatown; Ana Pla Catala's GSD Barcelona Project; Neil Leach's Swarm Urbanism; Vicente Guallart's Cristobal de Moura Street.

Projects identified criteria as parameters to develop new planning tools for the district based on research identifying the purpose, importance and whom it served within the mixed IT worker and residential neighborhood. Projects included Historic/Open Space, with students Ben Prager & Ivan Kostic; Food Market @22, with students Jeffrey Stattler & Yin Yu and Cultural Use of Open Space with students Jared Barak and Casey Hagerman.

The approach used Grasshopper parametric plugin to develop sets of relationships. Data including zoning use, protected built fabric, transportation, open space and street types were extracted from the 22@ district Ajuntament de Barcelona plan provided to us by the 22@ district planning office. Research also relied on data from the 22@ Ten Years (22@ 2011) analytical research done by the Universidad Politécnica de Catalunya, UPC.

4 PARAMETRIC PLACES

Projects strived to develop an adaptive unit block first analog and then digitally. Conventional planning relies on two dimensional plans and written guidelines. The research tests the possibility of using parametric models as zoning use tools. The research projects looked to close a research gap of organization while building on the data collection and qualitative solutions of the previous l|c|a:BCN Urban Design programs including: in 2012, University of Oregon's 'Connectivity' and independent program 'Materiality as Identity research'; in 2011 'Cultural Events' and 'Public Food Market Networks'. In these programs zoning uses and other criteria were mapped. Previous research understandings of pueblos in Barcelona were explored in the paper (Speranza 2013) and place branding value (Speranza 2012) supported this research.

4.1 *Historic/Open Space, Kostic Prager*

The project Historic/Open Space investigates the relationship between historic *modernisme* presence and open space. The importance of this idea was based on the place branding and sustainable value latent in such a relationship. The relationship of the public open space to *modernisme* place branding would benefit both the business identity and residential *pueblo* identity of the district.

Human spatial experience is the real world objective of the project. The identification and protection of city texture via details and sensory experience can play an import role in how we recognize and remember a unique place (Vitiello and Willcocks 2006). Historic/Open Space seeks to optimize the latent value of the small-scale *pueblo* experience (Speranza 2013) of the 22@ blocks.

The Francisco Franco regime's strict enforcement of no new residential use resulted in limited residential construction between 1935 and 1975, with distinguishable modernism buildings from 1880 to 1935. As an industrially zoned Poblenou area, open spaces were not provided as parks or as interior block patios, an exemplary aspect of typical Barcelona Exiample blocks.

The urban analysis tool Historic/Open Space thusly has the objective of relating public open space with the place branding value of historic built fabric, giving this limited but evenly distributed public space the *modernisme* identity of Barcelona, and in a similar though lesser degree the economic value of the Sagrada Familia example listed earlier. This relationship is important because it increases the places branding value of the space and also reinforces the value of both the protected and subsequently unprotected historic built fabric of the block. The reuse of these buildings in turn is a sustainable strategy by reducing the energy of new construction, conserving natural resources and reinforcing the cultural value for people to live and work in the district.

Criteria: Location, Size and Number.

To abstract and quantify this qualitative experience of associating historic form with open space, three criteria were identified as indicators of this general relationship: location, size and number. While the parametric system focuses on the unit of one block, the relationships in fact operate across blocks, namely across the public right-of-ways of streets. For this purpose parametric definitions, the term for scripting interface in Grasshopper, were developed at first within one block.

It was important to be reminded that the analytical value of this tool was not to make suggestions of holistic architecture nor urban design but rather a tool to reveal patterns based only on the parametric criteria built into the system.

Location: a simple parametric relationship was set up to attract open space to historic building.

Size: a simple parametric relationship was setup to match the size of the open space with the historic building.

Number: Given the 10% total open space but not knowing its number of distribution, the last parameter relates the number of evaluated historic building massing with the number of open spaces generated to equal 10% of the total block area.

These three simple and abstract relationships were tested within a one block system. To converge all three parametric criteria and use a real world block to test the system, the use of optimization was used to recommend an ideal solution to with real world external force data and the given block unit/organization. The Galapagos plugin for

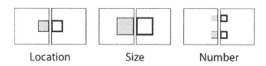

Figure 1. Historic/Open Space, three criteria, students I. Kostic & B. Prager.

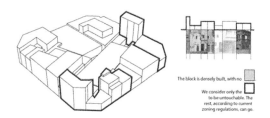

Figure 2. Historic/Open Space, protected historic fabric, students I. Kostic & B. Prager.

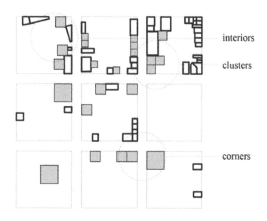

Figure 3. Historic/Open Space, Galapagos Optimization, students I. Kostic & B. Prager.

Grasshopper computes and visualizes over time the possible solutions, arriving at a final optimal solution.

The project team used a given site bound by Carrer de Pujades to the north and Carrer de Lull to the south, and Carrer de Avila to the west and Carrer de Badajoz to the east. The irregular buildings were regularized for the exercise. A total nine-block set was used including one adjacent block in all directions.

4.2 Emergent patterns

The resulting optimization proposal highlighted three emergent patterns: 1) clustering spaces; 2) corner spaces and 3) interiors (block patios). The analysis also made more observable new relationships such as the scale of walkability between open spaces and whether this scale would be more supportive of an emergent organization of a networking of connective nodes of open space or isolated islands of open space that are more disconnected across the district.

5 ADDITIONAL URBAN ANALYSIS TOOLS

Parametric Places tools were also developed with respect to place branding identity of food and cultural use of open space. The following tools demonstrate similar parametric criteria identification methods as the Historic/Open Space tool. In each project qualitative cultural identity is analyzed via quantifiable indicators characteristic of the place.

5.1 Food market @22

Food, like *modernisme*, is a central feature to the place brand identity of Barcelona. The MMBL Mercats de Barcelona network of thirty-nine public food markets provides fresh local and distant food to walkable neighborhoods in the City of Barcelona (Mercats 2013). As a previously industrial district area, 22@ has no public food markets, with an area that could support up to three such markets. Previous research within the l|c|a:BCN Public Food Markets Network investigated how food distribution may provide connectivity between 10% open spaces.

Trends in food activities types are shifting in Barcelona due changes in business hours, commuting times and immigration patterns from the traditional siesta two-hour lunch period and five-meal day, to a more business focused one-hour lunch and three major meals. The Food Markets 22@ tool visualizes different types of food activities at any chosen location. Food activity types include bakery, cooked/prepared foods, meat shop, wine store, fish story, grocery, fruits and vegetables, restaurants and bars. The tool uses food activity at store hours and real-time data in visualizations.

The food visualization tool also identifies parametric criteria of walkable scale. The radius of a 500 meter walking distance would be adjustable by individual users based on comfort. The radius may also be adjusted for topography. Qualitative cultural design objectives are quantified through more quantifiable indicators, in this case food establishment types, time, distance and topography. The project highlights Bruno Latour's idea of the attachment of design in real-time to its site (Latour & Yaneva 2008). Like other tools of the

Figure 4. Food market @22, students J. Stattler & Y. Yu.

Parametric Places research this tool provides visual analysis of existing behaviors allowing urban design to better respond to changing patterns of use and place brand over time.

5.2 Cultural use of public space

The use of public space in Barcelona has been attributed to a long history of cultural events, pueblo participation and climactic accessibility to exterior space. The tool Cultural Use of Public Space relates the size of the 10% open space to the adjacent distance to large public spaces such as large parks and the water front recreational spaces in Barcelona.

The tool first analyzes the relation between the sizes of public open space the cultural use that it supports. Based on this understand the tool builds a relationship between distance and size: the closer the 10% open space is to a large public space, the most dispersed the 10% unit block open space breaks up to support smaller activities. The farther the 10% open space is the more that space attempt to whole and large to support larger activities. Other criteria and complexity would be needed to match the complexity of the qualities of cultural use of public space but this tools provides a possible missing analysis in the guidelines of how the 10% open space me be used in regard to cultural use.

5.3 Place branding value and intelligent manufacturing

The three tools under the Parametric Places research use intelligent design methods to support the historic manufacturing context. They reveal synergistic values to the place brand of a location. Consider that 'the important thing to realize about branding a (place) is that it must be an amplification of what it is already there and not a fabrication'

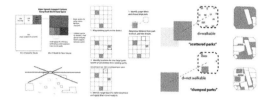

Figure 5. Cultural use of public space, students J. Barak & C. Hagerman.

(Gilmore 202). These tools do not create new place brands or new economic and sustainable value but reveal the latent value that preexists in a site. The complexity of urban systems, the variety of criteria and the large data sets at the scale of districts and cities have provided relative resistance when compared to the use of parametric design and qualitative indicators at the scale of individual buildings.

If these tools can identify and provide a metric for analyzing historic materiality at the scale of the block and district, then with the right metric they may also serve to consider the inclusion of materiality of new construction as well in urban design tools. The focus on protected historic buildings has provided an advantage in this specific research for energy and material conservation based on existing cataloged building data but this would not likewise preclude the additional inclusion of performative criteria at the scale of districts as is currently being researched by others.

6 CONCLUSIONS

Urban analysis tools used for measuring place branding and sustainability value are presented. These tools rapidly evaluate urban design relationships with regard to existing and future building material manufacturing methods and augment existing non-traditional guideline planning strategies such as the 22@ guidelines for open space, protected residential use, services use and protected historic buildings.

With increased complexity of systems understandings of cities and the increased recognition of place branding value of identity of districts, urban analysis tools should be developed that inform designers during the planning criteria process and during the individual lot build out negotiated during the block and district development period.

It is envisioned that Parametric Places research tools such as the Open Space/History, Food Market @22 and Cultural Use of Open Space tools could guide city politicians, planners and designers toward synergy-oriented design of new urban spaces and the reuse of existing districts in a way that captures latent place branding value of a location.

The use of parametric design in urban design at the scale of the district and block enhances the holistic design of urban space. Once existing manufacturing methods are destroyed, they are often lost forever. Multi-objective optimization criteria could be used for complex urban design conditions including reuse. These intelligent tools would support a new generation of sustainable design, where values of the past inform environments of the future.

REFERENCES

22@ Barcelona: 10 Years of Urban Renewal. 2010. Barcelona: Ajuntament de Barcelona. 30.

Barcelo, M. 2001. La Ciutat Digital, Pacte Industrial de la Region Metropolitana de Barcelona. Barcelona: Beta Editorial. 193.

Barcelo, M. 2001. La Ciutat Digital, Pacte Industrial de la Region Metropolitana de Barcelona. Barcelona: Beta Editorial. 193.

Gilmore, F. 2002. A Country—Can it be repositioned? Spain—the success story of country branding. In *Brand Management* Vol 9, No 4–5. Henry Stewart Publications: London. 281–293.

Latour, B & Yaneva, A. 2008. Give Me a Gun and I Will Make All Buildings Move: An Ant's View of Architecture. In R. Geiser (ed.), *Explorations in Architecture: Teaching, Design, Research*. Basel: Birkhäuser. 80.

Mercats de Barcelona, 2013. *www.bcn.cat/mercatsmunicipals*. Ajuntament de Barcelona.

Monza & Brianza 2012. Monza & Brianza Chamber of Commerce & Anholt Brand Index.

Rutten, D. 2007 (1st ed.). *Grasshopper & Galapagos*. Seattle: Robert McNeel & Associates.

Speranza, P. 2012. Place Branding from the Bottom Up: Strengthening Cultural Identity through Small-Scaled Connectivity. *Conference Proceedings: Roots-Politics-Place*. Berlin: inPolis. 237–252.

Speranza, P. 2013 Catalan Nation Building and Bottom-Up Placemaking: Placemaking and Barcelona Coastal Redevelopment in the 22@. *International Association for the Study of Traditional Environments*, The Myth of Tradition, Volume 242, 2013. Traditional Dwellings and Settlements Working Paper Series. Berkeley: IASTE.

Vitiello, R. and Willcocks, M. 2006. The Difference is in the Detail. *Place Branding Vol. 2*, 3, 248–262.

Sustainable construction

Green Design, Materials and Manufacturing Processes – Bártolo et al. (eds)
© *2013 Taylor & Francis Group, London, ISBN 978-1-138-00046-9*

Sustainable design of prefabricated solutions for the rehabilitation of ancient buildings

A.C. Coelho, J.M. Branco & P.B. Lourenço
ISISE, Department of Civil Engineering, University of Minho, Guimarães, Portugal

H. Gervásio
ISISE, Department of Civil Engineering, University of Coimbra, Coimbra, Portugal

ABSTRACT: A wood-based "kit-of-parts" for the rehabilitation of existing buildings is under development. The aim is to merge the benefits of the standardized manufacture, with the flexibility needed to suit the specific requirements of the built heritage. The proposed system should be reversible, flexible and adaptable, while ensuring the adequate structural, thermal and acoustic performances. Mechanical joints, reversible if possible, should be used for the connections with the existing construction. Light solutions, allowing on site adjustment and future material separation are preferable.

1 INTRODUCTION

Intervention in ancient buildings aiming to preserve the heritage is highly challenger, with regard to the difficulty to understand the existing construction and to propose an adequate retrofit, in comparison with the design of a new building over a blank sheet. However, the rehabilitation needs will increase as the built heritage ages and the economic and social conditions will favour the return to the urban centres.

Rehabilitation and retrofit are currently major issues in the construction industry, focusing the attention of many researchers in an international level. The research motivations may vary among heritage preservation, economical concerns, environmental management or urban revitalization.

Sustainability issues in buildings were initially concerned with thermal efficiency, due to the long use phase, that represented the greatest share of energy consumption in the whole life cycle of the building. A great effort has been done in recent years to promote energy efficiency in buildings, with the increase in mandatory insulation of the outer shell, as well as the use of more efficient HVAC (Heating, Ventilation and Air Conditioning) systems. It is actually possible to design "zero energy buildings", or even buildings that are featured to produce energy, therefore having a positive energy balance. To keep improving the environmental profile of buildings after this achievement, the focus should now be shifted to the construction phase and the energy embodied in the construction materials and processes (Rossi et al., 2012).

2 SUSTAINABILITY

Sustainability may be assessed throughout three different perspectives, which should work in combination: the economical, the environmental and the social impacts.

2.1 Rehabilitation

Rehabilitation of ancient buildings certainly produces impacts in the three categories, being some of them more remarkable than the others.

On the economical point of view, the rehabilitation of ancient buildings in city centres may be important to the city's economical revitalization. On the other hand, in case the building presents an average good condition, its retrofit may be cheaper than the construction of a new one.

Concerning environmental issues, rehabilitation is a means of reuse existing structures, thus extending its materials life span. Reducing the demolition needs means limiting the residues production and also limiting the demand for new products.

On the social perspective, rehabilitation of ancient buildings may be analysed under two major highlights: the valorisation of a region and the creation of jobs due to the fact that working in ancient buildings, normally requires more man work than new building construction.

2.2 Prefabrication and standardization

Abdallah (2007) argues that the use of prefabricated methods provides two main benefits for the construction industry: the first of them is the lower cost of the structures, with quality improvement over the life cycle; the second one is the improvement of the maintainability of the product, by means of an advanced manufacturing process, which balances present needs with long-term needs. Some other advantages of prefabrication also pointed by Abdallah (2007) are: the effectiveness of cost (first cost and life-cycle cost); earlier occupancy, reduced financing costs allowed by fast construction and the possibility of winter construction, avoiding weather delays. On the other hand, the same study refers to some special requirements of prefabricated methods: the additional effort on planning the work, comparing with on-site building methods; the additional engineering effort, in order to detail the prefabricated components and to develop construction sequences, optimizing the design; the need of specialized workmanship and equipment.

Prefabrication is actually an effective process on the improvement of the environmental performance of construction works, as well as on the promotion of self-construction.

2.3 *Wood*

Many centuries of wood-based construction made up a significant collection of wood building know-how, which preservation is important for local communities and may perform an important role in social sustainability issues.

When performing a Life-Cycle Analysis (LCA) of wooden buildings, it is usually considered that trees store carbon dioxide in their tissues, which will only be released by the decay or combustion of wood. This wood feature is highlighted on long lifespan wood-based products, likewise most of the construction materials (Borjesson & Gustavsson, 2000). In this discussion, Buchanan & Levine (1999) point out that the ability of wood to store carbon is not significant when compared to the total carbon emissions of building products manufacturing. As all the wood products have a finite life, the carbon storage balance will remain constant over time, considering that the amount of wood in use will eventually reach a steady state. For those two reasons, the carbon storage of wood products cannot offset the manufacturing emissions in the long term. In any case, Buchanan & Levine (1999) concludes that wood products require small amounts of energy in its manufacture, comparing with bricks, aluminium, steel and concrete. Therefore, when wood products are replacing energy-intensive ones, its lower embodied energy is significant towards the aim of carbon emission reduction in the long term, which is a permanent gain.

It can be assumed that the transformation process of wood produces virtually no waste,

since all the wastage can be used for production of wood-based products or fuel, decreasing the demand for fossil fuels (Lippke *et al.*, 2010, Sathre & Gustavsson, 2009). Although wooden constructions will need maintenance throughout its lifetime, the traditional wooden building systems allows partial replacement of modules or damaged elements, without affecting the structure as a whole. The use of wood also contributes to the energy efficiency of buildings, since it is a material with low thermal conductivity (Branco, 2003) particularly when compared with steel.

When dismantling a wooden building, the recovered wood can be directly reused in another building or used as raw material for wood-based products, either by extending its useful life or simply used as biofuel, avoiding the need for fossil fuels.

3 PREFABRICATION AND STANDARDIZATION IN REHABILITATION

Based on the fact that large sets of buildings have been erected under a common architecture and constructive system, usually even using prefabricated elements (as far as prefabrication was developed back in the 18th and 19th centuries, as a consequence of the industrial revolution), that have been kept under similar conditions (environment and use), therefore presenting common pathologies, it becomes clear that the use of a prefabricated system for its rehabilitation is a suitable option. According to Teixeira & Póvoas (2010), in the past centuries, the industrialization did not change the characteristics of the existing construction materials, but only increased its availability. Nowadays, it seems to be rational to use the same principle of increasing material availability and improving quality control on rehabilitation specific materials, which should present special features, necessarily different from the currently available new construction materials.

Assuming that, in these buildings, granite masonry structures usually present a good condition, being the timber elements more frequently damaged due to the lack of maintenance throughout several decades, the focus should be on repairing or replacing those elements. The slabs are suitable for prefabrication, due to the relatively regular span between structural supports (the gable masonry walls), provided that some adjustments may be executed on site, to offset any irregularities of the building. The choice of the material is critical to assure this on-site flexibility: wood plays an important role because it allows easy cut on-site. The interior partitions may be supplied with the right height (that is roughly repeated on these buildings), but some on-site adjustments in the extension of prefabricated boards may be required and should be enabled.

Concerning the roof, its main structure is made of timber trusses that are harder to prefabricate, due to its specific geometry and variety of shapes between neighbours buildings. Nevertheless, there is still the chance to prefabricate the sheeting boards to apply over the trusses, providing the necessary thermal and acoustic insulation, as well as the base to apply the waterproofing membrane and the ceramic tiles.

4 EXAMPLES OF AVAILABLE PRODUCTS

There are some prefabricated wood structural products available on market, for example:

- *ET³ slab and wall panel* (Pequeno & Cruz, 2009): is a composite panel, combining timber and glass; besides being structural, it also performs a thermal and an aesthetical function;
- *Wenus slab and wall panel* (www.cbs-cbt.com): was designed under the same principle of the corrugated cardboard, as a folded plate element with a "w" shape;
- *O'portune slab* (www.cbs-cbt.com): prefabricated slabs made with solid wood planks nailed or screwed together. There is the option to complement the slab with further layers to add finishing or to increase acoustic, thermal and fire resistance performance;
- *Kielsteg slab* (www.kielsteg.com): is a honeycomb slab, 100% made of wood, designed to perform a high structural performance and to dismiss additional finishing.

5 REQUIREMENTS TO FULFIL

The system under development in the current research project should merge, in a single product, the concerns related with sustainability and the concerns related to old-building rehabilitation. Some general requirements were defined.

5.1 General requirements

According to Couto *et al.* (2006), some of the recommendations in order to promote recycling in construction are: the use of recycled materials, to help the development of this market segment; the minimization of different types of materials on the same element; the avoidance of toxic and hazardous materials; the allowance of future separation of different recycling-potential materials; the avoidance of secondary finishing; the permanent

identification of the different materials in use; the minimization of different parts on the same element; the use of mechanical connections, or, when necessary, chemical connections weaker than the connected parts, enabling future separation; the design of easy-handling parts; the design with realistic tolerances, allowing necessary movements in assembling and disassembling operations; the limit of different connectors, to allow disassembling with minimum different tools; and also the design taking account of disassembling order, related to the expected lifespan of components, assuring easier access to less durable parts.

The size of the components is also a major issue, considering the need to deliver the parts into the right place in the building (most of the times keeping the masonry envelope of the building), as well as the practicability of the dismantling process at the end-of-life.

5.2 Ventilation

Timber needs ventilation in order to increase its useful lifetime. In the specific case of masonry buildings, ventilation is particularly important in the connection between timber and stone, due to the probability that stone presents some capillarity moisture, arising from its contact with the soil.

Also the roof should be carefully designed in order to avoid any moisture to get into the building, while allowing the necessary ventilation.

6 PROPOSAL DESIGN

In the first phase of the research project, a prefabricated slab was proposed, according to Figure 1.

It respects the same principles of the traditional Portuguese timber structures, starting with

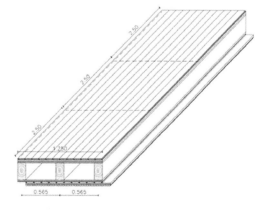

Figure 1. General prefabricated slab.

structural beams that cover the span between masonry gable walls, which may vary between 5 and 7 meters, depending on the age of the building. On the 18th century buildings, the spans were around 5 or 6 meters, arising to 7 meters in the 19th century. Nevertheless, in order to suit the needs of the vast majority of the buildings, avoiding the need to manufacture a special product, a span up to 7,5 meters is considered by default. As long as the slab's composition is the same over its entire area (considering a transversal section), the slab may be cropped on site, to fit the desired geometry of the building. For this reason, all the materials in use are timber and wood products, to allow workability with simple tools. The only exception is the waterproofing membrane.

The solid wood beams are sheeted in both faces by OSB panels. In order to avoid any waste of material, the OSB standard size of 2500 × 1250 × 22 mm is fully used. Therefore, each panel's width is 1250 mm and its length vary in multiples of 2500 mm, so that the slabs may be manufactured with the span length of 5 or 7.5 meters, to suit the needs of smaller (and older) buildings, as well as the needs of more recent and bigger ones. In any case, it is assured that an existing timber floor-structure may always be replaced by this system, because the traditional span allowed by the use of timber is never higher than 7.5 meters.

Above the OSB up layer, there is a cork layer (25 mm thickness), to assure the necessary acoustic performance, both for air and percussion noise. The floor finishing is made with a solid wood flooring with 20 mm thickness, likewise the traditional flooring used in this kind of building (Fig. 2).

Below the timber beams, there is also an OSB layer (22 mm thickness), coated on the bottom side with a waterproofing membrane, to assure the sealing between floors. The prefabricated slabs are supplied with the standard width of 1.25 mm, to allow easy handling on site. The slab has an indentation, in order to perfectly associate it with the contiguous parts, assuring the visual continuity. For continuity reasons, both the top and bottom finishes follow a stereotomy that is parallel to the structure and to the assembling direction (Fig. 2).

The finishing of the bottom side of the slab, which corresponds to the ceiling of the floor below, is made of solid wood slats, designed to assure the continuity of the ceiling, dismissing the need to a on site additional finishing process (Fig. 2).

The first and last prefabricated slab to apply in each floor is likely to have a special geometry, in order to fit with a third masonry wall (the front and the back facade). Anticipating this issue, a special slab was designed, replacing the solid wood beams by a solid structural layer of solid wood, made of several solid wood beams joined together (Fig. 3). This assures that this slab may be cropped in any shape and size, keeping its structural behaviour, specially in case the necessary area to cover demands a slab which width is less than the regular spacing of 565 mm between structural solid wood beams.

The on site assembling should start with the fixing of a continuous steel L shaped profile to the gable masonry walls that will support the slab's loads (Fig. 4, step 1). The prefabricated slab is supplied in two parts, to enable the assembling process. The top part (structure, acoustic insulation, waterproofing and floor finishing) should stand directly over the steel perimeter beam, allowing some geometric tolerance (Fig. 4, step 2) that will be later offset by the addition of wall panels to improve the performance of the external walls. At last, the bottom part (ceiling finishing) should be screwed to the slatted wood on the top part, that is designed in such a way that the steel beam keeps hidden inside the slab (Fig. 4, step 3). The final section is shown on Figure 5.

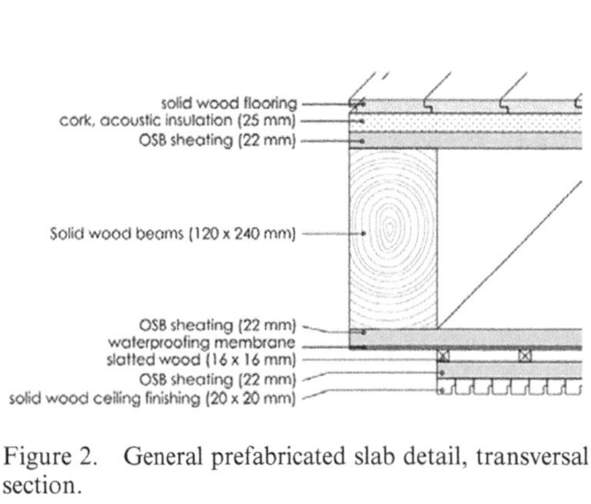

Figure 2. General prefabricated slab detail, transversal section.

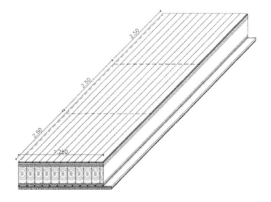

Figure 3. Special slab to be used in the tops.

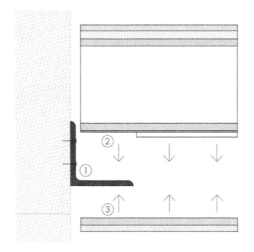

Figure 4. Assembling order.

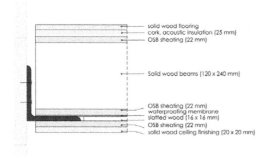

solid wood flooring
cork, acoustic insulation (25 mm)
OSB sheating (22 mm)

Solid wood beams (120 x 240 mm)

OSB sheating (22 mm)
waterproofing membrane
slatted wood (16 x 16 mm)
OSB sheating (22 mm)
solid wood ceiling finishing (20 x 20 mm)

Figure 5. Detail after assembling, connection to the masonry wall.

7 CONCLUSIONS AND FURTHER DEVELOPMENTS

It has been discussed in this paper that there are some actual opportunities to develop an existing building intervention strategy that makes it cheaper, faster, safer and more reliable than the current processes. One of the avenues in order to achieve these goals may be the introduction of prefabricated products and systems, specially featured for the use in the rehabilitation of existing buildings. Nevertheless, the use of a prefabricated system is only possible to apply in case the building is not protected heritage, when new elements may be a suitable option to replace its damaged parts. The starting research presented in this paper aims to propose prefabricated and industrialized wood-based systems for the intervention in existing buildings in Portugal. The solutions, inspired and respectful to the traditional materials and techniques, aimed to be fully compatible with the existing structures and materials. Despite the local characteristics and requirements, it is believed that those construction systems can be used in the traditional buildings of different European countries.

In this paper, a wood-based slab has been proposed. The further development of the slab design includes its economical and sustainable optimization, as well as its real scale prototype and testing.

The research project includes the development of some other parts to be used in the same sort of projects, mainly a versatile wall, that should fulfil the requirements for an interior partition or the upgrade of the external walls thermal performance, being attached to their internal face.

In order to fully assess the sustainability of the proposed system, in comparison to the current means of intervention on these buildings, a detailed LCA (Life Cycle Assessment) of all the proposed prefabricated products will be performed and published, to allow the environmental labelling of each product. The aim of labelling each product is the inclusion of sustainability criteria early in the project phase, allowing architects and engineers to make the project options based not only in economical and structural performance of products, but from now on including the environmental and social performance as well, for a complete sustainability characterization.

ACKNOWLEDGEMENT

The research project presented in this paper is part of "WoodenQuark n.º 2011/21635" from the Portuguese financing programme "Quadro de Referência Estratégico Nacional".

The authors also acknowledge the cooperation of "SRU—Sociedade de Reabilitação Urbana do Porto Vivo".

REFERENCES

Abdallah, A. (2007). *Managerial and economic optimizations for prefabricated building systems.* Technological and Economic Development of Economy (13:1), 83–91.

Borjesson, P., & Gustavsson, L. (2000). Greenhouse gas balances in building construction: wood versus concrete from life-cycle and forest land-use perspectives. Energy Policy (28), 575–588.

Branco, J. (2003). *Comportamento das ligações tipo cavilha em estruturas mistas madeira-betão.* Guimarães: Dissertação apresentada à Universidade do Minho para obtenção de grau de Mestre em Engenharia Civil, Especialização em Estruturas, Geotecnia e Fundações. (*in Portuguese*).

Buchanan, A., Levine, S. (1999). *Wood Based Building Materials and Atmospheric Carbon Emissions*. In: Environmental Science & Policy, 2, pp. 427–437.

CBS-CBT, Engineering, Technology, Timber construction—website "www.cbs-cbt.com", last visited on 1st March 2013.

Couto, A., Couto, J., & Teixeira, J. (2006). Desconstrução—Uma ferramenta para sustentabilidade da construção. NUTAU 2006. (*in Portuguese*).

Kielsteg Bauelemente, innovative smart wood structures—website "www.kielsteg.com", last visited on 1st March 2013.

Lippke, B., Wilson, J., Meil, J., & Taylor, A. (2010). *Characterizing the importance of carbon stored in wood products*. Wood and Fiber Science (42, Corrim Special Issue), 5–14.

Rossi, B., Marique, A.-F., Glaumann, M., & Reiter, S. (2012). *Life-cycle assessment of residential buildings in three different European locations, basic tool*. Building and Environment (51), 395–401.

Sathre, R., & Gustavsson, L. (2009). Using wood products to mitigate climate changes: External costs and structural change. Applied Energy (89), 251–257.

Teixeira, J., & Póvoas, R. (2010). Proposta de metodologia de intervenção para a reabilitação do património urbano edificado—as casas burguesas do Porto, coberturas. *Reabilitar 2010—Encontro Nacional Conservação e Reabilitação de Estruturas*, 1–11 (*in Portuguese*).

Green Design, Materials and Manufacturing Processes – Bártolo et al. (eds)
© *2013 Taylor & Francis Group, London, ISBN 978-1-138-00046-9*

Smart technology for the passive house

S. Vattano

Department of Architecture, University of Palermo, Palermo, Italy

ABSTRACT: Planning for sustainable policies aimed at identifying *Smart* urban spaces is determined by the parameters of *Smart Cities* model that encourage the production of intelligent building for construction energy efficient. The proposed article analyzes the effectiveness of the passive house in the Mediterranean area, offered by traditional building techniques, proposing a re-reading technology for the detection of a smart design methodology that can lead to the use of technologies with high energy efficiency.

1   INTRODUCTION

The European *Smart Cities and Communities* initiative, proposed under the *SET-Plan* (*Strategic Energy Technology Plan*), is affecting a very broad context, not only in terms of the parameters used for the definition of the *Smart City*, but also refer to the type of urban realities involved (Komninos, 2008). Social awareness of the environmental problem and necessity to preserve the balance of the biological system in which we live, lead to design solutions, technologies and use of materials which are sometimes different from "traditional" ones. The aim is to adopt models of production and less impact on consumption of exhaustible resources of the planet. In the complexity of the construction process is therefore required a balanced way that manages together innovative technologies and traditional knowledge in a reasonable architecture, capable of adapting different solutions depending on the needs to be met, the environmental context and the regulations to be observed. The buildings have a profound impact on the ecosystem, economy, health and productivity. The knowledge of this impact in the fields of building science, technology and operation, management and maintenance are available to designers, builders, contractors, operators, and owners who want to build green buildings and maximize both economic and environmental aspect. The solutions of sedimented architecture over the centuries, are characterized by a close relationship with the natural elements of the site and the daily and seasonal weather changes. The fact that these have persisted over time shows basically the correctness of their principles. Architectures then conditioned by the specific climate of the places in which they were built and characterized by a careful use of resources and technologies that enhance the principle of maximum efficiency with minimum expenditure

of energy. The traditional architecture, in fact, was the result of all the knowledge and processes that were competing to increase man's ability to adapt to the environment and to evolve in it.

2   A SMART TECHNOLOGY FOR PASSIVE HOUSE MODEL

When we talk about super-technological cities we can consider urban planning that even before the presentation of European *Smart Cities* model, were built in order to reduce CO_2 emissions and improve the energy performance of buildings through the implementation of innovative technologies (*ICT—Information and Communications Technologies*), like, for example, the city of Abu Dhabi, capital of United Arab Emirates. When talking about the model of *Smart City* we follow the tendency to idealize a composition of factors whose feasibility is closely linked to the rapid progress of the strategies that have new goals in the field of climate and energy (Gibson, D.V. *et al.* 1992). Above all, the attempt is to involve citizens and also a natural propensity towards improving the quality of life and local economies. It deals with investments in favor both of the efficient use of energy and of the reduction of CO_2 emissions (considering 1990 as the base year) by 2020, necessary to support cities and regions with ambitious and pioneering measures. Another objective is to achieve a 40% reduction of greenhouse gas emissions through a sustainable use of available resources and systems approaches related to new measures on buildings, on networks of local energy and transport. The initiative allows to structure national and international proposals for programming and planning *Grid Solar Energy Grid* and *Smart Buildings and Green Cars*. Therefore the objectives of this European initiate concern the implementation of best practices related to sustainable energy especially in relation to the redevelopment of existing heritage. To move in this direction it is necessary to integrate the most appropriate technologies and policy measures best suited through the consideration of important structural intervention, in the field of existing buildings. All this is necessary to reach certain parameters on the isolation of external and internal horizontal structures and on the casing of the building, in order to avoid heat losses. In this way it will aim at identifying ambitious measures guide for the construction *Smart* of buildings, above all re-thinking the existing buildings (Chesbrough, 2003).

Specifically, we can define *Smart Buildings*, zero energy buildings, or zero emissions ones, taking into account the requirements on the energy performance (*EPBD, Energy Performance of Buildings*) through the use of innovative materials (nanomaterials, solid insulating, vacuum insulating, cold roofs, etc.), which have to respond to lower levels of energy consumption. The technologies of *Smart Building* implement an automation system for buildings which allows you to manage in real-time safety, energy conservation, control of the whole structure and integration with innovative monitoring systems. The logical structure of a *Smart Building* therefore locates the following levels: the first one is physical and includes all the sensors/actuators (humidity, temperature, fire, intruder, etc.) and systems (HVAC, elevators, etc.) that innervate the structure of the building (Tselentis *et al.* 2010). The second level is characterized by the so-called *Building Management Systems* (Fig. 1) that have the purpose of supervising the systems and implement some automation. The third one is constituted by a system able to collect events from all systems and to correlate with each other

(*Building Control Room*). The *Smart Building* also includes the typology of *Green Building*, because it uses materials with a low environmental impact and systems for the efficient management of heat. It is an active building, which produces and distributes energy through the implementation of *Renewable Energy Sources* (*RES*), of *ICT* technologies and of materials that contribute to energy savings for the improvement of living comfort and safety (Komninos, 2002).

Smart Building can be considered as communication systems and data sources in a more integrated urban eco-system; for this reason the design approach to the entire building process has become a methodology of integrated design (*Whole Building Design*), founded on the principles of synergy and interconnection. But to go from design of sustainable energy building to *Smart Buildings* it must innovate the individual phases of the building process starting from the holistic point of view (costs, flexibility, energy efficiency and overall environmental impact, etc.) of building which will have be able to better respond to the demands of productivity, creativity, quality of life of the occupants, that is, the *Whole Building Design*. One of the Italian emblematic examples of *Smart Building* was presented by the prototype *biosPHera*, example of *passivhaus* model, designed and built by *ZEPHIR* (*Zero Energy and Passive House Institute for Research*), with the aim of showing how a passive building born and lives with an almost energy zero. The building consists of two blocks whose walls screens provide information about the inner well-being: temperature, humidity, percentage of CO_2, air-conditioned environment and surface temperature of the building envelop. It was necessary the three-dimensional modeling of the house, with the study of the optimal orientation and shading optimized in three different climates:

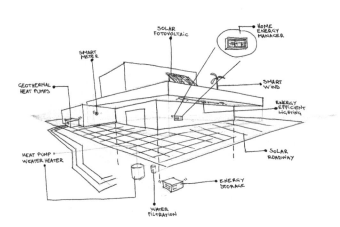

Figure 1. Example of home automation associated with the concept of *Smart Building*. (by Starlight Vattano).

northern, central and southern Italy. The use of free solar energy, derived from the environment, together with an energy performance envelope and from an efficient plant system, have allowed to develop a near-zero energy building component. In addition, the proper arrangement of the openings ensures the house a free energy intake, which is maintained and exploited thanks to an efficient heat casing, equipped with opaque components and limited thermal transmittance. The energy obtained by free solar heat is distributed through a system of mechanical ventilation with heat recovery, high efficiency, which minimizes dispersion and waste; the limited portion of thermal power, still needed to achieve an adequate indoor comfort during the coldest days of the year, is ensured by electric radiant plates positioned inside the walls and connected to the PV system placed on the roof. The building is also equipped with an innovative plant called "aggregate compact" which incorporates, in a single machine, the production of heating air, cooling air, controlled mechanical ventilation, heat recovery, production of domestic hot water, having a high overall efficiency. Wood was the natural element used in the construction of building both limited environmental impacts, due to the production, transport and disposal, and its capacities hygrometer and of healthy indoor air: the structure of the house is made entirely of wood panels; design attention has been devoted to the study of the casing, optimizing the quality of thermal insulation of each component and assessing regularly the absence of thermal bridges at the overlaps of different components, using innovative materials (airgel panels, nano-technological superinsulating, from the aerospace industry; insulating expanded panels extruded with high insulating power; rock wool panels phono-thermal-insulation at low content of aldehydes; fiber-plaster biocompatible slabs with thermal-phono-insulation function and high absorption capacity of volatile pollutants). The design attention is then focused around windows, active element in the energy needs of the building; starting from the choice of the transparent element, a triple-glazed low emissivity is able to give at the same time a high thermal insulation and a suitable sun transmission in the winter; were studied and simulated, at dynamic level, the conformations of three types of frames: wood, wood and aluminum, aluminum. For the opaque elements of the housing, both vertical and horizontal, were composed stratigraphy consist of numerous elements, able to meet the requirements of thermal insulation, hygrometric migration of steam and air tightness. We have to questioning if already existed a precursor of the *Smart Building*. How you can approach the concept of contemporary intelligent

building to traditional construction technologies? To associate the idea of *Smart Building* to a building belonging to a particular historical period and of which we can recognize a certain standard of energy performance, it means being able to read in the tradition of that building a smart technology to integrate with the contemporary innovation (Ishida, 2000). It is very interesting to watch the bioclimatic aspects of traditional architecture, the immense heritage of knowledge that has led to ingenious design solutions from the functional point of view and often very significant in terms of aesthetic and symbolic. This heritage, made by culture and history of building, of symbiosis between local climate, physical environment and local building materials, is accorded to a regionalism that varies from place to place. Types, shapes and construction techniques, components and used materials, show formal solutions of an architecture that has developed without architects, as result of human intelligence and of the deep knowledge of environmental factors. The nature of the soil, exposure to sun and wind, the presence of vegetation, the moisture content are some of the factors that have influenced the settlements in every time and place. A study conducted on traditional building of popular architecture caved in the rock, leads to the identification of some of the most interesting examples for the analysis of methods of environmental control over bioclimatic architecture of the past. The most explicit examples can be seen in areas where climatic factors take on extreme values, because the first need is the protection. In Mesa Verde (Colorado), several hundred houses are carved into the rock faces south and use ingeniously the physical and environmental resources for the air conditioning (Fig. 2).

Mesa Verde is located within a large cave with south exposure, away from direct high sunlight

Figure 2. Group of houses carved in the rock of Mesa Verde in Colorado.

of the summer and enjoys ventilation of summer breezes coming from the valley. At the same time it is well exposed to the south to collect the lowest winter rays, so that the heat of solar radiation can be accumulated from the rocky bottom of the cavern. The heat is stored by the rock of the cave, that has a great thermal inertia, and by land bricks of the buildings in *adobi*, that is made of earth and sun-baked. The accumulated heat is transferred gradually during the night, to create a microclimate constantly comfortable. In southern Tunisia in the middle of the desert, at 7–12 meters deep, living for two thousand years *Matmata*, Berber troglodyte diggers whose agglomerations are a model in the field of bioclimatic architecture. Following only the intuition, the ancient builders realized with a small number of techniques and materials available, comfortable and functional works based on the exploitation of microclimate and local resources, for thermal comfort and shelter from the external agents. And with a wide range of solutions, because each site has its own characteristics, even at short distances. In Islamic countries you protect yourself against heat and temperature range night-day forming communities by continuous complex tissue, such as the *casbah*, urban systems characterized by buildings huddled together, deep porches and walls thick, sometimes culminating in the so-called wind-towers. The buildings are adjacent so as to minimize the surfaces exposed to the south and the thermal exchanges with the outside air excessively hot. The typical form of Islamic *casbah* is closed, with a few outdoor narrow spaces and shaded street. In coastal regions the road tissue is often oriented in order to facilitate the channeling of cool and moist breezes from the sea and to avoid dry-warm winds from the desert surroundings. In Iran and Pakistan there are, since sec. X, *baud-geers* (wind catcher), kind of towers or chimneys, which collect the air flow at a height and through the vacuum created by a careful study of the prevailing winds, the channels through the house. The implementation of these ducts and hoods is well known in traditional architecture of the Middle East: it is possible to obtain an internal temperature cooler than ten degrees outside. The architecture of the past decades totally ignores the natural techniques of environmental control related to the structure; the new energy technologies are introduced with difficulty, and we tend to think of them as additional elements that apply to the building without formal integration or conceal as much as possible. In many regions are hidden technological and traditional aspects and territorial morphologies that would allow a natural construction of urban settlements under bioclimatic principles: underground, houses carved into the rock, ducts, pipes, etc. To interpret

bioclimatic architecture of the past through the *Smart* key, means to deep the functioning, the performance and technology of a building system, so that brings new benefits to existing buildings and provide the best solution to save energy. The reality of urban settlements of the Islamic tradition, for example, shows examples of spaces constructed according to the principles of passive. The design of *Masdar City* (Fig. 3) represents the first city in the world to be designed and conceived as *Carbon Neutral* (zero emissions and completely ecological) will be built in the desert about 20 kilometers from Abu Dhabi.

The proposed project focusing fields such as thermal insulation of buildings, lighting, low energy consumption, the percentage of glass surface, the optimization of natural light, the installation of smart appliances, smart meters and intelligent systems for the management of buildings, through an integrated distribution grid throughout the city and an energy management system that interacts with the electrical load on the grid base or *Smart Grid*. The energy required it will be produced by a solar power station, as well as photovoltaic, wind, geothermal and hydrogen; the water produced by a desalination plant powered by the sun and 99% of waste will be reused, recycled, end up in composting and energy plants. In *Masdar City*, the form of the city is dictated by climatic criteria clearly defined: the heating of the walls of the buildings are negated by the very close distances between the facades, the spaces between buildings favored the ventilation of the roads, which also contributes the wind tower, at the center of the square. It is a conical structure which exploits the air currents above the buildings, to ventilate urban

Figure 3. Courtyard that exploits the principles of bio-architecture in the project of *Masdar City*.

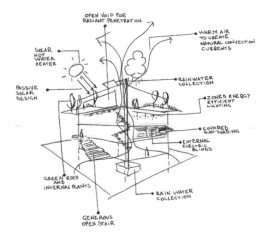

Figure 4. Study sketch based on the principles of passive architecture synthesized through the study of traditional building techniques.

environments. The walls of residential buildings are covered with corrugated panels similar to clay with the function of solar shading (*brise-soleil*) and elements which facilitate ventilation. Were adopted design criteria of the typical houses Yemen, with large thickness and perimeter natural ventilation chimneys. It was possible to develop and apply a wide range of environmentally friendly materials, such as low carbon concrete; recycled aluminum; timber.

3 CONCLUSIONS

It was recognized that the greatest environmental benefits come from some of the tools more passive and less expensive: the city (and buildings), the

orientation (with regard to the sun and the prevailing winds) and its shape. Respecting tradition means to observe the local culture, transpose messages, check the local availability of materials, find a form of functional and aesthetic integration into the landscape and in the existing buildings that do not break the harmony. The *Smart Building* today proposed must take into account the new reading of bioclimatic architecture of the past (Fig. 4); only in this way you can adapt your project to the reality of the place, trying to learn from existing local, from constructive, typological and climatic changes, and the use of the materials available on site.

The rules of traditional construction are discovered and cleverly adapted to current needs, supplemented by the current opportunity to express the best results in the balance between tradition and innovation.

REFERENCES

Chesbrough, H.W. 2003. Open Innovation: The New Imperative for Creating and Profiting from Technology. Boston: Harvard Business School Press.
Gibson, D.V., Kozmetsky, G. and Smilor, R.W. 1992. The Technopolis Phenomenon: Smart Cities, Fast Systems, Global Networks. New York: Rowman & Littlefield.
Ishida, T. 2000. Understanding digital cities. In Ishida T. & Isbister K. Digital Cities: Technologies, Experiences, and Future Perspectives. Berlin: Springer.
Komninos, N. 2002. Intelligent Cities: Innovation, knowledge systems and digital spaces. London and New York: Taylor and Francis.
Komninos, N. 2008. Intelligent Cities and Globalisation of Innovation Networks. London and New York: Routledge.
Tselentis, G., Galis, A., Gavras, A., Krco, S., Lotz, V., Simperl, E., Stiller, B., Zahariadis, T. 2010. Towards the Future Internet—Emerging Trends from European Research. Amsterdam: IOS Press.

Green Design, Materials and Manufacturing Processes – Bártolo et al. (eds)
© *2013 Taylor & Francis Group, London, ISBN 978-1-138-00046-9*

On-demand post-disaster emergency shelters

M. Tafahomi

Shelterexpert.org, Delft, The Netherlands
Aalborg University, Aalborg, Denmark

ABSTRACT: In post-disaster emergency shelter aid, various relief agencies and other actors with different backgrounds need to cooperate in a tremendously short amount of time and under very different circumstances. Lives depend on the quality of aid and, specifically, on the quality of shelter aid.

In this article we describe a research project that develops and tests a new approach in which the demands for shelters in a specific situation are methodically connected with available and innovative, sustainable shelter solutions, a Decision Support System (DSS).

1 INTRODUCTION

1.1 *Post-disaster shelters*

In post-disaster emergency shelter aid, various relief agencies and other actors with different backgrounds need to cooperate in a tremendously short amount of time and under very different circumstances. Lives depend on the quality of aid and, specifically, on the quality of shelter aid.

Relief organizations including the International Red Cross and the UNHCR, have recently expressed a need for more variety when choosing shelters in the immediate aftermath of a disaster (Geneva, shelter meeting 2009 & 2010). When choosing or designing a post-disaster emergency shelter, it is essential to connect the needs for post-disaster emergency shelters in a specific situation with the available shelter solutions and innovations. There is a need for on-demand provision of post-disaster emergency shelters and shelter items. Knowledge of the earlier performed post-disaster shelter relief, accurate evidence and the understanding of the whole process play an important role in this complex and hectic process.

1.2 *Innovation*

The increasing emphasis on innovation, resilience and preparedness in humanitarian actions, and the need for optimal intersectoral communication, lead to new approaches where data gathering and data quality become major elements that affect the results of relief activities.

The recent initiatives to gather data on the needs directly from the beneficiaries after a disaster, using mobile phones is an example of the inevitable role of new technologies in post-disaster emergency aid.

Connecting the gathered data to the available solutions for providing tailored (shelter) aid can be realized with a Decision Support System (DSS). The needs are to be systematically connected to the available solutions. Objective decision making can be facilitated to enable the beneficiaries and the shelter experts, to choose the most optimal shelter solutions in each situation.

In addition, this approach addresses the transparency and accountability in post-disaster emergency aid that lead to a more sustainable process of shelter aid.

Investigation of the possibility of using a systematic method, such as a Decision Support System (DSS), for providing on-demand shelters in immediate post-disaster, can result in more sustainable post-disaster shelter designs, as tailored solutions mean less waste, optimal use of energy and materials and less transportation costs.

2 SUSTAINABLE POST-DISASTER EMERGENCY SHELTERS

2.1 *Needs and solutions*

Stainable development is 'Development that meets the needs of the present without compromising the ability of future generations to meet their own needs' according to the report our common future.

For creating sustainable post-disaster emergency shelters that meet the needs without compromising the ability of future generations to meet their own needs' (1987, Our common future), thinking in advance and considering the long-term effects of the shelter aid in each specific context are the key elements.

According to Corsselis and Vitale (2008, shelter Centre) a 'shelter' is a habitable covered living space, providing a secure, healthy living environment with privacy and dignity to those within it. In specific case of post-disaster shelter aid, in many occasions the shelters are used for years, even for generations, while according to the international protocol, maximal use of an emergency shelter ought to be between 18 and 36 months. The necessity (semi)-permanent shelters is indicated by experiences as in Darfur and Bam. In addition the transformation of the emergency shelters into transitional and permanent habitats leads to more sustainable shelters and optimal use of materials.

2.2 Long-term effects of post-disaster emergency aid

Interviews with relief specialists as graham Saunders head of the IFRC and UN reports as Pakistan 2004 earthquake suggest that the first 24–48 hours are crucial for saving lives. Enduring 24 hours without shelter can be life-threatening in extreme climatic conditions, which may include severe cold, high altitudes, extreme heat or exposure to dust storms. For a large scale emergency, shelters are often delivered internationally from pre-positioned stockpiles. In the first weeks, aid packages including shelter items are requested based on initial assessments and consultation. A pipeline for aid is then set up. Immediate needs are to be met in this phase.

The shelters provided in the immediate phase after a disaster (post-disaster emergency shelters) have long-term effect on the living conditions. The recent focus on the long-term effects of shelter and the strategy of using disasters as opportunities to improve people's living conditions in a sustainable manner has resulted in greater emphasis on transitional housing solutions.

We need to bring together needs and solutions in the immediate aftermath of a disaster, in a manner that the long term effects are positive, to provide sustainable post-disaster shelters. The ability to provide optimal, immediate post-disaster shelter while simultaneously taking into account the long-term effects and factors such as sustainability, cultural acceptance, costs, interaction with beneficiaries are to be incorporated in the DSS.

2.3 Data/evidence

Reliable data gathering, data storage, data analysis, data connection, and data security are the key words in the DSS. As service calls to external sources can be realized, data gathering is not limited to the data that the DSS can gather. However, realizing service calls to external data gathering

systems need additional reliability and security steps and causes that the DSS is dependent of external factors for optimal functioning.

Master data are gathered and administrated by the data manager.

Currently the DSS performs service calls to the KNMI, CIP and Google maps. The service calls can be expanded. Regarding the solutions Data gathering involves data from the materials and solutions that are available on the market, including local resellers.

2.4 Evidence gathering and implementation

Optimal choice of shelter solutions in post-disaster emergency situations depends on deep understanding of needs, contexts of the needs, practical knowledge of the available solutions, and smart strategies to connect the two optimally.

In this article we investigate a new approach in which the needs for shelters in a specific situation are methodically connected with available and innovative shelter solutions by using a Decision Support System.

Evidence of needs can be submitted both by the beneficiaries and relief specialists as NGO's and the UN. The solutions can be entered by local or international suppliers. The questions asked by the DSS can be adapted to each type of user.

Figure 1 presents a summary of the first steps in the interface of the current DSS.

The DSS presents the specific questions for each type of user. The following steps will then be performed:

1. Log on: Type of user
2. A Enter Product specifications OR
   B Enter New Disaster
3. B Select the disaster-location (continent/ country/region) and disaster-type
4. B Enter disaster specifications (different questions for beneficiaries and relief specialists)
5. B3 The DSS presents the location data and a link to relevant sites as Google maps
6. B The DSS provides a first analysis of the disaster and a first set of advice

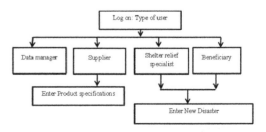

Figure 1. Log on and user identification.

7. B Based on this advice and personal judgment, the user selects the shelter specifications
8. B The DSS presents a list of 5 shelters that meet the specifications the most with ranking
9. B Choose the cheapest or the fastest
10. B The DSS produces a final report of the number and price for the shelters.

2.5 *Final report*

Currently writing reports is one of the tasks of senior shelter relief specialists. Writing reports offered shelter aid, is a time consuming administrative activity needed to fulfill the accountability aspect as well as documentation of the experiences.

In addition to advice for the optimal shelter, the DSS produces a final report of the offered shelters. The current DSS provides three chapters:

1. Problems caused by the disaster
2. General advices as: Earlier disasters/learned lessons/advice in Shelter related sectors
3. Report of the provided shelters-shelter advice including the costs.

3 SHELTER SOLUTIONS

3.1 *Sustainable post-disaster emergency shelters*

Regarding the total shelter solutions, when a suitable shelter is not available, an innovative design can be suggested by the DSS.

Successful choice of shelter solutions in post-disaster emergency situations depends on deep understanding of the needs; contexts of the needs; practical knowledge of the available solutions and smart strategies to connect the two optimally.

Post-disaster emergency shelter designers are as any other designer concerned with the ecological, social, cultural, and spatial application of technologies to meet specific human needs after each disaster and in each location. The choice for an on-demand designed shelter contributes to optimal post-disaster shelter relief process and sustainable post-disaster emergency shelters.

3.2 *Meeting the standards*

The first initiative for standardization of sheltering process, the sphere, was launched in 1997 by a group of humanitarian NGOs and the Red Cross and Red Crescent movement. They formulated a Humanitarian Charter and identified Minimum Standards to be attained in disaster assistance, in each of five sectors of aid: water supply and sanitation, nutrition, food aid, shelter and health

services. This process led to the publication of the Sphere handbook in 2000.

The Sphere Project was to develop a common framework and improve accountability for humanitarian aid. The Sphere Project is continuously being up to dated and developed. The humanitarian community mostly uses Sphere standards inconsequently. In some cases, Sphere indicators and standards have been dismissed, as in case of Pakistan where was decided that "Sphere standards will not be met". An insufficient shelter supply was the motive not to meet the Sphere-approved quality levels in order to provide aid for more affected population.

Currently various guidelines for shelters are available. Each organization has specific standards as the UNHCR handbook and the IFRC catalogue.

In the complex environment of post-disaster shelter-relief however, the task of creating optimal designs needs a systematic framework. The framework should take the following aspects into account:

- context-sensitivity
- crisis-sensitivity
- 'performance standard' paradigm.

Performance standards lead to flexible design environment that leaves room for flexible sustainable shelter designs.

3.3 *On-demand sustainable post-disaster emergency shelters*

Designing a post-disaster emergency shelter with the aid of a DSS needs adaptive design thinking. The basic design is to be flexible.

For this reason we developed a basic design for a sustainable shelter that can be realized using the parameters of a DSS and can be adapted to each specific situation with the aid of the DSS. A central column in the shelter can include additional part solutions as energy, water and sanitation.

Figure 2a and 2b illustrate two possible shapes of the designed shelters. The frame and the central column provide a freedom for designs and flexibility to create culturally accepted shapes for the shelters. This means more acceptance and longer use of the shelters, thus more sustainable shelters.

Usage of local materials, energy and cost effective designs are the consequences of this approach.

3.4 *Total concept*

The design contains a central column that is a part of the construction, the construction and the skin including the floor.

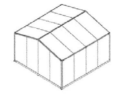

Figure 2a. and b.

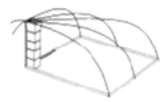

Figure 3. The position of the column in the shelter construction.

Depending on the needs and the available solutions the content of the central column will be chosen. Sanitation, heating, cooking facilities can be included in the column. There are various options for the skin and the construction. Two examples are:

A. The rapid response: Lightweight emergency solution with flexible tubes as construction and lightweight fabric (Fig. 2a). The flexible elements can be replaced by Aluminum Profiles or locally available materials for long-term usage.

B. Transitional solution: with construction profiles and the possibility of filling the surface with panels and extra accessories with the locally available materials (Fig. 2b).

As the central column provides flexible content, tailored solutions for each post-disaster situation can be offered. Figure 3 illustrates the position of the column in the shelter.

Based on the gathered evidence on needs and solutions the DSS can provide advice for tailored shelter solutions. The advice can be materialized with this shelter design approach. This approach will lead to more sustainable and tailored shelter solutions that meet the needs of the beneficiaries the most (Fig. 3).

The DSS can provide detailed advice regarding used materials and insulation. An optimum will be calculated and presented in diagrams as Figure 4 shows.

To meet the specific needs of the beneficiaries after each disaster, the DSS needs to provide advice for shelter parts and integral shelter designs. Water, sanitation and energy can be included in the shelter advice. Figure 5a illustrates the design for the central water and sanitation unit, developed in this research project.

An energy unite that can be adapted to the needs in each situation and meet the maintenance criteria is being investigated currently. Figure 5b illustrates the basic mobile energy unit that is developed and tested. We need to keep in mind that the choice of water, sanitation and energy facilities is made by shelter experts. The task of the DSS is the provision of advice, by connecting the evidence on needs, location and solutions

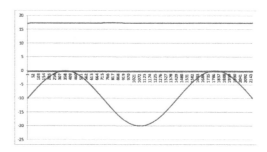

Figure 4. Realizing constant temperature inside in various outside conditions (source template Tonny Grimberg, Saxion).

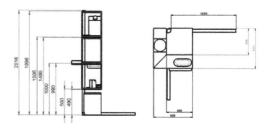

Figure 5a. The central column, can include energy, water and sanitation units.

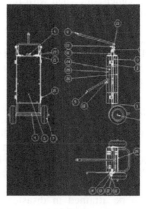

Figure 5b. The energy section can be delivered as a separate unite (Johan Kok, Iindustrialdesign.nl).

724

and presenting the possible solutions that meet the needs of the beneficiaries the most, for each situation.

Regarding energy provision, the advice provided by the DSS is to be based on long term effects of the usage of wood that results in deforestation, versus durable energy sources as solar energy or the fire sensitivity when using kerosene versus the usage of alternative energy sources that are currently less cost effective and need maintenance and technical know-how. The latter can mean educating the local population and growing local economy.

4 CONCLUSIONS

This research project can be summarized in a set of characteristics for a Decision Support System for post-disaster emergency shelter aid. The optimal DSS for post-disaster emergency shelters has the ability to provide uniform advice.

The test in this research project indicates that users including relief specialists and the beneficiaries need to trust the DSS and to believe in accuracy and reliability of the advice provided. In addition, participation of the users in the creation of the DSS, a flexible framework and a reliable and secure infrastructure are needed for a DSS that will be used by the involved parties.

According to our findings, a DSS that meets these characteristics leads to the provision of sustainable shelters that meet the needs of the users and therefor have higher performance in long-term.

The DSS is a digital tool that translates the gathered evidence on needs, solutions and location including long term effects of shelter provision into advice for the provision of sustainable, integral broader shelter solutions and designs.

The current DSS that is a prototype needs to be optimized and further developed, in cooperation with the users, e.g. the beneficiaries, the shelter experts and the designers and suppliers.

REFERENCES

Architecture for Humanity 2006. design like you give a damn.
Ashmore, J. & Fowler, J., & Kennedy, J. (2008). Shelter projects, 2008: Un-habitat. BARADAN, B. (2006).

Barenstein, J.D. & Pittet, D. 2007. Post-disaster housing reconstruction Current trends and sustainable alternatives for tsunami-affected communities in coastal Tamil Nadu.
Corsellis, T., & Vitale, A. 2005. Transitional Settlement: Displaced Populations: Oxfam.
Cuny, F. 1972, shelter documentation, UN archive.
Cutter, S.L. 1996. Vulnerability to environmental hazards. Progress in human geography.
da Silva, J. 2010. Lessons from Aceh. Key Considerations in Post-Disaster Reconstruction.
Davis, I, UNDRO 1978. Shelter after Disaster.
Howard, J., & Spice, R. 1989. Plastic Sheeting: Its use for emergency shelter and other purposes: Oxfam GB.
Howard, J., & Spice, R. 1989. Plastic Sheeting Oxfam GB.
Humanitarian Accountability Partnership 2005. Humanitarian Accountability.
ICRC/IFRC, catalogue 2002. ICRC, Geneva.
Leon, E. UN HABITAT 2009. Local Estimate of Needs for Shelter and Settlement, LENSS.
Manfield, P. 2000. A Comparative Study of Temporary Shelters used in Cold Climates.
Medcins Sans Frontiers, shade nets: use and deployment in humanitarian relief environments. Médecins sans Frontières, (MSF) Catalogue, 2002.
OCHA, 2005. Humanitarian Response Review.
Oxford university Press, 1987. Our Common Future.
Procurement catalogues and specifications IAPSO 2012.
Project, S. 2003. Sphere Project Training Package: Humanitarian Charter and Minimum Standards in Disaster Response: Oxfam.
Saunders, G. 2004. Dilemmas and challenges for the shelter sector: lessons learned from the sphere revision process.
Street, A. Action Aid, Parihar, G. independent consultant HPN 2007. The UN Cluster Approach in Pakistan earthquake response: an NGO perspective.
Tafahomi, Egyedi, 2008. Defining Flexible Standards for Post-Disaster Emergency Sheltering.
UN HABITAT, IASC, 2008, 2009, 2010. shelter projects.
UN report 1976 Guatemala earthquake & 2004 Pakistan earthquake.
UNHCR, 2002. Cooking Options in Refugee Situations. UNHCR, Geneva.
UNHCR, 2009. Handbook Tents.
United Nations, 2008. Transitional settlement and reconstruction after natural disasters.
USAID, 2002. Demographic Assessment Techniques in Complex Humanitarian Emergencies: Wisner, B et al. 2004. At Risk: Natural Hazards, People's Vulnerability and Disasters.
World Bank, 2010. Handbook for reconstructing after natural disasters, chapter 1, Safer Homes, Stronger Communities, the Six Options for Displaced Populations.

Green Design, Materials and Manufacturing Processes – Bártolo et al. (eds)
© *2013 Taylor & Francis Group, London, ISBN 978-1-138-00046-9*

Towards zero energy house integrated design: Parametric formwork for capillary systems in precast concrete panels

Christopher Beorkrem & Mona Azarbayjani
UNC Charlotte, North Carolina, USA

ABSTRACT: This paper investigates current research in the design of a parametric facade system for a prototype solar-powered house being developed for the 2013 Solar Decathlon to promote high performance building while using traditional passive strategies. The house design employs a precast carbon geopolymer concrete panel system, which is inlaid with a capillary system capable of circulating cooling and heating liquid to provide efficient comfort for occupants. To improve the efficiency of the concrete panel system, we have introduced a complex parametric formwork, which was developed through an analysis of various computer controlled router bit profiles and curve logics, in three dimensions. The parameters of this system are based on earlier conclusions about the effectiveness of our combination of geopolymer concrete and capillary system, while attempting to improve the system by increasing the surface area of the wall and through the creation of more expressive surface.

1 INTRODUCTION

The 2013 U.S. Department of Energy Solar Decathlon is a competition between twenty teams of students from colleges and universities across the globe. Contending teams are challenged to design and build an attractive, energy-efficient house that is powered entirely by the sun. The competition is sponsored by the U.S. Department of Energy and will be held in October in Irvine, California. Interest in clean and renewable energy sources is increasing as recognition of the limits of non-renewable energy continues to grow.

The University of North Carolina at Charlotte's solar decathlon house team has worked to introduce fly ash geopolymers into their concrete mix to help reduce the carbon footprint of the house. During the generation of electricity, we burn a billion tons of coal every year, leaving significant quantities of coal ash. Rather than sending this ash to the landfill, or storing it in costly and potentially dangerous pools, this recycled material can be used as an additive to concrete. The geopolymer concrete uses fly ash as part of the binding mixture within concrete, which accounts for approximately seven percent of fly ash being diverted from landfills each year. Fly ash can replace Portland cement as the binding agent in concrete. Since the production of Portland cement is estimated to generate between two and five percent of the world's GreenHouse Gas (GHG) emissions, 4 percent according to the EPA, the substitution of fly ash for cement is often cited as a means to reduce carbon footprint.

The geopolymer concrete used results in a nearly similar condition to traditional materials, but using geopolymer concrete results in a four percent reduction in carbon output.

Secondly, the exterior wall system we specified for the house includes a capillary system, which is intended to further enhance the performance of the wall and further diminish its impact on the environment. The capillary system is a thin series of tubes, which allow for the environment of the house to be mediated more quickly using the mass of the concrete, by circulating liquid through the interior of the wall.

We wanted to further enhance the performance of the wall by increasing the surface area of the wall condition, in essence creating a radiator like effect. Using the constraints of the capillary system, insulation core, and geopolymer concrete mix, we established a set of parameters to constrain the options for exposing more surface area without compromising the strength of the wall or creating components which were too thin and fragile. We designed a parametric model to delimit these options and to test out a series of options.

The design of the model accounted for the necessary thicknesses, which need to be maintained for the optimum use of the capillary system while attempting to increase the surface area of each façade. We determined a set of limitations for maintaining the strength and durability of the façade. The model effectively was attempting to find the maximum amount of surface area that we could produce while minimizing increases in volume of concrete. Once we had found a workable model which both increased the surface area of the façade design, but also created a compelling aesthetic to the façade, we needed to develop a method for creating the customized formwork necessary to manufacture the concrete panels.

Ideally, the formwork design needed to optimize the form and dimensions of the formwork while increasing the efficiency of the system, by saving time and reducing expenses. This paper discusses the role of parametric modeling for exploring the design goals and performative applications of complex formwork for the geopolymer concrete facade. Through a series of full-scale mockups, we attempted to find a balance between the effectiveness of various profiles, dimensions and patterns, formwork efficiency, and surface area to volume ratios to optimize heating and cooling transfer. The parameters of this system are based on earlier conclusions about the effectiveness of our combination of geopolymer concrete and capillary system, while attempting to improve the system through the creation of a more expressive surface.

2 METHODOLOGY

2.1 *Assumptions—parametric limitations on the system*

The use of a capillary system effectively creates a wall which functions as a radiator, interfacing with the interior of the house to both heat and cool the interior volume. The heating-cooling system of the house is well integrated to produce optimum comfort while using minimal resources. The insulated geopolymer walls contain a series of tubes embedded in the concrete, which are circulating water that flows warm in the winter.

As shown in the picture below, during the summer time warm water will transfer heat to the night sky and cooled water will then circulate during the day to reduce the temperature of the thermal mass geopolymer concrete walls. Step 1, the heat that is exchanged from the space to the wall will be stored in the mass and step 2, the heat is exchanged to the water running in the capillary tubes. Step 3, the heat will be transferred to the roof. During the night, the heat that is absorbed into the mass during the day will be sent to the roof to be transferred to the sky via radiant mode heat transfer. Increasing the interior wall surface area, therefore will increase the heat transfer rates from the space to the mass and subsequently to the water in the capillary tubes.

2.2 *Parameters*

We defined three parameters, which could effect the geometry of the façade. The goal is to optimize the formwork while increasing the surface area to volume ratio. Increasing the surface area will increase the heat transfer rate. The concept is similar to a car's radiator, which has thousands of fins that increase the surface area to increase heat transfer between the air and water. Adding surface area to the concrete interior will increase the heat exchange rate 3–4 times faster. The volume has not change, though the surface area has increased

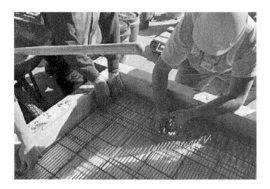

Figure 2. Concrete formwork with capillary tube system.

Figure 3. Geopolymer concrete and capillary tube system.

Figure 1. Section rendering of house design.

by adding the undulations. The main driver of increasing the surface area is to increase the heat dissipation from the air to the mass (heat sink).

We were however aware of the possibilities of creating elements on the façade, which were so thin that they could result in an increase in the fragility of the surface. Therefore, the bands on the walls could not be less that 2 1/4″ thick and not protrude from the façade more than 3/4″.

Once the correct proportion of undulating geometry was determined, we needed to develop a system, which was capable of producing a large variety of geometric options without increasing the amount or cost of the formwork. Plastic formwork

Figure 4. Geopolymer concrete wall with insulation and capillary system.

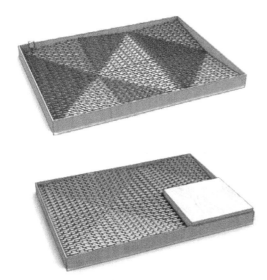

Figure 5. SHoP Formwork, and formwork with blocking.

and the molds necessary for their production constitute a significant proportion of the overall cost of generating any non-uniform precast panel. Following the precedent set by SHoP Architects in their 290 Mulberry project, we considered how we might create a single large-scale mold, out of which the portions could be blocked to create variations in the form. SHoP used this to great effect by creating an entire subset of panel shapes, forming around windows, creating corner conditions, and allowing for geometric shifts to seamlessly blend between panels, with only a single piece of custom formwork.

Though the SHoP method would work to minimize the amount of custom fabrication, it still created strong vertical lines between panels of exposed concrete. However, we desired to create the aesthetic of a single surface across the rear wall of the house, which is approximately thirty feet long.

3 MODEL DESCRIPTION

The design intent of our system was simplified to create a series of waves to give a dynamic effect on the building while providing shade through the day, which also increases the effect of the capillary tubes laid in the wall. The defined a texture with increasing density in the middle which would fade towards the edges, to allow for easy transitions in the formwork, while not requiring any other custom componentry to manufactured for each of those conditions (though it should be noted that using vertically cut halves of components would allow for us to create corner conditions as well).

3.1 Formwork

Typical precast formwork is created using a vacuum formed piece of plastic, which is pulled onto a wooded mold, fabricated using either traditional wood fabrication methods, or CNC tooling. Typically, this is done as a single large sheet, which can be used multiple times. Sometimes however, multiple versions of the same mold can be made in plastic as with each pour there is wear and tear on the plastic mold. The mold needs to be created with a draft angle along any edge, which is perpendicular to the surface to ensure that the formwork will pull away from concrete once it is cured. This draft angle is typically approximately 5 degrees.

3.2 Hexagonal panels

We worked with our precast formwork manufacturer who offered us the option of using smaller components to create larger panels. This included the possibility of creating a series of molds that could be flipped and mirrored. Each mold could be used to create many versions of a single shape. We quickly realized that a rectangular shaped mold would only produce two real variations right-side-up and upside-down. If we were to rotate the panel 90 degrees the grain of the form would not work with our desired outcomes. What we did discover is that by using a hexagonal panel, oriented vertically (with the parallel edges moving horizontally, we could get four different variations of each panel by rotating and flipping the geometry). This would give us the flexibility to minimize the number of molds we needed to create our intended geometry, while still having an enormous variety of design options.

While in search of a method for breaking down the vertical lines in the surface and maintaining a significant amount of variation, we used two particular references. Our primary precedents were *P_Wall* by Andrew Kudless (MATSYS) in 2009, and the Spanish Pavilion by Foreign Office Architects (FOA) in 2005, both of which demonstrated how variations in hexagonal panel shapes could help to break down strong grid lines in a panelized system. *P_Wall* created variation in the widths of hexagonal panels to breakdown any diagonal lines, which would develop across a hexagonal pattern. Kudless' variations allowed for each angled portion of the hexagonal geometry to remain at the needed angle while only creating variations in the lengths of the horizontal lines.

The *Spanish Pavilion* by Foreign Office Architects, also created variations in the hexagonal pattern, but used a series of panels, which maintained relational shifts between one another. FOA developed a panel system of six different hexagonal forms, whose perimeter still makes perfect hexagonal edges, but whose internal relationships vary from the balanced angles, which are typical for this shape. This allowed for many more variations in the system while creating only six different mold shapes.

The initial parameter to determine for the panels is the appropriate scale for the hexagonal shapes. Different proportions across the twelve foot tall wall could result in more molds having to be manufactured, but also could provide more variation. A hexagon of equal sides has equal 120-degree angles all around. The height of a hexagon is the square root of three times one of its sides. We determined that simplest method for creating the largest variety was to create a series of ridges, which are centered on one half of a hexagonal panel. The width of these ridges could vary but needed to be centered on one half or the other, so as always to align back to a baseline condition. Therefore the size of each hexagon, in proportion to the wall, would result in the possible densities and scales of ridges.

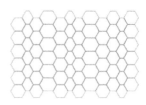

Figure 6. Hexagonal variations in *P_Wall*- Kudless.

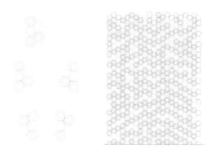

Figure 7. Hexagonal pattern on *Spanish Pavilion* (FOA).

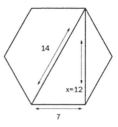

Figure 8. Spanish pavilion.

Figure 9. Hexagon proportions.

3.3 Router process

We determined that if we delimit the widths of possible ridges to two varieties, then we could minimize the number of molds that to produce. The variety was further minimized, so that each hexagon could do one of only three things: slope up into a ridge, continue a ridge, and slope out of a ridge. To simplify, we called these three pieces, "the on-ramp", "the highway", and "the off-ramp." As

Figure 10. Highway metaphor ("on-ramp", "highway", "off-ramp") Highway component can be flipped and rotated.

Figure 11. Fifteen unique mold shapes.

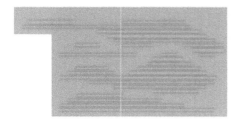

Figure 12. Variations on pattern configurations.

Figure 13. Façade.

we have two varieties of widths, we also had to create a panel, which was a continuous component for two different widths, and for doubles of each variety (two of the same width on the same panel).

This constraint resulted in nine unique molds, which could result in a dense variety of different patterns.

Additionally, this system allowed us to easily, within a panel and half, to terminate or fade out more gently all of the ridges on the system. As previously mentioned, a flat concrete panel condition at each corner and each perforation (window, door, etc...) would result in fewer complex panels and shapes.

3.4 Fabrication

To further minimize cost, we wanted to fabricate each of the molds for the system, using our in-house three-axis CNC router. Each mold was to be cut from Medium Density Fiberboard (MDF). Though our router is capable of cutting the slopes of each of the "on/off ramps" in the system, those cuts would result in a scalloped shape if we were to use a ball nose-cutting tool, or a stair stepped shape if we used a square-tip cutting tool. To avoid the imprecision that would result from having to sand down each of those surfaces, we elected to create a custom jig, allowing for us to run a square tip bit parallel to the ground while cutting the material to

Figure 14. Jig profile.

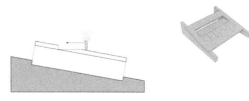

Figure 15. Angle relief cut on jig.

Figure 16. Tapered bit profile.

create a slope. This would result in a very clean cut-surface. Additionally, we use a tapered bit with a 5-degree angle to create the necessary slope for the concrete to release properly from the formwork.

Once complete, the molds could be used to create sets of hexagonal tiles, which were die cut by the formwork manufacturer to create perfect edges. Each of these panels is seamed together using only tape. The formwork was supported from below using MDF blocking of the same depth as the cut, supporting each of the tiles to prevent any deformations under the load of the concrete.

4 CONCLUSIONS

This paper discussed parametric approaches as a generative tool in order to provide balance between the forces driving the design and the constraints of the material. Parameters of the design, aesthetics, efficiency of materials and increasing the heat transfer rate through capillary tubes provide constraints that form the design process, which proves to be useful, especially during design exploration. Moreover, parametric manipulation of form is particularly important in performance-driven design processes, facilitating the combination of analysis and design synthesis into an integrated process.

Tightly parameterizing a novel system for the creation of both expressive wall systems, while also improving the efficiency of advanced wall compositions, including capillary heating and cooling, can be done in cost effective ways. Using technology, which is now more common than ever, while computationally deriving methods for expression in the system, can result from simple yet compelling shapes and forms. We are now able to iterate much more quickly using the tight constraints of a material system, and all of its parameters as a guide.

REFERENCES

Burry, Jane, Burry Mark, (2010). The New Mathematics of Architecture, (pp. 108–111). London: Thames and Hudson.

Hanle, Lisa, (2002). "CO2 Emissions Profile of the U.S. Cement Industry," U.S. Environmental Protection Agency, 2002.

Kosnya Jan, Kossechab, Elizabeth, (2002) "Multi-dimensional heat transfer through complex building envelope assemblies in hourly energy simulation programs" Energy and buildings, 34(5), (pp. 445–454). Retrived from: http://www.sciencedirect.com/science/journal/03787788/34/5.

Kudless, Andrew (2011). "Bodies in Formation: The Material Evolution of Flexible Formworks." Proceedings of the 31st Annual Conference ACADIA, (p. 98). Stoughton, WI: The Printing House.

Kudless, Andrew (2012). "Bodies In Formation: The Material Evolution of Flexible Formworks," In Gail Peter Borden & Michael Meredith (eds) Matter: Material Processes in Architectural Production (pp. 475–487). New York: Routledge.

Green Design, Materials and Manufacturing Processes – Bártolo et al. (eds)
© 2013 Taylor & Francis Group, London, ISBN 978-1-138-00046-9

Architectural rehabilitation and NZEB: The expansion of the Library of FDUL

A.P. Pinheiro
CIAUD Faculty of Architecture, Technical University of Lisbon, Portugal

ABSTRACT: With the introduction of the new Energy Performance of Buildings Directive recast challenges, in 2010, whose aim Europe "EU 20-20-20" for 2020 is to be a 20% reduction in GHG emissions, the introduction of 20% energy from renewable sources and a 20% increase in energy efficiency, were additionally implemented other measures. There is another more ambitious goal for new buildings construction, NZEB Nearly zero-energy building, which applies to new public buildings from 2018 and all new buildings from 2020. For case study, it was chosen one university building as an example where we can verify the hypothesis that it is possible to intervene in Architectural Rehabilitation in a sustainable way—passive and active—favored by a minimalist design. Thus, we selected the expansion of the Library of the Faculty of Law of UL as case study, which aimed to create a synthesis between architecture, sustainability and design.

1 INTRODUCTION

The sustainability concern is vital to the continued health of the planet, existing a need to redefine the acts of designing, thus the energy impact in buildings is critical. However, it is essential to find a balance between the solutions to adopt, articulating them conveniently between each other, questioning in time lengths the impact that they produce in construction, so that the interdisciplinary integration takes place in a productive manner. Therefore, the validity of the principles will only be achievable if they constitute contributions articulated in time, assuring accuracy and objectivity.

A diverse set of policies, regulations and laws have taken place towards building constraints, in order to optimize their energy performance throughout their existence.

The introduction of the new challenges resultant of the recast of the Energy Performance of Buildings Directive, EPBD, in 2010, defines the European goal "EU 20-20-20" for 2020, which predicts three situations: 20% reduction in GEE emissions, the introduction of 20% energy from renewable sources and a 20% increase in energy efficiency.

According to Directive 2010/31/EU of 19 May 2010, the building—with high energy performance and which requires nearly zero or very small energy amounts of energy—covered by energy from renewable sources produced on-site or nearby is called a Nearly Zero-Energy Building (NZEB).

In Portugal, the goal is that all public buildings are NZEB from 2018 on, and from 2020, all new buildings are as well.

As Case Study, it was chosen the expansion of an university building that sets an example of where the hypothesis can be verified: that it is possible to intervene in Architectural Rehabilitation in a sustainable manner—passive and active—favored by a minimalist design—the building of the Faculty of Law, University of Lisbon (FDUL), by Porfirio Pardal Monteiro, completed in 1958 (Fig. 1).

The choice of this building results from the dual reason of being an architectural heritage and of corresponding to an intervention as a coauthor with Rui Barreiros Duarte, on the Conservation, Restoration and Rehabilitation (Fig. 2) with architectural Expansion (Fig. 3). The building, with an area of 27,659 m², is developed through four floors plus parking and was opened in 2001.

Currently, the same team of architects is working on the new rehabilitation, presently in the

Figure 1. Faculty of Law by Porfirio Pardal Monteiro. Photo by Ana Paula Pinheiro, 1994.

Figure 2. Faculty of Law: rehabilitation by Ana Paula Pinheiro and Rui Barreiros Duarte. Photo by Sérgio Mah, 2001.

Figure 3. Expansion of Faculty of Law (FDUL) by Ana Paula Pinheiro and Rui Barreiros Duarte. Southeast view. Photo by Rui Barreiros Duarte, 2012.

Figure 4. Expansion of the Library of FDUL by Ana Paula Pinheiro and Rui Barreiros Duarte. Southeast view. 2013.

design phase of the renovation and expansion of the Library (Fig. 4).

With this project, it was intended to create a synthesis between Architecture, Design and Sustainability, that justifies an overall design with simplicity and that minimizes the environmental

impact, considering all phases of the building's lifecycle—design/construction/deconstruction.

The interaction between the various specialties and specificities of each one constitutes a fundamental technical contribution in order to optimize the aimed results.

The procedures recurrent from the work progress, lead to an articulate coordination of technical and aesthetic principles, design quality and optimization of the set, being the result an integrated system. The whole project was thought according to sustainable construction principles, using zero-emission materials. (Kibert, 2008).

The building design has considered the possibility of, in the future, being extended in height.

2 EXPANSION OF THE LIBRARY OF FDUL

2.1 Energy from renewable sources

Photovoltaic panels were used for energy production, and geothermal energy for heating and cooling.

2.2 Photovoltaic solar modules

The creation of a photovoltaic solar production central electricity for self-consumption, in accordance with Decree-Law 34/2011 of 8 March, which defines mini-production as an activity with decentralized production of electricity on a small scale through renewable resources, allows selling electricity to the public, with subsidized rates, provided there is demand on the same site. The Decree also allows the installation to be based on a single technology as long as the maximum power for central's grid connection is 250 kW.

According to the J. Oliveira's Project, the proposed central of power production is constituted by photovoltaic solar modules consisting of 60 square cells of 152.4 mm side, made of monocrystalline silicon, connected in series in order to obtain the nominal power of 245 Wp.

2.3 Installation functioning

"Photovoltaic conversion" is the name given to the process that converts the solar energy which insides on the modules from photovoltaic panels.

The materials used to guarantee this process are special and are called semiconductors. This process optimization implies that even at the diode junction of each photovoltaic cell, it establishes a different electric potential.

It is from the external union of each cell that an amount of electrons is released that, once produced, will create an electric current as long as there is solar incidence.

Thus, each module contributes in a dynamic way towards the production of electric current—with a defined potential—although dependent on the following parameters: azimuthal orientation, solar radiation intensity; perpendicularity of the module's solar incidence; environmental temperature; junction temperature.

The modules in series combination allow obtaining greater tension values, the greater the number of modules there is.

This module unit associated in series is designated as Series or "string". Therefore, the "strings" once joined in parallel, allow obtaining the desired nominal intensity of electric current.

The energy produced can have two purposes: be directly consumed or injected into the public network. To measure the energy (power) in any of the mentioned cases, a proper energy meter is set within the connection. (Oliveira, 2012).

2.4 Optimization factors

In addition to the taken options and the interdisciplinary articulations, one should optimize the performance of all secondary operations: acknowledge the inclination of the solar rays, optimize sunlight, major the function criteria.

It is vital to avoid shading. If a panel is partially shadowed, the entire assembly to which it is connected ceases to function.

2.5 Location of photovoltaic panels

The initial option to place photovoltaic panels on the exterior, in order not to interfere with the architecture, was abandoned due to vandalism.

Thus, the design of the south facade was thought to include high efficiency solar photovoltaic modules with crystalline silicon technology (mono or poly). These modules extend through the roof with the green areas' metric bands, combining technology with design.

The photovoltaic panels act as the building's loose skin and can be removed afterwards if there is need to expand the library in height.

This skin develops first horizontally, functioning as shading, then vertically, at the opaque area of the construction, and finally on the roof with inclination—about 6%–corresponding to the vegetation strips, always avoiding shadowing over one another.

The metric of the facade is 1.75 m wide, relating with the existing building's structure and enabling the possibility of placing various brand panels, not conditioning to a single market offer. This metric extends through the roof's green coverage, each strip corresponding to a different type of vegetation.

On the roof, different sets of lengths were associated with 7 m and 5 m (5 and 7 panels specifically) with slightly different inclinations because the height of the last module was defined as being 0.40 m above the low walls of the roof, as the maximum quota of the ramp plan.

Technically, the sets where associated in order to allow side access to the drainage boxes of the roof for downpipe maintenance.

The "string" design also took into account the need to avoid waste accumulation and the growth of plants beneath, which would end up damaging the several sets.

2.6 Geothermal energy

The Geothermal Energy Accumulators are elements of the system which work in order to accumulate the thermal energy collected for posterior usage (environmental heating and cooling).

However, the capacity of the Geothermal Energy Accumulators is limited, normally corresponding to a fraction of the energy needed for the building. Nevertheless, the accumulation and transference capacity may vary depending on the type, compaction and amount of water of soil. Also the accumulated energy accounting can substantially vary depending on its usage, specifically within thermal differentials.

The deposits have to be isolated within their entire periphery and at the top, in order to accumulate energy.

Its location will be made in the building implantation area, having the captors—composed by flexible pipes—to adapt to the infrastructure.

The pipes will be connected to similar collectors as the under floor heating, which in return will be connected to the central systems of hot and cold water production.

Collectors will be embedded in the pavement, creating access boxes.

A traditional system for hot and cold water acclimatization will always exist, being the heating performed by GTI (Group Thermal Integral). Whenever this type of chiller is producing hot water for acclimatization, it will simultaneously be filling the cooling deposits for the next season, functioning more efficiently. (Teles, 2013).

2.7 Green roofs

The usage of green roofs is a sustainable practice that has been acquiring more and more importance in order to optimize the energy consumption in buildings.

Apart from aesthetic qualities and aromatic green areas and flowery, green roofs also allow the annulment of the built, both from a pedestrian point of view as well as from seen from above.

Other advantages can also be described: increased thermal insulation, micro and macro external environment temperature diminish; contribution towards the building's durability, reduction of temperature range, reduction of heating and cooling costs; rainwater filter, contributing towards pollution reduction, improving air quality; fixing air dust; fixing solvents; reduction of carbon emissions, air pollution mitigation; increase sound insulation; recreating ecosystems. (Pinheiro, 2012).

It is predicted a vegetal coating with 3 to 4 times a year maintenance.

2.8 Water

In the Library expansion program, sanitary facilities were not provided or any other areas that require hot water as the ones existent within the support built building will be used. For this reason, solar panels for water heating are not contemplate.

The recovery of rainwater for irrigation is anticipated and to restrain water consumption drip irrigation was used.

2.9 Shading

Being an ancient practice, picked up by Le Corbusier in Modern Architecture, it acquires cultural variables and highly diverse design through its expression and language, having been used in this case study shading in a more pragmatic sense, related with the surroundings: the photovoltaic panels also serve as shading on the south facade; trees create shade in the western façade increasing passive cooling (Fig. 5); interior blinds.

Although it was designed to complement the exterior cooling, these references cannot be considered within thermal calculations as legislation only allows assessment of equipment accounted for. The remaining only matters as an additional benefit of the building's energy performance.

2.10 Natural light

Being an essential architecture component, focus was taken on natural lighting throughout the building, reducing energy consumption.

2.11 Lighting

High-efficiency lighting and daylight-dimming systems reduce power requirements. Occupancy sensors automatically lower the heat when spaces have been vacant for a period of time.

The lighting equipment includes energy efficient luminaires, lamps, and electronic ballasts along with automatic controls to maximize use of available daylight.

2.12 Thermal insulation

As thermal insulation, ETICS (External Thermal Insulation Composite System) was used, composed by HCFC free rigid polystyrene foam that has zero Ozone Depletion Potential. This rigid blue core is reinforced on both sides with glass fibre matting coated with smooth cement finish.

An insulation with 40 mm thick slab in the basement and 50 mm in intermediate slab and exterior walls was anticipated. The covers had initially 80 mm of thermal insulation, but it was reduced to 50 mm as the vegetable coating also acts as thermal and acoustic insulation.

Apart from avoiding thermal bridges with this type of insulation, aluminum window frames were also chosen with thermal cutting with doubled glass with solar factor 0.40 and Coefficient U 2.

3 CONCLUSIONS

The design of the building took into account environmental and sustainable considerations.

The project was developed focusing on energy efficiency, taking advantage of the maximum of the available resources, but with the goal of reducing the building energy needs.

Green roofs filter rainfall as part of a natural storm-water management system, reduce heating and cooling costs, and improve air quality by trapping air-borne dust and dirt.

The building uses natural light as much as possible.

The building meets the criteria of sustainable construction and, therefore, can be considered a green building.

Figure 5. Expansion of the Library of FDUL. West view. 2013.

Indeed, it is a Nearly Zero-Energy Building (NZEB).

REFERENCES

Kibert, Charles 2008. *Sustainable Construction: Green Building Design and Delivery*. 2nd ed. New Jersey: John Wiley & Sons, Inc.

Oliveira, João 2012. Faculdade de Direito da Universidade de Lisboa, Ampliação da Biblioteca: Minigeração de Energia Eléctrica (fotovoltaico). In Estudo de Viabilidade Técnica, Económica e Ambiental.

Pinheiro, A.P. 2012. Reabilitação Arquitetónica Verde e Design. In Edição Academia de Escolas de Arquitetura e Urbanismo de Língua Portuguesa—Vol. I, *Palcos da Arquitetura*, Faculdade de Arquitectura da Utl de 5 a 7 de novembro de 2012. Lisboa: pp. 232–240.

Teles, Miguel 2013. Faculdade de Direito da Universidade de Lisboa, Ampliação da Biblioteca. In Memória Descritiva e Justificativa.

Green Design, Materials and Manufacturing Processes – Bártolo et al. (eds)
© 2013 Taylor & Francis Group, London, ISBN 978-1-138-00046-9

New paradigms in post-hydrological disaster shelter practice; hydrological disasters & informal urban settlements—an insoluble problem?

J. Lacey & A. Read
Institute of Technology Carlow, Co. Carlow, Republic of Ireland

ABSTRACT: Informal urban settlements within burgeoning cities in the developing world are becoming increasingly vulnerable to the effects of hydrological disasters. Informality of tenure and poor land availability can cause difficulty for the humanitarian sector when providing transitional shelter, a necessity for recovery; furthermore the majority of experience within the sector primarily lies in the rural context. Primary data collected from an on-going rural flood resilient shelter strategy in SW Bangladesh revealed the sustainable aspects of local low-tech prefabrication and identified land ownership and site modification as necessary to achieve substantial flood resilience. Ultimately, to provide support to landless communities during the recovery phase, fabrication of re-locatable transitional shelters locally could accord with their landless status and remove them temporarily from the disaster zone while supporting the local economy.

1 INTRODUCTION

Migration towards urban centres within developing countries has led to the rapid and unsustainable urbanisation of cities experiencing growth and economic development. The inability of local governments to accommodate the large population influx results in restricted access to affordable housing, thus driving lower income groups to pursue their own methods of acquiring shelter (Davidson et al., 2007; Pantuliano et al., 2012). Informal settlements are inevitably established illegally on the periphery of urban centres on available land generally not suitable for residential development, i.e. flood plains, tidal flats or reclaimed swamps (Douglas et al., 2008), thus leaving the settlement dwellers highly susceptible to the effects of routine flooding from hydro-metrological events. Humanitarian crises based in the informal urban environment present aid agencies with difficulty when providing shelter due to the numerous issues attached to the illegality of tenure (Fan, 2012), addtionally the bulk of the expertise exists in rural shelter provision.

The purpose of this paper is to investigate the flood resilient shelter strategy adopted by a consortium of implementing agencies in SW Bangladesh to examine if this process of providing shelter in the rural post hydrological disaster context can be adapted and applied to the informal urban environment.

2 BACKGROUND

2.1 *Unplanned urbanisation*

95% of urban population growth over the next 30 years will occur in small to medium sized cities (500,000 inhabitants) in low to middle income countries (Haub, 2009). Economic development of these cities is often characterised by informality, illegality and unplanned settlements and is above all strongly associated with urban poverty. Many researchers subscribe to the theory that urban growth is synonymous with the growth of informal settlements (UNHABITAT, 2011). Asian cities host the largest portion (an estimated 61%) of the global slum population, with Southern Asian cities alone accounting for approx. 35% (Ibid).

Shelter construction within these rapidly growing settlements is extremely rudimentary. Tenure insecurity and the lack of access to formal funding mechanisms are primarily responsible for lower income groups taking the informal process of attaining housing, an almost inverted version of the formal process (Greene & Rojas, 2008). Additionally this dictates the construction technology adopted for shelter in slums where the risk of eviction is imminent, thus enthusiasim to upgrade the shelter fabric is typically absent (Jabeen et al., 2010). Impermanent/demountable materials such as Corrugated Galvanised Iron sheeting (CGI) or straightened out oils drums are commonly fixed to lightweight timber framed structures, essentially these settlments represent a modern day vernacular architecture (Oliver, 2006).

2.2 Hydrological hazards and the urban context

In 2011, natural phenomena reported were responsible for the deaths of approx. 30,733 people and the displacement of 244.7 million (Guha-Sapir et al., 2012), hydrological disasters accounted for 57.1% of those affeced and 20.4% of those killed (Ibid). Flooding within the urban context typically stems from a combination of complex causes, urban geomorphological hazards can be divided into two broad groups—hazards associated with settlement location and those created or accentuated by accelerated resource utilisation and city metabolism (Gupta & Ahmad, 1999), the latter severely increasing susceptability to flooding from routine metrological events (Jha et al., 2012; Douglas et al., 2008). The direct impacts of flooding may only affect those settled along water bodies or culverts, yet the indirect impacts i.e. the increased spread of communicable waterbourne diseases associated with flooding such as intestinal disease, malaria etc. propagate far beyond (Adelekan, 2010).

Some cultures have developed vernacular preventive and impact minimising approaches to mitigate the effects on flooding on their shelter, whether it be the slum dwellers of Indore, India 'fixing' their CGI roof sheets with rocks to allow quick disassembly in the event of a flood (Twigg, 2006) or the simple stilting and raising of furniture by residents of the Korail slum, Dhaka (Jabeen et al., 2010). Cultural significance plays a major role in the selection of technology e.g. the Assamese community living within the flood prone Brahmaputra basin, Assam, India must build on the earth, as stilting their dwelling for flood protection contradicts their beliefs (Living with FLOODS, 2010).

2.3 Shelter in the post disaster context

Providing shelter in any post disaster situation entails an element of confusion, placing shelter alongside displacement activates contradictory meanings, while shelter by definition in most contexts is both a durable commodity and developmental in its impact (Zetter, 1995), its associated with grounding and finitude while displacement is associated with up rooting, mobility and transience (Boano & Hunter, 2012). Providing the affected population with a durable shelter solution may only proceed after legal issues over adequate land for reconstruction have been resolved, a period which can span 2–15 years (Shelter Centre, 2012).

Shelter items and materials distributed during the relief phases only possess a lifespan of 1 year or less depending on contextual conditions (Barakat, 2003), thus the affected population must then be provided with interim housing.

Alternatively the transitional shelter (T-shelter) approach, an incremental shelter process, can be implemented to support families while reconstruction proceeds parallel (Shelter Centre, 2012). T-shelter is defined as a shelter which provides a habitable covered living space and a secure, healthy living environment, with privacy and dignity, to those within it, during the period between a conflict or natural disaster and the achievement of a durable shelter solution (Corsellis & Vitale, 2005). While there is no generic form or design for a T-shelter, it corresponds to a specific set of principles and can be characteristically defined as upgradable, re-useable, re-locatable, re-sellable or recyclable (Shelter Centre, 2012). Impermanent, light weight framed structures are commonly specified, typically using timber, bamboo or steel—an approach which supports incremental upgrading of the shelter's fabric using materials sourced locally (Shelter Centre, 2012). The general consensus from academics presently active within the sector (Setchell, 2001; Barakat, 2003; Twigg, 2006) advocate the procurement of local construction materials and labour to the extent where it is sustainable, as reliance on locally produced materials will aid economic recovery (Sheppard et al., 2005). Furthermore utilisation of locally available resources supports the ethos of sustainable architecture—recognising that the product may wear out over time but the process remains (Norton, 1999).

However prefabrication must be considered as it offers the ability to mass produce shelter components to meet the quantity demands for shelter in the post disaster context (Davidson et al., 2008), conversely it may be perceived as being exclusive as opposed to the much more supported concept of inclusive shelter solutions. Participation of the affected community within the response has always been highlighted as key to the success of a shelter strategy, as top down approaches tend to often ignore the complexity of the built environment, the local conditions and the user's needs and potentials (El-Masri & Kellett, 2001).

The response to recent urban disasters demonstrates some of the difficulties and obstacles encountered by implementing agencies in the post disaster urban context, particularly when dealing with landless communities i.e. renters, squatters, informal settlement dwellers (Clermont et al., 2011; Davis, 2012), e.g. complex land laws can require NGOs to attain land rights for T-shelters which exhibit a robust permanent aesthetic (Levine et al., 2012), causing major delay while shelter strategies

Table 1. Basic profile of candidates interviewed.

	Technical	Environmental	Social	Financial
Project manager (x4)	•	•	•	
Focus group (x5)				
Beneficiaries	•		•	•
Non-beneficiaries	•		•	•
Production centres				
Pillars (x7)	•	•		•
Latrines (x6)	•	•		•
Bamboo (x3)	•			•

with a less permanent presence can reduce fears of it potentially becoming a permanent settlement (de Oro & Hirano, 2012). There is recognition among the humanitarian community that a deeper understanding of the specific challenges of shelter in urban humaniatarian crises is required to shape appropriate and effective responses (Fan, 2012). Experience providing shelter predominantly lies within rural context, cities remove the humanitarian sector from their comfort zone as the post disaster informal urban environment contrasts to the traditional transparent environment of the Internally Displaced Person (IDP) camp (Crisp et al., 2012).

3 RESEARCH METHODOLOGY

The methodology for developing a prototype transitional shelter solution for the informal urban context involved a secondary data review and primary data collection from an on-going shelter programme. Factors derived from the secondary data review formed the basis of enquiry for the primary research questions which revolved around the sustainability of an active rural shelter programme and whether the key aspects of the strategy could be applied to the informal urban context. Sustainability is a nebulous term that often generates many definitions, for the purpose of the research sustainability will be investigated under 4 key areas defined by Norton (1999) namely technical, environmental, social, and financial sustainability.

Primary data was collected on a field study during September 2012 with Oxfam GB (OGB) specifically on the FRESH project, a rural shelter strategy constructing 11,092 flood resilient shelters and latrines for those worst affected by prolonged waterlogging in SW Bangladesh. Approximately 64% of the global population exposed to flooding live in Southern Asia (UNISDR, 2011). Moreover numerous flood resilient shelter strategies have been implemented in response to large scale disasters in the past, particularly in SW Bangladesh

(Cyclone's Sidr 2007 and Alia 2009). Although the FRESH project may not be classified as an exemplary case study as it is still on-going, it is a typical representation of a rural flood resilient shelter strategy. The opportunity to observe the process of providing shelter, and engage with all stake-holders involved contributed significantly to the research Table 1 represents the interviewees and participants and highlights their contribution.

A combination of both conventional researcher guided, inflexible, top down methods to collect quantitative data alongside participatory research methods which were community led and are typically used by NGOs when collecting qualitative data in the field (Rahman et al., 2010) were implemented.

4 RESULTS AND DISCUSSION

During the end of July and August 2012 heavier than usual monsoon rains in conjunction with numerous human induced factors led to the prolonged flooding situation in SW Bangladesh which destroyed approx. 19,534 shelters in the region. To aid those worst affected, the flood resistant shelter and latrine programme for SW Bangladesh (FRESH) was initiated by members of the NARRI consortium and implemented by a coalition of 6 INGOs and 5 PNGOs.

Based on a review of flood resilient shelter strategies executed in the past in conjunction with a participatory design process, the FRESH shelter design attempted to improve on the vernacular approach to mitigating the effects of prolonged periods of water logging. The indigenous practice of raising the homestead using earth was retained and improvement in the form of a rendered brick wall surrounding the plinth provided increased resilience. The structure comprised of both prefabricated Reinforced Concrete (RC) pillars, braced using timber and a timber kingpost roof construction. Although prefabricated RC is new to the region, its use is gaining popularity since being introduced by NGOs implementing shelter strategies in the recent past (Flinn & Beresford, 2009). Prefabricated bamboo panels were selected as a walling material due to their climatic compatibility and CGI roof sheeting over indigenous roofing materials—burnt clay tiles and thatch, as a result of beneficiaries condemning them due to their poor performance in extreme metrological events.

4.1 Technical sustainability

Technical sustainability was determined through investigating the skills required to construct shelter and what measures were taken to build technical capacity within the community to seek and ensure good quality in the built environment (Norton, 1999). The FRESH shelter design utilised skills and materials which existed in the project region, best practice was applied to traditional methods to ensure structural integrity during the next period of prolonged waterlogging. Particular modification was made to the manufacturing of precast RC products, as those available on the market were not subject to quality control thus homogeneity was an issue. Enhancement of this low-tech construction technology allowed concrete to perform more satisfactorily in future flooding situations i.e. more concrete coverage of reinforcement. However the sustainability of this production method is questionable, due to the extra cost associated with higher spec pillars, of manufacturers surveyed, 20% cited they would produce a cheaper alternative if requested. Off-site production in a controlled environment allowed for homogenous production to proceed despite in situ works being halted as a result of heavy rains, also the dry assembly of the predetermined kit of parts on site could be carried out in adverse conditions. The arrival of the components as a kit of parts eased the burden of transportation to restricted sites.

Quality control of the shelter construction was primarily the responsibility of the beneficiaries, technical training was provided to aid in the monitoring process. Only 1 of 15 shelters surveyed displayed any signs of failure, faults observed were predominantly associated with construction activities carried out in situ e.g. cracking was observed in 60% of the shelters' plinth construction, yet 83% were crudely repaired by the beneficiary using sand and cement procured locally.

4.2 Environmental sustainability

Environmental sustainability is determined through investigating the use of local natural resources for shelter construction and whether the approach avoids contaminating the local environment (Norton, 1999). Deforestation was cited as being a

major issue in the region and a contributory factor to the prolonged water logging. The shelter strategy was successful in mitigating the immediate effects of deforestation through specifying a hybrid frame of RC pillars and timber bracing, thus reducing the quantity of timber required. Timber procured locally was poor in condition primarily as a result of a rushed, inadequate curing process, indicating a market struggling to meet current demands. Regarding long term environmental sustainability, materials with a high embodied energy (CGI sheeting, RC products) are not environmentally sustainable within the global sense. Durability is often put forward as the rationale for use of these products, but this is entirely dependent on the quality and workmanship of the finished product; inferior production is not durable or sustainable.

Generous consideration was given to the development of an open pit latrine solution that satisfied all criteria—societal requirements dictated that latrines remain within close proximity to the shelter, regardless of proximity to water bodies, yet to prevent contamination a minimum of 15 m is required (Sphere Project, 2011).

4.3 Social sustainability

Respecting local customs and cultural requirements is crucial to acceptance of a shelter strategy (Norton, 1999) as it is often the reason cited for low occupancy. The beneficiary's selection of non-indigenous shelter materials indicates both disbelief in traditional methods due to the failure of their shelters in the flooding and possibly their aspiration towards more modernised construction methods. Adaptation and investment was witnessed in 90% of shelters surveyed and is a strong indicator of ownership and cultural acceptance of the shelter strategy.

4.4 Financial sustainability

Financial sustainability is determined through the ability of the general population to financially access materials and services within the locality (Norton, 1999). The FRESH shelter design utilised skills and materials which existed within the project region, figures taken from internal OGB data suggest indirect economic gains in the OGB project area alone (quota of 1086 shelters) of approx. BDT 4,958,910 (US $62,838.81). While craftsmen cited they were familiar with the design and construction technology needed to mitigate the effects of prolonged water logging, it was simply beyond the financial means of lower income groups; average monthly earnings of those within the FRESH project area are approx. BDT 3000 ($38.02), while the entire cost of the flood resilient shelter came to BDT 78,052 (US $989.72). However the flood resistant element of the shelter i.e. the plinth, costs approx. BDT 20,059 (US $254.19) is slightly more feasible.

5 CONCLUSIONS

Findings are subject to review upon analysis of the prototype and completion of the current research programme, yet certain aspects become apparent.

The FRESH shelter strategy offers an insight into the numerous criteria implementing agencies must satisfy when providing sustainable shelter solutions which underpin development while ensuring there is no negative effect on the local environment and society. It's evident that empowering beneficiaries in the earlier stages and monitoring of the project proved successful through an initial high occupancy rate and the level of financial investment made by beneficiaries. Simply not exploiting them as labour proved to be an effective strategy, thus suggesting that the low-tech prefabrication process isn't necessarily a top down approach and can be inclusive and sustainable.

Prefabrication of the primary construction components off site played an instrumental role within the FRESH shelter strategy, highly contributing to efficiency and quality control. Removing skill from the construction process through the introduction of a low tech component based construction method which utilises universally accessible materials can potentially aid the humanitarian sector achieve their quantifiable targets while supporting an inclusive process. To permanently mitigate the effects of flooding, site modification is required whether it's through a risk reduction infrastructure programme or done on an individual basis and typically consumes a large portion of the shelter's budget. Informality of the tenure infers modification of the landscape is not viable, therefore supporting disassembly and the re-locatable aspect of a T-shelter is fundamental to allowing lower income groups to relocate in the locality and remain within proximity of their livelihood activities until flood waters recede.

ACKNOWLEDGEMENTS

This paper forms some of the findings from on-going research being carried out in the Institute of Technology, Carlow. The research is funded as part of the Enterprise Partnership Scheme, a scholarship which is co-funded by the Irish Research Council and Edenshelters Ltd. The authors are grateful to the staff from Oxfam GB and the FRESH consortium for accommodating the research.

REFERENCES

Adelekan, I.O., 2010. Vulnerability of poor urban coastal communities to flooding in Lagos, Nigeria. *Environment and Urbanization,* Volume 22(2), 433–450.

Barakat, S., 2003. Housing reconstruction after conflict and Disaster, Humanitarian Policy Group, Network Papers, 43, 1–40.

Boano, C. & Hunter W., 2012. Architecture at Risk(?): The Ambivalent Nature of Post-disaster Practice. Architectoni.ca,1(1) 1–13.

Clermont, C., Sanderson, D., Sharma, A. & Spraos, H., 2011. Urban disasters—Lessons from Haiti. Study of member agencies' responses to the earthquake in Port-au-Prince, Haiti, January 2010, London: Disasters Emergency Committee.

Corsellis, T. & Vitale, A., 2005. Transitional settlement—Displaced populations. Oxford, Oxfam.

Crisp et al, 2012. Displacement in urban areas: new challenges, new partnerships. *Disasters,* 36(1), pp. 23–42.

Davidson, C., G. Lizarralde, and C. Johnson, 2008. "Myths and realities of prefabrication for post-disaster reconstruction." 4th international i-Rec conference on post-disaster reconstruction: building resilience, achieving effective post-disaster reconstruction.

Davis, 2012. What is the vision for shelter and housing in Haiti Summary Observations of Reconstruction Progress following the Haiti Earthquake of January 12th 2010, London. UNHABITAT.

de Oro, C., & Hirano, S., 2012. Learning from urban transitional settlement response in the Phillipines: Housing, Land and property issues, Catholic Relief Services.

Douglas, I., Alam, K., Maghenda, M., Mcdonnell, Y., McLean, L., & Campbell, J. (2008). Unjust waters: climate change, flooding and the urban poor in Africa. *Environment and Urbanization, 20*(1), 187–205.

El-Masri, M. & Kellett, P., 2001. Post-war reconstruction: Participatory approaches to rebuilding the damaged villages of Lebanon, a case study of al-Burjain. *Habitat International,* Issue 25, 535–557.

Fan, L., 2012. Shelter strategies, humanitarian praxis and critical urban theory in post crisis reconstruction. *Disasters,* 36(1), 64–86.

Flinn, B. & Beresford, P., 2009. Post-Sidr Family Shelter Reconstruction Bangladesh, UNDP.

Greene and Rojas, 2008. Incremental construction: a strategy to facilitate access to housing. *Environment and Urbanization,* 20(1), 89–108.

Guha-Sapir, D., Vos, F., Below, R. with Ponserre, S., 2012. Annual Disaster Statistical Review 2011: Brussels, Belgium. The Numbers and Trends, Centre for Research on the Epidemiology of Disasters (CRED).

Gupta, A. & Ahmad, R., 1999. Geomorphology and the urban tropics: building an interface between research and usage. *Geomorphology,* Volume 31, pp. 133–149.

Haub, C. 2009. Demographic Trends and their Humanitarian Impacts. Humanitarian Horizons Project, Washington, Feinstein International Center.

Islam, 2001. The open approach to flood control: the way to the future in Bangladesh. *Futures,* Issue 33, pp. 783–802.

Jabeen, H., Johnson, C. & Allen, A., 2010. Built-in resilience: Learning from grassroots coping strategies to climate variability. *Environment and Urbanisation,* 2(22), 415–432.

Jha, A.K., Bloch, R. & Lamond, J., 2012. Cities and Flooding: A guide to integrated urban flood risk managment for the 21st century, Washington D.C.: World Bank Publications.

Levine, S., Bailey, S., Boyer, B. & Mehu, C., 2012. Avoiding reality: Land, institutions and humanitarian action in post-earthquake Haiti, London: Humanitarian Policy Group.

Living with FLOODS. 2010. ICIMOD [Online video]. Nepal; ICIMOD. Available at; https://www.youtube.com/watch?v=t-soVznnYKs [Accessed February 27, 2013].

Norton, J. (1999). Sustainable architecture: a definition. Habitat Debate, 5(2), 10–11.

Oliver, 2006. Why Study Vernacular Architecture?. In: Build to Meet Needs: Cultural Issues in Vernacular Architecture. Italy. Architectural Press, pp. 3–17.

Pantuliano, S., Metcalfe, V., Haysom, S. & Davey, E., 2012. Urban vulnerability and displacement: a review of current issues. *Hazards,* 1(36(s1)), pp. 1–22.

Rahman, M.A., Rahman, T., & Mondal, M.S., 2010. *A* Comparative Analsis of Different Typesof Flood Shelters in Bangladesh, Dhaka, DIPECHO.

Setchell, C., 2001. Reducing vulnerability through livelihoods promotion in shelter sector activities: An initial examination for potential mitigation and post disaster application, Feinstein International Famine Center Working Paper No 5 Shelter Centre, 2012. Transitional Shelter Guidelines, Geneva, Shelter Centre.

Sheppard, S., Hill, R., Tal, Y., Patsi, A., & Mullen, K. 2005. The economic impact of shelter assistance in post-disaster settings. *CHF International Report. Washington: CHF.*

Sphere Project, 2011. Humanitarian Charter and Minimum Standards in Humanitarian Response *3rd Edition,* Geneva, The Sphere project.

Twigg, 2006. Technology, Post disaster housing reconstruction and livelihood security—*Benefit Hazard Research Centre* Disaster Studies, London, Working Paper No. 15.

UNHABITAT, 2003. The Challenge of Slums, Global report on Human settlements2003., Earthscan Publications.

UNHABITAT, 2011. State of the Worlds Cities 2010/2012—Bridging the Urban Divide, Earthscan Publications.

UNISDR, 2011. *2011* Global Assessment Report on Disaster Risk Reduction: Revealing Risk, Redefining Development, Geneva: United Nations International Strategy for Disaster Reduction.

Zetter, R. (1995). Shelter provision and settlement policies for refugees: a state of the art review. *Studies on emergencies and disaster relief*, (2), 51–71.

Green Design, Materials and Manufacturing Processes – Bártolo et al. (eds)
© 2013 Taylor & Francis Group, London, ISBN 978-1-138-00046-9

Improving urban environment with green roofs

G. Darázs & I. Hajdu
Corvinus University of Budapest, Budapest, Hungary

ABSTRACT: What makes our topic relevant is the fact that more and more proportion of the population lives nowadays densely built cities, where practically everything is artificial except for some public parks and smaller greeneries.

Green roofs as building constructions serve good examples of how growing passive building surfaces can be utilized and bring nature back into people's life. Moreover, they are also capable of reducing the anticipated effects of climate change and urbanization.

In our study, we assess the appropriate buildings to be supplied with green roofs in Budapest. Next we analyze the opportunities of green roof enlargement in the city, and a detailed research on assessing the economical, social and environmental role of green roofs will be done. Based on economic calculations, we show why and how we can encourage investors to build green roofs.

Our case study could serve as a good example for building green roofs in similarly structured cities to Budapest.

1 INTRODUCTION

The rate of urban areas all over the world is less than 1 percent but it contributes significantly to climate change. More than half of the population lives in cities, where 75 percent of power consumption and 80 percent of greenhouse gas emission of the world is concentrated, which significantly worsens the negative effects of climate change (urban heat island effect, change ventilation conditions in connection with decreasing air quality). (WHO Statistic, Vág, 2008).

For the interest of stopping or turning this process as soon as possible, we have to pay more attention to take more effective power generation and utilization of energy (using renewable energy sources, suitable insulation for buildings, constructing bypasses in transport, and creating more and more green areas). (MUT, 2011).

Because of packed buildings and concentration of transport it is increasingly unhealthy to live in the city. (Jaffal, 2012) Most of the air is used by cars and heating systems which have an adverse effect on the atmosphere too. Mainly in the summer hot the city air is overheated by lots of asphalt jungles and upswing air takes settled dirt and pollution. (FLL, 2002).

In order to decrease these effects, the urban climate could be improved in large measure by greening roofs and facades. Building green roofs is an eco-friendly and economic building method. Due to its architectural structure and decreasing proportion of green areas, Budapest is the most appropriate city to examine where and which tools could encourage green roofs constructions. The elemental basis for constructing green roof is to find areas which are suitable for architectural conditions and could be a good solution for environmental problems of the area.

Not only in Budapest, but also in the whole country it is difficult to estimate how many square meters of green roofs were built in the last decades. According to last years' data of air pollution and smog alerts in Budapest it is easy to admit the great need for the solutions which could make better essential condition of the urban population.

2 GREEN ROOF STRUCTURE

When building green roofs, there are four substantial viewpoints which could be taken into consideration in order that the roof-integrated vegetation subsists as long as possible. The first one is the choice of vegetation which suitable for local conditions and for the type of the green roof. The next one is the appropriate growing medium which is suitable for the selected vegetation and its thickness and composition for type of green roof.

The third one is to prevent the slur of the growing medium. Finally, an important factor function is that green roof could be able to retain maximum amount of rain water which is necessary for growing vegetation and healthy microclimate.

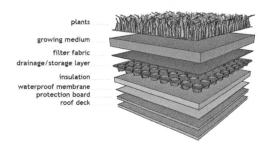

plants
growing medium
filter fabric
drainage/storage layer
insulation
waterproof membrane
protection board
roof deck

Figure 1. Green roof structure (http://newyork.thecity atlas.org/lifestyle/green-roofs-101/).

Table 1. Types of green roofs (based on Minke 2009).

	Intensive	Extensive
Slope of roofing	Max. 10°	0–25°
Load of roofing	2,5–15 kN/m²	0,5–2 kN/m²
Thickness of structure	Min 25 cm	Max 15 cm
Type of watering	Irrigation plant	Rain water
Vegetation	Ornamentals, crop, tree	Perennial, succulents, xerophilous plants
Utilization	Public park	Impassable

In addition it should take surplus water quickly away on the surface of the drainage layer to protect roofing from leaching and dilapidation in the future. (Minke, 2009).

2.1 Structural elements

In spite of the different features of roofs and environment, different type of green roofs could be applied, but basically green roof has a defined structures with consideration of the above mentioned viewpoints (Fig. 1).

3 TYPES OF GREEN ROOFS

Basically, two types of green roofs can be distinguished. (Table 1). From the point of the horticultural view, green roofs have two distinct types; the extensive and the intensive green roofs. The differences between the two types are based upon slope and maximum load of roofing, thickness of growing medium, type of watering, and the type of vegetation. Every type of roof has innumerable opportunities for forming vegetation with running into each other, and different kinds of habitat. (FLL, 2002).

4 FUNCTION AND EFFECT OF GREEN ROOFS

In the case of green roofs, we can talk about lots of effects which have significant environmental and physiological roles, mainly in cities. (Vestvik, 2012). The application of green roofs is useful part of economical and eco-friendly building method, since they

– reduce the rate of built-up and unbuilt areas,
– produce oxygen and set carbon dioxide,
– filter/swallow dust and pollutants from air,
– reduce warming up of roofing thus moderate swirling up the dust,
– moderate changes in temperatures between day and night,
– function as heat insulation,
– arrest rain water runoff, thus relieve sewerage system,
– increase the value of the building,
– create recreation value since vegetation pleases one's eyes and smells beautiful.(Minke, 2009).

5 SUITABLE AREAS FOR GREEN ROOF CONSTRUCTION IN BUDAPEST

Comparing the environmental circumstances, the junctions and the most polluted areas with location of flat roofs, it seems to be outlined that where are the areas suitable for green roof constructions and the advantages of green roofs affect for the environmental conditions of the given area.

In the inner districts of Budapest (Fig. 2), where high-roofed blocks of flats dominate and in the

Figure 2. Zone system of Budapest http://hg.hu/cikk/epiteszet/13479-a-strategia-meg-egyseges-budapesthez-keszul/nyomtatas.

suburb there are not flat roofs. Inspite of this in the transition zone on the ring around the inner city, there are owner-occupied blocks, industry centers, office buildings and residentials which have flat roof. In the case of these buildings the extensive type of green roof is recommended, since these roofs could not be able to carry weight of intensive roofs. In the transition zone, the renewing of old-fashioned buildings could be linked with creating green roofs.

If we talk about newly built flat roofs the situation is easier. In the case of office buildings, residentials, and shopping malls, it is worth building them with green roof along with choosing the suitable type for the conditions given. These types of buildings are basically in the transition zone, but in different parts of it due to their different kinds of demand.

While residential are rather in the external part of the transition zone making the most of nearness to the city center and have bigger rate of green surfaces, the office buildings are rather in the internal part of a zone near the city center and the busy road junctions. In the case of newly built constructions it could be necessary to make the building extensive types of green roofs compulsory or to use incentive shemes to make investors interested in green roof constructions.

Regarding the exact amount of rate of green roofs, there have not yet accurate surveys made. According to an in-depth study dealing with flat roofs suitable for green roof construction, in Budapest there is about 6.472.000 square meter flat roof which means 3 to 5percent of the built-up area. Unfortunately, in the case of static, the whole amount of flat roof is not suitable for building green roof. In the case of buildings in the one-time industrial area and ten-storey panel blocks, it is questionable to build green roofs and in most cases it would be an expensive solution. The surplus 2.95 square meter area with another type of flat roof is suitable for green roof construction. This is the 2 to 3 percent of city area. (Szabó, 2009).

From the point of view of horticultural viewpoints, it is desirable that green roofs are built where they can display their positive effects such as the hold back of rain water, urban heat island effect, decrease air pollution and the recreation value. (Minke, 2009).

In the capital the following areas are suitable for green roof constructions thus the incentive systems could be the most effective in these areas:

– air pollution I., II, VII., VIII. and XI. district
– buildings for green roof construction IX., XI., XIII. district, where there are the most newly built office buildings and residentials
– in the one-time industrial area of district IX., X., XI. where buildings will be renovated

Figure 3. Districts of Budapest http://www.buda pestetbekescsabatszeretok.eoldal.hu/cikkek/budapest/ keruletek.html.

– recreation value districts II. and XII. where old-fashioned buildings could be joined with green roof construction. (Fig. 3).

6 CONCLUSIONS

Unfortunately, not the lack of flat roofs causes the largest obstacle for green roof constructions, since more and more residential, panel blocks, office buildings with flat roof are built, which are the most suitable forms to built green roofs. The obstacle of wide spreading them is caused by large costs of investment and repairing caused by slipshod work, the appropriate incentive and regulation system or being afraid of the unknown. Compared to West-European countries, willingness for building green roofs is low in Hungary, which could be increased by developing suitable incentive systems. For instance, there are no laws or regulations regarding building green roofs in Hungary upon which constructors could rely. Due to this, to some extent, there are a lot of negative instances which impair the already low reputation of green roofs in Hungary.

Some possibilities to map these reasons are using questionnaires and in-depth interview methods; by means of these the attitude of inhabitants and investors as well as the impediment of spreading and building can be mapped.

After the determination of critical points economic calculation is necessary to encourage investors to build green roofs.

With full knowledge of international praxis as well as the incentive and regulation systems, it would be expedient to offer recommended procedures for both decision-makers and investors to construct green roofs in the similarly structured cities.

REFERENCES

Darázs G. 2012., A zöldtetőépítés ösztönzésének lehetősége Budapesten és más hazai nagyvárosokban Agora 9. 21–32. (Green roof promoting possibilities in Budapest and in other big cities in Hungary).

Forschungsgesellschaft Landschaftsentwicklung Landschafts 2002 (FLL): Zöldtetők tervezési, kivitelezési és fenntartási irányelve—Zöldtetők irányelve—12–15.old.

Jaffal I.,Ouldboukhitine S., Belarbi R, 2012., A comprehensive study of the impact of green roofs on building energy performance Renewable Energy 43. 157–164.

Minke G. 2009, Zöldtetők (Green roofs), Cser Kiadó, Budapest 6–48.

Magyar Urbanisztikai Társaság 2011., Városklíma kalauz, (Magyar Urbanisztikai Tudásközpont Nonprofit Kft.) 4–11. (Hungarian Society for Urban Planning, Urban climate quide).

Szabó L., 2009., A zöldtetőépítés hazai kilátásainak vizsgálata Budapest példáján 23–59. (Examination of prospects for green roof construction in Budapest).

Vág A., 2008., Kutatási keret a klímaváltozás városi hatásainak és az adaptáció európai gyakorlatainak tanulmányozásához—Társadalomkutatás 26(2.) 209–224.

Vestvik M. 2012: Green roofs, National Institute for Consumer Research (SIFO) WHOStatistics http://www.who.int/gho/urban_health/situation_trends/urban_population_growth_text/en/index.html.

Green Design, Materials and Manufacturing Processes – Bártolo et al. (eds)
© 2013 Taylor & Francis Group, London, ISBN 978-1-138-00046-9

Self-sustaining home: Design liaises technology

Jan-Michael Werner

University of Applied Sciences, Joanneum GmbH, Graz, Austria

ABSTRACT: This abstract briefly outlines the three-pillar-structure of ecology, economy and sociality displayed in the practical approach of designing zero-emission residential buildings. In short, the concept is focused on the intersection of architecture and technology. Furthermore, it describes the holistic point of view which is necessary to further develop a passive house into an active house. Additionally, it is essential to put the terms 'identification', 'ambiance' and 'convenience' into relation with the three-pillar-structure. The goal of this approach would be a self-sustaining home. This autonomy can be observed from the technical side in the electrical standard, green heating systems and water supply and from the architectural side in the inclusive design and sustainable construction method.

On the one hand, the abstract highlights the technical aspects and talks about the possibilities of making a building self-sustaining by putting not just PV on the roof, but also by saving its energy in large batteries or using geothermal energy and infrared panels for the heating or small, local wastewater treatment plants. On the other hand, it shows the importance of the architectural design which is able to make a building cheaper in the technical configuration and to convert it into a functional green building.

Hence, it is all about smart-design related to intelligent building technologies to reduce the CO_2 emissions and to handle our ecological resources with care. The talk is primarily about the integration of these technologies into architectural design—not about the collision or the dualism of these. Eventually, it is not exclusively about zero-emission and lots of insulation. The focus is on the interaction of smart design, using renewable energy resources and low life cycle costs to create and use buildings not just in a green but in a blue way.

1 INTRODUCTION

1.1 *The aim of sustainability*

Sustainability has been the keyword over the last years. The term 'sustainability' in general refers to people and to preserve sustainable human habitat. Meanwhile, the awareness of sustainability is present to many people. On the one hand, the industry and the economy are working hard on solutions and services, on the other hand, more and more users buy sustainable products mainly to benefit financially on a long-term basis.

Sustainability has also left its footprint in the building industry. Passive houses, new dampening technologies, the use of geothermal, solar and wind-driven energy controlled residential and office space vents are state of the art and a sign of sustainability in the construction industry. Ecologically improved heating systems, such as geothermal units, condensing units and infrared heaters, etc. have become more popular.

Long-distance heating as the long-time favored central energy supply has become less important in the energy use. It is well known that that the central energy supply is connected with very high losses.

On the one hand, the permanent stand-by in energy results in great losses for a high proportion of the connected consumers and on the other hand, the enormous conduits and transport routes release a lot of energy.

Technical achievements have been increasingly integrated into buildings for some time and have been pooled in the use of labels and certificates, which reflect the high level of technology and design in these structures in form of certificates. The energy performance certificate has been for years a proof of standardization and energy consumption. It has also become a tool to assess the sustainability of buildings.

2 ANALYSIS

2.1 *Status quo*

All these instruments only mirror the current situation. At the time of the assessment of the building an inspection takes place which solely places its focus on the material and technical component of the building. The performance of materials is assessed ecologically and economically using variables such as U-value, heating demand, type and quality of materials used, ventilation, among others and standardized.

To exclude the self-interest of contractors and producers to a certain extend passive houses are certified by independent institutions, which confirm the quality of the execution. The so-called Green Building Standard is now required by law for new buildings in many countries.

As this standard only refers to an examination of the building at a specific time, it is aimed at assessing the building ecologically and economically over the whole life cycle. The entire life cycle of a building is observed and classified while integrating the aspect of sustainability.

For this reason new labels, certification bodies and processes have been established. A new field of business has been created which deals with the ongoing investigation and measurement of quality of buildings. Similarly to the recurring technical examination of motor vehicles a recurring inspection of buildings has to be expected in the future. This procedure is sensible in contrast to the green building standard as it assesses and classifies the ecological and economic standards: The Blue Building Standard.

2.2 *Human needs*

Despite all these considerations and developments, the primary interest is focused on the measurement of technical qualities and the classification of different systems for energy savings in operation. How sustainable the used techniques are, is considered in the energy supply in the production of materials and in technological development. Their recyclability is included into the analysis as well. The 'Blue Building Standard' aims at creating an eco-balance of buildings. The starting point is a holistic approach which includes apart from building technology, building materiality and the life cycle economic and human needs. Human needs in this context are defined as socio-cultural aspects in general and the careful confrontation with the environment and energy. So to speak: the Blue Building Standard gives shape to a three-pillar-structure of ecology, economy and sociality.

A standardization, check and classification of buildings are difficult as soon as people are involved. People are heterogeneous as their behavior cannot be standardized and monitored. Eventually, it is clear that all of the previously described efforts were undertaken to offer a sustainable life to people. This leads to fundamental questions: Can the living environment of heterogeneous individuals—the built environment—by standardization and classification of materials and techniques

become sustainable? Can an imposed regulation of individual life processes cause a personal rethinking and intrinsically promote a sustainable way of life? Or does a forced mechanization and the compliance of competition-motivated rules driven by economic means not rather lead to antipathy than to a quiet revolution?

It is striking that the human being and their needs in their direct contact with buildings are excluded from the development of test methods and the legally laid down technical qualities. The factor human being that is primarily responsible for the smooth organization of the systems is due to being uncontrollable excluded from the development of these systems. As a result it is tried afterwards to handle the characteristics of individuality and to categorize the human being while taking negative conditioning and extrinsic motivation—such as regulations and penalties for non-compliance with the guidelines into consideration. This procedure is not successful. Increasing ignorance and lack of motivation are the logical consequences. The user cannot be forced into a functional diagram, which demands daily supervision and acceptance of existing building technology. Even building designers and planners give up with regard to the technical performance that is demanded for buildings. The necessary co-operation turns out to be a trap. Through negative conditioning over a longer period of time it cannot be expected that people start a rethinking process and change their habits as this might limit their personal freedom. Nevertheless, a rethinking in the motivation of heterogeneous individuality is desirable.

2.3 *Consumption*

Is the constantly increasing use of technology the truly decisive way to sustainable management of our planet? Must energy be saved on principle? Or is it not rather a question of the kind of energy that is used? In general, sufficient energy is available on our planet. Energy is not consumed, it is merely transformed into another energy state. Thus, it seems reasonable to discuss the ways of energy conversion and to define new ways.

Currently, sustainability is often associated with restrictions and compromise in the construction industry.

– Sustainable buildings are more expensive,
– Life in sustainable buildings has to follow predetermined rules
– Individual needs are restricted, etc ....

If you want to build sustainably, the ultimate goal is that both investor and user of a building feel comfortable and experience personally the benefits of sustainability.

A building is built for a specific purpose. This purpose has to be the ultimate goal, it may not become the building itself by its construction to the highest goal and purpose.

In general, the purpose of a building is to provide space and protection against environmental influences for people. It should help the user to facilitate daily things and processes and not create additional tasks and difficulties through highly complex building services and maintenance during operation (see Fig. 1). The use of technology is only reasonable and accepted to this extent by people as long as it supports the habits of the user. Once technology turns out to be a compulsion or restricts the habits of the users, they will try to get rid of them or develop an antipathy (see Fig. 2). Doubtlessly, they will try to get their personal liberty back by a deconstructive, negative attitude towards the building.

So it is essential to put the terms 'identification', 'ambiance' and 'convenience' into relation with the three-pillar-structure to support a positive attitude of the user. The identification with the environment—that is the building—is necessary that the interaction between user and building works. The user is willing to be confronted with the functioning of the building. The user of a building identifies with the building as soon as he feels comfortable in it.

Figure 1.   Complex building services.
Photo: DI Arch. Jan Werner, www.arch-vision.at.

Figure 2.   Massive wiring.
Photo: DI Arch. Jan Werner, www.arch-vision.at.

To feel comfortable in a building the spatial effect and not the technical performance is relevant. The spatial effect has direct influence on the perception of the people. Technical support can have an indirect effect e.g. in the form of less heating energy or better indoor air.

A very important component that contributes to the identification and well-being in a building is, therefore, the essence of a building. The importance of design, and thus the architectural coefficient of a building have in recent years become very important again. Architecture is the philosophy of effect. This formulation is not 'the quick gesture', meaning the 'big show', but rather the effect of a building on a user, even in its smallest detail. The essence of architecture is the spatial effect, which could be about pure construction technology far beyond reaching the ability to create enclosed space with well-being, but also just dislike, discomfort and displeasure. Interior design beyond subjective terms such a 'beautiful' or 'ugly' is to a large extent responsible by its conscious use of specific materials, surface design, basic shapes, volumes, etc. to ensure that users and viewers of the works are emotionally touched. The range of emotions is responsible for good and bad, and thus an essential part of human life. Architecture in its abstract form is co-responsible for the quality of human life. Only by the convincing incorruptibility of the functional design principles and the emotional effect the immense value and a true sustainability of a building are created: Broad acceptance beyond personal and subjective taste.

Since human beings differ from each other, the room configuration that is necessary for the highest possible level of comfort also differs and hard to reach over a given solution. From this the need for more flexibility individuality in interior design directly derives!

Space must be subordinated to the user—buildings should serve the user, not vice versa!

For the user a 'high-tech performance' is useful and acceptable, as long as he does not feel it, does not have a higher support and maintenance, and he is not confronted with higher cost in the construction and operation of the building but lowers them in the long term. Of course, among the individualists there will always be people who love high tech appliances but the majority will reject a prescribed discussion with technology.

3   CONCLUSIONS

3.1   *Become flexible*

A high level of technical equipment makes a building stiff and inflexible (see Fig. 3). Many technical

Figure 3. Air ventilation duct.
Photo: DI Arch. Jan Werner, www.arch-vision.at.

interventions require many cable ducts and interfaces and a detailed technical planning in advance. An adaptation of the technical building equipment during the utilization phase, let alone a complete change of this, is connected according to today's construction methods with huge effort. A flexible floor plan and interior design is restricted from the very beginning by a built-up technique. Therefore, the sustainability of the building by flexible re-design is excluded.

The goal of building development should therefore be: Minimize technology and powerful materials but maximize the flexibility and adaptability of buildings. We can by intelligent use of 'low-tech' concepts convert from consumer to user and from user to finisher. We are encouraged to create self-sufficient buildings, keep the technical standard as low as possible and only use it as a backup. Architecture itself must become an important element in the sustainability assessment. Architecture should be seen as a tool for harmonic networking/fusion with the environment and the supporting technology.

The highest amount of sustainability in the building industry can only be generated by a high flexibility of the structure itself. All the technical equipment does not contribute to sustainability if there is a constant need for maintenance, or after a few years it needs to be replaced by a newer, better technology. In this case, the technical equipment lowers rather than improves the sustainability performance of a building.

If a building can adapt easily to new requirements, it is probably the most sustainable and efficient and therefore the most ecological and economical building.

When we think about improving the sustainability of buildings, the way towards more insulation, more controlled ventilation systems and defined, manageable units is a dead end. Just as a few years ago, the centralization of power generation and supply was favored and turns out to be critical

now, the classification of buildings in sustainability scales based on technical factors will prove to be insufficient. Today, a building classified as 'good' building will be overtaken tomorrow by new technology and then already be classified as 'bad'. This cannot be the target of a Blue-building-policy.

If sustainability is primarily measured by the biggest amount of flexibility, then the assessment of sustainability must not be focused on technical equipment that makes a building necessarily inflexible. New systems to improve the flexibility of buildings are needed. An efficient building envelope with a decent basic insulation combined with a CO^2 neutral heating/cooling system is the basis for this. Systems such as geothermal, PV panels, infrared heating systems, solar thermal energy, etc. can provide the necessary energy supply to heat and cool the building. A configuration of the building envelope according to the Passive House Guidelines is no longer necessary when the provision of the required energy does not generate CO^2 emissions. A saving on heating and cooling energy is no longer necessary if the energy does not come from fossil fuels, but is produced locally and renewably.

3.2 *An example: Active life—active house*

The purpose of this planning task was not only to equip a building with sustainable technology and thus implement the ecological thought, but to design the building itself sustainably and ecologically. Thus, the structural and architectural conditions were superimposed to find a common solution. This is a solution that already solves the technical conditions in the planning stage and respects the conditions by choosing the right technology to implement.

The result of this task is a diamond-like structure, the south providing a roof surface at an angle of 40° to accommodate a roof integrated PV system (see Fig. 4). A corresponding elevation of the PV modules can therefore be dispensed with. At the same time in this roof area on each floor,

Figure 4. Visualization: Commercial and residential building 'Goller'.
Source: Martin Ernst, www.vizu-me.com.

a roof terrace is incised, which forms the threshold between inside and outside. Also integrated into the roof a staircase provides direct access to the private garden to the south. Through the mix of uses from residential and commercial space, the building is in building class 3 according to OIB in Austria and therefore requires increased fire protection. These fire safety conditions are met by an 18 cm solid wood wall. This wall is also due to the building design capable of spanning back a 9metre long overhang in the north of the building. An additional structural measure is not necessary. In the semi-public sector this design results in the overhang, which covers the entrance, supply and delivery area extensively and thus forms a canopy without being an independent canopy. This shelter is from a technical point of view a waste product. The design condition of the overhang originates by choosing the right material for a technical side aspect (see Fig. 5).

The result is a compact structure with a clearly recognizable design. Additionally, it has a high recognition value and presents because of the materials used a sustainable, ecological building that is typical for the region. Through the combination of geothermal heating and photovoltaic, the entire building is built and operated in an energy—self-sufficient and CO_2 neutral form. Excessive insulation is not necessary as passive houses due to their CO_2 neutrality do not need it. The basic insulation for low energy houses is wood wool and thus is ecological. The only renewable resource that the building needs is the sun.

The responsibility for sustainability in the construction industry is focused on the combination of architecture and technology. Only this can provide a holistic sustainability and also recognize the spatial effect, aesthetics, appearance, and the materiality of an acceptable status. Thus, sustainability is partly achieved by technical, physical building design principles, but never without functional, user-oriented design processes that provide the emotional value of a building, placing the room into the focus. This can only be achieved by the acceptance of the building in its environment, which is essential for the user to feel comfortable in the building. A technically perfect building has little value if it is not accepted, not being used, or asks too much of the user. An aesthetically, physically perfect building has little value if it does not follow the design principles of sustainability and does not function properly in a technical respect. So both areas are inseparably related to each other. It is precisely this mutual condition, this driving force of development, which equally forces the planner and the construction industry to take over responsibility. A discussion and advancement of evaluation processes, and measurement models in only one of the two areas alone is never useful and will always fail.

It is essential to re-think the development, the evaluation and mapping of sustainable buildings on a regular basis as both the technical basis that is subject to constant change due to further development and improvement and the concept of space and effect that is subject to permanent sociological and cultural changes request this measure.

The technical equipment, such as ventilation systems, installation systems, electrical power supply, etc., should be minimized.

The room configuration must become sustainably flexible. New partitioning systems that are reusable, adaptable and easily displaceable are necessary on the market. This allows a building to be sustainably flexible.

3.3 *Fewer, but better*

A building envelope with high quality and ecological building materials with a flexible inner structure without complex, technical building equipment combined with an ecological heating and cooling system, minimized installation ducts and a water treatment plant that makes rainwater usable, treats grey water already in the building, and returns the used water in usable condition back to the environment, represents a permanently sustainable building.

REFERENCES

Haus der Zukunft 2013. www.hausderzukunft.at.
Martin Putschögl, 2012. BlueBuild statt Real Vienna— ein Idee In Der Standard 9. März 2012.
ÖGNI 2012. http://www.ogni.at/de/unternehmenszerti fizierung/.
Österreichisches Institut für Bautechnik (OIB) 2012: *OIB-Richtlinie 2007*, www.oib.or.at.
Passiv Hause Institut 2013. www.passiv.de.
Wolfgang Feist 2007. Passivhaus Projektierungs Paket 2007. In PHPP 2007. Passivhaus Institut. 7. Auflage. Darmstadt 2007.
Wolfgang Feist 2012: Zertifizierungskriterien für Passivhäuser mit Wohnnutzung. *Zertifiziertes Passivhaus.* In PHPP 2012. Passivhaus Institut. Darmstadt 2012.

Figure 5. Commercial and residential building 'Goller'. *Photo:* DI Arch. Jan Werner, www.arch-vision.at.

Green Design, Materials and Manufacturing Processes – Bártolo et al. (eds)
© 2013 Taylor & Francis Group, London, ISBN 978-1-138-00046-9

Harnessing advances in eco-innovation to achieve resource efficient cities

P. Vandergert, S. Sandland & D. Newport
University of East London, London, UK

P. van den Abeele
IBGE-BIM, Brussels, Belgium

ABSTRACT: This paper analyses institutional approaches to embed eco-innovation in the urban realm, contrasting radical step change with incremental change. It firstly reviews three different eco-innovation models: CleanTech, Eco-design and Product Service Systems. The paper then summarises different institutional approaches that can be adopted by local governments. It reviews three cities' practices to embedding eco-innovation: London, Copenhagen and Stockholm. It concludes by identifying some of the opportunities and challenges that governance models need to overcome in order to achieve a step change to business as usual.

1 INTRODUCTION

1.1 Context

Cities are acknowledged to be at the forefront of the quest for sustainable development. Resource efficiency is one key aspect of sustainable development and cities provide a particularly appropriate scale at which to address this. This is because the confluence of business activity, local governance and consumption patterns provide both the challenge of unsustainable resource consumption patterns and the opportunity to harness eco-entrepreneurship with enlightened institutional frameworks and consumer demand (Fischer-Kowalski & Harberl, 1998).

1.2 TURAS research programme

In order to achieve resource efficient cities, local city governments are faced with a number of challenges. This paper reflects ongoing research conducted as part of the project Towards Urban Resilience And Sustainability (TURAS), funded under the 7th Framework Programme of the European Commission. The TURAS programme is designed to develop transition strategies to enable European cities to strengthen their resilience and sustainability. Our research investigates how local governments can support and facilitate innovative sustainable business activities through a range of measures, thus harnessing the power of sustainable industry, in particular Small and Medium Enterprises (SMEs), to contribute to more resource

efficient cities. This paper focuses on three sustainable business models: Clean technologies (CleanTech); Eco-design; Product Service Systems (PSS). Alongside analyzing support for these business models we also consider how to evaluate effectiveness and promote adoption of innovation.

The TURAS emerging research suggests that: innovative products, processes and practices; emerging technology; urban design; infrastructure transformation; behaviour change; and enlightened governance will all play a significant role in achieving step change with regard to resource efficiency and eco-innovation. In addition, prioritizing resource efficiency has a positive effect on the linked environmental priorities of carbon emission reduction and waste minimization. The research team will use action research to evaluate how to harness eco-innovation to delivering a sustainable and resilient urban vision.

2 ECO-INNOVATION MODELS

2.1 CleanTech

The definition of CleanTech adopted in this research is—products, processes and services helping to reduce negative ecological impact and improve the responsible and efficient use of natural resources.

The CleanTech business sector is developing its own taxonomy(ies) of CleanTech, with the company responsible for the Global CleanTech 100 listing of top companies proposing 13 categories,

Category
Advanced Materials
Agriculture and Forestry
Air
Biofuels & Biochemicals
Conventional Fuels
Smart Grid
Energy Efficiency
Energy Storage
Fuel Cells & Hydrogen
Recycling & Waste
Solar
Transportation
Water & Wastewater
Deloitte Development LLC, 2011

Figure 1. CleanTech taxonomy for the Cleantech Global 100.

see Figure 1. Whilst a definitive definition has not been arrived at, this provides a good representation of what is generally meant by those in the CleanTech sector.

CleanTech is maturing as a defined business sector, and cities provide an appropriate scale for developing and testing these new technologies.

2.2 *Eco-design*

Eco-design means taking into account the environmental impacts of a product in the early stages of design. Eco-design is based on a life-cycle approach: the environmental impact of the product is analysed throughout its life-cycle, from cradle to grave, or cradle to cradle (Braungart and McDonough, 2002). It is seen not just as a tool for increased efficiency but also as a way to promote more radical innovative approaches to new product and service development. The European Commission Eco-design directive establishes a framework for eco-design of energy-related products and states that 'designed to optimise the environmental performance of products, while maintaining their functional qualities, it provides genuine new opportunities for manufacturers, consumers and society as a whole' (EC Directive 2009/125).

2.3 *Product service systems*

Product Service Systems (PSS) provide an opportunity for a profound shift in the dominant production/consumption model. PSS 'focuses on value, especially value in use, rather than product or service characteristics, and thus provides designers with a higher degree of freedom' (Sakao, 2011). If PSS also embeds eco-design thinking at the start, and is able to harness CleanTech, it could become a

powerful tool for delivering eco-innovation within broader sustainable development goals. In other words, PSS looks critically at whether a service can be provided instead of a product, thus potentially reducing the resource and carbon intensity of production and replacing it with value-added, sustainable service. This can provide social and economic benefits, in terms of employment and cohesion.

3 INSTITUTIONAL TOOLS FOR LOCAL GOVERNMENTS

3.1 *Economic measures*

Comprising a mixture of taxes, subsidies and loans, the most effective economic measures are those that provide competitive advantage to eco-innovation and help compensate for the higher prices of products and services that are new on the market and unable to enjoy economies of scale (Bowers, 1997; Osawa & Miyazaki, 2006). Where products are concerned, taxation at this level is beyond the remit of most local governments, although they can seek to fund or find loans or subsidies that compensate for the higher cost of eco-innovation. The most effective loans are those where repayments are calculated to replicate the savings gained, thus imposing no extra cost on the business, and repayments are used to fund further loans.

Services offered within the city can be priced in a manner that encourages the use of eco-innovation. Whilst the public may perceive the pricing of services as additional taxation, it can gain acceptance when transfer to the cleaner technology is facilitated, encouraging take up. By separating the charge from a general local tax, the user is made more aware of the environmental choices open to them.

3.2 *Regulation*

There are three commonly used forms of regulation, that of permits, tradable pollution permits and the banning of substances. General permits to trade could be used with more rigour. Currently, in London, the heavy polluters are monitored centrally and local authorities issue permits to smaller organizations that could also create an environmental hazard, including animal treatment or printing processes. A licencing system could be required for all businesses, requiring renewal, which would enable local government to set conditions, such as obtaining an environmental audit or adopting a clean activity during the life of the permit. The terms of the permit could be on an escalator for some industries in order to encourage continual improvement.

When deciding on this manner of control, local governments need to ensure that they maintain a competitive environment and the ability to attract industry. This can be achieved if regional/national government make it a blanket requirement, or the local government creates a package that includes subsidies or loans to compensate for costs.

Although there is evidence that banning substances at district level can be achieved, such as in the case of banning the use of smoke-producing fuel in cities, it is somewhat limited in scope because removing items from sale in one area does not prevent purchasers going elsewhere to buy the product or purchasing online. Therefore, to be successful, educating the market and availability of viable alternatives is required.

Similarly, tradable pollution permits are most appropriately used at national or even international level because the tradable nature circumnavigates the desired end and removes the ability to ensure that emissions within a specific area are controlled.

3.3 *Urban design*

The design and management of physical infrastructure within a city provides a clear opportunity to embed eco-innovation at scale, through infrastructure and buildings development and management. A lot of good practice is carried out in this respect and councils can clearly learn from the practice of others. Whilst Scandinavian cities have specialized in developing clean urban design infrastructure including the use of district heating and cooling, green roofs and walls, effective transport systems and waste management, London has established low carbon zones that also educate the public and seek to improve the housing stock.

It is clear that good, sustainable urban design should address the local geography and governance structures of the city. It is also clear that some practices are not sustainable, such as Copenhagen's use of taxation to encourage the adoption of the electric car. Copenhagen only produces 30% of its electricity demand using renewable sources. This means that any increase in electricity demand must be met through non renewable means. In addition to this, the only benefit of the electricity is that it is clean at the point of use. In terms of efficiency, electricity uses far more fuel to deliver the same mileage.

3.4 *Public procurement*

Public procurement within the EU is worth 19% of EU GDP and therefore is a potentially powerful tool in encouraging resource efficiency and eco-innovation. Under EU public procurement rules, contracting authorities may take multiple aspects into account (on a voluntary basis), provided they are relevant to the product, service or work authorities want to procure, appropriate in scope, and within the overarching principles of value for money, transparency and fairness. These other aspects include protecting the environment, social considerations and fostering innovation. Also, in recognition of the important role SMEs play in the EU economy in terms of their potential for job creation, growth and innovation, making it easier for SMEs to access public procurement is an EU priority.

Within the EU, business and product environmental certification tools are available on a voluntary basis and these can be used as part of the public procurement process by local governments to achieve improved environmental performance, both of products and businesses. They can be useful tools to pull eco-innovation towards the mainstream, but tend not to be as helpful in pushing eco-innovation. They are currently less accessible to SMEs and as part of the TURAS programme the aim is to develop a more inclusive rating system.

3.5 *Knowledge exchange*

The authors have found that knowledge exchange is critical for the diffusion of eco-innovation. Research suggest that for this to be done successfully there needs to be a network of organizations that can offer expertise and facilities. Thus the first step for local governments that have not engaged in this area should be to develop such a network. A number of European cities have established CleanTech clusters, which are informal networking and match-making opportunities for investors, manufacturers and entrepreneurs to identify mutually beneficial business opportunities.

4 ANALYSIS FROM THREE CITIES

4.1 *London*

With a population of 8.5 million, London is the second largest urban district in the EU. London is administered by 32 London boroughs councils plus the City of London, and these 33 local governments are responsible for anything between 140,000 and 250,000 residents. The councils are responsible for maintaining most of the services in their borough, including planning, waste collection and roads. However the scope of control varies because some boroughs share facilities and operations to gain economies of scale and others are run by the regional Greater London Authority, for example transport. This latter control has seen the introduction of congestion charging, smart ticketing, and a cycle hire scheme. London borough councils seek to support a healthy business environment in order to underpin the prosperity of the borough. Thus individual boroughs would be reluctant to impose standards that run counter to business interests. As part of the TURAS programme, researchers undertook a review to evaluate the performance of London borough councils in assisting businesses, in particular SMEs, in developing clean and sustainable business practices.

Council action fell into one of three categories. Firstly, some councils appeared to deliver a minimum standard of service, sometimes aligned with central government. Secondly, some councils provided businesses with links to external agencies and resources. Finally, some councils, invariably with the support of third party organizations such as charities or universities, and supplemented with funding from the European Regional Development Fund (ERDF), ran projects that instigated the development of significant activity.

The research found that London's borough councils are generally active in the adoption of clean technologies. Most of this is in the form of knowledge exchange, and is supported by the provision of resources at a regional level, the issuing of licences and permits to control the density of some of the more polluting activities and a few grants and subsidies. Support is varied geographically, with some councils doing little more than the legal minimum. In contrast, others have forged links with or developed some excellent agencies and charities that support activities across a wide range of environmental challenges. However, researchers consider that the full extent of the activities is masked by an inconsistent use of terminology and technology.

4.2 *Copenhagen*

The municipality of Copenhagen has set itself the target to become one of Europe's first carbon neutral cities, and has in recent years systematically reviewed the provision of services to its residents. Some activities have benefited a variety of sectors. Electricity: Denmark has a history of leading the world in wind power generation. Danes produced the first megawatt turbine that could be used commercially enabling them to dominate the market and, supported the by economies of scale that followed, develop the technology further. Their large coast line also facilitated the development of windfarms in advance of other countries. This is supported by a target to produce 50% of Denmark's electricity through wind power by 2030.

Heating: Denmark started to adopt district heating schemes as early as 1925 and was developed

further during the fuel crisis in the 1970s. The systems have become increasingly clean tech to extract maximum efficiency and transfer to low carbon fuels. Research is being undertaken to assess the viability of geothermal heat production. End of pipe technologies clean flue emissions and the extracted matter is used in other products. CO_2 emissions are 40% lower than individual gas boilers and 50% lower than individual oil boilers. In producing electricity and heat together, fuel use is cut by 30% and the cost to the consumer is reduced by about 50%.

Cooling: The city has recognised that the demand for air cooling is likely to rise as a result of climate change and planners have sought a centralised solution. Areas where there is a sufficient concentration of demand were identified and a cooling plant developed using sea water. Surplus heat and heat exchange technology provides further cooling. This saves 67% of the CO_2 compared with traditional methods of cooling and, using less fuel, reduces SO_2 and NO emissions between 62% and 69% respectively. Businesses are allowed to develop networks provided their viability is proven.

Waste and recycling: The first environmental legislation introduced was related to waste, in 1973. Because of landfill capacity becoming scarce, waste in Copenhagen systematic recycling of a wide range of materials was introduced. Residues that cannot be recycled provide 30% of Copenhagen's demand for heating. Taxation on fossil fuel makes the renewable alternatives a viable option and legislation prohibits sending combustible waste to landfill. Efforts are being made to create a competitive market for waste. Only 1.8% of Copenhagen's waste goes to land fill.

Water: Copenhagen authorities recognised that the demand for water was outstripping supply. This was addressed by strict planning regulation covering aspects such as the sewerage of water, ensuring that rain water at roof level and road level and black waste water are handled in three separate streams. This has provided the additional benefit of reduced flooding risks, and preserves groundwater for uses where its purity is essential. Within the planning regulations there is mandatory green roof policy. This applies to all new buildings with a roof pitch below a specified angle. Its principle advantage is that of absorption of roof water, and ground level water being channelled into a wide range of outlets, including filtering plants for non-potable use. Software has been developed to control water pressure, which in conjunction with leak detection technology has reduced loss through leaks to 6%–7%, in comparison with 40%–50% in other cities. This has been accompanied by financial incentives for businesses and residents in the form of payments if they reduce their sewerage usage by collecting and re-using run off from their properties, and a high cost of drinking water. The average water savings are 26%.

Transport: A variety of steps have been taken to encourage commuters out of their cars. These involved firstly developing a company that served to integrate the different modes of transport through smart ticketing. Secondly, information systems were improved, providing users with integrated route planning facilities and access to real-time data. A network of designated cycle tracks was developed alongside roads, and speed was facilitated by "greenwaves", a technology previously used for cars; the cyclist who travels at 20 kmh will always arrive at traffic lights when they are green. Some roads and bridges are exclusively for bicycle use, and routes have been shortened by taking them through parks and open space whenever possible. Information is provided to other road users to heighten awareness of cyclists and "family bikes" have been designed and promoted. Use of bicycles has been integrated into the public transport system. As a result of these measures, carbon emissions have been reduced by 90,000 tonnes per annum, 88% of users find it the fastest way to work and others appreciate the associated health benefits.

4.3 Stockholm

The City of Stockholm is the largest and most densely populated area in Sweden. Named European Green Capital 2010, Stockholm has a strong record on integrating environment into city planning.

Transport: There has been a greater provision of cycle tracks and 75% public transport runs on clean fuel, either in the form of ethanol or biogas, and trains run on wind power-sourced electricity. Information systems are used to aid journey planning. A congestion charge introduced in 2007 deters some car users. Together, these initiatives have seen an 18% rise in the use of public transport and cut traffic to and from the city centre by 20% and significantly sped up transit through the city. Campaigns were conducted to educate residents about efficient driving practice. Stockholm has worked with manufacturers, retailers and owners of fleets of vehicles to promote the purchase of clean vehicles which will run on biofuels; it is also promoting the development of renewable fuels including biogas which will reduce emissions by 85%.

District Heating and Cooling: Steps are being taken to convert these processes to biofuels, using the same heat pumps to distribute both. Using much the same approach as Copenhagen with co-production of heat and electricity, Stockholm is

seeing the same benefits as well as the reduction of sulphur emissions.

Stockholm Climate Pact: Stockholm has developed a pact with nearly 100 local businesses in which they publish their environmental targets and commit to match Stockholm City's reduction targets.

Waste management: This is a highly regulated activity at national level, placing obligations on producers and users. Stockholm charges for waste removal and requires producers to facilitate the recycling of all their packaging, including ensuring the end user has access to a collection point.

5 CONCLUSION

In considering the instruments available to local governments, and the practices in three European cities, research undertaken to-date by the TURAS team suggests two themes that are relevant in identifying ways for cities to harness eco-innovation.

The first theme is the appropriate scale for intervention, both spatially and in terms of governance. The Copenhagen and Stockholm cases show how effective it is to intervene at the scale of a city's infrastructure in terms of driving and delivering resource efficiency. Across all three cities it seems that policies and actions to nurture and support (even subsidise) individual entrepreneurship is key to driving eco-innovation. This suggests that a nuanced governance approach based on a diagnostic of each city's strengths, opportunities, challenges and barriers is key to developing effective government interventions, crucially at different levels of government. The research team are developing a tool to do this. It appears that partnership working and collaboration is cruciall. For the TURAS team, a significant work stream will be action research to bring different stakeholders together to generate challenge-driven eco-innovation at the city scale.

The second theme is distinguishing between resource efficiency and more radical eco-innovation. Whilst much of the business and governance focus to date has been on promoting CleanTech to deliver resource efficiency and carbon reduction, the research team are investigating how more radical eco-design approaches can move away from unsustainable production and consumption models. One avenue being explored is using eco-design to integrate sustainable Clean-Tech with product service systems, in a way that supports a city's social and economic values and aspirations. Local governments who have the tools to identify partners to deliver this step change will be at the forefront of embedding eco-innovation. Given the acute environmental, economic and social challenges facing urban areas, it is within innovative governance approaches that we are likely to find opportunities to promote the step change needed.

REFERENCES

Bowers, J. 1997. *Sustainability and Environmental Economics. An alternative text*. Harlow: Prentice Hall.

Deloitte Development LLC. 2011. *Global Cleantech100 Report 2012*.

Fischer-Kowalski, M. & Harberl, H. 1998. Sustainability problems and historical transitions in Hamm, B. & Muttagi, P. (eds) *Sustainable development and the future of cities*. London; IDTG Publishing: 57–76.

McDonough, W. & Braungart, M. 2002. *Cradle to Cradle: Remaking the Way We Make Things*. New York: North Point Press.

Osawa, Y. & Miyazaki, K. 2006. An Empirical Analysis of the Valley of Death: Large-scale R&D Project Performance in a Japanese Diversified Company. *Asian journal of technology innovation 14(2): 93–11*.

Sakao, T. 2011. What is PSS design?—Explained with two industrial case studies. *Procedia—Social and Behavioral Sciences 25: 403–407*.

Green Design, Materials and Manufacturing Processes – Bártolo et al. (eds)
© 2013 Taylor & Francis Group, London, ISBN 978-1-138-00046-9

Opportunity to complement Vitoria-Gasteiz's green network, using the roofs of public facility buildings

Elena Basanta Carmona

Department of Urban and Regional Planning, School of Architecture,
Polytechnic University of Madrid, Spain

ABSTRACT: One of the main problems of today's cities is heavy air pollution caused largely by emissions of GreenHouse Gases (GHG) onto the atmosphere. The roofs of the buildings are presented in this context, as constructive elements which include plant surfaces that help offset these emissions. The green roofs, as part of the urban green infrastructure, act as sinks for the CO_2.

This article discusses the benefits and consequences of introducing green roofs in public facilities, turning them into places of opportunity, focusing on improving the quantification of CO_2 emissions and the role they play in the green network configuration inside the city of Vitoria-Gasteiz.

1 INTRODUCTION

One of the major challenges facing humanity in the 21st century is climate change, which has gone from a virtual threat to becoming a reality. Among its main consequences is global warming, caused by the increased emissions of GreenHouse Gases (GHGs) onto the atmosphere. The international measure of these gases is carbon dioxide or CO_2.

The developed countries are the main emitters of GHG from energy combustion, which highlights the importance of energy efficiency as a means of exchange and as one of the main goals of sustainability in cities. To achieve this it is necessary, among other things, to improve the energy efficiency of many existing buildings.

The energy efficiency renovation of buildings mainly focuses on their shells, i.e. in the facades and roofs. But roofs, because of their specific location and configuration, become particularly relevant when introducing green elements that reduce GHGs in cities because they act as sinks for CO_2. Furthermore, green roofs act as temperature regulators, providing greater insulation for the building. This results in a reduction in fuel consumption and consequently in a reduction in CO_2 emissions.

2 OBJECTIVES AND HYPOTHESIS

The overall objective of this study is to find out the level of decrease in urban pollution achieved when green roofs are installed in public facility buildings in a city. Public facility buildings were chosen because they are the biggest publicly-owned urban buildings.

Specifically, the main objective is to study the installation of green roofs in large public facilities in order to assess and quantify the reductions in emissions that this action would produce at an urban scale.

The hypothesis that we will try to verify is that the installation of green roofs in facility buildings of Vitoria-Gasteiz will reduce, in at least 10%, the emissions of CO_2 in the city.

3 METHODOLOGY

3.1 *Phase 1*

Firstly, we will produce a catalog of good practices including the most significant examples of the implementation of green roofs internationally in recent years and the existing public mechanisms for implementation and support.

The main objective is to learn about the latest and best solutions available in terms of green roof construction in large public buildings.

3.2 *Phase 2*

We will study the importance of green roofs as part of the green infrastructure of the city, where many of its strengths and benefits derive from.

Secondly, there will be an evaluation of the benefits of the green elements installed in buildings at an urban scale.

3.3 *Phase 3*

We will study the case of Vitoria-Gasteiz, highlighting the following points:

- urban green spaces, because they form an infrastructure to which green roofs belong.
- existing city planning, with special emphasis on plans related to Climate Change and CO_2 emissions.
- facility buildings, elaborating a database which allows for the classification of all buildings as suitable or unsuitable for a green roof.

3.4 *Phase 4*

We will analyze the results of the study in order to prove the hypothesis and draw conclusions.

4 CASE STUDY: VITORIA-GASTEIZ

4.1 *Study of urban green spaces*

The municipality of Vitoria-Gasteiz is notable for its biodiversity, which is the result of a historical tradition, a privileged location and currently, urban planning focused on the conditions of the city. This was reinforced with the ratification of the Aalborg Charter in 1995 and derived from the Agenda 21.

Urban green is of great importance for the city, and authorities are well aware of this.

The city has about 13 m²/inhabitant of green surface, but if you include the green ring around the city, this percentage rises to about 40 m²/inhabitant. Although it would seem that there is quantitatively sufficient green space in the city, it would be necessary first to analyze in depth the quality and configuration of these green areas, and then consider the introduction of green roofs on buildings in Vitoria-Gasteiz from energy rehabilitation and linked to their role as a sink for CO_2.

In fact, today there are several actions being carried out to improve biodiversity in the town, mostly through the Green Belt project. This project consists of a set of suburban parks of great ecological and landscape value, linked by eco-recreational corridors that surround the city. Its realization has generated numerous benefits to the city at an environmental, social and economic level.

This has influenced the City Council to propose in 2012 the creation of another green belt in the city interior.

4.2 *Study of public facility buildings*

For the study of Vitoria-Gasteiz public facilities those proposed by the General Plan of the city have been studied and the steps outlined below have been taken.

4.2.1 *Division of the city in areas*
The city has been divided into five zones by grouping districts as follows.

- Central Zone: Casco Viejo, Coronación, Lovaina, Ensanche and Desamparados.
- North Zone: Arriaga-Lakua, El Pilar and Zaramaga.
- South Zone: San Cristobal, Adurtza, Mendizorrotza and Ariznabarra.
- East Zone: Salburua, Anglo, Santiago, Arana, Santa Lucía, Aranbizkarra, Arantzabela and Judizmendi.
- West Zone: Gazalbide, Txagorritxu, Sansomendi, San Martin, Ali-Gobeo and Zabalgana.

4.2.2 *Selection criteria of the public facilities*
Subsequently, we have established several criteria to classify the public facilities as suitable or unsuitable for the installation of green roofs. These are basically the following:

- Location: We have considered only the facilities located within the Green Belt which surrounds the city.
- Property: We studied the facilities of a public nature, focusing on a possible action plan carried out by local municipalities.
- Existing roofs: existing roofs were considered at the time of the study (2012). Empty parts of plots or the vacant lots were not taken into account.
- Existing green roofs: there are none in public facilities, but a pilot proposal is being developed at the Europe Conference Centre.
- Roof type: some facility buildings, based on the specific characteristics of their roofs were considered unsuitable. For that purpose we have four major considerations:

1. Building Protection: buildings protected by the Master Plan in the categories of "full protection" and "structural conservation" have been ruled out because regulations do not allow the modification of the characteristics of their roofs.
2. Building System: buildings considered have either flat roofs or a slope of less than 30°, because these characteristics are optimal for the installation of a green roof of any type (extensive or intensive). Buildings covered with traditional systems such as tile or slate, were ruled out, in order to respect the aesthetic conditions. Moreover, domes, vaults and extremely divided roofs have not been considered because of their complexity.

3. Size: We have computed the roofs over 400 m²
   which are the ones with enough relevance to
   influence the overall study.
4. Others: In this case, mainly vacant plots have
   been included, plots which have not yet been
   built. Some of these sites have become green
   areas and, therefore, they are included in the network of green spaces in the city used by citizens.
   In other cases, parking areas or outdoor sports
   courts have been considered. Also included here
   are plots that have changed their use due to
   planning revisions. For instance, some of them
   have been reclassified as residential. Without
   more information, all these sites are considered
   public but unsuitable.

4.2.3 *Classification of equipment in suitable and unsuitable*

Finally, there were five tables for the five zones of
study. The classification of the buildings is shown
graphically, with a map of the entire city. After
the analysis, 122 facility buildings were considered
suitable of a total of 423 buildings found. This
amounts to 28.8% of suitable roofs versus 71.2%
of unsuitable ones.

4.3 *Surface calculation*

After classifying all the buildings studied as suitable or unsuitable to host a green roof, we calculated
the surfaces of those roofs that were considered
suitable. The result: 266,501 m².

4.4 *Energy and emissions savings*

The values of the energy savings and CO_2 emissions of suitable facility buildings can be known
starting with their roof surfaces.

Let's keep in mind that the CO_2 absorption of
green roofs depends on the plants on them. The
choice of plants depends on the type of green roof
considered, intensive or extensive. If you need
know exactly how much CO_2 will be absorbed
by the plants, it is necessary to design a specific
project for each roof.

The current study considers two different cases
to know the variation that would exist in the
choice between the two types of green roofs. It is
important to note that the placement of an intensive green roof requires a previous analysis of the
structure of the building, because the load considerably increases (and can reach 1200 kg/m² versus
the 100 kg/m² of the extensive green roofs). If the
structure needs to be reinforced, installation costs
can be excessive.

The two cases considered are:

– Case 1, an extensive green roof, 10-centimeters
  thick and various sedum species. The absorption

rate of this kind of plants is of about 0.02 kg of
CO_2 eq /m² per year.
– Case 2, an intensive green roof, 60-centimeters
  m thick and plants that can absorb an average
  of 30 kg CO_2 eq/m² per year.

With respect to energy savings, the study was
done using a virtual system, consisting of a natural gas heating boiler. To get accurate results, we
should also analyze each of the buildings in detail.

4.4.1 *Energy saving*

In general, adding a green roof provides an additional layer of insulation which can be quantified
from the *thermal transmittance* (U) of the roof.

$$U = 1/\Sigma R^T = 1/\Sigma e/\lambda$$

where R^T = total thermal resistance of constructive component (m²C/W); which is defined by
the expression $R = e/\lambda$; where e = thickness of the
layer (m); and λ = thermal conductivity of the
material.

For the sand layer of green roof are considered:

– e = 0.1 m (case 1); and e = 0.6 m (case 2)
– λ = 0.6 W/m² °C

When calculating the thermal transmittance of
a standard accessible roof and a green roof, we get
Table 1.

For the total energy savings of any roof, the difference obtained for thermal transmittance (U),
the annual heating demand of the building (S), and
the surface of the roof (A), have multiplied.

$$\text{Energy saving} = U \times S \times A$$

$$[kWh] = [(W/(m^2 \, °C)] \times [°C \, h] \times [m^2]$$

where U = 0.12 W/m² °C for case 1; and U = 0.43
W/m² °C for case 2.

For annual heating demand (S) of a building
type has taken 21,409 °C h.

4.4.2 *Emissions saving*

To find the emission reduction (AD) that this
energy saving represents, it has been associated to
two variables, the reduction of natural gas consumption of the boiler and the reduction of CO_2
absorption by plants.

$$AD = AE \times E + AB \times A$$

$$[t \, CO_2 \, eq] = [kWh] \times [t \, CO_2 \, eq/kWh] \\ + [t \, CO_2 \, eq/m^2] \times [m^2]$$

where AE = energy saving calculated above;
E = t CO_2 eq emission into the atmosphere of a

Table 1. Differences in the thermal transmittance between a standard accessible roof and a green roof.

Roofing type	U roof (W/m² °C)	U humid sand (W/m² °C)	U total (W/m² °C)	U difference (W/m² °C)
Standard roof	0.90	–	0.90	0.00
Green roof case 1	0.90	6	0.78	0.12
Green roof case 2	0.90	1	0.47	0.43

Table 2. Summary.

Surface m²	AE case 1 KW h	AE case 2 KW h	AD case1 tCO_2	AD case 2 tCO_2
266,501	684,678	2453,431	144.32	8493.08

Table 3. Comparative analysis.

Surface (m²)	Energy consumptions		Emissions	
	Total (MWh)	Per inhabitant (MWh/in h)	Total (tCO_2)	Per inhabitant (tCO_2/in h)
City				
–	2858,580	12,090	841,068	3557
Facility buildings (current)				
277,951	77,540	327	19,964	84
Facility buildings (Climate change plan savings)				
42,600	307	1.30	63	0.27
Facility buildings (Study case 1)				
266,501	684	2.90	144	0.61
Facility buildings (Study case 2)				
266,501	2454	10.38	8493	35.92

Table 4. Percentages.

	City	Facility buildings
Climate change plan	0.01%	0.32%
Study case 1	0.02%	0.72%
Study case 2	1.01%	42.54%

boiler type; AB = absorption of tCO_2 eq of a plant per m² of surface; and A = roof surface considered.

We have taken as values for E and AB the following data: E = 2.03 × 10^{-4} [tCO_2 eq/kWh]; AB = 2.00 × 10^{-5} [tCO_2 eq/m²] for case 1; and AB = 30.0 × 10^{-3} [tCO_2 eq/m²] for case 2.

Then, there is a Table 2, containing the results, with the summary of surface data, energy savings and emission obtained for the two cases considered.

5 CONCLUSIONS

To compare the data obtained with the generals of the city, the Table 3 was prepared. It presents the total emissions and energy consumptions of the city and for the facility buildings, according to the Climate Change Plan of Vitoria-Gasteiz.

To find out if the hypothesis is verified, percentages that these values represent were calculated.

The Table 4 shows that the hypothesis is not verified, since according to the research, the installation of green roofs in public facilities in Vitoria-Gasteiz would not reduce CO_2 emissions in the city by 10%.

Moreover, in the same table other important conclusions can be drawn. It should be noted that the results refer to the city of study, Vitoria-Gasteiz, and conclusions therefore cannot be directly extrapolated to other cities without completing the corresponding study.

1. In the case of introducing an extensive green roof (case 1), the emissions savings are more than twice those calculated by the Climate Change Plan. This is because in the present work a much more comprehensive study of the existing buildings has been carried out in order to know the real surface of the roofs that could be replaced.

2. When considering an intensive greenroof (case 2), saving CO_2 emissions to the atmosphere grows significantly, accounting for 1% of total emissions in the city.

Therefore, if you install an extensive green roof, savings emissions are much less significant than if you install an intensive one. Logically, the emissions savings would increase if the study added other plant species that would absorb more CO_2 or by adding a thicker green layer. As already mentioned, for this kind of green roof a structural analysis of the building would be required to assess the feasibility of the proposal.

3. The CO2 saving is less important compared to the city, but it is very significant when calculated with respect to the emissions produced by the facilities themselves, representing 42.54% of these emissions. You could add this measure to others in order to reach the goal of zero-emissions buildings, if not for the whole of the city, at least for the facility buildings.

4. The facility buildings, therefore, are an important source of suitable surfaces to install green roofs. In the case study, the number of those considered suitable (122, 28.8%) is less than the number of the unsuitable ones (301, 71.2%), but the absolute number of suitable surfaces that they represent is high (266,501 m^2). We must bear in mind that the city considered has a large number of facilities compared to other cities. In addition, one of the main causes of the high number of unsuitable buildings is the large number of them that are yet to be built in the new large developments planned (132 of the 301 as unsuitable, and 30% of the total), which may somewhat distort the relationship between them (suitable and unsuitable).

5. The introduction of green roofs in the facilities can help reduce CO_2 emissions in the atmosphere, but if the goal is to achieve a carbon neutral city, the measure itself has not sufficient entity to be applied and must be accompanied by other measures.

REFERENCES

Monograph

Bettini. V. 1998. *Elements of urban ecology*. Torino: Einaudi Editore, s.p.a ana.—2010 *Spain Global Change España 2020/2050: Building Sector*. Madrid: Centro Complutense de Estudios e Información Medioambiental, Green Building Council España, Asociación Sostenibilidad y Arquitectura. 248 p.

Fariña Tojo, J. 2007. *The city and the natural environment*. 3ª ed. Madrid: Ediciones Akal, S.A. 342 p.

Figueroa Clemente, M.E. & Redondo Gómez, S. 2007. *Natural CO_2 sinks: a sustainable strategy between climate change and the Kyoto Protocol from urban and regional perspectives*. Sevilla: Universidad de Sevilla, secretariado de publicaciones. 207 p.

Higueras García, E. 2006. *Urban bioclimatic*. Barcelona: Gustavo Gili. 241 p.

Hough, M. 1995. *Cities and Natural Process*. London: Routledge. 315 p.

Minke, G. 2005. *Green roofs: planning and practical advises* Teruel: EcoHabitar. 85 [1ª ed (2000): Dächer begrünen einfach und wirkungsvoll; planung, ausführungshinweise und stippss. Alemania: Ökobuch].

Neila Gonzalez, J. 2004. *Bioclimatic architecture in a sustainable environment*. Madrid: Munilla-Lería, DL 443 p.

Salvador Palomo, P.J. 200.3 *Green planning in cities*. Barcelona: Gustavo Gili, S.A. 326 p.

Articles

Chanampa, M. et al. 2010. Green technologies as tools for architectural restoration. In: *Sustainable Building Conference* (SB10mad). Madrid. 12 p.

Currie, B.A. & Bass, B. 2008. Estimates of air pollution mitigation with green plantas and green roofs using the UFORE mode.l In: *Urban Ecosyst* nº 11. Pág. 409–422.

Roehr, D. & Laurenz, J. 2008. Greening the Urban Fabric: contribution of green surfaces in reducing CO_2 emissions. Algarve, Portugal: *1st WSEAS International Conference on Landscape Architecture* 7p.

Reports and regulations

Air Quality Directive: on ambient air quality and cleaner air for Europe. [on line]. Eupean Comission. 2008/50/EC [ref. 28/07/2012]. web: http://ec.europa.eu/environment/air/review_air_policy.htm.

Ensuring quality of life in Europe's cities and towns: Tackling the environmental challenges driven by European and global change. Luxembourg: European Environment Agency Report. Nº5 (2009) 110 p.

Spain. Código Técnico de la Edificación: Ley 38/1999, de 5 de noviembre, de ordenación de la edificación, Real decreto 314/2006, 17 de marzo. Ministerio de Fomento. (2010). 1229 p.

Spain. Master Plan of Vitoria-Gasteiz. Servicio de Planeamiento y Gestión Urbanística del Ayuntamiento de Vitoria-Gasteiz. Departamento de Urbanismo, (2007). 1883 p.

Spain. Climate Change Plan of Vitoria-Gasteiz. (2010–2020). Vitoria-Gasteiz: Ayuntamiento y Agencia de Ecología Urbana de Barcelona. (2010) 243 p.

Spain. Vitoria-Gasteiz: carbon neutral city. 2020–2050. Vitoria-Gasteiz: Ayuntamiento y Agencia de Ecología Urbana de Barcelona. (2010). 101 p.

Green Design, Materials and Manufacturing Processes – Bártolo et al. (eds)
© 2013 Taylor & Francis Group, London, ISBN 978-1-138-00046-9

3 Scales of repurposed disposability—Construction, Renovation, and Demolition (CRD)

Wendy W. Fok
Digital Media & Design Program, Gerald D Hines College of Architecture, University of Houston, atelier//studio WF, New York, USA

ABSTRACT: Project managers and construction contractors have long recognized the importance of reducing waste and salvaging high value construction and demolition materials such as copper and other metals. Contractors are usually careful about the quantity of materials ordered, how materials are used and how to carefully de-construct valuable materials. In most cases however, materials that are more difficult to separate and that are worth less per unit weight are still going to landfill, even when they are present in large quantities. This represents an inefficient use of natural resources and uses up landfill capacity unnecessarily.

Unfortunately, some contractors do not realize that there are new opportunities for waste minimization, while others are reluctant to implement environmental practices because they believe these practices will increase their project costs. Most contractors are concerned about the cost of the labour that is needed to deconstruct materials for reuse or recycling. However, it has been shown that effective waste management during CRD projects not only helps protect the environment, but can also generate significant economic savings. Demonstration projects have shown that the diversion of waste from landfill can reduce waste disposal costs by up to 30%. This is accomplished through reduced tipping and haulage fees and the sale of reusable and recyclable materials.

Various projects from within our practice and within our academic curriculum will be brought into the attention of this paper. Specifics of modularity, form/fit/analysis, fabrication, and off-site production, will be demonstrated within the larger discussion through the focus onto three case studies, ranging from three different scales.

1 INTRODUCTION

The premise of the paper will discuss three scaled prototypes of which utilized the idea of how repurposed waste materials are sourced within the field of architecture, and the developmental nature of the design utility based within the parameters of repurposing. The basic use of these prototypes range from 1) an architectural modular developed as an attempt to address the issue of construction waste produced by demolished project sites for unskilled workers within developing countries, 2) an art installation piece which addresses overstocked and over-ordered materials and EPS foam, originally used for insulation, by contractors for an existing institutional construction project in Houston, Texas, and 3) an academic design-research project that was further pursued as an art installation sculpture, made for the HKSZ (Hong Kong Shen Zhen) Biennale, which uses localized wood materials, pre-fabricated through CNC, tried-and-tested to be constructed offsite, to limit the onsite production, delivery, and man-power.

Each of these projects instigate the a range of developing processes of computational design-research and the parallel intentions of diversion within the construction, repurpose and demolition processes inherit to the production of architectural design. All three case studies utilize and appropriate construction waste, and consider the repercussions of the construction waste within a novel form of fabrication and processing for the built environment. Incorporating the 3Rs (reduce, reuse and recycle) into construction, renovation and demolition waste management creates a closed-loop manufacturing and purchasing cycle. This significantly reduces the need to extract raw materials, reduces the amount of materials going to landfill sites and reduces the life-cycle costs of buildings and building materials.

2 BASIS OF DEVELOPMENT

According to the *United States Environmental Protection Agency* for Region 8, the qualifying basis of Construction and Demolition (C&D) materials are as follows:

> *Construction and Demolition (C&D) materials consist of the debris generated during the construction, renovation, and demolition of buildings, roads, and bridges. C&D materials often contain bulky, heavy materials, such as concrete, wood, metals, glass, and salvaged building components. Reducing and recycling C&D materials conserves natural resources and landfill space, reduces the environmental impact of producing new materials, creates jobs, and can reduce overall building project expenses through avoided purchase/disposal costs.*
>
> *In recent years, numerous efforts have been underway to reduce the environmental impacts of construction and demolition projects. EPA Region 8 helps promote and facilitate the recycling and reuse of these materials by providing useful information and grants, tools, and resources. The following provides links to these resources through frequently asked questions.*

Given the basis of the classification of materials, which consist as construction and demolition materials, the following projects were researched and developed according to the nature of the use and need. Most importantly, each prototype and design-research project dived into the computational aspects, in both high-tech and low-tech opportunities, and investigated the fundamental requirements, computational rigour, and intrinsic opportunistic possibilities of computational tooling in both analogue and digital testing as a means of understanding the material opportunity provided as a CRD material, while looked upon the use of the material as a hybridized retainer for the larger development of the premise.

3 RESILIENT MODULAR SYSTEM (RMS)

Resilient Modular System (RMS) is a continued collaborative academic project that is working towards a regional grant approval. RMS is a multi-disciplinary research proposal to forge the synergy and efforts between three different colleges/departments within the University of Houston: College of Architecture, Department of Industrial Engineering, and Department of Material Studies and Engineering.

Figure 1 demonstrates the different scales and iterative opportunities, topics, and fields of research will include, but will not be limited to: Architecture/Design, Industrial Engineering and Prototyping (Digital/Analogue), Patents, and the Material Sciences. Within the larger understanding of the design-research, all conducted research will require a high level of computational science

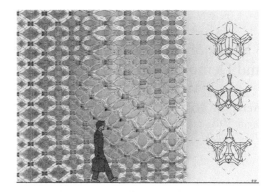

Figure 1. Scale, iteration, and tessellation of the RMS Modular.

and bio-engineering support. Each collaborator/Faculty member is a key asset to the development of this project and is experts within their respective fields.

The division of research and development will be as follows: **Prof Wendy W Fok** (Architecture/Design/Prototyping), a graduate student assistant (Architecture/Design/Prototyping), **Prof Ali Kamrani** (Patent/Industrial Engineering/Modular design aspects for form and fit analysis), and **Prof Ramanan Krishnamoorti** (Material Sciences/Bio-related engineering). The developmental nature of the RMS is to design-research into a modular system that could be applied as urban interventions within the context of temporary and permanent settings.

Using both eco-intelligent architectural design objectives, as demonstrated on Figure 2 displayed the dualistic approach between the knowledge and technique of manipulating sustainable materials ultimately pursues a positive impact on the planet as a growth opportunity and engenders a focus on enhancing benefits (not only reducing costs) through its decision-making and actions—taking an approach of optimization rather than minimization. This Project can understand the perspective of "people, planet and profit," as expansionist and enabling leadership through the achievement of advanced success metrics. For example, the concept of effective design of products and services should move beyond typical measures of quality—cost, performance and aesthetics—to integrate and apply additional objectives addressing the environment and social responsibility.

Through both digital and analogue (physical) prototyping in both architectural and design scales and migrating the opportunity of a full-cycle cradle-to-cradle design process into a Design-Fabrication project—with real-world contextual testing, and use of both repurposed construction

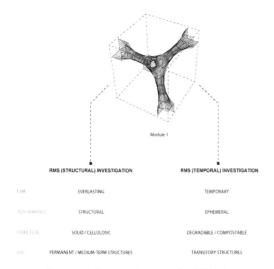

RMS (STRUCTURAL) INVESTIGATION		RMS (TEMPORAL) INVESTIGATION
TIME	EVERLASTING	TEMPORARY
PERFORMANCE	STRUCTURAL	EPHEMERAL
STRUCTURE	SOLID / CELLULOSIC	DEGRADABLE / COMPOSTABLE
USE	PERMANENT / MEDIUM-TERM STRUCTURES	TRANSITORY STRUCTURES

Figure 2. Dualistic research approach of RMS.

waste and biodegradable materials (specifically, biodegradable soy-based polyurethanes, ceramic fillers, and composite plastics)—RMS (temporal + structural) is to find a dualistic opportunity into sourcing ecological solutions of constructing temporary structures within the built environment in locations of need.

The idea of the RMS (temporal) is the ability of it to become an ecological and resilient modular construct for the built environment that could be subsequently dissolved, yet, in an effort of full-cycle design, also contributes to nourishing the natural landscape.

The temporary proposal is that one of these structures could be possibly constructed as a retaining wall system—similar to the ones that are seen along the side of the highway or a landslide retention wall. The composite within the mixture of this will consist of ceramic filler, broken down glass, and biodegradable plastic as the main composite material. The process of this works as follows: 1) a landslide retaining wall is constructed with the RMS module, 2) due to exposure and UV tested breakdown, when the biodegradable plastic comes to the end-life, 3) the plastic will degrade and dissolve. 4) Since the plastic is made with a mixture of ceramic filler, 5) when the plastic dissolves, 6) the ceramic filler will be left, and since the ceramic filler itself retains moister, 7) when the ceramic filler is deposited into the soil, it would provide itself as a form of nourishment for plantation and development for agricultural growth.

The primary material research for the RMS (temporal), ephemeral structure, will be based on agricultural or soy-based biodegradable polymers have been in research since the late 90s and have

been improved, bought out and carried forward, by some of the world's largest companies, like food and agricultural giant Cargill, who in 2008, spent over 22 million USD on developing a method to research and use polyols that can replace petroleum-based chemicals. The most effective method is to blend soy protein plastic with biodegradable polymer to form soy protein based biodegradable plastic, and forming the material with the method of extrusion and injection-molding to form useable pieces of plastic. Therefore, using the same traditional methods of constructing plastics, the same design fabricated parts would be used for applying similar 'thermoforming' or 'vacuum' forming techniques into constructing the prototypes.

While the secondary research for the RMS (structural) will be research for repurposing construction waste, as a mixture for the remediation of the structural testing and joint detailing, the same modular structure will be utilized to further the innovate on studying the structural form/fix/analysis of the RMS (structural) modular.

4 GEO-COGNITION

Geo-Cognition is based on the geometric concepts of projective geometry (duality principal) and the convergence theory, and the fusion of the four main geographic locations that had the most significant impact within the artist's career and life. The supervening confluences, which occur through transitional developments between the cities, are formalized by utilizing a form of projective geometry, and attach itself within an underlying cognitive geometrics theory.

The confluences of the cities, through its linearity and dynamics, are representations of both durational and formal natures of the transitions. These factors are carefully developed and linked to the artist's respective influences and the relative time spent within the period of that city, resulting in the dynamic affects which transition between the axioms of the different skylines and planes. Formally speaking, the different skylines merge (converge) from one into another, creating a morphogenesis between the planes.

Projective Geometry, shown on Figure 3, operates as a method of design appropriation that playfully interprets the original definition, which is a branch of geometry dealing with the properties and invariants of geometric figures under projection. In older literature, projective geometry is sometimes called "higher geometry," "geometry of position," or "descriptive geometry" (Cremona 1960, pp. v-vi). The most amazing result arising in projective geometry is the duality principle, which states that a duality exists between theorems such

Figure 3. Singular modular example of the Geo-cognition installation.
*Note: Shadows produce Cityscape.

as Pascal's theorem and Brianchon's theorem, which allows one to be instantly transformed into the other. More generally, all the propositions in projective geometry occur in dual pairs, which have the property that, starting from either proposition of a pair, the other can be immediately inferred by interchanging the parts played by the words "point" and "line."

The material exploration of GEO-COGNITION was made possible by utilising the byproduct of an architectural construction site. Fact shows that most US construction sites and construction managers overstock on more than 30% of building materials for construction. Over 70 million tonnes (155 000 million pounds) of waste is produced in the construction industry each year. This amounts to 55 LBS per week for every person, about four times the rate of household waste production.

The EPS DOW insulation foam used to CNC mill, produce, and fabricate GEO-COGNITION was made possible through donations from an actual construction site (a commercial building on the campus of the University of Houston), whereby the site manager offered us to take the overstocked material.

Through reusing the overstocked insulation material, it provided us (the artist) complimentary materials for production, while also lessening the waste creation and dumping cost for the contractor. This type of cradle to cradle/grave approach to design allows an innovation of creation, and amalgamation between art and architecture. Different material explorations, including the use of HIPS and MDF, and several prototypes were made before finalising on the EPS DOW insulation.

5 TETRA V2

Tetra V2 an urban sculptural installation created for the HKSZ (Hong Kong Shen Zhen) Biennale 2012 that provided evidence of offsite production (4 days of CNC and production work), and less than eight (5) hours of on-site installation, with the assistance of three workers.

The intention of the Tetra V2 computational process was developed through rhino as an overall procedure to expedite the installation process by devising an offsite pre-fabrication, manufactured, and construction system, using localised and repurposed MDF materials within the region of Hong Kong. The installation, as illustrated on Figure 4, was developed as an academic project at the Chinese University of Hong Kong for a summer 2011studio, which was subsequently furthered as an installation commission for the Hong Kong Shen Zhen Biennale. The larger intention of the piece was to understand the load bearing materials of repurposed wood materials, and understanding the manufacturing process of the CNC for offsite assembly. The design of the efficiency for offsite transportability and onsite construction, therefore, became a key asset into the umbrella premise of designing the sculpture itself.

Given the minimal budget and constraints of the design itself, the continuous production of utilising the CNC in an innovative flat-cut 2-axis process, rather than the typical 3-axis production, made this structure an assembly project rather than an innovation of the tooling itself. Each arm of the tetra-pod is composed of two pieces of 2-axis flat-cut MDF, whereby each tetra-pod itself is composed of six arms, whereby each pod is comprised of twelve pieces. The ability to construct a three-dimensional structure is therefore, played into both the computational tooling of the piece, and also the innovation of the assembly.

The construction and demolition of the piece was, therefore, an innovation of repurposed materials,

Figure 4. Onsite installation of the Tetra V2 at the HKSZ Biennale.

however, the hybridized approach of offsite assembly and onsite installation also expedited the de-installation of the structure. Whereby lead to the ease of transportability of the piece to be later become part of the permanent collection of the BGCA Foundation in Sai Kung, Hong Kong.

6 PERFORMATIVE CRITERIA

The basis of these prototypes and design-research projects are prospective projects in hopes to deliver a performative criteria and incentive to produce a continued effort into generating the material appreciation, and conscious approach to the continued discussion of generated waste production and management within the construction industry. While a large part of the debate is to better fulfil full-cycle design and cradle-to-cradle full loop development, in the case of the Resilient Modular System (RMS), the larger discussion is to provide a model of research those circumstances to allow the cradle-to-end result of construction waste to be repurposed rather than to have been disposed. All three instances functionally outsources to the utilitarian approach to further the results of the architectural state of the material, and transposes the traditional expectations of the end-result of the produced product or design—by creating a viable and creative method to the end-product.

The current environmental crisis and diminished natural resources has challenged the practice of Architecture to re-think its outdated processes of design and construction. New processes that act as full regenerative cycle systems are replacing existing wasteful construction models. The scope of this work focuses on the understanding and development of minimal surfaces specifically of those that are Triply periodic (i.e., Periodic in three direction) as an efficient modular building component fabricated out of high content recycle/salvaged construction solid waste. Each building component will be designed utilizing computational generative strategies to find the most optimal performance. Rapid prototyping and digital fabrication methods will be utilized in order to find efficient and economical modular structure systems that perform at three levels: **structurally, environmentally and socio-economically.**

7 CONCLUSION

The diversion of Construction, Renovation and Demolition (CRD) waste from landfill sites is an issue that has been gaining attention within both the public and private sectors. Surveys have indicated that as much as one third of the 20 million tonnes of solid waste of municipal waste streams is generated by construction, renovation and demolition activities. Many of our landfill sites are reaching capacity. In addition, CRD waste is sometimes illegally dumped or burned, causing land, air and water pollution. The increasing costs of disposal are ultimately reflected in project costs, as contractors must incorporate anticipated disposal costs in their bid costing. Realities such as these emphasize the need for initiatives that focus on reducing and diverting as much waste as possible from CRD activities.

With the rise of computer aided technology, the vast amount of rapid prototyping tools prompts designers to question how our visions of objectivity diverge into the tendency to push and understand the limits of different material properties to further the development of architectural design. The premise of this research proposal is to achieve speculative studies within a project framework, which will be presented through a quad-fold process of: design-research, fabrication-construction, exhibition-publication, and international distribution (including patents).

Design—the larger function of the term inclusive of Research and Development of Applied Sciences, Engineering, Technology and Architecture—today could perhaps be described as the relational equations mediated by digital techniques assisted with production and knowledge of fabrication. Like many fields in the modern culture, it strives to be truly integrated wherein the designer can move seamlessly from concept to production in a single, contained process.

The much larger discussion is less of how the demolition technique is developed, however, is the greater control of the material that is processed, where with demolished materials are reused. Part of the problematic debate within the construction, renovation and demolition argument originates from the structural integrity of reused materials, and the incentives provided by the localised governments for the repurposing of the materials. Whether computational techniques are required as a means to further the research, computation should however, be viewed as a means to test and further the potentials for opportunistic developments, rather as purely as means to digitize the technique of building.

According to the *United States Environmental Protection Agency* a large part of the initiatives for repurposing materials within the field of construction is dedicated to the reduction and reuse. Rather than looking into the means of solely researching with the on-going problems that end-materials produce, perhaps the larger research and development should be to look into potential re-establishments of the materials into cradle-to-end results, which place a larger affect—whether cultural, social, or economical—into the societies of which architecture and construction place an importance in.

ACKNOWLEDGEMENTS

This research is made possible by the generous support of several graduate students and faculty members at the Chinese University of Hong Kong and the Gerald D Hines College of Architecture at the University of Houston. Grant support and tooling were provided by local Hong Kong and Houston research lab facilities.

REFERENCES

Baerlecken, Daniel. and Duncan, David.: 2012, CAADRIA 2012—Beyond Codes and Pixels (eds.), *Junk—Design Build Studio,* CAADRIA, Hong Kong, 305–313.

Bechthold, Martin.: 2008, Innovative Surface Structure—Technologies and Applications, Taylor and Francis, *United Kingdom.*

Grobman, Y.J. and Neuman, Eran. 2012, Performalism—Form and Performance in Digital Architecture.

Oxman, Rivka., and Oxman, Robert.: 2010, The New Structuralism, *Architectural Design,* John Wiley & Sons Ltd, United Kingdom.

Peinovich, Ella. and Fernandez, John.: 2012, CAADRIA 2012—Beyond Codes and Pixels (eds.), *Localised Design-Manufacture for Developing Countries,* CAADRIA, Hong Kong, 285–294.

Sheil, Bob.: 2012, Manufacturing the Bespoke, AD Reader, *Architectural Design,* John Wiley & Sons Ltd, United Kingdom.

ENDNOTE

Reduction

Techniques for reducing the amount of material used in construction without any harmful consequences to the structure are still being developed. One of the best debris reduction techniques, Advanced Framing, can greatly reduce the amount of lumber used in wood framing for houses.

Reducing the amount of C&D materials disposed of in landfills or combustion facilities provides numerous benefits.

- Less waste can lead to fewer disposal facilities, potentially reducing associated environmental issues including methane gas emissions which contribute to global climate change.
- Reducing, reusing, and recycling C&D materials offsets the need to extract and consume virgin resources, which also reduces greenhouse gas emissions.
- Deconstruction and selective demolition methods divert large amounts of materials from disposal and provide business opportunities within the local community.
- Recovered materials can be donated to qualified 501(c)(3) charities, resulting in a tax benefit.

Source: *United States Environmental Protection Agency.*

Green Design, Materials and Manufacturing Processes – Bártolo et al. (eds)
© 2013 Taylor & Francis Group, London, ISBN 978-1-138-00046-9

Author index

Printed and bound by CPI Group (UK) Ltd, Croydon, CR0 4YY

18/10/2024

01776251-0005